International Conference on Disaster Mitigation and Management (ICDMM 2021)

E3S Web of Conferences Volume 331 (2021)

Padang, Malaysia
30 September – 1 October 2021

Editors:

L. Comfort
S. Saravanan
I.W. Sengara
Fauzan

ISBN: 978-1-7138-4176-0

Printed from e-media with permission by:

Curran Associates, Inc.
57 Morehouse Lane
Red Hook, NY 12571

Some format issues inherent in the e-media version may also appear in this print version.

For additional information, please contact EDP Sciences – Web of Conferences
at the address below.

EDP Sciences – Web of Conferences
17, Avenue du Hoggar
Parc d'Activité de Courtabœuf
BP 112
F-91944 Les Ulis Cedex A
France

Phone: +33 (0) 1 69 18 75 75
Fax: +33 (0) 1 69 28 84 91

contact-edps@webofconferences.org

Additional copies of this publication are available from:

Curran Associates, Inc.
57 Morehouse Lane
Red Hook, NY 12571 USA
Phone: 845-758-0400
Fax: 845-758-2633
Email: curran@proceedings.com
Web: www.proceedings.com

TABLE OF CONTENTS

COVID-19

ENHANCING FRAMEWORK FOR DISASTER PREPAREDNESS

LESSON LEARNT IN DISASTER MANAGEMENT

STRUCTURAL ENGINEERING

TRANSPORTATION AND LOGISTIC

TSUNAMI AND SEISMIC ENGINEERING

WATER AND RESOURCES ENGINEERING

Statement of Peer review

In submitting conference proceedings to *Web of Conferences*, the editors of the proceedings certify to the Publisher that

1. They adhere to its **Policy on Publishing Integrity** in order to safeguard good scientific practice in publishing.
2. All articles have been subjected to peer review administered by the proceedings editors.
3. Reviews have been conducted by expert referees, who have been requested to provide unbiased and constructive comments aimed, whenever possible, at improving the work.
4. Proceedings editors have taken all reasonable steps to ensure the quality of the materials they publish and their decision to accept or reject a paper for publication has been based only on the merits of the work and the relevance to the journal.

Title, date and place of the conference :

The International Conference on Disaster Mitigation and Management (ICDMM), September 30th – October 1st, 2021 at Convention Hall of Andalas University, Padang, West Sumatra, Indonesia

Proceeding editors:

Prof. Dr. Louise Comfort
Assoc. Prof. Dr. Subbarayan Saravanan
Prof. Dr. I Wayan Sengara
Prof. Dr. Eng. Fauzan

Date and editors' signature : November 25th, 2021

Prof. Dr. Louise Comfort

Assoc. Prof. Dr. Subbarayan Saravanan

Prof. Dr. I Wayan Sengara

Prof. Dr. Eng Fauzan

Factors associated with COVID-19 test willingness in Indonesia: tele-survey study after 'new normal' policy in risk groups

Mutiara Widawati [1,2*], *Firda* Yanuar Pradani[1], *M. Ezza* Azmi Fuadiyah [1], *Hubullah* Fuadzy [1], and *Rohmansyah* W. Nurindra [1]

[1]Pangandaran Health Research and Development Center, National Institute of Health Research and Development, Ministry of Health of Indonesia, Pangandaran KM. 3, Babakan, Kabupaten Pangandaran
[2]School of Public Health, Edward Ford Building, A27 Fisher Rd, University of Sydney NSW 2006, Australia

Abstract: Data about testing for COVID-19 is important to be recognized since disease diagnostic tests are useful as a foundation for various purposes. Therefore, to obtain comprehensive data regarding the willingness of the Indonesian people to test for COVID-19, a study was carried out. This study is a population-based quantitative study with a cross-sectional design conducted on risk groups located in Pangandaran Regency, Indonesia in 2020. Data collection used a tele-survey technique (telephone interview) with a closed electronic questionnaire instrument. and 400 people as samples. A multivariate analysis was applied between factors such as age group, gender, income, confirmed COVID-19 environment, intensity of information seeking, social media exposure, and willingness to test for COVID-19. The results showed that there were more respondents who stated they did not want to take the COVID-19 test than those who did. The majority of respondents stated that the reason for not wanting to take the COVID-19 test was because they felt it was unnecessary and considered expensive. This study also shows that factors such as the latest level of education, monthly income before the new normal, confirmed COVID-19 environment, intensity of information seeking, and exposure to social media have the potential to affect willingness to test for COVID-19.

1 Introduction

COVID-19, a disease caused by the novel coronavirus (Severe Acute Respiratory Syndrome Coronavirus 2, [SARS-CoV-2]) is rapidly spreading and affecting 216 countries worldwide. This disease already infected 98 million people and caused death to 1.9 million people. [1] Approximately all world's population is at risk of COVID-19 infection, and Indonesia is one of them. This country already has 836,718 positive COVID19 cases and 24,343 deaths because of it. [2] Indonesia is one of the countries with the fastest COVID-19 infection rate in Asia. During this situation, the World Health Organization (WHO) advised every country to implement multiplication of the COVID-19 confirmation test [3] to produce more appropriate and efficient disease control.

The increasing number of COVID-19 cases in Indonesia shows that the control and prevention implemented in Indonesia are not effective enough to overcome the fast rate of this disease spread. Recently, the COVID-19 vaccine has been approved to be included in several countries' health programs including Indonesia. The government has launched a gradual provision of vaccines for all Indonesian citizens. [4] Despite the vaccine launching, vaccines do not necessarily eliminate this disease in a short time, especially by taking into account the uncertainty of public acceptance of vaccines due to various media attacks published by anti-vaccines and distrust spread by social media. [5]

Besides the vaccine, another effort implemented by the Indonesian government to deal with this disease is the implementation of the Large-Scale Social Restriction (PSBB) policy in early 2020. This effort is aimed to reduce the spread of COVID-19 by maintaining distance between individuals. This effort is thought to be more realistic to be implemented in Indonesia compared to the 'lockdown' efforts which have been done in other countries because these efforts still allow people's economic activities to be held as long as they applied health protocol. [6] After several months of implementing the PSBB, the Indonesian government began implementing a new policy called AKB (New Normal Adaptation). During this period, the government permitted people to perform several activities (which previously were not allowed) under certain conditions. [7] The implementation of a new, "weaker" policy more

* Corresponding author: mutiarawidawati@litbang.kemkes.go.id

https://doi.org/10.1051/e3sconf/202133101001

likely gave an impact on people's willingness to be tested and if it goes wrong, then it could lead to the second wave of disease spread. [8]

The COVID-19 test is an initial step to control the spread of this disease. [9] The awareness of individuals who knows that they have COVID-19 will make them take more active prevention of transmission than individuals who do not know about it. Furthermore, this test can also be used as an asymptomatic transmission in population overview for the government and the community. [10] The delay in public and Indonesian government awareness of the COVID-19 test at the start of the pandemic is probably one of the reasons why some studies doubt the incidence of COVID-19 in Indonesia. [11,12] Although now Indonesia has performed widespread PCR testing by utilizing the coordination of various institutions, the total number of tests performed is not sufficient to describe the actual conditions occurring in Indonesia. [12]

Until recently, the most accurate test recommended determining a person's COVID-19 status is still the PCR test. [13] A person's willingness for a PCR test can be influenced by many things. Indonesia and other lower-middle-income countries residents would not think that taking a paid PCR test is such an easy matter. The high cost is likely to affect the willingness of both the government and the community, let alone wishing that the community will carry out this test independently without government assistance. Studies conducted by Grossman and Van der Weele show that ignorance in society might emerge towards positive actions when those actions are high cost. [14]

Other possibilities can also influence people's willingness to take the PCR test as a confirmation test for COVID-19. Individual characteristics are one of them. Characteristics of a person can influence their behaviour and their viewpoint on a disease. Information circulating in the community is also one of the factors that can influence society's willingness. The government needs to develop a strategy to optimize communication with the public regarding the risk of infectious diseases so that the various factors that cause negative behaviour towards diseases circulating in the community can be mitigated. [10]

So far, there are very limited articles discussing the willingness to take a disease prevention effort that involve the willingness to pay in Indonesia [15] and as far as we know, there has been no study that specifically addresses the willingness for a COVID-19 disease test in Indonesian residents. It is important to acknowledge that data about disease diagnostic tests are useful as a foundation for various purposes. Therefore, to acquire comprehensive data regarding the COVID-19 test willingness of Indonesian people, a study has been conducted. This study underlines factors that might influence a person, who are at high risk of COVID-19, to be willing to test for COVID-19 independently in a developing country like Indonesia. The data in this study will be very useful as a foundation for the government to create policies that can support the widespread COVID-19 test for all Indonesians. In the future, this study can also be used as a basis for policies that can support initial diagnostic testing to prevent other diseases.

2 Methodology

This study is a part of the project titled "Information Seeking Behaviour and Covid-19 Prevention Measures in Risk Groups during AKB (Adaptation to New Normal)". [16] This research is a quantitative population-based study with a cross-sectional design. The research was conducted from October to December 2020 and was conducted in Pangandaran district, west java province, Indonesia.

The study population was taken from the data of people who were at high risk of contracting COVID-19 listed in the Pangandaran District Health Office and the Pangandaran Regional Health Laboratory. The high-risk community population consists of tourism personnel (Pangandaran residents who are involved in the tourism sector), crowd subjects (Pangandaran residents taken in crowded places), health workers (Pangandaran residents who are health workers), religious leaders (Pangandaran residents who are religious figures such as the head of the *pesantren* (moslem's religious school) or the teacher of the *pesantren*), and people who had close contact with the patient confirmed positive for Covid-19. The total population obtained was 3923 people. The sample size is calculated using the sample calculation formula for the randomized stratified sampling method. [17] The sampling method is used considering the proportion of each risk group. The sample size obtained was 400 people. To anticipate sample shortages due to exclusion criteria, the same sample calculation was used to select 150 people as a sample reserve.

Ethical approval was provided by the Ethical Committee for Health Research, National Institute of Health Research and Development, the Ministry of Health of Indonesia (Number: LB.02.01/2/KE.553/2020). The data was collected using a tele-survey technique (telephone interview) with a closed-ended electronic questionnaire as an instrument. Interviews were conducted by trained data collectors. Every participant was informed about and understood the purpose of our investigation before entering the study. Before being interviewed, each participant was asked for their consent to participate in this research. Approval is obtained from each respondent without coercion and the respondent can pull out their consent at any time. Respondents who refused, resigned, disconnected, could not be contacted again during telephone interviews, came from outside the Pangandaran district, as well as respondents who were less than 17 years old were excluded from this study sample and replaced with respondents who had been prepared from the reserve. Consent from parents or guardians of those participants aged under 18 was not required. This has been approved by the Ethical Committee for Health Research, National Institute of Health Research and Development, the Ministry of Health of Indonesia.

This research instrument was made based on the framework of the CMIS (Comprehensive Model of

Information Seeking) model [18] and Theory of Planned Behaviour. [19] The questionnaire contains close-ended and multiple-choice questions which are designed to be as simple as possible to be filled. Before being used in data collection, the reliability test was carried out on 30 people using the questionnaire prepared. The questionnaire was revised and retested until the reliability coefficient (Cronbach alpha) was 0.655 (> 0.5). The data in the preliminary test were not included in the final analysis.

To acquire comprehensive data regarding the description of respondents who did not want the COVID-19 to test independently, respondents were also given open questions regarding the reasons why they did not want to take the COVID-19 test independently. The same answers are grouped and then the percentage of all respondents is made. Multivariate analysis was performed using logistic regression with backward Wald method to determine the relationship of the response variable (willingness to test) and each explanatory variable (age group, gender, education, income before the 'new normal' period, intensity of information seeking, and social media exposure). The relationship between variables is described by the value of the Odd ratio (OR) along with the value of the confidence interval (95% CI). The analysis is carried out to find the most optimal equation model. The difference in OR > 10% between the previous model and the model after the variable was removed is used to determine whether the variable can be included in the final model or not. All data were analysed using SPSS version 15 (IBM Inc., Chicago, IL.USA).

3 Findings

The results of the logistic regression analysis quantifying the effect of various factors on willingness to do tests independently are presented in table 1. Based on multivariable analysis, predictors of willingness to do test independently were last education level, income per month before new normal, positive COVID-19 neighbourhood, Information seeking intensity, and social media exposure. While age group and gender did not significantly affect the response variable.

Table 1. Factors that influence people to get vaccinated

Factors	Test willingness		OR	95%CI	p-value
	Yes (%)	No (%)			
Age groups					0.178
< 21 years old	0.8	3.5	NA	NA	NA
21 - 40 years old	13.3	45.8	NA	NA	NA
41 - 60 years old	6.3	25.5	NA	NA	NA
> 60 years old	1.5	3.5	NA	NA	NA
Gender					0.125
Men	11.3	50.0	NA	NA	NA
Women	10.5	28.3	NA	NA	NA
Last education level					0.033
Elementary	1.5	10.3	0.480	0.172-1.336	0.160
High school	7.8	47.5	0.455	0.248-0.835	0.011
University/college	12.5	20.5	1		
Income per month					0.009
< Rp 2 million	7.3	46.5	0.311	0.117-0.826	0.160
Rp 2 –5 million	11.3	28.0	0.696	0.274-1.772	0.448
> Rp 5 million	3.3	3.8	1		
Positive COVID-19 neighborhood					<0.001
Yes	5.0	5.8	4.369	2.055-9.289	< 0.001
No	16.8	72.5	1		
Information seeking intensity					0.005
Rarely	7.5	47.5	0.354	0.189-0.663	0.001
Sometimes	3.3	11.8	0.465	0.204-1.058	0.068
Often	11.0	19.0	1		
Social media exposure					0,016
Often	12.3	35.5	2.035	1.143-3.625	0,016
Not often	9.5	42.8	1		

NA: Not Applicable (were not included in the multivariate analysis)

Table 1 showed that respondents who have university/college degree education tend to be more willing to do the test independently compared to the ones who have lower degrees of education. High school level of education respondents is 0.455 times less likely to be willing to do the test compared to the one who has university level of education (CI= 0.248-0.835; p=0.011). Elementary school level of education respondents is 0.480 times less likely to be willing to do so as well (CI= 0.172-1.336; p=0.160).

The result also showed that respondents who have an income higher than 5 million IDR tend to be more willing to do the test independently compared to the ones who have a lower monthly income. Respondents who have a monthly income of less than 2 million IDR are 0.311 times less likely to be willing to do the test compared to the ones who have the highest salary (CI= 0,117-0.826; p=0.160). Similar to the lower salary, respondents who have a monthly income between 2 million and 5 million are 0.696 times less likely to be willing to do the test compared to the ones who have the highest salary (CI=0.274-1.772; p=0.448).

Respondents who have positive COVID-19 neighbours are 4.369 times more likely to be willing to do the test compared to the one who does not have it (CI= 2.055-9.289; p<0.001). Similar to this behaviour, respondents who often got information from social media are 2.035 times more likely to be willing to do the test independently compared to those who do not often get it (CI=1.143-3.625; p=0.016).

The results also showed that respondents who rarely seek information about COVID-19 tend to have lower test willingness (0.354 times) compared to the ones who often seek it (CI=0.189-0.663; p=0.001). Similar to this, respondents who seek information 'sometimes' also tend to have lower test willingness (0.465 times) compared to the ones who often seek it (CI=0.204-1.058; p=0.068).

Table 2 shows an overview of the reasons for respondents who did not want to take the COVID-19 test independently. More respondents stated that they did not want to take the COVID-19 test (78.5%) than those who did. The majority of respondents answered that the reason they did not want to take the COVID-19 test was that they felt it is unnecessary (49.75%). In addition, the second largest answer related to reasons for not wanting to be tested was because the COVID-19 test was considered expensive (17.5%).

Tabel 2. Reasons that influence people to get vaccinated

Reasons	Total	Percentage of all respondents (%)
Unnecessary	199	49.75
Expensive	70	17.5
Afraid to be tested	33	8.25
Don't understand anything about the COVID-19 test	10	2.5
The test place distance is too far	1	0.25
Later, if I got close contact with the COVID-19 confirmed case	1	0.25
Total	314	78.5

4 Discussion

This study is inseparable from several limitations. This research was conducted using a closed questionnaire so that the answers given by respondents were not explored further. There is a possibility that there will be hypothesis bias because the analysis is based on survey data. Respondents can answer the survey with a different answer from what they did. This bias is common in studies that measure a person's positive behaviour [20] Besides hypothetical bias, this study also does not describe Indonesia as a whole. This study was conducted in only one district in Indonesia and does not necessarily represent the whole of Indonesian society, especially since this study was conducted in the Pangandaran district that still consists of many villages. Certainly, the individual characteristics will be somewhat different from urban communities such as Jakarta (the capital city of Indonesia).

A person's willingness to take a COVID-19 test is probably caused by various factors, similar to other behavioural changes that can be influenced by several factors. Those factors include the expensive PCR test for COVID-19, changes in policies implemented by the government, anxiety, concern, boredom due to continuous exposure to information [21], socio-demographic characteristics [22], social norms [23], information-seeking behaviour, facility availability, and other factors. [10,24]

The results in this study indicate that the majority of respondents do not want to take the COVID-19 test because they feel it is unnecessary and they feel that the COVID-19 test is expensive. The minimum costs incurred for taking one test is 900,000 IDR (equivalent to 60 USD) [25], in addition, costs can also be incurred during self-isolation which is a direct effect of the COVID-19 test if the test results are positive. [10] This research shows that the majority of respondents have an income of less than 2 million IDR (equivalent to around 140 USD) per month, and indeed to spend 60 USD per test will be quite difficult for people with this income. As shown in the results, income is one of the factors of individual characteristics that has a significant effect on a person's willingness to take a COVID-19 test. Respondents who have a monthly income of less than 5 million IDR (equivalent to 340 USD) are more likely to refuse to do a COVID-19 test than respondents who have an income of more than 5 million.

Another factor that can influence behavioural change is individual characteristics. [22,26] This study result showed that the individual characteristics, which are the level of education and income, influenced the willingness of the COVID-19 test. Similar conclusions have been drawn from other studies in Pakistan. [22] This is probably because individuals with higher education have a greater tendency to increase knowledge related to disease risk and have a higher probability to digest information from health promotion. Therefore, prevention efforts can be more adapted to their daily lives. The insignificance association between the willingness to test for COVID-19 and ages or genders is probably because the will tends to be random across all ages and genders, so these factors have no effect on this behaviour.

The low percentage of respondents who want to take a COVID-19 test after the new normal period might also be because the community is bored and tired of taking prolonged prevention. Several studies state that prolonged intervention is one of the challenges that may occur in a community and can cause psychological fatigue of the individual that affects the practice they would take. [21,27] The low willingness for a COVID-19 test may also be because respondents choose not to care and choose to avoid information by hardly seeking it. This information-seeking ignorance behaviour is also demonstrated by a study done by Grossman. [28] The rarity of individuals who seeks information about COVID-19 in this study might also be explained by previous research that suggests if a person already knows their medical diagnosis and knows that this diagnosis is important as their basis for improving their health behavior, there is still the possibility that they will consciously do not care about it. [10,29]

The level of public compliance to health recommendations in several places in Indonesia is still in question. Several studies conducted in various regions in Indonesia have shown that the level of community compliance tends to be less than good, moreover, there is one study that reported that the disinfection activities carried out by people who should know health protocols is have not met the correct standard of health protocols. [30,31] In contrast to several countries that have implemented penalties due to violations of COVID-19 prevention protocol standard [32], penalties due to COVID-19 prevention violations are not discussed both in PSBB and AKB policies. The light penalty is only applied by a few regions depending on the local policies applied there, and still, it did not show a significant impact on public compliance in Indonesia. [33]

This study shows that although the number of individuals who frequently seek information is less than those who rarely do it, the ones who frequently seek information are more likely to be willing to take the COVID-19 test compared to the ones who rarely seek it. It might be due to their concern which was affected by the high intensity of information seeking that they performed. [34]

Respondents who frequently access social media are also more likely to be willing to take tests than those who do not. A large number of social media usage by people as a form of their concern for COVID-19 was also reported by descriptive research conducted in China. [35] Probably people feel that it is a necessity to look for news related to cases of an infectious disease. News related to the number of cases is a topic that can help people to be more aware of the disease. Furthermore, information related to the topic of disease epidemiology can also help people in promoting disease prevention and control for their families and their social life. [36] This is also supported by the results which show that respondents whose aware of their positive COVID-19 neighbours were more likely to have a will in the COVID-19 test compared to respondents who do not. This study also shows that the more exposed someone is to information, the more prudent his subconscious will be, which will lead to curiosity to get more information. The amount of information obtained can cause concern and a willingness to change behaviour.

Hopefully, the result of this study is going to be useful for the government in developing countries such as Indonesia. It can be used as a foundation and complement for the body of literature in formulating policies to support COVID-19 widespread testing for all populations. The various factors presented can be used as the base data to achieve a focused and compatible connection between health promotion intervention and the population targets. Both individuals who have a will for testing and those who have not might be directed toward more positive behaviour.

5 Conclusion

This study concludes that the number of people who did not want to be tested for COVID-19 is greater than the number of people who did want to be tested for COVID-19. The majority of respondents' reason why they did not want to take the COVID-19 test was that they felt it was unnecessary and the test was considered expensive. This research also shows that risk factors such as last education level, income per month before new-normal policy, positive COVID-19 neighbourhood, information seeking intensity, and social media exposure are related and have the potential to influence the willingness for a COVID-19 test.

The author would like to thank the Pangandaran District Health Office, Pangandaran Health Laboratory, and all staff of the Pangandaran Health Research and Development Unit for their assistance in conducting this research. I would also like to thank Firda Yanuar Pradani for her guidance in this research and in the process of writing this paper.

References

1. World Health Organization (WHO). Coronavirus disease (COVID-2019) situation reports. https://www.who.int/docs/default-source/coronaviruse/situation-reports (accessed on 10 of January 2021). (2021).
2. Task Force for the Acceleration of Handling COVID-19. The situation of the COVID-19 virus in Indonesia. Indonesia; (2020).
3. World Health Organization (WHO). WHO Director-General's opening remarks at the media briefing on COVID-19. Media briefing on COVID-19. (2020).
4. Presidential Decree No. 14 of 2021 concerning Vaccine Procurement and Vaccination Implementation in the Context of Combating the 2019 Corona Virus Disease (Covid-19) Pandemic. (2021).
5. R. Rahayu. COVID19 Vaccine in Indonesia: Analysis of Hoax News. J Econo Sauce Hum [Internet]. 2021;**2**(7):39–49. Available from: https://www.jurnalintelektiva.com/index.php/jurnal/article/view/422. (2021).
6. A Roziqin, Mas'udi S, Sihidi I. An analysis of Indonesian government policies against COVID-19. Public Adm Policy An Asia-Pacific J. 2021;
7. WHO Indonesia. COVID-19 Indonesia Situation Reports. 2020. (2020).

https://doi.org/10.1051/e3sconf/202133101001

8. Xu S, Li Y. Beware of the second wave of COVID-19. Lancet. 2020;395(10233):1321–2.
9. MW. Seeger. Best practices in crisis communication: an expert panel process. J Appl Commun Res.**34**(3):232–44. (2006).
10. L. Thunström, M. Ashworth, J. Shogren, S. Newbold, Finnoff D. Testing for COVID-19: Willful ignorance or selfless behavior? Behav Public Policy.1–8. (2020).
11. P. De Salazar, R. Niehus, A. Taylor, C. Buckee, M Lipsitch. Using predicted imports of 2019-nCoV Cases to determine locations that may not be identifying all imported cases. medRxiv. (2020).
12. R. Djalante, J. Lassa, D. Setiamarga, A. Sudjatma, M .Indrawan, B. Haryanto, et al. Review and analysis of current responses to COVID-19 in Indonesia: Period of January to March 2020.
13. Centers for Disease Control and Prevention. Guidance for Health Departments about COVID-19 Testing in the Community [Internet]. 2019 [cited 2021 Apr 21]. Available from: https://www.cdc.gov/coronavirus/2019-ncov/php/testing.html
14. Z. Grossman, J. van der Weele. Self-image and willful ignorance in social decisions. J Eur Econ Assoc.**15**(1):173–217. (2017).
15. H. Harapan, S. Anwar, A. Bustamam, A. Radiansyah, P. Angraini, R. Fasli, et al. Willingness to pay for a dengue vaccine and its associated determinants in Indonesia: a community-based, cross-sectional survey in Aceh. Acta Trop.;**166**:249–56. (2017).
16. Pradani FY, Widawati M, Fuadiyah MEA, et al. Research Report Information seeking behavior and COVID-19 preventive measures in at-risk groups during new normal period (Adaptation of New Habits). Pangandaran; (2020).
17. ER. Ziegel, Lemeshow S, D. Hosmer, J. Klar, S. Luanga. Adequacy of Sample Size in Health Studies [Internet].**36**, Technometrics. 232. Available from: https://apps.who.int/iris/handle/10665/41607.(1994).
18. JD. Johnson, J. David Johnson, WA. Donohue, CK. Atkin,S. Johnson. A Comprehensive Model of Information Seeking. 1**6**, Science Communication. 274–303. (1995).
19. NA Flanders, M Fishbein, I Ajzen. Belief, Attitude, Intention, and Behavior: An Introduction to Theory and Research. Addison-Wesley; (1975).
20. N Jacquemet, R V Joule., S Luchini, JF Shogren. Preference elicitation under oath. J Environ Econ Manage.**65**(1):110–32. (2013).
21. O Ilesanmi, A Bello, A Afolabi. COVID-19 pandemic response fatigue in Africa: causes, consequences, and counter-measures. Pan Afr Med J.37(Suppl 1). (2020).
22. M Yaseen, A Saif, T Khan, M Yaseen. Public knowledge and adherence to hand hygienic guidelines for the prevention of SARS-CoV-2 transmission: A cross-sectional survey from Pakistan 2021. Disaster Med Public Health Prep.;Mar:1–6. (2021).
23. RB Cialdini, NJ Goldstein. Social influence: compliance and conformity. Annu Rev Psychol.**55**:591–621. (2004).
24. N Nguyen, T Hoang, V Tran, Vu C, Siewe FJ, Colebunders R, et al. Preventive behavior of Vietnamese people in response to the COVID-19 pandemic. PLoS One. 2020**15**(9):e0238830.
25. Indonesian Ministry of Health. Circular of the Ministry of Health of the Republic of Indonesia No. HK.02.02/I/3713/2020 concerning the Highest Tariff Limit for Real Time Polymerase Chain Reaction (RT-PCR) Examination. HK.02.02/I/3713/2020. (2020)
26. S Uddin, T Imam, M Khushi, A Khan, Moni M. How did socio-demographic status and personal attributes influence compliance to COVID-19 preventive behaviours during the early outbreak in Japan? Lessons for pandemic management. Pers Individ Dif.;175(June):110692. (2021).
27. M Roberts, V Andreasen, A Lloyd, L Pellis. Nine challenges for deterministic epidemic models. Epidemics.10:49–53. (2015).
28. Z Grossman. Strategic Ignorance and the Robustness of Social Preferences. Manage Sci.**60**(11):2659–2665. (2014).
29. T Sharot, CR Sunstein. How people decide what they want to know. Nat Hum Behav.;4:14–9. (2020).
30. W Wiranti, A Sriatmi, W Kusumastuti. Determinants of Depok City community compliance with large-scale social restriction policies in prevention. J Kebijak Kesehat Indones JKKI. **9**(3):117–24. (2020).
31. A Athena, E Laelasari, T Puspita. Implementation of Disinfection in Prevention of Covid-19 Transmission and Potential Health Risks in Indonesia. J Ekol Kesehat.;**19**(1):1–20. (2020).
32. S Rab, M Javaid, A Haleem, R Vaishya. Face masks are new normal after COVID-19 pandemic. Diabetes Metab Syndr Clin Res Rev.;**14**(6):1617–9. (2020).
33. Y Farhan. Implementation of Sanctions Against Violations of the Use of Masks in the Context of Preventing the Spread of COVID-19 (Analysis of the Mayor of Banda Aceh City Regulation No. 25 of 2020). ETD Unsyiah. Maret. (2021).
34. P Liu. COVID-19 information seeking on digital media and preventive behaviors: the mediation role of worry. Oct 1;23(10):677-82. Cyberpsychology, Behav Soc Netw. 2020;**23**(10):677–82. (2020).
35. Y Zhao, S Cheng, Yu X, Xu H. Chinese Public's Attention to the {COVID-19} Epidemic on Social Media: Observational Descriptive Study (Preprint).
36. Gu H, Chen B, Zhu H, Jiang T, Wang X, Chen L, et al. Importance of Internet Surveillance in Public Health Emergency Control and Prevention: Evidence From a Digital Epidemiologic Study During Avian Influenza A {H7N9} Outbreaks.**16**, 20 (2021)

The performance of implementation movement restrictions during COVID-19 pandemic in West Sumatra and its impact

Purnawan[1*] and *Efendi* Heru[1]

[1]Department of Civil Engineering, University of Andalas, Padang, Indonesia

Abstract. The COVID-19 outbreak is currently spreading to all regions of Indonesia. This virus was suspected to spread from Wuhan at the end of December 2019. Restrictions on people's movements are one of the government policies that aim to reduce the spread of COVID-19. To find out the performance of government policy, a study on the implementation of restrictions policy to prevent the spread of COVID-19 in West Sumatra has been conducted. This study used the before-after method as data analysis, then the data was examined using the Chi-Square test. Based on the results of data analysis, it was found that the movement of people after the implementation of restrictions movement policy was decreased significantly. The decrease in the need to leave the house was average 57.7%, frequency leaving the house more than 5 times decreased by 80%. Time of leaving the houses in the day, in the afternoon and at night decreased more than 50%, the use of vehicle modes decreased by 14.9% and the number of people traveling outside the house decreased by 23.2%. The performance of restrictions on people's movement policy in West Sumatra was still not optimal, because many people did not know the detail of policies.

1 Introduction

At present, all regions in Indonesia are still busy dealing with the COVID outbreak. This virus was suspected from Wuhan City China, the virus spread at the end of December 2019 [1]. COVID-19 at that time was known as pneumonia due to the novel coronavirus, or a new coronavirus that had never been found before. Wuhan has a transportation hub in the form of the Hankou train station. Every day, tens of thousands of Chinese citizens pass through this station. This station was the starting place for dozens of COVID-19 cases. One of the factors supporting the spread of COVID-19 is the high flow of travel in the context of the Chinese New Year. Hundreds of millions of Chinese citizens then traveled to visit relatives' homes. Mostly concentrated in Beijing, Shanghai, and Guangzhou. The flow of travel from Wuhan continues to increase. In early 2020, about seven million Wuhan residents traveled to various regions. Thousands of people are thought to have contracted COVID-19. When the Chinese government realized the risk of human-to-human transmission, the local spread of COVID-19 was already occurring in Shanghai, Beijing, and several other big cities. The government has also implemented lockdowns in several cities and restricted travel in China. However, the plague has already spread widely. Health Sector: Many people affected by COVID-19 are scattered in various countries. At that time, the countries most affected by the pandemic were China, the USA, and Iran [2].

In mid-January 2020, thousands of Wuhan residents were still traveling to cities in various countries. The first case appeared in Thailand. Indonesia also reported its first two cases in Depok. Three weeks later, the Figure soared to 514 cases with 49 deaths. Cases in Indonesia have increased significantly and this continues and spreads throughout Indonesia. The COVID-19 pandemic has created chaos in Indonesia. This pandemic has turned the stability aspect of the country into chaos. Starting from the economy, health, social, culture, and also transportation aspects. The Indonesian government has reviewed various policies to minimize the spread of the COVID-19 pandemic. Starting from examining citizens who have just traveled from abroad, carrying out travel restrictions, and imposing regional quarantine. Parmet. and Sinha [3] reveal the importance of quarantine and travel restriction to reduce the spread of COVID-19.

Regional quarantine is a government policy that aims to reduce the spread of COVID-19, so as not to be exposed to more positive victims. One implementation is to limit the movement of the community and the desire of the people to gather in a crowd. The community is not allowed to leave the house if there is no need for an emergency. Government policies are enforced in various regions in Indonesia, including West Sumatra province. This policy is part of a traffic management measure that is implemented by local government or local people. Traffic management itself is a process of regulating using transport system conditions to meet with required [4]. Tanariboon reveals that transport demand management

* Corresponding Author: purnawan@eng.unand.ac.id

https://doi.org/10.1051/e3sconf/202133101002

can be implemented by utilizing all existing transportation facilities and means more efficiently, it can use regulation to minimize the use of vehicles [5], another type of transport demand management is using physical restriction, this can be implemented to obstruct the movement of vehicles to reduce congestion [6]. To reduce the spread of COVID-19, the Indonesian government finally implemented large-scale social restrictions (PSBB), this implemented depending on the facility and other resources of the city that is prepared. This restriction was implemented on an area suspected of being infected with COVID-19, this was conducted to prevent the possible spread of COVID-19 [7]. This policy is a combination of traffic management and quarantine scenario. This paper reveals the result of the implementation of several types of movement restrictions implemented by local government and local people in West Sumatra. Identifying people's movements before and during movement restrictions was conducted, then the impact of restriction movement is explained.

2 Research Method

This research was conducted based on the information obtained from the respondent that living in West Sumatra. The survey was carried out using an online system from 10 April 2020 until 20 June 2020, during this period, there are restriction movement policies in several cities. From the survey, 247 respondents have been collected. These respondents were obtained from a selected WhatsApp group, an online questionnaire was sent to several WhatsApp groups. The data collected include a profile of respondents, characteristics of movement restriction, and impact of its restriction movement. To obtain the impact of restriction movement, a comparison of people movement before COVID 19 epidemic and during COVID 19 epidemic was compared. Statistical analysis using the chi-square test was used to identify the difference between those conditions.

3 Finding and Discussion

3.1 Characteristic of Respondent

The result of the research is greatly influenced by the respondent characteristic, Tabel 1 below displays the characteristic of the respondent.

Table 1 indicates that the majority of the respondents are from West Sumatra, and most of them are from Padang (20%), Padang is a city in West Sumatra that has the highest *COVID*-19 patient. Most of the respondent is active people that usually have activity outside the house. Based on information from respondents it is known that most of the cities in West Sumatra implement movement restriction, only 55% of respondents say that their city implements the movement restriction policy. At that time, positive COVID-19 patients still have not spread evenly at all cities in West Sumatra.

Table 1. The characteristic of respondents

Characteristic	Percentage of respondent
The origin of the respondent	West Sumatra (85.8%), Others (14.2%)
Gender	Male (50.5%), Female (59.5%)
Respondent of district	Padang (20%), Padang Pariaman (1.4%), Pariaman (1%), Solok (2.4%), Solok City (2.4%), Tanah Datar (30%), Padang Panjang (1.4%), Agam (3.8%), Bukittinggi (3.3%), Lima Puluh Kota (1.4%), Payakumbuh (3.8%), Sijunjung (13.8%), Sawahlunto (2.4%), Pasaman (1.4%), Pasaman Barat (1.4%), Dharmasraya (1%), Solok Selatan (0.5%), Pesisir Selatan (8.1%), Mentawai (0.5%)
Age	< 20 years (5.7%), 20-29 years (41.2%), 30-39 years (26.5%), 40-49 years (8.5%), >50 years (18%)
Education	Primary School (0.5%), Yunior High School (0.0%), Senior High School (17.5%), Diploma (10%), Bachelor (62.6%), Master of Science (9.0%), Doctoral (0.5%)
Occupation	Government Official (29.1%), Bussinessmen (1.9%), Housewife (6.7%), Student (29.1%), Retired (2.4%), Others (31%)
Implementation movement restriction in the respondent area	Yes (55%), No (44.1%)

3.2 Characteristic of Movement Restrictions During COVID-19 Pandemic

3.2.1 Quarantine and social restriction.

During COVID-19 hits Indonesia, several movement restriction types were implemented by the local government or local people. At that time, local government officials and local people has limited knowledge about how COVID-19 spreads. To avoid COVID-19, some quarantine type has been implemented, Fig. 1 shows the type of quarantine experienced by the respondent in that area, then the type of social restriction implemented at the respondent area is shown in Fig. 2.

From Fig. 1, it is shown that the highest respondent experienced type quarantine is large scale social restriction, it is because the high sample respondent is from Padang, however, some people have experience house and district quarantine. During local government or local people implement quarantine to save the people from COVID-19, they implement social restrictions, this restriction to hamper people to leave a certain area. Fig 2 shows the type of social restrictions experienced by respondents, the highest percentage is closing access in and out of district/city area, and other respondents did not know about those implementations of social restrictions.

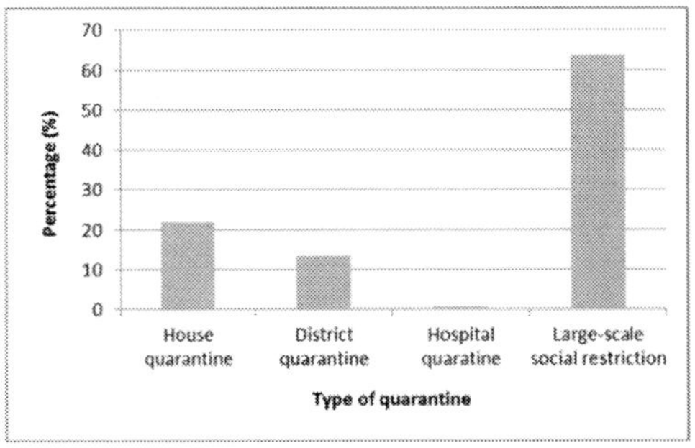

Fig. 1. Type of quarantine that known by respondent

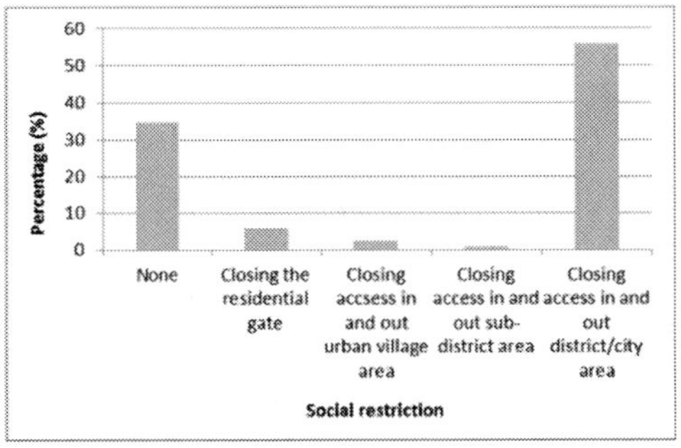

Fig. 2. Type of social restriction that implemented

From Fig. 1, it is shown that the highest respondent experienced type quarantine is large scale social restriction, it is because the high sample respondent is from Padang, however, some people have experience house and district quarantine. During local government or local people implement quarantine to save the people from COVID-19, they implement social restrictions, this restriction to hamper people to leave a certain area. Fig. 2 shows the type of social restrictions experienced by respondents, the highest percentage is closing access in and out of district/city area, and other respondents did not know about those implementations of social restrictions.

3.2.2 Area restriction and limitation leaving the house.

At the beginning of the COVID-19 spread, the local government and local people have implemented several policies without coordination with a higher level of government. Therefore, several types of restriction areas and time limitations to leave the house, Fig. 3 and Fig. 4 show the result of data analysis.

From Fig. 3, it is shown that restriction movement experienced by the respondent was implemented in district or city area, and in Fig. 4 is depicted that most respondent has the experience that there is no time limitation to leave the house (75%), but there is an area that implemented time limitation to leave the house, this is about 30%.

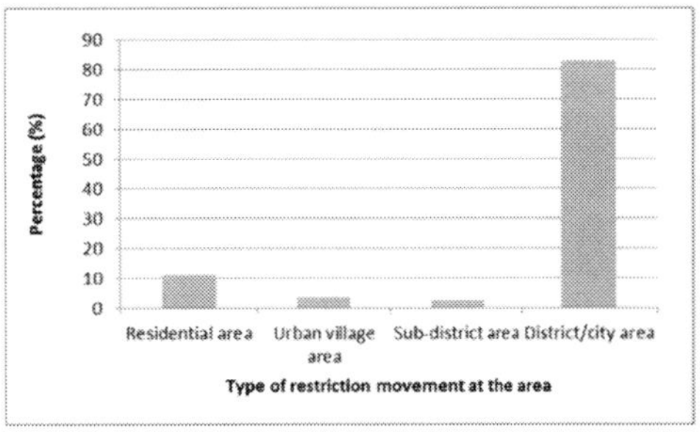

Fig. 3. Type of restriction movement at the area of respondent

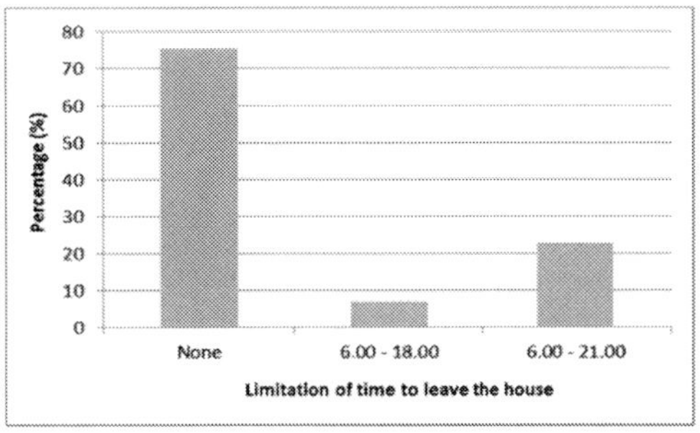

Fig. 4. Limitation of time to leave the house at a respondent area

3.2.3 Socialization and sanction of movement restriction policy.

Before the implementation of the restriction movement policy, there are socialization to the people. Fig. 5 and 6 shows the type of socialization and sanction to people who violate the policies

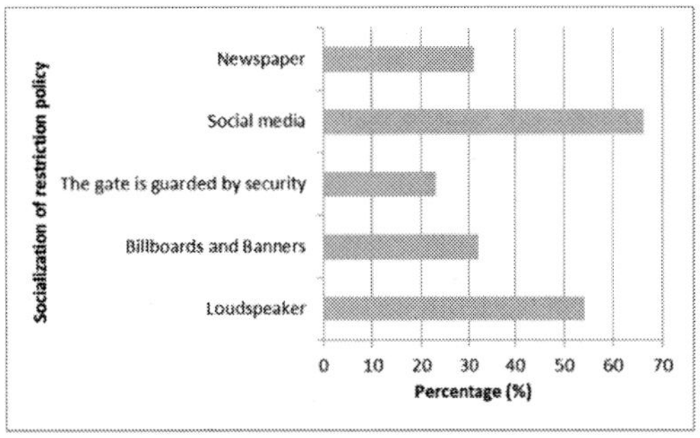

Fig. 5. Socialization of restriction policy that respondent known

From the information of respondents, most of them know the movement restriction policy from social media (67%) and loudspeaker (55%), the other know from the newspaper, billboard, and gate that guarded by security in

Fig. 5, this means that people always have COVID information from social media. This data also indicates that most of the people did not know there is a sanction for the violator and most of the policies were implemented without sanction, this condition that affect movement restriction policies was ineffectively implemented.

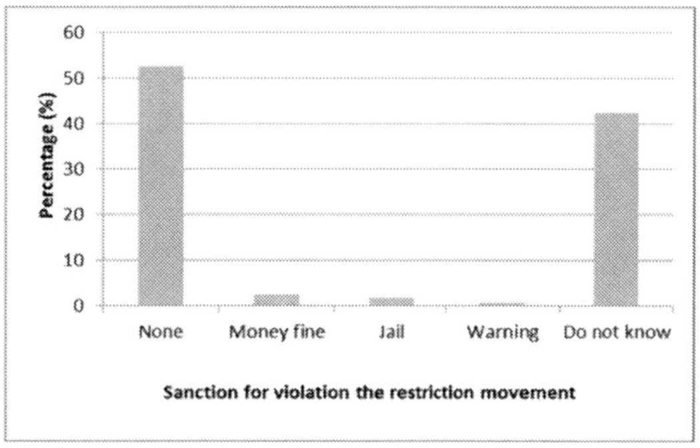

Fig. 6. The sanction for violation of restriction movement that respondent known

3.3 Performance and Impact of Movement Restrictions During Pandemic COVID-19

The government policy to reduce the spread of the pandemic COVID-19 has an impact on people's movement. The performance of movement restrictions is indicated by the condition people's movements before and during pandemic COVID-19 occur. Then, the impact of movement restriction policy is determined based on the value of percentage increase or decrease condition before and during pandemic COVID-19. Four variables have been analyzed for the impact of movement restriction policy, i.e. type of the need to leave the house, the use of transport mode, the frequency leaving the house, and the number of people leaving the house.

3.3.1 Type of the need to leave the house

The government has implemented the movement restriction policy, however, people still need to leave the house to fulfill the family need. Based on this condition, the type of people leaving the house before and during the COVID epidemic is identified. Fig. 7 show the different condition of people leaving the house for two conditions, and Fig. 8 indicates the impact of both condition that shown in the value of percentage decrease or increase.

Fig. 7 shows that the percentage of people leaving the house to fulfill daily need increase, the other type of need to leave the house decreased. This condition indicates that people obey the government policy, they stay at home and leave the house majority to buy necessities.

Based on the results of the Chi-Square test using α = 0.05, the calculated Chi-Square value is 42.5, this value is greater than the Chi-Square value from the table (18,307), it can be concluded that Ho is rejected and H_1 is accepted, meaning that there is a significant decrease in the needs to leave the house. Fig. 8 indicates that the percentage increase people leave the house to buy daily necessities is about 28%., the high decrease is on the activity social gathering and recreation, it decreases about 100% and 90%. This condition would affect the tourist business and social gathering facilities business, such as cafe, gathering room, restaurant, etc.

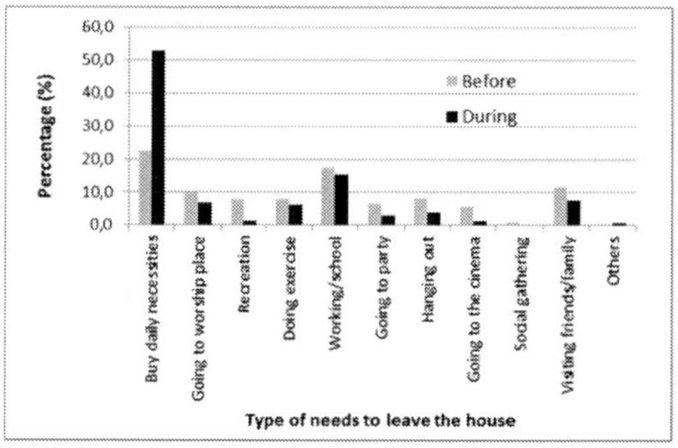

Fig. 7. Type of the need to leave the house before and during the COVID epidemic

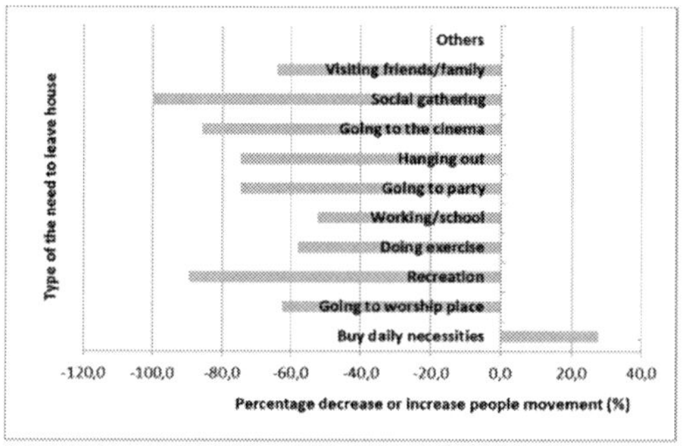

Fig. 8. Percentage decrease or increase the need to leave the house because of the COVID epidemic

3.3.2 The use of transport mode

During people leave the house, they use a transport mode, the transport mode that is used by people depending on the distance from the house. Fig. 9 give information on the condition of people using transportation mode before and during the COVID-19 epidemic, and Fig. 10 indicate the impact of both condition that is in the percentage decrease of increase of transport mode that used by people during the COVID-19 epidemic.

People tend to walk during the COVID-19 epidemic, if they should use the vehicle, they use the motorcycle, public transport, and car was avoided by people, this condition is indicated in Fig. 9. The impact of this condition is depicted in Fig. 10, it would decrease the use of public transport by about 66%, and increase the use of motorcycles by 16%, and increase people to walk by about 59%. This condition will make it difficult for public transport divers and owners because their income reduces significantly. Chi-Square test has been conducted using α = 0.05, the result of calculated Chi-Square value is 7.920, this value is greater than 7.779 the value from the table, it can be concluded that Ho is rejected and H_1 is accepted, meaning that there is a significant difference the transport mode that used by people before dan during COVID-19 pandemic.

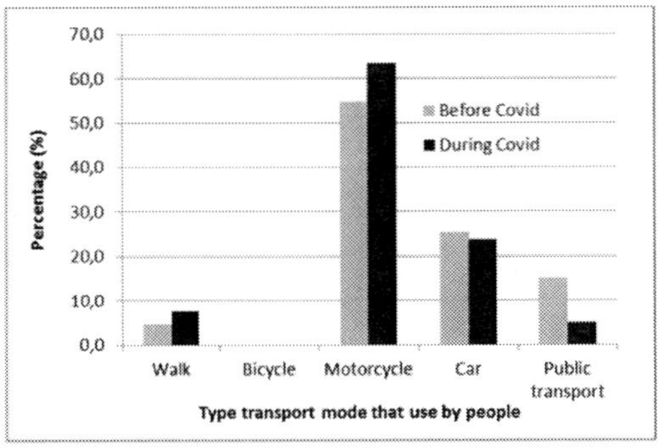

Fig. 9. The use of transport mode before and during the COVID epidemic

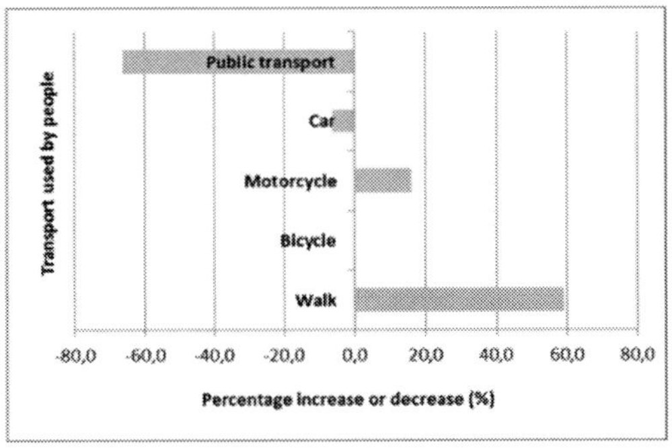

Fig. 10. Percentage decrease or increase the use of transport mode because of COVID epidemic

3.3.3 The frequency leaving the house

The performance of the movement restriction policy also could be identified from the frequency of people leaving the house before and during the COVID epidemic. The different number frequency of people leaving the house has been compared, the result is shown in Fig. 11. Then, the percentage difference between conditions before and during pandemic COVID-19 could be seen in Fig. 12.

Fig. 11 shows the reduction of frequency leaving the house, this is indicated that the frequency of leaving the house 1-2 times increase during the COVID-19 pandemic, and the frequency leaving the house more than 3 times/day decrease. Analysis using Chi-Square test with $\alpha = 0.05$, the calculated Chi-Square value is 35.438, this value is greater than the Chi-Square value from the table (7.815), this can be concluded that Ho is rejected and H_1 is accepted, there is a significant difference between the frequencies of the needs to leave the house before and during COVID-19. The impact of this condition is shown

In Fig. 12, the percentage decrease for leaving the house more than 4 times is 400%, and the increase of leaving the house for 2 times or less was increased about 80%.

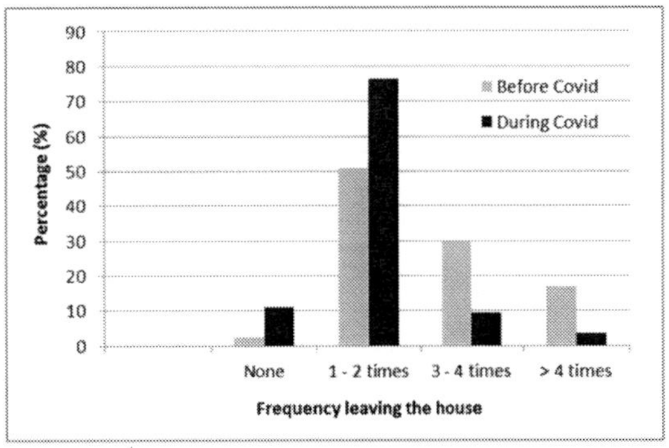

Fig. 11. The frequency of leaving the house before and during the COVID epidemic

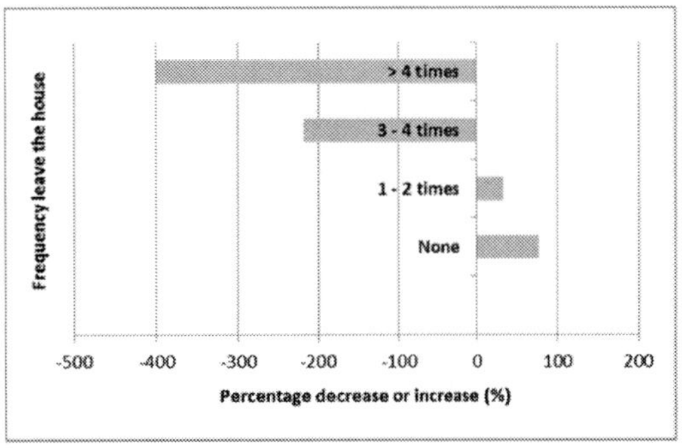

Fig. 12. Percentage decrease or increase the frequency because of COVID epidemic

3.3.4 The number of people leaving the house

To reduce the effect of spreading COVID-19, Government also produce a policy that the people were prohibited to near crowd people in the public area. Fig. 13 depicts the condition the number of people leaving the house before and during the COVID epidemic, for Fig. 14 indicates percentage increasing or decreasing the number

It is shown in Fig. 13 that during the COVID-19 pandemic, people tend to leave the house alone, the number of people leaving the house more than 2 was decreasing. Analysis Chi-Square test using $\alpha = 0.05$, the calculated Chi-Square value is 6.947, this value is less than the Chi-Square value from the table (7.815), this can be concluded that Ho is accepted and H_1 is rejected, there is not a significant difference between the number of people leaving the house before and during COVID-19. The highest percentage decreasing number of people leaving the house for more than 5 people is about 66%, and the percentage of people leaving alone increase by 40%, this is shown in Fig. 14. This can be concluded that during the pandemic COVID-19 people obey Government policy.

Although most people obey Government policies, however, the policy is still not optimum implemented. Several people are still violating Government regulations, some people still leave the house to public facilities that crowd by people, and there are still available people leave the house more than 2 people. Therefore, Government needs to improve the policy. The following policy should be implemented to increase the performance of movement restrictions in West Sumatra.

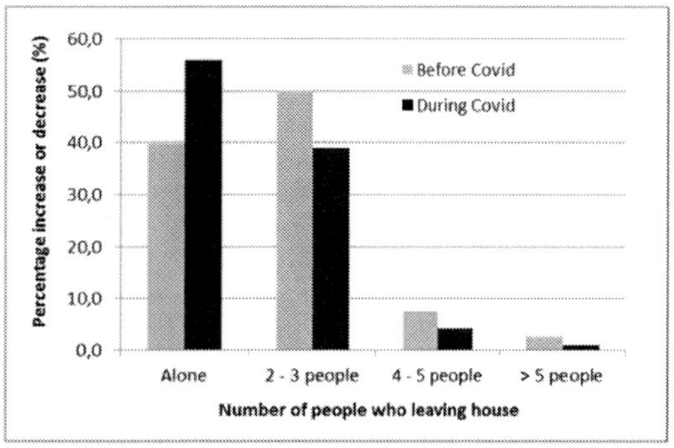

Fig. 13. The number of people who leaves the house before and during the COVID epidemic

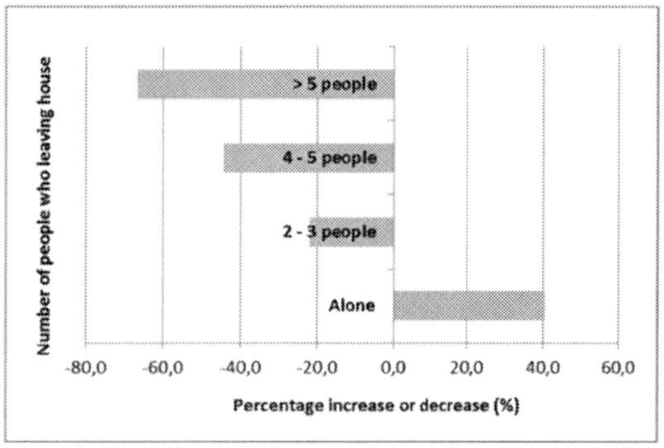

Fig. 14. Percentage decrease or increase the number of people who leave the house because of the COVID epidemic

Apply physical restrictions on guard by officers, so the movement of people is not easily carried out in the area. If the people have an emergency need to leave the area, that person must be free from COVID-19 or equipped with health protocol facilities.

- Control the time to leave the house for each area, so there are no crowds when meeting their daily needs in public facilities.
- Imposition of fines consistently and continuously by officers for violators.

4 Conclusions

Based on the analysis of respondent perception, it can be concluded that the movement of people after the implementation of restrictions on movement has decreased. This is because, before the implementation of movement restrictions, the people could fulfill their needs to leave the house freely, but after the movement restrictions were implemented by the Government or people, they are usually stayed at home and left the house only to fulfill basic needs such as buy daily food. This restriction movement policy has decreased by an average of 57.7% for all types of movement., The social gathering and tourism is the highest decrease. Statistical analysis using the Chi-square test shows that there is a significant decrease in all activity and all conditions. Although the people movement decrease, however, the performance of the people movement restriction policy is still not optimum, because people can leave the house and their area easily. Physical restriction with a guard by officers is required, control the time to leave the house should be implemented and the need of imposition the fines consistently and continuously by officers for violators.

The authors would like to thank the Engineering Faculty, University of Andalas. that supported this research, also Transportation laboratory assistants who help data collection.

References

1. Q. Li, X. Guan, , P. Wu, X. Wang, L. Zhou, Y. Tong, The New England Journal of Medicine (2020).
2. World Health Organization; Coronavirus (COVID-19) (Accessed on 2 March 2020)
3. https://who.sprinklr.com/ (2020)
4. W. Parmet, and M .Sinha, The New England Journal of Medicine (2020).
5. A. A. Alamsyah, Traffic Engineering Revised Edition. UPT UMM; (2008)
6. Y. Tanariboon, Regional Development Dialogue **13** , 3 (1992).
7. J. Luk, Road, and Transport Research, 3 (1992).
8. Government Regulation No. 21 of 2020 about Large Scale Social Restriction in acceleration handling of coronavirus disease 2019 (COVID-19)

Collaborative governance model in COVID-19 pandemic mitigation: a temporary unconditional cash transfer program

I Gede Eko Putra Sri Sentanu[1*], *Klara* Kumalasari[1], and *Ardian* Prabowo[1]

[1]Public Administration, University of Brawijaya, Indonesia

Abstract. The COVID-19 disaster has spread all over Indonesia, causing a sense of laziness for people in the economy by paying very large costs, limited activities, and even losing their jobs. The government through the direct cash assistance program as a mitigation measure in reducing the impact of the COVID-19 pandemic, but so far it has not been effective. The purpose of this study was to determine the ideal cooperative governance model for the cash transfer program during the COVID-19 pandemic. This study uses a qualitative method with a descriptive approach. The results of this study indicate that the cash transfer program still has many obstacles in distributing it to many people who are not on target, which is the result of collaboration between stakeholders such as the Social Service, Population and Civil Registry Office, Village Government, Village Community, Coordinator Team Field, and the COVID-19 Task Force Team. To increase cooperation between these stakeholders by strengthening on several sides such as: First, the government conducts data collection and systematic data collection between the recommendations of the names of beneficiaries from the village with the approval by the Population and Civil Registry Office and the Social Service Office. Second, the Government and the Community carry out monitoring and supervision. The desired impact is that the temporary unconditional cash transfer program can run effectively and benefit people in need. This program is a social network during the COVID-19 period.

1 Introduction

Currently, the outbreak of the COVID-19 infectious disease can spread rapidly throughout the world [1], including Indonesia [2, 3]. In addition to having an impact on the health sector, this pandemic has an impact on the economy. In the macroeconomy, the world economy lost + US$ 1 trillion and world economic growth was below 2%, far below the previous prediction of 2.5% [4]. As for the micro economy, the pandemic has had an impact on traders, especially since the call for restrictions on mobility. Shopping centers are limited in the number of visitors, so sellers have started to complain that their sales have dropped since the COVID-19 case. This impact also threatens workers or employees affected by the termination of employment (PHK) as a result of the heavy workload and reduced financial liquidity of the company [5]. Some of these problems have an impact on the decline in people's purchasing power in line with the low inflation rate [

According to the Social Health Insurance Administration Body (Badan Penyelenggara Jaminan Sosial Kesehatan/ BPJS), there were 2.1 million layoffs in Indonesia in 2020, while according to the Ministry of Manpower, there were 3.5 million formal and informal workers affected by COVID-19 [7]. One of the main sectors affected is tourism. The increasing number of layoffs in the tourism sector worsens the situation of an area. Batu City, a city that relies on the tourism sector, has felt the impact of the COVID-19 Pandemic. According to the Batu City Government, the tourism sector almost accounts for 40% of the total Original Local Government Revenue (OLGR) of Batu City [8]. The following are the numbers of employees who work at home and employees on layoff in Batu City as follows:

Table 1. Number of Employment Termination in the Malang area

No	City	Total Pegawai	
		Employees On Layoff	Employees at home
1.	Malang City	162	1.465
2.	Malang Regency	2.359	247
3.	Batu City	52	2.555

From these problems, government efforts need to be made in mitigating social networking from the COVID-19 pandemic disaster to overcome the economic problems felt by the community. One of the government programs for social networking by providing social assistance to people affected by COVID-19. The social assistance policy for people affected by COVID-19 is the government's response to the COVID-19 pandemic,

* Corresponding author: sentanu@ub.ac.id

 https://doi.org/10.1051/e3sconf/202133101003

especially for people who have decreased income or even lost their jobs [9]. However, in the process, there are still some problems in its distribution [10, 11, 12]. In the distribution of direct cash social assistance, there are several obstacles to the implementation of the policy. Several problems occurred during the COVID-19 pandemic, including the distribution of social assistance. the central and local governments, have several obstacles in distributing social assistance to the community [10]. The social assistance policy is still not effective, there are still many obstacles that must be resolved by the government. Therefore, the government needs to make improvements in the distribution of target recipients.

Collaborative governance theory is an important issue in the relationship between local governments and village governments, communities, and businesses. Collaborative Governance is one way to respond to the wishes of stakeholders to be involved in the implementation of development and to respond to the limited government funding that cannot keep up with the development of community demands for better government performance to obtain resources to carry out development according to the expectations of these stakeholders [13, 14]. In a pandemic emergency, it is not only the government's responsibility but also contributions from other stakeholders are needed through deliberation in forums, building understanding and commitment, and a sense of responsibility to immediately end this pandemic [15, 16].

2 Experimental Method

This study uses a qualitative descriptive approach. Descriptive research is research that provides a clear and systematic description of the object under study. This research also provides information and data following the phenomena in the field. The method used in this study uses qualitative research methods. According to Creswell [17], qualitative research is a method to explore and understand the meaning that some individuals or groups of people ascribe to social or humanitarian problems. This qualitative research process involves important efforts, such as asking questions and procedures, collecting specific data from participants, analyzing data inductively from specific themes to general themes, and interpreting the meaning of the data.

The location of this research is in Batu City, East Java. This research is sourced from primary data and secondary data, namely by data collection techniques consisting of interviews, observations, and documentation. This research consisted of one key informant, namely Deny Sumatra as an employee of the Social Service Secretariat of Batu City. Key informants should be people who are willing to share concepts and knowledge with researchers and are often used as a place to ask questions by researchers. For this reason, in collecting data, researchers should start from the key informants to get a complete and comprehensive picture of the problems observed [18]. The interview method in this study is a semi-structured interview. The interview starts from the issues covered in the interview guide, then the researcher can develop questions and decide how the issues are raised [19].

3 Result and Discussion

3.1 Factors

3.1.1 Starting Condition

Starting Condition is the initial condition when Collaborative Governance is implemented. The initial stages include the historical conditions of cooperation in the past, potential, and incentives expected from the Collaborative Governance process. Since the beginning of the establishment of social assistance policies for communities whose economies have been affected by COVID-19, the Social Service has actively collaborated with several actors who are considered competent in their fields, such as activities from planning to evaluation. Cooperation is also carried out during the implementation of the policy. However, in the implementation of the social assistance policy, there are several obstacles. It is explained by Sumatra, as follows:

"Kalau dalam pelaksanaan kebijakan bantuan sosial tersebut tentu ada kendala paling sering adalah kendala teknis". (In the implementation of the social assistance policy, of course there are obstacles, most often technical problems). Interview results on January 20, 2021.

The existence of these obstacles stems from technical constraints, for example, the distribution of social assistance is still incomplete, causing double data problems for prospective recipients of social assistance.

3.1.2 Institutional Design

Institutional Design is a cooperation forum attended by members, has clear rules, and is an open process. In the policy of distributing COVID-19 social assistance in Batu City, the social service has several times held focus group discussions with the actors involved in the policy. The focus group discussion is held once a month to discuss technical guidelines related to the distribution of social assistance and evaluation. This is explained by Sumatra, as follows:

"Dinas Sosial di Kota Batu melaksanaan beberapa kali FGD dengan tujuan seperti memusyahwarakan kendala-kendala pada kebijakan penyaluran bantuan sosial, dalam FGD yang kami ikutkan adalah aktor-aktor yang terlibat pada kebijakan itu misalnya Dinas Sosial, Dinas Kependudukan dan Catatan Sipil, Pemerintah Desa, Masyarakat Desa, Tim Koordinator Lapangan, dan Tim Satgas COVID-19". (The Social Office in Batu City held several FGDs with the aim of discussing the constraints on the policy of distributing social assistance, in the FGDs that we included were the actors involved in the policy, for example the Social Service, Population and Civil Registry Office, Village Government, Community Village, Field Coordinator Team, and COVID-19 Task Force). Interview results on January 20, 2021.

The social service in building this forum involves several actors who participate even though they have not been maximized in the forum. Every time a forum is held, the Batu City Social Service forms a closed committee, usually, the committee consists of employees from the Batu City Social Service.

3.1.3 Facilitative Leadership

Facilitative Leadership relates to deliberation by actors, setting clear ground rules, building trust, facilitating dialogue between stakeholders, and sharing mutual benefits. The social assistance policy for the COVID-19 community under the leadership of the head of the Batu City social service Ririk Mashuri, S.St, Mm, the head of the social service has a major influence on the COVID-19 social assistance policy for the economy of the community affected by COVID-19. In addition, the Head of the City Service is a leader who has charisma. The head of the Batu city social service has charisma in his leadership such as a competent educational background according to his field. This statement is supported by the statement of Sumatra, as follows:

"Kita sanggat terbantu dengan kepala dinas sosial yang menjabat sekarang dikarenakan latar belakang pendidikan nya sesuai dengan bidang dinas sosial sehingga sumbangsih ilmu diberikan kepada staffnya". (We are greatly helped by the current head of the social service because his educational background is in accordance with the field of social services so that knowledge contributions are given to his staff). Interview results on January 20, 2021.

3.1.4 Collaborative Governance Process

3.1.4.1 Face to face dialog

Face-to-face dialogue or face-to-face meeting is a meeting between various parties in the same location and time so that there is a direct or interactive dialogue between the various parties, to discuss a common interest. Direct dialogue with the Batu city social service has been well institutionalized. The form of face-to-face dialogue carried out by the Batu City social service includes meetings. These meetings are like focus group discussions. These meetings can be attended by all actors involved. However, the FGD was deliberation and consensus. This was conveyed by Sumatra as follows:

"FGD bersifat tertutup dan terbatas. Ada rapat yang dilakukan oleh beberapa orang dengan terbatas karena beberapa hal tidak perlu dilakukan dengan banyak orang. Namun ada rapat bersifat umum semua orang dapat berpartisipasi, biasanya sesuai pengaruh dari topik rapat tersebut". (FGDs are closed and limited. There are meetings that are held by a few people to a limited extent because some things don't need to be done with a lot of people. But there are general meetings everyone can participate, usually according to the influence of the topic of the meeting). Interview results on January 20, 2021.

3.1.4.2 Trust building

Trust building or building trust is a management strategy not only focused on now, but also for future cooperation. Building trust has the same intention to take the best policy for all parties. Trust Building begins by building communication between various actors involved in the policy of distributing COVID-19 social assistance. Building trust among actors has been carried out by the social service. Usually to keep a regular meeting at all times. In these meetings as a means of informing important matters and also given time to express their opinions in the next session. Disclosure of this information is key to building trust.

3.1.4.3 Commitment to Process

Commitment to Process is a commitment or agreement to carry out a certain process to achieve the desired common goal. Mutual commitment, means They depend on each other to solve problems and solutions, that this process is common property, and are open to each other for mutual benefits. Meanwhile, Mutual understanding is knowing the rules, goals, and limits of the pursuit. This is the rule of the game for every actor involved to know. Mutual understanding is a condition that is derived from deliberation, not left to unintentional thoughts and wishful thinking. Various understandings and understandings have been sought by the Batu City Social Service in several activities. The Batu City Social Service always tries to establish communication and provide information to other actors in a reciprocal manner and ask for input. D.Intermediate outcome.

3.1.4.4 Intermediate outcome

An intermediate outcome is a temporary result of an ongoing process. The temporary results have provided benefits and strategic value. The Batu City Social Service has implemented a policy for distributing social assistance, during the 2020 COVID-19 pandemic, the Batu City Office distributed cash social assistance seven times.

3.2 Stakeholders

The following are the actors involved in the distribution of social assistance programs to communities affected by COVID-19 in Batu City:

3.2.1 COVID-19 Taskforce

Actors of the COVID-19 task force unit have the role of policymakers or technical guidelines in the distribution of social assistance programs for communities affected by COVID-19 in Batu City. Actors of the COVID-19 task force unit have a low level of interest and power. the actor of the COVID-19 task force unit only establishes a network with the social service, this causes the actor of the COVID-19 task force unit to have minimal interaction with other actors

3.2.2 Social Office

The social service has a very crucial role in the implementation process, the role of the social service actor is as a facilitator or provider of facilities in the continuity. From the beginning of program development to the end of program implementation, it is the responsibility of the social service office, besides that the role of the social service also includes monitoring and evaluation, proposing technical guidelines. This actor has a high level of importance and power. So that it can be ensured that social service actors also interact with all actors to create a strong network

3.2.3 Village Government

Village actors act as proposers who propose candidates for social assistance recipients to the social service, the Village Head and/or Lurah report the results of data collection coordination in their area by knowing the Camat to the Mayor through the Head of the Batu City Social Service. In addition, the village is also an intermediary between the community and the social service. Actors with high importance but low power. Village actors have a low capacity to achieve goals but can become influential by forming alliances with other actors.

3.2.4 The head of Community Association (RT/RW)

The Community Association (RT/RW) is a communities organization that has a role to help social services and as representatives of the community in their area. as for the role of the actor. RT/RW such as data collection on prospective recipients of social assistance. In this case, the RT/RW actor cannot provide much assistance but fully support the implementation of the program

3.2.5 Field coordinator team

The field coordinator team actors are part of the volunteers from the social service. The field coordinator team actor consists of several communities selected by the social service. The role of the field coordinator team actor is believed to be to assist the social service for monitoring during program implementation such as validating potential recipients of social assistance, besides that the field coordinating team actor assists the RT/RW actor if there is a problem in collecting data on potential recipients of social assistance. Actors The field coordinator team does not have strong power but supports the program

3.2.6 Population and Civil Registry Office

The Population and Civil Registry Office is an actor from a government institution that is involved in the distribution of social assistance programs to communities affected by COVID-19 in Batu City. The role of the civil registry office of population actors is to verify the data of prospective recipients. Data verification is carried out so that there is no double data problem for recipients of COVID-19 assistance. Actors of Population and Civil Registry Office do not have high power because they only validate data on candidates for social assistance recipients for people affected by COVID-19

Discussion

From the previous data presentation, in general, the implementation of the direct cash transfer program has involved several stakeholders but still has obstacles in several ways, such as first, the government has not carried out data collection in a centralized and systematic manner between the recommendations for the names of beneficiaries from the village government and validation by the Population and Civil Registry Office and Social Office. This can provide opportunities for less targeted recipients of the assistance. Second, monitoring or supervision is carried out by the social service and the community. The community can supervise and report to the social service if something has not been received or the community receiving assistance is not appropriate. Third, to minimize the spread of COVID-19 as a result of crowds, it is necessary to carry out restrictions by the COVID-19 Taskforce and the police.

The general problems are the lack of cooperation between the actors involved, causing the implementation of the program not to run according to the needs of the community or other non-government stakeholders. The planning process is not carried out with optimal community involvement. For this reason, efforts are needed through collaborative governance. Next is the discussion about the ideal collaborative governance model in the direct cash assistance program for people affected by the COVID-19 pandemic.

Collaborative Governance models sometimes describe the gradual development of collaboration [20]. Collaborative Governance lies within the dynamics of a system context that can include resource, policy, and legal frameworks, socio-economic and cultural characteristics, network characteristics, political dynamics and power relations, and conflict history, among other dimensions [21].

Based on some previous data and referring to the Collaborative Governance Model [20] it can be seen as follows:

In particular, pre-established networks (relationships) and institutional contexts are important because they shape opportunities and constraints, and influence the if, how, and when of collaborative governance to be opened and implemented [22, 23, 24, 25]

In facilitating collaboration among stakeholders, the researchers made the coordination flow among stakeholders in implementing the cash transfer program as follows:

In Fig. 2., the researcher divides the actors by color, where each color represents the origin of government agencies, businesses, and the community. At the initial stage, community representatives, namely the head of the Community Association (RT/RW) recorded the community of potential recipients of direct cash assistance. This data collection is based on several indicators such as not receiving assistance from the Pre-Employment Card Holder, experiencing loss of livelihood (not having sufficient economic reserves to survive for the next three months), and having family members who are prone to chronic/chronic illness. After the data is obtained, the RT/RW provides the Village Government with this action so that the Village Government can validate the names of the prospective beneficiaries. Furthermore, the Village Government reports by inputting data on prospective beneficiaries of assistance to the Social Service. The Department of Social Affairs and the Department of Population and Civil Registry validate the data received from the Village Government. If the data is appropriate, the Social Service will coordinate with the Bank and the Field Coordinator Team to distribute assistance. In the distribution of aid distribution, it is necessary to coordinate with the Police and the COVID-19 Taskforce to ensure that the distribution process runs smoothly and under the applicable COVID-19 health protocol. In the process of collecting data on prospective beneficiaries to validation of beneficiaries, it is necessary to have a centralized system that is computerized or not done manually anymore, therefore minimizing errors in the manual data collection process. In addition, centralized computerization can simplify and speed up the processing of prospective beneficiaries.

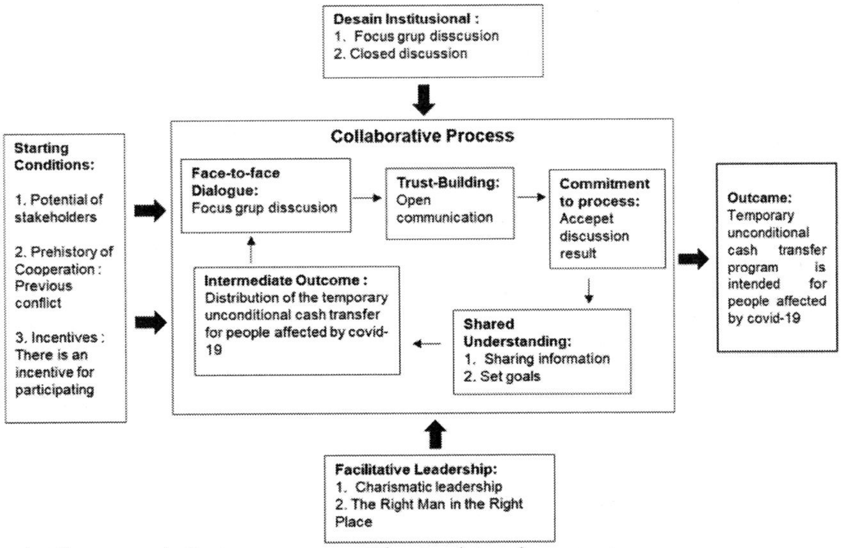

Fig. 1. Collaborative Governance in the temporary unconditional cash transfer program
Source: processed by researchers, 2021

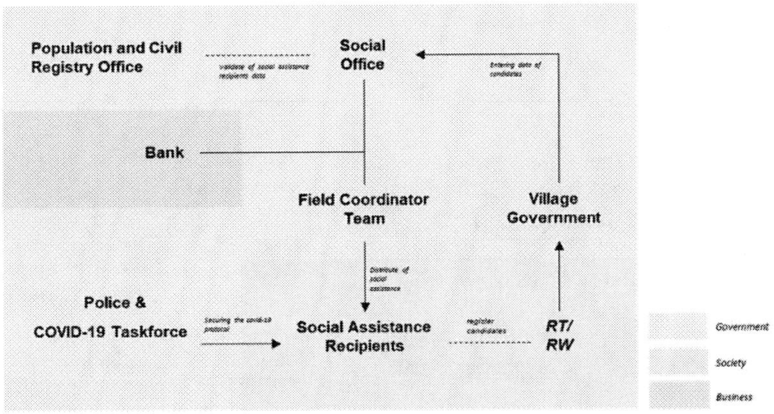

Fig. 2. Coordination Flow of the temporary unconditional cash transfer program
Source: Processed by Researchers, 2021

In the process of coordination between stakeholders in the temporary unconditional cash transfer program. Researchers adopted [20] starting with the principle of collaboration which consists of facilitative leadership, setting clear ground rules, building trust, facilitating dialogue between stakeholders, and sharing mutual benefits, procedures, and basic rules in collaboration for procedural legal collaboration processes, process transparency, participant inclusivity, forum exclusivity, and accountability. In this collaboration principle, especially accountability, it is necessary to involve all stakeholders in the village. The findings of this study, the government has not involved all segmentation actors. Because the involvement of the community and the village government alone is not enough to meet the needs of rural development. So that in this model the involvement of all stakeholders is needed. According to [26] limited stakeholder involvement causes the accountability process to not run smoothly. The involvement of external or wider stakeholders must be carried out immediately to play an important role in assisting the government in carrying out its duties and functions to distribute the temporary unconditional cash transfer program as a social safety net (SSN).

4 Conclusions

In the collaborative governance model, an important element of collaboration between local governments and village governments, the COVID-19 Taskforce and the Police in ensuring the process runs securing to the COVID-19 health protocol, the Bank helps facilitate the storage and provision of money to the community receiving direct cash social assistance. This study suggests that in the whole process of the direct cash transfer program by strengthening the role of local governments and village governments in conducting centralized and systematic data collection through computers to facilitate and speed up validation by the Population and Civil Registry Office and the Social Service Office. Supervision is carried out by the social service and the community. The community can supervise and report to the social service if something has not been received or the community receiving assistance is not appropriate. Furthermore, to minimize the spread of COVID-19 as a result of crowds, it is necessary to carry out restrictions made by the COVID-19 Task Force and the police. In addition, in the collaboration process, it is necessary to carry out facilitative leadership, establish clear ground rules, build trust, facilitate dialogue between stakeholders and share mutual benefits, procedures, and basic rules in collaboration for procedural legal collaboration processes, transparency of the process, inclusiveness of participants, forum exclusivity, and accountability. This process is needed at the time of cooperation to minimize problems in the field.

Thanks to Universitas Brawijaya and the Government of Batu City for facilitating this research

References

1. WHO, Report of the WHO-China Joint Mission on Coronavirus Disease 2019 (COVID-19) (2020)
2. H. P. Sari, *Indonesia Called Potentially New Epicenter of COVID-19, this is the response of the government spokesman* Retrieved from kompas .com on April 11 (2020)
3. L. Agustino, Analysis of COVID-19 Outbreak Handling Policy: The Experience of Indonesia, Jurnal Borneo Administrator, 16,2. (2020)
4. World Bank. Global Economic Prospects. A World Bank Group Flagship Report. 30th Anniversary edition (2021)
5. Purwanto, Agus, et al. "Impact of work from home (WFH) on Indonesian teachers performance during the COVID-19 pandemic: An exploratory study." International Journal of Advanced Science and Technology **29**.5. 6235-6244 (2020)
6. M. Yuniati, R. Amini, *Analysis of the Impact of COVID-19 on People's Purchasing Power NTB.* MPU PROCURATIO: Management Research Journal. **2**, 2. (2020)
7. Kurnia, Ade Miranti. *Due to Corona, more than 3.5 million workers have been laid off and laid off.* Retrieved from kompas.com on Septembre 1, 2021 (https://money.kompas.com/read/2020/08/04/163900726/imbas-corona-lebih-dari-3-5-juta-pekerja-kena-phk-dan-dirumahkan?page=all.)
8. Fizriyani, W. *Tourism Sector Becomes the Largest PAD Contributor in Batu City.* Retrieved from Republika.co.id on Septembre 1, 2021. (https://www.republika.co.id/berita/qmfajs384/sektor-wisata-jadi-penyumbang-pad-terbesar-kota-batu) (2021)
9. Darwadi MS, A. Rosihan, B. S. Utomo, *Mitigation of the COVID-19 Pandemic Disaster at Ogan Komering Ulu,* ETTISAL : Journal of Communication. **5**,2 (2020)
10. Ihsanuddin. Presiden Jokowi Teken Keppres *Determine the National Disaster COVID-19 Outbreak.* Retrieved from Kompas.com on Agust 1, 2021. (https://nasional.kompas.com/read/2020/04/13/18101841/presiden-jokowiteken-keppres-tetapkan-wabah-COVID-19-bencana-nasional)
11. Dewi, Wahyu Aji Fatma. *The impact of COVID-19 on the implementation of online learning in elementary schools. Educational: Journal of Educational Science* 2.1, 55-61 (2020)
12. Herdiana, Dian. " *Collaborative supervision in the implementation of social assistance policies affected by COVID-19."* Jdp (government dynamics journal) **3**.2, 85-99 (2020)
13. T.Sambodo, G, U Pribadi, U. Implementation of Collaborative Governance in the Brosot Cultural Village, Galur, Kulonprogo , *DI. Yogyakarta.* Journal of Governance and Public Policy, **3**(1) (2016)
14. A. Prabowo, M. R. K. Muluk, A. Hayat. Collaborative Governance Model in Village Development during the COVID-19 Period: A Study in South Lampung Regency,PUBLICIA: Journal of Public Administration. **6**,1 (2021)
15. S. Megawati, F. Niswah, T. A. Oktariyanda, *Collaborative Governance as Handling Efforts of Pandemic COVID-19 in Surabaya City,* in Proceedings of the 3rd International Conference on Social Sciences, ICSS, Atlantis Press (2020)
16. I. Wahidah, M. A. Septiadi, M. C. A. Rafqie, N. F. S. Hartono, R. Athallah, COVID-19 Pandemic: Analysis of Government and Community Planning in Various Prevention Measures, J. Manajemen dan Organisasi. **11**, 3. (2020)
17. Creswell, John W., and Cheryl N. Poth. Qualitative inquiry and research design: Choosing among five approaches. Sage publications (2016)
18. Martha, Evi, and Sudarti Kresno. " *Qualitative research methodology for the health sector."* (2016)
19. Rachmawati, Imami Nur. "Data collection in qualitative research: interviews." *Indonesian Journal of Nursing* **11**.1 35-40 (2007)
20. Ch, Ansell, and A. Gash. "Collaborative governance in theory and practice." *Journal of Public Administration Research and Theory* 18 :548-571 (2007)

21. K. Emerson, & T. Nabatchi, Collaborative governance regimes. Georgetown: Georgetown University Press (2015)
22. J. M. Bryson, B. C. Crosby, & M. M. Stone, Designing and strategically managing cross-sector collaborations, John M. Bryson: The Us-India Relationship (2015)
23. J. Sandfort, & S. Moulton, Effective implementation in practice: Integrating public policy and management, USA: John Wiley & Sons (2014)
24. H. Dickinson, & J. Glasby, 'Why partnership working doesn't work' pitfalls, problems, and possibilities in English health and social care. Public Manage Rev, 12 (2010)
25. M. McGuire, & R. Agranoff, The limitations of public management networks. Public Adm 89 (2011)
26. F. H. Zeho, A. Prabowo, R. A. Estiningtyas, Mahadiansar, & I. G. E. P. S. Sentanu, Stakeholder collaboration to support accountability in village fund management and rural development. Journal of Socioeconomics and Development, **3**(2), 89-100 (2020)

Health management of the college student during the COVID-19 pandemic

Sri Siswati[1]*, *Elsa* Giatri[2], *Yolanda* Safitri[1], and *Ice* Yolanda Puri[3]

[1]Health Policy and Administration Departement, Faculty of Public Health, Andalas University, Padang, Indonesia
[2]West Sumatra Senses Health Center, Padang,Indonesia
[3]Departement of Dietetics, Faculty of Medicine and Health Science, University of Putra Malaysia

Abstract. In early 2020, the world was hit by the COVID-19 outbreak This virus was originated in Wuhan, China, and hit 216 countries. In Indonesia, there is an increasing trend of data cases on June 11, 2020, as many as 35,295 positive people. This study aims to look at learning and health situations during the pandemic and the health impacts on students. As a result of the pandemic, lecture activities that were originally on campus were moved to homes with a distance learning system. The descriptive research method is authoritative with a sample of 96 students. The results of this study show that during the COVID-19 pandemic, some students experienced health problems as a result of the transition from offline to online learning, such as anxiety, headaches, and fatigue caused by staring at screens for too long and just doing too many assignments. Even though, students are aware of the importance of health protocols and use them, such as social distancing, wearing a mask when outside, and washing their hands. Students affirm that there is an adequate supply of nutritious food. It can be concluded that changes in learning activities have an impact on student's health, but this can be mitigated with good management and adherence to health protocols.

1 Introduction

The emergence of a new virus, a new strain of Corona (SARS-CoV-2), a condition known as Coronavirus Disease 2019, shook the world in early 2020. (COVID-19). The virus's origins were discovered in Wuhan, China, at the end of December 2019 [1].

A mystery incidence of pneumonia was initially reported in Wuhan, Hubei Province, in December 2019. new coronavirus. The disease was initially known as the 2019 novel coronavirus (2019-nCoV), but on February 11, 2020, WHO published a new name for it: Coronavirus Disease (COVID-19), which is caused by the Severe Acute Respiratory Syndrome Coronavirus-2 (SARS-CoV-2).

This virus can be transmitted from human to human and has spread widely to 216 countries and is designated as a Pandemic. The Pandemic effect makes all fields experience changes, especially in the field of higher education. According to WHO data, on June 24, 2020, globally, there are 9.296.202 cases and 479.133 deaths, whereas according to Indonesian task force data, on June 24, there are 49.009 cases and 2.573 deaths [2].

A new policy has been implemented by the Ministry of Education and Culture's Directorate of Higher Education No.1 of 2020 concerning the prevention of the spread of COVID-19 in universities, which asks all universities in Indonesia to begin distance learning, or in other words, students must do lessons at home [3].

Many students return to their homes and districts after the university is closed. The socioeconomic diversity, the existence of little resources, regional culture, and an epidemic-plagued atmosphere that is dense with fears is not favorable to learning. The epidemic's current stage, which shows an upward trend in positive cases and deaths, appears to be tense psychologically. Students with particular qualities may be affected to adapt to the learning models and outbreak environment's conditions. Students from low socioeconomic backgrounds, for example, may have restricted housing alternatives for distance learning, limited access to high-speed internet, and significant study requirements. It is not only the facilities that are criticized during online learning, but also the students' health in this Pandemic condition.

Student health and safety are very important. Every day students have to face their laptops or android mobile phones to learn from one lesson to another. Anticipation of being impacted is also influenced by vigilance to wash hands, wear masks and maintain distance if you are to locations that provide internet, electricity, and unstable networks. How can students manage their learning in this distant learning, so that learning runs well and student health and safety are maintained?

* Corresponding author: srisiswati@gmail.com

Based on research by Pustikasari, show that online learning uses zoom media. Stress and fatigue in students. Fatigue and stress occur more learning using zoom applications than Goggle class and Whatsapp. Learning media using zoom applications risk 2 times experiencing fatigue in students and 4 times at risk of experiencing stress compared to the use of learning media using Google class and Whatsapp [4].

Research conducted by Rahardjo 2020, that social media as a medium learning that should be used as a stress coping in his research Become a new source of stress and fatigue. Because of social media fatigue. experienced by students who study at home and the role of internet activities has been a lot Take many aspects of an individual's personal life, especially the stress of anxiety related to COVID-19 [5].

The health and safety of students are critical. Students must face their computers or Android mobile phones every day to learn from one lesson to the next. The likelihood of being impacted is also controlled by vigilance in washing hands, wearing masks, and keeping a safe distance from places that offer internet, energy, or unstable networks. How can students manage their learning in this distance learning environment such that learning continues to function smoothly while maintaining student health and safety?

2 Methods

This study uses the quantitative approach to descriptive research. This research took place in Padang, Indonesia, in May 2020. The population in this study is made up of students that study online from home using a variety of programs, but the actual population number is unknown. Based on the formula's calculation, the sample size is 96 respondents. The following are some sample criteria: Bachelor, Master, or Doctorate students from public universities or private colleges in Padang City are eligible to apply. Willing to be a respondent, currently learning "study from home," and the exclusion in a preliminary study, criteria are not a sample.

In this study, primary data was collected by filling out a questionnaire that was provided by researchers and related to the questions on the questionnaire. The secondary data has been collected through document review, which also involves gathering information from reports, documents, journals, seminars, and data relating to researchers. related to COVID-19 issues. Data were analyzed using univariate analysis techniques in IBM SPSS Statistic Version 25 to determine the description of learning models as well as the health of specific students. There are a few issues that stand out.

3 Results and Discussion

According to the table, the majority of respondents in this study were 90.6 percent women and 9.4 percent men. This condition could be caused by the fact that there are more female students than male college students who fill out the questionnaire.

Table 1. Characteristics of Respondents

Variable	Category	Frequency (n)	Percent age (%)
Gender	Male	9	9.4
	Female	87	90.6
Aged	15-20	67	69.8
	21-30	29	30.2

The majority of respondents aged 15-20 years were 69.8 percent, while those aged 20-30 years were 30.2 percent. At the age of 15-20 years, the age condition is relatively severe. It is estimated that students who completed the questionnaire had undergraduate education, with one-third of the respondents being undergraduate, postgraduate, or doctoral students. Filling out circulated questionnaires shows the respondents' interest in participating in online or online learning.

Table 2. Condition of a college student during COVID-19 pandemic

Variable	Category	Frequency (n)	Percentage (%)
Anxiety	Anxiety	82	85.5
	No Anxiety	14	14.5
Health Condition	Healthy	87	91
	Unwell	8	8
	People under surveilence	1	1
Headaches and Fatigue	Yes	60	62
	No	36	38

According to the data table above, 85.5 percent of those who learned online were anxious, while 14.5 percent were not. Anxiety in learning can be caused by a variety of issues, including the COVID 19 pandemic situation, the existence of multiple tasks, and the difficulties of online learning that must be attempted. According to research, anxiety arises as a result of uncertainty about the possibility of future threats, affecting a person's ability to avoid or manage the effects of these threats. This also applies to students. It's a good way to begin the day.

According to the data in the table above, 91 percent of the students feel physically and mentally healthy. Students may be able to learn as a result of this condition. People who are asymptomatic during a pandemic may be uncertain of the severity of the COVID 19 outbreak. Compliance with health protocols is an important measurement. Students' health is raised by medical personnel who take their body temperature, blood pressure and perform a general examination.

The table above showed whether or not students suffer from headaches. The majority of students, 62 percent, experienced headaches and fatigue, while 38 percent did not experience headaches or fatigue. According to Pustikasari and Fitriyani's research, the relationship between fatigue and online learning media using zoom has a p-value of 0.019. Using the OR value of 0.019 The study concluded that there is a significant relationship between fatigue and online learning media using zoom.

https://doi.org/10.1051/e3sconf/202133101004

When using an online learning application, a student is twice as likely to experience fatigue as when learning with Gcr+Wa media [4].

Table 3. Adequacy of nutritious food

Variable	Category	Frequency (n)	Percentage (%)
Nutrition Food	Enough	92	95.8
	Not Enough	3	4.2

The nutritional condition of students can be seen from the adequacy of food, with 95.8 percent believing it was sufficient and less than 5 percent believing it was insufficient. Food adequacy does not represent balanced nutrition, because an adequate balance of carbohydrates, animal and vegetable protein, fruits, and vegetables is needed.

According to research, who studied changes in body weight and eating habits during the COVID-19 pandemic, 1.5 kg is affected by lack of exercise boredom/loneliness/anxiety/depression, increased eating, more snack consumption, and do not consume highly nutritious foods. Diet and dietary habits produce results. There is a significant relationship between body mass index and energy intake, as well as the nutritional status of students who eat processed foods [6].

Table 4. Health protocols

Variable	Category	Frequency (n)	Percentage (%)
Social Distancing	Implemented	87	90.6
	Not yet	9	9.4
Using Mask When Outside the House	Yes	90	91
	No	6	8
Wash hands with soap	Yes	61	63.6
	No	35	36.4
Adequacy of Sleep Times	Enough	66	68.8
	Not Enough	30	31.2
Vitamins/Supplements	Available	61	63.3
	Inavailable	35	38.7

To prevent the spread of the COVID 19 outbreak, three indicators are used. Most students follow health protocols, which include 90.6 percent social distancing, 86 percent wearing masks, and 63.6 percent washing their hands with soap under certain conditions. When students don't get a good signal in their place of residence, they wear a mask and look for a signal network outside their home or residence. It is critical to use face masks correctly. The entire face, from the bridge of the nose to the chin, must be covered by a face mask. Before applying and removing the face mask, wash your hands with soap and water or an alcohol-based hand sanitizer. Remove the face mask from the back, avoiding touching the skin, and Throw away the face mask safely once used.

During this period of the pandemic, the Indonesian government's social distancing policy is expected to reduce and control the spread of the COVID-19 virus. Learning and working from home, staying at home and avoiding contact, prohibiting excessive activity, and limiting operating hours in public places are all policies that promote social distancing. To reduce COVID-19 cases, the government, which previously promoted efforts to promote social distancing (restrictions on social distancing), is now tightening its grip on physical distancing or maintaining physical distance from other people [7,8]. Many factors affected compliance, including knowledge, motivation, perception, and belief in disease control and prevention efforts, environmental variables, the quality of health instruction, and the ability to access available resources.

Adequate sleep is one sign of a healthy body. The majority of respondents, 68.8 percent, reported getting enough sleep for at least 7 hours, while 31.2 reported not getting enough sleep for 7 hours. This is consistent with the findings of a survey conducted by the Biology Center for Chronobiology, Psychiatric Hospital of the University of Basel, which found that respondents, on average, slept 15 minutes longer at night during the pandemic, and while there is a decrease in sleep quality, it is recommended that students' sleep patterns improve during the pandemic. Improving sleep patterns in students during the pandemic COVID-19 is beneficial to their health. The reason for this is a lack of sleep and an irregular sleep schedule can lead to a variety of health problems varying from weight gain to diabetes, depression, and anxiety disorders [9].

According to the survey results of participants who participated in the online learning process while working from home during the COVID 19 pandemic, the majority had already implemented it. Part of it has not been addressed and may be the result of a lack of careful planning for safety and health during the learning process. The number of courses taken does not have a plan for health protection in the event of a COVID-19 outbreak.

Preparation of medicine or vitamin at home is necessary during this COVID-19 due to diseases that can be overcome without going to a health care facility, as it reduces mobilization and crowding, resulting in a limited spread of COVID-19. During this pandemic, the following medications must be prepared at home: paracetamol, antihistamines, decongestant, cold medicine, and digestive medicine.

During the COVID-19 pandemic, nutritional adequacy, especially vitamins and minerals, is needed in maintaining 65 optimal immune systems, vegetables and fruits are sources of vitamins, minerals, and fibre. Vitamins and minerals contained in vegetables and fruits act as antioxidants and help increase immunity. By increasing immunity the body can help in the prevention of COVID-19 outbreaks [10].

Since the outbreak of the COVID-19 pandemic, so many recommendations to consume supplements or vitamins for endurance improvement. A supplement is a product that contains one or more vitamins, minerals, amino acids, fatty acids, and fiber. Supplements are not a complete food substitute but we still need to consume a wide variety of healthy foods to meet our daily needs.[10]

COVID-19 pandemic and supplement use endurance maintenance, coefficient numbers These findings show a positive correlation. 0.175, indicating that the correlation is unidirectional. As a result, there is little chance that COVID-19 is now better understood. The utilization of power maintenance supplements The resistance of the body is increasing [10]. Six factors influence supplement use and they are as follows: Level of education, experience, age of accumulation, beliefs, information, and income [11].

The authors would to thank college students as respondents to this research, thanks to the reviewer for their valuable comments to revise the paper.

4 Conclusions and Recommendations

Online learning is one alternative that can be done when the situation is not possible to be on campus and face to face. Efforts to improve online learning capacity have been carried out by various universities as evidenced by the existence of an online learning platform developed by each university with different names and models. In addition, several companies have also developed this online learning platform and can be accessed easily. But not all the technology that is.

References

1. Y Yuliana. Corona virus disease (COVID-19): A literature. Health Heal Ma, **2**(1):187–92 (2020)
2. COVID Control Cluster Team. Latest COVID Data (2020)
3 Ministry of Education and Culture of the Republic of Indonesia. Circular of the Minister of Education and Culture of the Republic of Indonesia Number 36962/MPK.A/HK/2020. Minister of Education and Culture of the Republic of Indonesia [Internet]. 2020;1–2. Available from: https://www.kemdikbud.go.id/main/index.php/files/download/c5d9f0ec9ff40c6 (2020)
4. A. Pustikasari, L. Fitriyanti. Stress and Zoom Fatigue on Students During Online Learning During the COVID-19 Pandemic. J Health Sciences, **13**(1):25–37 (202)
5. W. Rahardjo, N. Qomariyah, I. Mulyani, I. Andriani , Social media fatigue in college students during the COVID-19 pandemic: The role of neuroticism, information overload, life invasion, anxiety, and gender. Social Psychology, **18**(59):12 (2020)
6. C.J.E. Rambing, A.S.L. Bolang, S.E.S. Kawengian, Energy Intake and Nutritional Status of Students During the COVID-19 Pandemic J Biomedicine Jbm, **13**(2):175–9 (2021)
7. Yanti B, Wahyudi E, Wahiduddin W, Novika RGH, Arina YMD, Martani NS, et al. People's Knowledge, Attitudes, and Behaviors towards Social Distancing Policies as a Prevention of COVID-19 Transmission in Indonesia. J Adm Indonesian Health **8**(2):4 (2020)
8. Lathifa AR, Kamalia F, Putra FP, ... Student Compliance in Implementing Health Protocols during the COVID-19 Pandemic Period. Processing Inter [Internet].**1**(1):1–8. Available from: https://press.umsida.ac.id/index.php/iiucp/article/view/598 (2021)
9. C. Blume. Since January 2020 Elsevier has created a COVID-19 resource center with free information in English and Chinese about the novel coronavirus COVID-19. The COVID-19 resource center is hosted on Elsevier Connect, the company's news and public information, (January). (2020)
10. K. Lidia, E. Levina, S. Setianingrum, C. Folamauk, M. Riwu, A. Lidesna, et al. Health Improvement with Supplements and Balanced Nutrition in the Era of the COVID-19 Pandemic J Lemb Service to Masy Undana [Internet]. **14**(2):63–8. Available from: http://ejurnal.undana.ac.id/jlppm/article/view/3445 (2020)
11. S. Notoatmodjo. Health promotion and health behavior. Jakarta: Rineka Cipta (2012)

Balla ewako : community response to the COVID-19 pandemic based on local wisdom

Ahmad Fatkul Fikri[1*], *Syamsul* Maarif[1], *Dody* Ruswandi[1], *Titisari* Haruming Tyas[1], and *Islamia* Kharimah[1]

[1]Disaster Management Study Program, National Security Faculty, Republic of Indonesia Defense University. IPSC Sentul Area, Bogor, West Java, Indonesia

Abstract. The COVID-19 pandemic has entered a new phase with the discovery of new variants from several countries entering Indonesia. This epidemic is capable of destroying the joints of people's lives, such as health, education, politics, economics, social, culture, politics to the point of disrupting regional resilience. Various efforts have been made by the government in the context of dealing with COVID-19 which cannot be ascertained to end. This research has the aim of analyzing the other side of the COVID-19 pandemic in the form of local wisdom. This study uses descriptive qualitative data obtained through interviews, observation, and documentation. The analytical technique used is the Miles, Huberman, and Saldana model. The results of this study indicate that the local wisdom of *Balla Ewako* has the meaning of a tough house in the face of the COVID-19 pandemic. *Balla Ewako* is a tough village program by the South Sulawesi Regional Police which later became a movement for the community in dealing with the COVID-19 pandemic based on local wisdom. The *Balla Ewako* program has the goal of a strong village for health, food resilience. and security with direct community involvement. *Balla Ewako* has proven to be effective in tackling the COVID-19 pandemic by mobilizing all elements of the grassroots community.

1 Introduction

In the current era of technological progress, countries in the world are focused on fighting for the safety of mankind, including Indonesia. As a country, Indonesia is required to protect the entire nation and the entire homeland of Indonesia. This mandate has been stated in the Preamble to the 1945 Constitution of the Republic of Indonesia. The statement is the state's promise to its people to be able to live in prosperity and safety from various kinds of threats. The real threat facing the Indonesian nation today is a disaster.

Disasters come in various forms, ranging from natural disasters to man-made disasters that affect people's lives. One of the factors that cause disasters is a pandemic. World history records epidemics that have been experienced by health, for example, the Peloponnesian Pandemic (430 BC), the Justinian Plague (527-565 AD), the Black Death pandemic (1347-1351 AD), smallpox which occurred in 1492, cholera around 1961, the Spanish Flu in 1918-1919, the 2003 SARS outbreak, the 2009 Swine Flu, the 2014 Ebola pandemic and most recently the COVID-19 pandemic [1, 2, 3]. While the pandemic in Indonesia was recorded during the reign of King Airlangga in 1006-1042 AD, and what is currently happening is the COVID-19 pandemic [4].

To date, the struggle of countries in the world to face the COVID-19 pandemic has not ended, including in Indonesia. Corona Virus Disease 2019 (COVID-19) is a disease caused by a virus called Severe Acute Respiratory Syndrome CoronaVirus-2 (SARS-CoV-2) [5]. COVID-19 is called a zoonotic virus, which is a disease that is transmitted from animals to humans, with the first case in a wet market selling wild animals in Wuhan City, China [6]. COVID-19 spreads faster and more massively than the SARS and MERS outbreaks [7]. Based on this, on March 11, 2020, the World Health Organization (WHO) declared COVID-19 a global pandemic. Countries in the world, whether they are developed or underdeveloped, rich or poor, have all been affected by COVID-19.

Based on data from the World Health Organization (WHO) until June 24, 2021, there were more than 179 million confirmed cases of COVID-19, 3.8 million more of whom died spread over 223 countries in the world. While in Indonesia, on June 24, 2021, there were 2.053.995 confirmed cases of COVID-19, 55.949 people died from the virus. Not only taking lives, but the effects of the COVID-19 pandemic have also extended to aspects of social, political, economic, cultural, tourism, education to defense and security and the worst thing is the household [8, 9, 10, 11]. This condition has prompted the central and regional governments to work hard in dealing with the COVID-19 emergency. The COVID-19 emergency period in Indonesia was established through Presidential Decree No. 12 of 2020 concerning the Status of COVID-19 Non-Natural Disasters as National Disasters. A disaster emergency has the meaning of a condition that threatens and disrupts the order of people's

* Corresponding Author: ahmadfatkulfikri02@gmail.com

lives due to natural, non-natural causes that result in casualties, damage, and psychological disorders.

To achieve success in dealing with the COVID-19 pandemic, good policies from the government and public awareness are needed. Various efforts have been made by the government to suppress the spread of COVID-19 by involving health workers to volunteers. However, all the government's efforts and strategies have encountered obstacles when people don't care about this virus. Low literacy skills encourage people to be indifferent to the handling of COVID-19 so that public understanding of the dangers of COVID-19 is very low [12, 13]. The existence of a vaccine does not guarantee that this pandemic will end soon. The active participation of the community in building awareness and responsibility is the key to success in ending this pandemic. However, the diverse and particularistic conditions of Indonesian society are not easy for the government to do. A cultural approach is needed as an alternative route with the complex social conditions of the Indonesian nation.

Culture is an important foundation for present and future people's lives [14]. Local wisdom is in the form of knowledge, myths, and ancestral messages containing prohibitions, invitations, and sanctions in environmental management. Local wisdom includes life values that apply to the community in being wise, full of wisdom, and local ideas that are embedded and followed by residents to protect a sustainable environment [15]. Local wisdom also consists of a view of life and knowledge, consisting of various livelihood strategies to solve life's problems [16]. Local wisdom is important to be preserved to maintain the balance and sustainability of the environment. Having local knowledge of adapting and interacting with the environment to find adaptation strategies to the environment at the local level [17]. Positive things in culture This is called local wisdom. Koentjaraningrat, Spradley, Taylor, and Suparlan, have categorized the human culture which is the container of local wisdom to ideas, social activities, artifacts [18].

As a country that has a variety of cultures, there are various kinds of unique ethnic customs that are different for each tribe. Most of the tribes in Indonesia have local wisdom that can be used to fight the COVID-19 pandemic as it is today. One solution to overcome the problems that occur in the community to overcome COVID-19 is cultural power [14]. One of the local wisdom to fight the COVID-19 pandemic is *Balla Ewako* who is in the community of South Sulawesi. *Balla Ewako* is an activity to help each other between communities in terms of health, economy, and security during the COVID-19 pandemic.

As an area with a strong culture, Palopo City has an alternative in dealing with the COVID-19 pandemic. Based on this background, it is deemed necessary to research to determine the application of local wisdom in the form of *Balla Ewako* to the people of Palopo City as a risk reduction for the COVID-19 pandemic based on local wisdom. Broadly speaking, this research will discuss the other side of dealing with the COVID-19 pandemic, namely the local wisdom approach.

2 Method

This study uses a qualitative method with a descriptive approach. Researchers describe the situation under study by enriching the information and looking for relationships to answer why and how the phenomenon occurs. It is similar to the view of Nazir that descriptive methods are used to examine the status of a group of people, an object, a set of conditions, a system of thought, or a class of events in the present [19]. The subjects of this research are parties who are directly involved as resource persons and understand the research topic. The research subjects were selected by purposive sampling, meaning that the specified resource persons had met the specified criteria. The resource persons in question are the Lurah, Babhinkamtibmas, and Babinsa in Palopo City.

In this study, the data needed are primary and secondary. Techniques for providing data using interviews, observation, and documentation. The data analysis technique used is an interactive model as proposed by Miles, Hubberman, and Saldana (2014) which is an analytical technique that is carried out continuously during data collection in the field until data collection is completed so that the data obtained is saturated. The interactive model data analysis technique as proposed by Miles, Hubberman, and Saldana (2014) consists of data collection, data condensation, data presentation [20], and conclusion. While the location in this study is in Palopo City as shown in Fig. 1.

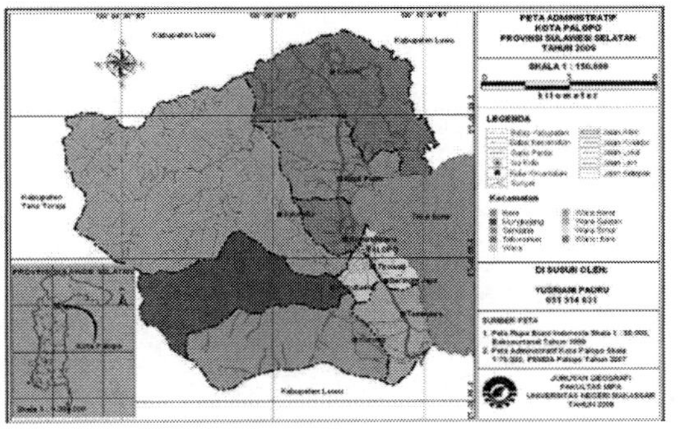

Fig. 1. Location of Area Study

3 Result and Discussion

3.1 Spread of COVID-19 in Palopo City

Coronavirus or better known as Corona Virus Disease 2019 (COVID-19) is a new virus that infects the respiratory tract [21]. This virus was first discovered in December 2019 in Wuhan City, Hubei Province, China [22]. China reported 44 patients with severe pneumonia whose cause was unknown [23]. The initial suspicion of this virus originating from bats sold in the Wuhan wet market, although animal-to-human transmission has not been confirmed, the process of human-to-human transmission has been extensively proven [24, 25]. This virus spread to almost all corners of the world until the World Health Organization (WHO) declared it a global pandemic on March 11, 2020, including Indonesia [26, 27].

The Indonesian government officially announced two cases of COVID-19 infection on March 2, 2020 [28]. The spread of COVID-19 occurred massively throughout Indonesia, including Palopo City, South Sulawesi. COVID-19 first entered Palopo City on April 29, 2020. The first patient was confirmed after traveling from Makassar, which is the epicenter of the COVID-19 pandemic in the South Sulawesi region. The rate of spread of COVID-19 in Palopo City continues to occur and is experiencing an increasing trend. The rate of development of positive confirmed cases can be seen in Fig. 2.

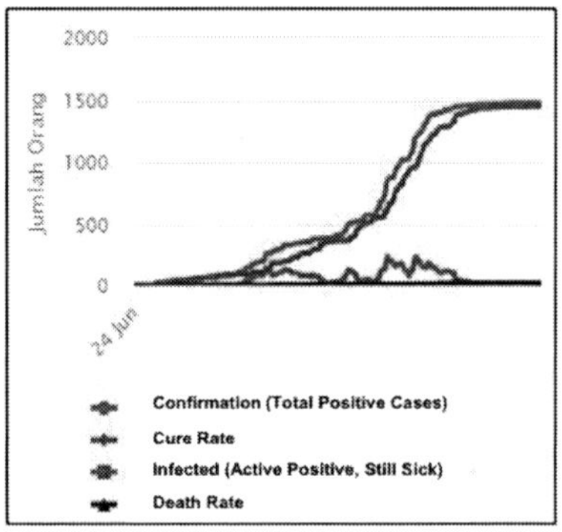

Fig. 2. The Development of Cumulative Cases of COVID-19 in Palopo City

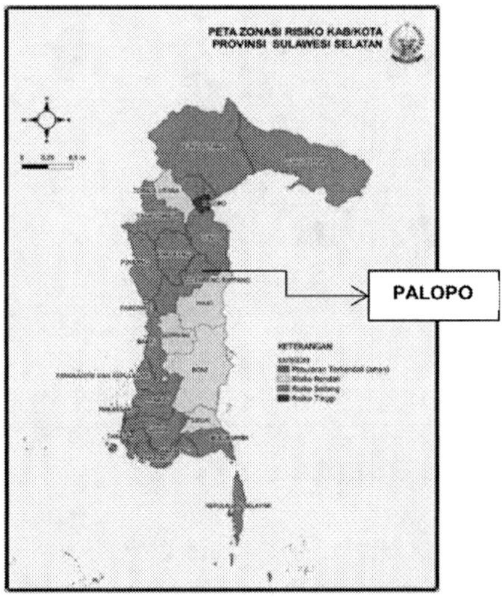

Fig. 3. Map of District/City Risk Zoning of South Sulawesi Province

Based on Fig. 2. shows that the total number of confirmed cases of COVID-19 experienced an upward trend in February 2021 and experienced a decrease in the number of confirmed cases until June 3, 2021. The decrease in the number of confirmed cases of COVID-19 was due to massive vaccination and the implementation of health protocols. While patients who died due to COVID-19 reached 54 people or 3.92%. This figure is higher than the death rate in South Sulawesi Province of 1.5%. The high mortality rate and low recovery rate of COVID-19 patients have pushed Palopo City as the area with the most vulnerable spread rate on October 30, 2020, as shown in Fig. 3.

On October 30, 2020, apart from the relatively low recovery rate in Palopo City, another thing that makes Palopo City still in the red zone is the high-density level. A high level of population density will have an impact on high interactions between individuals or groups in the region [29]. An area with a high population density, the higher the COVID-19 cases especially in urban areas [30]. Cities that are urban areas become catalysts in the context of the spread of infectious diseases [31]. History records that the Spanish Flu pandemic that occurred in 1918, positive confirmed cases and high deaths occurred in countries with high-density levels [32].

3.2 Concept and Implementation of *Balla Ewako*

The *Balla Ewako* concept is a program launched by the South Sulawesi Regional Police in the form of a tough village to accelerate the handling of the COVID-19 pandemic. *Balla Ewako* comes from the Bugis-Makassar language which consists of two syllables, namely *Balla* and *Ewako*. *Balla* means house and *Ewako* or *Rewako* means brave. In addition, the word *Ewako* is also used to provide encouragement or support and intimidate opponents. This program is based on the consideration that COVID-19 is contagious, namely through the breathing of people who have been infected [33, 34] thus placing the community at the forefront of handling the COVID-19 pandemic with the government as a supporter. As a conception, local wisdom is more easily understood by traditional communities because this community has a strong link between experience and the ability to read real phenomena [35]. This is evidenced by Prasetyo that local wisdom in traditional communities in Indonesia is an alternative in preventing disasters or in post-disaster management so that local wisdom has an important role in mitigating disasters that occur in Indonesia [36]. So that the Government can take advantage of the values of local wisdom as a step in disaster mitigation in the country.

The *Balla Ewako* concept is implemented in collaboration with the Lurah, Babinsa, Bhabinkabtimas, and the family community. The *Balla Ewako* concept is implemented by the Neighborhood Association (NA) and the Residents Association (RA) by involving the community directly overseeing the families and neighbors affected by COVID-19. This program was implemented before the implementation of the Micro-Community Activity Restriction policy in South Sulawesi. A community-based approach from the grassroots is a rational choice in dealing with the COVID-19 pandemic. Movements originating from the community in dealing with the COVID-19 pandemic, especially in their respective environments, are the main focus of the government.

The government implemented *Balla Ewako* as a response to the COVID-19 pandemic so that families obey the government's appeal. *Balla Ewako* has proven to be effective in the government's efforts to respond to the COVID-19 pandemic by collaborating with the community. Although not all sub-districts are aware of and understand the *Balla Ewako* concept, this program has proven to be able to reduce the positive number of confirmed COVID-19 in Palopo City. Since this program was issued, the community has responded well to the *Balla Ewako* program to respond to the COVID-19 pandemic. However, the public's adherence to the health protocols set by the government can fade, for example, the pressure to meet basic needs. This happens especially in people with middle to lower economic conditions [37]. Through the *Balla Ewako* movement, the government wants to achieve a resilient village in terms of health, food, and security.

First, after the *Balla Ewako* movement was implemented, data were found regarding the emergence of community movements to work together independently to maintain health. These activities include distributing masks for free, awareness of washing hands, awareness of using masks, checking body temperature, reporting by neighbors when there are residents who travel, and so on. The movement emerged from the community to maintain the health of every family member. Each resident takes care of the vulnerable groups in their respective homes. Vulnerable groups include those who have chronic health problems, such as heart disease, asthma, diabetes, and the elderly [38]. Older people have a higher susceptibility to coronavirus infection compared to younger people [39]. This shows that to achieve health resilience, community involvement is needed.

One of the main priorities in health empowerment is initiating the concept of work by the government that follows the current national development [40] to control the COVID-19 pandemic. The way to empower the community is to strengthen the community as a subject, not an object. The formulation and the National Health Promotion Policy states that the global health promotion strategy consists of three things, namely advocacy, atmosphere building, and community empowerment. This strategy can generally be used to address health problems and prevent health problems. Community empowerment is one of the strategies in health promotion that involves directly the participation of the community to maintain and improve their health [41].

Second, COVID-19 has had an impact on food insecurity where people experience food shortages and have to work harder to get food because some kiosks and markets have to be closed from operation [42]. The United Nations Food and Agriculture Organization (FAO) estimates that the number of people who are malnourished will increase to 132 million this year. One effort to be able to maintain the family's economic situation is to optimize food security by empowering the potential of each region [43].

During the COVID-19 pandemic, people's lives experienced difficulties, but life had to go on, therefore the government through the *Balla Ewako* program made donations to the community to meet the needs of daily life. To obtain food resilience, the *Balla Ewako* program requires the community through the kelurahan to cultivate crops on vacant land around their homes. Based on the results of interviews, the program was able to reduce public panic over the scarcity of food sources.

Fig. 4. Planting Process for Food Security

Food security during the COVID-19 pandemic also comes from donations from neighbors. When people are self-isolating, neighbors donate food to the neighbors. This shows that *Balla Ewako* brings back social capital that has been formed for a very long time. *Balla Ewako* generates existing social capital to survive, maintain survival and meet common needs. Social capital that has existed so far is mobilized for togetherness, establishing mutual trust and benefit, and to maintain life together.

In addition, social capital is built on mutual trust, interaction and relations between citizens, social solidarity, and norms between communities. This instrument provides community strength in dealing with the COVID-19 pandemic. Through this social capital, government programs, steps, strategies, and policies can be implemented synergistically. Social capital facilitates mutually beneficial cooperation in society because it contains networks, norms, and social beliefs that exist and is very strong in society [44, 45, 46]. Social capital can be utilized by the people of South Celebes, especially Palopo City when facing the crisis caused by the COVID-19 pandemic. Social capital is believed to be one of the main components in moving togetherness, mobility of ideas, mutual trust, and mutual benefit to achieve common goals [47].

Third, the sluggish economy due to the COVID-19 pandemic has triggered problems in public security and order. On the other hand, the government with the reason of reducing the risk of the spread of COVID-19 in detention homes freed prisoners. During the COVID-19 pandemic that has occurred in Palopo City since the beginning of 2020, the environmental security system must be strengthened. Every member of the community must carry out personal security, then maintain security in the surrounding environment. The *Balla Ewako* movement raises public awareness to maintain security and order in the surrounding environment. Awareness to protect the environment is done by taking care of the neighbor's house when the occupants of the house are doing maintenance in the hospital. In addition, the community independently began to limit territorial areas by using entrance portals to the area, collecting data on incoming guests, imposing curfews, and so on to be reported to the tough *Balla Ewako* village post.

Fig. 5. quarantine portal

The environmental security system has been regulated through the Regulation of the Chief of the Police of the Republic of Indonesia Number 23 of 2007. The regulation explains that a mobile security system is a unit that includes components that are interdependent and interconnected and influence each other which results in the ability to be used as an effort to meet the demands for security and order conditions in the environment. The environmental safety system during the COVID-19 pandemic is directed to manage so that the community obeys the health protocols set by WHO and the government. The system that needs to be built is a system that is fast and precise in preventing, socializing, and reporting mechanisms when there are symptoms that are similar to the symptoms of people infected with COVID-19 [48].

4 Conclusion and Suggestion

Balla Ewako is a tough village program by the South Sulawesi Regional Police which later became a community movement in dealing with the COVID-19 pandemic based on local wisdom. This program uses Bugis-Makassarese phrases to provide understanding to the community. The *Balla Ewako* program has the goal of a strong village for health, food resilience, and security with direct community involvement. *Balla Ewako* has proven to be effective in tackling the COVID-19 pandemic by mobilizing all elements of the grassroots community. So far, *Balla Ewako* has proven to be capable of tackling the COVID-19 pandemic, because its implementation is driven by social capital that already exists and is strong in society.

References

1. Dowling, S. How global outbreaks are contained. Retrieved From https://www.bbc.com/future/article/20200422-how-disease-outbreaks-are-contained. 2020. accessed 23 Mei 2021.
2. Putri, R. H. Mass Killing Disease Outbreaks. Retrieved from https://historia.id/sains/articles/wabah-wabah-disease-killer-massal-P7eL5. 2020. accessed 23 May 2021.
3. Rusdi. Diachronics, 20(1), (2020).
4. Harriyadi. Disease Outbreaks in Historical Records in Indonesia. National Archaeological Center. Retrieved from http://arkenas.kemdikbud.go.id/contents/read/article/67ihzv_1586426994/wabah Disease-dalam-Note-Sejarah-di-Indonesia#gsc.tab=0. (2020). accessed 23 May 2021.
5. World Health Organization. *Critical preparedness, readiness, and response actions for COVID-19.* (World Health Organization, Geneva 2020).
6. T. Liu, J. Hu, M. Kang, L. Lin, H. Zhong, J. Xiao, G. He, T. Song, Q. Huang, Z. Rong, A. Deng, W. Zeng, X. Tan, S. Zeng, Z. Zhu, L. Li, D. Wan, J. Lu, H. Deng, J. He, and M. Ma, bioRxiv preprint. (2020).
7. M. N. K. Boulos, and E. M. Geraghty, Int'l J of Health Geo, **19**(8). (2020)
8. Y. Estriyanto, DEDIKASI: Commun Service Reports, **2**(2), (2020)
9. E. G. Samudro, and M. A. Madjid, J National Resilience, **26** (2), (2020)
10. Y. Sariwaty, and D. Rahmawati, J Syntax Transformation, **1**(291–296), (2020)
11. S. Susilawati, R. Falefi, and A. Purwoko, Budapest International Research and Critics Institute (BIRCI-Journal): Humanities and Social Sciences, **3**(2), (2020)
12. D. R. Buana, GREETINGS: J Sos and Syar-i Culture, 7(3), (2020).
13. B. Yanti, H. Priyanto, and T. Zulfikar, Martabe: J Community Service, 3(1), (2020)
14. M. F. Shodiq, SALAM: J Sos dan Budaya Syar-i, **8**(2), (2021)
15. Takiddin, Sos Didaktika: Social Science Education J **1**(2), (2014)
16. S. Mawaddahni, Local Wisdom, **9**(1), (2017)
17. D. R. Manrique, T. Völker, J. Zoghbi, and A. G. Pereira, *Arctic: Traditional Knowledge, Livelihoods and Community Engagement.* (Publications Office of the European Union, Luxembourg, 2018)
18. Fajarini, U, Sos Didaktika: Social Science Education J, **1**(2), (2014).
19. M. Nazir, *Metode Penelitian*, (Ghalia Indonesia, Bogor, 2014)
20. M. B. Miles, A. M. Huberman, and J. Saldana, *Qualitative Data Analysis, A Methods Sourcebook, Edition 3,* (Sage Publications, USA, 2014)
21. C. C. Lai, T. P. Shih, W. C. Ko, H. J. Tang, and P. R. Hsueh, Int'l J Of Antimicrobial Agents, **55**(3), (2020)
22. H. Lu, C. W. Stratton, Y. W. Tang, J of Medical Virology, **92**(4), (2020)
23. D. Handayani, D. R. Hadi, F. Isbaniyah, E. Burhan, and H. Agustin, J Respirologi Indonesia, **40** (2), (2020)

24. F. K. Ayittey, C. Dzuvor, M. K. Ayittey, N. B. Chiwero, and A. Habib, J of Medical Virology, **92** (4), (2020)
25. M. A. Shereen, S. Khan, A. Kazmi, N. Bashir, and R. Siddique, J of Advanced Research, **24**, (2020)
26. I. Wahidah, N. F. S. Hartono, M. A. Septiadi, R. Athallah, and M. C. A. Rafqie, J Manajemen dan Organisasi, **11**(3), (2020)
27. Karyono, Rohadin, and D. Indriyani, J Kolaborasi Resolusi Konflik, **2**(2), (2020)
28. R. Djalante, J. Lassa, D. Setiamarga, A. Sudjatma, M. Indrawan, B. Haryanto, C. Mahfud, M. S. Sinapoy, S. Djalante, I. Rafiana, L. A. Gunawan, G. A. K. Surtiari, and H. Warsilah, Progress in Disaster Science, **6**, (2020)
29. T. S. Edriani, A. Rahmadani, D. M. M. Noor, Indonesian J of Applied Mathematics, **1**(2), (2021)
30. J. E. Nelwan, J of Public Health and Commun Medicine, **1** (2), (2020)
31. C. J. Neiderud, and MD. Infection Ecology & Epidemiology, **5** (1), (2020)
32 E. Alirol, L. Getaz, B. Stoll, F. Chappuis, and L. Loutan, The Lancet Infectious Diseases, **11** (2), (2011)
33 H. Harapan, N. Itoh, A. Yufika, W. Winardi, S. Keam, H. Te, D. Megawati, Z. Hayat, A. L. Wagner, and Mudatsir, J Infect Public Health. **13** (5), (2020)
34. N. Mona, J Sos Humaniora Terapan, **2** (2), (2020)
35 N. K. Agusintadewi, *Building a Culture of Disaster Response Based on Local Wisdom in Indonesia, in National Seminar on Architecture and Spatial Planning, Samarta, 2019, Bali, Indonesia* (2019)
36. B. Prasetyo, *Kearifan Lokal sebagai Basis Mitigasi Bencana.* Seminar Nasional FST-UT 2019, Jakarta, Indonesia, (2019)
37. C. T. Mukuruva, *Community-Based Emergency Management: A Case Study on a Cholera Outbreak in Zimbabwe,* Thesis, (Auckland University, Auckland, 2012).
38. K. J. Farkas, and J. R. Romaniuk, Society Register **4** (2), (2020)
39. B. Yanti, H. Priyanto, and T. Zulfikar, Martabe: J Pengabdian Kepada Masyarakat, **3**(1), (2020).
40. S. Lardo, J Pertahanan & Bela Negara, **10**(1), (2020)
41. S. Notoatmodjo, Health Promotion and Health Behavior (Revised Edition), (Rineka Cipta, Bandung, 2012)
42. Bahagia, B. Hudayana, R. Wibowo, and Z. Anna, Forum Geografi, **34** (2), (2020)
43. Z. Aidha, Z. and R. A. Harahap, Tropical Public Health J, **1** (1), (2021)
44. R. Putnam, Bowling Alone: The Collapse and Revival of American Community, (Simon and Schurster, New York, 2000)
45 J. S. Coleman, Social Capital in Creation of Human Capital, (University of Chicago Press, Chicago, 1989)
46. J. Field, Social Capital (translated by Nurhadi), (Kreasi Discourse, Bantul, 2010)
47. Hasbullah. Social Capital (Towards Excellence in Indonesian Human Culture). (United Press. Jakarta, 2006)
48. S. Personal, J National Security, 7(2), (2020)

The COVID-19 pandemic in the perspective of Indonesia's national security

Mir'atul Azizah[1*], *Adi* Subiyanto[1], *Sugeng* Triutomo[1], and *Lexi* Jalu Aji[1]

[1] Disaster Management Study Program, National Security Faculty, Republic of Indonesia Defense University, Indonesia

Abstract. Corona Virus Diseases 2019 or COVID-19 is a global problem and become a health problem in Indonesia seriously. Currently, in Indonesia, till June 27, 2021, the number of cases has reached 2,115,304, with 57,138 deaths and 1,850,481 healing. Every day the issues of COVID-19 tend to increase. The COVID-19 pandemic affects all sectors. It is the political, economic, socio-cultural, defense and security sectors. Generally, this study analyzes COVID-19 from the national security perspective, in which one of the elements is human security. This study used a qualitative research method with a descriptive analysis approach to answering COVID-19 research problems from a national security perspective. This study indicated that COVID-19 impacts national security, military, political, economic, social, environmental, and human security. The government has released the regulations that arrange the community to mitigate the spread of COVID-19. The government has issued a vaccination policy to protect the public from COVID-19. The conclusion of this study, COVID-19 affects national security, both general security, and human security. The entirety handling of COVID-19 will enhance the sense of security of each citizen, especially in terms of human security.

1 Introduction

Corona Virus Diseases 2019 or COVID-19 is a global problem, including in Indonesia. The Indonesian government has even designated COVID-19 as a non-natural national disaster by Presidential Decree Number 12 of 2020 [1].

The number of world COVID-19 cases until June 27, 2021, reached 181,870,356, with the number of cases dying 3,939,471 and recovering 166,382,287. Indonesia is one of 17 countries with the highest COVID-19 cases in the world. The number of cases in Indonesia currently reaches 2,115,304 with 57,138 deaths and 1,850,481 recoveries [2]. The rate of increase in COVID-19 cases cannot be said to be able to be suppressed. The trend of COVID-19 cases that continues to increase is an indication.

The impact of COVID-19 is vast, occurring in various fields of life [3] both in the political, economic, socio-cultural, defense and security areas. It will undoubtedly affect the stability of the national.

The conception of national security is a condition that creates a sense of security, comfort, peace, tranquillity, and order. Security refers to state security in the face of military threats, public security, and safety from non-military threaten [4]. Security is currently developing, not only military security but also non-military security, one of which is human security.

This study aims to analyze the response to the COVID-19 pandemic from a national security perspective. Then one of the elements is human security that will explain how this element influence life because of COVID-19.

2 Methods and Data

This study uses a qualitative method with a qualitative descriptive design. The source of the data used is primary data with in-depth interview techniques to the sources/informants. The study used secondary data in data/information from official reports from related agencies and literature studies.

3 COVID-19 in Indonesia

The first COVID-19 entered Indonesia, starting when two Indonesian citizens were exposed to COVID-19 after direct contact with foreign nationals visiting Indonesia. The Indonesian government also officially determined that COVID-19 had entered Indonesia on March 2, 2020 [5]. The government has developed strategies to protect the public from COVID-19, although the spread and transmission of COVID-19 in Indonesia are unstoppable. As previously stated, Indonesia is on the list of 17 countries with the highest COVID-19 cases globally, even number 1 in the Southeast Asia Region. Indonesia's

* Corresponding Author: miartul.azizah.sumitro@gmail.com

territory and high population are factors in the high number of COVID-19 cases in Indonesia. The wider the area and the more residents, the higher the risk of the COVID-19 case.

Table 1. Number of COVID-19 casesin every Province of Indonesia as of June 6, 2021.

Province	Number of Cases
Jakarta	435,135
West Java	319,761
Central Java	205,232
East java	156,326
East Kalimantan	72,114
Riau	63,145
South Sulawesi	62,347
Banten	50,042
Bali	47,472
Yogyakarta	46,183
West Sumatera	46,005
South Kalimantan	35,043
North Sumatera	32,528
South Sumatera	25,213
Central Kalimantan	23,090
Papua	20,473
Bangka Belitung Islands	19,126
Lampung	18,735
Riau Island	18,467
East Nusa Tenggara	16,694
Aceh	16,065
North Sulawesi	15,816
Central Sulawesi	12,971
North Kalimantan	12,376
West Nusa Tenggara	12,353
West Kalimantan	11,504
Southeast Sulawesi	10,538
Jambi	10,104
West Papua	9,397
Bengkulu	8,340
Maluku	7,793
West Sulawesi	5,579
Gorontalo	5,523
North Maluku	4,548

The cases of COVID-19 are increasing day by day, even new variants are emerging. It causes the number of COVID-19 instances to increase. The World Health Organization has also established a SARS-CoV-2 Risk Monitoring and Evaluation Framework to identify, monitor, and assess evolving variants. Surveillance, research on variants of concern, and impact evaluation on diagnostics, therapy, and performing vaccines [6].

The distribution of the number of COVID-19 cases in Indonesia in each province is different. The number of COVID-19 instances tends to be proportional to the population density of each area. The number of COVID-19 cases in each province in Indonesia as of June 6, 2021, is shown in Table 1.

Based on Table 1, it can be seen that the province with the most COVID-19 cases in Jakarta with 435,135 cases, followed by West Java with 319,761 and Central Java with 205,232. The area with the lowest case was North Maluku, which was 4,548. It is proportional to the total population density in the province. The population density in 2019 in Jakarta was 15,900 people/km^2, West Java was 1,394 people/km^2, Central Java was 1,058 people/km^2, and North Maluku was 39 people/km^2 [8].

COVID-19 is transmitted through humans, so interactions between humans will increase the risk of COVID-19 transmission. Jakarta Province has the highest population density compared to other provinces in Indonesia, so it is only natural that the highest COVID-19 cases are in the Jakarta area.

A virus that circulates widely in a population and causes infection, the probability of viral mutation increases. The opportunity for the virus to spread is even more and more and more the virus is replicating and undergoing changes [6].

4 Policies for Handling COVID-19 in Indonesia

The Indonesian government issued various policies. The policies made are regulations that regulate and limit the public to prevent and reduce the risk of transmission and spread of COVID-19, including health services and vaccinations to all levels of society.

4.1 Regulation

The Indonesian government has issued various regulations since the emergence of COVID-19. The existing rules are shown in Table 2 based on CHCNER [9].

Table 2 shows that the government has issued product regulations since COVID-19 came to Indonesia. These regulations relate to health protocols, health services, education, economy, culture, politics, transportation, and so on. The government is concerned about the community's health and safety, so most of everything is regulated in regulations related to COVID-19. Regulation will be optimal if supported by people who understand the function and implement the rule correctly.

Table 2. Indonesian Government regulations related to COVID-19 in 2020

No	Regulation	Date
1	The President of the Republic of Indonesia Decree Number 7 of 2020 concerning the Task Force for the Acceleration of Handling COVID-19	March 19, 2020
2	The President of the Republic of Indonesia Decree Number 9 of 2020 concerning Amendments to the Decree of the President of the Republic of Indonesia Number 7 of 2020	March 20, 2020
3	Government Regulation Number 21 of 2020 concerning Large-Scale Social Restrictions in the Context of Accelerating Handling of COVID-19	March 31, 2020
4	Presidential Regulation of the Republic of Indonesia Number 54 of 2020 and its Attachments regarding Changes in Posture and the 2020 Draft State Budget	April 3, 2020
5	Regulation of the Minister of Health of the Republic of Indonesia Number 9 of 2020 concerning Guidelines for Large-Scale Social Restrictions in the Context of Accelerating Handling of COVID-19	April 3, 2020
6	Regulation of the Minister of Transportation of the Republic of Indonesia Number 18 of 2020 concerning Transportation Control in the Context of Preventing and Preventing the Spread of COVID-19	April 9, 2020
7	Decree of the President of the Republic of Indonesia Number 12 of 2020 concerning the Designation of Non-Natural Disasters COVID-19 as a National Disaster	April 13, 2020
8	Regulation of the Minister of Transportation of the Republic of Indonesia Number 25 of 2020 concerning Transportation Control During Eid Al-Fitr Homecoming Year 1441 H in the Context of Prevention and Prevention of the Spread of COVID-19	April 23, 2020
9	Presidential Regulation of the Republic of Indonesia Number 82 of 2020 concerning the Committee for Handling COVID-19 and National Economic Recovery	July 20, 2020
10	Presidential Regulation of the Republic of Indonesia Number 99 of 2020 concerning Vaccine Procurement and Vaccination Implementation in the Context of Coping with COVID-19	October 6, 2020
11	Regulation of the Minister of Health of the Republic of Indonesia Number 84 of 2020 concerning the Implementation of Vaccination in the Context of Coping with COVID-19	December 14, 2020

The regulatory function [10] controls social [11]. The regulatory function will run well if the human resources who perform the regulatory function understand and are skilled [12]. Local governments also adjust to the regulations made by the central government. They socialize with the public either directly, the media, or other parties so that all the rules reach the community.

4.2 Health Services

Health services during the COVID-19 pandemic are very important. Through the Ministry of Health, the government has appointed several hospitals in each region as referral hospitals for COVID-19 patients. The government wants the people who need health services, especially those related to COVID-19, can be handled quickly and increase their chances of recovering.

Jakarta currently has 106 referral hospitals for COVID-19 [13]. At the end of last year, there were 90 referral hospitals [14]. It is an indicator that more and more COVID-19 patients in Jakarta need treatment. As the area with the most COVID-19 cases, Jakarta is trying hard to keep its people safe from COVID-19. One of them is by adding referral hospital facilities for COVID-19.

The government has prepared to allocate a special budget for free treatment for COVID-19 patients in government-owned hospitals. The government hopes that people from all walks of life will have the same opportunity to recover with this policy.

Doctors and nurses are at the forefront of direct contact in treating patients [15]. COVID-19 attacks everyone, both the general public and medical personnel. Medical personnel directly deal with patients, so they are also at high risk of contracting COVID-19. One by one, medical personnel died due to exposure to COVID-19. They are the saviors of the patients, but on the other hand, they also cannot refuse to be exposed to COVID-19

4.3 Vaccination

Vaccination is one of the Indonesian government's efforts to protect the public from COVID-19. The vaccines used are not produced by themselves. It is also a challenge for Indonesia, which has been independent for 75 years.

The COVID-19 vaccine is expected to protect against new virus variants. The development of vaccine effectiveness is continually monitored [6]. The success of a vaccination program is the responsibility of all, whether individuals, health professionals, government, or industry [16].

Indonesians have started receiving the COVID-19 vaccine. The number who have implemented the first vaccine is 17,775,918 of the 181,554,465 targeted or 9.79%. The second vaccination was 11,197,069 people from the target of 40,349,049 targets or 27.75%. This amount is vaccination as of June 7, 2021 [7]. Vaccine priority is given in stages, starting from health workers, public service workers, and the community.

Vaccine effectiveness is affected by its coverage, access to health centers, costs, and other factors [16]. The government provides free COVID-19 vaccines to the public. Exceptions are given to pregnant women, the elderly, and people with comorbidities. The implementation of vaccination is expected to achieve all that has been targeted.

Vaccination is not a sure guarantee of being safe from COVID-19. People who have received the vaccine must still carry out health protocols, namely wearing masks, washing hands with clean water and soap, maintaining distance, not crowding and reducing mobility, and consuming nutritious food.

5 Impact of COVID-19 on National Security

One of the state's goals based on the preamble to the Constitution of the Republic of Indonesia is to protect the entire Indonesian nation. The word save has a broad meaning, including protecting its citizens from all threats, including the danger of non-natural disasters COVID-19

The COVID-19 pandemic has affected the economic sectors of countries in the world, including Indonesia. The health sector is challenged to solve this condition. Other sectors were also affected, such as politics, society, and culture. It has an impact on national defense and security.

Security is one of the focuses of attention to be achieved by all parties. It is what makes the issue of national security a vital thing in a country. State behavior in anticipating threats to its national security by using a military approach states behavior in anticipation threats to its national security by using a military approach dominates the security. Now, the concept of national security developed. Security includes five main sectors, namely (a) military security, (b) political security, (c) economic security, (d) societal security, and (e) environmental security [17].

5.1 The Military Security

Military security includes the management of the state's weapons capabilities both offensively and defensively and the state's perception of the intensity of one another [17]. The military is the primary national security and defense force, both in human resources, defense equipment, systems, and other resources.

COVID-19 is a non-military threat, so the government appoints non-military institutions. The National Disaster Management Agency (NDMA) and the Ministry of Health are the leading sectors in dealing with the COVID-19 pandemic. NDMA and the Ministry of Health are working hard and working smart to protect the public from COVID-19.

COVID-19 is not just a non-military affair. The military is still involved in dealing with COVID-19. The military's involvement in handling COVID-19 includes contributing to the provision of humanitarian assistance.

As part of the military, Indonesian National Army (INA) is at the forefront of mitigating COVID-19. It can be used as an indicator in measuring Indonesia's defense strength against the threat of disease outbreaks, including COVID-19. The TNI provides health services for COVID-19 through the deployment of health units, including the Army Strategic Reserve Command Health Battalion, Marine Health Battalion, and the Air Force Health Battalion Headquarters [18].

Indonesian National Army is also involved in enforcing discipline in the implementation of health protocols. The military's involvement in handling COVID-19 brings people's trust in the government. INA, as the main component in security and defense, makes people feel more secure.

5.2 The Political Security

Political security focuses on the stability of the state organization, the ideological system that gives legitimacy to the government [17]. At this level, the government, through the National Disaster Management Agency and the Ministry of Health, is trying very hard to free the Indonesian people from COVID-19 immediately.

Indonesian politics is still within safe limits. All Ministries/Agencies cooperate to overcome COVID-19 and its impacts.

5.3 The Economic Security

Economic security includes access to resources, finance, and markets to support good welfare and state power [17]. COVID-19 has caused a decline in economic activity such as household consumption, investment, and financial institutions in Indonesia [3].

The Indonesian government has spent a large amount of money on efforts to prevent and deal with COVID-19. The government also revised the budget several times to meet these needs.

Significant impacts occur on the global economy, including economic stability in Indonesia [19]. COVID-19 pandemic has been running for more than a year, the Indonesian economy has weakened and is now starting to rise

Indonesia's economic recovery began to be seen in the second semester of 2020 and is predicted to continue in 2021. Progress in handling COVID-19, including vaccines, global economic recovery, macroeconomic policy stimulus policies, and various policy sharpening efforts that support the optimization of the economic recovery [20].

5.4 The Societal Security

Societal security includes traditional language, culture, religion, national identity, and customs [17]. COVID-19 is spread by human intermediaries, namely through close contact, droplets when coughing, sneezing, or talking [21]. High community mobility will lead to a higher risk of COVID-19 transmission. It makes people more aware of both themselves and the environment.

The social life of the community has been affected by COVID-19. Societies that previously interacted freely changed, limiting interactions. The education system also adapts, namely online learning. Silaturahmi (meeting each other) during Ramadhan (fasting month) and Eid Al-Fitr (holy days) that are usually done in person, is now turning online. Large-scale social restrictions were also implemented in almost all regions in Indonesia.

At the beginning of the emergence of COVID-19, people were apprehensive if there were neighbors in the same area who experienced COVID-19. They are afraid of getting infected, but over time people begin to understand and even help each other to get well soon

Sufficient social support from the environment affects the desired goals [22], in this case recovering from COVID-19. Indonesian people have local wisdom "gotong royong" or a form of cooperation. The community helps each other when COVID-19 infects a neighbor or family. They provide support in the form of motivation to recover and help provide food for daily needs during self-isolation

5.5 The Environmental Security

Environmental security focuses on maintaining the environment both locally and globally as essential support for the system in which humans live [17]. COVID-19 affects the environment, both in the environment where you live, where you work, and the environment in general.

The COVID-19 pandemic has forced people to wear masks and hand sanitizers. The impact of using masks and hand sanitizers is the accumulation of medical waste in the form of used masks and hand sanitizers.

Medical waste management is critical because it can pose a high risk to human health and the environment [23]. Medical waste processing in hospitals or other medical services has become commonplace because they usually have a third party to manage. How is the waste processing, both in the household, village or sub-district, sub-district and even at the national level? It is something that needs further attention. Medical waste that is not managed correctly will certainly cause environmental damage and other impacts.

5.6 The Human Security

The five security sectors described above are military, political, economic, social, and environmental security contain human security. Human security is part of national security.

Security means being free from insecurity and threats. Safe is free from fear and lack. Anxiety is due to physical, psychological violence, abuse, or death while lacking work, food, and poor health [24]. Human security, including freedom from threats and fear of COVID-19.

Human security is the right of every citizen. They have the right to feel safe, not afraid of COVID-19 or its effects. They also have the right to feel adequate in food, health services, including work. COVID-19 has frightened people. They are scared of contracting and transmitting it. They are also scared of the worst risk if they get COVID-19, which is death. They see the updating of COVID-19 cases increasing, and the number of deaths is increasing every day. It added their fear.

Households are the sector most affected by COVID-19 [19]. Some people have lost their jobs because the places where they work have decreased demand and even lost money. Those who rely on their income from trading also experience a decrease in income or even none at all. Examples of traders in the canteen or around schools and campuses stop because education is currently carried out online.

6 Conclusion

COVID-19 has affected all aspects of people's lives globally, including in Indonesia. The impact of COVID-19 has occurred in almost all sectors. National security is one of the most important things because it is related to national stability. COVID-19 affects military, political, social, environmental, and human security. The human security aspect of the community will be threatened. COVID-19 has been affecting their lives. Society must fight harder to survive and be safer from COVID-19. The Indonesian government issued policies in regulations and vaccinations to protect the public from COVID-19 to be safer. The government has also mobilized non-military and military institutions to deal with COVID-19 jointly. These are government efforts in realizing Indonesia's national goal, which is to protect all citizens, especially from the threat of COVID-19. National security is a shared responsibility, both government and society. The government is preparing all its strategies so that the community is safe, including safe from COVID-19. We, as citizens, are supposed to comply with all government policies. One thing that needs more attention is environmental safety in managing medical waste from used masks and bottles of hand sanitizer used by the general public. Safe in the political, military, economic, social, and environmental is the dream of all of us.

References

1. Decree of the President of the Republic of Indonesia Number 12, in *2020* (n.d.)
2. Worldometers, (2021)
3. P. B. Astuti and A. S. Mahardhika, J. Inov. Ekon. **5**, 85 (2020)
4. B. Darmono, J. Ketahanan Nas. **15**, 1 (2010)
5. M. H. Sukur, B. K. Haris, and R. N. Faradillahisari, Inicio Legis **1**, 4 (2020)
6. World Health Organization, (2021)
7. Committee for Handling COVID-19 and National Economic Recovery, (2021)
8. Bureau of Statistics Center, (2021)
9. Committee for Handling COVID-19 and National Economic Recovery, (2021)
10. A. Fitri, L. Trisnantoro, and D. H. Sulistyo, J. Kebijak. Kesehat. Indones. **03**, 3 (2014)
11. A. Shleife, Eur. Financ. Manag. **11**, 442 (2005)
12. T. Sukarma, *Designing a Health Service Regulatory System in the Provinces of DIY and West Java* (2005)
13. Kompas, (2021)

14. Decree Of The Governor Of Jakarta The Capital Of Region, (2020)
15. R. N. Putri, J. Ilm. Univ. Batanghari Jambi **20**, 708 (2020)
16. M. Doherty, P. Buchy, B. Standaert, C. Giaquinto, and D. Prado-Cohrs, Vaccine **34**, 6707 (2016)
17. B. Buzan, *People, State, and Fear: An Agenda for International Security Studies in the Post-Cold War* (1991)
18. A. Octavian, IOP Conf. Ser. Earth Environ. Sci. **708**, (2021)
19. S. Susilawati, R. Falefi, and A. Purwoko, Budapest Int. Res. Critics Inst. Humanit. Soc. Sci. **3**, 1147 (2020)
20. Bank of Indonesia, *Indonesia Economic Report 2020 "Synergy to Build Optimism for Economic Recovery"* (2021)
21. M. Khan, H. Khan, S. Khan, and M. Nawaz, J. Med. Microbiol. **69**, 1114 (2020)
22. E. Tezci, F. Sezer, U. Gurgan, and S. Aktan, Anthropologist **22**, 284 (2015)
23. Z. M. Shareefdeen, J. Environ. Prot. (Irvine, Calif). **03**, 1625 (2012)
24. S. Tadjbakhsh, *Human Security: Concepts and Implications with an Application to Post-Intervention Challenges in Afghanista* (2005)

Natural disasters and COVID-19: health worker preparedness and response

Dewi Wahyuni[1*], *Syamsunasir[1]*, *Adi* Subiyanto[1], *Anissa* Nurur Rachmatika[1], and *Tri* Winugroho[1]

[1]Disaster Management Study Program, National Security Faculty, Republic of Indonesia Defense University. IPSC Sentul Area, Bogor, West Java, Indonesia

Abstract. The purpose of this study was to determine the preparedness of health workers individually and in the workplace to deal with natural disasters during COVID-19. Data collection was carried out by distributing online questionnaires through social media with a total number of respondents, namely 68 health workers. The research questionnaire was adapted from the COVID-19 Preparedness Checklist for Rural Primary Health Care and Community Settings. During the COVID-19 pandemic, health workers' personal readiness to deal with natural disasters is still lacking. Primary health care and hospitals do not yet have adequate health facilities to deal with COVID-19, and in the case of a natural disaster, the situation will worsen.

1 Introduction

As the number of natural disasters rises, so does the demand for qualified health workers. Competence in the context of disaster management activities should be acquired by education and training so that health personnel is better prepared to deal with potential disasters [1]. COVID-19 and other natural disasters are causing a lot of damage in Indonesia right now, thus steps are needed to lessen the impact in the future, such as strengthening disaster management and raising preparedness [2].

Several studies have been undertaken to determine how well health personnel are equipped to deal with natural catastrophes and COVID-19. As a way to improve disaster preparedness, disaster simulations have been conducted for health students [3]. Nurses in Hong Kong are still frequently underprepared and incompetent [4]. The same thing happened in Indonesia, where disaster preparedness and understanding of roles are still lacking. Continuous education is needed to strengthen the capacity of health workers in disaster preparedness, response, recovery, and evaluation [5].

A country requires the readiness of human resources, standard operating procedures, and health facility infrastructure to deal with the COVID-19 cases [2]. Although mitigation and readiness initiatives to combat the disease have been adopted in several countries, the procedures in place to deal with COVID-19 cases are still inadequate. Affected countries have requested assistance in dealing with COVID-19 in their respective jurisdictions [6]. This study assesses the preparedness of health workers personally and at work to deal with natural disasters during the COVID-19 pandemic.

2 Methods

The preparation of health workers to deal with natural catastrophes during the COVID-19 pandemic is investigated using a descriptive cross-sectional research approach. Data was gathered via social media-distributed online questionnaires. The questionnaire is in the form of a link that respondents can fill out. Respondents in this study were health care professionals who worked in primary health care and hospitals and were willing to take part in the research. A questionnaire adapted from the COVID-19 Preparedness Checklist for Rural Primary Health Care and Community Settings was used to collect data [7].

3 Results

This study included a total of 68 respondents from various primary health care settings and hospitals, as shown in Table 1.

Table 1 shows that 51.5% of respondents are under 30 years old, 82.4% are women, 55.9% have a bachelor's degree last education, and 69.1% work in primary health care. Health workers in this study have different educational backgrounds

* Corresponding author : dewiwhyunii@gmail.com

https://doi.org/10.1051/e3sconf/202133101007

Table 1. Demographic characteristics of study participants (N=68)

Characteristics	n	%
Age (years		
- < 30	35	51.5
- > 30	33	48.5
Gender		
- Male	12	17.6
- Female	56	82.4
Educational level		
- Diploma	30	44.1
- Bachelor	38	55.9
Profession		
- Doctor	5	7.4
- Nurse	22	32.4
- Public Health	13	19.1
- Others	28	41.1
Agency		
- Primary Health Care	47	69.1
- Hospital	21	30.9

Table 2. Health Workers' Preparedness

Survey items	Yes		No	
	n	%	n	%
Are you aware of the disaster risks that may occur in the area where you live and work?	63	92.6	5	7.4
Have you prepared a survival kit (a bag containing items needed in case of a disaster) at home?	21	30.9	47	69.1
Have you prepared a survival kit (a bag containing items needed in case of a disaster) at work?	41	60.3	27	39.7
Have you planned a gathering point for family members in the event of a disaster?	29	42.6	39	57.4
Have you kept your valuable documents in a safe place?	57	83.8	11	16.2
Do you already have copies of essential documents that you can bring along with your survival kit in case of a disaster?	34	50	34	50
Have you ever attended disaster training?	22	32.4	46	67.6
Have you ever assisted disaster victims?	28	41.2	40	58.8
If the agency you work for holds disaster preparedness training, would you be interested in participating?	67	98.5	1	1.5

Table 2 demonstrates that 92.6% of respondents already know the risk of disasters that may occur in their region of residence and work. More than half of the respondents (69.1%) had not prepared a disaster survival kit at home in the event of a disaster, but 60.3% had prepared a survival kit at work. As many as 57.4% of respondents have not arranged a spot to congregate with their families in the event of a disaster. Respondents who have kept valuables in safety storage are 83.8%. A total of 67.6% of respondents have never participated in disaster training, 58.8% have never assisted disaster victims, and 98.5% are willing to participate in disaster training. As many as 62% of respondents will contact their families in the event of a disaster, and most of the respondents, 92.6%, stated that disaster information could be obtained through the internet (social media, official website).

Table 3. Health Facilities Preparedness in Facing COVID-19

Survey Item	n	%
Type of health facilities		
- Primary Health Care	47	69.1
- Hospital	21	30.9
Availability of disaster management guidelines		
- Yes	48	70.6
- No	20	29.4
Availability of patient handwashing facilities		
- Yes	68	100
- No	0	0
Separate waiting room for patients with COVID-19 symptoms		
- Yes	51	75
- No	17	25
Separate examination room for patients with COVID-19 symptoms		
- Yes	51	75
- No	17	25
The practice of physical distancing	48	70.6
- Yes	20	29.4
- No		
Availability of adequate PPE		
- Yes	56	82.4
- No	12	17.6
Ownership of guidelines for handling COVID-19		
- Yes	64	94.1
- No	4	5.9
Routine disinfection		
- Yes	60	88.2
- No	8	11.8

Table 3 summarizes that 69.1% of respondents work in health centers, 70.6% of respondents stated that in health facilities there are disaster management guidelines available, all health facilities have provided handwashing facilities for patients, as many as 75% of respondents stated that waiting rooms and examination rooms are between patients with COVID-19 symptoms and other patients have been separated, 70.6% of respondents stated that patients can implement physical distancing in health facilities, 82.4% stated that the available PPE was adequate, 94.1 already had guidelines for handling

COVID-19, and 88.2% had carried out routine disinfection.

Table 4 shows that 72.1% of respondents have attended seminars related to COVID-19, 67.6% have attended PPE training, 94.1% stated that they already have PPE guidelines in the workplace, 89.7% said that masks and handscoons have been facilitated adequately, 83.8% can implement physical distancing in their workplace, and 94.1% have carried out routine COVID-19 symptom checks.

Tabel 4. Health Worker Safety

Survey Item	n	%
Experience attending seminars related to COVID-19		
- Yes	49	72.1
- No	19	27.9
Experience in PPE training		
- Yes	46	67.6
- No	22	32.4
Availability of PPE guideline		
- Yes	64	94.1
- No	4	5.9
Availability of adequate masks and handscoons		
- Yes	61	89.7
- No	7	10.3
The practice of physical distancing in the health facility		
- Yes	57	83.8
- No	11	16.2
Routine COVID-19 symptom detection		
- Yes	64	94.1
- No	4	5.9

4 Discussion

Disasters are complex problems that require emergency health services immediately and in the long term can trigger the emergence of public health and psychological problems for victims [8]. Health services are the essential sector during pre-disaster, emergency response, and post-disaster because humans are the most important element to be saved in the event of a disaster [9]. In areas especially prone to disasters, it is necessary to prepare professional health personnel and supporting medical equipment to increase the capacity to deal with disasters, both in hospitals and community health centers [10]. Hospitals and primary health care are health service providers who have an important role in the event of a disaster. The earthquake in Nepal in 2015 damaged health care centers, which were needed at the time, making basic health services difficult to access [11].

The results of this study indicate that most health workers already know the risk of disasters that may occur in their area. Less than half of the respondents have prepared survival kits at home, but most have prepared them at work. Most of the respondents have not determined the point of gathering with their families, more than half have stored valuables in a safe place. Less than half have attended disaster training and assisted disaster victims. It can be concluded that the personal preparedness of health workers in this study is still low.

Health workers and other support staff in a health facility are not fully prepared to deal with disasters. Until now, the most effective method for preparing health workers for disasters is not known and evaluation is still needed regarding their willingness to assist in the event of a disaster [12].

Khalaileh et al., (2012) stated that only 42% of nurses were aware of the disaster management plan in their workplace. Most of the respondents are not aware of the level of preparedness of the health system in their workplace to deal with disasters. There is no Jordanian national action to prepare nurses for disasters. Workshops have been held and several universities have included disaster preparedness in the curriculum but no evaluation has been carried out on the extent of nurses' understanding of disaster management [13]. The same results were found in the research of Rassin et al., (2007), the level of disaster preparedness is still low, while early preparation significantly affects the health care system to respond to disasters [14].

The capacity of the primary health care system in rural Vietnam is inadequate to respond to health problems caused by storms and floods [15]. In central Vietnam as many as nine out of ten primary health facilities have a plan for dealing with floods, and four facilities receive regular flood preparedness training. Six facilities reported a lack of support for disaster response [16].

Currently, Indonesia is faced with an increase in COVID-19 cases and ranks fourth with the most positive confirmed cases in Asia [17]. Data from the Indonesian Medical Association shows that until March 2, 2021, the number of health workers who died from COVID-19 was 718 people [18]. The ratio of the number of doctors to the total population of Indonesia is 0.38 per 1000 population. Availability of hospital beds in Indonesia 1.2 per 1000 population in 2015 [19]. Based on this condition, it can be concluded that Indonesia does not yet have adequate health facilities to deal with COVID-19 and it will be even more difficult if another natural disaster occurs.

Not all health facilities provide separate rooms for patients with COVID-19 symptoms. There are still health facilities that find it hard to implement physical distancing, both patients and health workers. This is possible because of the limited area of health facilities. Until now, there are still health facilities that do not have adequate PPE and not all officers have attended PPE training. This condition is a factor causing the high number of positive confirmed cases among health workers.

5 Conclusions

This study explored the preparedness of health workers personally and at work to deal with natural disasters during the COVID-19 pandemic. The study shows Indonesia does not yet have adequate health facilities and health workers to deal with COVID-19 and it will be even more difficult if another natural disaster occurs.

 https://doi.org/10.1051/e3sconf/202133101007

References

1. A. Y. Loke, O. W. M. Fung, Nurses competencies in disaster nursing: Implications for curriculum development and public health. Int J Environ Res Public Health, **11**, 3289-303 (2014).
2. M. I. Firmansyah, F. Rahmanto, D. Setyawan, The preparedness for the COVID-19 pandemic management in Indonesia, JAKI, **8**(2), 188-201 (2020).
3. B. G. Kaplan, A. Connor, E.P. Ferranti, L. Holmes, L. Spencer, Use of an emergency preparedness disaster simulation with undergraduate nursing students, Public Health Nurs, **29**(1), 44-51 (2012).
4. O. W. M. Fung, A. Y. Loke, C. K. Y Lai, Disaster preparedness among Hong Kong nurses, J Adv Nurs, **62**(6), 698-703 (2008).
5. Martono, Satino, Nursalam, F. Efendi, A. Bushy, Indonesian nurses perception of disaster management preparedness, Chinese Journal of Traumatology, **22**, 41-6 (2019).
6. A. A. Thobaity, F. Alshammari, Nurses on the frontline against the COVID-19 pandemic: An integrative review, **3**, 87-92 (Dubai Med J. 2020).
7. COVID-19 PHC Action Group, COVID-19 preparedness checklist for rural primary health care and community settings (2020).
8. D. E. Hogan, J. L. Burstein, General concepts: basic perspectives on disasters. In Hogan, D.E., Burstein, J.L. (Eds.), Disaster Medicine. second ed. Philadelphia: Wolters Kluwer & Lippincott Williams & Wilkins (2010).
9. United Nations. Health and Disaster Risk Reduction http://www.wcdrr.org/uploads/HEALTH.pdf (2014)
10. A. Boyd, N. Chambers, S. French, D. Shaw, R. King, A. Whitehead, Emergency planning and management in health care: priority research topics, Health Systems, **3**, 83-92 (2014)
11. M. L. Hall, A. C. K. Lee, C. Cartwright, S. Marahatta, J. Karki, P. Simkhada, The 2015 Nepal earthquake disaster: lessons learned one year on, Public Health, **145**, 39-44 (2017).
12. J. R. Gowing, K. N. Walker, S. L. Elmer, E. A Cummings, Disasters preparedness among health professionals and support staff: what is effective? An integrative literature review, Prehospital and Disaster Medicine, **32**(3), 1-8 (2017).
13. M. A. A. Khalaileh, E. Bond, J. A. Alasad, Jordanian nurses' perceptions of their preparedness for disaster management. International Emergency Nursing, **20**, 14-23 (2012).
14. M. Rassin, M. Avraham, A. Nasi -Bashari, S. Idelman, Y. Peretz, S. Morag, et al, Emergency department staff preparedness for mass casualty events involving children, Disaster Management and Response, **5**(1), 36-44 (2007).
15. H. V. Minh, T.T. Anh, J. Rocklov, K. B. Giang, L. Q. Trang, K. G. Sahlen, et al, Primary healthcare system capacities for responding to storm and flood-related health problems: a case study from a rural district in central Vietnam, Global Health Action, **7** (2014).
16. A. Alga, T. A. T. Dang, D. D. Saulnier, G. T. Nguyen, Jv. Schreeb, Hope for the best. prepare for the worst-an assessment of flood preparedness at primary health care facilities in Central Vietnam Int J Environ Res Public Health, **15**, 1-8 (2018).
17. WorldoMeters. Reported Cases and Deaths by Country or Territory [Available from: https://www.worldometers.info/coronavirus/ (2021) [Accessed June 18, 2021].
18. D. Prajogo, A. Sudiarno, Sohal A, Maryani A, Rahman A, Dewi DS, et al. Discrepancies in facilities and services among hospitals in Indonesia increase COVID-19 risks among health workers [Available from: https://theconversation.com/discrepancies-in-facilities-and-services-among-hospitals-in-indonesia-increase-covid-19-risks-among-health-workers-157336 (2021) [Accessed June 18, 2021].
19. Deloitte. Rising to the COVID-19 health care challenge in Indonesia [Available from: https://www2.deloitte.com/id/en/pages/life-sciences-and-healthcare/articles/rising-to-covid-19-health-care-challenge-in-indonesia.html (2020) [Accessed June 18, 2021].

Positive impacts among the negative impacts of the COVID-19 pandemic for community life

Aam Amirudin[1]*, *M.* Syamsul[1], *Christin Sri* Marnani[1], *Nadiva Awalia* Rahmah[1], and *Wilopo*[1]

[1]Disaster Management Study Program, National Security Faculty, Republic of Indonesia Defense University. IPSC Sentul Area, Bogor, West Java, Indonesia

Abstract. The Covid-19 pandemic is a global disease outbreak that quickly spreads throughout the world and transmits from one person to another so easily and is a form of danger that has the potential to threaten all aspects of people's lives. These negative impacts include social, economic, health, and psychological impacts, even to the point of threatening national defense and security, due to the limited space for human movement in carrying out daily activities. But there are also positive influences that we unconsciously admit, not only producing clean air but making individuals mentally strong and changing personalities for a better life. Adaptation of new life patterns from before makes people learn about the meaning of maintaining cleanliness in the face of a pandemic with new habits in maintaining personal hygiene, family, and the surrounding environment. The long-term impact on society is to produce people who are tough in facing life by taking advantage of existing opportunities into innovations in running life with the emergence of creative businesses as a new source of income in meeting the needs of families today and in the future.

1 Introduction

The Covid-19 pandemic has greatly affected most of the world's people's lives, all aspects of life such as aspects of ideology, politics, economy, social, culture, defense, and national security with the Covid-19 pandemic. All of this happened because the spread of Covid-19 was very fast and massive, so the transmission rate was so large with a very worrying effect. as evidenced by the positive confirmation numbers and large mortality rates in each affected country, both developed countries, developing countries, especially poor countries with low levels of knowledge, sanitation, and care, thus adding to the very high vulnerability of the spread of Covid-19.

Countries in the world are so intense in tackling the Covid-19 pandemic, all ways are done so that their people avoid the spread of the Covid-19 pandemic, the methods used are based on pathogenic characteristics, epidemiological characteristics, and clinical characteristics. These three characteristics can be used as a reference to find out if the pandemic is spreading to healthy humans and in the end, it will be known how to deal with it.

Good response efforts will also have a significant impact on people's lives. Good cooperation is needed between all parties, the Government as policymakers and the community as policy implementers will be interrelated to the success of overcoming the Covid-19 Pandemic in a country. Policy Strict health protocols are very popular policies in every country in the world. Wearing masks, keeping a distance by staying away from crowds and contact with sick individuals and washing hands are real efforts that are highly recommended in efforts to overcome the Covid-19 Pandemic [1].

The application of strict health protocols in each country has an impact on the pattern of people's daily lives, efforts to maintain a distance of 1-2 meters between fellow humans lead to restrictions on activities outside the home. This is done to avoid interaction between humans because droplets from the human nose or mouth as a nesting place for Covid-19 are feared to spread to various places, either directly or indirectly, that will reach the surrounding humans [2].

Restrictions on community activities in daily activities lead to a decrease in work productivity, especially in business fields that depend on workers, the impact is that many employees are laid off. Many conventional traders lose out because of the lack of buyers and many other business fields are affected so that the income they usually get is reduced or even non-existent. This causes many in the community to become the new poor [3].

* Corresponding author: aamonusco130519@gmail.com

The Covid-19 pandemic has also caused community resilience to be high because difficulties become a challenge for people to survive, causing innovations to emerging in all respects, including in creating their business world. Supported by advances in science and technology with the internet which makes it easy for people to access various kinds of information. This is very helpful to get out of the difficult zone in the era of the Covid-19 Pandemic

From the explanation above, the author is interested in raising the theme "Positive impacts among the negative impacts of the Covid-19 Pandemic on people's lives" with the hope that this paper can be a motivation for the wider community that in trouble there is always wisdom and a way to make us at ease.

2 Methods

The research method used is a descriptive qualitative approach based on a literature review and analysis of several references. The references used in this study serve as a basis for providing valid information to discuss and draw conclusions on the positive impacts among the negative impacts of the Covid-19 Pandemic on people's lives. This study uses the qualitative data analysis technique of Miles and Huberman (1994), whose process consists of data reduction, data presentation, and conclusion drawing and verification [4]. The data reduction process in this study was carried out by making choices about which parts of the data were coded and discarded. In the data reduction process, qualitative data can be simplified and transformed in various ways, such as through strict selection, through a summary or brief description, through data classification in a broader pattern, and so on. The presentation of data is done by developing a structured set of information that will lead or produce conclusions. This process is also carried out to further improve the understanding of this research and serve as a reference for taking action. By analyzing the data that has been displayed, we will be able to understand what is going on and what to do based on our understanding of the data.

3 Results and Discussion

Global Covid-19 Pandemic cases are still showing a significant increase, data shows that there is still a high rate of increase in the number of confirmed patients as of May 17, 2021. According to data from Worldometers, the total number of confirmed cases of Covid-19 worldwide has reached positive numbers. as many as 163,694,333 cases, 3,392,634 people died and 142,144,554 people were declared recovered. From the data of 10 countries with the highest number of Covid-19 in the world, the USA is the country with the highest number of infection cases, followed by India, Brazil, France, Turkey, Russia, UK, Italy, Spain, and Germany [5].

The Covid-19 pandemic has a very complex impact on every individual life or relationship between individuals. The very fast transmission of the virus requires that efforts be made to limit activities or isolate from the outside world to break the chain of transmission, causing depression, anxiety conditions which are also associated with instability, severe emotional trauma, and economic recession because people cannot carry out production, distribution, and distribution activities. consumption and has a major impact on the structure of society that causes a social change in society [6]. UNCTAD predicts the global economy is expected to contract a staggering 4.3% in 2020. Millions of jobs have been lost, millions of livelihoods are threatened, and it is estimated that an additional 130 million people will live in extreme poverty if the crisis continues [6].

A Covid-19 pandemic is an unexpected event that comes suddenly beyond the reach or ability of humans. unintended social consequences. The phenomenon of Covid-19 certainly has a very complex impact on every life or relationship between individuals in humans. The psychological and social impact is influenced by people's fear of becoming sick. People's suspicion will lead to an uncertain social situation, people's attitudes become more protective of the surrounding environment, they are more suspicious of people who cough, sneeze, or people who look sick. From this stigma, finally, a discriminatory attitude emerged by not wanting to help other people by direct physical contact with people suspected of contracting the coronavirus [8].

Public suspicion will lead to an uncertain social situation, people's attitudes become more over-protective of their surroundings, they are more suspicious of people who are coughing, sneezing or people who look sick. From this stigma, a discriminatory attitude finally emerged, for example not wanting to help others in direct physical contact with people suspected of having the coronavirus, refusal of a nurse in the neighborhood because the nurse working in a hospital that accepts Covid-19 patients, refusal to bury the bodies of Covid-19 victims. -19 by several residents, even throwing stones at the medical officers who are tasked with carrying the bodies of Covid-19 victims by ambulance.

From an economic aspect, it is closely related to the economic adequacy of the community, which will be achieved if the income can cover their daily needs, with the Covid-19 Pandemic many business sectors are experiencing shocks, according to Small Business Report data by Facebook, OECD, and the World Bank, which surveyed nearly 30,000 businesses in 50 countries, said 61% of small and medium enterprises (SMEs) in the Asia Pacific had reported a drop in sales due to the Covid-19 pandemic [9].

The industrial, tourism, trade, and other services sectors are also greatly affected because these sectors are one of the forums that provide job opportunities for the community, but due to the Covid-19 pandemic, many people who depend on work in these sectors are missing, the impact is Layoffs (Termination of Employment). Data from the UNWTO showed that tourist arrivals in March fell by 57% as lockdowns began in many countries, and extensive travel, airport closures, and national borders. This means 67 million international arrivals and around US$ 80 billion in revenue, 100 to 120 million jobs will be at risk [10].

In addition, the Covid-19 pandemic has a positive impact on some citizens in the world, many countries in the world impose lockdowns or restrictions on community activities so that all activities ranging from work, school, shopping, to other activities become more at home, it makes us become spending more time with family, people in various circles are also competing to donate to help people in need. This is social solidarity that is unconsciously embedded in society [11].

Besides that, the condition of the earth is getting better due to reduced human activity so that the level of pollution is reduced due to the decrease in fossil fuel consumption by 435,000 barrels globally in the first 3 months of 2020 [12], for example, the PRC can cut 1,600 metric tons of CO2 or the equivalent of above 4%. of total Global CO2 production in 2019 [13].

The Covid-19 pandemic has affected business growth in various sectors, although some of them have lost money and even gone bankrupt but several successful business opportunities have made big profits. Launching from CNN Business, there are several lists of businesses that thrive in the USA during 2020 and are predicted to grow rapidly in the future, namely: 1) Housing increases 23%; 2) Energy based on solar power increased by 300%; 3) Bitcoin increases 3x; 4) Video Games increased by 100 billion game content broadcast hours; 5) Large retailers increased by 19%; 6) Technologies like Zoom skyrocketed by 307%; 7) Digital content increased by 134%; 8) Stream like Warner Bros. that thrives with its films [14].

Disruption or drastic changes due to the COVID-19 pandemic has forced the public, business actors, and the government to carry out various digital innovations. These measures are carried out, among others, through distance learning, online public services, the development of online surveys, and telemedicine development [15].

Changes in lifestyle habits due to the Covid-19 Pandemic have caused a shift in the behavior of the world community, due to limited movement, this is an opportunity to create new innovative business fields as according to the Board of Innovation (BoI) there are several changes in human behavior as a result of the Covid-19 Pandemic, namely: 1). Excessive anxiety; 2). Does not believe in cleanliness outside of itself; 3) Limiting traveling; 4). Work from home; 5) escalation of conflict increases; 6). Unemployment is increasing; 7). Accustomed to sending goods; 8). Limited interaction with others; 9). Self-identity is more important than job identity; 10). Consumer Immunity as Priority [16].

From this pattern of behavior change, people use it to become business opportunities, especially online businesses, which have increased greatly during the Covid-19 Pandemic, considering that online business can minimize interactions between humans without leaving their essence to earn income to support daily needs, according to data from the Ministry of Communication and Information. that the number of online shop transactions has increased by 400% and is predicted to continue to increase as the Covid-19 pandemic has no signs of ending [16]. In addition, thanks to the Covid-19 pandemic the world gave birth to new billionaires, according to data from the Swiss bank, UBS, during the April to July 2020 period, the wealth of the number of these billionaires increased by the US $ 10.2 trillion or around Rp144,078 trillion, UBS data also shows, the number of billionaires increased from 2,158 in 2017 to 2,189 [17].

4 Conclusion

The Covid-19 pandemic has had a major impact on the lives of people around the world in general. This influence certainly leads to a negative influence with various accompanying impacts from various aspects of human life, such as increased mortality rates, increasing unemployment rates, decreasing the number of visits to the tourism sector, and causing an economic recession because people cannot carry out production, distribution, consumption activities such as at normal times.

On the other hand, there is a positive influence with the emergence of new creativity and innovation as an adjustment effort from humans. The Covid-19 pandemic is still developing, forcing people to adapt to the current situation, which can allow us to continue to survive during the Covid-19 pandemic. Various habits that were carried out when conditions were still normal can no longer be carried out because of the fear of the Covid-19 attack. This is felt by some parties with the increasing standard of living by taking advantage of the existing situation by finding new business methods in the form of online businesses and other businesses. Besides that, with the demands of activities at home, it creates a lot of time together with family. In terms of changing behavior towards the environment, Covid-19 forces people to get used to healthy and clean living behaviors, in addition to the environment because of the limited movement of vehicles so that the air quality becomes clean and healthy too. It turns out that we can all learn that disasters make humans survive so that they give birth to resilient people who can solve all problems and get out of these problems to a better life.

References

1 Öner Özdemir, "Coronavirus Disease 2019 (COVID-19): Diagnosis and Management (narrative review)," *Erciyes Medical Journal* **42**, 3, 242–247 (2020)

2 WHO, "Rational use of personal protective equipment (PPE) for coronavirus disease (COVID-19)," last modified 2020, diakses Mei 19, 2021, https://apps.who.int/iris/bitstream/handle/10665/331498/WHO-2019-nCoV-IPCPPE_use-2020.2-eng.pdf (2020)

3 Maryla Maliszewska, "The Potential Impact of COVID-19 on GDP and Trade," *Policy Research Working Paper* (2020).

4 M. B. Miles, A.M Huberman, Qualitative Data Analysis: An Expanded Sourcebook. New York: Sage (1994).

5 Jawahir Gustav Rizal, "*Update Corona Dunia 17 Mei 3,3 Juta Orang Meninggal Dunia karena Covid-19*"(World Corona Update May 17, 3.3

Million People Died due to Covid-19). https://www.kompas.com/tren/read/2021/05/17/093407265/update-corona-dunia-17-mei-33-juta-orang-meninggal-dunia-karena-covid-19.

6 R. Perween, "Global Impact of COVID-19 Pandemic," *International Journal of Natural Disasters & Health Security*, July, 53–64, (2020)

7 UNCTAD, *Impact of the COVID-19 pandemic on trade and development: Transitioning to a New Normal., United Nations Conference on Trade and Development* (2020)

8 D. Joy Osofsky, J. Howard, Osofsky, Y. L. Mamon, "Psychological and Social Impact of COVID-19," *Psychological Trauma: Theory, Research, Practice, and Policy* 12, no. 5 (2020): 468–469.

9 Cloudhost, "Dampak Pandemi (COVID-19) Terhadap Bisnis UKM dan Startup," *Cloudhost*, last modified 2020, diakses Mei 19, 2021, https://idcloudhost.com/dampak-pandemi-covid-19-terhadap-bisnis-ukm-dan-startup/.

10 T Haryanto, "Editorial: Covid-19 Pandemic and International Tourism Demand," *Journal of Developing Economies* 5, no. 1 (2020): 1.

11 K. Dwari, "Covid-19 Pandemic: A Positive Influence on Lifestyle, Economy, Administration & Environment; An Overview," *Indian Journal of Forensic Medicine & Toxicology* 14, no. 4 (2020).

12 IEA, "Oil Market Report - March 2020," *IEA*, last modified 2020, diakses Mei 19, 2021, https://www.iea.org/reports/oil-market-report-march-2020.

13 Simon Evans, "Analysis: Coronavirus set to cause largest ever annual fall in CO2 emissions," *CarbonBrief*, last modified 2020, [Accessed May 19, 2020] https://www.carbonbrief.org/analysis-coronavirus-set-to-cause-largest-ever-annual-fall-in-co2-emissions (2020)

14 CNN Indonesia, "Daftar Bisnis Untung dan Buntung di AS Sepanjang 2020," *CNN*, last mo, https://www.cnnindonesia.com/ekonomi/20210104114852-92-589177/daftar-bisnis-untung-dan-buntung-di-as-sepanjang-2020/1. [Accessed May 19, 2020],

15 Widyati, IDK Kertawidana, " Positive impacts of Covid-19 pandemic to strengthen National Health Defense" Technium Social Sciences Journal **24**, 562-570, October (2021)

16 Bentangku, "14 Jenis Usaha Ini Diramalkan Naik Daun di tahun 2021,"(*This Business is Forecasted to Rise in 2021*), Qasir [Accessed May 19, 2020] https://www.qasir.id/inspirasi/14-jenis-usaha-ini-diramalkan-naik-daun-di-tahun-2021.

17 Gusti, "Transaksi Penjualan Online Naik 400 Persen di Masa Pandemi Covid-19," *UGM*, last modified 2020, diakses Mei 19, 2021, https://www.ugm.ac.id/id/berita/19452-transaksi-penjualan-online-naik-400-persen-di-masa-pandemi-covid-19.

18 BBC, "*Kekayaan para miliarder bertambah Rp144.078 triliun selama pandemi Covid-19, Jeff Bezos bertambah Rp1.200 triliun*,"(The wealth of billionaires increased by Rp144,078 trillion during the Covid-19 pandemic, Jeff Bezos increased by Rp1,200 trillion) *BBC News*, https://www.bbc.com/indonesia/majalah-55465990 [Accessed May 19, 2020]

Capabilities of the city government in combating COVID-19 in the city of Padang

Nia Audia[1*], *Roni Ekha* Putera[1], *Desna* Aromatica[1], and *Mizan B. F.* Bisri[2]

[1]Department of Public Administration, Faculty of Social and Political Science, Andalas University
[2] GSICS, Kobe University

Abstract. The purpose of this research is to explain the capability of the Padang city government in dealing with Covid-19, where Padang City is one of the cities in West Sumatra whose level of spread of Covid-19 is quite worrying. This is based on the increase in the number of Covid-19 cases in the city of Padang when the New Normal policy began to be implemented. The low awareness of the community to implement health protocols is considered one of the factors for the soaring positive case of Covid-19 in the city of Padang. The theory that I use in this research is Disaster Management and Local Government Capability by Bevaola Kusumasari. The data collection method is qualitative, in the form of interviews and documentation. The results of this study indicate that the capability of the Padang City government in dealing with Covid-19 shows that the City of Padang has the ability to suppress the rate of positive Covid-19 numbers. However, Padang City is still weak in terms of law enforcement in dealing with Covid-19, because there are still people who violate health protocols.

1 Introduction

The global pandemic of the corona virus disease (Covid-19) is changing the lives of everyone in the world. The disease is caused by a virus that is thought to have plagued Wuhan, Hubei Province, China in December 2019, (Adhikari et al., 2020). In Indonesia, Covid-19 began to enter on March 2, 2020, which was revealed by President Joko Widodo with the discovery of three positive cases of Covid-19 infection, (*Kompas.Com*, n.d. 2020). Almost all regions in Indonesia have been exposed to Covid-19, (*Covid19.Go.Id*, n.d.). The very high level of transmission causes this disease to have a much greater impact than diseases caused by previous corona viruses such as SARS and MERS, (Titik, 2020).

Taking into account the increase in cases of patients exposed to Covid-19, and the attitude of the public who are indifferent to the threat of Covid-19, the government of Indonesia has prepared various regulatory frameworks, institutional frameworks, and financing frameworks, which are not only in response to health emergency response conditions with the outbreak of the Covid-19 pandemic, but also in the preparation of social safety nets and the economic impact of this multidimensional disaster, (Hadi, 2020).

Based on the Covid-19 data update on February 7, 2021, it is known that positive cases of Covid-19 per province in Indonesia shows that DKI Jakarta province is the area with the highest ratio of positive cases of Covid-19. In addition to the Jabodetabek area which is the epicenter of cases of sufferers and victims due to Covid-19, several provinces on the islands of Sumatra, Bali and Banten also show quite a lot of victims infected with Covid-19. The continued increase in the number of Covid-19 patients is also due to the very high and easy transmission of the virus.

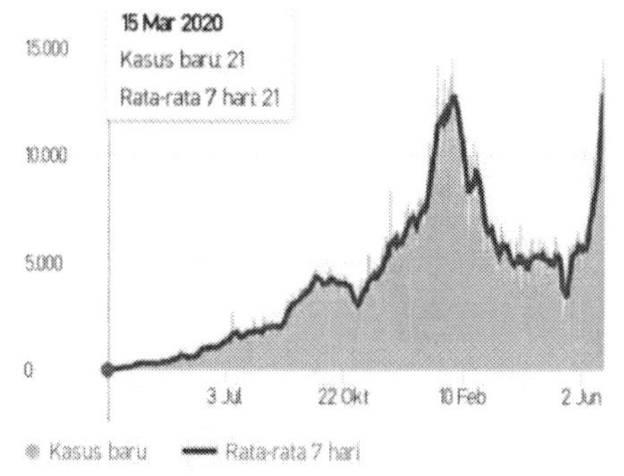

Fig. 1. Graph of Covid-19 Developments in Indonesia [6]

West Sumatra is one of the provinces that has a high number of confirmed positive cases of Covid-19 in Indonesia. Based on the corona.sumbarprov.go.id data update on September 14, 2020, it is known that the city of Padang is the area that has the highest number of Covid-19 infections. As one of the cities most affected by Covid-19 in West Sumatra Province, Padang City also has a high

* Corresponding author: niaaudia24@gmail.com

comorbidity rate. Based on the health profile of the City of Padang in 2019, the distribution of cases of comorbid patients was found in all areas in the City of Padang, so the risk of being exposed to Covid-19 was higher, (Yuniarti et al., 2020).

Table 1. Positive Number of Covid-19 in the Province of Indonesia [4]

No	Province	Positive Covid-19	Percentage
1.	DKI Jakarta	280,261	25.2%
2.	West Java	157,611	14.2%
3.	Central Java	129,228	11.6%
4.	East Java	115.071	10.4%
5.	South Sulawesi	49,166	4.4%
6.	East Kalimantan	42,924	3.9%
7.	Riau	29,211	2.6%
8.	West Sumatra	27.211	2.4%
9.	Bali	26,975	2.4%
10.	Banten	26,404	2.4%

The city of Padang is currently listed as the largest cluster of positive Covid-19 cases in West Sumatra Province, with a total of 13,819 confirmed positive cases. In addition, the number of deaths due to Covid-19 is also quite high in the city of Padang. The increase in the number of Covid-19 cases was also experienced by the City of Padang when the New Normal policy began to be implemented. The low awareness of the community to implement health protocols is considered one of the factors for the soaring positive case of Covid-19 in the city of Padang. Even though community participation is also needed in overcoming this Covid-19, which has been regulated by the Padang City government policy as a guideline for complying with health protocols when doing activities outside the home. The lack of public awareness and also the reduced role of government agencies in carrying out the Covid-19 health protocol during the new normal, resulted in the graph of positive cases of Covid-19 continuing to increase. Researchers assume that the Padang City Government does not utilize the capacity it has in dealing with Covid-19 in the new normal period.

Capability Covid-19 task forceThe city of Padang is highly demanded in controlling the increase in the number of positive cases and deaths due to being infected with Covid-19. ExistenceWest Sumatra Provincial Regulation Number 6 of 2020 concerning the Adaptation of New Habits in the Prevention and Control of Corona Virus Disease 2019, has the aim of reducing the risk and wider impact of Covid-19 in the community. This New Habit Adaptation Regulation (AKB) is mandatory, that is, it applies to districts/cities in West Sumatra. Therefore, the Padang City Government through the Satpol PP and the Padang City Covid-19 Task Force team conducted socialization to the public regarding the IMR Perda.

With the enactment of this regulation, it is hoped that the community will be disciplined in implementing health protocols, so that those who violate will be given sanctions. However, there are still many people who violate health protocols by not wearing masks at the Padang Raya market, (*langgam.id*, n.d.). This shows that a good capability is needed by the Padang City Government to control positive cases of Covid-19. Researchers assume that local governments have the greatest responsibility for dealing with Covid-19. This responsibility demands a good capability from the Padang City government.

Capabilities are a form of capability that the Padang City Government must possess in dealing with challenges and problems that occur in the dynamics and changes in the development of the Covid-19 cluster. The high number of cases infected with Covid-19 indicates that the Padang City Government has not been able to suppress the rate of positive Covid-19 numbers, which can be seen from the development of Covid-19 cases to date.

Capability regional government one of which can be seen from the institutional. To focus on tackling Covid-19, the Padang City government formed a Covid-19 Task Force. The Padang City Regional Disaster Management Agency (BPBD) as the leading sector in Padang City Disaster Risk Reduction, is part of the Padang City Covid-19 Task Force. The Padang City BPBD carries out its duties in dealing with Covid-19 as a member of the Padang City Covid-19 Task Force. Therefore kability will be a benchmark in the understanding of the Covid-19 Task Force in the response that has been implemented so far, in accordance with the conditions and challenges that are always developing in the city of Padang.

As for other than institutional, that for capability The Covid-19 Task Force must have clear roles and responsibilities that must be carried out with good coordination. Then, technicality is one of the variables in assessing the capability of the government, which in this case is regarding information systems and communication networks in dealing with Covid-19.Based on the background and problem formulation above, the research questions that arise are as follows: How is the capability city government in dealing with covid-19 in the city of Padang?, and the purpose of this study is to explain the capabilities regional government in dealing with COVID-19 in Padang City.

2 literature Review and Hypothesis Development

In various literatures, studies or the concept of organizational capability have attracted a lot of interest, especially in management research. Ability is also called the same as competence which is not only limited to having skills, but in more detail so that they really master their abilities from weak points to how to overcome them. Another opinion says that capability is the ability to exploit both the resources that are owned within and within the organization, as well as self-potential to carry out certain activities or a series of activities, (Aisyah & Purwanda, n.d.). Therefore, capability is an assessment or measure of what is done by the government to be capable.

Observing capabilities (capabilities) is perhaps the most significant structural problem in the management of todays complex organizations (Kusumasari & Alam, 2012). Where capability does not represent one resource only, without showing other resources, such as financial assets, technology or labor, but rather is a distinctive and superior way of allocating resources. In relation to disaster events, it is very important to identify the demands and demands (dynamic and developing conditions, role uncertainties and situational constraints) that characterize the disaster response environment and develop the management capabilities needed to deal with disasters.

The following are some studies that researchers consider relevant to the research topic that the researcher will do, which can be used as research reference material. The first is the research conducted by Bevaola Kusumasari and Quamrul Alam with the research title "Bridging the gaps: the role of local government capability and the management of a natural disaster in Bantul, Indonesia". This research uses descriptive qualitative research method. The results of this study are the ability of local governments and the necessity of disaster management is very important to explore the important role of local governments in disaster mitigation, preparedness, response and recovery activities, especially those related to the characteristics of local governments in developing countries. The ability of local governments to deal with disasters is defined as the function of institutions, human resources, policies for effective implementation, finance, technical resources and leadership.

The second is research conducted by Resa Vio Vani and Mayarni with the research title "The Dynamic Governance Capability of the Pekanbaru City Government in the Enforcement of New Normal During the Covid19 Pandemic" (Mayarni, 2020). The method in this study uses qualitative research methods based on post positivist or interpretive philosophy to examine the condition of natural objects. The results of the study conclude that the concept of dynamic capability is the most essential aspect in the Pekanbaru City government's efforts to achieve a competitive advantage that is able to anticipate the Covid-19 pandemic and take advantage of opportunities to support the achievement of the expected goals, so as to produce and implement adaptive policies that used for the development of dynamic capabilities which include the ability to think ahead, think again and think across, especially in the process of implementing the new normal during the Covid19 pandemic.

The third is research conducted by Made Hendriek Prasetia and Gusti Ngurah Darma Paramartha with the research title "The Effectiveness of the Role of the Traditional Village-Based Mutual Assistance Task Force (Satgas) for Outbreak Prevention Corona Virus Disease 2019 (Covid-19) In Padangsambian Village, Denpasar City" (Padangsambian et al., 2020). This study uses a qualitative method with a descriptive approach.Data collection methods in this study are through the internet as well as interviews, and observations. The results of the study concluded that The low quality of Human Resources (HR) for members of the Special Task Force related to legal knowledge and Covid-19 health protocols has hampered the operational tasks in the field. In addition, obstacles in the aspect of legal culture are also an important concern for researchers considering that public compliance with a special regulation or government regulation related to Covid-19 is still considered very minimal. This condition is also expected to be a response to the prolonged Covid-19 outbreak so that there is economic pressure that encourages people to ignore health protocols in order to continue to meet their needs.

Fourth, research conducted by Awang Darumurti dwith the research title "Human Resource Development in Disaster Management" (DWulansari et al., 2017). This research uses descriptive qualitative research method. The result of this research is that informal development is carried out by employees by reading and studying books/literature. While formal development is carried out with employees attending education or training. With the development of human resources both informally and formally, it can help BPBD in carrying out the stages of disaster management. From the stages of disaster management above, there are still 3 stages that have not been carried out by training, namely Mitigation, Rehabilitation, and Reconstruction. Inhibiting factors in HR development at BPBD Wonosobo Regency include, different types of organizations, limited facilities and infrastructure, heterogeneity of employees. While the supporting factors include high employee motivation, government policy support,

The fifth is research conducted byRisyal Hardiyanto Hidayat, with the research title "Strategic Steps to Prevent the Covid-19 Pandemic in Indonesian Correctional Institutions" (Hidayat, 2020). This research uses the method qualitative descriptive with data collection using library research. The results of the study concluded that the Ministry of Law and Human Rights had taken steps to prevent Covid-19 in Correctional Institutions through ministerial decisions such as; WFH, implementation of strict health protocols, delays, transfer of prisoners and the use of video conferencing in the online trial system as well as prisoner release programs through assimilation and integration rights.

The sixth is research conducted by I Wayan Putra Yasa, with the research title "*Tri Hita Karana* For the Prevention of Covid-19 in Bali" (Yasa, 2020). This research uses descriptive qualitative research method. The results of this study are the involvement of traditional villages as a task force team for handling Covid-19 began with a meeting between the Bali provincial government and the Bali Province Traditional Village Council (MDA) with the issuance of decree numbers 472/1571/PPDA/DPMA and Number 05/SK/MDA-ProvBali/III/ 2020 dated March 28, 2020. The form of implementation of the policy is carried out based on the concept *Tri Hita Karana* consisting of bad luck by performing religious rituals such as limiting traditional activities and religious approaches. *Pawongan*by monitoring the entry and exit of the community and distributing masks through pecalang (customary police), and *weak* provide a place for washing hands and spraying disinfectants in the traditional village environment. handling the spread of Covid-19 based on custom by applying local wisdom such as *Tri Hita Karana* Bali can be used as a role model for controlling the Covid-19 pandemic. This is important to do in order to welcome the adaptation of new habits. Involving the community holistically will raise collective awareness in preventing the massive spread of COVID-19.

This study will discuss the Capabilities of Local Governments in Combating Covid-19 in the City of Padang. There is still little previous research that has focused on the capability of local governments in handling Covid-19 in the city of Padang. The difference between this research and previous research is that the researcher wants to examine the aspect of capability regional government Padang City in handling Covid-19 based on the Regional Regulation of the Province of West Sumatra Number 6 of 2020 concerning the Adaptation of New Habits in the Prevention and Control of Corona Virus Disease 2019.

3 Research Methodology

This study uses a qualitative research approach with a descriptive type. The data collection method in this research is by conducting interviews, and documentation. Informants in this study were selected through purposive sampling technique.

4. Results and Discussions

Capability is an organization's ability to change and develop the essential prerequisites quickly to maintain a competitive advantage. The complexity of capabilities reflects an internal need to master complex tasks, (Schreyögg & Kliesch-Eberl, 2007). Connecting government capabilities and important factors in implementation, can be assessed from institutions, human resources, policy implementationijwill, financial, technical and leadership, (Kusumasari, 2014). Therefore, it is very necessary to have the capability Covid-19 task force in controlling Covid-19 in the region.

4.1 Institutional

According to Kusumasari, an effective institution must have: clear organizational structure, roles, duties, responsibilities and be able to establish networking with all levels of government (Kusumasari, 2014). The structure within the organization can be interpreted as a formal organizational framework, in which the work tasks are divided, grouped, and coordinated, (Robbins, S and Coulter, M, 2007). Role is a dynamic aspect of position (status), if a person carries out his rights and obligations according to his position, then he carries out a role. Status is a set of rights and obligations that a person has if he performs the rights and obligations according to his position, then he carries out a function (Soerjono Soekanto, 2002). A task is a part, element or component of a position which is a combination of two or more elements so that it becomes a complete activity.Moekijat, 1998). Therefore increasing the level of networking between organizational and government levels has been seen as essential to optimizing the flow of resources between institutions and increasing the accountability and effectiveness of disaster management.

The Padang City Covid-19 Handling Task Force has various roles with members from various institutions and OPDs including disaster volunteers. The Padang City Covid-19 Handling Task Force has coordinated in carrying out its duties with a clear division of roles and tasks within the structure of the Covid-19 Handling Task Force for the Covid-19 response. So that in carrying out its function as a chain breaker for the spread of Covid-19, the Covid-19 Handling Task Force has the capability in terms of the division of tasks and its role in carrying out the Covid-19 response. This is in accordance with what Moe & Pathranarakul said, namely that institutions are a very important determining factor in the success of disaster management, especially in conditions when the responsible government has no authority or can lead to unclear distribution of authority and delays in decision making, especially in the case of emergency assistance (Bevaola Kusumasari, 2014). A clear division of tasks and establishing networking with parties outside the government by the Padang City Covid-19 handling task force is carried out in order to avoid overlapping activities in carrying out Covid-19 prevention by institutions or OPD. So that with the division of tasks in dealing with Covid-19 within the Covid-19 Task Force, it is hoped that it can accelerate the response in order to break the chain of the spread of Covid-19. A clear division of tasks and establishing networking with parties outside the government by the Padang City Covid-19 handling task force is carried out in order to avoid overlapping activities in carrying out Covid-19 prevention by institutions or OPD. So that with the division of tasks in dealing with Covid-19 within the Covid-19 Task Force, it is hoped that it can accelerate the response in order to break the chain of the spread of Covid-19. A clear division of tasks and establishing networking with parties outside the government by the Padang City Covid-19 handling task force is carried out in order to avoid overlapping activities in carrying out Covid-19 prevention by institutions or OPD. So that with the division of tasks in dealing with Covid-19 within the Covid-19 Task Force, it is hoped that it can accelerate the response in order to break the chain of the spread of Covid-19.

4.2 Human Resources

The task of disaster management requires the cooperation of human resources based on problems and technical efforts related to the program of each work unit. In order to increase the efficiency and effectiveness of disaster recovery efforts, it is necessary to integrate several related programs and sectors which can be achieved through intensive periodic meetings. SHuman resources in question must have sufficient resources accompanied by division of work and sufficient delegation (Kusumasari, 2014). Human resources include staff who are trained in disaster emergency management, the ability to disseminate warning information, assist with evacuations, carry out emergency response activities, make needs assessments and can manage the distribution of relief items, (John Twigg; Humanitarian Practice Network, 2015). Human resources in the Padang City Covid-19

Handling Task Force in handling Covid-19 are adequate in terms of quantity. This is indicated by the sufficient number of personnel within the Padang City Covid-19 Task Force to carry out their roles and duties in accordance with their respective functions. For law enforcement to deal with the Covid-19 in Padang City, Padang City TNI personnel, Padang City Police and also Padang City Satpol PP are coordinating, so that the Padang City Covid-19 Handling Task Force does not lack personnel in carrying out public order, which is called the judicial operation. However, for the existing resources at the Padang City Regional Hospital, namely RS. Rasidin needs to maximize its health workforce resources. This is evidenced by the lack of services provided by health workers at the hospital as stated by one of the Covid-19 survivors who had received health services at the hospital. Rasidin City of Padang City.

Meanwhile, in terms of the quality of its human resources, the Padang City Covid-19 Handling Task Force is still not optimal. However, this is due to the lack of adequate facilities and infrastructure resources in carrying out the task of handling Covid-19 patients. So that the Padang City Covid-19 Handling Task Force cannot provide maximum Padang City Covid-19 handling facilities for Covid-19 patients. In addition, this condition occurs because Covid-19 is a disaster that has just been experienced in Indonesia, causing the completeness of tools and facilities and infrastructure in an effort to overcome Covid-19 is very minimal and not prepared beforehand.

4.3 Policy Implementation

According to Kusumasari, the implementation of policies in the government's capability model in disaster management is the availability of laws, policies and regulations as the basis for decision making, establishing relationships with other institutions and for resource mobilization. The basic rules for making policies to overcome Covid-19 in the City of Padang refer to national regulations specifically related to handling Covid-19. The implementation was carried out to strengthen capacity and increase regional resilience in suppressing the increase in the rate of positive confirmed cases of Covid-19 in the city of Padang. The implementation of the policy in this study is the availability of policies that regulate the handling of Covid-19 in the city of Padang.

The Indonesian government has issued a policy in dealing with Covid-19. As an effort to break the chain of the spread of Covid-19, the West Sumatra Provincial Government has also made a policy, namely Regional Regulation Number 6 of 2020 concerning the Adaptation of New Habits.In the Prevention and Control of Corona Virus Disease 2019. This Perda IMR can be applied in all Regencies/Cities in West Sumatra, one of which is implemented by the City of Padang.

In response to this, the Padang City Covid-19 Handling Task Force has also carried out socialization to the community, especially in crowded areas, one of which was carried out at the Padang City Raya Market. However, the socialization that was carried out before the implementation of the Perda AKB was considered not to have been carried out optimally, so that at the beginning of the implementation of the Perda AKB, there were still many people who did not know about the matters regulated in the Perda AKB. Therefore, there are still many people who do not wear masks such as traders in the Padang City market. However, there are also many people who think that the existence of this IMR Regulation will discipline the community more. This is because people feel anxious if they are subject to fines when they do not wear masks when doing activities outside.

Fig. 2. Dissemination of the Regional Regulation on IMR by the Padang City Covid-19 Handling Task Force, (Padang City Satpol PP in 2020).

Regarding the implementation of the local regulation on IMR, only some people are aware of the importance of wearing a mask when active outside the home. However, there are still many people who do not use masks when in public places or in crowds, such as at the Padang market even though the Perda AKB has been enacted. Thus, the researchers considered that some of the community would obey the prokes or wear masks if they knew that the Yustisi Operation was carried out by the law enforcement field of the Padang City Covid-19 Handling Task Force.

4.3 Finance

According to Kusumasari in the government capability model, the financial means that local governments must have adequate financial support to support all activities in disaster management (Kusumasari, 2014). Basically, the management funds needed for the implementation of the Decision of the Corona Virus Disease Handling Committee 2019 (Covid-19) and the National Economic Recovery are charged to the State Budget for the National Disaster Management Agency, as well as other legal sources in accordance with the provisions of the legislation. .

In dealing with Covid-19 in Padang City,The budget provided by the Padang City Government is adequate. The existing budget has been able to support all activities in the implementation of efforts to overcome Covid-19. In addition, to prevent a shortage of funds, the Covid-19 Task Force has distributed funds based on the priority scale and needs of each OPD. In addition, the Padang City Covid-19 Task Force has also received assistance from PPE (Personal Protective Equipment) from parties outside the government and NGOs. However, the Padang City Covid-19 response efforts are still focused on handling, so this will drain a lot of the Padang City Covid-19 budget later.

4.5 Technical

In the model of government capability in disaster management according to Kusumasari, the technical referred to is having a management logistics system and an effective information technology system to be able to communicate and establish networks with various stakeholders (Kusumasari, 2014). Based on the Regulation of the Head of the National Disaster Management Agency Number 13 of 2008, the logistics management system and disaster management equipment is a system that explains the logistics and equipment needed to cope with disasters during pre-disaster, during disaster and post-disaster. While the information technology system is a useful system to support and improve the quality of information for the community.

Based on the discussion above, it is known that in handling the Covid-19 of Padang City, the Padang City Covid-19 Handling Task Force related to technical matters need to be maximized. The technical matters in this research are related to logistics and information systems. First, for logistics, the Covid-19 Handling Task Force, in this case under the responsibility of the Padang City Social Service, provides logistical assistance to people who are self-isolating Covid-19 at home. This is done so that the community does not carry out activities outside the home, so that the Padang City Social Service Office helps in fulfilling their basic needs. However, when the logistics are borne by the Padang City urban village, there are still people who self-isolate at home who have not been recorded and do not receive logistical assistance. Therefore, the Padang City Covid-19 Handling Task Force needs to maximize coordination with urban villages so that Padang City people who need logistical assistance when self-isolating Covid-19 receive assistance. Because it is very risky for people who are confirmed positive for Covid-19 if they carry out activities outside the home to fulfill their basic needs. In addition, through the Padang City Social Service, the community also received BST assistance for the economic recovery of the community during the Covid-19 Pandemic. Therefore, the Padang City Social Service as the Padang City Covid-19 Handling Task Force is quite good in terms of logistics management to the community during this Covid-19 period. The community also received BST assistance for the community's economic recovery during the Covid-19 Pandemic. Therefore, the Padang City Social Service Office as the Padang City Covid-19 Handling Task Force is quite good in terms of logistics management to the community during this Covid-19 period. The community also received BST assistance for the community's economic recovery during the Covid-19 Pandemic. Therefore, the Padang City Social Service Office as the Padang City Covid-19 Handling Task Force is quite good in terms of logistics management to the community during this Covid-19 period.

Furthermore, in terms of technical information regarding the Covid-19 information system to the public, which in this case is carried out by the Padang City Communication and Information Office, it is necessary to further improve coordination with the Padang City Health Office. This is because Diskominfo is still having problems with updating data on confirmed Covid-19 in the city of Padang. So that an evaluation is needed at the Covid-19 Handling Task Force regarding the delivery of Covid-19 news to the public so that it can be managed properly. Because this will also have an impact on the Covid-19 data that will be uploaded by the West Sumatra Provincial and National Task Force, so that later it will not cause confusion in the community due to the data found on the official website of the Padang City Covid-19 Handling Task Force which does not match the numbers.

4.5 Leadership

The government capability model according to Kusumasari states that the leadership in question is having the leadership capacity that can make quick and appropriate decisions (Kusumasari, 2014). Efforts to overcome Covid-19 are not only the responsibility of local governments, but now it is a shared responsibility. Therefore, an explanation of leadership and disaster needs to focus on the extent to which leadership is able to direct and manage the ecological adaptation pattern of its community in collective action (with the community and stakeholders) for disaster reduction. In this case, disaster risk reduction can also be determined by the extent to which leaders (national and local) are able to drive a change together between institutional organizations. Based on the discussion above, in dealing with Covid-19 in Padang City, the Mayor of Padang, Mr. Mahyeldi Ansharullah, has shown a responsive and responsive attitude.

Fig. 3. Mayor of Padang Distributes Masks at Pasar Raya Padang.

The distribution of masks made by Mahyeldi is one of the efforts to prevent and protect the public from the spread of Covid-19 in the city of Padang. In addition, Mahyeldi also appealed to the public to wear masks.

This is in accordance with what Kusumasari said, namely capable leadership has the leadership capacity to make quick and appropriate decisions. Mahyeldi has issued several policies in dealing with Covid-19 in the city of Padang. In addition, the Mayor of Padang also went directly to distribute masks to the public in the context of dealing with Covid-19. In addition, the Chair of the Padang City Covid-19 Task Force also appealed to his subordinates to contribute, such as setting aside a small portion of the salary received for the Covid-19 response. Although a spike in the positive number of Covid-19 has occurred in Padang City, by the end of 2020, the spread of Covid-19 numbers can be suppressed, and the fluctuating Covid-19 number in Padang City tends to decrease.

5. Conclusion

The capability of the Padang City Government in dealing with Covid-19 shows that the City of Padang has the ability to suppress the rate of positive Covid-19 numbers. However, Padang City is still weak in terms of law enforcement in dealing with Covid-19, because there are still people who violate health protocols.Meanwhile, the Padang City Covid-19 response fund has also been sufficient, in addition to the Covid-19 Handling Task Force using Covid-19 funds based on a priority scale. In addition, the attitude of the Head of the Covid-19 Handling Task Force can also be categorized as a responsive and responsive leader. Although the number of Covid-19 spreads is quite high and has experienced a spike, this can be overcome by the evidence that the graph of the spread of Covid-19 at the end of 2020 has been sloping or also known as fluctuating which tends to decrease.

In this study, researchers describe the capabilities of the Padang City Covid-19 Handling Task Force. This research is expected to be able to contribute to the Padang City government, especially the Padang City Covid-19 Handling Task Force to improve its performance in an effort to break the chain of the Padang City Covid-19 spread.

Praise and gratitude we pray to Allah SWT. God Almighty, because of His blessings and grace, I was able to complete this scientific paper on time.We would like to thank those who have assisted in the completion of the research. In addition, we would like to thank the Padang City Covid-19 Handling Task Force and Covid-19 survivors in Padang City for being willing to become informants in this research.I hereby present this scientific paper with great gratitude and I hope that this scientific paper can provide benefits in the future.

References

1 S. P Adhikari, S Meng, Y. J Wu, Y. P Mao, R. X Ye, Q. Z Wang, C. Sun, S. Sylvia, S. Rozelle, H. Raat, , H. Zhou,. Epidemiology, causes, clinical manifestation and diagnosis, prevention and control of coronavirus disease (COVID-19) during the early outbreak period: A scoping review. In Infectious Diseases of Poverty , 9, 1, (2020).

2 S. Aisyah, E. Purwanda, (n.d.). Capabilities Analysis of PPIC Employees at PT. Idola Selaras Abadi. Retrieved September 11, 2021, from

3 Indonesia COVID-19 Data Analysis (Update As of 7 February 2021) - Latest News | Covid19.go.id. (n.d.). Retrieved September 11, 2021, from https://covid19.go.id/p/berita/analisis-data-covid-19-indonesia-update-7-februari-2021

4 Home | Covid19.go.id. (n.d.). Retrieved September 11, 2021, from https://covid19.go.id/

5 D. Wulansari, A. Darumurti, D. H. A. P Eldo,.. Human Resource Development in Disaster Management. Journal of Governance and Public Policy, **4**(3), 407–421, (2017).

6 GitHub CSSEGISandData/COVID-19: Novel Coronavirus (COVID-19) Cases, provided by JHU CSSE. (n.d.). Retrieved September 11, 2021, from https://github.com/CSSEGISandData/COVID-19

7 S. Hadi, Participatory Reducing the Risk of the Covid-19 Pandemic: An Overview of National Resilience to Disasters. Journal of Development Planning: The Indonesian Journal of Development Planning, **4**(2), 177–190, (2020).

8 Hidayat, risyal hardiyanto. Strategic Steps to Prevent the COVID-19 Pandemic in Indonesian Correctional Institutionsa. Journal of Health Education (e-Journal), **9**(1), 43–55, (2020).

9 John Twigg; Humanitarian Practice Network. (2015). Disaster Risk Reduction- Good Practice Review 9. Humanitarian Policy Network. http://odihpn.org/wp-content/uploads/2011/06/GPR-9-web-string-1.pdf

10 The Padang Raya Market Trader said about the IMR Perda whose sanctions are starting to take effect. (n.d.). Retrieved September 11, 2021, from https://langgam.id/kata-pedagang-pasar-raya-padang-soal-perda-akb-yang-sanksinya-mulai-berlaku/

11 B. Kusumasari, Q. Alam, Bridging the gaps: The role of local government capability and the management of a natural disaster in Bantul, Indonesia. Natural Hazards, **60**(2), 761–779, (2012).

12 B. Kusumasari, Disaster management and local government capabilities. Gava Media, (2014).

13 M. Mayarni, The Dynamic Governance Capability of the Pekanbaru City Government in the New Normal Enforcement During the Covid19 Pandemic. Journal of Aggregation: Government Reform Action in Democracy, **8**(2), 145–167, (2020).

14 C.-D. I. K Padangsambian, K Denpasar, I. M. H. Prasetia, I. G. Ngurah, D. Paramartha, Parta: Journal of Community Service in Indigenous Villages in the Framework of Preventing Corona Virus Disease Outbreak 2019 Introduction. Journal of Community Service, **1**(1), 36–43. (2020).

15 Jokowi's Sudden Announcement Surprises Corona Positive Patients Page all - Kompas.com. (n.d.). Retrieved September 11, 2021, from https://nasional.kompas.com/read/2020/03/04/08051

361/pengumuman-mendadak-jokowi-yang-kejutkan-pasien-positif-corona?page=all

16 S. P. Robbins, M Coulter,. Principles of management. Translated by Seyyed Mohammad Arabi and Mohammed Ali Hamid Rafiee and Behrouz Asrari Ershad, Fourth Edition, Tehran: Office of Cultural Studies, (2007).

17 S. Soekanto, Teori peranan. Jakarta, Bumi Aksara, (2002).

18 G. Schreyögg, M. Kliesch-Eberl, How dynamic can organizational capabilities be? Towards a dual-process model of capability dynamization. Strategic Management Journal, **28**(9), 913–93, (2007).

19 R. H. S. R. Titik, Bunga Rampai Artikel Penyakit Virus Korona (COVID-19) Editor : Titik Respati. Kopidpedia, 203–215. http://repository.unisba.ac.id:8080/xmlui/bitstream/handle/123456789/26743/fulltext_bc_16_feriandi_kopidpedia_fk_p2u_unisba_2020.pdf?sequence=1%0Ahttp://repository.unisba.ac.id, (2020).

20 I. W. P. Yasa, Tri Hita Karana for COVID-19 Prevention in Bali. Socius Journal: Journal of Sociology Research and Education, **7**(1), 54–66, (2020).

21 E .Yuniarti, P. M Indika, I. Dewata, E. Barlian,. Komorbidity Mapping of Covid-19 Events in West Sumatera. Geography and Geography Education, **4**(1), 2580–1775. http://sjdgge.ppj.unp.ac.id/index.php/, (2020).

Forecast of COVID-19 cemetery land needs with system dynamics modeling

Bondan Prakoso[1*], *I Dewa* Ketut Kerta Widana[1], *Rudy* Pramono[1], *Ernalem* Bangun[1], and *Hafizh* Surya Islami[1]

[1]Disaster Management Study Program, National Security Faculty, Republic of Indonesia Defense University. IPSC Sentul Area, Bogor, West Java, Indonesia

Abstract. As of June 30, 2021, Indonesia ranks third in Asia with the highest number of deaths due to COVID-19. One of the provinces in Indonesia, Central Java, is ranked third nationally in the accumulation of positive cases, recoveries, and deaths. Fulfilling the need for burial grounds for COVID-19 positive bodies needs attention because several countries and regions in Indonesia were experiencing a COVID-19 public cemetery land crisis. Sragen Regency is one area in Central Java designated as a red zone. The number of death cases is relatively high needs to ensure the availability of COVID-19 burial grounds. Therefore, this study aims to model the forecast for COVID-19 burial grounds. The approach used in this research is systems thinking, while the method used is system dynamics with Powersim software. Reference data for modeling is obtained from the *http://corona.sragenkab.go.id* page. The simulation results show the AME value of 4.7% or less than 5% so that the model is declared valid. Based on the Business as Usual (BAU) simulation, in the 20th week, it is estimated that there will be 320 deaths with a burial area of 961.13 m². Until the 20th week, the increase in COVID-19 Positive Population, Death, and Cemetery Land Needs on the graph of the BAU simulation results shows an exponential growth trend.

1 Introduction

One of Indonesia's national goals stated in the preamble to the 1945 Constitution is to protect the entire Indonesian nation and the entire homeland of Indonesia. This condition cannot be separated from the development of the strategic environment, which generally shows increasingly escalating and complex symptoms on a global scale [1]. These conditions allow the arrival of various threats, military threats (armed and unarmed), non-military threats, and hybrid threats[1]. Furthermore, the source of the threat can come from within or outside the country and is carried out by state and non-state actors, whether national, regional, or international. One of the non-military threats that hit Indonesia in disaster.

Indonesia is the largest archipelagic country in the world, consisting of more than 17,508 islands. With a beach length of 81,000 km, this area has a population of about 268 million people, making it the fourth most populous country in the world [2]. Known as the emerald of the equator, Indonesia has extraordinary beauty and natural wealth. However, Indonesians must also be aware that they live in areas that are at high risk of disaster threats [3].

In addition to being surrounded by three main active tectonic plates, Indonesia also has 127 active volcanoes known as the ring of fire. These active tectonic and volcanic processes cause Indonesia to frequently experience earthquakes, tsunamis, volcanic eruptions, and other natural phenomena. For example, the major disaster that hit Indonesia was the 2004 Aceh earthquake and tsunami. In addition, hydrometeorological factors also triggered other natural disasters, such as flash floods, landslides, hurricanes, and tidal waves. The large Population and the large potential for multiple disasters are quite a challenge for Indonesia.

In addition to natural disasters, other disaster risks hit Indonesia, namely social and non-natural disasters. At the beginning of 2020, the world has been hit by a non-natural disaster caused by a virus, namely COVID-19, and Indonesia is no exception. The first COVID-19 was reported in Indonesia on March 2, 2020, in several two cases announced by President Jokowi and Minister of Health Terawan Agus Putranto. With the issuance of Presidential Decree No. 12 of 2020, on April 13, 2020, the Government of Indonesia declared COVID-19 a national disaster because its scale was so massive. As of June 29, 2021, cumulatively, there were 2,156,465 positive cases, 1,869,606 recovered cases, and 58,024 deaths (*Covid19.go.id*), with a significant increase in cases occurring after Lebaran 2021. This increase in cases cannot be separated from the inclusion of several new variants introduced from other countries, for example, the

* Corresponding author : bondanhootowl@gmail.com

Delta variant from India. Compared to other less populous countries, the dense Population in Indonesia is expected to impact and suffer for a longer time [4]. The trend of positive confirmed cases of COVID-19 in Indonesia can be seen in Fig. 1.

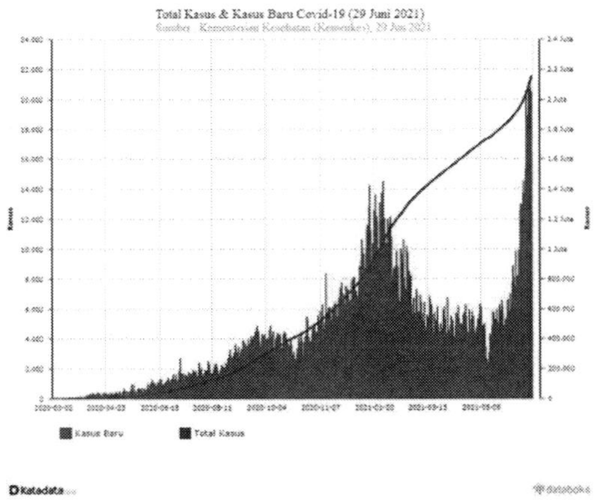

Fig. 1. Trends in COVID-19 Positive Confirmed Cases in Indonesia [5]

COVID-19 is a serious disruption to the functioning of multidimensional aspects of Indonesian people's lives. COVID-19 has disrupted the state of Indonesia, national security, and human security. Therefore, COVID-19 is designated as a non-natural disaster because it negatively impacts the form of death and other adverse impacts. When compared to COVID-19 deaths in Asia as of June 30, 2021, Indonesia ranks third after India and Iran. The comparison graph of these mortality trends is illustrated in Fig. 2.

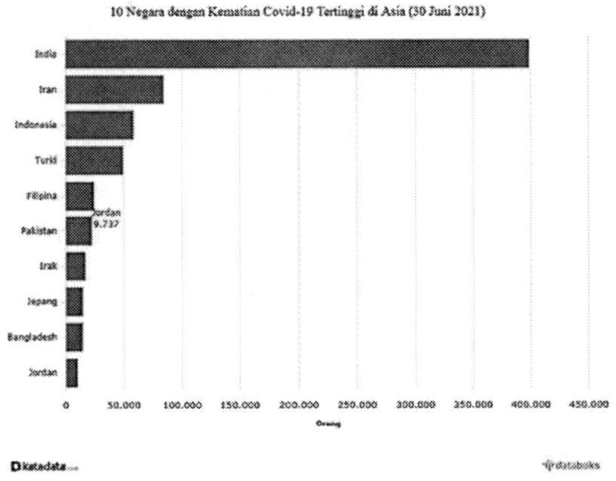

Fig. 2. Comparison of the 10 Countries with the Highest COVID-19 Deaths in Asia [5]

One of the provinces in Indonesia that have been affected by COVID-19 is Central Java. Based on data obtained from the https://corona.jatengprov.go.id page, the number of COVID-19 cases in Central Java Province as of June 29, 2021, was 251,604 positive cases, 212,386 recovered cases, and 16,082 deaths. These data put Central Java Province in the third-highest rank after DKI Jakarta and West Java. In addition, almost all districts/cities in Central Java have a high risk of transmission of COVID-19 or red zones. The areas in question are Brebes Regency, Cilacap Regency, Tegal Regency, Pemalang Regency, Pekalongan City, Pekalongan Regency, Batang Regency, Banjarnegara Regency, Kebumen Regency, Kendal Regency, Semarang City, Semarang Regency, Purworejo Regency, Magelang Regency, Demak Regency, Jepara Regency, Kudus Regency, Pati Regency, Grobogan Regency, Sragen Regency, Sukoharjo Regency, Karanganyar Regency, Wonogiri Regency, Rembang Regency, Blora Regency, and Sragen Regency. The map of the risk of COVID-19 transmission in Central Java can be seen in Fig. 3.

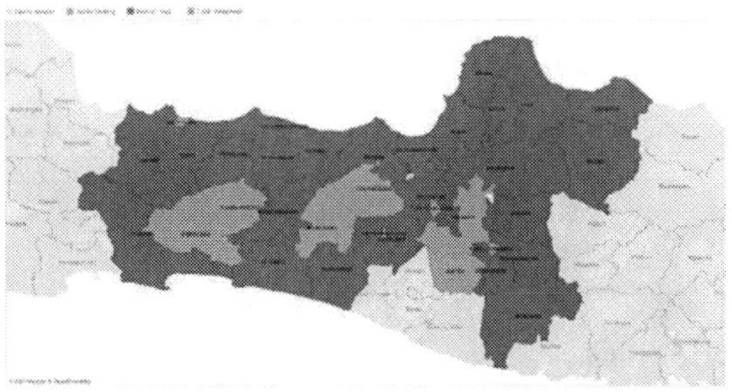

Fig. 3. Map of COVID-19 Transmission Risk in Central Java

Sragen Regency is one of the areas in Central Java that is designated as a red zone. Based on data as of June 29, 2021, Sragen Regency is ranked sixth in Central Java with an accumulation of 9,059 positive confirmed cases, 1,106 active cases, 7,268 recovered cases, and 685 cases died.

Fig. 4. Conditions of the Pondok Ranggon COVID-19 Public Cemetery in Jakarta

An interesting thing to research related to COVID-19 is the fulfillment of the need for burial grounds. For example, India is experiencing a crisis of cemeteries and crematoriums being overwhelmed by the number of deaths. Cemeteries are running out of space in many cities and funeral cremation fires burn through the night [6]For example, in Indonesia, the Pondok Ranggon COVID-19 public cemetery has experienced a severe crisis since

November 8, 2020, and all COVID-19 graves are full. To accommodate more bodies, funeral parlors began placing more than one body into graves. This policy only applies if the corpse is a close family member recorded on the same Family Card (KK) [7]. The condition of the Pondok Ranggon COVID-19 public cemetery can be seen in Fig. 4.

As the province with the third-highest number of positive cases in Indonesia, Central Java Province also needs to prepare burial grounds for cases of deaths due to COVID-19. Therefore, the Circular Letter of the Governor of Central Java Number 443.5/0007521 was issued on Thursday, April 17, 2020. The circular letter aims to ensure the availability of burial grounds for the bodies of victims of COVID-19. Sragen Regency, as one of the areas in Central Java with a relatively high number of deaths, needs to ensure the availability of COVID-19 burial grounds. In ensuring the availability of burial grounds, local governments need to make fast, precise, and effective policies. Do not let there be a crisis related to the availability of burial grounds due to unplanned policies. Aspects of policy and governance are a serious concern because COVID-19 has resulted in crises in both aspects in various countries.

Specifically in Indonesia, when COVID-19 emerged, policymakers showed a non-serious attitude and instead made humor. This resulted in minimal anticipation efforts. In addition, various preventive equipment, especially detection equipment and personal protective equipment, were lacking when COVID-19 had not yet spread. Rapid data assessment or analysis is needed because knowledge about COVID-19 is not sufficient [8]. As an anticipatory measure, an assessment of the basis for policymaking in handling COVID-19, especially for the provision of burial grounds, needs to be carried out. Therefore, this study aims to estimate the need for COVID-19 burial grounds.

To obtain an overview of the estimated demand for COVID-19 burial areas, system dynamics modeling is used to obtain trends in death cases and the required burial area. The data needed for the modeling is data on positive cases and data on causes of death. The data on positive cases and death cases used in this study is data on the development of COVID-19 in Sragen Regency, which was taken from the *http://corona.sragenkab.go.id* page with the accumulation of cases in units each week.

2 Research Methods

The approach used in this research is systems thinking. Systems thinking is an effort to think by placing an event or problem that does not stand alone but has a relationship with other events or problems [9]. There are three processes in systems thinking, namely synthesis, dynamic, and stochastic. Meanwhile, the method used is system dynamics. In systems thinking, the dynamic character means that all elements or sub-systems will experience changes based on time.

There are five stages in the systems thinking method. First, map the problem. Second, create a causal loop model. Third, perform dynamic modeling. Fourth, modeling and designing scenarios. Fifth, implementation and organizational learning [10].

Modeling in system dynamics requires software. By using software, the behavior of the model being created can be quickly identified. In this study, the software used is the Powersim program. Powersim is a program that can be used to simulate a dynamic model. In addition, Powersim can also simulate complex models in the real world [9].

3 Results and Discussion

Based on the method that has been described, there are five stages in the systems thinking method, including system dynamics. Therefore, the results and discussion in this study are described in 11 sub-chapters which are divided into these five stages. Problem mapping is described in-story modeling, modeling objectives, variables used in modeling, and definitions of operational variables. After the problem is mapped, it is then made into a causal loop diagram. Then, proceed with dynamic modeling in variable equations, stock-flow diagrams, and modeling assumptions. Scenario modeling and design are realized in simulations, validations, and business as usual simulations. Finally, implementation and organizational learning are carried out by making presentations in seminars or publishing in scientific journals. In more detail, the following is an explanation.

3.1 Modeling Story

The unit of time used in this modeling is per week, while the unit used in the population aspect is people. The need for COVID-19 Cemetery in Sragen Regency can be calculated by multiplying the number of deaths by the area of the cemetery per person. Death is obtained by multiplying the COVID-19 Positive Population with the COVID-19 Death Rate. The Positive COVID-19 Population is obtained from the multiplication between the Weekly Population with the COVID-19 Positive Rate. Meanwhile, the number of residents per week is obtained from the number of residents multiplied by the population growth rate. Population data is obtained from Sragen Regency data in figures for [11]. The Population Growth Rate is obtained from the division of the population growth rate of Sragen Regency in Figures in 2020 into units per week.

Table 1. Reference Data

Week-	Residents Positive COVID-19	Death
1	313	20
2	493	14
3	303	17
4	202	16
5	219	13

The COVID-19 Death Rate is obtained from comparing the average reference data for deaths due to COVID-19 and the average reference data for COVID-19 positive residents. Meanwhile, the COVID-19 Positive

Number was obtained from comparing the Total Population and the average COVID-19 Positive Population reference data per week obtained from the official website *http://corona.sragenkab.go.id.* The reference data for COVID-19 Positive Population and COVID-19 Deaths used in this study can be seen in Table 1.

3.2 Modeling Purpose

The purpose of this modeling is to find out the need for burial land in Sragen Regency for the next 20 weeks using system dynamics modeling with Powersim software.

3.3 Variables Used in Modeling

Based on the problem story, the main components or variables used for system dynamics modeling are:

- Total Population;
- Population Growth Numbers;
- Population Growth Rate;
- Total Population Per Week;
- COVID-19 Positive Numbers;
- COVID-19 Positive Rate;
- Residents Positive COVID-19;
- COVID-19 Death Rate;
- Death;
- Cemetery Land Needs; and
- Cemetery Area Per Person.

3.4 Definition Operational of Variable

Based on the main components or variables used for system dynamics modeling, the operational definition of each variable is explained as follows.

- Total Population is the number of residents calculated through the population census in years. The Population of the Sragen Regency based on the 2020 Census is 890.518 people.
- Population Growth Numbers is a number that shows the percentage of population growth in one week. In this modeling, the Population Growth Rate is 0,00583% per week.
- Population Growth Rate is a number that shows the development of the Population and is obtained from the multiplication of the Population with the population growth rate.
- Total Population Per Week is the number of residents calculated through the population census in the year plus the increase in Population per week. Total Population Per Week is obtained from the multiplication between Total Population and Population Growth Rate.
- COVID-19 Positive Number is data in the form of several positive cases of COVID-19 per week, as shown in Table 1.
- COVID-19 Positive Rate is a number that shows the development of positive cases of COVID-19 and is obtained from the multiplication of the Number of Population Per Week with the Positive Number of COVID-19.
- Residents Positive COVID-19 are residents who are confirmed positive for COVID-19 based on PCR testing. The positive Population is obtained from the multiplication of the Number of Population Per Week with the Positive Rate of COVID-19.
- COVID-19 Death Rate is the death rate due to COVID-19 based on comparing the number of deaths per week with the COVID-19 Positive Population.
- Death is COVID-19 Positive Resident who died. Death is obtained by multiplying the COVID-19 Positive Population with the COVID-19 Death Rate.
- Cemetery Land Needs is the size of the tomb required for a certain period. In this model, the unit of the time interval is weeks.
- Cemetery Area Per Person is the standard length and width of the tomb per person. In this model, the area of the tomb is 1.5 meters multiplied by 2.5 meters or $3.75m^2$/person.

3.5 Causal Loop Diagram

Based on the variables that have been defined, the relationship between variables is explained in the form of a Causal Loop Diagram (CLD) structure as can be seen in Fig. 6.

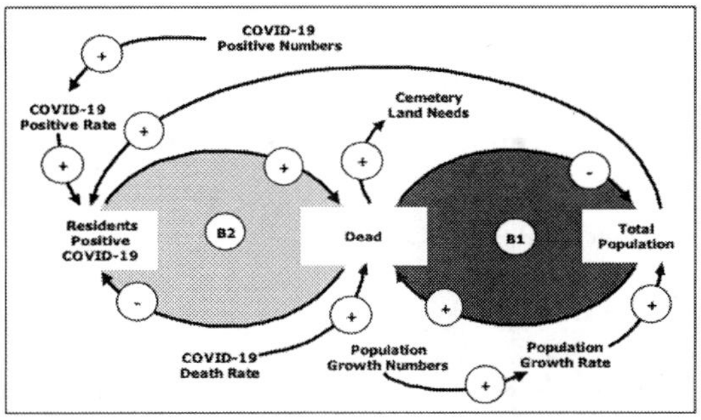

Fig. 6. Causal Loop Diagram

Based on Fig. 1, there are loops in the CLD structure. The loop in question, i.e. Population increases Death, while Death reduces Population. This loop is included in the type of balancing (B) because the multiplication between the relationship (+) with (-) produces the form (-). The second loop, namely COVID-19 Positive Population adds Death, while Death reduces COVID-19 Positive Population. The loop type of the second case is the same as the first loop, namely balancing.

3.6 Variable Equation

The variables contained in the CLD are then realized in the form of a Stock Flow Diagram (SFD). However, before making SFD, it is necessary to group the types of variables based on their symbols, values, and units. The grouping of variables based on the three models is shown in Table 2.

Table 2. Equation of Variables in Stock Flow Diagrams

Variable	Symbol	Value	Unit
Total Population		890518	People
Population Addition Rate	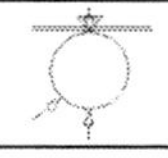	0,00583	%/week
Population Growth Rate		('Total Population'* 'Population Growth Numbers')- Death	People/ week
Weekly Population	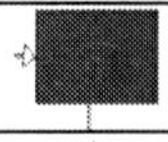	890518	People
COVID-19 Positive Numbers	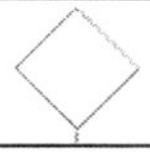	0,034362	%/week
COVID-19 Positive Rate	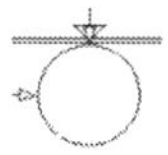	'Weekly Population'* ' COVID-19 Positive Numbers'	People/ week
COVID-19 Positive Residents		313	People
COVID-19 Death Rate		5,2287582	%/week
Death	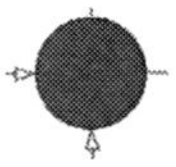	'Population Positive COVID-19'*' COVID-19 Death Rate'	People/ week
Cemetery Land Needs	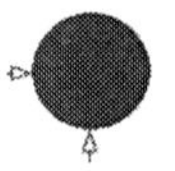	Death*' Cemetery Area Per Person'	m²/week
Cemetery Area Per Person	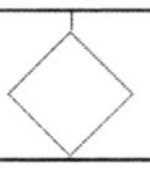	3,75	m²/Peop le

3.7 Stock Flow Diagram

Based on the CLD along with the variable equations, symbols, values, and units, a flow chart or Stock Flow Diagram (SFD) is drawn up. The SFD that will be used to simulate COVID-19 Positive Population, Mortality, and Cemetery Land Needs is shown in Fig. 7.

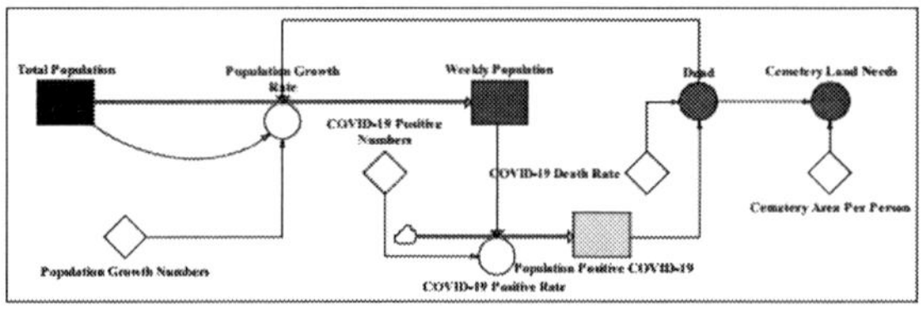

Fig. 7. Stock Flow Diagram

The variables contained in the SFD are more than the variables contained in the CLD because the calculation of Cemetery Land Needs requires additional other variables so that the results are valid, namely Population Each Week and Cemetery Area Per Person.

3.8 Modeling Assumptions

To clarify the model to be made, model assumptions are needed. The assumptions of the model in question include the following.

- The cemetery area per person includes the distance between graves.
- All people in the Death Count are the same size, which is 1.5x2.5 meters or 3.75 m².
- Cemetery Land Needs in the form of accumulation of one district so that it is not broken down by sub-district/village/hamlet/RW/RT.
- All cases of Death are carried out by burial, not cremation or other means.
- The causative factors are not taken into consideration in determining the Active Rate, Recovery Rate, and Death Rate.
- Positive cases of COVID-19 are actively treated and the number of cures is not simulated.

3.9 Simulation Results

Based on the input values and units to the SFD, a simulation is carried out to produce timetables and graphs. The simulation results of the number of deaths and the need for burial ground can be seen in Table 3.

Table 3. Simulation Data

Week-	Residents Positive COVID-19	Death
1	313	16
2	619	32
3	925	48
4	1231	64
5	1537	80

3.10 Validation

Table 4. Reference Data and Simulation for Validation

Week-	Death (Reference) (A)	Death (Simulation) (B)
1	20	16
2	34	32
3	51	48
4	67	64
5	80	80
Mean	50,4	48

To measure the suitability between the simulation data and the reference data, validation is needed. Validation was carried out by calculating the average mortality from the simulation results (B) and the average mortality from the reference data (A). The data and the average of the two types of Death Case data can be seen in Table 4.

Testing the validity of the simulation results can use the Absolute Means Error (AME) formula. The maximum acceptable fault tolerance limit with the AME test. The model is valid if the AME is less than 5% [12]. The AME formula in question is as follows.

$$AME = \frac{|B - A|}{A} x100\% = \cdots \%$$

$$AME = \frac{|48 - 50{,}4|}{50{,}4} x100\% = 4{,}7\%$$

The reference and simulation data contained in Table 4 are then entered into the AME formula. Based on the calculation, the AME value obtained is 4.7% so that it is greater than the maximum error tolerance of 5%. Therefore, the model is declared valid. The following is the calculation of AME against the data in Table 4.

3.11 Business as Usual Simulation

Based on the simulation results that have been declared valid, it is necessary to continue with the Business as Usual (BAU) simulation activity for policy analysis [13]. In this case, a BAU simulation was carried out on Death and Cemetery Land Needs. According to the research objectives, the Cemetery Land Needs is modeled until the 20th week to show future trends. BAU results can be seen in Table 5.

Table 5. Simulation of Business as Usual Cemetery Land Needs

Week -	Death (People)	Cemetery Land Needs (m²/Week)
1	16	61,37
2	32	121,37
3	48	181,37
4	**64**	**241,38**
5	80	301,38
6	96	361,39
7	112	421,39
8	**128**	**481,38**
9	144	541,38
10	160	601,37
11	176	661,35
12	**192**	**721,33**
13	208	781,29
14	224	841,25
15	240	901,20
16	**256**	**961,13**
17	272	1.021,06
18	288	1.080,96
19	304	1.140,86
20	**320**	**1.200,74**

Table 5 shows that in week 4 the number of deaths due to COVID-19 is predicted to be 64 people, so the Cemetery Land Needs is 241.38 m². In the next interval, namely the 8th week, the estimated number of deaths is 128 people, so it requires a burial area of 481.38 m². Furthermore, in the 12th week or third month, the number of deaths is estimated at 192 people and requires a burial ground of 721.33 m². Then, in the 16th week, it is estimated that there will be 256 deaths, requiring a burial area of 961.13 m². Finally, in the 20th week, it is estimated that the death toll will reach 320 people with a burial area of up to 961.13 m². In addition to being in tabular form, the trend of COVID-19 Positive Population along with Death and Cemetery Land Needs can be illustrated in Fig. 8.

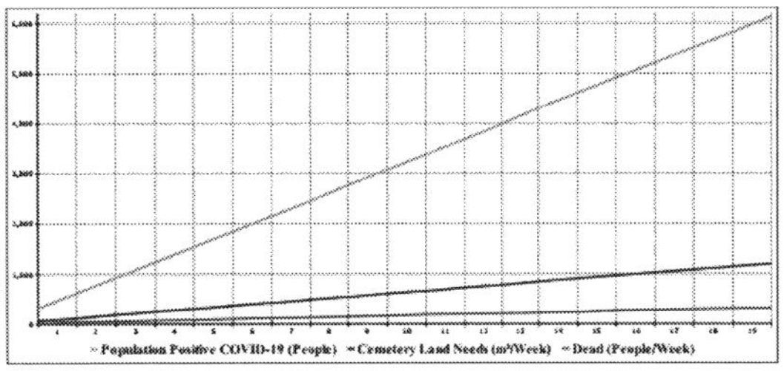

Fig. 8. Comparison of COVID-19 Positive Populations, Deaths, and Cemetery Land Needs

Fig. 6 shows a comparison between the COVID-19 Positive Population, Mortality, and Cemetery Land Needs based on BAU results. The BAU results show that the COVID-19 Positive Population always increases every week. This also has an impact on the increase in Burial Land Demand (blue line) which appears to be twice as high as Death (red line). The graph of the BAU simulation results shows that from week 1 to week 20 it always shows an increase or exponential growth. This shows that this model is in the form of a scenario without government policy or a model that works properly. That is, the value of the intervention model is 0 percent [9].

4 Conclusion

Based on the analysis and discussion, it can be concluded that the simulation results show an AME value of 4.7% or less than 5% so that the model is declared valid. After being declared valid, a Business as Usual simulation was carried out to model Death and Cemetery Land Needs. The results of the BAU modeling in the 4th week or first month show that the number of deaths due to COVID-19 is estimated at 64 people so that the Cemetery Land Needs is 241.38 m². Meanwhile, in the 20th week, it is estimated that the death toll will reach 320 people with a burial area of up to 961.13 m². Trends in Positive COVID-19 Population, Death, and Cemetery Land Needs from week 1 to week 20 based on the graph of the BAU simulation results show that there is always an increase or exponential growth.

Policymakers can use the model to be considered in decision-making because it is valid or feasible. In particular, this model can be used as a reference in preparing Cemetery Land Needs based on the development of COVID-19 Positive Population and Death. Thus, security related to Cemetery Land Needs will ultimately improve national security to prevent conflicts due to fighting over places or the practice of buying and selling burial land for COVID-19 positive bodies.

References

1. K. Defense, Jakarta Ministry. Defense of the Republic of Indonesia. (2015)
2. BPS, *Statistik Indonesia* (Central Bureau of Statistics of the Republic of Indonesia (BPS), 2020)
3. S. Maarif, Thoughts and Ideas: Disaster Management in Indonesia, National Agency for Disaster Management (2012)
4. R. Djalante, J. Lassa, D. Setiamarga, A. Sudjatma, M. Indrawan, B. Haryanto, C. Mahfud, M. S. Sinapoy, S. Djalante, and I. Rafliana, Prog. Disaster Sci. **6**, 100091 (2020)
5. T. P. Katadata, Kata Data (2020)
6. Apnews, (2020)
7. J. Post, (2020)
8. W. Mas' udi and P. S. Winanti, Governance for Handling COVID-19 in Indonesia: A Preliminary Study (Gadjah Mada University Press, 2020)
9. I. Umar and I. Dewata, (2017)
10. Hardjosoekarto, Soft Systems Methodology: (All Software Systems Method) (University of Indonesia Publisher (UI-Press) (2012)
11. B. P. Statistik, Sragen BPS (2020)
12. D. D. Suryono and E. S. Pujilestari, J. Segara **15**, 179 (2019)
13. R. Z. Surya and M. Gasali, J. Ind. Eng. Manag. Syst. **7**, (2017)

COVID-19 pandemic conditions: the role of sustainable environment in intentions to give zakat, infaq, and sadaqah

Alin Fithor[1*], *Dewi* Muliana[2] *Teuku* Afrizal[3] , and *Jumadil* Saputra[4]

[1]Diponegoro University, 50275 Semarang, Indonesia
[2]Pancasakti University, 52121 Tegal, Indonesia
[3]National University of Malaysia, 43600, Bangi, Selangor, Malaysia
[4]University of Malaysia, Terengganu, 21300 Kuala Terengganu, Terengganu, Malaysia

Abstract. The purpose of this paper is to determine the factors that influence the intention to give zakat, infaq, and sadaqah of Indonesian Muslim communities using the Theory of Planned Behavior (TPB) model. The sample data used in this study were 128 respondents and processed with SPSS to test the significance of the regression model. The independent variables used are religious beliefs, attitudes, subjective norms, and perceived behavioral control, while the intention is the dependent variable. The results showed that the idea of a sustainable environment had a positive and significant effect on attitudes. In addition, attitudes, subjective norms, and perceived behavioral control also have a positive and significant impact on intentions to the surrounding community. Amil institutions can predict the behavior of muzak through this paper. As for academics, this paper is used for further research by integrating perceived behavioral control as a predictor of Indonesian Muslim intentions. This study is an initial study that investigates the determinants of muzak behavior through the TPB model. Therefore, this paper will be helpful to a contribution from academics regarding the behavior Muslim community in the conditions COVID-19 pandemic in Indonesia and the world.

1 Introduction

The pandemic continues to provide bad news for countries in the world, including Indonesia. Indonesia's COVID-19 case fatality rate (CFR) reached 2.7% on Monday (June 21, 2021). COVID-19 has claimed 54,956 lives. Indonesia's CFR percentage is higher than the global 2.2% [1]. The increasing trend of deaths, but more than that, the COVID-19 outbreak has also impacted the economy of Indonesian millennial families. Millennial families do not have the savings to survive this pandemic. Most of them are also in debt. They also began to complain about their family's economic condition, so they were not satisfied with their family's economic [2].

The Indonesian Survey Institute (LSI) released a survey related to Indonesia's economic condition during a pandemic on July 18, 2021. As a result, most respondents rated the current national economy as bad; as many as 44.9% of respondents said the Indonesian economy was bad. People feel that their lives have become worse due to COVID-19. The majority, 70.9% of them, feel very/quite likely that their lives will worsen due to the economic impact of the COVID-19 outbreak. Public pressure for the government to prioritize economic problems is 50.7% compared to prioritizing health problems (46.2%). People also think that the blockades carried out by the government should be stopped. The majority of the people, 57.1%, feel that the PSBB is sufficient and can be stopped so that the economy can start immediately. However, 38.8% thought that PSBB should be continued so that the spread of the COVID-19 virus could be contained [3].

COVID-19 has had a huge domino effect on human life globally. The health crisis that initially occurred eventually created other crises, such as economic and social crises. One of the efforts to reduce poverty and economic inequality is to maximize zakat collection during the current COVID-19 pandemic. The potential for zakat, infaq, and sadaqah (ZIS) is reliable. The study of the Zakat Potential Mapping Index (IPPZ) conducted by the BAZNAS Puskas noted that in 2018 the potential for zakat in Indonesia reached 233 trillion rupiahs [4].

This great potential turned out to have obstacles, and in fact, there was a significant decrease in zakat during the pandemic was due to 48.4 percent of BAZNAS muzak experiencing a decrease in income (Sari et al., 2020). This condition will certainly affect the assets that can be muzak, especially from income zakat, which has a zakat base based on income or regular income from work that does not violate sharia. Not only that, the large potential for zakat, infaq, and sadaqah (ZIS) has not been well organized during the pandemic. The COVID-19 condition has made Zakat Management Organizations (OPZ) have difficulty collecting zakat conservatively by opening outlets in crowds and shopping centers [4]. The collection channels that previously could be utilized cannot be used optimally due to the Large-Scale Social Restriction policy

* Corresponding author: afithor@gmail.com

or the COVID-19 health protocol. These two things can have a negative influence on people's behavior in issuing ZIS in Indonesia.

The behavioral approach has been described by Ajzen 1991 through the theory of planned behavior (TPB). TPB is a development of the theory of rational action (TRA). TRA is a theory that is only intended for behaviors easily carried out by individuals without causing obstacles to do so [5]. Because the individual determines behavior, but there is a controlling factor in performing a behavior, the perceived behavioral control variable is added [5] TPB is a theory that aims to direct behavior into real action. TPB can reveal the reality of determining the variables that affect behavior [5].

Research on zakat, infaq, and sadaqah (ZIS) during a pandemic is very important, but there are very few references. In previous research, several factors influenced the intention of someone who issued his ZIS, such as religious beliefs, attitudes, subjective norms, and perceived behavioral control [6]. However, other studies have shown different things, such as [7], proving that subjective norms cannot form zakat intentions. Likewise, [8] research proves that perceived behavioral control cannot form a Muslim's intention to tithe.

2 Material and Methods

This research is included in the field research using a quantitative approach. The data in this study were taken using a survey method. The survey was conducted using a questionnaire as a data collection technique, and it aims to take a sample from the population [9]. This study collects information from a population of entities, aiming to compile a quantitative description of the attributes possessed by a larger population of which the entity is also a part [10]

A purposive sampling technique was used in the sampling process. *Purposive sampling* is a sampling technique carried out using certain criteria [11].

2.1 The Theory of Religious Beliefs

Religious beliefs affect attitudes [12], so the first hypothesis can be built:

H10: There is no effect of religious beliefs on attitudes.

H1a: There is an influence on religious beliefs on attitudes.

2.2 Attitude Theory

Attitudes affect the intention to tithe [12], so the first hypothesis can be built:

H10: There is no effect on attitudes towards the intention to give ZIS.

H1a: There is an effect on attitudes towards the intention to give ZIS.

2.3 Subjective Norm Theory

Subjective norms affect the intention to tithe [12], so the first hypothesis can be built:

H20: There is no effect on subjective norms on the intention to give ZIS.

H2a: There is an effect on subjective norms on the intention to give ZIS.

2.4 Perceived Behavioral Control Theory

Perceived behavioral control affects the intention to tithe [12], so the first hypothesis can be built:

H30: There is no effect on perceived behavioral control on the intention to give ZIS.

H3a: There is an effect on perceived behavioral control on the intention to give ZIS.

The research instrument used in this study was a questionnaire distributed online. The measurement scale used in this research instrument is a Likert scale with a range of 1 to 5. Value 1 = strongly disagree, number 2 = disagree, number 3 = neutral, number 4 = agree, and number 5 = strongly agree. Demographically, respondents are shown in Table 1.

3 Result and Discussion

Table 1 shows the demographic recapitulation of the respondents taken in this study. The table shows that the gender of the respondents who participated in this study 52.3% was male, and the remaining 47.7% were female. Respondents are also in productive age; it is proven that most respondents are aged between 17-40, namely 88.3%, the rest are spread in the age range between 41-50 years. Most of the respondents came from Central Java, namely 66.4%, the remaining 33.6% were spread in the provinces of Aceh, Banten, DIY, DKI Jakarta, West Java, East Java, Jambi, West Kalimantan, NTB, NTT, South Sulawesi, and West Sumatra.

The respondents' background is quite heterogeneous, as seen from the distribution of high school graduates as much as 38.3%, undergraduate 35.2%, master's degree 24.2%, and diploma 2.3%. When viewed from the background of the respondents' work, most work in the field of education, which is as much as 75%, the remaining 25% work in other fields. The income of 62.5% of respondents is less than 2 million, 28.9% is 2-5 million, 5.5% is 6-10 million, and 3.1% is more than 10 million per month

From the KMO and Bartlett test of sphericity, it can be seen that the KMO value is 0.834, which means the value is making better. While the Bartlett of Spherecity value is 0.000, which means the test results are significant. From the validity test, it can also be seen that the research instrument is valid.

Table 3 shows the value of the instrument reliability test by showing the calculation of the Cronbach alpha value. From these calculations, it can be seen that the Cronbach alpha value for the religious belief variable is 0.870, the attitude is 0.745, the subjective norm variable is 0.798, the perceived behavioral control variable is

0.844, and the intention variable is 0.870. From the Cronbach alpha value that exceeds 0.6, it can be concluded that the research instrument is reliable. In addition, all indicators used in this study are valid; it is proven that the Pearson correlation value of each indicator is greater than the r table value of 0.146.

Table 1. Information on Demographic

Demographic	Percentage
Gender	
Male	52.3
Female	47.7
Age	
17-20	38.3
21-25	12.5
26-30	11.7
31-35	15.6
36-40	10.2
41-45	7.0
46-50	2.3
Over 50 year	2.3
Province	
Aceh	1.6
Banten	1.6
Yogyakarta	1.6
Jakarta	7.8
West Java	7.0
Jambi	1.6
Central Java	66.4
East Java	6.2
West Kalimantan	2.3
West Nusa Tenggara	0.8
East Nusa Tenggara	0.8
South Sulawesi	1.6
West Sumatera	0.8
Education	
Senior High School	38.3
Diploma	2.3
Bachelor	35.2
Master	24.2
Profession	
Financial Services Sector	1.6
Personal Services	1.6
Tourism Services Sector	2.3
General Services Sector	5.5
Health Sector	0.8
Construction Area	1.6
Manufacturing Field	0.8
Government Sector	2.3
Education	75.0
Trade Area	4.7
Telecommunications Field	0.8
Others	3.2
Income	
Less than 2 million	62.5
2-5 million	28.9
6-10 million	5.5
More than 10 million	3.1

Table 2. The Result of KMO and Bartlett of Sphericity

Testing	Value
Sample Adequacy KMO	0.834
Bartlett of Sphericity	0.000*

*significant at $\alpha<0.05$

Table 3. Convergent Validity and Reliability Test

Item	P-Correlation
Religious Beliefs (CR=0.870)**	
KA1	0.486*
KA2	0.690*
KA3	0.883*
Attitude (CR=0.745)**	
SK1	0.676*
SK2	0.899*
SK3	0.856*
Subjective Norm (CR=0.798)**	
NS1	0.712*
NS2	0.860*
NS3	0.875*
NS4	0.747*
Perceived Behavioral Control* (CR=0.844)*	
PBC1	0.697*
PBC2	0.629*
PBC3	0.542*
ZIS intent (CR=0.870)**	
NZ1	0.842*
NZ2	0.821*
NZ3	0.906*
NZ4	0.809*

*valid (Pearson Correlation>r table 0.146)
**reliable (Cronbach alpha>0.6)

From the results of testing the first hypothesis, the R-value is 0.386, the R Square value is 0.149, and the Adjusted R Square value is 0.143. From the value of R Square, it can be interpreted that the variable of religious belief forms 15% of the variation in attitude. In other words, 65% of attitudes are formed by variables other than religious beliefs.

Religious beliefs have a positive and significant influence on attitudes, which can be seen from the regression test results of 0.149, with an at-count value of 4.704 and a significance value of 0.000. A significance value that is smaller than 0.05 means that if the religious beliefs of the Muslim community are strong, then the attitude in giving zakat, infaq, and sadaqah is getting stronger.

From testing the second hypothesis, the R-value is 0.603, the R Square value is 0.364, and the Adjusted R Square value is 0.359. From the R Square value, it can be interpreted that the attitude variable forms 36.4% of the variation in intention to give ZIS. In other words, 63.6%

of intentions to give ZIS were formed by variables other than attitude.

Attitude has a positive and significant effect on the intention to give zakat, infaq, and sadaqah can be seen from the results of the regression test of 0.364, with an at-count value of 8.493 and a significance value of 0.000. A significance value that is smaller than 0.05 means that if the attitude of the Muslim community is positive when the COVID-19 pandemic occurs, then the intention to give zakat, infaq, and sadaqah is even greater.

Furthermore, the results of testing the third hypothesis obtained an R-value of 0.613, an R Square value of 0.376, and an Adjusted R Square value of 0.371. From the value of R Square, it can be interpreted that the subjective norm variable forms 37.6% of the variation in intention to give ZIS. In other words, 62.4% of intentions to give ZIS were formed by variables other than subjective norms.

Subjective norms have a positive and significant effect on giving zakat, infaq, and sadaqah can be seen from the results of the regression test of 0.376, with an at-count value of 8.719 significance value of 0.000. A significance value that is smaller than 0.05 means that if the subjective norms developed in the Muslim community are positive during the COVID-19 pandemic, the intention to give zakat, infaq and sadaqah is even greater.

Tabel 4. Test Results of Direct Effect Determinant Coefficient and Adjusted R Square, t-test, and F dan test

DeterminantCoefficient & Adjusted R	**KA→SK**	**SK→NZ**	**NS→NZ**	**PCB→NZ**
R	0.386	0.603	0.613	0.369
R2	0.149	0.364	0.376	0.136
Adjusted R Square	0.143	0.359	0.371	0.129
Standardized Beta Coefficient				
T	4.704	8.493	8.719	4.450
Sig.	0.000	0.000	0.000	0.000
Result of F test				
F	22.126	72.132	76.025	19.805
Sig.	0.000	0.000	0.000	0.000

*Significance (Sig.<0.05)

Then the results of testing the fourth hypothesis obtained an R-value of 0.369, an R Square value of 0.136, and an Adjusted R Square value of 0.129. From the R Square value, it can be interpreted that the perceived behavioral control variable forms 13.6% of the variation in intention to give ZIS. In other words, 86.4% of intentions to give ZIS were formed by variables other than perceived behavioral control.

Perceived behavioral control has a positive and significant influence on giving zakat, infaq, and sadaqah can be seen from the results of the regression test of 0.136, with an at-count value of 4.450 and a significance value of 0.000. A significance value that is smaller than 0.05 means that if the perceived behavioral control formed on each Muslim is strengthened, then the intention to give zakat, infaq, and sadaqah is greater.

The study of intention is very familiar in behavioral theories such as Theory Plan of Behavior (TPB), Theory of Reason Action (TRA), Theory of Acceptance Model (TAM). By assessing intentions within the scope implementation of zakat, infaq, and sadaqah, this research has contributed to the development of these theories.

This research contributes to the amil zakat, infaq, and sadaqah to create a ZIS management strategy during a pandemic outbreak. The study of behavioral aspects of the Muslim community proposed in this study results from empirical research in the field.

The limitation in this study is that the variables used as a measure in predicting intention are limited to religious beliefs, attitudes, subjective norms, and perceived behavioral control. This study opens the opportunity for further research to add variables that predict the intention variable. In addition, this study also provides an opportunity to elaborate on the theory of intention.

4 Conclusion

The attitude of Muslims in giving zakat, infaq, and sadaqah can be formed by their religious beliefs. Meanwhile, to shape Muslim intentions in giving zakat, infaq, and sadaqah, it can be built by looking at several components in attitudes, subjective norms, and perceived behavioral control. In this study, attitudes, subjective norms, and perceived behavioral control positively affect the intention to give zakat, infaq, and sadaqah during the COVID-19 pandemic. A well-managed intention will be a strong predictor in the behavior of the Muslim community in giving zakat, infaq, and sadaqah.

References

1. R. R. Forradellas, S. N. Alonso, J. J. Vázquez, M. Á. E. Fernández, and N. V. Miró, "Entrepreneurship, sport, sustainability and integration: A business model in the low-season tourism sector," *Soc. Sci.,*. **10**, 4, 10040117, (2021).
2. M. L. Bustos *et al.*, "Tourism in the COVID-19 context in mesotidal beaches: Carrying capacity for the 2020/2021 summer season in Pehuén Co,

Argentina," *Ocean Coast. Manag.*, **206**, 27209, 105584, (2021).

3. Y. XIE, "社會利用與動員遣散 作為國家行動者的律師 Social Appropriation and Demobilisation," *Hongkong J. Soc. Sci.*, 51–88, (2020).
4. J. A. Erubami, P. Bebenimibo, and E. U. Ohaja, "Newspaper Exposure , Efficacy Feeling and Political Apathy among Youths in South-East Nigeria," *Pertanika J. Soc. Sci. Humanit.*, **29**, 3,. 1611–1629, (2021).
5. M. F. Othman and V. V. Okpe, "Assessment of Political Participation and Democratic Governance in Nigeria ' s Fourth Republic," *Pertanika J. Soc. Sci. Humanit.*, **29**, 1, 547–563, (2021).
6. L. Ode, A. Safar, H. Kusuma, D. A. Harjito, and Z. Arifin, "香港社會科學學報 Political Patronage and Capital Structure of Companies in Indonesia," *Hongkong J. Soc. Sci.* **57**, (2021).
7. C. Yee-hong, "檢視馬來西亞在中國 的戰略槓桿機遇與挑戰 Opportunities and challenges," *Hongkong J. Soc. Sci.*, 3–34, (2018).
8. R. K. Lai, "兩種港式粵語論述 民間中文 的語言政治 Hong Kong-Style Cantonese The Linguistic Politics of ' Chinese Textbook ' by Civil Society," *Hongkong J. Soc. Sci.*, 65–86, (2017).
9. G. Mjema, "Linear Programming Models for Estimating Weights in Analytic Hierarcy Process and For Optimization of Human Resource Allocation," *Int. J. Anal. Hierarchy Process*. **8**,. 2, 299–317, (2015).
10. G. Liu *et al.*, "A method to nourished beach stability assessment: The case of China," *Ocean Coast. Manag.*, **177**, 178, 166–178, (2019).
11. M. Yang and Y. Chen, "兩岸合憲性審查的展望 共和擺設抑或 憲政門神 ? Development and Prospect of Taiwan and Chinese Mainland Furnishings of Republic or Guardian of Constitution ?," *Hongkong J. Soc. Sci.*, 89–123, (2020).
12. S. Wang, "香港青年 教育 、 生涯 、 政治 的全球脈絡 Youth in Hong Kong," *Hongkong J. Soc. Sci.*, 3–25, (2017).

COVID-19 threat projection and Its influence toward national defense system

Amarulla Octavian[1], Syamsunasir[2*], and Nadiva Awalia Rahmah[2]

[1]Maritime Security Study Program, National Security Faculty, Republic of Indonesia Defense University, Sentul IPSC area, Tangkil Village, Citeureup District, Bogor 16810, Indonesia
[2] Disaster Management Study Program, National Security Faculty, Republic of Indonesia Defense University, Sentul IPSC area, Tangkil Village, Citeureup District, Bogor 16810, Indonesia

Abstract. The covid-19 pandemic that is currently experienced throughout the world has led to great uncertainty upon public health and the national economy. The impact of the Covid-19 pandemic has shaped the strategic environment landscape with four distinct attributes that might threaten a nation's survival, namely volatility, uncertainty, complexity, and ambiguity (VUCA). Thus, the government is expected to be able to make a swift and timely decision in responding to various threats in the VUCA era. Against the background, this paper discusses issues on global threat projection (especially the Covid-19 pandemic) and its influence on the national defense system, the re-actualization of the Total Defense System in the VUCA era, and the implementation of the Total Defense System in responding to the threats dynamic. To discuss the topic, this study analyzed the influence of global threats resulting from the Covid-19 pandemic on the national defense system, and how the Total Defense System is re-actualized and implemented in the VUCA era. The theoretical foundations employed in this study are securitization, neo-Malthusian, food security, dan society empowerment theories. The changing shape of threats has to be paid careful attention in formulating policy on the national defense system. In line with the complexity of threat characteristics in the VUCA era, the "re-actualization" of the Total Defense System has become an urgency given the current context. Against the dynamics of the current threats, the implementation of the Total Defense System could be examined from national food resilience and society empowerment factors. Total Defense System has to put people's roles at the forefront by guaranteeing the availability of necessities. By doing so, the people's loyalty in both national defense and the Total Defense System will improve along with people's welfare improvement.

1 Introduction

Before discussing the Covid-19 pandemic and its impact on the national defense system in depth, it is necessary to understand the concept of defense and security that Indonesia adopts. First, Article 30 Paragraphs 1 and 2 of the 1945 Constitution of the Republic of Indonesia stated that every citizen has the right and obligation to participate in the defense and security of the state, which is implemented through the Total Defense and Security System (Sishankamrata). The Indonesian Armed Forces and the National Police of the Republic of Indonesia are the main components in Sishankamrata, who are supported by the people. On the other hand, the Law of the Republic of Indonesia Number 3 of 2002 on National Defense and Law of the Republic of Indonesia Number 34 of 2004 on the TNI emphasizes that the national defense system is a Total Defense System that involves all people and other national resources.

This paper will elaborate on issues related to the projection of global threats (especially the Covid-19 pandemic) and their effects on the national defense system, the re-actualization of Sishanta in the VUCA era, and its implementation to face the dynamic threat landscape. Some of these problems are analyzed based on the relevant laws and regulations, along with several theories as to the analytical tools, that comprise securitization theory, neo-Malthusian theory, food security theory, and community empowerment theory.

2 Global Threat Projection and National Defense System

In the context of international relations, a state has to consider every changing situation and issue within its national, regional, and global strategic environment. Issues in the spotlight are part of a state's strategy to weaken other states or strengthen itself and to show that it

* Corresponding author: syamsunasir@yahoo.com

can respond to various threats and safeguard its existence as a sovereign state. On the other hand, globalization has had a great impact on every dimension of life. In addition to military issues, the state has to pay careful attention to issues in ideology, politics, economy, socio-culture, technology, public safety, and legislation dimensions.

There are several shapes of strategic environment related to the national defense which has interrelated roles. The correlation between the environment and the national defense system is very important since national defense analyzes strategic environment at the national, regional, and global levels [1]. Thus, analytical skills are required to understand whether changes in the strategic environment that occur at the national, regional, and international levels could be considered threats

2.1 Global Threat Projection (Covid-19 Pandemic)

Based on the 2020 World Economic Forum (WEF) report [2], the Covid-19 pandemic (infectious diseases) had not been categorized as a serious threat compared to several other threats at the global level until the beginning of 2020. Meanwhile, the 2021 WEF report [3] reconsidered the risks related to the global pandemic, economic downturn, political turmoil, and climate crisis that continued to worsen throughout 2020. One of the major changes in the 2021 report compared to 2020 is the risks caused by the Covid-19 pandemic.

The risk posed by Covid-19 is ranked first in terms of its impact in 2021 Fig. 1, whereas the risk of infectious diseases was only ranked 10th in 2020 [3]. The WEF 2021 report also reveals that in the next few years there will be mass unemployment, social inequality, and a slowdown, and even a prolonged economic downturn that might lead to mass riots. This condition will lead to national stability disruption in all states, including Indonesia. A prediction also said that a sharp economic downturn will occur and trigger a global recession.

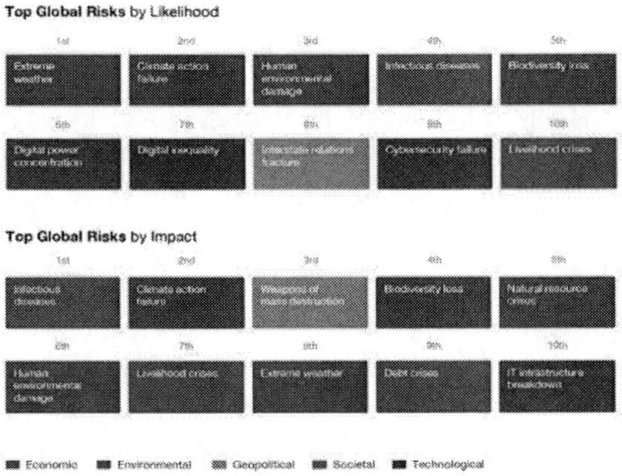

Fig. 1. 2021 Top Global Risk in terms of Likelihood and Impact [3].

A study of the risks posed by the Covid-19 pandemic showed that it is ranked as the 4th highest global risk in terms of likelihood and the 1st highest global risk in terms of impacts. The global Covid-19 pandemic does not only affect public health and economic sectors, but also the political and defense/security sectors. Nevertheless, WEF still places environmental issues, such as extreme weather, failures in adaptation and mitigation, and natural disasters as the main threats in the long term. In other words, the Covid-19 pandemic and environmental issues are real threats to the sustainability of life on earth.

According to Fig. 1, interstate relations fracture is ranked as the 8th highest global risk, while the impact of weapons of mass destruction is ranked in the 3rd place. Then, extreme weather and its effects are classified as a major threat. Therefore, states need to carry out international cooperation to mitigate those threats.

2.2 The Influence of Global Covid-19 Pandemic Threat toward National Defense System

Threats of military or geopolitical nature are currently predicted to be relatively less likely to occur than environmental and social threats. Therefore, a redefinition of the defense or security system concept is required. Against this need, securitization theory could be employed as a new framework for analyzing (non-military) security issues [4]. Securitization theory criticizes the traditional security paradigm which is too state-centric and military-centric by suggesting four sectors, namely: military, environmental, economic, social, and political [5]. Securitization can simply be understood as a process that places a non-security issue as a security issue.

In this case, the first-order effects of the Covid-19 pandemic and its derivative impacts such as mass unemployment, massive demonstrations, and food scarcity are issues that threaten national defense/security. As a response to these issues, states can no longer prioritize the use of militaristic methods or the deployment of armed forces. On the other hand, if a policy is implemented using a securitization approach, the government will treat ordinary-classified issues as threats to public safety and the nation's survival. These issues/threats only require reasonable action without any repressive military action.

Indonesia adopts a Total Defense System that involves all citizens and other national resources. In the context of the Covid-19 pandemic, the Total Defense System could no longer only use the power of the people as the last layer of defense. On the contrary, people who experience difficulties in life as a result of the rapid changes in the VUCA era might be a threat to the survival of the nation. Therefore, Sishanta must be oriented towards the development of people's capability to anticipate and face various socio-economic and political problems in the future. Thus, the threat to the nation's survival has become broader in scope.

Threats beyond the military dimension have even become a serious concern, where many countries have undergone securitization. Securitization itself operates in three spectrums: First, a non-politicized spectrum in which state does not attempt to address an issue; Second, the spectrum of politicization (politicized) in which state manages issues using ordinary means; and Third, the spectrum of securitization (securitized) where an issue is perceived as a security issue. The securitization spectrum depends on the securitizing actors who have the authority to declare an existential threat through a speech act against particular referent objects [6].

Securitization on non-military threats (such as the Covid-19 pandemic and climate change) can be classified as real threats. In its Defense White Paper or what is known as the Quadrennial Defense Review (QDR) 2014, The United States classified the threat of climate change as a threat to national security [7]. In responding to threats against the national defense, Indonesia uses a multi-agency approach. The government mandates ministries and agencies other than the defense sector as the leading sector or the main element to deal with non-military threats (Law of the Republic of Indonesia Number 3 of 2002 Article 7 paragraph 3). This indicates that the state has realized that threats to national defense do not only come from military threats, but also non-military threats.

Changes in the global threats landscape also indirectly affect decision-makers in various countries in determining their internal threats since changes in the nature and form of global threats could trigger conditions of uncertainty and disruption. Disruption does not only mean the phenomenon of changes that occur today but also reflects the meaning of the future change [8]. The Covid-19 pandemic has become a global threat and disrupted various sectors of life. Thus, it needs to be responded to by placing it as a threat to national defense.

Indonesia has gradually adopted the concept of human security. In this concept, the scope of security is no longer narrowed to state security that is monopolized by certain apparatus (actors) but becomes wider by taking into account the security of the people. Issues related to security are becoming more multi-sectoral and comprehensive, such as food and energy security. Thus, security sector management becomes more participatory because it involves many institutions. Therefore, synergy and solidity from various actors such as the armed forces, police, intelligence agency, immigration agency, academics, and other stakeholders are the building block of national defense against this kind of threat.

The spectrum of security also needs to be broadened. If so far security issues categorized as "domestic security" only include social unrest, inter-group conflict, armed insurgency, and armed separatist movements; then the category has to be broadened up to include poverty, difficulties to access clean water and food sources, disease outbreaks, increasing unemployment, and other social problems. These issues are considered threats to human life and therefore must be seen as threats to the state. The state is obliged to protect its people from various threats that might endanger their lives.

Against the aforementioned argument, efforts to build security have shifted from the concept of "security against" to "security with." Thus, cooperative security, confidence-building measures, and preventive diplomacy carried out both at the bilateral and multilateral levels are also parts of the effort to answer this problem. As a consequence, Sishanneg also needs to be prepared in a framework to deal with various global threats, such as the Covid-19 pandemic, climate change, natural disasters, and environmental damage.

3 Re-actualization of Total Defense System in VUCA Era

For the Indonesian people, the Total Defense System is the embodiment of the national defense system because it involves the armed forces as the main component as well as the people and other national resources as supporting components. Along with the development of the strategic environment, it is necessary to review the Sishanta concept. The validity of the Total Defense System concept has been proven in resisting colonialism in the pre and post-independence eras. However, in line with the changing dynamics of threats, especially in the VUCA era which is characterized by uncertainty and disruption, it is necessary to re-actualize the Total Defense System concept.

3.1 Re-actualization of Total Defense System

Re-actualization of the Total Defense System is very important, not only for decision-makers but also for all stakeholders. In the past, only certain elites understood defense/security issues. Today, the monopoly of defense/security issues is no longer appropriate to be practiced. The current democratization system in organizing the wheels of government has become a necessity, and defense issues have become the public domain. Moreover, issues related to defense systems cannot be in a vacuum or constant, since they are also closely linked to environmental changes. National defense is largely determined by various factors surrounding it [9].

The defense system cannot be separated from the problem of modernizing the Main Weapons and Equipment System. Nevertheless, another issue that is no less important, especially when the non-military defense is increasingly decisive in building a country's defense posture, is the involvement of other defense components, namely human resources, natural and artificial resources, and other national facilities and infrastructure. Therefore, the attention of defense planners has changed from the calculation of armed forces strength (hard power) to non-armed forces strength (soft power).

Furthermore, to defend the nation from various threats, Sishankamrata and the Total Defense System are also original and genuine concepts from Indonesia. Therefore, there are two choices that Indonesia might take to deal with the dynamics of the threats, namely: redefinition or re-actualization. If redefinition is chosen, then the mandate in the 1945 Constitution of the Republic of Indonesia has to be amended along with Law Number 3 of 2002 and Law Number 34 of 2004. In the end, re-actualization is an option that needs to be chosen, considering that every country has its differences in historical and cultural aspects.

When change has become a certainty (on the one hand) and the desire to maintain intrinsic values is very strong (on the other hand), then the best middle way is to

https://doi.org/10.1051/e3sconf/202133101012

re-actualize. In principle, re-actualization is to ensure the relevance of what is considered as a value in a system and ensure that it is not obsolete over time. History has also proven that Sishankamrata, which is supported by the people, was enabled Indonesia to gain its independence and defeat the colonials.

As a system that underlies defense and security policies, the concept of Sishankamrata still needs to be preserved despite the rapid changes in the legal and operational environments. In terms of spirit, the concept of Sishankamrata at least cannot be separated from the historical characteristics of the defense politics of the Republic of Indonesia. When most of the articles in the 1945 Constitution were amended, the concept of Sishankamrata was still maintained as stated in Article 30 Paragraph 2 of the 1945 Constitution.

The Total Defense System Doctrine places the people as the last resort of national defense to deal with various threats. People should be at the forefront of national defense since they are important parts of the national defense system. The Covid-19 pandemic is a real and actual example in which civilians are the ones who possess the competencies to deal with this threat. On the other hand, people still face various problems in their lives so they need to be more empowered by understanding them from a contemporary (contextual) perspective.

Besides calculating the need to increase military defense capabilities in dealing with threats, the evolution of threats dynamics in the fast-paced (disruptive) and uncertain VUCA era also requires changes in military strategy. Threats to national defense change rapidly and become increasingly uncertain like the "Volatility" and "Uncertainty" nature in VUCA. Science and technology development have also driven the change in defense characteristics of a nation. If defense threats in the past were always aimed at external enemies that possess certain military capabilities, now defense threats could be in the form of disease outbreaks, food scarcity, environmental damage, economic recession, and others. These non-traditional threats have also resulted in the development of various unconventional weapons types.

3.2 Total Defense System in VUCA Era

The changing forms of threats require a state to emphasize the role of supporting components in the Sishanta concept, which consists of defense components other than the military. In other words, it is not only military readiness that a state must prepare, but also various aspects beyond the military defense sector, ranging from ideological to economic matters. This means that the battlefield is getting broader, which ranges from physical war (hard power) to non-physical war (soft power). Due to the increasing complexity of threats toward national defense and the insufficient capability of the armed force, defense components other than the military have to play a bigger role in national defense.

This necessity is also in line with Article 7 Paragraph 3 of Law Number 3 of 2002 which stated that national defense system in dealing with non-military threats places government institutions other than the defense sector as the main element according to the form and nature of the threats, with the support of other national power elements. Strengthening government institutions other than the defense sector has also become one of the general defense policies, as stated in Article 2 Paragraph g of Presidential Regulation Number 8 of 2021 on the National Defense General Policy for 2020-2024. The document stated that one of the mid-term national defense priorities is to improve non-military defense capabilities carried out by ministries/agencies other than the defense sector.

Sishankamrata and Total Defense System is a total defense concept that is equal to the concept of total war in terms of the purpose and force mobilization. It is total, meaning that it uses all elements of national power, including natural and artificial resources, human resources, as well as national infrastructure and facilities. The people's role in national defense is also non-negotiable. National defense is essentially all defense efforts that are universal in nature and implemented based on the awareness of the rights/duties of citizens and belief in the nation's power (Article 2 of Law Number 3 of 2002).

For this reason, the law emphasizes the need for Reserve Components with high readiness be deployed anytime and anywhere. Reserve Components are national resources that are prepared and deployed through mobilization to reinforce the main components' capabilities (Article 1 paragraph (6) of Law Number 3 of 2002). Currently, no law specifically regulates Reserve Components and Supporting Components, but Law Number 23 of 2019 on National Resources Management for National Defense could be used as the legal basis for preparing Reserve Components and Supporting Components to carry out national defense duties.

The bill for Reserve Components and Supporting Components has also been prepared and has been discussed several times, but has not been passed into law due to rejections from several elements of society. By far, the law on Reserve Components has not been prioritized to be enacted in the near time. Another issue that hinders the law's enactment is the magnitude of globalization and the free market's influence in current national affairs. The strong market logic has shifted the logic of the national interest of each economic actor, both within and outside the government. As a result, the rhetoric about the importance of preparing national defense stands out more than reality.

Nevertheless, Law Number 23 of 2019 still can be used as the legal basis for preparing the Reserve Components and Supporting Components [10] as national resources. Komcad which consists of human resources, natural and artificial resources, and national infrastructure and facilities are spread throughout Indonesia. This condition urges the necessity to develop those national resources in an integrated manner to be reliable Reserve Components with capabilities to defend the entire territory of the Republic of Indonesia as a defense unit.

Indonesia's national defense capability can be measured based on the level of community welfare, which is viewed from three criteria, namely: poverty, education, and unemployment. If these criteria are in a good state, it will indirectly improve Indonesia's national defense

capabilities. In other words, improving people's welfare is directly in line with the improvement in national defense capability. Thus, both aspects have to be prepared simultaneously, and the fulfillment of people's basic needs is the priority. By doing so, people's loyalty to participate in national defense efforts will grow.
Efforts to improve people's capabilities in the Total Defense System are part of the national development goals. In the current VUCA era, people are not able to participate as national defense components in the Total Defense System if their lives are still far from a decent standard of living. Furthermore, the neo-Malthusian theory argued that the increase in population is in line with a decrease in food reserves. Such conditions can trigger communal conflicts to fight over scarce resources, especially necessities such as food. The scarcity of natural resources might potentially increase the risk of conflict [11]. Therefore, the re-actualization of Sishanta in the VUCA era is very important, especially the state's ability to prepare people as national defense components through the improvement of their living standards.

4 Total Defense System Implementation Against the Threat Dynamics

In the previous discussion, the Covid-19 pandemic that threatens national life could be categorized as securitization. According to the Copenhagen School [4], securitization is an extraordinary measure that turns a (non-security) issue into a security issue. The Covid-19 pandemic, which has hit all countries, is expected to last for a long period and has implications on the global food resources availability. For this reason, Indonesia needs to prioritize efforts to improve food resources available to ensure national food security.

The importance of food resources availability and provision was also expressed by Basundoro and Sulaeman in their research "Meninjau Pengembangan Food Estate Sebagai Strategi Ketahanan Nasional pada Era Pandemi Covid-19" (Review of Food Estates Development as a National Security Strategy in the Covid-19 Pandemic Era). The study concluded that the Covid-19 pandemic was a multi-dimensional phenomenon that could disrupt various strategic sectors, including the food security of a nation. The lack of food reserves availability owned by the state could lead to communal conflict that might threaten national stability. The food estate project is an embodiment of the national food security strategy [12]. Similar to this study, both studies view the Covid-19 pandemic as a threat that could disrupt various vital sectors related to people's lives, especially the national food security. In this case, the author views the necessity to develop national food security through the implementation of a protracted war concept to fulfill logistical support (especially food) in all regions during a long period.

For this reason, the implementation of the Total Defense System is closely related to the condition of national food security, the urgency of the food estate project, and community empowerment. These factors represent the core aspects of Total Defense System implementation to deal with various threats. The threat in question is related to the Covid-19 pandemic and several other threats that have been mentioned in the 2021 WEF Report as discussed in the previous section.

4.1 Food Security

Determining development priorities to achieve food security is very important since global level issues with high impact and likelihood of occurrence are water crises, food crises, and infectious diseases [3]. For the Indonesian context, the availability of water has not become a crucial or critical issue. On the other hand, infectious disease (especially Covid-19) has become a crucial issue because it has turned into an epidemic that caused many people to suffer and die. The COVID-19 pandemic has hit and disrupted all sectors of people's lives, including one of the most strategic, namely food security. In the end, the production and distribution of community food are disrupted. The Food and Agriculture Organization (FAO) even said that the COVID-19 pandemic impact could lead to a food crisis [13]. Likewise, national food security experiences ups and downs from time to time, so the Covid-19 pandemic can become a serious threat multiplier to national food security.

The covid-19 pandemic has disrupted the national food logistics system due to limited food logistics activities during the pandemic. The disruption of the food supply chain causes difficulties in people's food accessibility, thus threatening their lives. It is also feared that the uneven food distribution could lead to a disparity, where one or few regions experiences an excess of food commodities while the other experiences shortage.

Food security is closely related to national defense/security. This view has been recognized since ancient Egypt. In the case of Sishanta, Indonesia relies on its national resources to support national defense, including national food availability. The four pillars of food security include availability, accessibility, utilization, and stability [14]. For this reason, the Total Defense System will be effective if the state of national food resources fulfills these four pillars. Indonesia itself is in the category of a country with "serious risk" in terms of national food security. This condition indirectly reduces the nation's defense capability based on the support of people's power.

The Covid-19 pandemic does not directly disrupt the availability of national food, but it affects food distribution and even bans the import/export of food resources at the global level. Indonesia's food security is one of the emerging domestic (geopolitical) issues, where the management is still far from perfect. On the other hand, Indonesia has a large population (more than 270 million), which has large demands for food resources.

Without food security, especially in terms of availability and accessibility, people within a community will experience insecurity in terms of inadequate nutrient intake, which might potentially cause mass starvation. Such conditions could drive people to conduct social movements or large-scale demonstrations demanding changes to the government. In other words, the government could be deemed to fail if it is unable to fulfill the basic food necessary for its people. General Thucydides, military leaders, and historians from the Ancient Greek era explained that food security is a crucial issue and even becomes an important aspect for the continuity of a government [15]. Until now, the issue of food security remains a major concern for policymakers around the world, especially during the Covid-19 pandemic that is being experienced globally.

4.2 Food Security

As a supporting component in Sishanta, it is necessary to improve people's life and skills through empowerment measures to enable them to participate in national defense. In essence, community empowerment is not only aimed at individuals, but also in the community. For this reason, the people/community can be used as a normative benchmark, which places the concept of community empowerment as a part of efforts to build the existence of individuals, families, and even the nation as a realization of just and civilized humanity. Community empowerment activities require an understanding of human nature to enrich insight in applying various concepts or empowerment programs to the people.

History has proven that a nation that only relies on its natural resources without developing its human resources will never become a great nation. On the other hand, nations with limited natural resources but has qualified human resources were able to become developed and independent nations. Thus, improving human resources quality is a necessary condition for national development. If this is fulfilled, our nation can be looked at and visited by foreign experts.

Community empowerment programs or activities will not run properly without the support of other national resources, especially the availability of natural resources as the agricultural and maritime nation's greatest wealth. Indonesia possesses renewable and nonrenewable natural resources, such as oil, minerals, gas, and other mining commodities. Because these resources are one of the main factors to support community empowerment programs, these potentials should be explored in a controlled manner with strict adherence to sustainable development principles to ensure environmental balance and sustainability. One of the principles of community empowerment is based on the regional potentials and a healthy environment. Thus, the management and development activities must strictly adhere to the principles of sustainable development and environmental conservation for the benefit of all Indonesian people in the present and the future.

5 Conclusion

Based on the discussion on the projected threat of the Covid-19 pandemic against the Total Defense System, several conclusions were drawn. First, global threats are always changing from time to time. The Covid-19 pandemic has caused various negative impacts on a global scale. Changes in global threats also indirectly affect decision-makers in defining internal threats within each state. Threats changes have to become a concern in national defense system policies formulation. Second, in line with the increasingly complex nature of threats in the VUCA era, the "re-actualization" of the Total Defense System has become an urgent need in the current context. If in the past, the so-called threats to national defense always came from external enemies that possess a certain level of military capabilities, the current forms of threat are much broader than that. Threats to national defense in the current context could be in the form of disease outbreaks, food scarcity, environmental damage, economic recession, and other nontraditional threats. Third, the implementation of the Total Defense System to deal with the dynamics of the threats could be viewed from the national food security and community empowerment aspects. With an adequate level of national food security and social welfare, people will have a high level of resilience to improve their understanding of their roles in national defense as well as their capacity to perform their roles in the Total Defense System.

References

1. B. Bandoro. *States Choice of Strategies*. Yogyakarta: Graha Ilmu (2014).
2. WEF. *The Global Risks Report 2020.* http://www3.weforum.org/ docs/WEF_Global_Risk_Report_2020.pdf (2020).
3. WEF. *These are the world's greatest threats in 2021.* https://www. weforum.org/agenda/2021/01/these-are-the-worlds-greatest-threats-2021/ (2021).
4. B. Buzan, O. Waever. *Regions and Powers: The Structure of International Security*, Cambridge: Cambridge University Press, (2003).
5. A. Alfajri, Setiawan, H. Wahyudi. Synergy of Regional Defense Spatial Development in Facing Non-Military Threats in Indonesia, Jurnal Global and Strategis. **13** (2019).
6. 6. R M. Lutfi. Non-military Defense Against a Pandemic. Indonesian Defense White Paper (2020).
7. QDR. *Quadrennial Defense Review*. USA: Department of Defense (2014).
8. R. Kasali. *Disruption*. Jakarta: Gramedia Pustaka Utama (2017).
9. I. Samego. Contextualization of 'Sishanneg': Empowerment of Defense Areas in a Changing Perspective. Journal of Defense **5**, 1, 1-14 (2015).
10. B. T. Sofyan. *Development of Aerospace Potential in the Context of Empowering Air Defense Areas. Potdirga National Seminar, Bidakara Ballroom, Pancoran Jakarta* (2020).
11. T. Bernauer, T. Böhmelt, V. Koubi. *Environmental changes and violent conflict.* Environmental Research Letters **7**, 1, 015601. (2012).
12. A. F. Basundoro & F. H. Sulaeman. *Mininjau Pengembangan Food Estate Sebagai Strategi Ketahanan Nasional Pada Era Pandemi Covid-19.* Conference. Jakarta: Lemhannas (2020).

13. SDGs-Indonesia. *Pentingnya Menjaga Ketahanan Pangan di Masa Pandemi COVID-19.* https://localisesdgs-indonesia.org/beranda/v/pentingnya-menjaga-ketahanan-pangan-di-masa-pandemi-covid-19 (2020).
14. Food and Agriculture Organization (FAO). *Impacts of coronavirus on food security and nutrition in Asia and the Pacific: Building more resilient food systems*, 9. Bangkok. (2020).
15. A. Burn. *The Pelican History of Greece*, 297-299, London: Penguin (1982)

Strengthening social capital of urban community during COVID-19 disaster

Dwi Kristanti[1], *Edison* Edison[2], *Mohammad Kus* Yunanto[3], *Alfiandri* Alfiandri[2], *Diah Siti* Utari[4], *Tri* Samnuzulsari[5], *Dhani* Akbar[6], *Suyito* Suyito[7], *Emmy* Solina[5], and *Wayu Eko* Yudiatmaja[2*]

[1]Department of Public Administration, Open University, SouthTangerang, Indonesia
[2]Department of Public Administration, University of Maritim Raja Ali Haji, Tanjungpinang, Indonesia
[3]College of Administrative Sciences, Yogyakarta, Indonesia, Yogyakarta, Indonesia
[4]Department of Public Administration, Stisipol Raja Haji, Tanjungpinang, Indonesia
[5]Department of Sociology, University of Maritim Raja Ali Haji, Tanjungpinang, Indonesia
[6]Department of International Relations, University of Maritim Raja Ali Haji, Tanjungpinang, Indonesia.
[7]Faculty of Applied Social Sciences, University of Sultan Zainal Abidin (UniSZA), Trengganu, Malaysia

Abstract. COVID-19 disaster has destroyed many facets of societal and economic circumstances. However, it remains a crucial debate how the effect of COVID-19 on community social capital. This study aims to examine the relationship between a community's risk perception on COVID-19 and social capital. In addition, we also investigate the mediating role of social collectivism in the association. This research used a survey approach by performing an online questionnaire. Data were gathered from 156 respondents at the residence in an urban community in Sukoharjo, Central Java. A partial-least square structural equation modeling (PLS-SEM) was applied to analyze the data. The results show that social capital enhances during COVID 19 because of the increasing risk perceptions on COVID-19 and social collectivism. The impact of risk perceptions on social capital is mediated by social collectivism. The present study adds our understanding of social capital at the time of the COVID-19 outbreak, especially what factors drive it. The research also has practical implications for the government and urban community to anticipate risks of COVID-19 by enhancing social capital and collectivism.

1 Introduction

Since March 2020, Indonesia has officially been attacked by the Novel Corona Virus (COVID-19) disaster. COVID-19 has been recognized as a national disaster because of the significant effect of the COVID-19 on community health and well-being. Although the government has shown serious commitment and effort to beat COVID-19, the pandemic has not been left until now. Instead, the COVID-19 has gone, the spread of the virus becomes exceptionally uncontrollable. Based on the data of the COVID-19 task force, there were 2,072,867 people infected by the COVID-19, and 56,317 people died by June 25th, 2021 [1]. These numbers grew threefold than the beginning of the year.

The COVID-19 brings many impacts on human life. The social and economic effects of COVID-19 have become worst and massive. Suryahadi et al. reported that household expenditure decreased during COVID-19 because of various restrictions of economic activities in many regions in Indonesia. These problems contribute to an increase in the number of poverties in Indonesia [2]. COVID-19 also disrupts the social life of people around the world. Because of many restriction policies, many people can close interact and communicate than before. As a consequence, social bonding declines during the COVID-19 outbreak.

Many people avoid the impact of COVID-19 risk by using a variety of strategies. One of the strategies is to reinforce social capital collectivism. The people realize that they cannot face COVID-19 alone. If the COVID-19 and self-isolated infect them, they need the other and vice versa to fulfill their utilities because there are many forbidden copied during their sick. The people with COVID-19 also need others to strengthen them emotionally and psychologically. The COVID-19 can also increase social collectivism. If the persons are attacked by COVID-19, many people who have sympathy and empathy intentionally tend to help them. The effort to help others is ordinarily delivered sincerely and genuinely. In other words, the increasing social capital and collectivism can be influenced by COVID-19 risks.

Previous studies have shown the importance of social capital during a disaster [3, 4]. Social capital is considered as the community's strength in facing a disease. Diverse dimensions of social capital, such as trust, norms, and networks, are highly believed to predict social cohesion in the time of disaster. Although several studies have sought

* Corresponding author: wayuguci@umrah.ac.id

the impact of COVID-19 on economic and social life, the research highlighting what factors affect social capital during COVID-19 is still rare. Much of the study sheds light on natural disasters in portraying social capital, such as earthquakes, floods, tsunami, and volcanic eruptions [5]. Therefore, we need an obvious explanation of how social capital in the era of the COVID-19 disaster.

To fill these research gaps, this study explicitly seeks the association between community's risk perception on COVID-19, social collectivism, and social capital of urban communities in Indonesia. The present research contributes to the current literature in three routes. First, this study analyses the factors determining social capital during COVID-19. Second, we also elucidate the mediating role of social collectivism in the impact of COVID-19 risk perceptions. Lastly, instead of investigating rural communities recognized as high social capital by past studies [3], this work focuses on the urban community in Indonesia.

The structure of this paper is proposed in four sections. In the first part, we highlight the theoretical underpinnings of this study. This section clearly explains the concept used in this study--COVID-19 risk perceptions, social collectivism, and social capital. It is followed by the relationship among the variables, hypothetical statements, and research model. The second section contains methodological issues of the study, covering research method and instrument, sampling, and analytical strategy. The key results of the study are placed in the next section. It also includes the discussion and explanation about the data. Finally, the conclusion provides a summary of the research, including its key findings.

1.1 Theoretical Framework

1.1.1 Risk Perception on COVID-19, Social Collectivism, and Social Capital

The concept of risk perception is applied to investigate why people take (or do not take) precautions against external threats. To effectively communicate risk and build increased catastrophe risk mitigation techniques, it is necessary to understand risk perception [6]. In the context of COVID-19, individuals who have a higher perception of the risk of the coronavirus affect the way they act. If the perception of the possibility of being exposed to COVID-19 is small, people will tend to ignore health protocols. Public disobedience to health protocols and public distrust of the pandemic tend to have a negative impact on the high spread of COVID-19 cases.

Social collectivism is understood as a set of attitudes, beliefs, behavioral goals, and behaviors associated with solidarity and concern for others [7]. Social collectivism is the antithesis of individualism. It is related to the expected movement of the people socially connected in which they enable to develop social relationships intensively in the community. The process of social collectivism extensively generates added value for common interests, such as fulfilling general needs. Social collectivism can be directed to social actions in the community, specifically helping each other, working together, and social charity, strongly required in disaster.

Social capital refers to a series of informal norms and values collectively owned by the community facilitating cooperation and coordination into society [3]. Social capital becomes an important aspect in the time of the COVID-19 pandemic in Indonesia. Indonesian people are accustomed to living communally, both formally and informally. The group is one of the crucial social capital in Indonesia. The power of groups can encourage solidarity for problem-solving, including the corona disaster. Group participation is required to build collective strength against the COVID-19 outbreak.

Most scholars regard that social collectivism and social capital are similar. However, social capital is quietly different from social collectivism. If social capital is directly implemented in real action, social collectivism is only on the idea and artificial support. Social collectivism is underpinned by cooperation, social ties, and mutual respect. Individuals who are aware of the member of their community will improve community social capital. In sum, social collectivism is an antecedent of social capital. It cannot exist without social collectivism.

1.1.2 The Relationship among Studied Variables

This paper assumes that the three latent variables, risk perception on COVID-19, social collectivism, and social capital, are connected. Risk perceptions on COVID-19 positively influence social collectivism. The COVID-19 risk perception is related to the impact of COVID-19 in human life, such as economic, social, and psychological effects. In the economic aspect, the people infected by the COVID-19 have a high vulnerability to the loss of their job and income. They potentially become impoverished because of COVID-19. In terms of the social dimension, the people should apply physical and social distancing to comply with health protocol, implying the people are entirely separated from their social environment.

On the other hand, the people infected by the COVID-19 should be self-isolated. It means that they are also detached from their social life. In the end, it can make people more stressed and mentally pressure. All risks make the people have a similar sense that the COVID-19 is dangerous for their humanity. The feeling makes the people more connected and develops social collectivism, mainly for the people who are ever infected by the COVID-19 [8, 9]. Thus, risk perceptions on COVID-19 will enhance community social collectivism (Hypothesis 1).

The perceptions of COVID-19 risk tend to care with others because it can be a serious impact on the infected people. In addition, individuals also realize that COVID-19 are a danger for not infected people if the infected people do not follow health protocol, especially staying at home. Those infected ordinarily will get sick and cannot carry out their activities normally, such as working, cooking, and praying. The neighbors voluntarily will help provide daily needs of infected people, such as groceries or some foods, medicines, and vitamins, to get well soon

[10]. The people are willing to help infected people because of two reasons, respectively; it is to help infected people quickly recover and prevent the spreading of the virus. Furthermore, the neighbor will help the people who are losing their job because of COVID-19 by providing meals and donations. Therefore, COVID-19 risk perceptions will improve social capital (Hypothesis 2).

The community with high social collectivism ordinarily also has high social capital [11, 12]. Thereby, social collectivism can positively enhance social capital (Hypothesis 3). The COVID-19 risk perception can influence a community's social capital through social collectivism. High community social collectivism [13, 14]. Therefore, social collectivism positively mediates the relationship between COVID-19 risk perceptions and social capital (Hypothesis 4).

1.2 Hypotheses and Research Model

This research has three variables, predicted (risk perceptions on COVID-19), mediated (social collectivism), and dependent variable (social capital). From the explanation of theoretical framework and relationships among studied variables, the hypotheses of this research can be formulated as follows: *H1*: Risk perceptions on COVID-19 have a significant impact on social collectivism, *H2*: Risk perceptions on COVID-19 have a significant impact on social capital, *H3*: Social collectivism has a significant impact on social capital, and *H4*: The effect of risk perceptions on COVID-19 is mediated by social collectivism.

Based on these hypotheses, the research model can be drawn in Fig. 1.

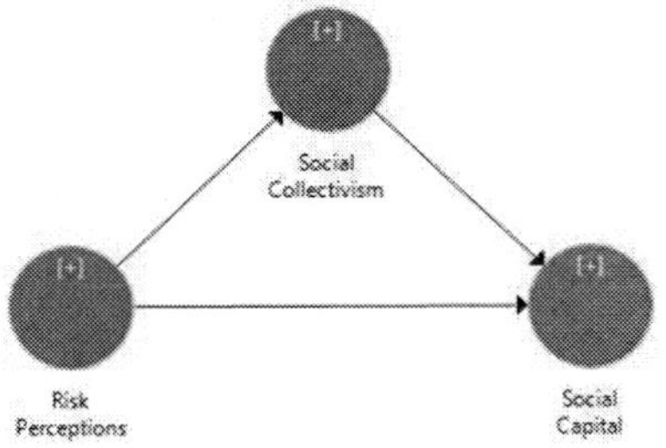

Fig. 1. Research Model

2 Methods

This research used quantitative inquiry to achieve the study purpose. A survey approach was applied in this research. This study examined the relationships between risk perceptions on COVID-19, social collectivism, and social capital during the COVID-19 pandemic.

2.1 Sampling Procedure

This study used a cross-sectional survey approach. A non-probability sampling procedure was applied in this research. Data were gathered using a self-administered questionnaire. Because of the COVID-19 pandemic condition, we distributed the questionnaire through an online form (google docs) in the WhatsApp Group (WAG) of the residence. The participants came from households of Korpri residence at Gayam, Sukoharjo, Central Java. One hundred eighty respondents participated in filling the questionnaire. After screening the data, only 156 questionnaires were valid and complete.

The demographic characteristics of the respondents were quietly different. The majority of the respondents were female (58%) than male (42%). Most of the respondents attained a Bachelor's degree (56), senior high school (42%), and graduates (2%), indicating the respondents were well educated. In terms of occupation, there was a mother household (35%), civil servant (26%), teacher (13%), entrepreneur (12%), student (10%), and doctor or health workers (4%). Around 3% of respondents had monthly income greater than IDR 10.000.000, 52% IDR 5.000.000-10.000.000, 38% IDR 3.000.000-5.000.000, and following 2% less than IDR 3.0000.000.

2.2 Measures

The constructs included in the questionnaire consisted of three variables, namely risk perceptions, social collectivism, and social capital. The constructs of this research were measured using tested and validated items developed by prior studies. Risk perceptions were assessed utilizing six items expanded by de Bruin [15]. The sample statement for the item was, "I worried about the loss of my job because of COVID-19". Social collectivism was evaluated using five items developed by Yamaguchi [16]. The sample statement for the construct was, "I would like to support my neighbor infected by COVID-19". Social capital was rated using six items built up by Onyx & Bullen [17] and validated by Yudiatmaja et al. [3, 18]. The sample statement for the construct was, "Participated local emergency action during COVID-19". All items were assessed using a five-point Likert's scale, 1 (strongly disagree), 2 (disagree), 3 (neutral), 4 (agree), and 5 (strongly agree).

2.3 Statistical Analysis

After completing the questionnaire, data in this research were analyzed employing partial least square structural equation modeling (PLS-SEM) [19]. It was used because of two reasons related to the usage of PLS. First, although the respondent of this research was only 156 persons, it was adequate to process with PLS-SEM because the minimum standard of the respondent was 100. Second, the use of PLS-SEM was quietly suitable for non-parametric statistics with non-normality data [20]. A SmartPLS software was performed as a research tool. Using SmartPLS, the data were analyzed using the PLS algorithm and bootstrapping techniques. PLS algorithm was applied to check the validity and reliability of the data. To test the studied hypotheses, this research employed bootstrapping using 5000 resample. The decision was made by identifying path coefficient or beta (β), t-value, and p-value.

3 Results and Discussion

3.1 Measurement Model

Measurement model assessment is aimed to inspect the validity and reliability of the data. Validity was assessed using convergent and discriminant validity. Convergent validity is evaluated by identifying the loading factor of the items and the average variance extracted (AVE). Based on the data in Table 1, it can see that all items have a loading factor exceeding 0.70; thereby, all items are valid and acceptable [21]. To assess reliability, composite (CR) and construct reliability. CR was evaluated by identifying CR value, and construct reliability of all constructs was inspected by checking Cronbach's alpha (α). The results indicate that CR and Cronbach's alpha is greater than 0.70, implying the data are reliable and acceptable [22].

Table 1. Validity and Reliability

Variable/Item	Item	Loading	CR	AVE
Risk Perceptions	RPC1	0.876	0.953	0.772
	RPC2	0.853		
	RPC3	0.903		
	RPC4	0.888		
$\alpha = 0.914$	RPC5	0.891		
	RPC6	0.870		
Social Collectivism	SOC1	0.876	0.949	0.788
	SOC2	0.888		
	SOC3	0.904		
$\alpha = 0.933$	SOC4	0.889		
	SOC5	0.882		
Social Capital	SCP1	0.818	0.936	0.709
	SCP2	0.823		
	SCP3	0.819		
	SCP4	0.879		
$\alpha = 0.918$	SCP5	0.856		
	SCP6	0.855		

In assessing discriminant validity, we apply Fornell & Larcker's (FL) and HTMT criteria. Table 2 exhibits that all square roots of AVE (in bold and diagonal) are higher than the association of respective rows and columns [23]. Accordingly, discriminant validity was fulfilled by the measurement model. HTMT criterion also was employed to estimate discriminant validity. As suggested by Kline, the threshold of HTMT values is 0.90 [24]. As can be seen in Table 2, the values of HTMT are lower than 0.90, indicating adequate discriminant validity.

Table 2. Discriminant Validity

FL Criterion Construct	1	2	3
Risk Perception	**0.879**		
Social Collectivism	0.685	**0.888**	
Social Capital	0.633	0.662	**0.842**
HTMT Criterion Construct	1	2	3
Risk Perception			
Social Collectivism	**0.726**		
Social Capital	**0.673**	**0.708**	

3.2 Structural Model

The structural model was assessed after the measurement model is fit. To test the multicollinearity, the values of all variance inflation factors (VIF) before testing the hypotheses. Based on the result of SmartPLS, VIF values are little than a rule of thumb, 3. Therefore, the measurement model is free from multicollinearity issues.

Table 3 and Fig. 2 summarize the results of hypothesis testing. The findings show that the community's risk perceptions on COVID-19 positively and significantly affect social collectivism ($\beta = 0.685$, $t = 11.38$, $p < 0.001$), and H1 is supported by the data. Community's risk perceptions on COVID-19 also positively and significantly impact social capital ($\beta = 0.338$, $t = 3.817$, $p < 0.001$) and H2 is accepted. Community's social collectivism positively and significantly influences social capital ($\beta = 0.431$, $t = 5.131$, $p < 0.001$) and H3 is accepted. Lastly, the effect of community's risk perceptions on COVID-19 on social capital is mediated by social collectivism ($\beta = 0.295$, $t = 4.430$, $p < 0.001$), and H4 is supported.

Table 3. Hypotheses Testing

Path	β	T Value	Sig.	Decision
RPC -> SOC	0.685	11.38	***	Supported
RPC -> SCP	0.338	3.817	***	Supported
SOC -> SCP	0.431	5.131	***	Supported
RPC -> SOC -> SCP	0.295	4.430	***	Supported

Note: Sig. *** = $p < 0.001$

The predictive power of the model is identified by analyzing R-square. Fig. 2 also reveals the R-square of social collectivism and social capital. The data show that R-square for social collectivism is 0.469, followed by social capital 0.500. The results indicate that the variance of social collectivism was shaped about 46.9% by risk perceptions. In addition, about 50% of the variance of risk perceptions and social collectivism can establish social capital.

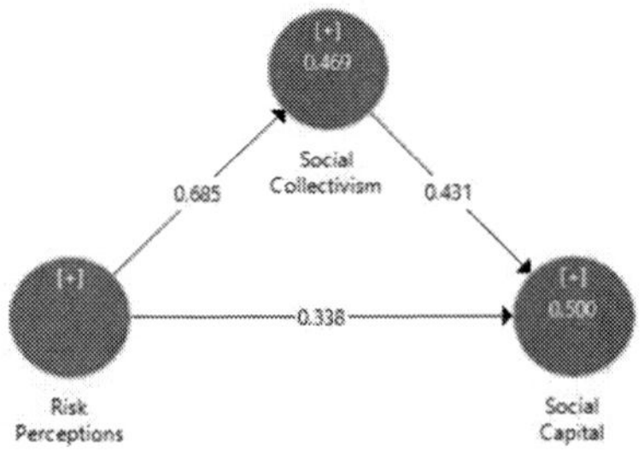

Fig. 2. Structural Model

3.3 Discussions

This work shed light on the impact of the community's risk perceptions on COVID-19 and social collectivism on social capital. After processing the data, we reveal that there is a positive and significant relationship between risk perceptions on COVID-19 and social collectivism.

There is a significant influence of the community's risk perceptions on COVID-19 in predicting social collectivism. The findings show the positive connection between the urban community's COVID-19 risk perceptions and social collectivism. Residents realizing COVID-19 risks can encourage social collectivism in their community. The result confirms the study conducted by Germani et al. stressing individualism and collectivism in reducing perceived risks of COVID-19 [25].

The research claims that COVID-19 risk perceptions positively and significantly influence social capital. The individuals understanding the impact of COVID-19 on their life tend to strengthen social capital. It is applied because they realize the significant effect of COVID-19 on the social, economic, psychological, and health of persons. Then, they have a consciousness to help others or neighbor through various actions, such as providing some foods, medicine, and vitamins. If the efforts are conducted by many people in the community, it can enhance community social capital.

This investigation notes that social collectivism has an essential role in strengthening social capital. There are five statements of social collectivism, such as "I would like to support my neighbor infected by COVID-19", "I would like to assist my neighbor infected by COVID-19", and "I feel the pain felt by my neighbors infected by COVID-19", who has been identified to increase social capital. Espirit of the corps of the urban community enhance social capital during COVID-19. The findings are supported by previous research pointing out the emergence of collectivism on social capital [26].

Moreover, this research also highlights the mediating effect of social collectivism. Social capitalism can be a mediator in the relationship between COVID-19 risk perceptions and social capital. Social collectivism can improve the impact of the community's perceptions on COVID-19 risk. The positive relationship between risk perceptions on COVID-19 and community social capital is partially mediated by social collectivism. The result corroborates previous research predicting individualism and collectivism on social capital in OECD member countries [27].

The findings of this study have several implications for the study and practice of social capital. Regarding the theory, our research contributes to the debate of social capital during COVID-19 in an urban community. We consider risk perceptions on COVID-19 and social collectivism as predictors of the increasing social capital at the time of the COVID-19 pandemic. These variables may lead to a better understanding of how social capital can work during a disaster. This study also provides a valuable suggestion for the government in socially handling COVID-19 impact on society. The government should offer intensive understanding to the community about the significant effect and risks of COVID-19 not only to human health and economy but also to social circumstances.

4 Conclusion

In this study, we investigate the effect of the COVID-19 disaster on the social capital of the urban community and whether the impact is mediated by social collectivism. Four hypotheses related to the connections among variables are tested. The findings of the research support all hypotheses by which community's risk perceptions on COVID-19 and social collectivism can predict social capital. This study also finds the mediating role of social collectivism in associating risk perceptions on COVID-19 and social capital. Theoretically, the current research enriches the study of social capital by recognizing COVID-19 risk perceptions and social collectivism of urban communities. Practically, our findings can be worthy policy recommendations to the government in reducing the impact of the COVID-19 pandemic by strengthening the community's social capital.

References

1. "Covid19.go.id," [Online]. Available: https://covid19.go.id/. (2021)
2. A. Suryahadi, R. Al Izzati, and D. Suryadarma, "Estimating the impact of Covid-19 on poverty in Indonesia," *Bull. Indonesia. Econ. Stud.*, **56**, 2, 175–192 (2020).
3. W. E. Yudiatmaja, Y. Yudithia, T. Samnuzulsari, S. Suyito, E. Edison, Social capital of local communities in the water resources management: An insight from Kepulauan Riau," in *IOP Conference Series: Materials Science and Engineering*, **771**, 1 (2020)
4. N. E. Ganapati, "Downsides of social capital for women during disaster recovery: Toward a more critical approach," *Adm. Soc.*, **45**, 1, 72–96 (2013).
5. S. Partelow, "Social capital and community disaster resilience: post-earthquake tourism recovery on Gili Trawangan, Indonesia," *Sustain. Sci.*, **16**, 1, 203–220 (2021).
6. I. A. Rana, S. S. Bhatti, A. B. Aslam, A. Jamshed, J. Ahmad, and A. A. Shah, "COVID-19 risk perception and coping mechanisms: Does gender make a difference?" *Int. J. Disaster Risk Reduct.*, **55**, 1–10 (2021).
7. C. H. Hui, "Measurement of individualism-collectivism," *J. Res. Pers.*, **22**, 17–36 (1988).
8. M. Yıldırım, E. Geçer, and Ö. Akgül, "The impacts of vulnerability, perceived risk, and fear on preventive behaviors against COVID-19," *Psychol. Heal. Med.*, **26**, 1, 35–43 (2021).
9. S. Suyito and N. A. Ghani, "The Pattern of People Participation of Island Society of Dendun Village, Bintan Regency," *Policy Soc. Rev.*, **1**, 1, 1–6 (2021).
10. Y. Yudithia, E. Edison, D. Kristanti, T. Samnuzulsari, S. Suyito, and W. E. Yudiatmaja, "Evaluating the Impact of Technological Adoption Policy for Rural Coastal Communities," *E3S Web Conf.*, **202**, 1–7 (2020).
11. J. Allik and A. Realo, "Individualism-collectivism and social capital," *J. Cross. Cult. Psychol.*, **35**, 1, 29–49 (2004).

12. D. Kristanti *et al.*, "Network governance in addressing climate change: a case study of the Asian Cities Climate Change Resilience Network (ACCCRN) in Indonesia," *IOP Conf. Ser. Earth Environ. Sci.*, **724**, 1, 012091 (2021).
13. N. Plohl and B. Musil, "Modeling compliance with COVID-19 prevention guidelines: the critical role of trust in science," *Psychol. Heal. Med.*, **26**, 1, 1–12 (2021).
14. G. G. Alrubaiee, T. A. H. Al-Qalah, and M. S. A. Al-Aawar, "Knowledge, attitudes, anxiety, and preventive behaviors towards COVID-19 among health care providers in Yemen: An online cross-sectional survey," *BMC Public Health*, **20**, 1, 1–11 (2020).
15. W. B. de Bruin, "Age differences in COVID-19 risk perceptions and mental health: Evidence from a national U.S. survey conducted in March 2020," *J. Gerontol. B. Psychol. Sci. Soc. Sci.*, **76**, 2, e24–e29, (2021).
16. S. Yamaguchi, "Collectivism among the Japanese: A perspective from the self," in *individualism and collectivism: Theory, method, and applications*, no. Cross-cultural research and methodology series, U. Kim, H. C. Triandis, C. Kagitcibasi, S.-C. Choi, and G. Yoon, Eds. Newbury Park, CA: Sage, **18**, 175–188 (1994).
17. J. Onyx and P. Bullen, "Measuring social capital in five communities," *J. Appl. Behav. Sci.*, **36**, 1, 23–42 (2000).
18. W. E. Yudiatmaja *et al.*, "Reducing solid waste through waste banks: an empirical study in Kepulauan Riau, Indonesia," *IOP Conf. Ser. Earth Environ. Sci.*, **755**, 1, 012076 (2021).
19. I. Y. Prastya, S. D. Meilinda, R. Sulistiowati, A. Alfiandri, and W. E. Yudiatmaja, "Application of SEM: an analysis of Depok City's website," *IOP Conf. Ser. Mater. Sci. Eng.*, **1088**, 1, 012028 (2021).
20. J. F. Hair, G. T. M. Hult, C. M. Ringle, and M. Sarstedt, *A primer on partial least squares structural equation modeling (PLS-SEM)*, 2nd Ed. Thousand Oaks, CA: Sage Publications (2017).
21. J. F. Hair, J. J. Risher, M. Sarstedt, and C. M. Ringle, "When to use and how to report the results of PLS-SEM," *Eur. Bus. Rev.*, **31**, 1, 2–24 (2019).
22. W. W. Chin, "How to write up and report PLS analyses," in *Handbook of partial least squares*, V. E. Vinzi, W. W. Chin, J. Henseler, and H. Wang, Eds. Berlin: Springer, 655–690 (2010).
23. C. Fornell and D. F. Larcker, "Evaluating structural equation models with unobservable variables and measurement error," *J. Mark. Res.*, **18**, 1, 39–50 (1981).
24. R. B. Kline, *Principles and practice of structural equation modeling*, 4th Ed. New York: The Guilford Press (2015).
25. A. Germani, L. Buratta, E. Delvecchio, and C. Mazzeschi, "Emerging adults and covid-19: The role of individualism-collectivism on perceived risks and psychological maladjustment," *Int. J. Environ. Res. Public Health*, **17**, 10, 1–15 (2020).
26. W. E. Yudiatmaja *et al.*, "Social policy on the rural coastal communities: Why the implementation fails?," *E3S Web Conf.*, **232**, 1–7 (2021).
27. K.-I. Yoon, "The cultural effects of individualism and collectivism on social capital," *Int. Area Rev.*, **13**, 2, 187–212 (2010).

COVID-19 pandemic impact: an identification of the cause of cost overrun in construction project

Yervi Hesna[1*], *Jati* Sunaryati[1], and *Ayu* Hidayati[1]

[1]Department of Civil Engineering, Faculty of Engineering, Andalas University, Indonesia

Abstract. The paper provides a thorough examination of the impact of COVID-19 on increasing the cost of the construction project. The pandemic of COVID-19 had disrupted the financial and operational of the construction industry. Many projects have been closed and suspended, and this situation has caused great uncertainty in the construction industry. Even though the Instruction of the Minister of Public Work No. 2/2020 issued March 27, 2020, stated that construction activities in Indonesia could continue work. However, it still believes COVID-19 can lead to poor project cost performance. The study used a literature review to explore the factors contributing to the increase in project cost. A case study in the hotel construction project is used to test the factors. The research found six factors have a frequency of occurrence for each cause of cost overrun, namely labor, material, implementation time, project financial aspects, estimated cost, and occupational health and safety. The case study's two factors are not the cause of the cost overrun based on the interviewee's answer: heavy equipment and overhead. This paper also explains the mechanism to address cost overrun continue in construction work.

1 Introduction

The pandemic of COVID-19 had arisen globally. COVID-19 has not only had an impact on the health sector but has had a profound impact on various sectors, ranging from the tourism, economic, manufacturing, social, to construction sectors. The operational of companies, construction works, and finances are the primary aspect of the construction business industry. Esa [1] argued that operational and financial are the top issue caused by COVID-19.

Sihombing [2] believed that the construction phase is the most impacted phase under the pandemic COVID-19. Project stakeholders across the construction industry lacked understanding in managing the project while COVID-19 happened [3]. It is resulting deficiency of cost performance.

Cost overrun is considered a persistent phenomenon in construction projects [4]. As one of the project success indicators, cost deviations from the project baseline reflect the project's health. The cost overrun comprises all the relevant factors for the time overrun and or vice versa.

On the other side, Al Hazim [5] argued that natural disaster is undoubtedly one of the significant factors contributing to cost overrun in construction projects. Natural disasters are unpredictable incidents that may cause cost overrun due to delay or suspension in work. As a natural disaster, the pandemic COVID-19 makes people confused because there is no previous experience.

Many scholars have identified cost escalation as an impact of the pandemic COVID-19 in the operational and financial that construction owners and contractors can have never imagined earlier. These unexpected circumstances can create contract applications and interpretive contests for the unwary construction partner.

Several scholars in cost overrun during the pandemic COVID-19 show that the area of study focuses on methodological approaches such as identifying cost overrun factors and finding the most significant factors [3]. Zamani [6] strengthened the problems the construction industry when facing by COVID-19. As an impact of COVID-19 in the construction industry, Ullah [7] identified the cause of cost overrun based on the party involved in project construction. Annamalaisami [3] arranges the cause of cost overrun based on finance, design, execution, quality, and professional skill-related factors.

The area of research in cost overrun based on input or resources of the construction project is still limited. This study is conducted to identify the cause of cost overrun based on the impact of the pandemic COVID-19 on the input or resources of the construction project. And also, this study will determine mechanisms that can control cost overrun on building construction.

The study begins with a literature review followed by collecting primary data through a study case. Finally, the recommendations are given to minimize the cost overruns

* Corresponding author: yervi@eng.unand.ac.id

in the future caused by a natural disaster so that the construction projects will not stop midway

2 Literature Review

2.1 Effect and Prospect of COVID-19 Pandemic to The Construction Industry

COVID-19 outbreak disrupted businesses globally, including built asset procurement and facility management. The pandemic COVID-19 is also having an impact on the construction industry. Biswas [8] identified some problems in the construction sector. The supply chain has been disrupted, which makes various construction materials are not reaching the construction site. All transportations have been disrupted caused by the lockdown. Labor shortage because workers cannot reach their workplaces and many workers are not willing to come to work. The company has a financial problem, and the party also has contractual implication problems.

The cost overrun occurs during the monitoring phase of construction as the effect of COVID-19. Cost variance between the planning and market value strongly affects the performance of the project cost. Cantarelli [9] argued the correlation between the project size and the cost overrun indicates the percentage of cost overrun in the larger construction projects.

The effect of this pandemic has both positive and negative impacts. Ogunnusi [10] explained the first positive impact is the opportunity to improve managing the project virtually. The situation has encouraged the usage of virtual communication to attend to operational needs such as correspondence, meetings, which could have been done face to face previously, as stated by Khan [11]. The project manager can expand technological tools to re-organize Workplan virtually and increase off-site working or use precast elements to minimize the worker in the construction site.

Despite the positive experiences, the pandemic also raises uncertainty in project construction. Even without COVID-19, time overrun and cost overruns are some of the significant defeats the construction industry has been battling. Now with this pandemic, the case has become extraordinarily complex.

One of the causes of the large impact on construction work is the issuance of regulations by the Occupational Safety and Health Administration (OSHA). Regulations issued is to carry out construction work must pay attention to the potential risk of being exposed to the coronavirus. Steps that must be taken in controlling and preventing COVID-19 in the sector construction, are to analyze the hazard risk by mapping the potential risk of exposure to each work activity.

By conducting this hazard risk analysis, it can be seen which activities are included in the hazard high, medium, and low risk so that action can be taken according to the risk. If the risk is high, the work must be postponed or even temporarily suspended until it can be carried out with the maximum level of risk being a moderate level.

To avoid risk high, workers who are suspected of COVID-19 will be prohibited from entering the field. And to can minimize the level of moderate risk, workers must be reduced at a time at the same location, which reduces the productivity of construction work.

3 Research Methodology

This study focuses on determining the cause of cost overrun in construction projects and test whether there is a significant difference in project prices between the before and during this pandemic period.

To fulfill the research purpose, the study first explores the factors contributing to the increase in the project cost from much literature. Following a review of the literature concerning cost overrun, this research developed a list of 31 major causes of cost overruns in construction projects.

A summary of factors that can affect construction cost overrun is shown in Table 1 below. This table is arranged by collecting the probable factors based on the input/resources of the project. All of the input factors of the project are material, labor, heavy equipment, occupational health and safety, overhead, project financial, cost estimation, and execution time.

Then, a case study was used to test whether the factors have been affected in terms of the COVID-19 pandemic. This study also identifies the magnitude of the change in cost.

The Hotel Santika Premiere Construction Project in Padang City, Indonesia, was explored as qualitative research. Interviews with the contractor and consultant conducted the study. A questionnaire list was designed to collect general information on respondents' and interviewees' perceptions following Table 1 to find out the critical elements causing cost overruns.

A yes/no answer was provided for each cost overrun factor to collect responses on the possibility of occurrence of the factors in this case study. The use of open-ended questions is enforced to enable the interviewees to provide a wide range of needed information. on how to determine the impact of the COVID-19 on the project cost.

4 Result and Discussion

4.1 Case Study Findings

PT. Central Tetha Jaya owns the project of Hotel Santika Premiere. With a contract value of IDR 128,000,000,000, this project is planned for 425 calendar days with a contract date starting 19 June 2019. Having 11 floors, the construction of this hotel involves seven consultants consisting of construction management consultants, architectural planning consultants, structural, interior consultants, Mechanical Electrical Plumbing (MEP) consultants, and quantity surveyors consultants. This project also has four specialist contractors who handle foundation work, STP, lifts, and heat pumps.

Table 1. Factors Affecting Construction Cost Overrun

Factors Affecting Construction Cost Overrun			Reference
A. Materials	A.1	Materials are not following the specification	[12]
	A.2	Material prices tend to rise	[12]
	A.3	Poor material preparation	[12]
	A.4	Shortage of material	[10]
	A.5	Delivery problem	[13], [14], [15]
	A.6	Transportation cost	[16]
B. Labor	B.1	Shortage of material	[13], [15]
	B.2	Low quality of labor	[12]
	B.3	Weaknesses in negotiating prices with labor foremen	[12]
	B.4	Increase in labor prices/wages	[15], [17], [18], [5]
	B.5	Reducing the number of workers on-site to comply with social distancing	[5], [6], [13], [14]
C. Heavy equipment	C.1	High rental price of equipment	[15], [19]
	C.2	High cost of equipment transportation	[12], [15]
	C.3	Lack of supervision causes equipment productivity is not to be maximized	[5], [12]
D. Occupational health and safety	D.1	Lack of PPE availability causes a decrease in productivity levels	[14]
	D.2	Due to strict health protocols (masks, hand gloves, disinfectants, hand sanitizers, soap) increased project costs during the COVID-19 pandemic.	[16], [13], [20], [1]
	D.3	There is a cost for the COVID-19 test procedure	[6]
	D.4	Provide Health/COVID-19 Supervisors on site	[1]
	D.5	Adding a mess for social distancing	[16]
	D.6	Location quarantine for positive workers and close contacts	[16]
	D.7	Project delays during the COVID-19 pandemic	[14]
E. Overhead	E.1	Additional working hours and overtime	[21]
	E.2	Fees for field meetings	[22], [20]
F. Project Financial	F.1	Poor control in the field	[23], [15], [24], [19]
	F.2	Funding from owner	[15], [18], [19]
G. Cost Estimation	G.1	It does not take into account contingencies	[19], [22]
	G.2	Inaccuracy of cost estimation	[19], [22], [5]
H. Execution Time	H.1	Frequent job delays	[25], [22], [19], [10], [17]
	H.2	Overtime schedule requirements	[18]
	H.3	Poor scheduling	[26], [18]
	H.4	Consideration of allowed working hours during the COVID-19 pandemic	[27]

4.2 Result and Discussion

Thirty-one factors have been identified from the literature study as the cause of cost overrun. Based on the interviewee's opinion, the percentage of the causative factors on cost overruns to the Santika Premiere Hotel Development Project under the pandemic COVID-19, the results are shown in Table 2.

Table 2. The Frequency of Occurrence for Each Factor

No	Factors Affecting Construction Cost Overrun	Percentage of each answer	
		Yes	No
1	Materials	50%	50%
2	Labor	67%	33%
3	Heavy equipment	0%	100%
4	Occupational health and safety	43%	57%
5	Overhead	0%	100%
6	Project Financial	50%	50%
7	Cost Estimation	50%	50%
8	Schedule	50%	50%

Six factors have a frequency of occurrence for each cost overrun cause: labor by 67%, material by 50%, implementation time by 50%, project financial aspects by 50%, estimated cost by 50%, and occupational health and safety by 43%. Two factors are not the cause of the cost overrun in the Hotel Santika project based on the interviewee's answer, namely heavy equipment and overhead.

In the Hotel Santika project, inappropriate materials can cause cost overruns. It usually happens due to an order error and or an error in material estimation during the tender. Regarding the increase in material prices, cost overruns occurred due to escalation prices for steel, iron, hollow iron, and glass materials. Steel has a significant increase of up to 35%, while hollow iron and iron increased between 10%-15%. Likewise, glass experienced a 10% increase in price.

The delivery problem also leads to some material shortages in the construction project. During the large-scale social restriction phase, the brick usually had a delivery problem. It would lead to delay in-wall work.

The shortage of labor and low quality of labor can cause cost overrun for this case study. The shortage of labor occurred at the beginning of the implementation of large-scale social restrictions in Padang City. Caused labor from the island of Java is not able to back after the Eid holiday.

Meanwhile, the low quality of the labor is considered as the cause of cost overrun. Low quality of labor is the fault of the foreman in finding labor. That is why labor can not achieve the job target. This causes an increase in project overhead. Likewise, an increase in labor prices/wages by 5% - 10% can cause cost overruns.

The increase in expenses is also due to the strict health protocol implemented in the Hotel Santika project. Contractors provide masks, hand gloves, disinfectants, hand sanitizers, hand soap, and thermometer guns during the COVID-19 pandemic. Masks are only given once, namely cloth masks, after which workers buy their masks. However, if workers do not carry masks, they will be given masks. The proportion of increase due to the completeness of Personal Protective Equipment is also not too large, namely 0.01%. But the contractor provides a special location for quarantine for labor. Labor who had positive with COVID-19 and or for the close contact should be quarantined. The contractor cost almost 10 million rupiahs.

The suspension of work also happened in this project two times in Mei 2020 and July 2021. The labor is free for fourteen days. Even though the labor did not work, the contractor still had to provide food with a total of 60 million rupiahs.

The contractor should be maintained properly cost overrun. Cost overrun can be because of inaccurate cost estimates and Cost overruns during execution. But the contractor still has the chance to overcome the situation. King [13] explore mechanisms for reducing COVID-19's negative impact on the construction industry. He pointed that one of the proposed solutions is highering productivity. The method used in this case study to overcome cost overrun is by tightening the supervision of the foreman to the worker. Tightening the supervision can control the productivity target and also complete the pending work within the allotted time. This is in line with Fahadila [19], who pointed out consistency to control the cost, work schedule, materials, and labor as a mechanism to address the cost overrun.

Riedl [28] showed 87% of budget overruns, and delays cases included identification to poor planning. As cost overrun is the impact of COVID-19, so the project manager should revise the master plan.

5 Conclusion

The COVID-19 pandemic has brought fundamental disruption at the state, community, and construction industry levels. Although construction activities in Indonesia are still allowed to continue operating in the social restriction phase, through the Minister of PUPR's Ministerial Decree Number 2/2020, its implementation brings some operational and financial problems for the project.

This research was conducted based on a case study at the Hotel Santika Padang Construction Project by conducting interviews with the contractor and the Construction Management Consultant. This study aims to identify the cause of cost overrun by COVID-19 in the construction industry. And also, this study will determine mechanisms that can reduce COVID-19's negative impact on building construction.

Of the eight indicators studied, six factors influence the cost overrun of the project under the COVID-19 pandemic. Revision in the project management plan is made as a response of the project manager to make the adaptation of the contractor in achieving project progress that disturbs while social restriction phase.

Publication of this article is supported by the Civil Engineering Department of Andalas University Grant No 133/UN16.09.D/PL/2021.

References

1. M. B. Esa, F. S. B. Ibrahim, and E. B. M. Kamal, *Adv. Sci. Technol. Eng. Syst.*, **5**, 5, (2020).
2. L. B. Sihombing, (2021).
3. C. D. Annamalaisami and A. Kuppuswamy, ASCE-ASME J. Risk Uncertain. Eng. Syst. Part A Civ. Eng., **7**, 2, (2021).
4. K. Ullah, M. S. Khan, M. T. Lakhiar, A. A. Vighio, and S. Sohu, J. Appl. Eng. Sci., **8**, 1, (2018).
5. N. Al-Hazim, Z. A. Salem, and H. Ahmad, Procedia Eng., **182**, (2017).
6. S. H. Zamani, R. A. Rahman, M. A. Fauzi, and L. M. Yusof, "Effect of COVID-19 on building construction projects: Impact and response mechanisms," IOP Conf. Ser. Earth Environ. Sci., **682**, 1, (2021).
7. K. Ullah, A. H. Abdullah, S. Nagapan, S. Suhoo, and M. S. Khan, "Theoretical framework of the causes of construction time and cost overruns," IOP Conf. Ser. Mater. Sci. Eng., **271**, 1, (2017).
8. A. Biswas, A. Ghosh, A. Kar, T. Mondal, B. Ghosh, and P. K. Bardhan, J. Phys. Conf. Ser., **1797**, 1, (2021).
9. C. C. Cantarelli, B. Flyvbjerg, E. J. E. Molin, and B. van Wee, Eur. J. Transp. Infrastruct. Res., **10**, 1, (2010).
10. M. Ogunnusi, M. Hamma-Adama, H. Salman, and T. Kouider, Int. J. Real Estate Stud., **2**, 2, (2020).
11. R. A. Khan, R. Yazdani, G. Azeem, and M. Mirmozaffari, Int. J. Appl. Decis. Sci., **1**, 1, (2022).
12. A. . G. A. Sumadi, I Wayan Edi., Wiranata, A.A., Asmara, A Sci. J. Civ. Eng., **20**, 1, (2016).
13. S. S. King, R. A. Rahman, M. A. Fauzi, and A. T. Haron, "Mechanisms for addressing the impact of COVID-19 on infrastructure projects," IOP Conf. Ser. Earth Environ. Sci., **682**, 1, (2021).
14. A. Alsharef, S. Banerjee, S. M. J. Uddin, A. Albert, and E. Jaselskis, Int. J. Environ. Res. Public Health, **18**,. 4, (2021).
15. T. Sahusilawane, J. Rekayasa Sipil,. 5, 2, (2011).
16. H. Mohamed, A. L. Mansoori, A. B. Alsaud, and H. Yas, Int. J. Pharm. Res., **13**, 01, (2021).
17. C. Sun and S. Xu, "Analysis of the impact of the COVID-19 epidemic on the construction engineering epc projects and claims," IOP Conf. Ser. Earth Environ. Sci., **676**,1, (2021).
18. S. Sohu, A. H. Abdullah, S. Nagapan, N. A. Memon, R. Yunus, and M. F. Hasmori, Int. J. Integr. Eng., **10**, 9, (2018).
19. R. Fahadila F, J. Tek. Mesin, **06**, 2017.
20. K. J. Santoso, Dimens. Pratama Tek. Sipil, **10**, 1, (2021).
21. S. Lavidaloka, "Adaptasi Pelaku Industri Konstruksi Dalam Masa Pandemi COVID-19," (2021)
22. A. J. Mangaline, Henry Pascal., Haryono, Survei Mengenai Biaya Overhead Serta Faktor-Faktor Yang Mempengaruhinya. Andi.
23. F. Fahira, Smartek, vol. **03**, 03, (2005).
24. N. A. Memon and I. A. Rahman, Mod. Appl. Sci., **07**, 08, (2013).
25. I. Santoso, Dimens. Tek. Sipil, **1**, 1, (1999).
26. Wattimury, Sipil Statik, **03**, 04, (2015).
27. D. S. T. and P. C. G. P. Mr. Moses Nyathi, Connections, **19**, 2, (2020).
28. E. Riedel and A. Al-Keim, "Strategies to Reduce Cost Overruns and Schedule Delays in Construction Projects," Walden University, (2017).

The analysis of problems, causes, and efforts to resolve problems in the service of covid-19 patients at the army hospital in Padang City

Adila Kasni Astiena[1]*, *Mohamad* Reza[2], *Fitri Sari* Susanty[2], and *Jonsriwanto*[3]

[1]Departement of Administration and Public Health Policy, Faculty of Public Health, Andalas University, Indonesia
[2]Public Health Study Program Faculty of Medicine, Andalas University, Indonesia
[3]Reksodiwiryo Hospital, Padang, Indonesia

Abstract. During the adaptation period to new habits, every hospital needs to make some changes to prevent the transmission of covid infections in the hospitals. The Ministry of Health has issued a technical guide that should be a guide for every hospital. The Army Hospital in Padang City has a different communication flow and chain of command from hospitals owned by health institutions. This study aims to analyze efforts to optimize services for COVID-19 patients during the adaptation period for new habits. The research design is qualitative. The study was conducted in October-December 2020. The unit of analysis is an emergency department for the Covid and non-covid services. Data were collected by interview, observation, and document review. The results showed that emergency room services were not following the technical guidelines for hospital services during the adaptation period for new habits. Some of the causes are seen in terms of input, ie. personnel, room and equipment, funding, and patient behavior as well as from the screening process. Steps that can be taken to optimize services are to create a service flow, install zone barriers, move screening rooms and improve patient education and carry out periodic monitoring and evaluation.

1 Introduction

Every sector of life has changed since the declaration of Covid-19 as a global pandemic by WHO in March 2020. After the announcement of the first case of Covid-19 in March 2020, the government has issued Presidential Decree Number 12 of 2020 concerning Determination of Non-Natural Disasters for the Spread of Corona Virus Disease. 2019 (COVID-19) as a National Disaster [1]. Coronavirus is spreading rapidly in Indonesia. Indonesia is a country with the highest number of deaths due to Covid-19 in the world. Health worker mortality in Indonesia is also the highest in the world. Until August 2020, the death of health workers in Indonesia reached 350 people.

The Covid-19 pandemic has changed every activity in people's lives, including changes in health care organizations. This change aims to minimize the risk of spreading Covid-19 by implementing the Adaptation of New Habits in the community. Health facilities such as health centers, clinics, and hospitals are very at risk of becoming clusters of the spread of the Sar Cov-2 virus. On the one hand, health care institutions are trying to save Covid-19 patients, but on the other hand health workers and visitors are faced with a high risk of contracting Sar Cov-2.

To anticipate this, the government has issued guidelines for handling COVID-19 for local governments, health service facilities such as puskesmas, and hospitals [2]. Especially for the prevention of transmission of Sar Cov-2 in hospitals, the government has issued a technical guide for hospital services during the adaptation period of new habits [3]. Through the implementation of this guide, it is hoped that hospitals will optimize service delivery to COVID-19 patients and otherwise be able to prevent transmission to other patients, health workers, and visitors.

The city of Padang is currently in the Covid-19 red zone due to a significant increase in cases. The Army Hospital is a reference for COVID-19 patients in the city of Padang. This hospital is under the command of the Ministry of Defense and Security, unlike other hospitals under the Ministry of Health. This is a challenge in providing health services, where on the one hand hospitals follow the military chain of command but on the other hand need to follow service standards from the Ministry of Health. This hospital not only serves soldiers and their families but also serves civilians. Even the number of patients from civil society exceeds that of patients from

* Corresponding author: adila@ph.unand.ac.id

https://doi.org/10.1051/e3sconf/202133101015

the army. If not bridged properly, this can have an impact on the quality of health services.

Assessment of the quality of emergency unit services is a crucial part of every hospital. Patients enter the door of inpatient services after going through the Emergency Department (ER) or outpatient services. The ER is the main gate of community assessment of the high or low quality of service at a hospital. Thus, the ER has several requirements and quality indicators that must be met, so that a hospital can provide quality health services. However, with the pandemic, the ER service requirements have several modifications that aim to prevent the risk of Covid-19 transmission in hospitals. In reality, there are several problems in the implementation of emergency services during the pandemic. Service adaptation must adapt to the pandemic situation so that not all hospitals have the readiness to adapt to the pandemic situation. This creates problems in the quality of health services that can endanger the safety of patients, health workers, and visitors.

During this Covid-19 pandemic, efforts to identify problems, analyze the causes of problems, and find solutions are very crucial in efforts to improve service quality. The Army Hospital in providing health services to Covid-19 patients also guides government regulations and the Technical Guidelines for Hospital Operations during the adaptation period of new habits. From the preliminary study, it is known that Covid-19 services are still not optimally implemented with several service problems in the emergency unit and inpatient units.

The purpose of this study was to analyze the problems of service for Covid-19 patients at the Army Hospital in Padang City by identifying problems, looking for causes of problems, exploring problem-solving steps, and making a model plan for monitoring and evaluating patient services. Problem identification models, problem-causing analysis, problem-solving steps, and problem-solving monitoring and evaluation plans can be used by management as tools that can be used in decision-making, especially related to handling pandemics at home that are across the Ministry, namely the Ministry of Defense and Security.

The name of the author concerned. Please ensure that affiliations are as complete as possible and include the country.

2 Literature Review

Hospital visitors are screened first before entering the ER or Outpatient Installation. Screening is a process of evaluating and screening patients and visitors to separate suspected Covid-19 patients from non-suspect Covid-19 patients, health workers, and visitors. Patients are declared suspect of Covid-19 if they meet at least 1 criteria for an epidemiological history or 1 clinical symptom. Screening aims to reduce the risk of exposure to Covid-19 to patients and health workers. Requirements include building standards, location, and flow of patients so that the goals and objectives of screening can be achieved [2].

Buildings for Covid-19 screening can be in the form of temporary buildings that already exist, can be simple tents, but must have adequate natural ventilation. The screening location should be as close as possible to the main entrance so that it can be accessed by all visitors, avoiding the accumulation of queues and ensuring the safety and security of visitors. The flow of visitor movement should be one-way with separate entrances before screening and exits after screening to avoid the risk of COVID-19 transmission. Patients who are declared Suspected Covid-19 are then triaged to determine whether they require immediate medical intervention, wait, or be referred to another hospital [2].

The requirement for the location of the triage room is to be at the entrance before entering the Covid-19 ER waiting room. At the triage location, it is necessary to have hand hygiene facilities, in the form of a sink, soap, and running water. In addition, there is a need for information and education media in the form of brochures and posters related to Covid-19 prevention. Officers need to be aware of visitors with symptoms of cough, fever, shortness of breath, and difficulty breathing [4]. Hospital services in the pandemic era also need to pay attention to space zoning, namely the division or grouping of service rooms based on the similarity of activity function characteristics. This is done to prevent the spread of COVID-19 in hospitals. The Covid-19 zone includes the Covid special service area and the Covid Special support area. Among the COVID-19 special service, areas are the emergency room and the treatment room.

According to the Regulation of the Minister of Health Number 47 of 2018 concerning Emergency Services, emergency services are services provided to patients in an immediate time to save lives and prevent disability [5]. The physical requirements of the emergency room building are related to the building area, location, entrance, and exit, vehicle entry and exit routes, special areas for ambulance parking, arrangement of room flows, and room requirements. The emergency room must also have a triage room, which is room to carry out a process of sorting patients based on the severity of the injury or disease and determining the type of treatment [6].

There are 3 options in optimizing patient care during the adaptation period to new habits [3].

1. First, the Covid and Non-Covid rooms are in separate locations, but if there are problems with limited room availability, this can be done by providing temporary or permanent barriers marked with clear special markers and adhering to a one-way lane system.
2. Second, dividing the number of special human resources for Covid-19 and Non-Covid-19 services. However, if the number of human resources is not sufficient, then a schedule/division of picket shifts is carried out between ordinary services and COVID-19 special services.
3. Third, if the availability of the room does not allow for zone separation at all, then arrangements are made for service schedules, division of service shift hours or service days followed by decontamination and sterilization measures for both rooms and

medical devices after providing services to COVID-19 patients according to the rules applicable.

Emergency room patients who have symptoms of COVID-19 must be prioritized to wait separately in the room provided. Patients must remain in this waiting room and must not visit other parts of the hospital such as the treatment room, cafeteria, and so on [5]. The COVID-19 special ER area is an observation room or action room for emergency room patients with COVID-19 symptoms. This area is separated from the non-COVID-19 emergency room area through a permanent or temporary boundary. After entering the COVID-19 ER area, patients are not allowed to return to the non-COVID-19 ER area and officers are only allowed to enter and leave this area through the dressing room/ doning/doffing [2]. The arrangement of Covid-19 services that are carried out properly during a pandemic situation is expected to improve the quality of Covid-19 services to the community to heal and prevent transmission to officers and fellow patients and visitors.

3 Method

The design of this research is qualitative. The study was conducted for 4 months from October 2020 to February 2021. The research variables were Problem Identification, Causes of Problems, Problem Solving, Monitoring and Evaluation, and Follow-up Plans for Optimizing Emergency Room Services during the adaptation period of new habits. Data were collected by interview, observation, document review, and FGD methods. The research informants were 7 people, namely the Head of the Hospital, Head of Outpatient Installation, Head of the Emergency Room, 1 doctor, 1 nurse, and 2 patients. Data analysis was carried out thematically. The validity of the data was tested by the source triangulation method and the triangulation method.

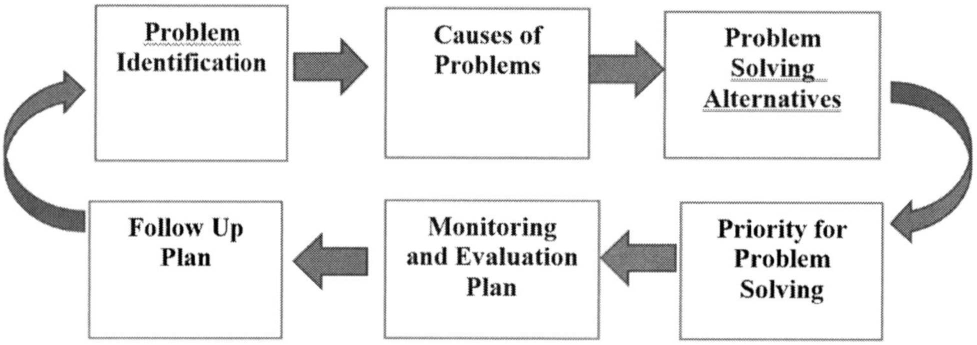

Fig. 1. Research Frameworks

4 Results

4.1 Problem Identification

From the results of observations, interviews, and document review, the results of problem identification were obtained

4.1.1 Covid patient services in the ER are not up to standard.

From the results of direct observations and interviews with ER officers, data showed that the patient screening room was located in one room with the ER patient registration counter which was separated by a barrier made of glass. This screening room has 2 beds, but there is no barrier between the beds, so there is a high risk of causing Covid transmission to non-Covid patients.

Screening is carried out by a nurse or doctor on duty using a rapid assessment form. After screening, the patient is placed in the appropriate room, namely the Covid isolation room for Covid patients and the triage room for non-covid patients. The Covid isolation room and triage room are located next to the screening room. The Covid isolation room and the triage room have the same entry route.

Officers enter the Covid isolation room using PPE. If the patient is examined in the laboratory, the sample will be taken by the labor officer in the Covid isolation room. However, if a Covid patient requires a radiological examination, the patient will be taken to a radiology room using an isolation chamber. The Army Hospital does not yet have a portable X-ray, so Covid and Non-Covid patients must take turns to carry out examinations. After X-ray examinations are carried out on COVID patients, the radiology room will be decontaminated and non-COVID patients may not conduct examinations. This decontamination process takes about 1 hour. If there are non-COVID patients who need X-rays, they have to wait. Procurement for this portable X-ray has been proposed but has not yet been realized.

Some patients who come to the ER behave dishonestly regarding the complaints and symptoms of Covid. Patients who should have been placed in the Covid isolation room were finally placed in the triage room. This increases the risk of transmitting Covid to non-Covid patients. After laboratory and radiological examinations, patients with suspected Covid-19 are given education for self-isolation if they only experience mild symptoms and do not require treatment. Patients who need Covid treatment will be transferred to the Covid treatment room using an isolation chamber.

Several Covid patients are recommended to be treated but are not willing to be treated and choose to go home of their own free will. Patients who agree to be treated will be escorted by officers to the Covid treatment room. There is no special route to the Covid special treatment room. Access from the ER to the radiology examination room has the same path as Non-Covid patients, visitors, and hospital staff.

The hospital already has an isolation chamber, which is a means of transporting patients to carry out radiological examinations and move to the COVID-19 isolation treatment room. There are no posters, banners, brochures, or so on about Covid in the emergency room so that patients and visitors get limited information.
There are 3 service shifts in the ER where the number of doctors on duty is 3 people per shift and 5 nurses per shift. The effective working hours of Covid services are 140 hours per month. Based on the Ministry of Health, the minimum need for general practitioners is calculated by the formula = (3 shifts x24 hours) x30 days divided by 140 hours of optimal service = 16 people. The number of nurses on one shift is 5 people. The minimum need for nurses is calculated using the formula = (5x24) x 30 days divided by 140 hours = 26 people.

4.1.2 Covid Isolation Treatment Is Not Up To Standard

The capacity of the Covid patient treatment room at the Army Hospital is 68 beds divided into 2 locations, namely Room V and the Pavilion Building. The Covid treatment room is not equipped with negative pressure or HEPA filters as air filters. All COVID patients treated at the Padang City Army Hospital are patients with mild to moderate symptoms. Patients with severe symptoms to M Djamil Hospital. Because the Army Hospital does not yet have a ventilator.

The division of service zones at RST Padang already exists but is not informed to patients/visitors. There is no barrier between the red zone and other zones. No officers are guarding the red zone, or warning signs so that non-Coivid patients and visitors enter the zone without knowing it.

4.2 Prioritizing Problems

Priority problems using the CARL method with a weighting of 1-5 values. CARL itself has the meaning:

1. C (Capability), namely the availability of resources (funds, facilities, equipment)
2. An (Acceptability), namely convenience, the existing problems are easy/can be overcome with the availability of methods/methods/technology as well as implementation support.
3. R (Readiness) which is the readiness of the implementing staff as well as the readiness of targets such as expertise or ability and motivation
4. L (Leverage) that is how much influence one criterion with another in problem-solving.

Furthermore, to determine the priority of the problems identified by making a CARL criteria table and scoring with a range of 1 to 5, where 5 = very important, 4 = important, 3 = quite important, 2 = not important, 1 = very unimportant. And the priority is the one with the highest value after each criterion is multiplied (CxAxRxL = Value). The value is determined based on the results of the FGD with the management with the results shown in table 1. Based on table 1, the priority problem is "Covid patient services in the emergency room are not up to standard".

Tabel 1. The Problem Priority in the Covid Services in The Army Hospital

No	Problem	C	A	R	L	Score	Priority Number
1	The service for Covid patients in the ER is not up to standard	5	5	4	5	500	I
2	Patient care in the Covid isolation ward is not up to standard	4	5	4	5	400	II

4.3 Cause of Problem

To help determine the possible causes of the problem, the Ischikawa diagram is used. Ischikawa diagram is a tool to describe several causes of problems in detail based on input aspects consisting of man, money, material, method, and machine. The cause of the problem can be described as follows:

4.3.1 Man

The cause of the problem in the Covid service come from Man Aspect, due to ;

1. The patient is not honest in providing information about the complaints experienced. Patients who suffer from COVID will be treated like ordinary patients in an ordinary room and after the results of laboratory and radiology examinations are obtained, it will be known that COVID-19. This patient is at risk of transmitting the disease to staff and other patients in the room.
2. Rejection of patients who do not want to be diagnosed as COVID patients.
3. Patients who have been diagnosed as COVID-19 patients sometimes deny their illness and even accuse

https://doi.org/10.1051/e3sconf/202133101015

the hospital of only looking for profit so that they refuse to be treated and ask to go home at their request. Some patients argue that they have the right to refuse to perform a service or action.

4.3.2 Method

1. The arrangement of the screening room in the ER is not up to standard
2. The location of the screening room is in the ER and there are no barriers between beds in the ER. This poses a high risk of transmitting COVID to non-covid patients.
3. The arrangement of the ER isolation room is not up to standard.
4. The IGD Covid isolation room is inside the ER and has the same route as the non-Covid patient triage room. There should be a different path between the isolation room and the patient triage room to avoid the transmission of Covid.
5. The flow of services for COVID patients in the ER uses the old flow
6. The flow of patient care in the ER still uses the old flow that does not include screening and triage in the ER. The channel is still asking for travel history to China, while the transmission of Covid in Padang City has been going on by local transmission.

4.3.3 Material

1. There is no portable X-ray device yet
   There is no portable X-ray device so that the examination for COVID and Non-COVID patients is carried out alternately. This causes the waiting time for ER radiology services to increase, thereby reducing the quality of health services.
2. No media information related to Covid was found
   There were no information media such as banners, billboards, or brochures that explained Covid, so other patients/visitors did not have sufficient understanding.
3. There is no warning for those who hinder the performance of the Health Officer
   No information media was found, it could be in the form of banners containing laws and sanctions for those who try to hinder the performance of health workers in p. for patients who hinder the performance of health workers in handling Covid.

4.3.4 Money

Limited funds for the procurement of portable X-rays. The budget for handling Covid comes from the Army Hospital. There has been no financial assistance from other parties, both from the Ministry of Health and from the Regional Government regarding the handling of Covid.

4.3.5 Market

An increase in the number of COVID patients compared to the beginning of the pandemic. There has been a significant increase in Covid patients in Indonesia, including in West Sumatra Province

The description of the causes of the problem is described as follows; This problem is searched for the cause of the problem by describing it on the Ischikawa Diagram as shown in Fig. 2.

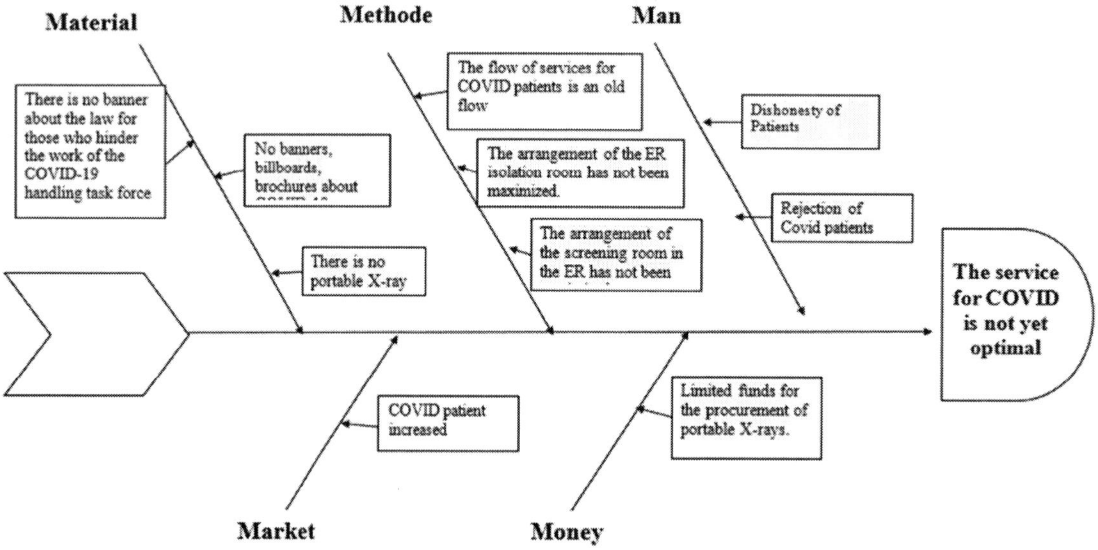

Fig. 2. Causes of the Problem

4.4 Troubleshooting Alternatives

Table 2. Causes of Problem and Troubleshooting Alternative

Subject	Causes of	Alternative Problem Solving
Man	Patients are dishonest in providing information about their complaints	Provide a statement letter during screening for COVID that the information provided is correct and if there is an error, the patient is willing to be prosecuted
	Rejection patients who do not want to be diagnosed as COVID patients.	Provide an explanation that COVID is not a disgrace that must be covered up.
Method	The arrangement of the screening room in the ER has not been maximized.	Screening at the ER entrance by officers (security) who will then place patients according to their illness, namely COVID or Non-COVID
	The arrangement of the ER isolation room has not been maximized.	Moving the isolation room IGD into space screening IGD present and in the bulkhead permanently so patients COVID not come into contact with the patient Non COVID
	Chronology of service Patients COVID a groove longer	Creating groove patient care covering new screening measures and triage
Material	There are no X-ray portable	Accelerate procurement portable x-ray device for COVID patients
	There are no banners, billboards, brochures about COVID-19	Making banners, billboards and brochures about COVID-19
	There are no banners about the law for those who hinder the work of the COVID-19 handling task force	Make a banner containing the law for people who hinder the work of health workers.
Funds (Money)	Limited funds for the provision of portable x-rays.	Requesting financial assistance from the Ministry of Health or the Provincial
Market	An increase in the number of COVID patients compared to the beginning of the pandemic.	Maintaining the quality of patient care by implementing health protocols

4.5 Priority for Problem Solving

The method used in determining the root of the problem is to continue the analysis of the fishbone diagram using the effectiveness/efficient method:

1. M (Magnitude): the magnitude of the problem that can be overcome
2. I (Important): the importance of a way out to solve the problem
3. V (Vulnerability): sensitivity (path accuracy) out for problems)
4. C (Cost) : Efficiency (Costs incurred)

$$\text{Priority (P)} = \frac{M \times I \times V}{C}$$

Scoring for MIV :
1: Least effective
2: Ineffective
3: Less effective
4: Effective
5: Most effective

Scoring for C :
1: Most efficient
2: Efficient
3: Less efficient
4: Inefficient
5: Least efficient

Table 3. Problem Solving Priority

No	Activities	M	I	V	C	Score	Priority
1	Provide a statement letter	5	5	5	1	125	I
2	Provide an explanation that COVID is not a disgrace which must be covered	5	5	5	1	125	I
3	Patient Screening at the entrance	5	5	5	2	62.5	II
4	Moving the ER isolation room to an existing IGD screening room	5	5	5	2	62.5	II
5	Create a new patient service flow including screening and patient triage measures	5	5	5	1	125	I
6	Accelerating the procurement of portable X-rays for COVID patients	4	4	4	3	21.33	IV
7	Making banners, billboards, brochures and others regarding COVID-19	5	5	5	2	62.5	II
8	Making banners about the law and sanctions if preventing officers from carrying out COVID response	5	5	5	2	62.5	II
9	Requesting financial assistance from the Ministry of Health or the Province	4	4	4	4	16	V
10	Maintaining the quality of patient care by implementing health protocols	4	5	4	2	40	III

Based on table 3, the authors take the top two priority scales for alternative problem solving as follows:

1. Provide a statement letter at the time of screening for COVID that the information provided is correct and if there is an error, the patient is willing to be prosecuted

2. Explain that COVID is not a disgrace that must be covered
3. Creating a new patient care flow including screening and patient triage tindakan
4. Screening at the ER entrance by officers (security) who will then place the patient according to their illness, namely COVID or Non-COVID
5. Move the ER isolation room to the current ER screening room and in a permanent partition so that COVID patients do not come into contact with Non-COVID patients
6. Make banners, billboards, brochures, and others about COVID-19
7. Make a banner regarding the law and sanctions if it prevents officers from handling COVID

4.6 Activity and Budget Plan of Action

After the alternative solutions to the problem are determined, a work plan (the plan of action/POA) is prepared according to table 4

Table 4. Activity and Budget Plan of Action

NO	Activity	Objectives	Elements Involved		Time/Place	Person in Charge	Budget Plan (Rp)
			HR	Facility			
1	Providing a statement letter	Prividing the correct information about covid	Heads of Outpatient room, Head of Room, Doctor on duty, Nurse, Screening officer	Statement Letter	Every service hour/ER	Head of Outpatient Installation, Head of Room	Does not incur additional costs
2	Providing the explanation that COVID is not a disgrace which must be covered	Understanding that COVID is not a disgrace	Doctor on duty	*Informed consent*	Every service hour/ER	Doctor on duty	Does not incur additional costs
3	Creating a new service flow including screening and patient triage measures	The flow of patient care is clear	Head of Outpatient Installation, Head of Room,	Flow of patient care	December er 2020/ER	Head of Outpatient Installation, Head of Room	Does not incur additional costs
4	Conducting screening for patient at the entrance	Avoiding transmission of COVID to non-COVID patients	Head of Installation Outpatient	Blank Screening	Every service hour /ER	Screening officer	Does not incur additional costs
5	Moving the ER isolation room to the current ER screening room and making permanent partition	Avoid transmission of COVID to Non COVID patients	Head Outpatient Installation, Head of Room, Doctor on duty, Nurse, Screening officer	Isolation room does not pass the triage room (Non COVID)	December 2020/ER	Head of Outpatient Installation, Head of Room	10,000,00 0
6	Making banners, billboards, brochures regarding COVID-19	Provide information to patients and visitors about COVID	Head of Service Affairs Medical, Head of Finance, Head of Outpatient Installation, Head ofRoom	Banner, Billboards, brochure	December 2020/ER	Head of Outpatient Installation, Head of Room	2,500,000
7	Making banners about the law and sanctions if obstructing officers in handling COVID	Informing the Act and sanctions against persons who hinder the work of the	Head of Medical Service Affairs, Head of Finance, Head of Outpatient Installation, Head of Room,	Banner	December 2020/ER	Head of Outpatient Installation, Head of Room	1,500,000

Plan of Action (POA) for number 1- 4 do not incure costs, meanwhile for POA number 4 costs 10 million rupiahs, for POA number 5, costs two and a half million rupiahs and for POA number 7 costs 1,5 million rupiahs.

4.7 Monitoring and Evaluation

In guiding the implementation of activities in the field later, monitoring and evaluation activities are planned as shown in table 5.

Table 5. Monitoring and Evaluation Plan

No	Activities	Input	Process	Output
1	Providing a statement letter	Head of Outpatient Installation, Head of Room, Screening Officer, Statement Letter The	officer provides a statement letter to all patients Filling in the statement by the patient	The statement letter is filled out honestly by the patient
2	Providing an explanation that	Doctor Inform Consent	on dutyExplanation from the doctor on duty to the	Patient understands 100% that COVID is not
3	Creating a new service flow including	Head of Outpatient Installation, Head of Room,	Making a newnew	patient service flow There is apatient service flow
4	Conducting screening for patient at the	Officers, screening forms	Officers screen at the ER entrance by filling in the form	Patients have been placed based on COVID or Non COVID
5	Moving the ER isolation room to the current ER screening	Head Outpatient Installation, Head of Room,	Moving the isolation room to the current screening room and given a permanent partition	Separation of paths for COVID and Non-COVID patients
6	Making banners, billboards, brochures and	Banners, billboards, brochures, equipment officers	Patients/visitors can see and read the information on banners, billboards and brochures	Understanding about COVID-19
7	Making banners about the law and sanctions	Head of Hospital, Head of Outpatient Installation, Head of Room, Banner	Patients/visitors read the banner provided	Understanding about about the sanctions if obstructing officers in handling COVID

5 Conclusion

The service for COVID-19 patients in the ER at the Army Hospital has not been following the technical guidelines for hospital services during the adaptation period for new habits issued by the Ministry of Health. This is caused by several underlying factors, namely from the 5 M aspect (Man, Method, Material, Money, Market). First, the patient is dishonest with the complaints he experiences and the rejection of the patient diagnosed with COVID (man). Second, the arrangement of the screening room in the ER is not following the service technical guidelines for handling COVID and there is no new COVID Patient service flow (method). Third, there is no portable x-ray; there are no banners, billboards, brochures about COVID-19, and no banners about laws containing sanctions for people who hinder officers from working (materials). Fourth, limited funds for procuring portable X-rays (money), and fifth, the increase in the number of COVID patients is very fast compared to the beginning of the pandemic (the market).

The priority of alternative problem solving taken is by 1). Provide a statement letter for honesty of information from patients, 2) Educate the public that Covid is not a disgrace and does not need to be stigmatized, 3) Create a flow of services for Covid patients in the form of screening and triage actions 4) Conduct screening at the entrance and place COVID or Non-COVID patients 5) Making permanent insulation in the ER so that COVID patients do not come into contact with non-COVID patients. 6) Provide health promotion media in the form of banners, billboards, brochures related to Covid-19, and laws and sanctions for those who hinder officers in handling COVID.

References

1. Decree of the President of the Republic of Indonesia. Presidential Decree No. 12 of 2020 concerning the Determination of Non-Natural Disasters for the Spread of Corona Virus Disease 2019 as National Disasters. Fundam Nurs.;(1):1–2. Available https://peraturan.bpk.go.id/Home/Details/135718/keppres-no-12-tahun-2020 (2020)
2. Ministry of Health. Guidelines for Prevention and Control of COVID-19 5th Available https://covid19.go.id/storage/app/media/Protokol/2020/Juli/REV-05_Pedoman_P2_COVID-19_13_Juli_2020.pdf (2020)
3. Ministry of Health. Technical Guidelines for Hospital Services During the Adaptation of New Habits. **5**, ARSI Journal. Available: https://arxiv.org/pdf/1707.06526.pdf%0Ahttps://www.yrpri.org%0Ahttp://weekly.cnbnews.com/news/article.html?no=124000%0Ahttps://www.fordfoundation.org/%0Ahttp://bibliotecavirtual.clacso.org.ar/Republica_Dominicana/ccp/20120731051903/prep%0Ahttp://webpc.cia (2020)
4. World Health Organization. COVID-19 Risk Communication Materials for Health Service Facilities. WorldHealing Organs.Available from:https://covid19.go.id/p/protokol/pedoman-pencegahan-dan-pengendalian-coronavirus-disease-covid-19-revisi-ke-5 (2020).
5. Ministry of Health. Regulation of the Minister of Health of the Republic of Indonesia Number 47 of 2018 concerning Emergency Services. Available from: http://ieeeauthorcenter.ieee.org/wp-

https://doi.org/10.1051/e3sconf/202133101015

content/uploads/IEEE-Reference-Guide.pdf%0Ahttp://wwwlib.murdoch.edu.au/find/citation/ieee.html%0Ahttps://doi.org/10.1016/j.cie.2019.07.022%0Ahttps://github.com/ethereum/wiki/wiki/White-Paper%0Ahttps://tore.tuhh.de/handle/11420 (2018).

6. Indonesian Ministry of Health. Hospital Emergency Room (IGD) Standards. Minister of Health of the Republic of Indonesia Regulation of the Minister of Health of the Republic of Indonesia. 1–29. Available from: http://hukor.kemkes.go.id/uploads/produk_hukum/PMK_No__47_Th_2018_ttg_Emergency_Services.pdf (2009).
7. Health England P. COVID-19: Guidance for maintaining services within health and care settings Infection prevention and control recommendations Version 1.2.;1–49. Available https://assets.publishing.service.gov.uk/government/uploads/system/uploads/attachment_data/file/990923/20210602_Infection_Prevention_and_Control_Guidance_for_maintaining_services_with_H_and_C_settings__1_.pdf.(2021).

Analysis of the government regulations on COVID-19 pandemics in Indonesia: implementation and challenges

Benny Hidayat[1*], *Ade Suzana* Putri[2], *Taufika* Ophiyandri[1], *Bambang* Istijono[1], Dilanthi Amaratunga[3], *Richard* Haigh[3], and *Asitha* de Silva[3]

[1]Civil Engineering Dept., Faculty of Engineering, Andalas University, Padang, West Sumatra, Indonesia
[2]Faculty of Public Health, Andalas University. Padang, West Sumatra, Indonesia
[3]Global Disaster Resilience Centre, University of Huddersfield, United Kingdom

Abstract. COVID-19 started in China and then spread to other countries. The Indonesian government officially announced the first case of COVID-19 on 2nd March 2020. Previously, on 4th February, the Ministry of Health issued a Decree of the Minister of Health number HK.01.07 of 2020 regarding "Determination of Novel Coronavirus Infection (2019-Ncov Infection) as a Disease That Can Cause Outbreaks And Efforts to Overcome it". WHO then declared COVID-19 as a pandemic on 11th March 2021. After that, the central government issued various regulations related to the handling of COVID-19. This paper aims to study the regulations issued by the Indonesian government regarding the handling of COVID-19. The research method uses a content analysis method using nVivo software. The study results show that the regulations are quantified, consisting of 30% related to social, 10% related to the budget, 25% related to health, 20% related to COVID-19 handling organizations, and 15% related to economic recovery. This paper also presents the challenges related to the implementation of these regulations

1 Introduction

The World Health Organization (WHO) on 11th March 2020 declared the coronavirus outbreak or COVID-19 as a global pandemic. This outbreak was initially discovered in Wuhan (China) and then spread to other countries. The determination of the status of this pandemic is due to the rapid and widespread to areas far from the center of the outbreak. With this pandemic status, WHO asks all countries to increase their response to the COVID-19 emergency.

In modern human history, this ongoing outbreak of COVID-19 is unprecedented. WHO has declared COVID-19 as a pandemic, but the factors that cause, vulnerability, and impact of COVID-19 are not only limited to health factors This pandemic has had a very hard impact on people who are already in a vulnerable condition, such as women, children, migrant workers, and people with disabilities. This pandemic is an example of systemic risk where the existing hazard not only harms parts of the system but also has the potential to fail for all systems. The normal approach that the government has taken so far for disasters does not seem to be able to answer the challenges in handling the COVID-19 pandemic. It is particularly important to better align the prevention and response efforts of health ministries and disaster management authorities, from national to local. COVID-19 has underlined that response mechanisms require much strengthening. As this pandemic is unprecedented, most countries do not have practical experience in handling events resulting from a combination of natural hazards and biological hazards.

The COVID-19 pandemic has been declared a national disaster in Indonesia. Its massive impact, particularly on health and socioeconomic aspects, requires quick and effective responses from every level of government to administer the pandemic control measures and to enhance community capacity to cope with the pandemic impact. Some literature explores the government responses from countries at the regional or national level [1–4]. However, rarely are discussing the preparedness at the local level, particularly in decentralized government countries, and exploring how COVID-19 pandemic enhance the local government to take into account the epidemic or pandemic as an important biological hazard in their contingency policy and embedding health into their disaster risk management.

Padang city is the capital of the West Sumatra Province of Indonesia and has a population of around one million people. It is situated on the very flat liquefiable ground, and 50% of residents live in tsunami inundation areas. Indonesia is one of the most vulnerable countries to disasters globally, frequently exposed to various hazards. Indonesia has a high risk of flooding, as more than 60%

* Corresponding author: bennyhidayat@eng.unand.ac.id

of Indonesia's districts are exposed to flood hazards. Indonesia also faces high seismic, tsunami, and volcanic risk, as situated in the Pacific Ring of Fire with 127 active volcanoes. With approximately five major events per year, Padang is a metropolitan area that is one of the most prone to geophysical disasters in Indonesia. The disasters that occurred in Indonesia not only had an impact on humans but also had an impact on the economy. Living in hazardous areas, having limited financial resources, and limited access to basic services make the poor and vulnerable the most affected by disasters.

Like many other cities across the globe, Padang could not escape from the current COVID-19 pandemic. As of 28th December 2020, in Indonesia, there were a total of 765,350 confirmed cases with 22,734 deaths [5]. The numbers increased significantly to 3,967,048 confirmed cases and 125,342 deaths per 21st August 2021 [6]. In Padang, there are 40,016 total confirmed cases and 511 total deaths as of 22nd August 2021 [7]. A significant increase of new cases and deaths were notified in June and July 2021, and Padang has been identified as one of the cities with the highest level of risk by the Ministry of Health [8].

With the condition of the city of Padang, which is very prone to disasters, and the condition of the ongoing COVID pandemic, the need for disaster risk reduction policies that embed health factors appears. As an initial step to achieve this, a study is carried out on how the handling of COVID-19 is still ongoing. This paper aims to examine the government's response in handling COVID-19 in the form of reviewing key documents regarding regulations related to COVID-19. Problems and challenges were also identified in the implementation of these regulations. Then a study was conducted on the implementation of regulations at the local government level, taking the case in the city of Padang.

2 Research Method

This paper is an initial part of research entitled 'Embedding COVID-19 Preparedness Into Local Disaster Risk Reduction, which was carried out by a team of researchers from Universitas Andalas and Huddersfield University. This paper is an initial study conducted by the team from Universitas Andalas.

At this initial stage, a study was conducted on the response that had been carried out by the Indonesian government in dealing with COVID-19, in the form of issuing regulations at the central government and local government levels. Also identified are the obstacles and challenges in implementing these regulations.

The study in this paper uses a literature review by applying the content analysis method from regulatory documents and research paper documents related to COVID-19 in Indonesia. Regulatory documents related to COVID-19 issued by the government are traced from the National Legal Documentation and Information Network (JDIHN) system with the website address at https://jdihn.go.id/. Meanwhile, research papers on COVID-19 were traced from the Garuda Portal, a portal containing articles published in journals in Indonesia. Because this is about COVID-19, the selected paper was published in 2020 and 2021.

The regulatory and research papers are then downloaded in the file. Then using the nVivo software (Fig.1), content analysis was carried out on these papers. This nVivo software will help determine the themes of problems in implementing regulations related to COVID-19. Additional data and information were also obtained by conducting interviews with key officials in the Padang city government and the Padang City Regional Disaster Management Agency (BPBD).

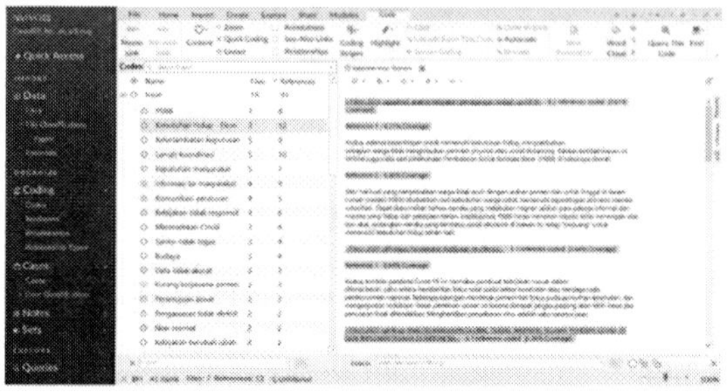

Fig. 1. nVivo software to analyze regulations and papers

3 Results and Discussions

The government officially announced the first case of COVID-19 on 2nd March 2020. Previously, on 4th February, the Ministry of Health issued a Decree of the Minister of Health number HK.01.07 of 2020 regarding "Determination of Novel Coronavirus Infection (2019-Ncov Infection) as a Disease That Can Cause Outbreaks And Efforts to Overcome it". WHO then declared COVID-19 a pandemic on 11th March 2021.

After that, the central government issued various regulations related to COVID-19 (see figure 2). On 20th March, the President issued Presidential Instruction Number 4 of 2020 regarding "refocusing of activities, reallocation of budgets and procurement of goods and services in the context of accelerating the handling of Corona Virus Disease 2019 (COVID-19)". This instruction calls for prioritizing the use of existing budget allocations for activities that accelerate the handling of COVID-19 in ministries, institutions, and government units. The Presidential Instruction also aims to accelerate the procurement of goods and services related to the handling of COVID.

On 20th March 2020, Presidential Decree No. 9 of 2020 was also issued regarding the Task Force for the Acceleration of Handling COVID-19. This decision improves the previous decision, increasing the membership of the task force from various institutions, because the spread of COVID-19 has spread and has an impact on social, economic, economic, health and welfare aspects of the people, including WHO has declared COVID-19 a pandemic.

At the end of March, the government again issued several regulations. Law Number 2020 concerning state financial policies and financial stability for handling COVID-19. Government Regulation Number 21 of 2020 was also issued regarding Large-Scale Social Restrictions (PSBB) in the context of accelerating the handling of COVID-19. Also, on the same date, Presidential Decree No. 11 of 2020 concerning the Determination of the COVID-19 Public Health Emergency was issued, which declared COVID-19 as a public health emergency that must be taken care of. At the same time, Presidential Regulation Number 52 of 2020 was issued, which assigned the Ministry of Public Works and Public Housing to carry out the construction of observation and shelter facilities in the fight against COVID-19 along with public infrastructure, facilities, and utilities on Galang Island, Batam City, Riau Islands Province.

Following up on PP No. 21 of 2020 concerning PSBB, the Ministry of Health issued guidelines for large-scale social restrictions (abbreviated in Indonesian as PSBB) with health regulation number 9 of 2020 on 3rd April 2020. Large-Scale Social Restrictions are restrictions on certain activities of residents in an area suspected of being infected with Corona Virus Disease. Implementation of Large-Scale Social Restrictions includes school and workplace holidays, restrictions on religious activities, restrictions on activities in public places or facilities, restrictions on social and cultural activities, restrictions on transportation modes, and restrictions on other activities specifically related to defense and security aspects.

Then on 13th April 2020, Presidential Decree No. 12 of 2020 was issued regarding the Determination of Non-Natural Disasters that Cause Corona Virus Disease 2019 (COVID-19) as a National Disaster. The consideration is that this non-natural disaster caused by the spread of COVID-19 has impacted increasing the number of victims and property losses, expanding the coverage area affected by the disaster, and having implications for broad socioeconomic aspects in Indonesia. This presidential decree also orders governors, regents, and mayors as the Chair of the Task Force for the Acceleration of Handling COVID-19 in the regions; in setting policies in their respective regions, they must pay attention to the policies of the Central Government.

In May and the following month, the government issued a series of government regulations in the economic field related to COVID-19. On 11th May, Government Regulation No. 23 of 2020 was issued for the National Economic Recovery Program (PEN). This program is a series of activities for the recovery of the national economy, which is part of the state financial policy implemented by the government to accelerate the handling of the pandemic and face threats that endanger the national economy and financial system stability and save the national economy. On 10th June, PP No. 29 of 2020 was issued regarding income tax facilities in the context of handling COVID-19. In August, following the issuance of PP No. 49 of 2020 regarding the adjustment of contributions for the social security program for workers during the non-natural disaster of COVID-19.

In September 2020, a series of regulations related to the COVID-19 vaccine was issued. Presidential Decree No. 18 of 2020, dated 3rd September 2020, regarding the acceleration of the development of the COVID-19 vaccine. Then on 5th October 2020, Presidential Regulation No. 99 of 2020 was issued regarding the procurement and implementation of vaccinations in response to the COVID-19 pandemic. Then on 9th February 2021, Presidential Regulation number 14 of 2021 was issued regarding the procurement and implementation of vaccinations, amending PP Number 99 of 2020.

Of the various regulations related to COVID-19 issued by the central government, if grouped based on the theme of the regulation, the quantification consists of 30% related to social society, 10% related to the budget, 25% related to health, 20% related to organizations handling COVID-19, and 15 % related to economic recovery.

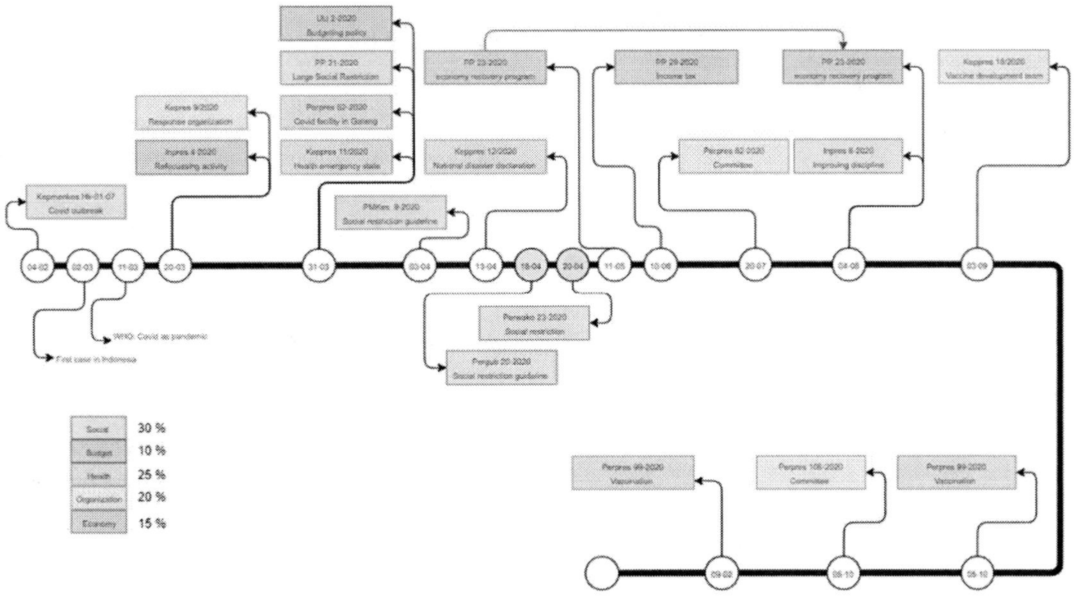

Fig. 2. Timeline of COVID-19 related regulations

3.1 Challenges

The first case of COVID-19 in Indonesia was announced by the government on 2nd March 2020. After that, the government issued a series of regulations related to the handling of COVID-19, forming an organization to handle COVID-19, regulations related to social society, related to health, related to the economy, and related to budgeting.

One government policy that is directly related to the spread of the COVID virus is to implement a large-scale social restriction (called PSBB, Pembatasan Sosial Berskala Besar). The implementation of PSBB is based on epidemiological considerations, the magnitude of the threat, effectiveness, resource support, operational technical, economic, social, cultural, and security considerations. This policy is in the form of Government Regulation Number 21 of 2020 concerning PSBB in the Context of Accelerating Handling of COVID-19. PSBB is a restriction on certain activities of residents in an area suspected of being infected with COVID-19 to prevent the possible spread of the virus. Furthermore, through the Regulation of the Minister of Health Number 9 of 2020, the government issued guidelines for implementing PSBB to reduce the number of COVID-19 cases that have increased.

The implementation of this PSBB is one of the government regulations that are directly related to the general public. The government banned the holding of mass activities and crowds, and companies were closed and schools and campuses. Offices were asked to lay off their employees. Ports, airports, stations, terminals, and other transportation infrastructure are restricted. Business places such as hotels, shopping malls, and tourist attractions, and sports stadiums are closed. Houses of worship are also not allowed to hold mass activities that invite crowds.

In the implementation of these government regulations, there are several issues and challenges in their implementation. Several issues and challenges can be identified from the content analysis of journal articles regarding COVID-19 in Indonesia. Issues and challenges in implementing this regulation are interrelated.

3.1.1 The government's initial response was slow.

At the beginning of the COVID-19 outbreak, the government was considered slow in responding [9], [10]. The President issued Presidential Decree Number 7 of 2020 concerning the Task Force for the Acceleration of Handling the 2019 Corona Virus Disease after 11 days of announcing the first case in Indonesia. On 20th March 2020, the President issued Presidential Decree No. 9 of 2020, which gave the governor the authority to direct and evaluate the handling of COVID-19 in their respective regions.

3.1.2 The economic needs of the community.

The implementation of the regulations is faced with the community's basic needs, and many people violate the rules to stay at home (during PSBB) because of the pressure of household economic needs. Most of those affected are informal workers who depend on their daily income. Generally, they have no savings, and when their movement is restricted, then their income becomes non-existent.

One strong thing that causes residents to be indifferent to the government's directive to stay at home (through PSBB) is the need for residents to fulfill their daily economic interests. It can be assumed that those who migrate are informal workers and those who make a living from daily work. The implication is that the PSBB only suppresses the migration of the middle-upper and upper classes, while those with lower socioeconomic status continue to struggle to meet their daily needs.

Data from the Central Statistics Agency (BPS) recorded that formal workers in Indonesia amounted to 55,272,968 people while informal workers numbered 74,093,224. This large number of informal workers experience social vulnerability because they are faced with the choice to comply with government advice or maintain family economic resilience [11].

The research of Mujani and Irvani [12] noted that there was a relationship between income and community compliance. They noted that residents who used to work before COVID-19 tended to violate PSBB. Residents who have a negative attitude towards the PSBB are more likely to have educational backgrounds and lower incomes. Because people's incomes are low, they cannot live without working regularly outside the home unless they get social security while staying at home and not working.

3.1.3 Weak coordination.

Weak coordination between stakeholders is also a problem in controlling the spread of the coronavirus in Indonesia [13]. The government system in Indonesia is a unitary state, placing the central government as the main stakeholder in terms of disasters, both natural and non-natural disasters, including health. The government system in Indonesia also recognizes the concept of regional autonomy, which mandates that regional governments can carry out all affairs except foreign policy, defense, security, judicial, monetary and fiscal, and religion.

The lack of transparency of the central government in providing valid data regarding COVID-19 has forced several local governments to lock down or quarantine their territories independently. The central government opposed this decision because the regional quarantine decision was the authority of the central government. This example shows the weak coordination between the central and local governments, which has caused the spread of COVID-19 in almost all provinces in Indonesia to move quickly. The "authority of affairs" issue between the central and local governments was also a debate in the early days of COVID-19 attacking Indonesia.

3.1.4 Community non-compliance and sanctions.

Many people have not complied with the regulations made by the government, such as the PSBB. In addition to meeting economic needs, people are also bored staying at home for a long time. People are also fed up with the restrictions on work and activities and choose to keep doing their routines.

In the implementation, officers are also considered less alert, have unclear sanctions, and have low discipline, which is why the rules are not successfully implemented. This is also exacerbated by the lack of oversight of the implementation of regulations.

Government firmness is also needed in implementing these regulations, in the form of sanctions for those who violate them, so that people become more obedient. The government must have the courage to impose sanctions on citizens who do not comply with the government's appeal or direction. These sanctions not only deter people but also make people consent to do things regulated by the government.

3.1.5 Culture and politics.

The implementation of regulations related to COVID-19 is also faced with Indonesian culture. The implementation of PBBB becomes ineffective when the cultural activity is being done. For example, 'mudik' or traveling to the place of origin is considered something that cannot be abandoned. Religious activities are also something that cannot be abandoned, so that restrictions on movement by the government are also in some places ineffective.

Mujani & Irvani's research [12] also found that their political preferences also influenced people's compliance with COVID-19 regulations. People who support government parties tend to agree to regulations, and on the other hand, people who are not pro to government parties tend to not comply with government regulations.

3.1.6 Information to the public.

Communications conveyed by the government regarding the handling of COVID-19 are less effective, including the lack of accurate data and information submitted by the government to the public, the lack of socialization by the government on several issues, low public trust in the government, to the ineffectiveness of government communication between institutions. The Indonesian government has issued many public policies, which in the end confuses the public with the many synchronized policies issued.

3.2 Local Government Response

Through the Ministry of Health, the Indonesian government responded to the state of public health emergency of international concern due to 2019-nCoV infection by assigning several control measures that health authorities should take at the national, province, or district/city level. The government formally regulated the control measure through a ministerial decree issued in February 2020, while no cases had been reported in Indonesia. The responsibilities of the governments stated in the decree included improving the communication, information, and education to the community, particularly those who planned to travel to high-risk places; the preparedness and early detection in the entrance area; health care capacity, laboratories, and referral system; and inter-sector coordination.

Responding to the state of emergency of the COVID-19 pandemic and following the identification of the first case of COVID-19 in Indonesia, the central government established a national task force to handle the COVID-19 pandemic through inter-sector involvement. The Task Force for the Acceleration of the Response to COVID-19 was formed in March 2020 to improve the national health system, accelerate the government response, anticipate COVID-19 escalation, and make operational decisions. Then, the subnational government (provincial and district/city) was required to establish a similar task force at the local level.

The governor of West Sumatra Province responded to the decree by issuing an official letter on 16th March 2020 stating that the local government leaders agreed to take prompt preventive and control measures since no COVID-19 case was reported at the time. They coincided with establishing the COVID-19 Local Task Force at the province, district, or city level. In addition to establishing the task force, the local leaders would enhance the implementation of preventive protocols at public places, postpone any public meeting except a strategic plan attended by limited participants, and deliver official information regarding the pandemic situation in the province. Particularly in Padang City, a Local COVID-19 Task Force was formed through the mayor decree 149/2020, issued on 27th March 2020. This regulation underlined the organization, responsibilities, coordination mechanism, and financing of the task force. The role of the local task force is identical to the duties of the national task force mentioned in the presidential decree.

The Indonesian government implemented a large-scale social restriction (PSBB) to restrain public mobilization rather than a total lockdown. PSBB limits certain public activities, including education, formal sectors, religious activities, social and cultural events in the high-risk province or district/city. PSBB in West Sumatera Province was implemented at all 19 districts and cities on 18th April 2020, extending to 19th May 2020. The city mayor of Padang reinforced the policy by issuing legal instruments regulating the public activities during PSBB. City regulations regarding PSBB included the arrangement of PSBB in Padang City, the decree on penalties for any violation during PSBB, and a law on the restriction of domestic travels out or into Padang City.

The negative impact of PSBB on social and economic aspects in Indonesia led the government to discontinue the implementation of PSBB. The government adopted a micro-level of public activities restriction (PPKM). PPKM differed from PSBB in terms of the flexibility of social limitations. The government could decide to implement PPKM in a certain subdistrict, village, office, or even neighborhood only rather than the whole area of a district or city during PSBB to minimize more significant economic and social impact. However, occasional restrictions related to specific public events were still maintained in PPKM, such as shortening religious celebrations days and national holidays. The Ministry of Transportation also contributed by issuing several regulations to be implemented nationally regarding domestic and international travel requirements for all modes of transport.

The government of Padang City showed a quick response in issuing local regulations during the implementation of PPKM. Since the PPKM was initiated in April 2021, the Ministry of Health (MoH) categorized Padang City several times at level four (the highest level). The local regulation arranged education activities at school and educational institutions, working arrangements for formal sectors, the operational time of restaurants and other public facilities, travel procedures, and other public events. The lower the level of PPKM, the more activities are allowed.

The government of Indonesia called new-normal for a new life with the adaptation of behavioral change since the emergence of COVID-19 following the official statement from MoH regarding national movement on prevention protocols of COVID-19, including mask use, washing hand, and physical distancing. Then, the terms of adaptation of behavioral change (*Adaptasi Kebiasaan Baru*, AKB) were first stated in the official letter of the Minister of Religion, which regulates the religious activities during the pandemic in May 2020. Adopting prevention protocols in public activities was then encouraged to reduce COVID-19 morbidity and mortality and economic recovery. West Sumatera government responded to this initiative by issuing a regulation of the AKB in COVID-19 prevention and control on 30th September 2020, following the provincial law on travel supervision out from and into West Sumatera. The local house of legislative issued the derivative regulation on AKB in Padang City on 7th January 2021. As the PPKM has been implemented, more public activities are opened, and the law of AKB is relevant to guide the communities in COVID-19 prevention during their daily movements.

The pandemic of COVID-19 challenges the function of local government to maintain the people's health, particularly in Indonesia, with 34 provinces spread in more than 17,000 islands. In a country such as Indonesia, strengthening inter-sector coordination and improving local government capacity are two essential strategies in addressing an extreme disaster such as the COVID-19 pandemic. The subnational or local government is at the forefront for COVID-19 control in a country with a decentralized governmental system. Therefore, decision-making and regulation-making are crucial to the legal justification for the program financing and resource allocation.

The local government of the City of Padang showed a quick response in terms of regulation-making. The local regulations are issued immediately once the national law is acted. The local law arranged the unit in-charged of each program for pandemic control, the organization, the responsibilities of each structure, the coordination mechanism, and the financing arrangement.

The regulation issued by the city government covers the whole aspects of public needs and activities. It included education, transportation, social and religious events. There were also regulations in the health sector, including improving tracking, testing, treatment, expanding vaccination coverage, and the resource allocation for COVID-19 control. In terms of mitigating the impact of a pandemic, the local regulations covered government subsidies for the communities and informal sectors.

4 Conclusions

The study has shown that the Government of Indonesia has responded to COVID-19 pandemics by issuing many regulations at the national and local levels. The regulations are 30% related to social society, 10% related to the budget, 25% related to health, 20% related to organizations handling COVID-19, and 15 % related to economic recovery. The local government has responded quickly with the national regulation by issuing a local regulation at the province, district, and city-level as legal justification for resource allocation.

The challenges on implementing those regulations are the slow initial response by the government, fulfillment of the community's economic needs, weak coordination between stakeholders, community non-compliance, culture and politics, and ineffective public information.

This paper is a preliminary review of an ongoing study. To understand how health factors can be embedded into disaster risk reduction, a series of interviews are being conducted with key officials from the National Disaster Management Agency (BNPB), the COVID-19 taskforce, and the Ministry of Health. It is also planned that direct observations will be made to locations that are experiencing disasters during the COVID-19 pandemic; and also planned a series of focused group discussions with key stakeholders in the city of Padang.

The authors would like to thank the Newton Fund Institutional Link British Council for funding this research, which collaborates between the Universitas Andalas (Indonesia) and the University of Huddersfield (UK). Universitas Andalas is funded by the Basic Partnership Research (RKD) scheme from the Ministry of Education, Culture, Research, and Technology with a derivative contract number T/2/UN.16.17/PT.01.03/RKD-Soshum/2021.

References

1. M. Coccia, "Preparedness of countries to face a COVID-19 pandemic crisis: Strategic positioning and underlying structural factors to support strategies of prevention of pandemic threats," Environ. Res., (2021).
2. R. Djalante et al., "Review and analysis of current responses to COVID-19 in Indonesia: Period of January to March 2020," Prog. Disaster Sci. **6**, p. 100091, (2020).
3. B. W. Kulohoma, "COVID-19 risk factors: The curious case of Africa's governance and preparedness," Sci. African, **13**, p. e00948, (2021).
4. B. M. Vallejo Jr and R. A. C. Ong, "Policy responses and government science advice for the COVID 19 pandemic in the Philippines: January to April 2020," Prog. disaster Sci., **7**, p. 100115, (2020).

5. WHO, "Indonesian Situation," 2021. https://COVID19.who.int/region/searo/country/id (accessed 22nd August 2021).
6. Indonesian COVID-19 Taskforce, "Distribution Map," 2021. https://COVID19.go.id/peta-sebaran (accessed 22nd August 2021).
7. Diskominfo Padang, "Padang City Condition," 2021. http://corona.padang.go.id (accessed Aug. 22, 2021).
8. Indonesian COVID-19 Taskforce, "Risk Map," 2021. https://COVID19.go.id/peta-risiko (accessed 22nd August 2021).
9. L. K. Alfirdaus and T. Yuwono, "Pandemi COVID-19 dan Pendekatan Kebijakan Multikrisis: Sebuah Refleksi Teoritis," JIIP J. Ilm. Ilmu Pemerintah., **5**, 2, 206–216, (2020).
10. A. A. Sitorus, " Synchronized Indonesian Government Policies in Handling COVID-19," J. Renaiss., **6**, 1, 721–732, (2021).
11. Z. Harirah and A. Rizaldi, " Responding to State Policy Reasons in Dealing with the COVID 19 Pandemic in Indonesia," J. Econ. and Policy. Indonesian public., **7**, 1, (2020).
12. S. Mujani and D. Irvani, " Citizens' Attitudes and Behaviors towards Policies for Handling the COVID-19 Outbreak," Polit. J. Ilmu Polit., **11**, 2, 219–238, (2020).
13. L. Agustino, "a Analysis Of COVID-19 Outbreak Handling Policy: The Experience Of Indonesia," Junal Borneo Adm., **16**, 2, 253–270, (2020).

https://doi.org/10.1051/e3sconf/202133102001

Designing blood supply policy using simulation approach

Eri Wirdianto[1*], *Putri* Ramadhani[1], and *Milana*[2]

[1]Industrial Engineering Department, Faculty of Engineering, Andalas University, Indonesia
[2]Automotive Engineering Department, Faculty of Engineering, Padang State University, Indonesia

Abstract. Blood transfusion is one of the critical requirements of many medical treatments. To ensure timely blood availability, the Indonesian Red Cross (Palang Merah Indonesia/PMI) has to tackle the blood supply challenges of stockouts and overdates. This research aims to develop a simulation model to determine the blood supply policy to reduce stockouts and overdates. The simulation model was developed according to the existing system at UDD PMI Padang. Simulation experiments were then attempted with some sequential scenarios and carried out on variables that affect the blood supply. The simulation result recommends the blood supply policy to reduce blood stockout and overdate at a higher service level, at conditions of the blood demand is assumed to be constant or increased.

1 Introduction

Blood is needed in the human body in certain portions to function correctly. However, there are times when a person experiences a lack of blood that requires blood donation from others. For example, the need for blood can be caused by traffic accidents, natural disasters, surgeries, or other conditions. In this situation, blood donation is needed to replenish the amount of blood in the body.

The hospitals in Indonesia usually get blood supply from the nearest branch of the Indonesian Red Cross (Palang Merah Indonesia/PMI). In the City of Padang, the hospitals collaborate with a particular unit of PMI to manage blood supply called Blood Supply Unit (Unit Donor Darah/UDD). In the rest of this article, this unit is called PMI UDD Padang. PMI UDD Padang works to supply blood by collecting blood from the volunteers in the communities.

PMI UDD Padang obtains blood through volunteers who donate their blood. The blood supply system mechanism includes blood collection and qualification, blood processing, blood storage, and blood distribution. The blood obtained will go through an initial examination, doctor examination, followed by taking blood samples for quality and serological test, respectively. Damaged blood (containing disease) will be destroyed, while good blood is processed and separated into several blood components and stored in the blood bank as a stock.

The blood is commonly classified into some blood groups: A, B, AB, and O. Each blood group has to be stocked at certain levels to meet the blood demand of different blood groups [1]. Each of these blood groups can be synthesized into several blood components, which are Whole Blood (WB), Pack Red Cell (PRC), Fresh Frozen Plasma (FFP), Trombosit Concentrate (TC), and Cryo-Precipitate/Anti-Haemophilic Factor (C-P/AHF). The blood managed by PMI UDD Padang was presented in Fig. 1.

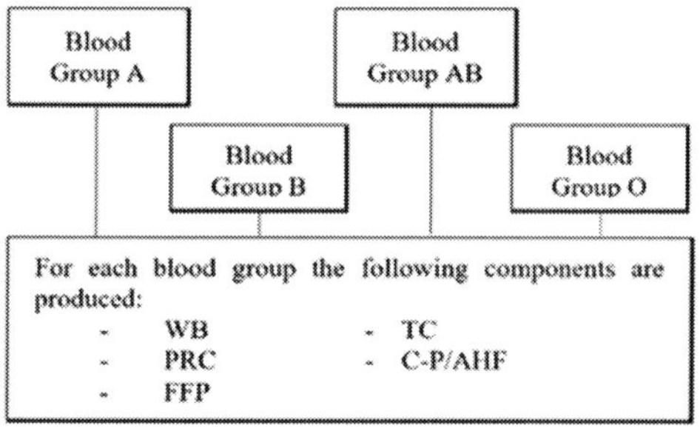

Fig. 1. Blood groups and types of blood components.

Based on a preliminary survey conducted at PMI UDD Padang, it is known that there was a condition where PMI UDD Padang was unable to meet the demand for some blood components due to stockouts. In contrast, some other blood components had excess stock and passed their expiration date. This circumstance is caused by inaccuracy estimation of the amount of blood needed at different components. The amount of blood supply shortage of PMI UDD Padang at a particular year can be seen in Fig. 2, while the amount of blood that has expired during that year can be seen in Fig. 3. Fig. 2 shows that the blood shortage for some blood components is very high, dominated by the lack of TC, followed by PRC. It indicates a significant difference between the availability of blood at PMI UDD Padang and the blood demand in specific blood components. Fig. 3 represents that the overdate blood for all blood products is still high. It happens because there is an excess stock of specific blood components, which causes such blood components to be

* Corresponding author: eri_wirdianto@eng.unand.ac.id

 https://doi.org/10.1051/e3sconf/202133102001

stored until their expiration date, so they must be destroyed.

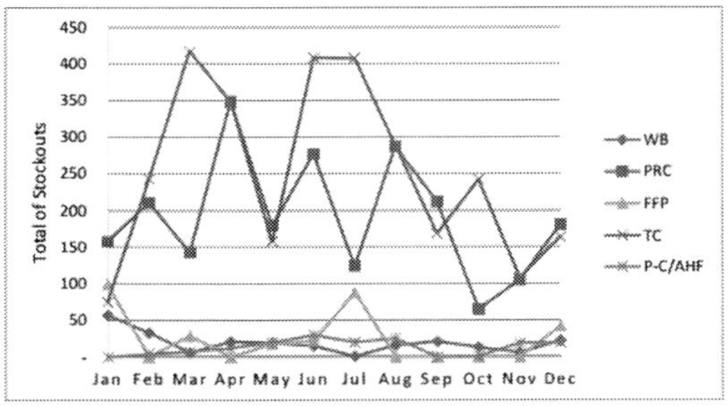

Fig. 2. Total shortage of blood.

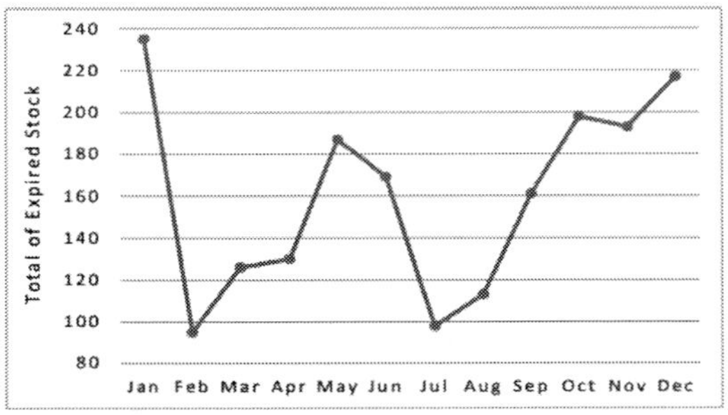

Fig. 3. Total of expired blood.

Considering that blood supply, production, and demand are stochastic and having complex interactions among system entities, a discrete-event simulation is determined as a suitable method to mimic the real system [2]. The use of discrete-event simulation on health services has been grown significantly for the last four decades [3,4]. However, from 483 journal papers examined, only 13 addressed the logistics model of health service products where blood is one of its topics [3]. Moreover, in another review article in health care, only 1 out of 211 papers used the discrete-event simulation model to study the blood center [4].

The recent researches related to blood supply have been conducted from different perspectives. The availability of blood at the right time and specified quality have made this topic interesting to discuss. Onggo [5] modeled a hybrid simulation model for the blood supply chain and found that the shortages of blood supply in Low- and Middle-Income countries have caused a high unavoidable death rate, especially for women and children, compared to High-Income countries. In the blood collection process, the blood supply unit could provide two different ways of collection. Therefore, the blood collection strategy has been modeled and simulated by considering fixed and mobile blood collection [6]. Moreover, different characteristics of the blood supply chain in different countries also become challenging issues to solve [5-8].

In this research, a discrete-event simulation model is developed by considering the complex interaction of blood supply, production, and demand in order to determine the relevant blood supply policy to reduce the number of stockouts and overdates to increase service level. The development of a simulation model on the particular system is novel and beneficial to improving PMI UDD Padang's performance.

2 Research Method

The main activity of this research was developing the simulation model of blood supply at PMI UDD Padang. Before developing the simulation model, it was essential to understand the system description, conceptual model, and model specification.

2.1 System Description

PMI UDD Padang carries out several stages to obtain blood before it is distributed to the recipient. These stages are blood collection and qualification, blood processing, blood storage, and blood distribution. The illustration is presented in Fig. 4.

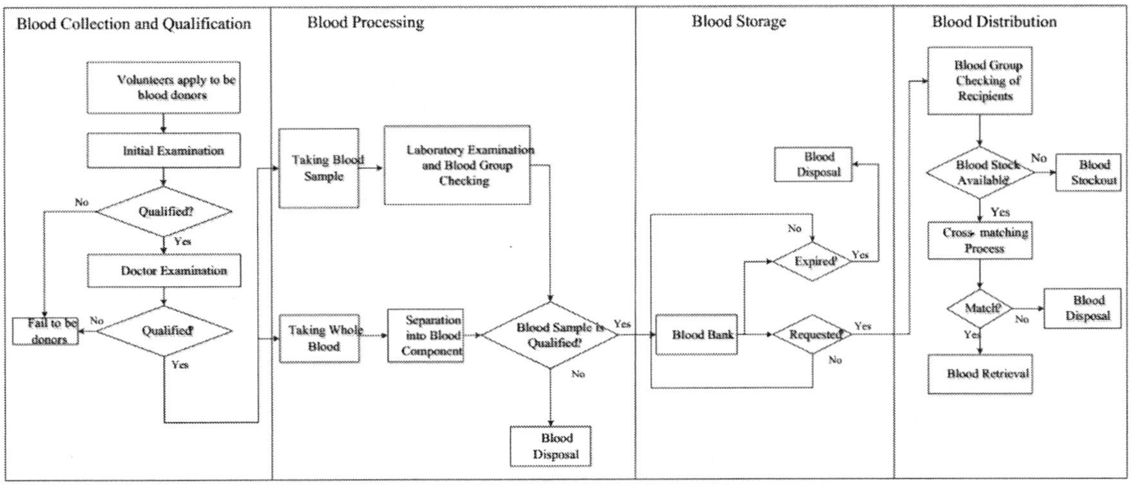

Fig. 4. System description of blood inventory at PMI UDD Padang.

https://doi.org/10.1051/e3sconf/202133102001

2.2 Conceptual Model

The simulation model for the blood supply system used the software of Microsoft Excel and Arena. Microsoft Excel was used in the initialization process, while Arena was used in developing simulation models of the existing system [9,10]. The conceptual model of this research can be seen in Fig. 5.

2.3 Model Specification

Model specification is a more specific description of the model than the conceptual model. The model specification of this research describes the events experienced by the blood throughout the system. An event graph is a graph that describes the events in the simulation. The event graph model of the blood supply system can be seen in Fig. 6.

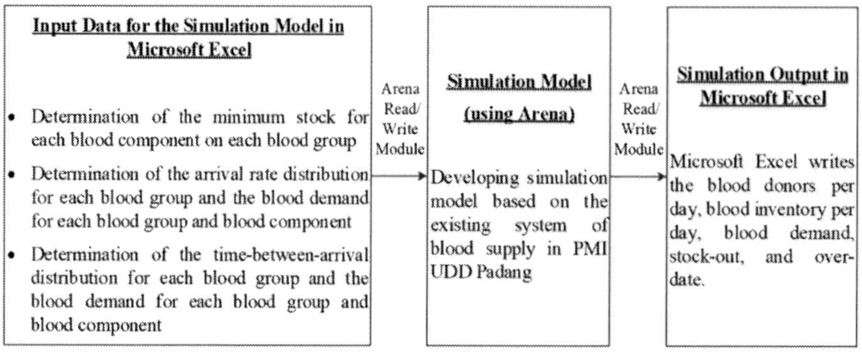

Fig. 5. Conceptual model of the research.

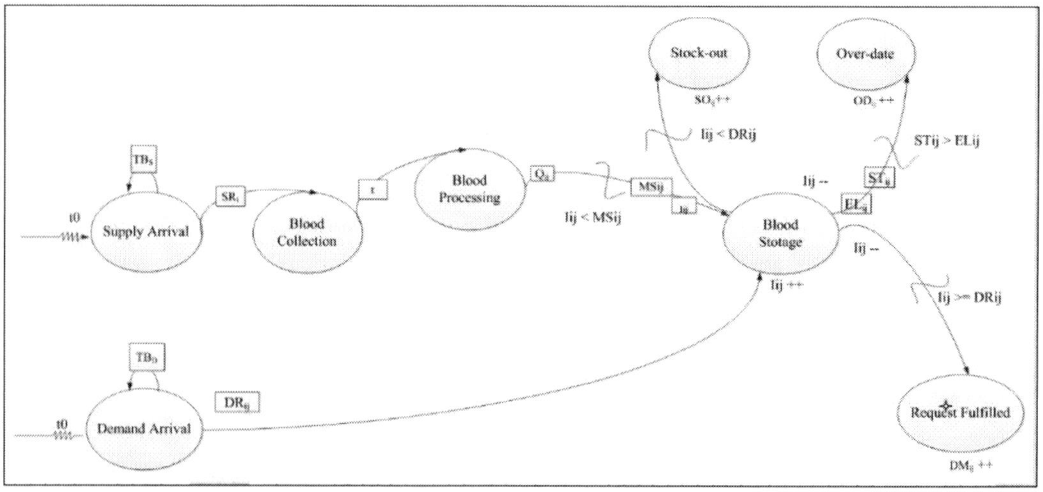

Fig. 6. Event graph of blood supply system at PMI UDD Padang.

Where:

TBS = Time between arrivals of donors
TBD = Time between arrivals of demand
I_{ij} = Total inventory of each blood group and component
MS_{ij} = Minimum stock of each blood group and component
Q_{ij} = Total production of each blood type and component
DR_{ij} = Demanded quantity for each blood type and component
SR_{ij} = Total donation of each blood type
ST_{ij} = Storage time of each blood group and component
EL_{ij} = Expiry limit for each blood type and component
So_{ij} = Total stockout of each blood group and component
Od_{ij} = Number of overdates for each blood group and component
DM_{ij} = Total fulfilled demand for each blood group and component

2.4 Development of Simulation Model

Based on the existing condition, a computer simulation model was created using the Arena software to develop the simulation model of the blood supply system at PMI

UDD Padang. The blood supply system was observed from blood collection and qualification, blood processing, blood storage, and blood distribution. Determination of the blood supply is based on the minimum stock that must be available for each blood component of each blood group. In addition, each blood component also has an expiration date which will be immediately destroyed if it has passed that limit.

2.4.1 Input Data

The input data used in the simulation model are the arrival rate of blood, the time between arrivals of blood, the arrival rate of blood demand, and the time between arrivals of blood demand. The distribution used for the two data is discrete. It is because the time interval used in the system is discrete.

As an example, based on historical data, the demand for Blood Component WB of Blood Group A is following DISC (0.6438 , 0 , 0.7370 , 1 , 0.8548 , 2 , 0.8959 , 3 , 0.9370 , 4 , 0.9589 , 5 , 0.9671 , 6 , 0.9808 , 7 , 0.9890 , 8 , 0.9918 , 9 , 0.9973 , 10 , 1 , 11), which means that for a particular day probability of demand equals to 0 unit is 0.6438 or demand equals to 1 unit is 0.0932 and so on.

2.4.2 Simulation Input and Output

The input of the simulation model read by Microsoft Excel is the minimum stock of each blood component in each blood group. The simulation results of the blood supply model of PMI UDD Padang can also be seen in Microsoft Excel using the Read/Write module. The Read/Write module is used to view the simulation results for all the variables needed for each blood group and component. The variables displayed in the module are the day, the number of incoming blood donations per day, the number of blood that has been processed, the demanded quantity, the fulfilled demand, the number of stockouts, the amount of final inventory, and the number of overdates for each blood component of each blood type. The logic model in this process can be seen in Fig. 7.

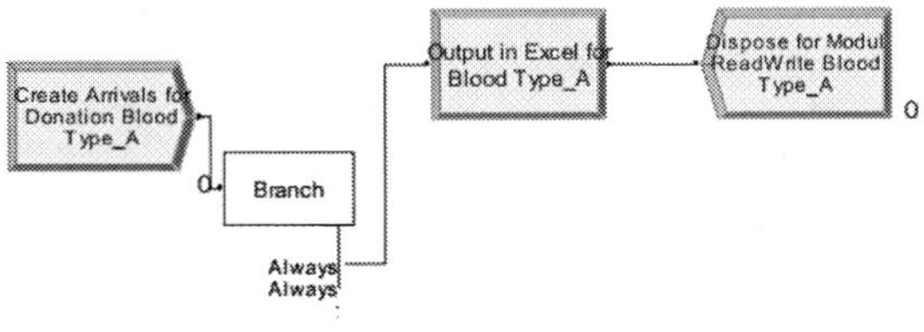

Fig. 7. Logic model of read/write module.

2.4.3 Blood Arrival Module

The process that occurs in this module is the process of generating the number of blood arrivals. There is a decision to use a blood bag (triple/double). Every incoming blood will use a triple bag of blood until the triple bag stock requirement is met. Then proceed with the use of a double bag. The modules used in the blood acquirement process are Create – Assign – Route – Station – Batch. The logic model and modules in this process can be seen in Fig. 8.

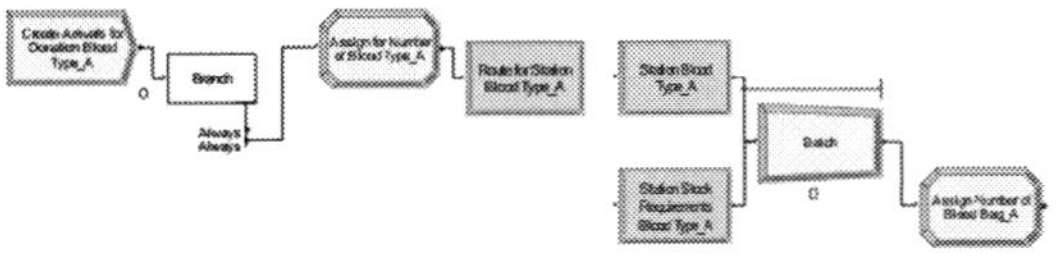

Fig. 8. Logic model of blood arrival module.

2.4.4 Blood Stock Determination Module

This requirement of blood is based on the number of stock available and the minimum stock. The formula for determining the need for this blood component are as follows:

1) WB Requirement = Max (0, (Minimum Stock A WB - WB Final Stock Amount)). This method is the same for calculating the number of other blood components needed.

2) Amount of triple bag requirement = Max ((Total FFP Stock Requirement + AHF Stock Requirement Amount), (TC Stock Requirement + AHF Stock Requirement Amount))

3) Number of double bag needs = Max (Number of WB Stock Requirements, (WB Stock Requirements Quantity + (PRC Stock Requirements Amount - Triple Bag Stock Requirements Total))))

The modules used to model this process are Create – Assign – Route. The logic model of this process can be seen in Fig. 9.

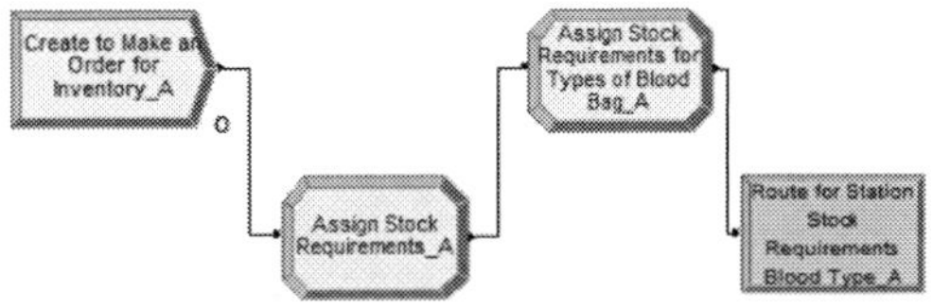

Fig. 9. Logic model of stock determination module.

2.4.5 Blood Processing Module

The process that occurs in this section is the separation of whole blood into several blood components. This process is based on the number of stock requirements provided by the blood storage department based on the amount of blood stock in the blood bank. The modules used to model this process are Branch – Assign – Separate – Route. The logic model of this process can be seen in Fig. 10.

2.4.6 Blood Storage Module

The processes that occurs are storing production results and determining the final inventory value at the end of each day. The modules used to model this process are Station – Hold – Assign – Branch – Dispose. The logic model of this process can be seen in Fig. 11.

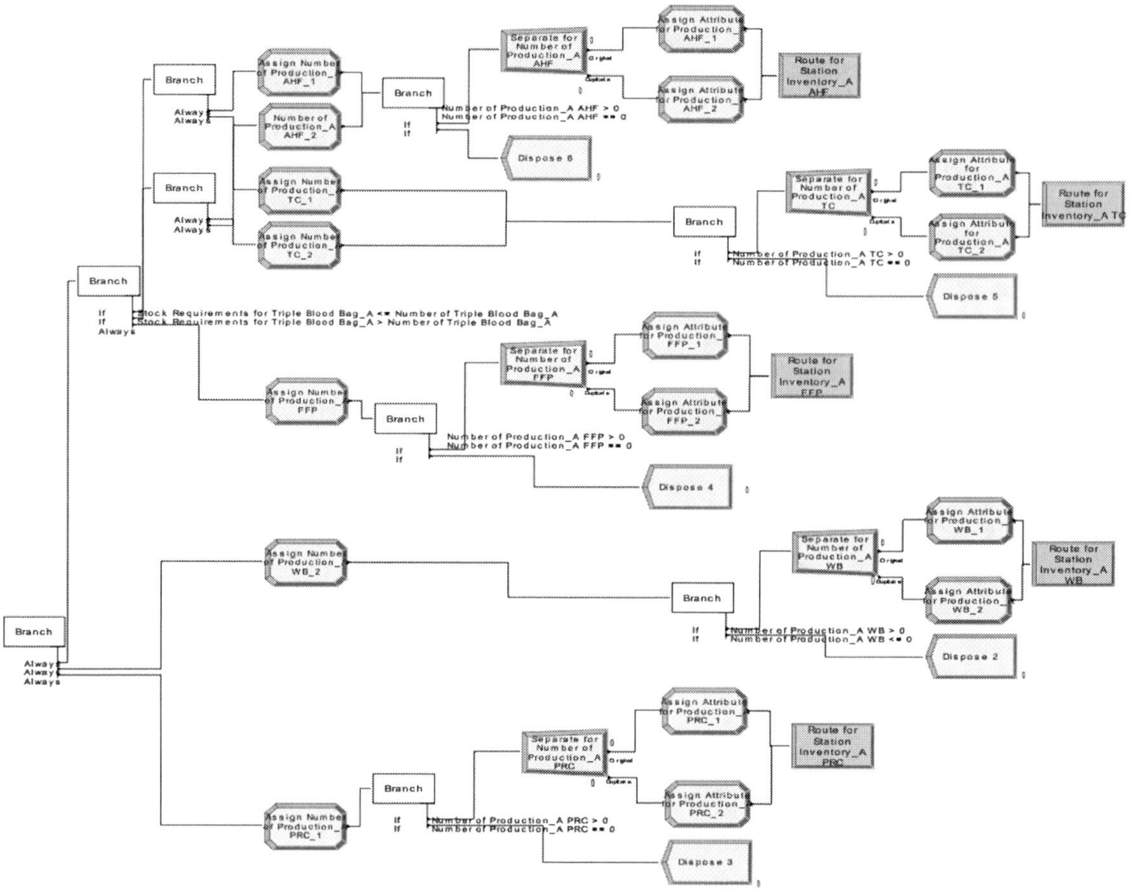

Fig. 10. Logic model of the blood processing module.

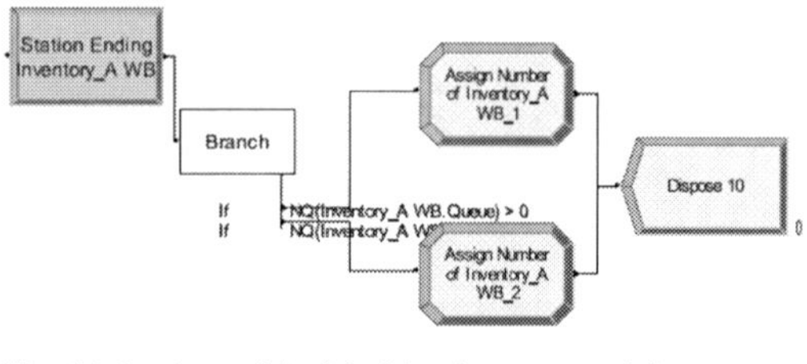

Fig. 11. Logic model of the blood storage module.

2.4.7 Blood Demand Module

This module works when the demand for blood from the recipient occurs at the blood distribution section. The modules used to model this process are Create – Assign. The logic model and modules in this process can be seen in Fig. 12.

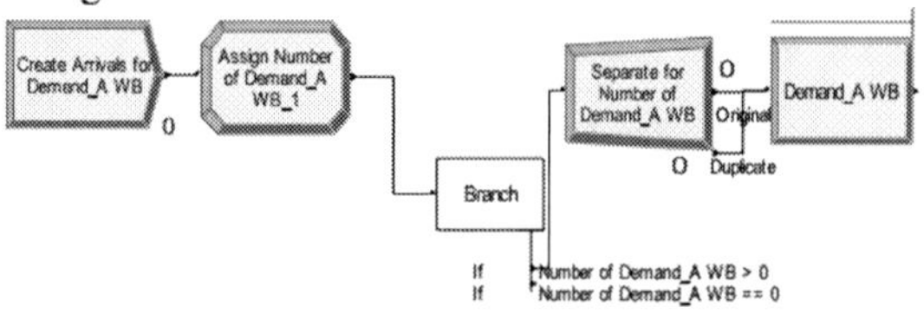

Fig. 12. Logic model of blood demand module.

2.4.8 Blood Disposal Module

The process of disposing of expired blood is a process that occurs in the storage section by removing blood from storage if the blood has passed the expired time. The formula for determining storage time and the number of overdates in this process are:

1) Storage time= TNOW - The first time the blood entered the storage
2) Number of overdates = Number of overdates + 1

The modules used to model this process are Assign – Branch – Route – Station – Dispose. The logic model of this process can be seen in Fig. 13.

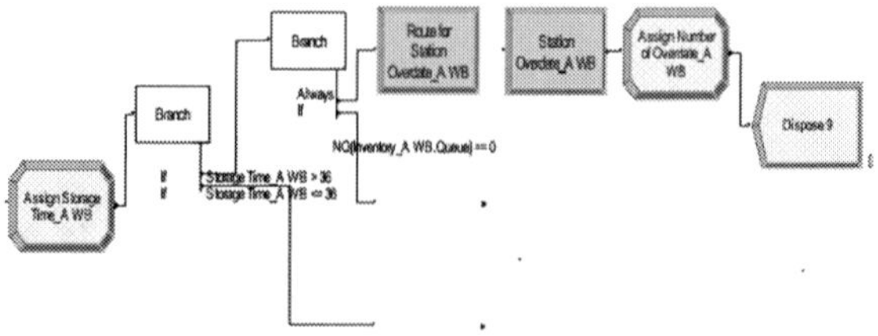

Fig. 13. Logic model of blood disposal module.

2.4.9 Fulfilled Demand Module

Fulfilled demand is a continuation of blood demand. If the stock is more than the demanded quantity, then the blood will be issued from the storage section, and the calculation of fulfilled demand for each blood component will be carried out. The blood demand will be fulfilled until the end of the day, so the model is also designed to fulfill the demand at the end of the day. The fulfilled demand formula in this process are:

1. Demand is fulfilled if = Demand > 0 andTotal Inventory > 0
2. Fulfilled demand = Fulfilled demand + 1

The modules used to model this process are Branch – Batch – Route – Station – Assign. The logic model of this process can be seen in Fig. 14.

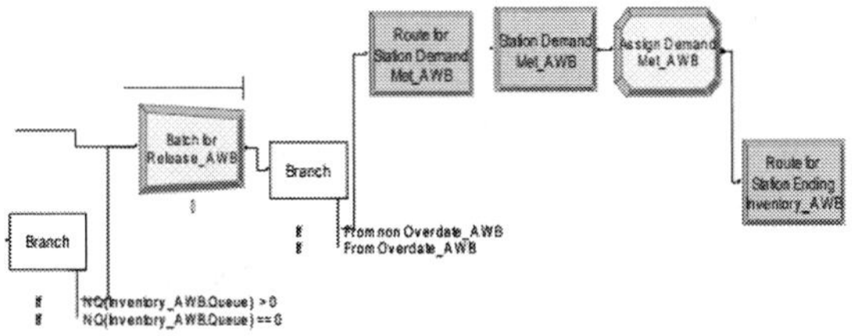

Fig. 14. Logic model of fulfilled demand module.

2.4.10 Stockout (Demand Not Fulfilled) Module

Stockout is a continuation of the blood demand. If the stock is less than the demanded quantity, the blood will be issued as much as the number of available stock, while the demand quantity that cannot be fulfilled will be calculated for each blood component.

The formula for determining the decision and calculating the amount of stockout blood in this process are:

1) Stockout occurs if = Demand > 0 and total inventory = 0 or Demand > 0 and total inventory > 0 (overdate blood)
2) Total stockout = Total stockout + 1

The modules used to model this process are Branch – Assign – Dispose. The logic model of this process can be seen in Fig. 15.

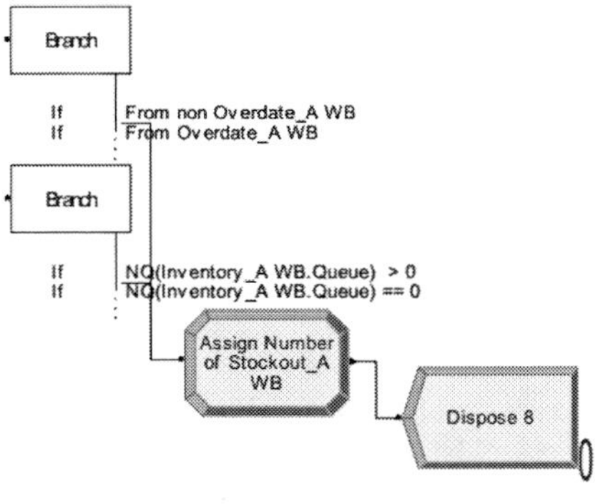

Fig. 15. Logic model of stockout module.

3 Result and Discussion

3.1 Simulation Model Verification and Validation

Verification is the stage to determine whether the computer simulation program has been made as expected [11]. The verification of this simulation model has been done by ensuring that the model has all the components specified during the system identification phase and does not experience errors when running.

Validation is a process to see whether the model is appropriate or similar to the actual system conditions to represent the whole system [12]. The model is validated using the black box method, which compares the simulation results with the actual system. The statistical test was carried out using SPSS software. The results of the paired sample t-tests are shown in Table 1.

Table 1. The result of paired sample t-test using SPSS.

Simulation - Actual		Paired Differences					t	df	Sig. (2-tailed)
		Mean	Std. Dev.	Std. Err. Mean	95% Confidence				
					Lower	Upper			
Pair-1	Donors Arrival	- 11,65	372,79	53,81	- 119,89	96,60	-0,216	47	0,830
Pair-2	Production	- 10,41	252,80	16,32	-42,56	21,73	-0,638	239	0,524
Pair-3	Stock-out	10,18	130,39	16,83	-23,50	43,87	0,605	59	0,548
Pair-4	Over-date	19,25	115,14	33,24	-53,90	92,40	0,579	11	0,574
Pair-5	Demand	-1.56	264.54	17.08	-35.2	32,08	-0.092	239	0.927

Table 1 shows that there is no significant difference between the simulation results and the actual system, so it is concluded that the simulation model is valid and can be used to represent the actual system.

3.2 Initial Simulation Result

The verified and validated simulation model was run for 366 days with ten replications based on the normal approximation method. The initial minimum stock policy decided by PMI UDD Padang is shown in Table 2.

The simulation output is based on actual conditions. The number of stockouts (SO) and overdates (OD) for each blood component of each blood group can be seen in Table 3.

Table 2. Initial minimum stock policy at PMI UDD Padang.

Bl. Comp. / Bl. Group	WB	PRC	FFP	TC	AHF
A	5	30	20	30	30
B	30	5	30	10	30
AB	30	30	5	30	20
O	20	30	10	5	30

Table 3. Output of initial simulation model.

Bl. Comp.	WB		PRC		FFP		TC		AHF		Total
Bl. Group	SO	OD	SO	OD	SO	OD	SO	OD	SO	OD	
A	10	103	353	0	1	0	1654	4	3	0	2128
B	5	695	137	0	0	0	1072	2	1	0	1912
AB	32	14	408	0	0	0	482	93	1	9	1039
O	4	1235	103	0	2	0	2347	0	0	0	3691
Total	51	2047	1001	0	3	0	5555	99	5	9	8770
Total SO	6615										
Total OD	2155										

3.3 Experimentation Scenarios

The large numbers of stockouts and overdates are the problem at PMI UDD Padang. Therefore, the solution is to determine the production quantity for each blood component to minimize it. The minimum stock policy influences the production quantity of each blood component at PMI UDD Padang and the number of blood collected every day. In this study, the experiment was divided into two scenarios: when the blood demand is assumed to be the same as historical data and when the blood demand increases at the population growth rate.

Based on data from the Central Bureau of Statistics of the City of Padang, the population increases by 1.38% on average annually. So that, it is concluded that the blood demand will also increase in that percentage. The simulation output with increased demand under initial minimum stock policy can be seen in Table 4.

Table 4. Output of simulation model with increased demand under initial minimum stock policy.

Bl. Comp.	WB		PRC		FFP		TC		AHF		Total
Bl. Group	SO	OD	SO	OD	SO	OD	SO	OD	SO	OD	
A	11	83	427	0	1	0	1684	4	3	0	2213
B	5	658	147	0	0	0	1088	2	1	0	1901
AB	32	14	409	0	0	0	482	93	1	9	1040
O	4	1101	125	0	2	0	2400	0	0	0	3632
Total	52	1856	1108	0	3	0	5654	99	5	9	8786
Total SO	6822										
Total OD	1964										

In this simulation model, there are three things that will affect the blood supply system, which are blood demand, minimum stock policy, and the number of donors. So the scenarios were prepared based on these three things to obtain alternative policies related to the minimum stock policy and efforts to increase the number of donors when the condition of blood demand is the same as historical data and the condition of blood demand increases by population growth. The criteria used to select policy alternatives are the smallest stockout and overdate generated by the designed scenarios, which are shown in Fig. 16.

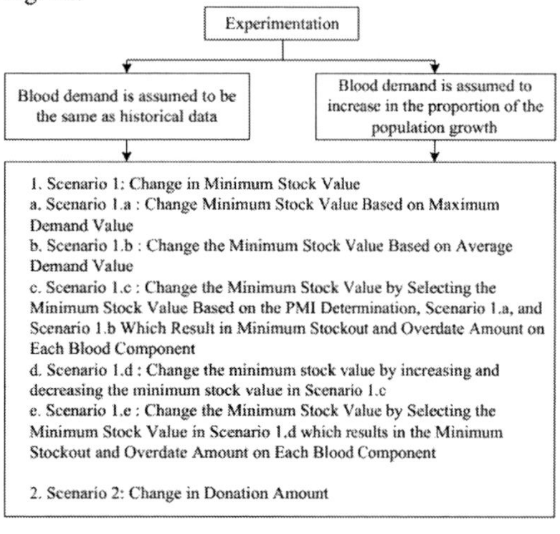

Fig. 16. The designed scenario of the simulation.

The results obtain from running these scenarios are explained below:

1. Under the condition of constant blood demand.

The selected minimum stock policy is shown in Table 5. Table 6 shows SO and OD consequences for that policy under the conditions that (i) the number of donors similar to that of the last year and (ii) the number of donors increases by 10% from last year (this 10% was chosen as the best percentage obtained).

Table 5. The chosen stock minimum policy under the condition of constant blood demand.

Bl. Comp. / Bl. Group	WB	PRC	FFP	TC	AHF
A	1	3	30	56	27
B	1	3	30	51	30
AB	1	1	30	21	13
O	1	15	32	70	15

Table 6. Output of simulation model under the condition of constant blood demand.

Bl. Comp.	WB		PRC		FFP		TC		AHF		Total
Bl. Group	SO	OD	SO	OD	SO	OD	SO	OD	SO	OD	
A	17	75	332	0	1	0	194	102	5	0	726
A + 10%	6	889	76	0	1	0	148	106	4	0	1230
B	8	531	111	0	0	0	117	113	1	0	881
B + 10%	4	1531	25	0	0	0	85	118	1	0	1764
AB	37	7	384	0	0	0	205	234	4	2	873
AB + 10%	12	164	103	0	0	0	175	254	3	2	713
O	8	1259	104	0	0	0	118	88	11	0	1588
O + 10%	5	2720	40	0	0	0	81	92	9	0	2947

From Table 6, it can be concluded that if the amount of blood demand remains constant, then the minimum stock policy used is as in Table 5, and the number of donors for blood group AB should be increased by 10% while for other blood groups remain the same. This proposed scenario will reduce stockout and overdate on all blood groups.

The comparison of stockout and overdate before and after scenarios designed for constant blood demand are illustrated in Fig. 17.

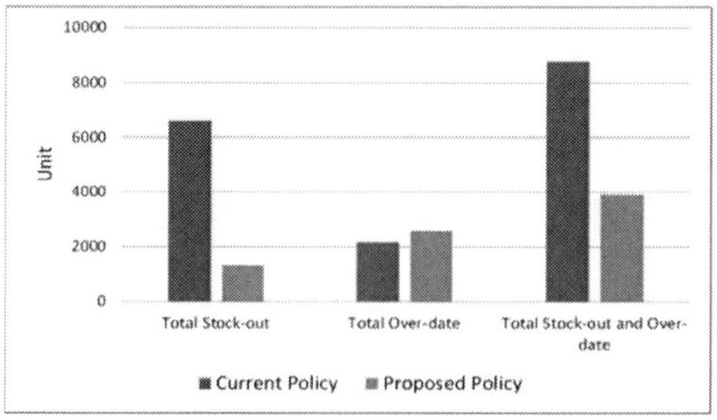

Fig. 17. The comparison of stockout and overdate before and after scenarios designed for constant blood demand.

2. Under the condition of increased blood demand.

The selected minimum stock policy is shown in Table 7. Table 8 shows SO and OD consequences for that policy under the conditions that (i) the number of donors similar to that of the last year and (ii) the number of donors increases by 10% from last year (this 10% was chosen as the best percentage obtained).

Table 7. The chosen stock minimum policy under the condition of increased blood demand.

Bl. Comp. / Bl. Group	WB	PRC	FFP	TC	AHF
A	1	3	30	57	27
B	1	3	30	47	30
AB	1	1	30	21	13
O	1	12	32	71	15

Table 8. Output of simulation model under the condition of increased blood demand.

Bl. Comp.	WB		PRC		FFP		TC		AHF		Total
Bl. Group	SO	OD	SO	OD	SO	OD	SO	OD	SO	OD	
A	20	54	339	0	1	0	198	109	5	0	726
A + 10%	6	785	81	0	1	0	151	114	4	0	1142
B	8	491	120	0	0	0	158	82	1	0	860
A + 10%	4	1478	26	0	0	0	118	85	1	0	1712
AB	37	7	384	0	0	0	205	234	4	2	873
A + 10%	12	164	103	0	0	0	175	254	3	2	713
O	9	1123	127	0	0	0	124	94	7	0	1484
O + 10%	5	2556	44	0	0	0	86	99	9	0	2799

From Table 8, it can be concluded that if the amount of blood demand remains constant, then the minimum stock policy used is as in Table 7, and the number of donors for blood group AB should be increased by 10% while for other blood groups remain the same as last year. This proposed scenario will reduce stockout and overdate on all blood groups.

The comparison of stockout and overdate before and after scenarios designed for increased blood demand are illustrated in Fig. 18.

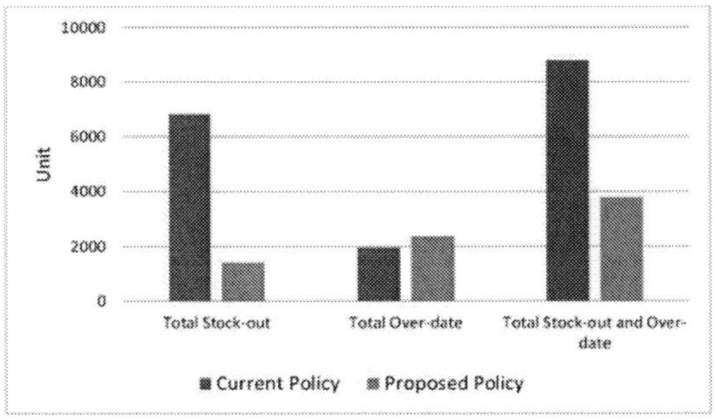

Fig. 18. The comparison of stockout and overdate before and after scenarios designed for increased blood demand.

Various scenarios can be generated using this developed simulation model to obtain feasible alternatives of blood supply systems in various conditions.

According to results obtained from model simulations, it is suggested that PMI UDD Padang increases the stock of blood group AB by 10%, which has been consistently stockout. Considering that the number of donors of blood group AB is rare, PMI UDD Padang needs to actively identify people who have that blood group in the community. They need to be socialized about the importance of blood availability and encouraged to donate their blood regularly to enhance health service quality in the community.

4 Conclusion

This research produces a simulation model that can represent the actual system at PMI UDD Padang. This simulation model is designed by considering the processes that occur at PMI UDD Padang. The processes are the donation arrival process, the blood demand arrival process, the process of determining the need for the amount of blood production, the blood processing, the blood storage, the blood distribution to fulfill the demand, the blood disposal, and the process of calculating the stockout. The proposed blood supply policies that can increase service levels are found at the minimum stock policy obtained and the increase donations to blood group AB by 10%.

The blood supply is closely related to the availability of disposable medical equipment to support the blood supply process. It is recommended to design an inventory planning policy to ensure the availability of disposable medical equipment at minimum inventory cost.

The authors would like to thank the Industrial Engineering Department, Faculty of Engineering, Andalas University for supporting this publication through the 2021 Publication Grant. They also express their gratitude to the LPPM of Andalas University, Indonesia.

References

1. M. Asadpour, O. Boyerhasani, and R. Tavakkoli-Moghaddam, in Proceeding of the 1st International Conference on Challenges and New Solutions in Industrial Engineering and Management and Accounting, 16 July 2020, Adib Mazandaran, Iran, (2020)
2. M. Doneda, S. Yalçındag, I. Marques, and E. Lanzarone, Vox Sang. (2021)
3. S. Liu, Y. Li, K. P. Triantis, H. Xue, and Y. Wang, MDM Plcy. & Pract. **5** (1), 1 (2020)
4. X. Zhang, BMC Health Serv. Res. **18** (1), 687 (2018)
5. B. S. Onggo, in Proceeding of the Winter Simulation Conference, 7-10 December 2014, Savannah, GA, USA, (2014)
6. E. Alfonso, X. Xie, V. Augusto, and O. Garraud, Health Care Manage. Sci. **15**, 63 (2012)
7. J. Blake and K. McTaggart, in Proceeding of the 6th International Conference on Simulation and Modeling Methodologies, Technologies and

Applications, 29-31 July 2016, Lisbon, Portugal, (2016)

8. A. Mansur, I. Vanany, and N. I. Arvitrida, in Proceeding of the 9th International Conference on Industrial Engineering and Operations Management, 5-7 March 2019, Bangkok, Thailand, (2019)
9. W. D. Kelton, R. P. Sadowski, and N. B. Zupick, *Simulation with Arena*, 6th ed. (McGraw-Hill Education, New York, 2015)
10. T. Altiok and B. Melamed, *Simulation Modeling and Analysis with Arena*. (Elsevier, Burlington, 2007)
11. C. A. Chung, *Simulation Modeling Handbook: A Practical Approach*. (CRC Press LLC, Boca Raton, 2004)
12. M. D. Rossetti, *Simulation Modeling and Arena*, 2nd ed. (John Willey & Sons, Inc., New Jersey, 2016).

https://doi.org/10.1051/e3sconf/202133102002

Governing sustainable tourism in time of COVID-19 disaster: empirical evidence from Bintan, Kepulauan Riau

Mahadiansar Mahadiansar[1*], *Andy Fefta* Wijaya[1], *Alfi Haris* Wanto[1], *Wayu Eko* Yudiatmaja[2], and *Ramadhani* Setiawan[2]

[1]Department of Public Administration, Brawijaya University, Malang, Indonesia
[2]Department of Public Administration, University of Maritim Raja Ali Haji, Tanjungpinang, Indonesia

Abstract. The COVID-19 disaster has brought various negative effects on human activities, including the tourism sector. However, only a few studies have conducted to the development and governance of tourism by the stakeholders during the COVID-19 pandemic. The purpose of this study is to explain the governance of sustainable tourism during the COVID-19 outbreak. The research was conducted in Bintan Regency, Kepulauan Riau, by using a qualitative approach. The data were collected using an interview with several tourism stakeholders, involving business actors, resort owners, government, and village officials. The results suggest that the governing process of tourism in Bintan focused on formal policy and implementation. The government and business actors become crucial actors in helping the sustainability of tourism during the COVID-19 pandemic. The study can be used as a valuable recommendation to the government in managing tourism in the era of COVID-19.

1 Introduction

Bintan Regency has very strategic tourism potential, since it has direct borders with neighboring countries, that is, Singapore and Malaysia. Thus, the number of tourist visits, both domestic and foreign tourists is the main contributor to Regional Original Income (PAD). Based on data from the Central Statistics Agency (BPS) of Bintan Regency, the increase in the number of tourists in 2016 was 305,404 people, in 2017 was 368,592 people and in 2019 was 522,399 people [1].

An understanding of sustainable tourism policy that depends on institutional sustainability is essential. Practically, in the sustainable development process, the completion of the basics of sustainable governance is always missed it is to ensure equity in the tourism development policy process[2]. However, not much known about the types of governance processes and instruments that can effectively synergize economic, ecological, and social interests in the tourism sector [3]. Sustainable tourism is an interconnected concept involving all sectors, including economic, social, cultural and political [4].

Sustainable tourism is not about forecasting or a case study of what will happen now and in the future. The development of tourism in Bintan Regency socially is very helpful as an additional income besides the profession as fishermen. The impacts in the implementation of tourism policies will involve stakeholder including communities around areas that have sustainable tourism potentials to be developed.

Implementation development of sustainable tourism policies has a problem or phenomenon occur both formally and informally, such as studies by Prasetyanti et al.[5]. Ceballos-Lascuráin highlights the basics of sustainable tourism policies managed by multiple stakeholders [6]. This study aims to analyze sustainable tourism policies during the COVID-19 pandemic. The questions in this study can be formulated as follows: (1) How is the implementation and policy of sustainable tourism during the COVID-19 pandemic? (2) How is Tourism Governance during the COVID-19 pandemic in Bintan Regency?

2 Research Methods

This research is qualitative research with a case study approach [7]. Case study research is conducted to explore unique and interesting cases in certain limited systems. Case studies are also more detailed in describing a phenomenon by collecting data directly [8]. Primary data were obtained through interviews with several informants, which included the Head of the Bintan Regency Culture and Tourism Office, the Head of the Bintan Regency Village and Empowerment Agency, the Head of the Bintan Regency Malay Customary Institution, the Head of Teluk Bakau Village, the BBC Tourism Manager (Bintan Blue Coral), and Raja Ali Haji Maritime University Academic (Tourism Policy Expert Staff).

* Corresponding author: mahadiansar@student.ub.ac.id

The instrument in this study adopts the concept of implementation of sustainable tourism development policies with indicators covering several aspects, namely: the content of policy and context of implementation [9]. Content of policy consists of interest that is affected, the type of benefits, the change of degree, the point of the decision-making, implementation of the program, and used resources. The context of implementation consists of the power, importance, and strategies of the actors involved, the characteristics of institutions and regimes, compliance, and responsibility. The analytical method used in this research is the qualitative analysis by analyzing the problem by looking for solutions and strategies through using case study analysis. Data analysis consists of examining, grouping, tabulating, testing, or recombining evidence, to generate findings based empirically using the Nvivo 12 Pro application [10].

3 Results

3.1 Bintan Tourism Condition

Bintan Regency is one of 4 regencies and 2 cities in the Riau Islands Province based on Law no. 25/2002 (BPS Kepri, 2020). Based on data from the Riau Islands BPS, (2020) there are around 241 islands in the Bintan Regency area. With a large number of islands and a strategic location with foreign countries, namely Singapore and Malaysia, Bintan has become one of the destinations for foreign tourists to vacation. Bintan has 61 tourist attractions, 31 tours & travels, 24 tour groups, and 51 hotels and resorts spread across the Bintan Regency (Bintan Culture and Tourism Office, 2021). Here are some locations for tourism development in Bintan, namely:

1. Sebong Pereh as an Eco-Tourism area (Mangrove tourism and Tourism Village),
2. Bintan Bunyu as an Eco-Tourism area (Mount Bintan, Tourism Village, and camping location),
3. Bintan Integrated Resort (Beach, Golf, Mall, etc.),
4. Trikora Beach (White Sand Beach, Cottage, Water Tourism, Snorkeling, Diving, Fishing Locations, and Fishing Village), and
5. Batu Licin as a new Resort development.

Due to a large number of tourist sites and accommodations, Bintan Regency in 2019 attracted 1,094,442 visitors both local and foreign tourists. However, the pandemic hit the tourism industry in Bintan. Data shows that during the COVID-19 pandemic the number of tourist visits to Bintan Regency decreased drastically by 81.49%, in 2020 the number of visits to Bintan Regency was only 202,583 visitors (Bintan Culture and Tourism Office, 2021).

Based on the table of coding results above, shows that some of the sources of tourism activities in Bintan Regency have stopped globally. However, sustainable tourism governance is only focused on local communities while maintaining health protocols. Bintan Regency Tourism currently only expects government policies to open border access, but the impact of the break of global tourism in the district has an impact at a loss of employment as if they are laid off or laid off en masse.

Table 1. Bintan Regency Tourism Coding Results During the COVID-19 Pandemic

Interviewees	Statement
BBC Tourism Manager	When COVID tourism activities fall.
	Since the *border* was closed during pandemics (all tourists 90% come from *Singapore*), the tourist comes from *Singapore*, so no tourists. The number of employees on weekdays Monday - Friday (8 employees) / previously used to be around 100 more employees. Currently, expecting only local tourists
Malay Customary Institution	No cultural events after COVID.
	After COVID there are no integrated cultural and tourism activities. However, the youth of Bintan initiate small-scale cultural and tourism activities in the villages by obeying the health protocols.
Head of the Mangrove Bay Village	There are plans such as combining culture and tourism through dance and culture. This program was only running before COVID. Still not sure because I'm still afraid of the stigma of COVID.
	People are afraid of the stigma of COVID, there are pros and cons during the pandemic. They were afraid that COVID would spread in their area, but after the Regent, Governor, and health experts arrived. After socialization, the community understands this.
	During COVID, some foreign investors, businessmen, keep visiting to ask for coastal land. But, most of the coastal land has been run out. Some Japanese investors/entrepreneurs want to open investments and build, but there is no land available.
Head of the Department of Culture and tourism	During COVID, the Lagoi ferry only operates once a week but no one enters because the border is still closed.

3.2 Content of Policy

As pandemic COVID-19 takes place, there are several rules and decisions issued by the government in the form of government regulation, Kementerian, Institutions, and the Force in an attempt to prevent the spread of COVID-19, the following is a regulation that affects Bintan tourism:

1. Regulation of the Minister of Law and Human Rights No. 11 of 2020 concerning the Prohibition of Foreigners from Entering the Territory of the Archipelago State of the Republic of Indonesia,
2. Circular Letter (SE) Task Force of the COVID-19 No. 12 of 2021 concerning the Extension of Domestic Travel Provisions, and
3. Guidelines for the Implementation of the Ministry of Tourism and Creative Economy policies in the form of the implementation of CHSE (Cleanliness, Health, Safety, and Environmental Sustainability) in the Tourism Industry which refers to the Decree of the Minister of Health of the Republic Indonesia Number HK.01.07/MENKES/382/2020 regarding Public Health

Protocols in Places and Facilities General In the Context of Prevention and Control of COVID-19.

Based on the explanation, the content of policy focuses on rules and regulations that must be obeyed by local governments and other stakeholders in handling the COVID-19 pandemic in Bintan Regency tourism activities in the implementation of CHSE. The latest data shows the number of tourism industries in Bintan Regency that have a satisfactory CHSE category is 47 (forty-seven), the tourism industry can be seen in Table 2.

Table 2. Bintan Tourism Industry with Satisfactory CHSE

No	Tourism Industry	Amount
1	Tourist Transport	1
2	Tourist attraction	2
3	Home *stay* / Pondok Wisata	9
4	Hotel	23
5	Restaurant	10
6	Golf Business	2
Amount		47

The CHSE certification policy was established by the Ministry of Tourism and Creative Economy of the Republic of Indonesia as a guide for the implementation of CHSE which includes Hygiene, Health, Safety, and Environmental Sustainability. There are 11 guidelines in the implementation of CHSE [11], including the Tourist Attraction Guide, Hotel Guide, Homestay Guide, MICE Guide (Meetings, Incentives, Conferencing, Exhibitions), Event Organizing Guide, and so on. The implementation of the CHSE certification is intended as one of the efforts in preventing and handling COVID-19 in the tourism sector [12][13]. The guidelines it is described in great detail as an example in the Hotel Guide about tools of Health Protocols, optimization of air circulation, the implementation of the cleaning and disinfecting, entrances, rooms, meeting rooms, dining room, swimming pool, fitness center, mosque, and other public facilities in the business area.

3.3 Context of Implementation

The implementation of the Health Protocol and CHSE certification goes well because humans need to stay alive. The implementation of CHSE and the Health Protocol received a positive response based on the respondents' results. In addition, the Government in this case assists in the form of providing the tools of Health Protocol, implementing certification, and the implementing of vaccines is welcomed. Some expectations from respondents are the implementation of the vaccine will be accomplished immediately for all groups not only for tourism employees thus tourists feel safe when doing tourism activities. The CHSE certification is expected to build the confidence of potential tourists by knowing that the locations visited have followed the CHSE standards implemented by the Government.

Fig. 1. The Implementation of CHSE in Bintan Regency.

4 Discussion

Tourism governance aims to take strategic actions for the public interest in general so that during the COVID-19 pandemic it can be well controlled, this aims to ensure that every stakeholder can carry out their duties for tourism in Bintan Regency adjusts to new adaptations. By looking at governance in the grindle concept, here are the results of the connection map analysis in Content of Policy and Context of Implementation using NVivo 12 Pro

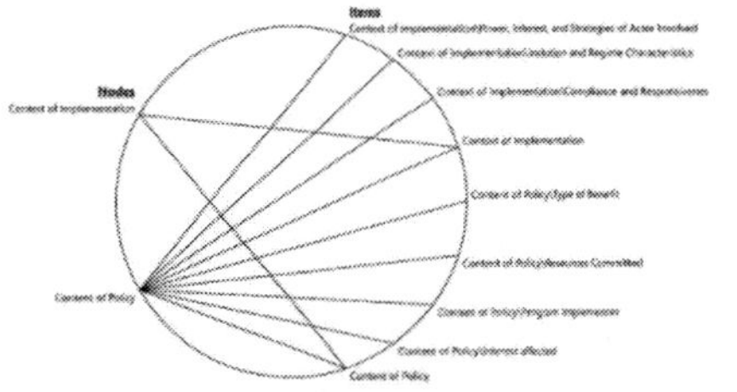

Fig. 2. Connection Map of Tourism Governance on the Grindle Concept during the COVID-19 pandemic.

Based on the results of interviews with the interviewees, the actors of interest show that the connection map of governance in the tourism sector of Bintan Regency during the COVID-19 pandemic is not all well connected, wherein Content of Implementation only leads to the implementation of the COVID-19 program as of the connection map line remains constant on Content of Implementation. Therefore, the Content of Implementation based on the interview results of the interviewees still provides an explanation of Content of Policy in general, which is not able to describe the explanation in terms of Content of Policy dimensions. Furthermore, in the Content of Policy based on the results of interviews with interviewees, it can be concluded the dimensions of Content of Policy influence each other's involvement in the COVID-19 pandemic in the tourism sector in Bintan Regency.

https://doi.org/10.1051/e3sconf/202133102002

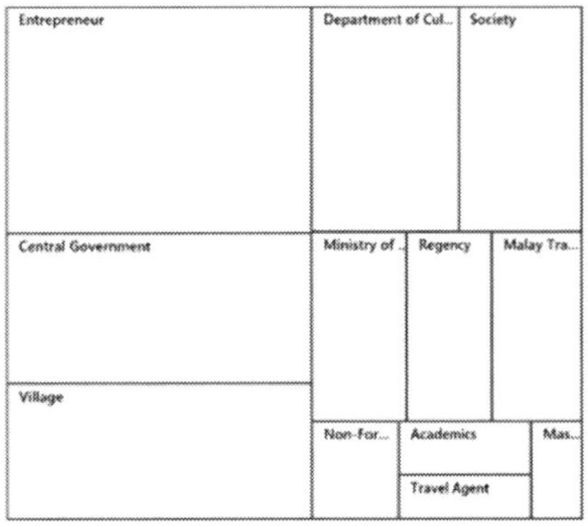

Fig 3. Hierarchy of Tourism Governance during the COVID-19 Pandemic Based on the results of interviewees.

Based on the results of interviewee interviews using NVivo 12 Pro, it shows that the entrepreneur hierarchy is the actor who has the most influence in the development of sustainable tourism during the COVID-19 pandemic. This is the most widely discussed businessman in the interview session on tourism developments during the COVID-19 pandemic. Furthermore, by strengthening tourism governance, the central government positions itself under the entrepreneurs, which means that the central government's attention to entrepreneurs in the tourism sector can be a special concern.

This is due to the Bintan Regency based on BPS Bintan data, Regional Original Income (PAD) in Bintan Regency in the amount to 87% moves the tourism sector and the income of tourism entrepreneurs dominate in Bintan regency [14]. Therefore, the smallest hierarchy in the mass media, the contribution of the mass media in tourism governance is not significantly involved because the mass media only focuses on tourism conditions during the COVID-19 pandemic and there is no tourism activity in Bintan Regency which is covered by the mass media.

5 Conclusion

Based on the results and discussions as well as secondary and primary data, tourism governance in Bintan Regency during the COVID-19 pandemic still focuses on implementation and formal policies. This is the absence of a technical application to the actualization of tourism governance in standard operating procedures (SOP), necessarily tourism governance in the scope of the content of policy and context of implementation dimensions should prioritize the application of tourism governance to formal policies. Thus, technically it presents policy innovations that can make Bintan Regency the concept of sustainable tourism during the COVID-19 pandemic.

References

1. Oktaviana, R. Feni, A. S. Muhammad, F. Kurnianingsih, Mahadiansar, Internal condition analysis on tourism development of Bintan Regency 2019. *Indonesian Journal of Tourism and Leisure* 2: **51** (2021).
2. Patterson, Carol *Sustainable tourism: business development, operations, and management*. II. Human Kinetics (2016).
3. Schroeder, Kent. 2015. Cultural values and sustainable tourism governance in Bhutan. *Sustainability (Switzerland)* 7. MDPI AG: 16616–16630 (2015)
4. Sharpley, Richard, D. J. Telfer *Tourism and Development: Concepts and Issues*. Bristol: Channel View Publications (2002).
5. Prasetyanti, Retnayu, I. A. Nugroho Governance, network in sustainable Tourism development: A case of thematic village tourism in Malang, Indonesia. In *Proceedings of the Annual International Conference of Business and Public Administration*. Poor (2018)
6. Ceballos-Lascuráin, Héctor. 1996. *Tourism, ecotourism, and protected areas . Tourism, ecotourism, and protected areas* (1996)
7. Miles, Matthew, M. Huberman, J. Saldana *Qualitative data analysis*. New York: SAGE Publications (2014).
8. Yin, K. Robert, 2018. *Case study research and applications: Design and methods*. Sixth. Sage Publications (2018).
9. Grindle, M S *Politics and policy implementation in the third world*. Princeton University Press (2017).
10. Bazeley, Patricia, *Qualitative data analysis with NVivo*. London: SAGE Publications (2007).
11. Maemunah, Ita. 2021. Implementation of Cleanliness, Health and Environmental Sustainability Guidelines in Restaurants around the Southern Java Crossing Route in Ciamis Regency. *Budapest International Research and Critics Institute (BIRCI-Journal): Humanities and Social Sciences* **4**: 2635–2642 (2021)
12. Syahrin, Alfi *Health Protocol Certification for tourism business due to Covid-19 outbreaks in Bali. Journal of Geography of Tropical Environments*.**5** (2021)
13. Yudiatmaja, Wayu Eko, Dwi Kristanti, Imam Yudhi Prastya, Yudithia Yudithia, Tri Samnuzulsari, Suyito Suyito, and Dian Prima Safitri Social policy on the rural coastal communities: why the implementation fails?. In *E3S Web of Conferences* **232**(1): 02006 (2021).
14. Akbar, Dhani, A. Setiawan, R. Prayuda, A. Putra, A. Aznor, W. E. Yudiatmaja, Community preparedness on transboundary oil spill governance in Bintan island. *Journal of Physics: Conference Series*, **1655**(1): 012144 (2020)

Forms of government collaboration at the local level in handling the COVID-19 disaster

Cici Safitri[1*], *Tengku Rika* Valentina[1], and *Ario* Wicaksono[2]

[1]Departmen of Political Science, Faculty Social Science and Political Science, Andalas University, Padang, Indonesia
[2]Institude for Governance and Policy Analysis, University of Canberra, Australia

Abstract. The current COVID-19 disaster requires serious handling, with the involvement of various parties. A different approach is shown by *Nagari* Pasie Laweh which is one of the in West Sumatra, namely by using a people-based approach which means that the low-level government *Nagari* collaborates with non-government local actors to handle and prevent COVID-19 with a different approach. clans (people united by tribes). The purpose of this research is to see how the efforts or forms of collaboration made by the *Nagari* Government with local actors deal with the COVID-19 disaster. This study uses a qualitative method with a descriptive approach and data collection techniques using literature studies from various articles/journals and related sources that help answer research questions. The results of this study show that the collaboration carried out by the *Nagari* government has made the handling of the COVID-19 disaster more optimal. Disaster management using a people-based method involves local non-governmental actors called the *tigo tungku sajarangan (ninik mamak, alim ulama, cadiak pandai).* The involvement of these actors helps the handling of men j a near to people's needs, this effort also facilitates communication between the government *Nagari* with society, so that collaboration raises public confidence towards the policy applied in *Nagari*.

1 Introduction

The pandemic has not ended, the number of cases has soared, with around more than 1 million Indonesians already exposed to the virus, which has made many countries lose their citizens. The government's efforts to recover the national economy in the second quarter are still experiencing shocks with the current spike in cases in June 2021. Following are daily case updates in Indonesia on 7 June 2021

Table 1. Daily COVID-19 cases in Indonesia until 7 June 2021

Confirmed	**1.863.031**+6.993 cases
Active Case	**99.663**+1.208
get well	**1.711.565**+5.594
Died	**51.803**+191

From the table above, it can be seen that the number of cases that occurred until June 2021 continues to increase, requires special attention from the government for the sake of national security and the impact that will be caused and felt by the community. Not to mention the government's efforts to control the virus and its impact on the community are still not optimal in reducing cases until now. This makes the government have to tighten policies and collaborate with local governments so that case by case does not experience a significant increase. This unrest is experienced not only by the central government but also by local governments. The regional government must make efforts so that the region does not experience a downturn during this pandemic, including West Sumatra.

Nagari is a leading government that has the power to interact directly with the community. *Nagari* can also make various efforts in dealing with disasters. Collaboration between the central-regional government and all elements of society is the main key in controlling the virus. West Sumatra, with its local wisdom and unique culture, has made the handling of the COVID-19 disaster in collaboration between the *Nagari* government and local non-governmental actors consisting of *Tigo tungku sajarangan (ninik Mamak, alim ulama, cadiak pandai).* This makes the efforts of the *Nagari* government easier and touches the needs of the people in *Nagari*. To control the virus, the *Nagari* government has implemented policies by the direction of the central government, such as establishing a COVID-19 post in each region. However, the establishment of a post without support and collaboration with local actors will not work well. The synergy built by the *Nagari* government through central elements in society is also important for the smooth running of the policies carried out.

* Corresponding author: cicisaafitri@gmail.com

This form of collaboration built by the *Nagari* Government with local actors is called a people-based approach in handling the COVID-19 disaster in *Nagari* in West Sumatra, one of which is *Nagari* Pasie Laweh who is the locus of this research. This method or approach is used in all aspects, both in the form of virus control as well as handling and preventing the impacts caused by the COVID-19 disaster. Not infrequently, many studies related to the handling of COVID-19 carried out in various regions in Indonesia are also carried out following local norms and culture so that it is easier to reach the needs of the community. This synergy can also increase the community's legitimacy to the *Nagari* government in handling COVID-19. The involvement of all elements of society is an important aspect of the success of the policies to be implemented.

Many previous studies serve as references for researchers who have conducted similar research including [4, 1, 7, 3, 2] this category of research talks about the readiness of local governments in dealing with the COVID-19 pandemic outbreak and the efforts made. The research above can help the authors find that to see the readiness and how the local government deals with this outbreak, it has not been seen from the lowest government which also makes efforts to prevent and handle the virus, but has not been explained more fundamentally by the four studies above. The involvement of the *Nagari* government as the lowest government can also be seen in its efforts to help deal with the outbreak. This is what the author wants to examine in this research. Disaster management efforts carried out by the *Nagari* government have a relationship with the government's success in handling this disaster.

2 Research Question

This people-based method is an important factor to see how the *Nagari* government is handling this disaster, both in controlling the virus, preventing and handling COVID-19 which has been regulated in the policies of the central government, but how is the form of collaboration between the *Nagari* government and local actors in handling this disaster? the COVID-19 disaster at the local level? The purpose of this study is to further see and analyze how the efforts or forms of collaboration made by the *Nagari* Government with local actors to deal with the COVID-19 disaster.

3 Research Method

This study uses descriptive qualitative methods to analyze how the form of collaboration carried out by the *Nagari* with local actors in dealing with the COVID-19 disaster. Descriptive analysis method to collect data and information on phenomena or actual problems through the process of data collection, compilation, processing, and concluding, which seeks to describe an objective empirical situation on the phenomenon or problem being studied. The data collection technique the researcher uses is a literature study, due to the limitations of the researcher to conduct direct interviews with the informants in this study.

A literature review is a critical and in-depth evaluation of previous research, also a good literature review is done by evaluating new findings from scientific articles/journals. It does not stop at the reading process alone; the researcher also summarizes making analysis and critical and in-depth synthesis of the articles/journals reviewed. Furthermore, the data is analyzed by the principles of the exploratory postpositive method, besides the researcher will obtain data from reports (articles/journals and documentation related to research) a large amount of data will be reduced by researchers by making abstractions as a core summary.

4 Literatur Review

Many scholars discuss the handling of the COVID-19 disaster at the domestic and national levels. Several categories of group researchers start from research on handling COVID-19 at the National level and the local level. Various forms of policies that were born from the central government must of course be in line with the policies carried out by the local government so that efforts to overcome and handle COVID-19 are optimal to break the chain of the spread of the COVID-19 virus. Like the research conducted by [1] regarding policy analysis for handling the COVID-19 outbreak: Indonesia's experience. The efforts of the Indonesian government in handling cases of the spread of the coronavirus can be understood by the efforts, steps, and policies implemented by the government so that they can see the handling strategy. Using a deliberative policy approach, this research shows that the government's response to the spread of COVID-19 is still slow because, at the beginning of the first case entering Indonesia, there was still no feeling of a crisis that endangered national security. Furthermore, the lack of coordination between stakeholders, both the central and local governments, the lack of synchronicity of the coordination has resulted in very slow virus control. When viewed from the aspect of society, they also disobeyed the government's appeal, which in the end did not synergize these three elements, causing the handling of the COVID-19 disaster to be slow at the beginning.

The research revealed [1] is an evaluation for both central and regional governments, as well as the community to carry out various policy efforts in handling this disaster. From time to time, there are many breakthroughs made by the central and regional governments such as research conducted by [10] which says that communication in disasters is an important part of handling every disaster, including the COVID-19 pandemic. His research entitled "Disaster Communications for Handling Coronavirus Diseases. 2019 (COVID-19) in Indonesia" revealed that communication is needed to build public trust to prevent public panic. This study explains the various steps and policies of the Indonesian government in dealing with COVID-19, especially in managing the flow of information and maintaining good communication to the public to maintain transparency and prevent mass hysteria. Communication is an important component in response and recovery, communication can also contact all elements, agencies, communities that work together to stop the spread of the virus. The availability of reliable information and communication is important in disaster management, the government must be able to ensure that the public can easily get information about the development of COVID-19. So that the availability of clear information and communication also helps all elements to act by policies and needs for handling COVID-19.

If this can be built, then all policies and collaborations between these elements can work. Research from [5] also talks about the harmonization of the authority to handle the COVID-19 pandemic between the central and local governments, seen from the health quarantine law policy. The law stipulates that COVID-19 is designated by the central government as a health emergency. The central government and local governments work together to make efforts to ensure the availability of the necessary resources in the implementation of health quarantine, namely through the establishment of a task force for handling COVID-19, providing health facilities, providing education on the dangers of COVID-19 to the entire community by mobilizing all stakeholders to participate in dealing with COVID-19. -19, provide social safety nets and deal with economic impacts. There are still problems such as the lack of coordination, communication, and synergy that was felt at the beginning of the pandemic. the government should respond quickly, which ultimately greatly affects the handling process. However, it has been seen that the government is increasingly focused and realizes that coordination is an important aspect, this can be seen from the existence of task forces and studies of the status of various regions in Indonesia.

The next category of researchers looks at the handling of COVID-19 from the domestic aspect, namely how the government at the local level can handle the COVID-19 disaster, such as research conducted by [6] talking about local leadership during the COVID-19 pandemic: Responses, Policy, and the Electoral Stage. This research was conducted in four provinces, namely DKI Jakarta, West Java, Central Java, and East Java in the face of the COVID-19 pandemic. This study uses literature review and descriptive analysis which reveals that local leaders in four provinces not only have a role in controlling the virus but also have a role in maintaining many aspects, such as health, social and economic. To deal with this dangerous threat, local leaders coordinate with central leaders, issue public policies in the regions, and encourage active community participation. In addition, the role of leaders at the local level also raises the view that the performance of leaders related to controlling the COVID-19 virus at the subnational level will push the prospect of political leadership to the national political stage.

This is in line with the research conducted by [9] which talks about local leadership issues in the COVID-19 pandemic. His research, entitled building cooperation and activating the role of local leadership: The Penta helix strategy for handling the impact of COVID-19. His study [9] revealed that cooperation and local leadership are the dominant characteristics of social capital practices in Indonesia. In an organic or systematic sense, it becomes a thread of synergy and integration between stakeholders. Therefore, packaging and strengthening social capital in policies with the Penta helix framework is the ultimate arrow in handling the COVID-19 pandemic. With Penta helix, the community's social capital can be used to deal with COVID-19. This makes the community more creative in building and modifying social capital for the prevention and handling of the pandemic.

Social capital is also positioned appropriately in government policies and programs, the vision, and mission of business actors, evidence-based policy formulation activities (evidence) from academic groups, being part of public campaigns that can be carried out in the media, and a new spirit in building cooperation and activating the role of local leadership for pandemic response mechanisms at all stages of the disaster management cycle.

5 Result

5.1 Forms of collaboration with local governments in dealing with the COVID-19 disaster

One of the *Nagari* that is the focus of this research is *Nagari* Pasie Laweh, Palupuh District, Lima Puluh Kota Regency, West Sumatra Province. This *Nagari* has innovations to carry out COVID-19 disaster management at the *Nagari* level (local level). The policies that have been implemented by the center to the regions are managed by the *Nagari* Pasie Laweh government based on local wisdom and local culture. Such as regional quarantine policies, isolation houses, handling and prevention, handwashing assistance facilities can be managed and run properly through people-based methods.

This method means that the applied policies are applied based on local values, numbers, and local cultural wisdom, by utilizing local actors and the community who can help optimize policies. Kaum is defined as a group, or tribe living in one *Nagari*. So, the activities of the *Nagari* or the *Nagari* Government must involve the tribes in the *Nagari* as an effort to synergize in carrying out the goals to be achieved together. The involvement of local non-government actors is an important synergy and collaboration to gain legitimacy from all elements of society. As revealed by [10] that communication is an important aspect in disaster management, communication can increase synergy and collaboration between elements of both government and society as well as other institutions that create public trust. The involvement of local non-governmental actors consists of *Ninik Mamak* from each clan in *Nagari* Pasie Laweh.

The government of *Nagari* Pasie Laweh uses people-based methods to prevent and manage the COVID-19 disaster and also makes village fund management successful during the COVID-19 pandemic so that it is in line with government policies and the needs of the *Nagari* people. The government of *Nagari* Pasie Laweh is synergizing and collaborating with 100 *Ninik Mamak* in the *Nagari* to jointly build the *Nagari* and avoid the impacts caused by the current pandemic, starting from the health, economy, and social community.

The form of synergy built by the *Nagari* government with this local actor(100 *Ninik Mamak*) can be seen through the policies of the *Nagari* Pasie Laweh government in line with the direction of the central government, namely as follows

1. The establishment of the COVID-19 handling post
   By ministerial directives and circulars, the *Nagari* Pasie Laweh government established a COVID-19 post which aims to be an initial screening location, for the public or anyone who enters *Nagari* must pass through the post. At this post, there are screenings such as recording and requesting a report on travel history and destinations to *Nagari* Pasie Laweh. The command post is provided with medical equipment, such as a body temperature checker, a place to wash hands, and a supply of masks. The guarding of this post consists of staff from the *Nagari* Government, *Ninik Mamak*, Youth, and Community (in rotation) to carry out monitoring.
2. Forming a COVID-19 education team, this education team consists of all elements in *Nagari* Pasie Laweh consisting of Midwives, Babinsa (Village Trustees), Bhabinkamtibmas (Bhayangkara Pembina Community Security and Order), MUI *Nagari*, *Ninik Mamak*, Youth, and *Nagari* staff who also brought questionnaires. to measure people's knowledge of COVID-19, as well as what are the impacts on community activities and the economy. Furthermore, this activity also asked for advice from the community for handling COVID-19 to the needs of the community.
3. Establishment of data and information center. Here we are connected with various data that has been provided by the *Nagari* government related to the COVID-19 case. How many people entered *Nagari*, Until April 18, 2020, based on the data that researchers got, 75 immigrants, entered *Nagari* with distribution areas of Jakarta, Riau, and Malaysia. We can find these data at the *Wali Nagari* office; there are also available data on recipients of PKH assistance, social assistance, and other assistance so that no more people receive two aids at once. In addition, such as Village Cash Direct Assistance (BLT-DD), the *Nagari* government also cooperates with *ninik Mamak* to formulate and determine at the *Nagari Musrebang* who is entitled to receive the assistance. This involvement is also reasonable because *ninik mamak*'s closeness to their people, and it is felt that they know more about the family's economic situation during society are also taken into consideration by the *Nagari* Government.

The collaboration can be seen on the chart handling agency COVID-19 in *Nagari* Pasie Laweh in this Fig.1
This form of synergy between the Government at the local level and non-government actors seems to greatly facilitate the *Nagari* government in preventing and overcoming COVID-19. The collaboration between the two has resulted in good collaboration so that it gains legitimacy and active participation from the community to support efforts to reduce the impact of the COVID-19 virus. When viewed from Fig 1. looks actor local (*ninik Mamak*) has a line parallel with the *Wali Jurong* to synergize to handle efforts nan close to the community. resulting collaboration and cooperation as formation posts COVID-19, team educate the COVID-19, the establishment of data centers and information, as well as the warehouse logistics.

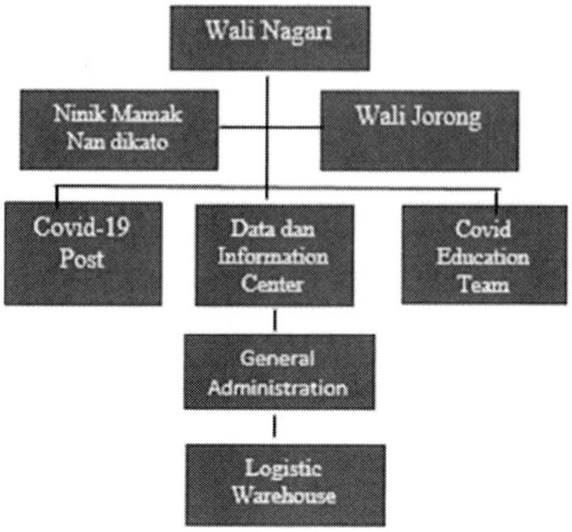

Fig. 1. Chart of institutions handling COVID-19 *Nagari* Pasie Laweh

Not only that, another innovation that can be utilized by *Nagari* Pasie Laweh in collaboration with *ninik Mamak* outside of central government policies or regulations is the innovation of using the *Rumah gadang* as an isolation house, with the agreement of the people and *ninik mamak*. The *Nagari* government has succeeded in cooperating to establish this isolation house. The isolation house used is an unoccupied house belonging to the people that can be used for people who are exposed to the virus. Furthermore, people exposed to the virus who are in isolation at home are the responsibility of their people. So, the community or *ninik Mamak* helps the needs of the people who are in isolation, such as providing food and vitamin intake.

In addition, to empower affected communities such as those in PHK, or not working, WFH, and people who have lost their income due to the pandemic are empowered through the management of vacant communal lands. The *ninik Mamak* of each clan allowed the *Nagari* government to be managed temporarily to be able to help the community and meet their daily basic needs. This collaboration between the *Nagari* government and *ninik Mamak* helps all elements of society to reduce the economic impact. The *Nagari* government provided funds to buy fish seeds, and plantation seeds to be managed in the vacant land. The results of these seeds can be distributed to the community later. The *Nagari* government and *ninik Mamak* invite and take turns empowering the community to manage the land.

This people-based approach or method was successfully applied by the *Nagari* Pasie Laweh

Government to cope with the COVID-19 disaster. This synergy and collaboration between the *Nagari* government and local non-government actors can make disaster management more optimal. So that the role of both becomes important to achieve the needs and welfare of the people in *Nagari* Pasie Laweh. Collaborative cannot always be done either between the government and the government alone or with institutions/institutions. however, it is also necessary to collaborate with actors or community groups who are deemed to be able to build public trust so that there is legitimacy to carry out optimal activities or policies.

6 Conclusion

The government's collaboration at the local level in dealing with the COVID-19 disaster in *Nagari* Pasie Laweh is quite effective. The collaboration built by the *Nagari* government with local non-government actors, namely *Ninik Mamak*, has made the handling and prevention of the COVID-19 disaster run well. Utilization through aspects of local wisdom and cultural norms is also an important aspect in disaster management at the local level. This is proven by the several forms of collaboration that give birth to innovations that are closer and by the needs of the community. It is important to know that making the community feel involved is also an aspect to build public trust in the policies that will be implemented.

The support of elements of the community at the local level for the policy makes this response closer to the community. The people-based approach in *Nagari* Pasie Laweh can also be used as a reference for regions that have local wisdom and culture that can help the *Nagari* government to be stronger in carrying out the policies that are implemented. The needs and welfare of the community in *Nagari* become more important with various approaches that can be taken and various forms of synergy and collaboration with actors or elements of the community.

References

1. L. Agustino, Analyses of the Policy for handling the COVID-19 outbreak: the experience of Indonesia. Analyses of the COVID-19 outbreak handling policy: the experience of Indonesia. Journal of Borneo Administrator, **16** (2), 253–270 (2020)
2. RM. Amin, RF. Envelope, B Wicaksono, Handling COVID-19 from a Collaborative Governance Perspective in Pekanbaru City. 13, 1–13 (2021)
3. A. Arie, S. Dharma. The COVID-19 Infodemic in an Open Government Perspective : A Literature Review. **6** (1) (2021)
4. D. Aromatica,. Strengthening the Capacity of the Solok District Local Government in Handling COVID-19 Through Collaborative Governance. 2008, 138–141 (2020)
5. S. Khadijah, Harmonization of the authorities for handling the COVID-19 pandemic between the central and regional governments. Journal of Kertha Semaya, **8** (6), 858–866 (2020)
6. RH Imawan, I Syafii, MR Surya, P Prayoga,. Local Leadership in the Time of the COVID-19 Pandemic: Response, Policy, and the Electoral Stage. Journal of Political Research, **17** (2), 195–281 (2020)
7. DR. Hizbaron, D. Ruslanjari, D. Mardiatno. Amidst COVID-19 Pandemic: An Adaptive Disaster Governance (2021)
9. MA. Humaedi, SS Purwaningsih, LV Sunday, R. Fathy,. Building cooperation and activating the role of local leadership : Penta helix strategy for handling the impact of the COVID-19 role of local leadership : pentahelix's strategy for handling the impact of COVID-19. **23** (1), 39–58 (2021)
10. F. Noor, F. Ayuningtyas, W Prihatiningsih. International Journal of Multicultural and Multireligious Understanding Disaster Communications for Handling Coronavirus Disease 2019 (COVID-19) in Indonesia. Journal of Multicultural and Multireligious Understanding (IJMMU), **7** (4), 25–35 (2020)
11. Tribunsumbar.com, 2020. Paste Laweh Manages Village Funds and Hands I COVID-19 Base on People. In Access (https://www.tribunsumbar.com/pasie-laweh-kelola-dana-desa-dan-tangani-COVID-19-berbasiskan-kaum/) On 15 May 2021
12. Binews.id, 2020. Pasi e Laweh Manages Village Funds and Handles COVID-19 Always Baiyo Batido with *Ninik mamak.* accessed at (https://www.binews.id/berita/3209/pasie-laweh-kelola-dana-desa-dan-tangani-COVID-19-berbasis-kaum.html) On May 15 2021

https://doi.org/10.1051/e3sconf/202133102004

Improving food security through the urban farming workshop for disaster risk reduction: evidence from Jogoyudan Village, Yogyakarta as a Resilient City

Lexi Jalu Aji[1*], *Siswo* Hadi Sumantri[1] *Yuli* Subiyakto[2], *Mir'atul* Azizah[1], and *Annisa* Rahmawati[1]

[1] Disaster Management Study Program, National Security Faculty, Republic of Indonesia Defense University. IPSC Sentul Area, Bogor, West Java, Indonesia
[2] Management Department, Gadjah Mada University, Yogyakarta, Indonesia

Abstract. This study aims to determine the effect of the implementation of urban farming workshops on food security of Jogoyudan Village residents supporting the concept of Yogyakarta as a Resilient City as a disaster risk reduction. This study is a quasi-experimental study using a nonequivalent pretest-posttest control group design. This study involved 200 residents in 4 RWs (Neighborhoods) in Jogoyudan Village as the research sample selected by cluster random sampling. RW 08 and RW 10 as the experimental group, while RW 11 and RW 12 as the control group. The research data were collected through assessment instruments in the form of pretest and posttest that were analyzed quantitatively, as well as using the method of observation and documentation. The result of this study indicates that the implementation of the urban farming workshop has a positive effect on improving the understanding level of food security of the people in Jogoyudan Village as evidenced by the value of the Wilcoxon test of 0.000 ($\alpha > 0.05$) and the Mann-Witney test of 0.000 ($\alpha > 0.05$). In addition to that, the urban farming workshop can also provide participants with farming skills, which can facilitate greater and easier access for food which is the main indicator of food security, as attested by the experimental group in Jogoyudan Village who was able to independently make hydroponic installations at their homes. This shows that the urban farming workshop can be one of the ways toward disaster risk reduction supporting the realization of Yogyakarta as a Resilient City.

1 Introduction

Population density is a challenge that has to be addressed in the consideration of disaster mitigation. The emergence of potential disasters due to population density is caused by several triggers, one of them being dense settlements that extend to riverbanks [1]. The village on the banks of the Code River is an example of a disaster-prone settlement located in the middle of Yogyakarta City. Every year, a flood occurs from the Code River due to heavy rains that overtops the riverbanks. In 2010, there was even an eruption of Mount Merapi which spewed cold lava into the river and caused massive flooding. One of the affected villages was Jogoyudan Village [2].

Flood may cause various kinds of derivative disasters, one of them being food disasters. When a disaster occurs, the food availability decreases while the community needs post-disaster increase [3]. The consequences of floods might immobilize daily activities and might cause economic losses. Furthermore, disrupted food production will increase the price of food due to a decrease in food supply [4]. The need for food will arise along with the increasing population growth in a specific area [5]. This may escalate the community's vulnerability to disasters. To prevent this, public awareness is needed to maintain food security.

Food security is a highly complex issue in national development and is mainly a top priority in development [6], it even has become one of the main agendas in the *Sustainable Development Goals (SDGs)* forum [7]. According to Law No. 18 of 2012 concerning Food, food security is the fulfillment of food for households, carried out by the availability of sufficient food, both in quantity and quality, safe and affordable in a sustainable manner [8]. Based on this concept, there are 3 aspects of food security, namely, 1) *availability*, 2) *affordability* and 3) *quality and safety* [9].

Based on data collected by the *Global Food Security Index (GFSI) 2020*, Indonesia's food security level is ranked 65 out of 113 countries. It is ranked 34th for availability, 55th for affordability, and 89th in terms of quality and safety [9]. Given the fact that Indonesia is an agricultural country with vast and abundant fertile soils,

* Corresponding Author: lexijalu@gmail.com

Indonesia's food security rank in terms of affordability, as well as quality and safety is considered low.

As stated by the report of the Ministry of Agriculture Republic Indonesia (2021), part of the reason for Indonesia's low ranking in two aspects of food security was caused by natural disasters and environmental damage (non-natural disasters) that occurred in Indonesia. The occurrences of natural disasters in Indonesia contributed to the disruption in the food distribution chain and the damage to agricultural land, resulting in the community's affordability aspect diminished. The unbounded environmental pollution and excessive use of chemicals in agriculture caused the low quality and safety aspects of food security in Indonesia [10]. On top of that, based on research [11], the Covid-19 pandemic has impacted food security as it hampered the distribution, availability, affordability, and stability of food.

As one of the cities proclaimed as a resilient city, Yogyakarta City needs to fix this food problem. The concept of a *Resilient City* is a city that can recover quickly in the face of disasters (Adiyoso, 2018). The disaster management paradigm has evolved from "*fatalistic responsive*" that is oriented to emergency responses to "*proactive preparedness*" that focuses early on thorough preparation. This paradigm requires both the government and the community to jointly undertake disaster risk reduction efforts [12]. Consequently, the government and the people of Yogyakarta City are obliged to put efforts to lower disaster risk concerning the threat of food disasters.

The *affordability* aspect has an assessment indicator in the form of the community's ability to access food. If the community can produce food independently, the community's access to food will be easier, assessing the affordability aspect better. The affordability aspect also has an indicator of the people's understanding of food diversity. The better the community's understanding of food diversity, the higher the affordability aspect's value [13]. Meanwhile, the quality and safety aspect is related to the nutritional quality of food and its safety. The better the nutritional quality and safety of the food consumed by the community, the higher the quality and safety aspect will be. In addition to that, this same aspect also has indicators of community understanding regarding nutrition and food safety [9]. Based on this explanation, assistance that can improve skills, as well as people's understanding of food security, is needed.

A workshop is an educational program designed to teach students theoretical and practical skills that can be used in everyday life functions [14]. The results of research conducted by Lahami (2021) show that workshops can improve students' competence in conceptual and practical skills in agricultural management [15].

Jogoyudan Village is a dense urban settlement located on the banks of Code River in the middle of Yogyakarta. Many people living in this village are not owner-occupiers and almost the entire land surface has been covered by buildings and concrete, making it difficult for residents to garden and generate livestock [16]. One farming method that can be applied in urban areas is Hydroponics. Hydroponics is one of the *urban farming* systems that does not require fertile soils and can be applied in narrow areas [17].

In reality, the Gowongan Village Government has developed programs related to food counseling for the community, but the evaluation of these activities has not shown significant results on the level of food security in Jogoyudan Village. The counseling given to the community focuses more on food nutrition and health education; it has not yet reached the field of food security [18].

Based on the information obtained in the above description, we conducted this study using the title "Improving Food Security Through Urban Farming Workshop for Disaster Risk Reduction: Evidence from Hogoyudan Village, Yogyakarta, Indonesia". The purpose of this study was to measure the effect of the urban-farming workshop on food security in Jogoyudan Village for disaster risk reduction. The formulated hypothesis of this research is that the Urban Farming Workshop can improve food security in Jogoyudan Village for disaster risk reduction.

2 Method and Data

This quantitative research used a *Quasi Experiment* research model with a *Nonequivalent Pretest-Posttest Control Group Design.* The sampling technique used in this study was *Cluster Random Sampling* [19].
This research was conducted in 2019 in Jogoyudan Village, Gowongan Sub-district, Jetis District, Yogyakarta City. The population of the study was the residents of Jogoyudan Village and using a sample of 4 (four) RWs (Neighborhoods) whose areas are on the banks of the Code River, namely RW 08 and RW 12 as the experimental group, while RW 10 and RW 11 as the control group. For the determination of sample size, this study referred to Isaac and Michael's reference table [20]. The independent variable in this study was the urban farming workshop which was applied to the experimental group, while the control group used the Gowongan Village government's food training program. The dependent variable in this study was the food resilience level of the residents of Jogoyudan Village.

The data of this study included data on the understanding level of food security as well as urban-farming skills. Data on urban-farming skills were collected through observation and documentation that were analyzed descriptively. Meanwhile, data on the understanding level of food security were collected through pretest and posttest assessment instruments that were analyzed with a statistically inferential method.

The prerequisite tests carried out by this study conducted a normality test of the data using the Shapiro Wilk test, while the homogeneity test was used to determine the data using the Levene test. Based on the results of the prerequisite test, the Mann-Whitney U test and the Wilcoxon test were used for the inferential statistical test [21]. The Wilcoxon test was used to test the value of the level of understanding of food security before and after the workshop, while the Mann-Witney test was used to test the difference in the value of the level of understanding of food resilient in the experimental group and the control group.

3 Result and Discussion

3.1 Description of Research Activities

Jogoyudan Village is located at the banks of Code River in Gowongan Sub-district, Jetis District, Yogyakarta City. Jogoyudan Village has 5 RWs (Neighborhoods), namely RW 08, RW 09, RW 10, RW 11, and RW 12. In this study, RW 08 and RW 10 were used as the experimental group, while RW 11 and RW 12 as the control group.

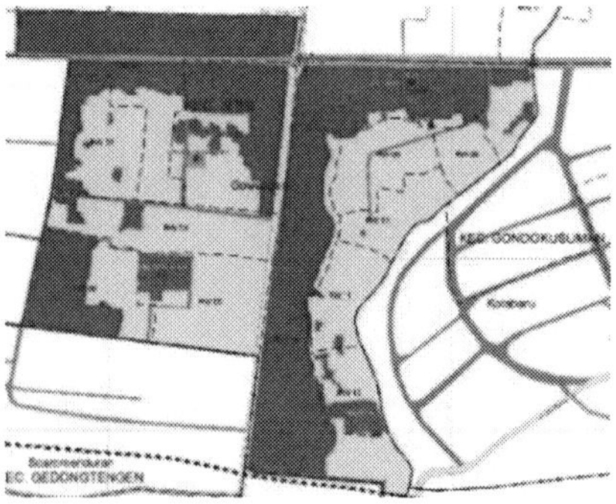

Fig. 1. Gowongan Village Administration Map

The treatment for the experimental group was given in the form of an urban farming workshop. The urban farming workshop activity consists of 3 activities, specifically 1) teaching activity, 2) hands-on/practical activity, and 3) a study visit to PT. Indmira. RW 08 carried out activity 1 and activity 2 on Friday, July 15th, 2019, while RW 10 carried out the activities on Friday, July 22nd, 2019. The teaching materials for the experimental group started by providing a brief introduction to urban farming and the concept of food security through food diversification, then followed by wick-hydroponic practice activities.

Fig. 2. Documentation of Urban Farming Workshop Activities

After the participants were able to practice hydroponics, we continued activity 3 by performing a study visit to PT. Indmira, which was held all together on Saturday, July 23rd, 2019. PT Indmira is a research company in the agri-complex field engaged in the business of organic vegetable products using the hydroponic method.

In the control group, the training activities were successfully carried out without the intervention of the researcher. The training activities were conducted by the Gowongan Village Government following the village program plan in the form of food nutrition counseling related to public health.

Fig. 3. Documentation of Industrial Visits

3.2 Understanding Level of Food Security

The total number of families in RW 08 consists of 132 families, while RW 10 consisting a total of 136 families. In this study, one person from each family was selected as a representative for the research sample. Based on Isaac and Michael's table (Jakni, 2017), the minimum sample size with a significance of 5% is 100.

Table 1. Result of Homogeneity Test Value of Understanding Level of Food Security.

Descriptive Analysis	Experimental Class		Control Class	
	Pretest	*Posttest*	*Pretest*	*Posttest*
Number of Participants	100	100	100	100
Minimum	25,00	65,00	20,00	50,00
Maximum	55,00	100,00	60,00	100,00
Mean	39,75	83,80	40,70	76,75

Based on the results of descriptive analysis, the average value of the level of understanding of food security in the experimental group increased from 39.75 (pretest) to 83.80 (posttest) with a difference of 44.05 increase in value. The average value of the level of understanding of the control group also increased from 40.70 (pretest) to 76.75 (posttest) with a difference of 36.05 increase in value. This shows that there is a difference in the increasing deviation value between the experimental group and the control group.

Table 2. Normality Test Results Values of Understanding Level of Food Security.

Tests of Normality

Class	Kolmogorov-Smirnov			Shapiro-Wilk		
	Statistic	df	Sig.	Statistic	df	Sig.
Postest Exp.	.160	100	.000	.950	100	.001
Postest Control	.212	100	.000	.929	100	.000

For the normality test, the above table showed that the significance value of the posttest results in the experimental and control groups were both less than 0.05 ($\alpha < 0.05$), this tells that the research data were not normally distributed. Since the research data were not normally distributed, non-parametric tests were used for inferential tests. The inferential tests performed as mentioned used the Wilcoxon test and the Mann-Witney test.

Table 3. Wilcoxon Test Results in Value of understanding Level of Food Security.

Test Statistics[a]

	PostestEx - PretestEx
Z	-8.810[b]
Asymp. Sig. (2-tailed)	.000

a. Wilcoxon Signed Ranks Test
b. Based on negative ranks.

The Wilcoxon test was applied to see if there was a significant difference in the average value between the average pretest value and the average posttest value of the experimental group (Creswell, 2017). The results of the Wilcoxon test showed a significance value of 0.000 ($\alpha > 0.05$), which means that there is a significant difference in the average value between the average pretest value and the average posttest value of the experimental group.

Table 4. Mann-Whitney Test Results Value of Understanding Level of Food Security.

Test Statistics[a]

	PostestResult
Mann-Whitney U	2886.000
Wilcoxon W	7936.000
Z	-5.264
Asymp. Sig. (2-tailed)	.000

a. Grouping Variable: Postest

After the Wilcoxon test, we conducted the Mann-Witney test to find a significant difference in the average score between the average posttest scores of the experimental group and the control group [21]. The results of the Mann-Witney test showed a significance value of 0.000 ($\alpha > 0.05$), meaning that there was indeed a significant difference in the average value between the average posttest value of the experimental group and the average posttest value of the control group.

3.3 Urban farming Skills

Fig. 4. Documentation of Urban Farming Workshop Activities

Based on the results of observations made in the field, residents of RW 08 and RW 10 as the experimental group who received treatment with the urban farming workshop were able to practice the given training. The urban farming workshop in the form of hydroponic training was successfully carried out by some residents of RW 08 and RW 10 as evidenced by the following documentation

The residents demonstrated great enthusiasm during the training, as seen from the number of participants in the urban farming workshop from RW 08 and RW 10 that reached the total amount of 124 people, exceeding the minimum target required of 100 participants. In addition to being introduced to hydroponics, the residents were also brought into the farming techniques of vertical gardens and fruit plants in pots.

The urban farming workshop effectively developed the residents' skills in managing plants. One reason was that after the training, of the 124 participants who attended the workshop, 76 participants were succeeded and continued to make hydroponics independently in their respective homes. Although constrained by limited land, residents took the initiative to use the walls of their houses and yard fences to put hydroponics. There was even one resident who could build hydroponic installation in their yard, with documentary evidence as follows:

Fig. 5. Residents' Hydroponic Installation Documentation

3.4 Discussion

Population density is one of the challenges in disaster risk management [1]. The dense settlements on the banks of the Code River in Yogyakarta City are one example of disaster-prone settlements due to overcrowding. Every year in that place, a flood occurs due to the overflowing banks of the Code River. The most severe flooding happened during the 2010 eruption of Mount Merapi, the material from the eruption of Mount Merapi filled the river and caused massive flooding in the settlements around the Code River. One of the affected villages was Jogoyudan Village [2].

Floods can cause various kinds of derivative disasters, one of them being food disasters [3]. Food disasters may occur due to the immobilization of people's daily activities when a flood disaster occurs. The inhibition of food distribution and production activities causes scarcity and increases in food prices [4]. In addition, food problems can arise along with an increase in population that is not accompanied by a rising in food production. The decrease in agricultural land also decreases the quantity of food. Therefore, it is necessary to increase the capacity and understanding of the community to maintain food security [5].

It is such an irony that Indonesia's food security was ranked 65 out of 113 countries, knowing that Indonesia is an agricultural country with abundant fertile soils and a tropical climate. Indonesia's low ranking in food security is due to Indonesia's low score in terms of food accessibility and food quality and safety [9].

https://doi.org/10.1051/e3sconf/202133102004

The Ministry of Agriculture of Indonesia reported that Indonesia's low ranking in two aspects of food security was caused by natural disasters and environmental damage. The occurrences of natural disasters in Indonesia contributed to the disruption in the food distribution chain and the damage to agricultural land, resulting in the community's affordability aspect diminished. Furthermore, the widespread environmental damage caused the food quality and safety aspects to below [10].

As one of the cities proclaimed as a resilient city, Yogyakarta City needs to fix the potential for foods disasters. The disaster management paradigm has evolved from fatalistic responsiveness oriented to emergency response to proactive preparedness that focuses early on efforts to reduce disaster risk [12]. This shifting paradigm requires both the government and the community to jointly undertake disaster risk reduction efforts.

Training using a workshop method is considered effective in improving skills. A workshop is an educational program designed to teach students theoretical and practical skills that can be used in everyday life functions [14]. Providing workshops with material on urban farming has the potential to improve community skills for farming in urban areas, this will affect aspects of affordability and quality as well as food security in the community.

The indicators for improving the aspect of food affordability consist of two things, namely 1) the level of community ability in accessing food and 2) the level of public understanding of the diversity of food. The aspect of food quality and safety also has two indicators, namely 1) the level of quality of food nutrition and food safety obtained by the community and 2) the level of public understanding about nutrition and food safety [9].

In this study, the urban farming workshop given to the experimental group consisted of 3 activities, namely 1) teaching activity, 2) hands-on/practical activity, and 3) a study visit. In the teaching activity, the material given was an introduction to urban farming and hydroponics as a method of urban farming. Workshop participants were also given the material on the concept of food security, introduction to food nutrition, and food diversification programs as a solution to food problems and disasters.

Based on research data obtained using pretest and posttest instruments, the implementation of an urban farming workshop can increase the average value of the level of understanding of food security in the experimental group. The average score increased from 39.75 (pretest) to 83.80 (posttest) with a difference of 44.05 increase in value. The increase in value was significant, as evidenced by the results of the Wilcoxon test on the pretest and posttest scores of the experimental group which obtained a significance value of 0.000 ($\alpha > 0.05$). From this result, it is concluded that the increase in the average value of the level of understanding in the experimental group shows that the urban farming workshop can improve indicator number 2 on aspects of affordability and quality, and safety.

The urban farming workshop has proven to be more effective in increasing the level of understanding of community food security, compared to the training provided by the Gowongan Village government. This is supported by the result of the average posttest score obtained by the experimental group (83.80) which is higher than the control group (76.75) with an average difference of 7.05 points. The difference in average values is significant, validated by the results of the Mann-Witney test that obtained a significance value of 0.000 ($\alpha > 0.05$).

Apart from being able to increase the level of understanding of food security, the urban farming workshop was also expected to build urban farming skills in the community. Accordingly, it was expected to increase the first indicator on the availability aspect in terms of increasing the ability of the community to access food, along with the third indicator on quality and safety aspect, in terms of increasing the level of quality of food nutrition and food safety obtained by the community.

The practical activity was implemented by providing a hands-on experience and a study visit to workshop participants. The practical activity was conducted by practicing making hydroponics. The hydroponic method used in this study was the wick technique. The wick technique is one of the simplest hydroponic techniques that does not require electricity and complicated installation. The materials for conducting the wick technique only require used goods, namely bottles and used cloth. In addition to low production costs, practicing hydroponics using the wick technique is more environmentally friendly because it can reduce waste that may pollute the environment.

Fig. 6. Hydroponics with Wick Technique

The study visit was carried out by visiting PT. Indmira, a company engaged in the organic vegetable industry, let the participants have in-depth learning about hydroponics. The participants were permitted to carry out several activities. For starters, they were introduced to various hydroponic techniques other than Wick Technique, such as the Floating Raft Technique and the Nutrient Film Technique (NFT) that were documented in Fig. 3. We also provided the know-how on how to market hydroponic products to participants that were conducted with the aim that after the workshop, apart from being able to grow crops using hydroponics, participants can also grow their crops independently and can market their products.

The results of observations and research documentation showed that the urban farming workshop can improve the community's skills in producing food independently, meaning that people may be able to gain greater and easier access to food sources. This was attested by all 124 participants in the urban farming workshop who could make hydroponics using the wick technique as documented in Fig. 4.

Based on the results of observations and research documentation post-follow up the urban farming workshop, there were 76 people out of 124 workshop participants who succeeded and continued to make hydroponics independently at home. There was even one resident who was able to build hydroponic installations using Floating Raft Techniques and NFT techniques that could be seen on the documentation in Fig. 5. This also proves that limited land does not become a limiting factor in agricultural farming.

Based on the data provided and analysis conveyed, this leads to the conclusion that the urban farming workshop can improve the first and second indicators of food security. For the first indicator, namely the affordability aspect, people can produce food independently resulting in them getting greater and easier access to food. As for the second indicator, the community can have better quality and safety of the food they produce themselves, which can increase the quality and safety aspect of food security.

The two value increases in terms of affordability and quality and safety aspects regarding food security prove that the implementation of urban farming workshops can effectively improve the food security of the people in Jogoyudan Village. This will furthermore reduce vulnerability and increase community capacity concerning the threat of food disasters. In addition to that, this research also supports the realization of Resilient City which is a program of the Yogyakarta City government. By conducting the urban farming workshop, it has become one of the steps to reduce disaster risk that can be applied to people living in urban areas and dense settlements.

4 Conclusion

Based on the analysis conveyed, it can be concluded that the urban farming workshop has a positive effect on improving the understanding level of food security of the people in Jogoyudan Village as validated by the results of the Wilcoxon test of 0.000 ($\alpha > 0.05$) and the Mann-Witney test of 0.000 ($\alpha > 0.05$). In addition to that, the urban farming workshop has also provided participants with farming skills, that will furthermore facilitate greater and easier access to food which is the main indicator of food security, as attested by the experimental group in the Jogoyudan Village who were able to construct hydroponic installations independently at their homes. Finally, this study shows that the urban farming workshop can be one of the ways toward disaster risk reduction supporting the realization of the Resilient City concept in Yogyakarta.

References

1. L. J. Aji, D. P. Meiliasari, R. K. Apriyadi, S. Maarif, H. Sumantri, PENDIPA J. Sci. Educ. **6**, 64 (2022).
2. T. Harahap, Sudayono, D. Kristiadi, J. Tekno Glob. **5**, 29 (2016).
3. U. Azmiyati, J. Social Science and Education. **3**, (2019).
4. Y. Oktari, J. Social Science and Education. **2**, (2018).
5. Y. Purwaningsih, J. Ekon. Pembang. **9**, 1 (2008).
6. A. R. Quisumbing, L. R. Brown, H. S. Feldstein, L. Haddad, C. Peña, Food Nutr. Bull. **17**, 1 (1996).
7. A. Suryana, *Toward Sustainable Indonesian Food Security 2025: Challenges and Its Responses Achmad* (2014).
8. Ministry, of Health, Bul. Data And Inf. Window. health **Semester I**, (2015).
9. GFSI, Regional Report: Asia Pacific (2021).
10. Ministry of Agriculture, Performance Report Food Availability and Insecurity (2021).
11. N. A. Fitriah, Impact of the Covid-19 Pandemic on Indonesian Food Security: An Exploratory Research, Bogor Agricultural University (2021).
12. Z. D. G. Raja, H. Hendarmawan, S. Sunardi, J.L Environment. And the Geol Disaster G. **8**, 103 (2017).
13. I. Rural, L. Through, B. Productive, Nature **544**, 1 (2008).
14. R. H. Kurniawan, Manag. Account. Res. J. **04**, 16 (2019).
15. M. A. G. Lahami, J. Educ. Technol. Curriculum, Learn. Commun. **1**, 13 (2021).
16. N. R. P. Ayodiya, J. Pembang. Wil. Kota **10**, 22 (2014).
17. E. Sugianto, Hydroponics, Altitude Requirements, Temperature, Relative Humidity, Sunlight 1st ed. (Gramedia, Jakarta, 2008)
18. T. Rahayu, L. J. Aji, A. Suta, A. Hergani, A. Fitria, R. N. Utami, S. V. Vidyaningsih, Int. Inventory. Innov. Competence. **2**, 54 (2017)
19. Sugiyono, Educational Research Mrthods (Qualitative, Quantitative And R&D) (Alfabeta, Bandung, 2019)
20. Jakni, Experimental Research Methodology in Education (Alfabeta, Bandung, 2017)
21. J. Creswell, *Riset* Educational Research: Planning, Implementation, and Evaluation of Qualitative & Quantitative Research (Learning Library, Yogyakarta, 2017)

Enhancing pandemic resilience: planning and institutional resilience, a learning model for Indonesia

Marnani. Christine S[1*], *Hartanto*[2], and *Sobar* Sutisna[1]

[1]Disaster Management Study Program, National Security Faculty, Republic of Indonesia Defense University. IPSC Sentul Area, Bogor, West Java, Indonesia
[2]Department of International Relations, Respati Yogyakarta University

Abstract. The COVID-19 pandemic is one of the most expensive and destructive threats facing millions of people around the world. Policymakers generally have a linear worldview and can help the economy and society to track epidemic shocks and crises by pulling the right levers. For example, in recent years, Ebola, MERS, Zika, and SARS have demonstrated a limited willingness to respond promptly and responsibly. The pandemic emphasizes the need to be more proactive. Yet, existing epidemic response frameworks are either disease-specific, category-specific, or non-specific, with no comprehensive framework encompassing all components of institutional resilience. In addition, the COVID-19 pandemic requires a holistic approach to meet the challenges. In this study, the goal of resilience to the outbreak of the epidemic is to develop an overall framework by identifying institutional departments to strengthen resilience planning. The framework can guide evidence-based decision-making and prioritization. This article encapsulates a framework that focuses on the interactions between processes and their system characteristics, which leads to an overemphasis on a limited number of features, significantly enhancing the flexibility of the organization

1 Introduction

Environmental emergencies were rarely prioritized in catastrophe planning, development sustainability, or business continuity. Anthropogenic, climate-change-related, or technological accidents involving ecological or biological systems have taken a significant amount of the risk-space we live in recent years. Health catastrophes of the magnitude that the COVID-19 pandemic has brought to the world's attention necessitate a rigorous study of the processes and dangers of such environmental and health outbreaks. This is crucial to comprehending the emerging issues surrounding low-frequency high-intensity disasters such as pandemics, tsunamis, floods, liquefaction, water contamination emergencies, landslide, or volcanic eruptions causing death and food shortages in specific areas.

The topic was written in response to the necessity of learning from the COVID-19 health catastrophe, which turned into a massive global calamity. The lessons learned from this pandemic calamity are helpful not only for health and social systems but also for the economy and business sectors, environmental protection, and the pursuit of sustainable development goals. The study summarises and recommends lessons learned from cross-domain assessments of worldwide virus outbreaks, and it will be beneficial in creating disease-resistant populations. Scholars in humanities, management, and public policy intend to use this study as a knowledge support system and reading reference, especially for Indonesian scholars.

2 Pandemic: An Overview

Pandemics have shaped human history [1, 2, 3], but the name "pandemic" is new. Modern cultures have the dubious fortune of experiencing near-simultaneous epidemics over the world as a result of several centuries of relentlessly growing globalization. An increase in the number of developing diseases over the last two decades has fueled fears of a new devastating pandemic on the horizon. Nonetheless, much of the globe was caught off guard when the World Health Organisation (WHO) proclaimed the developing coronavirus SARS-CoV-2 a pandemic on March 11, 2020.

Since the first coronavirus disease of 2019 (COVID-19) was discovered in the Chinese city of Wuhan in December 2019, the globe has been facing a crisis on numerous fronts, including public health, social welfare, and economic development. This global problem does not appear to be slowing down, and the disease's effects are posing new challenges to governments worldwide. According to a general understanding of similar diseases in the past, the evolution, spread, and fatality of such

* Corresponding Author: christinemarnani@yahoo.com

epidemics are dependent on both natural and human variables. Many investigations have been undertaken to determine the disease's complex link with natural and anthropogenic variables.

For example, Zhou revealed in a significant study that individuals with pre-existing medical conditions are more susceptible to COVID-19 [4]. People with severe disease conditions are more vulnerable to high health repercussions [5]. These elements, particularly in developing nations where health infrastructure is inadequate and socioeconomic conditions are insufficient to deal with the pandemic's effects, play a critical influence in determining fatality rates. According to a study of African countries' healthcare systems, the COVID-19 pandemic poses a significant threat to the continent on numerous levels [6].

Not only is the continent ill-prepared to combat such a sickness, but it has also been shown to be vulnerable to the effects of such a large-scale outbreak. Similarly, several scholars have looked into the pandemic's economic repercussions. For example, Zhang discovered that the COVID-19 has proven to be one of the most important contributors to increasing global financial market volatility [7]. The study of several micro and macro variables in various disciplines found that significant efforts and resources will be required to limit the amount of harm, particularly in developing countries with higher population density [8].

Qu has suggested that environmental variables play a role in the pandemic spread. Although several studies have been conducted to explain the complex interaction between environmental factors and COVID-19, there is still a lack of consensus among researchers and scientists on the subject [9]. For example, Wu found that a rise in temperature and humidity could help to restrict the COVID-19 pandemic to some extent, based on data from more than 150 nations [10].

However, Zhu and Xie found that an increase of one degree Celsius in the mean temperature was positively connected with roughly 5% of the infected cases after studying the growth of instances of more than 100 in China. They concluded that there was no compelling evidence of COVID-19 confirmed cases associated with possible warming scenarios [11]. However, earlier research has shown that the wide range of diseases caused by the 'severe acute respiratory syndrome (SARS) virus family are linked to environmental factors such as pollution, temperature, relative humidity, and various other climatic parameters [12, 13, 14]. Given the above debates, these studies call decision-makers' attention to the need to increase the population's capacity, particularly in rural areas. The density of educated people is also a factor in pandemic vulnerability. This will help policymakers create more effective risk measuring, readiness, response, and mitigation programs.

3 Pandemic's Effects

Even though pandemics are rare occurrences, they have a significant impact on the world's shaping. The consequences necessitate well-thought-out contingency plans to prevent disease spread and death. While containment methods are implemented in stages, and medical therapies take time, the pandemic has significant economic, financial, and social consequences.

In many ways, the COVID-19 pandemic is similar to earlier pandemics, according to an analysis of their economic implications. Except for the Spanish flu, which spread through military networks, trade and travel by land, air, or sea propagated other pandemics. As in the case of swine flu, the greater the extent of global trade, the faster the virus spreads. Because of the rising demand for healthcare items, there has been a surge in healthcare service businesses. Higher pricing and strains on healthcare spending have resulted because of this. Due to the lack of products, prices were higher during the Ebola outbreak. Some pandemics, such as the Black Death and the Spanish Flu (both had a high death rate), resulting in a labor shortage and increased salaries.

Work has been disrupted, schools and companies have closed, and as a result, productivity has decreased. Hotels and the entertainment business have suffered as a result of the downturn in the service sector. In more recent pandemics, there had been a drop in tourism and travel earnings (2000 onwards). The effect on productivity, on the other hand, is debatable. It was discovered that the Spanish flu resulted in higher capital per worker and thus higher productivity. However, another study found low long-term productivity due to lost skills (due to unemployment) and a negative impact on the productivity of children born to moms who were pregnant during the epidemic. In the aftermath of Ebola, there was a drop in productivity and participation in the labor force.

GDP, investment, and consumer confidence have all been negatively impacted. For countries that rely heavily on tourism, the impact on GDP was larger. Pandemics that occurred before the twenty-first century had little effect on trade, especially in nations like Guinea, Liberia, and Sierra Leone, which had a small share of the global economy. The news of the pandemic has a significant impact on market sentiment and fear.

Though it may appear that societal implications are boosting short-term stress levels and posing issues in terms of women's abuse, child labor, and domestic violence, these can have far-reaching consequences that have the potential to worsen conditions in the future. Because there are significant interdependencies between social features and economic, political, and financial aspects, it is critical to limit these negative impacts on the economy's bright future.

In terms of international coverage and transmission speed, COVID-19 stands out. "Preliminary medical investigations suggest that COVID-19 is less lethal than SARS but more transmissible [15]. Because the world is now much more integrated in terms of trade and travel, it has absorbed far more countries than earlier pandemics. Because previous pandemics spread primarily through trade and travel routes, it is obvious that growing economic integration has made countries more vulnerable to the severe economic effects of such pandemics.

The list of the ten countries severely afflicted by COVID-19 is nearly equal to the list of the world's ten largest economies (Iran and India are the exceptions) [16].

The United States, China, Japan, Germany, the United Kingdom, France, and Italy are among the top 10 countries impacted by the disease. Just the United States, China, Japan, Germany, the United Kingdom, France, and Italy account for 60% of global supply and demand (GDP), 65 percent of global manufacturing, and 41% global manufacturing exports". These factors indicate that a modern-day pandemic such as COVID-19 will have far more significant economic implications for the entire planet than any other pandemic previously witnessed.

4 Institutional Resilience

The pandemic of COVID-19 is among the most expensive and devastating threats to millions of people worldwide. On June 4, 2020, the highest death rate of 315.54 per million was being faced by the United States of America (USA), whereas a 190.93 death rate per million people for Canada and a lower death rate of 0.027 per million for Indonesia. The first confirmed case of COVID-19 virus infection was in Jakarta. She had been in contact with a Japanese citizen who tested positive in Malaysia on February 27, 2020. Then onwards, Indonesia is facing an exponential increase in the number of COVID-19 patients with symptomatic and asymptomatic symptoms.

Infectious illness data-driven frameworks are increasingly being utilized to provide real-time or near-real-time situational awareness during disease outbreaks. Despite the inherent constraints of predicting the size and magnitude of complex processes, mathematical and computational models have been employed to anticipate epidemic size and scale. This aids in calculating the danger of case importation around the world and the risk of disease outbreaks that go unchecked [17]. Despite differing viewpoints on using modeling tools in epidemiological studies, many studies have been conducted in recent years to evaluate infection containment and mitigation tactics and develop pandemic preparation warning systems. In most cases, model-based epidemic scenarios concentrate on the magnitude and length of the epidemic or pandemic [18].

In addition, the modeling community is presently evaluating resilience, mitigation, and containment policies by lowering the attack rate per hundreds, thousands, or tens of thousands of persons in the population. The evaluation of epidemic risks and consequences is a multifaceted challenge that must incorporate societal issues such as infrastructure and service interruption, lost output, inflated prices, crisis-induced budget deficits, and poverty [19]. As a result, it's critical to expand the framework-based approach to epidemic analysis. This framework will include measurements that can be used to assess systemic and institutional resilience, such as the overall system's response to disruptions, the implications of those disruptions, the outcome of mitigation, and the recovery and retention of system functionalities [20].

COVID-19 outbreaks have a multi-cascade effect, affecting many industries and perhaps resulting in worldwide institutional crises and poor economic development. Understanding these areas is vital for human survival, and their resilience will serve as important indicators. Institutional performance is partly dependent on robust growth, but mostly on resilience: the ability to bounce back, develop workflows that absorb losses and deficiencies, be aware of and participate in changing conditions and feedback, and deal methodically with the pattern of problems that will inevitably arise along the way [21].

Although there has been a lot of research on evaluating the resistance of intricate processes to unfavorable events, its integration into the analysis technique for epidemic hazards is still mainly unexplored. Most crucially, institutional resilience must include the temporal horizon or the recovery and stability of functionality in the case of an epidemic [22, 23, 24]. As a result, identifying and managing institutional resilience to epidemics requires assessing the temporal profile of how essential functions of the system are kept or managed to recover in response to pandemic events.

Mitigation, preparedness, response, and recovery are well-known disaster phases; however, they don't work well as frameworks for Institutional Resilience. Social media impacts the intricacies of national and international institutions such as banking, trade, and communication rather than communities [25]. The different activities of epidemic response, the links between them, and potential avenues for an impact have been developed for an epidemic response to this review to determine the different activities of epidemic response, the links between them, and potential avenues for an impact.

COVID-19 pandemics, which differ from epidemics in that they have more defined policy aims, pose similar issues in terms of establishing understanding frameworks. Many frameworks, for example, were created to develop appropriate interventions, set priorities during the implementation phase, or assess specific aspects of performance in many circumstances. A comprehensive action encompassing all approaches to examine all stages of a life cycle intervention was not taken. As a result, an improved framework consisting of preparedness, coping mechanisms, adaptation, awareness, and transformation has been identified as a means of improving institutional resilience.

4.1 Preparedness

Five EU member states have been affected by Middle East Respiratory Syndrome (MERS) and poliomyelitis, including the UK, Greece, Spain, Poland, and Cyprus. Policies and tactics exist in all nations, and valuable lessons have been learned and institutionalized as a result of previous public health catastrophes such as Ebola (2014–2016), the H1N1 pandemic (2009), and SARS (2003). [26, 27, 28]. There were also significant lasting benefits from large meetings like the Olympics, held in Greece in 2004, and the United Kingdom in 2012. However, new technology, knowledge, and experience gained from these events are subject to political and policy priorities shifting away from securing effective core public health capacities in some participating countries [29]. There was no preparedness for COVID-19, but the lessons acquired from earlier outbreaks have presented institutions with strategies to improve their preparedness.

There are crucial measures to ensuring that European public health preparedness programs are ready to conduct evaluations, document outcomes, and communicate them broadly, such as frequent training of all necessary staff personnel [30]. A similar methodology can be used in poorer countries to improve institutional preparation. More catastrophe preparedness planning is needed, and institutional processes must be integrated. Multi-scalar integration and governance structures that enable cohesive collaboration and cooperation within the system are part of this readiness for national and local institutional systems with shorter supply chains.

4.2 Coping Mechanism

Resilience is more challenging and necessitates acts that may appear contradictory to logic at times. A literature review indicates three critical measures they may take to strengthen the resilience of their personnel, teams, and institutions [31]. The ability to manage with the consequences of economic losses, no employment, job recession, and insecurity in the demand and supply chain emphasizes the importance of coping needs in improving the institutional resilience of businesses and governmental structures [32]. It needs to Adopt a systemic approach to health policy to create flexible, resilient, sustainable, and equitable resilient systems.

4.3 Adaptation

Over the last 30 years, the incidence of epidemics has led to increasing adaptation measures for dealing with them. Outbreaks of such diseases and pandemics impair transportation, food, health, tourism, education, and services, resulting in lost output, inflated prices, a crisis-induced budget deficit, and poverty [33]. The supply chain suffers the greatest, affecting all areas of society. A systemic shift is also being witnessed around the globe, both in developed and developing countries, from spending on luxury items to spending on necessities such as life-saving pharmaceuticals, which is improving resilience structures.

4.4 Awareness and Transformation

Increased awareness is a necessary component of improving institutional resilience. During the Ebola outbreak, promoting awareness served as a strategic means of overcoming the crisis. It aided in improving Liberia's health system's resilience to the Ebola epidemic [34]. Furthermore, improving social media networks to ensure their viability would assist in raising awareness. The importance of knowledge sharing will improve epidemic response. The sharing of value is needed as a means of disseminating best practices and lessons learned as well as fostering solidarity from the communities to elevate assistance. Science must be measured in terms of various types of learning. Small farmers, indigenous peoples, women, and young people must actively raise their voices.

The national-local link is critical for guaranteeing policy and implementation continuity. The incorporation of readiness, scoping, adaptation, and awareness into the system results in modifying several concepts, designs, and analyses [35]. Transformation in all kinds of structural reformation and prioritizing strengthen the establishment of a resilient foundation. These are country-specific because each has a unique collection of resources and capacity to deal with epidemics.

5 Institutional Planning

With rising concerns about environmental challenges/climate change and developing disease scenarios, it is critical to consider the creation of sustainable production systems. The need for planning in an organization for any business or activity is expanding to society's constant logical progress today [36]. When it comes to corporate activity planning, the function of management tools is crucial. Effective management tool use is a possible actor for long-term planning and production practices [37]. This section discusses some significant instruments that are useful when it comes to institutional planning.

5.1 Evaluation of The Life Cycle

It is a system for examining the environmental consequences of all processes connected with all stages of product, service, and process development. This system is a critical tool for industries, businesses, and policymakers to properly strategize their decisions and operations. It employs a cradle-to-grave method that provides a full review of the entire organizational structure, including up and down chain parts and third-party elements involved in the system. Doing so decreases the impact of external demand on organizational performance and action approaches, boosting organizational resilience.

5.2 Cost-Benefit Analysis

It is a system for assessing the environmental impacts of all processes associated with all stages of product, service, and process development. The Cost-Benefit Analysis is an essential tool for industries, businesses, and regulators to incorrectly strategize their decisions and operations. It uses a cradle-to-grave approach to give a comprehensive examination of the complete organizational structure, encompassing up and down chain portions and third-party elements involved in the system. It reduces the impact of external demand on organizational performance and action approaches, hence increasing organizational resilience.

5.3 On-Site and Off-Site Planning

Emergencies necessitate urgent action, and adequate preparation is required to carry out these measures.

Emergency preparedness is needed for any organization to respond to emergencies both on and off-site [37]. On-site planning is specific to dealing with the consequences of any emergency event that disrupts internal operations, structures, or workforce. The emergency response needs of the areas surrounding the organization are addressed via site planning. Before generating operational plans, on-site and off-site studies are valuable for prioritizing emergency response planning.

5.4 Exercises in Impact Dependence

Businesses, organizations, and institutions rely on biodiversity (through ecosystems and associated ecosystem services) to run their operations and activities smoothly. Any disturbance in the critical ecosystems and related services on which organizations rely will significantly impact their (structural and functional) operations. Conducting impact dependency exercises assists businesses in future-proofing their operations by [38]: ensuring the future supply of vital resources, improving resource efficiency, securing future operating licenses, should plan better business decisions and activities. It also aids in spotting company risks and identifying new opportunities.

5.5 Assessing Vulnerability

Environmental issues, such as natural catastrophes or epidemics, impede organizational processes by disrupting services and causing infrastructure damage. Vulnerability assessment is a tool that aids in finding hotspots or zones/areas that are most vulnerable to risks. Conducting a vulnerability assessment helps identify the required adaptive responses—whether the organization's structure and processes have the necessary coping and adaptive capability to take on or move forward with specific activities [39].

5.6 Continuity Preparation

Business continuity planning is a proactive strategy to understand an organization's or institution's weaknesses and risks when a crisis arises. The fundamental strategy is to maintain critical functions operating and eventually recover with minimal downtime, preventing systems from failing [39]. Effective planning enables organizations to respond quickly and efficiently in the face of unexpected events, building risk management systems to avoid disruptions in operations. A business continuity plan is built on three major components: Resilience: Considering many adversities during operation planning and design improves organizational resilience for multiple risks. Recovery: During a crisis, focusing only on operations that include critical/essential services should be preferred, followed by establishing a recovery time aim for the rest of the functions. Contingency Plan: Ensures that proper procedures are in place for various internal work plans and value chain points [40].

6 Conclusion

The current coronavirus or COVID-19 pandemic has taught many important lessons, and many people are now discussing how to deal with crisis inside crisis. Incidences of crisis during a crisis are not new; instead, they have a long history. However, it is critical to consider what lessons we have gained from previous situations and if they have been incorporated into policies and management practices. As a result, the emerging field of resilience assessment and administration, as well as its implementation, could assess cross-domain alternatives to identify a policy design that improves the system's ability to (1) plan adverse events, (2) absorb stress, (3) recover, and (4) prevent and prepare for future stressors by appropriately adapting them. System planning should be inclusive, considering the entire value chain and evaluating life cycle components into planning procedures. To summarise, the paradigm we described may help balance reduced risk with disruption to important activities linked with public health interventions to optimize policy outcomes.

References

1. WH. McNeill, Plagues and peoples (Garden City, New York, 1976)
2. JM. Diamond, Guns, Germs, and Steel: The Fates of Human Societies (W.W. Norton & Cole, New York, 1999)
3. A. Crosby, The Columbian exchange: biological and cultural consequences of 1492 (Greenwood, Westport, 1972).
4. F. Zhou, T. Yu, R. Du, G. Fan, Y. Liu, Z Liu, J. Xiang, Y. Wang, B. Song, X. Gu. L **395**(10229):1054–1062 (2020)
5. L. Fang, G. Karakiulakis, M. Roth. LRM **8** (4):e21(2020)
6. JN. Nkengasong, W Mankoula (2020). L **395**,:841–842 (2020)
7. D. Zhang, M. Hu, Q. Ji, FRL:101528 (2020)
8. WJ. McKibbin, R. Fernando, The global macroeconomic impacts of COVID-19: seven scenarios (March 2, 2020). CAMA Working Paper No. 19/2020. (2020)
9. G. Qu, X. Li, L. Hu, G. Jiang. EST **54** (7), :3730–3732 (2020).
10. Y. Wu, W. Jing, J. Liu, Q. Ma, J. Yuan, Y. Wang, M. Du, M. Liu, STE **729**,:139051 (2020).
11. Y. Zhu, J. Xie. STE **724**,:138201 (2020).
12. J. Bao, Z. Wang, C. Yu, X. Li. BMC Public Health **16**:375 (2016).
13. Y. Cui, Z-F. Zhang, J. Froines, J. Zhao, H. Wang, S-Z. Yu, R. Detels. EH **2**:15 (2003).
14. K. Lin, DY-T. Fong, B. Zhu, Karlberg. **134**,:223–230 (2006).

15. R. Baldwin, di Mauro BW, Economics in the time of COVID-19. (CEPR Press, Washington, DC, 2020).
16. Q. Zhang, H. Deng, C. Zhang. The value of early response by surrounding areas of epidemic center Hubei during COVID-2019 outbreak in China: a quasi-experiment analysis. Available at SSRN 3548372 (2020).
17. E. Massaro, A. Ganin, N. Perra, I. Linkov, A. Vespignani, SR **8**(1):1–9 (2018).
18. AA. Ganin, E. Massaro, A. Gutfraind, N. Steen, JM. Keisler, Kott A et al. SR **6**(1):1–12 (2016)
19. G. O'Brien, UK emergency preparedness: a holistic response? DPM **17**(2):232–243 (2008).
20. E. Massaro, A. Ganin, N. Perra, I. Linkov, A. Vespignani, SR **8**(1):1–9 (2018).
21. J. Wang, Z. Wang, Strengths, weaknesses, opportunities and threats (SWOT) analysis of China's prevention and control strategy for the COVID-19 epidemic. IJERPH **17**(7):2235 (2020).
22. J. Romaní, E. Baselga, O. Mitjà, N. Riera-Martí, P. Garbayo, Vicente A et al, AD **111**(5):426–429 (2020).
23. CA. Bevc, MC. Simon, TA. Montoya, JA. Horney, PHR **129**:35–41 (2014)
24. DJ. Dausey, M. Moore, BMC Res Notes **7**(1):474 (2014)
25. J. Kinsman, J. Angrén, F. Elgh, M. Furberg, PA. Mosquera, L. Otero-García, R. Snacken, T. Derrough, PC. Santisteve, M. Ciotti, S. Tsolova, BMC Health Serv Res **18**(1):528 (2018)
26. B. Aguirre, RR. Dynes, J. Kendra, R. Connell, JHSEM **2**(2) (2005)
27. MA. Barton, M. Christianson, CG. Myers, K. Sutcliffe. Resilience in action: leading for resilience in response to COVID-19. BMJ Leader, pp leader-(2020)
28. A. Warsame, K. Blanchet, F. Checchi, BMJ Glob Health **5**(1):e002109 (2020)
29. EJ. Ling, E. Larson, RJ. Macauley, Y. Kodl, B. VanDeBogert, S. Baawo, ME. Kruk, HPP **32**, iii40–iii47 (2017)
30. AA. Ganin, E. Massaro, A. Gutfraind, N. Steen, JM. Keisler, Kott A et al. SR **6**(1):1–12 (2016)
31. UNEP, Benefits of life cycle approaches. Life Cycle Initiative (UNEP, New York, 2020)
32. LfN I (2012) Why follow the natural capital roadmap? Leaders for Nature. http://leadersfornature.in/get-started. [Accessed June 3, 2020]
33. Bindal, Sonal, Pritha Acharya, Anil Kumar Gupta, and Jugal Kishore. "Enhancing Epidemic Resilience: Planning and Institutional Resilience." In Integrated Risk of Pandemic: COVID-19 Impacts, Resilience, and Recommendations, 463-480. (Springer, Singapore, 2020)

Socio-economic resilience of Pidie Jaya to against disaster in the terms of national security

Riskina Tri Januarti[1*] *Heridadi[1]*, *Achmed* Sukendro[1], *Rio* Khoirudin Apriyadi[1], and *Dandung* Ruskar[1]

[1] Disaster Management Study Program, National Security Faculty, Republic of Indonesia Defense University. IPSC Sentul Area, Bogor, West Java, Indonesia

Abstract. Pidie Jaya (Piday) District of Aceh Province is a newly formed district in 2007. In addition to the COVID-19 pandemic, the Pidie Jaya district also experienced a series of disasters throughout 2020. Nevertheless, the Piday District Human Development Index (HDI) in 2017 exceeded the National achievement. But piday district poverty ranks 3rd in the Province. This has the potential to make the Piday district prone to disasters in addition to geographical location factors. Therefore, the existence of such gaps in this study will be discussed. This study uses qualitative methods through descriptive narrative approaches. Sources of information and data were obtained through Focus Group Decision (FGD) from several sources and studies of some literature. The results stated that the gap is due to economic development being less evenly distributed and less felt by lower-level people. This is due to low education factors and lack of capital for farmers and fishermen as the majority of workers in Piday District. The research suggests increasing community capacity and resilience by 1) Improving the quality of human resources for farmers and fishermen in the form of education and training based on improving the economy and living standards of farming communities wrapped in local wisdom; 2) Infrastructure development both facilities and infrastructure; 3) Provision of capital and ease of access in business and insurance coverage against disasters by taking into account local wisdom; 4) Empowerment of the role and involvement of local community leaders in the formulation of policies and local government institutions.

1 Introduction

As a country with the most islands in the world, Indonesia has abundant natural beauty and wealth. However, behind the beauty and natural wealth it has, Indonesia turns out to have a potential threat to natural disasters that are quite risky. When inaugurating the Opening of the 2021 National Coordination Meeting for Disaster Management on March 3, 2021, the President of the Republic of Indonesia Joko Widodo said that Indonesia was ranked 35th in the world as the country most prone to disaster risk. It was also emphasized that mitigation efforts are the main focus of disaster management [1].

According to data from BNPB, throughout 2021 Indonesia has experienced 1.472 disaster events from January to June 24 [2]. The most dominant disasters are disasters caused by hydrometeorological factors, especially floods, as shown in the infographic on the distribution of natural disasters in Indonesia in 2021 in Fig. 1.

Various threats of disasters, both natural and man-made, exist in the life cycle of the entire Indonesian nation, which can directly hinder the smooth progress of development and the results achieved [3].

Fig. 1. Infographics on the distribution of natural Disasters, January 1 – June 24, 2021 [2].

President of the Republic of Indonesia in Law 24/2007 states that anyone has the right to a sense of security, especially those affected by disasters [4]. This shows that human security against the threat of disaster is a major concern. Given the nature of Indonesia's defense is that national defense is universal, and its implementation is based on an understanding of the rights and obligations as citizens and trust in their fighting power [5]. National

* Corresponding Author: riskina.tri@gmail.com

security is no longer the content of war, nor is it a threat or defense of combatants, and non-combatants against all sectors of the state or foreign military attacks, but related to the security of all sectors that threaten human survival [6].

Geographically, Aceh Province is on the island of Sumatra which is also a meeting area between the Eurasian and Indo-Australian plates, and the existence of a living fault "The Great Sumatra Fault" or the Sumatran Fault cuts Sumatra from Semangko Bay to Aceh Province and extends to the Andaman Sea [7]. Some faults divide Aceh Province, both active and inactive, making Aceh Province a disaster-prone area [8]. At the end of 2016, the province of Aceh was again shaken by the occurrence of two earthquakes in the same month. The disaster resulted in 104 deaths, distributed in 3 (three) districts, namely Piday, Bireue, and Pidie districts. The most victims were in Piday Regency, which was 97 people and also caused tremendous damage to community housing, infrastructure, and socio-economic community [9].

The entire area of Piday Regency which includes 8 sub-districts, 222 villages, and 34 settlements is a disaster-prone area, especially earthquakes. This is because Sumatra is located at the intersection of three major faults in the world, namely Australia, Eurasia, and the Pacific Plate [10]. The same thing was conveyed by the Aceh Social Service in the 2020 Psychosocial Support Service Training of Trainer event in Banda Aceh on Wednesday 11 March 2020 which stated that Aceh Province is a disaster-prone province, due to the geological and geographical environment of Aceh Province which is in the middle of a disaster. Ring of fire, making several volcanoes and subduction zones the epicenters of earthquakes and tsunamis. Aceh is also prone to flash floods, landslides, floods, and droughts. Others pointed out that the disasters that caused the most serious casualties were floods, landslides, and flash floods [7].

This incident finally placed Piday District as the third rank in Aceh Province in terms of poverty. However, from all regencies in Aceh Province, it turns out that Piday Regency has an HDI that continues to increase even above the Aceh Province and National HDI. It was recorded that in 2017, the HDI of Piday Regency was at an index of 71.73 while the Provincial HDI was 70.60 and the National HDI was 70.81 [9].

Based on the above background, the researcher finally conducted a study on the causes of the gap between poverty and HDI in the Piday Regency community in this COVID-19 pandemic situation. According to disaster risk analysis, the vulnerability index is exponentially in line with risk, which means that the greater the vulnerability index, the potential risk also increases and vice versa. And the HDI value, which is categorized by researchers as capacity, has an inverse relationship to threats. Where the higher the HDI, the smaller the vulnerability of an area, in this case, is poverty in Piday Regency.

2 Method and Data

This research is qualitative research using a descriptive narrative sentence approach based on the results of seminars and questions and answers with resource persons in the form of Focus Group Decisions (FGD). The resource persons in this study came from the Aceh Besar Geophysics Station, Piday Kodim, Regional Disaster Management Agency (BPBD), Regional Central Statistics Agency (BPS), Regional Development Agency, Community Social Institutions, and from academics of disaster experts at Syiah Kuala University, Piday Regency. Meanwhile, the data used in this study are intended as supporting material in providing an overview and evaluation of the social and economic vulnerability index and its relationship to the HDI. Social and economic vulnerability data based on information from informants will focus on poverty levels, types of work, and local wisdom because the index has a large influence. Literature review of several kinds of literature is intended to provide a more realistic picture and explanation to strengthen and deepen information from the informants.

3 Result and Discussion

3.1 Socio-Economic Life of the People of Piday Regency

Piday Regency which is located in the Aceh Province is one of the regencies that can be said to be still new. It is a division of Pidie Regency and was formed through Law 7/2007 on January 2, 2007. The area of the Regency is. Piday is 1,162.84 km2 (952 km2 land and 210.84 km2 ocean). The Pidie Regency government oversees 8 sub-districts, 34 settlements, 222 *gampongs*, and 600 hamlets [11].

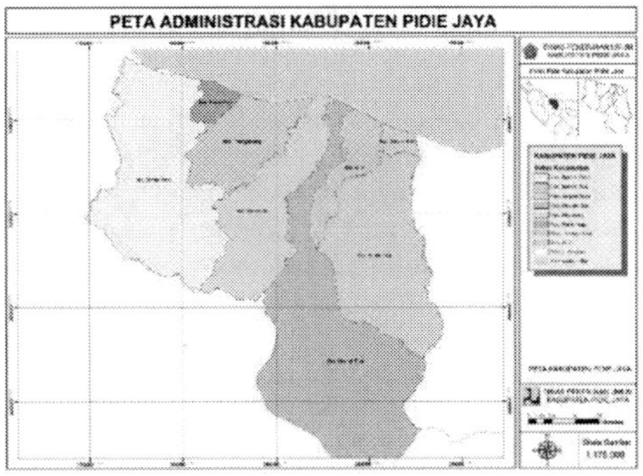

Fig. 2. Administration Map of Pidie Jaya Regency [12]

Through FGD seminars with sources from the Social Service of Piday Regency, it can be seen that the profession of work, poverty, and local wisdom have a great influence on the socio-economic vulnerability of the community. In addition, other information provided is that the majority of the Piday community depends on agriculture, fisheries, and forestry for their livelihoods. Along with this information, the Piday Regency Disdukcepatl data also explains that indeed the majority of the Piday Regency people's livelihood is working as farmers and fishermen in fulfilling their daily needs as

https://doi.org/10.1051/e3sconf/202133102006

shown in Table 1. This is due to the geographical conditions of the area that support the profession [9].

Table 1. The Majority of Professions/Livelihoods of the Population of Pidie Jaya Regency 2018 [9].

Profession/Livelihood	Total	%
Farmers/Growers	16.913	10,56
Fisherman	7.907	4,94
Marketer	2.686	1,68
Entrepreneur	15.510	9,69
Civil Servants	3.461	2,16
TNI/POLRI	509	0,32
Employee/Honorary	1.864	1,16
Others	5.767	3,60
Housewife, Students Toddler (Unemployed)	105.498	65,89
Total Population	**160.115**	**100,00**

The following information obtained from the FGDs is that the poverty rate in Piday District is 19%, ranks third in Aceh Province, and is still considered high. The Central Statistics Agency (BPS) noted that since 2002, Aceh Province has been the poorest area in Sumatra [13]. Sharp poverty began to be felt in 2000 - 2004 due to the occurrence of armed conflict and the 2004 Tsunami [14].

Eka Agustina (2018) concludes that poverty in Aceh province is influenced by the level of education and unemployment rate while the population does not affect poverty so that the government must improve the education and skills of the workforce through training to improve the quality of human capacity so that they are competitive to get jobs. , so as not to fall into poverty [15]. The difficulty of opening a business field and being constrained by capital and marketing capabilities are the main factors causing regional unemployment [9].

The 2004 Aceh tsunami disaster greatly affected the socio-economic life of the Acehnese. The Acehnese were forced to rebuild their lives after the tsunami hit. Specifically, in Pidie Jaya on December 7, 2016, an earthquake occurred centered in Pidie Jaya. The disaster left hundreds of people dead and thousands of people affected. Economic losses are estimated at billions of rupiah [16]. The geographical area of Pidie Jaya Regency is prone to disasters, resulting in a lack of interest from investors to invest in Pidie Jaya Regency. So that in terms of trade economy, the Pidie Jaya Regency area is still classified as underdeveloped [17].

In general, 70% of Piday people in rural areas work in agriculture. In the aftermath of the 2004 tsunami, restoration and reconstruction work had been carried out for nearly a decade with a budget of nearly one trillion, including the construction of housing and public buildings for the people affected by the tsunami. The improvement of infrastructure and the community's economy (community livelihoods) with the restoration of the agricultural and fisheries sectors has led to various improvements in Aceh's fisheries and agricultural potential in many ways. Although the impact on the wider community is not significant and does not provide a stronger role for economic growth, income, and welfare, as well as the maintenance of people's lives [18].

In the context of regional economic development, Aceh's coastal and maritime capacity management requires strategies and controls on ecosystem sustainability. This is reflected in local wisdom in marine protection. The wisdom of Acehnese fishermen in protecting the sea is included in the customary rules of the Aceh fishing community. The Commander of the navy is responsible for leading the customary associations that govern maritime customary law. Maritime customary law is formulated under Islamic law which guarantees regulations for the exploitation of maritime wealth *(meupayang)*, equitable distribution of profits, determines the schedule for catching marine fish, enforces customary rules and rituals, raises fishermen, and resolves problems among fishermen. Rules in resolving maritime disputes and violations, warnings and sanctions for seafarers who violate maritime rules, and relations between fishermen and local governments. [19].

In addition, Piday Regency also has local wisdom called "Meureukon" which is a method of religious education or da'wah about religious knowledge based on the Koran. Meureukon is one of the traditions and local wisdom of the Piday community. There are no restrictions on the number of members, age, and gender in Meureukon. The special feature of Meureukon is the use of the Koran in the explanation of religious knowledge with the material as a whole adapted and associated with Islamic religious law. Meureukon activities are usually held in the framework *(pendopo)*, so it is also said to be a debate event in the Tengku *Rangang* style. However, it usually starts after the Maghrib prayer and stops at the meunasah *(surau)* before the dawn prayer. Meureukon activities are not just a means of broadcasting or a means of Islamic da'wah, but there are values of religious education, unity, society, and aesthetics in Meureukon which have been applied in people's lives. Meureukon also has a function, namely as a means of expression, ritual, communication, and entertainment [20].

The people of Pidie Jaya Regency also have local wisdom in the form of the Peusijeuk Tradition, which is a tradition in the form of a tradition to ask for safety, peace, and happiness in life [21]. This tradition is carried out during wedding ceremonies, performing Hajj, about to migrate, commemorating the birth of a child, or to commemorate something happiness that is obtained, such as when a family member earns a bachelor's degree. This tradition can strengthen the social ties that exist in the Pidie Jaya community which is expected to be able to increase the socio-economic level of the community [22].

3.2 Piday District Human Development Index

Indonesia's HDI in the early days of the COVID-19 pandemic compared to previous years grew by 0.03 percent (an increase of 0.02 points). This slowdown in growth is strongly influenced by the decline in per capita expenditure due to the COVID-19 pandemic. Based on the performance achievements of the Piday district government since 2011 that the HDI has reached 68.69 and in 2015 it increased to 70.49 in 2016 and 2017 by 71.73 with the calculation of the new method. The HDI of the Piday Regency is increasing every year. Similarly, when compared to the Provincial HDI in 2015 of 69.45, it increased to 70.00 in 2016 and 2017 to 70.60 (table 2). In general, it can be seen that the HDI of the Piday Regency is always above the Aceh and National HDI [9].

Table 2. Pidie Jaya District HDI 2011-2017 [9]

No	Descrip-tion	Year					
		2012	2013	2014	2015	2016	2017
1	Pidie Jaya District	68,9	69,3	69,9	70,5	71,1	71,7
2	Aceh Province	67,8	68,3	68,8	69,4	70,0	70,6
3	National	67,7	68,3	68,9	69,5	70,2	71,8

HDI includes three basic human components, namely life expectancy (Longevity), education (Knowledge), and decent living (Decent living). The achievement of the HDI that continues to increase is a positive sign of continuous improvement in the quality of the Piday community in terms of health, education, and economy. The high and low value of the human development index cannot be separated from the development plans implemented by the government. The achievement or increase in the number of HDI does not happen instantly. Human development is a method that cannot be measured right away. It is difficult to determine the health outcomes of education and development in the short term, which is generally the opposite of economic development. Communities with higher human development, especially those with good levels of education and health, are more resilient than those with less nutrition and less education. So that these vulnerable people will face difficulties in changing their fate when shocks or disasters occur. Human development will shape the strength of individuals and communities to reduce vulnerability and build resilience socially and economically [10].

3.3 The HDI Gap and Piday Community Poverty

With a high HDI value even higher nationally, it should be able to suppress and reduce poverty in Pidie Jaya and at the same time increase resilience to natural disasters that often occur. The entire area of Piday Regency is a disaster-prone area. Various disasters ranging from earthquakes, tsunamis, floods, strong winds, erosion, volcanic eruptions, landslides, forest fires, and abrasion will always be a threat. On February 16, 2017, at 02:53 WIB, there was an earthquake with a value of 5.4 on the Richter Scale at a depth of 12 km, located at latitude 5.21-96.13 East Longitude, 9 km northwest of Piday with 97 fatalities. Another disaster that recently occurred was the flood disaster in April 2020 amid the Covid-19 pandemic which resulted in material losses [9]. And based on information from sources, disasters that often occur are floods. One of the high negative impacts of disasters is believed to be related to the vulnerability of low-income communities [23].

However, the high HDI value of Pidie Jaya is not in line with the existing poverty rate. Through research conducted by Suliswanto (2010) concluded that the HDI variable has the greatest influence on poverty [24]. A high HDI value will reduce the poverty rate. The high HDI which is not in line with the poverty rate shows that economic growth still does not provide many benefits for lower-income people. However, a surge in the quality of human resource capacity will have a major impact on reducing poverty. For this matter, the solution is to form a just and equitable economy for the poor to ensure the fulfillment of basic needs and to encourage secondary and tertiary needs. Based on the experience of economic development from developed countries in Asia such as Taiwan, Singapore, South Korea, and Hong Kong, it has been proven that the direction of investment towards improving human resources (through the education process) is more "effective" than physical investment (buildings, other infrastructure) [24]. This happens because the poverty of an area can also be caused by the quality of the population. Even with the construction of road materials, factories, hospitals, and others, if the quality of the population is not high, then the material capital cannot be utilized optimally [25]. This shows that there is a relationship between economic growth and poverty with HDI. In addition to the quality of human resources, another factor that affects poverty according to Susilowati (2015) is the debt factor. It is concluded that there is a mutually influencing relationship between debt factors, economic growth, and poverty. So that in carrying out economic development, the Government should pay attention to these three variables [26].

Judging from the geography of Pidie Jaya Regency which has abundant natural resources, poverty should not be a problem in the people of Piday Regency where the majority of their livelihoods are farmers and fishermen. Based on the results of research conducted by Usman (2020) it is concluded that there are obstacles in the maritime sector to be able to develop. The fishing communities, toke, and crew fishermen in Pidie Jaya have lost their jobs sharply. This causes unemployment in a very short time because the income of fishermen only depends on the sea as the sole employment sector. Based on the results of interviews conducted, it turned out that the fishermen asked to be allowed to re-install FADs in the areas that had been cut. In addition, payments for FADs that have been cut are expected to be paid as soon as possible so that they can be used for preparation for the manufacture of new FADs. Fishermen also ask for compensation in the form of empowerment, necessities, and side dishes [27].

Other obstacles faced include natural factors due to climate change, high prices of fuel oil (BBM), and shallow fish-selling points (TPI), which make large vessels unable to dock so they cannot directly carry out fish trading transactions. Meanwhile, from the work of farmers, Adhiana (2017) concludes that the factors causing the vulnerability of farming communities in Aceh province are the level of education and access to financial credit. Other factors such as the number of productive age, health, culture, and the role of traditional institutions have a significant influence on the resilience of farmers. This is because these factors also affect poverty. The higher the index of these factors, the more it can suppress and reduce the poverty rate so that the resilience of farmers in dealing with disasters will increase [28].

4 Conclusion

The influential socio-economic vulnerabilities in the people of Piday Regency are the high poverty rate and type of work. Cultural factors, religion, and local wisdom reduce the level of socio-economic vulnerability. Poverty and type of work have a big influence on the impact of disasters both during pre-disaster to post-disaster considering that the majority of types of work are based on existing geographical conditions.

In general, the poverty of the Piday community is caused by the quality of the human resource capacity which is minimal in terms of knowledge. In addition, the existence of debt and economic growth that is not under the existing physical development is considered to be of little benefit to the community. So based on this, there is a relationship between poverty and HDI and there is still a gap between the high HDI in Piday Regency and the existing economic growth. Theoretically, with the increase in HDI, economic growth will also increase and it is expected to reduce poverty, but this does not apply in Pidie Jaya Regency. The poverty that occurs in fishermen is making the fishing profession a single profession, inadequate market access, and lack of business capital so that it increases the number of unemployed. An increase in the number of unemployed will eventually increase the poverty rate and increase the needs of fishermen for basic needs. From the agricultural sector, farmers experience poverty vulnerability at the level of education and access to business credit. It can be concluded that farmers and fishermen collectively have poverty vulnerabilities in terms of education and business capital.

5 Suggestion

As input material that can be given by the author to build the socio-economic life of the community in Piday Regency so that it is in line with economic development so that it will increase community capacity and be resilient in facing disasters, namely: 1) Infrastructure development, both facilities and infrastructure, and economic distribution must consider the prosperity the surrounding community, especially farmers and fishermen as well as local wisdom that exists in the community; 2) Improving the quality of human resources for farmers and fishermen in the form of education and training based on improving the economy and standard of living of the community wrapped in local wisdom; 3) Provision of capital and ease of access in business and insurance coverage against disasters by taking into account local wisdom; 4) Empowerment of the role and involvement of local community leaders in policy formulation and local government institutions.

References

1. Humas, (2021)
2. BNPB, (2021)
3. D. Putra, *Disaster Management and National Defense* (2012)
4. RI, *Law Number 24 of 2007 Disaster Management* (Government of the Republic of Indonesia, Jakarta, 2007)
5. RI, *Law Number 23 of 2019 Management of National Resources for National Defense* (Government of the Republic of Indonesia, Jakarta, 2019)
6. A. Mack, Die Friedens-Warte **80**, 177 (2005)
7. I. Rusydy, (2013)
8. M. Affan, M. Syukri, L. Wahyuna, and H. Sofyan, Aceh Int. J. Sci. Technol. **5**, 54 (2016)
9. Pidie-Jaya, *Qanun of Pidie Jaya Regency Number 3 of 2020 Concerning the Medium Term Development Plan (RPJM) of Pidie Jaya Regency for 2019-2024* (Pidie Jaya Regency Government, Aceh, 2020)
10. Pidie-Jaya, *Qanun of Pidie Jaya Regency Number 1 of 2018 Concerning the Long-Term Development Plan (RPJP) of Pidie Jaya Regency for 2015-2035* (Pidie Jaya Regency Government, Aceh, 2015)
11. N. Nurjanah, J. Pengabdi. Kpd. Masy. **19**, 68 (n.d.)
12. D. P. Pidie Jaya, *Pidie Jaya District Administration Map* (2011)
13. C. G. Asmara, (2021)
14. Safrina, (2021)
15. E. Agustina, M. N. Syechalad, and A. Hamzah, J. Perspekt. Ekon. Darussalam **4**, 265 (2018)
16. M. Fadri, M., & Zuhri, J. Ilm. Mhs. Bid. Huk. Kenegaraan **2**, 336 (2018)
17. L. J. Aji, D. P. Meiliasari, R. K. Apriyadi, S. Maarif, and H. Sumantri, PENDIPA J. Sci. Educ. **6**, 64 (2022)
18 A. Graf, S. Schroter, and E. P, *Aceh: History, Politics, and Culture* (Institute of Southeast Asian Studies, Singapore, 2010)
19. E. Apriana, Serambi Saintia J. Sains Dan Apl. **4**, (2016)
20. F. Fitria, R. Ramdiana, and I. Ismawan, J. Ilm. Mhs. Pendidik. Seni, Drama, Tari Musik **4**, (2019)
21 T. Rahayu, L. J. Aji, A. Suta, A. Hergani, A. Fitria, R. N. Utami, and S. V. Vidyaningsih, Int. Invent. Innov. Compet. **2**, 54 (2017)
22. N. Noviana, DESKOVI Art Des. J. **1**, 29 (2018)
23. W. Adiyoso, *Disaster Management: Introduction and Strategic Issues* (Bumi Aksara, Jakarta, 2018)
24. M. S. W. Suliswanto, J. Ekon. Pembang. **8**, 357 (2010)
25. M. L. Jhingan, (2016)
26. D. Susilowati and M. S. W. Suliswanto, J. Innov. Bus. Econ. **6**, 89 (2015)
27. U. Usman and M. Nur, J. Serambi Akad. **8**, 1671 (2020)
28. A. Adhiana, Agrifo J. Agribisnis Univ. Malikussaleh **2**, 34 (2017)

The impacts of the policies implementation to handle the COVID-19 pandemic in the field of employment

Ernalem Bangun[1], *Muhammad* Rifqi[2*], and *Bondan* Prokoso[1]

[1] Disaster Management Study Program, National Security Faculty, Republic of Indonesia Defense University. IPS Sentul Area, Bogor, West Java, Indonesia.
[2] Indraprasta PGRI University, Jakarta, Indonesia

Abstract. The implementation of policies to handle the COVID-19 pandemic that has been taking place in Indonesia since the beginning of 2020 has resulted in various impacts, especially in the field of employment. This paper aims to analyze the positive and negative impacts of implementing those policies. This study uses a qualitative method, where data collection was conducted through interviews. The implementation of pandemic-related policies in PT Freeport Indonesia was selected as a case study. The findings of this study indicate that related to employment issues, the positive and negative impacts of the implementation of the COVID-19 handling policy were felt to be most significant on the issues of Occupational Health and Safety (OHS) and income. As for social protection (insurance), employment contracts, and severance pay, the impact was relatively not felt by workers.

1 Introduction

Since its first appearance in the city of Wuhan, Hubei Province, China, at the end of 2019, to date Coronavirus Disease 2019 (COVID-19) caused by SARS-CoV-2 has infected 142,017,767 people worldwide, of which 120,554,767 (98%) of them have been declared cured, while 3,033,003 (2%) have died (Worldometers.info, 2021). In Indonesia, the first positive case of COVID-19 was detected on March 2, 2020, namely two cases in Depok City, West Java Province. To date, the number of positive confirmed cases throughout Indonesia has reached 1,604,348 cases, of which 1,455,065 (90.7%) have been declared cured and 43,424 (2.7%) have died. Thus, there were 105,859 (6.6%) active cases (COVID19.go.id, 2021). During a period of approximately eleven months, the absence of a vaccine made social distancing the main strategy, if not the only strategy, to deal with the threat of transmission and spread of the coronavirus. It was only at the end of 2020 that various researches to find vaccines in various countries, including Indonesia, came to fruition. Finally, since early 2021 vaccination has become the main strategy in almost all countries in the world, in addition to social restrictions.

The COVID-19 pandemic situation certainly has an impact on almost all areas of life, including health, social, culture, economy, education, and other fields, although the two most frequently highlighted are the health and socio-economic aspects. The employment sector was one of the social institutions that cannot be separated from the influence of COVID-19. The impact of COVID-19 on employment institutions was very significant. The results of the KSBSI research in May 2020 showed that the COVID-19 pandemic had an indirect impact on workers by the implementation of the rules set by the government on the business fields where the workers devoted themselves. The results show that there were workers who continue to work with full wages, continue to work but not full wages, were laid off with full wages, sent home with not full wages, layoffs with severance pay, layoffs without severance pay, work from home with full wages, work from home with not full wages. This condition resulted in the emergence of social, economic, cultural, and psychological problems for workers (KSBSI, 2020).

In reality, the pandemic situation was a complicated situation that must be addressed and overcome with complex steps as well. Policies and countermeasures that were taken and implemented, cannot be denied, also resulted in impacts that must be considered. The problem examined in this study was the impact of the implementation of the COVID-19 handling policy itself. The impacts in question were more about the impact on labor conditions, especially in the mining sector.

2 Method

This study uses a qualitative approach. PT Freeport Indonesia (PTFI), which operates in Mimika Regency, Papua Province, is used as a case study in this study. However, because in carrying out its operations PTFI is

* Corresponding Author: muhammad.rifqi.pensyarah@gmail.com

supported by subcontracting and privatization companies in its supply chain (SC) network, some of the informants in this study also came from these subcontracting companies. According to PTFI's 2018 list of privatizations and contractors, there are a total of 351 organizations in the SC network, with the total number of human resources involved reaching 14,777 people.

The research was carried out for approximately three months, namely between January and March 2021. The informants of this research consisted of workers/laborers, trade union officials/laborers, and company management, both of whom have authority as policymakers and are tasked with implementing policies on handling the COVID-19 pandemic in the work environment. During the research period, the Research Team conducted interviews with 49 informants, consisting of 11 informants from the management of PT Freeport Indonesia, 19 people from the management of trade unions/labor, and 19 informants from the workers/laborers.

Informants were selected based on several considerations. For the category of company management, the main consideration is the involvement of informants in the implementation of policies for handling the COVID-19 pandemic in the work environment, both as policymakers and policy implementers. Informants in this category have various positions, ranging from the superintendent, manager, to vice president. For the category of workers/labor union administrators, the main consideration in selecting informants is their involvement in the implementation of policies for handling the COVID-19 pandemic in their capacity as workers/labor union administrators. For the category of workers/labor, informants were selected to represent different work units. This is based on the consideration that different work units have different operational characteristics, so they must implement measures to handle the COVID-19 pandemic under their activities. In addition, the informants did not only come from PTFI, but also several other companies within PTFI's operational network, such as PTPJP, PTKPI, and PTPU.

The research was conducted in the context of PTFI's production activities during the COVID-19 pandemic. However, the COVID-19 pandemic situation did not allow field research to be carried out in Mimika Regency, where PT Freeport Indonesia operates. Therefore, almost all data collection activities are carried out by relying on the intermediary of technology.

First, data collection through structured interviews. In this activity, the Research Team has prepared a list of questions submitted to the informants through the Google Form application. The list of questions includes personal information from informants, as well as closed and open questions regarding the impact of implementing policies to handle the COVID-19 pandemic, especially for workers.

Second, collecting data through in-depth semi-structured interviews. In this activity, the Research Team conducted interviews in two ways, namely face-to-face interviews and telephone interviews. Direct, face-to-face interviews were conducted with informants from the management of the company who were in Jakarta during the research period. Given the pandemic situation that requires social distancing. Meanwhile, telephone interviews were conducted with informants from the management of the trade/labor union who were in the Mimika Regency during the research period.

Third, primary data collection is in the form of documents, including policy documents such as Government Regulations, Minister of Health Regulations, Circulars, as well as in PTFI's internal context, the Interoffice-Memorandum (IOM). Finally, secondary data collection through literature study.

3 Result and Analysis

3.1 An Update on the Situation of COVID-19 Pandemic and the Impacts of Policies Implementation to Handle It

In Mimika Regency, the first positive case of COVID-19 was detected on March 29, 2020. The latest data as of February 22, 2021, shows that the total number of positive cases in the district reached 4,606 cases, of which 4,162 (90.4%) were declared cured and 40 (40). 0.9%) died (COVID19.MimikaKab.go.id, 2021).

During a period of approximately eleven months, social restrictions, the implementation of health protocols, and the application of a Clean and Healthy Lifestyle (PHBS) became the main strategy, it could even be said to be the only strategy, to deal with the threat of transmission and spread of the coronavirus. This is because as a new disease, there is no vaccine for COVID-19. It was only at the end of 2020 that various studies to find the vaccine in several countries, including Indonesia, showed encouraging results. Furthermore, in January 2021 the COVID-19 vaccination in Indonesia can begin.

To achieve herd immunity to stop the transmission and spread of the coronavirus, at least 70% (182 million people) of the Indonesian population must receive vaccinations. For this reason, the government plans a vaccination program in stages, in which the first phase is prioritized for 1.6 million health workers throughout Indonesia, and is targeted for completion in February 2021. The next phase will be for 17.4 million public service personnel and 21.,5 million elderly people (Kompas.com, 2021).

The COVID-19 pandemic has had a tremendous impact on the socio-economic sector. During 2020 the world economy experienced a sluggish, with slowing economic growth. Even in several countries, including Indonesia, the government officially announced a recession. However, this impact does not come from the COVID-19 disease itself. This impact is precisely the price that must be paid from the implementation of policies to handle the pandemic that relies on social restrictions.

Various economic activities that rely on the gathering of many people are the worst affected sectors, such as entertainment and tourism venues, and shopping centers. In addition, various manufacturing activities that use a large number of workers to carry out their production activities (labor-intensive) are also experiencing serious impacts. Social restrictions forced factories to utilize only a small part of the number of resources and labor they had so that the utilization of production capacity experienced a significant reduction, and productivity fell drastically. In contrast to office workers who can still carry out their work from home (work from home), more workers in these sectors are then laid off, even experiencing layoffs (PHK).

The next economic sector that is also significantly affected by the social restriction policy is the service sector in the transportation sector. With the appeal not to travel—even in some countries, travel bans—most people no longer travel, either for sightseeing and recreation, visiting friends and relatives, or for official purposes. The number of people traveling, whether by land, sea, or air, has thus fallen sharply.

Many experts from various countries have highlighted economic conditions that have experienced a slowdown, stagnation, and even negative growth due to the COVID-19 pandemic. Not a few of these experts have underlined various arguments why the current economic recession is more serious than the previous economic recessions, such as in 2008, there are even those who argue that the situation is worse than the Great Depression in the 1930s. One of the things that distinguish this recession from previous ones, is the fact that it was caused by a deliberate, even programmed, cause, namely social distancing to slow the spread of the pandemic. Economic activity seems to have been stopped on purpose.

At the same time, supporting the existing health system to survive is a very expensive undertaking. Most of the resources owned by the state are unavoidably reallocated to the health sector, both to support the treatment of the sick, especially COVID-19 patients, as well as to conduct research to produce vaccines that can be used to fight the rampant epidemic.

However, this does not mean that all sectors of the economy are paralyzed. Even certain sectors are experiencing growth during the emergency of the pandemic. The digital economy is one sector that is gaining momentum for rapid growth. A drastic reduction in people's travel means an increase in the need for online buying and selling and thus an increase in the delivery of goods. Furthermore, teaching and learning activities in the education sector, ranging from the level of playgroups and kindergartens, primary and secondary education, to higher education at various institutes, colleges, and universities, must be carried out online. This is an opportunity for the emergence and development of various digital companies, including start-up companies, which provide applications to facilitate distance learning. Likewise with the implementation of the work from the home system; the need for technological facilitation is also increasing, both for coordination purposes such as short message services (Whatsapp, Telegram, etc.) and video conferencing (Zoom, Webex, Google Meet, Microsoft Team, etc.), as well as sending documents (electronic mail/ e-mail).

3.2 Evaluation on the Impacts of the Policies Implementation to Handle the COVID-19 Pandemic

In this subsection, we will discuss the impact of implementing policies for handling the COVID-19 pandemic from the point of view of workers/laborers and, in particular, trade/labor unions. The discussion will be divided into several subtopics. First, the perception of workers/labor and trade/labor unions on the significance of implementing policies on their lives in general. In this case, they will also discuss what areas of life they think are most affected by the implementation of the existing policies. Second, subtopics on the positive and negative impacts of policy implementation on various employment issues, including Occupational Health and Safety (OHS), income, social protection (insurance), employment contracts, and severance pay.

3.2.1 The Perception of Labor Union on the Significance of the Implementation of Policies to Handle COVID-19 Pandemic

Rationally, the implementation of policies to handle the COVID-19 pandemic, especially in the form of implementing health protocols and social restrictions to keep the company's activities operating, of course, I have an impact on the lives of workers/laborers. However, to confirm this assumption, we feel it is necessary to categorize the perceptions of informants among those who believe the implementation of the policy has a significant effect on the lives of those who believe otherwise. As a result, the majority of informants (86%) are of the view that the implementation of the policy significantly affects the lives of workers/laborers. The results of the quantification of the informants' perceptions can be seen in Fig. 1:

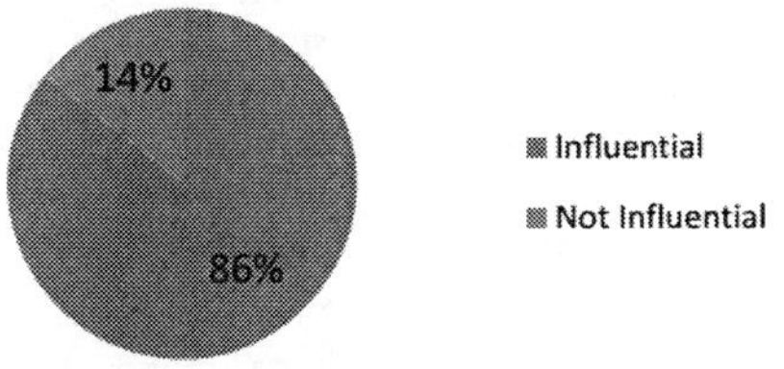

Fig. 1. The informants' perception concerning the significance of the implementation of policies to handle the COVID-19 pandemic.

Based on the results of the interviews, we also identified areas of life that were felt to be most affected by the implementation of the COVID-19 handling policy. These fields include economics, health, social, lifestyle, food, quality of life, family, and production. Of all these areas of life, economic, health, and social are the three areas that are felt to be most affected by the implementation of existing policies.

https://doi.org/10.1051/e3sconf/202133102007

In detail, the informants' perceptions of the areas affected by the implementation of the COVID-19 pandemic handling policy can be seen in Fig. 2:

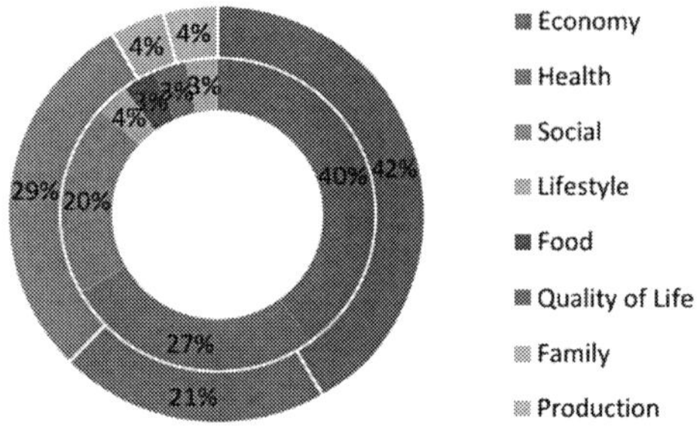

Fig. 2. The areas of life that were significantly influenced by the implementation of the policies to handle the COVID-19 pandemic.

In Fig. 2, we present a diagram that compares the perceptions of trade unions/labor and employees in viewing the areas most affected by the implementation of policies to handle the COVID-19 pandemic. As a result, these two categories of informants have similar perceptions, that the impact of policy implementation is greatest in the economic, health, and social fields.

In the economic field, based on policies issued by the government (in this case the Ministry of Manpower) and companies, there are no salary cuts—employees still receive the full amount of salary. However, the rearrangement of work mechanisms and patterns still affects the income received by employees. This is due to the reduced number of benefits and bonuses given to employees. Regarding economic conditions, at the household level of workers/laborers, a positive impact is felt in the form of a frugal lifestyle driven by prohibitions or restrictions on activities outside the home and in crowded places. However, in a broader context, this causes economic activity to experience sluggishness because most people do not consume through shopping, recreation, tourism, and the like.

In the health sector, of course, the implementation of the COVID-19 handling policy has an impact on the lives of employees. With this policy, particularly related to social restrictions and health protocols, employees can be sufficiently protected from the transmission of COVID-19. In this case, existing policies tend to have a positive impact. However, social restrictions to prevent transmission also have undesirable effects related to the psychological health of employees. The results of interviews with several informants showed that reduced contact and time to gather with family had implications for increasing employee stress levels.

In the social sector, the impact of the implementation of the COVID-19 handling policy can be seen in several ways. Social restrictions cause interaction and socialization among employees to decrease, especially because they cannot gather to do activities that are usually done in a social or friendship environment. Reduction of employee travel from the Highlands (Tembagapura) to the Lowlands (Timika City) and vice versa also affects the decrease in the quantity and quality of time spent by employees with their families. This creates a psychological burden for them. Some of the informants also complained about the boredom that is generally felt by employees because of the necessity to do an antigen swab test every time they want to leave and enter Tembagapura. Nevertheless, they are still aware and understand the importance of the test both to monitor their health condition, protect the health of their families, and prevent uncontrolled local transmission in the work area. In other words, the willingness to always carry out an antigen swab test every time they want to enter and exit Tembagapura is a 'form of sacrifice' that they must do.

3.2.2 The positive and negative impacts of the implementation of the policies to handle the COVID-19 pandemic

Based on interviews with informants, we tried to identify the positive and negative impacts of implementing policies to handle COVID-19 on various employment issues, including Occupational Health and Safety (OHS), income, social protection (insurance), employment contracts, and severance pay. The following is an explanation of each of these issues.

a. Occupational Health and Safety

Based on interviews with informants, we identified several positive impacts of implementing the COVID-19 handling policy on Occupational Health and Safety (OHS). These positive impacts include (1) educating workers/laborers so that their awareness of OHS increases, (2) workers/laborers are protected from the threat of coronavirus transmission, (3) reduced concerns of workers/laborers to go to the work area because generally everyone is aware of the health protocols that must be adhered to, as well as (4) increasing cleanliness in the work environment.

In addition to the positive impacts above, we also identified some negative impacts expressed by the informants. These negative impacts include (1) increased awareness and mutual suspicion among workers/laborers because everyone can be a carrier and transmit the coronavirus, (2) the necessity to work individually and not be able to work in groups can lead to reduced effectiveness and performance, (3) there is a lot of fatigue (fatigue) in the workplace, both physically and psychologically, due to a reduction in crew/personnel working at the same time, (4) delays in certain jobs because they must always carry out health protocols, and (5) the decline in the achievement of OHS performance targets that have been set by the company.

In one interview with an informant, it was also found that there is a link between OHS both with the productivity of the company and the income earned by the worker/laborer. He explained that the implementation of the COVID-19 handling policy could help fulfill the availability of human resources (manpower) needed. With

https://doi.org/10.1051/e3sconf/202133102007

a large number of employees who are healthy and safe, it will be more guaranteed to fulfill the number of human resources who work under the needs of the job. The availability of human resources is very necessary for all operational activities of the company to keep it running smoothly and achieve monthly and annual production targets. For workers/laborers themselves, good OHS quality will directly affect the amount of income received, especially from production bonuses, because OHS performance is one of the important indicators or requirements in calculating bonuses given by the company. Thus, in a broader context, the implementation of policies for handling COVID-19 to ensure the quality of OHS also contributes to building mutually beneficial industrial relations during the pandemic.

There are also problems in implementing policies related to OSH, such as technical constraints, especially the lack of masks so that some workers/laborers wear masks that do not meet the standards, work safety violations, and negligence in using masks due to lack of information and awareness, there are still employees who lack concerned with health protocols, there are still employees who lack discipline in the work environment (such as smoking in the workspace), decreased focus and concentration of workers/laborers due to anxiety. On the other hand, some factors make it easier to implement policies for handling COVID-19, one of which is the habit of workers/laborers who have been sufficiently developed in complying with OHS rules.

To optimize the implementation of policies, the company's management always cooperates with trade unions to carry out communication and coordination (social dialogue), as well as monitoring and evaluation (monev) regularly to minimize the negative impact of implementing policies on handling COVID-19, both on the condition of workers/laborers. and their families, as well as the company's operational and production activities.

b. *Income*

The results of interviews with informants also helped us in identifying some of the positive and negative impacts of the implementation of the COVID-19 handling policy on the income of workers/laborers. The positive impacts that have been identified include (1) the maintenance of income for workers/laborers due to the provision of incentives, especially for those who work in the Highlands, and (2) the emergence of an increasingly frugal lifestyle due to the prohibition/restrictions on carrying out activities outside the home and in public places.

The negative impacts that have been identified include (1) reduced employee rights, especially in the form of bonuses, so that their income is reduced, (2) there is discrimination in the provision of incentives, and (3) reduced benefits received by employees if they are confirmed. positive for COVID-19 or quarantined.

The results of interviews with some informants indicate that the impact of implementing the COVID-19 handling policy on employee income, whether it will be positive or negative, is largely determined by the employee's health condition. If the employee is in good health, the income tends to remain good; but if they are sick, their income will decrease. This is because salary payments and various forms of benefits are still calculated based on employee productivity. As a result, employees who are sick or exposed to COVID-19, experience a decrease in income because they have to undergo quarantine/isolation and cannot work. Employees often complain about this. According to the trade/labor union, the circular letter from the Minister of Manpower stipulates that employees who cannot work because they have to undergo quarantine/isolation should still receive their full salary. However, in reality, this policy has not been fully implemented by the company for reasons of performance and productivity.

If we look closely, the problems above are not naturally rooted in the implementation of policies to handle the COVID-19 pandemic, but rather on the provisions that have been imposed by the company in normal times. In this case, one form of bonus received by employees is a safety bonus (a bonus based on performance in carrying out OHS)—if the employee's safety performance is good, the bonus he receives is good, and vice versa. Therefore, an informant explained, when a work accident occurs in a certain division, for example, the relevant division will experience a reduction in the safety bonus, and thus, the employees' annual production bonus will also be affected. Based on this way of thinking, employees who are infected with COVID-19 are considered to have decreased OHS performance so that they experience a reduction in bonuses, especially if this happens due to negligence in carrying out health protocols.

Considering the above conditions, where employee complaints arise regarding declining income on the one hand, and on the other hand, the calculation of bonuses by the company which is still based on the provisions of OHS performance in normal situations, it is necessary to have a dialogue between employees represented by trade unions. with the company to achieve a common perception.

Regarding the impact of implementing this policy on income, there is one interesting finding that needs to be highlighted in particular. In the interviews we conducted with informants from the workers/labor and trade unions, we found that some of the informants stated that the positive impact was more pronounced because there were additional incentives and bonuses provided by the company. However, at the same time, some informants stated the opposite, that their income was decreasing. At first glance, there are contradictions in the answers and descriptions among the informants. To get a better understanding of this 'contradiction', we tried to map the statements of the informants.

The decrease in income, based on the results of the interview, was caused mainly by the absence of the bonuses they would normally get under normal circumstances. In addition, a more crucial reason is that some employees cannot work because they have to undergo quarantine/isolation, either because they have been diagnosed positive for COVID-19, become a suspect, or because they have physical contact with a positive COVID-19 patient. As for those who state that

https://doi.org/10.1051/e3sconf/202133102007

incentives exist, these incentives are generally 'conditional'; incentives are given only to employees who can continue to work productively, as well as to those who work in certain areas, for example in the Highlands. Thus, in this case, COVID-19 does not directly affect the income of workers/laborers; however, its spread does drastically increase the risk of workers/laborers getting sick and losing some of the incentives or bonuses that they should have been able to get.

The above situation then creates a new problem in the form of social jealousy among fellow workers/laborers. There is a perception among some workers/laborers that discrimination has occurred because of the unequal distribution of incentives. This is also exacerbated by the emergence of the perception that there is no clarity in determining employee criteria, which are important (and still employed normally) and which are less important (thus working from home or even being laid off). A similar perception was expressed by other informants who viewed unjust discrimination in the application of work presence rules.

In a pandemic, several external factors affect the family's economic condition. Government policies such as cutting/discounting PLN electricity payments for 900 W power consumers that have been running for one year, for example, ease the burden felt by worker/labor households. However, several other external factors added to the burden they felt, such as rising prices for various basic needs and transportation costs.

c. Social Protection (Insurance)

In general, based on the results of interviews with informants, the policy for handling the COVID-19 pandemic and its implementation has not had any impact on the social protection/insurance program participated by the workers/laborers. However, several things are of concern to workers/ laborers and trade unions/ laborers.

First, if the worker/laborer remains healthy and continues to work, his saving plan can continue to run so that its value increases; subsidies from the government are very helpful in this regard. On the other hand, if the worker/laborer is exposed to COVID-19 and cannot work for days during the quarantine/isolation period, then the saving plan may also be disrupted. In other words, social protection which should eliminate or at least reduce the concerns of workers/laborers during the pandemic, especially when exposed to the coronavirus and have to undergo quarantine or treatment, actually becomes a source of anxiety because of the possibility that workers/laborers cannot maintain the value of their savings. their plan is due to illness.

Lack of socialization regarding insurance is also a problem because it causes many employees who do not know well the ins and outs of the insurance they follow. In such a situation, some workers/laborers are compelled to perceive injustice in terms of insurance, so that in general the quality of the implementation of the insurance program is considered poor.

d. Employment Contract

In general, the COVID-19 handling policy does not directly affect the work contract, because there are no changes to the work contract or the articles in the Collective Labor Agreement (PKB) due to the implementation of the COVID-19 handling policy. However, during the pandemic, the company did not hold any new employee recruitment activities. On the one hand, from the employee's point of view, this can be seen as a positive impact because it means that their employment contract will be able to be extended; but on the other hand, from the point of view of people who hope to work at PTFI, this can also be seen as a negative impact because there is very little chance of applying for a job.

So far, PTFI itself has not experienced layoffs due to the impact of the COVID-19 pandemic. However, several informants revealed that many employees were laid off, even laid off, in several subcontractor companies that were included in the PTFI supply chain (SC). Some subcontracting companies also did not renew the contracts of some of their workers/laborers because these workers/laborers were not considered an 'essential' unit during the pandemic. In a situation like this, it is not surprising that there is widespread concern among workers/laborers, both at PTFI itself and in its various subcontracting companies, that they may be laid off or even laid off.

e. Severance Pay

In general, there is also no impact whatsoever due to the implementation of the COVID-19 pandemic handling policy on severance pay. This is because PTFI so far has not taken a policy to lay off its employees.

However, some informants revealed that the amount of severance pay has decreased, so it is considered unfair for dismissed employees. However, there is no definite information on whether this happened to some of PTFI's employees, or to employees who work in subcontracting companies within the PTFI supply chain network. If it is true that PTFI did not lay off employees during the pandemic, then it is very likely that what the informants said happened to PTFI's subcontracting companies.

4 Conclusion

Based on the discussion that has been carried out in the previous chapter, it can be concluded that the impact of the implementation of the COVID-19 handling policy is considered to be the most significant in the economic, health, and social fields. In the economic field, changes in work patterns affect the income received by employees. Some employees feel an increase in income from incentives provided by the company related to COVID-19, but others feel a decrease. In the health sector, social restrictions and health protocols have proven to be effective in keeping employees from transmitting the coronavirus. However, unwanted effects appear in the form of stress and work fatigue. In the social sector, social restrictions and travel restrictions cause a decrease in the intensity and frequency of interaction and socialization among employees, as well as the quantity and quality of time with family. This creates a psychological burden for employees.

Regarding employment issues, the positive and negative impacts of the implementation of the COVID-19 handling policy are felt most significantly on Occupational Health and Safety (OHS) and income issues. As for social protection (insurance), employment contracts, and severance pay, the impact is relatively insignificant. Regarding OHS, the positive impacts of policies for handling the pandemic include (i) more educating workers/laborers about OHS, (ii) safeguarding workers/laborers from the threat of coronavirus transmission, (iii) reducing concerns of workers/laborers to go to work areas, and (iv) increasing cleanliness in the work environment. The negative impacts include (i) increased awareness and mutual suspicion among workers/laborers, (ii) unable to work in groups, (iii) there is a lot of fatigue in the workplace, both physically and psychologically, because of reduction of crew/personnel, (iv) delays in certain jobs because they must always implement health protocols and (v) decrease in the achievement of OHS performance targets that have been set by the company. Regarding income, the positive impacts of the pandemic handling policy include (i) providing incentives, and (ii) increasing frugal lifestyles. The negative impact includes (i) reduced employee rights in the form of bonuses, (ii) discrimination in the provision of incentives, and (3) reduced benefits if employees are quarantined.

This research was funded by *Federasi Pertambangan dan Energi Konfederasi Serikat Buruh Seluruh Indonesia* (FPE KSBSI). We would like to thank the management of FPE KSBSI at the national, regional, commissariat, and PTFI management whose have supported us during this research.

References

1. Public Relations Bureau of the Ministry of Manpower, Minister of Manpower Issues Circular on Labor Protection and Business Continuity Related to COVID-19. Retrieved December 9, 2020, from the Indonesian Ministry of Manpower: https://kemnaker.go.id/news/detail/menaker-terbitkan-surat-edaran-about-pelindungan-buruh-dan-kelanjutan-usaha-terkait-COVID-19 [Accessed March 17, 2020]
2. COVID19.go.id., Distribution Map. Retrieved April 19, 2021, from COVID19.go.id: https://COVID19.go.id/peta-sebaran [Accessed April 19, 2021]
3. COVID19. MimikaKab.go.id, Mimika COVID19 Monitoring Data. Retrieved April 19, 2021, from COVID19.MimikaKab.go.id: https://COVID19.mimikakab.go.id/page/data[Accessed February 22, 2021]
4. Febriana, E, Basic Food Program for Vulnerable Communities to Face the COVID 19 Pandemic. Retrieved 12 8, 2020, from Social Counseling Center: https://puspensos.kemsos.go.id/program-sembako-untuk- Masyarakat-rentan-hadapi-pandemi-COVID-19. [Accessed April 16, 2020]
5. Indika Energy. Kuala Pelabuhan Indonesia. Retrieved March 15, 2021, from indikaenergy.co.id: https://www.indikaenergy.co.id/energy-infrastructure/kuala-pelabuhan-indonesia/ (2021).
6. Emerging Infections.. Five Handling Protocols Become the Main Guidelines for Preventing and Controlling COVID-19. Retrieved 12 8, 2020, from the Ministry of Health: https://COVID19.kemkes.go.id/situation-infection-emerging/info-corona-virus/ [Accessed March 12, 2020]
7. Kompas.com. (2021, January 13). Seconds Before the First COVID-19 Vaccination Starting from Jokowi. Retrieved March 19, 2021, from Kompas.com: https://nasional.kompas.com/read/2021/01/13/09164581/detik-detik-jelang-vaccinasi-COVID-19-pertama-yang-dimulai-dari -jokowi?page=all [Accessed January 13, 2021]
8. Minister of Manpower of the Republic of Indonesia. (2020, March 17). SE Minister of Manpower RI No. M/3/HK.04/III/2020. Indonesia. [Accessed March 17, 2020]
9. Nurachman, A. (2020, April 25). The Ministry of Social Affairs Gives Cash Social Assistance for 102,727 KPM in Tangerang Regency. Retrieved 12 8, 2020, from Family Hope Program: https://pkh.kemsos.go.id/ [Accessed April 25, 2020]
10. CNN Editor. (2020, April 14). Jokowi's Incentive List for Victims of Layoffs due to Corona. Retrieved December 9, 2020, from CNN Indonesia: https://www.cnnindonesia.com/ekonomi/20200414084421-532-493304/List-insentif-jokowi-untuk-korban-phk-karena-corona [Accessed April 14, 2020]
11. Republika Editor. (2020, May 1). Jokowi Promises to Protect Workers During the COVID-19 Pandemic Retrieved December 9, 2020, from Republika.co.id: https://republika.co.id/berita/q9nfj1409/jokowi-promise-tetap-lindungi-buruh-saat-pandemi-COVID19 [Accessed May 1, 2020]
12. Wiyono, A. (2020, April 20). Social Safety Net Program in the Midst of a Pandemic. Retrieved 12 8, 2020, from Social Counseling Center: https://puspensos.kemsos.go.id/program-jaring-pengaman-social-di-tengah-pandemi [Accessed May 1, 2020]
13. Worldometers.info. (2021, April 19). COVID-19 Coronavirus Pandemic. Retrieved April 19, 2021, from Worldometers.info: https://www.worldometers.info/coronavirus/ [Accessed April 19, 2021]

https://doi.org/10.1051/e3sconf/202133102008

Pentahelix synergity natural and non-natural disaster management in Pidie Jaya District Aceh Province to support national security

Sugeng Yulianto[1]* *Fauzi* Bahar[1], *Sugimin* Pranoto[1], and *Aam* Amirudin[1]

[1]Disaster Management Study Program, National Security Faculty, Republic of Indonesia Defense University. IPSC Sentul Area, Bogor, West Java, Indonesia

Abstract. Geographically, Indonesia is located on disaster prone area. Natural disasters such as earthquakes, tsunamis, floods, landslides, volcanic eruptions and non-natural disaster such as Covid-19 Pandemic often occur in various places in Indonesia including in Pidie Jaya, Aceh Province. These disasters have a big influence on many aspects of the socio-economic life of the affected communities. Managing disaster properly will reduce the risk so that it will provide security and resilience community that can anticipate all the impacts of disasters. Collaboration in the form of the Pentahelix Synergy concept involving elements of the Government, Society, Academics, and the Business Industries is one of the important aspect in disaster management. Furthemore, the mass media and the private sector will provide great energy to solve the problem during disaster. This paper discuss about the synergy of Pentahelix in dealing with natural and non-natural disasters in Pidie Jaya Regency, Aceh Province. The results will be useful as lesson learned to support National Security program in Indonesia.

1 Introduction

In the second paragraph of the Preamble to the 1945 Constitution of the Republic of Indonesia, it is stated that the noble goal of the nation is to realize the Unitary State of the Republic of Indonesia which is independent, united, sovereign, just, and fair. In addition, the national goals are also stated in the fourth paragraph of the Preamble to the 1945 Constitution of the Republic of Indonesia, namely to protect the entire Indonesian nation and the entire homeland of Indonesia, promote public welfare, educate the nation's life, and participate in implementing world order based on independence, eternal peace, and social justice.

The existing national goals and visions are then implemented in the nation-building program by involving all Indonesian citizens. All of these efforts must be built on the foundation of national security, namely to protect the entire Indonesian nation and the entire homeland of Indonesia.

This condition cannot be separated from the development of the strategic environment which generally shows increasingly escalative and complex symptoms in various parts of the world [1]. These conditions allow the incoming threats, both military threats (armed and unarmed), non-military threats, and hybrid threats [2]. Sources of the threats can come from within or outside the country and are carried out by state and non-state actors, whether national, regional, or international.[3].

From prehistoric times to the present time, disasters are an inseparable part of human civilization. Disasters affect human life and livelihoods in livelihoods, customs, and culture. Rapid industrialization and urbanization lead to global warming and climate change that increasing frequency of disasters. However, the development of science and technology can manage disasters by reducing the losses and damages that accompany disasters.

Due to Indonesia's geographical location, from the past, present, and future, Indonesia is faced with the threat of geophysical and hydro-meteorological disasters. Currently, the threat of these disasters is magnified by the impact of global warming and the effects of climate change. Geophysical disaster threats include earthquakes, tsunamis, volcanic eruptions, and landslides. Meanwhile, the threats of hydro-meteorological disasters include droughts, floods, storms, and forest fires.

The total number of natural disasters in Indonesia from January 1 to September 3 2020 was 1,944 incidents. Floods still dominate with 730 events, followed by tornadoes with 523 events, landslides with 368 events, and forest and land fires with 263 events. The most severely affected provinces were Central Java Province with 356 incidents, followed by West Java Province with 336 incidents, and East Java Province with 267 incidents.

* Corresponding author: yulianto_s79@yahoo.co.id

Of the 1,944 total natural disasters, 272 people died, 24 people were missing, 422 people were injured, and 3,871,567 people suffered and were displaced.[4].

The mention disaster events do not include the impact of non-natural disasters which are still ongoing. In April 14, 2020, the Government of Indonesia has designated Covid-19 as a non-natural national disaster.[5]. The number of people exposed to Covid-19 as of June 9, 2021, spread across 34 provinces and 493 regencies/cities. It is confirmed 1,877,050 positive, 1,723,253 people cured and 52,162 dead [6].

With so many natural and non-natural disasters in Indonesia and tending to increase every year, the impact is not only taking human lives and livelihoods, but also the damage of the infrastructures. For example, the impact of natural disasters in 2020, there are 29,964 houses, 1,317 facilities (555 education, 637 worship, and 125 health), as well as 126 offices and 330 bridges are damaged [7]. Likewise, the Covid-19 pandemic, the impact felt not only on the health sector but also on other sectors, including the economy and employment sectors. According to the Central Statistics Agency (BPS), there has been a significant decline in Indonesia's economic growth from the first quarter of 2019 by 5.02%, to the first quarter of 2020 by 2.97 percent, and to the second quarter of 2020 by minus 5.32%.[8].

Furthermore, according to data from the Ministry of Manpower (Kemnaker) it was recorded that from the formal sector a total of 1,010,579 (873,090 workers from 17,224 companies were laid off, and 137,489 workers were laid off from 22,753 companies). Meanwhile, the number of companies and workers affected in the informal sector is 34,453 companies and 189,452 workers[9].

Aceh Province is the land closest to the epicenter of the Indian Ocean earthquake which has the potential for a tsunami disaster. This is influenced by the geographical location of Aceh Province in the north by the Malacca Strait and the Andaman Sea, in the east by the Malacca Strait, in the south by North Sumatra Province, and in the west by the Indonesian Ocean.[10].

According to DIBI, the history of disasters in Aceh Province from 1815 to 2015 places flood disasters as the most impactful events. While other potential disasters are flash floods, extreme waves and abrasion, earthquakes, tsunamis, forest and land fires, droughts, epidemics and disease outbreaks, volcanic eruptions, extreme weather, and landslides.

Pidie Jaya Regency is one of the areas in Aceh Province where the disaster risk index still occupies a high class. There are four disasters with different risk classes that threaten eight sub-districts. A tsunami is a natural disaster threat that mostly covers almost all sub-districts, especially the coastal areas, namely the sub-districts of Meureudu, Meurah Dua, Bandar Dua, Jangka Buya, Ulim, Trienggadeng, and Panteraja. Then moderate flooding inthe districts of Bandar Dua, Meureudu, Meurah Dua, and Bandar Dua. Strong winds with high risk in Bandar Dua sub-district, Meureudu, Bandar Baru. As well as lava and volcanic ash disasters in the Meurah Dua sub-district.

In addition to the above disaster, the spread of the Covid-19 Pandemic case in Aceh Province as of February 24, 2021, was reported by the Covid-19 Handling Task Force as being in the yellow zone or low-risk zone, but has expanded from 11 to 14 districts covering Southeast Aceh Regency, Aceh East, North Aceh, Bireuen, Central Aceh, Gayo Lues, Pidie Jaya, Pidie, West Aceh, Southwest Aceh, Nagan Raya, South Aceh, Simeulue, and Aceh Singkil Districts [11].

The latest development of Covid-19 infections in Pidie Jaya Regency until March is still showing increasing. The number of infected victims has reached 229 people, 16 people died, 3 people are still being treated (positively active), and 210 people have been declared cured.

The paper aims to discuss more about the synergy of Pentahelix in dealing with natural and non-natural disasters in Pidie Jaya Regency, Aceh Province to support National Security.

Based on the above background, some problems occur in disaster management efforts in Pidie Jaya Aceh Regency, so the author arranges the questions asked about the pentahelix synergy to support this research, namely:

1. How is the communication between Pentahelix in disaster management in Pidie Jaya Aceh Regency?
2. How is the coordination between Pentahelix in disaster management in Pidie Jaya District, Aceh?
3. How is the cooperation between Pentahelix in disaster management in Pidie Jaya District, Aceh?

From the formulation of the problem above, the objectives of this research are:

1. Analyzing the communication between Pentahelix in disaster management in Pidie Jaya District, Aceh
2. Analyzing the coordination between Pentahelix in disaster management in Pidie Jaya District, Aceh
3. Analyzing the cooperation between Pentahelix in disaster management in Pidie Jaya District, Aceh

2 Study Method

This research uses qualitative analysis methods that is *library research*. The qualitative rationale in this paper is a real condition in describing Pentahelix Synergy in handling natural and non-natural disasters in Pidie Jaya Aceh Regency, where the data obtained were obtained from the results of the Group Discussion Forum (FGD) with the Pidie Jaya Regional Government Apparatus and other sources. Others are also who supported the 2021 Domestic Work Lecture (*Kuliah Kerja Dalam Negeri* / KKDN) for the Defense University Disaster Management Study Program which took place from 22 to 26 March 2021

3 Results and Discussion

Disaster outcomes are a phenomenon whose impact will be felt by all affected parties in an area, therefore in its management, it is necessary to have good cooperation between all components in the form of a pentahelix, namely parties or helix that have roles, interests, and characters consisting of the government, society, the business world, academics or experts and the mass media. Disaster prevention and management can be done by one party. Pentahelix is an answer, just adapted to the pentahelix stance for pre-disaster, emergency response, and post-disaster. Because the character of the problem will be different and take into account local aspects as well[12].

In addition, the value of synergy and cooperation is a value that should be upheld in the archipelago. This is inseparable from the fact that Indonesia is a disaster-prone area. The value of synergy taken means that disaster management in Indonesia will take place effectively because the value of cooperation is purely the identity of the Indonesian nation. Through the pentahelix, this value must be pursued by various parties according to their functions and capacities, so that the concept of pentahelix synergism will go hand in hand considering that disaster is a shared business.

Disaster management in Pidie Jaya District, Aceh Province, obtained data and information about pentahelix synergy in the form of communication, coordination, and cooperation between relevant stakeholders which are important points in achieving disaster management success.

a. Communication

BPBD and BMKG have communicated related to disaster mitigation by studying and analyzing potential tectonic earthquake zones in Pidie Jaya Regency which is presented in Seismicity Data originating from grove bands sensor earthquake detection equipment installed at the BPBD office and the Pidie Jaya Regent's Office. From April to August 2020, data obtained that there have been very significant tectonic earthquakes 10 (ten) times a day in Pidie Jaya Regency, this has resulted in Pidie Jaya Regency being a prone zone with the potential for tectonic earthquakes. All of this information is disseminated in the form of WhatsApp Groups to relevant agencies within the Pidie Jaya Regency Government.

BPBD and BMKG have also provided education to the public in the form of earthquake and tsunami simulations so that when the coast recedes, within 30 (thirty) minutes people can run for shelter to avoid the tsunami disaster. BMKG also has a special program with other countries periodically every 2 (two) years to conduct simulations related to community preparedness for earthquakes that cause tsunamis.

Information dissemination has also been carried out by BMKG to the wider community, both in the form of SMS, email, printed and electronic media information. BMKG has an android application that can be downloaded by the public, and there are books that BMKG publishes every month as a form of disaster information to the public. The Go to School disaster education program has been carried out by BMKG for elementary to high school students as public information. BMKG also has a Tsunami evacuation guide in the Covid-19 situation that can be applied to the tsunami emergency response during the Covid-19 pandemic.

BPBD has carried out socialization and education to the public regarding the handling of the Covid-19 pandemic in the form of a disaster management information system (SIGUNA) application that can be accessed by anyone on the website. https://sigunapijay.com/index.php

Conducted a coordination meeting for the Regional Leadership Communication Forum (FORKOMPINDA) involving elements of Pentahelix in Pidie Jaya Regency including Unsyiah who has breadth in the aspect of communication networks as well as disseminating education about disaster management to internal and external, to create community resilience/resilience. Therefore, the disaster communication carried out by Unsyiah can help the Pidie Jaya Regency Government to disseminate disaster information.

BPBD in disaster communication is also assisted by the Disaster Risk Reduction Forum (FPRB) which is a combination of NGOs, community leaders, and the media who are directly in the field so that FPRB can provide information to each other with BPBD.

The synergistic relationship between stakeholders in disaster management efforts can be seen through the communication relationship between related parties. The ability to communicate that is owned by the Regional Government is a key skill to create good synergy. One form of good communication in disaster management in Pidie Jaya Regency is an information system that is communicated with the WhatsApp Group and related forum coordination meetings. This communication has an impact on increasing the participation of all existing elements of Pentahelix, to build better preparedness.

In essence, communication is present whenever and wherever humans are, including during disasters. Pre-disaster, disaster emergency, or post-disaster both involve communication so that the media element of Pentahelix is so important in disaster management, it's just that it is necessary to understand how the communication function is present in each condition. In other words, in disaster management, communication can be present as a function of socialization and information dissemination, management and coordination functions, and counseling and rehabilitation functions.

b. Coordination

BPBD coordination with BMKG has been carried out regarding socialization to the public on the dangers of earthquakes and tsunamis, installation of earthquake and tsunami equipment as an early warning system (EWS) to the community as well as how to read these tools and interpret them, as well as installation of ground motion detectors (Acelorograph).

BPBD has also coordinated with the Pidie Jaya Regency Government including Bappeda and Public Works regarding the mapping and preparation of a master plan for disaster-prone areas, revetment of rivers and beaches, and installing breakwaters, building evacuation routes, revising/reviewing the RTRW, and building earthquake-resistant houses in this activity. post-earthquake rehabilitation and reconstruction.

The form of BPBD coordination by involving elements of Pentahelix on the threat of natural disasters is implemented, among others, in the coordination meeting of the Disaster Risk Reduction Forum between the leaders of Pidie Jaya Regency; coordination meeting with the SAR Task Force of Pidie Jaya Regency; and BPBD as a coordinating measure also held an assembly for the title of preparedness forces to deal with natural disasters.

In handling the Covid-19 pandemic as a non-natural disaster in Pidie Jaya Regency, BPBD has strengthened the coordination function between Pentahelix and there is a command function that has been formed, namely the Covid-19 Task Force (Satgas Covid-19), the Covid-19 Task Force has built post The command post (posko), the Covid-19 Task Force and BPBD have held a coordination meeting for the Regional Leadership Communication Forum (Forkompinda) in Pidie Jaya Regency and a coordination meeting for the Covid-19 Task Force Team.

Coordination was also carried out by the Pidie Jaya Regency Government with academics, who were appointed, namely Syiah Kuala University (Unsyiah). Unsyiah was asked by the Pidie Jaya Regency Government through its Chancellor a few hours after the 2016 Earthquake to immediately conduct a rapid assessment, the results of data analysis of data information can be used by policymakers in disaster management efforts.

In structural mitigation, BPBD strengthens riverbanks and coastal abrasion which is carried out with a budget sourced from the APBD so that structural mitigation efforts are carried out at several priority points, it is necessary to coordinate the budget with the Pidie Jaya Regency Government and coordinate the implementation of development along rivers and streams. coast with PUPR as the authority holder.

The coordination process has been adapted from the Rietjens (2008) model, there is a need for operational establishment in the field, and evaluation. The formation of an organizational structure in the coordination of Pentahelix to synergize is needed to see who does what and as a control in achieving goals. In handling non-natural disasters such as Covid-19, coordination has been established between Pentahelix who does what, and the line of command is led by the Covid-19 Task Force, and evaluation of coordination in the implementation of the Covid-19 response has been carried out through the Covid-19 Task Force forum meeting with Pentahelix elements. This coordination can be considered quite synergistic because all elements of Pentahelix are involved even though in its implementation it has not been synergized, such as overlapping authorities or throwing each other's authority, especially related to a limited budget[13].

While the coordination of the elements of Pentahelix for natural disaster management has not yet been established, it can be seen from the absence of a Crisis Center, so that if there is a natural disaster such as an earthquake there is no leading center as a command line that leads the coordination line between Pentahelix, it is necessary to form a team and command leadership for implementation disaster management, especially for emergency response, clear programs as operational coordination in the field, and evaluation in the implementation of coordination between Pentahelix regularly.

c. Cooperation

BPBD has collaborated and coordinated among stakeholders, one of which is Bappeda and PU in the construction of earthquake-friendly houses for rehabilitation and reconstruction after the 2016 earthquake.

BPBD together with the Disaster Risk Reduction Forum (FPRB) assisted by the TNI/Polri conducted a study on the identification of flood-prone areas. From the data collected, there are around 56 villages that are very vulnerable with high intensity because they are located on the river flow line. Then 16 villages with medium risk and 7 villages with low risk.

In dealing with Covid-19, BPBD has collaborated with the Health Office, Social Service, Dandim, Polri, Bappeda, PUPR, and other pentahelix elements in several Covid-19 prevention activities, including implementing 3T efforts (tracking, tracing, treatment); Implementation of Judicial Operations / Community Disciplinary Raids under Pidie Jaya Regent Regulation Number 27 of 2020; Providing BLT assistance to the poor; and the establishment of a resilient village for the Covid-19 disaster for economic resilience during a pandemic.

The cooperation between the Pentahelix elements in dealing with Covid-19 and its program has been established and the Covid-19 Task Force has been appointed as the command line and the front line in the fight against Covid-19.

The collaboration between the Pidie Jaya Regency Government and the community as elements of Pentahelix through an approach through religious leaders in handling Covid-19 is carried out as figures who become role models for the people of Pidie Jaya Regency. Religious leaders can lead public opinion by conducting outreach and education to the public to be able to comply with health protocols for the spread of Covid-19 and the implementation of vaccines for all the people of Pidie Jaya Regency.

The government of Pidie Jaya has also collaborated with the people of areas prone to natural disasters, with the construction of two disaster-resilient villages which are expected when there is a disaster, the village will be able to carry out emergency response and recovery efforts quickly.

Academics appointed by the Pidie Jaya Regency Government a few hours after the 2016 Earthquake, namely Syah Kuala University (Unsyiah) by conducting a Rapid Assessment as a form of collaboration to increase the effectiveness and efficiency of initial humanitarian actions taken to ensure the survival of the affected communities. disaster.

The results of the Rapid Assessment are given to the Pidie Jaya Regency Government policymakers as advice in carrying out emergency response efforts quickly, precisely, and coordinated. In addition, it can also be useful as a reference in preparing a contingency plan for Pidie Jaya Regency on earthquake prevention measures in the future. According to Unsyiah's contingency plan in an interview with Mr. Agus, there were constraints in resources, causing delays in updating the contingency plan document.

The concept of disaster management, at first glance, looks quite simple, but, in disaster management, it is quite complex. The things that are always a problem are communication, information, coordination, and cooperation[14].

Disaster management at the stage of disaster management, both in the pre-disaster, emergency response, and post-disaster phases cannot all be done alone. There needs to be cooperation between each element of Pentahelix starting from good communication so that they can synergize in providing useful information for the community, coordination and cooperation can also run synergy between Pentahelix if there is a clear leader or line of command, and commitment between Pentahelix, especially elements of the Government towards efforts risk reduction from planning to reconstruction.

Based on information obtained from resource persons in FGDs, cooperation in natural disaster management in Pidie Jaya Regency such as during the 2016 earthquake is still not good. Obstacles faced at that time included limited resources, limited physical facilities, and all components in Pidie Jaya Regency were affected or became victims of disasters which caused cooperation between agencies to be not optimal. Cooperation with business institutions has not yet been carried out by the Pidie Jaya Regency Government.

Pidie Jaya Regency is the 3rd (three) poorest district in Aceh Province, plus a limited disaster budget taken from APBN sources, collaboration with business institutions can be a solution to solving poverty problems and limited disaster budgets in Pidie Jaya Regency.

In making cooperation what needs to be agreed is the appointment of a leader or line of command in the implementation of disaster management cooperation so that the lines of communication and coordination between Pentahelix are clear so that trust can be created between Pentahelix so that in the implementation of disaster management there is a commitment from all elements of Pentahelix.

Cooperation in handling the Covid-19 pandemic has been carried out well, with the formation of a Covid-19 task force that can keep Pidie Jaya Regency in the green zone. However, there are obstacles faced by the COVID-19 task force, including the low level of compliance with the use of masks and patient tracing.

The culture of the Acehnese people who are obedient to religious leaders has become a shortage as well as an opportunity for the Covid-19 Task Force in socializing the Covid-19 response. This is where the role of cooperation between the Covid-19 Task Force and religious leaders is needed so that exposure to Covid-19 in the Pidie Jaya community can still be controlled. Religious leaders in Pidie Jaya Regency have a very important role which is a guide and role model for the community. So that religious leaders and other elements of Pentahelix need to work together in handling disasters, both natural and non-natural disasters. This will create common ground and mutually agreed-upon initiatives.

BPBD of Pidie Jaya Regency as an official agency in disaster management should have planning, control, supervision, and evaluation in the implementation of disaster management. Contingency plans and action plans are important to always exist and be updated regularly as guidelines for the cooperation program of all elements of Pentahelix as a preventive measure in disaster risk reduction to create human security for all communities in Pidie Jaya Regency.

4 Conclusion

Synergy of natural and non-natural disaster management in Pidie Jaya Regency is carried out by taking into account three aspects, namely Communication, Coordination and Cooperation. Communication is carried out by holding forum coordination meetings for natural and non-natural disaster management which also involves the mass media in delivering the results of coordination to the public. Coordination is carried out well in the management of non-natural disasters. However, it is different from the coordination carried out during natural disaster management. There is still a need for coordination between elements of the pentahelix so that the stuttering that occurred in the 2016 earthquake disaster management does not recur. Cooperation carried out during natural disaster management is still not good. There are constraints on resources and limited physical facilities in Pidie Jaya Regency experiencing stuttering during the 2016 Earthquake.

The synergy that exists in handling the Covid-19 pandemic is very good. This can be seen in the seriousness of various parties such as the Pidie Jaya Regional Government in collaboration with several institutions such as religious leaders, community leaders, TNI/Polri, government, and private institutions, as well as involving academics who can provide education to the public about the transmission of Covid 19, implementing health protocols with good and right and the people of Pidie Jaya support all the efforts that have been carried out by the Government to overcome the Covid 19 pandemic.

Efforts to increase the synergy of natural and non-natural disaster management in Pidie Jaya Regency can be carried out in several ways, such as a). Strengthening cooperation between ministries, community institutions, and other government agencies, including the TNI where one of the main tasks in Military Operations Other than War (OMSP) is to assist in tackling natural disasters; b). Implement existing regulations, which will regulate disaster management in the hope that it will provide a legal basis for stakeholders to work under their duties and responsibilities and authorities so that all efforts to be carried out can run and provide optimal results.

References

1. M. Markum, W. Winarno, Efforts to Build Nation's Character to Prevent Radicals in order to Build Nation's Character to Prevent Radicalism and Terrorism in Indonesia, J. PPKn Researcher and Thinker. Educator. Pancasila and Citizenship, **8** (1), 38–61, (2020).
2. A. Muchaqqi, M. Halkis, A. A. Mundayat, Impact of Socio-Economic Inequality on the Defense and Security System of the Universal People (Case Study in Muara Angke Area), Ekon. Defense, **5** (2) (2020).
3. A. Alfajri, A. Setiawan, and H. Wahyudi, "Synergy of Regional Defense Spatial Development in Facing

Non-Military Threats in Indonesia, J. Glob. Strategy, **13**, (1), 103, 2019, 0.20473/jgs.13.1.2019.103-122 (2019).

4. R. Jati, 2,925 Natural Disasters Occurred in 2020 in Indonesia, Hydrometeorological Disasters Dominate - BNPB, BNBPB, 2020. https://www.bnpb.go.id/berita/sebanyak-2-925-bencana- nature-happening-2020-in-land-water-disaster-hydrometeorology-dominate (accessed June 10, 2021).
5. BNPB, President Designates COVID-19 as a National Disaster, (BNPB, 2020)
6. Home Covid19.go.id." https://covid19.go.id/ (accessed June 10, 2021).
7. More than 42,000 Residents' Houses Damaged Throughout 2020 - BNPB. https://bnpb.go.id/berita/more-dari-42-000-rumah-warga-rusak-sepanjang-tahun-2020 (accessed June 10, 2021).
8. Ministry of Finance of the Republic of Indonesia, Facing the Pandemic, the Indonesian Economy in the First Quarter of 2020 Grows by 2.97% (2021).
9. A. M. Kurnia, As a result of Corona, more than 3.5 million workers have been laid off and laid off, (Kompas.commpas.com, 2020).
10. National Disaster Management Agency, Aceh Disaster Risk Assessment 2016 - 2020, Disaster Risk Management Agency, **43** (2015)
11. D. Aceh, Yellow Zone Expands in Aceh, New Cases of Covid-19 Increase by 26 People, (dinkes.acehprov.go.id, 2021).
12. Disaster and the Pentahelix in Disaster Management in Indonesia - BNPB. https://bnpb.go.id/berita/kebencanaan-dan-pentahelix-dalam-penanggulangan-bencana-di-indonesia#: (accessed Jun. 10, 2021).
13. S. P. Nugroho, Civil-Military Cooperation in Disaster Management, Dialogue on Disaster Management, **7**, 2016, 103–110 (2015).
14. B. Setio, Disaster Communication: System Aspects (Coordination, Information and Cooperation), **1** (2012).

https://doi.org/10.1051/e3sconf/202133102009

Comparison of air pollutant standard index before and during the COVID-19 pandemic in DKI Jakarta

Dedy Hartono[1]* *Syamsul* Maarif[1], *Deffi* Ayu Puspitosari[1], *Iko* Saikanti Ponangsera[1], *Jihan* Fitriyanti[1]

[1] Disaster Management Study Program, National Security Faculty, Republic of Indonesia Defense University. IPSC Sentul Area, Bogor, West Java, Indonesia

Abstract. Clean air is a basic thing that is needed by living things. This makes clean air is very important for the continuity of life. Therefore, an air pollution standard index is needed to determine air quality based on the concentration of gas contained in it. As the State Capital which is the center of the economy, DKI Jakarta Province has an important air quality issue that needs to be controlled regularly. This research is a qualitative descriptive study of secondary data on the air pollutant standard index to compare the gas concentration parameters in the air before and during the COVID-19 pandemic. The results of the analysis show that during the COVID-19 pandemic there was an increase in the "good and moderate" category days and a decrease in the "unhealthy and very unhealthy" category days compared to before the COVID-19 pandemic.

1 Introduction

DKI Jakarta Province is the economic center and the capital city of the Republic of Indonesia, which consists of five administrative cities with a dense population [1][2]. The high daily use of motorized vehicles, factory smoke, and other pollutants that tend to increase due to the impact of infrastructure development have resulted in low air quality in DKI Jakarta Province [3][4]. Modern urban life cannot be separated from various technologies and goods production systems to support people's welfare. But most of these things are the main causes of air pollution that we are not aware of.

Based on the air quality observation site aqicn.org on June 1, 2021, at 15:38, DKI Jakarta ranks 3 out of the top 100 countries with the worst air pollution [5]. The air quality in DKI Jakarta reaches the level of 134 US Air quality index which indicates the air in DKI Jakarta is unhealthy [5].

Air quality is a measure of the level of good or bad of a mixture of gases contained in the troposphere that is needed and affects the health of humans, living things, and other elements of the environment whose composition is not always constant [6][7]. Air quality is a basic need that is very important and needs to be considered because humans need oxygen to be able to continue to live [8], especially during the COVID-19 pandemic.

Data shows that several regions in Indonesia, including DKI Jakarta Province, have experienced an increase in positive cases of COVID-19 since it was first announced, both in the number of sufferers and the number of deaths [9]. COVID-19 As a new pandemic that is so complex and we do not fully understand, both medically such as viral compounds and their spread, health protocols, mitigation, vaccines, and treatment, but various other non-medical dimensions that we need to understand regarding the characteristics of this virus, such as air quality.

Based on the description above, this paper will try to analyze the comparison of the air pollutant standard index before and during the COVID-19 pandemic in DKI Jakarta Province. This needs to be analyzed further to provide a better understanding of the dynamics of handling and the impact of COVID-19 on air quality in DKI Jakarta Province.

2 Study Method

This study uses a qualitative research method with a descriptive approach in answering research problems. The research answers will be described based on secondary data on the air pollutant standard index [10][11][12] to compare the gas concentration parameters in the air before and during the COVID-19 pandemic in the discussion section, then conclusions will be drawn based on the results of the analysis that has been carried out by researchers.

3 Result and Discussion

3.1 Concentration in 2019 and 2020

Particulate Concentration ($PM_{2.5}$) is the concentration of

* Corresponding author: dedyhartonoaal@gmail.com

airborne particles smaller than 2.5 microns (micrometers) [13]. In simple terms PM2.5 is a very small dust particle, even when compared to a strand of our hair split into 20.

Because of its very small size, $PM_{2.5}$ makes it dangerous, because it can enter the body to the deepest organs through the respiratory tract, and can even reach the lungs directly. More than that, $PM_{2.5}$ can also reach into the blood vessels because of its very small size. So if the buildup continues to occur in the blood vessels, $PM_{2.5}$ can also cause diseases such as heart disease and stroke. Another impact of $PM_{2.5}$ is that this dust can affect pregnant women and the fetus they contain [14][15].

This is what makes $PM_{2.5}$ concentration one of the mandatory benchmarks in looking at air quality in the environment. $PM_{2.5}$ assessment can be categorized based on the threshold value. Threshold Limit Value (NAV) is the air pollution concentration limit that is allowed to be in the ambient air. NAV $PM_{2.5}$ = 65 gram/m3.

$PM_{2.5}$ concentration in DKI Jakarta Province was carried out at 5 points representing the five administrative city areas of DKI Jakarta, namely Bundaran HI, Central Jakarta (DKI1), Kelapa Gading, North Jakarta (DKI2), Jagakarsa, South Jakarta (DKI3), Lubang Buaya, East Jakarta (DKI4), Kebon Jeruk, West Jakarta (DKI5). The results of the $PM_{2.5}$ concentration in 2019 and 2020 in DKI Jakarta Province can be seen in Fig. 1.

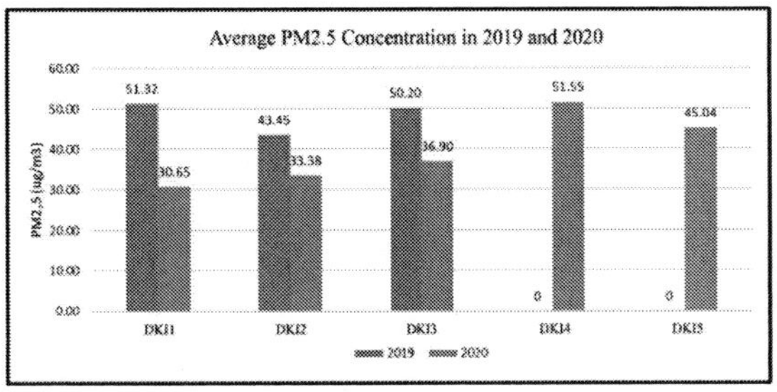

Fig. 1. Average $PM_{2.5}$ concentration in 2019 and 2020

$PM_{2.5}$ concentration in DKI Jakarta Province was carried out at 5 points representing the five administrative city areas of DKI Jakarta, namely Bundaran HI, Central Jakarta (DKI1), Kelapa Gading, North Jakarta (DKI2), Jagakarsa, South Jakarta (DKI3), Lubang Buaya, East Jakarta (DKI4), Kebon Jeruk, West Jakarta (DKI5). The results of the $PM_{2.5}$ concentration in 2019 and 2020 in DKI Jakarta Province can be seen in Table 1.

Table 1. Average $PM_{2.5}$ decline in 2020

Station	Decrease	
	$\mu g/m^3$	%
DKI1	20,67	40,28
DKI2	10,07	23,19
DKI3	13,30	26,49
DKI4	-	-
DKI5	-	-

Table 1 informs that there was a decrease in the annual $PM_{2.5}$ concentration in 2020 compared to 2019. The decrease in $PM_{2.5}$ concentration in DKI1 was 20.67 g/m3 (40.28 %), DKI2 was 10.07 g/m3 (23, 19 %), DKI3 by 13.30 g/m3 (26.49 %). However, the comparison of $PM_{2.5}$ concentrations in 2019 and 2020 in DKI4 and DKI5 cannot be seen because there is no $PM_{2.5}$ concentration data found in 2019.

In general, the decline in $PM_{2.5}$ concentrations in 2020 occurred due to the work from home (WFH) program and large-scale social restrictions (PSBB) in DKI Jakarta Province, as shown in Fig. 2.

Fig.. 2. Comparison graph of the monthly average $PM_{2.5}$ in 2019 and 2020

Fig. 2 explains that overall the monthly average concentration of $PM_{2.5}$ in 2020 is lower than in 2019. When calculated during the COVID-19 pandemic in the period March to December 2020, the average $PM_{2.5}$ at each monitoring station air quality can be seen in Table 2.

Table 2. Average $PM_{2.5}$ for the period March to December 2019 and 2020

Station	Decrease	
	Ug/m3	%
DKI1	22,96	42,25
DKI2	7,56	16,83
DKI3	14,31	27,45

Table 2 informs that in the period March to December 2019 and 2020 in DKI1 there was a decrease in $PM_{2.5}$ concentration by 22.96 g/m3 (42.25%), DKI2 by 7.56 g/m3 (16.83%), DKI3 of 14.31 g/m3 (27.45 %).

3.2 CO Concentration in 2019 and 2020

Carbon monoxide (CO) is a gas produced by a variety of processes, including the burning of coal, wood, and the use of fuel in motor vehicles. This gas is odorless, colorless, and cannot be tasted [16][17][18].

When a person is exposed to or inhales carbon monoxide gas, the blood's ability to bind oxygen is reduced. This is because CO gas is more easily bound to hemoglobin and then forms carboxyhemoglobin (COHb). The more COHb that is formed, the less oxygen will be circulated throughout the body. As a result, the body will experience a lack of oxygen (hypoxia). Complications of carbon monoxide poisoning can cause brain damage, heart disease, and fetal compromise [19][20][21].

This is what makes the concentration of CO in the air one of the mandatory benchmarks in looking at air quality in the environment. CO assessment can be categorized based on the threshold value. Threshold Limit Value (NAV) is the air pollution concentration limit that is allowed to be in the ambient air. NAV CO = 25 gram/m3.

The concentration of CO in DKI Jakarta Province was carried out at 5 points representing the five administrative city areas of DKI Jakarta, namely Bundaran HI, Central Jakarta (DKI1), Kelapa Gading, North Jakarta (DKI2), Jagakarsa, South Jakarta (DKI3), Lubang Buaya, East Jakarta. (DKI4), Kebon Jeruk, West Jakarta (DKI5). The results of the 2019 and 2020 CO concentrations in DKI Jakarta Province can be seen in Fig. 3.

 https://doi.org/10.1051/e3sconf/202133102009

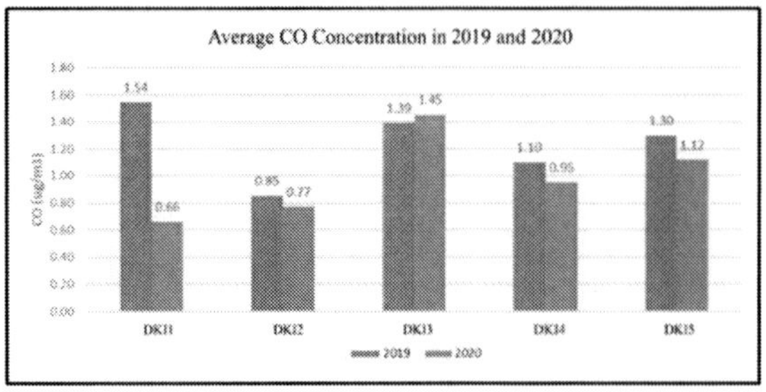

Fig. 3. Average CO concentration in 2019 and 2020

The result of CO concentration in DKI1 in 2019 was 1.54 g/m3 and 2020 was 0.66 g/m3, CO concentration in DKI2 in 2019 was 0.85 g/m3 and 2020 was 0.77 g/m3, the concentration CO in DKI3 in 2019 was 1.39 g/m3 and 2020 was 1.45 g/m3, CO concentration in DKI4 in 2019 was 1.10 g/m3 and 2020 was 0.95 g/m3, the concentration CO in DKI5 in 2019 was 1.30 g/m3 and in 2020 it was 1.12 g/m3. The decrease in the average concentration of CO in 2019 and 2020 can be seen in Table 4.

Table 4. 2020 average CO reduction

Station	Decrease	
	$\mu g/m^3$	%
DKI1	0,88	57,24
DKI2	0,08	9,59
DKI3	-0,06	-4,32
DKI4	0,15	13,64
DKI5	0,18	13,70

Table 4 informs that there was a decrease in the annual CO concentration in 2020 compared to 2019. The decrease in CO concentration in DKI1 was 0.88 g/m3 (57.24 %), DKI2 was 0.08 g/m3 (9.59 %), DKI3 of -0.06 g/m3 (4.32%), DKI4 of 0.15 g/m3 (13.64 %), DKI5 of 0.18 g/m3 (13.70%). However, the CO concentration in DKI4 in 2020 has increased by 0.06 g/m3 (4.32%) compared to the previous year.

In general, the decline in CO concentrations in 2020 occurred due to the work from home (WFH) program and large-scale social restrictions (PSBB) in DKI Jakarta Province, as shown in Fig. 5.

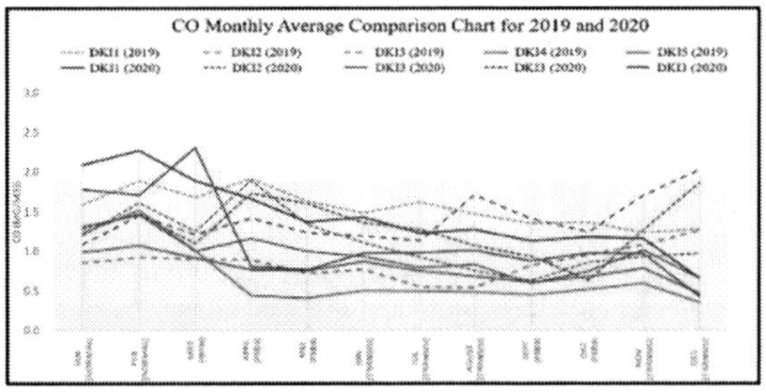

Fig. 5. Comparison graph of monthly average CO in 2019 and 2020

Overall, it can be seen that the average monthly CO concentration in 2020 is lower than in 2019. When calculated during the COVID-19 pandemic for the March to December 2020 period, the average CO at each air quality monitoring station can be seen in Table 5.

Table 5. Average CO reduction for the period March to December 2019 and 2020

Station	Decrease	
	Ug/m3	%
DKI1	0,98	65,34
DKI2	0,13	14,35
DKI3	0,13	9,21
DKI4	0,19	18,20
DKI5	0,27	21,56

Table 5 informs that in the period March to December 2019 and 2020 in DKI1 there was a decrease in CO concentration by 0.98 g/m3 (65.34 %), DKI2 by 0.13 g/m3 (14.35 %), DKI3 of 0.13 g/m3 (9.21 %), DKI4 of 0.19 g/m3 (18.20 %), and DKI5 of 0.27 g/m3 (21.56%).

3.3 Trend ISPU Year 2019 and 2020

In 2020 there was an increase in the number of days in the "Good and Moderate" category and a decrease in the days in the "Unhealthy and Very Unhealthy" category compared to the previous year (2019), as shown in Table 6.

Table 6. Comparison of ISPU in 2019 and 2020

Air Pollutant Standard Index	2019		2020	
	∑	%	∑	%
Good	2	1	29	8
Moderate	172	50	244	67
Not healthy	183	48	90	24
Very Unhealthy	8	2	3	1
Dangerous	0	0	0	0

In 2020 there was an increase in the number of days in the "Good and Moderate" category and a decrease in the number of days in the "Unhealthy and Very Unhealthy" category compared to the previous year (2019). Table 6 above informs that the Air Pollution Standard Index in 2019 with a good category of 2 days (1 %), moderate 172 days (50 %), unhealthy 183 days (48%), very unhealthy 8 days (2%), and dangerous 0 days (%). In addition, the Air Pollutant Standard Index in 2020 has a good category of 29 days (8%), moderate 244 days (67%), unhealthy 90 days (24%), very unhealthy 3 days (1%), and dangerous 0 days (0%).

4 Conclusion

The conclusion from the results that have been carried out shows that the COVID-19 pandemic has had a good impact on air quality in DKI Jakarta Province. This is evidenced by the decrease in PM2.5 concentration in the period March to December 2019 and 2020 in DKI1 by 22.96 g/m3 (42.25%), DKI2 by 7.56 g/m3 (16.83 %), and DKI3 of 14.31 g/m3 (27.45 %). The CO concentration in the same period showed a decrease in DKI1 by 0.98 g/m3 (65.34 %), DKI2 by 0.13 g/m3 (14.35 %), DKI3 by 0.13 g/m3 (9, 21 %), DKI4 was 0.19 g/m3 (18.20 %), and

DKI5 was 0.27 g/m3 (21.56%). The decrease in PM2.5 and CO concentrations has provided benefits to the air pollutant standard index in DKI Jakarta Province in the form of increasing the number of days in the "Good and Moderate" category and decreasing the days in the "Unhealthy and Very Unhealthy" category.

References

1. W. Salim, B. Kombaitan, The rise and challenge of a capital *City,* **13** ,120–8, (Jakarta: 2009).
2. R. Setiowati, H. S. Hasibuan, R. H. Koestoer, Green open space masterplan at Jakarta Capital City, Indonesia for climate change mitigation *IOP Conference Series: Earth and Environmental Science,* **200,** (IOP Publishing), 12042 (2018).
3. L. Dsikowitzky, W. S. A. Van der, F. Ariyani, K. J. Hesse, A. Damar, J. Schwarzbauer, Transport of pollution from the megacity Jakarta into the ocean: Insights from organic pollutant mass fluxes along the Ciliwung River *Estuar, Coast. Shelf Sci.* **215,** 219–28 (2018).
4. G. Baum, H. I. Januar, S. C. A. Ferse, A. Kunzmann, Local and regional impacts of pollution on coral reefs along the Thousand Islands north of the megacity Jakarta, Indonesia *PLoS One* **10** e0138271 (2015).
5. IQAIR, World Air Quality Index Ranking | Air Visual (2021)
6. K. Mommens, N. Brusselaers, L. T. Van, C. Macharis, A dynamic approach to measuring the impact of freight transport on air quality in cities *J. Clean. Prod.* **240** 118192 (2019).
7. S. Schmitz, L. Weiand, S. Becker, N. Niehoff, F. Schwartzbach, V. Schneidemesser, An assessment of perceptions of air quality surrounding the implementation of a traffic-reduction measure in a local urban environment *Sustain. Cities Soc,* **41,** 525–37 (2018).
8. R. M. Harrison, *Pollution: causes, effects, and control* (Royal society of chemistry, 2015)
9. I. S. Ponangsera, R. K. Apriyadi, D. Hartono, W. Wilopo, Identification of the Characteristics of COVID-19 on the Perception of the Number of Positive Cases, Recovered and Deaths Due to COVID-19 in Indonesia PENDIPA *J. Sci. Educ,* **5,** 277–83 (2021).
10. Diskominfotik-Jakarta, Air Pollution Standard Index (ISPU) 2019 - Open Data Jakarta
11. K. Midander, K. Elihn, A Wallén, L. Belova, A. K. B. Karlsson, I. O. Wallinder, Characterisation of nano-and micron-sized airborne and collected subway particles, a multi-analytical approach *Sci, Total Environ,* **427**, 390–400 (2012).
12. J. S. Apte, J. D. Marshall, A. J. Cohen, M. Brauer, Addressing global mortality from ambient PM2, 5 *Environ. Sci. Technol,* **49,** 8057–66 (2015).
13. X. Ji, Y. Yao, X. Long, What causes PM2. 5 pollutions? Cross-economy empirical analysis from socioeconomic perspective *Energy Policy*, **119**, 458–72 (2018).
14. S. W. Ryter, D. Morse, A. M. K. Choi, Carbon monoxide: to boldly go where NO has gone before *Sci. STKE* **2004** re6–re6 (2004).
15. S. Mahajan, S. Jagtap, Metal-oxide semiconductors for carbon monoxide (CO) gas sensing: A review *Appl. Mater. Today*, **18**, 100483 (2020).
16. A. Paliwal, A. Sharma, M. Tomar, V. Gupta, Carbon monoxide (CO) optical gas sensor based on ZnO thin films *Sensors Actuators B Chem,* **250,** 679–85 (2017).
17. M. S. Al-Moamary, A. S. Al-Shammary, A. A. Al-Shimemeri, M. M. Ali, H. H. Al-Jahdali, A. A. Awada, Complications of carbon monoxide poisoning *Saudi Med. J,* **21,** 361–3, (2000).
18. C-P. Lo, S-Y. Chen, K-W. Lee, W-L. Chen, C-Y. Chen, C-J. Hsueh, G-S. Huang, Brain injury after acute carbon monoxide poisoning: early and late complications *Am. J. Roentgenol,* **189,** W205–11, (2007).
19. K. Betterman, S. Patel, Neurologic complications of carbon monoxide intoxication *Handb. Clin, Neurol.* **120,** 971–9, (2014)

Risk analysis of production process for food SMEs using FMEA method: a case study

Prima Fithri[1*], *Muhammad* Rafi[1], Pawenary[2], and *A. S.* Prabuwono[3]

[1]Department of Industrial Engineering Faculty of Engineering, Andalas University, Indonesia
[2]Electrical Engineering, Faculty of Engineering, ITPLN, Indonesia
[3]Department of Information Technology, Faculty of Computing and Information Technology in Rabigh, King Abdulaziz University, Saudi Arabia

Abstract. The increasing development of the industry makes every industry have to compete with other competitors to gain an edge. The advantages of competition are influenced by several factors, one of which is good human resource management. Where if a company has good human resources, it will increase profits indirectly and can increase productivity. This research discusses case studies about the potential dangers of IKM Heppy Bakery's potential dangers that can harm workers in bread production. The method used is Failure Mode And Effect Analysis (FMEA). Later, the data will be filled and given a rating distinguished into three parts: severity, occurrence, and detection. The data were obtained through questionnaires given to 3 workers at IKM Heppy Bakery and filled in rating values based on the provisions that have been given to the questionnaire. This Value helps determine the Risk Priority Number (RPN) obtained from multiplication between severity, occurrence, and detection. After processing the RPN multiplication data, the highest RPN value was obtained by 193 with the danger factor of the operator overheating and dehydrating due to high temperatures. Furthermore, the calculation of critical Value was obtained by 109. Based on the critical Value obtained seven hazard factors above the critical value, these seven hazard factors need to be improved so that workers do not avoid accidents when conducting the production process.

1 Introduction

Currently, the government makes Small and Medium Industries (IKM) a focus in development. Therefore, Small and Medium Industries are one of the pillars of the country's economy, which is strong enough to fight the world economy that can use many resources to create jobs for organizations. In addition, Small and Medium Industries need training or socialization, especially those related to the safety and health of workers. Most workers who work in Small and Medium Industries have low jobs and irregular working hours, and the wages given are not sufficient for workers' needs. Some hazards found in the workplace are high work environment temperatures, poor workplace lighting, inhaled dust, and exposure to chemicals used. A form of business that can increase its productivity by having good resources, the company must pay attention to the safety and health of the workers as well as possible.

The safety and health of workers is an obligation that the company must consider (BBC NEWS, 2010; Ishimaru et al., 2020). Occupational safety and health must be considered because it will cause work accidents that impact worker productivity [3]. Small and Medium Industries (IKM) know the importance of Safety, Health, and Environment (SHE) in the business sector. They lack the SHE field that company owners must consider, which can be seen in IKM Happy Bakery. These SMEs pay less[†] attention to the application of SHE in the production process carried out by workers.

Potential hazards are often as operators who work often experience overheating and dehydration due to a narrow working space; operators are fatigued due to the position of items that are not ergonomic; operators do not wear protection when mixing mashing dough which can result in pinched hands. Based on the problems in the Happy Bakery IKM, the authors complete the case study to identify what risks can be encountered in the IKM during production. This is done so that IKM Heppy Bakery can apply the importance of Safety, Health, and Environment (SHE) in the work environment, increasing worker productivity and providing benefits for the company. This case study was completed by observing the potential work hazards in the production process to prevent work hazards arising from hazards during the production process at IKM Heppy Bakery.

The problem raised comes from a previous study that discusses determining priorities for overcoming work

* Corresponding author: primafithri@eng.unand.ac.id

accidents using the FMEA method [4]. The result of this study is that 12 kinds of work accidents occur at the research site. In addition, a previous study was conducted regarding the dangers of the machines using the FMEA method [9]. This research shows that the body plate component is the main hazard that affects the engine most. Based on previous research conducted by Paciarotti C et al. (2014), which discussed the A revised FMEA application to the quality control management, the results obtained were cost savings time efficiency (Paciarotti et al., 2014). In addition, based on previous research conducted by Doshi, J., and Desai, D. which discussed the application of Failure Mode & Effects Analysis (FMEA) For Continuous Quality Improvement Multiple Case Studies in Automobile, it was found that to identify potential failures that occurs in the production process and product quality can use the FMEA method [2].

From that previous study, this method can be used for this case study to know the potential priorities danger in this IKM.

2 Research Methodology

The stages of observing problem objects and methods used in case studies start from preliminary studies, literature studies, problem identification and formulation, data collection, data processing, analysis, and conclusions and suggestions.

A preliminary study is to identify the actual circumstances of the event at the company. This stage is a stage that conducts direct observations about the process of making bread at the IKM Heppy Bakery. This direct observation found a problem about the potential risk of harm to workers during the production process.

Literature studies are done hand in hand with field studies. This study is also helpful for solving problems found. In this literature study, all references related to field studies or problems will be resolved. References help formulate problems, data processing, and analysis. Some theories related to the problem are Risk, Risk Management, Damage, Maintenance, Failure Mode, and Effect Analysis (FMEA).

The procedure of problem identification is done by paying attention to the circumstances when processing bread production at IKM Heppy Bakery. The observations showed a risk that could potentially occur hazards when conducting the production process

The data was collected regarding potential hazards that can occur when workers during the production process. This potential hazard data is obtained from questionnaires filled by three factory workers. These three workers work on the grinding machine operator, dough mixer, and roasting, where they later fill out the questionnaire given. This data is needed to calculate the FMEA value and find the main Risk Priority Number (RPN) sequence that has the potential for hazards in doing the job.

After the data is collected, it will be done against the data. This research uses the FMEA method to analyze what risks are potentially dangerous when doing the production process. A rating does this FMEA method based on *severity, occurrence,* and detection *rating.* Furthermore, the RPN value is calculated to determine which priority is first improved. The formula used in calculating RPN Value.

$$RPN = S \times O \times D \tag{1}$$

Formula description:

RPN = Risk Priority Number

S = Severnity

O = Occurance

D = Detection

When you have obtained the RPN value, further calculation of critical Risk Priority Number (RPN) is used to see how many failure modes are above RPN critical Value and provide solutions to factors that risk the occurrence of hazards in work. The formula of the critical value is.

$$Critical\ Value = \frac{RPN\ Total}{Number\ of\ Risks} \tag{2}$$

The results of data processing are analyzed to evaluate the problems that have been solved. First, the analysis can be done by looking at the main potential that causes the hazard, then the sequence of potential hazards that can occur, then the critical value of the Risk Priority Number and suggestions for problem-solving against the risk of potential hazards exist.

3 Results and Discussion

The results and discussion obtained in this study are based on a data questionnaire given to 3 workers who do the process of bread production.

Respondent 1: Ana Laila (dough mixer machine operator)
Respondent 2: Khairul Azmi (grinding machine operator)
Respondent 3: Junior (Roasting Machine operator)

This questionnaire data contains rankings on factors that allow the occurrence of hazard risk when workers do the process of bread production in IKM Heppy Bakery is distinguished into three parts, namely *Severity, Occurrence,* and *Detection* that have been averaged can be seen in Table 1

After obtaining severity, occurrence, and detection values from 3 respondents, RPN calculation is done by multiplying three variables in the FMEA method that have been obtained and then done the process of sequencing RPN value to see the order of the highest causes of potential hazards in workers can be seen in Table 1.

Table 1. RPN Scores from3 Respondents

No.	Potential Hazards	Severity	Occurance	Detection	RPN
1	Operator overheating and dehydration due to high temperatures	7	7	4	196
2	Narrow work areas cause accidents while working	7	4	4	112
3	Hearing loss due to noise	5	6	5	150
4	Improper body position as the operator fatigues	6	5	5	150
5	Operator At Risk of burns in the grill	5	4	5	100
6	The operator is concerned when using the seat as a tool	5	5	6	150
7	Operator slips due to slippery floor	6	5	4	120
8	May be electrocuted due to irregular cable position	5	3	3	45
9	Inhaled dust causes disturbance to the smell	3	6	5	90
10	The operator's hand is at risk of getting pinched when smoothing the dough	4	2	6	48
11	Crash due to load stack overriding operator	6	4	6	144
12	Potential gas leakage due to improper gas location	7	2	2	28

Here's an example of a calculation of the RPN value:

$$RPN\ Scores\ 1 = S\ x\ O\ x\ D$$
$$= 7\ x\ 7\ x\ 4$$
$$= 193$$

Based on the sequence of RPN values of each factor that has the potential to cause harm to workers, further determining the critical Value of RPN to show what danger factors are above the critical value so that it needs to be evaluated improved quickly. Based on Table 1, as many as 12 factors are causing potential hazards with a total RPN value of 1313; based on the total value can be determined critical value RPN with the calculation using the formula:

$$\text{Critical Value} = \frac{Total\ RPN}{Number\ of\ Potential\ Hazards}$$
$$\text{Critical Value} = \frac{1313}{12}$$
$$\text{Critical Value} = 109$$

Based on the order of RPN value results in Table 1, seven factors have the potential for danger in work above the critical Value, namely with the values of 193, 151, 149, 142, 139,131, and 115, respectively. Therefore, seven factors above the critical Value need to be resolved, and a solution is sought in advance to overcome the occurrence of accidents at work.

4 Conclusions and Suggestions

The results of this research are 12 potential hazards identified in the process of bread production in IKM Heppy Bakery using the FMEA method consisting of three variables, namely severity, occurrence, detection, and calculation of critical risk level (RPN) that is prioritized in repair. The result of data processing that has been obtained is the order of RPN values from the most considerable Value to the smallest Value. The enormous value obtained was 193, with the potential danger of overheating and dehydration operators. At the same time, the smallest value of 31 with the potential danger of gas leakage due to the location of gas is not appropriate. After obtaining the RPN value sequence, then determined the critical value of 12 potential hazard factors. The critical Value obtained is 109, which indicates seven factors of potential hazards above the critical value. So, improvements need to be made in the form of solutions to overcome the seven potential hazards.

The solution that can be done to the potential dangers is to rearrange the layout of the machines so that in doing the work, the workers do not feel too overheated and the items placed to be more ergonomic so that workers do not easily experience fatigue when picking up goods. Workers should also pay attention to the condition of the floor of the production site so as not to slip during the production process.

Recommended suggestions for future research are: further research into identifying potential hazards can use methods other than the FMEA method. It should be necessary to make Standard Operating Procedures (SOP) related to the production work area at IKM Heppy Bakery. There need to be countermeasures to the repair solutions provided in IKM Heppy Bakery to increase productivity during bread production.

References

1. BBC NEWS.. *Policy-by-policy: The coalition government's plans*. BBC News Channel. (2010).
2. J. Doshi,., & D. Desai,.. Application of failure mode & effect analysis (FMEA) for continuous quality improvement - multiple case studies in automobile SMEs. *International Journal for QualityResearch*.02-07. (2017).
3. P. Fithri, N. A. Riva, L Susanti, & B. Yuliandra,. Safety analysis at weaving department of PT. X Bogor

using Failure Mode and Effect Analysis (FMEA) and Fault Tree Analysis (FTA). *2018 5th International Conference on Industrial Engineering and Applications, ICIEA 2018*,382–385. (2018).

4. G. Halvani., M. Mohammadnia, M. Zare Mehrjardi, & M. Keshtvarz. Risk assessment with intervention of gas pipelines repair using FMEA method and determination of safety performance monitoring: a case for a gas and oil operating company in 2012-13. *Occupational Medicine*.(2012).
5. G. B. HM. Analysis of Occupational Safety and Health in the Production Process Section Using the Failure Mode And Effect Analysis (FMEA) Method at PT Mitra Mandiri Perkasa. *Jurnal IPTEK*, *25*(1). (2021).
6. T. Ishimar, H. Hirosato, T. Mori, N. Ito, K. Hiraoka, R. Minohara, S. Kajiki, M. Uehara, Y. Kobayashi, & K. Mori. Systems and human resource development for occupational health in India: Effective occupational health management for Japanese enterprises with overseas branches. *Sangyo Eiseigaku Zasshi = Journal of Occupational Health*, **62**(3).(2020).
7. M. Mangeli, A. Shahraki, & F. H. Saljooghi, Improvement of risk assessment in the FMEA using nonlinear model, revised fuzzy TOPSIS, and support vector machine. *International Journal of Industrial Ergonomics*, *69*. (2019).
8. Paciarotti, C., Mazzuto, G., & D'Ettorre, D. A revised FMEA application to the quality control management. *International Journal of Quality and Reliability Management*, **31**(7). (2014).
9. R. F. Ramadhan, E. Widowati, & M. Mardiana. Failure Mode and Effect Analysis (FMEA) Application for Safety Risk Assessment Design of "X" Bakery. *Unnes Journal of Public Health*, **8**(1). (2019).

Managing policies for stunting prevention during the COVID-19 disaster: the case of West Sumatra Province

Azwar[1*], *Muhammad* Ichsan Kabullah[2,3], *Lucky* Zamzami[4], and *Rifki* Dermawan[5]

[1] Department of Sociology, Faculty of Social and Political Sciences, Andalas University
[2] Department of Public Administration, Faculty of Social and Political Sciences, Andalas University
[3] Centre of Constitutional Studies (PUSaKO), Faculty of Law, Andalas University,
[4] Department of Anthropology, Faculty of Social and Political Sciences, Andalas University
[5] Department of International Relations, Faculty of Social and Political Sciences, Andalas University

Abstract. The Covid-19 pandemic has had major implications for government administration. Policymakers set policies for handling Covid-19 as a priority agenda at this time. However, these measures have reduced other problems such as stunting which still requires government attention. This article explores policy management of stunting by the local government amid the Covid-19 pandemic with the West Sumatra province case. The findings in the field, show that the implementation of stunting prevention policies is not easy considering the powerlessness caused by budget refocusing and overlapping authorities between institutions. Meanwhile, the meaninglessness is indicated by the low level of public support for the handling of stunting. This is exacerbated by a bureaucratic culture that is slow to implement policies, including in managing stunting-related programs during the Covid-19 crisis. So that there is alienation or what we call as alienation in the bureaucracy in implementing stunting prevention policies.In the end, the management of stunting policies during the Covid-19 pandemic should receive strong support from all involved parties.

1 Introduction

The Covid-19 pandemic has put heavy pressure on all aspects of life. Although the Covid-19 pandemic is classified as a non-natural disaster, the implications it causes are also experienced in government administration where handling Covid-19 has become a policy priority. This is inseparable from the trend of Covid-19 cases, which continues to increase where there are 253,286 people affected by Covid-19 with a death toll of 58,995 people on July 1, 2021 [1]. The government has done its best in handling Covid-19. This is shown through a series of massive regulations and policies. At the central government level, President Joko Widodo has carried out a series of policies such as the establishment of the Task Force for the Acceleration of Handling the Corona Virus through Presidential Decree (Keppres) Number 7 of 2020, which is headed by the Head of the National Disaster Management Agency. The government has also designated Covid-19 as a non-natural national disaster) [2]. In addition, the government has also implemented lockdown in temporary (PKMM) to contain the spread of Covid-19 cases. In the local government, several regional heads are encouraged to do their best to handle Covid-19, starting from refocusing the budget, increasing social spending, to trimming the regional government's routine budget such as the official travel budget [3].

It should be realized that disasters such as the Covid-19 pandemic are situations that will always be faced by the bureaucracy and cannot be avoided considering that no one knows when the Covid-19 pandemic will end. This is termed VUCA (Volatile, Uncertain, Complex, and Ambiguous) a terminology that describes the dynamic, fluctuating, uncertain, complex, and ambiguous multilateral world conditions [4]. It is just that in the discussions that developed, some parties began to realize that an excessive response to Covid-19 could reduce the government's presence in providing guarantees for fulfilling people's welfare in other sectors. Moreover, by looking at the characteristics of the public bureaucracy that has been imaged so far which is full of rigidity, stuck with routines, and procedural, the task of the bureaucracy to accommodate issues other than the Covid-19 pandemic is becoming increasingly difficult [5].

For West Sumatra province itself, one of the problems that have not been resolved by the local government is stunting. Based on the results of the Nutrition Status Monitoring (PSG) in 2015-2017, there was an increase in the prevalence of stunting from 25.5% in 2016 to 30.6% in 2017 [6]. In 2018, based on Basic Health Research (Riskesdas) found that the stunting prevalence rate in

* Corresponding author: azwar@soc.unand.ac.id

West Sumatra Province increased to 30.8% [7]. The three-year trend shows that there has been no progress made by local governments in implementing stunting prevention policies. Indeed, it appears that the numbers stunting prevalence rate in West Sumatra are just equal to the national average of stunting prevalence 30.8% in 2018. Therefore, stunting prevention policies must be carried out continuously. All parties, especially policymakers in the government can interpret the stunting issue should receive similar attention with the Covid-19 pandemic.

From the government's point of West Sumatra province, much pressure was put on the bureaucracy in policy implementation of stunting prevention during a pandemic Covid-19. Excessive reporting on Covid-19 has eliminated the problem of stunting as a priority issue in government administration. Even if stunting is to be put forward as a priority agenda by the provincial government, there is a tendency to reject it from the public. This can be seen in the labeling of the Governor of West Sumatra Province as a figure who is considered to have less concern in handling Covid-19 [8]. This indicates that there is great pressure experienced by policymakers if they want to prioritize issues other than Covid-19. Whereas the problem of stunting is acute and has not been completely resolved by the local government. With the enormous pressure faced by policymakers in managing stunting prevention policies, there should be an adaptation in the bureaucracy.

In this article, we would like to identify the factors that hinder the implementation of stunting prevention policies during the Covid-19 disaster. The conceptualization built departs from the theory of policy alienation, which allows us to identify the level of alienation of policymakers in implementing stunting policies during the Covid-19 pandemic. This is theoretically considered relevant considering that the incremental stunting prevention policy is often seen only in a limited way. Even though several other factors create the alienation of stunting policies, which should also be the subject of a thorough analysis by policymakers. In terms of method, the author uses a qualitative method with a case study approach at the West Sumatra Provincial Health Office, by looking in detail at regulatory factors, budgets, and policies

2 Problem Statement

What are the factors that hamper the West Sumatra Provincial Government from implementing stunting prevention policies during the Covid-19 disaster?

3 Research Methods

To answer the problem statement above, this study will use qualitative methods. The reason for using qualitative methods is the existence of standardized research procedures, thus, they can assist in finding descriptive data both written and oral from the observed phenomena [9]. Qualitative methods are also effective in exploring phenomenon about what is experienced by research subjects such as behavior, perception, motivation, action, and others. For these reasons, qualitative methods are very relevant in identifying the factors that hamper the West Sumatra Provincial Government in implementing stunting prevention policies during the Covid-19 disaster. The data collection was carried out at institutions directly related to stunting prevention policies, including the National Population and Family Planning Agency (BKKBN) of West Sumatra Province, West Sumatra Province Health Office, and West Sumatra Province Social Service. The duration of data collection itself is carried out over a period of two months, from July to August 2021.

4 Theoretical Framework

Various policy studies have been conducted in analyzing the problem of stunting in Indonesia. A study from Nisa [10] explains that although stunting prevention is regulated by many regulations, its implementation is not followed by follow-up at the local level to the village level. Another study conducted by Azwar [11] and colleagues describes various social policy packages made by the government as a model for the welfare state that can have an impact on reducing stunting cases. These are the various policies contained in social policy packages such as the Family Hope Program (PKH), Community Health Insurance (Jamkesmas), Smart Indonesia Card (KIP), Home Renovation Assistance Program, and others become part of the indirect intervention in stunting. Tampubolon's [12] also revealed that coordination, funding, and public participation, especially from village government officials, influenced the implementation of stunting prevention policies. Based on these studies, it is clear that stunting prevention policies are still running partially and have not been able to bind many parties.

In this article, we will analyze the factors that hinder the West Sumatra Provincial Government in the implementation of stunting prevention policies using the conceptual framework alienation. The conceptualization of policy alienation departs from the philosophy of a cognitive state faced by implementers who feel alienated from the policy program being implemented [13]. This usually occurs in a complex issue and there is strong pressure from many parties. At the same time, the alienation policy theory is also relevant for policymakers because they can psychologically identify the effectiveness of policy [14]. This is exacerbated by the absence of an accurate database from the local government for people who deserve to receive assistance, a situation that indicates alienation from stakeholders in implementing policies.

In the first variable, powerlessness can be defined as a person's lack of control over the events that are happening in front of them [15]. Seeman [16] defines powerlessness as "the expectation or probability held by the individual that his behavior couldn't determine the outcome of what he has done". In the realm of policy implementation, powerlessness is related to the level of influence that bureaucrats have in shaping policy programs. While the second variable in policy alienation is meaninglessness [17]. Sarros and colleagues [18] define meaninglessness as "the inability to understand the relationship of one's contribution to a larger goal". Further explanation of the meaninglessness in policy implementation can be seen from the level of support at the community and client/ customer levels. Through the use of policy alienation as a conceptual framework, this study is expected to provide another perspective in analyzing the implementation of stunting prevention policies.

5 Results

5.1. Overview of Stunting in West Sumatra Province

The issue of stunting requires great attention for policymakers considering the negative impacts it causes. According to a study by the Health Research and Development Agency (Balitbangkes) of the Ministry of Health, stunting in other words kerdil or stunted children has real implications for poor educational achievement and low income as adults [19]. Another study conducted by Cahyono and colleagues [20] also revealed that stunting has an impact on reducing the IQ of children in Indonesia by 10 to 15 points, the lack of academic achievement of children and in the long term will exacerbate poverty due to low levels of competence in entering the workforce. Stunting is also likened to a hidden tragedy where the impact does not only affect low intelligence levels but also morbidity and mortality in toddlers, low intellectual and cognitive abilities, and causes degenerative disease problems [21].

Table.1.Statistics of stunted toddlers based on regency/city in West Sumatra province Year 2007 - 2019

No	District/City	Riskesdas 2007	Riskesdas 2013	Riskesdas 2018	SSGBI 2019
1	Mentawai Islands	32.6	40.9	32.4	43.3
2	Pesisir Selatan	37.4	50.0	26.5	30.56
3	Kab. Solok	36.2	42.0	41.5	35.45
4	Sijunjung	46.1	37.2	35.0	30.62
5	Tanah Datar	39.9	38.8	26.3	23.31
6	Padang Pariaman	34.9	41.0	26.1	21.29
7	Agam	36.1	30.0	30.0	26.64
8	Limapuluh Kota	29.8	28.8	40.1	29.94
9	Pasaman	48.6	55.2	40.6	35.67
10	Solok Selatan	34.7	48.6	27.7	34.99
11	Dharmasraya	30.4	40.4	30.3	26.37
12	Pasaman Barat	45.2	51.5	35.1	31.66
13	Padang	30.9	33.7	22.4	20.92
14	Solok	39.4	30.7	23.6	20.73
15	Sawahlunto	31.2	22.5	20.6	23.43
16	Padang Panjang	34.4	28.5	27.2	13.55
17	Bukittinggi	31.0	28.5	17.2	23.46
18	Payakumbuh	44.4	30.4	19.2	22.3
19	Pariaman	23.3	44.0	17.8	20.78
	West Sumatra	36.5	39.2	29.9	27.47
	National	36.8	37.2	30.8	27.67

Categories of Public Health Problems stunting Prevalence according to WHO	
1. Low (Prevalence <20%)	
2. Moderate (Prevalence between 20 - 29.9%)	
3. High (Prevalence between 30 - 39.9%)	
4. Very High (Prevalence above 40%)	

The negative impacts of stunting will be difficult to manage amid the catastrophic pandemic Covid-19 experienced at this time. With the increasing economic burden on families due to the Covid-19 pandemic, it is believed that the poverty experienced by many families can directly affect the increasing malnutrition that children receive. When a child is malnourished, it gradually fails to achieve a normal and healthy height according to the child's age. Therefore, the problem of stunting cannot be intervened locally from a health point of view alone, but as a complex issue that needs to be tackled together.

At the practical level, stunting prevention policies have been carried out across sectors and involve various stakeholders. Stunting prevention efforts implemented by the central government are usually conducted at the macro level with massive scale such as the provision of vitamins and supplementary food, free immunization for toddlers, education about nutrition, and provision of National Health Insurance (JKN), Healthy Indonesia Card (KIS) to the implementation of the Family Hope Program (PKH) [22]. At the local government level, stunting management is carried out through complementary interventions such as optimizing Posyandu in promoting environmental health, reducing poverty, and empowering women [23]. From these various policy interventions, it can be seen that the government in a structured manner starting from the central government and local governments have taken various steps in preventing stunting.

Unfortunately, the stunting problem has not been completely resolved by the government in the West Sumatra Province. This is illustrated by the data on stunting under five in the last thirteen years as shown in table 1

`5.2. Alienation in Stunting Policies

It is interesting to see the role of local governments in particular the Government of West Sumatra province in implementing prevention policies of stunting. This departs from the reality that the budget capacity of the provincial government is stronger than that of the regencies/municipalities. In addition, with a smaller number of affairs, as stated in Law Number 23 of 2014, the provincial government should be able to fill the resource gap of regencies/ municipalities in the implementation of stunting prevention policies during the Covid-19 disaster. However, this is not easy to realize given the powerlessness and meaninglessness experienced by policymakers.

Powerlessness in policy can be interpreted as the inability of policymakers to control events that occur [24]. In the case of stunting prevention policies in West Sumatra Province, the inability of policymakers in the Provincial Government to accelerate stunting prevention policies is due to budget changes and overlapping authorities between institutions. On the budget side, pandemic Covid-19 has led to changes in budget allocation of every government agencies President Joko Widodo has ordered every provincial government and regencies/municipalities to participate in helping to overcome Covid-19 in their regions, such as by refocusing and reallocating the local budget (APBD), forming regional task forces and increasing social assistance. In the West Sumatra Province case, the provincial government also followed the president's instruction in refocusing and reallocating the budget. Several regulations that require provincial governments to refocus their budgets include Government Regulation in place of Law (Perppu) Number 1 of 2020, Decree of the President of the Republic of Indonesia Number 11 of 2020, Regulation of the Minister of Finance Number 19/PMK.07/2020, Minister of Home Affairs Regulation Ministry of Home Affairs Number 20 of 2020, and Instruction of the Minister of Home Affairs Number 5 of 2020. Through those changes, every government area including West Sumatra Provincial Government needs to do refocusing and re-allocation the budget to support the handling of Covid-19. The provincial government also responded by refocusing the budget for handling Covid-19 to reach Rp. 541.2 billion [25]. In other words, as a percentage of the budget, programs/activities for handling Covid-19 received a proportion of 7.3% of the 2020 West Sumatra Provincial Budget.

In 2021, Provincial Budget still refocusing on the handling of Covid-19 as mandated by the Minister of Home Affairs Regulation Number 64 of 2021. This results in limited budget support for stunting prevention policies. In 2021, the budget allocated West Sumatra Provincial Government for the prevention of stunting only Rp. 148,030,000. With this budget amount, it is difficult to fulfill the target of reducing the stunting prevalence rate nationally by 14% in 2024. The stunting prevalence rate in West Sumatra Province in 2019 still reached 27.47%, so a 13% reduction is needed in the next five years.

As for the overlapping authority between institutions, it can be seen from the lack of regulation in determining the main actors who implement stunting prevention policies in West Sumatra Province. At this moment, there is no regional regulation (Perda) at the provincial level that specifically contains the stunting issue. The existence of the Regional Regulation can be a reference for policymakers in implementing stunting prevention policies. The existing regulations related to stunting are the Decree of the Governor of West Sumatra Province Number 440-380-2020 concerning the Establishment of the Team for the Acceleration of Prevention and Handling of stunting in the Province of West Sumatra in 2020, where the Head of the Regional Development Planning Agency (Bappeda) of West Sumatra Province is appointed as the head of the stunting prevention management team. By choosing Bappeda as the leading sector makes no guarantee of the successful implementation of prevention policies stunting in the province of West Sumatra. Bappeda does not have the power of direct execution as owned by other local government organizations such as the health service office and social services office.

The problem of overlapping authority between institutions is also due to the unclear regulations that regulate the authority of institutions. This was explained by one of the informants as follows:

"... We have a lot of work in this government and not only dealing with Covid. We are also obliged to take care of stunting which is still high... There is still a transition between the health office and the BKKBN, and for the time being there is only a Governor's Decree regarding the acceleration of stunting handling. In the decree, it is stated which OPD is involved in planning stunting activities. " (NS Interview, West Sumatra Provincial Health Office, 22/7/2021)

Based on the interview above shows that BKKBN transition and the health department can not be separated from the discourse of the President in accelerating prevention policy stunting where BKKBN gave a greater role in the management of stunting. The Government of West Sumatra province does not only do the authority of stunting prevention but most of the authority is also taken by the BKKBN. Unfortunately, it was not followed up through more concrete rules. The result is a lack of clarity in the division of responsibility between the provincial governments and the BKKBN so that it can trigger a conflict that has implications for incremental policy.

The meaninglessness can be interpreted as the level of individual understanding of a policy [26]. If the level of individual understanding of a policy is low, it will naturally give birth to resistance in the form of omission and rejection of a policy. Conversely, if the level of individual understanding of a policy is high, support for a policy will appear.

In the context of the stunting prevention policy in West Sumatra Province, meaninglessness is indicated by the low public attention to the stunting issue because it is more diverted to the Covid-19 issue. The Governor of West Sumatra Province, who was considered less responsive in handling Covid-19, as has been reported in the mass media and social media recently, also experienced this. The Governor of West Sumatra Province further stated as follows:

"... We have a lot of work in this government and not only dealing with Covid. We are also obliged to take care of stunting which is still high... There is one area, most of the residents are under 150 cm. In the holiday (Eid al-Adha), they have never seen a buffalo. Bringing the buffalo three days and three nights there from Lubuk Sikaping because there is no road access. If through water need three hours. I and my friends from Terabas went to those areas." (Interview with MA, the Governor of West Sumatra province, 29/07/2021)

The statement from the Governor of West Sumatra Province emphasized that the provincial government still have attention in handling stunting during the Covid-19 period. Unfortunately, the attention of policymakers on the stunting issue can be eliminated given the great pressure from the public in the Covid-19 crisis. Concerns arise over the implementation of stunting prevention policies during the Covid-19 situation, forcing every policymaker to be prepared to face resistance that arises from the public when fighting for stunting as an unpopular agenda. Therefore, it is important to build awareness of all parties that the government's task is not only to deal with Covid-19. It takes tactical steps as impartial news and campaigns in the mass media.

The meaninglessness of stunting prevention policies during the Covid-19 pandemic is also influenced by the characteristics of the bureaucracy running very slowly or termed decision-making delays, namely the time lag between the formulation of a policy and its implementation [27]. In the case of stunting, the meaninglessness faced by provincial government policymakers is also due to the inadequacy of government officials to maintain the sustainability of stunting prevention policies. The bureaucracy seems to be run slowly and rigidly, which represents the ongoing Weberian bureaucracy model [28]. At the same time, meaninglessness is indicated by the lack of space for public participation, especially for women. Although the Minang as the biggest ethnic in West Sumatra adheres to a matrilineal system by prioritizing the role of women, the various gender agendas are difficult to implement due to the domination of patriarchy in the bureaucracy [29]. Whereas the position of women in the issue of stunting can be both an object and a subject that can support policy changes. With the participation of women, it can be a force in encouraging the acceleration of the implementation of stunting prevention policies.

6 Conclusion

For every policymaker, it is not easy to escape from public pressure if an issue is deemed unpopular. In the context of stunting in West Sumatra, stunting prevention policy stunting is hard to be implemented since the Covid-19 pandemic has become a major issue, and crucial at this time. The alienation of stunting prevention policies is also marked by the powerlessness of the provincial government, especially in terms of budget and inter-agency coordination and the existence of meaninglessness, especially the minimal support from the wider community. Therefore, the West Sumatra Provincial Government must build a strong will to collaborate with relevant parties to implement stunting prevention policies. In addition, efforts are needed to build a more adaptive bureaucratic management in responding to the stunting problem.

This article is one of the research outputs for the Acceleration of Professorship (PGB) of the Faculty of Social and Political Sciences, Andalas University in 2021. We especially thank all informants for this research. We also express our deepest gratitude to the lecturers and students at the Faculty of Social and Political Sciences, Andalas University who have supported this research. We also give appreciation to reviewers for their constructive input in improving the article. Lastly, we also expressed appreciation to the committee organizing the 2nd International Conference on Disaster Mitigation and Management (ICDMM) who still held the conference during the Covid-19 pandemic.

References

1. https://www.covid19.go.id/situasi-virus-corona/
2. https://tirto.id/update-corona-indonesia-covid-19bencana-nasional-case-reach-117-eFq1

3. M.I. Kabullah, S. Sudarmoko, H. Koeswara, Rahman, F., & Azwar, A. *Dysfunction of Collaborative Governance in The Handling Policy of Covid-19* at *Jambi Province.* Jurnal Sosial Humaniora, **14** ,1, 99-111 (2021)
4. E. A. Purwanto, 2019. Agile and Innovative Public Policy in Winning Competition in the VUCA Era (Volatile, Uncertain, Complex and Ambiguous). Inauguration Speech for Professor of Public Policy Gadjah Mada University: Yogyakarta.
5. A. Codreanu, A VUCA Action Framework for A VUCA Environment: Leadership Challenges and Solutions. Journal of Defense Resources Management, **7**, 2, 31–38 (2016)
6. M. Syafrina, M. Masrul, & F. Firdawati, Analysis of the Padang Pariaman Regency Government's Commitment in Overcoming Stunting Problems Based on the Nutrition Commitment Index. Andalas Health Journal, **8** (2), 233-244 (2019)
7. https://www.harianhaluan.com/news/detail/72257/stunting-sumbar-masih-di-atas-20-persen
8. https://langgam.id/dokter-dan-pengamat-kritik-sikap-gubernur-sumbar-hadapi-covid-19/
9. W. John, Qualitative Inquiry & Research Design: Choosing Among Five Approaches. SAGE: Thousand Oaks, California (2013).
10. L. S. Nisa, Stunting Prevention Policy in Indonesia. Journal of Development Policy, **13**, 2, 173-179 (2018)
11. A. Azwar., Chatra, E., & Zuldesni, Z. 2019. Welfare State Policy Model as A Poverty Reduction Strategy in the West Sumatra Districts and Cities. Journal of Anthropology: Socio-Cultural Issues, **21**, 2, 227-235
12. D. Tampubolon, Kebijakan Intervensi Penanganan Stunting Terintegrasi. Jurnal Kebijakan Publik, **11**. 1, 25-32. (2020)
13. L. G. Tummers, V. J. J. M. Bekkers, & A. J. Steijn, Policy alienation of public professionals: Application in a new public management context. Public Management Review, **11**, 5, 685-706. (2009)
14. M. I. Kabullah, H. Koeswara, & D. Rahmadi, Alienasi Kebijakan Anggaran Provinsi Riau dalam Pencegahan dan Pengendalian Kebakaran Hutan dan Lahan. Jurnal Administrasi Dan Kebijakan Publik, **5**, 2, 114-133. (2020)
15. L. Tummers, Policy Alienation: Analyzing the Experiences of Public Professionals with New Policies. Ph.D. Thesis Erasmus University Rotterdam: Rotterdam. (2012)
16. M. Seeman, On the meaning of alienation. American Sociological Review, **24**, 6, 783-791. (1959)
17. L. Tummers, Policy Alienation: Analyzing the Experiences of Public Professionals with New Policies. Ph.D. Thesis Erasmus University Rotterdam: Rotterdam. (2012)
18. J. C. Sarros, G. A. Tanewski, R. P. Winter, J. C. Santora, & I. L. Densten, Work alienation and organizational leadership. British Journal of Management, **13**, 4, 285-304 (2002)
19. T. Trihono, A. Atmarita, D.H. Tjandrarini, A. Irawati, I. Nurlinawati, N. H. Utami, & T. Tejayanti, Short (stunting) in Indonesia, problems and solutions. Jakarta: Publishing Agency for Research and Development Agency. (2015)
20. F. Cahyono, S. P. Manongga, & I. Picauly, Determinants of stunting in children under five in various ecosystem zones in Kupang Regency. Journal of Nutrition and Food. **11**. 1. (2016)
21. N. K. Aryastami, Policy studies and overcoming stunting nutrition problems in Indonesia. Indonesian Bulletin of Health Research, **45**, 4, 233-240 (2017)
22. N. F. Rahmawati, N. A. Fajar, & H. Idris, Social factors, economics, and utilization of posyandu with the incidence of stunting under five of poor families receiving PKH in Palembang. Indonesian Journal of Clinical Nutrition, **17**, 1, 23-33 (2020)
23. L. S. Nisa, Stunting Prevention Policy in Indonesia. Journal of Development Policy, **13**, 2, 173-179. (2018)
24. L. Tummers, Policy Alienation: Analyzing the Experiences of Public Professionals with New Policies. Ph.D. Thesis Erasmus University Rotterdam: Rotterdam. (2012)
25. https://dprd.sumbarprov.go.id/home/berita/1/1270
26. L. Tummers, Policy Alienation: Analyzing the Experiences of Public Professionals with New Policies. Ph.D. Thesis Erasmus University Rotterdam: Rotterdam
27. T. Jovanovski, & M. Muric, The Phenomenon of Lag in Application of the Measures of Monetary Policy. Economic Research-Ekonomska Istraživanja, **24**, 2, 154–163. (2012)
28. M. I. Kabullah, The Limits of Weberian on Anti-Corruption Approaches in the Indonesian Municipalities. IOP Conference Series: Earth and Environmental Science, **175**. (2018)
29. M.I. Kabullah, & M. N. Fajri, Neo-ibuism in Indonesian politics: election campaigns of wives of regional heads in West Sumatra in 2019 Journal of Current Southeast Asian Affairs **40** (1): 136-155. (2021)

Removal of nitrate using modified pumice as adsorbent for reducing groundwater pollution

Shinta Indah[1*], *Denny* Helard[2], *Gusti* Lusiani[1], and Diana Hapsari[2]

[1]Department of Environmental Engineering, Faculty of Engineering, Andalas University, Padang 25163
[2]Departemen of Applied Life Science, Faculty of Applied Biological Science, Gifu University

Abstract. Nitrate adsorption onto the physically and chemically modified pumice was investigated as an effort for reducing groundwater pollution. The treatments were heating at temperatures of 300°C, 450°C, and 600°C for physical and soaking in acid solutions (HCl, H_2SO_4, and HNO_3) for chemical treatments. The adsorption was performed in a batch system at room temperature (25±1°C) with the optimum condition (pH 4; 0.3 g/L of adsorbent dose; <63 μm of adsorbent diameters and 30 minutes of contact time). The results indicated that the physically and chemically modified pumice resulted in increasing removal efficiency and nitrate uptake compared to the natural pumice. The highest removal efficiency and nitrate uptake were achieved from 300°C of heating temperatures (62.04% and 155.09 mg/g) and H_2SO_4 for the acid solution (83.30% and 208.25 mg/g), while by using the natural pumice only reached 57.02 % and 142.55 mg/g. The SEM images of the modified pumice confirmed the change in the surface morphology of pumice including the pore structure and surface area which can be proper sites for adsorption of pollutants. This study demonstrated that physical and chemical modification could be the potential treatment to increase the removal efficiency and nitrate uptake of the natural pumice, thus can solve the problem of groundwater pollution.

1 Introduction

Due to its harmful effect, contamination of nitrate in groundwater has become a serious environmental problem. In water resources, the most common sources of nitrate are the discharge of raw wastewater, biodegradation of nitro-organic compounds, septic tank effluents, fertilizers, and pharmaceuticals. In groundwater, nitrate is possibly the most widespread contaminant, causing a serious threat to drinking water supplies due to its high water solubility [1-3]. High nitrate concentrations in drinking water cause health problems, such as cancer due to the formation of nitrosamine and cyanosis among children. Therefore, the US Environmental Protection Agency (EPA) and World Health Organization (WHO) set the maximum acceptable contaminant level of nitrate for drinking water to be 45 mg/L and 50 mg/L of nitrate, respectively [4].

Adsorption is an effective technique for nitrate removal, as it is an effective alternative for removing ions from water, economical, easy design, and operation as well as low costs [5-7]. For nitrate removal, different adsorbents have been studied [8] such as geomaterials like kaolinite [9], and pumice [10, 11]. Since in several cases the geomaterials have lower adsorption capacity, various methods have been proposed for the modification of geomaterials to improve their adsorption capacity. Physical treatment by heating process and chemical treatment by using chemicals including acids, bases, minerals, and the oxidizing agent has been evaluated for adsorbent modification. Those treatments proved could improve the adsorption capacities of the adsorbent due to the changes in the pore structure, surface chemical functional groups, and specific surface area of the adsorbent. [12].

Pumice is one of natural pozzolan which has low weight, porous structure, and a large surface area. In addition to its applications in the construction industry, researchers in the field of water and wastewater treatments have been studied a possible extension of the pumice to become an adsorbent [5, 13-15]. The objective of the present study is to evaluate the adsorption capacity of physically and chemically modified pumice on nitrate removal. This is one of the efforts for reducing or solving the groundwater pollution caused by nitrate.

2 Material and Methods

2.1 Reagent

Nitrate stock solution was provided by dissolving potassium nitrate (KNO_3) in distilled water. The acid solutions i.e. HCl, H_2SO_4, and HNO_3 were used for chemical modification of the pumice

* Corresponding author: shintaindah@eng.unand.ac.id

2.2 Modification of Pumice

The pumice was collected from the riverside of Sungai Pasak, West Sumatera, Indonesia. To remove any impurities, pumice was washed with distilled water and dried at room temperature (25±10°C). The pumice was ground and sieved to obtain particle size fractions of <63 μm before modification. A physical modification was conducted by heating at a temperature of 300°C, 450°C, and 600°C for 3 h. Chemical modification was carried out by soaking and stirring the pumice HCl 1 M, H_2SO_4 1 M, and HNO_3 1 M for 4 h and then washing by distilled water and drying at 130°C for 3 h.

2.3 Batch adsorption experiment

Batch adsorption experiment was conducted at room temperature (25±1°C) with the optimum condition: pH 4; <63 μm of adsorbent diameters; 0.3 g/L of adsorbent dose and 30 minutes of contact time. These conditions were obtained from the previous research. In each experiment, 100 mL of nitrate solutions with 75 mg/L of initial concentration were treated with seven kinds of adsorbents in a set of Erlenmeyer flasks. A shaker machine was used to shake them at a speed of 100 rpm for 30 minutes of contact time. Then, the mixture was filtered through a 0.45 μm membrane filter to separate the adsorbents and supernatant. The concentration of nitrate in the supernatants was measured using spectrophotometer UV-Vis (Shimadzu, UV-2600). The difference between the initial and final nitrate concentrations of the solutions was calculated as the number of nitrate ions adsorbed by the pumice. All experiments were conducted in triplicate and the averaged values of replicate tests were presented as the results.

The following mass-balance equations were used to calculate the removal efficiency and the nitrate uptake (q_e, mg/g) on natural and modified pumice:

$$\boldsymbol{Removal}\ (\%) = \frac{C_0 - C_e}{C_0} \times \mathbf{100\%} \ .... (1)$$

$$q_e = \frac{C_0 - C_e}{m} \times \boldsymbol{V} \ .... (2)$$

where C_0 is the nitrate's initial concentration (mg/L), C_e is the nitrate's equilibrium concentration (mg/L), V is the volume of the solution (L), and m is the mass of the pumice (g).

2.4 Analysis of surface morphology of pumice

a scanning electron microscopy (SEM, model S-3400N, Hitachi, Japan) was employed to observe the surface morphology of pumice.

3 Results dan discussion

3.1 Physically modified pumice

The removal of efficiencies and nitrate uptakes of natural and physically modified pumice are shown in Fig. 1. Physical modification of pumice was carried out by heat treatment at a temperature of 300°C, 450°C, and 600°C. The nitrate removal efficiency by using the natural pumice was 57.02%, while by the modified heating pumice, the removal efficiency reached 62.04%, 57.24%, and 29.86%, respectively at 300°C, 450°C, and 600°C. The nitrate uptake obtained by using the natural pumice was 142.55 mg/g, whereas by modified heating pumice at 300°C, 450°C, and 600°C were 155.09 mg/g, 143.11 mg/g, and 74.64 mg/g, respectively. It was observed that the removal efficiency and nitrate uptake increased in using modified pumice by heating at 300°C and 450°C. However, the lowest removal efficiency and nitrate uptake were found at the modified heating pumice at 600°C.

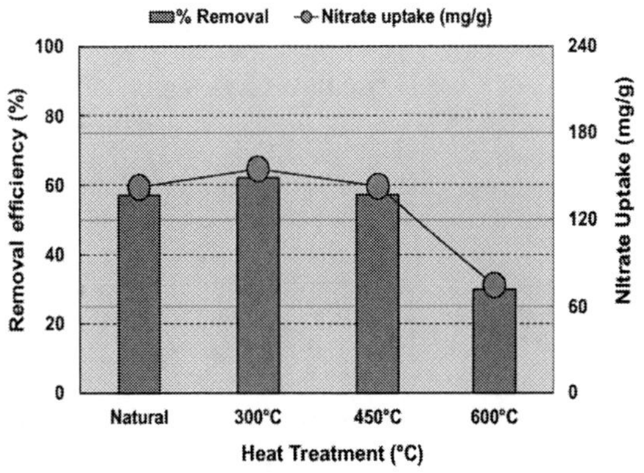

Fig 1. Removal efficiencies and nitrate uptakes of natural and physically modified pumice.

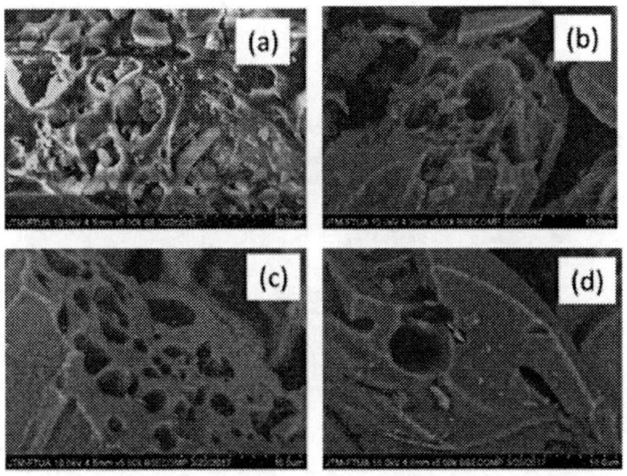

Fig 2. The SEM images (5000x) of natural pumice (a) and physically modified pumice by heat treatment at 300°C (b), 450°C (c), and 600°C (d).

Generally, physical modification results in the enhancement of physical characteristics (total pore volume and BET area) of the adsorbent. The calcining process or heat treatment causes a change in the porosity and the chemical composition of the adsorbent surface. The heat treatment also revealed a significant increase in pore volume and specific surface area of the adsorbent and may increase its adsorption ability [16]. Nevertheless, the higher temperature leads to damage in the adsorbent structure and affects the adsorption process, as observed from the experiment with 600°C of heating temperature.

Fig 2 presents the SEM images of the natural and physically modified pumice. A relatively large number of pores is shown in the natural pumice with some of the pores are covered by other compounds (Fig 2a). The physically modified pumice heated at 300°C and 450°C showed a greater number of open pores (Fig 2b and 2c). It demonstrates that heat treatment increased the pore volume and surface area of the pumice, which also increased its adsorption capacity. Yet, the modified heating pumice at 600°C has fewer pores, as shown in Fig 2d, due to damage in the structure of the pumice pores caused by overheating temperatures

3.2 Chemically modified pumice

Chemical modification of pumice was performed by acid immersion of pumice. An increase in removal efficiency was observed by using the chemically modified pumice than the natural pumice, as shown in Fig 3. Using the natural pumice, 57.02% of nitrate removal efficiency was obtained while by modified pumice soaked in HCl, H_2SO_4 and HNO_3 removal efficiencies increased to 72.15%, 83.30%, and 57.87%, respectively. The increase in removal efficiency was followed by an increase in nitrate uptake. The nitrate uptake observed by using the natural pumice was 142.55 mg/g while using modified pumice soaked in HCl, H_2SO_4 and HNO_3 reached 180.37 mg/g, 208.25 mg/g, and 144.67 mg/g, respectively.

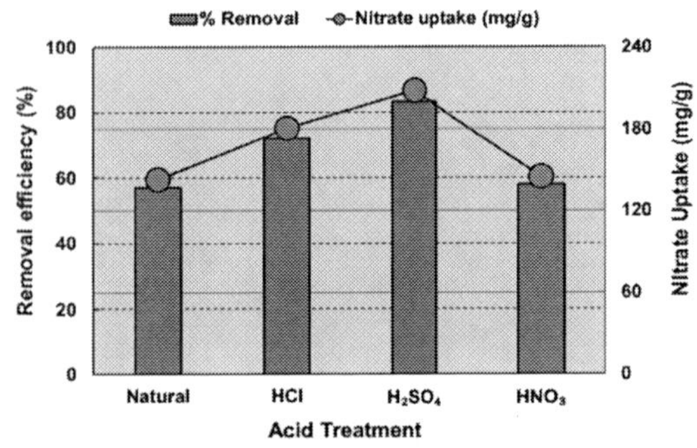

Fig 3. Removal efficiencies and nitrate uptakes of natural and chemically modified pumice

Acid treatment of adsorbent is applied to oxidize the porous surface of the adsorbent. It caused increases in the acidic property and also removes the mineral elements, as well as improved the hydrophilic nature of the surface [17]. Sulfuric and nitric acid is the most widely studied ones besides some other acids used for the purpose. The purpose of the acidic treatment of pumice was to improve the positive surface charge of the adsorbent and improve its sorption capacity [18]. Besides, soaking the adsorbent in acid solution may dissolve the impurities on the surface of the pores of the adsorbent, increasing the pore volume, surface area, and adsorption capability of the adsorbent [19].

The results obtained that the highest removal efficiency and nitrate uptake were found using modified pumice by soaking in H_2SO_4 1 M, followed by HCl and HNO_3. It indicated the different interactions between the acid solution and the pumice. This finding is supported by the result obtained by Soleimani et al (2019) in investigating the efficacy of acid-modified pumice for the removal of humic acid (HA) from aqueous solutions. They evaluated five different acids i.e. hydrochloric acid, acetic acid, sulfuric acid, phosphoric acid, and nitric acid. the result of this study, indicated that the modified pumice increased the adsorbent efficiency and the effectiveness of acids on the pumice adsorbent was in the order of H_2SO_4> HNO_3> H_3PO_4> HCOOH> HCl [7].

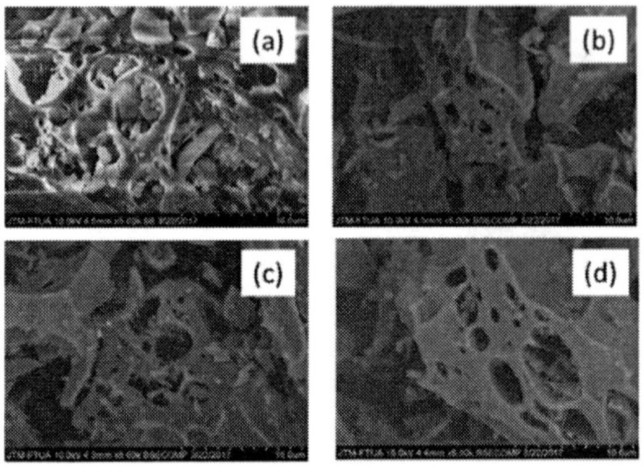

Fig 4. The SEM images (5000x) of natural pumice (a) and chemically modified pumice by acid solution: HCl (b), HNO_3 (c), and H_2SO_4 (d).

The effect of acidic treatment on the surface morphology of pumice is shown in the SEM images of natural and modified pumice, as displayed in Fig 4. Fig 4a shows the pores of natural pumice were largely in number and relatively smooth surface but were covered by other compounds or impurities, as mentioned previously. The uncovered pores and cleaner surfaces were found at the acid-modified pumice, as shown in Fig 4b-4d. The surface of acid-modified pumice also became rugged and more porosity as well as deep cavities, which can be proper sites for adsorption of pollutants. The surface morphology of modified pumice with sulfuric acid has more porosity and heterogeneous than the other samples, as shown in Fig 4d. This condition generated more active adsorption sites and increased removal efficiency as well as the adsorption capacity of pumice. Similar results were also obtained by Soleimani et al. (2019).

The comparison of removal efficiencies and nitrate uptakes obtained by using the natural pumice and modified pumice by heating at 300°C and soaking in H_2SO_4 1 M is summarized in Table 2. These two treatments are the best treatment for physical and chemical modification of pumice which produced the highest removal efficiency and nitrate uptake in the batch adsorption experiment. The physically and chemically modified pumice demonstrated an increase in the removal efficiency and nitrate uptake. The increase was caused by the change in the surface area, pores, and chemical functional groups on the pumice surfaces due to the physical and chemical treatment or modification. Based on the experimental results, it was revealed that various treatments of modification could be applied to increase the removal efficiency and nitrate uptake of the pumice. Further investigations should be conducted to find out various treatments for the best technique in the modification of pumice.

Table 2 Comparison of removal efficiencies and nitrate uptakes by using the natural pumice and the physically and chemically modified pumice.

Kind of adsorbent	Removal Efficiency (%)	Nitrate Uptake (mg/g)
Natural Pumice	57.02	142.55
Physically modified pumice (Heating at 300°C)	62.04	155.09
Chemically modified pumice (Soaking in H_2SO_4 1M)	83.30	208.25

4 Conclusion

As one of the efforts for reducing groundwater pollution, the nitrate adsorption onto the physically and chemically modified pumice was studied in the batch system. The physical modification was conducted by heat treatment at 300°C, 450°C, and 600°C, while for chemical modification, soaking the pumice in three kinds of acid solution namely HCl, H_2SO_4, and HNO_3 with 1 M of concentration were performed. The result shows that the physical and chemical modifications proved to increase the removal efficiency and nitrate uptake. The performance of the heat treatment at 300°C was the best temperature for enhancing the removal efficiency and nitrate uptake, compared to natural pumice and heat treatment at 450°C and 600°C. The removal efficiency increases from 57.02% by the natural pumice to 62.04%, 57.24%, and 29.86% using modified heating pumice, while nitrate uptake was 142.55 mg/g by the natural pumice and 155.09 mg/g, 143.11 mg/g, and 74.64 mg/g by modified heating pumice at 300°C, 450°C, and 600°C, respectively. For chemical modification, the highest removal efficiency and nitrate uptake (80.30% and 208.25 mg/g) were achieved using modified pumice by soaking in H_2SO_4 1 M. The results of SEM analysis demonstrated that physical and chemical treatment modified the pore structure of the natural pumice that could change the surface area and chemical functional group on the surface of the pumice. Overall results were suggested that modification by physical and chemical treatment increased the adsorption capability of pumice in the nitrate removal

The authors are thankful to Universitas Andalas, Indonesia (Grand No. 45/UN.16.17/PP.PGB/LPPM/2018) for supporting this research.

References

1. A. Bhatnagar and M. Sillanpa, Chem Eng J. **168** 493504. (2011)
2. C. H. Liao, S. F. Kang and Y. W. Hsu., Wat Res. **37**. 41094118. (2003)
3. D. Helard, S. Indah and M. Oktavia, Wat Sci and Tech: Wat Sup. **20**(4). 1484-1492. (2020)
4. M. H. Ward, T. M. deKok, P. Levallois, J. Brender, G. Gulis, B. T. Nolan and J.Van Derslice, Environmental Health Perspective. **113** 1607–1614 (2005).
5. M. N. Sepehr, A B. Amrane, K. A. Karimaian, M. Zarrabi, H. R. Ghaffari, J Taiwan Inst Chem Eng. **45** 635–647 (2014).
6. S. Indah, D. Helard and A. Binuwara. Wat Sci. Tech. Bonus Issue **2** 509-515. (2017).
7. H. Soleimani, A. H. Mahvi, K. Yaghmaeian, A. Abbasnia, K. Sharafi, M. Alimohammadi and M. Zamanzadeh, J Mol Liq. **290** 111181 (2019).
8. M. Islam, P. C. Mishra and R. Patel, J Enviro Manage. **91** 1883–1891 (2010).
9. L. Tribe, R. Hinrichs and J. D. Kubicki, J Phys Chem B. **116** (36) 1266–11273 (2012)
10. J.Y. Kim, M. S. Balathanigaimani and H. Moon, Wat Air Soil Pollut. **226** 431. (2015)
11. D.Helard, S. Indah and D. P. Kiflia, IOP Conf Ser: Mat Sci Eng. **846**(1) 012059 (2020)
12. A. H. Mahvi and B. Heibati. Res. J. Chem. Enviro. **16** 62–68 (2012)
13. S. Indah, D. Helard, T. Edwin and R Pratiwi. AIP Conf Proc. **1823** 020072 (2017b)
14. S. Indah, D. Helard and F..Hudawaty, Int J Adv Sci Eng Inform Tech. 9 (5) (2019)
15. H. Shayesteh, A. Rahbar-Kelishami and R. Norouzbeigi, J of Mol Liq. **221** 1–11 (2016)
16. G M. Walker, L. Hansen, J. A. Hanna and S. Allen, J. Wat Res. 37(9) 2081-9 (2003)
17. W. Shen, Z. Li, and Y. Liu, Rec. Patents Chem. Eng. **1** 27–40 (2008).
18. M. R. Samarghandi, M. Zarrabi, M. N. Sepehr, A. Amrane, G. H. Safari, and A. Bashiri, Iran Environ Health Sci Eng. **9** 9 (2012)
19. S. D. Gisi, G. Lofrano, M. Grassi and M. Notarnicola, Sustain Mat Tech J. **9** 10-40 (2016)

Evaluation of reproductive health program during disaster in West Sumatra Province

Azrimaidaliza[1*] *Qori* Andayani Putri[2], and *Yessy* Markolinda[2]

[1]Departement of Nutrition, Faculty of Public Health, Andalas University, West Sumatra, Indonesia
[2]Departement of Public Health Science, Faculty of Public Health, Andalas University, West Sumatra, Indonesia

Abstract. Indonesia is one of the countries with the highest disaster risk. Disaster could affect all aspects of human life including reproductive health. The study aims to evaluate reproductive health programs during disasters. The study was conducted with a qualitative study regarding the system approach. There were 19 informants from health centers, disaster management agencies, and social institutions in the South of West Sumatra Province. Data was collected through in-depth interviews, focus group discussions, and document reviews. Data analysis used content analysis. In the input component, there has been no policy for reproductive health, double responsibility on human resources, and the budget from disaster management for existing facilities. In the process component, a plan for the reproductive health programs does not exist yet and no specific team of health reproductive. The implementation of reproductive health services was data collection, assessment of needs, reproductive health care, and monitoring will be done after a disaster. In the output component, the coverage of indicators for women of childbearing age has been well-served, but not for young women. Management of reproductive health programs is not optimal yet. It is recommended to improve the management of reproductive health programs before, during, and after disasters.

1 Introduction

A disaster is an event or series of events that can threaten and disrupt people's lives caused by natural and non-natural factors. Disasters can result in human casualties, natural damage, property losses, and also psychological impacts [1]. Data from 128 developing countries in the world found that the average number of deaths from natural disasters and armed conflict in East Asia and the Pacific region was not significantly different from the rest of the developing world [2].

Indonesia is amongst the countries with the highest disaster risk in the world [3]. Demographic, geological, and geographical of Indonesian situation are the factors that contribute to the risk of disaster happened. Based on the Indonesian Disaster Information Data (DIBI), the number of disaster events that occurred in Indonesia has increased over the last three years, namely, there were 1,694 disasters in 2016, 2,306 disasters in 2017, and 2,397 disasters in 2018 throughout Indonesia [4].

According to the Regional Disaster Management Authority report, West Sumatra Province has four main potential disasters, namely the potential for landslides, floods, earthquakes, and tsunamis [5]. Several areas in West Sumatra Province are areas prone to natural disasters, especially flash floods, and have a high impact on disaster events [5].

Disaster cases that occur will cause health crises, including paralysis of health services, fatalities, injured victims, refugees, nutritional problems, problems with the availability of clean water, environmental sanitation problems, infectious diseases, and psychiatric disorders, and disruption of reproductive health services, especially on women's groups [6, 7]. Fact, reproductive health services during disasters are often not available because they are not considered a priority, even though there are always pregnant women, mothers in labor, and newborns who need help. Reproductive health services that are not available during disasters such as Antenatal Care (ANC) services and services for adolescent girls [8].

Reproductive health services in overcoming the health crisis are implemented through the Minimum Initial Service Package (MISP) for reproductive health. The MISP will be able to prevent reproductive health problems that will arise, such as preventing and dealing with sexual violence, reducing HIV transmission, preventing increased maternal and neonatal morbidity and mortality [9].

Based on the research of Siti Nuruniyah (2014), it is found that there were 3 cases of sexual harassment in the refugee camps resulting from reproductive health services during the disaster that was not optimal in terms of input. It might have happened because there were no operational policies and organizational structures and infrastructure

* Corresponding Author: azrimaidaliza@ph.unand.ac.id

facilities were not widely available such as intimate booths, tools, and equipment. Besides that there was no contraception and bathrooms were not separated between men and women. The study has also identified that process of monitoring and evaluation has not been carried out in a structured manner. Meanwhile, in terms of output, the responsible team has not yet been formed and not all officers and the community can use the existing inputs [10].

In February 2018, there was a flash flood that hit one area in the south of West Sumatra Province. The disaster caused one of the villages to feel the impact. The disaster resulted in the death of one person, 3 units of heavy houses, and 4,000 housing units submerged by floods, and as many as 781 families were affected and evacuated. Then there was one area that was worst hit by the flash flood disaster and the people were evacuated to the evacuation post. Based on the results of an interview with the head of the area, it was stated that the community affected by the disaster had received general health services such as health checks and additional food for babies from the health center. As for the provision of assistance, It was conducted by a local disaster management agency and social service [11, 12].

According to the data and the previous studies, the study aimed to describe and evaluate reproductive health programs during a disaster in one of the areas in the South of West Sumatra Province which had experienced a disaster. The research used a systems approach consisting of input, process, and output.

2 Methodology

This type of research is a qualitative study using a systems approach to know in detail about health reproductive programs consisting of input, process, and output components.

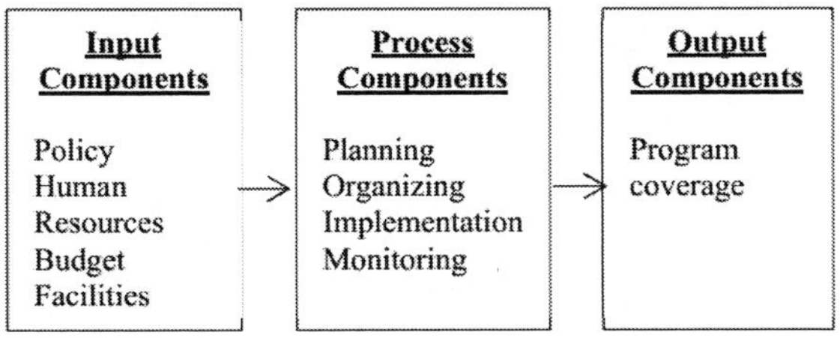

Fig. 1. System Approach in Research of Reproductive Health Program Evaluation

The research was held in one region in the South of West Sumatra Province, specifically in health institutions, local disaster management agencies, and Social Institutions. 19 informants were carried out by purposive sampling and agreed to participate in the study.

Data was collected through in-depth interviews with the standard of interview guidelines, Focus Group Discussion or FGD, and document review. The input data component consisted of policy, human resources, budget, and facilities. Process component consisted of planning, organizing, implementation, and monitoring. Then, the output component was coverage of the reproductive health program. After that, data were analyzed by the content analysis method.

3 Analysis and Discussion

We evaluated the input components which consisted of policies, Human Resources (HR), funds, and infrastructures or facilities. Based on interviews and document review, there was no policy regarding special programs for reproductive health during disasters. But the reference was based on disaster guidelines, namely Law Number 24 of 2007 concerning disaster management and guidelines for the MISP. Human resources for reproductive health at the time of a disaster at the Health Office did not yet exist, but when a disaster occurred related to reproductive health, it was assigned to village midwives who were assisted by nurses. There is no special budget for reproductive health during the disaster, but it is combined with the budget for disaster management sourced from the budget income expenditure. Availability of facilities or infrastructure for reproductive health during a disaster, namely utilizing existing facilities and infrastructure.

The process component consists of planning, organizing, implementing, and monitoring. Based on the results of the in-depth interview, FGD, and document review that there was no specific reproductive health program planning during the disaster. The activities were included with all disaster-related programs and coordinated with the cross-sector of disaster. Those involved in the management of reproductive health programs were in the field of public health, especially the section on family health and nutrition. Meanwhile, at the health center, it is conducted by the worker of the maternal-child health program. Reproductive health services provided during a disaster are the provision of blood-added tablets and general health checks. The community received assistance in the form of food, clothing, blankets, kitchen utensils, towels, and toiletries, but disaster victims, especially women of childbearing age and teenagers, did not receive assistance in the form of menstrual pas. Supervision of reproductive health programs during a disaster is carried out directly by visiting the refugee camps for disaster victims and those who carry out supervision can be the head of the health center and the health office.

The output component is based on the results of in-depth interviews, FGD, and document review that the Health Service, Health Center, local disaster management agency, and social service do not yet have specific indicators for reproductive health but for disaster indicators in general they have been implemented well. Reproductive health services and needs for reproductive women and young women were still not fulfilled.

3.1 Input Component

3.1.1. Policy

Based on the results of the research that policies for reproductive health during disasters in the Regency had not yet been written because policies regarding disasters were still the responsibility of the Provincial Health Office. The guidelines that have been used as a reference during disasters so far are general disaster guidelines, namely Law Number 24 of 2007 concerning disaster management and guidelines for the MISP of reproductive health in health crises.

The finding was also in line with the research of Siti Nuruniyah (2014) in her research on reproductive health services for refugees stating that the policy basis used is based on Law Number 24 of 2007 concerning disaster management and local regulation number 31 of 2010 concerning the state of emergency response, but policies in the form of decrees and standard operating procedures were not yet existed [13].

3.1.2. Human Resources

The results showed that there were no specific health workers which were involved in reproductive health programs during the disaster. The performance of the health workers was under the competence and educational background. Only a few health workers received special training for reproductive health during the disaster, namely only 1 worker who had received training twice. The staff also knew the MISP guidelines and the reproductive health kit during the training. In this health center, there had not been a special team for reproductive health during the disaster.

This result was consistent with research by Siti Nuruniyah (2014). The current research stated that human resources were not yet available specifically for reproductive health services for refugees. Health human resources took care of all health problems as a whole in the refugee camps [13]. Based on the MISP manual, a coordinator for reproductive health services in emergency response situations must be appointed to coordinate across programs, across sectors, local and international institutions [14].

3.1.3. Budget

The results of the study showed that there was no special budget for reproductive health during the disaster, but it was combined with the budget for disaster management sourced from state revenue funding or APBN and regional expenditure budget or APBD. The overall availability of budget during the disaster at the Health Service, Social Service, and Regional Disaster Management Agency was not sufficient, but the health center was sufficient.

Based on the results of Sandra Krause's (2013) research on reproductive health services for refugees in Jordan, it was stated that there was a gap in funding and equipment as an obstacle in implementing the MISP in Jordan [15]. According to Decree of the Minister of Health of the Republic of Indonesia No. 145 of 2007 concerned to guidelines for disaster management in the health sector that the budget for disaster management use disaster budgets allocated by each Regency, City, or Province. If there are deficiencies, they can propose in stages from the Regency, City, Provincial, and Central levels.

Determining how much financial loss results from a natural disaster is not an easy task and budgeting in advance for a disaster could be far more complicated. The need for a budget according to disaster depends on expenditure for pre-disaster, emergency response, and post-disaster. Heru Fahlevi's study (2019) found that the total budget of the local governments has a significant positive correlation with the disaster budget in Aceh Province, Northern Indonesia. The surveys also identify the correlation between the total budget of the local government and disasters as the same budgeting procedure is applied [16].

3.1.4. Facilities

The results showed that the availability of logistics for reproductive health during a disaster was partially available. Meanwhile, at Regional Disaster Management Authority and the Social Service, there were also infrastructure facilities such as logistics warehouses, evacuation equipment, and equipment needed during disaster management. The existing infrastructure when a disaster occurs was 1 unit of an ambulance and a special tent for health checks for victims.

Based on the results of the study that the lack of facilities and infrastructure were operational vehicles that only had 1 unit of an ambulance for patient referrals during a disaster. The situation was made it difficult for health officers in the health center to go back and forth to refer patients from the disaster scene to the health center.

The results of research from Siti Nuruniyah (2014) stated that reproductive health infrastructure such as sterile delivery kits for maternal and neonatal emergency medicines was not yet available in evacuation centers and referral cars to limited health care facilities [13].

3.2 Process Component

3.2.1 Planning

Based on the results of the study, it was shown that there was no specific planning for reproductive health programs during the disaster, but the plan is still joined with all programs and coordinated with cross-sectors related to disasters.

The result of the study was in line with research conducted by Sanaz Sohrabizadeh (2018) that there was no pre-disaster planning for reproductive health services during a disaster. This can be seen from the lack of coordination between local, regional, and national health systems. Another consequence of the lack of planning is the lack of equipment and limited human resources faced by health centers in providing reproductive health services in the post-disaster period [17].

3.2.2 Organizing

The results showed that there was no specific organizational structure for reproductive health during the disaster, but the overall structure was used. Those involved in the management of reproductive health programs are in the field of public health, especially the section on family health and nutrition. While at the health center is done of the Maternal and Child Health program.

According to the MISP Guidelines for dealing with reproductive health during a disaster, an organization consisting of a reproductive health coordinator as the Chairperson and team members who are responsible for the components of gender-based violence, prevention of HIV transmission, maternal and neonatal health, and logistics is formed [9].

The study was in line with current the study in East Lombok. the management of reproductive health post-earthquake was not well organized. The study also identified that the organizational structure for managing the district-level health care program, both during the emergency response and rehabilitation phase, has not yet been formed [18].

3.2.3 Implementation

The program of reproductive health was for during disaster, and there is no program of reproductive health before and after the disaster. Reproductive health services during the disaster were provided by giving tablets from the health center to add blood and general health checks. During a disaster emergency response situation, the Health Service had also coordinated with a local disaster management agency, social service, armed forces, and police to evacuate disaster victims and provide first aid for victims' needs. When the situation stabilized, the victim was given food and clothing assistance, but the victim did not receive assistance related to reproductive health such as menstrual pads, and was also not given food assistance for mothers who have children under five.

The result of the study is consistent with the research conducted by Suryati (2013), The study found that the needs of young women during menstruation in disaster emergencies, 95.10% need a menstrual pad, 55.24% need underwear, and 19% need medicine, but during a disaster, these needs have not been met [8]. Disasters have a significant impact on the reproductive health conditions of the affected residents, especially women of childbearing age, children, and young women, these impacts include increased maternal and neonatal deaths, increased risk of cases of sexual violence and further complications, increased transmission of Sexually Transmitted Infections (STIs), the occurrence of unwanted pregnancies and unsafe abortions, as well as the increasing spread of HIV [8].

A study by Nurhidayati in 2020 found that health workers' types of service activities after the earthquake are midwives, nurses, and doctors from the unaffected health center. The activities routinely include antenatal care, delivery assistance, high-risk detection, and case referral. Some services were not provided, namely family planning services and contraceptives. The study also found that there was an increase in mothers' and neonates' high-risk cases and an increase in pregnant women in the first trimester during the rehabilitation period. Provision of reproductive health tent for delivery services, pregnant women, and maternity services is available in collaboration with overseas organizations, but there is only one for the entire district in East Lombok. Limited facilities, logistics, sanitation, and health workers are also less sensitive to other basic needs [18].

Reproductive health program for women is not only conducted during the disaster but also after the disaster. The program which provides for the identified sexual and reproductive health needs is conducted in terms to follow the guidance of the Minimum Initial Service Package by the global Inter-Agency Working Group. Reproductive health services are included the distribution of medical camp kits, the deployment of nurses with birth attendance skills, and the organization of outreach reproductive health camps. The service is also the provision of emergency reproductive health kits and midwifery kits to health facilities and the psycho-social counseling support provided to maternity health workers. Shelter and transition homes were needed for pregnant and post-partum mothers and their newborns, the distribution of dignity kits, motivational kits for affected women and girls, and female community health volunteers [19].

3.2.4 Monitoring

The results showed that the supervision of reproductive health programs during the disaster was carried out directly by visiting the refugee camps for disaster victims and supervising the head of the Community Health Center in the Health Office.

This finding is not in line with the results of research conducted by Sanaz Sohrabizadeh year 2018. The study identified that the lack of supervision of pregnancy care for women living in disaster areas, especially for women as heads of families, resulted in these women not getting reproductive health services during disasters [17].

Based on MISP, monitoring during the health crisis emergency response is carried out periodically after one or two weeks of implementing the health MISP. Meanwhile, in the post-crisis stage, monitoring is carried out using existing mechanisms and is used in normal situations, namely the Maternal and Child Health Area Monitoring mechanism which is carried out routinely every month [9].

3.3 Output Component

Based on the results, the coverage of reproductive health program indicators during the disaster has not been implemented properly. The implementation of reproductive health programs during a disaster for Women of Childbearing Age is still not fulfilled because this vulnerable group only gets a blood pressure check and is given blood-added tablets. Meanwhile, young women are only given blood-added tablets and do not receive reproductive health services related to reproductive health.

One indicator that has not been achieved in the reproductive health program during the disaster of a health institution is health services and the needs of young women. Reproductive health services during a disaster for young women do not yet exist, while the needs of adolescents have not been fully met and some adolescents also do not know about the services and needs for reproductive health during disasters. For this reason, it is expected that health workers can improve reproductive health services and needs during disasters, including priorities for young women. As we know that a woman's reproductive system is delicate and complex in the body. So, it is important to take steps to protect it from infections and injury and prevent problems—including some long-term health problems

The input component of the evaluation of reproductive health program management during disasters is no policy in reproductive health programs during disasters in the health center but is guided by Law No. 24 in the year 2007 about disaster management and MISP. Human Resources in reproductive health programs during disasters that follow the training of crisis emergency response were still limited. Funds for the implementation of reproductive health programs during disasters come from general program funds sourced from the APBN and APBD. Some of the supporting facilities and infrastructure for reproductive health programs during the disaster were already available and in coordination with the head office and local disaster management agency [12].

The process component for reproductive health programs during a disaster does not yet exist specifically. No special organizational structure has been established for reproductive health during the disaster. The implementation of reproductive health programs during disasters is the provision of services and needs for disaster victims. Supervision is carried out during a disaster, namely directly after a disaster event.

The output component is that the coverage of reproductive health program indicators during the disaster. Mainly, this program is specifically for vulnerable groups, such as women of childbearing age and young women who have not yet received good services and needs.

4 Conclusion

Programs of reproductive health during a disaster in the regions are not optimal yet in terms of inputs, processes, and outputs. It is recommended to improve the management of reproductive health programs during disasters regarding specific divisions on health reproductive teams. The Program is not only conducted during a disaster but also before and after a disaster. The program should include such a social empowerment approach provides a broad strategy within which they may acquire resources and procedural means within which they can enhance their own lives or reduce existing vulnerabilities and health outcomes.

References

1. RI Law No. 24 of 2007 concerning 2007 Disaster Management
2. RJ Swatzyna, VK Pillai: The Effects of Disaster on Women's Reproductive Health in Developing Countries. *Global Journal of Health Science* 2013, **5**(4).
3. R Djalante , M Garschagen , F Thomalla , R Shaw : *Disaster Risk Reduction in Indonesia: Progress, Challenges, and Issues. Springer International Publishing* (2017).
4. Regional Disaster Management Agency (BPBD): Indonesian Disaster Information Data (DIBI). Jakarta: Regional Disaster Management Agency; (2017)
5. Regional Disaster Management Agency (BPBD): Indonesia's Disaster Hazard Index. Jakarta: Regional Disaster Management Agency; (2018)
6. .The Ministry of Health of the Republic of Indonesia: Technical Guidelines for Managing Health Crisis Due to Disasters. Jakarta: Ministry of Health of the Republic of Indonesia; (2007)
7. Azrimaidaliza: Socio-economic and nutritional status of mothers in disaster Prone areas. *Andalas Journal of Public Health* 2014, **9**(1).
8. Suryati: Description of the Needs and Behavior of Young Women during Menstruation in Disaster Emergency Situations in West Sumatra. Journal of Public Health 2013, **7**(2):60-65
9. Ministry of Health of the Republic of Indonesia: Guidelines for the Implementation of the Minimum Initial Service Package (MISP) in a Health Crisis. Jakarta: UNFPA (2017)
10. S I Munauwarah, Hermansyah, Strategy for Reproductive Health Services in Disaster Emergency Situations at the Pidie District Health Office. Aceh Health Ministry Health Polytechnic 2014, **2**(1):1-10
11. South Solok District Health Office: South Solok District Health Office Profile Report 2016. South Solok: South Solok District Health Office; (2016)
12. South Solok District Health Office: Profile of South Solok District Health Office 2016. South Solok: South Solok District Health Office; (2016)
13. S Nuruniyah: Evaluation of Reproductive Health Services Prone to Merapi Eruption. Journal of NERS and Midwifery Indonesia 2014, **2**(2):51-61.
14. The Ministry of Health of the Republic of Indonesia: Guidelines for the Implementation of the Minimum Initial Service Package (MISP) for Reproductive Health in a Health Crisis. Jakarta: Indonesian Ministry of Health; 2007.
15. S Krause, H Williams, MA Onyango, S Sami, W Doedens, N Giga, et al: Reproductive health services for Syrian refugees in Zaatri Camp and Irbid City, Hashemite Kingdom of Jordan: an evaluation of the Minimum Initial Services Package. *Journal Conflict and Health* 2015, **9**(1).
16. H Fahlevi, M Indriani, RS Oktari: Is the Indonesian disaster response budget correlated with disaster risk? *Journal of Disaster Risk Studies* 2019, **11**(1):759.

17. S Sohrabizadeh, K Jahangiri, RK Jazani: Reproductive health in the recent disasters of Iran: a management perspective *Journal BMC Public Health* 2018, **18**(389):1-8.
18. Nurhidayati, RS Padmawati, A Wahab: Organizing reproductive health services for post-earthquake refugees in East Lombok. *Berita Kedokteran Masyarakat* 2020, **36**(6).
19. P Chaudhary, G Vallese, M Thapa, VB Alvarez, LM Pradhan, K Bajracharya, K Sekine, S Adhikari, R Samuel, S Goyet: Humanitarian response to reproductive and sexual health needs in a disaster: the Nepal Earthquake 2015 case study. *An international journal on sexual and reproductive health and rights* 2017, **25**(51):25–39

Implementation of non-linear controller on photovoltaic maximum power point tracker for energy storage equipment charger

Muhammad Imran Hamid[1*], *Abdul* Rajab[1], *Andi* Pawawoi[1] and *Rezy* Chaniago[1]

[1]Electrical Eng. Dept. Andalas University – Padang, West Sumatera, Indonesia

Abstract. The portable energy storage system is an infrastructure for providing electrical energy needed to support the recovery process after a natural disaster. This system is a battery arrangement that can be recharged using locally available primary energy sources such as photovoltaic. The main problem in using photovoltaic as the power source for this equipment is to increase the efficiency of power extraction (energy harvesting) during the recharging process. Traditionally, to obtain maximum extraction conditions, conventional linear maximum power point tracker (MPPT) mechanisms such as PID-based MPPT and the like are used. However, if the PV and the storage system works at various locations with environmental conditions behave unusually, the conventional MPPT cannot work accurately and optimally. In this paper, the Fuzzy method for constructing a nonlinear controller-based MPPT was studied. The step size of the tracking process in the conventional MPPT P&O method is modified by involving the fuzzy algorithm. This algorithm then is applied to a DC-DC converter to test the performance criteria such as the response and efficiency of the resulting power extraction. The testing and computer simulations show that the conventional MPPT mechanism can provide prospective results through modification and application of a non-linear controller.

1 Introduction

In the application of PV as part of an electric power supply system, such as a charger for a portable energy storage system (ESS), Fig.1, the efficiency of energy conversion, is the main focus of various development and research efforts.

Energy storage system refers to the device of converting electrical energy from power systems/sources into a form that can be stored for converting back to electrical energy when needed. In disaster conditions, energy storage role to support the critical infrastructure and services. It is providing electrical power. Energy storage systems of all shapes, types, and sizes reduce the time it takes for first responders to begin recovery efforts. ESS must be recharged. If ESS uses renewable, locally primary sources, such as photovoltaic, it must be able to extract the available energy optimally, in high-efficiency mode, and accurately (because the source in the area is limited). All of these requirements can be achieved by implementing advanced, innovative control on ESS components, processes, and systems.

In a PV-based source system, to get maximum efficiency, a control technique is used so that the operating point of the photovoltaic module or array always works at its maximum power condition. This technique is called maximum power point tracking, and it plays a key role in the functioning and effectiveness of the system [1].

The basic conventional maximum power point tracking (MPPT) technique consists of the *incremental conductance* (IC) technique, the *fractional open voltage, and short circuit* technique, and the *perturb & observe (P&O)* technique [2]. These techniques are widely used because of their simplicity, ease of implementation, and cost-effectiveness. However, these techniques have their respective drawbacks. The IC and P&O techniques experience oscillations around the maximum power point (MPP) due to fluctuations in the operating point when the PV system approaches the MPPT [3]. Furthermore, this becomes a steady-state error which ultimately causes a loss of energy produced or should be extracted. In addition, P&O is also unable to track MPP in rapidly changing environmental conditions due to slow convergence. Then, in applying the P&O algorithm, it is difficult to determine the number of steps for each updating step.

Faced with this, the researchers developed MPPT techniques based purely on non-linear or artificial intelligence such as Artificial Neural Network (ANN) [4], fuzzy logic [5], and metaheuristics such as Ant Colony Optimization, Particle Swarm Optimization (PSO), and Cuckoo Search Algorithm [6]. These non-linear and artificial intelligence-based MPPT techniques also have drawbacks, including The ANN-based MPPT technique requires a large amount of data (different irradiation,

* Corresponding author: imran@eng.unand.ac.id

temperature, and partial shading conditions) for proper training, [7]. Metaheuristic-based methods show extreme transient behavior during the search for optimal parameters (duty cycle or reference voltage). This causes convergence to be achieved more slowly and may result in early convergence under rapidly changing environmental conditions [8]. Meanwhile, Fuzzy Logic Control (FLC) offers moderate computational complexity and has an efficiency comparable to ANN. However, FLC design requires designers to have complete knowledge of PV operations and FLC rule tables can lead to computational improvements, if not optimized properly [9]. To overcome this weakness, various combinations of intelligent techniques are proposed, for example, the Particle Swarm Optimization (PSO) technique [10], is used to optimize the membership function and FLC rules. Then in the literature [11], FLC combined with Hopfield ANN is used to calculate and determine fuzzy rules (fuzzy rules table). The intelligence-based MPPT technique combined with these other techniques provides an advantage in dealing with inaccuracies and parameter variations [12].

In this paper, fuzzy logic as the non-linear controller is integrated into the conventional MPPT P&O algorithm, by exploiting the advantages of the MPPT P&O technique, while the shortcomings of artificial intelligence Fuzzy techniques that require expertise in planning are relatively accommodated, considering that knowledge of PV module behaviour as a condition for fuzzy implementation is sufficiently mastered by designer. The MPPT P&O technique in operation controls the PV array output power through a PWM switching mechanism in a photovoltaic generation system (Fig. 1). The P&O MPPT control algorithm is based on the calculation of the PV output power and the power change. These signals are acquired by sampling the voltage and current of the module or array. The power and voltage change is detected by comparing their present and previous levels and used to change the reference voltage by adding or subtracting a voltage change (ΔV). Furthermore, fuzzy logic is used to vary the magnitude of this ΔV according to its position concerning the MPP point. With this combination, the problem of oscillation around the MPP which is the main weakness of the MPPT P&O method can be reduced.

In applying this Fuzzy and P&O combination method, the next problem faced is to increase the MPPT response performance so that conversion efficiency can be optimized. The addition of fuzzy logic to the MPPT algorithm however provides additional processes that must be passed to produce the final output of the controller signal. This must be compensated by a response acceleration strategy that can be applied to the membership function design or fuzzy logic membership used. For this reason, a strategy to improve MPPT performance is proposed by adjusting the level of ΔV in tracking the optimum working point of the PV module. It takes into account the wide range of dP/dV oscillation values in the area to the left and right of the maximum operating point on the PV characteristic curve. The range of each membership function is different between the two so that the membership curve profile is not symmetrical.

2 PV Module Characteristic and Fuzzy P&O MPP Tracker

A simple diagram of an ESS is shown in Fig. 1. It consists of main components such as primary energy sources, in this case, a module or array of photovoltaic arrays, energy storage components in the form of a battery bank, DC-DC converter for charging/discharging, and DC-AC converter as the interface between this energy storage system and the load. A transformer is also sometimes needed to adjust the system voltage level. In this system, the MPPT mechanism in the form of a control algorithm is placed on the DC-DC charging/discharging converter.

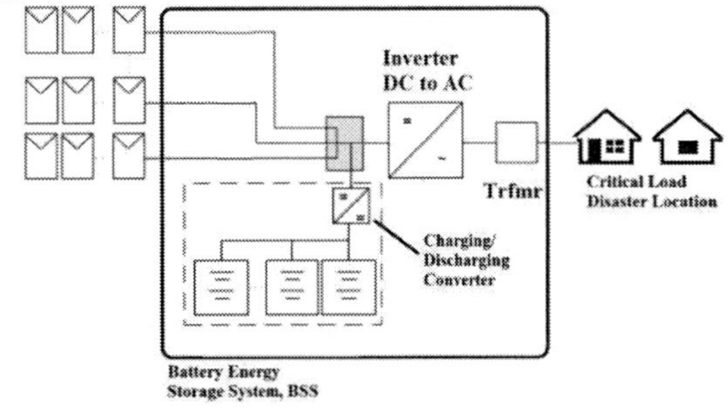

Fig. 1. An energy storage system with photovoltaic as a primary energy source.

PV module or array as Fig. 1 can be modeled according to [13], from [14] and [15],

$$I = I_{LG} - I_{os}\left\{exp\left[\frac{q}{AkT}(V - IR_s)\right] - 1\right\} - \frac{V - IR_s}{R_{sh}} \quad (1)$$

Where

$$I_{os} = I_{or}\left[\frac{T}{T_r}\right]^3 exp\left[\frac{qE_{Go}}{B_k}\left(\frac{1}{T_r} - \frac{1}{T}\right)\right]$$

$$I_{LG} = [I_{SCR} + K_I(T - 25)]\frac{\lambda}{100}$$

and

I and V	cell output current and voltage;
I_{os}	cell reverse saturation current;
T	cell temperature in C;
k	Boltzmann's constant;
q	electronic charge;
K_I	short circuit current temperature coefficient at ISCR;
λ	solar irradiation in W/m
I_{SCR}	short-circuit current at 25 C and 1000 W/m;
I_{LG}	light-generated current;
E_{GO}	band gap for silicon;
B = A	ideality factors, 1.92;
T_r	reference temperature;
I_{or}	cell saturation current at ;
R_{sh}	shunt resistance;
R_{se}	series resistance.

 https://doi.org/10.1051/e3sconf/202133102014

By multiplying both sides of equation (1) by Voltage V, we get the power equation of the PV module P(V). The plot of the power as a function of module voltage, at a given temperature and irradiance condition, is shown in Fig. 2., where MPP points are seen at the top of the curve. The figure also shows the dP/dV parameter, which is the rate of change of the extracted power to the voltage of the module. These three-parameter positions are shown on the P(V) curve, namely the position on the curve where the slope (gradient) dP/dV > 0 is to the left of the MPP and the slope (gradient) dP/dV < 0 is to the right of the MPP.

The traditional P&O algorithm is depicted with a flowchart as shown in Fig. 3. The MPP point can be tracked. This algorithm continuously calculates the value of module power from voltage and current signal that is sensed at the PV module output. If there is a change in power, then the P&O algorithm will take action to increase or decrease the reference voltage Vref on the DC-DC converter operating control system. Furthermore, in the converter control system, this reference voltage change is used to adjust the duty ratio of the PWM signal as the gating of the converter's switching components.

The change in the converter reference voltage Vref is done by adding or subtracting a voltage change ΔV to the converter's operating reference voltage in the previous sample. The value of ΔV is predetermined by the designer. P&O algorithm will change the converter output voltage reference along with the condition of the dP/dV gradient position [3]. If the gradient is dP/dV>0 then ΔV will be subtracted from the previous reference, and if the gradient is dP/dV<0 then ΔV will be added until dP/dV = 0, i.e. when the operating point of the module is at its optimum.

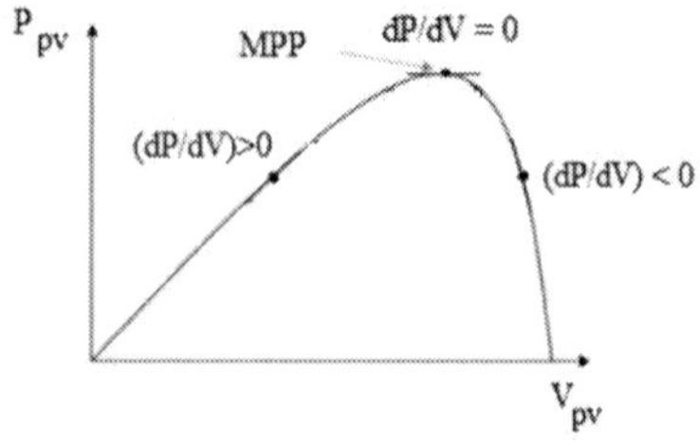

Fig. 2. P(V) characteristic curve of a photovoltaic module.

In the development of the P&O algorithm into Fuzzy-P&O, fuzzy logic is used to adjust the ΔV voltage which was fixed in the traditional P&O. Fuzzy logic makes the gradient or rate of change of the extracted power to the voltage, dP/dV, as the input *error* signal E(n), and changes in the dP/dV as input signal *delta-error*, dE(n). *error* signal (E(n)) serves to determine the location of the photovoltaic power gradient while *delta-error* (dE(n)) serves to determine the level of error change.

The value of ΔV is varied, which is determined based on the previously compiled Fuzzy rule. This can be seen in the flowchart in Fig. 4

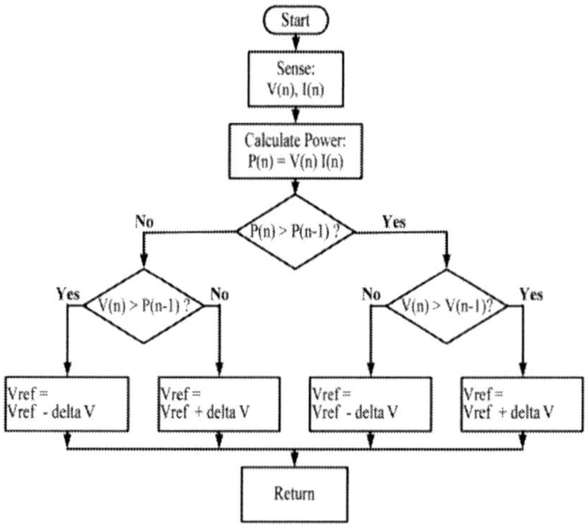

Fig. 3. Flowchart of P&O algorithm to track the position of MPP on the P(V) characteristic curve of PV module.

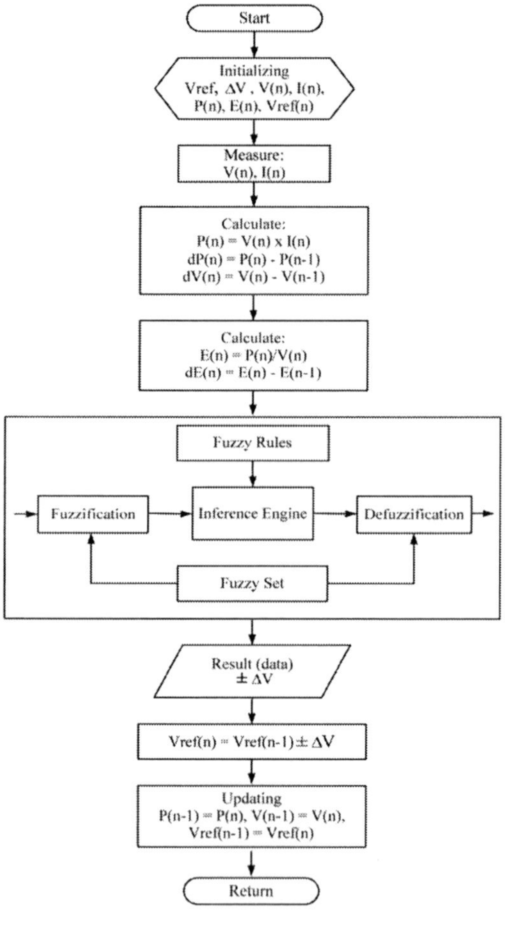

Fig. 4. Flowchart of Fuzzy-P&O algorithm to track the position of MPP on the P(V) characteristic curve of PV module.

3 Proposed Fuzzy Set Design for Fuzzy-P&O MPPT Algorithm

In this section, we will discuss the methods in membership function design, testing, and simulation. Based on Fig. 4., it can be seen that the focus of fuzzy controller design is aimed at the following three aspects: the preparation of fuzzy sets in the process of fuzzyfication and defuzzification, and the determination of the rule-based fuzzy.

The fuzzy set in the fuzzyfication step is formed based on the photovoltaic power slope characteristic data obtained from the PV module test to obtain the P(V) characteristic curve. Furthermore, for each segment on this curve, the value of dP/dV and its changes can be obtained which are used as input error signals E(n) and delta error dE(n) for fuzzy blocks. As a test case data from a PV module with data as Table 1.

From the simulation, the range of dP/dV is between (-206.72) on the right side of MPP to (15.2) on the left side. The PV power characteristic curve has an asymmetrical parabolic shape. The right and left sides of the MPP have different power gradients, the steeper the curve, the greater the power gradient. This condition causes the power gradient range of the right side to be larger than the left side of the MPP and of course, the power gradient to the right side of the MPP is also negative because the direction of the curve is linearly decreasing. With this range, six fuzzy sets for E(n) are formed. Also, five fuzzy sets for delta error dE(n) and ΔV. The boundaries for each fuzzy set are determined as follows:

Error, E(n):
NB : [-247 -206.7 -126.7 -86.44]
NS : [-126.7 -47.5 -15.41]
NK : [-47.5 -9 0]
M : [-9 0 2.5]
PK : [0 2.5 9]
PB : [6.28 9 15.2 17.92]

delta error, dE(n):
NB : [-0.024 -0.02 -0.008 -0.004]
NK : [-0.008 -0.004 0]
M : [-0.0015 0 0.0015]
PK : [0 0.004 0.008]
PB : [0.004 0.008 0.02 0.024]

Voltage reference change, ΔV(n):
NB : [-1.3 -1 -0.6 -0.3]
NK : [-0.6 -0.3 0]
M : [-0.1 0 0.1]
PK : [0 0.3 0.6]
PB : [0.3 0.6 1 1.3]

The rule base table that connects the resulting input and output fuzzy ΔV is shown in Fig 5. The base rules in the table use if-then rules. Defuzzyfication to produce a crisp value of V is carried out through the fuzzy inference system using the Mamdani method. The range that will be used to change the ΔV level is ± 1.3 Volts.

Table 1. Parameter of the PV Array at G = 1000W/m^2.

Parameters	Value
Maximum Power (P_{max})	6224 Watt
Voltage Maximum Power (V_{mp})	444.6 Volt
Current Maximum Power (I_{mp})	14 A
Short Circuit Current (I_{sc})	15.2 A
Open Circuit Voltage (V_{oc})	527.5 Volt

Table 2. Fuzzy rule base table.

dE \ E	NB	NS	NK	M	PK	PB
NB	PB	PB	PB	PK	M	NK
NK	PB	PB	PK	M	NK	NK
M	PB	PK	PK	M	NK	NB
PK	PB	PK	PK	M	NK	NB
PB	PB	PK	M	NK	NB	NB

4 Results

To evaluate the performance of the Fuzzy-P&O MPPT development, testing was carried out using the Simulink software. A simulation circuit to execute MPPT algorithm on a DC to DC converter type interleaved DC-DC converter is built. Furthermore, this circuit is operated by input power from the photovoltaic module where the amount of power towards the converter circuit and the MPPT algorithm can be adjusted by varying the level of irradiance received by the PV module. Furthermore, the extraction efficiency and power response on the converter side are acquired and analyzed. Three simulation conditions were carried out, namely the operation of the DC-DC converter system in distributing power from the PV module without applying the MPPT algorithm; implementation of conventional P&O MPPT, and application of Fuzzy-P&O MPPT. Fig. 5. shows the full system converter simulation diagram on the Matlab/Simulink software (upper); PV module model diagram (middle), and Fuzzy P&O implementation diagram (lower).

Based on the power generated in the simulation, it can be seen that the Fuzzy Perturb and Observe can track the MPP of the PV array faster: 14 ms, than the conventional Perturb and Observe which takes 156 ms, this shows that Fuzzy technology can improve the quality of MPPT P&O in tracking the photovoltaic MPP. In addition, the power of the PV Array that can be generated by Fuzzy Perturb and Observe is also greater, when the Irradiance 1000W/m^2 Fuzzy P&O produces 6207.83 W of power while the PO is only able to generate 6078.91W of power. Table 2 shows the output power value generated for each irradiance level

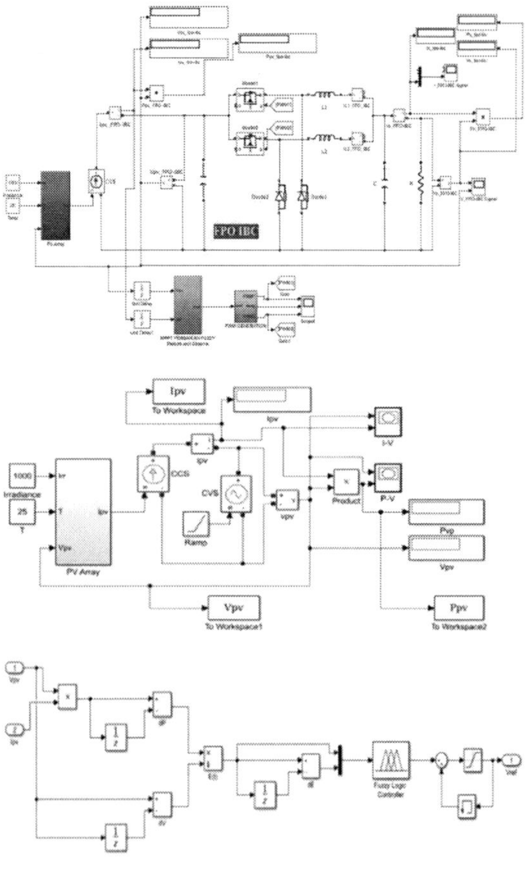

Fig. 5. Simulation diagram of Fuzzy-P&O MPPT on Interleaved DC-DC converter

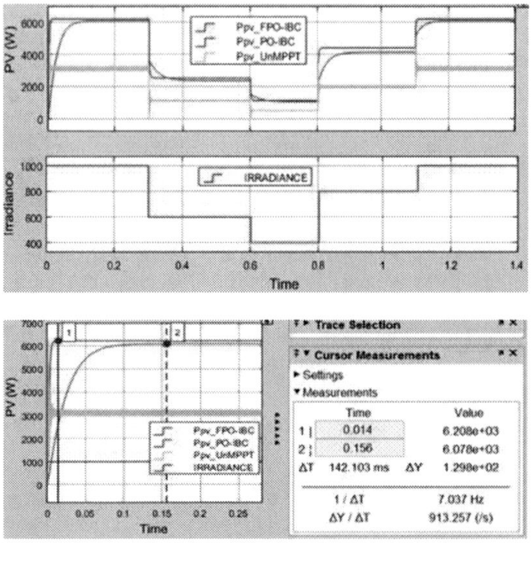

Fig. 6. PV system output power at Irradiance 1000 W/m^2, (a) Response of the MPPT to the irradiance, and (b) detail response.

Table 2. System output power for various operation schemes

Irradiance (W/m^2)	PV System output		
	Without MPPT	FPO-IBC	PO-IBC
400	497.91W	1128.33W	1048.18W
600	1119.16W	2525.25W	2347.77W
800	1988.52W	4396.14W	4126.68W
1000	3105.74W	6207.83W	6078.91W

From the simulation carried out by applying the same irradiation pattern and level: 1000W/m2, an analysis of the power extraction efficiency was carried out, using the equation:

$$\eta = \frac{P_{MPPT-x}}{P_{Max}} \times 100\% \quad (2)$$

Where

η efficiency of power extraction
P_{MPPT-X} output power using the x-type MPPT
P_{Max} output power from PV array

The resulting power extraction efficiency conditions as outlined in Table 3.

Table 3. Power extraction efficiency

MPPT Implementation		
None	P&O	Fuzzy P&O
44.89	99.74	97.66

MPPT Perturb and Observe have high power efficiency in optimizing photovoltaic power compared to MPPT with other algorithms, with the application of Fuzzy technology in Perturb and Observe, MPPT P&O on photovoltaic systems has a faster response in tracking with a large power efficiency of 99.74%.

5 Conclusion

The performance of the traditional MPPT development of P&O type by utilizing a non-linear Fuzzy algorithm into a Fuzzy-P&O algorithm has been described. In this paper, a fuzzy set method has also been introduced which is based on the condition of the dP/dV value of the power-to-voltage characteristic curve of the photovoltaic module. The fuzzy set is arranged in an asymmetrical shape that resembles the power function characteristic curve of the PV module voltage. The speed of response and also the increase in the efficiency of the resulting power extraction are the advantages of this algorithm. With these results, it can be said that the development of MPPT that combines traditional methods and non-linear techniques such as fuzzy perspective is to be developed.

The authors would like to express gratitude to Universitas Andalas for providing financial support under Fundamental Research Grant (RD), Contract No. T/9/UN.16.17/PT.01.03/Energi-RD/2021.

References

1. M. I. Hamid, M Anwari, Single-Phase Photovoltaic-Inverter Operation Characteristic in Distributed Generation System, Distributed Generation, D N Gaonkar, IntechOpen,(2010)
2. A. R. Jordehi., Maximum power point tracking in photovoltaic (PV) systems: A review of different approaches. Renewable and Sustainable Energy Reviews; 65:1127-1138 (2016)
3. S. K. Kollimalla, M. K. Mishra, A novel adaptive P&O MPPT algorithm considering sudden changes in the irradiance. IEEE Transactions on Energy conversion; **29** (3), 602-610 (2014)
4. S. Saravanan, N. R. Babu, RBFN based MPPT algorithm for PV system with high step up converter. Energy Conversion and Management; 122:239-251 (2016)
5. N. Karami, N. Moubayed, R. Outbib General review and classification of different MPPT Techniques. Renewable and Sustainable Energy Reviews; 68:1-18 (2017)
6. J. P. Ram, T. S. Babu, N. A Rajasekar comprehensive review on solar PV maximum power point tracking techniques. Renewable and Sustainable Energy Reviews; 67:826-847 (2017)
7. Z. Salam, J. Ahmed J, B. S. Merugu, The application of soft computing methods for MPPT of PV system: A technological and status review. Applied Energy; 107:135-148 (2013)
8. A. Khare, S. Rangnekar, A review of particle swarm optimization and its applications in the solar photovoltaic system. Applied Soft Computing; **13** (5):2997-3006 (2013)
9. N. Karami, N. Moubayed, R. Outbib, General review and classification of different MPPT Techniques. Renewable and Sustainable Energy Reviews; 68:1-18 (2017)
10. Y. Soufi, M. Bechouat, S. Kahla. Fuzzy-PSO controller design for maximum power point tracking in the photovoltaic system. International Journal of Hydrogen Energy; 42(13):8680-8688 (2017)
11. R. Arulmurugan, N. Suthanthiravanitha. Model and design of a fuzzy-based Hopfield NN tracking controller for standalone PV applications. Electric Power Systems Research; 120:184-193 (2015)
12. L. Suganthi, S. Iniyan, A. A. Samuel, Applications of fuzzy logic in renewable energy systems–a review. Renewable and Sustainable Energy Reviews; 48: 585-607 (2015)
13. E. Koutroulis, K. Kalaitzakis, "Development of a Microcontroller-Based Photovoltaic Maximum Power Point Tracking Control System, *IEEE Trans. On Power Electronics.*, **16** (1) (2001)
14. G. Vachtsevanos and K. Kalaitzakis, "A hybrid photovoltaic simulator for utility-interactive studies," *IEEE Trans. Energy Conv.*, **EC-2**, 227–231, (1987)
15. M. G. Jaboori, M. M. Saied, and A. A. Hanafy, "A contribution to the simulation and design optimization of photovoltaic systems," *IEEE Trans. Energy Conv.*, **6**, 401–406 (1991)

Juridical review of disaster management policy in ASEAN and the correlation with the domestic management system of disaster in Indonesia

Delfiyanti[1*] and *Magdariza*[1]

[1] Departement of International Law, Faculty of Law, Andalas University, West Sumatera, Indonesia

Abstract. Southeast Asia was in a natural disaster thus the management was supposed to be a priority to the existing states in this territory. It is the most vulnerable to disaster in the world. By then, the member states of ASEAN agree to issue the regulation for disaster management, ASEAN Agreement on Disaster Management and Emergency Response (AADMER) that in forwarding established ASEAN Coordinating Centre for Humanitarian Assistance on Disaster Management (AHA Centre). It is facilitating cooperation and coordination for disaster management in the ASEAN territory. The organization was established in Indonesia as a member state with potential disaster. To reinforce the regulation and disaster management system, the government issues Act No.24 of 2007 on disaster management as the base and manual. The policy refers to the activities implemented immediately for an accident in control arising worst impact, involving rescue and evacuation of the victim, properties, compliance of demand, shelter, refugees handling, and facilities-infrastructure restoration. Moreover, the Act regulating disaster mitigation-based layout system set in an attempt to improve safety and living comfort.

1 Introduction

Southeast Asia was in potential natural disaster thus the management was supposed to be a priority to all states in this territory. It is the most vulnerable to all kinds of natural disasters in the world. The causing factor is that geographically Southeast Asia span between potential tectonic layers resulting in an earthquake, volcano eruption, and tsunami. Moreover, the area is located among the ocean, Indian and Pacific triggering hurricanes.[1]

One of the most horrible natural disasters of the Tsunami of Aceh in December 2004 caused by seismic in the Indian Ocean killed more than 200.000 people. Hurricane in Myanmar of 2008 with 130.000 victims. The natural disaster in Asia does not only cause fatalities but material also. According to the chief of Badan Nasional Penanggulangan Bencana (BNPB), Willem Rampangilei in the speech of the 29th ASEAN Committee on Disaster Management (ACDM), the member states of ASEAN experiences the high-risk natural disaster with the potential loss of material about US$4.6 billion or 0.2 percent of gross domestic income annually. Based on data from World Bank since 2000, more than 100 million people were impacted by natural disasters and almost all kinds potentially occurred in ASEAN such as earthquakes, volcano eruptions, tsunami, floods, cyclones, landslides, and drought.[2]

Under the situation, as the institution for member-states cooperation in Southeast Asia, ASEAN must take an action to minimize the risk as low as possible. The community already has an instrument to reinforce the cooperation and unity for natural disaster management, but the effectiveness must be improved in the implementation.

In the ASEAN Ministerial Meeting on Disaster Management (AMMDM) summit was issued ASEAN Agreement on Disaster Management and Emergency Response (AADMER) by ASEAN Committee on Disaster Management (ACDM) as the coordination center. AADMER is the regional framework with proactive for cooperation, coordination, technical assistance, and mobilization of resources in the whole aspect of disaster management.

Moreover, the ASEAN community focused on disaster management policy under the ASEAN Regional Forum (ARF) and ASEAN Defense Ministerial Meeting Plus (ADMM Plus) with all the members including the speech partner state of ASEAN. The focus of institutions is cooperation in support of humanitarian and assistance as a part of disaster management in Southeast Asia. One of the most important components in the AADMERs agreement is the establishment of the ASEAN

* Corresponding Author: defi_anti@yahoo.com

Coordinating Centre for Humanitarian Assistance on Disaster Management (AHA Centre) in 2011. This intergovernmental organization was established by ten member states of ASEAN with the purpose to facilitate the cooperation and coordination of disaster management for member states of ASEAN.

As a member of ASEAN, Indonesia is the island's state located in tectonic plates geographically, Asia Continental, Australia, India Ocean, and Pacific. There is a volcanic belt in the south and east span from Sumatera-Java-Nusa Tenggara-Sulawesi with the side of old volcanic mountains and lowland dominated by swamps. It is vulnerable to all kinds of a disaster such as eruptions, earthquakes, tsunami, and landslides.[3]

Data of Badan Nasional Penanggulangan Bencana (BNPB, 20005) shows that more than 2000 disaster occurred since 2002 to 2005. It is consisting of 743 floods, 615 droughts, 222 landslides, and 217 fires. The total of victims since the period is 165.945 people who died by earthquake and tsunami, and 2.223 of social conflicts. Furthermore, the flood caused 2.665.697 people to lose their houses and become refugees. According to the Bappenas mapping, Indonesia was suffering from seven huge disasters since 2004-2007 with enormous damages caused by the Tsunami of Aceh (2004), Avian influenza (2004-2005), volcano eruption (2006), the earthquake of Yogyakarta (2006), Lapindo (2006), the tsunami of south Java (2006) and the flood of Jabodetabek (2007). It is predicted that the total loss of material is 110.4 trillion.

Indonesia realizes that the disaster must be handled seriously since an earthquake followed by a tsunami occurred in Aceh in 2004. Disaster is the most comprehensive discussion and multidimensional. In response to the increasing disaster annually, consideration to the disaster management must be understood and implemented by all concerned parties. The disaster is a concern for all parties. Consequently, periodically Indonesia establishes the national system of disaster management.[4]

Therefore, the required system and policy of disaster management provide manual or standard when disasters happen by nature, engineering, or social factors. The regulation required the base of law to take any action efficiently and immediately to minimize the loss impact. Another important factor is the appointment of a responsible party for coordination when a disaster has happened. Therefore, issuing Act No.24 of 2007 on disaster management has important meaning as the manual of concerned parties for disaster management much occurred in Indonesia.

On the other side, it is important to consider the international regulation of disaster to compare the system or policy for disaster management such as ASEAN Agreement on Disaster Management and Emergency Response implemented in ASEAN and the other one.

2 Methods

Dominantly, a method used in this research is "normative juridical" applied by literature, and the observed data is secondary with primary law material, secondary, tertiary, etc.[5] In correlation with the research, use published laws relate to the research objects consisting of regulation on disaster applied in ASEAN and Indonesia, textbook, journal, dictionary, any comments, and judge's decision. Moreover, there is no non-law material used in relevant and supporting research.

The study use technique and processing method and "qualitative data analysis". In general, processing and analysis consist of (1) data reduction, (2) simplification and presentation of data, and (3) verification of result and conclusion. The data analysis was arranged simultaneously by data processing, even it was begun since the data collection process. As the legal research, the qualitative analysis used is juridical. Therefore, the phases of data analysis are the simplification of the data processing phase only.

3 Analysis and Discussion

3.1 The Legal Aspect of Disaster Management Policy System In ASEAN

There was established AHA Centre in ASEAN or Humilitarian Assistance of ASEAN as the institution o international organization in 2007 as the implementation result of AADMER from ASAN Special Leader's Meeting on Aftermath of Earthquake and Tsunami Disaster known as Tsunami Summit in Jakarta of 2005. The vision of AADMER is the state survives against disaster and social safe in ASEAN. The establishment of AADMER intends to reduce casualties and loss of property influencing the economy, society, physic, and environment of member states of ASEAN in natural or caused by human error through AHA Centre.

Establishment of AHA Centre easy monitoring on natural disasters in Southeast Asia. The organization has a disaster monitoring system when a disaster is detected in the member state. In this case, AHA Centre could give a fast response to the disaster in coordination with the member states for reducing impact. Additionally, set up AHA Centre is become the base of coordination for humanitarian assistance and disaster management, facilitating cooperation and coordination among the parties, such as United Nation and concerned international organizations to promote regional cooperation. Under the AHA Centre, ASEAN could give a fast response to the member states of ASEAN suffering disaster, especially in emergency response.

Especially, to implement the mandate, AHA Centre collaboration with disaster management national organization of ASEAN member states. There is two high-level commitment to reaffirmation the role of the AHA Centre as the main regional coordination agent of ASEAN in disaster management and emergency response, the declaration of ASEAN on the response of disaster in and out of the territory and the vision of ASEAN 2025 on disaster management. The headquarter of AHA Centre is in Indonesia and the mechanism of work and policies is as follows :

1. Risk assessment, early warning, and monitoring.

The strategic instrument of the working program intends to reduce casualties, property through disaster identification and risk before occurring by increasing the warning period. The activities described in three sub-components will bring proper assessment of regional risk and early warning through a focus on cross-border issues requiring interstates collaboration, thus it will bring benefit to the area, most inclusive-restoration plan, and attempt to mitigation effort and target management response and recovery.

a. Risk Assessment
   In constant progress and environmental change, the risk growing in the new area and among the population. Identification of high-risk populations and geographical areas and understanding the principle risk, vulnerability, and capability of disaster management is important to be effective in design, management, and recovery. The instrument of disaster management, also, intends to develop and implement the assessment program of regional risk that establishes the existing protocol and best practice to accomplish the initiative of member states. Furthermore, the development and implementation of policy and procedural management are required to ensure that the AHA Centre accepts and consolidates the data of National Focal Point (NFP) and publish the identifiable and analyzed data and risk to the concerned parties.
b. Early Warning
   Loss by disaster could reduce by early detection before occurring and coordinated early warning. It is intended to ensure that the member states can develop, maintain, and reviews temporarily on national early warning information system, communication network, public awareness, and preparedness, as well as developing an existing ASEAN system and competence such as Disaster Information Sharing and Communications Network (DISCNet), Online Southeast Asia Disaster Inventory (OSADI), Online Southeast Asia Disaster Map(OSA -Map).
c. Monitoring
   Advanced monitoring on risk, vulnerability, and capability of disaster management in juridiction is potential for effective implementation of preparedness activity and response required to the early warning system. Article 5 and 7 of AADMER require control that implements continuously in support of risk identification and early warning activity.

2. Recovery
   The particular instrument of the working program in Article 17 of AADMER is rehabilitation. It is called "recovery" as the program includes all spectrums of the recovery process including rehabilitation and reconstruction. Moreover, it is secure the consistency with the term used internationally. Article 17 of AADMER summarizes that the member states, individually or collectively, should develop the strategy, implementation of the program, and increase cooperation (bilateral, regional, and international) for recovery of disaster.
3. Emergency Logistic Assistance
   The humanitarian assistance coordination center of disaster management in ASEAN, AHA Centre, issuing emergency logistic system. It is arranged in conjunction with the first anniversary of the AHA Centre in Malaysia Airbase at Subang of Malaysia on December 7th, 2012.
4. Logistic Distribution by AHA Centre
   Since its establishment, the organization showed its existence in response to the earthquake that occurred in Myanmar. AHA Centre distributes 250 tents, and 70 tarpaulins in response to immediate requirements confirmed by the government through NFP. The powerful earthquake of 6.8 Richter Scale hit Myanmar on November 19th of 2012. AHA Centre coordinating with RRD of the social department of Myanmar to implement the rescue mission. The main purpose of the team is to understand the impact and assessment for emergency response. Led by the executive director of the organization, the team is accompanied by the Senior Emergency Preparedness and Response Officer and Senior Programme Officer of Japan - ASEAN Integration Fund (JAIF) Management Team. Moreover, the team was supported by a member of ASEAN-ERAT of Myanmar and successfully reach the affected areas such as Kyauk Myaung and Shwe Bo city. By the mission, identified the logistic requirement must be immediate. The first distribution of tent arranged on November 23th from the UNHRD-WEP warehouse in Subang, Kuala Lumpur of Malaysia. Distribution of 250 tents was accomplished on November 26th of 2012, meanwhile, 70 tarpaulins were supplied by Myanmar locally. Handover from AHA Centre to RRD Social Department of Myanmar was carried out on Tuesday 27 November 2012 in Yangon, Myanmar.

3.2 Legal Aspect of Disaster Management Policy System In Indonesia

Indonesia is the largest archipelago in the world with 17.508 islands. It is a site among the Asia and Australia continentals and India and Pacific Oceans. Even though having abundant natural resources and wonderful islands, however, powerful concern must give as it has 129 active volcanos, known as the ring of fire, and it is located in the confluence of three active tectonic plates in the world. On another side, Indonesia was in a tropical area and hydrologic situation triggering other natural disasters such as storms, extreme rainfall, flood, landslide, and drought. Moreover, there are other threats besides natural disasters such as forest fires, social conflict, as well as technological failure.

The tragedy of Tsunami Aceh in 2004 was paid serious attention by the government of Indonesia and international for disaster management. In follow up the situation, issued the Decree of President No.83 of 2005 on National Coordination Agency of Disaster Management as there is no specific legal instrument in concern to the disaster management.

In 2007, therefore, the government establish a policy of disaster management system based on Act No. 24 of 2007 on disaster management with 85 Articles ruling on the mechanism. According to Article 1, described that disaster is an accident or sequel threatening and disturb of social living caused by, whether in natural or human error thus result in fatalities, environmental destruction, properties, and psychological impact. This Act classifies disaster according to the natural, engineering, and social. The system of disaster management policy of Indonesia is arranged by four aspects as follows:[6]

a. Social, economic, and culture.
b. Environment sustainability.
c. Useful and effective.
d. Scope of area

In the implementation of disaster management, the government can set the vulnerable area to forbidden one for settlement and/or repeal in the partial or comprehensive right, ownership on the property according to the regulation [7] The principle of disaster management involves humanitarian, justice, equality, position before the law and government, balancing, synchronization, and harmonization, order and legal certainty, togetherness, environmental sustainability, and science technology (Article 3). Meanwhile, implementation of disaster management consists of 3 phases including pre-disaster, in time-emergency response, and post-disaster.[8] Temporarily, in the potential situation of a disaster, there are preparation, early warning, and mitigation.

Additionally, the central and local governments are responsible for the implementation of disaster management (Article 5). According to Article 6 the government (central) is responsible for the implementation of disaster management including a) reduction of risk and combination between risk and development program, b) protection of people, c) properly right indemnity, d) recovery, e) allocation of state budgeting, f) ready-used fund budgeting, g) credible data and authentic document maintenance from disaster risk.

In the meantime, local government is responsible according to Article 8 as follows: a) assured social rights and refugees properly, b) social protection, c) risk reduction and combination between reduction and development program, and d) budgeting of local government; the Act No. 24 of 2007 was described to establish the Disaster Management National Agent follow up by Decree of President No. 8 of 2008.

Act No. 24 of 2007 resulted in alterations significantly on disaster management policy in Indonesia, whether at the national or local level, that in general, it is capable to protect people by management in characteristic, frequency, and understanding of vulnerability and risk. According to Article 8 of the Act mandated to establish a new unit, the Disaster Management National Agent (called, *BNPB*) substitutes the Disaster Management National Coordination Agent and Local Disaster Management Agent substitute the Implementing Coordination Unit and Local Coordination Unit. Based on Article 13 the Agent implements the responsibility as follows:

a. Formulate and establish the regulation of refugees-disaster management.
b. Fast response is effective and efficient.
c. Coordination in implementation of disaster management on schedule, integrated, and comprehensively.

In the meantime, duties of BNPB (Article 12) are as follows: a) provides the manual and instruction on disaster management effort including prevention, emergency response, recovery, and reconstruction in equal, b) establish the standard and requirement of disaster management according to the regulation, c) publishing information of disaster management in public, d) reports the implementation of management to the President monthly in a normal situation and all the times in emergency, e) uses and responsible the national-international assistances, f) responsible the received national budgeting, g) implement another responsibility according to the regulation, and h) arrange the establishment manual of Local Disaster Management Agent.

Moreover, the government exposed the chance to the third party to sustain the disaster management according to the Articles 28 and 30 that the business institution and international agent have an opportunity to implement disaster management, whether collective or individual is required to minimize the occurred disaster.

4 Conclusion

a. In the ASEAN Ministerial Meeting on Disaster Management (AMMDM) summit was issued ASEAN Agreement on Disaster Management and Emergency Response (AADMER) by ASEAN Committee on Disaster Management (ACDM) as the coordination center. AADMER is the regional framework with proactive for cooperation, coordination, technical assistance, and mobilization of resources in the whole aspect of disaster management. Any attempts of AHA Centre in disaster management in Southeast Asia is implementing the function in establishing the emergency response system starting from risk assessment, early warning and monitoring, recovery, and establish the logistic assistance system insecure the material availability required in proper, as a part of response system after the disaster in middle and high level.

b. Act No. 24 of 2007 on disaster management result in any alteration significantly to the disaster management in Indonesia, whether at the national and local level in general; the regulation can provide the safety for people in Indonesia in a characteristic manner, frequency, and understanding of vulnerability and risks. In this Act, mandated to establish new agent, Disaster Management National Agent (called, *BNPB*) substitutes the Disaster Management National Coordination Agent and Local Disaster Management Agent substitute the Implementing Coordination Unit and Local Coordination Unit. Article 23 of Act No. 24

of 2007 was described that the government (central) responsible to implement disaster management. Therefore, the policy of the system was ruled comprehensively by the regulation.

References

1. S. Zakiah, The Role of ASEAN Coordinating Centre For Humanitarian Assistance On Disaster Management (AHA Centre) for Disaster Management In Philipine, (2016).
2. M. Feybe,BNPB: The Risk Disaster in ASEAN Country, accessed from https://www.beritasatu.com/nasional/392135-bnpb-negara-asean-berisiko-bencana-yang-sangat-tinggi, (2021).
3. Supriyadi, Disaster Management in Indonesia, (2008).
4. Disaster Centre of IPB, Bogor, (2016).
5. Lili Rasjidi and Liza Sonia Rasjidi, Monograf : Philosopy of Science and Research Methode Bandung, 5. (2005)
6. Article 31 of Act No. 24, 2007 on Disaster Management.
7. Article 32 of Act No. 24, 2007 on Disaster Management.
8. Article 33 of Act No. 24, 2007 on Disaster Management.

Analysis of wind power potential as an electricity source for safety light and sign on limestone mining hauling road

Ikhsan Purnama[1] and *Muhammad Imran* Hamid[1*]

[1]Andalas University, Electrical Engineering Department, 25163 Padang, West Sumatera - Indonesia

Abstract. A work accident can occur anytime and anywhere to the worker if they take unsafe action and are in an unsafe condition. Mining areas located in the hills have unsafe conditions for operating heavy equipment, such as steep terrain conditions, slippery roads, and foggy weather, and lack of lighting. Installation of light assign is the right step in reducing work accidents at night, foggy and rainy weather. Measurement of wind data and calculation of wind potential statistically using Weibull distribution. The parameter values of scale and shape are 1.67 and 1,71 respectively, with an average wind speed of 1,49 m/s and the wind power potential of 3.14 W/m^2. Based on measurement and analysis results, this quarry always gets wind gusts both day and night, so it has the potential to take advantage of wind energy either on a small turbine scale or in small and medium electric power.

1 Introduction

Limestone is a raw material for cement production. In the mining of limestone, there are stages of the process, they are land clearing, drilling, blasting, loading, hauling, crushing, and conveying. Each of them has a high risk of work accidents. The mining process cannot be separated from the worker factor must also be considered, by guaranteeing health and safety and preventing accidents.

Work accidents can occur to workers anytime and anywhere if they take unsafe actions and they are in unsafe conditions. On the other hand, a work accident is undesirable even and often unpredictable, which can cause loss of a good time, possessions, or property or loss of life that occurred in a work mining process or related to it. Therefore, as soon as possible work accidents are minimized to make the whole mining process run smoothly and the safety of workers is protected [1].

The hauling process has a high level of risk of work accidents in the mining world which have a high impact on death and injury, as well as financial losses in the form of repairs, replacement parts, and downtime [2]. Mining areas located in the hills have unsafe conditions for operating heavy equipment, such as steep terrain conditions, slippery roads, and foggy weather, and lack of lighting. The performance of a truck driver or operator depends on fatigue, concentration, skills, training experience [2][3][4]. The ability to drive is often impaired by the failure of the work system on the machine and the environment such as heavy fog, dust, and rain [2][3]. Installing safety lights as a sign and powered by the wind as electricity is one of the right solutions in reducing work accidents at night, foggy, dusty, and rainy weather on hauling roads. Besides that, the safety light can be a warning sign of an earthquake for vehicles while going to pass through cliffs and ravines.

The wind is renewable energy, abundant, and easy to obtain. Wind energy can be converted into mechanical and electrical energy by a wind turbine. Wind intensity is different at each location on the earth, depending on time and position toward sea level. Wind dynamics can be studied by applying the probability density function or probability distribution [5]. This distribution can be used for modeling wind speed in various regions.

In probability theory and statistics, the Weibull distribution is a continuous distribution that was first introduced by the Swedish physicist Walodi Weibull in 1939 [6]. The Weibull distribution is widely used in wind energy studies because of its approach which is considered suitable and accurate to describe wind speed variations even with very small samples [7]. The potential use of wind energy as a power plant can be realized if it meets sufficient wind speed analysis data using certain methods [8].

The purpose of this study is to determine the characteristics of wind speed as a power plant expressed in the Weibull probability distribution curve. As well as knowing the form factor (k) and scale factor (c) of the Weibull probability curve as a feasibility reference and turbine selection, then the wind can be utilized as green energy to power safety lights and sign on the site.

2 Methodology

2.1 Location Overview and Measurement

Wind speed measurements are carried out on limestone mining quarry namely Karang Putih Hill, there are about

* Corresponding author: imran@eng.unand.ac.id

10 kilometers of existing hauling road on different altitudes. In this study, the wind speed was measured at 260 meters above sea level altitude, at the waist of the hill. Not far from the beach, about 13 kilometers away, topologically Karang Putih Hill located in Padang City, West Sumatera Province, Indonesia at (0°58'23.88" S and 100°28'15.88" E) and imaged by Google Earth software in Fig. 1.

Fig. 1. Karang Putih Hill imaged by Google Earth

The data needed are wind speed, daily air temperature, and air pressure. they were measured over 30 days (February 2021 - March 2021) at the one-minute interval. Wind speed was measured using a 3-cup anemometer. The resolution of the wind speed measurement is 0.1 m/s. The accuracy according to the manufacturer's specifications is ±3% over a speed range of 0 - 30 m/s. For temperature and humidity were measured using the DHT11 temperature sensor module. These sensors are connected to the data logger system using Arduino Uno microcontroller, depicted by Fig. 2, and then placed at 10 meters from the ground. Furthermore, data analysis was carried out.

Fig. 2. Hardware for wind measurement

2.2 Wind Data Assessment

Besides wind speed, air density also affects wind power. First, by measuring the altitude of the site, can get vertical air pressure. The higher elevations have lower air pressure because air has mass and is influenced by the earth's gravity-based on Torricelli's Law. In addition, horizontal air pressure by air temperature also affects air density, the cold air temperature has solid mass and the hot air temperature has floating mass. The air density, air temperature, and air pressure values are related and can be obtained from the following equation [9]:

$$\rho = \frac{P}{RT}(kg / m^3) \quad (1)$$

$$P = 760 - \frac{h}{100} x133(Pa) \quad (2)$$

(P), (R), (T), and (h) are air pressure, specific gas constant, ambient air temperature in Kelvin, and location altitude above sea level, respectively.

The wind is a developing atmospheric condition, its parameters change with time, therefore it is suitable to be calculated using probability density function (PDF). The probability density functions include; Weibull, Reyleigh, Gamma, Lognormal, Exponential, and Gausian [8][10]. In various literature, Weibull is the best method in determining wind energy potential [5], as follows:

$$f(v) = \frac{k}{c}\left(\frac{v}{c}\right)^{k-1} exp\left[-\left(\frac{v}{c}\right)^{k}\right] \quad (3)$$

Where $f(v)$, (k), and (c) are Weibull probability density functions in wind speed, shape parameter, and scale parameter respectively. A greater value of (k), will make a higher wind amount or wind speed duration and vice versa. The value of (c) affects the maximum and minimum values of wind speed. A greater value of (c), will make the curve shows high wind speed [11]. While the cumulative density function (CDF) of the Weibull distribution is as follows [9]:

$$F(v) = 1 - exp\left[-\left(\frac{v}{c}\right)^{k}\right] \quad (4)$$

The Weibull distribution is a useful function in analyzing reliability and maintenance. The value of the shape and scale parameters according to the Weibull distribution is very important in choosing a wind turbine installation location [12]. There are several methods for determining the parameter (k) and (c) of the Weibull distribution, namely empirical method, energy pattern factor method, graphic method, moment method, maximum likelihood method, maximum likelihood modification, and equivalent energy method [13]. This research uses an empirical method in which (k) value can be determined directly only by requiring the value of the average wind speed and standard deviation [14], after that we can get the value of (c) as in the following equation:

$$k = \left(\frac{\sigma_V}{v_m}\right)^{-1,086} \quad (5)$$

$$c = \frac{v_m}{\Gamma\left(1 + \frac{1}{k}\right)} \quad (6)$$

Where σ_v and v_m are the standard deviation of the wind speed data and the average wind speed respectively. The ratio v_m represents the average turbulence intensity of the area. For Γ() is Stirling's approximation of a gamma function [15] and can be calculated by the equation [16]:

$$\Gamma(y) = \int_0^{\infty} e^{-x} x^{y-1} dx \qquad (7)$$

The wind is air movement and has kinetic energy [17][18]. The power from the wind per unit area can be obtained by the equation (8) Meanwhile the instantaneous power generated from a wind turbine depends on the kinetic energy of the wind and the efficiency of the turbine by equation (9) as follows:

$$P(v) = \frac{1}{2}\rho A v^3 \qquad (8)$$

$$P_{turb}(v) = C_p \frac{\rho A v^3}{2} = C_p \frac{\pi \rho d^2 v^3}{8} \qquad (9)$$

Where A, ρ, and P_{turb} are turbine sweep area, air density, and turbine output power at wind speed (v). Then C_p is the overall wind turbine power coefficient, consisting of electrical efficiency, mechanical efficiency, and aerodynamic efficiency (Betz coefficient: 0.593) and d is propeller diameter [19].

To analyze the wind energy potential in a region at a period, two parameters should be known;
they are power density and energy density respectively [20]. The wind power density (WPD) is the amount of power or power produced per turbine sweep area and wind energy density (WED) is an estimation of average wind turbine output power over the period, for example, one year or one month [21][22][23]. They are obtained from the equation (10) and (12). The average WPD for the Weibull distribution can be obtained from equation (11).

$$WPD = \frac{P(v)}{A} = \frac{1}{2}\rho v^3 \qquad (10)$$

$$WPD = \int_0^{\infty} P_{out}(v) f(v) dv \qquad (11)$$

$$= \int_0^{\infty} \frac{1}{2}\rho v^3 f(v) dv$$

$$= \frac{1}{2}\rho c^3 \Gamma\left[1 + \frac{3}{k}\right]$$

$$WED = T\int_0^{\infty} P(v) f(v) dv \qquad (12)$$

$$WED = \frac{1}{2}\rho c^3 \Gamma\left[1 + \frac{3}{k}\right] T \qquad (13)$$

Where $P_{out}(v)$ is the instantaneous output power at wind speed (v), $f(v)$ is the frequency of the wind speed (v) (from the appropriate wind speed distribution measurement in the form of a PDF value for a certain period). (T) here is the period for which the average power output of P_{turb} is estimated, for example, one month T = 730 hours or one year T = 8760 hours.

Assessment of the speed and potential wind energy parameters by considering the reference from LAPAN [21] class/scale table as the minimum and maximum limits of wind energy that can be utilized for electrical energy, resulting in the assumption of modified parameters as shown in Table 1.

Table 1. Turbine scale based on minimum wind speed

Turbine Scale	Wind Speed Minimum (m/s)	Blade Diameter (m)
Very Small	1,5	2
Small	2,6	5
Medium	4,1	20
Big	5,1	50

3 Result and Discussion

Three points of analysis will determine the wind potential value of a site, they are air density, wind speed variation, wind power, and wind energy density.

3.1 Air Density

Using equations (1) and (2), the average daily air density is between 1.19 kg/m³ to 1.14 kg/m³. Fig. 3 is the comparison of air density with ambient air temperature. Higher air temperature results in lower air density.

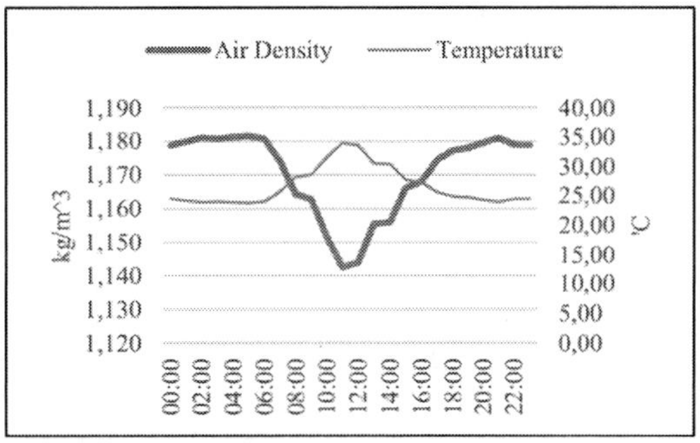

Fig. 3. Daily air density based on daily air temperature.

3.2 Wind Speed Variation

Wind data were collected for one month in February 2021. The whole daily average and daily powerful wind speed are graphed in Fig. 4 and Fig. 5. They show the highest average daily wind speed is at 10:00 to 15:00, and the lowest average daily wind speed is at 19:00 evening and 02:00 midnight. It means this site often gets wind gusts in the nighttime more than 2 m/s.

Table 2 is the result of measuring the number of wind speed frequencies that occur during one month. The obtained speed range of 1.5 m/s is the most frequent wind speed. Statistically from the data can be obtained the average wind speed is 1,49 m/s. Furthermore, to determine the Weibull distribution, it takes the value of the shape parameter (k) and the scale parameter (c). These parameters will be used to determine the potential for wind power and energy in the site. But these parameters will vary according to time and location, as well as wind speed over time.

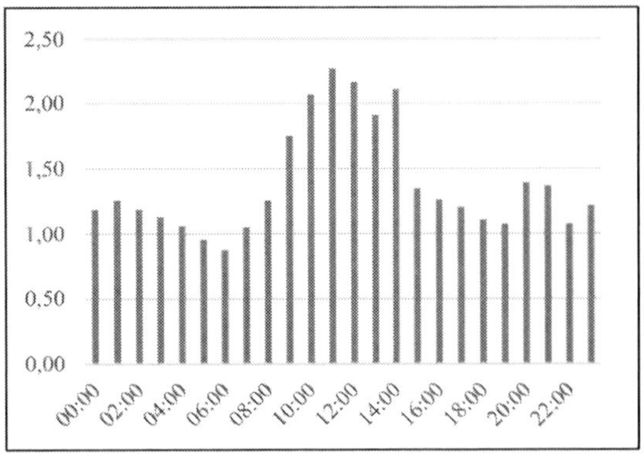

Fig. 4. Daily average wind speed in February 2021.

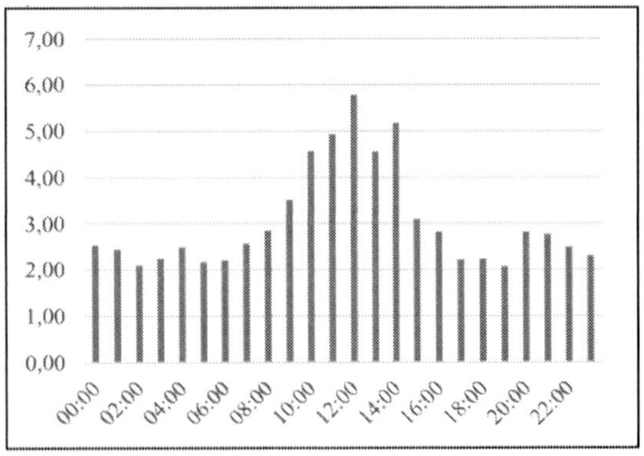

Fig. 5. Daily powerful wind speed in February 2021.

By using equations (5) and (6) the parameters of the Weibull distribution are obtained. They are 1,71 for (k) value and 1,67 for (c) value. Then by using equations (3) and (4), the probability density function (PDF) and the cumulative density function (CDF) can be determined which is graphed in Fig. 5.

Table 2. Weibull distribution calculation results.

Speed Range (m/s)	Freq	(%)	PDF	CDF
0	79	0,18	0,00	0,00
0 - 0,5	4775	11,16	0,33	0,04
0,5 - 1	8952	20,92	0,44	0,22
1 - 1,5	12277	28,68	0,40	0,46
1,5 - 2	8031	18,76	0,31	0,66
2 - 2,5	3922	9,16	0,22	0,81
2,5 - 3	2078	4,86	0,14	0,9
3 - 3,5	1166	2,72	0,08	0,96
3,5 - 4	624	1,46	0,05	0,98
4 - 4,5	361	0,84	0,03	0,99
4,5 - 5	227	0,53	0,02	1,00
5 - 5,5	167	0,39	0,01	1,00
5,5 - 6	40	0,09	0,01	1,00
6 - 6,5	72	0,17	0,00	1,00
> 6,5	29	0,07	0,00	1,00

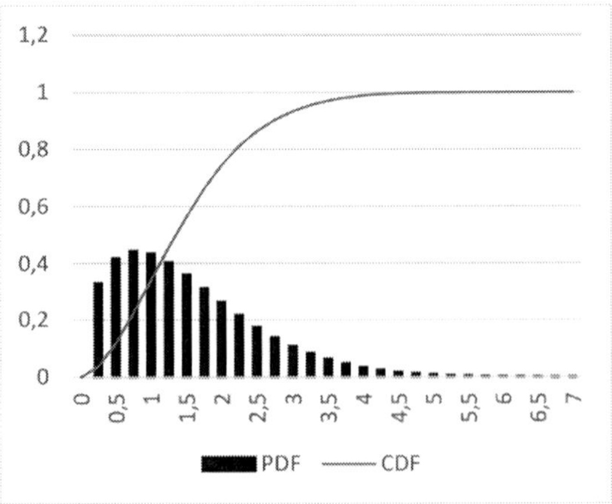

Fig. 5. PDF and CDF chart.

3.3 Wind Power and Energy Density

Based on the data obtained from the wind speed distribution, then the calculation of the potential power and wind energy per month is carried out with equations (10) and (12) so that the results are obtained in Table 3. The potential for wind power and energy at this location is 3,14 W/m^2 and 2,26 kWh/m^2 when using a turbine with a 1 m^2 sweep area. Based on Table 1 and the analysis result, the wind potential of this site can be utilized by using very small and small wind turbine scales. As a result, monthly wind energy can be obtained effectively of 7,10 to 44,37 kWh/m^2. Table 4 gives example for using various wind turbine blades or rotor diameters.

Table 3. Analysis result of wind power potential.

ρ (kg/m^3)	Parameter		Mean (m/s)	WPD (W/m^2)	WED (kWh/m^2)
	k	c			
1,17	1,71	1,67	1,49	3,14	2,26

Table 4. Wind power density and wind energy density monthly achieved based on various wind turbine blade diameter

Rotor Diameter (m)	WPD (W/m^2)	WED (kWh/m^2)
1	2,46	1,77
2	9,86	7,10
3	22,18	15,97
4	39,44	28,40
5	61,62	44,37

4 Conclusion

This study focuses on wind potential assessment used for safety lights and signs on mining hauling roads in Karang Putih Hill. Based on shape and scale parameters, can we infer that wind characteristic in the hauling road at 260 meters altitude has regular wind gust with low to medium wind speed variation? So that the analysis results of air density, wind variation, and wind power density, this site has the potential to utilize wind energy by using up to a small turbine scale.

The authors would like to express gratitude to the Electrical Engineering Department of Andalas University for providing publication support under Publication Grant Scheme (No. 79/UN.16.09.D/PL/2021)

References

1. Institute of Electrical and Electronics Engineers, "*2016 2nd International Conference of Industrial, Mechanical, Electrical, and Chemical Engineering (ICIMECE) : October, 6-7, 2016, Royal Ambarrukmo Hotel, Yogyakarta, Indonesia.,* "196–200 (2016)
2. E. Sun and X. Zhang, "3D assisted driving system for haul trucks in surface mining," *Proc. 2011 Int. Conf. Transp. Mech. Electr. Eng. TMEE 2011*, 363–366, (2011)
3. Y. Nishimura, T. Osafune, S. Kato, A. Hiromori, H. Yamaguchi, and T. Higashino, "Vehicle Proximity Awareness by Inter-Vehicle Communication for Surface Mine Operation Safety," *IEEE Veh. Technol. Conf.*, **2017**-June (2017)
4. E. Sun, "The fuzzy neural network based haul truck driver fatigue detection in surface mining," *ICNC-FSKD 2017 - 13th Int. Conf. Nat. Comput. Fuzzy Syst. Knowl. Discov.*, no. 2016, pp. 1522–1526 (2018)
5. M. B. Hemanth Kumar, S. Balasubramaniyan, S. Padmanaban, and J. B. Holm-Nielsen, "Wind energy potential assessment by Weibull parameter estimation using multiverse optimization method: A case study of Tirumala region in India," *Energies*, **12**, 11 (2019)
6. L. G. Otaya, "Distribusi Probabilitas Weibull Dan Aplikasinya," *J. Manaj. Pendidik. Islam*, **4**, 2, pp. 44–66 (2016)
7. E. Parameter and I. Tsani, "Estimasi Parameter ... (Indria Tsani),"27–35.
8. J. Waewsak, C. Chancham, M. Landry, and Y. Gagnon, "An Analysis of Wind Speed Distribution at Thasala, Nakhon Si Thammarat, Thailand," *J. Sustain. Energy Environ.*, **2**, no. January, 51–55, (2011)
9. Z. H. Hulio, W. Jiang, and S. Rehman, "Techno-Economic Assessment of the wind power potential of Hawke's Bay using Weibull parameter: A review," *Energy Strateg. Rev.*, **26**, no. February, 100375 (2019)
10. O. Douglas Queiroz de Oliveira Filho *et al.*, "Analysis of Weibull parameters for wind power generation," no. March 2018 (2018)
11. Ilyas, Suriadi, and M. Gapy, "Analisis Potensi Tenaga Angin Menggunakan Metode Distribusi Weibull di Kawasan Banda Aceh," *Semin. Nas. dan Expo Tek. Elektro*, no. February, 66–68 (2017)
12. P. Bhattacharya, "Weibull Distribution for Estimating the Parameters," *Wind Energy Manag.*, no. 1 (2011)
13. S. Deep, A. Sarkar, M. Ghawat, and M. K. Rajak, "Estimation of the wind energy potential for coastal locations in India using the Weibull model," *Renew. Energy*, **161**, 319–339 (2020)
14. C. F. De Andrade, H. F. Maia Neto, P. A. Costa Rocha, and M. E. Vieira Da Silva, "An efficiency comparison of numerical methods for determining Weibull parameters for wind energy applications: A new approach applied to the northeast region of Brazil," *Energy Convers. Manag.*, **86**, 801–808 (2014)
15. S. L. Walker, "Building mounted wind turbines and their suitability for the urban scale-A review of methods of estimating urban wind resource," *Energy Build.*, **43**, 8, 1852–1862 (2011)
16. R. Maouedj *et al.*, "Wind energy resource assessment at three sites in the Algerian highlands," *Proc. 2018 6th Int. Renew. Sustain. Energy Conf. IRSEC 2018*, pp. 0–5 (2018)
17. A. Serban, L. S. Paraschiv, and S. Paraschiv, "Assessment of wind energy potential based on Weibull and Rayleigh distribution models," *Energy Reports*, **6**, April, 250–267 (2020)
18. F. Oral, I. S. Ekmekçi, and N. Onat, "Weibull distribution for determination of wind analysis and energy production," *World J. Eng.*, **12**, no. 3, pp. 215–220 (2015)
19. R. D. Prasad, R. C. Bansal, and M. Sauturaga, "Wind modeling based on wind input data conditions using Weibull distribution," *Int. J. Glob. Energy Issues*, **32**, 3, 227–240 (2009)
20. C. Paper *et al.*, "Wind energy potential assessment to estimate the performance of selected wind turbine in the northern coastal region of Semarang-Indonesia," no. January (2017)
21. A. Fadlullah, S. Riyanto, and S. Sudirman, "Pengembangan Aplikasi Analisis Potensi Angin Sebagai Sumber Energi Terbarukan Menggunakan Estimasi Parameter Weibull Berbasis Metode Power Density (Studi Kasus: Kota Tarakan)," *Inovtek Polbeng*, **9**, 1, 129 (2019)
22. B. S. Premono, D. D. D. P. Tjahjana, and S. Hadi, "Wind energy potential assessment to estimate the performance of selected wind turbine in the northern coastal region of Semarang-Indonesia," *AIP Conf. Proc.*, **1788**, no. January (2017)
23. M. A. Hyams, "Wind Energy in the built environment," *Metrop. Sustain. Underst. Improv. Urban Environ.*, pp. 457–499 (2012)

Vertical and horizontal accountability on disaster preparedness and response of two highly urbanized cities in Metro Manila, Philippines

Marlon Era[1]*

[1]De La Salle University, Manila, Philippines

Abstract. This paper focuses on the accountability of disaster preparedness and response of two cities in Metro Manila, Philippines. The paper draws a lesson from the experiences of two highly urbanized cities in Metro Manila and how coordinated efforts in disaster preparedness and response among the members of the City Disaster Risk Reduction and Management Councils (CDRRMC) were accounted for. This paper provides a deeper understanding of how city governments fulfill their commitments and responsibilities to disaster victims during disaster preparedness and response. This study utilized key informants interviews and focus group discussions among key members of CDRRMC and review of existing laws and relevant policies. The upward vertical accountability is well established as this involves members of the City DDRM Councils as the council members are accountable to their respective City Mayors. Downward vertical accountability is not well defined as this refers to the accountability of DRRM frontline service workers to the disaster victims. The horizontal accountability among members of the City DRRM Councils is yet to be established as this involves different department heads who are at the same level of power and authority. The paper provides fresh insights on the vertical and horizontal accountability in disaster governance of cities in Metro Manila.

1 Introduction

Vertical and horizontal accountability are forms of accountability and are facets of democracy and good governance. These two forms of accountability exist in disaster risk reduction and management (DRRM). Due to the severe impact of disasters experienced by the affected countries, it is imperative to establish accountability in disaster governance as it promotes the stakeholders' fulfilment of responsibilities.

Globally, there have been significant achievements in DRRM. However, there remain certain issues that nations encounter in disaster governance. First, some countries fail to demonstrate good governance, which hampers the effective implementation of DRRM. Second, there can be challenges brought by having multiple disaster governance paradigms in the processes of DRRM.

Firstly, countries become vulnerable to disasters when they suffer from having bad governance. National development and disaster risk reduction efforts are cultivated when features of good governance, i.e., "participation, consensus, accountability, transparency, responsiveness, effectiveness, efficiency, equity, inclusivity and rule of law" are present in a country [1] (Khotami, 2017). Conversely, [2] Van Niekerk (2015) pointed out that carrying out effective disaster governance can be difficult when such features of good governance are non-existent in a country. The susceptibility to disasters results from institutional failure—which is associated with bad governance. Institutions are comprised of various state and non-state actors that make up the governance structure that formulates public policies and provides the needs of the citizenry. When such a governance structure fails to deliver services to the public, consequently, it results in institutional failure, and ultimately exposes the nation to disaster risk [3] (Ahrens & Rudolph, 2006).

Secondly, governments can encounter difficulties in the emergence of multiple paradigms and paradigm shifts in DRRM. Disaster governance paradigms, which guide plans and policies, have changed and shifted during the course of nations' plethora of disaster experiences. However, some countries struggle in handling multiple and shifting disaster governance paradigms. For instance, international institutions advocate for a shift in disaster management paradigm—from "'response and recovery' to 'prevention and preparedness'"; this paradigm shift is represented through Disaster Risk Reduction (DRR). [4] Jones et al. (2015) explained the difficulty in mainstreaming DRR into national development activities,

* Corresponding author: marlon@dlsu.edu.ph

which requires shifting the paradigm to prevention and preparedness. This is because disaster managements are still response-oriented than prevention-oriented. This can also be difficult for nations that lack resources to achieve this [5] (Jones et al., 2015). Another example is the case of Nepal where various state and non-state actors uphold their paradigms in disaster governance. The differences in paradigms among these actors result in the division of resources, competition among them, and struggles in campaigning for their paradigm [6] (Vij et al., 2020).

Specifically, some developing countries in Asia and some parts of the West, also deal with issues in DRRM. The local governments, who are a fundamental unit in disaster governance, lack the capacities to administer management in disasters. Likewise, some countries suffer the poor legislation that supports DRRM. Also, there have been ambiguous hierarchical structures and delegation of responsibilities among stakeholders in some nations, which entail difficulties in DRRM.

The local government units play a great role in disaster governance. Local or regional areas of a country are the ones generally hit by disasters, and therefore they typically carry out the activities to handle disasters [7] (Jones et al., 2015). However, there is a profound lack of capacities in administrative, organizational, financial, and political aspects in developing countries [8] (Ahrens & Rudolph, 2006). In the case of Nepal, the organizations at the local level struggled with a deficiency in financial and technical resources. This resulted in the inability to deal with large-scale disasters and to implement disaster-related activities. Such insufficient capacity was attributed to a lack of autonomy as they were restricted to making their own decisions and mostly depended on assistance from higher-level organizations [9] (Choudhury et al., 2019). The issue of lack of capacity at the local level is also evident in India where human, material and financial resources were inadequate. However, in contrast, in their case, decentralization or having autonomy empowered the local government and general population [10] (Pal & Shaw, 2018). Furthermore, lack of capacity is also suffered by other countries such as Nepal and Timor-Leste [11] (Jones et al., 2014 & [12] Cook et al., 2019).

Legislation is an integral part of disaster governance, as it "provides the framework around which strategies to build risk reduction into development and reconstruction activities can be empowered" [13] (Pal & Shaw, 2018). However, legislation for disaster governance aiming for public development tends to not be prioritized due to other matters, such as socioeconomic issues being prioritized instead [14] (Samutereko, 2021) as well as deficiency in financial resources [15] (Pal & Shaw, 2018). [16] Samutereko (2021) explained that, in Zimbabwe, they have a weak institutional basis for DRR at the local levels. Disaster legislation has not been prioritized in the country. Thus, he pointed out the need to revise the country's disaster legislative framework in which vulnerability reduction and community resilience have to be taken into account to tackle disasters effectively. Similarly, in Pakistan, [17] Deen (2015) called for the government to formulate policies that help decrease vulnerability to disasters as well as to prioritize disaster preparedness and response at the national level.

The difficulties exist among state and non-state actors in terms of hierarchical structure and distribution of responsibilities. These actors include government institutions, non-governmental organizations (NGOs), international non-governmental organizations (INGOs), and other civil society actors. [18] Jones et al. (2015) discussed the evolution of disaster risk governance wherein delegation of influence and responsibility among different stakeholders has been "reoriented upward, outward, and downward." It then creates vagueness as to "who governs DRR?" This is somewhat reflected in the case of Uganda where people "were lost in the complex hierarchical structure" and there was a lack of clarity of the responsibilities and roles vested on the actors involved in disaster management [19] (Maes, 2018). This issue is also evident in the case of Nepal where it was "difficult to locate centers of power or identify who is leading the [disaster risk reduction] process" [20] (Jones et al., 2014).

The Philippines also experiences certain problems concerning the implementation of DRRM. In particular, issues in the local government in the Philippines have been present. [21] Sonny & Manejar (2018) explained these issues. For instance, the DRRM was not mainstreamed within the local government as manifested by certain misalignments between LGUs' local plans with the major planning documents for DRRM. In addition, the number of permanent appointees does not meet the minimum staff number per LGU, as suggested in the policy. Also, the mandated 5% funds of DRRM allocated for the LGUs were underutilized.

In the case of Typhoon Yolanda, Sonny & Manejar (2018) discussed several problems that occurred as the government tackled the disaster. First, the local government units (LGUs) were "sidelined three years after RA 10121 was passed in the wake of Yolanda." Second, command responsibility and the persons who were supposed to be in charge of decision-making were ambiguous. Third, despite the issue of ambiguity at hand, multiple actors who were involved did not coordinate among themselves. Lastly, funds from the government were not granted to municipal mayors, and funds from NGOs were delayed due to the bureaucratic processes [22] (Sonny & Manejar, 2018). Furthermore, the LGUs themselves were ravaged by the impact of the typhoon and were immobilized at the time and the government officials at the higher level presided. However, during the disaster, there had been "too much politicking" as those officials who took over had arguments among themselves and were sensationalized publicly [23] (Blanco, 2015).

1.1 Accountability

Critical to that of a developing democracy is the delivery of much-needed services to its citizenry. Hence, accountability connotes both the responsiveness of elected officials to citizens' demands and the responsibility that government officials have to act upon those preferences [24] (Acosta et al., 2010). [25] Amaratunga et al. (2016) stated that "[accountability]

https://doi.org/10.1051/e3sconf/202133102017

rather than being a bureaucratic or legal term, is about improving democratic processes, challenging power and claiming citizenship." It is also instrumental in establishing public trust towards the performance of the government [26] (Khotami, 2017). Accountability is also a key feature in building good governance [27] (Khotami, 2017 & [28] Bovens, 2007). Governance refers to the decision-making and implementing processes among governments and public or private institutions [29] (Yap, 2007). As stated earlier, good governance includes "participation, consensus, accountability, transparency, responsiveness, effectiveness, efficiency, equity, inclusivity and rule of law" [30] (Khotami, 2017).

The concept of accountability indicates a relationship between, at least two types of actors, "a principal" who entrusts (a task or responsibility) to "an agent" for the sake of the principal [31] (Acosta et al., 2010). Accountability takes various forms, two of them being—vertical accountability and horizontal accountability. Vertical accountability is a principal-agent relationship in which the principal (state's citizenry) holds the agents (governments) accountable for their actions. Horizontal accountability involves various institutions of different government branches that hold each other accountable and monitor so that no institution exhibits abuse of power and such [32] (O'Donnell, 1998 as cited in Acosta et al., 2010). In addition, accountability entails concurrence among organizations and individuals on their assigned roles, tasks, and responsibilities as well as reporting on relevant actions [33] (Amaratunga et al., 2019).

There are two mechanisms in pursuing accountability in relationships between citizenry and government and government performance. First, Political Mechanism—referring to a specific form of relationship between elected government officials being directly responsible for their public action, including but not limited to the provision of public goods. This presents an opportunity for voters to reward or sanction their government officials by granting them the right and duty to re-elect or not their political representatives out of office due to their vertical linkage. This is regardless of whether these officials have complied or defected with the electoral mandate [34] (Manin et al., 1999). Second, Social Mechanism—involves social dynamics and whose processes that may not take place via formal electoral procedure but through street protests, mobilizations, and social media. The main feature of this mechanism is to achieve direct influence on government officials to acquire the desired result of effective government action. Albeit its effectivity is short-termed and not binding, it applies to services that cover, health, sanitation, water, transportation, wages, etc. [35] (Acosta et al., 2010).

1.2 Dimensions of Accountability

Accountability encompasses financial, legal, and social dimensions. These dimensions of accountability are necessary as a part of the democratic venture and on the account of good governance. First, financial accountability is the provision of relevant information and reports on financial activities and the performance of the stakeholders [36] (Schiavo-Campo and Tomasi, 1999 as cited in Zeyn, 2018). It also connotes the obligation of the stakeholders to manage the funds responsibly [37] (Madiasmo, 2002 as cited in Zeyn, 2018). In the context of public financial accountability, at the core of it is "the checking and balancing of potential abuse of power by public officials to limit the potential for corruption of public offices and officials" [38] (Khaile & Davids, 2020).

In addition, an important activity that strengthens financial accountability is a financial audit. It is the examination of financial statements to ensure transparency of stakeholder actions [39] (Zeyn, 2018). [40] Dewi et al. (2019) also pointed out that the information quality of financial statements influences financial accountability. Therefore, a financial audit is an important aspect of the institutional management function.

Second, legal accountability involves processes that "seek to ensure that government delivers on electoral promises, fulfills the public trust, aggregates and represents citizens' interests, and responds to ongoing and emerging societal needs and concerns" [41] (Brinkerhoff, 2017). It involves elected public officials' fulfillment of their duties, such as providing public needs [42] (Acosta et al., 2010). In addition, within such dynamics, citizens observe the officials' performance in terms of establishing and executing policies, building public confidence, and carrying out societal rules [43] (Brinkerhoff, 2017).

Third, [44] Amaratunga et al. (2016) argued that accountability in DRR is "more social than political." Social accountability drives increased participation of citizens in checking the governments and their actions in fulfilling their responsibilities [45] (Amaratunga et al., 2016). Social accountability moves beyond formal ways, i.e., elections, (contrary to political accountability) but rather it is "a continuous relationship of citizen demands' through street protests and mobilizations, *public naming and shaming*, signing of petitions, etc." [46] (Acosta et al., 2010). Its salient feature is to get direct answers from the government as demanded by the citizens [47] (Acosta et al., 2010).

1.3 Accountability in Disaster Risk Reduction and Management

Accountability is invaluable in ensuring the government's provision of public goods and services and is a key feature of good governance. Its role extends to DRRM. The stakeholders who are in charge of governance obligations are held accountable to guarantee successful DRRM [48] (Amaratunga et al., 2019). Accountability in DRRM aims to enable scrutiny and understanding of the measures taken by involved actors and the actors themselves [49] (Amaratunga et al., 2016).

Due to the severe losses brought about by disasters, the key stakeholders' performance in DRRM must be enhanced. This can be achieved by establishing systematised effective accountability mechanisms. These mechanisms are ought to be conducted during the various phases within the cycle of disaster management (i.e., response, recovery, rehabilitation, reconstruction, prevention, mitigation, and preparedness) [50] (Amaratunga et al., 2016). Moreover, according to [51] Amaratunga et al. (2019), accountability can be ensured through: mainstreaming DRR into overall national policies, enabling legislation, institutional development, acquiring adequate resources, empowerment of stakeholders, and regular monitoring, evaluation, and review.

There are international frameworks that guide the governments in formulating policies and mechanisms for DRRM. Some of the widely-recognized frameworks globally are Hyogo Framework for Action (HFA), Sendai Framework for Disaster Risk Reduction (SFDRR) 2015-2030, Climate Change Adaptation (CCA), among others. These frameworks emphasize the importance of accountability. For example, the SFDRR acknowledges the significant role of the government in fulfilling their responsibilities, particularly, in dealing with disaster risks and thus it recognizes the importance of holding them accountable for their actions [52] (Amaratunga et al., 2019). In the Philippines, Republic Act 10121 mandates the governments to practice transparency and accountability as attributes of good governance [53] (Republic Act No. 10121, 2010).

1.4 Policy on DRRM Accountability

The knowledge and experiences gained from venturing to prevent, mitigate, prepare and respond to disasters have been shared among international entities and nations [54] (Amaratunga et al., 2019). Frameworks containing lessons learned and guidelines have been created to direct governments in establishing effective DRRM. The Sendai Framework for Disaster Risk Reduction (SFDRR) 2015-2030 recognizes the State's principal role in achieving disaster risk prevention and reduction. Together with the State, it emphasizes the role of other stakeholders, stating that:

Disaster risk reduction and management depend on coordination mechanisms within and across sectors and with relevant stakeholders at all levels, and it requires the full engagement of all State institutions of an executive and legislative nature at national and local levels and a clear articulation of responsibilities across public and private stakeholders, including business and academia, to ensure mutual outreach, partnership, complementarity in roles and accountability and follow-up [55] (Sendai Framework for Disaster Risk Reduction 2015-2030, 2015).

In the Philippines, the Philippine Disaster Risk Reduction and Management Act of 2010 (Republic Act No. 10121) is the national law that lays out the policies, plans, and programs that are to be taken when dealing with disasters (Republic Act No. 10121, 2010). The law also assigns the government departments with responsibilities and respective agencies to be supervised. The highest-level government agency that is organized by the law is the National Disaster Risk Reduction Management Council (NDRRMC) or National Council that supervises disaster governance [56] (Republic Act No. 10121, 2010 & Opulencia-Calub et al., 2017). It is administered by the Office of the Civil Defense under the Department of National Defense (DND) which oversees the planning, training, coordination, emergency response, and operational support in DRRM [57] (Opulencia-Calub et al., 2017). The Council is mandated to create "an enabling environment for government action at all levels, and embed a culture of continued accountability to and communication with people affected by a disaster" [58] (Opulencia-Calub et al., 2017). Other departments include those who supervise the four thematic areas of disaster in RA 10121, namely, the Department of Interior and Local Government for preparedness, Department of Social Welfare and Development for a response, Department of Science and Technology for prevention and mitigation, and the National Economic and Development Authority for rehabilitation and recovery [59] (Republic Act No. 10121, 2010).

Furthermore, the law provides directives for the NDRRMC, in particular, the civil defense offices, to assist the regional and local councils in disaster governance (Opulencia-Calub et al., 2017 & Blanco, 2015). The law mandates the organization of offices, that is the Local Disaster Risk Reduction and Management Office (LDRRMO), at all local government levels: provincial, city, municipal, and barangay (Republic Act No. 10121, 2010). These local offices are to administer the DRRM activities and plans and capacity-building and knowledge management for DRRM [60] (Opulencia-Calub et al., 2017).

2 Objectives and Method

The study initially documented the horizontal and vertical accountability of Cities of Quezon and Muntinlupa to appraise if certain provisions (Sections 12, 19 & 20) from Republic Act 10121 were enforced by the Local Disaster Risk Reduction Management Offices (LDRRMOs).

Table 1 FGDs Conducted in Muntinlupa and Quezon City

Date	Participants	Venue
August 30, 2017	City Disaster Risk Reduction and Management Office; City Health Office; Department of Education; City Engineering Office; City Social Services	Muntinlupa City Hall
September 2017	City Disaster Risk Reduction and Management Office; City Health Office; City Environment and Natural Resource Office; Gender and Development Office and City Planning and Development Office	Quezon City Hall

Given that much of the LDRRMO's performance presented and much-publicized are those derived from its upward and downward accountability. Study findings were also derived from the Focus Group Discussion (FGD) conducted in both cities in August and September 2017 respectively which were participated by the members of the Disaster Risk Reduction Management Councils (Table 1). Likewise, selected representatives of the two cities were interviewed as key informants. All data about the research inquiry were enumerated according to questions posed by the research. Questions ranging from their role identification to their accountability were documented to appraise the existing condition and status in both cities regarding their respective DRRM councils. However, the last set of three (3) questions revealed the emerging pattern/s from which to address the concerns and issues derived from both study areas.

3 Findings

Particular findings revolved around identifying the thematic groupings involved in the DRRMC; their roles and responsibilities; performance rating of their respective committees and council in the implementation of the disaster mitigation and response in the last five (5) years; the major accomplishments as a committee in said implementation; the issues and challenges encountered by their committees in said implementation; level or degree of accountability with regard to the respective committees actual duties and responsibilities; the feedback mechanism of each committee in the city disaster response and mitigation; presence of a feedback mechanism of DRR in the city; factors facilitating accountability in the performance of disaster response and mitigation in the city; factors that hindered accountability in the performance of disaster response and mitigation in the city; and what the recommendations are to improve accountability in the DRRM council especially in the implementation of disaster response and mitigation.

3.1 Accountability in Disaster Preparedness and Responses

The existing conditions in the two (2) cities relative to their accountability in Disaster Preparedness and Response (DPR) were determined. The DPR includes how the City Disaster Risk Reduction Management Office (C/DRRMO) accounts to their Superior or Donor in the execution of their functions. For both Quezon City and Muntinlupa City, DRRM Council representatives claimed that they are accountable to the Office of the Mayor since that office is the one making final approval of the DRRM funds release and utilization. However, the City Health Office (CHO) of Muntinlupa clarifies that in the case of in-kind donations, CHO representatives view no need for accountability for such type of donation.

Other responses elicited from the same FGDs regarding accountability albeit are concerning disaster victims and their families. There were different sets of responses from both Quezon City and Muntinlupa, as documented by the research team. Representatives from two (2) (Environment and Engineering) of the three (3) respondents from the Quezon City DRRMO claimed no accountability as they provide only services to DRRMO responders and are not front liners. The CHO, by contrast, concur being accountable towards disaster victims as they are in direct contact with the beneficiaries of their health services.

The majority of the participants from Muntinlupa declare that the DRRMO, like in any LGUs, aims for zero casualties, making the office accountable to every victim. This was affirmed by the Medical Controller and Physician of the DRRMO who backed up his claim that all barangay personnel of 9 barangays of Muntinlupa was trained to respond to emergency cases as an initial action towards zero casualties.

Concerning the accountability with other members of the City DRRM Council, all respondents from Muntinlupa FGD reveal that they have yet to establish accountability with other members of the City DRRM Council. Participants declare that the issuance of accountability within said council is subject to further discussion and without prejudice being that the establishment of the DRRMO is still in its nascent stage.

However, the Philippine National Police and the Department of Education, both members of the DRRM council, mention the various modalities in which they can be held accountable. For example, an administrative case may be filed against the Police Officer of the Philippine National Police (PNP) at their headquarters should they encounter an erring police officer. For the Department of Education, their modality is to rely mostly on their Brigada Eskwela,(school brigade) which operates before the school openings only, and on the teachers for the rest of the year. The City Health Office claims open lines of communication between the CHO staff city and barangay concerning the DRRM program. Meanwhile, the City Engineering Office has yet to determine their DRRM accountability having recently joined the Council.

Quezon City DRRM council members were asked about responsibilities during disaster preparedness and response. The Head of the City Engineering Office cites no accountability within the DRRM Council, while the DILG representative notes its accountability including the Bureau of Fire Protection and the Quezon City Council Secretary.

The explanation is to ensure that information given to the Department of Interior and Local Government, coming from regional and national channels, is expected to be coordinated with local services. A representative from the Gender and Development (GAD) department confirms its accountability within the council but it is not proactive. The City Health Office (CHO) reports feedback among members of the council and works very closely with each member. In addition, they feel accountable to the Council and free and open to suggestions.

3.2 Vertical Accountability among City DRRM Councils

Results of the study indicated the various conditions which substantiate the two (2) types of accountability. Beginning with the vertical accountability for the Quezon

City Disaster Risk Reduction Management Council, all FGD participants signified their familiarity with their respective committees in the City's DRRM Council namely: Disaster Preparedness, Search and Rescue, and Prevention and Mitigation. The Department of Environment and Natural Resources (DENR) representatives are included as part of the operating team handling relocation emergency, relocation and rehabilitation, and emergency and rapid assessment operating team. Adding the participation of the Engineering Department–National Capital Region (NCR) that provides the services of intervention, restoration, and post-rehabilitation of facilities, aside from the Technical Assistance provided to the DRRMO, has also been documented.

3.3 Horizontal Accountability among City DRRM Councils

The existing horizontal accountability noted in this particular study is according to set standards by the government; the main to this is its observance to prescribed national standards. Horizontal accountability has a basic core of standards set by the Commission on Audit (COA) and those prescribed by government awarding bodies i.e. KALASAG and Seal of Good Local Governance as well as standards as suggested concerning internal operations (i.e. Annual Investment Plan) of its local government units. This means all government officials and employees under each department, division, down to its local units are cognizant of their accountability towards their respective department heads, mayors, headquarters, and the like.

On the matter of horizontal accountability for the City of Muntinlupa, all members of the city DRRM council are aware of their accountability to the City Mayor and the modalities in which this is enforced. Regarding accountability to disaster victims, who are the clients or beneficiaries of various disaster response interventions, during the FGD, many of the participants declared their aim for Zero casualty—making the City DRRMC accountable for every victim. Meanwhile, from the Quezon City DRRMC, most of those who recognized accountability are those teams in direct contact with beneficiaries or front-liners such as the Officer-in-Charge for Disaster, Community Health Office, and the DRRMO. The rest of the DRRMC members claimed to not view accountability to the victims as they do not have direct contact with the disaster victims.

Based on the documented responses from both Focus Group Discussion of Quezon City and Muntinlupa, both groups are unclear of their accounts within their respective councils. What is mostly indicated from their responses is their accountability in the upward pattern and that is leading towards their City Mayors and their respective department heads (i.e. DILG Secretary, PNP Chief, etc.). There is also the lack of appreciation for downward vertical accountability among the council members for their target clients which is seemingly treated as part of the demographics. The quality of feedback needs to be evaluated to provide better service to its constituents with a set of standards with regards to their feedback mechanism to change the culture of demography/statistics of its constituents and provide quality government service.

4 Conclusion

Good governance recognizes the fact that accountability aims to ensure an improvement in the delivery of public services, measuring performance, and providing incentives to achieve targets and sanctions in case of non-performance among institutions. It forms a part of a whole among the main characteristics of good governance to which the desired results is to promote community confidence assuring the community that its local government will act and pursue the community's overall interest regardless of dissenting opinions among members. This also serves as a reminder for local governments that they are acting on behalf of their community to help them understand the importance of having open and ethical processes which adhere to the law and stand up to scrutiny.

Just as the Philippine government struggles to evolve into a new democracy, paradigm shifts relevant to government and government service continue to occur. In the context of disaster preparedness and response, certain improvements could be introduced in the bureaucracy particularly in the sector of government service. This is particularly in the areas of downward vertical accountability in terms of how the government direct service workers can be accountable to the disaster victims and families. In the same token, horizontal accountability has yet to happen including among members of the different local special bodies like the City Disaster Risk Reduction and Management Councils wherein, there is no clear accountability mechanism or procedure is available to make members more accountable. Unless a better understanding of the cause of horizontal accountability is better appreciated between and among the main actors in fulfilling their respective agencies mandate, institution blaming will continue without leading to likewise the maturity for government bureaucracy and service delivery. Having both vertical and horizontal accountability established and properly observed by both parties in the disaster preparedness and response platform, we may expect to see more transparent and accountable governance and happier and more resilient citizens

References

1. S. C Opulencia-Calub, E. M. Luna, J. G. J., Molina, Pontejos, G. G., & Chico, J. S. (2017). *Tools, platforms, and mechanisms to support accountability to disaster-* Khotami, M. (Eds.). *The Concept of Accountability in Good Governance*. Atlantis Press. (2017)
2. Van Niekerk, Disaster risk governance in Africa: a retrospective assessment of progress against the Hyogo Framework for Action (2000-2012). *Disaster Prevention and Management,* **24**(4), 397-416. (2015).

3. J. Ahrens, & P. M. Rudolph, The importance of governance in risk reduction and disaster management. *Journal of Contingencies and Crisis Management*, **14**(4), 207-220 (2006).
4. S.Jones, B. Manyena, & S. Walsh, Disaster risk governance: evolution and influences. In J. F. Shroder, A. E. Collins, J. Jayawickrama (Eds.), *Hazards, risks and disasters in society* (pp. 45-61). Academic Press. (2015).
5. S. Jones, B. Manyena, & S. Walsh, Disaster risk governance: evolution and influences. In J. F. Shroder, A. E. Collins, J. Jayawickrama (Eds.), *Hazards, risks and disasters in society*, 45-61. Academic Press. (2015).
6. Vij, S., Russell, C., Clark, J., Parajuli, B. P., Shakya, P., & Dewulf, A. Evolving disaster governance paradigms in Nepal. *International Journal of Disaster Risk Reduction*, *50*, 1-8 (2020).
7. S. Jones, B. Manyena, & S. Walsh, Disaster risk governance: evolution and influences. In J. F. Shroder, A. E. Collins, J. Jayawickrama (Eds.), *Hazards, risks and disasters in society* 45-61. Academic Press. (2015).
8. J. Ahrens, & P. Rudolph, M. The importance of governance in risk reduction and disaster management. *Journal of Contingencies and Crisis Management*, **14** (4), 207-220.
9. M. U. I. Choudhury, M. S. Uddin, & C. E. Haque, "Nature brings us extreme events, some people cause us prolonged sufferings": the role of good governance in building community resilience to natural disasters in Bangladesh. *Journal of Environmental Planning and Management*, **62** (10), 1761-1781 (2019).
10. I. Pal, & R. Shaw, Disaster governance and its relevance. In I. Pal & R. Shaw (Eds.), *Disaster Risk Governance in India and Cross-Cutting Issues* 3-22. Springer. (2018).
11. S. Jones, K. J. Oven, B. Manyena, & K. Aryal, Governance struggles and policy processes in disaster risk reduction: a case study from Nepal. *Geoforum*, **57**, 78-90 (2014).
12. A. D. Cook, V. Suresh, T. Nair, & Y. N. Foo, Integrating disaster governance in Timor-Leste Opportunities and challenges. *International Journal of Disaster Risk Reduction*, **35**, 1-41 (2019).
13. I. Pal, & R. Shaw, Disaster governance and its relevance. In I. Pal & R. Shaw (Eds.), *Disaster Risk Governance in India and Cross-Cutting Issues*, 3-22 Springer. (2018).
14. E. Samutereko, disaster preparedness, and Response in Zimbabwe the case of Cyclone Idai of March 2019 in Chipinge and Chimanimani. *Technium Social Sciences Journal*, *21*, 877-885. (2021).
15. I. Pal, & R Shaw, Disaster governance and its relevance. In I. Pal & R. Shaw (Eds.), *Disaster Risk Governance in India and Cross-Cutting Issues* 3-22. Springer. (2018).
16. E. Z. Samutereko, Disaster preparedness, and Response in Zimbabwe: the case of Cyclone Idai of March 2019 in Chipinge and Chimanimani. *Technium Social Sciences Journal*, *21*, 877-885. (2021).
17. S. Deen, Pakistan 2010 floods. Policy gaps in disaster preparedness and response. *International Journal of Disaster Risk Reduction*, **12**, 341-349. (2015).
18. S. Jones, B. Manyena, & S. Walsh, Disaster risk governance evolution and influences. In J. F. Shroder, A. E. Collins, ... J. Jayawickrama (Eds.), *Hazards, risks and disasters in society*, 45-61. Academic Press. (2015).
19. J. Maes, C. Parra, K. Mertens, B. Bwambale, L. Jacobs, J. Poesen, & M. Kervyn, Questioning network governance for disaster risk management lessons learned from landslide risk management in Uganda. *Environmental Science & Policy*, **85**, 163-171. (2018).
20. S. Jones, K. J. Oven, B. Manyena, & K. Aryal, Governance struggles and policy processes in disaster risk reduction: a case study from Nepal. *Geoforum*, **57**, 78-90. (2014).
21. D. N. Sonny, & A. J. Manejar, Disaster preparedness and local governance in the Philippines (PIDS Discussion Paper Series, 2018-52 (2018).
22. D. N. Sonny, & A. J. Manejar,. Disaster preparedness and local governance in the Philippines (PIDS Discussion Paper Series, No. 2018-52 (2018).
23. D. V. Blanco, Disaster governance in the Philippines issues, lessons learned, and future directions in the post-Yolanda super typhoon aftermath. *International Journal of Public Administration*, **38** (10), 743-756 (2015).
24. A. M. Acosta, A. Joshi, & G. Ramshaw, *Democratic accountability and service delivery: A desk review*. International Institute for Democracy and Electoral Assistance. (2010).
25. D. Amaratunga, R. Haigh, & S. Hettige, (Eds.). *The Role of Accountability Within Disaster Risk Governance*. Massey University. (2016).
26. M. Khotami, (Eds.). *The Concept of Accountability in Good Governance*. Atlantis Press. (2017).
27. M. Khotami, (Eds.). *The Concept of Accountability in Good Governance*. Atlantis Press. (2017).
28. M. Bovens, Analysing and assessing accountability: a conceptual framework. *European Law Journal*, **13** (4), 447-468 (2007).
29. K. S. Yap, *What is Good Governance?*. Bangkok United Nations Economic and Social Commission for Asia and the Pacific (UNESCAP). https://www.unescap.org/sites/default/files/good-governance.pdf (2007).
30. M. Khotami, (Eds.). (2017). *The Concept of Accountability in Good Governance*. Atlantis Press.
31. A. M. Acosta, A. Joshi, & G. Ramshaw, *Democratic accountability and service delivery: A desk review.*

International Institute for Democracy and Electoral Assistance. (2010).

32. A. M. Acosta, A. Joshi, & G. Ramshaw, *Democratic accountability and service delivery: A desk review*. International Institute for Democracy and Electoral Assistance (2010).
33. D. Amaratunga, R. Haigh, & S. Hettige, (Ed.). *Accountability in the context of disaster risk governance*. UNDRR. (2019).
34. B. Manin, A. Przeworski, & S. Stokes, Elections and representation. In A. Przeworski, S. Stokes, & B. Manin (Eds.), Democracy, accountability, and representation, 29-54 (1999).
35. A. M. Acosta, A. Joshi, & G. Ramshaw, *Democratic accountability and service delivery: A desk review*. International Institute for Democracy and Electoral Assistance. (2010).
36. E. Zeyn, The effect of internal audit quality on financial accountability quality at local government. *Journal of Accounting, Business and Finance Research*, **4** (2), 74-82. (2018).
37. E. Zeyn. The effect of internal audit quality on financial accountability quality at local government. *Journal of Accounting, Business and Finance Research*, **4** (2), 74-82 (2018).
38. F. T., Khaile, & G. J Davids,. The role of council committees in promoting financial accountability. *Public Policy and Administration Research,* **10** (8), 66-75 (2020).
39. E. Zeyn, The effect of internal audit quality on financial accountability quality at local government. *Journal of Accounting, Business and Finance Research*, **4** (2), 74-82. (2018)
40. N. Dewi, S. Azam,.& S. Yusoff, Factors influencing the information quality of local government financial statement and financial accountability. *Management Science Letters*, **9** (9), 1373-1384 (2019).
41. D. W. Brinkerhoff, Accountability and good governance: concepts and issues. In A Huque & H. Zafarullah (Eds.), *International Development Governance* 269-287. Routledge (2017).
42. A. M. Acosta, Joshi, A., & Ramshaw, G. (2010). *Democratic accountability and service delivery: A desk review*. International Institute for Democracy and Electoral Assistance.
43. D. W. Brinkerhoff,. Accountability and good governance: concepts and issues. In A Huque & H. Zafarullah (Eds.), *International Development Governance*, 269-287. Routledge. (2017)
44. D. Amaratunga, R. Haigh, & S. Hettige, (Eds.). (2016). *The Role of Accountability Within Disaster Risk Governance*. Massey University.
45. D. Amaratunga, R. Haigh, & S. Hettige,. (Eds.). (2016). *The Role of Accountability Within Disaster Risk Governance*. Massey University.
46. A. M. Acosta, A. Joshi, & G. Ramshaw, *Democratic accountability and service delivery: A desk review*. International Institute for Democracy and Electoral Assistance (2010).
47. A. M. Acosta, A. Joshi, & G. Ramshaw. *Democratic accountability and service delivery: A desk review*. International Institute for Democracy and Electoral Assistance. (2010).
48. D. Amaratunga, R. Haigh, & S. Hettige, (Ed.). *Accountability in the context of disaster risk governance*. UNDRR. (2019).
49. D. Amaratunga, R. Haigh, & S. Hettige (Eds.). *The Role of Accountability Within Disaster Risk Governance*. Massey University. (2016).
50. D. Amaratunga, R. Haigh, & S. Hettige (Eds.). *The Role of Accountability Within Disaster Risk Governance*. Massey University. (2016).
51. D. Amaratunga, R. Haigh, & S. Hettige (Ed.). *Accountability in the context of disaster risk governance*. UNDRR (2019).
52. D. Amaratunga, R. Haigh, & S. Hettige. (Ed.). *Accountability in the context of disaster risk governance*. UNDRR. (2019).
53. Philippine Disaster Risk Reduction and Management Act of 2010, Republic Act No. 10121 https://pcw.gov.ph/republic-act-10121-philippine-disaster-risk-reduction-and-management-act-of-2010/ (2010).
54. D. Amaratunga, R. Haigh, & S. Hettige. (Ed.). *Accountability in the context of disaster risk governance*. UNDRR. (2019).
55. Sendai Framework for Disaster Risk Reduction 2015-2030 United Nations Office for Disaster Risk Reduction. https://www.preventionweb.net/files/43291_sendaiframeworkfordrren.pdf (2015).
56. Philippine Disaster Risk Reduction and Management Act of 2010, Republic Act No. 10121 https://pcw.gov.ph/republic-act-10121-philippine-disaster-risk-reduction-and-management-act-of-2010/ (2010).
57. S. C. Opulencia-Calub, E. M., Luna, J. G. J. Molina, G. G. Pontejos, & J. S. Chico, *Tools, platforms and mechanisms to support accountability to disaster-affected populations in the Philippines*. Institute of Development Studies. (2017).
58. S. C. Opulencia-Calub, E. M., Luna, J. G. J. Molina, G. G. Pontejos, & J. S. Chico. S*Tools, platforms and mechanisms to support accountability to disaster-affected populations in the Philippines*. Institute of Development Studies. . (2017).
59. Philippine Disaster Risk Reduction and Management Act of 2010, Republic Act No. 10121 https://pcw.gov.ph/republic-act-10121-philippine-disaster-risk-reduction-and-management-act-of-2010/ (2010).
60. *affected populations in the Philippines*. Institute of Development Studies.

Proposed solutions of accidents resulting use of 3D coordinate technology in the civil engineering industry

Amr Salah H. R. Ali[1,4], *Lilik* Ariyanto[2], and *Salah H. R. Ali*[3,4*]

[1]Business Info Systems Program, Helwan University, Cairo, Egypt
[2]Civil Engineering Department, Faculty of Engineering, Sang Bumi Ruwa Jurai University
[3]Precision Engineering Division, National Institute for Standards (NIS), Giza, Egypt
[4]Ministry of Higher Education and Scientific Research, Egypt

Abstract. Mechanical engineers have adapted 3D coordinate technology in the design and implementation of cranes to accelerate the production of construction industries, especially tall structures such as buildings and bridges. Structural engineering projects such as buildings, infrastructure, and bridges are central to the development of human civilization. However, it is also a source of some types of accidents and even disasters, which will be referred to in this paper as structural engineering disasters. This paper discusses the concept, types, and characteristics of structural engineering disasters, especially those that talk about the use of high-level 3D coordinate equipment (tower or mobile cranes) and ways to mitigate them. The accidents in practice occur due to dynamic balance defects during equipment operation, which is usually due to incorrect selection or behavior on the construction site, installation, maintenance, operation, or ambient environmental conditions. Structural engineering disasters occur frequently and globally because they are closely related to human behavior and performance. This paper confirms that solutions can be found entrepreneurial to manage the situation to reach the appropriate operation to mitigate the accidents, through implementing the proposed administrative and engineering recommendations related to an accurate scientific method to meet these challenges. The engineering recommendations are based on ensuring the proper implementation of maintenance, operation, and calibration under the engineering standards. While the administrative recommendations focus on ensuring the follow-up of the safety procedures for implementing maintenance programs and the appropriate behavior of operator and scientifically signal person to meet these challenges.

1 Introduction

Advances in cosmic civilization made 3D coordinate technology play an important role in engineering industries, especially the civil and construction industry. The civil and construction industry is one of the axes of the development plan of the world's countries. There are many engineering machines and equipment are designed by mechanical designer engineers to serve the construction industry based on the idea of 3D coordinate technology. This equipment has played a huge role in the progress of the construction industry, especially cranes. The operation of the crane is generally based on wire ropes or chains and beams. The efficient and safe operation of the cranes is of paramount importance to the safety of workers and the execution of works according to the schedule of the project program. Of course, the safe operation of the cranes brings the overall success of the project [1]. Cranes are the lifeline of the construction industry, while their accidents are still too frequent. Cranes are the lifeline of the construction industry, while their accidents are still very frequent. This is because they're operating at high altitudes, which contributes to the highest mortality rate in the construction industry as a result of bad human behaviors. Therefore, human behavior is an influential factor, whether in technical or administrative work. However, without proper safety management of cranes on construction sites, accidents that could be catastrophic for some people can occur due to loss of life and property.

There are many studies concerned with the design, manufacture, and testing of cranes [2, 7]. Many studies have been concerned with human behavior [4, 8], as operator and signalman and technician. However, there is a lack of systematic research on the relationship between the personality and unsafe behavior of construction workers. In general, the causes of construction risks can be categorized into three main factors: the customer, the contractor (the builder), and the designer [9]. Estimates suggest that the crane's operation is involved in up to one-third of fatalities in all construction and maintenance [10,

* Corresponding author: SalahAli20@yahoo.com

13]. Not to mention the number of injuries or damage to the cranes, equipment, and buildings around them. In light of this danger, there is always an urgent necessity to study the sources of accidents to find suitable solutions for them. Although some cranes are equipped with advanced safety technologies, accidents such as blind lifting and poor contact can lead to machine failure and death [7].

On the other hand, few studies have concerned the development of crane layout planning (CLP) in construction projects of modular integrated construction (MiC). In construction operations, the movement of the crane is repeated to suspend the load and transport it to the specified location. Undoubtedly, this requires careful planning such as selecting a suitable crane and choosing the position of the crane in terms of productivity and safety so that the performance is carried out accurately and thus the transfer process is successful. These studies can be categorized into mobile crane optimization [14,16], investigation of potential working areas of mobile cranes [17, 18], and mobile crane lifting trajectory optimization [19, 22]. It becomes clear that the studies related to CLP development of MiC in construction projects are still very few [23, 24].

The objectives of this article aim to search for sources of errors that lead to crane accidents during operation in construction projects. In addition to providing entrepreneurial solutions proposals on two grounds: the risks resulting from human errors, whether administrative, from lack of follow-up and technical errors in light of international Standards of CLP for MiC.

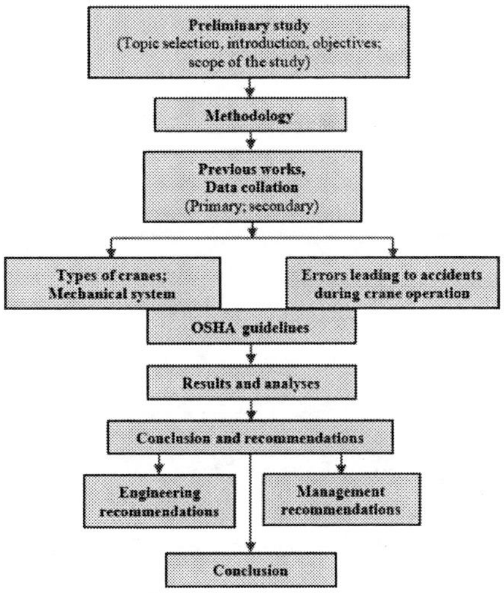

Fig. 1. Schematic flow diagram of the proposed methodology

The methodology used in this paper is presented in detail as mentioned in Fig.1. The objectives were set, the results of questionnaires, case studies, and the result of a personal interview with expert committees were read to compile information related to crane safety in the construction industry. Some secondary data was also collected from several research articles, books, online cranes companies, and previous research.

2 Entrepreneurial Management for Mechanical System

The relationship between entrepreneurial management and industrial enterprises will present in this section. This includes presenting a brief overview of the concept and history of entrepreneurship and the ambitions we aspire to in this work. Moreover, we will provide a summary of the common commercial types of cranes used in construction engineering projects, in addition to some design factors that must be taken into account for the cranes to operate in a safe dynamic manner.

2.1 Entrepreneurship in the Mechanical System for Sustainable Development

Entrepreneurship plays an important role in the safety management of sustainable engineering equipment and machinery in the industry, particularly when operating cranes. Entrepreneurial thinking is the way to operate and maintain cranes and equipment. This ensures successful construction projects and outstanding profits. The term "entrepreneurship" appeared with the need to make institutional projects for commercial industrial activities during the 1960s in European countries. Entrepreneurship in its early days was influenced by a range of humanities such as economics, marketing, sociology, behavioral psychology, management, and history.

Currently, without a doubt, information and communication technology (I&CTs) has contributed to maximizing the role of pioneering businesses for the success of many projects [25]. Studies have confirmed that the modern concept of entrepreneurship depends on a management's willingness to work rationally to organize the work of projects in light of the opportunities and challenges to achieve their ambitions, then serve the community and achieve profits [26, 28]. This makes entrepreneurship the main requirement as a step for the success of projects following its objectives to participate in the development of the economy. Hence, the importance of studying the behavior of employees' performance has become an important requirement that must be focused on for the success of projects. This leads to the continued implementation of engineering construction projects. Hence, in this paper, we were keen to present pioneering recommendations suggested as solutions to reduce the problems of crane damage and accidents. In addition, we would like to inform the community and engineers of the importance and power of entrepreneurship in the success of projects.

2.2 Types of Cranes and Mechanical System

Dynamic balance is considered one of the main causes of breakdown for carinas during the operation of equipment used in construction engineering projects. So, this section includes a review most common types of cranes used in the construction and civil industry, as well as presents a description detailed from the view of the mechanical design of one of the commonly used crane structures.

2.2.1 Types of construction cranes

Practically, all construction projects have goals and strategies for implementing the goals and end with a feasibility study to achieve profits. Many types of cranes are used in construction and industrial projects [29, 31]. There are also many factors that a project manager should consider in choosing the right type of crane to be recommended, including but not limited to the following: Safety and properly use, Weather, Project terrain, Power and torque, Loads, Heights, The area and size of the project, Project time and duration, productivity, and cost, etc. Therefore, one or more of these types of levers can be selected to achieve the objectives and feasibility of the project. More information about common types of cranes used in construction projects can be found in the following reference [29].

Table 1: Common types of commercial cranes

No.	Name of Crane	Description (Key feature, Uses)
1-	Carry Deck	Versatility and movability, Loading and moving materials on job sites.
2-	Crawler	Tracks instead of wheels, Jobs with soft/uneven terrain.
3-	Floating	Ability to work in water, Projects on the sea like ports or oil rigs.
4-	Rough Terrain	Four rubber tires (four-wheel drive), Pick and carry operations on rough terrain.
5-	Truck-mounted	Ability to travel on the highway, For loading and moving equipment on job site.
6-	Bridge/Overhead	Reliable stability and efficiency, Typically used for industrial environments.
7-	Bulk-handling	The large grabbing-like mechanism, Lifting & moving bulk cargo like coal or scrap metal.
8-	Hammerhead	Racking-ability for the trolley to move back and forth, One of the most commonly used crane types.
9-	Stacker	The forklift-like mechanism, Retrieval of cargo in warehouse storage.
10-	Telescopic	Hydraulic cylinder for changing length, Highly adaptable for a variety of mobile situations.
11-	Tower	Height and weight-lifting capabilities, Building large structures.

Once all the data and technical specifications for the project are available, then the technical manager can decide what type of crane need for successful construction work as shown in Table 1.

Experimentally, a tower crane is an essential commonly used tool in the construction projects such as bridges and tall buildings, due to the ability to lift heavy materials and various features. This is due to its amazing lifting capabilities. Due to the big size of the tower crane, it is equipped with a high operating cab that controls all motions. The Tower crane is also having 6 different types as follows: Luffing tower crane, an A-frame tower crane, a Flat top tower crane, a Self-erecting tower crane, a Portal tower crane, and a Remote control crane [31]. Whereas the idea and the working theory of all tower cranes are the same, but the deferent is manly in the height and power, weight. Thus, we will review the design, structure, and dynamic safety of the tower crane in the next section.

2.2.2 Mechanical design and dynamic safety of cranes

In the late 1940s, Hans Liebherr group (German design engineers) invented the first mobile tower crane. While as a result of advances in various life sciences, cranes are made up of the parts shown in Fig. 2 [32]. The specific definition of a tower crane is a mechanical system designed and equipped with electrical motors to lift and transport loads or materials through a hook suspended from a movable boom. More information about the technical development throughout the history of the cranes can be found in the following reference [33]. Fig. 3 shows the importance of considering the role of geometric dimensions in achieving dynamic balance during operation as a necessity for safety [32, 34].

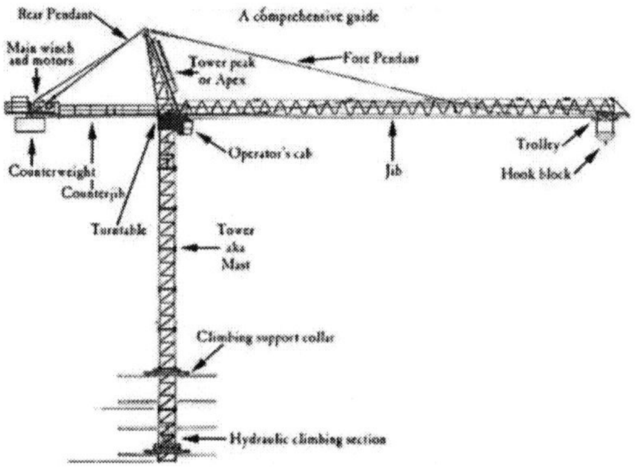

Fig. 2. Skeleton of a tower crane

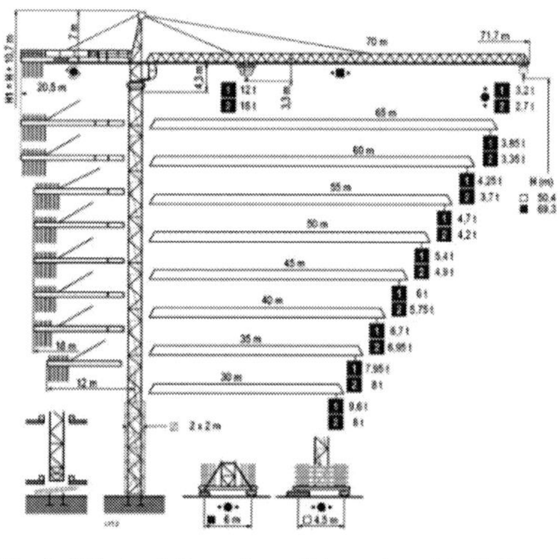

Fig. 3. Effects of dimensions of dynamic safety [34]

Many scientific kinds of research have been concerned with studying the design, construction, and dynamic control of cranes operation in construction projects [35]-[38]. Most of the research has focused on the study of the dynamic to insuring and achieving safe operation [37]-[39]. Fig. 4 shows some results of a study that demonstrated that the trajectory of the load carried H the working conditions are the least favorable as there is a risk of a loss of the crane system stability using three different variants [38]. Where t is the operation time.

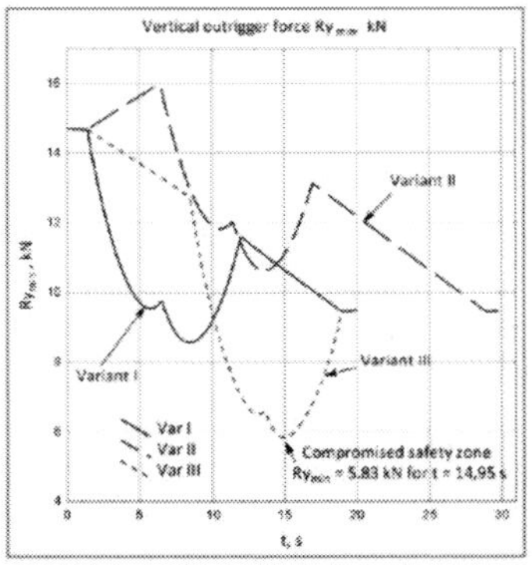

a)

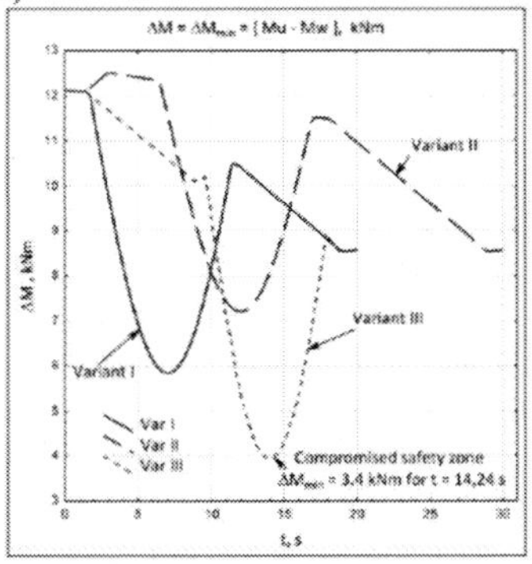

b)

Fig.4. Stability indicator for safety due to operation conditions during displacement of the cargo for the 3 different variants of handling assignment at values of a) and b)

The overturning torque *Mw* is counteracted by the stabilizing torque *Mu* with an opposite direction that is dependent on the mass and the location of the mass center of the crane elements, in the absolute coordinate system Oxyz. There are indicators necessary to ensure the safety of cranes in operation during operation in the construction industry. On the other hand, the tower and mobile cranes should include the safety devices or indicators according to EN 17076:2020 standards [40, 42] such as Obstacle lights, Height limiter, Slewing limiter, Range/trolley limiter, Weight limiter, Moment imiter and Anemometer and block box (optional). Fig. 5 shows several practical device indicators are equipped in a tower crane to ensure good vision, monitoring, and follow-up performance to ensure safety during operating [42, 43].

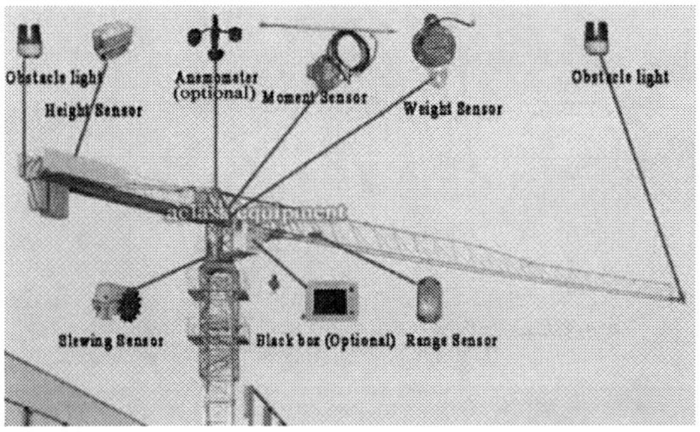

Fig. 5. Safety devices used in tower cranes according to Standards [42]

3 Errors Leading to Accidents during Tower Crane Operation (Structural Engineering Disasters)

Structural engineering disasters due to crane accidents lead to a variety of causes. Many errors lead to accidents during the operation of the cranes such as technical errors, and administrative or behavioral errors. When cranes fail, the results are often disastrous due to the height, size, and weight of crane structures (cab, boom, and tower) and their loads. This is mainly due to the dynamic instability during operation due to technical and management errors. Where studies have shown that the imbalance is caused by various errors such as crane bending of excessive loads, crane failure due to wind and weather conditions, incorrect assembly, unused proper safety procedures, improper operator training, operator error, poor maintenance, mechanical fatigue and wear, connect with electrical power lines, or failure to perform regular or appropriate safety checks [44, 47].

This section will focus on errors percentage caused by technical and administrative errors. Fig. 6 illustrates two of the crane collapse accidents in the projects sites that occurred in New York around 2015 and New Jersey and claimed the lives of several people and damaged property, which could amount to disasters.

a)At New York, the US, 205 [45]

https://doi.org/10.1051/e3sconf/202133102018

b) At New Jersey (BECHT Co), US [47]

Fig. 6. Pictures from real accidents that happened for some tower cranes in America

The facts have confirmed that crane accidents are relatively more serious than other construction equipment accidents, and also the most costly of all in terms of insurance claims and lawsuits. The study has shown that accidents caused by the use of mobile cranes in the construction industry are due to many technical and behavioral reasons, environmental conditions, and location. While a recent study [46] confirmed that the mortality rate when operating mobile cranes is due to the following reasons: loads 14%, drilling equipment 15%, electrocution 44%, operator error overloading 7%, and 7% due to boom breakage, 5% failure Wire rope, miscellaneous 3%, the result of hitting the worker with the winch is 5%. The study also attributed the main causes of mobile crane accidents to high winds 1.6%, machinery and structure failure 11.2%, outrigger failure 22.6%, support failure 31.5%, and operator error 33.1 %. Technical and administrative errors along with behavioral errors are the most important sources of mobile crane accidents.

4 Results and Analyses

Of course for each type of crane, they have a different type of failure depending on the design and operation. For example, the tower and mobile crane fail due to vibration-induced poor dynamic balance in operation, and the tower crane fails due to the straightened failure of the crane's steel support structure [9]. While, the examination and study of questionnaires related to the topic of the research, the results were presented in this section. Table 2 shows some of some important recommendations results to reduce tower crane failure and accidents as a function of error sources in the construction industry [9].

Analyzes have shown that crane failure considers Ahmed the most dangerous factor of all, as it can lead to permanent disabilities or even death. Therefore, the Occupational Safety and Health Administration (OSHA) has guidelines several workplace safety standards within CLP of MiC to reduce these risks or accidents, so site workers should follow these instructions to increase the level of safety.

Table 2: Some recommendation factors to reduce tower crane failure and accidents

Recommendation Factors	Error Source
Tower crane operator must be eligible and have a license (competent)	Human
Every 4 hrs. the crane operator needs to be changed	Human
The loading of certain materials or stuff must be checked with the limitation of the crane	Human
should add a camera which can +zoom in so that the operator can see signalmen	Mechanical
All cranes must be inspected by the supervisor and safety officer before use	Human
Take serious whatever you are doing because it's involving life	Human
Competency in all aspects not just crane operator, but the rigger and lifting supervisor + compulsory in lifting plan	Human
Weekly must do an inspection and fully supervise for mobile crane	Human
Take into accounts the speed of the wind	Human
Pay extra attention when working	Human
Competency personnel including lifting supervisor & lifting plan for each lifting operation.	Human

5 Conclusion

The study confirmed that safety management in construction projects has tasks and solutions, in addition to developing the necessary plans and strategies. These tasks should ensure that administrative and technical procedures are always followed up by international standards of CLP to find solutions to maintain a dynamic balance of cranes to mitigate the accidents, as follows:

A) Regarding administrative and management solutions:

1) Follow-up records of test and choose the location and the fixation installation method of crane, taking into account the distribution of crane assignments during work
2) It is forbidden to speak with the crane operator or the signal operator during work.
3) Follow-up records of maintenance and repair and spare parts that have been replaced, as well as records of checking the safety of operation or what is sometimes called calibration.
4) Wind speed and temperature must be measured and carefully taken into account.
5) The importance of following up on the crane operator change dates.

B) Regarding engineering and technical solutions:

1) Prevent the crane from being overloaded for any reason, as it affects its dynamic balance.
2) Avoiding human error because it is the main factor as a result of unplanned movements, which leads to the collapse of the crane, so the crane must be managed entrepreneurial by a specialized person who has been trained, qualified, and experienced in this.

3) Taking into account the poor communication between the crane operator and the signal operator. The most common mechanical failures are commonly lead to the failure of the crane, especially the mobile crane, which results in the crane falling or overturning.
4) Failure to maintain the crane by an untrained and unspecialized person. Because mechanical failure is the most common cause of crane failure, especially boom failure.
5) Ensure that the crane is calibrated according to the agreed technical standards
6) The crane must be properly and periodically maintained by the competent mechanic under the supervision of the competent engineer under the principles of the crane industry.
7) The effect of wind speed, temperature; precipitation intensity, and rain direction on operation should be taken into account.

Following the precautionary principle and in light of the above, we affirm that all personnel involved in lifting activities must be adequately trained and prepared for the working environment, with the cranes properly maintained under the catalog of the manufacturer's and the industrial assets.

References

1. Hwang, Seokyon, Automation in Construction, Ultra-wide Band Technology Experiments for Real-time Prevention of Tower Crane Collisions, Automation in Construction, **22**, 545-553 (2012).
2. Md. S. Hasan, Interactive and Dynamic Integrated Module for Mobile Cranes Supporting System Design, MSc thesis, University of Alberta, 2008.
3. P. Swuste, A 'normal accident' with a tower crane? An accident analysis was conducted by the Dutch Safety Board, Safety Science, **57**, 276–282, (2013).
4. K. Dhalmahapatra, O. Pradhan, S. Das, K. Singh, and J. Maiti, Prioritization of human errors in EOT crane operations and its visualization using virtual simulation,4th Int. Conf. on Recent Advances in Information Technology (RAIT-2018) (2018).
5. M.F. Milazzo & G. Ancione, V. Spasojevic Brkic, D. Vališ, Investigation of crane operation safety by analyzing main accident causes, Risk, Reliability, and Safety: Innovating Theory and Practice – Walls, Revie & Bedford (Eds), Taylor & Francis Group, London, ISBN 978-1-138-02997-2 (2017).
6. Y. Fang a and Y. K. Chob, Measuring Operator's Situation Awareness in Smart Operation of Cranes, 34th International Symposium on Automation and Robotics in Construction (ISARC 2017) (2017).
7. L. Ma; H. Guo; and Y. Fang, Analysis of Construction Workers' Safety Behavior Based on Myers-Briggs Type Indicator Personality Test in a Bridge Construction Project, Journal of Construction Engineering and Management, **147,** Issue 1, January 2021.
8. M. A H. Ramalan, M. F. Hasmori, A Study on Human Factor That Leads to Tower Crane Accident at Construction Site in Malaysia, Recent Trends in Civil Engineering and Built Environment **2**, 1, 852-861, Universiti Tun Hussein, Onn Malaysia Publisher's Office (2021).
9. M. K. Ali; M. I. Mohamad, Crane Failure and Accident in Construction, 141-154 (2016).
10. Neitzel, L. Richard, N. S. Seixas, and K. K. Ren, A Review of Crane Safety in the Construction Industry, 2001, **16**, 12, 1106–1117, Nov 2010 (2001).
11. A. Shapira, G. Lucko, & C. Schexnayder, Cranes for Building Construction Projects, (2007).
12. S. Mohamed, Safety Climate in Construction Site Environments, (2002).
13. J. E. Beavers, P.E. F.ASCE, Moore J. R., R. Rinehart, and W. R. Schriver, Crane-Related Fatalities in the Construction Industry, (2006).
14. A. Sawhney, A. Mund, Adaptive probabilistic neural network-based crane type selection system, J. Constr. Eng. Manag, **128**, 265–273 (2002).
15. D. Wu, Y. Lin, X. Wang, S. Gao, Algorithm of crane selection for heavy lifts, J. Comput. Civ. Eng., **25**, 57–65 (2021).
16. A. Alshibani, H. Elassir, M. Al-Najjar, H. Hamida, AHP based approach for crane selection of building construction in Saudi Arabia: A case study. In Proceedings of the Annual Conference–Canadian Society for Civil Engineering, Laval, QC, Canada, (2019).
17. H. Safouhi, M. Mouattamid, U. Hermann, A, Hendi, An algorithm for the calculation of feasible mobile crane position areas, Autom. Constr., **20**, 360–367 (2011).
18. K. Tantisevi, B. Akinci, Simulation-based identification of possible locations for mobile cranes on construction sites, J. Comput, Civ. Eng., **22**,21–30, 2008.
19. P.L. Sivakumar, K. Varghese, N.R. Babu, Automated path planning of cooperative crane lifts using heuristic search, J. Comput. Civ. Eng., **17**, 197–207 (2003).
20. S. Kang, E. Miranda, Planning and visualization for automated robotic crane erection processes in construction, Autom. Constr, **15**, 398–414, (2006).
21. W. Kacalak, Z. Budniak, M. Majewski, Optimization of the movement trajectory of mobile crane working elements. In Advances in Manufacturing; Advances in Manufacturing 475–484, Springer: Cham, Switzerland, (2018).
22. N. Kayhani, H. Taghaddos, A. Mousaei, S. Behzadipour, U. Hermann, Heavy mobile crane lift path planning in congested modular industrial plants using a robotics approach, Autom. Constr., **122**, 103508 (2021).
23. S. Tariq, M. Hussein, R. D. Wang, and T. Zayed, Trends and developments of on-site crane layout planning 1983–2020: bibliometric, scientometric,

and qualitative analyzes, On-site crane layout planning, (2021).

24. G. H. Seong, L. H. Chul, C. B. Yoon, and Y Mi, GA-Based Optimization Method for Mobile Crane Repositioning Route Planning, Appl. Sci. **11**, 6010, 1-18 (2021).
25. Amr S. H. R. Ali, Salah H. R. Ali, *Entrepreneurship and Advanced Technologies in Industry, Int. Conference on Advanced Information Tech. and Communication (IC-AITC), Insomnias*,(2021).
26. Amr. S.H.R. Ali; E. G. Khalil, Cloud Computing and E-Marketing in Entrepreneurship Development, 1st Int. Conference "Entrepreneurship and Socio-Economic Development" Women Empowerment and Youth Employment, MUST University, Faculty of Management, Economics and Information Systems,1-17, 6th of October City, Giza, Egypt, (2018).
27. E. M. M. M. Shuaib, The Impact of Cloud Computing Applications on Developing Technological Awareness, Journal of Specific Education Research, Faculty of Specific Education, Mansoura University, 126-169, (2017).
28. H. K. Al-Hadrawi, Leadership as an Entrance to Contemporary Business Organizations in the Light of Adopting the Concept of Intellectual Capital (A field study), University of Kufa, Iraq, Al-Ghari Journal of Economic and Administrative Sciences, **27**, 9, 85-128 (2013).
29. BigRentz, Inc, 11 Types of Cranes Commonly used in Construction, Oct. 2019. On the website at: https://www.bigrentz.com/blog/types-of-cranes
30. D. SDLS, Types of Cranes for Construction & Industrial Projects, February 2017. On the website at: https://brynthomascranes.com/types-of-cranes/
31. Cranes and Machinery. On the website at: https://www.gruasyaparejos.com/en/tower-crane/types-of-tower-cranes/
32. J. J. Rubio-Ávila, R. Alcántara-Ramírez, J. Jaimes-Ponce, I. I. Siller-Alcalá, Design, construction, and control of a novel tower crane, International Journal of Mathematics and Computers in Simulation, **1**, 2, 119 (2007).
33. W. Kacalak, & Z. Budniak, & M. Majewski, Computer-aided analysis of the mobile crane handling system using computational intelligence methods. Advances in Intelligent Systems and Computing, **662**, 250-261 (2018).
34. A. Trąbka, The influence of clearances in a drive system on dynamics and kinematics of a telescopic crane, Acta Mechanica et Automatica, **9** (1), 9-13 (2015).
35. W. Kacalak, Z. Budniak, M. Majewski, Crane Stability Assessment Method in the Operating Cycle, Transport Problems, **12**, Issue 4 (2017).
36. O. N. Anezirisa,& I. A. Papazoglou, & M. L. Mud, & et al. Towards risk assessment for crane activities, Safety Science, **46**, 6, 872-884 (2008).
37. EN 17076:2020(MAIN), Tower cranes - Anti-collision systems - Safety requirements, (2020).
38. Tower crane equipment, Professional Lift Building Tower Cranes Heavy Equipment Topless with ISO. On the website at: http://www.towercraneequipment.com/china-professional_lift_building_tower_cranes_heavy_equipment_topless_with_iso-6134691.html
39. R. Neitzel, N. Seixas, & K. Ren, Applied Occupational and Environmental Hygiene, A Review of Crane Safety in the Construction Industry, (1047-322X), (2010).
40. P. Swuste, A 'normal accident' with a tower crane? An accident analysis was conducted by the Dutch Safety Board, Safety Science, **57**, 276–282 (2013).
41. Posted by admin, Crane Accident, New York in (2021)
42. D. A. J. Smith, Crane Operations in Construction: Regulations, Standards, and Liability, Amundsen Smith LLC. On the website at: https://slidetodoc.com/crane-operations-in-construction-regulations-standards-and-liability/
43. J Collins, How To Manage A Crane Accident,. On the website at: https://becht.com/becht-blog/entry/how-to-manage-a-crane-accident (2019)

https://doi.org/10.1051/e3sconf/202133102019

Spatial assessment on health impact of atmospheric pollution in Makassar, Indonesia

Wesam Al Madhoun[1,2*], *Faheem* Ahmad Gul[3], *Faizah* Che Ros[4], *Hamza* Ahmad Isiyaka[5], *Anwar* Mallongi[6], and *Ashari* Rasyid[7]

[1]Gaza University, Gaza Strip, Palestina
[2]University of Tun Hussein Onn Malaysia, 86400 Parit Raja, Johor, Malaysia
[3]School of Civil and Architectural Engineering, Nanchang Institute of Technology, Nanchang 330099, China
[4]Malaysia Japan International Institute of Technology, University of Technology Malaysia Kuala Lumpur, 54100 Kuala Lumpur, Malaysia
[5]Kano University of Science and Technology, Wudil, Nigeria
[6] Hasanuddin University, Makassar, South Sulawesi 90245, Indonesia
[7]Polytechnic of Health Makassar, Makassar, South Sulawesi, Indonesia

Abstract. There has been little discussion to date on air pollution and its potential relationship with health in Makassar, Indonesia. This study aims to create a starting point for this discussion by investigating existing data points and the potential correlation between ambient air pollution and health in Makassar, Indonesia. Six months of air quality data (July-December, 2018) on CO, SO2, NO2, O3, PM10, and PM2.5 were provided by the city and were analyzed alongside tuberculosis and pneumonia data provided by the hospital and community health centers in Makassar. Data were analyzed using principal component analysis, dendrogram, and some GIS mapping. Quantitative data from the USAID-funded Building Health Cities project were also used to help explain some of the quantitative findings. Results show that principal component analysis (PCA) gave three statistics factors having eigenvalues exceeding one, which account for 83% of the total variance in the dataset. The three factors accounted for a strong impact by CO, O3, SO2, PM10, and PM2.5 attributed to the incomplete combustion of fuel from automobiles, bush burning, and industrial emission. Air pollution-related illnesses such as tuberculosis and pneumonia are found to prevail in the area. Real-time air quality monitoring is required to benchmark the health impact of extreme conditions. This study also encourages urgent intervention by decision-makers to tackle the level of tuberculosis and pneumonia occurrence that may be favored by the poor air quality in Makassar.

1 Introduction

Air pollution originates from a diverse source and consists of a mixture of substances found in chemical and physical states, many of which are considered harmful to human health. Air pollution was identified as a major environmental challenge associated with respiratory disease and reduced life expectancy [1]. To standardize the monitoring and quantification of air pollutants on a global scale, the World Health Organization (WHO), among others, has established a particular set of indicators, as the measurement of all possible air pollutants is not yet feasible worldwide. Therefore, WHO has defined four air pollutants that have severe impacts on the health of people including; Particulate Matter (PM), which is further divided into PM_{10}, or coarse particles which have an aerodynamic diameter of fewer than 10 μm, and $PM_{2.5}$, or fine particles with an aerodynamic diameter of fewer than 2.5 μm, sulfur dioxide, nitrogen dioxide, and ozone [2]. Focusing solely on these pollutants is not meant to undermine the impacts of other air pollutants on the environment and human health, rather it is meant to optimize the monitoring of the general state of air quality [2].

Ninety percent of the people across the world breathe polluted air that does not meet the WHO air quality guidelines [3-4]. Annually, ambient air pollution contributes to the death of three million people worldwide; however, South East Asia and the Western Pacific are the most affected regions [3]. Obstructive pulmonary infections such as pneumonia and tuberculosis have been reported to be among the chronic causes of death [5] and by 2020, it is estimated to be third among the leading cause of death in adults and children worldwide. Ambient air pollutions have proven to be

* Corresponding author: w.madhom@gu.edu.ps

associated with adverse cardiopulmonary health effects [6].

Indonesia, located in South East Asia, is a consortium of over 17000 islands and is the world's largest archipelagic State [7]. It is ranked fourth among the world's most populated countries with about 260 million people [7], and land-use change and deforestation make it a significant emitter of greenhouse gases [8]. Bush burning and fuel combustions are the primary determinants of climate change and air pollutions in Indonesia. Urban areas tend to be the most affected by air pollution. Pollution from industrial and domestic activities, forest fires, and the vast majority (about 80%) coming from the transportation sector may also play a significant role [9]. This is due to a fuel combination, poor infrastructures that result in traffic congestion, and long queues of vehicles emitting air pollutants which gravely impact human health, buildings, forests, the quantity and quality of crops, and surface water quality [9]. According to Haryanto and Franklin [10], 50% of the morbidity across the country is caused by poor air quality.

Especially in the last decade, there has been unusually high growth in the number of motor vehicles in major Indonesian cities, including Makassar [11]. Makassar is the capital of Indonesia's province of South Sulawesi and is a major and largest urban city in Eastern Indonesia. It has largely remained a trade hub of the country, especially Eastern Indonesia, albeit less developed than other major cities of Indonesia such as Jakarta, Bandung, and Serpong.

Makassar is a coastal city with complex local climatic conditions (land and sea breeze) which may result in the recirculation and accumulation of pollutants. However, the air quality indices still exceed the WHO guidelines. Generally, from October to March, Makassar experiences rainfall, and it is classified as the wet period; whereas, from April to September, it experiences dry season which is characterized by less rainfall and biomass burning, leading to high PM concentrations [12]. Previous studies as described in table 1 highlight the concentration of air pollution in Makassar, Indonesia.

Table 1. Studies on air pollution in Makassar

Period covered	Pollutant studied	Concentration	Ref
2006 - 2010	SO_2, CO, NO_2, O_3, Pb, TSP, and PM_{10}	76 μg/m³, 1041 μg/m³, 43.2 μg/m³, 54.5 μg/m³, 0.7 μg/m³, 188 μg/m³, 54.6 μg/m³, respectively	[13]
2010	$PM_{2.5}$	280 ug/m³	[14]
2006 - 2012	Particulate matter	80 ug/m³ – 380 ug/m³	[15]
2012 - 2013	PM_{10}	32.9 ug/m³	[12]

This study aims to start to fill this data gap in Makassar, by a) identifying what routine data are available in the city for air pollution and related health conditions; b) defining any limitations to those data; c) analyzing these data for any useful correlations or information that can help support a discussion on air pollution and health by Makassar decision-makers.

2 Materials and methods

2.1 Study Area

Makassar (Indonesia) is an urban city with high population density, traffic congestion, and industrial activities. It's the capital of the province of South Sulawesi. It is located between 119° 24' 38" east longitude and 5° 8' 19" south latitude covering an area of about 175,77 sq km with 14 districts [13]. This area is characterized by a tropical climate embedded with two major seasons (rainy and dry season). The rainy season is between September and January. The topographical and geology of the area is made up of hills, plains, and beaches. The average temperature is 27°C.

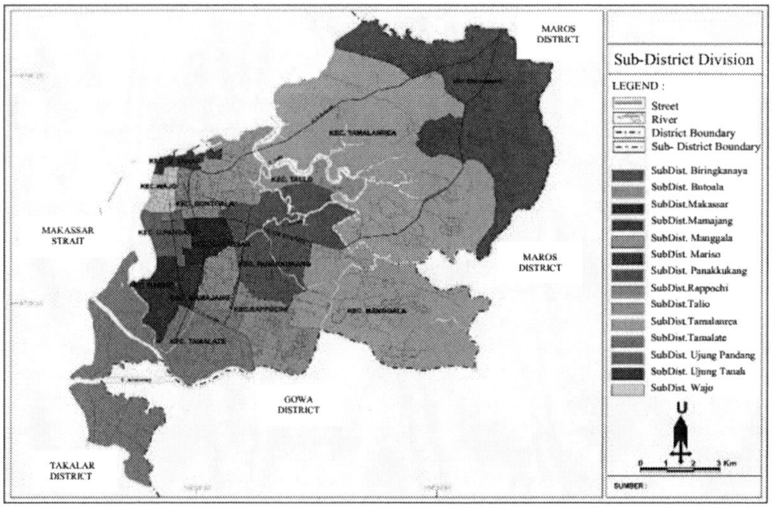

Fig. 1. Administrative Map of Makassar

https://doi.org/10.1051/e3sconf/202133102019

2.2 Data compilation and analyses procedures

Data compilation and analysis procedure comprise of the data collection, air quality assessment, and health risk assessment using descriptive statistics, multivariate statistical techniques, and geographical information system as described in Fig. 2.

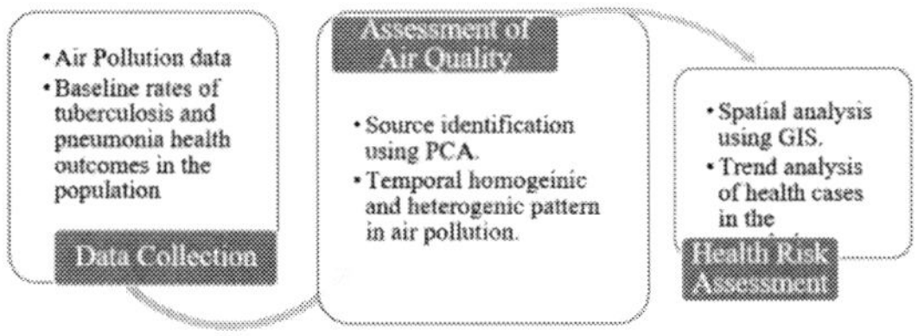

Fig. 2. Data collection and analysis procedures

2.3 Data screening procedure

Air quality monitoring data of Makassar, Indonesia comprising of CO, SO_2, NO_2, O_3, PM_{10}, and $PM_{2.5}$ were provided by the department of the environment while the health data (Tuberculosis and Pneumonia) were made available by the department of health Makassar from July 2018 to December 2018. The data were collected from a single available site and recorded by the department, which represents the period when measurements are monitored, recorded, and reported to assess extreme conditions and possible health impacts. This site represents the only functional station that monitors air quality status in the area with the availability of data. Except for July 2018 (where half of the data were collected at 5-minute intervals), all other air quality data (August-December) were collected at 30 minutes intervals, daily.

The data provided comprises some missing observations and was estimated using the nearest neighbor method in the XLSTAT 2019 software. The missing values amounted to 2231 (5.6%), which gave a total of 39684 (6614 observations x 6 parameters) data sets used for the analysis. The 2018 health data for tuberculosis and pneumonia in adults and children were collected from community health centers and hospitals in Makassar, Indonesia without any missing observations. A total of 464 pneumonia cases and 5485 tuberculosis data were also utilized. The data were used as provided due to their reliability to identify their relationship with the air quality data, their spatial occurrence, and temporal variations.

2.4 Method of analyses

Air quality and health data collected were subjected to different analyses to simplify, describe and interpret their complex and dynamic behavior. The following statistical techniques are proposed.

2.4.1 Principal component analyses (PCA)

PCA will be used to establish a minimum component that can be used to describe the maximum variance possible in a data set (Singh et al., 2005; Wang et al 2014). It is a pattern recognition technique that can analyze complex data [16]. It will be applied to extract the most significant parameters by eliminating the less significant parameters with minimal loss of the original variables [17]. The major possible source of pollution will be identified using PCA.

The equation is expressed as:

$$Z_{ij} = a_{i1}x_{1j} + a_{i2}x_{2j} + a_{i3}x_{3j} + ... + a_{im}x_{mj} \quad (1)$$

Where z represents the component score, a takes the component loading, x is used as the measured value of variables, the component number is represented by I, the sample number is represented by J and m is the total number of variables.

2.4.2 Geographical information system (GIS)

GIS provides a spatial visual presentation in maps. In this study, GIS will be used to provide the spatial occurrences of health incidences due to air pollution to identify areas of high, medium, and low effects. It will also be used to map out traffic flow and pollutant concentrations.

2.4.3 Hierarchical agglomerative cluster analyses

HACA is an unsupervised pattern recognition technique used to aggregate a similar group of entities that portrays a strong internal (within-class) homogeneity and strong external (between classes) heterogeneity [17]. It is used to aggregate a group with identical characteristics in a sequential manner with the aid of a dendrogram that measures the degree of risk homogeneity through Ward's method and Euclidean distance [18]. A dendrogram highlights a reduction in the dimensionality and complexity of the data sets. It also provides a visual representation of the level of uniformity in the pattern of observed parameters [19]. In this current study, HACA was used to extract the pattern of air pollutants that can be used as a guide for future sampling.

3 Results and discussion

3.1 Data availability and quality

Data availability is a priority in air pollution management and its impact on human health. Data provided from city offices are usually associated with outliers and noisy observations. These variations may require a series of data treatments and standardization to achieve a realistic result. The error may be due to the method of sensor calibration, staff skills, and other technical shortcomings. A better calibration technique may be adopted coupled with staff training to enhance skills in routine data collection. Deploying real-time sensors for air quality monitoring will improve quality data availability for proper analysis. It is also important to improve on the digitization of health data to enable easy accessibility for regularly monitored data. These will improve reliability and minimize errors.

3.2 The relative contribution of measured air pollutants to air quality

PCA was used to identify the factors with strong positive loading that can be used for source profile identification. Factor loading from 0.7 was selected as a threshold used to identify the possible spatial sources of the air pollutants. Three factors were obtained based on eigenvalues greater than one, which account for 83% of the total variance in the data set. Table 2 describes the factor loadings after varimax rotation, eigenvalues, and cumulative variance. Fig. 3 represents the PCA loading and scree plots where factor loadings are selected.

Varimax factor 1 (VF1) shows a strong positive loading for CO (0.773), NO_2 (0.793), and O_3 (0.802) which account for over 42% of the total variance in the data set as described in table 1. CO and SO_2 are by-products formed from the incomplete combustion of fuel in automobiles. Within the study area, these pollutants are emitted from automobiles especially during the early hours of the day and towards traffic peaks in the evening, as such these pollutants are associated with vehicle emission. The high association with CO and NO/VOC (via O3) may suggest that this is a factor associated with traffic pollution as vehicles will emit CO, NO, and VOCs. And then the NO/VOC would convert to NO_2 and O_3.

VF2 contributes 22% of the total variance in the data set indicating a strong impact contributed by $PM_{2.5}$ (0.708) and a moderate loading for PM_{10} (0.506). PM_{10} and PM2.5 may be associated with harmful trace elements that trigger respiratory, cardiovascular diseases and increase mortality rate by 1% for every increase in 10μg/m^3 (Yubero et al. 2011). These pollutants mostly originate from bush burning, industrial smoke, and urban dust (Shakir et al 2011).

VF3 has a strong positive loading for SO_2 (0.882) accounting for over 18% of the total variance in the data set. This could be attributed to industrial emissions due to gasoline and diesel fuel combustion during manufacturing processes.

Table 2: Factor loadings after varimax rotation

Parameters	VF1	VF2	VF3
CO	**0.773**	0.002	0.174
SO_2	0.009	0.001	**0.882**
NO_2	**0.793**	0.006	0.081
O_3	**0.802**	0.001	0.074
PM_{10}	0.057	0.506	0.074
PM_{25}	0.041	**0.708**	0.012
Eigenvalue	2.551	1.313	1.133
Variability (%)	42.516	21.885	18.879
Cumulative %	42.516	64.401	83.280

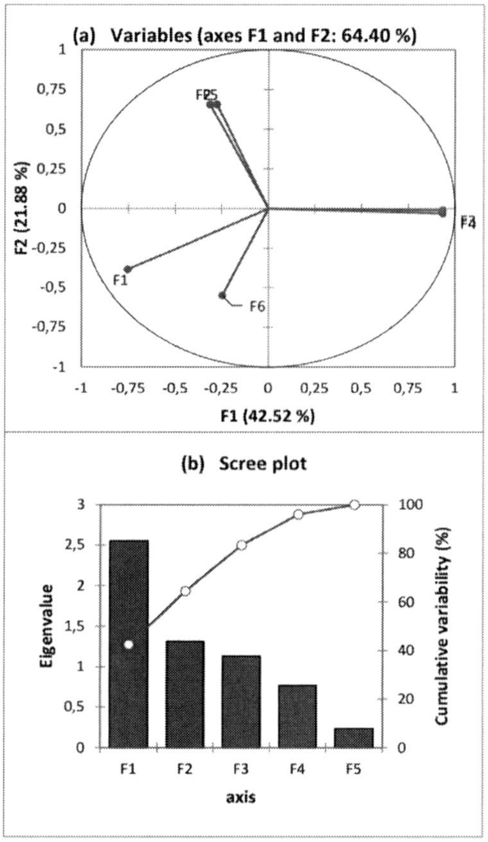

Fig. 3. (a) PCA loading and (b) scree plot diagram after varimax rotation

3.3 Spatial cases of tuberculosis and Pneumonia using GIS

The spatial variability and distribution in the pattern of tuberculosis (TB) and pneumonia (PE) in Makassar Indonesia were mapped out using geographical information system (GIS) in Fig. 4. The patter portrays a dominance in the occurrence of Pneumonia due to the impact of air pollution in the environment. Air pollution is a primary cause of TB and PN both in children and adults [6], [20].

Fig. 4. Spatial pattern of tuberculosis and pneumonia cases in Makassar

3.4 Similarities and dissimilarities among the sampled air pollutants

In this study, hierarchical agglomerative cluster analysis was used to identify the general and monthly pattern of the similarities and dissimilarities among the sampled parameters based on the dendrogram. The essence is to understand the parameters that fall within the same class of pollution level and those that have different characteristics in each month. Fig. 5 denotes a dendrogram cluster for sampled air quality data from July to December 2018 while Fig. 6 represents the monthly homogeneity and heterogeneity of the parameters based on their concentration. From Fig. 5, CO displays a different pattern from other pollutants due to its distinct emission sources especially from automobiles during the early traffic morning hours and late traffic evening. This is because CO is released from incomplete combustion that affects the air quality pattern from July to December within the study area.

NO_2 and O_3 have the same monthly pattern from July to December because O_3 is formed when solar Ultraviolet radiation reacts with nitrogen oxides and reactive hydrocarbon is the principal component of photochemical smog. NO_2 can act as a precursor in the formation of O_3 giving them a close interrelationship within the pollution spectrum. SO_2, PM_{10}, and $PM_{2.5}$ have a similar pattern within the same cluster due to their interrelated pollution sources. SO_2, PM_{10}, and $PM_{2.5}$ are emitted mostly from industrial activities, bush burning, construction sites, etc.

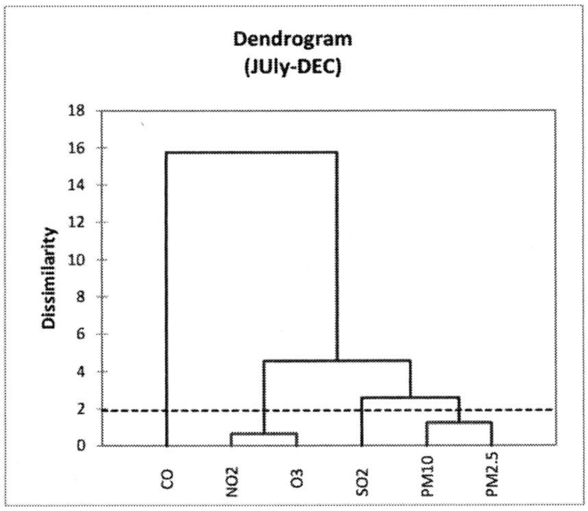

Fig. 5. Variation in the characteristic of the parameters

The monthly pattern of the studied pollutants was also described with a dendrogram in Fig. 6 which relates to the monthly climatic conditions observed in Makassar. The dendogram for July, August, and September 2018 shows similar behavior in the occurrence of NO_2, SO_2, PM_{10}, $PM_{2.5}$, O_3, and CO. This is because Makassar has a tropical climate with the driest months in July, August, and September. This makes the pollutants exhibit similar patter during this period of the year as described in dendrogram a, b, and c. Dry season experiences less rainfall which provides a favorable climatic condition for more pollutant concentration. More so, farmers engage more in bush burning in preparation for the rainy season to plant crops. This portrays a similar pattern in pollution level due to the homogenous bush burning activities.

The rainfall pattern in Makassar increases from the month of October to December. December experiences a high rain pour of about 552 mm within the last six months of the year. These change in rainfall and temperature pattern is responsible for modifying the pollutant behavior in Makassar especially towards the end of the year. Heavy rainfall may help to settle down the suspended pollutants in the atmosphere.

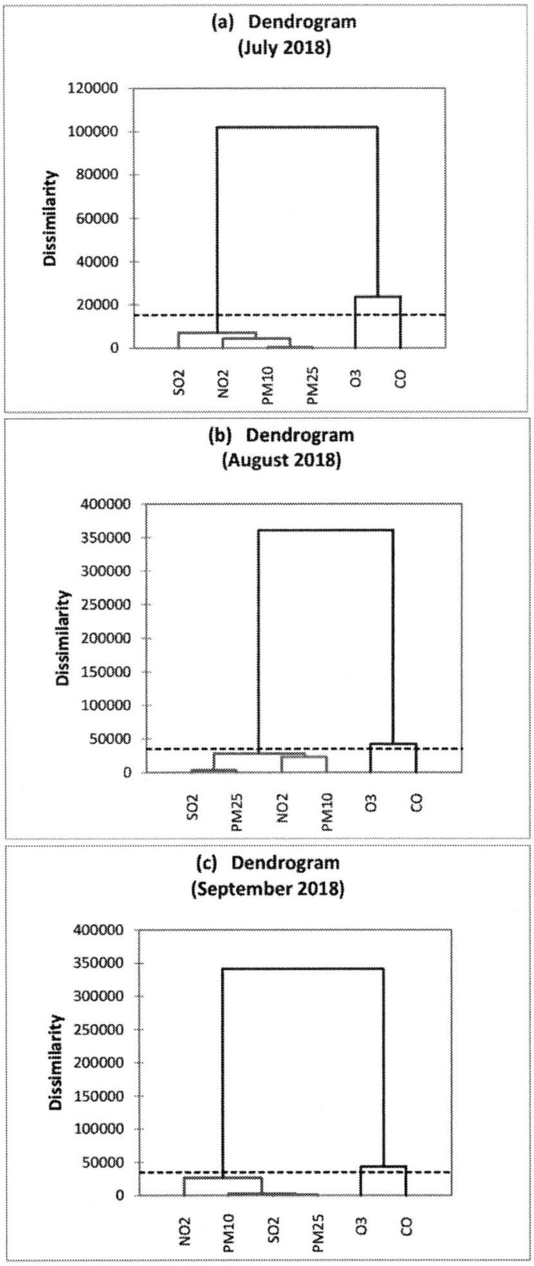

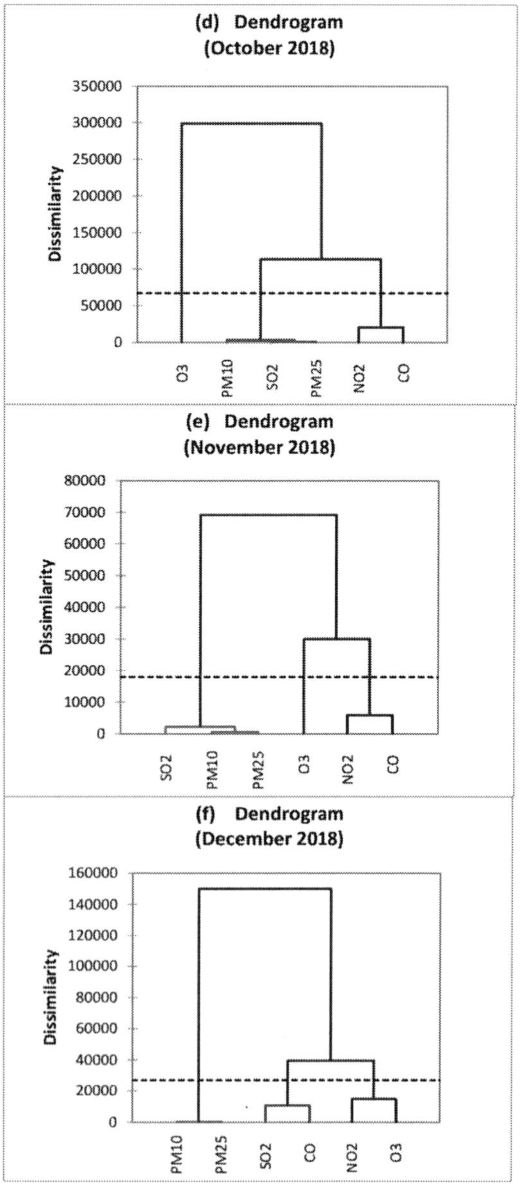

Fig. 6. Monthly variation in the characteristics of the sampled parameters

3.5 Air pollution trend analyses and variability in the pattern of health cases in Makassar

Ambient air pollutants that include PM_{10}, $PM_{2.5}$, O_3, CO, NO_2, and SO_2 have a direct or indirect effect on lung functions which can increase the occurrence of pleural effusion caused by the excess accumulation of fluid between the visceral and parietal flexural of the lungs. This shortfall inhibits lung expansion thereby causing serious health challenges that may lead to a series of hospital admission and finally death [21]. From Fig. 7, PM_{10} happens to be the most highly occurred pollutants followed by NO_2 due to bush burning and emission from industrial activities especially during the month of July, August, and September. This is evident in the cluster analysis displayed in the dendrogram in Fig. 6 which shows PM_{10} and NO_2 in the same cluster. This high concentration is controlled by the dry climatic season experienced in July, August, and September which provides favorable conditions for pollutant occurrence. During these months, farmers practice bush burning to clear their farmland in preparation for high rain pouring in November and December. However, less pollutant concentration is found in November and December due to the rainfall that beats and settles the pollutants. During this period there is less bush burning which reduces the number of pollutants in the environment.

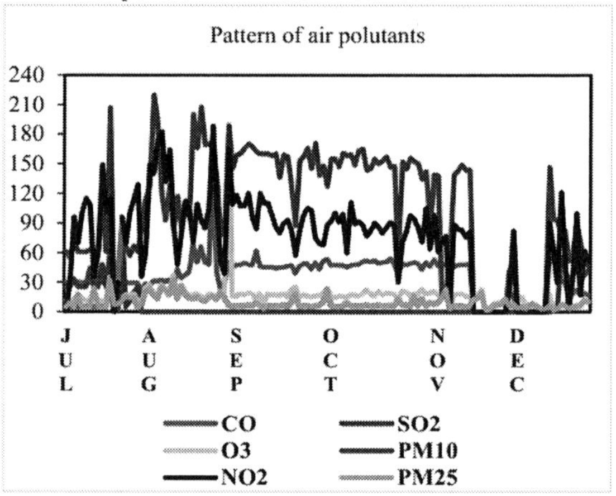

Fig. 7. Trend pattern of monthly air pollutant concentration

3.6 Cases of tuberculosis (TB) in adults and kids in Hospitals and community health centers

Ambient air pollution is among the frequent cause of respiratory infectious diseases globally. Despite the effort put towards reducing the trend of TB cases, it has continued to pose threat to the health of people in developing countries. Recent evidence has shown that exposure to ambient air pollution could increase the incidence of pulmonary TB [21-22]. This risk is obvious due to the dispersion of contaminated air, ventilation, and clinical cases in Makassar hospitals and community health centers as described in Figs 6 and 7. These pollutants may be associated with complex mixtures containing bacteria and other pathogens forming a composition that can affect the immune system and respiratory organs [23-24]. In 2018, the highest adult hospital TB cases in Makassar are 142, 134, 133, 131, and 130. For children's hospital cases, the highest incidence is recorded in 112, 92, 89, and 85. These numbers are of serious concern and require urgent interventions.

More so, community health centers have recorded several TB cases for adults and kids in 2018. For adults, the highest incidence is 215, 167, 159, and 148, while children's cases in different hospitals are in the range of 10, 8, 7, and 6.

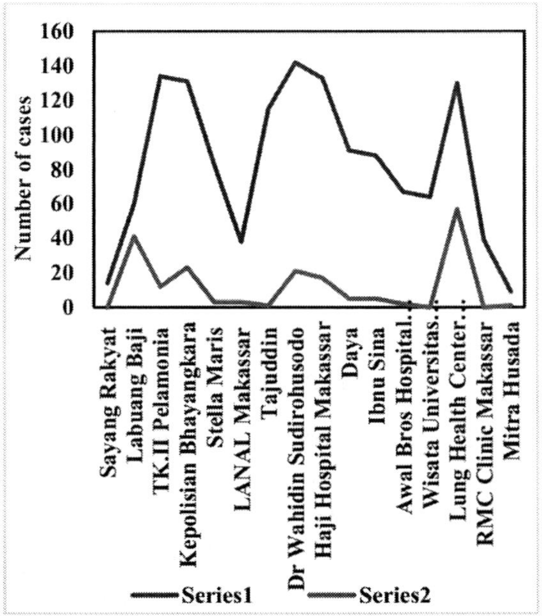

Fig. 8. Hospital Cases of TB in Kids and Adults

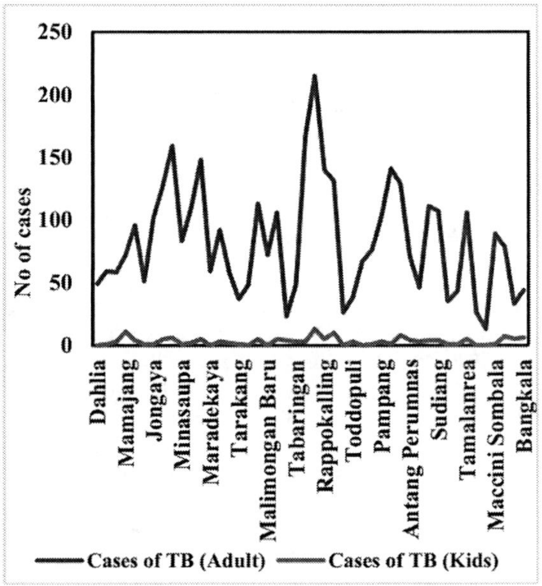

Fig. 9. Community Health Center Cases of TB in Kids and Adults

3.7 Pneumonia cases in Makassar Indonesia

The pattern of pneumonia cases in male and female children within the age of 0-4 years in Makassar was also analyzed. The result as described in Fig. 10 shows a similar but slight difference in trend between the male and female children. This indicates that air pollution affects pediatric pneumonia due to their weak immune system and other defense mechanisms [25]. From Fig. 10, the male children between the ages of 0-4 years have the highest occurrence case of 30, 29, and 28. Other cases display a similar effect for male and female children.

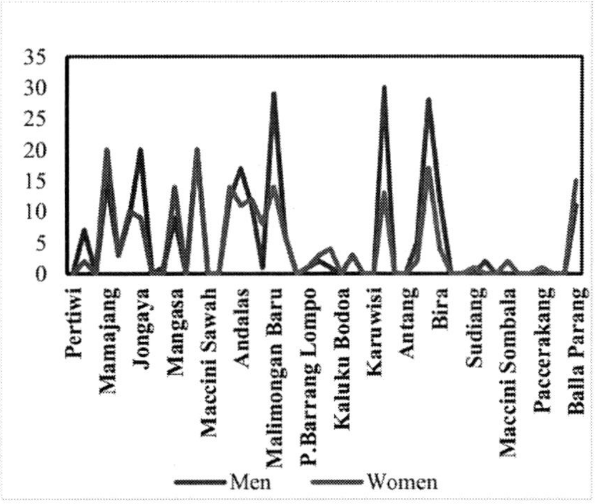

Fig. 10. Pneumonia case for male and female children (0-4 years)

4 Conclusion

This study aims to identify the major possible sources of air pollution, their trend, classification, and health impact in Makassar, Indonesia. Air quality data from July to December 2018 such as CO, SO_2, NO_2, O_3, PM_{10}, and $PM_{2.5}$ alongside tuberculosis and pneumonia data from hospital and community health centers were used. The result for the PCA produced three varifactors with eigenvalues greater than one which explains 83% of the total variance in the dataset. The first varifactor has a strong positive loading for CO, NO_2, and O_3 could be associated with vehicle emission. Strong positive loading of PM_{10} and $PM_{2.5}$ is also observed in VF2 accounting for 22% of the total variance in the dataset. The source of these pollutants is linked with bush burning, frequent urbanization and construction, industrial emission, vehicle emission, and re-suspension of soil dust. SO_2 has a strong positive loading in VF3 accounting for over 18% of the total variance in the data set. Industrial emissions from gasoline and diesel fuel combustion during manufacturing processes may contribute to this emission. July, August, and September are the months with the highest concentration of pollutants due to the presence of dry sky which provides a favorable condition. During these months, these pollutants are associated with complex mixtures that lead to several clinical cases of pneumonia and tuberculosis. Kaluku Bodoa, Jumpandang Baru and Rappokalling, Kassi-kassi, Tamamaung, Malimongan Baru and Antara areas are associated with highest clinical pulmonary cases. These locations have busy traffic, highway, and shipyard activities and some are also slum areas. GIS was also used to map out the spatial distribution in tuberculosis and pneumonia cases. The dendrogram was used to understand the general and monthly pattern of air pollution based on the climate. However, several hospitals and clinical admission cases in adults and children were presented. Urgent measures should be taken by the government to reduce the level of pollution which will, in turn, minimize the level of clinical cases in Makassar. The research findings show that serious and urgent interventions need to be taken to save public health by reducing transport and industry emissions. Stringent enforcement of policy and regulations is needed urgently as well.

The authors would like to acknowledge Ms. Ratna Dwi Puji Astuti and Ms. Annisa Utami Rauf for their contribution to this article, furthermore special thanks to the Department of Environment in Makassar for their cooperation.

References

1. Rauf, A. U. Mallongi, A. Daud, A. Hatta, M. Al Madhoun, W. Amiruddin, R. Abdul Rahman, S. Wahyu, A. Puji Astuti, R. D. Community Health Risk Assessment of Total Suspended Particulates near a Cement Plant in Maros Regency, Indonesia, J Health Pollution 30: (210616) (2021),
2. World Health Organization (WHO) (2018a) Ambient (Outdoor) Air Quality and Health Available:https://www.who.int/news-room/fact-sheets/detail/ambient-(outdoor)-air-quality-and-health, Accessed 20th May 2019 Google Scholar
3. World Health Organization, Geneva (WHO) (2018b), 9 out of 10 people worldwide breathe polluted air, but more countries are taking action Google Scholar.
4. World Health Organization (WHO) Global Health Observatory Available: https://www.who.int/data/gho/data/themes/topics/topic-details/GHO/ambient-air-pollution, Accessed 20th Sept 2021 (2021)
5. Vestbo, J., Hurd, S.S., Agustí, A.G., Jones, P.W., Vogelmeier, C., Anzueto, A., Barnes, P.J., Fabbri, L.M., Martinez, F.J., Nishimura, M., Stockley, R.A., Sin, D.D., Rodriguez-Roisin, R., Global strategy for the diagnosis, management, and prevention of chronic obstructive pulmonary disease. Am. J. Respir. Crit. Care Med. 187, 347–365 (2013).
6. Ho, S., Chuang, K., Lee, K., Chen, J., Wu, S., Chen, T., Lee, C., Chang, C., Feng, P., Su, K., Tsai, C., Chuang, H. Chronic obstructive pulmonary disease patients have a higher risk of occurrence of pneumonia by air pollution Science of the Total Environment 677, 524–529 (2019).
7. World Population Review at http://worldpopulation review.com/countries/indonesiapopul/ (2016)
8. World Resources Institute The wealth of the poor: Managing ecosystems to fight poverty. At http://www.wri.org. (2005).
9. Haryanto, B. Climate change and urban air pollution health impacts in Indonesia. In Climate Change and Air Pollution, 215-239. Springer, Cham (2018).
10. Haryanto B and Franklin P Air pollution: a tale of two countries. Rev Environ Health **26** (1):75–82, (2011)
11. Muchtar, F., & Ramli, M. I. Estimation of CO vehicular emission on arterial roads in Makassar City based on Caline-4. In IOP Conference Series: Earth and Environmental Science, **235** (1), 012056. IOP Publishing. (2019).
12. Rashid, M., Yunus, S., Mat, R., Baharun, S., & Lestari, P. PM10 black carbon and ionic species concentration of urban atmosphere in Makassar of South Sulawesi Province, Indonesia. Atmospheric Pollution Research, **5**(4), 610-615. (2014).
13. Sattar, A., Rashid, M., Mat, R., & Puji, L. A Preliminary Survey of Air Quality in Makassar City South Sulawesi Indonesia. Jurnal Teknologi, **57** (1) (2012).
14. Haryanto, B., Resosoedarmo, B., Utami, S. T. B., Hartono, B., & Hermawati, E. Effect of Ambient Particulate Matter ($PM_{2.5}$) to Prevalence of Impaired Lung Function and Asthma in Tangerang and Makassar. Kesmas: National Public Health Journal, **10** (4), 145-149 (2016).
15. Suryani, S. Status of Suspended Particulate Matters Pollution at Traditional Markets in Makassar City. In Journal of Physics: Conference Series **979**, 1, 012041. IOP Publishing (2018).
16. Al-Odaini, N. A., Zakaria, M. P., Zali, M. A., Juahir, H., Yaziz, M. I., & Surif, S. Application of chemometrics in understanding the spatial distribution of human pharmaceuticals in surface water. *Environmental monitoring and assessment*, **184** (11), 6735-6748 (2012).
17. Shrestha, S., & Kazama, F. (2007). Assessment of surface water quality using multivariate statistical techniques: A case study of the Fuji river basin, Japan. *Environmental Modelling & Software*, **22** (4), 464-475.
18. Lau, J., Hung, W. T., Cheung, C. S.. Interpretation of Air Quality with Monitoring Station's Surrounding. *Atmospheric Environmetric*. 43: 769-777. (2009)
19. Ogwueleka, T., C. Use of multivariate statistical techniques for the evaluation of temporal and spatial variations in water quality of the Kaduna River, Nigeria. *Environmental monitoring and assessment*, **187**(3), 1-17., 297-301 (2015).
20. Zhu, S., Xia, L., Wu, J., Chen, S., Chen, F., Zeng, F., Chen, X., Chen, C., Xia, Y., Zhao, X., Zhang, J. Ambient air pollutants are associated with newly diagnosed tuberculosis: a time-series study in Chengdu, China. Sci. Total Environ. 631-632, 47–55 (2018).
21. Zhou, S., Xu, B., Qi, L., Zhu, D., Liu, B., Wei, J., Next-generation sequencing reveals mutational accordance between cell-free DNA from plasma, malignant pleural effusion, and ascites and directs targeted therapy in a gastric cancer patient. Cancer Biol. Ther. 1–6 (2018).
22. Smith, G.S., Van Den Eeden, S.K., Garcia, C., Shan, J., Baxter, R., Herring, A.H., Richardson, D.B., Van Rie, A., Emch, M., Gammon, M.D. Air pollution and pulmonary tuberculosis: a nested case-control study

https://doi.org/10.1051/e3sconf/202133102019

among members of a northern California health plan. Environ. Health Perspect. **124**, 761–768 (2016).

23. Yan, D., Zhang, T., Su, J., Zhao, L.L., Wang, H., Fang, X.M., Zhang, Y.Q., Liu, H.Y., Yu, L.YStructural variation in the bacterial community associated with airborne particulate matter in Beijing, China, during hazy and nonhazy days. Appl. Environ. Microbiol. 84 (pii: e00004-18). (2018).
24. Sun, Y., Xu, S., Zheng, D., Li, J., Tian, H., Wang, Y. Effects of haze pollution on microbial community changes and correlation with chemical components in atmospheric particulate matter. Sci. Total Environ. 637-638, 507–516 (2018).
25. Nguyen, T.K.P., Tran, T.H., Roberts, C.L., Fox, G.J., Graham, S.M., Marais, B.J. (2017). Risk factors for child pneumonia - focus on the Western Pacific Region. Paediatric Respiratory Reviews, **21**; 95-101.

The earthquake of 1926 in Padang Panjang, West Sumatra

Yenny Narny [1*], *Robert* Cribb[2], *Yudhi* Andoni[1], and *Ifkar* Fikri[1]

[1]Department of History, Andalas University 25163 Padang, West Sumatera Indonesia
[2]Department of Political and Social Change, Coral Bell School of Asia-Pacific Affairs, Australian National University, Canberra-Australia

Abstract. This study explores contemporary newspaper narratives on the devastation caused by the earthquake of 1926 in Padang Panjang, West Sumatra. These narratives become crucial amid the difficulty of finding historical statistical data for disaster studies that depict the chaotic situation caused by natural disasters in the past. Historical methods and methodologies were used in this study to combine fragments of information found in contemporary newspapers, especially those discussing the situation of societies and the policies of the Dutch colonial government in dealing with the earthquake of 1926 in Padang Panjang. This study hopefully opens new insights for the development of disaster disciplines and provides access to the development of the study of the history of disasters in developing countries, particularly in Indonesia.

1 Introduction

An earthquake is a natural disaster that is characterized by the intense vibration or shaking of the earth's surface in the outermost layer. Severe earthquakes cause enormous damage to objects on the earth's surface, especially buildings which may be subjected to pressures they were not designed to resist. Earthquakes are commonly caused by the sudden movement of tectonic plates under the earth's surface. They can also be caused volcanically, as magma moves beneath the earth's surface, causing vibrations or tremors at the earth's surface [1]..

Earthquakes occur in various places around the world, but some areas have a greater earthquake potential than others due to local geographic conditions West Sumatra is one such region [2].. West Sumatra is vulnerable to volcanic earthquakes because of several active volcanoes in the region. Three volcanoes – Mount Marapi in Agam Regency, Mount Talang in Solok Regency, and Mount Tandikat in Padang Panjang City – are the main causes of volcanic earthquakes in the West Sumatra region in recent times.

Besides volcanic earthquakes, West Sumatra is also frequently hit by tectonic earthquakes. The region is flanked by two earthquake lines, namely the plate subduction zone in the western sea area and the Sumatran fault zone (Semangko fault), located along with the hill range (Bukit Barisan) [3].. Tectonic earthquakes are generally stronger and more dangerous than volcanic earthquakes [4].. Tectonic earthquakes occurring below sea level also have the potential to generate tsunamis that can lead to even greater damage.

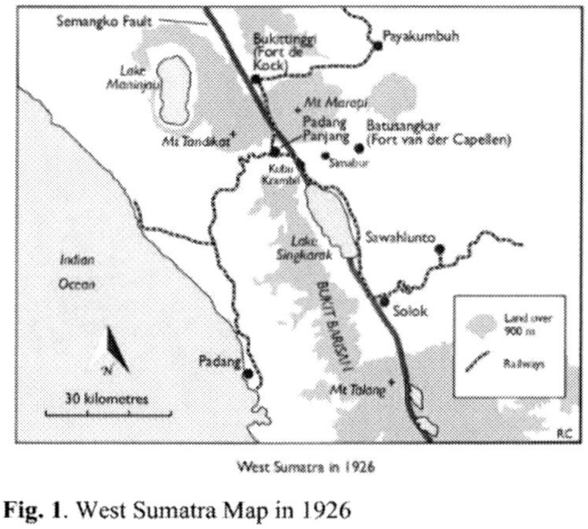

Fig. 1. West Sumatra Map in 1926

The oldest manuscript that records an earthquake in the West Sumatra region is the Du Puy report in 1845. James du Puy was the Dutch resident, or local governor, in Padang. In his report, he wrote about an earthquake in 1797 which hit the Padang City area. This earthquake occurred on 10 February 1797 at ten o'clock at night. The earthquake then triggered a tsunami wave that destroyed parts of the city of Padang. In another report in 1847, Du Puy also mentioned an earthquake forty years earlier in 1757 that shook the city of Padang causing massive destruction and resulting in the

* Corresponding author: yenny.narny01@gmail.com

land being torn apart [5]. This earthquake is the oldest disaster recorded in the historical records of West Sumatra, although not much information can be gathered for further study.

In the following years, various areas in the province of West Sumatra were repeatedly hit by earthquakes, causing the destruction of landscapes and buildings and claiming many lives. It was recorded that there were about 15 earthquakes with a significant intensity that caused devastation in various areas of West Sumatra from 1797 to 2010. The intensity of these earthquakes ranged between 5 and 8.5 on the Richter Scale. Some earthquakes triggered tsunami waves to cause even more enormous damage in this region6].

At the beginning of the 20th century, West Sumatra underwent both a socio-political transformation and a conflict with modernity. This transformation prevailed not only in big cities but also in remote villages. On the other hand, in the main cities in West Sumatra, political and social movements emerged. At this time, in the city of Padang Panjang, known as one of the centres of education in the region, many schools, and office buildings were built as well as public facilities such as hospitals and train stations. Padang Panjang was one of the important cities until an earthquake that struck in 1926 devastated the city. The earthquake not only claimed many lives but also caused enormous material losses.

The magnitude of the devastation caused by the earthquake attracted wide attention from newspapers circulating at that time. Their reports describing the devastation generated sympathy from both within and outside the colony. To restore the situation to normal, the colonial government then provided various forms of aid for earthquake victims. This study focuses on how the colonial government handled the social and economic consequences of the 1926 earthquake as well as the narratives about the earthquake that appeared in the local printed mass media.

2 Data Source

This paper utilizes various sources of print media published at the time of the earthquake. Other written sources were also obtained from the National Archives of the Republic of Indonesia. These sources were then studied to obtain data used to reconstruct historical events related to the earthquake that hit the Padang Panjang area in 1926. These data can be classified into two parts according to types and sources. There are four types of data. First is data on the causes of the earthquake obtained from the newspapers *De Maasbode* on 29 June 1926 and *Deli Courant* on 3 July 1926. Both newspapers were in the Dutch language.

The second type of data is on the damage caused. For this type of data, the sources are provided by newspapers and magazines written in both Dutch and Indonesian language. Among the newspapers and magazines were *Iqbalul Haq* magazine, June-July 1926 Edition, *Barito Minangkabau* magazine 10 July 1926, *Deli Courant* 3 July 1926, *Dagblad Radio* 29 June 1926, and *Soera Kota Gedang* 7 July 1926.

Third, data for casualties and material losses were obtained from magazines and archives such as *Dagblad Radio* 29 June 1926, *Soera Kota Gedang* 7 July 1926, and *Iqbalul Haq* magazine Edition June-July 1926 *Zutphensche Courant* 20 August 1926 and ANRI, AG archives July 3, 1926, No. 18679.

Fourth, for material losses and government policies, data are obtained from archives and newspapers such as; archive ANRI, AG 3 July 1926, No. 18679, *Oetoesan Sumatra* 11 November 1926, ANRI archives, AG 3 July 1926, No. 18679, *De Telegraaf* 3 July 1926, *Soeara Kota Gedang* 7 July 1926, ANRI, BGS July 2, 1926, No. 1224, ANRI archives, Tel 24 July 1926, No. 521[a], Grote Bundels.

3 Method

This paper employs the 4 stage- historical method, namely:

1. Heuristics or data collection stage. The data supporting this article were obtained from various libraries in West Sumatra, and the National Archives in Jakarta. The library collection included various contemporary newspaper articles and documents discussing the earthquake that occurred in Padang Panjang in 1926.
2. Criticism, namely the stage where all sources that have been collected are checked and verified for accucracy and correctness so that the data obtained are valid to be used as sources of writing.
3. Interpretation, which is the stage where the data that has been collected and criticized are combined into a unified understanding through one historical concept so that these data provide various historical facts about the earthquake of 1926.
4. Historiography, that is, the final stage of the historical research method, namely writing what is understood as a historical reality based on the facts found and explained as a valuable reconstruction effort of the past.

4 Discussion

The earthquake of Padang Panjang occurred on Monday, 28 June 1926. This earthquake coincided with the day of the city market day, with the consequence that most of the people at the time of the earthquake were in the *Pasar Serikat* of Padang Panjang Batipuh, X Koto. These market-goers were busy buying and selling or just looking around and sitting and relaxing at coffee shops[7].

The earthquake hit with two large quakes and many aftershocks of small earthquakes. The first earthquake had a strength of Ms = 6.75 while the second earthquake had a strength of Ms = 6.5[8]. The first earthquake was felt at 10 in the morning, amid the hustle and bustle of the city market, full of people buying and selling in the market. All of a sudden, there was a very strong quake that caused people to rush out of their homes or buildings to reach open spaces. The second earthquake occurred at around 1 in the afternoon with a strength almost the same as that of the first earthquake. The community continued to experience after shakes until 7 July. During this time, the sun was blocked by thick clouds, intermittently accompanied by drizzle. The sunlight entering the earth is blurred and dim. Cahaya matahari yang masuk ke bumi tidak sempurna, remang-remang. The dim light of the sun during these earthquakes also created a tense atmosphere and created a sense of fear that would never fade away from people's hearts and minds [9].

Muhammad Radjab, a well-known journalist from West Sumatra, describes in his autobiography how the community reacted when the earthquake hit the city. He reports that several people chanted the words *La ilaha illallah* ('There is none worthy of worship except Allah) continuously in the hope that if they died they would do so while chanting and reciting this sentence affirming tawhed (the oneness of Allah). He also described how most people chose to take refuge in a *surau*, an Islamic assembly building in some regions of Sumatra and the Malay Peninsula used for worship and religious instruction and usually smaller than a mosque. On the other hand, his father, who was an ulema, lectured the gathered people, "We must *tawakkal*, [the word for the Islamic concept of the reliance on God].. The world might end in a few days. Humans have been evil many times, therefore it is in the law of God."[10].

The earthquake destroyed several areas in West Sumatra and caused many casualties and enormous material losses. Padang Panjang and surrounding areas suffered the most from the earthquake. Nonetheless, the impact of this powerful earthquake was also felt in Padang, Solok, Agam, Bukittinggi, Payakumbuh, and Batu Sangkar [11].

The cause of the earthquake was initially thought to be volcanic activity from Mount Talang [12]. However, further investigation by Dr. S.W. Visser, a scientific collaborator at the Koninklijk Magnetisch en Meteorologisch Observatory (KMMO) led to a different insight. The KMMO was founded by the Netherlands government that to research meteorology, but its investigations later extended to the phenomenon of earthquakes. The colonial government, conscious that the Dutch East Indies region as a whole was vulnerable to earthquakes, both volcanic and tectonic. For this reason, they set up an institution in charge of researching various earthquake phenomena. On January 1, 1866, the Koninklijk Magnetischen Meteorologisch Observatorium te Batavia (KMMO) was established which was the forerunner of the birth of BMKG (Badan Meteorology, Climatology, and Geophysics). Initially, this institution researched the measurement of temperature, atmospheric pressure, wind speed and direction, duration and intensity of the sun, as well as measurements of rainfall [31].. However, in 1908 this institution began to research earthquakes, notably by installing horizontal components of the Wiechert seismograph in Jakarta, while the installation of vertical components was only carried out in 1928. In the previous year, 1907, the government had first placed these devices in the Ambon and Padang areas. However, these devices could not function properly so they were brought back to Batavia [32].. Visser found that the earthquake was tectonic. According to him, the earthquake was caused by fractures in the earth's crust that stretched along the Bukit Barisan [13]..

As a result of the earthquake, the city of Padang Panjang which had been a center of trade, education, railway transportation, and political and religious movements was destroyed in an instant. Impressive buildings that had adorned the city, ranging from shop buildings, houses, schools (both religious and government), to mosques and surau were destroyed [14].. Buildings made of stone were almost all destroyed. On the other hand, buildings made of wood were often in better condition, although dozens of them also collapsed in the earthquake. The train station was also badly damaged, the roof of the centre, as well as some walls, had collapsed, while large cracks could be seen everywhere [15]. The fact that the destruction was most severe in cities, where communists, reformist madrasas (Islamic schools), and colonial administrators lived, led traditional or conservative Muslim clergy to view the earthquake as divine retribution against impious modern institutions [16].

Losses due to the earthquake are estimated to have reached ten million guilders for the entire region of West Sumatra. Meanwhile, in downtown Padang Panjang, the loss was estimated at f [or *f*]. 574,000, while the total loss in the Padang Panjang area as a whole reached f1,175,000 due to damage and destruction of various community properties. Other areas outside Padang Panjang such as Fort de Kock (Bukittinggi) and Solok suffered losses of f 200,000 and f 632,000, respectively [17].

Not only did the earthquake destroy built infrastructure, but it also caused damage to the ecosystem and natural environment. The earthquake caused many roads to crack as wide as three to five centimeters. Besides that, the earthquake also caused landslides in Padang Panjang, Kubu Krambil, and Simabur. Some of the environment around Lake Singkarak also suffered severe damage because the earthquake caused the lake water to overflow which then flooded the surrounding area [18].

The earthquake also led to the complete paralysis of various economic activities of the community. To meet basic needs such as food, medicine, and clean water, the community heavily relied on assistance from various parties, both private and colonial government. Even though at that time there were also some people selling food ingredients, the prices were too high for people who had just survived the earthquake, so that the government and the private sector offered significant help with food assistance [19].

The number of people who died as a result of the earthquake is estimated to have reached a thousand victims in the entire region of West Sumatra [20]. In Padang

Panjang, the death toll from the earthquake was estimated at 195 people (other opinions put the death toll at around two hundred or three hundred). The toll may have been higher because the actual number of victims could not be identified due to limited information at the time. Of the 195 earthquake victims in Padang Panjang, 170 victims were from the Minangkabau community, 23 victims from the Chinese community, and one each from the Europeans and Bombay (British Indian) communities. The earthquake also caused many houses to be destroyed. It is estimated that as many as 7,000 people were forced to live in refugee camps set up by the government in a field. The earthquake victims later faced new problems such as lack of clean water, food, and medicine so that on 29 June the government immediately sent aid to ease the suffering of the earthquake victims [21].

The Chinese community was the second most disadvantaged party in the disaster. There was much destruction in the Chinese residential area in Padang Panjang, causing many casualties and enormous material losses. Meanwhile, only a single fatality from among the Europeans was reported in various newspapers (especially Dutch-language newspapers) at that time. The victim was an aide and pharmacist named Schmidt who died when he was buried under the rubble of a collapsed building. It was also reported that an infantry lieutenant named E.J. Donk suffered serious injuries and an officer named Kist suffered minor injuries. The other Europeans who managed to survive took refuge in emergency tents set up in the middle of the market [22].

In dealing with the earthquake, the Padang Panjang municipal government continued to send the latest information to the colonial central government in Bogor via telegram. These telegrams were then distributed to various relevant government institutions such as Binnenlands Bestuur (Department of Home Affairs), Leger (Army), BOW (Department of Public Works), Gouvernementsbedrijven or related government bodies [23]. To respond to the emergency arising from the earthquake in West Sumatra, the colonial central government immediately dispatched military troops to assist earthquake victims and secure the surrounding area on 3 July 1926. The government also sent food aid which was distributed to all areas affected by the earthquake [24].

Reporting about the earthquake and its aftermath in various newspapers attracted much attention and sympathy from within and outside the country, with the result that assistance also came from private parties. One of the private parties that assisted was Smeroefonds, a foundation that was established in 1911 in Batavia in response to a volcanic eruption in Java and which subsequently became the most important and private organization offering disaster relief. and specialized in assisting disaster victims [25]. Assistance also came from various Minangkabau associations located abroad such as Studyfonds Kota Gedang, Tjabang Betawi [26]. Foreign aid came from a visit by the Swedish Consul General [27]. and aid from Sydney of one million pounds which arrived on 24 July 1926 [28].

For the community to return to normal activities, the colonial government then carried out repairs to various infrastructures that had been damaged by the earthquake. The main repairs were carried out on the houses of government officials and public facilities; these repairs were targeted to be completed in 1927. The main targets for the repairs carried out by the government were the houses of officials destroyed by the earthquake so that they could immediately resume their state duties. For the construction of these official houses, the government received direct financial support from the Kingdom of the Netherlands in two payments of 162,500 guilders and 38,000 guilders. On the other hand, various public facilities such as railway stations and lines, hospitals, post offices, and so on were also a major focus of the government so that the city of Padang Panjang as a whole could rapidly return to normal activities. [29].

In the various policies taken by the colonial government, there was a tendency to prioritize the safety of the Europeans ahead of that of the natives. This priority was indicated in the reports made by Residents, in which they addressed the safety of their officials and employees rather than that of the indigenous people who had suffered more from the earthquake. The government also based its priority on the restoration of various infrastructures based on economic interests by prioritizing the improvement of transportation and communication routes. Restoring these lines of communication meant that the flow of income from the West Sumatra region could be restored for the benefit of the colonial government [30].

On the other hand, the KMMO prioritized scientific research and encouraged its employees to publish the results of their research. In 1911 KMMO began to publish its scientific journal, which was published every year and contained writings about the various studies they carried out [33]. However, the knowledge about earthquakes collected by this institution did not seem to have been widely disseminated so that most people were still very unaware of the earthquake that occurred. Lack of public understanding of the phenomenon of earthquakes contributed to the huge number of casualties and material losses as happened in Padang Panjang in 1926.

5 Conclusion

The large number of casualties and material losses caused by the earthquake of 1926 in Padang Panjang demonstrated the lack of public knowledge in dealing with the earthquake that occurred so that a high number of losses could not be prevented or minimized. Muslim society in West Sumatra at that time generally considered the earthquake as the coming of the apocalypse sent by God as a result of the sins of the community. As a result, in dealing with the earthquake, they could only surrender and pray for a miracle without making any preparations to prevent greater losses when the after-shakes occurred.

The colonial government played an important role in dealing with the earthquake and in restoring the facilities, infrastructures as well as the community who were suffering from the earthquake. Even though the policies taken were based on economic motives and the benefits that would be obtained by the colonial government, the humanitarian side was also seen from a large amount of government assistance to the community and infrastructure improvements to restore conditions in Padang Panjang to their original state. Moreover, the colonial government at that time also played a role in the development of science regarding the phenomenon of earthquakes in Indonesia in the long term through its institutions, so that losses due to earthquakes could be minimized in the future.

References

1. Sunarjo et.al, *Earthquake Popular Edition*, (Jakarta: Meteorology Climatology and Geophysics Council, 2012), 26, (2012)
2. S. Karim, Geography of West Sumatra and Earthquake Natural Disaster, Conference Paper Seminar on Natural Disasters, 2, (2011)
3. D. Surjo at.al, History of Earthquake Disaster in Sumatra, (Jakarta: Ministry of Education and Culture, 2016), 99, (2016)
4. Nur Hidayat dan Eko Widi Santoso, "Earthquake and its Mechanism", Natural Journal, **2**, 3, 50.
5. D Surjo et.al, op,cit, hlm. 101.
6. Ibid, 100.
7. Mestika Zed, Final Report of the Professor's Research: Inventory and Reconstruction of the History of the September 30, 2009 Earthquake in the City of Padang Through Documentary Films, (Padang: Padang State University, 2012), 36, (2012)
8. Untung et.al, Rupture along the Great Sumatran Fault, Indonesia during the Earthquake of 1926 and 1943, Bulletin of the Seismological Society of America, **75**, 1, 316.
9. Majalah Iqbalul Haq, June-July (1926)
10. Muhammad Radjab, Childhood in the Village 1913-1928: Autobiography of a Minangkabau Child, (Jakarta: Library, 1974), 128, (1974)
11. Magazine Berito Minangkabau 10 July 1926
12. De Maasbode 29 June 1926
13. Deli Courant 3 July 1926
14. Magazine Iqbalul Haq , op,cit.
15. Deli Courant, op,cit.
16. Jeffrey Hadler, The Endless Dispute: Matriarchy, Islamic Reformism, and Colonialism in Minangkabau, (Jakarta: Freedom Institut, 2010), 245, (2010)
17. Zutphensche courant 20 August 1926
18. Dagblad Radio 29 Juni 1926 and Soera Kota Gedang 7 July 1926
19. ANRI, AG 3 July 1926, No. 18679
20. Dagblad Radio dan Soeara Kota Gedang, op,cit.
21. Magazine Iqbalul Haq, op,cit.
22. Nieuwsblad van Friesland: Hepkema's courant 2 July 1926
23. ANRI, AG 30 June 1926, No. 18284
24. ANRI, AG 3 July 1926, No. 18679
25. De Telegraaf 3 July 1926; Sander Tetteroo, 'Java, 1919: Colonial Disaster Relief', Online Atlas on the History of Humanitarianism and Human Rights, https://hhr-atlas.ieg-mainz.de/articles/tetteroo-java
26. Soeara Kota Gedang, op,cit.,
27. ANRI, BGS 2 July 1926, No. 1224.
28. ANRI, Tel 24 July 1926, No. 521a, Grote Bundels.
29. Oetoesan Sumatra 11 November 1926.
30. Devi Riskianingrum, 'Disaster Management and Knowledge Transformation on Earthquake in the Colonial Period', Jurnal Paramita **23**, 1, 6.
31. K. van Berkel et.al, A History of Science in the Netherlands: Survey, Themes, and Reference, (Leiden: Brill, 1999), 223-224, (1999)
32. J. Paulus et.al, Encyclopaedie van Nederlandsch-Indie, Eerste Deel, (Leiden: E.J Brill, 1918), 55-59, (1918)
33. Van Berkel, op, cit, 224-225.

Modelling of coastal area management based on climate change adaptation in Bengkulu City

Muhammad Eko Ardiyanto[1*], *M Mustopa* Romdhon[2], and *Reflis*[2]

[1]Natural Resources Management Program, Faculty of Agriculture-Bengkulu University, Bengkulu, Indonesia
[2]Agricultural Socio-Economic Dept., Faculty of Agriculture-Bengkulu University, Bengkulu, Indonesia

Abstract. Climate change has had an impact on the environment and society In Indonesia. Delay in adaptation efforts will result in direct and indirect economic losses in 2100 of 2.5% of Gross Domestic Product (GDP). Climate change indicators such as surface temperature, rainfall (CH), sea surface temperature (SPL), sea level (TML), extreme climate events (ENSO, IOD/DMI, PIO/IPO), and extreme weather events (heavy rains, strong winds), and storm surges) will have potential impacts on related fields in the national development system, both in terms of economy, livelihoods, ecosystems, and special areas. Another impact is the increasing frequency of hydrometereological disasters. This study focuses on the vulnerability of coastal areas using multiple decision-making methods and geographic information systems and remote sensing. The relationship between demography and climate change is carried out through literature review and observation. Observations were made in 19 sub-districts An appropriate strategy is needed to adapt to vulnerabilities, especially in coastal areas.

1 Introduction

1.1. Background

Predicted changes in climate such as the increase in surface temperature will have a direct impact on the lives of humans, plants, and animals. Especially for countries that have a hot summer heatwave that can cause fatalities. Even though Indonesia does not have a distinct summer season, the increase in temperature during the day can result in local heating so that the use of air conditioning becomes more frequent and increases the consumption of electrical energy. The increase in temperature is also suspected to cause excessive evapotranspiration in plants, forest fires, and the proliferation of insects more quickly and widely. In a report issued by the IPCC (Intergovernmental Panel on Climate Change)[1], it is stated that if climate change including changes in land and ocean surface temperatures continues, there will be an increase in the frequency and duration of heatwaves in the oceans. The report also states that human-caused global warming will have an impact on increasing the frequency and intensity of rainfall and the risk of drought in the Mediterranean region. The IPCC predicts that if there is an increase in global temperature of 1.5 ^{c}o. then that the number of species that are expected to be lost if there is a temperature increase of 2 °C is 18% insects, 16% plants, 8% vertebrates and is projected to decrease to 6% insects, 8 % of plants and 4% of vertebrates if there is an increase in temperature of 1.5 °C. Other impacts include loss of biodiversity, increased forest fires, extreme weather, and the spread of invasive species, pests, and diseases.

Various studies have concluded that the areas most affected by climate change are urban and coastal areas. Muhammad's research [2] mentions several impacts of climate change on the sea and coast, namely the increase in seawater temperature, sea-level rise, increased CO^2 entering the sea, and frequent tidal waves. This will have further repercussions. In his research, Joesidawati [3], stated that due to sea-level rise, the Tuban district is vulnerable to damage to coastal resources and economic losses. Ismail [4] studied the socio-economic parameters of climate change and found that coastal areas of Malaysia have high a vulnerability by looking at indicators of quality of life, economic value, and infrastructure.

Bengkulu, Bengkulu Selatan Regency is ranked as the 30th most vulnerable district in Indonesia Meanwhile, other regencies/cities in Bengkulu Province are not included in the 50 vulnerable areas. Figure 1 shows the areas in Bengkulu Province that are vulnerable to climate change. The status of areas with high vulnerability is in the highlands. While the coastal areas tend to have low–medium vulnerability. Research from Zamdial [5], states that climate change is damaging the coastal area of North Bengkulu with impacts including coastal forest degradation, abrasion, and landslides, mining excavations, conversion of coastal forests, settlements,

* Corresponding author:ekokhanzalila@gmail.com

damage to river mouths, traditional gold panning, ponds, seawater intrusion. and port lanes. In another study by Zamdial [6], 14 observation sites in Bengkulu City generally experienced coastal degradation due to abrasion, land conversion, and pollution.

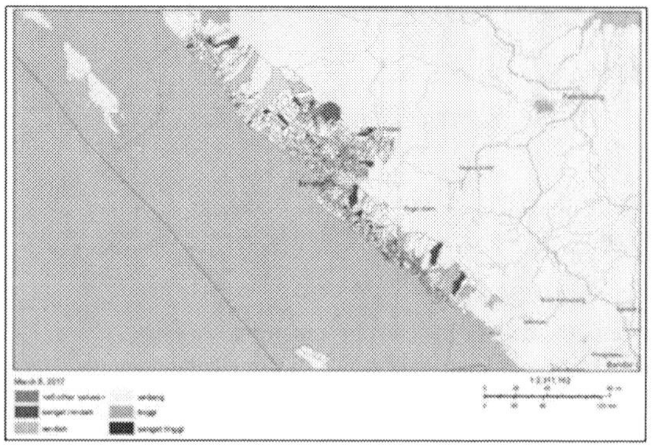

Fig 1. District/City Vulnerability Map in Bengkulu Province.

Vulnerability according to the Law of the Republic of Indonesia of 2007 concerning Disaster Management is a condition of a community or society that leads to or causes an inability to deal with the threat of disaster. Miladan [7] concludes that vulnerability assessment is an attempt to identify the impact of a disaster, both physically in the area and the socio-economic impact of the community. Joesidawati [3] defines vulnerability as the tendency of an entity to experience changes in the form of physical damage (marine ecosystems, coastlines, buildings, humans) or abstract concepts (society, economy, country, and so on). Vulnerability in coastal areas is a condition that will exacerbate the process of damage in coastal areas such as abrasion, sedimentation, or loss of coastal areas. Triatmodjo [8] also added the definition of vulnerability as damage to the residential environment, seawater quality, coral reefs, and mangrove forests.

1.2. Climate Change in Bengkulu City

Climate change in Bengkulu City will bring changes in sea surface temperature, surface temperature, and rainfall. Information from the National Aeronautics and Space Administration (LAPAN) through the SRIRAMA application has been assessed over a period of 30 years, between January 1989 and September 2021. In measurements from Malabero Fisherman Village with coordinates -3.79, 102.25 there was an increase in the increase in water surface temperature of the sea. In January 1989, the average temperature was around 301.83 K (28.68 °o.), in January 2010, it was around 301.74 K (29.1 °o.) and in January 2021, the temperature was at 302.28 K (29.13 °C). There is an increase in seawater temperature of 0.45 K (0.45 °C). For land surface temperatures in the Air Sebakul area (-3.83, 102.35, in 1991 it was around 299.5 K (26.35 °C) and there was an increase in temperature on the land surface by 0.67 K in 2021 so that the temperature became 300.17 K (27.02 °C).

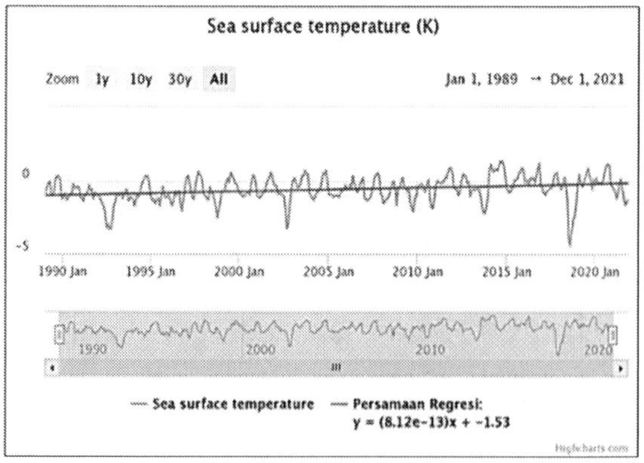

Fig. 2. Sea Surface Temperature (K) Period Jan 1989 – Sep 2021

In the rainfall parameter, there is a decrease in frequency. In January 1991, the average rainfall was 11.92 mm/day and this increased to 14.48 mm/day by 2021. Over a 30-year period, the highest rainfall observed in one day was 25.41 mm/day in October 2016. Anwar et al [9] concluded that some of the impacts of climate change are changes in rainfall patterns, shifts in seasons, and rising temperatures. Shifts also occur in the area affected by rain. This of course will disrupt cropping patterns for farmers.

Novianty, et al [10] found that the effects of climate change threaten the food security of coastal communities due to ocean acidification, physical damage to the shoreline (including building and infrastructure), and sea bleaching. This resulted in economic losses suffered by traditional fishermen reaching more than 73 trillion IDR. This condition can threaten the fisheries sector which has fishery potential consisting of demersal, pelagic, and tuna species [11]. Romdhon. MM [12] observed that in addition to fishing activities in the sea, there are also other activities, especially those carried out by women to supplement household income, including fish processing and fish marketing and trading around the house. From the form of this activity, it is concluded that fishermen's households in Bengkulu city are slightly above the poverty line. In Heryanti's research [13], fishermen start fishing at 16.00 (afternoon) and land at 06.00 (morning). It was concluded that there were various changes in the work of fishermen, especially in Pasar Bengkulu Village by collecting coal.

The process of change also occurs in the coastline in Bengkulu City. In Syukriani's study [14], changes in the shoreline caused abrasion of 19.41 Ha/year and sedimentation of 18.7 Ha/year. The areas that have experienced changes are mostly the meeting area between the river and the sea, namely the Kualo Estuary, Sungai Hitam Estuary, the Jenggalu River Estuary, and the Pulau Baai Harbour. From direct observation, there are several changes and damage that occurred on the coast of Bengkulu City. The worst damage was around the Taman Wisata Alam Pasir Putih Area and the Pulau Baai Area. The damage that occurs is in the form of damage to mangrove vegetation and coastal abrasion.

8,5Fig 3. Changes in the Bengkulu City Coastline in the Harbour Area of Baai Island Source Field Documentation, 2021

Looking at information from various sources, it is considered important to conduct further research on the vulnerability and capacity of the community in dealing with climate change in Bengkulu City considering the geographical location of Bengkulu City in the coastal area and dynamic demographic conditions. This is in line with Sinay's opinion [15] which states that demographic characteristics and cultural systems play an important role in adaptation.

1.3. Formulation of the problem

Climate change indicators of changes such as sea surface temperature, surface temperature, and rainfall will have an impact in the form of increasing the intensity of strong winds, storms, and extreme weather. Subsequent impacts will affect physical changes in coastal and urban areas. An analysis of vulnerability to climate change is very much needed down to the site/village/kelurahan level. Hence the formulation of the problem for this research: What is the impact of climate change on the coastal community of Bengkulu City?

2 Methodology

2.1. Description of the research area

Bengkulu City, which is located on the coast of the Indian Ocean, is the capital of Bengkulu Province, which is located between 3°45'-3°59' South Latitude and between 102°14'-102°22' East Longitude with an area of 151.7 km^2. Topographically, Bengkulu City is relatively flat and low lying with some areas located at a slope of 0-15% (14,224 ha) and 15-40% (228 ha). A total of 5 (five) sub-districts, namely Ratu Samban District, Ratu Agung District, Singaran Pati District, Sungai Serut District and Kampung Melayu District are areas that do not have slopes. In terms of an Indonesian total population of 265 million people in 2019, Bengkulu City is categorized as a medium city (100,000 to 500,000 people).

Fig. 4 Map of the Research Area in Bengkulu City

2.2. Source and data collection

The image data analysed were Landsat 5 TM images (acquired 27 July 1989, 25 May 1995,) and Landsat 8 OLI TIRS images (acquisition 8 August 2021) obtained through the website https://earthexplorer.usgs.gov/ and field observations using Garmin Handheld GPS assistance. Other data are taken from various literature and official documents such as the Bengkulu City Spatial Planning Document, the Bengkulu Province Coastal Zone Management Plan Document, the Maritime Area Action Plan Document, and the National Action Plan for Climate Change Adaptation.

2.3. GIS and Remote Sensing

Observation of shoreline changes in the coastal city of Bengkulu through remote sensing (Path 125/Row 63) using Landsat 5 TM imagery (acquired 27 July 1989, 25 May 1995, 22 May 2005, 9 September 2005, 31 March 2010) and Landsat 8 OLI TIRS (acquired June 3, 2015, and September 21, 2021) and extracted using the Normalized Difference Water Index Method [[16]–[19]The NDWI algorithm can be used to separate land and water. According to research [16], if the NDWI value is more than zero then it is categorized as water level while if it is less than zero then it is categorized as land.

$$NDWI = \frac{Green-NIR}{Green+NIR} \quad (1)$$

For NDWI processing, the Landsat 5 TM image has seven spectral channels as shown below:

1. Band 1 Visible (0.45 - 0.52 m), 30 m resolution, for soil/plant differentiation, coastal zone
2. Band 2 Visible (0.52 - 0.60 m), 30 m resolution, for vegetation
3. Band 3 Visible (0.63 - 0.69 m), 30 m resolution, for plant species differentiation
4. Band 4 Near-Infrared (0.76 - 0.90 m), 30 m resolution, for biomass

5. Band 5 Near-Infrared (1.55 - 1.75 m) 30 m resolution, for snow/cloud differentiation
6. Band 6 Thermal (10.40 - 12.50 m), 120 m resolution, thermal
7. Band 7 Mid-Infrared (2.08 - 2.35 m) 30 m resolution, for lithology

Meanwhile, the Landsat 8 OLI TIRS image, which was launched on February 11, 2011, has 2 sensors, namely the Operational Land Imager (OLI) and the Thermal Infrared Sensor (TIRS). Band details on OLI sensor:

1. Band 1 Coastal/Aerosol, (0.435 – 0.451 m), 30 m resolution
2. Band 2 Blue (0.452 – 0.512 m), 30 m resolution
3. Band 3 Green (0.533 – 0.590 m), 30 m resolution
4. Band 4 Red (0.636 – 0.673 m), 30 m resolution
5. Band 5 Near-Infrared (0.851 – 0.879 m) 30 m resolution
6. Band 6 SWIR-1 (1,566 – 1,651 m), 30 m . resolution
7. Band 7 SWIR-2 (2.107 – 2,294 m), 30 m resolution
8. Panchromatic Band, (0.503 – 0.676 m), 15 m . resolution
9. Cirrus band, (1,363 – 1,384 m), 30 m resolution

Band details on TIRS sensor:

1. Band 10 TIRS-1, (10.60 -11.19μm), 100 m resolution
2. Band 11 TIRS-2, (11.50 -12.51 m), 100 m resolution

In NDWI processing, different channels are used. Landsat 5 imagery uses channel 2 (Visible) and channel 4 (Near Infrared), Landsat 8 uses channel 3 (Green) and Channel 5 (Near Infrared).

3 Data Processing

Image data obtained are overlaid to get the area of abrasion and sedimentation. Processing is done by comparing changes every year. Change results are tabulated for analysis.

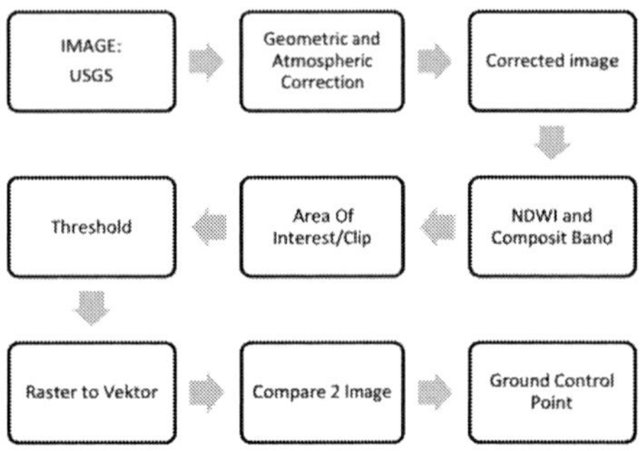

Fig 4. Data Processing

The image obtained from the USGS consists of various channels, then geometric and atmospheric corrections are carried out. The purpose of the correction is to change the reflectance value between objects on earth and satellite image objects due to the effects of absorption, scattering, and atmospheric reflection. This correction was made through the Quantum GIS 3.18 Zurich software with the Semi-Auto Classification Plugin (SCP) plugin. In addition to the core channel, Landsat 5 and Landsat 8 also include several useful files for geometric, radiometric, and atmospheric corrections in the form of files with MTL or MTL codes as file extensions.

Furthermore, as a comparison, all channels in raster form on Landsat 5 TM and Landsat 8 OLI TIRS are combined through the Build Menu Raster menu. After merging, the processed raster files are then stacked to bring up variations of the Red Green Blue (RGB) channel as needed. For this study, the composite channel on Landsat 5 is a combination of channel 543 [16] and channel combination 564 for Landsat 8 OLI TIRS [14]

4 Results and Discussion

The coastal and marine area of Bengkulu City is an area that is home to various activities such as coastal forest ecosystem services, marine ecosystems, fishery resources, Small islands, and potential marine tourism services.

The mangrove forest ecosystem is not too wide with different positions. Some of the locations include TWA Pantai Panjang, Kandang Village, Pelabuhan Pulau Baai Area, Padang Serai and the Jenggalu River. The condition of mangrove degradation is caused by pond activities, settlements, and area development [6].

Based on the Maritime Regional Action Plan document issued by the Bengkulu Province Bapelitbang in 2017, the condition of the coastal area of Bengkulu City has decreased in quality and quantity. Zamdial identification [6] concluded that there was damage to coastal areas, especially at 14 locations, namely Pondok Besi Beach, Bangkahulu River Bridge Beach, Pasar Bengkulu Beach, Teluk Sepang Beach, Sungai Hitam Beach, Muara Sungai Bangkahulu Beach, Jakat Beach, Beach Malabero, Sumur Meleleh Beach, Samudera Ujung Beach, Panjang Beach, Sumber Jaya Beach, Muara Lempuing and Pasir Putih Beach.

Fig. 5 Channel Combination 452 Landsat 5 the Year 1989 Bengkulu City.

Slightly different from the results of research with a focus on areas directly adjacent to the sea found that only six locations experienced severe degradation, namely Teluk Sepang Beach (sedimentation); Pantai Pelabuhan and Pelabuhan Area (abrasion); Pasir Putih Estuary (sedimentation and abrasion), Tapak Padri Beach sedimentation, Pasar Bengkulu Estuary sedimentation) and Sungai Hitam Beach Estuary (sedimentation).

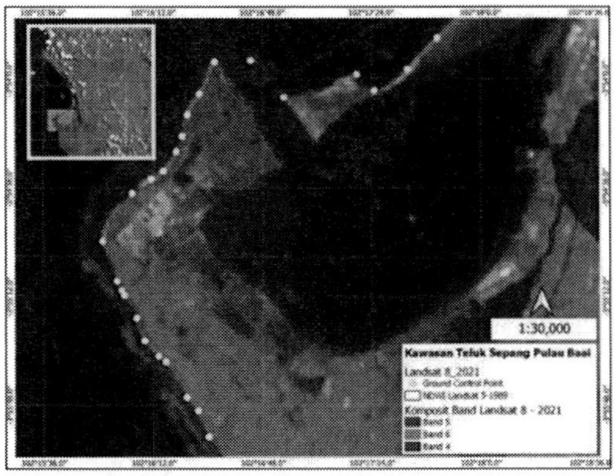

Fig 6. Overlay NDWI LS 5– 1989 with Composite LS8-2021 – Teluk Sepang Area, Pulau Baai

Table 2. Sedimentation and abrasion

Periode	Abrasion (Ha)	Sedimentation (Ha)	Difference (ha)
1989 – 1995	0	0	0
1995 – 2000	80.626	27.931	52.695
2000 – 2005	35.807	26.960	8.847
2005 – 2010	27.466	21.158	6.309
2010 – 2015	245.737	23.811	221.927
2015 - 2021	11.480	19.899	-8.419
Amount	401.117	119.758	281.358
Average per years	12.53	3.74	8.79

Teluk Sepang Area and the Pulau Baai Ocean/Port are areas with a high rate of change compared to other areas. The biggest change in the form of abrasion occurred in the 2010 – 2015 period in the Pulau Baai Port Area of 245,737 ha as shown in Figure 6. This area was under pressure from activities such as coal and plantation products and activities in the Baai Island port area.

Meanwhile, sedimentation has occurred since the initial period of observation but with an average area that is not too large. Sedimentation occurs in Muara Pasir Putih Period 2010-2015 as shown in Figure 6. This area is not too much pressure because there is not too much activity.

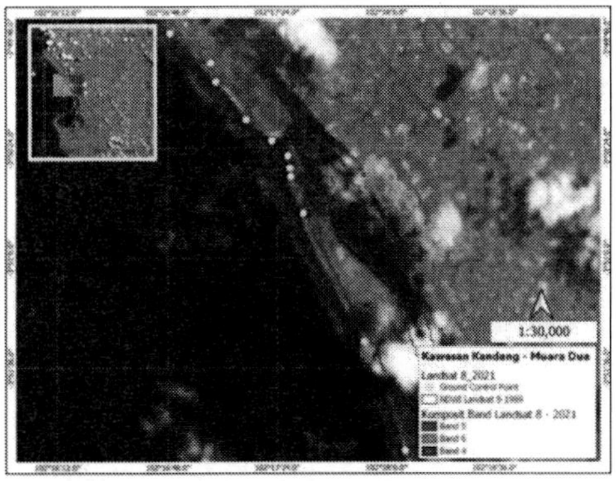

Fig 7. Overlay NDWI LS5 – 1989 with Composite LS8-2021 – Muara Pasir Putih Area

4.1. Vulnerability

Miladan [7] classifies vulnerability to climate change into five groups, namely physical vulnerability, socioeconomic vulnerability, population social vulnerability, environmental vulnerability, and regional economic vulnerability. A similar study was conducted by Choirunnisa [20] which classifies vulnerability into only three groups, namely physical vulnerability, social vulnerability, and economic vulnerability, while the environmental vulnerability is not included as vulnerability.

4.1.1. Physical aspect

The aspect of physical vulnerability is highly dependent on the existence of spatial management. If it is related to legality, the Regional Spatial Planning Document becomes appropriate in making adaptation efforts. Kelurahan Tengah Padang and Pondok Besi are the areas with the highest average population density of 1:80. Meanwhile, the lowest densities are Kelurahan Muara Dua and Kelurahan Teluk Sepang. This is because the two kelurahan have large areas and not too many residents. Conditions certainly create different vulnerabilities and solutions for each region.

In the flood event that affected Bengkulu Province in 2019, many infrastructures were not functioning due to floods and landslides, especially in coastal areas. The National Road, which connects Bengkulu City to West Sumatra, was inundated for some time by the water. As a result, all economic activities cannot run properly with an estimated loss of Rp. 144 billion [Source: https://bnpb.go.id/infografis/infografis-bencana-gunung-dan-longsor-benngkulu].

Residential buildings located in coastal areas may be vulnerable to climate change due to the selection of building materials and technology. The area of the Kampung Nelayan Sejahtera – Sumber Jaya Village is particularly vulnerable, it is classified as a medium slum area with an area of 47.42 ha and is an illegal area because it is owned by PT Pelindo [21]. Most of the buildings are permanent type and some are non-permanent buildings that border or are on the water's edge.

4.1.2. Aspects of socio-economic vulnerability

The total population in the study area is 80,869 people (Ministry of Home Affairs RI Data, accessed June 30, 2021) with a composition of 40,679 men (50.3%) and 40,190 women (49.7%). Based on the age group 0-14 years, there are 21,030 people, ages 15-69 years there are 58,082 people and those aged 70 years and over are 1,757 people. In addition, the number of people with disabilities for Bengkulu City in 2018 was 572 people (357 men and 215women) https://statistik.benngkuluprov.go.id/Sosial/disabilitas/2018.

In the education sector, the largest group are/not or are not yet in school with 23,642 people (29.23 %) of the total population. Followed by high school graduates with 20,793 people (25.71%). Wahyudi et all's research [22] state that there is a significant relationship between knowledge of climate change and education level. Education is considered to affect mindset, knowledge, and abilities.

In terms of employment status, those directly related to climate change are fishermen with a population of 2,032 people (2.51% of the total population of 19 urban villages). The impact felt by fishermen by climate change is that they cannot go to the sea. While the rest are not directly related but do not rule out business sector groups that have activities on the coast. Such as a group of traders on the beach who are unable to carry out their economic activities or a decrease in the number of buyers and the flow of goods and services is hampered. Data from the Ministry of Maritime Affairs and Fisheries of the Republic of Indonesia for 2010 – 2020 (https://statistik.kkp.go.id/home.php?m=prod_ikan_prov&i=2#panel-footer) states that in Bengkulu, there is an increase in the production of several fishery commodities. catch like snapper and grouper. The increase that occurred was quite significant as shown in Figure 8:

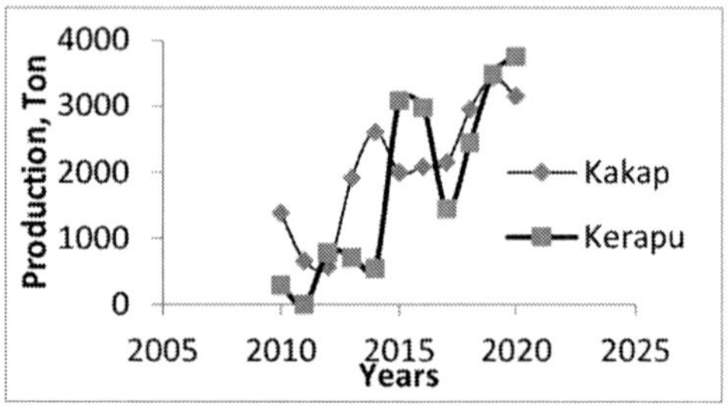

Fig. 8. Bengkulu Capture Fishery Production

As for Bengkulu City, snapper commodities in 2019 were 1,631 tons (47% of total provincial production) and grouper 2.451 tons (70% of total provincial production).

The Pulau Baai Port area is also a strategic area for the Bengkulu economy. Until May 2021, total recorded exports were US$ 49.9 million (Source: https://benngkulu.bps.go.id/indicator/8/45/1/value-eksport-provinsi-benngkulu-fob-free -onboard-port-island-baai-.html).

The economic improvement in Bengkulu City also contributes to the variety of forms of the economy, shifting the economy into secondary and tertiary sectors [23]. The economy is dominated by the accommodation sector, the food, and beverage sector, the real estate sector, and the education sector. So is the coastal area. The primary sector of fisheries, which is the backbone of the region, has begun to compete with tourism potential and its supporters. The decline in the primary sector also threatens the development of land conversion. Another study [24] states that the carrying capacity of the land for water supply is still in the conditionally sustainable category where the water demand does not exceed the availability of water.

4.1.3. Socio-cultural aspect

The results of Djafar's research in Laedyawati and Yuliani [25] stated that there was a change in attitudes and actions after various consulting was carried out to get information. The function of the extension is to mobilize the community (initiation), socialization (initial risk assessment), preparation, promotion, participation (support and community participation), and program dissemination. The condition of the community becomes vulnerable because they do not have the knowledge and there is no strategy because of the assumption that disasters are something that God decrees.

4.1.4. Environmental aspect

Mangrove vegetation has a function as an absorber of greenhouse gases (CO^2) from the atmosphere and is processed through photosynthesis into biomass in cells and tissues. In addition, it serves as a protector of coastal ecosystems from excessive accumulation of organic land [22]. The potential of mangroves in Bengkulu City is 214.62 ha (116.24 ha in the TWA Pantai Panjang and 98.38 ha outside the area) [23], this study also mentions tree vegetation and samplings that make up the mangrove ecosystem found only 9 species, namely *Rhizophora apiculata, Sonneratia alba, Bruguiera gymnoriza, Xylocarpus granatum, Avicennia alba, Hibiscus tiliaceus, Lumnitzera littoreae, Ceriops tagal and Acrostichum aureum* with carbon content stored in the mangrove ecosystem stands of 18.53 tons/ha.

Residential areas on the coast are also vulnerable to the possible spread of climate-related disease vectors. Although in the research of Chandra and Nugraheni [26], it was concluded that there was no relationship between the incidence of DHF in Bengkulu City and wind speed because the data used did not use the variables of human behaviour and environmental conditions

4.2. Adaptation Effort

4.2.1. Climate change policy

Indonesia's transition policy towards a low-emissions and climate-resilient future outlined in Indonesia's Nationally Determined Contribution (NDC) document, includes an affirmation of strengthening the role of ProKlim as a climate resilience program at the local level. In the figure below it can be seen that the components of adaptation cover the food sector, prevention, and preparedness including disease control.

Padang Serai Village, Kampung Melayu District is one example of a coastal area that follows the Proklim implementation initiated by the Ministry of Environment and Forestry of the Republic of Indonesia.

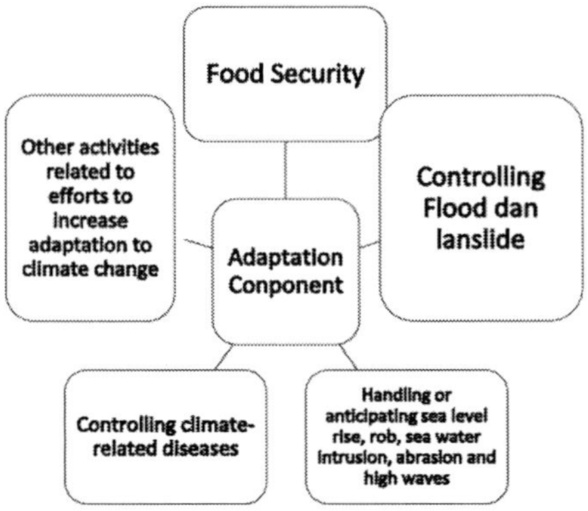

Fig 9. Components of Climate Adaptation Capacity

4.2.2. Strengthening local capacity

The diversification of catches in the Bengkulu city area is very diverse with a total of 108 varieties of fish and another marine biota [27] with export commodities in the form of tuna, skipjack, pomfret, grouper, snapper, white shrimp, tiger shrimp, lobster, and sea cucumber. Meanwhile, domestic marketing, includes tuna, mackerel, squid, octopus, squid, and others. Processed forms of fish include salted fish and fish meal. For this commodity, non-economic fish are generally used. Research from Romdhon [12] states that housewives can support their main income so that the average household in coastal areas is slightly above the poverty line.

The utilization of e-commerce is also a part of adaptation in the form of online marketing. Various platforms offered for free or paid can be an alternative in accelerating the distribution of fish catches and processed [28].

4.2.2. Knowledge management

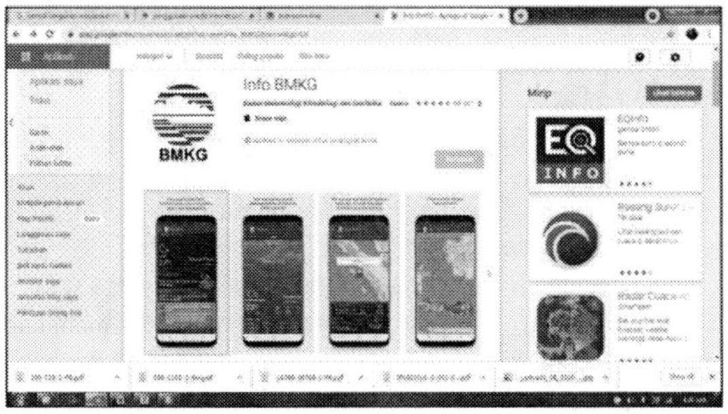

Fig 10. BMKG Apps

Information on climate and weather is a form of community preparedness and capacity. The role of BMKG as a data guardian in providing information on daily, monthly, and yearly weather conditions should be utilized by the community. Android-based applications can also access BMKG information through the Playstore. Information that can be accessed includes maritime weather, aviation weather, climate, air quality, earthquakes, and tsunamis. In addition, dissemination is also carried out via SMS or WA groups.

4.2.3. Institutional

Synergy in solving climate change issues requires the role of all regional apparatus organizations. At the national level, through NDC commitments, the Government of Indonesia established a Directorate General of Climate Change unit under the Ministry of Environment and Forestry. At the Bengkulu Provincial Government level, there are units related to climate change such as the Marine Spatial Management Sector – Provincial Marine and Fisheries Service, Bengkulu Province Regional Disaster Management Agency, Bengkulu, Bengkulu Province Bappeda, and Other Local Government Organizations. This is of course realized through various regulations and the implementation of programs and activities.

Not only government organizations, but the participation of universities and the community also is a strategy in implementing climate change adaptation. As done by Bank Indonesia in 2012 by providing training on salted fish processing. This training will provide opportunities for fishermen to market their catch if they are unable to go to sea due to the weather.

5 Conclusion

The impacts of climate change in Bengkulu City include a decrease in rainfall, an increase in surface temperature, and changes in the coastline. Areas that are vulnerable to being affected are the Teluk Sepang, Pulau Baai, Malabero and Pasar Bengkulu. The forms of morphological changes are sedimentation and abrasion with the highest abrasion process occurring in the 2010-2015 period in the Baai Island area. This area is an area with diverse economic activities and is the key access point for imports and exports in Bengkulu Province.

Community participation is needed to enable adaptation to climate change by paying attention to the local wisdom of coastal communities and appropriate technology. The form of programs and activities that can be carried out can be in the form of core adaptation activities or other activities based on environmental management.

References

1 IPPC, D. Roberts, R. Pidcock, Y. Chen, S. Connors, and M. Tignor, "Science Editor," (2018).

2 S. Muhammad, D. G. R. Wiadnya, and D. O. Sutjipto, " Adaptation of Coastal and Marine Area Management to the Impact of Global Climate Change," *Fak. Perikan.*, 1–6, (2009).

3 M. I. Joesidawati, "Study of climate change and damage to coastal resources in Tuban district," Inst. Teknol. Sepuluh Nop., 289, (2017).

4 I. Ismail, M. L. Husain, A. @ M. S. M. Muslim, and E. Saleh, "Assessment of Socio-Economic Vulnerability Index to Sea Level Rise along the East Coast of Peninsular Malaysia," Malaysian J. Soc. Sci. Humanit., **5**, 10, 21–33 , doi: 10.47405/mjssh.v5i10.497, (2020).

5 Z. et al Zamdial, "Study Of Identification of Coastal Damage in Bengkulu Utara Regency, Bengkulu Province," Enggano, **5** , no. October,2020, pp. 510–528, (2020).

6 Z. Zamdial, D. Hartono, D. Bakhtiar, and E. Nofridiansyah, "Study on Identification of Damage to Coastal Areas in Bengkulu City Study on Identification of Damage to Coastal Areas in Bengkulu City," J. Enggano, vol. **3**, 1,65–80, doi: 10.31186/jenggano.3.1.65-80, (2018,).

7 N. Miladan, "The Study of the Vulnerability of the Semarang City Coastal Area to Climate Change,", [Online]. Available: https://core.ac.uk/download/pdf /11723070.pdf, (2009).

8 B. Triatmodjo, Beach Building Planning, Cetakan Ke. Yogyakarta: Beta Offset Yogyakarta, (2011).

9 A. Anwar, S. Sudjatmiko, and M. F. Barchia, "Oldeman and Schmidth-Fergusson Climate Classification Shifts as a Basis for Natural Resource Management in Bengkulu," Nat. J. Penelit. Natural Resources and Environment Management., **7**, no. 1, pp. 59–68, , doi: 10.31186/naturalis.7.1.9261, (2019).

10 L. Ilmu and P. Indonesia, "Efforts to Formulate Vulnerability Indicators Related to Disasters," 185–200, 2016.

11 K. Sukiyono and M. M. Romdhon, "Assessing technical e ffi ciency for Bengkulu province catching fishery industries and determination of it ' s technical e ffi ciency," J. Int., **4**, 6, 168–174, [Online]. Available: www.fisherjournal.com, (2016).

12 M. M. Romdhon, "BENGKULU Women ' s Income Contribution And Food Access Fisherman ' s Household In Bengkulu City," 63–68.

13 M. Suryati and E. Solina, "Program Studi : Sosiologi," Jmm, **3**,. 2, p. 2019, (2019).

14 S. Sukriani, "Landsat Image Data Analysis Monitoring Bengkulu Coastline Changes," **2**, 1, 78–89, (2017).

15 L. Sinay and R. W. B. Carter, "Communities and Local Governments," Climate, **8**, 7, 1–15, (2020).

16 Komang Iwan Suniada, "Coastline Change Detection Using Remote Sensing Technology In Jembrana Bali Komang Iwan Suniada," J. Kelaut. Nas., **10**, 1, 13–20, (2015).

17 Z. Hidayah and A. Apriyanti, "Detection of Changes in the Coastline of the Eastern Jakarta Bay in 2003-2018", J. Kelaut. Indones. J. Mar. Sci. Technol., **13**, 2, 143–150,, doi: 10.21107/jk.v13i2.7980.

18 N. Anggraini, S. Marpaung, and M. Hartuti, "Analisis Perubahan Garis Pantai Ujung Pangkah Dengan Menggunakan Metode Edge Detection Dan Normalized Difference Water Index (Ujung Pangkah Shoreline Change Analysis Using Edge Detection Method and Normalized Difference Water Index)," J. Penginderaan Jauh dan Pengolah. Data Citra Digit., **14**, 2, 65–78, doi: 10.30536/j.pjpdcd. 1017.v14.a2545, (2018)

19 E. Özelkan, "Waterbody detection analysis using NDWI indices derived from Landsat-8 OLI," Polish J. Environ. Stud., **29**, 2, 1759–1769 , doi: 10.15244/pjoes/110447, (2020)

20 F. Bekele, D. Tolossa, and T. Woldeamanuel, "Local institutions and climate change adaptation: Appraising dysfunctional and functional roles of local institutions from the bilate basin agropastoral livelihood zone of Sidama, southern Ethiopia," Climate,**8**, 12, 1–21, doi: 10.3390/cli8120149, (2020)

21 M. Rijal and Ardiansyah, "The Slum Condition of the Prosperous Fisherman Village of Bengkulu City in Efforts to Prevent and Improve the Quality of Slum Settlements," Temu Ilm. IPLBI, 53, 61–66, (2016).

22 Wahyudi (LIPI), Sequestering Carbon-Coastal Ecosystem Services For Climate Change Mitigation, 1st ed. Yogyakarta: Gadjah Mada University Press, (2017).

23 B. Iin and E. Indraswanti, : "the journal of economic development," **2**, 2, 103–118, (2020).

24 H. Wiryono, Wiryono and Gunggung , Senoaji and Fajrin, "Study of Environmental Capacity and Supporting Capacity of Bengkulu Water Watershed Area," (2015).

25 Ledyawati and Fitri Yuliani, "Strategy to Improve Understanding and Resilience to Disasters in Bengkulu City," Talent. Conf. Ser. Local Wisdom, Soc. Arts, **2**, 1, 1–6,

26 C. G. Sihombing, E. Nugraheni, and W. Sudarsono, "the Relationship Between Rainfall, Air Temperature and Wind Speed Effects Dengue Hemorrhagic Fever Case in Bengkulu City At 2009-2014," Diponegoro Med. J. (Jurnal Kedokt. Diponegoro),**7**, 1,. 366–380, (2018).

27 Bank Indonesia, "Report on MSME Development Patterns for Dried Fish Processing Business in Bengkulu City,. [Online]. Available: https://www.bi.go.id /id/umkm/pembiayaan/Pages/Pola-Pembiayaan-Usaha-Kecil-Pengolahan-Ikan-Kering.aspx , (2012)

28 Y. Apridiansyah and A. Wijaya, "E – Commerce Application of Bengkulu Sea Products withApplication of Seo Algorithm Using Introduction to Literature Study Methodology," **1**, 5, 194–200, (2019)

https://doi.org/10.1051/e3sconf/202133102022

The effectiveness of the non-free plastic bag policy to reduce plastic waste in the community of Padang

Novia Wirna Putri[1*], *Septia* Pristi Rahmah[1], *Defriman* Djafri[1], *Irene* Sandra Olivia[1], and *Ulfah* Winanda Putri[1]

[1]Faculty of Public Health, Andalas University, Padang, Indonesia

Abstract. The waste generation of Padang City at the Final Processing Site (TPA) is 495.50 tons/day, and the amount of unmanaged waste is 62.42 tons/day, of which 40% is plastic waste. Efforts to reduce the use of plastic bags have begun in Padang, referring to Perwaliko number 36 of 2018. Shopping centers and modern stores must use plastic shopping bags that meet SNI no later than December 31, 2020, especially in various retail and contemporary shopping centers. This study aimed to analyze the relationship between the effectiveness of plastic use policies and the decrease in the use of plastic bags in people who shop at modern retailers in Padang City. This type of research is quantitative with a cross-sectional design. The number of samples is 116 respondents who shop at modern shopping centers, namely Matahari Dept. Store, Food Mart Padang, Ramayana Padang and SPAR Padang. Data analysis used univariate and bivariate analysis. The results showed that 46.6% of respondents stated that the plastic bag policy was not adequate, and 46.6% of respondents stated that there was no reduction in plastic bags. There is a significant relationship between the effectiveness of the plastic bag policy and the decrease in the use of plastic bags in the people of Padang City (p=0,029). With this paid plastic bag policy, government, retailers, and the public will be more concerned with the environment.

1 Introduction

Plastic bags are bags used as wrappers to carry shopping items in food, drinks, clothing, and other household items. The nature of cheap, practical, and compact plastic bags makes plastic bags increasingly out of control. The world's population can use 500 million to 1 billion plastic bags in one year. Everyone uses 150 plastic bags every year.2011 To make that much plastic requires 12 million barrels of oil and 14 million tree trunks that must be cut down as the basic raw materials for making it. Of course, the higher use of plastic bags will be followed by an increase in the amount of plastic waste. Plastic bags that are not used will be thrown away, even though it takes hundreds of years to decompose [2]. In the last ten years, the amount of plastic waste generated has continued to increase significantly. Indonesian people use around 9.8 billion pieces of plastic bags every year. Almost 95% of the use of plastic bags becomes plastic waste (KLHK, 2016).

The population of Padang City in 2018 reached 939,112 people, with a population density of 1,351 people/km2. This population is accompanied by the generation of Padang City waste at the Final Processing Site (TPA) as much as 495.50 tons/day and unmanaged waste as much as 62.42 tons/day. In 2019, the population of Padang City increased to 950,871 people, with a waste generation of 624.24 tons/day This shows that the increase in population goes hand in hand with the increase in the number of municipal solid waste generation.

Plastic bags take a long time, even thousands of years, to decompose completely. The decomposition of plastic in the soil into plastic particles can contaminate soil and groundwater. If burned, plastic will produce toxic fumes harmful to health because it will decompose in the air as dioxin compounds that are very dangerous if inhaled. Throwing plastic into rivers and seas can damage habitats in rivers and seas [2].

Therefore, it is crucial to manage plastic waste. To reduce environmental problems from the use of plastic bags, several countries such as Australia, Italy, the United States, Tanzania, and Ireland have used taxation or bans on the production and use of plastic bags for shopping or other purposes [5].

In Indonesia, the Ministry of Environment and Forestry (KLHK) has issued a policy aimed at reducing the generation of plastic waste through the Circular Letter of the Directorate General of Waste Management, Hazardous and Toxic Waste and Materials of the Ministry of Environment and Forestry Number: SE-06/PSLB3-PS/2015 concerning Anticipatory Measures Application of the Paid Plastic Bag Policy in Modern Retail Businesses and Circular Letter Number S. 1230/ PSLB3-PS/2016 concerning the price and mechanism for the application of paid plastic bags are the basis for the implementation of the paid plastic bag policy in the modern market (KLHK, 2016). Plastic bags are no longer

* Corresponding author: noviawina.skm@gmail.com

given free of charge by retailers to consumers. If the consumer needs a plastic bag, the consumer must buy a plastic bag from the shopping place. The Ministry of Environment and Forestry has reduced plastic waste with a paid plastic bag trial program. The program trial started on February 21, 2016, in conjunction with the commemoration of National Waste Care Day. With the existence of paid plastic bags, people will bring their shopping bags, not just throwing away plastic bags that have been given free of charge by supermarkets/other retail traders.

Padang City Government support for plastic waste management efforts in the form of the issuance of Padang Mayor Regulation No. 36 of 2018 concerning Control of the Use of Plastic Shopping Bags. This regulation explains that one of the government's policies in reducing waste, especially plastic waste, is applying a plastic bag policy that is not free in all shopping centers, modern shops, and traditional markets in Padang City. Modern shopping centers and shops must use plastic shopping bags that fully meet SNI no later than 31 December 2020. For this reason, it is necessary to research the effectiveness of the policy of using non-free plastic bags to reduce their use in people who shop at modern retail in the city of Padang.

2 Method

The type of research used is analytical research using a cross-sectional design. The population in this study were all people who shopped at modern retailers in Padang who applied the non-free shopping cart policy, namely Matahari Dept. Store Padang, Foodmart Padang, Ramayana Padang, and SPAR Padang, whose population is unknown. The number of samples in this study was 116 respondents. The location of this research is in a modern shopping center that has implemented a plastic bag policy that is not free, namely Matahari Dept. Store Padang, Foodmart Padang, Ramayana Padang, and SPAR Padang. The research variables are the effectiveness of the policy on the use of non-free plastic bags and the decrease in the use of single-use plastic bags. Primary data is collected directly through the distribution of questionnaires to respondents.

Univariate analysis was used on each of the variables studied with a frequency distribution table accompanied by an explanation, namely the frequency distribution of the effectiveness of the policy on the use of non-free plastic bags and the decrease in the use of plastic bags. Bivariate analysis was used to see the relationship between the independent and dependent variables using the chi-square statistical test at the 95% confidence level ($p < 0.05$).

3 Result

Table 1 shows an overview of the frequency distribution of respondent characteristics like gender, age, and education. Table 1 shows that female respondents (89.7 %) were more than male respondents (10,3 %). Based on age, more respondents were categorized as late adolescents aged 17-25 years (53,4 %). Based on education, most of the respondents have a senior high school education, namely 58.6%.

Table 1. Respondent Characteristic

Respondent Characteristics	F	%
Gender		
Male	12	10,3
Female	104	89,7
Age		
17-25 years	62	53,4
26-35 years	18	15,5
36-45 years	15	12,9
46-55 years	18	15,5
56-65 years	3	2,6
Education		
Elementary	2	1,7
Middle school	3	2,6
High school	68	58,6
Above college	43	37,1

Based on the results of the study, it can be described the frequency distribution of policy effectiveness in the people who shop at modern retailers in the city of Padang, can be seen in Table 2 below:

Table 2. Description of Policy Effectiveness Frequency Distribution to the Padang Community

No	Statement	Evaluation	
		Yes (%)	No (%)
1	There has been a notification from the government regarding the paid plastic bag policy	59,5	40,5
2	There has been a notification from the supermarket regarding the paid plastic bag policy	62,1	37,9
	Supermarkets offer alternative options for shopping bags other than plastic	58,6	41,4
3	The application of the paid plastic bag policy by the supermarket to buyers is correct	71,6	28,4
4	The price of a plastic bag is Rp. 200 - according to your current economic condition	89,7	10,3
	The paid plastic bag policy is very appropriate when the volume and types of waste are increasingly diverse today	87,1	12,9
5	Paid plastic bag policy can reduce the use of plastic bags	77,6	22,4
	After finishing shopping, the plastic bags have been recycled by business actors/supermarkets	56,0	44,0
6	Paid plastic bag policy has an impact on a clean and healthy environment	85,3	14,7
	Bring your shopping bag	78,4	21,6

Based on Table 2 shows that 89.7% of the people of Padang City agree that the price of plastic bags is Rp. 200 - according to the current economic condition of the community. As many as 87.1% of the people of Padang

City agree that the paid plastic bag policy is very appropriate when the volume increases and the current type of waste. As many as 85.3% of the community agree that the paid plastic bag policy impacts a clean and healthy environment.

The frequency distribution of the effectiveness of the policy on the use of plastic bags for people who shop at modern retail in Padang City can be seen in the following table:

Table 3. Frequency Distribution of Policy Effectiveness to the people of the city of Padang

The Effectiveness Policy	Frequency (f)	Percentage (%)
Ineffective	54	46,6
Effective	62	53,4
Total	116	100,0

Table 3 shows that as many as 54 people (46.6%) of respondents stated that the policy on paid plastic bags was not effective. The frequency distribution of the decrease in the use of plastic bags among people who shop at modern retail in Padang City (Table 4):

Table 4 Frequency Distribution of Decreased Use of Plastic Bags in Padang City Community

Decrease in the use of plastic bags	Frequency (f)	Percentage (%)
Bad	54	46,6
Good	62	53,4
Total	116	100,0

Table 4 shows that as many as 54 people (46.6%) respondents stated that the decrease in plastic bags was not good. The frequency distribution of the decrease in the use of plastic bags in people who shop at modern retailers in the city of Padang in Table 5:

Table 5 Description of Frequency Distribution of Decreased Bag Usage Plastic in Padang City

No	Statement	Evaluation	
		Agree(%)	Disagree(%)
1	Before there was a paid plastic bag policy, I didn't bring my shopping bag	80,2	19,8
2	After there was a policy and an appeal for paid plastic bags, I brought my shopping bag	70,7	29,3
3	Shopping bags that you bring are reusable, more durable, and more economical than buying plastic shopping bags every time you shop	85,3	14,7

Table 5 shows that 95.7% of respondents agree that supermarkets must provide types of shopping bags other than environmentally friendly plastic, 87.9% of respondents agree that supermarkets are the right party for implementing a paid plastic bag policy to reduce the use of plastic bags. As many as 86.2% of respondents agree that the selling price of plastic bags is worth Rp. 200, - is considered cheap. As many as 80.2% of respondents agreed that before there was a paid plastic bag policy, I did not bring my shopping bag, and as many as 70.2% of respondents agreed that after there was a policy and an appeal for paid plastic bags, I brought my shopping bag. There was a 50.9% decrease in single-use plastic bags after the policy of using non-free plastic bags.

3.1 The Relationship between the Effectiveness of the Plastic Bag Policy and the Reduction in the Use of Plastic Bags

Table 6 shows the relationship between the Effectiveness of the Plastic Bag Policy and the reduction in the use of plastic bags:

Table 6. The Relationship between the Effectiveness of the Plastic Bag Policy and the Reduction in the Use of Plastic Bags

The Effectiveness Policy	Reduce The Use of Plastic Bags				Total		p-*value*
	Bad		Good				
	f	%	F	%	f	%	
Ineffective	31	57,4	23	42,6	54	100,0	0,029
Effective	23	37,1	39	62,9	62	100,0	
Total	54	46,6	62	53,4	116	100,0	

Table 6 shows that more respondents stated that the policy of using plastic bags was not effective in reducing the use of plastic bags in daily life, namely 31 people (57.4%), compared to respondents who stated that the policy of using plastic bags was effective in reducing the use of plastic bags. Plastic in daily life has as many as 23 people (37.1%). The statistical analysis shows a relationship between the effectiveness of the plastic bag policy and a decrease in the use of plastic bags (p-value =0.029).

4 Discussion

4.1 The Effectiveness of Plastic Bag Policy

Based on the results of the univariate analysis on the effectiveness of the non-free plastic bag policy, 54 people (46.6%) of respondents stated that the policy on paid plastic bags was not effective, and 62, 9% said it was effective. Research by Asmuni (2015) states that the No Plastic Bag Day (NPBD) program in Malaysia has made consumers avoid the use of plastic bags by 52.3%. This program has influenced customers to use reusable shopping bags. (Asmuni, Hussin, Khalili, & Zain, 2015) .

The Malaysian government in 2011 launched the Day Without Plastic Bags Campaign. The goal is to reduce plastic bags that harm the environment. All retail, supermarkets, and hypermarkets participate in the program. Stores that do not provide plastic bags encourage customers to bring their bags or purchase eco-friendly bags. Some stores provide paper bags or synthetic fiber bags that can be reused several times. They strengthened the ban on plastic bags with a levy fee of MYR 0.20 or around 700 rupiahs (Asmuni, Hussin, Khalili, & Zain, 2015). The fee for the tax retribution is channeled to charity (Hong, 2011) or used for environmental conservation activities. However, the government is allowing the use of new plastic bags for traditional markets, restaurants, and night markets for hygiene when bringing groceries and wet food (Zen, Ahamad, & Omar, 2013).

Based on the description of the frequency distribution of respondents' statements, 89.7% of the people of Padang City agree that the price of plastic bags is Rp. 200 - according to the current economic condition of the community, as many as 87.1% of the people of Padang City agree that the paid plastic bag policy is very appropriate when it is getting worse. The current volume and type of waste increases, and as many as 85.3% of the community agree that the paid plastic bag policy impacts a clean and healthy environment.

The implementation of plastic bag policies in developed countries started in the early 1990s, namely by imposing taxes in Germany and Denmark. Although Germany and Denmark were the first countries to impose a plastic bag tax in 1991 and 1994, only Denmark can pioneer plastic bag policies. (Knoblauch, Linda, & Stein, 2018). In the southern states of Europe, the fact that plastic bag waste is much more visible and dangerous due to limited collection and recycling of waste has led policymakers to adopt plastic bag laws. (Knoblauch, Linda, & Stein, 2018)

The Government implements a paid plastic bag policy to reduce environmental pollution from plastic waste. During the trial of paid plastic bags, the Government, the National Consumer Protection Agency (BPKN), the Indonesian Consumers Foundation (YLKI), and the Indonesian Retail Entrepreneurs Association (APRINDO) agreed that the selling price of plastic bags was at least Rp. 200-, per bag including additional tax. Value (VAT). After the trial runs for at least three months, the price of plastic bags is evaluated by the Government, Local Government, and APRINDO. The agreement between retailers, the Government, BPKN, YLKI, and APRINDO that the plastic bags provided are chosen have the most negligible environmental impact and must meet national standards issued by the Government or independent institutions. APRINDO is committed to supporting the provision of incentives to consumers, environmental management through corporate social responsibility programs with respective regulations and waste management. This policy also applies to modern retail businesses that are not members of APRINDO (Bahagia, Yunita, & Susanti, 2019).

Applying the policy of not-free plastic bags has both positive and negative impacts from an environmental and economic perspective. From an environmental perspective, this policy has a positive value because it can preserve the environment by reducing plastics that can pollute the environment, water, soil, and air sources. From an economic point of view, this policy can be detrimental to the community because the public must pay a medium-sized plastic bag fee of between Rp. 200 – Rp. 250. If people spend a lot of money, of course, the money spent on buying plastic bags will increase (Bahagia, Yunita, & Susanti, 2019). This policy is straightforward to implement. It is enough for people to bring their shopping bags, then they can be involved in this policy and the consequences if they do not bring their shopping bags are not too burdensome, which only costs Rp. 200,- (Octavianti, Koswara, & Sari, 2016).

This policy is expected to change people's behavior to reduce the use of plastic bags starting from simple things, namely when shopping. This policy conveys a message to bring shopping bags from home. This policy is implemented as an innovation present in the community to shape and change people's behavior in a better direction so that environmental and ecosystem sustainability is maintained, disaster prevention, and reduction of the circulation of plastic waste (Octavianti, Koswara, & Sari, 2016). This paid plastic bag policy must, of course, be known by the broader community so that it will not cause polemics and question marks in its implementation. News of this policy is also communicated through many mass media, both traditional and electronic. The role of the media is still a very effective and favorite communication strategy because the reach of the media is extensive, not only on a local scale but also nationally, even internationally (Lusnita, 2019).

With this much news, it is hoped that the public will understand the intent and purpose of implementing this paid plastic bag policy and that the ultimate goal of reducing plastic waste can be achieved. After the communication from KLHK to Aprindo and Aprindo to its members, the most important thing is how retailers communicate with consumers. Therefore, the Ministry of Environment and Forestry requires retailers to install a form of communication at each outlet in an announcement or statement that plastic bags are no longer free. Following the method chosen by the Ministry of Environment in its communication strategy, the Ministry of Environment and Forestry created infographic information such as the one below, which provides education about waste in its announcements to provide an understanding to consumers why plastic bags are no longer given free of charge (Lusnita, 2019) (Asmuni, Hussin, Khalili, & Zain, 2015). One way to increase effectiveness and community participation is to foster a culture of carrying bags when shopping.

4.2 Decreasing the Use of Plastic Bags

Table 4 shows that 95.7% of respondents agree that supermarkets must provide types of shopping bags other than environmentally friendly plastic, 87.9% of respondents agree that supermarkets are the right party for implementing a paid plastic bag policy to reduce the use of plastic bags. As many as 86.2% of respondents agree that the selling price of plastic bags is worth Rp. 200, - is considered cheap. Many consumers prefer to buy plastic shopping bags when shopping. Therefore, the government needs to reconsider the price of plastic bags that are recommended to retailers. According to Novianti and Kartika, the recommended price to reduce the use of plastic bags is around Rp. 500 – Rp. 1000. The government should continue to socialize and provide an understanding of the paid plastic bag policy to strengthen consumers' desire to reduce the use of plastic bags (Astuti, 2016). As many as 80.2% of respondents agreed that before a paid plastic bag policy, they did not bring a shopping bag. As many as 70.2% of respondents agreed that they brought shopping bags after a policy and an appeal for paid plastic bags. There was a 50.9% decrease in single-use plastic bags after the policy of using non-free plastic bags. This change shows that implementing the plastic bag policy is changing consumer behavior.

The reduction of plastic waste in Indonesia cannot be calculated because it has only just begun, but there have been reports of its implementation in other countries. In China, the paid plastic bag policy was implemented in February 2008 in all regions of China. This plastic bag restriction covers the production, sale, and use of plastic bags. In addition, these plastic bags should not be distributed for free. Those who need it have to buy it. After three years of implementing a paid plastic bag policy, China reported that the use of plastic bags had decreased by 2/3 from the previous, which was 60 thousand tons of plastic. This amount of plastic is equivalent to 3.6 million tons of oil or 5 million tons of coal and cuts CO2 emissions by more than 10 million tons (Widhi, 2016). The decline in the use of plastic bags with the implementation of a paid plastic bag policy also occurred in South Africa. This policy was implemented in May 2003. In 2010, a study from the University of Cape Town stated that there had been a 44% decline in plastic bag consumption since the plastic bag restriction policy was implemented (Widhi, 2016). A similar thing has happened in the United States, which has implemented this policy since 2014. Some stores charge 10 US cents or the equivalent of Rp. 1385 for the use of pre-packaged, recyclable plastic bags (Tjahjono, 2016). In Hong Kong, people who shop and use plastic bags have to pay 50 cents. With a paid plastic bag program, these efforts can reduce plastic consumption by up to 73% (KLHK, 2016).

The UK has implemented a paid use of plastic bags in shops since October 2015. Shops charge 5 pence or equivalent to IDR 985 for each use of a plastic bag. The UK was the last country in the UK to implement the policy (Wales in 2011, Scotland and Northern Ireland in 2014). Within three years of its implementation, this paid plastic policy has reduced plastic consumption by up to 80% in Wales (Thomas, Sautkina, Wolstenholme, & Whitmarsh, 2019). Thus, the implementation of the paid plastic bag policy that has been implemented in several countries in the world has a positive impact on the environment. A significant decrease can see in the number of plastic bags used in these countries. In Indonesia, the implementation of the paid plastic bag policy is also expected to be able to contribute to reducing the number of plastic bags.

5 Conclusion

The generation of plastic waste in Indonesia is getting more and more out of control. It can pollute the environment because of the nature of plastic which is difficult to biodegrade in the long term. Plastic waste is also harmful to the health of living things. Plastics are made from materials that are carcinogenic if they enter the food eaten. Then, replacing plastic with biodegradable plastic and reducing plastic use by implementing a paid plastic bag policy. Applying paid plastic bags in Indonesia was carried out from February 21, 2016, to June 5, 2016. It is hoped that the implementation of this policy can reduce the use of plastic bags in Indonesia, as in other countries that have implemented this policy before.

The existence of a paid plastic bag policy is expected to encourage the government, retailers, the public, and the media to be more concerned about the environment. The government is expected to make policies and rules regarding reducing the use of plastic bags, starting from reducing production, storage, sales, distribution to the use of plastic. Retailers are expected to implement rules and systems to reduce the use of plastic bags and educate consumers. Retailers can provide non-plastic shopping bags with attractive designs and large sizes.

The community should implement a plastic diet lifestyle starting at home and participate in educating and socializing together as a community. The role of the media is needed to make this policy successful by disseminating the plastic bag diet solution and the reasons for doing so.

Thanks to the people of Padang City who have participated in this research. Thank you also to fellow lecturers in the K3 Kesling department, IKM Study Program, and the entire academic community of the Faculty of Public Health, Andalas University, for their support and cooperation.

References

1. S Asmuni, Hussin, N. B., Khalili, J. M., Z. M. Zain, Public Participation and Effectiveness of The No Plastic Bag Day Program in Malaysia. *Procedia - Social and Behavioral Sciences 168*, 328 - 340 (2015)
2. Astuti, D.A. Application of Paid Plastic Bags as an Effort to Reduce the Use of Plastic Bags. R&D, **12**, 32-40 (2016).
3. Bahagia, Yunita, I., & Susanti, R. M. Analysis of Paid Plastic Bag Policy Implementation at Modern Retail Banda Aceh. Porch Technique, 549 – 556 (2019).
4. Hong, L. C. (2011). Environmental tax laws in Malaysia today. *Malaysian Accountancy Research and Education (MAREF) Review* **1**(2), 9-10 (2011)
5. Jalil, M. A., Mian, M. N., & Rahman, M. K.. Using plastic bags and its damaging impact on environment and agriculture: An alternative proposal. *International Journal of Learning & Development,* **3**(4), 1-14 (2013)

6. Ministry of Environment.). 2019 National Waste Management Information System. Retrieved from http://.sipsn.menlhk.go.id (2019)
7. KLHK17 Cities Will Implement Paid Plastic Bags. Jakarta:Director General of Waste, Waste and B3 Management . (2016).
8. Knoblauch, D., Linda, M., & Stein, U. (2018). Developing Countries in The Lead - WHat Drives The Diffusion of Plastic Bag Policies? *Sustainability*, 1-24.
9. Lusnita, N. Plastic Waste Reduction Program Through Paid Plastic Bag Policy in Modern Retail. Servites **1**(1), 12 – 24 . (2019)
10. Octavianti, M., Koswara, I., & Sari, D. YCharacteristics of Plastic Bag Policy Innovation Not Free for Housewives in the City of Bandung. Communication **8**(2), 134-146 . (2016).
11. Safitri, P., Winda, S., & Mochamad, Z. (2018). *Statistik Lingkungan Hidup Indonesia*. Jakarta: Badan Pusat Statistik Indonesia.
12. Thomas, G. O., Sautkina, E., Wolstenholme, E., & Whitmarsh, L. (2019). The English Plastic Bag Charge Changed Behavior and Increased Support for Other Charges to Reduce Plastic Waste. *Frontiers in Psychology 10*, 266.
13. Tjahjono, S. (2016, Maret 16). *Pengelolaan Limbah Plastik "Fosil" Masa Depan*. Diambil kembali dari http://print.kompas.com.
14. Widhi, N. (2016, Oktober 10). *Tas Plastik Berbayar Rp. 200 di Indonesia. Bagaimana di Luar Negeri?* Diambil kembali dari http://news.detik.com/berita/31479.
15. Zen, L. S., Ahamad, R., & Omar, W. (2013). No Plastic Bag Campain Day in Malaysia and The Policy Implication. *Development and Sustainability* **15** (5), 1259 - 1269.

Hydraulic conductivity and calcium carbonate content of biocemented heavy-metal contaminated mine waste soil

Jodin Makinda[1, 2 *], *Khairul Anuar* Kassim[1], *Kamarudin* Ahmad[1], *Abubakar Sadiq* Muhammed[1,3], and *Muttaqa Uba* Zango[1,4]

[1]School of Civil Engineering, Faculty of Engineering, University of Technology Malaysia, 81310, Johor Bahru, Johor, Malaysia
[2]Faculty of Engineering, University of Malaysia Sabah, 88400, Kinabalu, Sabah, Malaysia
[3]Department of Civil and Water Resources Engineering, Unimaid, P.M.B. 1069 Maiduguri, Borno State, Nigeria
[4]Department of Civil Engineering, KUST, Wudil P.M.B. 3244 Kano State, Nigeria

Abstract. It is not uncommon for mining activity to generate wastes associated with negative engineering impacts include susceptibility to runoff due to the absence of vegetation, erosion, and sinkhole. Due to their high degree of permeability, movement of heavy metal contaminants in waste soils and aquifers occurs actively along with the hydrogeological parameters, and research on using biocementation methods such as enzyme induced calcite precipitation (EICP) to reduce the hydraulic conductivity is therefore beneficial. Mine waste soil collected from a copper mine in Lohan Dam, Sabah is characterized physically, morphologically, and chemically then treated with EICP under different operational parameters include curing duration (1,3,7 days), curing temperatures (5, 10, and 25 °C), and relative density (70 and 80 %). The hydraulic conductivity was then tested using a constant head permeability test and the calcium carbonate content (%) is determined using the HCL washing method. Properties of the Lohan Dam wastes are found to be predominantly coarse grain soil of low plasticity, high specific gravity, high permeability, acidic in nature, and low organic content. Morphologically, they are composed of powdered and hardened particles with dark brown color with high amount of irregular-shaped particles. Mineralogical, Lohan Dam soil wastes contain a high level of heavy metals beyond the safety level of the Department of Environmental Malaysia. EICP treatment had to change the degree of permeability from 'high and medium' to 'low' with a 94-97% reduction in hydraulic conductivity corresponding to the amount of calcium carbonate content produced ranging 6.94-9.63%. In conclusion, relative density shows the marginal effect, curing duration, and temperature shows a more significant impact on the treatment effectiveness.

1 Introduction

The mining process generates a large number of inert materials but the presence of toxic heavy metals such as mercury, arsenic, lead, zinc, cadmium is not uncommon. Hazardous chemicals may present naturally or due to addition during the extraction process. The potential threat of these contaminants to the environment sustainability may include alteration of soil composition and threatening the quality of water supply and groundwater. As a consequence, the living quality especially of the population living around the surrounding area is low, in some cases forcing resettlement and changing their economic structure.

In addition, negative engineering impacts include susceptibility to runoff due to the absence of vegetation, erosion, and sinkhole due to highly permeable soil, and landslide due to unstable grounds, which may increase because of the debilitated properties of soil. As a result, rupture of ponds, dams, or impoundments that can flood the adjacent lands is a real threat. As the movement of contaminants especially in soils and aquifers often occurs along with the hydrogeological parameters, study on remedial methods such as biocementation using enzyme-induced calcite precipitation (EICP) is necessary. Since EICP is considered relatively new, more inventive, and environmentally sustainable compared to the other methods, a wider application of this technique in geotechnical engineering should be explored.

Most of the current researches on EICP deals with improving the strength of either organic, clayey, or sandy soils. Limited available data on how the method can be applied in mine waste management, thus the necessity of this research. This paper reports the investigation of the performance of EICP in terms of reduction of the hydraulic conductivity of heavy-metal contaminated mine waste and the calcium carbonate content is produced.

2 Literature Review

Soils are the major end-recipient of the heavy metals released into the environment and unlike organic

* Corresponding author: jodin@ums.edu.my

contaminants that can be oxidized by microbial activity, most metals do not undergo chemical degradation, therefore their concentration stays in the soil long after their introduction [1]. In Sabah, Malaysia one of the mines that have been known to pose health and environmental risk is the Mamut Copper Mine (MCM) located in the interior part of Ranau District. The mine was in operation from 1975 to 1999 and during this period, it generated about 350 Mt of waste, of which 250 Mt are rocks and overburdened materials and over 100 Mt of tailings. The wastes were dumped at various sites near the mining areas but mostly at nearby Lohan Dam [2].

Investigation on the geotechnical properties of heavy metal-contaminated soil provided varied characteristics when different sources and localities are investigated. Researchers comparing iron mine in China and copper mine in Mexico reported the differences in their basic geotechnical properties, permeability, compression, and cyclic response [3], while a study on geotechnical properties of a clay soil from old tin mines in Pahang, Malaysia showed variances in compaction properties, consistency index, shear strength and permeability [4]. The differences in soil behaviors can be respectively attributed to their ore type, clay mineralogy and physical and chemical process used during extraction stages [3], and different percentages of tin tailings [5].

Ground improvement methods and techniques particularly on the enhancement of engineering properties of heavy metal contaminated soil can be divided into physical, chemical, or biological techniques. Physical improvement methods are soil replacement, thermal analysis, or electric repair, chemical treatments are either through passivation or leaching and biological remediation (or bioremediation) methods include pytoremediation and animal repair. However, these methods are considered costly for a small-scale project, not completely effective, or maybe cause secondary pollution, questionable long-term durability, limited heavy metal enrichment, slow and long treatment cycles [6]. Moreover, for bioremediation, different heavy metals may have different effects on the biochemical functions of the animals [7]. Therefore, the aforementioned limitations of these practices necessitate exploration for environmentally-friendly and sustainable technologies hence utilization of biological processes such as EICP, a biologically driven calcium carbonate (calcite or $CaCO_3$) precipitation technology, either controlled or induced.

The mechanism of the EICP is summarized in Eq 1-6 below [8],

$$CO(NH_2)_2 + H_2O \xrightarrow{urease\ enzyme} NH_3 + NH_2COOH \quad (1)$$

$$NH_2COOH + H_2O \rightarrow NH_3 + H_2CO_3 \quad (2)$$

$$2NH_3 + 2H_2O \rightarrow 2NH_4^+ + 2OH^- \quad (3)$$

$$2OH^- + H_2CO_3 \rightarrow CO_3^{2-} + 2H_2O \quad (4)$$

$$CO(NH_2)_2 + 2H_2O \rightarrow CO_3^{2-} + 2NH_4^+ \quad (5)$$

The urease enzyme supplied will hydrolyze the urea to produce ammonia and carbamate (Eq. 1) whereby carbamate then decomposes spontaneously into ammonia and carbonic acid (Eq. 2). The ammonia is then hydrolyzed into ammonium and hydroxide ions thus increasing the pH of the sample, making carbonate precipitation favorable (Eq. 3). The carbonic acid dissociates into inorganic carbonates (Eq. 4-5) and supplying calcium chloride anhydrous ($CaCl_2$) will provide calcium ions that combine with the carbonate ions to precipitate calcite, $CaCO_3$ (Eq.6).

$$Ca^{2+} + CO_2^{2-} \rightarrow CaCO_3 \quad (6)$$

The calcite generated from the chemical reaction is acting as the bonding and cementation agent for soil particles. The crystallographic forms of calcite affect the effectiveness of EICP treatment as different sizes, shapes, and distribution patterns can produce different strength responses of the biocemented soil [9]. Besides, factors such as type of biocementation solution, soil characteristics, treatment method, temperature, treatment concentration, pH level, and degree of saturation are also critical [6].

3 Research Methodology

3.1 Soil Sampling

Soil sampling was conducted at Lohan Dam where tailings were piped, routed, and deposited from the source, Mamut Copper Mine, located 8 km away [10]. The samples were collected at GPS coordinate of 6° 0' 45.936" N, 116° 44' 20" E from the depth of 0.5 m to 3.0 m of the soil profile.

3.2 Geotechnical and Geochemical Properties

The geotechnical properties determined were their natural moisture content, sieving analysis, Atterberg limits, pH, specific gravity, and organic matter. The dry and wet sieving method was used to determine the particle size distribution of coarser and finer soil respectively. pH was measured by pH meter and specific gravity, permeability, and organic matter were measured according to British Standard.

Following the tests, the mine wastes were classified based on their particle size, particle distribution, and texture using two major classification systems; AASHTO and USCS. All laboratory tests were performed according to BS 1377: Part 2: 1990. The air-dried and powdered sample were then made into pressure, digested, and analyzed using an inductively coupled plasma optical emission spectrophotometer (ICP-OES) to determine the concentration of heavy metals and major elements. A scanning electron microscope (SEM) with a maximum magnification of about 1000x was used to generate high-resolution images and precisely measures very small features on the surface of the sample.

3.3 Enzyme-Induced Calcite Precipitation Treatment

The EICP solutions used in this study are made up of urease enzyme, urea, and calcium chloride anhydrous. Cementation solution of 0.5 M concentration was made by mixing urea and calcium chloride anhydrous according to their molarity with distilled and deionized water until complete dissolution. The urease enzyme of 3g/L is added just before adding to the soil sample to avoid early precipitation of $CaCO_3$. The urease enzyme and the chemical reagents were sourced from Fischer Scientific Company, United States, and have a purity of 99%. The urease enzyme is plant-based, extracted from jack bean (*Canavalia ensiformis*) with the reported activity of 3500 U/g.

The cementation solutions were mixed with the samples, placed in permeability cells, and compacted at three-layer to achieve the densities relative to the optimum water content previously determined in the compaction test. Each layer of the mixture was gently tampered with to achieve the predetermined final height. The cell is then cured at different curing temperatures before a constant hydraulic conductivity test is performed according to BS1377. The series of the treatment are shown in Table 1 below;

Table 1. EICP Treatment Series

Sample	Relative Density (%)	Curing Temperature (°C)	Cementation Solution (M)
Control	-	-	-
EICP 1	70	5	0.5
EICP 2	70	15	0.5
EICP 3	70	25	0.5
EICP 4	80	5	0.5
EICP 5	80	15	0.5
EICP 6	80	25	0.5

3.4 Acid Digestion Test

After hydraulic conductivity is completed, the samples were then tested for the calcium carbonate content (CCC), using Acid Washing Method. Samples were collected at three points (top, middle, and bottom), mixed with HCL acid to dissolve the calcium carbonate, then all the solution and insoluble solids were washed by distilled water on filter paper with a coarse pore size sieve (No 200) for 10 minutes. The purpose of the washing process is to remove all soluble calcium from the soil particles. Then the remaining particles were oven-dried and weighed. The CCC was calculated using Eq. 7

$$CCC = 100 - (B/A) \times 100 \quad (7)$$

Where;

A = Weight of soil prewashing
B = Weight of soil post washing

4 Result and Discussion

4.1 Geotechnical Properties

The distribution curve of Lohan Dam soil in Fig. 1 showed the dominant amount of coarse-grain particles at 70%. This percentage is higher compared to wastes from other copper mines around the world but lower compared to the Mamut site where it originates [11]. For example, the study found that the average copper mine waste in the Philippines consists of 50% coarse [12], 30% coarse in British Columbia [13], 55% in Sarcheshmeh Iran [14], 20% coarse in Michigan [15], 60% coarse for average US slime [16] and 10% coarse for British Columbia slime [17]. This shows that the size of mine copper waste particles depends on the milling size, feeding materials, hydro cyclone pressure, slurry density, and presence of water during the extraction process.

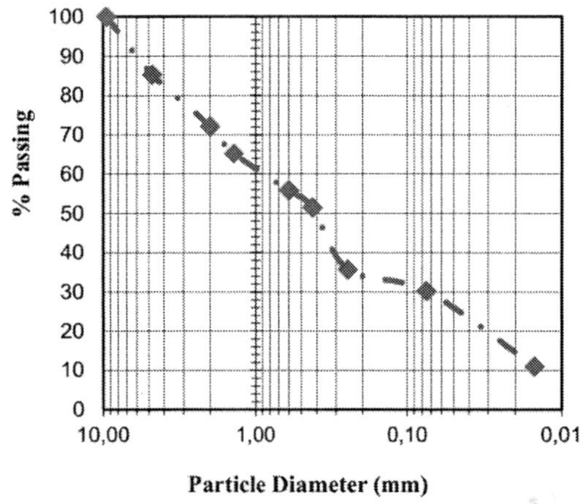

Fig. 1. Particle Distribution of Lohan Dam Soil

Classification by Unified Soil Classification System (USCS) indicates that Lohan Dam soil is SM with the presence of silt in a coarse grain that is predominantly sand. Using the American Association of State Highway and Transport Officials (AASHTO), it is classified as A-4, a fair to poor rate as road subgrade.

The consistency behavior of tailings can be grouped into coarser and finer than 0.074mm (Sieve No 200). The coarser group, also known as 'sand tailings' exhibits non-plastic (NP) behavior while the finer group, also known as 'tailings slimes' may exhibit low plasticity. The acidic nature of the mine is evident in the pH value recorded between 4.53-6.18. The average organic content is relatively low with the value ranging from 0.15-1.27%. As the organic content comes primarily comes from plant roots, this value may indicate that the soil is not conducive for vegetation and hence a challenge for stabilization method using phytoremediation method. This is supported by the fact that there are fewer high trees and plants observed here compared to the nearby area.

Specific gravity test showed that on average, mine wastes have Gs of 2.66-2.75, associated with heavy grained soil. Similar to the other processed mine waste soils, the Lohan Dam soil has a high maximum dry density and low cohesion but a high angle of friction. The engineering properties are shown in Table 2.

Table 2. Properties of Lohan Dam Soil

Properties	Values
Consistency Index	
Liquid Limit, LL (%)	27.0
Plastic Limit, PL (%)	23.5
Plasticity Index, PI (%)	3.5
Liquidity Index, LI (%)	0.91
Activity (A)	0.32
Average Natural water content (%)	26.68
Specific Gravity, Gs	2.66-2.75
pH	4.35- 6.18
Maximum Dry Density, kg/m^3	1640
Optimum Water content, %	16%
The angle of friction, °	6 - 37
Cohesion, kN/mm^2	0.04 – 0.1
Organic content, (%)	0..15-1.27

The SEM morphological analysis is shown in Fig. 2 (1000x) appears to be composed of powdered and hardened particles with dark brown color. Particles observed are well-rounded and spherical in shapes but some irregular-shaped are also observed.

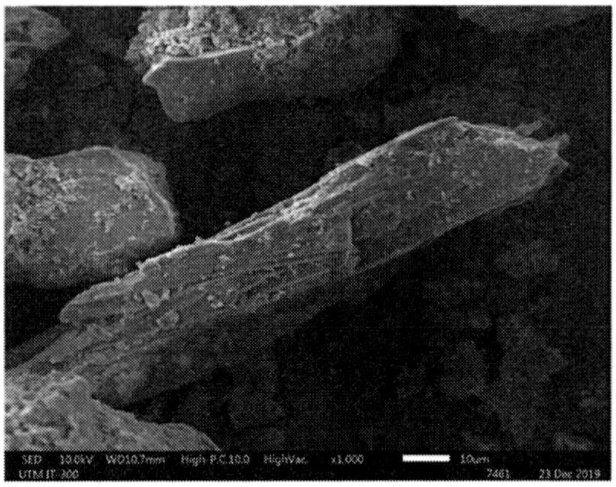

Fig. 2. Scanning Electron Magnetic of Lohan Dam Soil

4.2 Geochemical Properties

The level of heavy metals in soil samples collected from Lohan Dam are shown in Table 3. When compared to the standard of the Department of Environmental Malaysia (DOE) and Food and Agriculture Organization of the United Nations (FOA), the concentration (in mg/L) of arsenic, As (0.319), copper, Cp (4.337), and nickel, Ni (0.376) are beyond the safe levels indicate the necessitate for treatment.

Table 3. Heavy Metal Contaminant of Lohan Dam Soil

Heavy Metal Elements	Contaminant Level (mg/L)			
	Point 1	Point 2	Point 3	Average
Arsenic, As	0.315	0.319	0.323	0.319
Cadmium, Cd	-0.052	-0.051	-0.051	-0.051
Cobalt, Co	-0.026	-0.025	-0.024	-0.025
Chromium, Cr	0.578	0.588	0.591	0.586
Copper, Cp	4.287	4.327	4.396	4.337
Iron, Fe	2926.2	2951.1	2916.9	2931.4
Manganese, Mn	3.601	3.627	3.698	3.642
Nickel, Ni	0.372	0.378	0.378	0.376
Lead, Pb	0.508	0.554	0.541	0.535
Zinc, Zn	0.546	0.56	0.56	0.555

When the further chemical test is performed, it was found that the predominant compounds for the soil are quartz (SiO_2), orthoclase ($KAlSi_3O_8$), and zinc-aluminum hexanedycarboxylate hydrate (CH_3AlO_5Zn). The qualitative analysis results are shown in Fig. 3.

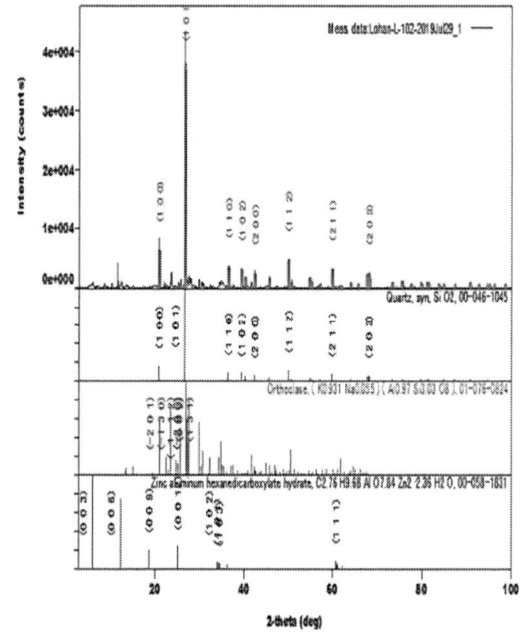

Fig. 3. Major Chemical Compound of Lahad Dam Soil

4.3 Effects of EICP Treatment to the Hydraulic Conductivity

The average hydraulic conductivity of the control sample indicates that soil has a 'medium' degree of permeability with a value between 10^{-3} and 10^{-1} cm/s, consistent with the soil description and classification in Section 4.1. Following 1, `3 and 7-day EICP treatment, almost all series measured an average value between 10^{-4} to 10^{-3}, a reduction between 94-97%. This indicates that the EICP treatment has changed the degree of permeability from

 https://doi.org/10.1051/e3sconf/202133103001

'medium' to 'low', corresponding to the permeability of soil that is classified as very fine sand and mixtures of sand, silt, and clay. The reduction is shown in Fig. 4.

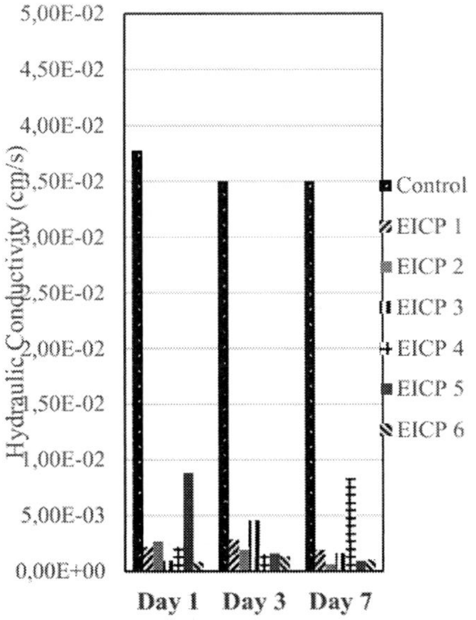

Fig. 4. Hydraulic conductivity of Lohan Dam soil

It is also noted that curing temperature is beneficial for a 1-day curing duration as shown by EICP 3 and EICP 6, where both samples were cured under temperature 25°C over the EICP 1 and EICP 4, where both samples were cured under temperature 5°C. This indicates that the biocementation reaction is more effective under higher temperatures. The findings are similar to a previous study on the rate of urea hydrolysis for uncontaminated soils which reported that they are marginally higher in temperature closer to 30°C [18]. The finding is also consistent with a previous study on semi-arid tropical soil which stated that the optimum urease activity lies at approximately 60° C, and the activity increased with increasing temperature from 10 until reaching the peak at 60° C [19].

On the other hand, the relative density of 70 % and 80% show smaller differences in permeability, probably due to the mixing method selected i.e premixing rather than injection. While premixing allows uniform distribution of cementation solution to all the soil grain particles, the injection method has major disadvantages i.e uneven distribution of bacteria and localized cementation around the injection points leads to pore-clogging at the area near the injection points [6]. Therefore, it is speculated that the soil particle in 70% and 80% relative density were both uniformly treated, hence smaller differences in permeability reduction is observed.

4.4 Effects of EICP Treatment on the Calcium Carbonate Content

Before the EICP treatment, the CCC of untreated mine waste is 3.66% but the percentage increases to a value ranging 6.38-6.95 % after 1 day and 6.94-9.63 after 3 and 7 days, as shown in Fig. 5. This is observed for all series except for EICP 5 for Day 1. Hence, it can be stated that the amount of improvement in EICP treated soil is much related to the amount of induced calcium carbonate content [20] and the reduction in permeability of mine waste is corresponding to the amount of calcium carbonate content.

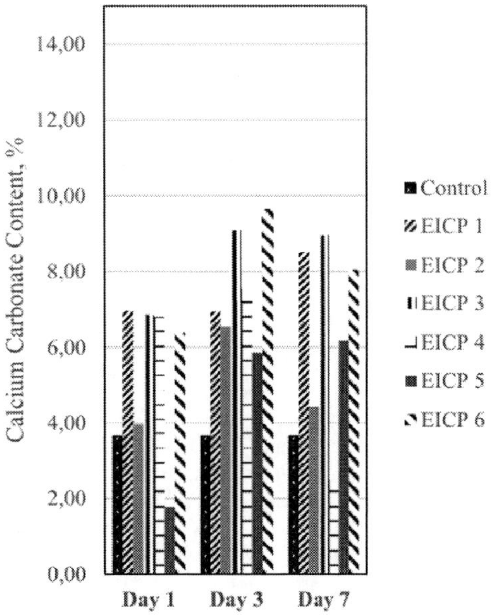

Fig. 5. Calcium Carbonate content of Lohan Dam soil

As is the case for the control sample, the lack of reduction in permeability can be explained by the low amount of calcium carbonate content, as shown in Figure. Before the treatment, the CCC is recorded as 3.60% but upon treatment, the range of CCC is between 1.46-4.10%, lower than the value of mine waste.

5 Conclusion

A series of geotechnical test EICP treatments have been performed on heavy-metal contaminated soil from Lohan, Dam Sabah, and the following conclusion can be made;

1. Geotechnical, Lohan Dam soil is predominantly coarse grain soil of low plasticity, high specific gravity, and high permeability making them susceptible to land contamination. Soil is acidic in nature and its low organic content may indicate that stabilisition by neutralization and phytho-stabilization methods will be costly and not conducive.
2. Morphologically, all wastes appear to be composed of powdered and hardened particles with dark brown color. They contain a higher amount of irregular-shaped particles but well-rounded and spherical are also observed. Their surfaces appear to be very smooth with no agglomeration between particles which indicates that without treatment, the strength value of

the soil is low due to the weak bonding between the loose grain structure.

3. In terms of mineralogy, Lohan Dam soil wastes contain a high level of heavy metals beyond the safety level of the Department of Environmental Malaysia (DOE) and Food and Agriculture Organization of the United Nations (FOA) making them good candidates for EICP stabilization.
4. EICP treatment has changed the degree of permeability of the mine waste from 'medium' to 'low' corresponding to the amount of calcium carbonate content. While relative density shows the marginal effect, curing duration and temperature show a more significant impact.

This work was supported by the Ministry of Education Malaysia's Fundamental Research Grant Scheme [R. J130000.7851.5F256] and the High Impact Research Grant [Q. J130000.2451.04G57].

References

1. A.V. Patel, *A Study on Geotechnical Properties of Heavy Metal Contaminated Soil.* Paripex - Indian Journal of Research, **3** (6), 2 (2014).
2. V. D. Ent, A. and M. Edraki, *Environmental Geochemistry of the Abandoned Mamut Copper Mine (Sabah) Malaysia.* Environ Geochem Health, **40** (1), 19 (2016).
3. L. Hu, et al., *Geotechnical Properties of Mine Tailings.* Journal of Materials in Civil Engineering, **29** (2), 04016220 (2017).
4. A. Z. Rahman, et al., *Influence of Amang (Tin Tailing) on Geotechnical Properties of Clay Soil.* Sains Malaysiana, **41**, 303-312 (2012)
5. W. Zuhairi, N.S. M. Pauzi, and H. A. Mutalib, *Acid mine drainage, and heavy metals contamination at abandoned and active mine sites in Pahang.* **55**, 15-20 (2009).
6. D. Mujah, M.A. Shahin, and L. Cheng, *State-of-the-Art Review of Biocementation by Microbially Induced Calcite Precipitation (MICP) for Soil Stabilization.* Geomicrobiology Journal, **34** (6), 524-537 (2017).
7. Y. Luo, *Study on the repair of heavy metal contaminated soil.* IOP Conference Series: Earth and Environmental Science, **300**, 032076 (2019).
8. A. S. Muhammed, et al., *Influence of multiple treatment cycles on the strength and microstructure of biocemented sandy soil.* International Journal of Environmental Science and Technology, (2021).
9. A. A. Qabany and K. Soga, *Effect of chemical treatment used in MICP on engineering properties of cemented soils.* Géotechnique, **63** (4), 9 (2013)
10. M. K. Azizli, , Y. T. Chee, and J. Birrel, *Design of the Lohan Tailings Dam, Mamut Copper Mining Sdn. Bhd., Malaysia.* Minerals Engineering, **8** (6), 705-712 (1995).
11. J. Makinda, et al., *Geotechnical properties of unprocessed soil from abandoned Mamut mine in Sabah, Borneo.* IOP Conference Series: Earth and Environmental Science, **708** (1), 012021 (2021)
12. R. G. R Salazar. *Design, construction, and operation of the tailings pipelines and underwater tailings disposal system of Atlas consolidated mining and Development Corporation in the Philippines.* in *Proceeding of the 1st International Tailing Symposium.* 1973. Miller Freeman, San Francisco.
13. E. Klohn, C. Maartman, *Construction of Sound Tailings Dam by Cycloning and Spigotting.* in *Proceeding of the 1st International Tailing Symposium.* 1973. Miller Freeman, San Francisco.
14. A. Shamsai, et al., *Geotechnical Characteristics of Copper Mine Tailings: A Case Study.* Geotechnical and Geological Engineering, **25** (5), 591-602 (2007).
15. F. Girucky, *New tailings dam construction at White Pine.* in *Proceeding of the 1st International Tailing Symposium.* 1973. Miller Freeman, San Fransisco.
16. R. Volpe, *Physical and engineering properties of copper tailings*, in *American Society of Civil Engineers*, ASCE: New York. p. 19. (1979).
17. K. H. Mittal, N.R.M., *Parameters for the Design of Tailings Dams.* Canadian Geotechnical Journal, **12** (2), 27 (1975)
18. M. Nemati, E. Greene, and G. Voordouw, *Permeability profile modification using bacterially formed calcium carbonate: Comparison with enzymic option.* **40**, 925-933 (2005)
19. K. Sahrawat, *Effects of temperature and moisture on urease activity in semi-arid tropical soil.* Plant and Soil, **78**, 401-408 (1984).
20. S. G. Choi, et al., *Methods for Calcium Carbonate Content Measurement of Biocemented Soils.* Journal of Materials in Civil Engineering, **29**, 06017015 (2017).

Understanding of subsurface conditions controlling flow liquefaction occurrence during the 2018 Palu earthquake based on resistivity profiles

Adrin Tohari[1*], *Dadan* Dani Wardhana[1], *Muhammad* Hanif[1], and *Keigo* Koizumi[2]

[1]Research Center for Geotechnology, National Research and Innovation Agency (BRIN), Bandung, West Java, Indonesia 40135
[2]Graduate School of Engineering, Division of Global Architecture, Osaka University, Osaka, Japan 565-0871

Abstract. The 7.4 Mw earthquake on 28th September 2018 in Palu City triggered a flow liquefaction phenomenon in the Balaroa and Petobo areas, contributing to significant casualties and building damage. This paper presents the results of a liquefaction study to map subsurface conditions in these areas using the multi-electrode resistivity method with the dipole-dipole configuration. The objective of this study is to understand factors controlling the flow liquefaction phenomenon. Based on the interpretation of 2-D resistivity images, the liquefied soil layers are characterized by lower resistivity values than the non-liquified layers. These contrasts of resistivity values form a gently sloping boundary between the liquefied and non-liquefied soil layers. The resistivity image perpendicular to the flow direction indicates the presence of a subsurface basinal morphology in the Balaroa area, suggesting that a shallow groundwater zone is present within the liquefiable soil layer. Thus, the subsurface topographical condition is the main governing factor of flow liquefaction phenomena during the 2018 Palu earthquake.

1 Introduction

The earthquake (7.4 Mw) that occurred in Palu City, Sigi, and Donggala Regencies on 28 September 2018 caused not only a tsunami but also a flow liquefaction phenomenon in several areas in Palu City and Sigi Regency. The occurrence of flow liquefaction resulted in significant casualties and extensive damage to residential areas and infrastructures in the Balaroa and Petobo areas in Palu City and Sigi Binomaru, Jono Oge, and the South Sibalaya areas in Sigi Regency (Fig. 1).

Fig. 1. Flow liquefaction phenomena in (a) Balaroa, (b) Petobo, (c) Ojo Oge and (d) South Sibalaya.

Flow liquefaction failure generally occurs in areas with a slope of more than 5% [1, 2] and is caused by the presence of a thin layer of sand beneath the dense soil layer [2]. As mentioned previously, the phenomenon of flow liquefaction in Palu City and Sigi Regency only occurred in some areas (see Fig. 1). This difference in liquefaction vulnerability is likely influenced by geological, hydrological, and subsurface soil structure conditions that vary from one region to another. Thus, it is essential to have good knowledge and understanding of the factors of geological conditions and subsurface structures that control flow liquefaction in Palu City and Sigi. Such knowledge is required to establish an effective strategical plan to mitigate this earthquake-collateral hazard in this region.

The understanding of liquefaction occurrences is commonly gained from field geotechnical investigations such as geotechnical drilling with standard penetration tests and cone penetration tests. Although these geotechnical methods provide detailed sub-surface data required for liquefaction analysis, such as soil types, groundwater level, and soil strength, these methods are costly and time-consuming.

Many previous studies have recently attempted to use geophysical methods, such as resistivity survey, to understand the subsurface condition in liquefaction-prone areas [3-5]. Based on the resistivity images, [3] established a correlation of resistivity values with liquefiable soil layers and liquefaction controlling

* Corresponding author: adrin.tohari@gmail.com

https://doi.org/10.1051/e3sconf/202133103002

fractures in San Carlo, Italy, due to the 20 May 2012 earthquake (Ml = 5.9). [4] found a good correlation of the thickness and depth of liquefiable soil layers determined from the resistivity image and the CPT data. [5] concluded that the liquefaction phenomenon in the Pawnee area due to the 2016 Pawnee, Oklahoma earthquake (Mw 5.8) is associated with the presence of isolated lenses of conductive sediments and subsurface fractures formed by the earthquake. The resistivity method has also been used to study the flow liquefaction phenomenon in Balaroa and Petobo areas after the 2018 Palu earthquake [6, 7]. However, the studies only focused on the identification of a shallow aquifer that controlled the liquefaction phenomenon. Thus, no previous studies have been attempted to evaluate subsurface factors controlling a flow liquefication phenomena.

This paper presents the results of a liquefaction study to evaluate subsurface conditions controlling flow liquefaction phenomenon in Balaroa and Petobo areas using the multi-electrode resistivity method. The objectives of this study are to (1) obtain the 2-D resistivity images and (2) evaluate the subsurface factors governing the flow liquefaction in Balaroa and Petobo.

2 Geology of Palu City

According to [8], Palu City is made up of two rock formations, namely alluvial and coastal deposits (Qap) and Celebes Molasse (QTms). The alluvial and coastal deposits consist of gravel, sand, silt, and coral limestone and are the youngest sediments in this area. The deposits were formed in rivers, deltas, and shallow seas and are likely Holocene in age.

The alluvial deposits can be differentiated into young and old alluvial fan deposits, alluvial, flood, and old river channel deposits [9]. The alluvial fan deposits are found on the hilly low plains that stretch on the west and east sides of the valley.

The Celebes Molasses are found at lower altitudes on both ridges, unconformably overlapping the Tinombo Formation (Tt) and metamorphic rock complex (km) [8]. The molasses contain debris originating from older formations and consists of conglomerates, sandstones, mudstones, coral limestones, and marls, which are weakly consolidated. Near the metamorphic rock complex in the western part of the eastern ridge, the molasse deposits consist mainly of coarse boulders and are presumably deposited near the fault. Toward the sea, the rocks turn into fine-grained clastic rocks.

The main fault controlling geological structure in the Palu City area is the well-known Palu-Koro Fault, an active fault with left-lateral movement and an NNW–SSE trend [10-14]. The strike-slip fault is observable within the basin, crosses the distal part of alluvial fans, and is expressed by intra-basin ridges in slightly uplifted places from the adjacent surface [15]. The 7.5 Mw earthquake in 2018 revealed the main Palu-Koro fault line in the ground surface. The surface fault line can be traced from Sigi Regency, Palu City, and Palu Bay [14, 16]. The maximum slip of the Palu-Koro fault is 6±0.5 m around Palu City, with an average of up to 4.7 m in the northern and southern areas of the City [17].

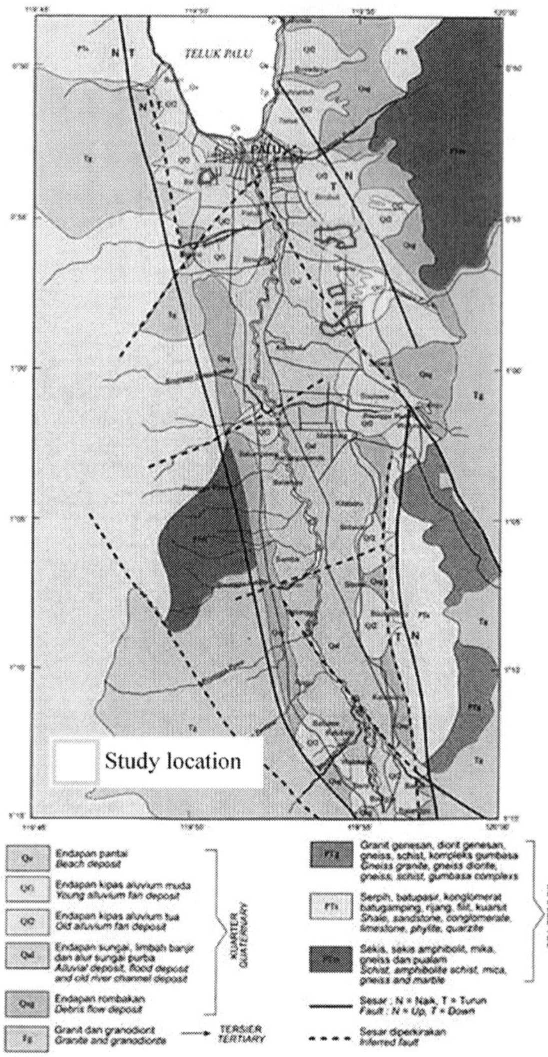

Fig. 2. Geological map of Palu City and its surrounding area [9]

3 Methodology

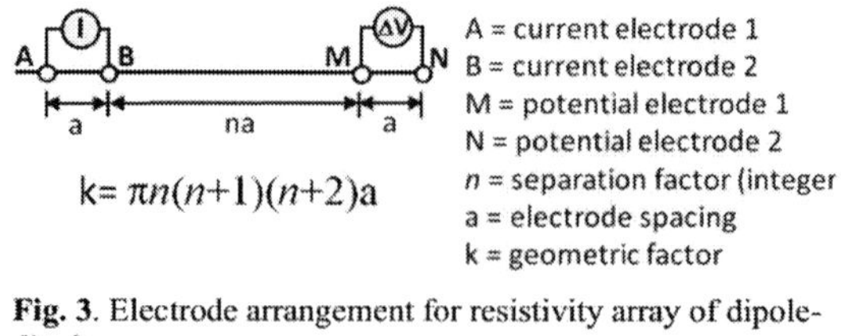

Fig. 3. Electrode arrangement for resistivity array of dipole-dipole.

Geoelectric measurements were carried out at flow liquefaction locations in the Balaroa and Petobo areas using Supersting R8/IP equipment to obtain two-dimensional resistivity profiles up to a depth of 70 m. The equipment was connected to 56 stainless steel electrodes, laid out in a straight line at equal distances apart using a multi-core cable. The dipole-dipole array was used for data protocol acquisition in providing comprehensive data resolutions (Fig. 3). Two survey lines crossing perpendicular to one another were selected at the Balaroa site. Only one line was selected at Petobo (Fig. 4). The length and direction of each line are summarized in Table 1. The elevation of each electrode was measured using GPS.

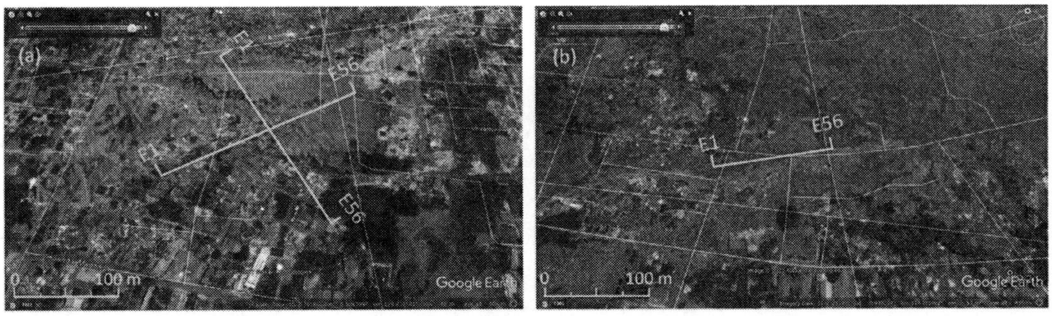

Fig. 4. Map of resistivity survey location showing the survey lines (yellow lines) in (a) Balaroa and (b) Petobo sites. The images of google earth were taken in January 2019. Notation E1 and E56 indicates the locations of the first and the last electrodes,respectively.

The measurement data were edited using AGI Administrator software. Then, inverse modeling was performed using AGI Earth Imager to obtain subsurface resistivity profiles.

Table 1. Resistivity survey locations.

Location	Coordinate (50M)		Spacing (m)	Length (m)
	X (mE)	Y (mS)		
Balaroa	816281.80	9899604.43	5.0	275
Balaroa	816281.80	9899604.43	5.0	275
Petobo	824241.00	9895970.00	2.5	140

4 Results and discussions

4.1 Resistivity profiles in the Balaroa site

Fig. 5 and 6 present the two-dimensional resistivity profiles for the Balaroa site in WSW-NNE and NNW-SSE directions, respectively. The resistivity values for the Balaroa site range from 25–105 Ωm with depth coverage of about 70 m. The resistivity profiles show that the soil layers near the surface have low resistivity values of less than 50 Ωm, which is likely to be associated with alluvial deposits. Meanwhile, the underlying high resistivity layers are probably associated with the old alluvial fan deposit.

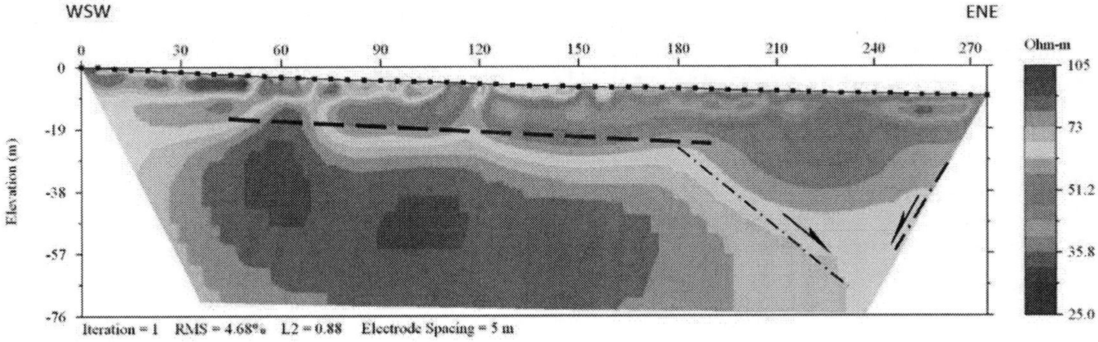

Fig. 5. Resistivity profile for WSW-ENE direction at the Balaroa site. The black dotted line indicates the inclined plane at the boundary between the soil layer with low resistivity and the soil/rock layer with high resistivity. Dashed-doted lines and arrows indicate the normal faulting. The profile map view is shown on Fig. 4a.

As shown in Fig. 5, the boundary between alluvial deposits and the old alluvial fan layer forms an inclined plane at a very gentle angle from west to east. Thus, this subsurface topography is likely to govern the initiation of movement of liquefied soil mass to turn into flow liquefaction during the earthquake. Fig. 5 also indicates the thickening of alluvial deposits to the east due to the existence of a basinal structure associated with the normal faulting component of the Palu-Koro Fault. [18] also found this basin geometry from the analysis of microtremor measurement data in this area.

Meanwhile, Fig. 6 indicates that the boundary between low and high resistivity seems to form a basin geometry associated with normal faulting components of the Palu-Koro Fault. This basin geometry will control groundwater flow to this liquefied area. Consequently, shallow

 https://doi.org/10.1051/e3sconf/202133103002

groundwater levels will easily develop in this area to cause the soil mass to liquefy during the earthquake. The old alluvial fan deposits seem to disappear, and the alluvial deposits become thicker in the southeast direction. Consequently, this area experienced less intense flow liquefaction than the main area during the earthquake, as evidenced in Fig. 4(a).

Furthermore, the examination of Fig. 5 and 6 clearly shows that the boundaries between alluvial deposits and alluvial fan deposits are still intact after the flow liquefaction event. It implies that the flow liquefaction only involved the alluvial deposits. The presence of high resistivity layers at the ground surface is associated with the remnants of alluvial fan deposits being displaced during the earthquake.

It is also evident that the alluvial deposits indicated by low resistivity of less than 30 Ωm at a depth of shallower than 10 m were locally developed after the liquefaction event. This finding suggests that the highly saturated soil layers are still present within the liquefied soils. Hence, the area is still prone to a liquefaction hazard in the future.

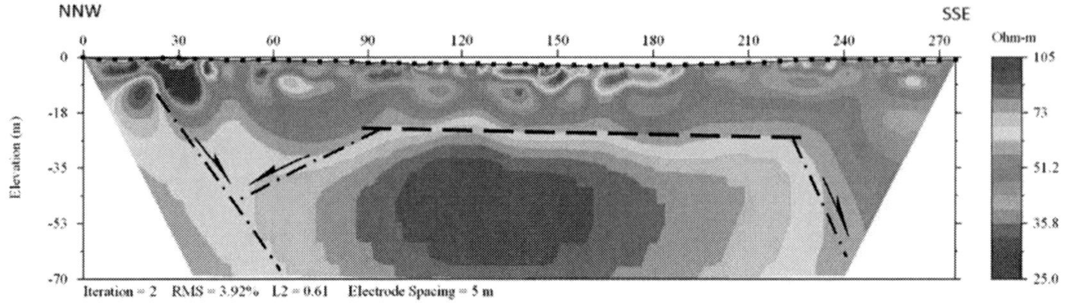

Fig. 6. Resistivity profile for NNW-SSE direction at the Balaroa site. The black dotted line indicates the inclined plane at the boundary between the soil layer with low resistivity and the soil/rock layer with high resistivity. Dashed-doted lines and arrows indicate the normal faulting. The profile map view is shown on Fig. 4a.

4.2 Resistivity profiles in the Petobo site

Fig. 7 presents the results of the geoelectric survey for the east-west direction at the flow liquefaction location in the Petobo area. The resistivity profile shows that the resistivity values range between 2 and 200 Ωm with a coverage depth of 36 m. The soil layer with a low resistivity profile (less than 40 Ωm) is probably associated with the alluvial deposits. In contrast, the high resistivity layers (>40 Ωm) are probably associated with the old alluvial fan deposits. Some fragments of alluvial fan deposits can also be found near the surface, as indicated by higher resistivity values, in most parts of the slope profile. The boundary between these different deposits forms an inclined plane at a gentle angle from east to west that controls the generation of flow liquefaction. Close examination of Fig 5 and 7 suggests that the inclined boundary in the Petobo site is steeper than in the Balaroa site. Consequently, the liquefied soils in the former flowed at a greater distance than in the latter.

Fig. 7 also shows the presence of basinal structures within the alluvial fan deposits, probably due to normal faulting associated with the movement of the Palu-Koro Fault. This basin geometry may control groundwater accumulation within the alluvial deposits, which leads this sediment mass to be highly saturated, and thus, highly liquefiable during the earthquake. Localized very low resistivity layers are still detected within the alluvial deposits, indicating saturated alluvial deposits are still present near the ground surface in the upper part of the slope profile. These pieces of evidence indicate that the Petobo liquefied area is still prone to flow liquefaction in the presence of a strong earthquake in the future.

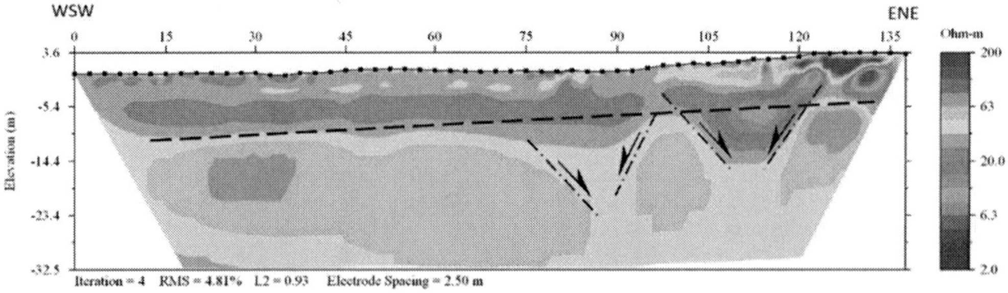

Fig. 7. Resistivity profile for WSW-ENE direction at the Petobo site. The black dotted line indicates the inclined plane at the boundary between the soil layer with low resistivity and the soil/rock layer with high resistivity. Dashed-doted lines and arrows indicate the normal faulting. The profile map view is shown on Fig. 4b.

4.3 The implication to flow liquefaction hazard risk assessment

The result of this current study indicates the critical role of local sub-surface topography as a governing factor for flow liquefaction. Even though thick loose sediments with shallow groundwater exist in most areas of Palu City, massive flow liquefaction only occurred in Balaroa and Petobo areas. In other words, the subsurface topographical condition in other areas with a gentle hillslope in Palu city is not favorable for flow liquefaction. Thus, this study signifies the importance of a comprehensive subsurface geological investigation, using a combination of geotechnical and geophysical methods to map sedimentary basin geometry for the potential flow liquefaction risk assessment in other cities prone to intense earthquake shaking. A massive flow liquefaction event, like in Balaroa and Petobo, may also occur in other coastal cities built on a sedimentary basin.

5 Conclusions

Based on the results of the geoelectric survey, the following conclusions can be drawn:

(a) Low resistivity sediment layers associated with alluvial deposits are mostly present above the higher resistivity sediments of alluvial fan deposits.

(b) The subsurface topographic features, such as a gently inclined boundary between the liquefiable soil and the underlying alluvial fan deposits and basin morphology associated with normal faulting of the Palu-Koro Fault, are the governing factors of the flow liquefaction phenomenon in the Balaroa and Petobo areas.

(c) The occurrence of flow liquefaction in Balaroa and Petobo signify the importance of a comprehensive subsurface geological study to assess the flow liquefaction risk in other earthquake-prone coastal cities.

This study was funded by the 2019 INSINAS research grant of the Ministry of Research, Technology, and Higher Education of the Republic of Indonesia.

References

1. T.L. Youd, Civil Eng. **48**, 4, 47 (1978)
2. T.L. Youd, *Geologic effects—liquefaction and associated ground failure*, in Proceedings of the Geologic and Hydrologic Hazards Training Program, 5-30 March 1984, Denver, Colorado (1984)
3. N.A Zeid, S. Bignardi, R. Caputo, G. Santarato, M. Stefani, Annals of Geophys. **55**, 4, 713 (2012)
4. M. Jinguuji, S. Toprak, Exploration Geophys. **48**, 28 (2017)
5. F. Kolawole, E.A. Atekwana, A. Ismail, Earthquake Seis. Res. Lett. **88**, 1017 (2017)
6. A. Arsyad, L. Samang, T. Harianto, A.B. Muhiddin, A.R. Djamaluddin, S. Aswad, *Reconstructing mechanism of large ground movement induced by 2018 Palu Earthquake 7.4 Mw*, in Proceedings of the 7th International Conference on Earthquake Geotechnical Engineering, 17-20 June 2019, Rome, Italy (2019)
7. S. Nurdin, T. Harianto, S. Aswad, A. Arsyad, S. Alexsander, *Liquefaction disaster mitigation and geohydrology conditions, lessons from the Palu earthquake magnitude 7.4 Mw 28*, in Proceeding of the 23rd Annual National Conference on Geotechnical Engineering, 12-13 November 2019, Jakarta, Indonesia (2019)
8. H. Sumadirdja, T. Suptandar, S. Hardjoprawiro, D. Sudana, *Reconnaissance geological map of the Palu quadrangle, Sulawesi*, Geological Research and Development Center Geologi, Bandung (1973)
9. A. Soehaimi, M. Firdaus., I. Effendi, *Earthquake susceptibility zonation map of Palu and its surrounding area*. Geological Research and Development Center, Bandung (2000) *in Indonesian*
10. J.A. Katili, Geol Rundsch. **59,** 581 (1970)
11. H.D. Tjia, Bull. Geol. Soc. Malaysia **10,** 73 (1978)
12. O. Bellier, M. Sébrier, D. Seward, T. Beaudouin, M. Villeneuve, E. Putranto, Tectonophys. **413**, 201 (2006)
13. M.R. Daryono, *Paleoseismologi Tropis Indonesia (With Case Studies of the Sumatran Fault, the Palu Koro-Matano Fault, and the Lembang Fault)*. Dissertation Doctoral Program. Institut Teknologi Bandung (2016). Unpublished
14. I.M. Watkinson, R. Hall, *Fault systems of the eastern Indonesian triple junction: evaluation of Quaternary activity and implications for seismic hazards Geohazards in Indonesia* in Earth Science for Disaster Risk Reduction. editor / P. Cummins; I. Meilano, Geol. Soc. London **441**, 71 (2017)
15. A. Patria, P. S. Putra, Geosci. Lett. **7**, 1 (2020)
16. A.I. Abdullah, Abdullah J. Phys.: Conference Series, **1434,** 012009 (2020)
17. H. Bao, J.P. Ampuero, L. Meng, E.J. Fiedling, C. Liang, C.W.D. Milliner, T. Feng, H. Huang, Nat. Geosci. **12**, 200 (2019)
18. A. Cipta, A. Rudyanto, H. Afif, R. Robiana, A. Solikhin, A. Omang, Supartoyo, S.Hidayati, Geol. Soc., London, Special Publications, **501**, 185 (2020)

Optimizing modified triaxial testing for small strain zone using local displacement transducers and bender element for cement-treated soft soil

Muhammad Akmal Putera[1*], *Noriyuki* Yasufuku[1], *Adel* Alowaisy[1], and Ahmad *Rifai*[2]

[1]Kyushu University, Department of Civil and Structural Engineering, 744 Motooka, Nishi-Ku, Fukuoka, Japan
[2]Gadjah Mada University, Department of Civil and Environmental Engineering, 55281 Yogyakarta, Indonesia

Abstract. The settlement behavior is a common problem on the railway structure that can be optimized by applying cement-treated soil as ground restoration. However, the application of a high cement mixing content needs a proper estimation that can be achieved by adjusting the element testing. The strain measurement devices can estimate the deformation characteristics, such as secant modulus, Poisson ratio, and shear modulus that can describe the settlement behavior and stiffness of cement-treated soil. This research is focused on a static analysis of triaxial consolidated undrained ($\overline{CU}$) testing that is improved by the axial and radial local displacement transducer (LDT) and bender element to increase the accuracy of measurement results. Furthermore, the secant modulus and shear modulus is more accurate when the combination of radial and axial LDT is used due to a small strain range. Lastly, the shear modulus measurement is improved by using a filler in the cement-treated soil for the bender element test. To conclude, this system of testing for the static condition can be utilized for the dynamic condition, because the measurement shows a reliable result for a small strain range which is the parameter of the dynamics condition.

1 Introduction

Recently, the railway structures industry has reported various technical problems related to deflection on ballastless structures. Ballastless structures consist of the slab track and cement composite with 50 cm thickness on average. The railway deflection problem is directly related to the underlying soil, where the excessive and long-lasting settlement in soft soil ground may induce significant damage and cracking to the structure. Therefore, proper control and evaluation of the soft soil settlement beneath railway structures is strongly needed, wherein standard, it should not exceed 2 mm/year.

Several techniques were developed to overcome the settlement of natural ground and can be generally divided into mechanical and chemical reinforcement. Under the chemical reinforcement group, cement treatment is widely adopted especially in developing countries due to its effectiveness and relatively low cost. Deformation characteristics associated with settlement behavior need to observe in soft soil and cement-treated soil is shear modulus, Young's modulus, and Poisson ratio.

To understand and provide effective solutions for the settlement problem, accurate determination of the soft soil's mechanical parameters is a key factor. Generally, those parameters are determined using laboratory element testing, where the Triaxial Consolidated Undrained ($\overline{CU}$) is commonly used when dealing with soft ground and cement-treated soils [1] [2]. The advantage of measuring the settlement parameter with Triaxial $\overline{CU}$ testing is the ability to determine the parameter only large strain (1 – 15%). However, using the conventional testing method includes difficulties associated with determining the soil behavior in a small strain zone (10^{-6} – 1%). For this reason, applying the local and external strain measurement devices can be used to determine strain range behavior along with the shear modulus.

Several studies have been carried out to investigate the local measurement using Local Displacement Transducer (LDT) and Bender Element (BE) in Triaxial [3] [4]. These studies focused primarily on the reliability of the LDT to measure the deformation characteristics of soil when subjected to monotonic and cyclic loadings. In addition, radial deformation determination is also crucial. Also, inevitable discrepancies in the axial strain orientation determined based on the local and external measurement in the strain range smaller than 1% that is used to determine the shear modulus especially when dealing with cement-treated soils were reported. In addition, the measurement is also affected by the shearing rate, where 0.005% is conventionally used, and the bedding phenomenon during the sample preparation. Bedding error has difficulties related to BE test which was the incorrect measurement from a large cavity on specimen [5] [6]. Consequently, to ensure the reliable determination of the deformation properties, the sample preparation, testing methodology, and measurement need to be optimized.

* Corresponding author: muhammad.akmal.putera@gmail.com

This study aims at optimizing the Triaxial $\overline{CU}$ testing system when combined with local and external strain measurement devices and BE system. Also, the accuracy of the adopted determination methodology of the deformation characteristic in cement-treated soil is discussed.

2 Experimental setup and methodology

In laboratory analyses focused on Triaxial $\overline{CU}$ to determine the deformation characteristic associated with the testing procedure. The testing procedure of the conventional method has been utilized for measuring the small strain ranges using LDTs and BE devices [7] [8]. Therefore, sample preparation has been explained by the optimization of cement-treated soil to ensure reliable measurement in a small strain range.

2.1 The testing procedure of Triaxial $\overline{CU}$

This research was using Triaxial $\overline{CU}$ in static condition test as shown in Fig. 1. The standard of testing in the conventional method on shear loading rate is about 0.005 mm/min, referred to as a clay material in JGS 0821. In the conventional method, there are critical measurement devices such as linear variable displacement transducers (LVDT) to measure the deformation characteristic from the outside triaxial chamber.

The coefficient of LVDT is about 0.005508 in 1 mm. Also, a pressure meter to determine volume change during the consolidation process. The coefficient of the pressure meter is about 0.0106 in 1 $\mu\varepsilon$. However, the limitation of using conventional tests cannot define the radial displacement during the shearing process. Also, the consequences for using the LVDT outside the chamber are the bedding error on the top and bottom on cement-treated soil. For this reason, the utilization of local strain measurement is needed.

The accuracy of strain measurement using LVDT is about 0.01 mm, this resolution is associated with strain range measurement. For instance, the measurement with LVDT has coverage of the strain range within 0.01 to 10% explained by Fig. 2. However, this strain range measurement needs to evaluate with the local strain measurement devices. Furthermore, the accuracies on deformation characteristics properties such as shear modulus (G) and secant modulus (E_{sec}) need to define by comparing the LDT and LVDT results.

2.1.1 Axial and Radial LDT

LDT specification made by bronze plate and the dimension is about 90 mm on length, 3.5 on width, and 0.2 on thickness. The dimension can be seen in Fig. 3. The LDT must be calibrated to keep the strain in every testing. The pseudo-hinged specification is made by a bronze plate with a specified angle to attach the axial LDT as illustrated by Fig. 4a. Therefore, the attachment method using the pseudo-hinged into membrane needs to reinforce with epoxy glue to ensure the detached problem during measurement, as described in Fig. 4b.

The coefficient of LDT can be defined by using the calibration of the deformation test. The calibration of axial LDT is used to describe the non-linear relationship of the output voltage with the displacement is shown in Fig. 5a. It describes that the limitation of the displacement is about 5 mm and 5% in the strain range. On the other hand, the radial LDT is used to describe the linear relationship. It can be seen in Fig. 5b. The Figure shows that the negative and the positive limitation is about 1.5 mm and 1.5 mm. From this result, the initial condition has been evaluated and the test condition saved in every axial and radial LDT testing.

The accuracy of strain measurement using LDT is about 0.0001 mm. This resolution is associated with the strain range measurement. The potential for using the LDT to measure the strain range between 1 to 0.0001% can be carried out from the initial test. The resolution is based on the calibration and limitation of the LDT initial measurement. However, the result of the calculation in LDT is used to evaluate the deformation characteristics properties during the shearing process.

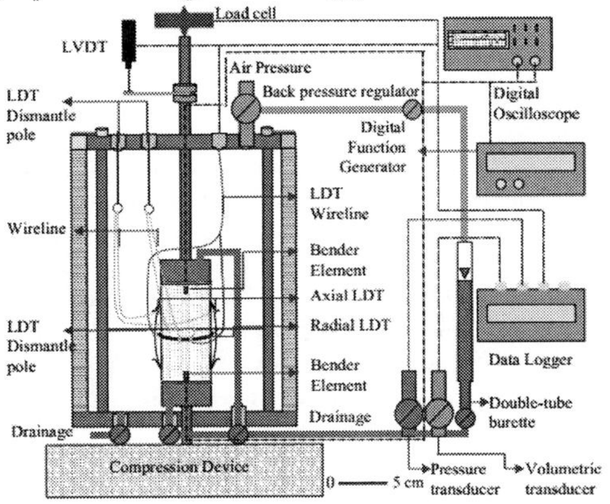

Fig. 1. Triaxial $\overline{CU}$ setup

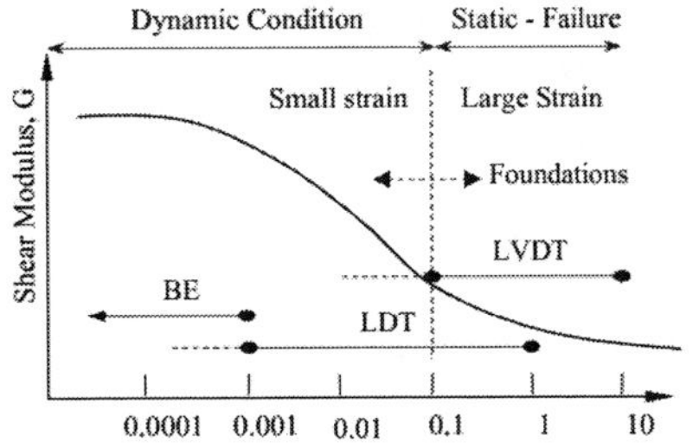

Fig. 2. Typical stiffness variation and variances of strain levels for laboratory tests and structures.

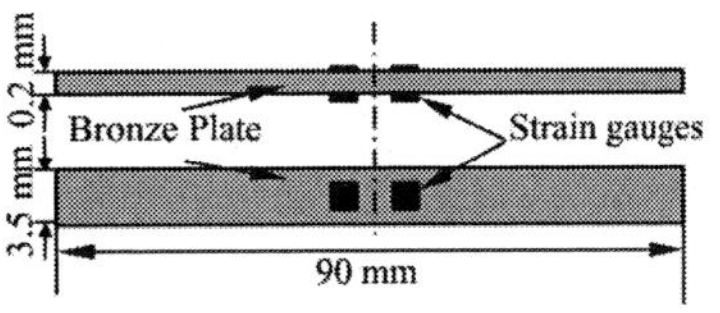

Fig. 3. LDT geometry specification

During the shearing process, the change of stress strain from the LDT and the load cell needs to be recognized. For instance, before the LDTs reach the limitation of displacement it needs to detach immediately from pseudo-hinged. Occasionally, the radial LDT was unable to be dismounted due to the pressure inside the triaxial chamber that pushed the LDT through the specimen. The dismounted problem in radial LDT has been solved by using dismantle pole with a wire connected to the axial and the radial LDT. Therefore, the system procedure of dismounted LDTs from pseudo-hinged has been explained with dismantling pole to maintain the LDTs lifespan.

2.1.2 BE devices

The material of BE is consists of silver, piezoelectric ceramic, and nickel as an electrode. The original dimension of BE that was used in the test is about 10-20 mm in length, 10 mm in width, and 1.5-2.5 mm in thickness. The dimension has been widened by coating the BE with epoxy glue to protect it from water. The dimension information in the test is the same as the curing mold shown in Fig. 6.

The shear modulus is obtained by analyzing the arrival time and amplitude in receiving the data from the oscilloscope. There are several methods to determine shear modulus, such as the start-to-start method, peak-to-peak method, and cross-correlation method. This research was used the peak-to-peak method to evaluate the shear modulus. The data from laboratories that used a single pulse of sine wave input. Shear wave velocity was observed from the peak position of an input wave to the first peak of the received wave.

There are different responses at shear modulus is from the effectiveness in sample preparation with filler and without filler. The filler needs to use from bedding errors on the top and bottom surface. The filler is consists of bentonite and gypsum mixture. Filler composition method that was used in test by ratio 1:2 of binder to water. The texture and structure can be explained by Fig. 7.

The result to analyze the reliable curve can be shown from Fig. 8a is used to determine the reliable results by using the bender element in the triaxial apparatus. Fig. 8b is used to identify the less reliable result due to bedding error and indicate the sample preparation in cement-treated soil. Furthermore, reliable results are used to obtain filler applications on both sides of the test object, then describe the effective results with filler on the test object. In the meantime, the filler proportion could be advantageous for some reason, and the future analysis should be enhancing the effectiveness of the filler types in several experimental conditions.

2.2 Sample Preparation

The specimen needs to be prepared with an appropriate condition to understand the reliable result condition to examine the system of testing with the triaxial $\overline{CU}$ test. The material used for this study is the Ariake clay from Saga Prefecture. Then the properties are shown in Table 1. The explanation of the preparation method for the specimen to mix soil cement from marine clay is depicted in Fig. 9. The steps of sample preparation and the installation procedure for the cement-treated soil-applied in the bender element test are as follow:

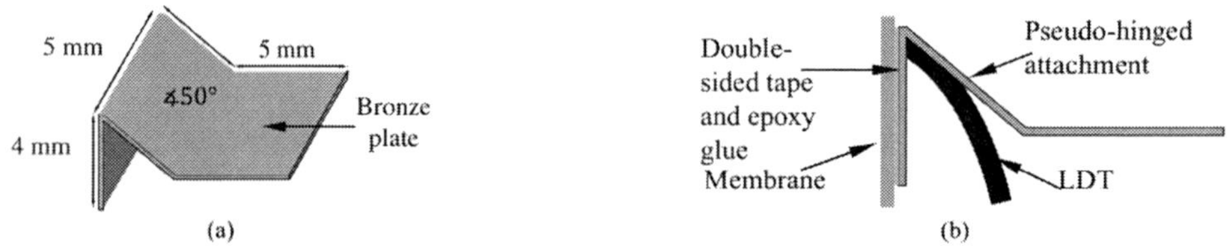

Fig. 4. (a) pseudo-hinged geometry specification and (b) pseudo-hinged technique attachment

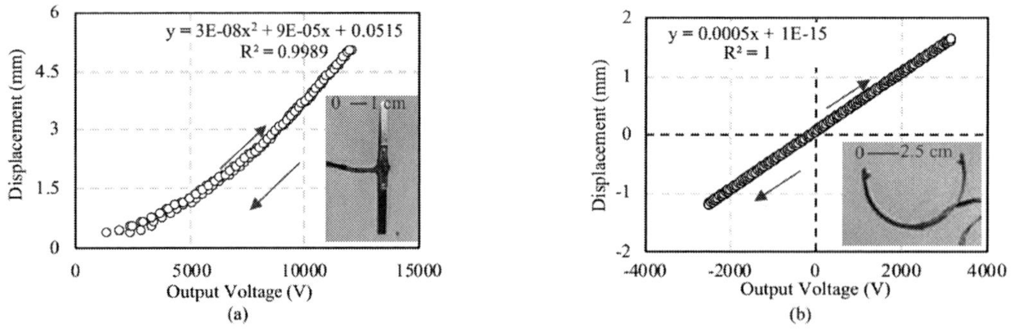

Fig. 5. (a) Calibration result of the Axial LDT and (b) radial LDT

 https://doi.org/10.1051/e3sconf/202133103003

Table 1. Soil properties

Soils	Natural Water Content (%)	Specific Gravity	425 μm size (%)	D_{50} (mm)	Liquid Limit (%)
Ariake Clay	240	2.547	100	8.59×10^{-4}	128.5

2.2.1 The mixing phase

The mixing process of cement with soil has been widely used by several researchers. Ariake Clay is characterized by a high water content which is about 240%. Due to the high water content, it can indicate that the use of mixing criteria is based on the liquidity index [2]. If the value of the liquidity index is twice the dry mass, the most suitable mixing method is the dry mixing method. Hence, no more water is needed. It is worth noting that the percentage of the cement addition is based on the ratio of the dry mass and the cement weight.

2.2.2 Pre-curing phase

In this phase, cement-mixed soil has been developed in a slurry condition. This condition is prepared by the modified mold and set up the slurry condition to the modified curing mold. The modified mold has been made using acrylic and a steel plate to create a casting hole which is used to penetrate the bender element. During the curing period, the modified curing mold has the advantage to created a proper casting for bender element plates from both directions (upper side and bottom side) to anticipate the cracking damage. Furthermore, to prevent any cracking damages at the post-curing phase, we need to rub the mold with grease in advance. These methods have been applied for mixing the specimen inside the curing mold by tamping the mold from the bottom side. The tamping process is carried out about 30 times for each layer. However, it is better to keep the slurry condition of the cement-treated soil before pouring the cement-mixed soil into the modified mold.

2.2.3 Post-curing phase

The next step is the post-curing phase. Inside the modified curing mold, there is a progressive reaction such as the hardened reaction inside the mold, and by the time the cement-treated soil structure then becomes concrete. The removal of the specimen from the curing mold is a common problem during this process. There was cracking damage from the specimen that can be caused by the looseness of the tape or the airtight seal that is attached to the specimen.

2.2.4 Specimen setup

The remaining phase is to set up the specimen into the bender element test. As the previous research encountered a problem and difficulties during handling a special procedure of the conventional testing method, so a slight modification is required. Conventional testing methods can be carried out in two conditions, static and dynamic conditions. Both have difficulty determining the incremental stress behavior with bedding error from surfaces related to the upper and bottom sides. For this reason, using the filler to cover the bedding surface can improve the shear modulus measurement. Furthermore, we need to carefully consider the sample preparation and installation procedures to evaluate the accuracy of the resulting deformation characteristics of the specimen, which in this study is a cement-treated soil with high cement content.

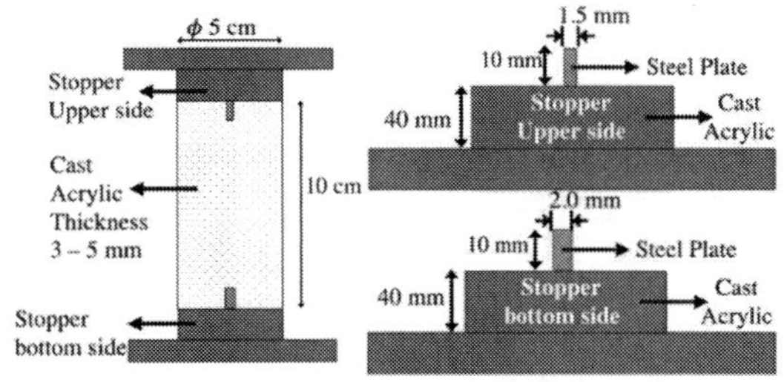

Fig. 6. Curing mold specification with bender element shape

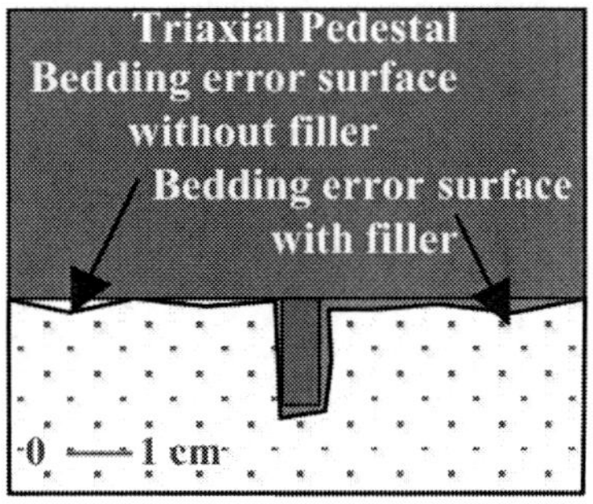

Fig. 7. The texture of structure of bedding surface with filler and without filler

3 Local and External strain measurement devices

3.1 Stress-strain relationship

The combination amongst the LDT and LVDT devices in Triaxial $\overline{CU}$ can evaluate the axial (ε_a) and radial (ε_r) strain parameters during the shearing process. The relationship between the strain and deviatoric stress ($\sigma_a - \sigma_r$) is described by a nonlinear curve as illustrated by Fig. 10a and b. During 7 days of the curing period, the result shows a small difference in the peak of deviatoric stress by using LDT and LVDT. These deformation characteristics can be seen in Fig. 10a. However, on day 28 of the curing process, the result shows a peak difference of about 2% in the axial strain. The differences in the deviatoric stress and axial strain are associated with the radial strain, which describes the peak of radial strain in the axial LDT as bigger than LVDT measurement. However, the results are related to the bedding error which increases as the stiffness of the specimen is increases. To improve the accuracy of stress-strain relationship using cement-treated soil needed to use the strain measurement such as axial and radial LDT.

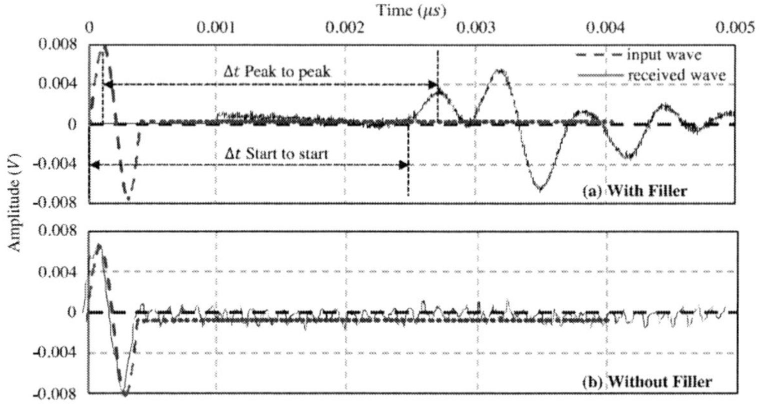

Fig. 8. Identification method using BE test (a) describe peak to peak and start to start method with filler and (b) describe the condition without filler

3.2 Poisson ratio

The deformation characteristics in the dynamic conditions such as Poisson ratio (υ) can be determined with the LDT by using the radial strain. The Poisson ratio can be defined by dividing the radial and axial strain.

$$\text{Poisson ratio} = -\frac{\text{change in radial strain}}{\text{change in axial strain}} = \frac{\Delta\varepsilon_r}{\Delta\varepsilon_a}$$

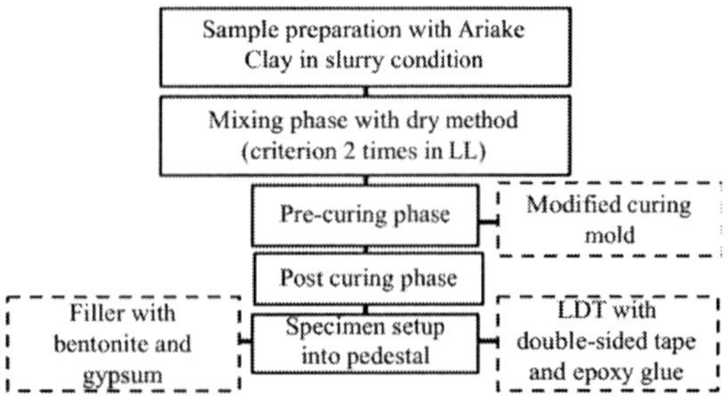

Fig. 9. Schematic diagram of the sample preparation procedure

The Poisson ratio can be determined during the shearing process associated with radiale strain and axial strain. The nonlinear occurred of the small strain range from 0.001 to 0.1%, which decreases of the Poisson ratio is observed by increasing the axial strain. The Poisson ratio from 28 days of curing is higher than the 7 days of curing as illustrated by Fig. 11a and b. The significant discrepancy is due to the axial strain and radial strain measurement of the 28 days of curing, which describes the stiffness higher than 7 days of curing. Fig. 11b shows the different result of the LVDT and LDT, where determine the significant discrepancy result start from 0.01% of the axial strain range with 28 days of curing. This can be interpreted as the differences of resolution between strain devices measurement and stiffness of specimen. The accuracy of system testing to measure the Poisson ratio in small strain is ranged from 0.001% to 1% by using radial LDT. Based on the result of this analysis, further testing using more varied samples is considered to be essential to enrich the strain level behavior in the radial LDT.

4 Accuracy of Secant Modulus and Shear Modulus

4.1 Secant modulus

The measurement of the stiffness modulus utilizing the conventional triaxial $\overline{CU}$ testing has been widely used for conventional testing to determine the strain range behavior. To understand the variety of the small strain range with stiffness modulus, the secant modulus must be determined. Furthermore, combining the LDT with conventional triaxial testing allows the determination of small strain ranges. The secant modulus can be defined by the stress-strain curve from zero value to 50% of the ultimate strength [9]. The advantage of using 50% of the ultimate strength level is that it can be applied in axial LDT, where the utilization of axial LDT has been recognized by measuring the variety of the small strain range.

$$\text{Secant Modulus} = \text{E} = \frac{\text{change in axial stress}}{\text{change in axial strain}} = \frac{\Delta\sigma_a}{\Delta\varepsilon_a}$$

The accuracy of the LDT testing in determining the secant modulus has been described in Fig. 12. The figure delineates the relationship between the axial strain and the secant modulus determined using conventional testing and the LDT. The results indicate that the soil stiffness decreases with the incremental increase of the axial strain. Fig. 12 (a) and (b) show the nonlinear relationship below the 1% axial strain, which describes the secant modulus for 7 days and 14 days of curing, respectively. Based on the results, it was found that the sample cured at 28 days has higher strength than the one cured at 7 days. Below the 1% axial strain for both samples in Fig. 12, the LDT and LVDT curves showed some differences. However, the accuracy of testing using the LDTs has been observed well in small strain range from 0.001% to 1%.

4.2 Shear Modulus

The development of the system of testing for determining the shear modulus has been widely used for the deformation characteristics in experimental testing. To describe the shear modulus effect, the starting point from the beginning of testing to determine the nonlinearity of small strain level below 0.001%. The system of testing must be considered before applying the stress from consolidation pressure and the shearing process. The result of measuring the shear modulus in a small strain range is used to determine a system of testing in the static condition with conventional testing and applied the same initial strength condition on a system of testing to the dynamic condition. The Shear modulus can be defined by the following equation.

$$\text{Shear Modulus} = -\frac{(E)}{2(1+\upsilon)}$$

The shear modulus can be determined during the shearing process based on the Poisson ratio and the secant modulus. The result of shear modulus was analyzed associated with strain measurements such as the axial strain in LDT and LVDT, as shown in Fig. 13. In 7 days of curing the small strain range below 0.1% is relatively linear, the condition can be seen in Fig. 13(a). On the other hand, the 28 days of curing has described the nonlinear condition. Therefore, it can be concluded that by increasing the axial strain, the shear modulus decreases.

The observation is made to evaluate the system of testing between LVDT and LDT from 0.01% to 1% in strain range as shown in Fig. 14. The result shows the shear modulus, which describes that the LVDT is underestimated from the axial LDT. Also, the capacity of LVDT cannot measure the value below 0.01%. The interpretation of initial shear modulus ($G_{\max}$) has been described in 0.001%, which is the condition based on measurement of LDT and BE before shearing process and consolidation process. Meanwhile, the BE test can measure the G_{max} that shows the linear condition similar with LDT in small strain 0.001% to 0.01%.

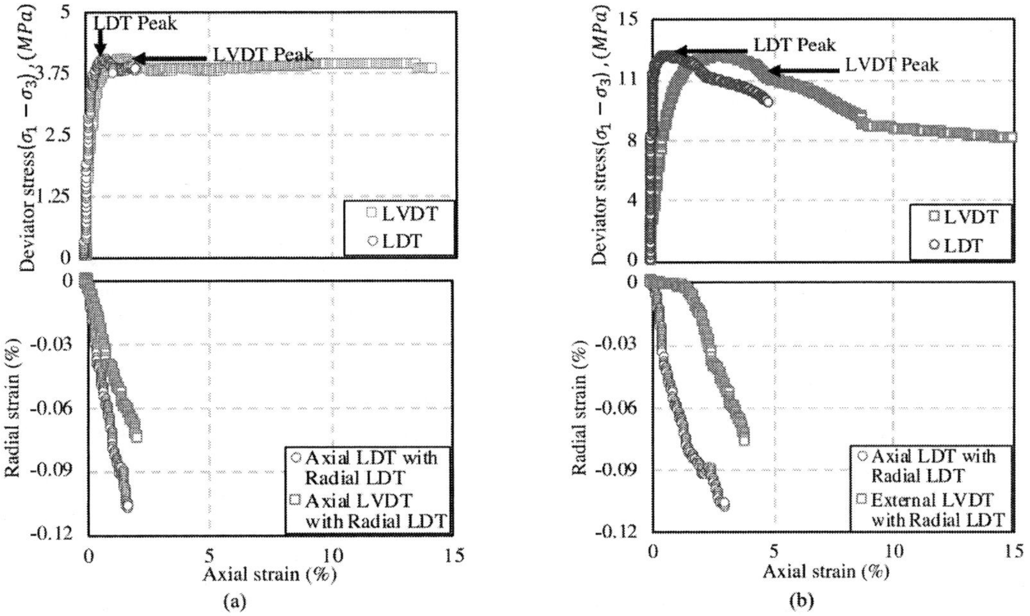

Fig. 10. Result of static loading test (a) 100 kPa of confining pressure with 50% cement content at 7D curing and (b) 100 kPa of confining pressure with 50% cement content at 28D curing

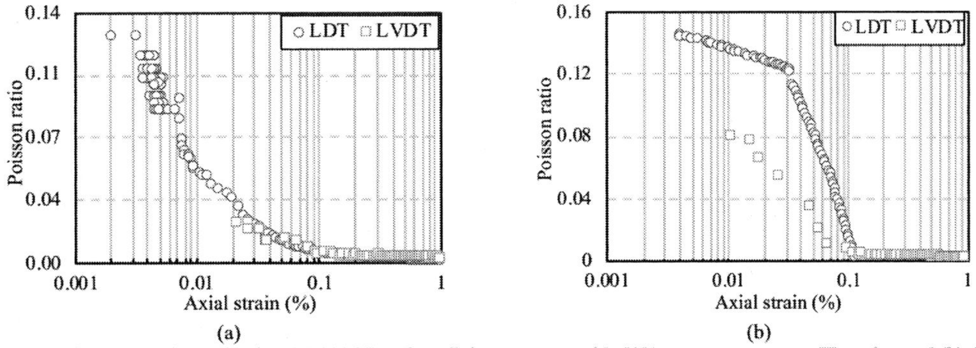

Fig. 11. Poisson ratio determination (a) 100 kPa of confining pressure with 50% cement content at 7D curing and (b) 100 kPa of confining pressure with 50% cement content at 28D curing

https://doi.org/10.1051/e3sconf/202133103003

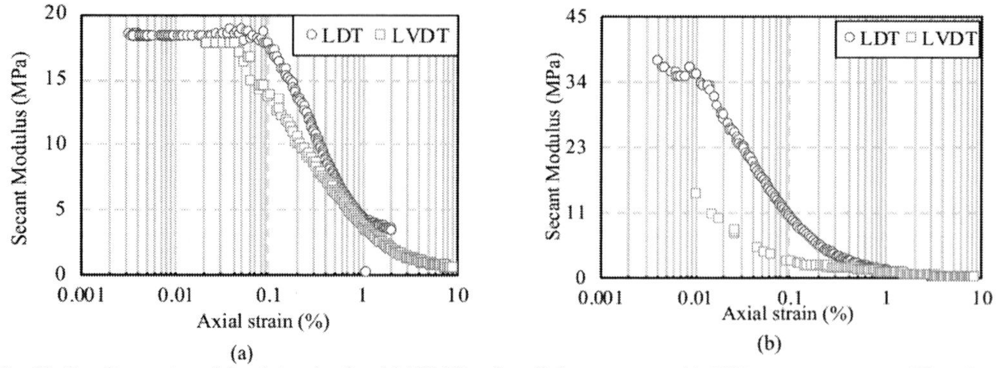

Fig. 12. Result secant modulus determination (a) 100 kPa of confining pressure with 50% cement content at 7D curing and (b) 100 kPa of confining pressure with 50% cement content at 28D curing

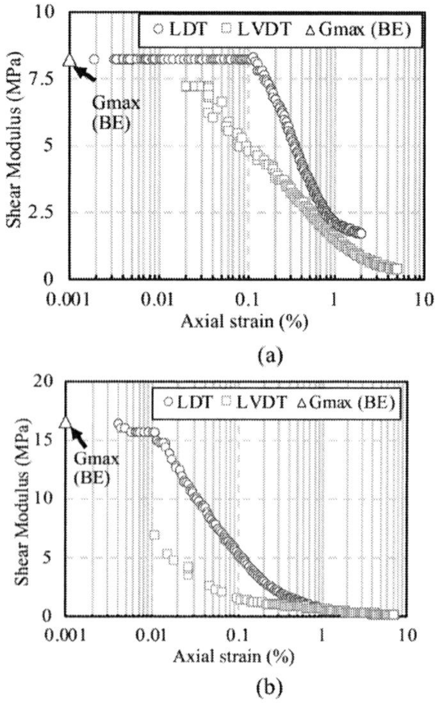

Fig. 13. The Shear Modulus result (a) 100 kPa of confining pressure with 50% cement content at 7D curing and (b) 100 kPa of confining pressure with 50% cement content at 28D curing

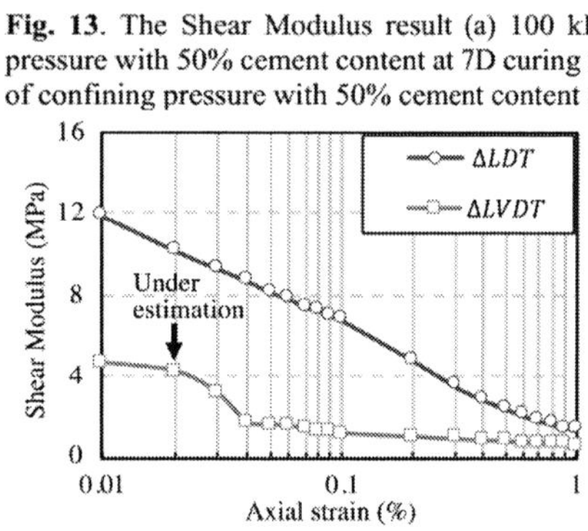

Fig. 14. Accuracy in shear modulus with axial strain using LDT and LVDT

5 Conclusion

Through this study, the CU bar testing system combined with local and external strain measurement devices and BE system was optimized for testing cement-treated soft soils focusing mainly on the small strain range. The main conclusions are delineated as follows:

1. To ensure reliable accurate measurement of the small strain using the axial and radial LDTs, it is recommended to fix the LDT's hinge to the outer surface of the membrane using epoxy glue and a double-sided tape to prevent the detachment of the hinge during shearing. On the other hand, it is necessary to use a filler paste to fill the cavity in the sample when installing the Bender Element to ensure reliable accurate measurement of the shear wave velocity.
2. A significant discrepancy of about 2% in the axial strain was measured using the LDT and the conventional LVDT. It was found that the difference became more significant after 28 days of curing, which can be related to the hardening of the cement-treated sample with time as confirmed from the BE results.
3. It was found that the shear modulus determined using the LDTs and the BE are in good agreement in the small strain zone (0.001 %), and differ significantly from the modulus determined using the LVDTs. The difference in the measured shear modulus determined using the LDTs and the LVDTs decreases towards larger strain ranges and converges to the same value for an axial strain equals or larger than 1%.
4. Finally, it can be concluded that to ensure the reliable determination of cement treated soft soils mechanical properties over the entire strain range, both LDTs for the small strain range (0.001 to 0.1) and LVDTs for the large strain range (> 1%) should be simultaneously used.

References

1 H. Atkinson, "Non-linear soil stiffness in routine design," G´eotechnique, **50,** 5,. 487-508 (2000)

2 K. Kasama, K. Zen, and K. Iwataki, "Undrained shear strength of cement-treated soils," Soils and Foundations, **46**, 2, 221-232 (2006)

3 S. Goto, F. Tatsuoka and S. Shibuya, "A simple gauge for local small strain measurements in the laboratory," Soils and Foundations, 31, 1, 169-180 (1991)

4 S. Yamashita, T. Kawaguchi, Y. Nakata, T. Mikami, T. Fujiwara, and S. Shibuya, "Interpretation of international parallel test on the measurement of Gmax using bender elements," Soils and Foundations,. 48,.4, 631-650 (2009)

5 K. Yao, Q. Chen., H. Xiao, Y. Liu, and F. H. Lee, "Small-strain shear modulus of cement-treated

6 marine clay," J. Mater. Civ. Eng., 32, 6 (2020)

7 H. Xiao, K. Yao, Y. Liu, S. Goh, and F. Lee, "Bender element measurement of small strain shear modulus of cement-treated marine clay – effect of test setup and methodology," Construction and building materials, 172,433-447 (2018)

8 D. Xu, L. Borana, and J. Yin, "Measurement of small strain behavior of a local soil by fiber Bragg grating-based local displacement transducers," Acta Geotechnica, 9, 935–943, (2013)

9 S. S. Kumar, A. M. Krishna, and A. Dey, "Local strain measurement in a triaxial test using an on-sample transducer," in Indian Geotechnical Conference, Chennai (2016)

10 P. Malkkowski, L. Ostrowski and J. Brodny, "Analysis of young's modulus for carboniferous sedimentary rocks and its relationship with uniaxial compressive strength using different methods of modulus determination," Journal of Sustainable Mining, 17, 146-157 (2018)

Effect of load on the shape of the slip surface and the velocity of the landslide on a mixture of clay sand

Julita Andrini Repadi[1*], *Fathol* Bari[1], *Junaidi*, and *Oscar* Fitrah Nur[1]

[1]Departement of Civil Engineering, Faculty of Engineering, Andalas University, West Sumatera, Indonesia

Abstract. The slip surface has an essential role in slope stability analysis. The slip surface is to be known to calculate the safety factor. The velocity of landslides needs to be done to mitigate when a landslide occurs to reduce the risk. This study aims to determine the shape of the slip surface and the velocity of the landslide with variations in load and 2 combination material. The slope is formed in a glass box measuring 110 x10x40 cm. The slope angle used is the same as the angle of internal friction. The slope is given a uniform load until the slope collapses. Loading is done by using a pressure device placed on the prooving ring at the bottom of the press. From both material variatn ions and load variations, it can be seen that the shape of the slip surface that occurs is almost the same, namely in the form of a slip surface. So it is concluded that the load does not affect the shape of the slip surface. The velocity that occurs in combination 1 is moderate-rapid, while combination 2 is moderate. Combination 2 clay is safer than combination 1 but not too significant.

1 Introduction

Landslides are known to occur on all continents, seas, and oceans. In many areas of the world, landslides are a serious threat. This has led to landslide modeling, and landslide risk evaluation becoming increasingly popular among scientists and practitioners. However, due to natural factors, landslides are difficult to predict, limiting the community's ability to reduce the risk of landslides [1].

Landslides with a high death toll are often the result of failures in land management. Based on BPS data in 2021, Indonesia's population continues to increase from 2018 to 2020, the significant increase in population is 1% per year or more than 2.6 million per year. The increasing number of residents has resulted in people having to build houses in hilly and mountainous areas. According to BPS 2021 data, in Indonesia the number of victims who died and went missing in 2020 due to landslides was 124 people, injured 87 people, affected and displaced as many as 23,375 people. Of the 34 provinces in Indonesia, West Sumatra is in the 4th position with the most deaths, 6th position with injuries and 11th position affected and displaced provinces.

According to the World Health Organization (WHO), the area most vulnerable to landslides is land that has changed due to human activities such as construction. The increase in population in Indonesia can trigger construction development, and construction development eventually triggers landslides.

Understanding the interrelationships between Earth's surface processes, ecological systems, land use, and human activities is key to reducing the risk of erosions. Due to limited land, people are no longer worried about the dangers that occur due to landslides. According to Hakam (2019) research, people in the Sianok Canyon area are no longer concerned about the threat of landslides due to limited land. They live at very high risk [2].

Analysis of slope stability has a crucial role in the planning of civil constructions. Calculating the minimum factor of safety of a slope is very important in geotechnical engineering. Calculating the minimum slope safety factor or finding the slip surface of a hill is a complex problem[3]. Slope stability analysis is related to safety factors. To calculate the safety factor of the slope, it is necessary to know the shape of the slip surface.

The various types of slip surfaces are translation, rotation, block, rock debris, soil creep, and debris. Researchers who have researched the shape of the slip surface numerically are in rotation [4-11].

Landslides occur when the driving force is greater than the resisting force. The retaining force is generally influenced by the strength of the rock and the density of the soil, while the driving force is one of the causes due to the load of the building on it.

This research is present by looking at the effect of the load on the slope with a decrease in the slope of 1 cm and 2.5 to see its impact on the shape of the slip surface

2 Slope Stability

Landslides occur when the driving force exceeds the resisting force. The stability of this slope is quantified in terms of the Factor of Safety (FS), as shown in Equation 1. part of the landslide as shown in Fig. 1.

* Corresponding author: julitaandrinirepadi@gmail.com

$$SF = \frac{\text{Resisting Force}}{\text{Driving Force}} \quad (1)$$

SF < Slope is unstable
SF >1 Slope is stable

Table 1. Classification of landslides based on Cruden and Varnes [13]

Velocitty class	Description	Velocity (mm/s)	Probable destructive significance
7	Extremely Rapid	$>5 \, . \, 10^3$	A major violent disaster, buildings destroyed due to the impact of abandoned materials, many deaths, society is impossible to escape.
6	Very Rapid	$<5 \, . \, 10^3$	Several lives were lost, the velocity was too great to support everyone to save themselves
5	Rapid	$<5 \, . \, 10^1$	Evacuation escape possible, structures, property and equipment destroyed
4	Moderate	$<5 \, . \, 10^{-1}$	Some temporary and insensitive structures can be preserved.
3	Slow	$<5 \, . \, 10^{-3}$	Construction improvements can be made. If landslide movement is not large, Construction can be maintained with frequent maintenance.
2	Very slow	$<5 \, . \, 10^{-5}$	Landslide did not damage some permanent buildings.
1	Extremely Slow	$<5 \, . \, 10^{-7}$	Imperceptible without instruments, construction possible with precautions

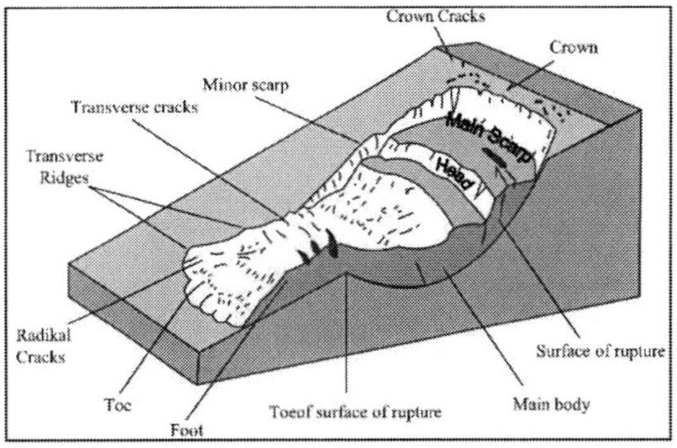

Fig 1. Part of the landslide

External Terzaghi classifies the occurrence of landslide into 2, namely external and internal factors, one of the causal factors is the load, in this study the cause of landslide. [12]

Understanding the interrelationships between Earth's surface processes, ecological systems, and land use, and human activities are key to reducing the risk of landslides. Knowing the velocity of landslides can also prevent and limit the number of human casualties and property damage. Landslide is the movement of rock, soil debris, etc., down a slope near the earth's surface due to the force of gravity. The rate of landslides can occur very slowly to very quickly down the slope, following the classification of slopes.[13]

3 Methodology

The steps taken in this research are as follows [14]:

1. Preparing material samples
   The material used is fine sand, passed the no.10 sieve and was retained by the no.200 sieve. The sand is first washed to remove the silt content and other organic matter, while the clay is dried to facilitate filtration with filter no. 200, Experiments were carried out with two combinations of materials, namely:

   a. The first combination is 95% sand+5% clay, as shown in Fig. 2

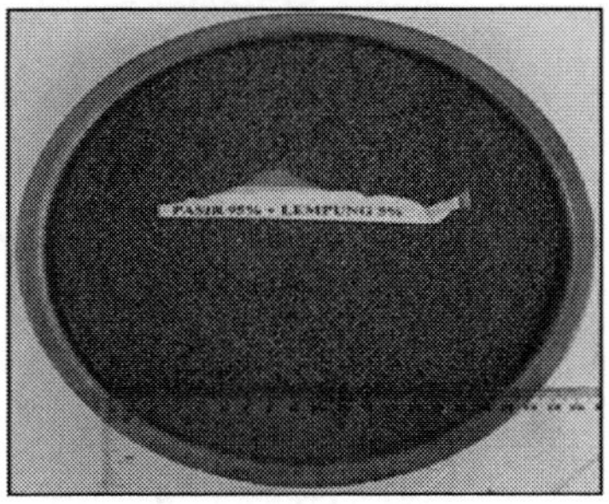

Fig. 2. Combination 1 (95% sand + 5% clay)

b. The second combination is 90% sand+10% clay, as shown in Fig 3.

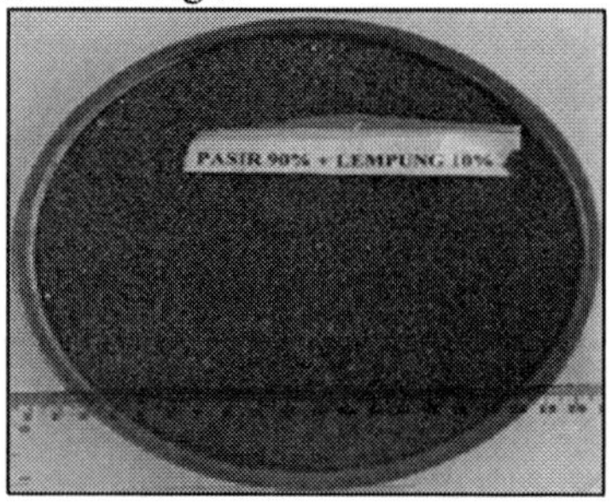

Fig. 3. Material mixture of sand 90% + clay 10%

2. Preparing equipment
The equipment used is a glass box measuring 110 cm x 40 cm x 10 cm, iron cut with a length of 9.8 cm as a static load, plate 9.8 x 21 cm x 1 cm, coloured markings, pressing tools, Proving rings,
3. The camera is mounted on a tripod placed in front of the slope and next to the slope.
4. Conducting soil parameter testing
Testing Total weight (γ), cohesion (c), and Angle of Internal Friction (Φ), Soil density (Dr) and Water content (w), as shown in Table 2

Table 2. Soil Parameters

	Combination 1	Combination 2
Total Weight (γ)	15,76 kN/m^3	15,89 kN/m^3
Cohesion (c)	0,09 kN/m^2	0,33 kN/m^2
Angle of Internal Friction (Φ)	35,06^0	33,69^0
Soil Density (Dr)	73 %	70 %
Water Content (w)	3,168%	3,766 %

5. Slope Making
Slopes are made with a slope angle equal to the shear angle in the slope. Then put a coloured mark 2.5 cm in the horizontal direction and 2.5 cm in the vertical direction.
6. Giving burden
7. Loading is done by arranging pieces of iron measuring 9.8 cm above the slope, then on top of the pieces of iron, a plate measuring 9.8 cm x 21 cm, 1 cm thick, is placed under the pressing device, then the proving ring is pressed with pressure tool. The proving ring is used to see the value of the load, and the dial gauge is used to see the value of the decrease in slope.Seeing the shape of the slope collapse,
8. Processing load data, time, settlement, slope failure form, velocity.
9. Giving conclusion

4 Analysis and Discussion

The first combination (95% sand + 5% clay) was carried out 3 times. The shape of the slip surface is observed, as shown in Fig. 3 where the shape of the slip surface from 3 experiments with a load of 1 cm and 2.5 cm, the shape of the slip surface is almost the same, namely rotation, so it can be concluded that the load does not affect the shape of the slip surface.

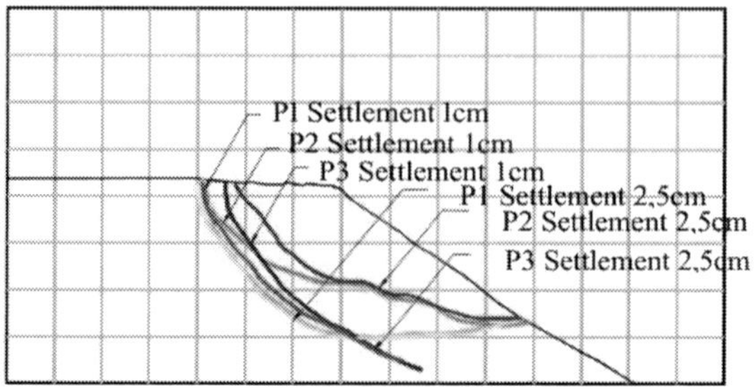

Fig. 3. Form of slip surface from the combination of 95% sand +5% clay

Furthermore, the load, time, and velocity were observed with a decrease of 1 cm and 2.5 cm. Experiment 1 are shown in Table 3, Experiment 2 is shown in Table 4, and Experiment 3 is shown in Table 5.

Table 3. Comparison of load, settlement, time, and velocity in experiment 1 material 95% sand +5% clay

Load(q) (ton/m2)	Settlement (cm)	Time (s)	Velocity (mm/s)	Class Velocity
0,00	0	0	0,000	
4,56	1	29	0,345	Moderate
9,41	2,5	65	0,385	Moderate

Table 4. Comparison of load, settlement, time, and velocity in experiment 2 materials 95% sand +5% clay

Load(q) (ton/m2)	Settlement (cm)	Time (s)	Velocity (mm/s)	Class Velocity
0,00	0	0	0,000	
4,56	1	36	0,278	Moderate
2,14	2,5	50	0,500	Rapid

Table 5. Comparison of load, descent, time and velocity in the 3 experiment materials 95% sand+5% clay

Load(q) (ton/m2)	Settlement (cm)	Time (s)	Velocity (mm/s)	Class Velocity
0,00	0	0	0,000	
8,20	1	19	0,526	Rapid
12,56	2,5	40	0,625	Rapid

Table 6. Comparison of load, descent, time and velocity in the 3 experiment materials 95% sand+5% clay

Load (q) (ton/m2)	Settlement (cm)	Time (s)	Velocity (mm/s)	Class Velocity
0,00	0	0	0,000	
5,77	1	28,00	0,38	moderate
8,03	2,5	51,67	0,50	moderate

From the 3 experiments, the average load and velocity were taken the results obtainec are as shown in Table 6

The results showed that with a combination of 95% sand + 5% clay, the velocity that occurred was between moderate and rapid. Landslide with Moderate class means that when a landslide occurs, some temporary and insensitive buildings can be maintained. Meanwhile, erosions with rapid velocity mean that evacuation is possible, property and equipment are destroyed. To take the average of 3 trials the velocity that occurs is moderate which means that if the landslide is not permanent the building can still be maintaned.

The shape of the slip surface was observed with a combination of 90% sand + 10% clay material. The slip surface that occurs with 3 trials with a decrease of 1 cm and 2.5 cm is rotational, as shown in Fig 4.

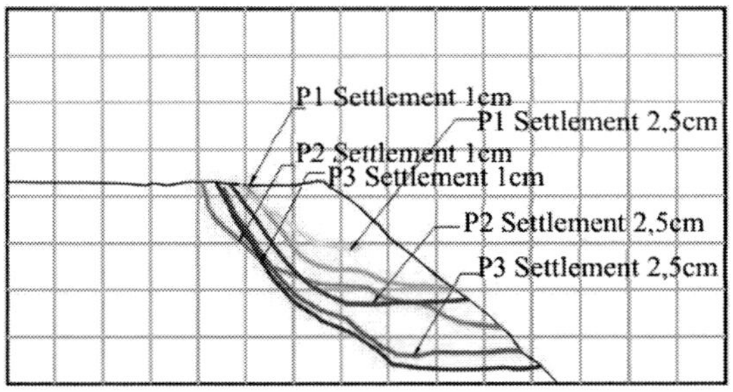

Fig. 4. Form of slip surface from the combination of 90% sand +10% clay

The second combination (90% sand + 10% clay) was also tested 3 times. Observing the load, time and velocity that occurs, with a decrease of 1 cm and 2.5 cm, the results are obtained as in Table 7, Table 8, and Table 9.

Table 7. Comparison of load, settlement, time and velocity in experiment 1 material 90% sand + 10% clay

Load(q) (ton/m2)	Settlement (cm)	Time (s)	Velocity (mm/s)	Class Velocity
0,00	0	0	0,000	
4,56	1	32	0,313	Moderate
3,35	2,5	58	0,431	Moderate

Table 8. Comparison of load, settlement, time and velocity in the experiment 2 materials 90% sand + 10% clay

Load(q) (ton/m2)	Settlement (cm)	Time (s)	Velocity (mm/s)	Class Velocity
0,00	0	0	0,000	
5,77	1	35	0,286	Moderate
3,35	2,5	65	0,385	Moderate

Table 9. Comparison of load, settlement, time and velocity in the experiment 2 materials 90% sand + 10% clay

Load(q) (ton/m2)	Settlement (cm)	Time (s)	Velocity (mm/s)	Class Velocity
0,00	0	0	0,000	
5,77	1	35	0,286	Moderate
5,77	2,5	60	0,417	Moderate

Table 10. Comparison of load, settlement, time and velocity in the experiment 2 materials 90% sand + 10% clay

Load (q) (ton/m2)	Settlement (cm)	Time (s)	Velocity (mm/s)	Class Velocity
0,00	0	0	0,000	
5,37	1	34,00	0,29	moderate
4,16	2,5	61,00	0,41	moderate

From Tables 7-10 it can be seen that the velocity that occurs is moderate, which means that some non-permanent and insensitive structures can be temporarily maintained.

5 Conclusion

The shape of the slip surface from 3 trials for combination 1 (95% sand + 5% clay) and combination 2 (90% sand + 10% clay) with a load of 1 cm and 25 cm has the same shape, which is rotational, so it can be concluded that the load and variation 2 materials do not affect the shape of the slip surface.

The velocity that occurs from combination 1 compared to combination 2, combination 1 (95% sand + 5% clay) has a greater velocity than combination 2 (90% sand + 10% clay) which is in the moderate-rapid range in velocity class 4 -5, which means that some non-permanent and insensitive structures can be temporarily maintained and it is still possible for self-evacuation, but property and equipment are destroyed. If the average velocity class is taken, it is taken, it is class 4 moderate, while combination 2 is in the range of 4 which is moderate, some temporary and insensitive structures can be temporarily maintained, but between combination 1 and combination 2 there is no significant difference

References

1. I. Machdi, *statistical year book of Indonesia*, (BPS-*Statistics Indonesia*, Jakarta, 2021)
2. A. Hakam, *Slope stability and ground retaining wall*, (Andalas University Press, Padang, 2010)
3. Li, S.H., Wu, L.Z. and Luo, X.H., A Novel Method for Locating the Critical Slip Surface of A Soil Slope, J. of Engineering Applications of Artificial Intelligence, **94**, 2020, 1-15 (2020)
4. W. Fellenius, Calculation of the stability of earth dams, In Proceedings of the second congress of large dams, 4,. 445–463, Wash- ington DC (1936)
5. A. W. Bishop, The Use of the Slip Circle in the Stability Analysis of Slopes, Geotechnique, 5, pp. 7-17 (1955)
6. N. R. Morgenstern, V. E. Price, The Analisys of the Stability of General Slip Surfaces, The Institution of Civil Engineers, 79-93, London (1965)
7. E. Spencer, A Method of Analisys of the Stability of Embankments Assuming Parallel Inter-Slice Forces, Manchester University Institut of Science and Technology, 11-26, Inggris (1967)

https://doi.org/10.1051/e3sconf/202133103004

8. S. K. Sarma, ICE, 23, pp. 323-433, London (1973)
9. M. Jafri, Iswan, M. Rizki, G.E. Susilo, CIVENSE, 3, 1 (2020)
10. Raj D and Yogendra S. Effect of Building Loads on the Stability of Hill Slopes. ASCE. 638-647 (2016).
11. Janbu, N. Slope Stability Computations. Embankment Dam Engineering - Casagrande *Volume*, R.C. Hirschfeld and S.J. Poulos, eds., John Wiley and Sons, New York, 47-86 (1973).
12. Terzaghi K Mechanisms of landslides. Geological Society of America, Berkley, 83–123 (1950).
13. S. P. Pradhan, V. Vishal, T.N. Singh, Springer, 50 (2019).
14. J. Andrini Repadi, F. Bari, Junaidi, F. Anas Ismail, Andriani, A. Hakam, *Slip surface in a clay-sand mixture slope test Using scaled laboratory models*, International Journal of GEOMATE, Sept 2021, Japan (2021).

https://doi.org/10.1051/e3sconf/202133103005

Soil classification at Muaro Baru beach of Padang City using CPT data

Rina Yuliet[1*], *Mas* Mera[1], and *Krismon* Hidayat[1]

[1] Department of Civil Engineering, Engineering Faculty, Andalas University, Indonesia

Abstract. Many semi-empiric correlations have been developed to estimate geotechnical parameters based on Cone Penetration Test (CPT) data for various types of soils. This paper aims to classify soil types based on CPT data with several semi-empiric correlations and compare the results of some of these correlations. In this study, the field CPT and the laboratory test were carried out on soil from two closely spaced locations in the estuary area of Muaro Baru, Padang city. The CPT data was used to determine the soil type using several semi-empirical correlations, namely; friction ratios, Schertmann (1978), Robertson and Campanella (1983), and Robertson et al. (1986), then updated by Robertson in 2010. Soil Behaviour Type (SBT) is based on the cone resistance (q_c), sleeve friction (q_s), and friction ratio (R_f). The Unified Soil Classification System (USCS) is also used to classify soils using sieve analysis. The results showed that from the several semi-empirical correlations obtained compatibility soil classification and soil profiles.

1 Introduction

The soil properties are very dependent on the size of the grain, so grain size is used as a basis for naming and classifying the soil-however, most types of soil consist of a mixture of particle sizes. Sand is described as non-cohesive and non-plastic soil. The soil properties are very dependent on the grain size, which is used as a basis for naming and classifying the soil. Investigation of soil properties is beneficial when evaluating building settlement risks, determining the flow rate of water passing through a soil structure, and evaluating the shear strength of a soil that allows the stability of a slope to be calculated.

Most soil classifications are very simple based on a sieving analysis and an Atterbergh limits test in the laboratory. Standard soil classification systems are the USCS and the AASHTO. These systems use soil properties such as grain size distribution, liquid limit, and plasticity index. Soil classification can also be done using a Cone Penetration Test (CPT). The CPT is one of the most common in-situ tests used to evaluate soil types. Much research has been done to identify and classify soils from CPT data [1, 2, 3].

Mohamed (2003) reports that no theoretical procedure can predict the actual soil type of a location. Cone resistance (q_c) and sleeve resistance (f_s) provide insufficient information to classify soil type. Factors such as the history of stress, water content, density, local inclusions of material such as gravel and lenses also affect the final value of q_c and f_s. At the same time, the method developed by previous researchers was only suitable for the area where the method was developed. Furthermore, a CPT test should also be supplemented with information from direct drilling. [3].[b]

Previous researchers have developed empirical equations to classify soil based on CPT data. However, this research has focused chiefly on sub-tropical and regions. Research on this subject for tropical regions like Indonesia is very limited [4]. These empirical equations are based on CPT data are inconsistent with laboratory tests [5].

The soil in the Nurul Haq Shelter area, 37 m from the coast of Padang city, West Sumatera, is known from previous Standard Penetration Test (SPT) data to be silty sand to a depth of 6 m [6-8]. In the present research, this soil was field-tested with CPT and a sieve analysis in the laboratory to determine the type of soil behavior and classify it according to the USCS system. The soil investigation is the coastal area of Muaro Baru close to the river mouth. The CPT test was carried out at 2 points 10 m apart, and the sieve analysis was carried out on shallow soil (0-60 cm).

1.1 Soil Behaviour Type (SBT) on CPT data

CPT can be used on all soil types, from soft soils and very stiff clay soils. It provides a quick soil profile, is economical, and provides repeatable data, and there is a solid theoretical basis for interpreting the results. The difference in friction ratio between side resistance and cone resistance is needed to interpret CPT data. The relative size of these variables depends on soil type. The cone resistance (q_c) in the sand is higher than in clay, and

* Corresponding author: rina@eng.unand.ac.id

https://doi.org/10.1051/e3sconf/202133103005

the friction ratio (R_f) in the sand is lower than in clay. CPT cannot determine soil grain size but can determine soil mechanical characteristics (strength, stiffness, compressibility) and SBT.

Many semi empiric correlations have been developed to estimate geotechnical parameters based on CPT data for various soils. The first CPT was made in 1932. After that, in 1953, soil classification based on CPT was first created by Begemann (Fig. 1) [9].

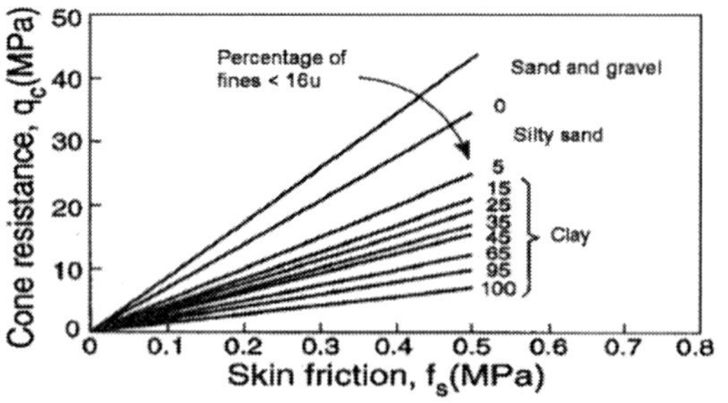

Fig. 1. Soil type based on Begemann (1953) method [9]

Another method for interpreting CPT data is the initial assessment using the friction ratio (R_f) shown in Table 1.

Table 1. Soil type based on friction ratios [10]

Friction Ratio, R_f(%)	Soil Type
<1	Coarse to medium sand
1-2	Fine sand, silty to clayey sands
2-5	Sandy clay, silty clays, clays, organic clays
>5	Peat

Variation between friction ratio and cone resistance is also needed in the final analysis. This correlation varies, and each has reliability in its application. Soil classification is based on a combination of both friction ratio and cone resistance according to Schertmann (1978), Robertson and Campanella (1983) are shown in Fig. 2 and Fig. 3, respectively.

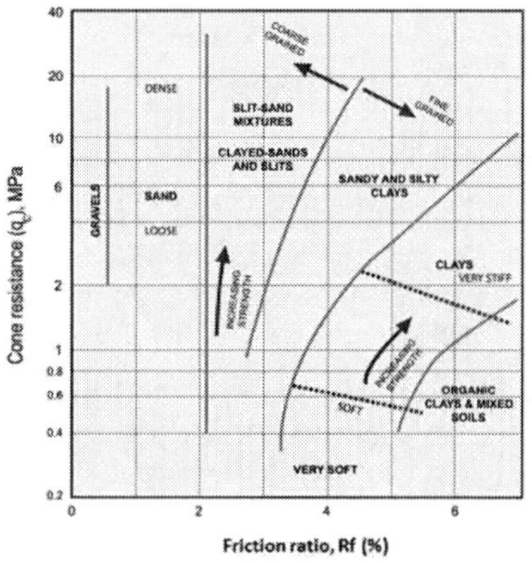

Fig. 2. Soil type based on Schertmann (1978) [10]

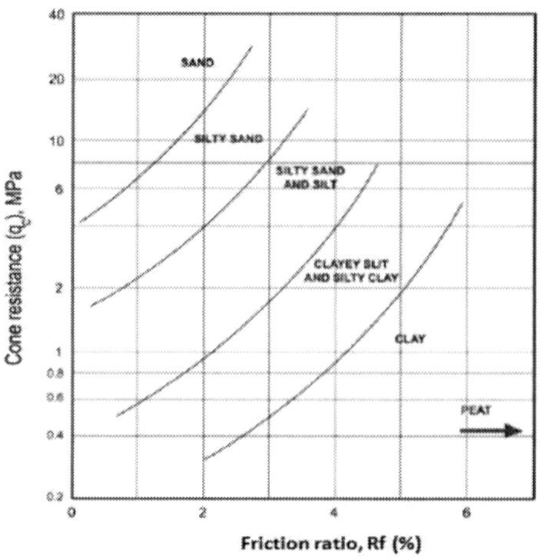

Fig. 3. Soil type based on Robertson and Campanella (1983) method [10]

Usually, based on CPT data, the cone resistance (q_c) in the sand is higher than in clay, and the sand's friction ratio (R_f=f_s/q_c) is lower than in clay. CPT cannot determine soil physical characteristics such as grain size but can determine soil mechanical characteristics (strength, stiffness, compressibility) and Soil Behaviour Type (SBT). The most common SBT chart using CPT data is suggested by Robertson et al. (1986) then updated by Robertson in 2010 in the dimensionless version (Fig. 4).

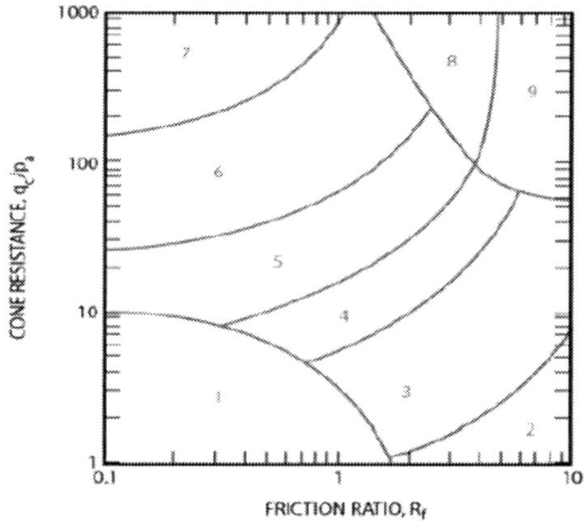

Fig. 4. Soil type based on Robertson et al. (1986) updated by Robertson (2010) [9]

Table 2. Soil Behaviour Type (SBT) [9]

Zone	*Soil Behavior Type*
1	*Sensitive, fine grained*
2	*Organic soils - clay*
3	*Clay – silty clay to clay*
4	*Silt mixtures – clayey silt to silty clay*
5	*Sand mixtures – silty sand to sandy silt*
6	*Sands – clean sand to silty sand*
7	*Gravelly sand to dense sand*
8	*Very stiff sand to clayey sand**
9	*Very stiff fine grained**

1.2 Soil classification based on USCS

The physical properties of the soil are influenced by the distribution of soil grain size and the amount of each mineral component in the soil which determines the soil plasticity. There are two commonly accepted soil classification systems; the AASHTO and the USCS classification systems. Both systems take into account the grain size distribution and the Atterbergh limit. Road engineers use the AASHTO classification system more often, while geotechnical engineers commonly use the Unified Classification System (USCS).

1.3 Clay strength from CPT

Table 3 uses the above relationships to establish the clay's likely strength.

Table 3. Correlation of q_c and cu [10]

Consistency	Cohesion Undrained, c_u (kPa)	q_c (MPa)	Assumption. Not corrected for overburden
Very Soft	0-12	<0.2	N_k=17
Soft	12-25	0.2-0.4	N_k=17
Firm	25-50	0.4-0.9	N_k=18
Stiff	50-100	0.9-2.0	N_k=18
Very stiff	100-200	2.0-4.2	N_k=19
Hard	≥200	>4.0	N_k=20

1.3 Sand strength from CPT

Meyerhoff (1956) presents the correlation of the q_c value of the CPT test results and the relative density (D_r) to the friction angle (ϕ) for sand (Table 4). Table 3 can be used to determine preliminary sand strength from CPT data [11].

Table 4. Correlation of q_c, D_r and ϕ (Meyerhoff, 1956) [11]

Consistency	Relative Density (D_r)	q_c (MPa)	Estimated friction angle value, ϕ (°)
Very Loose	<0.2	<2	<30
Loose	0.2-0.4	2-4	30-35
Medium Loose	0.4-0.6	4-12	35-40
Dense	0.6-0.8	12-20	40-45
Very Dense	>0.8	>20	>45

2 Methodology

In this study, the tests carried out were field tests (Cone Penetration Test/CPT, SNI 2827-2008) [12] and laboratory tests (grain size analysis test, SNI 3423-2008) [13]. Field testing and soil sampling were carried out at two adjacent points with a distance between the two points of 10 m in Muaro Baru Pasie Nan Tigo (Fig. 5). The two test points are coded, namely Muaro Baru 1 (MB1) and Muaro Baru 2 (MB2).

Fig. 5. Location map of research Muaro Baru beach of Padang City

To get the calculation parameters that are required, these two CPT data will then be plotted into the many semi empiric correlations to be used to estimate the soil type in the study area. The semi empiric correlations used to determine soil type are; friction ratios, Schertmann (1978), Robertson and Campanella (1983), and Robertson et al. (1986), updated by Robertson (2010). From CPT data, the cone resistance (q_c) and sleeve friction (f_s) values are obtained, then calculate the friction ratio ($R_f=f_s/q_c$).

3 Results and Analysis

3.1. Soil classification based on USCS

Grain size distribution graphs for the Muaro Baru area (MB1 and MB2) can be seen in Fig. 7. At 0-60 cm depth, the soil type is around 98% sand and 2% silt soil.

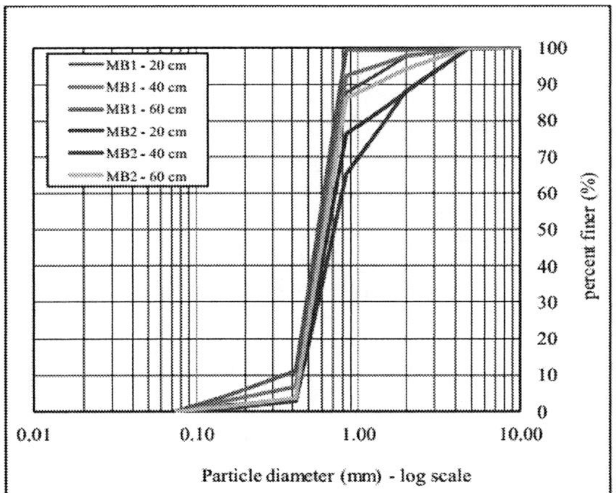

Fig. 7. Particle-size distribution of two locations

Based on Fig. 7, these soils be classified as Poorly graded Sand (SP). The results of soil classification based on USCS for MB1 and MB2 areas are shown in Table 4.

Table 4. Soil classification by USCS

Parameter	MB1-20 cm	MB1-40 cm	MB1-60 cm	MB2-20 cm	MB2-40 cm	MB2-60 cm	MB3-20 cm
% Passing No. 200 sieve	0	0	0	0	0	0	1.4
% Retained on No. 4 sieve	0	0	0	0	0	0	0
D10 (mm)	0.45	0.43	0.36	0.45	0.45	0.45	0.16
D30 (mm)	0.52	0.51	0.47	0.56	0.54	0.52	0.19
D60 (mm)	0.67	0.65	0.62	0.8	0.72	0.67	0.27
Cu	1.49	1.51	1.72	1.78	1.6	1.49	1.7
Cc	0.9	0.93	0.99	0.87	0.9	0.9	0.84
Group symbol	SP	SP	SP	SP	SP	SP	SP
Note : SP = Poorly Graded							

3.2. Interpretation results from CPT data

The results of the CPT data for the MB1 area can be seen in Fig. 8. The plotting data for the MB1 area can be seen in Fig. 9, Fig. 10, and Fig. 11, respectively. Recapitulation of the results of plotting data in the MB1 areas is shown in Table 5.

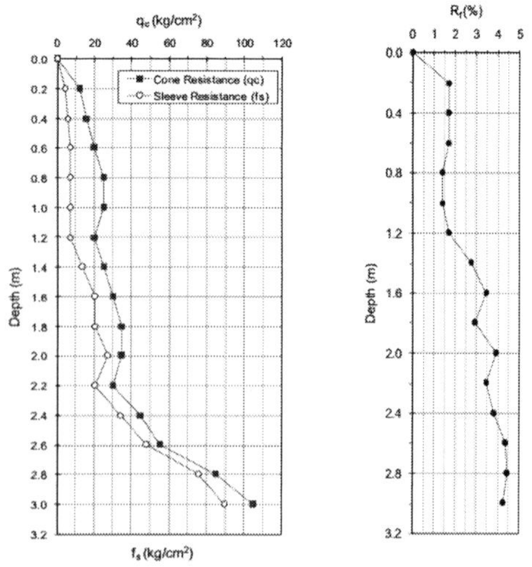

Fig. 8. CPT results for MB1

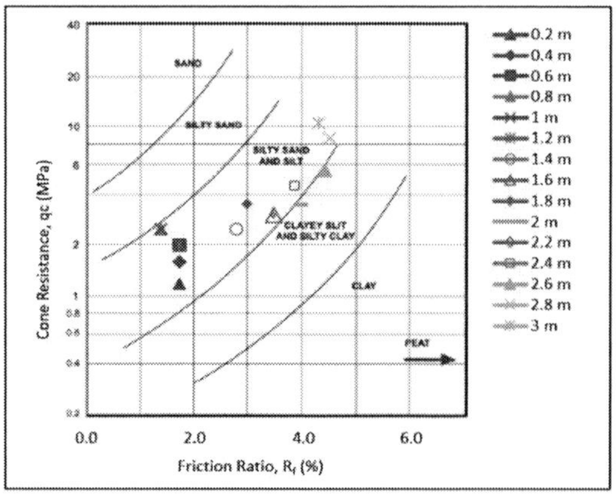

Fig. 9. Plotting CPT data for MB1(Robertson and Campanella, 1983)

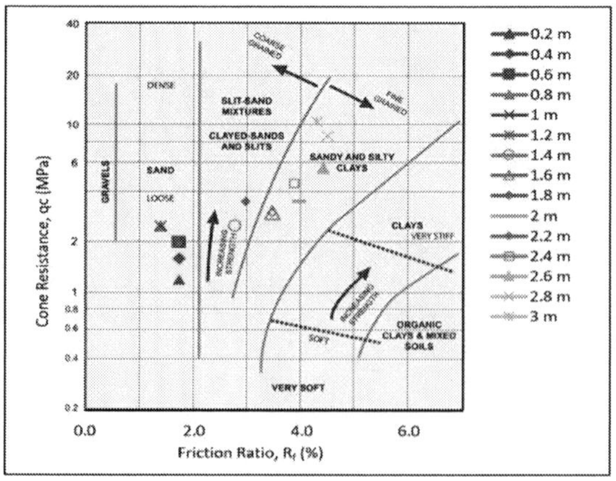

Fig. 10. Plotting CPT data for MB1 (Schertmann, 1978)

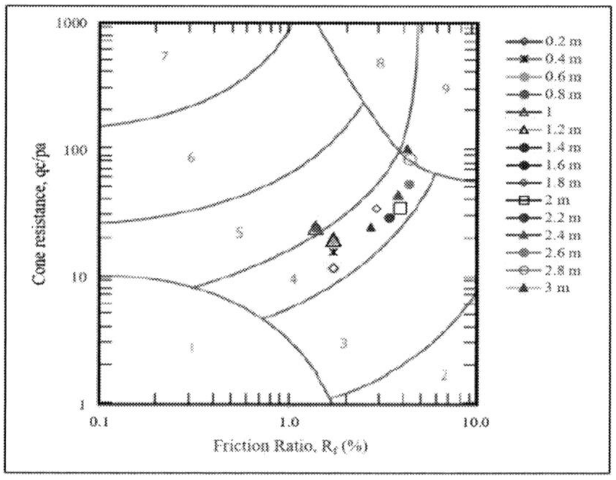

Fig. 11. Plotting CPT data for MB1 (Robertson et al., 1986, updated by Robertson, 2010)

From Table 5, it can be seen that the distribution of data for the semi-empirical correlation based on friction ratios at the MB1 areas obtained a soil profile consisting of two layers of soil. In the first layer, the soil type is fine sand, silty to clayey sands. In the second layer, the soil types are sandy clay, silty clays, clays, organic clays.

According to Robertson and Campanella (1983), the soil profile at the MB1 consists of five soil layers. The first, third, and fifth layers have the same soil type: silty sand and silt, while the second and fourth layers have soil types clayey silt and silty clay.

According to Schertmann (1978), the soil profile to MB1 areas consists of five soil layers; The first layer is loose sand; the second and fourth layers are clayed-sands and silts, the third and fifth layers are sandy and silty clay.

Robertson et al. (1986), updated by Robertson (2010), obtained four soil layers in the MB1 areas. The first and third layers are silty-mixtures clayey silt to silt clay, the second layer sandy-mixtures silty sand to sandy silt, the fourth layer is very stiff fine-grained.

 https://doi.org/10.1051/e3sconf/202133103005

Table 5. Soil Behaviour Type (SBT) MB1 area

Depth (m)	Friction ratios	Robertson and Campanella (1983)	Schertmann (1978)	Robertson et al., 1986, updated by Robertson, 2010
0	Fine sand, silty to clayey sands	Silty Sand and Silt	Loose Sand	silt mixtures-clayey silt to silt clay
0.2				
0.4				
0.6				
0.8				sand mixtures-silty sand to sandy silt
1				
1.2				silt mixtures-clayey silt to silt clay
1.4	sandy clay, silty clays, clays, organic clays		Clayed-sands and silts	
1.6			Sandy and silty clays	
1.8			Clayed-sands and silts	
2		clayey silt and silty clay	Sandy and silty clays	
2.2		Silty Sand and Silt		
2.4				
2.6		clayey silt and silty clay		
2.8		Silty Sand and Silt		
3				verry stiff fine grained

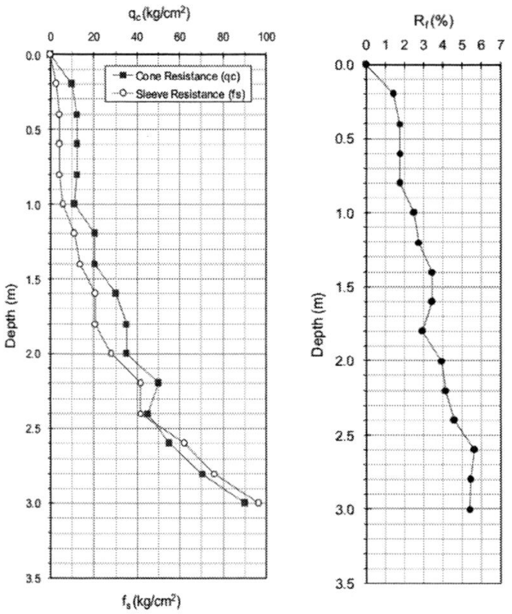

Fig. 12. CPT results for MB2

From Table 6, for the semi-empirical correlation based on the friction ratio in the MB2 area, a soil profile consisting of three soil layers is obtained; the first layer, the type of soil is fine sand, silty to clayey sands. In the second layer, the soil types are sandy clay, silty clays, clays, organic clays. The third layer, soil type, is peat. According to Robertson and Campanella (1983), the soil profile MB2 areas consists of eight soil layers.

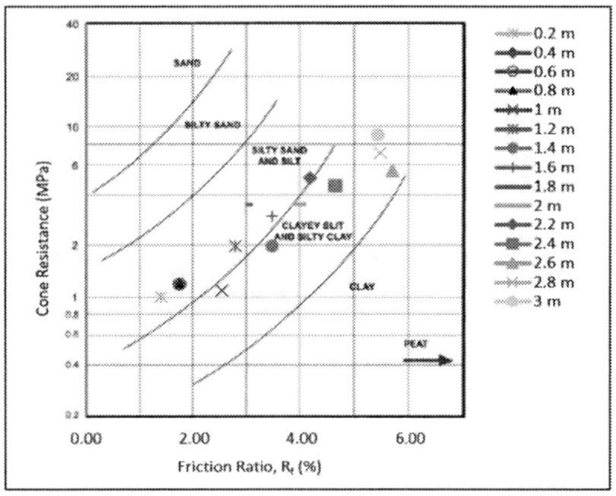

Fig. 13. Plotting CPT data for MB2 (Robertson and Campanella, 1983)

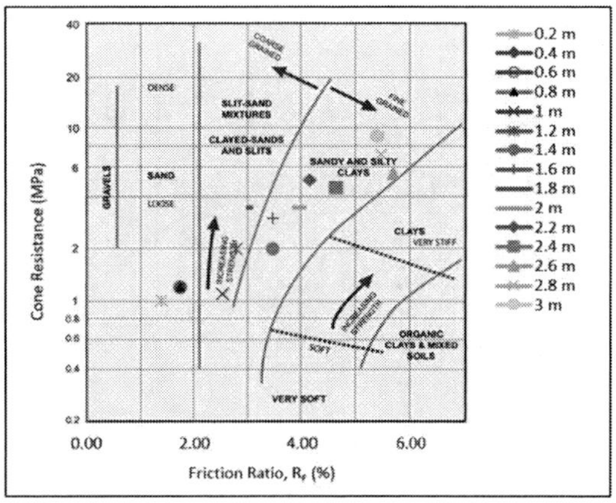

Fig. 14. Plotting CPT data for MB2 (Schertmann, 1978)

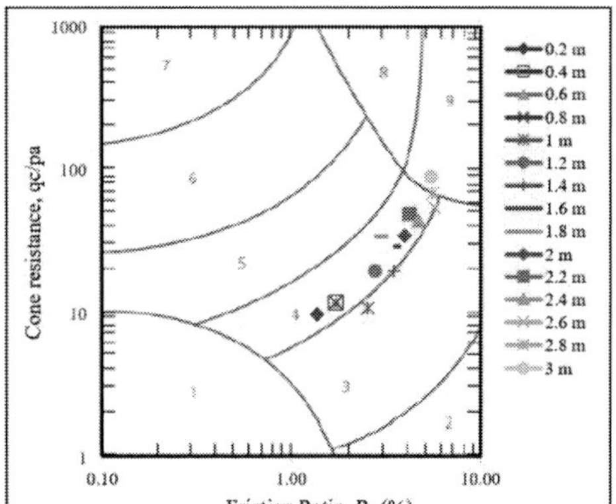

Fig. 15. Plotting CPT data for MB2 (Robertson et al., 1986, updated by Robertson, 2010)

The first, third, fifth, and seventh layers have the same soil type, silty sand, and silt. The second, fourth, sixth, and eighth layers are clayey silt and silty clay. According to Schertmann (1978), the soil type in the MB2 area is the same as the MB1 area, and only the depth is different.

Robertson et al. (1986), updated by Robertson (2010), obtained a soil type in the MB1 area similar to the MB2 area where the soil profile consists of four soil layers with the depth is different.

Table 6. Soil Behaviour Type (SBT) MB2 area

<table>
<tr><th>Depth (m)</th><th>Friction ratios</th><th>Robertson and Campanella (1983)</th><th>Schertmann (1978)</th><th>Robertson et al., 1986, updated by Robertson, 2010</th></tr>
<tr><td>0</td><td rowspan="5">Fine sand, silty to clayey sand</td><td rowspan="5">Silty sand and silt</td><td rowspan="5">Loose sand</td><td rowspan="5">silt mixtures-clayey silt to silt clay</td></tr>
<tr><td>0.2</td></tr>
<tr><td>0.4</td></tr>
<tr><td>0.6</td></tr>
<tr><td>0.8</td></tr>
<tr><td>1</td><td rowspan="8">sandy clay, silty clays, clays, organic clays</td><td>Clayey Silt and silty clay</td><td rowspan="2">Clayed-sands and silts</td><td>sand mixtures-silty sand to sandy silt</td></tr>
<tr><td>1.2</td><td>Silty sand and silt</td><td rowspan="9">silt mixtures-clayey silt to silt clay</td></tr>
<tr><td>1.4</td><td>Clayey Silt and silty clay</td><td rowspan="2">Sandy and silty clays</td></tr>
<tr><td>1.6</td><td rowspan="2">Silty sand and silt</td></tr>
<tr><td>1.8</td><td>Clayed-sands and silts</td></tr>
<tr><td>2</td><td>Clayey Silt and silty clay</td><td rowspan="6">Sandy and silty clays</td></tr>
<tr><td>2.2</td><td>Silty sand and silt</td></tr>
<tr><td>2.4</td><td rowspan="4">Clayey Silt and silty clay</td></tr>
<tr><td>2.6</td><td rowspan="3">peat</td></tr>
<tr><td>2.8</td></tr>
<tr><td>3</td><td>very stiff fine grained</td></tr>
</table>

4. Conclusions

Soil classification in the Muaro Baru area based on USCS is poorly graded sand (SP), where the soil is dominated by sand with little silt. Meanwhile, by using several semi-empirical correlations, it is found that the number of different soil layers but has almost the same soil type.

The author sincerely thanks the Department of Civil Engineering, Andalas University, for the Finacial Support with contract number 130/UN16.09.D/PL/2021.

References

1. H. K. S. Begemann, pp. 17-20, (1965).
2. P. K. Robertson, CAN. GEOTECH. J. **27**, (1990)
3. A. O. Mohamed and O. A. El Fatih, *J. Build. Road Res.* **5**(August), 37–46 (2003).
4. A. Apriyono, Yanto, P. B. Santoso and Sumiyanto, (EIC 2017) AIP Conference Proc.
5. N. Ecemis and M. Karaman, J. Eng.Geo. pp 48-57, (2014).
6. R. Yuliet, Fauzan, A. Hakam and H. Riani, J. GEOMATE, **17** (59), pp 106-114, (2019)
7. R. Yuliet, A. Hakam, M. Mera, and Fauzan, J. GEOMATE **19** (73), pp.14-19, (2020)
8. R. Yuliet, A. Hakam, M. Mera, and Fauzan, J. GEOMATE **21** (84), pp.41-47 (2021)
9. P. K. Robertson and K. L. Cabal, Gregg drilling and testing, Inc, (2016)
10. B. G. Look, *Geotechnical Investigation and Design Tables*, Second Edition, CRC Press, pp. 68-70, (2014)
11. Government of Indonesia, Ministry of Public Works and Housing (PWH), *Geotechnical Parameter Correlation Set and Foundation*, (2019)
12. BSN of Indonesia, *Standard test method for Mechanical Cone Penetration Testing of Soils*, SNI 2827-2008, Jakarta, Indonesia, (2008)
13. BSN of Indonesia, *Standard test method for Particle-size analysis of Soils*, SNI 3423-2008, Jakarta, Indonesia, (2008)

Planning of the foundation a three-story building constructed on potential liquefaction area in Air Tawar Estuary of Padang City

Rasyid Karami[1], *Rina* Yuliet[1*], and *Elsa Eka* Putri[1]

[1]Departement of Civil Engineering, Faculty of Engineering, Andalas University, West Sumatera, Indonesia

Abstract. The earthquake has become one of the most lethal problems in Indonesia, especially across the ring of the fire zone. On September 30th, 2009, an earthquake with a magnitude of 7.6 SR occurred, triggering tsunami and liquefaction. The liquefaction potential and tsunami forces need to be mitigated when an earthquake happens to reduce the risk. This study designs the building foundation that can endure the loads from an earthquake, tsunami, and liquefaction in the Air Tawar estuary of Padang City. The soil profile and liquefaction potential can be identified with CPT (Cone Penetration Test) data. After identifying the liquefaction potential, the upper structure was designed to consider the earthquake load and tsunami waves according to FEMA P-464 (2012). Afterward, design the foundation dimensions based on the ultimate load from the upper structure, wherein the amount of the settlement should be smaller than the tolerable soil settlement. The method used in the paper is taking the CPT data in the Air Tawar Estuary. After analyzing the liquefaction potential, making an upper structure design with ETABS 2016, recapitulate the maximum joint reaction to design the foundation that can withstand the ultimate load, and calculate the amount of soil settlement. Thus the design of the upper structure and lower structure take into account liquefaction. The amount of soil settlement obtained in the building design is still within the tolerance range of 9.79 mm, where the maximum limit of reduction is 32m.

1 Introduction

The building is known as one of the fundamental human needs. These days, the building can become the most dangerous place during natural disasters, especially earthquakes and tsunamis. Cyclic load from an earthquake can also trigger liquefaction, especially in the area around the coast and estuary river. The results of research conducted by Yuliet et al. at Nurul Haq shelter that is located close to the coastline, with a distance of 0.37 km towards the coastline of Padang, has a potential of liquefaction between the depths of 11.55 m and 33.55 m [1].

Putra et al. researched liquefaction potential at GOR Haji Agus Salim and Lapai, Padang City. On research conducted an analysis using CPT method, which focused on the correlation between cone penetration against cyclic stress ratio according to Seed and Idris chart which described the soil boundary whether the liquefaction occurs or not. Compared to a predetermined safety factor of 1.50. It is concluded that the research area has a considerable liquefaction potential [2]. Based on previous research in Padang city, in planning the foundation for high-rise buildings and considering the tsunami load, it is also necessary to consider soil conditions in areas with liquefaction potential.

2 Study Literature

2.1 Liquefaction Potential Analysis

Liquefaction is a natural disaster where the strength and stiffness of the soil are reduced due to earthquakes or other ground movements. It is a process of changing the nature of the soil from solid to liquid, caused by cyclic loads when an earthquake occurs so that the pore water pressure increases close to or exceeds the vertical stress. In this case, the soil undergoing liquefaction is sandy soil or contains much sand, which means that the soil is non-cohesive and saturated. Liquefaction occurs in loose sandy soil (not dense) and saturated with water. As the water pressure increases caused by earthquake shocks, the effective stress decreases [3].

During an earthquake, it is usually followed by a series of cyclic loads and sometimes a tsunami. One of the impacts of soil shaking and soil distribution in specific geological environments is liquefaction. This process can cause buildings to crack, damage, and even collapse; damage to buildings due to liquefaction is called soil failure. One way to evaluate liquefaction potential is to use CPT data, which in this equation compares the CRR value to the CSR value, which can be explained in equation (1) [4]:

* Corresponding author: rina@eng.unand.ac.id

https://doi.org/10.1051/e3sconf/202133103006

$$FS = \frac{CRR}{CSR} \quad (1)$$

FS < 1.50 Liquefaction not occurs
FS > 1.50 Liquefaction occurs

Before evaluating the liquefaction potential, the profile and soil properties of Air Tawar Estuary should be determined first, which will then be used to evaluate the liquefaction potential. In classifying soil, the Robertson and Wride methods are used [5], it can be seen in Fig. 1.

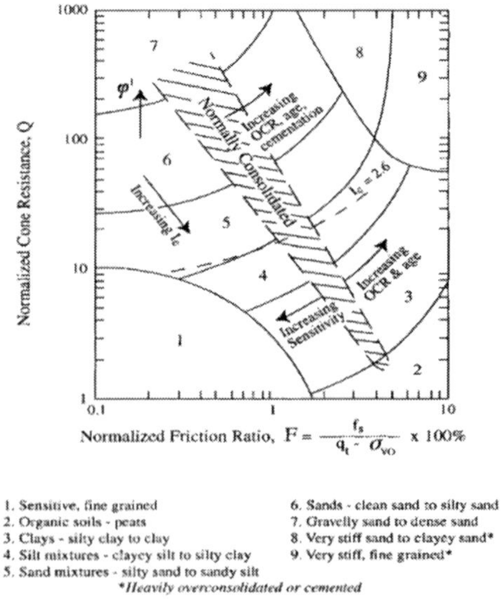

Fig 1. Soil Classification using Robertson and Wride methods [5].

To obtain other soil properties, we can use the correlation Table below [6]:

Table 1. Soil strength based on CPT data [6]

Soil classification		Estimated value qc (MPa)
Very Soft	Cu = 0-12 kPa	<0.2
Soft	Cu = 12-25 kPa	0.2-0.4
Firm	Cu = 25-50 kPa	0.4-0.9
Stiff	Cu = 50-100 kPa	0.9-2.0
Very stiff	Cu = 100-200 kPa	2.0-4.2
Hard	Cu > 200 kPa	> 4.0

Table 2. Preliminary sand strength from CPT [6]

Relative Compaction D_r (%)		q_c (MPa)	Internal Friction, ϕ (°)
Very loose	Dr < 15	<2.5	<30
Loose	Dr = 15-35	2.5 – 5.0	30 – 45
Med dense	Dr = 35-65	5.0 – 10.0	35 – 40
Dense	Dr = 65-85	10.0 – 20.0	40 – 45
Very dense	Dr >85	>20.0	>45

Table 3. A representative range of dry unit weight [6]

Type	Soil description	Unit weight range (kN/m³)	
		Dry	Saturated
Cohesion less Compacted broken rock	Soft sedimentary	12	18
	Hard sedimentary	14	19
	Metamorphic Igneous	18	20
Cohesion less; sands and gravel	Very loose	14	17
	Loose	15	18
	Medium dense	17	20
	Dense	19	21
	Very dense	21	22
Cohesion less sands	Loose-Uniformly graded	14	17
	Loose-Well graded	16	19
	Dense-Uniformly graded	18	20
	Dense-Well graded	19	21
Cohesive	Soft-organic	8	14
	Soft-non organic	12	16
	Stiff	16	18
	Hard	18	20

The Coefficient Stress Ratio (CSR) value (Seed and Idris, 1971) [5],

$$CSR = (\tau_{av}/\sigma'_{vo}) = 0.65\,(a_{max}/g)\,(\sigma_{vo}/\sigma'_{vo})\,r_d \quad (2)$$

Where the shear stress reduction coefficient (r_d) is, [7]:

$$r_d = \frac{(1.000 - 0.04113z^{0.5} + 0.04052z + 0.001753z^{1.5})}{(1.000 - 0.04177z^{0.5} + 0.05729z - 0.006205z^{1.5} + 0.001210z^2)} \quad (3)$$

To determine the Cyclic Resistance Ratio (CRR) value, use equation (4) and equation (5) [7],

$$\text{If } (q_{c1N})_{cs} < 211 \quad (4)$$

So, CRR= $\exp\,[((q_{c1N})_{cs} / 540) + ((q_{c1N})_{cs} / 67)^2 - ((q_{c1N})_{cs}\ 80)^3 + ((q_{c1N})_{cs} / 114)^4 - 3]$

$$\text{If } (qc1N)cs > 211 \quad (5)$$

So, CRR7.5= 2

2.2 Earthquake

According to the USGS earthquake, the term is used to describe the sudden slip of a fault. The resulting ground is shaking and emitting seismic energy caused by the slip, volcanic or magmatic activity, or other sudden changes in the earth's stress. The 2017 Indonesia Earthquake Source and Hazard Map were launched by the Ministry of Public Works and Public Housing (PUPR), This is done to increase preparation for the potential for earthquakes in the future to minimize losses. The 2017 earthquake map can be seen in Fig. 2

https://doi.org/10.1051/e3sconf/202133103006

Fig 2. Earthquake and hazard maps 2017 [8]

2.3 Tsunami

Tsunami comes from Japanese, composed of the words *tsu* which means harbor, and *nami* which means wave. Tsunamis are ocean waves that are formed due to impulsive disturbances in the ocean. The disturbance is formed due to a sudden change in seabed shape, either horizontal or vertical direction. These changes are caused by three primary sources: volcanic eruptions, tectonic earthquakes, or landslides on the seabed.

The tsunami loads are calculated based on FEMA P-464 2012 [9], as follows:

1. Hydrostatic loads
   The loads are applied to the center of the column as high as 1/3 of the maximum water height in the direction of the tsunami wave.
2. Buoyancy
   The buoyant force is concentrated at the center of mass of the floor plate submerged by the tsunami
3. Hydrodynamic loads
   The hydrodynamic loads located in the column as high as 1/2 of the maximum water height, in the direction of the tsunami wave
4. Impulsive forces
   The impulse forces are distributed on the walls of the structure at the height of the tsunami wave in the direction of the tsunami.
5. Debris impact forces
   The impact force from waterborne debris (e.g., floating driftwood, lumber, shipping containers, automobiles, boats) can cause significant building damage.
6. Damming of accumulated waterborne debris
   The force due to debris is uniformly applied to the structural element with a minimum width of 40 ft (12 m)
7. Uplift forces on elevated floors
   The uplift forces act evenly on the upper floors affected by the tsunami
8. Additional Retained Water Loading on Elevated Floors

2.4 Pile Foundation

The method used in the calculation of the pile foundation is a static analysis method. The static analysis method can be categorized as an analytical method that uses the compressibility and strength properties of the soil to determine the pile performance and capacity. Static pile capacity from the amount of soil/rock resistance along the sides of the pile and at the tip of the pile can be estimated from geotechnical engineering analysis using the following data:

a. Laboratory test data to determine the soil shear strength and rock around the pile
b. Standard Penetration Test /SPT data
c. Cone Penetration Test /CPT

The calculation of the foundation used the direct method proposed by Meyerhof (1956). The equation can be seen in equation (6) [10].

$$P_{allow} = \frac{q_c \times A_p}{3} + \frac{JHP \times K_{ll}}{5} \qquad (6)$$

3 Analysis and Discussion

3.1 Liquefaction Potential Analysis

The soil classification is obtained by using the Robertson and Wride method, as shown in Fig. 3.

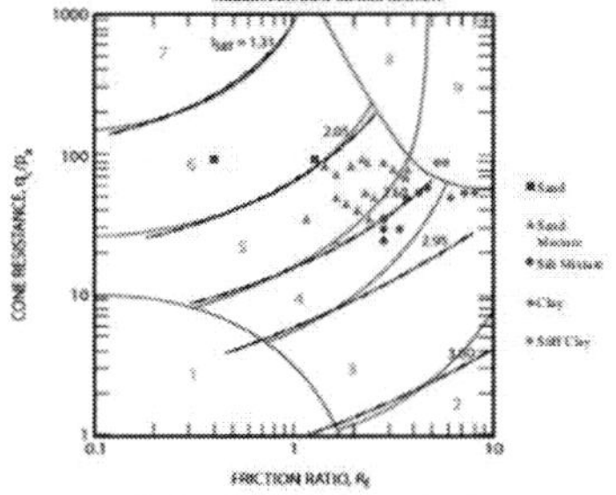

Fig 3 Soil classification in Air Tawar estuary

Analysis of liquefaction potential using the CPT method, while the CPT graph is as follows:

https://doi.org/10.1051/e3sconf/202133103006

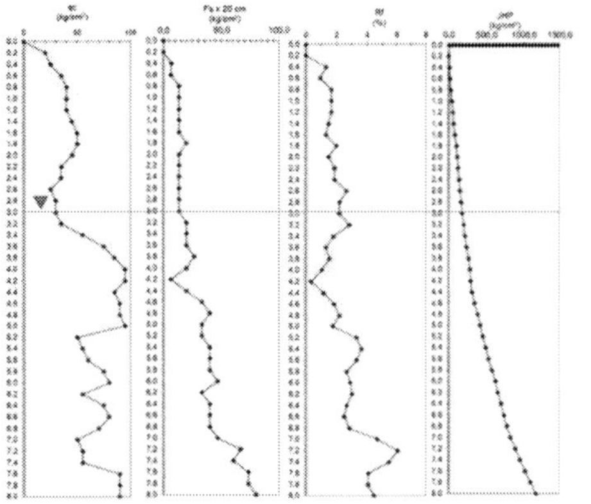

Fig 4. CPT graph in Air Tawar estuary

Table 4. Recapitulation of CSR, CRR, and FS values at each depth

Depth (m)	Soil type	CRR	CSR	FS	Description
0.0	Sand				
0.2					
0.4					
0.6					
0.8					
1.0					
1.2					
1.4	Sand mixture		Not evaluated		
1.6					
1.8					
2.0					
2.2					
2.4					
2.6					
2.8	Silt mixture				
3.0					
3.2		0.087	0.333	0.521	L
3.4		0.114	0.345	0.565	L
3.6	Sand mixture	0.156	0.356	0.622	L
3.8		0.179	0.367	0.794	L
4.0	Sand	0.206	0.376	0.713	L
4.2		0.200	0.385	0.520	L
4.4		0.166	0.393	0.566	L
4.6	Sand mixture	0.175	0.400	0.830	L
4.8		0.171	0.407	1.017	NL
5.0		0.180	0.413	0.792	L
5.2		0.110	0.420	0.942	L
5.4	Silt mixture	0.118	0.426	1.412	NL
5.6		0.126	0.432	1.269	NL
5.8	Sand mixture	0.127	0.437	0.762	L
6.0		0.134	0.441	0.999	L
6.2	Silt mixture	0.107	0.446	0.706	L
6.4		0.122	0.450	0.696	L
6.6	Sand mixture	0.129	0.453	0.682	L
6.8		0.111	0.456	0.662	L
7.0	Clay	0.091	0.460	1.405	NL
7.2		0.097	0.464	4.308	NL
7.4		0.095	0.468	4.037	NL
7.6		0.154	0.471	4.249	NL
7.8	Stiff clay	0.150	0.473	4.225	NL
8.0		0.147	0.476	4.203	NL

Note: L = Liquefaction, NL = Not Liquefaction

By using the correlation table in Table 1, Table 2, and Table 3, the soil profile in the Air Tawar Estuary is obtained, which is described in Table 4. After obtaining the properties and soil profile, CSR, CRR, and Safety of Factor are sought at each depth to evaluate the liquefaction potential, as shown in Table 4. From Table 45, it can be seen that the soil in the Air Tawar Estuary area has the potential for liquefaction where all Safety of Factor values are several depths smaller than 1.

3.2 Upper Structure Design

3.2.1 Building structure data

In conducting a structural analysis, the parameters used in the technical data of the building and the building design conditions are as follows:

Address	: Air Tawar Estuary, Padang	
Structure Type	: Reinforced Concrete	
Number of Floors	: 3	
Building Height	: 1st floor	(8 m)
	2nd, 3rd Floor	(4 m)
Building Area	: 320 m^2	
Building Function	: Shelter	
Floor Plate Thickness	: 14 cm	
Beam Dimensions	: 600 mm x 300 mm	
Column Dimensions:	Column K_1 (800 mm x 800 mm)	
	Column K_2 (600 mm x 600 mm)	

3.2.2Structural modelling

Fig. 5 shows a 3D model of the building designed using ETABS 2016, and the floor slabs are modeled as slabs, beams, and columns are modeled as frames.

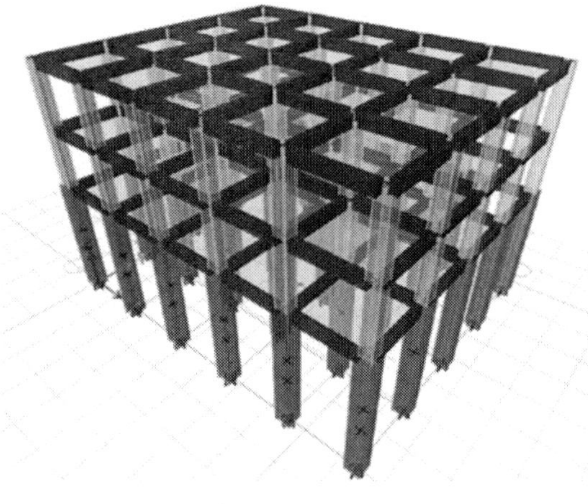

Fig. 5. Three-story building structure modelling

3.2.3 Load analysis

Load analysis refers to : PPIUG 1983 [11] ; SNI 1727-2013 [12] ; SNI 1726-2019 [13]. The living load of refugees used is 250 kg/m^2 (FEMA P-464 2012) [9]. SPT value was 6.15, which means the soil in the Air Tawar can be categorized as soft soil (SE) [14], where the N-SPT value is smaller than 15. The response spectrum can be seen in Fig. 6.

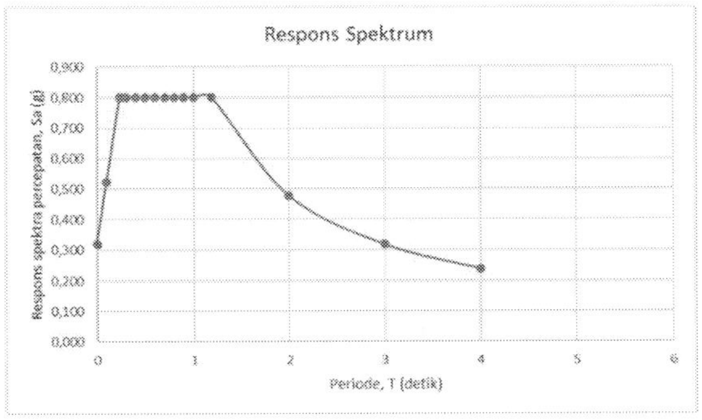

Fig. 6. Soil spectrum response

For tsunami loads recapitulation based on FEMA P-464, 2012 [10] can be seen in Table 6.

Table 6. Tsunami load recapitulation

No.	Tsunami Load	Load (kN)
1	Hydrostatic loads on the wall	236.40
	Hydrostatic loads on the front column	590.87
	Hydrostatic loads on a column	276.25
2	Floating force	10,860.06
3	1-floor hydrodynamic loads	101.00
	2-floor hydrodynamic loads	233.57
4	Impulsive force	151.51
5	Debris impact force	647.29
6	Damming of accumulated waterborne debris	5,908.72
7	Uplift forces on elevated floors	0
8	Additional retained water loading on elevated floors	39.93 kN/m^2

For the combination of loading using FEMA P-464 2012 [9] and SNI 1726:2012 [13], where the maximum reaction value can be seen in Table 7.

Table 7. Maximum reaction value

Force (kN)			Momen (kN)		
F_x	F_y	F_z	M_x	M_y	M_z
27.90	33.76	1,217.70	3,842.89	45.90	1.23

For the foundation calculation, only use the F_z value of 1,217.70 kN.

3.3 Foundation Design

The foundation design is carried out by trial and error using the direct cone method as described in equation (5). The foundation size design is obtained as follows:

a. Pile Cap (Fig. 7.):
- Thick (D) : 80 cm
- Length (P) : 200 cm
- Width (B) : 200 cm

b. Foundation (Fig. 8.):
- Number of Piles : 4
- Pile Height : 720 cm
- Pile Diameter : 30 cm
- Pile Material : Reinforced concrete

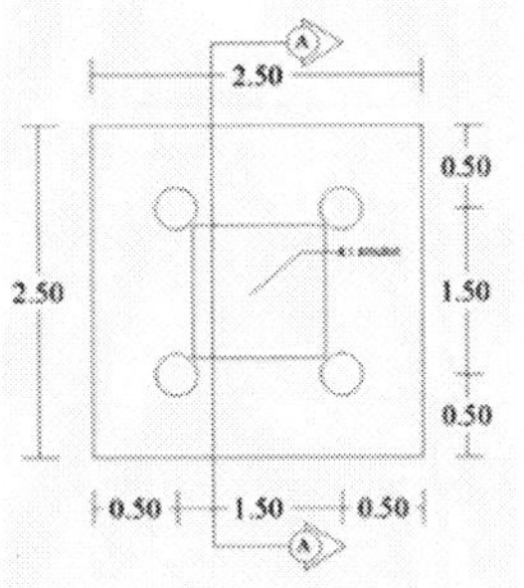

Fig. 7. Pile cap plan

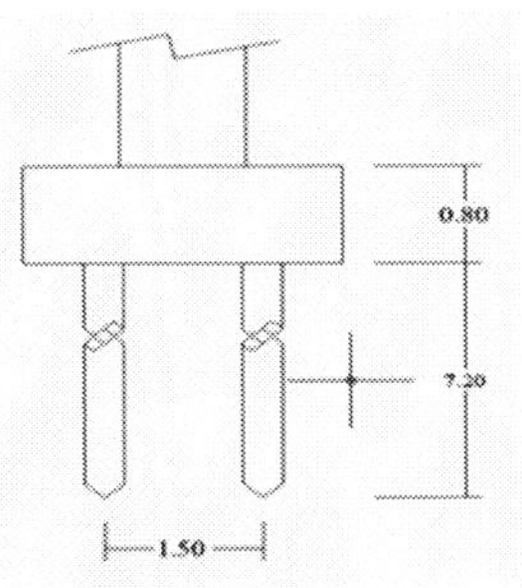

Fig. 8. A-A pile cap section

3.3.1 Foundation bearing capacity

In planning the bearing capacity of the foundation, only the combination of live load, dead load, and earthquake load is considered, which is the value of P_{max} borne by the foundation resulting from a combination of axial force and moment. At the same time, the value of the most significant force and moment obtained is 727.02 kN for axial force, 319.31 kN-m for a moment in the x-direction, and 147.75 kN-m in the y-direction. Then look for the P_{max} value with the following equation (7):

$$P_{max} = \frac{F_z}{n} + \frac{M_x.x_1}{n_x\,(2.x_1^2)} + \frac{M_y.y_1}{n_y\,(2.y_1^2)} \qquad (7)$$

$$P_{max} = \frac{727.02}{4} + \frac{319.31 \times 0.75}{2\,(2 \times 0.75^2)} + \frac{147.75 \times 0.75}{2\,(2 \times 0.75^2)} = 337.42 \text{ kN}$$

The maximum load (P_{max}) value for the building is 337.42 for one pile. The value of the single pile-soil bearing capacity uses equation (7) where Cross-sectional area (A) = 0.07 m^2 and Circumference (K_{ll}) = 0.94 m as follows:

$$P_{allow} = \frac{9{,}000 \times 0.07}{3} + \frac{673.33 \times 0.94}{5} = 338.98 \text{ kN}$$

It was found that the single pile bearing capacity is 338.98 kN, while the bearing capacity of the pile group is 1355.91 kN, of which the single pile bearing capacity of 338.98 kN could withstand the P_{max} of the superstructure of 337.42 kN. It can be concluded that four piles with a diameter of 30 cm can withstand the ultimate load from the building. The design of the foundation and soil profile can be seen in Fig. 9

3.3.2Soil settlement

In the construction of buildings, the magnitude of soil settlement should be reviewed to reduce the impact on the building itself. The amount of tolerance for pile settlement according to Skempton and McDonald in 1956 can be seen in Table 7, while the magnitude of the settlement in the pile group for consolidation settlement and quick settlement can be seen in equation (8) and equation (9) [15]:

$$S_c = \frac{C_c.H_c}{1+e_o} log \frac{\sigma_o'+\Delta\sigma'}{\sigma_o'} \quad (8)$$

$$S_i = H \frac{1}{C'} log \frac{\sigma'_o+\Delta\sigma'}{\sigma'_o} \quad (9)$$

Table 7. Pile group settlement tolerance [15]

Maximum settlement	
Sand	32 mm
Clay	45 mm

In equation (8), several soil properties have not been obtained. To find the value of C_c the Sanglerat method (1972) is used, which can be seen in equation (10) as follows [16]:

$$C_c = \frac{q_c-4,2}{6q_c-4,0} \quad (10)$$

After calculating the amount of soil settlement, the magnitude of the value of soil settlement for the equivalent layer under the foundation using the 4V:1H method can be seen in Table 8. The value of the pile group settlement is 9.79 mm, which meets the settlement limits in Table 7, which for the settlement tolerance is 32 mm.

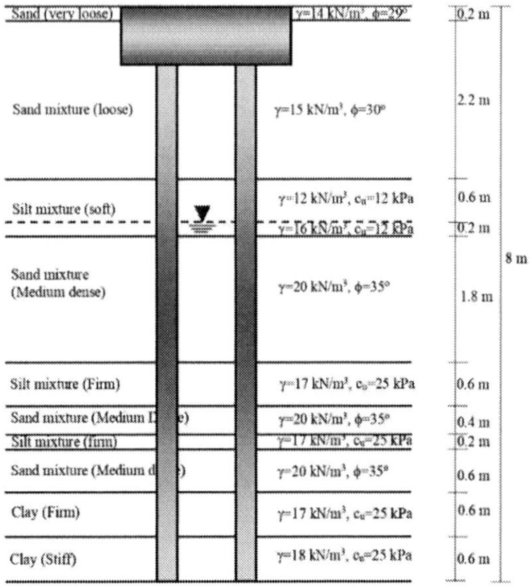

Fig. 9. Soil profile and pile design in Air Tawar Estuary

Table 8. The amount of soil settlement under the foundation is equivalent

Layer	Depth (m)	Settlement type	Total settlement (mm)
8	5.6 - 6.0	Quick settlement	1.1
9	6.0 - 6.2	Quick settlement	0.96
10	6.2 - 6.8	Quick settlement	1.21
11	6.8 - 7.4	Consolidation settlement	2.78
12	7.4 - 8.0	Consolidation settlement	3.74
The total settlement of the pile group			9.79

4 Conclusion

From the results of testing the CPT data, it was found that:

1. The soil profile in the Muara Air Tawar area is dominated by a mixture of sand consisting of 12 layers with the groundwater table at a depth of 3 m below the ground surface.
2. The analysis of the potential for liquefaction using CPT test data, it was found that the soil in the Muara Air Tawar area has the potential for liquefaction at a depth of 3 m - 4.6 m, 4.8 m - 5.2 m, and 5.6 m - 6.8 m with the lowest FS value of 0.52 at a depth of 4.2 m.
3. The planning design of a 3-story building in the Muara Air Tawar area with a height of the 1st floor as high as 8 m, the 2nd and third floors each 4 m, the beam dimensions are 600 x 300 mm, the column dimensions of the 1st floor are 800 x 800 mm and the dimensions of the 2nd and 3rd-floor columns are 600 x 600 mm, while the thickness of the floor slab is 140 mm thick.
4. The foundation design considers the liquefaction potential; the pile cap dimensions are 250 x 250 cm with a thickness of 80 cm, while the number of piles obtained is 4 per column point, with a length of 720 cm and a pile diameter of 30 cm.
5. The soil settlement that occurred was still in the safe category because the total settlement was 9.79 mm, where the maximum settlement limit on sandy soil is 32 mm.

References

1. R. Yuliet, Fauzan, A. Hakam, and H. Riani, J. GEOMATE, July 2019, **17**, 59, 106-114 (2019)
2. H. G. Putra, A. Hakam and D. Lastaruna, J. Rekayasa Sipil, **5** (1) (2009)
3. Tohwata, Ikuo, *Geotechnical Earthquake Engineering,* Springer-Verlag Berlin Heidelberg, (2008)
4. Idriss, I. M dan R.W. Boulanger, *Soil Liquefaction During Earthquake,* (2009)

https://doi.org/10.1051/e3sconf/202133103006

5. P. K. Robertson, C. E. Wride, *Evaluating cyclic liquefaction potential using cone penetration test.* Canadian Geotechnical Journal, (1998)
6. B. G Look, *Handbook of Geotechnical Investigation and Design Tables*. London, UK: Taylor and Francis Group, (2007)
7. T. F. Blake, Personal Communication as cited in Youd, et al. (2001). *Liquefaction Resistance of Soils*: Summary Report from the 1996 NCEER and 1998 NCEER/NSF Workshops on Evaluation of Liquefaction Resistance of Soils, (1996)
8. Ministry of Public Works and Housing (PUPR), *Indonesia Earthquake Source and Hazard Map*, (2017)
9. FEMA P646, *Guidelines for Design of Structure for Vertical Evacuation from Tsunami*. Washington D.C, USA: Federal Emergency Management Agency, (2012)
10. G. G. Meyerhof, *Penetration Test and Bearing Capacity of Cohesionless Soils.* ASCEJ Soil Mech Found Div 82:866-1019, (1956)
11. Ministry of Public Works, *Indonesian Loading Regulations for Buildings* (PPIUG), (1983)
12. BSN of Indonesia, *Minimum Load for The Design of Buildings and Other Structures*, SNI 1727:2013, Jakarta, Indonesia, (2013)
13. BSN of Indonesia, *Design Method of Earthquake Resistance for Buildings and Other Structures,* SNI 1726:2012, Jakarta, Indonesia, (2012)
14. BSN of Indonesian, *Geotechnical Design Requirements*, SNI 8460:2017, Jakarta, Indonesia, (2017)
15. B. M. Das, *Principles of Foundation Engineering 8th Edition,* Boston: Cengage Learning, (2016)
16. Sanglerat, *The Penetrometer and Soil Exploration 1st edition*. Elsevier Science, (1972)

https://doi.org/10.1051/e3sconf/202133103007

Analysis of landslide potential due to changes of land use/ land cover at the Kuranji watershed, Padang using normalized difference built-up index (NDBI)

Andriani Andriani[1*], *Geri Despita* Putra[1], *Salsabila* Ramadhani[1], *Ismael*[1], and *Hendri Gusti* Putra[1]

[1] Civil Engineering Departement, Andalas University, Padang, West Sumatera, Indonesia 25163

Abstract. The earthquake and tsunami predictions in the city of Padang have caused very rapid land-use changes, especially in the Kuranji watershed, where people tend to seek locations that are safe from tsunamis and liquefaction. Changes in environmental characteristics such as slope geometry conditions, vegetation density, and changes in land use will affect runoff coefficient and rainwater filtration, triggering a potential for landslides. This study aims to analyze the potential for landslides due to changes in land use in the Kuranji Watershed. The identification of land-use change is carried out using a remote sensing approach, namely the Normalized Difference Built-Up Index (NDBI). Landslide potential is determined based on the relationship between land use and runoff coefficient from 2007 to 2019. The results showed there had been an increase in the built-up area in the Kuranji watershed from 1602.212 ha (2007) to 2897.513 ha (2019). In contrast, the vegetation area has decreased. An increase in the runoff coefficient of 3.9% from 2007 to 2019. The final results of this study are thematic geospatial information obtained in the form of the relationship between changes in land use and the potential for landslides that occurred in the Kuranji watershed during the period 2007 to 2019.

1 Introduction

Landslides are disasters that often occur in Indonesia. Especially in hilly areas and rivers that have redundant cliff relief. One of the causes of frequent landslides in Indonesia is land-use change. Land-use change is caused by limited areas. West Sumatra is one of the provinces that often experience landslides/slope collapse disasters. The results of monitoring carried out by the Geological Agency, the Center for Volcanology and Geological Disaster Mitigation, in 2011 recorded 21 times the disasters of land movements in West Sumatra that are spread in several districts [1].

This occurs partly due to the factor of high rainfall, very steep slopes > 70 (100 - 150%), and is an area classified as prone to ground movement (red zone). One area that is vulnerable to slope collapse is Bungus Bay and surrounding areas. The vulnerability of land movements on the coast of Bungus Bay is indicated to be strongly influenced by physical factors such as geology, topography, land use, and rainfall [2]. Slope collapse causes huge economic losses, loss of life and property, environmental damage, and social life. Slope failure is a natural process due to the influence of the earth's gravitational force, unstable slopes, and impermeable rock as a slip plane for the soil or rock on it [3, 4,5]. The cause of slope collapse can be caused by natural and human factors, namely geological, geomorphological, slope, and climate conditions, including rainfall, humidity, and temperature. Meanwhile, factors caused by humans consist of land use, road network construction, mining activity, and others. Landslides are the mass of rock, soil, or material that makes up the slope (a mixture of soil and rock) moving down the slope. Factors causing natural landslides include earth surface morphology, land use, lithology, geological structure, rainfall, and seismicity [6,7]. Apart from natural factors, landslides are also caused by human activity factors that affect a landscape, such as agricultural activities, slope loading, slope cutting, and mining [8,9]. The main factors causing landslides include slopes, soil texture and type, rainfall, land use, vibrations, the material on cliffs, old landslides, deforestation, and landfill areas. The parameters that cause landslides and based on the weighting model used, and the most considerable influence on the level of landslide vulnerability are the parameters of rainfall and land slope; while rock types and land use have the same weight, then soil type is the parameter with the most influential small one [10,11].

* Corresponding author: andriani@eng.unand.ac.id

The Kuranji watershed is one of the areas experiencing rapid land-use change. The Kuranji River Basin is included in the green zone area from the danger of tsunamis and liquefaction. According to the BPBD of Padang City, from 2014-2019, five landslides occurred in the Batang Kuranji Watershed (DAS). Landslides arise because they are affected by several conditions, such as morphological conditions, geology, geological structures, hydrogeology, and land use. The use will affect the speed of water flow and the surface flow coefficient; the faster the water flow, and the greater the surface flow coefficient, the faster the erosion process will occur. High erosion will make the area prone to landslides [12]. To interpret land use, it can be done by remote sensing; one of the commonly used indexes is the Normalized Difference Built-up Index (NDBI) and the Normalized Difference Vegetation Index (NDVI) [13,14,15]. NDBI is an effective transformation to map built-up areas in cities automatically using Landsat 8 OLI imagery, and using NDBI; it is possible to map urban areas, such as industrial, commercial, and urban residential areas that cannot be separated. Besides, using NDBI can interpret rural areas [16,17]. NDBI is an index of built-up area, which is an algorithm to show the density of bare soil [18]. NDBI is very sensitive to developed or open land. This algorithm was chosen because it is the most frequently used transformation to assess the constructed land index. NDBI is an effective transformation to map the limited land area [19,20].

2 Materials And Methods

2.1 Study Area

The Kuranji watershed is an area that is experiencing rapid changes in land use. Geographically, the Kuranji watershed is located at 100°20'31"-100°33'50" East Longitude and 00°55'59.88"-00°47'24" South Latitude. While administratively located in Padang City which includes Pauh Subdistrict (120 km^2), Kuranji Subdistrict (53.65 km^2), Koto Tangah Subdistrict (29.41 km^2), Nanggalo Subdistrict (8.93 km2), and North Padang District (3.69 km^2). The total area of the Kuranji watershed is approximately 205.6 km^2.

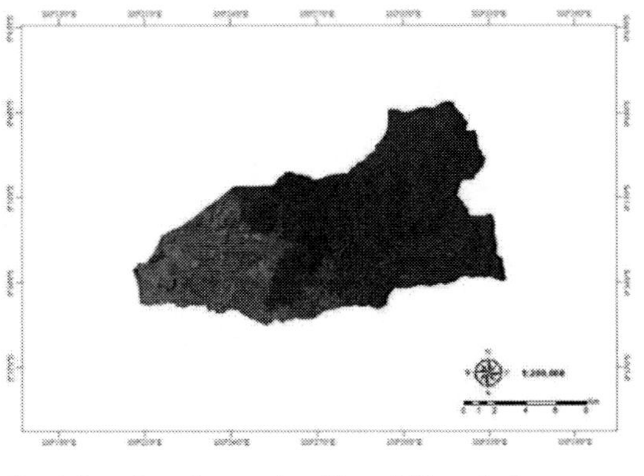

Fig. 1. Location of research at Kuranji Watershed.

The Batang Kuranji watershed flows from the upstream of the row hills with the highest elevation of ± 1,605 MSL on the Bukit Tinjau Laut and empties at the Padang beach with the main river length of ± 32.41 km and the total length along with all tributaries of 274.75 km. (Fig. 1).

2.2 Data Analysis

This research consists of three stages, land use analysis using NDBI, validating the data by conducting field observations and checking using google earth, and analyzing the potential for landslides based on the value of runoff coefficient on various land use.

The data used to identify land-use change is the Landsat 7 and Landsat 8 imagery data which is downloaded for free from https://earthexplorer.usgs.gov/.

The spatial resolution of the image is 30 x 30 m. The satellite image data used are Landsat 8 OLI imagery on path/row: 127/61, which was downloaded from the United States Geological Survey (USGS) website. Table 1 shows the types of data used in this study.

Table 1. The data criteria used in this study

No.	Type Data	Resolution (m)	Date of imagery data used
1.	Landsat 7	30	19 July 2007
2.	Landsat 8	30	27 April 2012 22 July 2017 12 July 2019
3	The boundary of Kuranji Watershed		

Data processing using Quantum Gis 3.14.0. Meanwhile, for data validation interpretation results, use google earth and direct observations in the field.

2.2.1 Landsat image data processing using the NDBI method

The data used are in the form of Landsat 7 and Landsat 8 image data. Analysis of Landsat image data was obtained using remote sensing methods, namely by using the NDBI method. The image data that is processed is image data from 2007, 2012, 2015, and 2019. The location of the image data used is the Kuranji watershed. The NDBI formula is as follows:

$$NDBI = \frac{SWIR - NIR}{SWIR + NIR} = \frac{B6 - B5}{B6 + B5} \quad (1)$$

Where :
SWIR = Shortwave Infrared or often referred to as band 6
NIR = Near Infrared or often referred to as band 5

2.2.2 Runoff coefficient analysis

The runoff coefficient or surface flow coefficient is the ratio between the peak rate of runoff and rain intensity.

Also referred to as the C value. The main factors influencing the C value are the soil infiltration rate, cover cropping, topography and geology, and rain intensity. The following is the runoff coefficient based on land use based on SNI 2415: 2016.

Table 2. Coefficient of runoff (C) per land use/ land cover (LULC).

No.	Land-Use/Land Cover	Coefficient of runoff
1.	Mixed Gardens	0.25
2.	Commercial	0.95
3.	Field	0.25
4.	Swap	0.20
5.	Rice fields	0.20
6.	bush	0.07
7.	Industry	0.90
8.	pond	0.20
9.	mangrove	0.01
10.	settlement	0.40
11.	bare land	0.50
12.	river	0.00
13.	education area	0.70
14.	garden	0.30
15.	tomb	0.25
16.	funeral	0.25

2.2.3 Validation of data

The results of the interpretation of land use are then validated using google earth and field observations, it is required that the results of the validation of the interpretation of image data meet 95% of observations in the field. Furthermore, the area in the Kuranji watershed that has experienced landslides has been checked by using data from the Padang City BPBD. The final result of this study is a thematic map of landslides due to land-use changes in the Kuranji watershed.

3 Results and Discussion

3.1 Interpretation of Land Use/ Land Cover using NDBI at Kuranji watershed

Analysis of land use/ land cover using NDBI is based on the reflectance value of each band. Furthermore, the guided classification is carried out from the identified map. The land use in this research is divided into 3 types, namely: built-up areas, vegetation areas, and bare-land. The built-up area consists of settlements, office areas, educational areas, and industrial areas. While the vegetation area consists of forests, rice fields, shrubs, rice fields, and dry-land agriculture.

Land use in the Kuranji watershed can be seen in Fig. 2. For vegetation areas are colored green, open areas are colored yellow, and built areas are colored red. From the figure, it can be seen that there is an increase in the built-up area and a decrease in the vegetation area.

3.1.1 Land Use/Land Cover at 2007

The area of vegetation is 15631.143 ha or 76.006% of the total area of the Batang Kuranji watershed. On the land use map, this vegetation class is colored light green; Included in this class are tropical forests, fields, and plantations, as well as areas with dense trees. This vegetated area is spread out in highland areas, including Pauh, Kuranji, and Koto Tangah sub-districts.

In the bare-land class, found an area of 3332.425 hectares, or 16.204% of the total area of the Kuranji watershed. On the land-use map, the distribution of this class is marked in yellow. Types of land included in this class are empty or dry land, dirt roads, and agricultural areas. This type of class of bare-land is spread across all sub-districts that are included in the Kuranji watershed, namely Pauh, Kuranji, Koto Tangah, Nanggalo, East Padang, and North Padang Districts. In the built-up land class, obtained an area of 1602.212 ha or 7.791% of the total area of the Kuranji watershed.

On the land-use map, the distribution of this class is marked in red. Types of land included in this class include residential areas, dense buildings, and asphalt roads. The types of classes of built-up land are mostly scattered in coastal areas such as Nanggalo, North Padang, Kuranji, and East Padang Districts.

3.1.2 Land Use/Land Cover at 2012

In the area of vegetation class marked with light green on the land use map, it was found that an area of 15768.401 Ha or 76.673% of the total area was obtained. The types of land included in this class are tropical forests, as well as fields or gardens. The distribution of this class is more focused on the upland and hilly areas, which include Kecamatan Pauh, Kuranji, and Koto Tangah. Bareland in yellow on the land use map has an area of 2954.891 hectares or 14.368% of the total land area. The types of an area that are included in the bare-land class are vacant land, dirt roads, and agricultural areas.

The distribution of this class is evenly spread across all sub-districts which are included in the Batang Kuranji watershed, namely Pauh, Kuranji, Koto Tangah, Nanggalo, East Padang, and North Padang Districts. Based on the land use map, in 2012, the area of built-up area was 1842.488 hectares or 8.959% of the total area. On the land-use map, this class is marked in red. Types of the area included in the built-up area class are settlements, solid buildings, asphalt roads, and places that have civilian constructions on them. The distribution of this class is more focused on the coastal areas, namely the Nanggalo, North Padang, Kuranji, and East Padang Districts. The value of Near-Infrared or Band 4 in 2012 is -0.0182003 up to 0.551485. Meanwhile, the value of Shortwave Infrared or Band 5 is -0.0163216 to 0.45015.

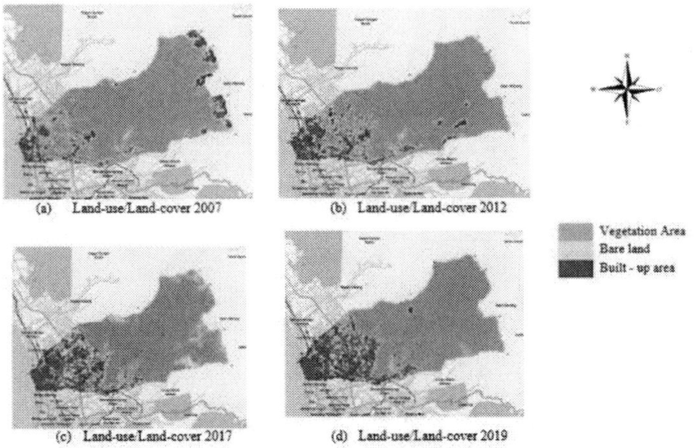

Fig. 2. Interpretation Land-use/ Land-cover using Normalized Difference Built-Up Index (NDBI)

3.1.3 Land-use/ Land-cover at 2017

The area of vegetation is 12651,451 hectares or 61.517% of the total area of the Batang Kuranji watershed. On the land use map, this area class is colored light green. The types of the area included in this class are tropical forests, fields, and plantations as well as areas that have dense trees. This type of vegetation class focuses on the spread of upland areas, which include Pauh, Kuranji, and Koto Tangah Districts. In the bare land class marked in yellow on the land use map, an area of 5315.226 Ha or 25.845% of the total land area is obtained.

The types of the area that are included in the bare land class are vacant land, dirt roads, and agricultural areas. The distribution of this class is evenly spread across all sub-districts which are included in the Batang Kuranji watershed, namely Pauh, Kuranji, Koto Tangah, Nanggalo, East Padang, and North Padang Districts. Meanwhile, the built-up area using NDBI was obtained for an area of 2599.103 hectares or 12.638% of the total area of the Batang Kuranji watershed. On the land use map, the distribution of this class is marked in red. Types of land included in this class include residential areas, dense buildings, and asphalt roads. The types of classes of the built-up area are mostly scattered in coastal areas such as Nanggalo, North Padang, Kuranji, and East Padang Districts.

3.1.4 Land-use/Land-cover at 2019

The area of vegetation class marked with light green on the land use map, it was found that an area of 14825.048 Ha or 72.086% of the total area was obtained. The types of the area included in this class are tropical forests, as well as fields or gardens. The distribution of this class is more focused on the upland and hilly areas, which include Kecamatan Pauh, Kuranji, and Koto Tangah. In the bare land class marked in yellow on the land use map, an area of 2843.329 ha or 13.825% of the total area is obtained. The types of land that are included in the bare-land class are vacant land, dirt roads, and agricultural areas. The distribution of this class is evenly spread across all sub-districts which are included in the Kuranji watershed, namely Pauh, Kuranji, Koto Tangah, Nanggalo, East Padang, and North Padang Districts.

Based on the land use map, the developed land class has an area of 2897.513 Ha or 14.08% of the total land area. On the land use map, this class is marked in red. Types of land included in the built-in land class are settlements, solid buildings, asphalt roads, and places that have civilian constructions on them. The distribution of this class is more focused on the coastal areas, namely the Nanggalo, North Padang, Kuranji, and East Padang Districts.

3.2 Changes in Land-use/ Land-cover at Kuranji Watershed in 2007-2019

The results of observations show changes in land use in 2007-2019 with using by the NDBI method in the form of Landsat 7 imagery and Landsat 8 imagery, it was obtained 3 class intervals, namely vegetation, bare land, and built-up area. This multitemporal observation resulted in a picture of change in land use/land cover at the Kuranji watershed in 2007, 2012, 2017, and 2019.

Based on Fig. 3, there have been various land use/land cover changes at the Kuranji watershed from 2007-2019. A noticeable change is in the built-up area class, which has increased each year in review. Whereas in the bare-land class and vegetation area, the graph fluctuates in each year of review. One of the factors that influence land-use change information is the satellite imagery used in this study. The use of satellite imagery, especially in Landsat

7 imagery in 2007 and 2012, has been damaged or is often called SLC-Off. This can be seen in the visual appearance of Landsat 7 in 2007 and 2012, which have black lines on their image. As result, the information provided was not optimal when identifying changes in land use in 2007 and 2012 in the Kuranji watershed.

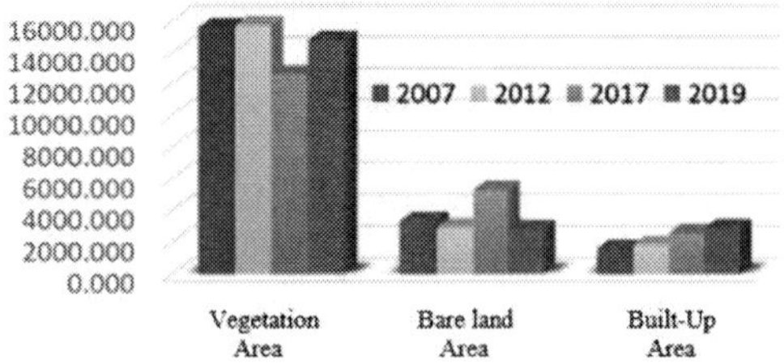

Fig. 3. Changes in Land-use/land-cover at Kuranji Watershed.

3.3. Relationship between Land-use/Land-cover and the Potential of Landslides at Kuranji Watershed

From Table 3 and Fig. 3, it can be seen that in 2007 the coefficient of runoff was 0.102, meaning that 10.2% of rainwater fell into the runoff. In 2012 there was an increase in runoff coefficient by 0.05% from 2007 to 0.107 or 10.7%. For 2012 there was another increase of 3.9% to 0.146 or 14.6%. Finally, in 2019 there was a slight decrease in the runoff coefficient by 0.5% to 0.141 or 14.1%. So can see that from 2007-2019 or for 12 years in the Kuranji watershed, the runoff coefficient has increased by 3.9%.

The increase in the runoff coefficient in the Kuranji watershed in the 12 years of review is still relatively safe because the average increase is only 0.325% per year. Changes in runoff coefficient values are strongly influenced by changes in land use in the Kuranji watershed. It can be seen in Fig. 3, in 2017 and 2019 that there have been significant changes in open-land and vegetation. Open land with a higher C value decreased in 2019, while vegetation land with a lower C value increased in 2019.

The resulted in a decrease in the value of the runoff coefficient from 2017 to 2019 by 0.5%. Table 3 and Fig.3, it showed that changes in the coefficient of runoff are strongly influenced by changes in land use within a certain time in the Kuranji watershed. Runoff coefficient values must be continuously monitored every year to know changes in coefficient values of occurs. If the runoff coefficient approaches the value of 1, further action needs to be taken to change land use in the Kuranji watershed to reduce the potential for landslides. Because it is increasing the value of the coefficient of runoff in an area, the potential for landslides is also increasing because more water becomes runoff; so the potential for surface erosion will also increase

Table 3. Changes in Runoff Coefficient from 2007-2019

Type of LULC	C Value	Area (Ha)				(C x A)/∑A			
		2007	**2012**	**2017**	**2019**	**2007**	**2012**	**2017**	**2019**
Vegetation Area	0.03	15631.143	1576.401	12651.451	14825.048	0.023	0.023	0.018	0.022
Bare-land Area	0.15	3332.425	2954.891	5315.226	2843.219	0.024	0.022	0.039	0.021
Built-up area	0.7	1602.212	1842.488	2599.103	2897.513	0.055	0.063	0.088	0.099
Total		20565.780	20565.780	20565.780	20565.780	0.102	0.107	0.146	0.141

4. Conclusions

The interpretation of land use/ land cover using NDBI shows that there has been an increase in the built-up area and bare-land in the Kuranji watershed area, while the vegetation area continues to decline. The increase in the built-up area and bare-land has resulted in a higher runoff coefficient, which has resulted in higher potential landslides in the Kuranji watershed.

This research was funded by Andalas University by following the Fundamental Research Scheme, Contract Number: T/22/UN.16.17/PT.01.03/IS-RD/2020 The Fiscal Year 2020.

References

1. Anonymous. Disaster Info (Actual Monthly Disaster Information), Bulletin National Disaster Management Agency (BNPB), Edition November (2018)
2. L.M., Highland, and P., Bobrowsky, U.S. Geological Survey Circular 1325 (2008)
3. D.G., Bayuaji, A.L., Nugraha, and A., Sukmono, J Geodesi UNDIP, **1** (2016)
4. V.B.S., Chandel, K.K., Brar, and Y., Chauhan, Int Journal of Geomatics and Geosciences, **2** (2011)

5. V.N., Mishra, P.K., Rai, and K., Mohan, Journal of the Geographical Institute, Jovan Cvijić" SASA (Serbia), **64**, 1 (2014)
6. D.M., Cruden, and D.J., Varnes, Transportation Research Board, U.S. National Academy of Sciences, Special Report, **247** (1996)
7. A.B.S.M., Arsjad, and S., Hartini, Majalah Ilmiah Globe, **16**, 2 (2004)
8. L.M., Highland, and P., Bobrowsky, United States Geological Survey Circular 1325 (2008)
9. D., Aringoli, B., Gentili, M., Materazzi, and G. Pambianchi, Nova Science Publishers, New York (2010)
10. A., Domenico, B., Marcello, M., Marco, G., Bernardino, and P., Gilberto, *Large landslides in sea-cliff areas of the central Adriatic coast (Italy),* in Proceedings of the Second World Landslide Forum – 3-7 October 2011, Rome (2011).
11. F.C., Dai, C.F., Lee, and Y.Y., Ngai, Engineering Geology, **64** (2002)
12. G.D., Bathrellos, H.D., Skilodimou, K., Chousianiti, A.M., Youssef, B., Pradhan, Sci. Total Environ., **575** (2017)
13. A., Andriani, E., Ibrahim, D.D.A., Putranto, and Affandy, A.K. , EEC, **6**, 2(2018)
14. P., Sinha, N.K., Verma, and E., Ayele, Int Journal of Advanced Remote Sensing and GIS, **5**, 8, (2016)
15. E., Li, P., Du, A., Samat, J., Xia, and M., Che, Int Journal of Remote Sensing, **36**, 24 (2015)
16. A., Purwanto, Jurnal Edukasi, **13**, 1 (2015)
17. S.S., Bhatti, and N.K., Tripathi, GIScience & Remote Sensing, **51**, 4 (2014)
18. X.B., Luo, and W., Li, Journal of Applied Remote Sensing, **8**, 9 (2016)
19. H., Xu, Photogrammetric Engineering & Remote Sensing, **73**, 12 (2007)
20. Y., Zha, J., Gao, J. and S., Ni, Int. Remote Sensing, **2**, 3 (2003)

Spatial model of land subsidence mitigation at lowland areas

Andriani Andriani[1*], *Hafizah* Novasari[1], *Masril* Syukur[1], and *M. Shubhi* Nurul Hadie[1]

[1]Civil Engineering Departement, Andalas University, Padang, West Sumatera, Indonesia 25163

Abstract. South Sumatera is one of the provinces with a large number of lowlands in Indonesia consisting of tidal land and lowland swamps, one of which is located in Tanjung Api-Api. The conversion of lowlands into built-up areas will cause physical, socio-economic, environmental, and public health impacts. One of the impacts arising from the development of settlements and infrastructure in lowland areas is land subsidence and flooding. The purpose of this research is to create a model for mitigating land subsidence in lowland areas spatially. The spatial model is carried out by weighting with Analytical Hierarchy Processes (AHP) and then mapping mitigation efforts that will be carried out based on the potential for land subsidence. The results show that if the lowland area is developed into an industrial and residential area, it will have an impact on land subsidence with a moderate to high potential level. Based on the zoning of the potential for land subsidence, the best pre-disaster mitigation efforts to do are choosing the type of construction according to the type of soil, conserving groundwater, maintaining infiltration areas and green open spaces, and regional spatial planning.

1 Introduction

The plan of development of a special economic zone in Tanjung Api-Api will change the environmental ecosystem, a large-scale landfill at the research site will affect environmental conditions, including the surrounding soil conditions. This will affect the lowland ecosystem in the region because lowlands have a very large environmental role. The development of residential and industrial areas on a large scale in the Tanjung Api-Api area will trigger land subsidence. Land subsidence is a problem faced by many big cities in the world and Indonesia. The impact of land subsidence will have direct and indirect impacts on human life; both physical, socio-economic, health, and environmental impacts [1]. To reduce the impact arising from land subsidence, it is necessary to carry out mitigation. Mitigation is an action to reduce the effects of natural disasters so that the resulting disasters do not cause a major impact. Disaster Mitigation Actions include the collection and analysis of disaster data to reduce the level of vulnerability and danger of a disaster, predict disaster events, preparedness, emergencies, and socialization of disasters through print media and lectures [2].

Natural disasters are a series of events caused by nature. Natural disasters can be caused by natural factors, such as geological factors, geomorphology, high rainfall, and others [3]. This will trigger natural disasters, such as floods in the rainy season and drought in the dry season, landslides and ground movements, land subsidence and liquefaction, earthquakes and tsunamis, and others. Disaster mitigation efforts are a process of anticipating disasters that will arise through urban planning and design. Disaster mitigation efforts can be displayed with a conceptual model. The conceptual model is a diagram of a set of relationships between certain factors that are believed to have an impact on or lead to a target condition [4]. Conceptual model of disaster mitigation that aims to avoid disasters, in two ways, namely: structural mitigation, namely efforts in the form of strengthening buildings or infrastructure and avoiding potential disaster areas and non-structural mitigation by not changing the natural environment that can protect against disasters, in addition to potential wisdom local communities can contribute to disaster mitigation [5].

Mitigation efforts to overcome land subsidence can be done technically and non-technically. The technical way to overcome land subsidence is to improve soil characteristics, make infiltration wells/ biopore, and others. To reduce subsidence and flooding problems, it is necessary to conserve underground water, by limiting the use of drilled wells and making injection wells [6]. Mitigation of land subsidence can be done by regulating the use of groundwater (limiting the construction of wells and excessive use of groundwater) [7]. Regulating the hydrological condition of peat soil by maintaining the water table and restoring the groundwater level is one way that can be used to prevent peat soil subsidence, peat soil decomposition, peat fires, and CO_2 emissions [8]. To minimize the impact due to the development of

* Corresponding author: andriani@eng.unand.ac.id

settlements in lowlands, it can be done by making infiltration wells, biopore, maintaining vegetation, and renaturalizing rivers [9].

Disaster mitigation actions can be carried out before and after a disaster occurs, to minimize casualties, preventive measures must be prepared as early as possible, preventive efforts are carried out to avoid large impacts on areas indicated by disasters. Natural disasters can be local but can also be widespread, causing macro impacts. One way that can be done to mitigate disasters is by spatial planning, to reduce physical and economic losses due to disasters [10]. Mitigation activities include risk assessment and determination of potential disaster impacts [11].

Geospatial information is one of the methods used to determine areas that are prone to land subsidence, then using spatial modeling can be done to mitigate land subsidence in lowland areas.

2 Research Methodology

2.1 Research Location

The Tanjung Api-Api area is part of the Banyuasin district which is located in the Banyuasin II sub-district and is located at coordinates 104° 45' to 104° 55' E Longitude and 2° 17' to 2° 24' S Latitude and is along the East coast of Sumatra Island. The existence of the Tanjung Api-Api port and the planned development of a Special Economic Zone has caused the Banyuasin district to have a very important role as a center for downstream industries. The Tanjung Api-Api area has been developed since 1969 through the transmigration program as an agricultural and residential area [12], currently, in the Tanjung Api-Api area, the Tanjung Api-Api port has been built. The existence of plans for the construction of the international port of Tanjung Carat and the Special Economic Zone in Tanjung Api-Api requires special handling because it is located on lowland and soft soil so that the potential for land degradation is likely to occur. Fig. 1 shows the planned location for the construction of the Special Economic Zone.

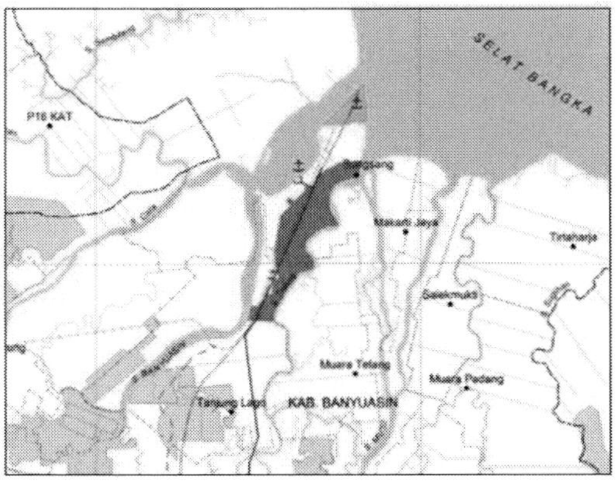

Fig. 1. The Plan of Special Economic Zone at Tanjung Api-Api) [13]

Fig. 2 shows the distribution of soft soil in Indonesia. From the map, it can be seen that the area along the east coast of Sumatra Island consists of soft soil, namely peat soil and soft clay [14] so that it has the potential to experience land subsidence when reclaimed.

Fig. 2. Map of Soft Soil Distribution in Indonesia

2.2 Data Analysis

To reduce the impact of land subsidence, mitigation efforts are carried out. In this study, mitigation efforts were carried out before the disaster occurred. There are many criteria used to determine disaster mitigation. To determine the magnitude of the weight of each criterion factor used the Analytic Hierarchy Process (AHP) method. AHP is a decision-making approach designed to help find solutions to complex multi-criteria problems [15], to determine decisions using criteria and alternative decisions in determining the final goal of a decision or decision suggestion. The AHP process is based on the assessment criteria for decisions from alternatives.

Assessment of mitigation efforts is carried out by processing questionnaire data obtained from experts using AHP. The experts used in this study amounted to 20 people who came from academics and practitioners who are experts in the fields of environment, disaster, and geotechnics. Furthermore, the validation of AHP results is carried out by conducting field observations and studies from previous journals. Mapping of mitigation efforts is carried out based on the results of the weighting obtained from the expert. The mapping was carried out using QGIS 3.12.2, the results of weighting and interviews from experts were displayed in the form of thematic geospatial information.

Fig. 3 shows the hierarchical structure of land subsidence mitigation. Mitigation is carried out technically and non-technically. This step aims to break down a complex problem arranged into a hierarchical form. A hierarchical structure itself consists of elements that are grouped into levels consisting of the goals or goals of decision making, assessment criteria, and choices.

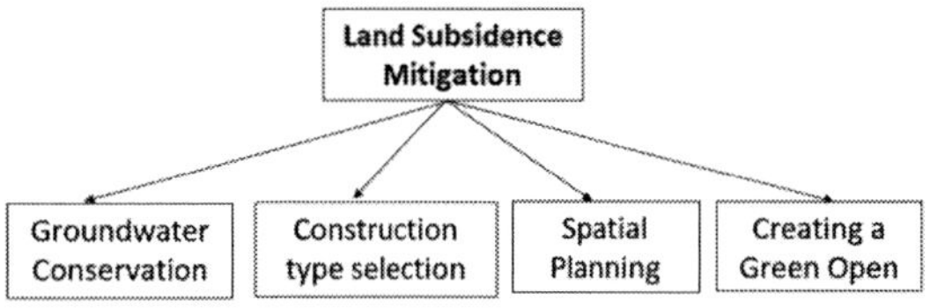

Fig. 3. The Hierarchical Arrangement of Land Subsidence Mitigation in Lowlands

https://doi.org/10.1051/e3sconf/202133103008

3 Results and Discussion

The results of previous studies indicate that the change in the Tanjung Api-Api area into a built area (residential and industrial areas) will trigger land subsidence. This occurs due to the embankment of the soil of more than 2 m and excessive groundwater extraction (deep wells). Currently, land use in the Tanjung Api-Api area is still dominated by forests, shrubs, agricultural land, and plantations, so the potential for land subsidence is still low. To anticipate high land subsidence due to area development, mitigation efforts need to be carried out to avoid and reduce the impacts that will occur due to land subsidence.

Mapping is one form of mitigation that can be done to reduce the impact that occurs. Based on the literature review, several things can be done to reduce the impact of land subsidence, namely selecting the appropriate type of construction to be applied in built-up areas and conserving groundwater. Groundwater conservation can be in the form of making infiltration wells, retention ponds, and biopore which function to fill groundwater (recharge) so that the pore does not empty.

Fig. 4 shows a conceptual model of land subsidence that can be carried out. The mitigation model consists of mitigation before land subsidence, during land subsidence, and after land subsidence. The conceptual model is an important set for mitigating soil subsidence in the lowland.

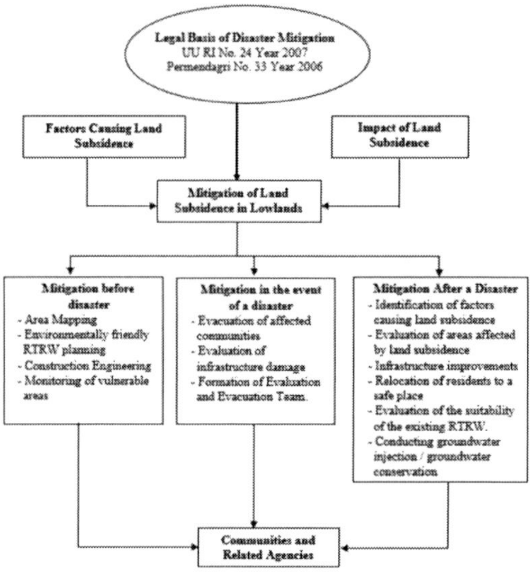

Fig. 4. Conceptual Model of Land Subsidence Mitigation Efforts.

The results of the questionnaire from the experts obtained weights for each criterion for mitigation of land subsidence.

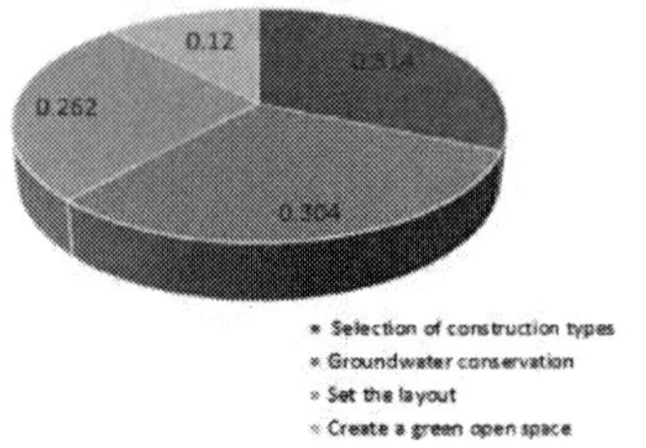

Fig. 5. Assessment of Land Subsidence Mitigation using by AHP

The selection of construction types weights 0.314; groundwater conservation with a weight of 0.304; set the layout with a weight of 0.262; and create a green open space with a weight of 0.12.

The selection of the right type of construction is very necessary to overcome the impact of damage to infrastructure, the selection of construction type is influenced by the type of soil, the depth of the soft soil, and the load. The depth of soft soil will also affect the bearing capacity of the soil, the deeper the soft soil is the type of foundation that can continue the load of the building is needed, for example, pile foundations / drilled piles are used, while for roads it is better to use geosynthetic or other types of soil improvement and rigid pavement. Large loading will cause the potential for land subsidence so that special construction is needed that can carry the working load. The choice of embankment material is also needed to accelerate the process of land subsidence so that the soil is quickly stable and does not form pores due to less dense soil.

Groundwater conservation is carried out to overcome the land subsidence that occurs. The purpose of groundwater conservation is to maintain the balance of groundwater so that there is no emptying of pores that can cause soil subsidence and instability in buildings. Groundwater conservation can be in the form of making infiltration wells, biopore, and groundwater injection and

https://doi.org/10.1051/e3sconf/202133103008

can also be in the form of government policies on groundwater extraction.

The regulation of land use and its designation in lowland is very important to maintain the balance of carrying and environmental capacity so that the development of lowland does not cause environmental degradation. The selection of the most appropriate mitigation efforts depends on the factors causing land subsidence and the most dominant impact due to land subsidence. Factors causing land subsidence will affect the mitigation methods that can be applied at the research site. Based on the results of the study, it was found that the main factors causing soil subsidence were soil type, the thickness of soft soil, land use, embankment height, and groundwater use. The most suitable mitigation effort is to choose the type of construction that is suitable to be applied in the research area to prevent or reduce the impact of infrastructure damage in the research area, while water conservation efforts need to be carried out to anticipate floods and droughts that often occur in the Tanjung Api-Api area so that the presence of groundwater conservation can prevent the formation of pores in the soil due to excessive groundwater extraction [16, 17, 18, 19, 20].

Fig. 6 shows the mitigation efforts for various land use based on the level of potential land subsidence at the stage before land subsidence occurs. The existence of a mapping of potential land subsidence in the research area will help to determine the direction of physical and non-physical mitigation in the disaster-affected area. The results of the mapping carried out due to the conversion of lowlands in Tanjung Api-Api into a built-up area, industrial areas (Special Economic Zones), and settlements have a high potential for land subsidence [21]. The high potential for land subsidence in this area is due to the embankment height of more than 2 meters and the extraction of deep groundwater (bore wells). With the thematic geospatial information, the vulnerability of the area to potential land subsidence can be evaluated and the most appropriate policy direction for mitigation efforts can be determined to be applied to the research location.

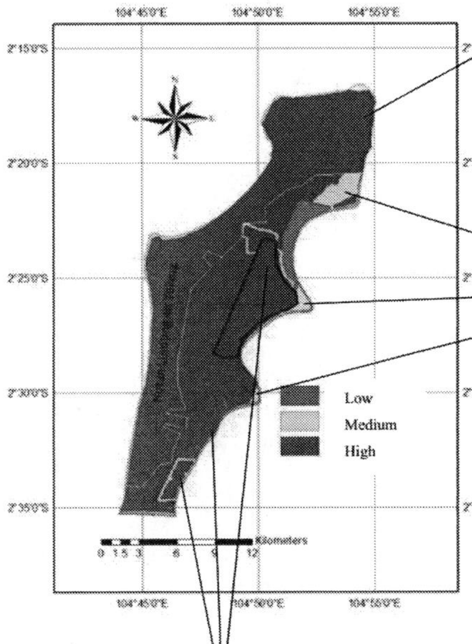

Low potential (forest and scrub)

- Maintaining protected forest areas as conservation areas
- Maintaining the balance of the lowland ecosystem

Medium potential (settlement area)

- Soil improvement using preloading and vertical drains.
- Using sand as heap material
- For vital buildings and more than 3 floors, it is better to use deep foundations (piles/drills)
- Using light building construction.
- Special design of utility buildings (water pipes, gas pipes, etc.) so that they are not broken/cracked when there is a change in ground elevation.
- Conducted subgrade stabilization
- The road network should use cast concrete/rigid pavement
- Providing clean water treatment buildings.
- Create a green open space and catchment area

High Potential (Industrial Area)

- Soil improvement using preloading and vertical drains.
- Using sand as heap material
- For vital buildings and more than 3 floors, it is better to use deep foundations (piles/drills)
- Using light building construction.
- Special design of utility buildings (water pipes, gas pipes, etc.) so that they are not broken/cracked when there is a change in ground elevation.
- Subgrade stabilization.
- The road network should use cast concrete/rigid pavement
- Providing clean water treatment buildings.
- Create a green open space and catchment area

Fig. 6. Mitigation Spatial Model in the Stage Before Land Subsidence Occurs

https://doi.org/10.1051/e3sconf/202133103008

4 Conclusions

Based on the results of the study, it can be concluded as follows:

a. Conceptually, disaster mitigation can be done before, during, and after a disaster occurs. Mapping of areas that have the potential for land subsidence is one form of mitigation before a disaster occurs. Regional mapping is divided into areas with low, medium, and high potential. To select the most suitable mitigation efforts, AHP analysis is carried out and then applied spatially to locations based on the level of potential disaster, the existence of a map of potential land subsidence is expected to reduce the impact of disasters.
b. AHP results show that mitigation of land subsidence can be done by selecting the type of construction (weight 0.314), conserving groundwater (weight 0.304), adjusting the spatial arrangement (0.262), and creating green open space (0.12). To obtain the most optimal mitigation model used depends on the regional development plan and land use.

This research was funded by a grant from Civil Engineering Department, Faculty of Engineering, Andalas University with Contract number: 119/UN16.09.D/PL/2021 for Fiscal Year 2021.

References

1. A. Andriani, E. Ibrahim, D.D.A. Putranto, A.K. Affandy, *Evaluation of Land Subsidence Impact at Lowland Areas Using Analytical Hierarchy Process (AHP) Method,* in Proceedings E3S Web Conf of the 1st The 1st Sriwijaya International Conference on Environmental Issues 2018 (1st SRICOENV 2018), 2-11-2018, Palembang, Indonesia
2. I.A. Sadisun, Disaster Management: Living Strategies in Potentially Disaster Areas, (2014)
3. Anonymous, Law no. 24 of 2007 concerning Disaster Management R. Wikantiyoso, P. Tutuko., Group Konservasi Arsitektur & Kota (2009)
4. R. Wikantiyoso, Local Wisdom, **2**, 1 (2010)
5. D. Murdohardono, Buletin Geologi Tata Lingkungan (Bulletin of Environmental Geology), **12**, 1 (2000)
6. M. Dehghani, A. Hooper, R. Hanssen, M.J.Z. Valadan, S.Saatchi, I. Entezam, *Hybrid conventional and Persistent Scatterers SAR Interferometry for Land Subsidence Monitoring in southwest Tehran, Iran.* Fringe Workshop, Italy (2009)
7. J. Jaenicke, H. Wosten, A. Budiman, Mitigation Adaptation Strategy Global Change, **15** (2010).
8. M.Rijal, P. Aldy, Jurnal Arsitektur UBL, **1**, 3 (2012)
9. D.P.E. Putra, S. Agung., K. Keokhampui, H. Fukuoka, *Land Subsidence Risk Asseessment in Karst Region, Case Study : Rongkop, Gunung Kidul, Yogyakarta-Indonesia. In: Mitteilungen zur Ingenieurgeologie und Hydrogeologie-Festschrift zu.* RWTH Aachen University, German (2011)
10. C. Tarhan, C. Aydin, V. Tecim, Urban Planning and Architecture Design for Sustainable Development, UPADSD (2015)
11. Ngudiantoro, Jurnal Penelitian Sains, **13**, 3(A) (2010)
12. BAPPEDA SUMSEL, Rencana Tata Ruang Wilayah Provinsi Sumatera Selatan (2012)
13. Anonim, *Panduan Geoteknik 2*, Puslitbang Prasarana Transportasi (1997)
14. T.L. Saaty, Int.J. Services Sciences, **1**, 1 (2008)
15. L.P. Sien, C.C. Shieng, C.F. Sheng, H.T.Yuan, *The Effect of Land Subsidence on the Embankment of the New Second Freeway in Taiwan.* Proceedings of the Fifth International Symposium on Land Subsidence, The Hague, October 1995 (1995)
16. J.C. Chai, S.L. Shen, H.H. Zhu, X.L. Zhang, Lowland Tech. Int., **7**,1 (2005)
17. D.L. Galloway, S. Michelle, Buletin DE LA Sociedad Geologica MEXICANA, **65** ,1 (2013)
18. F. Raspini, C. Loupasakis, D. Rozos, S. Moretti, Natural Hazards and Earth System Sciences, **13** (2013)
19. W. Yanlin, B. Xiangjun, F. Sheng, et al., Procedia Engineering, **15** (2011)
20. A. Andriani., E. Ibrahim., D.D.A. Putranto, A.K. Affandy, and H.G. Putra, *Spatial analysis of land subsidence potential due to lowland conversion (case study at Tanjung Api-Api)*, in Proceedings IOP Conference Series: Material Science and Engineering, **933**,012054, The 2nd AISCE, 18-19 September 2019, Banda Aceh, Aceh Province, Indonesia (2019)

https://doi.org/10.1051/e3sconf/202133103009

Changes in land cover to reduce erosion and peak discharge of sub-watershed of Danau Limau Manis

Taufika Ophiyandri[1], *Bambang* Istijono[1*], *Teguh* Haria Aditia Putra[2], *Aprisal*[3], and *Benny* Hidayat[1]

[1]Department of Civil Engineering, Andalas University, Indonesia
[2]Doctoral Program, Andalas University, Indonesia
[3]Department of Soil Science, Andalas University, Indonesia

Abstract. Land cover change is a strategic issue in the Subwatershed of Danau Limau Manis. The community was triggered by the big earthquake of September 30, 2009, so people moved to the upper watershed. The problem is the increasing number of people, the damage to the watershed ecosystem. This damage can be seen from the erosion of river water and the availability of river water during the rainy season and dry season. An appropriate solution is needed to reduce erosion by modeling land cover. This study aims to determine the value of soil erosion and peak discharge. Erosion using a rational formula using the modified universal soil loss equation model based on land units and peak discharge. The land unit is an overlay of land cover maps, slope maps, and soil maps. Land units that produce erosion and exceed the tolerance limit are carried out with a land cover change scenario using the spatial multi-criteria analysis model. Scenario determination is also based on slope maps, land cover maps, soil maps, river flow density maps, forest area maps, erosion values , and the Padang city spatial plan. The study results explain that the erosion and peak discharge resulting from the existing land cover is quite large, namely 47.89 tons/hectares/year and 152.81 m3/sec. After scenario modeling, erosion decreased by 11.91 tons/hectares/year and peak discharge 15.26 m3/sec.

1 Introduction

Watershed is seen as an area with a high integration between human interests in meeting the needs of life and the interests of ecosystem protection. This condition often causes problems for the environment. Where one side of the fulfillment of the needs of life that is done by the community often destroys the order of the ecosystem. The destruction of a watershed ecosystem is undoubtedly a big problem that will cause natural disasters, such as floods, landslides, and erosion.

The destruction of the watershed area is in line with the increasing number of people who need land for the built area. The destruction is due to the change in the vegetated land cover into built-up areas, resulting in erosion, sedimentation, and even on a large scale, therefore causing floods, landslides, and flash floods. Changes in land use into built-up areas will impact changes in flow characteristics, flooding in the rainy season, decreasing river discharge, and drought in the dry season [1].

The increasing population also occurs in the Subwatershed of Danau Limau Manis (DLM). This increase was driven by the occurrence of a huge earthquake in 2009 in the city of Padang. The impact of the earthquake prompted coastal communities to move to tsunami-safe locations. One of them is the sub-watershed of DLM in the middle. The threat of a megathrust earthquake that triggers a tsunami wave is still scary news among the public. This condition also encourages people to move to safer locations. So the subwatershed of DLM in the middle is experiencing high population pressure.

The increase in population has led to an increase in residential areas in the Subwatershed of DLM. Currently, the construction of residential areas has led to the upper watershed, which significantly affects the hydrology of the watershed area. The upper watershed of the DLM area, which is part of the Kuranji watershed, also converts forest land into agricultural land. In the Kuranji watershed, forest land cover decreased by 1,128.42 hectares (8.52%), and residential areas increased by 2,577.08 hectares (63.10%) [2]. In the Subwatershed of DLM, forest land cover decreased by 231.56 hectares (7.44%) and increased residential areas by 49.97 hectares (91.09%).

Judging from the condition of the slopes, the 50% wider Subwatershed of DLM has land with a high slope. This condition is hazardous if land cover changes are not appropriately controlled. The high slope conditions are also supported by high rainfall in the Subwatershed of DLM. This causes high runoff, increased soil erosion, and river discharge.

* Corresponding Author: bistijono@eng.unand.ac.id

 https://doi.org/10.1051/e3sconf/202133103009

Subwatershed of DLM also has high river density, which impacts high surface runoff and low infiltration. Subwatershed characteristics of DLM produce considerable erosion. This is caused by low soil infiltration and high surface flow, and the potential for landslides. This condition results in a large river discharge and the potential for flooding [3].

The problem in the Subwatershed of DLM can also be seen from the high fluctuation of river discharge during the rainy and dry seasons. The high fluctuation of river discharge indicates that the upper watershed area has begun to experience disturbances in vegetation land cover. Where vegetation should be able to reduce erosion and runoff so that infiltration increases. Increased infiltration makes groundwater reserves increase, and during the winter season, groundwater can fill river water. Based on the current Subwatershed of the DLM problem, the right solution is needed to restore the watershed area and keep it sustainable so that sustainability is maintained. Efforts are being made to optimize land cover in the upstream and middle areas of the Subwatershed of DLM.

This research has two objectives. First, calculate the erosion value based on land units and calculate the peak discharge. Second, the rational method will optimize land cover by predicting erosion using the Modified Universal Soil Loss Equation (MUSLE) and peak discharge. Next, a multi-criteria analysis (SMCA) spatial analysis was carried out to optimize land cover.

2 Research Significance

Changes in land cover Subwatershed of DLM is a current problem. The high rate of population growth is an indicator of land cover change. The high population encourages people to use the land for agriculture. This agricultural activity triggers illegal logging in this subwatershed, even in protected forest areas. The conversion of mixed garden lands on relatively steep slopes and paddy fields also occurs. This condition results in high erosion and high fluctuations in river discharge, reaching 191.83 m3/second. In the rainy season, the river discharge reaches a discharge of 409.94 m3/second with a high sediment concentration. In the dry season, the river discharge is shallow at 2.45 m3/second. Based on these problems, it is necessary to find an appropriate land cover model to reduce erosion and flood discharge.

3 Materials and Method

3.1 Research Location

The Subwatershed of DLM is administratively located in Padang City, West Sumatra Province. More details can be seen in Fig. 1.

Geographically, the Subwatershed of DLM is bordered to the north by the Subwatershed of Kuranji, the south by the Batang Arau watershed, and the east by the Indragiri watershed. Astronomically is located between 100^{0}26'24.74" E - 100^{0}33'10.96" E and 0^{0}51'23.93" S - 0^{0}55'17.79" S, has a maximum height of 1,865.75 meters above sea level.

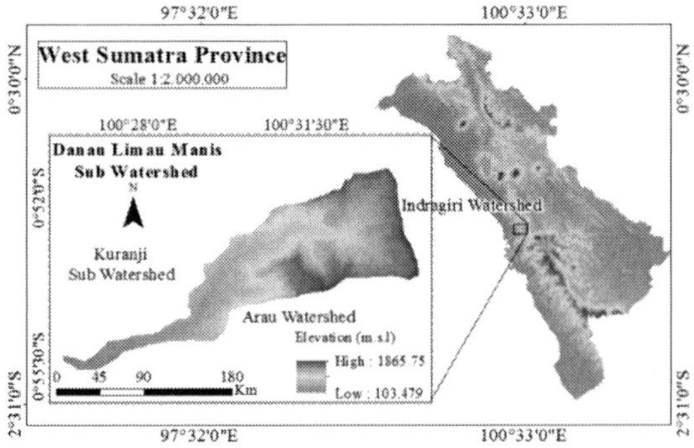

Fig. 1. Research location

3.2 Materials

The MUSLE model uses the highest daily rainfall in the last ten years, namely from 2009-2018 at 97.75 mm/day. Determination of peak discharge using the rational method.

The land cover map consists of forests (Fr), mixed gardens (Mg), rice fields (Rf), shrubs (Sc), and settlements (Sl). The land cover map is sourced from SAS Planet satellite imagery analysis released by Arc-GIS Imagery in 2018. This map uses a spatial resolution of 20x20 meters.

Soil sampling for the MUSLE model analysis used the land unit approach. The land unit map consists of a land cover map, a slope map, and a soil type map.
The scale of the soil map is 1:250,000 consisting of Entisols (Ent) and Inceptisols (Ept). The slope map was analyzed using a DEM map with a resolution of 8x8 meters. It consists of gentle (B), slightly steep (C), steep (D), and very steep (E). See Fig. 2.

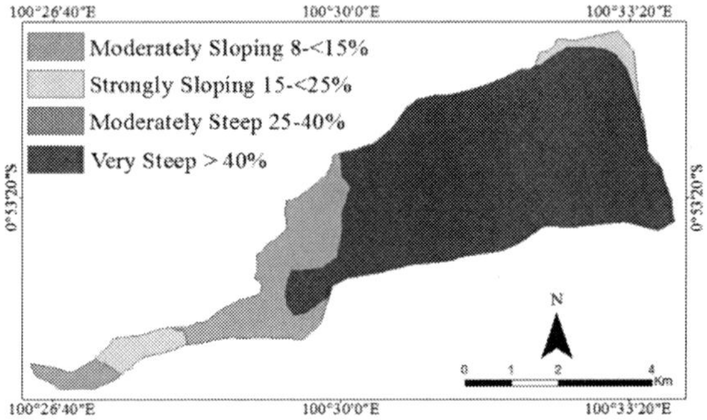

Fig. 2. Slope conditions in the DLM Sub Watershed

3.3 Method

The MUSLE equation with the formula:

$$Sed' = 11,8(Q_{surf} \times q_{peak})^{0,56} \times K \times LS \times C \times P \quad (1)$$

https://doi.org/10.1051/e3sconf/202133103009

Table 1. Soil erosion DLM Sub Watershed

No. LU	LU	RM	K	LS	CP	E
4	Ept-B-Sl	49245.56	0.15	4.41	1	10.29
5	Ept-B-Mg	33836.57	0.11	5.55	0.2	1.33
7	Ept-C-Rf	66341.94	0.18	9.45	0.02	0.70
8	Ept-C-Rf	65649.19	0.07	7.57	0.005	0.05
9	Ept-C-Sl	56563.47	0.15	6.52	1	17.81
11	Ept-C-Mg	47828.28	0.18	5.87	0.2	3.26
12	Ept-D-Sc	40838.80	0.13	9.45	0.3	4.82
13	Ept-D-Fr	168572.20	0.07	18.62	0.001	0.07
14	Ept-D-Mg	92315.84	0.12	10.37	0.2	7.17
15	Ept-E-Fr	464755.51	0.06	20.10	0.001	0.17
16	Ept-E-Sc	6060.51	0.20	18.24	0.3	2.15
22	Ept-B-Rf	56159.76	0.14	5.55	0.004	0.06

Lu is a land unit. RM is the erosivity of rain obtained from surface runoff and peak discharge. K is the value of soil erodibility, and LS is the value of the slope's

Length and slope. CP is the value of soil conservation measures. E is the erosion value of tons/hectares/year.

The surface flow equation uses the SCS (Soil Conservation Service) curve number (CN) method with the formula:

$$Qsurf = ((Rday - 0.2S)\char`^2)/((Rday + 0.8S)) \quad (2a)$$

$$S = 25.4\ 100/CN - 10 \quad (2b)$$

Calculating the peak discharge using the rational method with the formula:

$$q\ peak = 0{,}277\ C.I.A \quad (3a)$$

$$I = R/25 \times (24/Tc)^{2/3} \quad (3b)$$

$$Tc = (0{,}869 \times L^3)^{0{,}385}/H \quad (3c)$$

The optimization scenario model uses the SMCA model. There are five scenarios of land cover change in this study, where scenario 0 is the existing state. Scenarios 1-4 are based on the SMCA model using an overlay analysis technique of several criteria: slope maps, land cover maps, soil maps, river flow density maps, forest area maps, erosion values , and the Padang City Spatial Plan 2010-2030. Overley analysis was carried out for each map to determine the scenario model of land cover change. An analysis of the erosion value and morphometric characteristics of Subwatershed of DLM was also carried out. After that, the appropriate land cover can be determined in the area. Based on the data used, the only land cover was changed, while the other criteria remained.

4 Result and Discussion

4.1 Erosion

The subwatershed of DLM consists of 12 land units with an area of 3,158.90 hectares. In this subwatershed, there is one land unit whose erosion value exceeds the erosion tolerance, namely in sample 9 (Ept-C-Sl) of 17.81 tons/hectares/year. The high erosion is because the area is a residential area with a rather steep slope. Land units with low erosion values are found in sample 8 (Ept-C-Frs) of 0.05 tons/hectares/year. Although the RM value and land unit area of sample 8 are larger than sample 9, they do not increase erosion in sample 8. This is because this land unit is a forest area that has smaller K and CP values. Zema [4] explains that forest vegetation can reduce 45% of the total rainfall that falls on watershed areas. More details can be seen in Table 1.

Based on soil conditions, sample 9 has a clay texture with a clay percentage of 76.02%. Soil conditions with a dominant texture of clay coupled with a high rainfall intensity of 97.75 mm/hour impact low infiltration and high runoff. High runoff coupled with steep slope conditions will impact the strength of surface runoff in eroding the soil surface, resulting in significant erosion. Li [5] explained that erosion occurs due to runoff, and erosion increases as the slope conditions get steeper. Liu [6] added that increased rainfall and slope resulted in runoff and erosion increasing rapidly and getting bigger.

The volume weight (BV) value in sample 9 (Ept-C-Sl) is high at 1.29 g/cm3. The high density of the soil is also caused by the very low soil organic matter, which is 1.15%. The density has an impact on the soil pores. The higher the density of the soil, result the lower the pore space of the soil. The low soil pore space will result in small infiltration and impact high runoff and erosion. The overall erosion value in the Subwatershed of DLM is 47.89 tons/hectares/year.

The organic matter with a high percentage is found in forest land units with an average of 12.25%. Zeng [7] added that organic matter can improve soil aggregates so that the soil is not easily eroded. It is also added by Anderson [8] that the level of infiltration is positively correlated with soil organic matter. The more the amount of organic matter in the soil, the infiltration will increase. The high content of soil organic matter can give the shape of irregular aggregate pores, and this makes the infiltration bigger [9].

4.2 Land Cover Change Scenarios

The scenario model for the Subwatershed of DLM uses the SMCA model based on the erosion value of the land unit. The value of erosion on the land unit is the basis for optimizing land cover, whether it exceeds the erosion tolerance or has a relatively high erosion so that the amount of erosion and runoff can be reduced.

Scenario determination also looks at the status of protected and conservation forest areas or other use areas (APL). The scenario of the land unit being reforested if it is in a protected and conserved forest area. If it is in the APL area, it is directed to be a mixed garden and settlement.

The slope of the land is also a reference in determining residential areas. Cultivation areas located on slopes above 15% are used as mixed gardens and below 15% can be used as residential areas. Consideration of the designation of residential areas is also based on the Regional Regulation of the City of Padang No. 10 of 2011, which states the development of Padang City is directed to the east. Van Haaren [10] explained that GIS-based SCMA could assess large geographic areas and display the results on a map. Adiat [11] adds that the GIS-based spatial model in analyzing decision-making can be done very accurately and can be trusted if the criteria used are coherent. Vogdrup-Schmidt [12] explained that the model multi-criteria spatial in showing the most suitable area in analysis decision-making. For more details, the scenario for the Subwatershed of DLM is described below.

Subwatershed of DLM has an area of 3,158.90 ha consisting of forest land cover of 91.25%, mixed gardens of 3.92%, shrubs 0.60%, rice fields 2.5%, and settlements of 1.74%. With the condition of the existing land cover, erosion of 47.89 tons/hectares/year. The high erosion in this area is caused by high rainfall, conversion of forest land in the upper watershed, residential areas, and the weak application of soil and water conservation techniques. This land cover condition also produces high surface runoff and peak discharge of 1,456,209.33 m3 and 152.81 m3/sec.

Scenario 1 explains that there are shrubs in the Subwatershed of DLM due to illegal logging in conservation and protected areas with 102.89 hectares. This disturbed area produces high erosion of 14.14 tons/hectares/year. The forest area at this location should produce an average erosion of 0.10 tons/hectares/year; for more details, see Fig. 3.

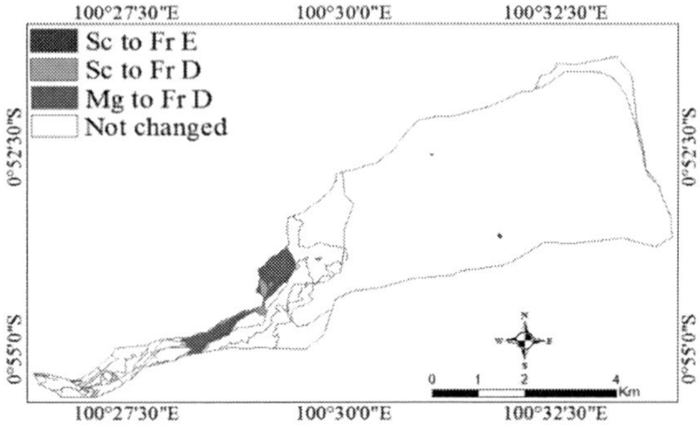

Fig. 3. Scenario 1 land cover change

The location of this area is on a steep slope. This condition causes erosion and runoff that occurs quite large due to the unprotected soil from the kinetic energy of rainfall. So that when it rains, soil particles are easily destroyed by the kinetic energy of rainfall and make it compressed so that it can clog the soil pores. If the soil pores are clogged, then infiltration is reduced, and surface runoff is increased. Impacted by the steep slope conditions, the surface runoff energy increases when transporting eroded soil particles. The mixed garden land cover was also found in protected forest areas of 83.98 hectares and resulted in erosion of 7.17 tons/hectares/year. This location is on a steep slope. If this condition is left unchecked, it will cause the land to be degraded and increase critical land in the upstream sub-watershed. Based on this, this location must be reforested. Tadesse [13] explained that increasing vegetation cover by 88% of the total area could reduce erosion by 4.80 tons/hectares/year.

Scenario 2 explains that in the downstream Subwatershed of DLM, mixed gardens are found on gentle to slightly steep slopes with an area of 39.94 hectares. The resulting erosion was 4.39 tons/hectares/year. This erosion is reduced by increasing the density of mixed gardens because this location is in the middle part of the subwatershed DLM, affecting erosion and surface runoff, increasing river discharge, and leading to flooding in the downstream watershed.

This condition is also hazardous because the Subwatershed has a high river density. When it rains, the surface runoff will quickly enter the river body carrying soil particles. If this condition is left unchecked, it will result in more significant erosion and runoff, leading to landslides, floods, and flash floods. Therefore, the density of mixed gardens needs to be increased. Based on the morphometric characteristics of the Subwatershed of DLM it does not allow land with low vegetation density.

The texture of the soil also influences the high erosion and runoff in the subwatershed. High clay content ranging from 71.17% - 86.23%. High clay content will reduce infiltration into the soil, especially soils with low vegetation density. The high volume of soil weight also influences erosion that occurs. The heavier the volume of the soil, the denser the soil will be and the less pore space the soil will have.

If these conditions are combined (there are shrubs, ex-illegal logging and low mixed garden density, steep slopes, high river density, and high rainfall), this condition will cause erosion, large runoff, and low infiltration. So that when it rains, surface runoff and soil particles will quickly enter the river and quickly increase sediment and river discharge. This condition makes the river discharge erode the river channel, thus potentially producing floods and flash floods.

The subwatershed also has dendritic river branches. River branches spread in all parts of the subwatershed and can produce large river discharges. This subwatershed also has a high river density, steep slopes in the upstream and middle, and rather steep at the outlet of the subwatershed. It has an extensive river gradient where the upstream has an average height of 1,747 m above sea level, and the outlet is 112 m above sea level with a distance of 17.50 km.

Based on these conditions, it can be explained that high river density will make surface runoff quickly enter the river and reduce the opportunity for water to enter the ground. The steep slope of the land further increases the

 https://doi.org/10.1051/e3sconf/202133103009

flow velocity of surface runoff, which can wash away more significant amounts of soil particles into the river. The large river gradient causes the rapid flow of the river downstream of the watershed. The rapid flow of this river has the potential to erode the river channel. This can cause extensive erosion, runoff that can erode and carry large amounts of soil particles and even cause landslides. This condition will result in flooding and flash floods. For details, see Fig. 4.

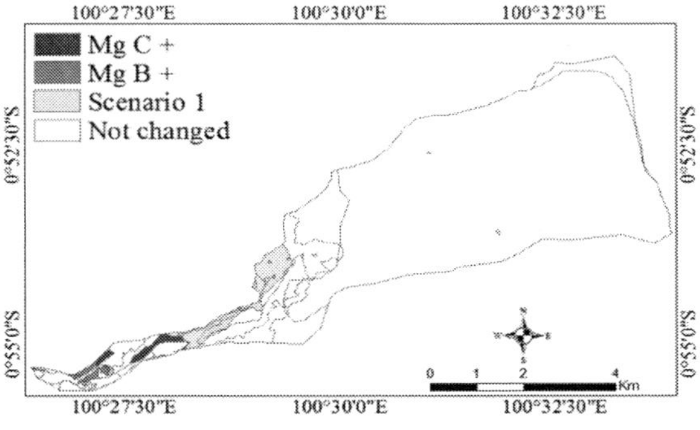

Fig. 4. Scenario 2 land cover change

Scenario 3 emphasizes increasing residential areas. This increase in residential areas needs to be done because the high population growth rate reached 2.83%. Population's need for residential is rising, and the right location is sought for the need. Handavu [14] states that increasing population growth will increase residential areas. The increase in residential areas is limited to the slope factor where the area with a rather steep to the very steep slope is very high, which is 3,084.75 hectares or 97.65%. In comparison, the area of land with a tiny slope is 74.15 hectares or 2.35%. The increase in residential areas can be done on slope B. The small area of land with gentle slopes makes a choice for residential areas very limited. The only option is to turn the mixed garden into a settlement. This condition will reduce the water catchment area in the Subwatershed of DLM.

In addition, the condition of the upstream area with high river density causes reduced infiltration so that the supply of groundwater decreases during the dry season. However, the erosion produced at this location is low, with a land area of 13.99 hectares.

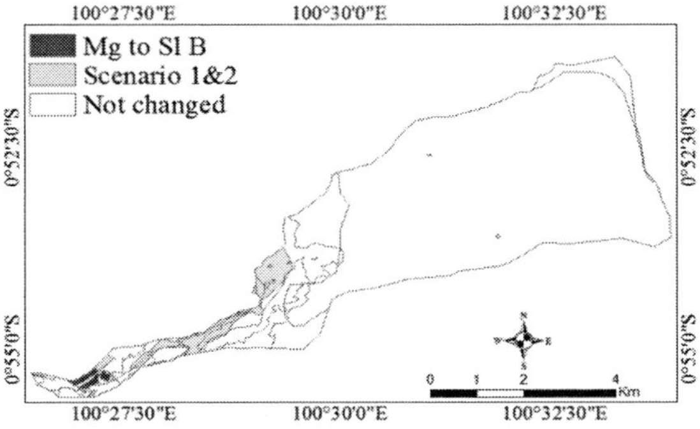

Fig. 5. Scenario 3 land cover change

The improvement of residential areas is carried out by applying soil and water conservation principles, one of which is by making soil pores in open residential areas. Soil pores serve to reduce runoff and increase the supply of water into the soil. For details, see Fig. 5.

In scenario 4, an attempt was made to increase the residential area on a mixed C-slope garden. However, it is not recommended because it is located on a rather steep slope and produces significant erosion of 16.65 tons/hectares/year. For details, see Fig. 6.

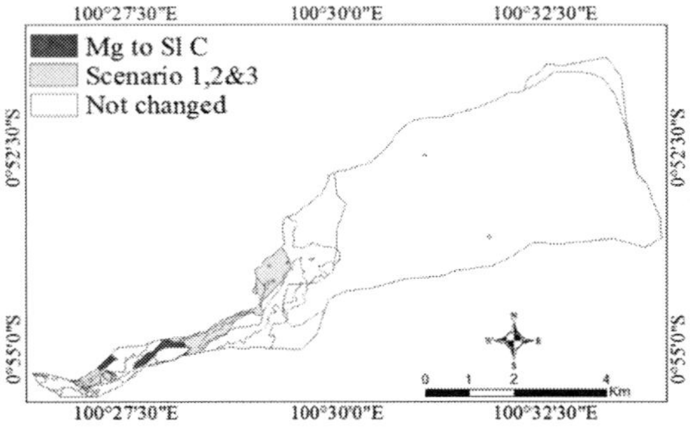

Fig. 6. Scenario 4 land cover change

This condition will have an impact on flooding in the downstream area. For this reason, an optimal scenario is needed so that the Subwatershed of DLM can be utilized optimally without damaging the quality of the environment. The scenario approach used for the subwatershed is based on the land unit's erosion value, which is the basis for optimizing land cover.

4.2.1 Erosion value based on land cover change scenario

Scenario 0 is the erosion value in the existing state. The resulting erosion was 47.89 tons/hectares/year. For more details, see Table 2.

Table 2. Erosion value based on land cover change scenario

No. LU	Lu	Erosion (tons/hectares/year)				
		Scenario 0	Scenario 1	Scenario 2	Scenario 3	Scenario 4
4	Ept-B-Sl	10.29	9.83	9.63	9.70	9.77
5	Ept-B-Mg	1.33	1.27	0.62	6.74	6.79
7	Ept-C-Rf	0.70	0.67	0.66	0.66	0.67
8	Ept-C-Rf	0.05	0.05	0.05	0.05	0.05
9	Ept-C-Sl	17.81	17.01	16.67	16.79	16.92
11	Ept-C-Mg	3.26	3.12	1.53	1.54	16.65
12	Ept-D-Sc	4.82	0.06	0.06	0.06	0.06
13	Ept-D-Fr	0.07	0.07	0.07	0.07	0.07
14	Ept-D-Mg	7.17	0.13	0.13	0.13	0.13
15	Ept-E-Fr	0.17	0.17	0.16	0.16	0.16
16	Ept-E-Sc	2.15	0.02	0.02	0.02	0.03
22	Ept-B-Sl	0.06	0.05	0.05	0.05	0.05
Total		47.89	31.79	29.64	35.98	51.35

In scenario 1, the land cover changes from shrubs to a forest in conservation and protected areas in samples 12 and 16. The original erosion from 4.82 and 2.15 tons/hectares/year changed to 0.06 and 0.02 tons/year. Hectares/year. Mixed gardens in a protected area of 83.98 hectares were also converted into the forest. The resulting erosion from 7.17 tons/hectares/year decreased to 0.13 tons/hectares/year or decreased by 98.19%. The total erosion produced in scenario 1 was 31.79 tons/hectares/year or decreased by 33.62% from scenario 0.

Scenario 2 is a continuation of scenario 1. However, in scenario 2, additional scenarios are carried out, namely the mixed gardens in sulfur B and C (samples 5 and 11) are increased in density. The resulting erosion was 1.33 and 3.26 tons/hectares/year to 0.62 and 1.53 tons/hectares/year, respectively. In scenario 2, the erosion is 29.64 tons/hectares/year or a decrease of 38.11%.

Scenario 3 is a continuation of scenarios 1 and 2. The addition is to change the sulfur mixture garden B (sample 5) into a settlement. Erosion resulted from 1.33 to 6.74 tons/hectares/year, or an increase of 80.27%. In scenario 3, erosion of 35.98 tons/hectares/year is generated or decreased by 24.87% from scenario 0.

Scenario 4 is also a continuation of the previous scenario. The difference is in sample no 11 by changing the mixed garden slope C into a settlement. The resulting erosion was 3.26 tons/hectares/year to 16.65 tons/hectares/year, an increase of 80.42%. In scenario 4, the erosion is 51.35 tons/hectares/year or an increase of 6.73% from scenario 0.

Scenario 2 produces a lower erosion value than the other scenarios. This scenario can reduce erosion by 18.25 tons/hectares/year or 38.11%. The disadvantage of this scenario is that it does not provide space for an increase in residential areas. The optimal scenario is in scenario 3. Where in this scenario, it provides space to increase the settlement area of 13.99 hectares. This scenario can reduce erosion by 11.91 tons/hectares/year or 24.87% compared to the existing erosion.

The increase in the settlement area on the Subwatershed of DLM can only be done on slope B. If it is increased on slope C, the resulting erosion will be substantial, more than 16.65 tons/hectares/year. From the morphometric conditions, increasing the settlement area on slopes above 15% is not recommended. It is due to the Subwatershed of DLM having a very high river density, large river gradient. This condition causes surface runoff that transports soil particles to enter the river and drastically increases river discharge quickly. This also results in erosion of river channels, resulting in floods, landslides, and even flash floods.

It can be concluded that the increase in a residential area can be done on slope B by changing the mixed garden. This change can be made if the need for residential areas is urgently needed. Because this location is located in the downstream Subwatershed of DLM and the middle part of the Kuranji watershed, the increase in residential areas is combined with the creation of pore holes.

4.2.2 Peak discharge

Calculation of peak discharge using a rational formula. Where A (km2) is in the area of the subwatershed, I (mm/hour) is the intensity of rain, Tc (hour) is the concentration-time, C is the runoff coefficient, and Qp (m3/sec) is the peak discharge. More details can be seen in Table 3.

Table 3. Peak discharge existing condition

A	I	Tc	C	Qp
31.59	25.79	1.51	0.67	152.81

The peak discharge of the existing land cover is higher than the other scenarios. The total peak discharge in the existing condition is 152.81 m3/sec. The high peak discharge is caused by many forest land cover being converted into shrubs and community plantations in the upper watershed. The mixed garden land cover also causes the high peak discharge, which is not yet optimal density, on a slope above 15%. For more details on changes in peak discharge based on land cover change scenarios, see Table 4.

Table 4. Peak discharge value based on land cover change scenarios

No	Subwatershed of DLM	Qp (m^3/sec)
1	Scenario 0	152.81
2	Scenario 1	135.70
3	Scenario 2	135.70
4	Scenario 3	137.55
5	Scenario 4	139.39

The high peak discharge in existing conditions will cause erosion of river channels and flooding downstream of the watershed. This condition can cause soil damage both physically and chemically. According to Aprisal [15], floods will affect the texture and organic matter of the soil. Flooded land will remove soil organic matter and damage soil structure.

5 Conclusion

The existing land cover resulted in erosion and peak discharge of 47.89 tons/hectares/year and 152.81 m3/sec. After the scenario of land cover change scenario, 1 was able to reduce erosion and peak discharge 33.62% and 11.20%, respectively. Scenario 2 can reduce erosion and peak discharge by 38.11% and 11.19%. Scenario 3 can reduce erosion and peak discharge by 24.87% and 9.99%. Scenario 4 increases erosion by 7.22% and decreases peak discharge by 8.78%.

The optimal scenario is in scenario 3 because it can increase the residential area by 13.99 hectares. This scenario can reduce erosion by 11.91 tons/hectares/year or 24.87% compared to the existing erosion, with a peak discharge of 137.55 m3/sec.

 https://doi.org/10.1051/e3sconf/202133103009

We want to express our sincere gratitude to the University of Andalas for the 2021 Research Grant no. T/17/UN16.17/PT01.03/IS-PTU-KRPGB-Unand/2021.

References

1. R. Permatasari, A. Sabar, D. K. Natakusumah, and H. Samaulah, "Effects Of Watershed Topography And Land Use On Baseflow Hydrology In Upstream Komering South Sumatera, Indonesia," Int. J. GEOMATE, **17**, 59, 28–33 (2019).
2. T. H. A. Putra, B. Istijono, Aprisal, B. Rusman, and T. Ophiyandri, "The Dynamics Of Land Cover Change And Causal Factors In The Kuranji Watershed," Int. J. GEOMATE, **21**, 84, 69–75 (2021).
3. T. H. A. Putra, B. Istijono, Aprisal, B. Rusman, Andriani, A. Hakam, Adrinal, and T. Ophiyandri, "Interrelations of rainfall and morphometric characteristics in generating geological disasters of Kuranji watershed Padang city," in IOP Conference Series: Earth and Environmental Science, **708**, 1 (2021)
4. D. A. Zema, A. E. L. Borja, B .G. Carra, P. Denisi, V. A. Rodrigues, M. Ranzini, F. C. S. Arcova, V. de Cicco, S. M. Zimbone, "Simulating the hydrological response of a small tropical forest watershed (Mata Atlantica, Brazil) by the ANNAGNPS model," Sci. Total Environ., **636**, 737–750 (2018).
5. T. Li, S. Li, C. Liang, B. He, and R. T. Bush, "Erosion Vulnerability Of Sandy Clay Loam Soil In Southwest China: Modeling Soil Detachment Capacity By Flume Simulation," Catena, **178**, January, 90–99 (2019).
6. Y. Liu, Y. Xin, Y. Xie, and W. Wang, "Effects Of Slope And Rainfall Intensity On Runoff And Soil Erosion From Furrow Diking Under Simulated Rainfall," Catena, **177**, 19, 92–100 (2019).
7. Q. Zeng, F. Darboux, C. Man, Z. Zhu, and S. An, "Soil aggregate stability under different rain conditions for three vegetation types on the Loess Plateau (China)," Catena, 167, May, 276–283 (2018).
8. R. L. Anderson, K. R. Brye, and L. S. Wood, "Landuse And Soil Property Effects On Infiltration Into Alfisols In The Lower Mississippi River Valley, USA," Geoderma Reg., 22, e00297 (2020).
9. Y. Zhao, X. Hu, and X. Li, "Analysis Of The Intra-Aggregate Pore Structures In Three Soil Types Using X-Ray Computed Tomography," Catena, **193**, 19, 104622 (2020).
10. R. Van Haaren and V. Fthenakis, "Gis-Based Wind Farm Site Selection Using Spatial Multi-Criteria Analysis (SMCA): Evaluating The Case For New York State," Renew. Sustain. Energy Rev., **15**, 7, 3332–3340 (2011).
11. K. A. N. Adiat, M. N. M. Nawawi, and K. Abdullah, "Assessing The Accuracy Of Gis-Based Elementary Multi-Criteria Decision Analysis As A Spatial Prediction Tool - A Case Of Predicting Potential Zones Of Sustainable Groundwater Resources," J. Hydrol., **440–441**, 75–89 (2012).
12. M. V. Schmidt, S. B. Olsen, A. Dubgaard, I. T. Kristensen, L. B. Jorgensen, B. Normander, C. Ege, T. Dalgaard, "Using Spatial Multi-Criteria Decision Analysis To Develop New And Sustainable Directions For The Future Use Of Agricultural Land In Denmark," Ecol. Indic., **103**, March, 34–42 (2019).
13. L. Tadesse, K. V. Suryabhagavan, G. Sridhar, and G. Legesse, "Land use and land cover changes and Soil erosion in Yezat Watershed, North Western Ethiopia," Int. Soil Water Conserv. Res., **5**, 2, 85–94 (2017).
14. F. Handavu, P. W. C. Chirwa, and S. Syampungani, "Socio-Economic Factors Influencing Land-Use And Land-Cover Changes In The Miombo Woodlands Of The Copperbelt Province In Zambia," For. Policy Econ., **100**, October 2018, 75–94 (2019).
15. Aprisal, B. Istijono, T. Ophiyandri, and Nurhamidah, "The Effect Of Flood To Quality Index Of Soil Physical Properties At The Downstream Of Kuranji River Watershed, Padang City," Int. J. GEOMATE, **16**, 54, 74–80 (2019)

Physical changes of low quality of clay bricks due to sandy-soil mixture under combustion performance

Muhammad Ridwan[1], *Ade* Indra[3], *Annisa* Febiana Yunera[1], and *Rudy* Kurniawan[2*]

[1]Department of Civil Engineering, Faculty of Engineering, Padang Institute of Technology, West Sumatera, Indonesia
[2]Department of Civil Engineering, Faculty of Engineering, Andalas University, West Sumatera, Indonesia
[3]Department of Mechanical Engineering, Faculty of Engineering, Padang Institute of Technology, West Sumatera, Indonesia

Abstract. This research studies the physical properties of the low quality of bricks mixed with sandy -soil in home industries located in Gunung Sarik of Padang City. The ratio mixture of the brick making material in the home industries was not measured with the proper measurement; as a result the quality of the brick was poor. The research used 5 x 5 x 5 cm and 180 cubes of bricks to see the physical changes. The ratio of mixed materials between sandy soil and clay are 0.11.025, 0.43,0.67,1.0, and 1.50. The procedure for making bricks was the same as a factory without changing the factory method. The process of burning bricks for three days and two nights using wood. From the test results, the density of the bricks did not change significantly between direct and indirect fire exposure but decreased in proportion to the addition of sandy soil. The addition of sandy soil reduces the percentage of damage and physical changes. The acceptable ratio of sandy -soil mixture is a range of 0.43.-0.6.

1 Introduction

Limestone West Sumatra is one of the province in Indonesia that is prone to earthquakes [1]. However, more than 80% of either residence or office buildings used bricks which are produced do not meet the Indonesian Industrial Standard (SNI) of 15-2094-2000.

Most of brick manufacturers in West Sumatra are home industries which are operated traditionally. They use sandy sand soil as the primary material to be mixed with clay. However, they do not have any standard amount of each material for the mixture. In other word, the portion of the material in each brick could be varied.

Some manufacturers use wood in the burning process of the bricks. Using wood in the process will allow more heat absorption in order to produce a good quality of the bricks. The ideal heat temperature in this process is in range of 800-1000 ^{O}C. However, they fail to have that range of temperature although having put the bricks close to the wood fire. [4].

Bricks are a mixture of sand (silica) and clay (alumina) mixed with a predetermined composition and added with water to become plasticity. In the making process the bricks should be easily printed, dried without shrinkage and produced without any cracks or curves. Investigating the physical changes in the making process of the bricks have been conducted by several researchers [1,2,3].

Producing strong bricks must use a good quality of clay which is added with other materials [5]. The location where the clay is excavated determines the quality of the clay itself [6]. Having a proper heat and fuel in the combustion also contributes greatly to the strength of the bricks. The properties of bricks depend on the nature of each brick unit, the quality of the raw materials used, and the technology in the manufacturing process. Therefore, it is essential to obtain information about the bricks' physical, chemical, and mechanical feasibility, the raw materials' characteristics, and the manufacturing process. Brick making initially often used a mixture of 30% sand and 70% clay [6]

This study examined the bricks produced by manufacturers in Gunung Sarik and find out how the changes of bricks that had been studied previously with materials and positions in different combustion processes. The similar making process was used in the production of the bricks. However, the material and combustion position were different. For this study, composition of sand and clay were at ratios of 0.11.025, 0.43,0.67,1.0, and 1.50. The burning process used firewood and was carried out with the brick position close to the fire and far from the fire. The distance of the fire was around ± 20 cm. It was expected that it could improve the quality and quantity which met the SNI 15-2094-2000. Then it could be a guideline for improving the method of making bricks in the home industries of West Sumatra

2 Clay Brick

* Corresponding author: ruddykurniawan@eng.unand.ac.id

According to SNI 15-2094-2000, a brick is a building or construction material made of a mixture of clay and other materials heated in high temperature.

According to the Indonesian Industrial Standard (SII-0021-1978), the compressive strength of clay bricks is divided into six strength classes. Table 1 shows the average compressive strength of bricks.

Table 1. The average compressive strength of bricks (*SII-0021-78)*

Class	fc'		Permit coefficient of variation
	Kg/cm²	N/mm²	
25	25	2,5	25%
50	50	5	22%
100	100	10	22%
150	150	15	15%
200	200	20	15%
250	250	25	15%

According to SNI 15-2094-2000, in term of its compressive strength, the quality of red clay bricks can be classified into three levels as presented in Table 2.

Table 2. The average compressive strength of bricks (SNI 15-2094-2000*)*

Quality	Class	Fc'		Permit coefficient of variation
		Kg/cm²	N/mm²	
Grade I	50	50	5	22%
Grade II	100	100	10	15%
Grade III	150	150	15	15%

Clay is the soil that has cohesion and plasticity. Carbonic acid and activity generated from geothermal energy will form clay during the weathering process of silica rocks. When mixed with water, the clay will produce plastic properties from the mineral particles contained in the clay [7]. On the other hand, clay production involving chemical reaction will compose groups of colloid-sized particles with tiny grain diameters of 0.002 mm.

The manufacturing of the bricks undergoes several processes; excavating molding, drying, and high temperature- burning, which result in color changes. [6] The color changes of the burnt clay depends on the substance it contains, such as aluminum, iron, carbon, manganese, and calcium.[7]. The colors produced by iron compounds are yellowish-white, yellow, red, brown, and black. On the other hand, Limonite compounds produce brown, yellowish-white, and yellow. The red color of the clay is due to the content of hematite compounds and the green color is due to the iron silicate compounds contained in the clay.

Meanwhile, manganese compounds produce brown carbon compounds that produce blue, gray, green, or brown colors. Determining the color of bricks during burning is quite challenging [8]. Table 3 shows the approximate color change of the raw clay after the burning

Table 3 Clay color changes after the combustion process [8]

Color of clay	Changes color after burning
Red	Red or brown
Dark yellow	Dark yellow, brown, or red
chocolate	Red or chocolate
White	White or yellowish-white
Gray or black	Red, dark yellow, or white
Green	Red
Red, Yellow, dark gray	Dark gray, early red then cream, dark yellow or greenish-yellow when melting

Fig. 1 shows that the changes in the bricks before and after being burned. Before burning, the bricks have two distinct surfaces close together, while the grains have only one boundary after being burned. The driving force from the combustion is a reduction in the surface area (which means a reduction in surface energy.[10]

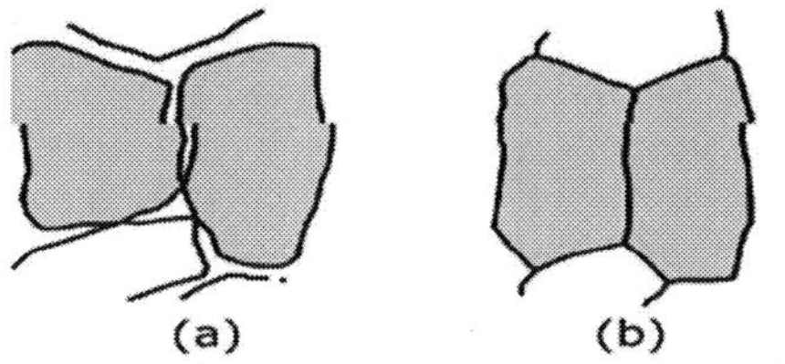

Fig. 1. (a) Bricks before burning and (b) Bricks after burning [10]

Good quality bricks are resulted from the perfect timing and temperature in the combustion. Silica crystals will re-crystallize and melt if they are at the precise temperature. If only using direct firewood without a furnace room, that temperature is impossible to be created. [9]

3 Brick Making Process

The stages in the brick-making process are excavation of soil or raw materials, material processing, brick formation, brick drying, brick burning, and brick cooling. The stages of brick making are as follows [6]

3.1.Raw Material Excavation

Excavation of brick raw materials should be carried out with less plastic soil and added with sand to prevent shrinkage. The excavation is still done manually with a hoe. After digging in the thickness of 40-50 cm on the top layer, the soil is cleaned of leaves, tree roots, plastic. After that, it is dug again to a depth of 1.5-2.5 meters or depending on soil conditions. The excavated soil is collected and stored in a protected place in order to break down the organisms contained in the clay. The longer the clay is stored, the more decayed it gets.

3.2.Raw Material Processing

When processing clay, raw materials must be mixed, which is referred to as pulverizing work. This work is still done manually by trampling the wet clay by either man power or animal power and can also be crushed by hand manually or by a tractor machine.All the mixed ingredients must be evenly mixed with the clay. Before being formed with a mold, this ready-made raw material is rested 2 to 3 days allowing the clay particles to absorb water in order to make them more stable resulting a whole shrinkage.

3.3.Formation of Bricks

After the raw material acquires the plastic properties in 2 to 3 days rested, it is printed with wood or glass molds according to the standard size of SNI 15-2094-1991 or SII-0021-78. So that the clay does not stick during the molding process, the wooden mold must first be soaked with water. The surface of the ground floor of the brick molding must be flat and sprinkled with rice husk ash. The first step in the brick molding process is placing the mold on the ground floor. The clay is ready to be stirred or mixed and it is then placed into the mold by pouring it, and after that, it is pressed by hand so that there are no cavities left in the mold. Next, the mold is flipped and exposed to the sun. Finally the raw bricks are then collected in a protected place to be ventilated.

3.4.Brick Drying

The brick drying process is still done manually using sunlight. The drying process of bricks would be better if done gradually. The raw bricks are coated with plastic to preven the direct exposure of the sunlight. The direct sunlight exposure will cause cracks on the bricks . The 1-day old bricks are then flipped. Furthermore, the dry bricks are arranged crosswise to be exposed to the wind so called fencing. This process only takes two days in a sunny weather and takes more or less a week in a cloudy or rainy weather.

3.5.Burning Bricks

Checking the heating and cooling temperature during the burning process is crucial. During the combustion process, the clay will undergo physical, chemical, and mineralogy changes. The process of burning bricks must be balanced with an increased temperature and a increased rate of temperature during the process. There are several stages; a. Stage of evaporation (drying). At this stage, the release of the forming water occurs up to a temperature of about 120°C. b. Oxidation stage. At this stage, the burning of plant remains (carbon) is contained in the clay. This process takes place at a temperature of 650°C-800°C. c. Full combustion stage. At this stage, the brick is burned until it is cooked and vitrification occurs until it becomes a solid brick. The ripening temperature varies between 920°C-1,020°C, depending on the nature of the clay used. d. containment stage. At this stage, the temperature should be kept stable for 1-2 hours. In stages 1, 2, and 3, the rise of the temperature should be too rapid to prevent the crack in the bricks, the black color, expansion, and etc.

Factors that affect bricks during the firing process are temperature, time, particle size, surface energy, and etc. Structural physical changes are caused by a reduction in the number and size of pores, grain growth, an increase in density, and shrinkage. Clay that has a large surface area and small size will produce great strength to bind both physically and chemically to the surrounding water. Water heated to temperatures above 1000°C will not be easily extracted from the clay.

During the burning process, wood or rice husks can be used as a fuel. To produce good quality of bricks, they should be burned at high temperatures, and it is better to use wood as the fuel since it contains carbon elements. Selecting the proper fuel ensure the quality of the bricks. [10]

4 Mechanical Properties of Bricks

The mechanical properties of bricks are significant for research purposes and are also used to determine the properties and capabilities by testing and analysis.

Porosity (Water absorption)

The strong bricks will have low absorption of water and humidity. The higher the level of the absorption to water and humidity is, the poorer the quality of the bricks is. Bricks are structural elements that have hygroscopic properties, which means they easily absorb water. The acceptable water absorption for clay bricks is less than 20% [11]. The amount of porosity in a material varies from 0% to 90%, depending on the type and quantity of the material. The strength of the object will increase if the porosity contained in the object is minor. On the contrary, if the porosity content in the object is significant, the strength will decrease. To calculate the porosity.[12]

$$Porosity\ (\%) = \frac{(Mb-Mk)}{Vb} \times \frac{1}{\text{pair}} \times 100\% \qquad (1)$$

Mb = Wet mass (g)
Mk = Dry mass (g)
Vb = Volume (cm3)
water = Density of water (g/cm3)

4.1.Burning Shrink

Burning shrink is a change in dimensions or changes in the volume of the material that has been burned. Shrinkage due to changes in microstructure (grain or grain boundaries) indicates that a sintering process or heating of brick has occured.

Bricks made of clay or clay with a good burn value is considered in a number of less than 2.5% in order to prevent bricks with crack or any defect. The equation is used to determine the amount of burn loss [13]

$$Shrinkage\ (\%) = \frac{(I_0 - I_1)}{I_0} \times 100\% \quad (2)$$

Information,
I_0 = Length of test sample before burning (cm)
I_1 = Length of the test sample after burning (cm)

4.2.Specific gravity

Density is defined as mass per unit volume with the following formula.

$$SG\ (\rho) = \frac{Massa\ (M)}{Volume\ (V)} = (gr/cm^3) \quad (3)$$

Information,
M = Normal mass (g)
V = Volume of object (cm3)

4.3.Compressive strength

The compressive strength of the red bricks needs to be measured by using a universal compression testing machine. It will assure the quality of the bricks is based on the acceptable standard. There are factors that affect the compressive strength of red bricks as follows;

- The properties of material used in the mixture
- The proportion of the material used in the mixture
- The mixing method
- The compressing method
- The hardening process
- The burning process

5 Methodology

To get the physical changes of the bricks due to burning using wood, 180 pieces of bricks in size of 50 x 50 x 50 mm were used. Each treatment used 15 pieces for different treatments. The samples were divided into two combustion conditions; the condition of 2-3 bricks from the fire burning position (F), and one brick from the burning (N). In these two conditions, 90 samples were made, each mix using sandy soil and clay in a ratio of 0.11.025, 0.43,0.67,1.0, and 1.50. The burning was performed for three days and two nights by using wood. Sandy soil was taken from Lubuk Alung, and clay was taken from Gunung Sarik. The image below is an example of the material used

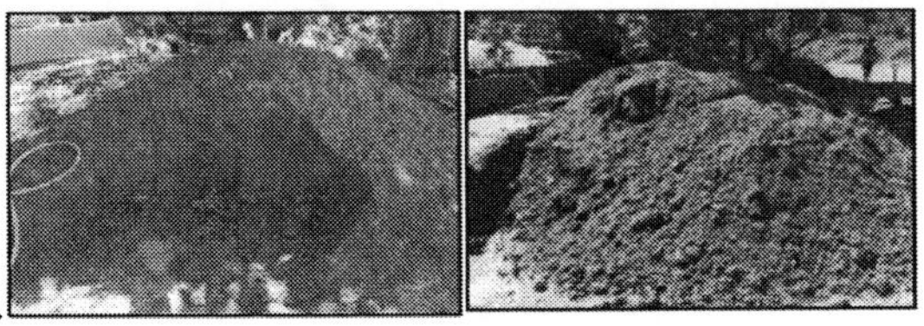

Fig. 2. Gunung Sarik clay and Lubuk Alung sandy-soil

Fig. 3. Sample of 50 x50 x50 mm of brick on factory

Fig. 3 is an example of a test object made in a factory according to the factory's manufacturing method

After the drying process, it underwent the burning process in the firewood for 3 days and 2 nights. The image below is the temperature data during the combustion.

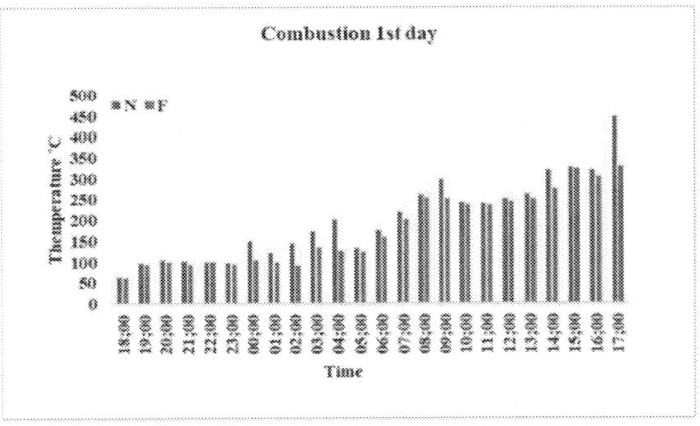

Fig. 4. The first day of combustion

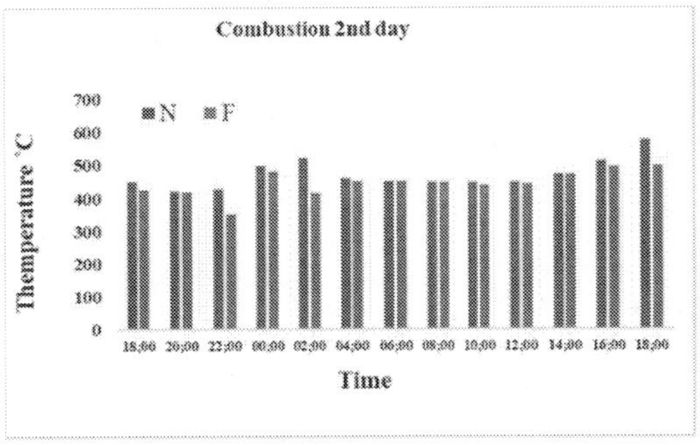

Fig. 5. The second day of combustion

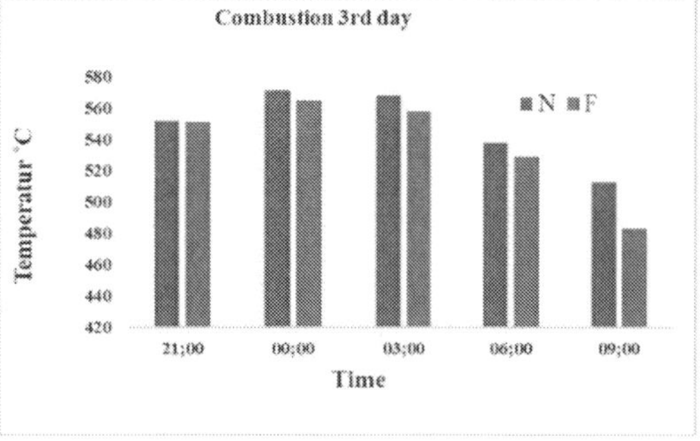

Fig. 6. The third day of combustion.

The temperatures value from low to peak temperatures in conditions near (N) and far from the fire (F) are shown in Figs. 4, 5, and 6. The burning temperature of the brick closest to the fire only reaches around 600 0C. It indicates that the temperature is low as in the temperature of 0- 800 0C the bricks only undergo the drying process not the forming of the bricks themselves.

6 Analysis and Discussion

Fig. 7. shows the relationship between specific gravity and compressive strength, including the ratio of the sandy-soil mixture. From the test results, the compressive strength value decreases as the ratio increases. In other word, there is no significant effect to the SG either the firing of the brick is close to the firewood or the firing of the brick is far from the firewood. The compressive strength will increase as the sandy-soil ratio increases, and the effect of strength is very different for bricks that experience combustion close to the fire. It indicates that the combustion temperature dramatically affects the strength of the bricks.

Fig. 8 shows that the percentage of samples that do not experience changing in shape due to the increase of sandy soil. It indicates that the addition of sandy soil will reduce the amount of clay in one mix. The effect of adding sandy soil is that the number of bricks with a cleft is less and less cleft will occur if the heat is not in maximum rate during the process

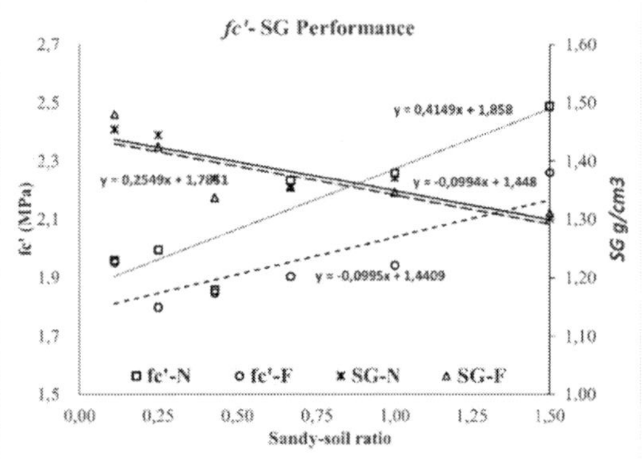

Fig. 7. Compressive strength vs Specific Gravity Performance

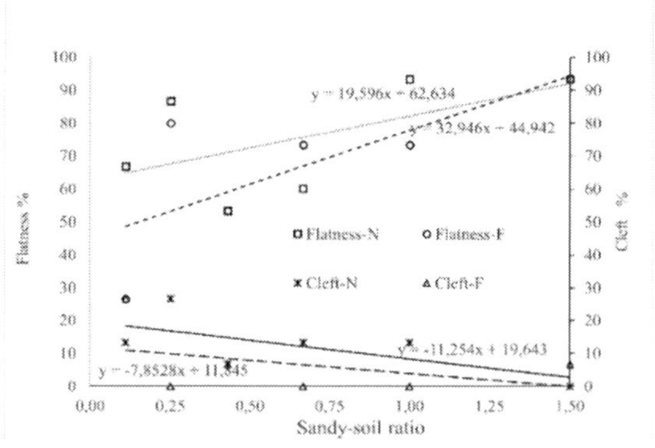

Fig. 8. Flatness and cleft performance of brick

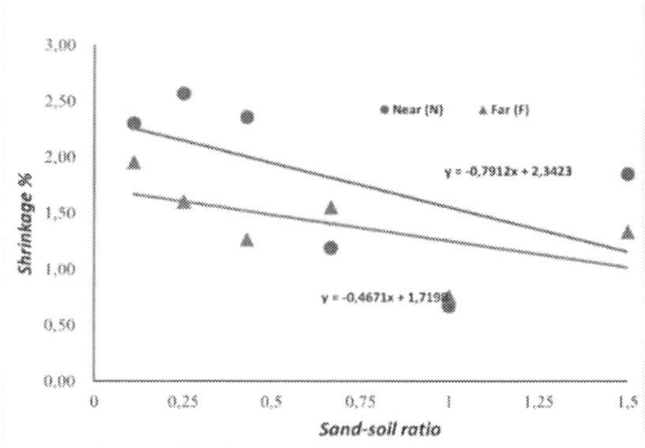

Fig. 9. Shrinkage percentages performance

Fig. 9 shows that the higher the sandy-clay ratio is, the less shrinkage occurs. However, the higher the amount of sandy clay used, the less the function of the clay, so the amount of sandy clay needs to be limited. From this research, the sandy-clay ratio is acceptable around 0.25-0.5. However, this highly depends on the temperature of the manufacturing and combustion processes. The figure 9 also shows that the bricks at that stage only undergo the drying process.

7 Conclusion

Physical changes due to burning using wood will give effect to the quality of the brick. Moreover, burning the bricks closed to firewood or not will also give impact to the brick quality. The addition of sandy soil in the brick mixture will increase the flatness of the bricks and it will reduce the cleft and shrinkage of the bricks.

The authors would like to thank Kemenristekdikti for the financial support through the PDUPT scheme by contract number 002/27.O10.5/PN/VII/2021.

References

1. M. Ridwan, I. Yoshitake, A. Nassif, Proposal of Design Formulae for Equivalent Elasticity of Masonry Structures Made with Bricks of Low

Modulus", Advances in Civil Engineering, **1**-11 (2017)

2. M. Ridwan, I. Yoshitake, A. Nassif, Two-dimensional fictitious truss method for estimation of out-of-plane strength of masonry walls",Construction and Building Materials, **152**, 24-38 (2017)
3. M. Ridwan, R. Kurniawan, Agus, An evaluation of mechanical properties of clay brick for masonry wall in Indonesia, MATEC Web of Conferences, **215**, 01034 (2018)
4. G. Pruthvi Raj, M. Ravula, K. Subramaniam, Failure in Clay Brick Masonry with Soft Brick under Compression: Experimental Investigation and Numerical Simulation, Key Engineering Materials, **747**, 472-479 (2017)
5. G. Kirkelund, L. Skevi, L. Ottosen, Electrodialytically treated MSWI fly ash use in clay bricks, Construction and Building Materials, **254**, 119286, (2020).
6. M. Erna Hastuti, "Effect of Combustion Temperature and Addition of Ash on Bricks Quality", Journal of Neutrino (2012).
7. M. Wilson, A Handbook Of Determinative Methods In Clay Mineralogy. Glasgow: Blackie (1987).
8. 8. M. Maulani, M. Mulyadi And S. Syakri, "Application Design To Identify Compressive Strength Of Bricks Based On Color Based Image Processing Using Statistical Feature Extraction Methods", Jurnal Infomedia,. 1, 1 (2016)
9. 9. I. Nur, "Improving the Quality of Clay as Raw Material for Making Bricks in Bukaka Village, Bone Regency, South Sulawesi", Correct Journal: Applied Technology Journal For Community Engagement And Services, 3 (2), 135-146 (2020)
10. 10. S. Pramono, "Waste as Raw Material for Brick Making", In Proceedings of Semnas Entrepreneurship, 275-294 (2014).
11. 11. A. Chandra, D. Djamas And R. Ramli, "The Effect of Annealing Temperature on Physical Properties of Natural Fiber Waste Bricks", Sainstek: Journal of Science and Technology, 8 (2), 150 (2017).
12. 12. V. Vlack And L. H, Materials Science and Technology. Jakarta: Erlangga (1992).
13. 13. Suwardono, Knowing the Making of Brick, Tile, and Glazed Tile, Cv. Yrama Widya, Bandung (2002)

https://doi.org/10.1051/e3sconf/202133103011

Implementation of earthquake mitigation at the engineering faculty, university of mataram

Teti Zubaidah[1,2*], *Rosmaliati*[1,2], *Cipta* Ramadhani[1,2], and *Dwi* Ratnasari[1,2]

[1]Center of Excellence in Science and Technology on Geomagnetic, University of Mataram, Indonesia
[2]Electrical Engineering Department, University of Mataram, Indonesia

Abstract. After the 2018 Lombok earthquakes, buildings within the Engineering Faculty of the University of Mataram (FT Unram), were severely damaged and there have been no comprehensive treatments. Facilities to support earthquake mitigation, such as signs for evacuation routes and assembly points (temporary evacuation sites), are still minimal, and even if they are available many of them are not designed and placed properly. Some parts of the buildings have been partially renovated, meanwhile, there are new buildings constructions that have not been investigated related to disaster mitigation. Anticipating the re-occurrence of big earthquakes in Lombok, it is very necessary to carry out activities for implementing earthquake mitigation at FT Unram. The activities are divided into three stages, namely *first*: conducting direct reviews for inventory and documentation of all vulnerable points, followed by making of maps and building plans; *second*: determining the proper locations of assembly points, creating labels and signs for evacuation routes and assembly points, and making tutorial videos on safety instructions; *third*: socialization to the policymakers (Faculties and Department's officials), and outreach as well as evacuation drill for all academic communities. The expected output of these activities is increasing of understanding and skills of the academic community in conducting disaster mitigation at their workplaces and application of technology as well as recommendations for disaster risk reduction policies at the faculty and university level.

1 Introduction

Lombok earthquakes were occurred with the mainshock on 5th August 2018, with M 6.9 as reported by USGS [1] and GEOFON [2], and M 6.8 by BMKG as a representation of national authority [3]. Fig. 1 shows the shake map generated with Did You Feel It (DYFI) as a product report of USGS [4]. According to the map, ground shaking reached MMI VIII to IX in the vicinity of the epicenter, which corresponds to severe to violent shakings and moderate to heavy damages. While in Mataram city (the distance around 80 km from epicenter) ground shakings reached MMI VI to VII, which correspond to strong to very strong shakings and light to moderate damages. However, at the University of Mataram heavy damages were also found and causing several buildings to collapse. At the Engineering Faculty of the University of Mataram (FT Unram), light to heavy damages was found, without collapse of any buildings.

Three years after the 2018 Lombok earthquakes, damaged buildings within FT Unram have not been treated comprehensively. Indeed a few months after the quakes, facilities to support earthquake mitigations, such as signs for evacuation routes and assembly (temporary evacuation) points, have been created. However, the available signs are still minimal, and many of them are not designed and placed properly. Some parts of the damaged buildings have been partially renovated, meanwhile, some new building constructions have not been studied related to disaster mitigations.

Anticipating the re-occurrence of big earthquakes in Lombok, it is very necessary to carry out activities for implementing earthquake mitigation at FT Unram. Careful investigations have been started in 2020 and re-investigations are conducted in 2021 in considering some partial renovations and additional buildings.

2 Methods

The methods are divided into 3 (three) stages which are expected to be completed over three years from 2021-2023, with the following stages:

Stage #1 (2021): Conducting a direct review for inventory and documentation of all vulnerable points, followed by making maps/buildings.

Stage #2 (2022): Determining the proper locations of assembly points (temporary evacuation sites), creating

* Corresponding Author: tetizubaidah@unram.ac.id

https://doi.org/10.1051/e3sconf/202133103011

labels and signs for evacuation routes and assembly points, and making tutorial videos on safety instructions.

Stage #3 (2023): Socialization to the policymakers (Faculties and Department's officials), and outreach as well as evacuation drill for all academic communities.

Because of pandemic situations, socialization for policymakers and outreach for all communities is considered possible to be conducted only in the third year.

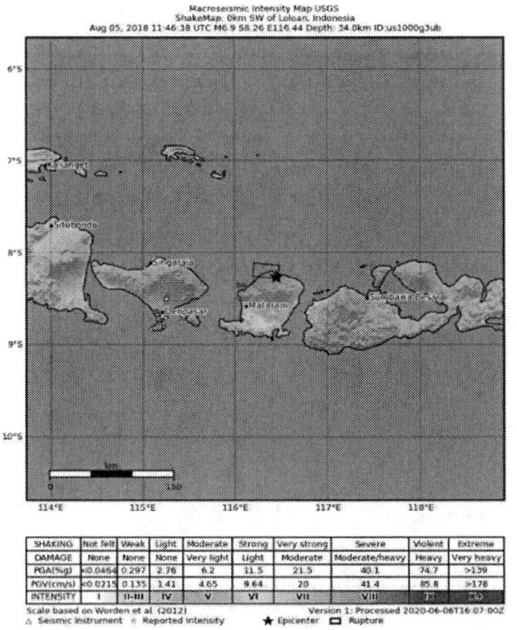

Fig. 1. Shakemap of the Lombok mainshock earthquake on 5th August 2018, generated with DYFI USGS [4].

2.1 Site Plan

The site plan of FT Unram is depicted in Fig. 2, with the positions of investigated buildings are marked as A, B, C1, C2, C3, and D. In all parts of these buildings careful investigations of actual conditions have been carried out to find out all points of vulnerability, that need to be anticipated for earthquake mitigations.

2.2 Investigations on Buildings

Investigations on all buildings have been carried out and continued. Here are descriptions of ongoing investigations up to 30th September 2021.

2.2.1 Building A

This building is the administrative center where the Dean and Vice Deans and faculty staffs have their offices on the 1st floor. There is a library and the Informatics Engineering study program on the 2nd floor, while a common meeting room is placed on the 3rd floor.

Fig. 3 shows the atmosphere on the 1st floor, where a corridor is narrow and blocked with several piles of goods because there is a warehouse at the end of the corridor. In Fig. 4 some staff occupied a narrow place under the stair, due to the lack of available spaces, while the condition of the walls in that room was quite dangerous. Fig. 5 shows the library with several reading shelves (some of them are made of glasses) are placed very closely together. Fig. 6 shows some cracks in the outside walls of the 3rd floor of the building. Since current renovations are only inside the building, these cracks will not be treated, because they can only be observed by seeing carefully through a window in the corner of the 3rd floor.

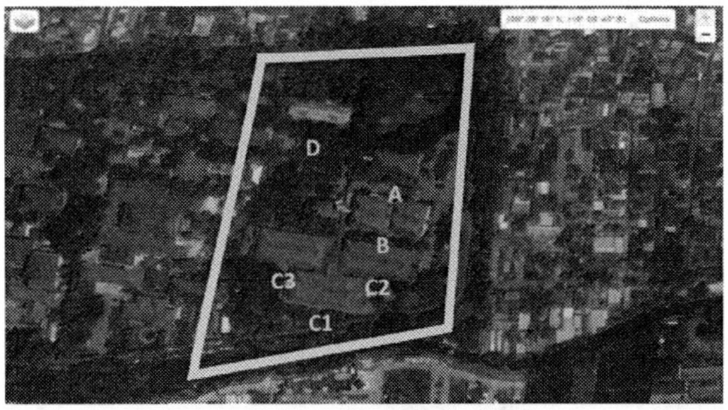

Fig. 2. Site plan of FT Unram with the investigated buildings (A, B, C1, C2, C3, and D), created with earth explorer [5].

Fig. 3. Daily situations of the 1st floor of Building A.

Fig. 4. Staff is working on a narrow place under stairs, and conditions of walls on the 1st floor of Building A.

https://doi.org/10.1051/e3sconf/202133103011

Fig. 5. The library with several reading shelves (some of them are made of glasses) is placed very closely together on the 2nd floor of Building A.

Fig. 6. Cracks in the outside walls of the 3rd floor of Building A.

Fig. 7. For an example of a misplaced and misused evacuation symbol on Building A, here a ladder symbol was used on a straight path.

Fig. 8. Example of a misused symbol on Building A, here the same symbol as evacuation one was used to give information of direction to a study program, which on a panic situation it might be misinterpreted as an evacuation route.

Fig. 9. A large crack between two columns was observed from the 2nd floor of Building B; the crack is initially from the ground floor.

Fig. 10. A large crack in the wall beside the hanging stair, connecting the 2nd and the 3rd floor of Building B.

Fig. 11. A large crack on the ceiling of the electrical laboratory on the 3rd floor of Building B.

https://doi.org/10.1051/e3sconf/202133103011

Fig. 12. Example of a misplaced evacuation symbol on Building B, which is quite impossible to be seen by the people who come out from the classroom.

Fig. 13. An example of a misplaced evacuation symbol on Building B, placed on a glass window which is most probably broken firstly on a large earthquake.

Fig. 14. A misplaced evacuation symbol on Building C1 might be misunderstood there is a staircase behind the door.

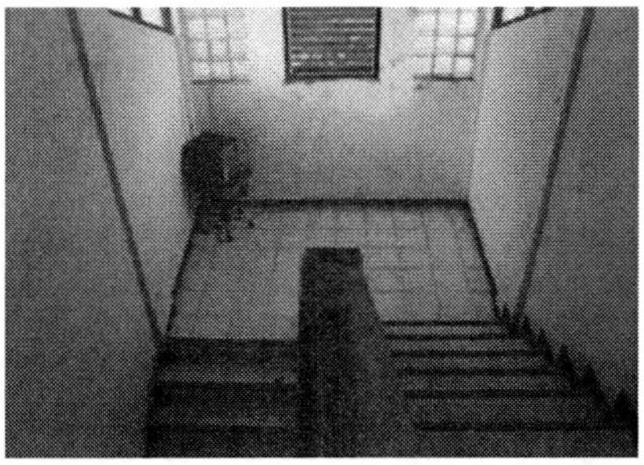

Fig. 15. Inaccessible evacuation route in Building C1, some items are placed in the stair and the next door is locked.

Fig. 16. Site plan of FT Unram with Buildings under investigation (A, B, C1, C2, C3, and D).

Fig. 17. Site plan of FT Unram with Buildings under investigation (A, B, C1, C2, C3, and D).

In this building, there were evacuation signs, but often installations and the symbols used are not correct. Fig. 7 is an example of an evacuation sign with a ladder symbol was placed, even though what is meant is only going straight. Fig. 8 shows a symbol that commonly resembles an evacuation sign, even though what is meant is showing the direction to reach the Informatic Engineering study program. In a panic situation, this sign might be misinterpreted. The placements of these evacuation signs are higher than the standard, which should be between 120 – 160 cm above the level [6].

2.2.2 Building B

This building is the administration room, lecture rooms, and laboratories of Mechanical Engineering and Electrical Engineering study programs. There are quite a large number of damages on this building, including heavy damage on the 1st, 2nd, and 3rd floors. Fig. 9 shows a large crack that can be observed on the 2nd floor. It is a connection between two columns, that are heavily fractured already from the ground floor. Some of the walls including one beside the hanging stair Fig. 10 had large cracks, and the ceiling of the laboratory on the 3rd floor was also broken in Fig. 11.

Many evacuation symbols are installed incorrectly, for example, it placed on the wall in the direction of the classroom door in Fig. 12, which is quite impossible for the people who come out of the room to see the sign. Several symbols are placed on the glass windows or glass doors in Fig. 13, which will most probably be broken firstly on a large earthquake.

2.2.3 Building C

Compared to the other two previous buildings, damage to this building was quite minimal, especially in Building C1. While in Buildings C2 dan C3 there were some minor damages on the 2nd floor; but the exact number of vulnerabilities could not be identified yet. The evacuation symbols are also in good availability in all important places. However, there was a misplaced symbol as depicted in Fig. 14. The symbol with ladder sign placed next to a closed glass door, which might be misunderstood that there is a staircase behind the door by guests; even though what is meant is an evacuation route and the stair is in the next position route (just after a large glass door). If we try to follow the route through the stair, it was inaccessible because the next door was locked and there were some items placed in the middle of the stairs, as depicted in Fig. 15.

2.2.4 Building D

In this building damages are not so visible, because the finished construction before the 2018 Lombok earthquakes was only for the 1st floor; and the rest three floors were only completed in the middle year of 2021. However, the condition of the corridors and stairs is very narrow (Fig. 17), which raises difficulties for people to evacuate. The wooden doors of some classrooms (Fig. 18) are difficult to open, which might cause some other vulnerabilities. Careful inspections on this building construction during the last few months show that there seems to be a slight slope in the concrete columns starting from the 2nd floor. Because of this, we must continue more detailed investigations.

2.3 Evacuation Plans

Further mapping is carried out to obtain the most probable position for the location of the assembly points and temporary evacuation points. The map was generated by using Google Earth Pro [7], which can be used also to identify the length of the curve and the areas of each shape.

3 On-Going Results

3.1 Identification and Mapping of Vulnerable Points

Fig. 18 is an example of a building sketch that shows all vulnerable points and levels of vulnerability in detail. This drawing is equipped with point codes and an accompanying document that describe the vulnerability of each point, also documentation in the form of photos and videos of each vulnerable point.

Results of calculation of the number of vulnerable points and their categories for each building and each floor as shown in Table 1. It appears that Buildings A and B are the most vulnerable and need immediate attention, while Building C is relatively safer. Meanwhile, in Building D, we have not finished collecting data and must continue more detailed investigations.

3.2 Proposal for Evacuation Plans

Assembly points and evacuation routes are proposed based on the Evacuation Plans in Fig. 19. The buildings and surrounding closest areas are considered not to be used for assembly points [8]; they are marked as areas with white colors. Permanent obstacles including a wind power turbine, a large power transformator, and semi-permanent buildings are considered not possible for assembly points; they are marked as areas with red colors. Temporary obstacles including parking areas are also considered not suitable for assembly points; they are marked as areas with blue colors.

The results of the calculation are depicted in Table 2, where we get free spaces of about 15,000 m^2. If it is assumed that only 10% of them can be used as assembly points, then the available spaces are about 1,500 m^2. Considering the capacity of 1000 people (according to the number of students, lecturers, staff, and guests), we will get approximately 1.5 m^2 area per person, which is still suitable for social distancing conditions (if an earthquake occurs while we are still in pandemic conditions). In a normal condition where people only need free space of about 0.3 m^2 according to standard [8], the available spaces are quite enough.

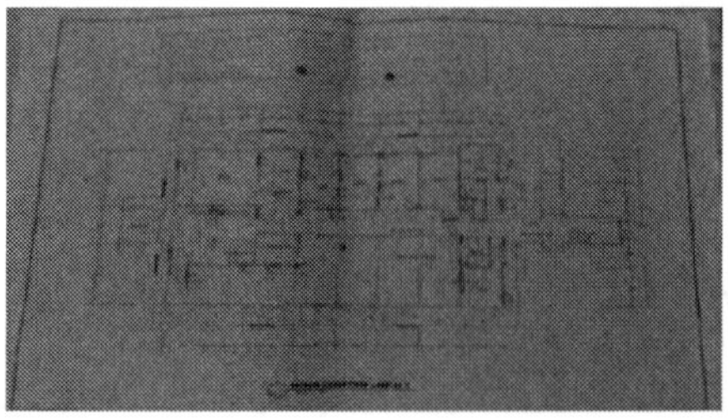

Fig. 18. Sketch of Building A, showing all vulnerable points.

Table 1. Vulnerable points and their categories for Buildings at Engineering Faculty of the University of Mataram.

Build-ing	Flo-or	Light	Medi-um	Heavy	Total
A	1	20	17	2	29
	2	5	9	1	15
	3	4	11	4	19
B	1	18	17	5	40
	2	30	10	4	44
	3	0	5	0	5
C1	1	1	0	0	1
	2	0	0	0	0
C2	1	0	0	0	0
	2	NA	NA	NA	NA
C3	1	0	0	0	0
	2	NA	NA	NA	NA
D	1	NA	NA	NA	NA
	2	NA	NA	NA	NA
	3	NA	NA	NA	NA
	4	NA	NA	NA	NA

Fig. 19. Evacuation plans of FT Unram, showing all available free space areas (green) which can be used as assembly points, while buildings (white) and permanent- (red) and temporary- (blue) obstacles are considered not suitable for assembly points. The map was generated by using Google Earth Pro [7].

Table 2. Proposal for assembly/temporary evacuation points at Engineering Faculty of the University of Mataram.

Considered Space	Area (m^2)
Total area	39,575
Buildings	15,481
Permanent Obstacles	5,116
Temporary Obstacles	3,019
Free spaces	15,959
10% availability	1,595
Capacity for 1,000 persons	1.5

4 Outputs and Final Results

The targeted outputs of activities are increasing of understanding and skills of the academic community in conducting disaster mitigation at their workplaces and application of technology as well as recommendations for disaster risk reduction policies at the faculty and university level. The results of these activities are:

1) Availability of documents and maps of vulnerability points,
2) Availability of earthquake mitigation facilities,
3) Implementation of research products, in the form of studies on vulnerability and risk levels due to earthquakes,
4) Increasing awareness and skills of the FT Unram academic community in mitigating earthquakes, so that will significantly reduce the level of vulnerability and risks,
5) Availability of policy recommendation documents to increase resilience to earthquake re-occurrences in the future.

5 Conclusions

Activities for implementation of earthquake mitigation have been conducted at FT Unram, and at this stage resulted in documents and maps of vulnerability points and proposals for assembly points. Considering the number of vulnerable points, Building B is considered the most vulnerable, while building C is the safest one. The available area of proposed assembly points is wide enough for 1000 people and suitable for social distancing (if an earthquake occurs during pandemic conditions).

These activities and related publications are funded by PNBP DIPA BLU UNRAM under contract number 1895/UN18.L1/PP/2021 entitled *"Implementasi Mitigasi Gempa Bumi di Fakultas Teknik Universitas Mataram"*. We thank to students who actively involved in these activities (Alif Adi Prasetya, I Made Wisnu Kusuma Putra, Ramdani Saputra, Muhammad Abulloh Hamzan, Irma Yunihad, Krisdian Ambarwati, Fitrianti Rukmana, Sekar Sari Bintang Solehah, and Khairunnisak).

References

1. USGS, https://earthquake.usgs.gov/earthquakes/eventpage/us1000g3ub/executive, accessed on 07 October 2021
2. GEOFON, http://geofon.gfz-potsdam.de/eqinfo/event.php?id=gfz2018pfnz, accessed on 07 October 2021
3. BMKG, http://repogempa.bmkg.go.id/repo_new/, accessed on 07 October 2021
4. USGS, https://earthquake.usgs.gov/earthquakes/eventpage/us1000g3ub/shakemap/intensity, accessed on 07 October 2021
5. USGS, https://earthexplorer.usgs.gov/, accessed 07 October 2021
6. Curtin University, *PDG Emergency Management Design Guidelines 000331*,

https://properties.curtin.edu.au/local/docs/guidelines/000331-Emergency-Management-Design-Guidelines-2020-02.pdf, accessed on 07 October 2021 (2020)

7. Google Earth Pro, https://www.google.com/earth/versions/#earth-pro, accessed on 07 October 2021 16:01 WITA
8. PUPR, *Permen PUPR No. 14 tentang Persyaratan Kemudahan Bangunan Gedung* (2017)

Slip surface in narrow backfill behind retaining wall

Abdul Hakam[1*] and *Deni Irda* Mazni[2]

[1]Andalas University, Civil Engineering Department, Padang, Indonesia
[2] Dharma Andalas University, Civil Engineering Department, Padang, Indonesia

Abstract. For a particular area in Geotechnical engineering, a soil slope is defined as a surface of soil mass which is inclined. It the slope is unstable or has insufficient factor of safety, then it needs to be strengthened by a retaining wall or a particular earth reinforcement to ensue slope failure does not occur. It has long been known that the pattern of slip failure is classified into two main types: translation and rotation. Other patterns of slip failure can be approached within the two mentioned types above. The main purpose of this classification is to assist the engineers in the process of the the stability analysis in purpose to obtain the safety factor of the slope and the reinforcement system if any. For the retaining wall reinforcement analysis, the developed method is generally in the form of soil pressure behind the wall. The pressure due to the self weight of the soil is generated by assuming the backfill is long enough, so that the slip failure can be fully described according to the two main types above. Then in cases where the backfill behind the wall is quite narrow, the method should be corrected or modified. These narrow areas are often found on roads that are built on relatively steep slopes. In this paper, the form of the slip failure behind a narrow retaining wall is presented. The results of this study are very useful for developing analytical methods for retaining soils that are built in narrow areas due to location limitations.

1 Introduction

A retaining wall is a man made construction that is built to restraint the soil behind and to avoid landslides. Retaining wall is also an artificial engineered construction as a countermeasure that restrains the moving soil mass. The type of retaining wall construction made must be adapted to the purpose. In addition, the shape and geometry of the retaining wall construction must be adapted to the site conditions, including slope geometry, hydrological conditions and soil conditions. For example, when a retaining wall is constructed adjacent to an exterior basement wall, it is necessary to refer the theoretical and model tests for understanding the failure mechanism and calculating the active earth pressure [1].

The decision to take the values of the acting force or pressure is the initial stage on stability analysis. This is a very important factor on obtaining the stability of the retaining wall in the actual conditions. A study on the collapse of retaining walls has been reported by [2], become good lesson for future stability analysis of retaining walls. The report has described the case description including the failure mechanism of the retaining wall. Based on past experience, it has been showed that the failure of retaining walls is usually caused by improper design of retaining and support systems as reported by [3].

Generally, built retaining walls are in terms of cantilever type made of reinforced concrete. Another type that is very often used is the gravity wall which is made of masonry or concrete. The two types of retaining walls are very commonly used to hold soil, so the general size of these walls has been conventionally mentioned in various literature since past [4,5]. Figure 1 shows a conventional dimension of a cantilever retaining wall.

Other failure cases of retaining walls have also been discussed by various researchers such as [6], where a cantilever retaining wall designed according to the procedures in the civil engineering handbook, failed shortly after construction was completed.

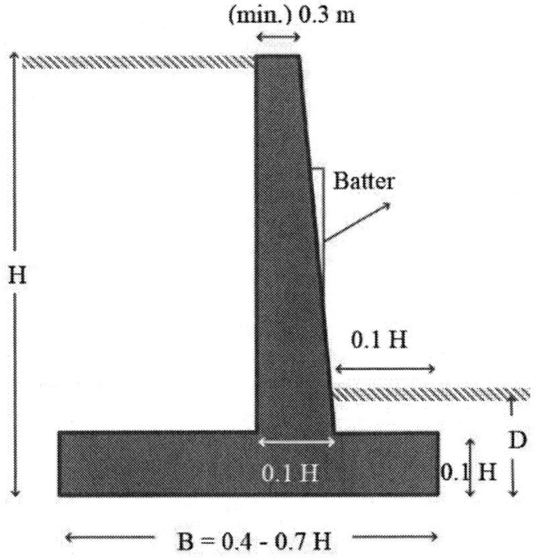

Fig. 1. Conventional dimension of a cantilever wall.

* Corresponding Author : abdulhakam2008@gmail.com

 https://doi.org/10.1051/e3sconf/202133103012

2 Methods

2.1 Lateral active pressure

Lateral earth pressure caused by the soil self weight is differentiated according to its direction and mode of action. The general procedure for estimating the stability of a retaining wall begins with the step of determining the main forces acting behind the wall. This is because the main purpose of the contracting retaining wall is to restraint the soil behind the wall against sliding due to the acting forces.

The main theory for calculating the lateral pressure of soil due to its self weight was originally introduced by [7]. Furthermore [8] proposed a relatively simpler calculation procedure for calculating the lateral forces. Rankine's theory is based on the stress state in the embankment of soil mass that reaches a plastic equilibrium boundary condition where all points in the soil mass are on the threshold of shear failure (Figure 2). This theory assumes that the lateral pressure at an angle that is parallel to the slope of the backfill behind the retaining wall. By using Mohr's circle, it can be shown that the lateral deformation produces boundary conditions for determining stress with the shear strength parameters of the embankment, c and ϕ.

Rankine's theory can be found in almost all soil mechanics reference books for estimating the lateral earth pressure. It is relatively easier to be understood and is often used in the stability analysis of retaining walls. Rankine's theory gives the lateral active stress due to soil weight (γ) increases linearly as a function of depth (z), and can be expressed by the equation:

$$pa = (\gamma . z)\ Ka \tag{1}$$

where the last term of the above equation is known as the active earth pressure coefficient, Ka:

$$Ka = \tan^2 (45 - \phi/2) \tag{2}$$

2.2 Laboratory experiments

In the design of retaining walls, the existing earth pressure theories are usually used where soil pressure calculated based on the coefficient of soil pressure theory. For example, in designing a retaining wall due to seismic forces, it is conventionally done using the Mononobe-Okabe theory. This theory is an adaptation of the dynamic conditions of the classical Coulomb solution for the evaluation of earth pressure under static conditions.

Based on the Mononobe-Okabe theory, the effect of earthquake motion on terms of the lateral pressure acting on the wall. It can be modeled by including in the boundary equilibrium equations of the collapsed soil mass. Furthermore, the inertial forces that exist on the ground due to seismic acceleration are assumed to be constant at one direction according to the direction of the earthquake being considered.

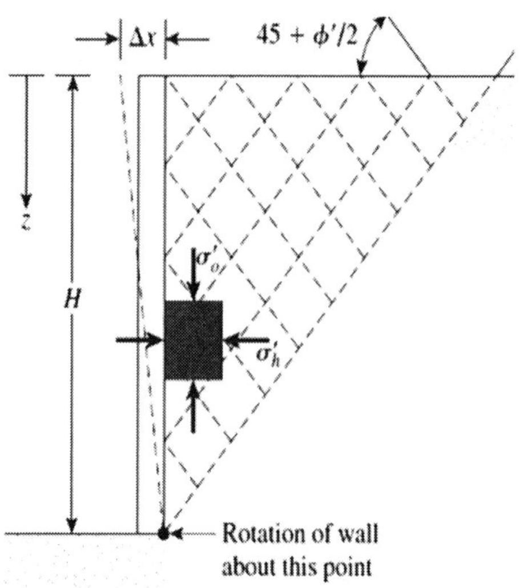

Fig. 2. Rakine's active pressure [5].

The explanation the conventional Coulomb on Mononobe-Okabe theoretical approach provides a linear seismic earth pressure distribution behind the retaining wall has been reported [9]. Furthermore, the pseudo-dynamic method can be used to calculate the seismically active earth pressure distribution in more realistic way. In the analysis of seismic effects, the effect of inertia on the mass of the walls and soil must be considered in the design of retaining walls.

Although the existing theories can also give quite good results, but the need for visual evidence through experimental and later test on the real cases also needs to be done. Thus the visualization resulted from the tests will become the documented evidence. Several tests to study the behavior of retaining wall systems on scaled laboratory models have been published. One example is a test on the vibrating table performs on an L-shaped retaining wall with dry sand fill, has been done by [10,11]. The experimental were carried out in a transparent box with records of acceleration and displacement during the tests. The experimental results are reported with the aim of providing visual and qualitative descriptions so it can be used to increase understanding of the important aspects of the behavior of retaining walls, especially due to dynamic loads. The test results of retaining wall system with horizontal embankments and inclined embankments has been carried out as shown in Fig. 3.

From these results, it can be seen that the failure line on the backfill behind the wall is in the form of a linear plane with a certain angle respect to the horizontal axis. Fig. 3b shows the initial failure pattern where the shows the failure pattern still in the form of single line. For the next vibration, the failure pattern involving another angle by including larger soil masses but the failure plane is still linear.

The results of a similar test have been reported that the failure plane behind the wall due to dynamic loads is equal to or greater than active loading with an angle of > 45+ϕ/2 degrees [12]. If the dynamic load continues to be increased, the failure plane moves slowly from a static state in terms of active stress (following the Rankine

https://doi.org/10.1051/e3sconf/202133103012

landslide plane) to a passive failure state with an angle of = 45-ϕ/2 degrees. In this state a complete sliding of the wall-

wall system has occurred.

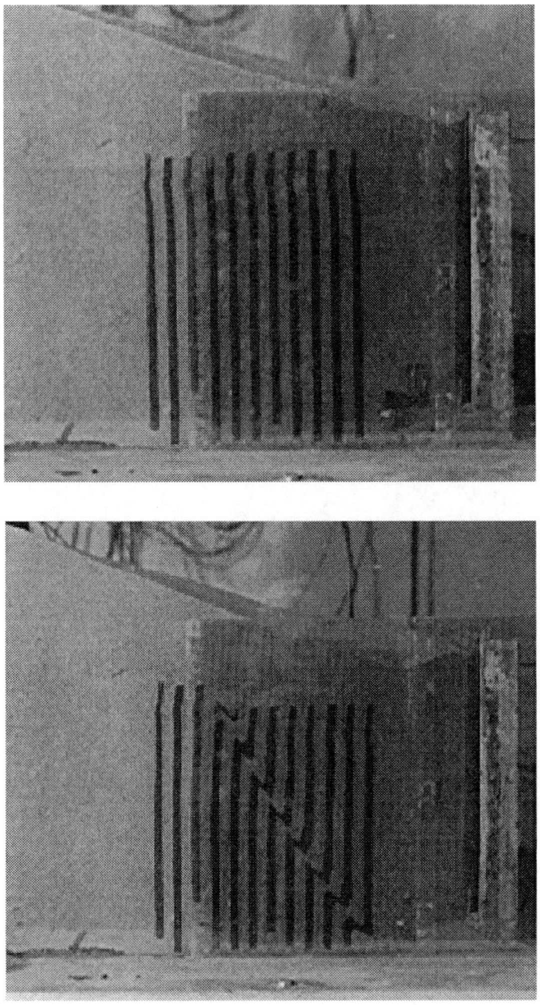

Fig. 3. Slip failure behind the wall [10].

3 Result and Discussion

The reported form of soil mass failure behind retaining walls generally refers to the failure of the backfill which is relatively far from the wall so that it has sufficient length to form a complete failure plane. To see the shape of the collapsed plane with a narrow fill area, an experimental retaining wall model has been carried out with the results as shown in Fig. 4.

Based on the above test results, it can be seen that the collapse occurred in the form of a linear slip failure plane. For long backfill, the collapse failure place is consistent with Rankine's theory. While the form of collapse plane on a narrow backfill, the angle of collapse is getting steeper. The failure plane on a narrow backfill will self adjust to the width of the top of the fill, so that the collapsed soil mass also decreases.

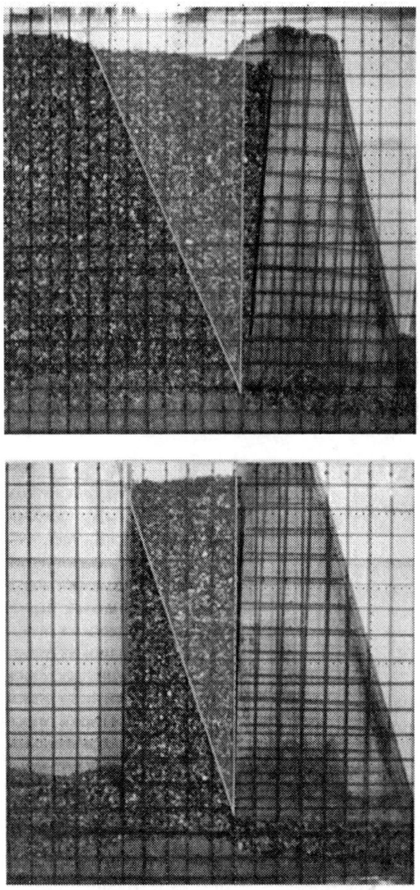

Fig. 4. Slip failure with scaled model

The form of failure plane of the soil mass behind the retaining wall to develop the theory of lateral earth pressure generally refers to the failure of the backfill which is relatively far from the wall so that it has sufficient length to build a perfect failure plane. However, for a relatively narrow backfill, then the collapse pattern will change. This was because the available area to form a Rankine-type collapse was insufficient.

It has also recently been reported that the limited use of classical failure theory for retaining walls with narrow backfill is due to the classical failure theory being developed on the assumption that there is a large space behind the wall [13]. The finite element simulation studies show that if the retaining wall moves in horizontal direction, one to three slip surfaces will be generated from the heel of the wall, and eventually expand to the surface along the wall (Figure 5).

Based on the description in the section above, it can then be derived the dominant forces acting on the retaining wall system. The values of these forces will be determined in the analysis of the stability of the retaining wall system. The values of the acting force must be considered against the dimensions of the moving mass, both under static loading conditions and due to seismic one.

The main thing that should be kept in mind is to pay attention to the generally accepted stability criteria. Some parameters that apply conventionally may can be shifted so that the expected construction planning can be completed with assured safety.

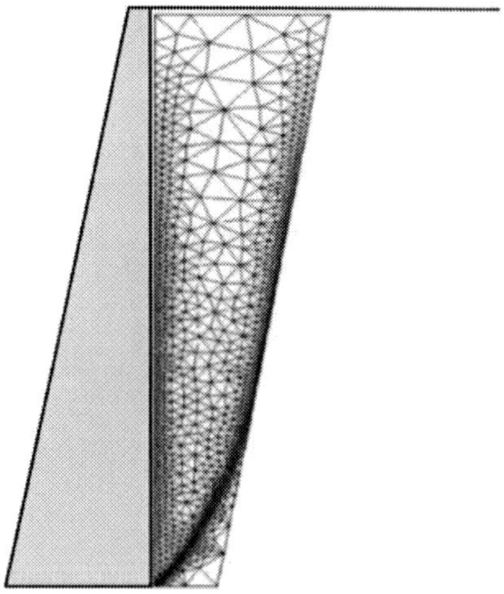

Fig. 5. Slip failure with numerical simulation [13].

4 Conclusion

Civil engineering practice is sometimes faced to the problems with various limitations due to the work location. The limitations of the construction site can be in terms of a small area for construction, very steep topography and sometimes unfavorable soil conditions. For this reason, to get the solution of these special cases, it is necessary to figure out the problem by considering the theories that have been developed, the studies that have been carried out, simulations in the form of numerical models as well as the experimental scaled physical models.

In this paper, a brief description of the formulation of the problem of location limitations resulting in a narrow fill, has been described. This problem arises later on in the design of retaining walls involving horizontal dimensional limitations as well as steep slopes. The solution is to figure out the problem while still considering the design procedure based on the studies that have been done and convinced by visual observations based on the results of laboratory tests.

The authors would like to thank the Department of Civil Engineering, Faculty of Engineering, Andalas University which has partially participated in providing financial support through Scopus Indexed International Publication Grants contract No. 117/UN16.09.D/PL/2021

References

1. H. Weidong, Z. Xinnian, L. Xiaohong, Z. Yongqing, and Z. Xiyu, Active Earth Pressure against Cantilever Retaining Wall Adjacent to Existing Basement Exterior Wall, Int. J. Geomech., **20**, 11, 04020207, (2020)
2. S. Sharma, Teaching Retaining Wall Design with Case Histories, Geo-Frontiers, ASCE, 2877-2886 (2011)
3. A. Lim, Lesson learned from retaining wall failures: A geotechnical disaster, MATEC Web Conf., **229** (2018)
4. J E., Bowles, Foundation Analysis and Design, 5th ed., McGraw Hill Inc, (1997)
5. B.M. Das, and K. Sobhan, Principles of Geotechnical Engineering, 8th ed., Cengage Learning (2014)
6. R.E. Olson, Failure of a Twenty-Foot High Retaining Wall, Proceedings, Missouri University of Science and Technology, Third International Conference on Case Histories in Geotechnical Engineering, **2**, 791-797 (1993)
7. C.A., Coulomb, Essai Sur Une Application DesRegles De Maximis Et Minimis A Quelques Problemes De Statique Relatifs A L'Architecture (Essay On Maximums And Minimums Of Rules To Some Static Problems Relating To Architecture), Mem. Acad. Roy,pres. Divers Savants, 7, Paris, 38. (1776)
8. W.J.M. Rankine, On the Stability of Loose Earth, Philosophical Transactions of the Royal Society of London (1776-1886). 1857-01-01. 147, 9–27 (1857)
9. S. Nimbalkar, A. Pain, S. Mohd Ahmad, S., and Q. Chen, Stability assessment of earth retaining structures under static and seismic conditions, Infrastructures, **4**, 2, (2019)
10. E., Cascone, A.S. Lo Grasso, and M. Maugeri, Dynamic Model Tests On L-Shaped Gravity Retaining Walls, Proc. WCEE 12th, paper no. 2445.(2000)
11. E. Cascone, A.S. Lo Grasso, and M Maugeri, Dynamic Model Tests on Gravity Retaining Walls with Various Surcharge Conditions, International Conferences on Recent Advances in Geotechnical Earthquake Engineering and Soil Dynamics. 9. (2001) https://scholarsmine.mst.edu/icrageesd/04icrageesd/session07/9
12. A. Hakam, Stabilitas Lereng and Dinding Penahan Tanah, Universitas Andalas, Padang (2010).
13. F. Chen, J. Yang, and Y. Lin,Active Earth Pressure of Narrow Granular Backfill against Rigid Retaining Wall Near Rock Face under Translation Mode, Int. J. Geomech., **19**, 12, p. 04019133 (2019)

Local wisdom of coastal communities in mitigating earthquake disaster risk reduction

Putri Febri Wialdi[1*], *Roni Ekha* Putera[1], *Kusdarini*[1], *Ria* Ariany[1], *Tengku Rika* Valentina[2], and *Rifki* Dermawan[3]

[1] Department of Public Administration, Faculty of Social and Political Science, Andalas University
[2] Department of Political Science, Faculty of Social and Political Science, Andalas Universitasy
[3] MSc International Relations, School of Sociology, Politics, And International Studies, Bristol University, The United Kingdom

Abstract. The purpose of this study was to describe how the application of the mitigation of earthquakes for coastal communities by indigenous communities coast of Padang. This study uses interviews and literature. The reduction is one of the measures for reducing disaster risks through physical development as well as awareness and increase the capacity of communities to deal with the threat of disaster. Disaster management can be viewed from different angles, one of them with mitigation based on local wisdom (Local Wisdom) to reduce the risk of earthquakes. Mitigation models based on local knowledge are the study of disaster by using the approach of positivism, where the application can not be separated from social and cultural methods in facing disasters and use natural signs as an early warning system. West Sumatra is one of the provinces which are in an area prone to natural disasters earthquakes. Where several districts/cities in West Sumatra are located in the area will be exposed to the phenomenon of earthquake megathrust, one of Padang. Therefore, it needs improvement in disaster management in the face of disaster.

1 Introduction

The National Agency for Disaster Management (BNPB) states that the trend of natural disasters in Indonesia, in general, has increased in recent years. Located in the Pacific ring of fire, or also known as the Pacific circumference, and flanked by three tectonic plates of the world, namely the Indo Australian, Eurasian, and Pacific plates, Indonesia is a disaster-prone country.[3]

In Indonesia, disaster is no longer considered an extraordinary event because it occurs almost at any time, whether it is a natural disaster such as an earthquake, volcano, flood, hurricane, and forest fire, or man-made disaster such as building fire and collapse. According to the result of the study by BNPB (2014), most of the disasters in Indonesia are caused by environmental damage, for example, flash floods in several regions such as Wasior in Papua and Pangkalan Subdistrict in Lima Puluh Kota District. Due to this high level of vulnerability and susceptibility to disasters, policy on disaster risk reduction is important to minimize property loss and casualties. Based on the data from BNPB (2011), West Sumatra Province is ranked 6th in the list of disaster vulnerabilities because of the geographical conditions of several districts and cities, located in coastal areas directly adjacent to the Indian Ocean. Therefore, 7 of 19 districts/cities have high levels of vulnerability and susceptibility to earthquake and tsunami threats, namely Padang Pariaman, Pasaman, Mentawai Islands, Pesisir Selatan, Pasaman Barat, Pariaman, and Padang. [3].

BNPB considers that the level of preparedness of the community and local government is facing and responding to natural disasters remains low and not optimal. This issue needs to be addressed to reduce disaster risk and minimize the worst possibility. Disaster management in the future needs to be conducted from now on.

Currently, Law No 24 of 2007 concerning Disaster Management is the legal basis for any activities of disaster risk reduction in Indonesia, wherein the regulation states that disaster management must be carried out comprehensively. Referring to this regulation, changes are made to institutions that handle and coordinate with each other regarding disaster management at both national and regional levels.
The province of West Sumatra, tectonically located in the western region of Indonesia, is a highly dynamic region. The subduction/interaction of 2 plates, namely the Indo-Australian plate and Eurasian plate, causes West Sumatra to be prone to disasters such as earthquakes and tsunami.

Currently, there is a heated discussion among the public regarding the megathrust earthquake that is estimated to have a magnitude of 8.9 SR and will hit several areas of West Sumatra, including Padang. Due to its location in a subduction zone, Padang has a source of destructive earthquakes. Furthermore, based on the disaster risk index graph compiled by BNPB from 34 provinces, Padang is a city on the island of Sumatra with a high level of risk of earthquake and tsunami.

Disaster management in West Sumatra Province has been regulated in West Sumatra Provincial Regulation

* Corresponding Author: putriwialdi@gmail.com

Number 5 of 2007 concerning Disaster Management. Meanwhile, disaster management in Padang has been regulated through the Regulation of the Mayor of Padang No. 25 of 2018 on Fixed Procedures for Disaster Management. Referring to the Presidential Regulation No 64 of 2010 concerning Disaster Mitigation in Coastal Areas and Small Islands, coastal areas and small islands have a high risk of earthquake and tsunami, thus requiring the government to carry out disaster management and include mitigation in the management of coastal areas and small islands that contains social, economic, and cultural aspects of the community.

Disaster mitigation must be carried out in a planned and comprehensive manner. Therefore, several approaches are necessary, namely technical, human, administrative, and cultural approaches. Local wisdom is the asset of the community that can be utilized to cope with disasters. Thus, the government as the leading sector in disaster mitigation can learn how to save the community, prevent disasters and reduce the risk or impact of disasters through local wisdom. Local wisdom has long existed and developed before the modern technology-based early warning system. The cultural approach through local wisdom has been proven to successfully help Acehnese people in reading coastal natural phenomena and save thousands of them from the tsunami hitting Simulue Island as well as residents around Mount Kelud in reading the initial signs of the eruption of Mount Kelud through animal behavior [15].

In addition, coastal communities generally have their own belief and tradition, including traditional ceremonies based on coastal cultures that have existed from generation to generation. In Padang, there are several areas with their tradition of warding off misfortune (*tolak bala*), namely Pantai Air Manis in Padang Selatan Subdistrict and Bungus Teluk Kabung Subdistrict. This ritual is carried out annually every Sha'ban month by the local community as a special means of asking for safety from tsunami and earthquake disasters.

By carrying out mitigation based on local wisdom, the government positions the community and community groups as the main actors in carrying out disaster management. According to Wiwik Ratna Kusumasari, community-based disaster management can be realized supposing there are motivators, clear concepts, the appropriate language of popular communication based on local wisdom, and easily accessible information networks. It is also in line with the current study by BNPB that intensively integrates local wisdom in disaster mitigation.

There are three actors or elements involved in the framework of comprehensive disaster management, namely the government, private sector, and the community. Their involvement is an important issue. As stipulated in Law No. 24 of 2007 Article 16 Paragraph (3), the activities of preparedness are the responsibility of the Government and local government and implemented together by the public and private sectors. Meanwhile, the Government Regulation No. 21 of 2008 on the Implementation of Disaster Management mentions that the implementation of disaster management is a series of efforts covering the determination of development policies at risk of disaster, the activities of disaster prevention, emergency response, and rehabilitation.

In the implementation of disaster management in a region, multi-stakeholders are involved, both from the government (from the lowest to the highest level of government, from the village to the center), the legislature, Non-Governmental Organizations (both local and foreign), and private sectors. Institutional institutions formed in the context of disaster risk reduction in Padang are well defined and complete because mostly all institutions engaged in the field of disaster are available in Padang. In this case, BPBDPK as the leading sector should be able to manage existing institutions.

The purpose of this study was to describe the implementation and process of Earth and Tsunami Disaster Mitigation Based on Local Wisdom for the Coastal Community of Padang City.

2 Literature Review and Hypothesis Development

2.1 Disaster Mitigation

Disaster mitigation is defined as an activity or an action to reduce the impact of a disaster as well as the efforts to minimalize casualties and property damage when the disaster occurs. In carrying out disaster mitigation, it is necessary to conduct a disaster risk assessment of the area To calculate the disaster risk of the area, we must know the hazard, the vulnerability, and the capacity of the area. Disaster mitigation is the obligation of various government parties, experts, and communities. In addition, disaster mitigation should be carried out continuously and routinely long before the disaster occurs [7] In addition, mitigation is also defined as an action aimed at reducing the impact of natural or man-made disasters on a nation or community.

Disaster mitigation must be carried out in a planned and comprehensive manner through various efforts and approaches, among others [8]

2.1.1 Technical Approach

Technically, disaster mitigation is carried out to reduce the impact of a disaster, for example by.

1. Creating a solid design of earthquake-resistant buildings.
2. Creating disaster-resistant materials, for example, fire-resistant materials.
3. Creating a technical safety design, for example, a flood embankment.

2.1.2 Human Approach

The human approach is aimed at forming people who understand and are aware of the dangers of disasters. Therefore, human behavior and way of life must be improved and adapted to environmental conditions and potential disasters faced.

2.1.3 Administrative Approach

The government or organizational leadership can take an administrative approach in disaster management, especially at the mitigation stage. For example, by conducting disaster risk analysis in making permits and conducting disaster studies.

2.1.4 Cultural Approach

There is still an opinion among the public that disaster is destiny, thus it must be accepted as it is. It is not entirely true, because with the ability to think and act, humans can attempt to avoid disasters as well as reduce the risks caused. Therefore, we need a cultural approach to raise awareness about disasters. Through a cultural approach, disaster prevention is adjusted to the local wisdom of the community that has been established for a long time.

Disaster risk management and reduction are adapted to the culture and traditions that develop in the community. This approach should be utilized by the government to develop local culture and traditions to build disaster awareness in the community.

In carrying out disaster mitigation, W.N Carter divides mitigation into two concepts, namely nonstructural and structural mitigation. Nonstructural mitigation is conducted as follows [1].

a) Creating a legal framework
b) Establishing institutions
c) Increasing public awareness
d) Conducting training and education
e) Providing incentives
f) Providing warning system

2.2 Local Wisdom

Local wisdom is all forms of knowledge, belief, understanding, or insight as well as customs or ethics that guide human behavior in living in an ecological community [2] Local wisdom that applies in a community has a positive impact on facing and responding to the coming disaster. Local wisdom is an extraction of various experiences of the ancestors who have experienced a disaster, passed down from generation to generation [11]. In addition, local wisdom is believed to have a contribution in terms of community participation that is expected to run sustainably by understanding local knowledge, local technology, local culture, and local traditions that can contribute to disaster mitigation [11].

Studies on disaster and disaster management can be carried out with a positivist approach. The studies on disasters using a positivism approach are inseparable from the theme of socio-culture and community, particularly about to with concerning community responses to disasters (Adaptive Capacity) and the ability to make adjustments and provide feedback in dealing with natural disasters [5].

The empirical experience of the community in a long and continuous period time on various kinds of processes, environmental dynamics with geomorphic processes and their results, and disaster events provide extensive and in-depth experience and knowledge to the community. This experience also plays a role in shaping and providing wisdom to the community in their interactions with the nature and the environment [5].

Local wisdom can play a role and provide a real contribution to overcome risks that may arise due to physical processes of (disasters). Moreover, it is believed to be able to support the sustainability of human life after a disaster [5]

3 Research Methodology

This research applies a qualitative approach. using techniques. Collecting data through observation, and interviews (structured or unstructured), documentation, and visual materials and designing protocols for recording information

4. Results And Discussions

Currently, there is a heated discussion among the public regarding the megathrust earthquake that is estimated to have a magnitude of 8.9 SR and will hit several areas of West Sumatra, including Padang. Due to its location in a subduction zone, Padang has a source of destructive earthquakes, tsunamis, and other natural disasters. It is evident from the disaster risk index graph in Indonesia presented in Fig. 1. Referring to the graph, Padang is a city in the island of Sumatra with a high level of risk for earthquakes and tsunami.

Fig. 1. Disaster Risk Index in Indonesia

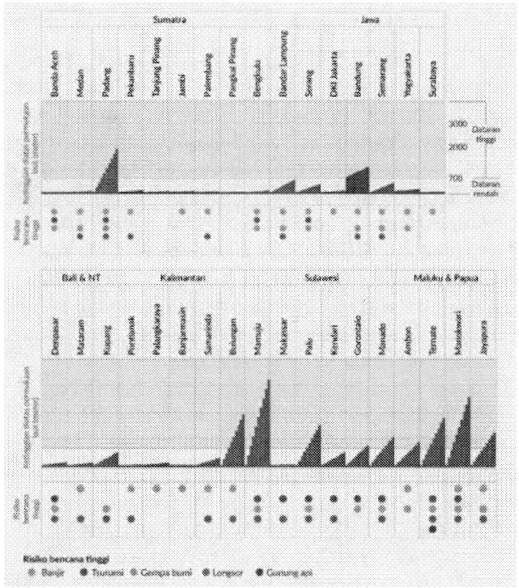

It is by following under in accordance with the statement conveyed by the head of BNPB, Lieutenant General Doni Manardo, in the agenda of the BNPB work visit to West Sumatra. The people of Padang, Mentawai, and several districts/cities located on the West Coast of Sumatra are supposed to be prepared to face the Megathrust phenomenon. It is also in compliance with the data compiled by the Regional Agency for Disaster Management (BPBD) of Padang regarding the population affected by the earthquake and tsunami in districts/cities located in the West Coast region of Sumatra, as presented in table 1.

Table 1. Affected Population in Districts/Cities in the West Coast of Sumatra, 2017

No	District/ Cities	Total Number			
		Subdistricts	Village/ Nagari	Popultion(People)	Population (%)
1	Pesisir Selatan District	15	182	245,916	43.26 %
2	City of Padang	11	104	508,804	58.38 %
3	Padang Pariaman District	17	103	24,861	5.23%
4	City of Pariaman	4	71	25,029	27.94 %
5	Agam District	16	82	20,644	3.81%
6	Pasaman Barat District	11	19	78,782	17.69 %
7	Mentawai District	10	43	17,313	16.40 %

Referring to Table 1, Padang, a city located along the west coastline, has the highest percentage of population among 7 districts or cities on the West Coast of Sumatra that is affected by the earthquake and tsunami. Furthermore, the research conducted by the Indonesian Association of Disaster Experts (IAKI) discovers that Padang is one of the cities prone to earthquakes and tsunamis (Radarmalang.id, 2019). Thus, to reduce disaster risk, the community must take mitigation and preparedness measures, starting with increasing public understanding and improving infrastructure, socialization, early warning system, and guidance to the community.

Disaster management in Padang has been regulated through the Regulation of the Mayor of Padang No. 25 of 2018 on Fixed Procedures for Disaster Management. This policy serves as the guideline for the government in improving disaster management in Padang. However, preparedness and mitigation measures from related parties and the community are also essential.

In compliance with the Presidential Regulation No 64 of 2010 concerning Disaster Mitigation in Coastal Areas and Small Islands, coastal areas and small islands have a high risk of earthquake and tsunami, thus requiring the government to carry out disaster management and include mitigation in the management of coastal areas and small islands that contains social, economic, and cultural aspects of the community [11].

Table . Potential Tsunami Hazard Areas in several sub-districts in Padang City in 2017

NO	Kecamatan	Luas Bahaya (HA)				Bahaya	
		Rendah	Sedang	Tinggi	Total	Indeks	Kelas
1	Bungus Teluk Kabung	109	507	807	1.423	0,944	Tinggi
2	Koto Tangah	271	692	2.346	3.309	0,778	Tinggi
3	Kuranji	298	307	16	621	0,476	Sedang
4	Lubuk Begalung	96	53	62	211	0,300	Rendah
5	Lubuk Kilangan	-	-	42	42	0,596	Sedang
6	Nanggalo	160	608	80	848	0,666	Sedang
7	Padang Barat	20	165	289	475	0,926	Tinggi
8	Padang Selatan	151	114	200	465	0,727	Tinggi
9	Padang Timur	141	405	34	580	0,566	Sedang
10	Padang Utara	44	287	415	747	0,905	Tinggi
11	Pauh	-	-	-	-	-	Rendah
KOTA PADANG		1.289,25	3.138,48	4.292,28	8.720,01	0,688	Tinggi

By improving the system and strengthening earthquake and tsunami disaster mitigation in coastal areas, it is hoped that the government and the community can carry out rescue independently and utilize the main infrastructure as a gathering place or a shelter when the 8.9 SR megathrust earthquake hits and causes a tsunami in Padang and other coastal areas. Padang has two areas located in the coastal area and the red zone of an earthquake that is at risk of tsunami, namely Padang Selatan and Bungus Teluk Kabung Subdistricts.

Disaster mitigation, according to Presidential Regulation No 21 of 2008 concerning the Implementation of Disaster Management, is one of the ways or actions to reduce disaster risk, either through physical development as well as awareness and increased capacity in dealing with disaster threats. Disaster mitigation is divided into two, namely structural and nonstructural mitigation. Nonstructural mitigation is a method or effort to reduce the risk of disaster by empowering and providing good knowledge to the community [10].

Meanwhile, according to W.N Carter, mitigation is part of the pre-disaster stage in the disaster management cycle. In disaster mitigation, there are efforts or actions taken by the government and the community to reduce the impact caused by disasters. Carter divides disaster mitigation into structural and nonstructural mitigation [1] . One example of structural mitigation carried out by the government is constructing buildings used as shelters. Currently, there are several buildings in Padang that are used as shelters, namely government buildings, banks, hotels, and other buildings whose construction is intended as shelters, one of which is located in Parupuk Tabing. However, several areas in Padang are still considered to have a shortage of shelters. It is confirmed by the Head of General Affairs Subsection of BPBD of Padang on initial observations.

"It is true that the number of tsunami shelters in several areas of Padang City is currently still limited. It is due to budget constraints and limited land since building a shelter requires location and in-depth study. Therefore, it is hoped that the community can utilize the natural shelters provided by their surrounding environment, such as in Bungus Teluk Kabung, Pantai Air Manis, and the hilly areas in Mount Pangilun." (Interview with Firdaus, the Head of General Affairs Subsection of BPBD of Padang, on May 6, 2019)

Meanwhile, W.N Carter divides nonstructural mitigation into several activities, one of which is the establishment of institutions or organizational groups that are aware of disasters [1]. Observed from this aspect, Padang currently has several organizational groups formed by the government whose duties and functions serve as an extension of the government in carrying out disaster management.

Based on the Decree of the Chief Executive of BPDB of Padang No 229 of 2017 concerning the Appointment of Volunteers Providing Disaster Information (Roll Call) in Padang, it is determined that the disaster personnel/members/volunteers are ready to assist BPDB of Padang in providing information regarding disaster in each subdistrict and village. BPBD of Padang has appointed Disaster Preparedness Groups comprising 2,080 members in 104 villages in 11 subdistricts in Padang (SK KP BPBD Padang City No 229, 2017).

According to W.N Carter, the next activity of nonstructural mitigation is effectively increasing disaster management and public awareness of disasters since disaster mitigation activities require community participation [1]. Based on the observation, the community living in the coastal areas in Padang Selatan and Bungus Teluk Kabung Subdistricts understand the risk of earthquake and tsunami disasters that may occur in their areas. They have an early warning system derived from local wisdom that has been ingrained among coastal communities, namely reading the natural conditions.

"... we already understand the risk of an earthquake that might have a potential for a tsunami. Therefore, in general, we have carried out government instructions to prepare disaster preparedness bags, considering we are located in the red zone of the tsunami." (Interview with Yelierni, the member of Air Manis Coastal community, on June 15, 2019)

Based on the interview above, it is obvious that the people living in Air Manis Coastal area are aware of the risk of disasters that may occur in their settlements, thus they have carried out the instructions of BPDB. Implementing good disaster mitigation does not only require community awareness, but also the knowledge related to disaster management, for example how to apply mitigation measures developing in the local culture of the community [1]. Therefore, training and education for the community regarding good disaster management are necessary, one of which is disaster mitigation. It is confirmed by the Head of General Affairs Subsection of BPBD of Padang on initial observations.

"Conducting training and education related to disaster mitigation for the community is the most important pillar in disaster mitigation. Therefore, by involving related parties currently, BPBD establishes several activities including socialization related to massive disaster mitigation, particularly to the community living in coastal areas, since, shelter is not the only main issue in reducing the risks of earthquake and tsunami disasters." (Interview with Firdaus, the Head of General Affairs Subsection of BPBD of Padang, on May 6, 2019)

Currently, there is still a lack of education and public understanding of disaster mitigation. Therefore, a method is needed to empower the community. For example, a more ingrained model that provides optimal results in shaping a culture of disaster mitigation in Indonesia, specifically Padang, considering that disaster mitigation has not been utilized as a local culture in the community at large. Disaster mitigation must be carried out in a planned and comprehensive manner. Therefore, several approaches are necessary, namely technical, human, administrative, and cultural approaches [8]. Local wisdom is the asset of the community that can be utilized to cope with disasters.

Thus, the government as the leading sector in disaster mitigation can learn how to save the community, prevent disasters and reduce the risk or impact of disasters through local wisdom. Local wisdom has long existed and developed prior to the modern technology-based early warning system.

The cultural approach through local wisdom has been proven to successfully help Acehnese people in reading coastal natural phenomena and save thousands of them from the tsunami hitting Simulue Island as well as residents around Mount Kelud in reading the initial signs of the eruption of Mount Kelud through animal behavior [15].

In addition, coastal communities generally have their own belief and tradition, including traditional ceremonies based on coastal cultures that have existed from generation to generation. In Padang, there are several areas with their tradition of warding off misfortune (*tolak bala*), namely Pantai Air Manis in Padang Selatan Subdistrict and Bungus Teluk Kabung Subdistrict. This ritual is carried out annually every Sha'ban month by the local community as a special means of asking for safety from the tsunami and earthquake disasters. By following the statement given by the Security and Order and Disaster Management Section in Bungus Teluk Kabung Subdistrict:

"As government officials of Bungus Teluk Kabung Subdistrict, we fully support every event or tradition that has existed for a long time, one of which is the annual tradition of warding off misfortune (*tolak bala*). Particularly in Bungus Teluk Kabung Subdistrict, this tradition is performed to encourage the community to return to the philosophy of *Adat Basandi Syarak, Syarak Basandi Kitabullah*. As community representatives, we believe that Bungus Teluk Kabung can be free from all threatening misfortune such as earthquake and tsunami disasters through this tradition." (Interview with Rustam, the Head of Welfare and Disaster Management Section on 02 July 2019)

In Bungus Teluk Kabung Subdistrict, the tradition of warding off misfortune involves religious scholars such as sheiks, subdistrict government officials, Mayor of Padang Mahyeldi, Al Quran Education Park (TPQ) or Madrasah Diniyah Awaliyah (MDA), mosque youth, and *majelis taklim*. The event begins with a joint parade, followed by dhikr before sunset and evening prayers in the congregation and closed by *al tarekat*.

Similarly, South Padang Subdistrict also has a tradition of warding off misfortune held in Pantai Air Manis Village. The traditional ceremony called Raba'akia is attended by several adat figures, the community, and government leaders such as the Head of the Tourism Office.

In Padang, earthquake and tsunami disaster mitigation by utilizing local wisdom has also been implemented or utilized by BPBD of Padang as the leading sector in carrying out disaster prevention and management to ensure the safety and security of the people of Padang from natural disasters such as earthquakes and tsunamis. It is confirmed by Firdaus, the Head of General Affairs Subsection of BPBD of Padang, as follows.

"The local wisdom of the community is the most important and prioritized method for mitigating natural disasters such as earthquakes, tsunamis, and floods since mitigation based on local wisdom is expected to be able to provide an early warning system to the community before to the disaster, considering the limitations of BPBD of Padang in providing early warnings to the community during disasters. Therefore, mitigation based on local wisdom needs to be highlighted in carrying out disaster mitigation." (Interview with Firdaus, the Head of General Affairs Subsection of BPBD of Padang, on May 6, 2019)

Implementing mitigation based on local wisdom obviously requires an understanding of the earthquake and tsunami disasters, followed by an understanding of methods or steps to be taken when the disasters occur. Therefore, the cooperation of every stakeholder involved is necessary as it is impossible to fully delegate these activities solely to the government or individuals. By following the statement of the Head of General Affairs Subsection of BPBD of Padang Firdaus that:

"To mitigate the earthquake and tsunami disasters, socialization has been carried out to make the public aware of the importance of mitigation and understand what actions to be taken when an earthquake with a potential for tsunami occurs in Padang, both at the community and at the education level. The importance of mitigation based on local wisdom has also been socialized through community groups, such as Disaster Preparedness Group (KSB) both at the village level and at the subdistrict level. In addition, several NGOs, such as Tsunami Preparedness Community (KOGAMI), are also established due to the limited ability of BPBD of Padang to reach the community directly." (Interview with Firdaus, the Head of General Affairs Subsection of BPBD of Padang, on May 6 2019, at 10.30 WIB.)

It is by following the findings of a study conducted by Zulfadrim, Toyoda, and Kanegae entitled "The implementation of Local Wisdom in Reducing Natural Disaster Risk: A Case Study Form West Sumatra". Yang states that the community an urban and rural areas in Minangkabau has developed ways to prevent or reduce disaster risk by well adaptation and preparation. Their ancestors had experienced a recurrent hydro-meteorological disasters and utilized their knowledge to anticipate the disaster. The most interesting feature about the Minangkabau community is their traditional house. Most people know about the traditional houses of West Sumatra with a unique triangular roof. This traditional house is called Rumah Gadang (Big House) because of its function as the main venue for several activities as well as a place for clan meetings and marriages. This house is built on stilts with high space from the ground to protect the community from flood, animal attack, and moisture and frequently to keep their stock of cattle. This traditional house is also able to withstand an earthquake since the column of this building is normally laid in stone foundation rather than inserted into the soil [16].

Referring to the research conducted by Zulfadrim, it is evident that disaster mitigation based on local wisdom has long been applied by the Minang community, observed from how their ancestors built Rumah Gadang equipped with *rangkiang* to store their food and livestock to prepare themselves for disasters.

Furthermore, based on the findings of this research, it is observed that people living in coastal areas such as Pantai Air Manis Village in South Padang Subdistrict and 6 villages in Padang Selatan Subdistrict, in general, have an early warning system by utilizing the megaphone in the mosque (*surau*) in their neighborhood. During long-last and severe disasters, the mosque keeper, also known as "*garin masjid*", will provide instructions for the community to evacuate to the hills in the two areas.

The existence of an early warning system based on local wisdom is considered capable of providing benefits in increasing activities for reducing the risk that will be caused by disasters, as it is known that there are limitations in the ability of the government to prepare and provide an easily accessible and understandable warning system for the community.

However, positioning the mosque keeper as the actor who will provide information related to disasters requires the mosque keeper to have good education and training related to disasters, for example by providing opportunities for them to be involved in disaster activities that have been carried out by the government. However, until 2019, it has not been carried out yet.

By carrying out mitigation based on local wisdom, the government positions the community and community groups as the main actors in carrying out disaster management. According to Wiwik Ratna Kusumasari, community-based disaster management can be realized supposing there are motivators, clear concepts, appropriate language of popular communication based on local wisdom, and easily accessible information networks. In the application of this model, the actors carrying out mitigation activities based on local wisdom are community groups, consisting of traditional leaders such as KAN and *niniak mamak*, religious leaders, and KSB of Subdistricts and Villages as the extension of BPBD.

5. Conclusion

In general, the government has provided support by making policies or regulations related to mitigation as guidelines in carrying out disaster management. However, these policies or regulations have only been studied in general. In addition, there are absolutely no clear rules regarding guidelines for mitigation by utilizing cultural aspects.

To facilitate the implementation of mitigation based on local wisdom, the government has formed community-based institutions to assist in implementing mitigation and increasing the role of the community in disaster mitigation. However, the government does not collaborate with outside organizations such as KAN and WALHI. Mitigation based on local wisdom will be more optimal supposing the government collaborates with these two institutions.

The community in the coastal area has understood the actual measures in disaster mitigation. However, in practice, people living in coastal areas tend to be more aware than people living in hilly areas. People living in coastal areas have carried out government instructions, one of which is preparing Disaster Preparedness Bags. This is inversely proportional to the people living far from the shoreline, feeling that they live in safer areas.
Therefore, it is necessary to increase training and education related to disaster mitigation for the community in this region. However, the education and training do not progress well and are optimal. Until 2019, training and education related to mitigation in Pantai Air Manis Village is carried out by house-to-house visits. Similar activity is carried out in 5 of 6 villages in Bungus Teluk Kabung Subdistrict. It leaves only 1 village, namely Teluk Kabung Tengah, as the only village holding training activities by involving stakeholders in the village.

It is caused by the limited budget provided by the government to the Forum for Disaster Preparedness Groups, both at the subdistrict and village levels. Can be overcome by utilizing the tradition of warding off misfortune as a forum for educating the public regarding disaster mitigation, considering that this tradition is carried annually in both regions and is attended by many people in the area. In addition, the government and FKSB can also take advantage of the existence of KAN and the culture of "returning to *surau*" to apply mitigation methods by utilizing local wisdom to optimize the implementation of mitigation.

6. Limitation And Study Forward

In this study, the researcher limits the mitigation study described or reviewed only from the nonstructural aspects of mitigation according to the field of public administration. This study is expected to provide contribute to the government of Padang to further appreciate the local wisdom of the community that can be used as a method for carrying out earthquake disaster management.

We would like to express our gratitude to the parties who contributed to the completion of this research. The completion of this research is inseparable from the contribution of several government agencies of Padang that have been willing to become the informants in this study. In addition, we would like to thank the informants from various circles of society who were willing to be interviewed as well as the parties involved in providing ideas, criticism, and input to complement this research.

References

1 W. N.. Carter, *Disaster management: A Disaster Manager's Handbook*. Mandaluyong City, Manila: Asian Development Bank (2008)

2 M. Desfandi, The urgency of local wisdom-based disaster education curriculum in Indonesia. Socio Didactics, 1(2). Dec 2014. Journal.uinjkt.ac.id/index.php/sosio-Fitk. (2014)

3 https://bnpb.go.id/

4 B. Kusumasari,. & Q. Alam,. Local wisdom-based disaster recovery model in Indonesia. Disaster Prevention and Management, **21**(3), 359-369 Law No 24 of 2007 concerning Disaster Management (2012)

5 .Marfai. Introduction to environmental ethics and local wisdom. Yogyakarta: Gajah Mada University Press. (2012)

6 N. Putera,, & Sugandi Synergizing stakeholder in reducing risk of earthquake and tsunami-disaster in the most vulnerable area. International Journal of Administrative Science & Organization, **23**(3) (2016)

7 R. E. Putera,. Mitigation of earthquake risk reduction from the perspective of policy implementation. PT RajaGrafindo Persada: Depok. (2018)

8 S. Ramli,. Practical guidelines for disaster management. Jakarta: PT Dian Rakyat. (2010)

9 The Decree of the Mayor of Padang No 122 of 2018 concerning the Forum for the Disaster Preparedness Group for the period 2018-2023.

10 The Presidential Regulation No 21 of 2008 concerning the Implementation of Disaster Management.

11 The Presidential Regulation No 64 of 2010 concerning Disaster Mitigation in Coastal Areas and Small Islands.

12 The Regulation of the Mayor of Padang No. 25 of 2018 on Fixed Procedures for Disaster Management.

13 Wikantiyoso, R. (2010). Disaster mitigation in urban areas; adaptation or anticipation of urban planning and design. The potential of local wisdom in urban planning and design for disaster mitigation efforts, local wisdom. **2**. 1 (2010)

14 www.bnbp.go.id/uploads/24/buku-rbi.pdf

15 L. Zamzami, & Hendrawati. The local cultural wisdom of the maritime community for disaster mitigation efforts in West Sumatra (Research Article, Andalas University, Padang). (2011)

16 Zulfadrim, Y .Toyoda, & H Kanegae. The implementation of local wisdom in reducing natural disaster risk: A case study from West Sumatera. Osaka, Japan. (2018)

https://doi.org/10.1051/e3sconf/202133104002

The preparedness of school community in facing tsunami disaster in Lhokseumawe City

Wheny Utariningsih[1*], *Rizka* Sofia[1], *Iffah* Farhana Abu Talib[2], and *T.* Saifullah[3]

[1]Medical Faculty of Malikussaleh University, Indonesia
[2]U-INSPIRE Malaysia, Malaysia
[3]Law Faculty of Malikussaleh University, Indonesia

Abstract. A tsunami in Aceh in 2004 resulted in many casualties and damage to educational facilities. The disaster taught about the importance of preparedness in anticipating disasters to humans. It will help in shaping and planning what actions need to be taken in the event of a disaster. This study aimed to see the level of preparedness of the school community in Lhokseumawe City in dealing with the tsunami disaster. School community preparedness was measured using preparedness parameters, namely knowledge, and attitudes; policies and guidelines; emergency response planning; disaster warning system; and resource mobilization. This research is descriptive quantitative research employing a survey method using a closed questionnaire. Utilizing the descriptive percentage method, the research shows that the school community's preparedness in Lhokseumawe City is in the almost ready category with an index value of 57.02. The index value is obtained by adding up the index values for each parameter by taking into account its weight. The index value of the knowledge and attitude parameter of the school community; and the level of school community policies and guidelines are in the ready category with index scores of 68,24 and 65. School community emergency planning parameters; the level of the school community disaster warning system and the level of resource mobilization are in the unprepared category; not yet ready and almost ready with index values of 35.6; 40; and 54.15 respectively.

1 Introduction

The location of Aceh Province is at the confluence of two plates (Eurasia and Indo-Australia) which makes Aceh Province a high-risk area for earthquake disasters. This can be seen in the earthquake that was followed by the tsunami in 2004. Based on the Indonesian Disaster Data and Information, the disaster has resulted in 165,791 deaths and 2,830 injuries. 518,450 people had to evacuate from their homes and 179,312 houses were seriously damaged. As many as 240 health facilities and 1,226 educational facilities were also reported to be damaged [1].

The data states that thousands of educational facilities were damaged by the disaster. BAPPENAS [2] added that in the education sector it was estimated that 45,000 students and 1,870 teachers were missing, around 1,962 schools were damaged and destroyed and the loss was estimated at Rp 1041 billion. The number of damaged educational facilities become a special concern for local governments as schools are a gathering place for students which makes them very vulnerable to disasters.

Communities, especially students who are dealing directly with disasters, are obliged to have preparedness in dealing with disasters. This will help in shaping and planning what actions need to be taken when a disaster occurs, especially a tsunami disaster. Preparedness in dealing with disasters is also very important for vulnerable groups, one of which is children. Reflecting on the 2004 tsunami which caused many educational facilities to be damaged, schools must be prepared in dealing with the tsunami disaster. This is because success in handling and evacuating students when a tsunami occurs is very dependent on the preparedness of the school itself.

Lhokseumawe is one of the cities in Aceh Province that has a high risk of the tsunami disaster. Based on the 2016-2020 Aceh Disaster Risk Study [3] in Lhoseumawe City 6,032 people were exposed to the tsunami disaster. The number is spread along the coast in the city of Lhokseumawe. A tsunami disaster is a disaster that takes the most lives and causes damage in a short time so that the paradigm of responsive disaster management must be changed to a preventive activity, namely disaster risk reduction [4]. One way to reduce disaster risk is preparedness. This preparedness will help the community, especially the school community, in planning what actions need to be taken in the event of a disaster.

2 Theoretical Framework

2.1 Tsunami Disaster

Tsunami comes from the Japanese language, namely "tsu" which means harbor, and "nami" which means wave. Both

* Corresponding author: whenyutari@unimal.ac.id

https://doi.org/10.1051/e3sconf/202133104002

words mean "big waves in the harbor". So in the simplest terms, a tsunami is a series of ocean waves that are generally often caused by violent movements on the seabed. Tsunamis are included in the ranks of the most complex and frightening physical phenomena. The tsunami was responsible for the huge loss of life and property destruction. Because of its ferocity, the tsunami had a significant impact on the humanitarian, social, and economic sectors of society [5].

The speed of a tsunami is determined by the depth of the water. In deep water, a tsunami can travel at speeds of 500 to 1000 kilometers per hour while in shallow waters the speed slows down to several tens of kilometers per hour. In addition to speed, the height of the tsunami also depends on the depth of the waters [6]. Some tsunamis, and nearly all tsunamis that travel across the ocean floor with destructive force, are caused by subductions beneath the ocean floor associated with major earthquakes [7]. In Indonesia, on average, earthquakes occur 15 times a day in all parts of Indonesia, causing a tsunami to frequently hit Indonesia. From 1600 until now, there have been 109 tsunamis in Indonesia. Even in the last fifteen years, the tsunami occurred on average once every two years. However, the recurrence of a tsunami in every place takes place over a long period [8].

Sumatra and Java are the most vulnerable islands to tsunamis because they lie directly in front of the Indo-Australian Plate [9]. One of the cities on the island of Sumatra that has a moderate to high risk of a tsunami disaster is Lhokseumawe City. This can be seen in the Tsunami Disaster Risk Map in Lhokseumawe City issued by BNPB.

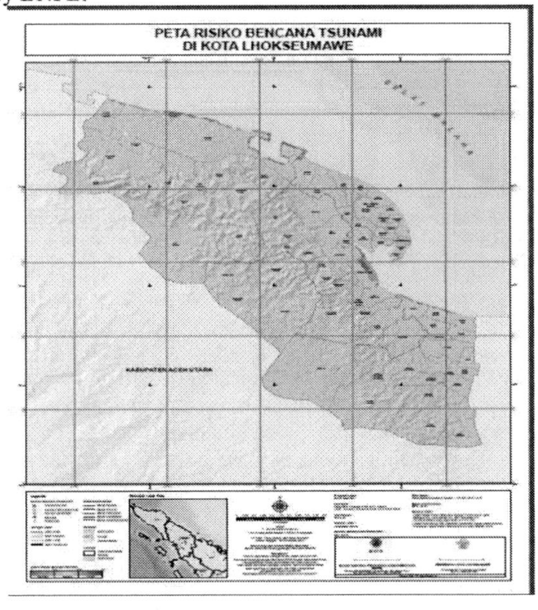

Fig. 1. Tsunami Disaster Risk Map in Lhokseumawe City.

Based on Fig. 1, it can be seen that the city of Lhokseumawe has a moderate to high tsunami risk index. This has become the basis for the Lhokseumawe community, especially the school community located on the coast, to increase preparedness in dealing with the tsunami disaster.

2.2 Preparedness

2.2.1 Definition of Preparedness

Preparedness is an effort made to get a better preparation to deal with existing disaster risks, including understanding and analyzing the relationship between threat risk and vulnerability and then using knowledge to take action to reduce the impact of disasters. Disaster preparedness is one of the dimensions in disaster management that is often associated with actions that enable individuals and communities to respond quickly and effectively to the event of disaster [10]. The change from responsive to preventive paradigm emphasizes that disaster preparedness is one of the important stages to reduce the number of losses that arise due to disasters. Preparedness is one part of disaster risk reduction which is in the period before a disaster occurs or non-disaster situations.

2.2.2 Aspects of Preparedness

Several aspects that require attention in the development of preparedness [11], namely:

1. Planning and organization: directives and policies related to planning for handling emergencies that are effective and continuously updated, as well as an adequate organizational structure for a response.
2. Resources: a complete inventory of all organizational resources, division of tasks and responsibilities
3. Coordination: strengthening coordination between institutions or organizations as well as eliminating friction and increasing cooperation between related institutions or organizations.
4. Readiness: disaster management organizational unit must be fully responsible for monitoring and maintaining preparedness standards for all elements.
5. Training and public awareness: the need for adequate training and public awareness and availability of accurate information.

In the analysis of this study, researchers used standard parameters to measure the preparedness of school communities according to LIPI – UNESCO/ISDR [11], namely: (1) Knowledge and Attitudes towards Disaster Risk; (2) Policies and Guidelines Relating to Preparedness to Anticipate Natural Disasters; (3) Plans for natural disaster emergencies; (4) Disaster warning system; (5) Mobilization of resources.

These five parameters are used to measure the preparedness of the school community. The preparedness study divides disaster preparedness stakeholders into 2, namely the main stakeholders and supporting stakeholders. Stakeholders included in the main stakeholder groups, namely: (1) Individuals and households; (2) Government and (3) School Community, while supporting stakeholders, namely: (1) Community institutions, such as PKK, youth organizations, taklim councils, KAN, mosque/church youth; (2) Non-

https://doi.org/10.1051/e3sconf/202133104002

Government Organizations (3) Professional groups (4) Private parties [11]. This study will focus on the preparedness of the school community which is divided into three namely school preparedness as an institution (S1), teacher preparedness (S2), and student preparedness (S3).

3 Research Method

This research is descriptive quantitative research utilizing the survey method. The survey was conducted using a closed questionnaire and using instructions for filling out. There are three series of questionnaires in this study, namely the S1 (school administrators), S2 (teachers), and S3 (students) questionnaires. The content used in the questionnaire refers to the preparedness study conducted by LIPI – UNESCO/ISDR [11], to assess the level of preparedness of the school community in dealing with disasters. The focus of this research is aimed at five things, namely the level of knowledge and attitudes, policies and guidelines, emergency planning, warning systems, and resource mobilization. The sampling technique used in this study is the purposive sampling method. The criteria used are under the research objectives, namely the distance of the school to the beach or the vulnerability of the school to the tsunami disaster. Based on the needs of the study, State Elementary School 5 Lhokseumawe, State Elementary School 9 Banda Sakti, State Junior High School 5 Lhokseumawe, and State Vocational High School 2 Lhokseumawe was selected.

In this study, preparedness was measured using preparedness index analysis. The index numbers in this study include all indices for each parameter, namely knowledge and attitude (KA); emergency planning (EP), Warning System (WS), and Resource Mobilization Capacity (RMC) at each source. The higher the index number, the higher the level of preparedness of the school community. After knowing the parameter index value of one respondent, be it students, teachers, or schools, the index value of the entire sample can then be determined.

4 Result and Discussion

School community preparedness was measured based on filling out questionnaires from schools represented by principals (S1), teachers (S2), and students (S3). The schools selected based on the distance between the schools and the coastline were SD N 5 Lhokseumawe, SD N 9 Banda Sakti, SMP N 5 Lhokseumawe and SMKN 2 Lhokseumawe.

4.1 Level of Knowledge and Attitude of School Community

4.1.1 Teacher

Teacher knowledge about disasters is important in school preparedness. It is the teacher who will provide or pass on this knowledge to his students. So that students' knowledge about disasters depends on the knowledge of their teachers at school, although this knowledge can also be obtained from other media such as TV, radio, and books. Based on the data obtained from the four schools as respondents, the level of knowledge and attitudes of teachers is included in the ready category, namely at index 67.

Utariningsih mentions that knowledge is the basis for carrying out the right activities in anticipating a disaster [4], 85% of teacher respondents know that earthquakes, tsunamis, floods, landslides, volcanic eruptions, and storms can cause disasters. Respondents also know the cause of the tsunami disaster even though they only know one cause, namely an earthquake under the sea. The questionnaire submitted was also about signs of a tsunami. Almost all respondents know that the signs of an impending tsunami are a strong earthquake that causes people to be unable to stand. The sign of an impending tsunami is not only an earthquake but also seawater that suddenly recedes, large waves on the horizon to a loud bang like an explosion. Only 20% of respondents knew all the signs of the tsunami.

Based on the data obtained, the knowledge about the tsunami disaster that was owned by the teachers was conveyed by 85% of the teachers. There are 20% of respondents who answered that they never talked about or were taught about the tsunami disaster. This can be caused by the lack of knowledge of teachers due to differences in teachers' educational backgrounds.

4.1.2 Student

Knowledge is the initial foundation of preparedness. A person will have an attitude of preparedness if they have good knowledge of the definition, types, signs, and ways to save themselves when a tsunami disaster occurs. Based on the data obtained, students' knowledge and attitudes have an index of 70.1 which is in the ready category. Knowledge about the tsunami disaster is the basis for awareness to carry out emergency planning, resource mobilization, and early warning [4]. Knowledge is also the basis for carrying out the right activities in anticipating a disaster. The basic question used to measure the parameters of knowledge and attitudes is the type of natural event that causes a disaster. 93% of the 84 respondents know that earthquakes, tsunamis, floods, landslides, volcanic eruptions, and storms can cause disasters. This shows that the respondents already know the basic concept of a disaster, namely hazard.

The data also shows that students' knowledge of tsunamis is already high. This is shown by 99% of respondents knowing that one of the causes of the tsunami disaster is an earthquake. This knowledge will be the capital for students in determining steps in the event of a tsunami disaster. This is because an earthquake is a natural warning for a tsunami disaster. The earthquakes, especially those that occur on the seabed, can cause a tsunami [7].

Knowledge about the causes of the tsunami disaster, which is an earthquake, is obtained from personal experience as well as the experience of people around the respondent. In 2004, Aceh was hit by a tsunami that caused nearly 200,000 deaths. The event began with an earthquake and receding seawater. Based on the data obtained, 83% of respondents know that the signs of a tsunami disaster are a strong earthquake and the seawater suddenly receding. 61% of respondents answered that when these signs occur, they should run away from the sea. In addition to gaining knowledge about disasters from experience, 81% of respondents answered that they gained knowledge from schools, print, and electronic media, as well as from books and television/radio. Among these media, the media that provides the most information to students about the tsunami disaster are radio and TV. This shows that the media, especially TV, plays an important role in increasing knowledge about disasters for both students and the general public.

Based on knowledge data from teacher (S2) and student (S3) respondents, the Knowledge and Attitudes index of School Community is:

KA Index (KS)
= 0,60*KA Index (S2) + 0,40* KA Index (S3)
= 0,60*(67) + 0,4*(70,1)
= 68,24

The level of knowledge and attitudes of the school community in Lhokseumawe City is in the category of ready with an index value of 68.24.

4.2 School Community Policy and Guidance Level

Since 2007, Indonesia has issued regulations on disaster management, namely Act no. 24 of 2007. Following up on this, the Ministry of Education and Culture developed a disaster risk reduction strategy in schools with teaching materials modules and training on integrating disaster risk reduction. The Ministry of Education and Culture has also issued Circular Letter 70a/MPN/2010 on disaster risk reduction in schools. This circular letter became the reference for the provincial and city/district education offices in carrying out disaster risk reduction activities in schools.

In a survey conducted to four schools in Lhokseumawe, school leaders/teachers/staff were aware of regulations and circulars regarding disaster risk reduction in schools. This is indicated by an index value of 65 and is included in the ready category. Even though they are aware of the policy, only 50% of respondents have formed a disaster preparedness group at school. In addition, only 50% of respondents answered that they had integrated preparedness into relevant subjects at school and only 62% had included it in extracurricular skills. The implementation of rules by schools, which is quite far from optimal, is influenced by the budget. 50% of the respondents mentioned that they did not have a special budget for school preparedness, so the school had never carried out activities to increase preparedness such as increasing knowledge and skills and training in evacuation simulations in schools.

The less than the optimal implementation of school policies is one indication that schools are starting to forget the tsunami disaster that occurred in 2004. A disaster is a recurring event. In addition, the Lhokseumawe City tsunami risk index is in the medium to high category. Therefore, it is necessary to re-implement policies on disaster risk reduction in schools so that the preparedness of the school community in anticipating the tsunami disaster is increased.

4.3 Level of School Community Emergency Planning

4.3.1 Teacher

Data obtained from a sample of 4 schools, namely two elementary schools, 1 secondary school, and 1 high school in the city of Lhokseumawe show that the level of emergency planning for teachers in dealing with the tsunami disaster is still relatively unprepared. Based on the emergency planning questionnaire, an index of 44 was obtained which was at the level of being unprepared. The data obtained shows that all respondents have anticipated the occurrence of a tsunami disaster by preparing copies of the class documents/lessons taught. However, only 5% of respondents have prepared several things to anticipate the tsunami disaster apart from preparing copies of documents. These include training students to save themselves, nailing/binding bookshelves to the wall, and placing heavy items on the floor.

Many school community activities are carried out in schools. Therefore, the preparedness of schools in anticipating the threat, especially the tsunami disaster, must be improved. Teacher preparation, when faced with disaster cases, is still low. 60% of teachers are only able to calm students down when a tsunami disaster occurs. When a tsunami occurs, teachers must be able to calm students down and guide students to immediately save themselves. However, 60% of respondents still do not know what they should do when it happens while teaching in class.

4.3.2 Student

The school community emergency planning parameter consists of 5 questions and an index value of 61.8 is obtained, which is in the almost ready category. Although 97% of respondents know that they need to save themselves when a tsunami occurs, the school does not support the evacuation route map. In addition, there are still many students, namely 27% of students who answered that they did not know what they should do to deal with the tsunami disaster. 92% said that they did not know an evacuation or rescue route. This is a problem because they are the ones who determine their safety in the event of a disaster. This is following a survey in Japan, in the 1995 Great Hanshin Awaji earthquake which showed that the percentage of survivors was caused by the actions of Self by 35%, Family Members 31.9%, Friends/Neighbors 28.1%, Passersby 2.60%, SAR Team 1.70%, and others 0.90% [12].

4.3.3 School

The data obtained from the 5 questions on the questionnaire show that the school's emergency response plan index is still in the unprepared category with an index value of 27.5. Although 75% have saved a copy of the data to anticipate the tsunami disaster, all respondents answered that there is no evacuation plan. The agreement on evacuation places, evacuation route maps, evacuation equipment/equipment as well as conducting evacuation training/simulation has not been carried out by all respondents.

Evacuation plan parameters are interconnected with other parameters. The knowledge possessed by students and teachers will be useless if it is not supported by evacuation facilities. If a tsunami disaster occurs during school hours, students/teachers will be confused about where they should save themselves because they have not done any evacuation training/simulation. In addition, this evacuation plan is also related to disaster warning parameters. Disaster warnings will be effective in helping the school community to save themselves when a disaster occurs if it is supported by the existence of evacuation sites, evacuation route maps, and training.

Based on the index obtained from the teacher (S2), student (S3) and school (S1) respondents, it can be calculated that the school community emergency planning index value is:

EP Index (KS)
= 0,61*EP index (S1) + 0,30* EP index (S2) + 0,09* EP index (S3)
= 0,61*27,5 + 0,3*44 + 0,9*61,8
= 35,6 (Unprepared)

The level of emergency planning for the school community in Lhokseumawe City is in the unprepared category with an index value of 35.6.

4.4 Level of School Community Disaster Warning System

4.4.1 Teacher

The level of teacher resource mobilization is at an index of 41. The index is included in the unprepared category. The level of teacher resource mobilization is seen based on seven questionnaires. On the question of tsunami warning signs/warnings in their area, 80% know the traditional tsunami warning signs/methods/local agreements and the national tsunami system. Indonesia has built a tsunami early warning system or often called the Indonesia Tsunami Early Warning System (Ina TEWS). The Indonesian Tsunami Early Warning System (Ina TEWS) serves to warn of a tsunami shortly after a major earthquake occurs at sea. The warning will help people living on the coast to immediately save themselves from the tsunami disaster [4]

In case when the teacher hears tsunami warning signs/warnings while teaching, only 10% of the respondents answer that they will not panic, guide students to evacuation sites, and immediately monitor the truth of tsunami news from the competent authorities. Based on the answers from the respondents, 80% of the teacher respondents did not know the signs of canceling the tsunami warning or the sign or information that the situation was safe after the tsunami was declared by the BPBD or the local government.

4.4.2 Student

Students' knowledge of the tsunami warning system is still in the unprepared category, namely at index 31.8. This can be seen in the question about tsunami signs, as many as 80% of respondents answered that they did not know the tsunami warning signs, both national and traditional warning systems. Early warning is a series of activities to give warnings as soon as possible to the community about the possibility of a disaster occurring in a place by the authorized institution [13]. This early warning system will notify people living in areas with a tsunami risk level to carry out the early evacuation. However, in this study, there are still many respondents who do not know about early warning even though 96% of respondents know what to do if there are signs of a tsunami danger, which is to stay away from the beach or run to a higher place.

4.4.3 School

The disaster warning index value for schools is still in the unprepared category, which is still below the number 40. Almost all questions in the questionnaire are answered with statements stating that there are none and have not been implemented by the school. Although schools have used a bell to signal the start/end of lessons, this bell has not yet been set by the school to be used as a broad disseminator of tsunami warnings in schools. In terms of disaster warning parameters, none of the schools that were sampled had the tools used for tsunami warnings.

WS Index (KS)
= 0,57*WS Index (S1) + 0,29* WS Index (S2) + 0,14* WS Index(S3)
= 0,57*40 + 0,29*41 + 0,14*38,1
= 40 (Unprepared)

The level of the school community disaster warning system in Lhokseumawe City is in the unprepared category with an index value of 40.

4.5 Level of School Community Resource Mobilization

4.5.1 Teacher

The level of mobilization of teacher resources in the city of Lhokeumawe based on data from 4 schools obtained an index number of 63. The index is in the almost ready category. On questions about whether or not they have attended training, workshops, seminars, lectures, discussions, or simulations, 70% of respondents answered that they have participated in disaster knowledge materials. Meanwhile, the material on early warning systems (equipment, sound signals, information dissemination) was only attended by 20% of respondents.

Knowledge about the tsunami disaster that was known by teacher respondents was conveyed by 85% of teachers to their students. The delivery is influenced by the educational background and the subjects taught by the teacher. However, 95% of teachers have never practiced disaster warning, first aid, and rescue/evacuation with their students.

4.5.2 Student

The level of mobilization of school community resources in Lhokseumawe City is still in the unprepared category, which is at an index of 40.5. The index is obtained from 2 questions about activities related to disasters that the respondents have participated in. Based on the data obtained, all respondents answered that they had never conducted disaster evacuation simulation training. This indicates that the school community in Lhokseumawe, as illustrated by the 4 educational institutions sampled in this study, has not optimally prepared for the tsunami disaster. Although students' knowledge of tsunamis is categorized as ready, evacuation training/simulation needs to be carried out so that students' preparedness in dealing with tsunami disasters increases. This needs to be done because Lhokseumawe is a coastal area that is in the moderate to a high category for tsunami disasters.

4.5.3 School

The mobilization of school community resources has an index score of 52 and is in the almost ready category. 75% of schools already have a School Health Unit, Scouts and First-aid trained students that can be used for school preparedness in dealing with disasters. However, only 25% of teachers/school staff have attended workshops/seminars about disaster knowledge, evacuation plans, first aid, early warning systems, and evacuation simulations.

Regarding the policy parameters, namely Circular Letter 70a/MPN/2010 from the Ministry of Education on mainstreaming disaster risk reduction in schools, only 25% of respondents have included preparedness materials in the relevant lesson materials. This can happen because there is no regulation from the local office for the implementation of the circular letter. The teacher's knowledge and the absence of guidance can also be the cause of the non-implementation of the circular letter. 62.5% of respondents answered that they had never received assistance/guidance related to preparedness.

Based on the analyzed data, the level of mobilization of school community resources is as follows:

RMC Index (KS)
= 0,60*RMC Index (S1) +0,30* RMC Index (S2) + 0,10* RMC Index (S3)
= 0,60 * 52 +0,30*63 + 0,10*40,5
= 54,15 (almost ready)

Based on the index value of 54.15, it can be seen that the level of mobilization of school community resources in Lhokseumawe City is in the almost ready category.

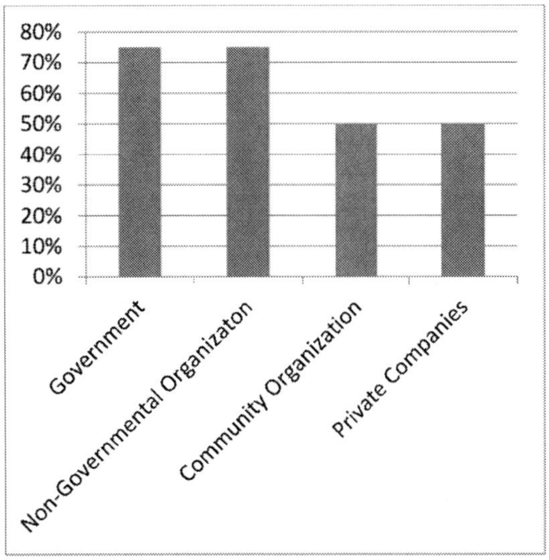

Fig. 2. Source of Aid/Guidance Related to Tsunami Disaster Preparedness

4.6 School Community Preparedness Level

A tsunami disaster is a type of disaster that can cause damage and loss of life. There is no tool to predict when a tsunami will occur. Therefore, it is necessary to carry out preventive activities. One of its important activities is to increase community preparedness for people who may be vulnerable to tsunami disasters. This study focuses on the preparedness of the school community in Lhokseumawe City. Based on the survey conducted, index numbers have been obtained for each parameter. The index values per community are as follows:

4.6.1 School Index (S1)

The school index value is a combination of the index value of school preparedness policy (PS), emergency response plans (EP), disaster warning (WS), and resource mobilization (RMC). The parameter index values are then added up according to their respective weights [11]. The following is the calculation of the school preparedness index value.

School Index (S1)
= 0,29 (PS Index) + 0,41 (EP Index) + 0,12 (WS Index) + 0,18 (RMC Index)
= 0,29 (65) + 0,41 (27,5) + 012 (4,1) + 0,18 (52)
= 30,6

Based on the sum of the index values of the school preparedness policy parameters (PS), emergency response plans (EP), disaster warnings (WS), and resource mobilization (RMC), school preparedness is in the category of unprepared with an index value of 30.6.

4.6.2 Teacher Index (S2)

The teacher index value is a combination of the index values of knowledge about disasters (KA), emergency response plans (EP), disaster warnings (WS), and resource mobilization (RMC). The parameter index values are then added up according to their respective weights [11]. The following is the calculation of the teacher preparedness index value.

Teacher Index (S2)
= 0,71 (KA Index) + 0,17 (EP Index) + 0,05 (WS Index) + 0,07 (RMC Index)
= 0,71 (69,7) + 0,17 (44,5) + 0,05 (40,6) + 0,07 (63)
= 61, 5

Based on the sum of the index values for disaster knowledge (KA), emergency response plans (EP), disaster warnings (WS), and resource mobilization (RMC), the level of teacher preparedness are in the almost ready category, with an index value of 61.5.

4.6.3 Student Index (S3)

The student's index score is a combination of the index values of the knowledge about disasters (K), emergency response plans (EP), disaster warnings (WS), and resource mobilization (RMC). The parameter index values are then added up according to their respective weights [11], The following is the calculation of the student preparedness index value.

Student Index (S3)
= 0,83 (KA Index) + 0,08 (EP Index) + 0,04 (WS Index) + 0,04 (RMC Index)
= 0,83 (70) + 0,08 (61,9) + 0,04 (31,8) + 0,04 (40,5)
= 65,9

Based on the sum of the index values for disaster knowledge (KA), emergency response plans (EP), disaster warnings (WS), and resource mobilization (RMC), the level of preparedness of students are in the ready category with an index value of 65.9.

The level of preparedness is the combined result of the school community knowledge index, the school community emergency planning index, the school community disaster warning system index, the school community policy, and guidance index, and the school community resource mobilization index using each weight in each parameter.

Total KS Index

= 0,50* KA Index (KS) + 0,10*PS Index (KS) + 0,23*EP Index (KS) +0,07*WS Index (KS) + 0,10*RMC Index (KS)
= 0,50* 68,24 + 0,10*65 + 0,23* 35,6 + 0,07*40 + 0,10 * 54,15
= 57,02

After summing up using their respective weights following the guidelines from LIPI-UNESCO [11], the school community preparedness index value in Lhokseumawe City is 57.02 which is in the almost ready category. The index value of each parameter can be seen in Fig. 3.

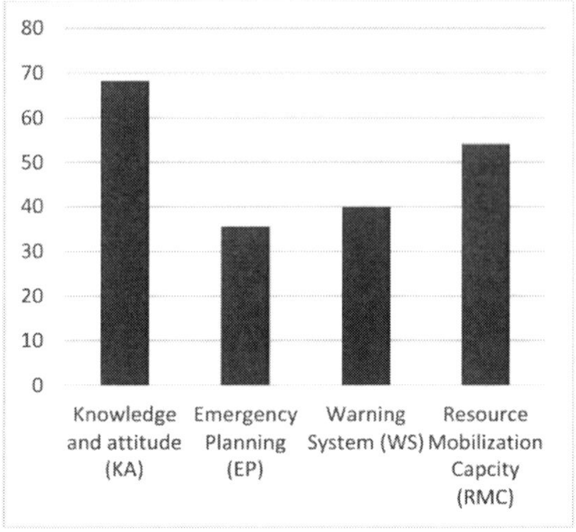

Fig. 3. Index Values for Each Parameter of School Community Preparedness

5 Conclusion

- The level of knowledge and attitudes of the school community in Lhokseumawe City is in the ready category with an index value of 68.24.
- The level of school community policies and guidelines is in the ready category with an index value of 65.
- The level of emergency planning for the school community in Lhokseumawe City is in the unprepared category with an index value of 35.6.
- The level of the school community disaster warning system in Lhokseumawe City is in the less ready category with an index value of 40.
- The level of mobilization of school community resources in Lhokseumawe City is in the almost ready category with an index value of 54.15.
- The level of preparedness of the school community in the city of Lhokseumawe is in the almost ready category with an index value of 57.02.

References

1. BNPB, Aceh Province Natural Disaster in December *2004* (DIBI, BNPB, 2021)
2. BAPPENAS, *Preliminary Damage and Loss Assessment* (Badan Perencanaan Pembangunan Nasional, 2005)
3. BNPB, *Aceh Disaster Risk Assesment 2016-2020* (BNPB, Jakarta, 2016)
4. W. Utariningsih, R. Sofia, Journal SPATIAL Wahana Komunikasi and Informasi Geografi, **20,** 22-27 (2020)
5. UNESCO-IOC, Tsunami Glossary:IOC Document Information No.1221 (UNESCO, Paris, 2006)

6. T. Puspito,Nanang, Managing Disaster Risk in Maritime Areas (ITB, Bandung, 2010)
7. Dudley, W.C, Min L. *Tsunami !* (Pakar Raya, Bandung, 2006)
8. E. Yulianto, F. Kusmayanto, N. Supriyatna, Dirhamsyah. Surviving the Tsunami Disaster, Learning from the Tsunami in Aceh and Pangandaran (JTIC-UNESCO House, Jakarta, 2008)
9. A. Y. Baeda, F. Husain, Journal Teknik Sipil Institus Teklogi Bandung, **19**, 1 (2012)
10. J. Sutton, K. Tierney, *Disaster Preparedness: Concept, Guidance, and Research* (University of Colorado, 2006)
11. LIPI-UNESCO/ISDR. Study of Community Preparedness in Anticipating Earthquake and Tsunami Disasters (LIPI, Jakarta, 2006)
12. BNPB, Disaster Preparedness Training Manual: Building Awareness of Alertness and Preparedness in Facing Disasters *(*BNPB, Jakarta, 2017)
13. W. Utariningsih, Disaster Management (Sefa Bumi Persada, Lhokseumawe, 2020)

The concept of an early warning system for the use of nuclear technology in Indonesia

Dewi Prima Meiliasari[1*], *Berton* Suar Panjaitan[2], *Toto* Heryanto[3], and *Wilopo*[1]

[1]Disaster Management Study Program, National Security Faculty, Republic of Indonesia Defense University, Indonesia
[2]The Disaster Management Training Center, National Disaster Management Agency, Indonesia
[3]Directorate of Technical Support and Nuclear Emergency Preparedness, Nuclear Energy Regulatory Agency, Indonesia

Abstract. The application of nuclear power plants in the world is increasing and has the potential for accidents, including nuclear weapons tests from outside the Nuclear Non-Proliferation Treaty, thus allowing radioactive releases to fall into Indonesian territory, which is a potential threat. Indonesia also utilizes nuclear power in many fields, thus also potentially becoming a national threat. To prevent the threat of radiation potential requires a nuclear disaster Early Warning System (EWS) that can be accessed by the community. This research analysed the EWS that Indonesia needed, using qualitative methods by describing the results of research obtained from interviews and secondary data. Results and research discussions starting from observation and monitoring of radiation exposure, warning services, information dissemination, and supported by countermeasures for the effectiveness of the EWS. The results showed that the development of an integrated nuclear disaster EWS that is easily accessible to the community quickly and continuously under the development of an emergency is a strategy as a decision to reduce the risk of nuclear disaster. The development of the strategy requires a legal basis that regulates coordination between ministries/institutions, from planning to information dissemination to ensure the safety of the public and Indonesian people's security.

1 Introduction

The use of nuclear power has been known since World War II as a weapon to destroy the cities of Hiroshima and Nagasaki in 1945, until now nuclear power is used by the world. According to data released by the World Nuclear Association in 2007 on the Indonesian National Nuclear Agency website, there were 436 uses of nuclear power used for Nuclear Power Plants (NPP) which operate in 33 countries around the world and can generate electricity as much as 16% of the world's electricity needs. Based on data from the International Atomic Energy Agency (IAEA) in 2018, the United States is the largest country operating nuclear power plants in the world with 99 units, followed by France in second place with the number of nuclear power plants reaching 58 units. The use of nuclear power as nuclear power plants worldwide is dominated by European countries, the United States, and several developed countries in Asia such as China, Japan, India, and South Korea, as well as 15 other countries that have just built nuclear power plants [1,2].

The operation of nuclear power plants around the world is not only beneficial but also has the potential for radioactive release to the environment, both under normal operating conditions and in accidental conditions that cause exposure to the public and the environment.

In the event of an accident, the release of fission product in large quantities with long half-lives, through the atmosphere through world global winds which have an altitude above 800 m will be very likely to be carried up to enter the Indonesian atmosphere and with high rainfall conditions, the slowing down wind speed at the time of the wind direction deflection and the high surface altitude will very likely allow the fission product radionuclides to fall into the surface of the region in Indonesia. Indonesia's geographical position, which is located between countries operating nuclear power plants around the world, has the potential to be exposed to radioactive release in the event of an accident so that it becomes a potential global threat given in Figure 1.

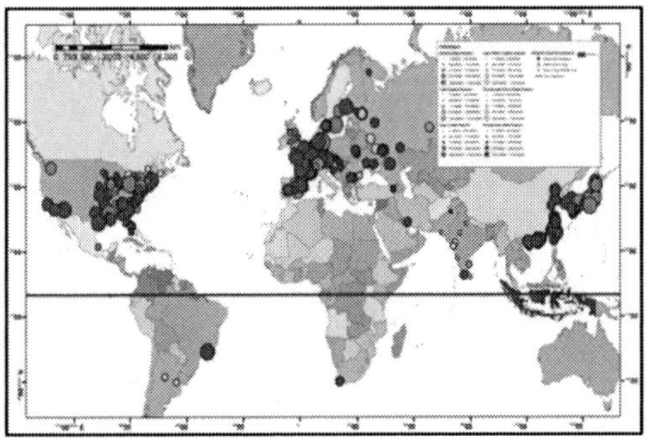

Fig. 1. The potential global radiation threat.

Indonesia as a country that ratified the Nuclear Non-Proliferation Treaty (NPT) has utilized nuclear technology in various fields, i.e health, industry,

* Corresponding author: d.meiliasari@bapeten.go.id

https://doi.org/10.1051/e3sconf/202133104003

agriculture, and research. In the fields of 1. Health, it is used as a *radiodiagnostic* and interventional tool, radiotherapy, and nuclear medicine; 2. Industry, as a tool for material analysis *(XRF, logging),* food preservation, sterilization of medical devices, non-destructive testing (radiography), measurement *(gauging),* Tracer, and Imaging *(checking baggage or containers)*; 3. Agriculture, for breeding rice seeds to produce superior rice varieties, to preserve ready-to-eat food, slowing down shoots in tubers, preserving fruits; and 4. Research carried out to support the development of the nuclear industry and preparation for the construction and operation of nuclear power plants in Indonesia carried out at the Triga 2000 (2 MW) reactor in Bandung, the Kartini Reactor (250 kW) in Yogyakarta, and the Siwabessy Multipurpose Reactor-GA (30 MW) in Serpong. In addition to being useful, the use of nuclear technology also has a great potential for the release of radioactive substances in the event of an accident. The location of the use of nuclear power in Indonesia as a potential threat is given in Figure 2.

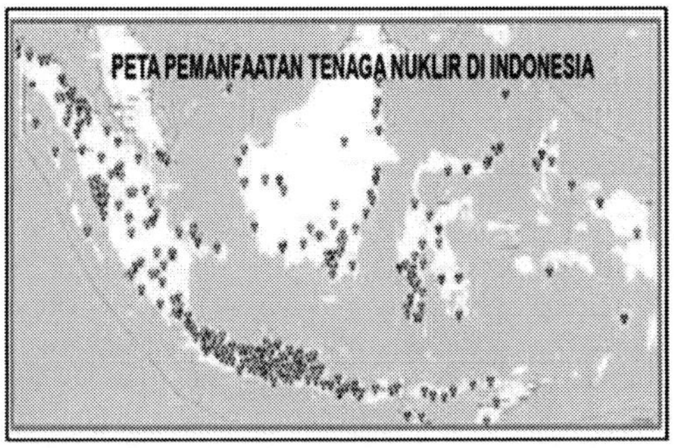

Fig. 2. The potential national threat.

The use and distribution of nuclear power plants and radioactive substances (ZRA) globally and nationally is a global, regional and national security threat, and the use of nuclear power for weapons testing increases the potential danger of radioactive release or contamination caused by nuclear emergencies from within or outside the territory of the Republic of Indonesia.

Nuclear weapons tests have been detected by several stations of the Meteorology, Climatology and Geophysics Agency (BMKG) on January 6, 2016, at 08.30.01 WIB, an unusual vibration on the earth's surface has been detected near the location of North Korea's nuclear test facility as reported on the website. BMKG official. The detection of vibrations caught by BMKG only identifies the location of the vibrations but has not been able to identify the possibility of radioactive fallout that may occur in the territory of Indonesia. The existence of equipment capable of detecting a fallout in Indonesia that is capable of functioning as a nuclear early-warning system is deemed necessary to be paired with the equipment already owned by BMKG to ensure the creation of protection and a sense of security for all Indonesian people.

An EWS for radioactive releases into the Indonesian environment is an effective nuclear emergency response monitoring system capable of monitoring the latest radiation levels in real-time, which is needed in several locations as an EWS, to enable the public to respond appropriately as a rescue and to avoid casualties, and to reduce the risk of disasters due to exposure to radioactive releases.

The Fukushima disaster in Japan that occurred in March 2011 taught a lesson that it is not enough just to have the equipment to detect radioactive releases with advanced technology, but also an EWS that can disseminate information quickly to the public in the early stages of an emergency, to communicate and coordinate, and to quickly report the status of an emergency to the IAEA [3]. Communication and coordination between authorized institutions that observe disaster symptoms and the National Disaster Management Agency (BNPB) and/or Regional Disaster Management Agency (BPBD) according to the location of the disaster level, as a basis for making decisions on disseminating information to the government, private broadcasters, and mass media for resource mobilization, so that the public can prepare themselves as an effort to reduce the risk of radioactive releases into the environment.

2 Methods and Data

The method in this study uses a qualitative method intending to get a detailed description of the EWS starting from observing the symptoms of radioactive release into the Indonesian environment so that the public can make efforts to reduce the risk of radioactive release. Data in the field were obtained through interviews and supported by secondary data collected by researchers. The research results are described qualitatively based on existing theories.

3 Theory

Early warning in Law Number 24 of 1997 concerning Disaster Management is a series of activities as an early warning received by the community that a disaster will occur. Early warning is a combination of technological capabilities and the ability of policymakers, as well as the ability of the community to follow up as an effort to reduce the risk of radioactive releases into the Indonesian environment. Early warning as part of disaster risk reduction begins with observing existing disaster symptoms, analysing the results of observing disaster symptoms, making decisions by the authorities, disseminating disaster warning information, and taking action by the community. Community action on radioactive releases into the environment is carried out immediately after an early warning is given following the amount of contamination in the environment. The community performs temporary protection at home (sheltering), evacuates, and takes iodine tablets [4]. If one of these components is not met, the EWS will not be successful as a whole.

The EWS is a specific link (critical link) between preparedness actions and emergency response activities. There are 2 (two) factors that play a role in the framework of the EWS, namely the Decision Makers and the Community. On the community side, there are 3 (three) elements that determine how the community reacts to the EWS. These elements consist of Knowledge, Attitude, and Skill. The first step in shaping the community's reaction to the EWS is to provide information about the EWS. For people who have obtained this information knowledge, it is hoped that there will be a positive attitude change towards the EWS. This change is expected to be able to make people behave positively towards the EWS. If the stages of change in the community's reaction to the EWS are as expected, the EWS can reach the community accurately [5].

In addition to community factors, other factors that play a role in the framework of the EWS are the decision-makers. In Indonesia, through the Government Regulation of the Republic of Indonesia Number 21 of 2008 concerning the Implementation of Disaster Management, the results of the analysis of observations of disaster symptoms from the EWS are submitted to BNPB and BPBD [6]. As we all know that disaster management and handling of refugees are coordinated by BNPB at the national level, regional management by BPBD. Through the existence of this institution, policies related to the EWS can be made, especially those related to the framework of the EWS, such as fixed procedure, Guidelines, and Work Mechanisms. Thus the EWS as an initial sub-segment in the preparedness stage can play a good role so that in the end when a disaster occurs, its severity can be controlled. With a good conceptual framework, the EWS as a link between preparedness actions and emergency response activities will result in response activities that lead to disaster management so that disaster victims can be reduced.

4 Observation of Disaster Symptoms

Risks can occur because of the threat of disaster risk (hazards), vulnerabilities, and levels of capacity [7]. The threat has been identified by researchers in preliminary research as described above, that the threat of radioactive release risk can come from global threats due to the use of nuclear power in the world as nuclear power plants and nuclear tests, as well as national threats from nuclear use in the fields of health, agriculture, industry, research in Indonesia [8].

The rapid increase in the population of Indonesia, increasing of housing land, offices and other public services office and approaching the area of nuclear technology utilization. Public education factors are still low, including disaster science, especially nuclear disaster risk reduction measures, and the lack of awareness of local governments in nuclear hazard risk preparedness [9]. This vulnerability is a threat to the risk of radiation exposure to the Indonesian environment.

Capacity building can be achieved by developing a nuclear disaster EWS in Indonesia. In developing a nuclear disaster EWS, it is based on Article 67 of Government Regulation Number 54 of 2012 concerning Safety and Security of Nuclear Installations which stipulates that early warning equipment is an element of infrastructure in the nuclear preparedness program, and in Article 16 of the Regulation of the Chairman of the Nuclear Energy Regulatory Agency Number 1 of 2010 concerning Nuclear Emergency Preparedness and Response, which regulates the authority of the licensee (PI) to develop an early detection system for nuclear emergencies, but has not regulated the authority of the nuclear energy regulatory agency (BAPETEN) in developing an EWS as a system that ensures the safety and security of the public from nuclear disasters [10–13].

In the Presidential Instruction of Indonesia, Number 4 of 2019 concerning Capacity Building in Preventing, Detecting, and Responding to Disease Outbreaks, Global Pandemics, and Nuclear, Biological, and Chemical Emergencies, the President of the Republic of Indonesia gave instructions to the Chairman of the Nuclear Energy Regulatory Agency to increase the ability to detect and respond to the occurrence of nuclear emergencies, whether arising from nuclear accidents in the use of nuclear energy for peaceful purposes, or all forms of use of nuclear energy for non-peaceful purposes, such as the misuse of nuclear materials or radiation sources by North Korea in nuclear weapon tests, and other activities. such as sabotage, and theft [3,14].

The nature of presidential instructions according to Jimly [15], only as "policy rules" that is a form of policy regulation that cannot be categorized as a form of legislation, formally it cannot be called an official regulation, because it is only limited to providing direction in terms of carrying out tasks. Seeing the importance of an EWS in anticipation of a nuclear disaster and as a mitigation effort to reduce the risk of nuclear threats. Therefore, it is necessary to have laws and regulations related to the EWS as a legal basis in the implementation of monitoring the detection and response of radioactive releases, which include regulations regarding the responsibilities and authorities of ministries, institutions, and local governments related to the nuclear disaster EWS, and governance how to implement an EWS.

The existence of laws and regulations regarding the nuclear disaster EWS can be the basis for making policies in developing an EWS, so that efforts to strengthen and develop a sustainable nuclear disaster EWS are created and synergize between institutions to improve the safety of life and property of the Indonesian people.

5 Monitoring and Warning Service

5.1 Analysis of Observational Data

To defend the Unitary State of the Republic of Indonesia (NKRI), Indonesia builds, develops, and operates a national EWS as integrated and efficient monitoring and supervision of radioactive releases to the environment in the territory of the Republic of Indonesia. Radioactive release EWS becomes real-time data that can be accessed online for monitoring radioactivity to the environment of the Republic of Indonesia.

Online data which in its planning can be accessed by all parties and the community based on social networks, has been able to be integrated with the International Radiation Monitoring Information System (IRMIS) owned by the IAEA. So that if an environmental radioactive release is detected into the territory of the Republic of Indonesia, the public can immediately take action to reduce the risk of a nuclear disaster, and the

 https://doi.org/10.1051/e3sconf/202133104003

IAEA will be immediately informed and will take action in emergency communication.

One of the detection equipment for the nuclear disaster early warning information system has been installed in several locations of vital objects, including 1. In the Serpong nuclear area, 6 detectors have been installed; 2. In the Bandung nuclear area, 1 detector has been installed; 3. In the Yogyakarta nuclear area, 1 detector has been installed; 4. In the palace of the independent state, 1 detector has been installed; 5. The BMKG CTBT station has been installed with 5 detectors; 6. The BMKG station has installed 10 detectors depicted in Figure 3; and the location of other vital objects that are still in the planning.

Fig. 3. Built-in IRDMS detector.

Radiation exposure is monitored in real-time that can be accessed online by the detector system supervisor officer every time. If there is an increase in radiation exposure in one region, the red alarm will appear at the BAPETEN - Nuclear Emergency Centre as shown in Fig. 4, it also displayed on the dashboard data on the NMC monitor web application given in Figure 5, the system will also notify the office by email and Short Message Service (SMS) and real-time data is monitored in the NMC Web application on Figure 6.

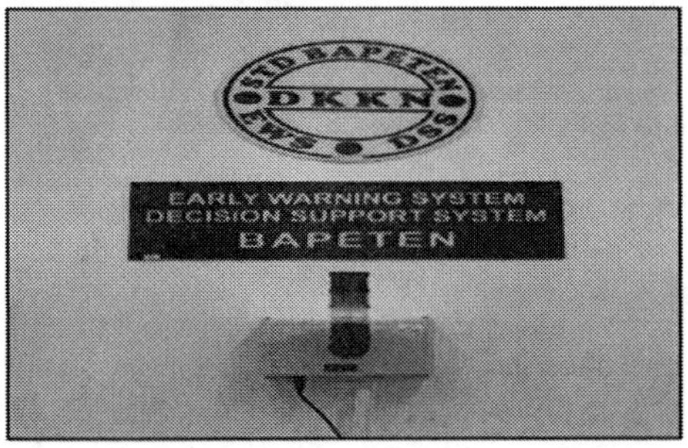

Fig. 4. EWS Alarm.

Fig. 5. Dashboard data.

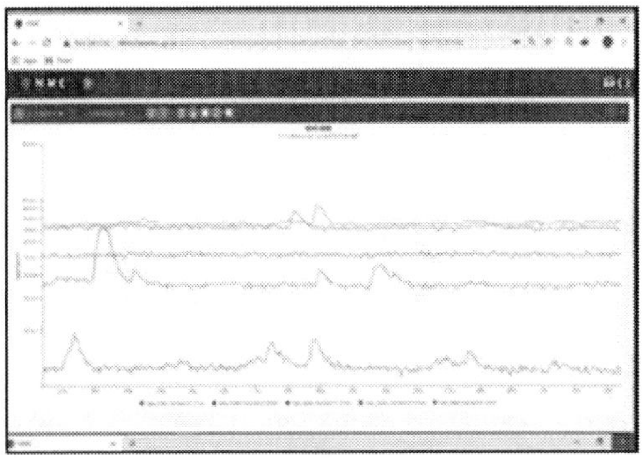

Fig. 6. NMC web app.

Figure 6 illustrates that there is an increase in radiation release which can be seen from the image of a wave that curves higher than the others, but the increase in radiation release in Figure 6 is still considered within normal limits because in abnormal conditions it appears in yellow or red.

5.2 Decision-Based on the Analysis Result

Nationally, the nuclear emergency management system is part of the National Disaster Management System. At the national level, this arrangement is coordinated by the National Disaster Management Agency. Currently, all Ministries and other institutions have been trying to build an integrated EWS at the national level, so that all efforts and capabilities in the field of disaster management can be implemented synergistically.

A synergistic radioactive release EWS is also part of the plan in the development of integration with the application of radiological impact analysis and geospatial information systems (GIS) named the National EWS (N-EWS). It is going to be a data integration system between the Central Statistics Agency (BPS), the National Institute of Aeronautics and Space (LAPAN), and BMKG. N-EWS detector as confirmatory data and also assessment for predictions of radiological failure. N-EWS can confirm the correctness of the prediction results if there is information from the licensee about an accident that results in a radioactive release, the N-EWS detector application can confirm the analysis of radiological impacts in certain areas with a certain distance from the analysis data based on the geospatial information system. Data that has been integrated into information on EWS for BNPB, BMKG, IRMIS, Association of Southeast Asian Nations Network of Regulatory Bodies on Atomic Energy (ASEANTOM). An integrated nuclear disaster EWS is given in Figure 7.

Currently, the implementation of the decision on the nuclear disaster EWS is based on the results of coordination between BAPETEN and the BMKG after analyzing the symptoms of a nuclear disaster from the early warning information systems that have been installed in various regions. The results of the decision on the occurrence of radioactive releases are submitted by the Chairman of BAPETEN to the Chairman of BNPB/Provincial BPBD/City BPBD following the mandate of Law Number 24 of 1994 and Government

Regulation Number 54 of 2012 and information on early warning of radioactive releases has been integrated with the IRMIS-IAEA system so that all Nuclear accident events that result in the release of radiation into the Indonesian environment can be known by the IAEA [5,13].

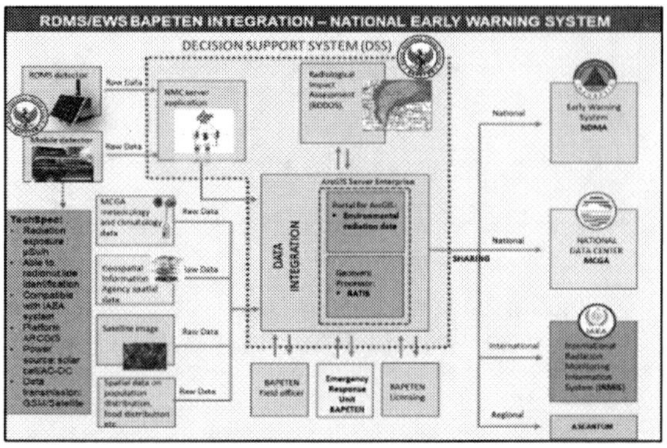

Fig. 7. Nuclear disaster EWS integration architecture.

BAPETEN together with BMKG has decided to verify an issue of nuclear radiation exposure in the South China Sea from online media on November 20, 2019, based on the early warning monitoring system of Radiological Data Monitoring Systems (RDMS) sensors which were installed side by side with the BMKG earthquake monitoring station network in Tarakan, Tanjung Pinang, and Gorontalo. From the results of RDMS no significant radiation level was detected compared to natural radiation levels, this symptom analysis is supported by the results of seismic recording analysis in two locations close to the estimated location of nuclear explosions in the South China Sea, namely in Vietnam and Taiwan that there is no seismic signal anomaly that shows a result of a nuclear explosion [9].

The results of the BAPETEN agreement with BMKG regarding the undetected nuclear disaster in the South China Sea can be accessed on social media, with the aim that all parties and the public can know and be able to reduce public concerns about the release of radiation hazards to the environment, especially people who live and earn income from the sea in the sea around the South China Sea.

6 Information dissemination

Dissemination of information to the public, mass media, and agencies related to the situation and condition of the facility-level nuclear emergency is carried out by the licensee based on the nuclear preparedness program of the facility that has been determined. The facility's nuclear preparedness program according to the IAEA (2015) in IAEA GSR Part 7 requirement 10 contains procedures for providing information, instructions, and warnings to the community around the facility or activity during an emergency [16]. Researchers still see that the facility's nuclear preparedness program has not been updated according to the latest references.

Nuclear emergency information to the public, mass media, and related agencies is sourced from information collection at the Control Command Center (PUSKODAL), information is provided by public relations at the facility. Dissemination of information to the public can also be carried out during nuclear emergency training and exercises, training is carried out once a year for facility-level nuclear emergencies, every two years for nuclear emergency training at the provincial level carried out by the provincial BPBD, and once every four years for training and exercises. Nuclear emergency at the national level organized by BNPB and involvement of the community around the reactor in nuclear emergency training and exercises need to be carried out to reduce the risk of nuclear disaster.

The Governor declares the status of a nuclear emergency at the provincial level to the public based on a recommendation from the Chairman of BAPETEN, in the event of a dose rate of 5 mSv/hour or more measured for ten minutes or more at the installation site boundary and in the event of abnormal radioactive release with a concentration of air activity equal to or exceeding a dose rate of 5mSv/hour at the installation site limit detected from the normal release level, while for a nuclear emergency at the national level, a declaration of an emergency status is made by the President based on a recommendation from the Chairman of BAPETEN because the dose rate is 500 mSv/hour or more which is measured for 10 minutes or more at the installation site boundary, and if abnormal radioactive release conditions with a concentration of air activity equal to or exceed a dose rate of 500 mSv/hour at the installation site boundary detected from the normal release path.

7 Countermeasures

Nuclear disaster management capability is implemented according to GSR Part 7 from a series of capabilities that are prepared systematically and planned based on nuclear emergency management guidelines or procedures, starting from preparedness, nuclear emergency response until the nuclear emergency response is declared ended [2].

Quick, precise, effective, and efficient handling of nuclear emergencies is carried out by anticipating nuclear emergencies through the provision of infrastructure and the capability of the response function [13]. The infrastructure prepared for the response capacity includes: 1. the establishment of a nuclear disaster emergency management organization so that the responsibilities and authorities between parties are seen to be integrated into carrying out nuclear emergency management, which consists of a chairman and implementer (operations controller, operations operator, radiology reviewer, public relations and Chairman of ministries and related institutions); 2. Coordination between organizations is an important pillar in carrying out integrated nuclear emergency management so that it can run following the nuclear disaster emergency management plan that has been made so that nuclear disaster management plans must be made and updated according to current conditions; 3. Facilities and equipment prepared, including early warning and alarm equipment as a means of detecting the early release of radioactive sources, either in the form of online data or in the coordination system for reporting from the nuclear emergency response team; 4. Procedures and work instructions as guidelines for the technical implementation of nuclear disaster emergency management, including reporting procedures and coordination of decisions on early warning information systems; emergency response procedures for handling victim decontamination; procedures for providing information to the public; evacuation procedures for victims which include gathering points, temporary shelters, Mashall yards, triage rooms and so on; 5. Nuclear emergency training and exercise are carried out periodically according to the level of the nuclear disaster category.

The countermeasures functions in carrying out nuclear emergency response measures include 1. Identification of nuclear disaster symptoms from the nuclear disaster EWS, reporting after disaster symptoms have occurred to BAPETEN as the institution that guarantees the safety and security of the use of nuclear energy, and activating notification systems and activating emergency management organizations. nuclear following procedures for dealing with nuclear emergencies; 2. Mitigation measures to reduce the escalation of radiological hazards, reduce the potential threat of radioactive release, and reduce the impact of the release of radioactive substances to the community; 3. Protection measures are taken immediately after the impact of the accident exceeds the intervention level by providing iodine tablets, sheltering, and evacuation immediately after getting the results of disaster symptoms from the nuclear disaster EWS; 4. Protection measures against nuclear emergencies for workers, the community, and the environment by providing iodine tablets, sheltering, evacuation, and using personal protective equipment for nuclear emergency response officers, monitoring dose acceptance, limiting dose acceptance, taking action against excess doses, and decontamination of personnel; and 5. continuous provision of information following the development of the emergency to the public in emergency response to a nuclear disaster is an important action so that the public immediately takes measures to protect themselves against the effects of radioactive releases, the instrument for disseminating information is also a concern so that the public can quickly receive the latest news on the actual situation by through print or electronic media.

8 Conclusion

It was concluded that EWS in real-time which was able to detect the threat of radioactive release that occurred in national and global in the prevention of nuclear disaster emergencies is needed so that people can receive information through print and electronic media quickly and continuously following the development of nuclear disaster emergencies.

The nuclear EWS integrated with the EWS owned by other ministries / Institutions is a strategy for the effectiveness of information on nuclear disaster early warning information systems. It needs coordination to produce an agreement according to the results of data analysis to achieve the effectiveness of information data. Coordination starts from planning to disseminating information continuously to the community.

EWS integrated with the detector system, able to coordinate systems with national and international and also related ministries/institutions, including BMKG, BPS, LAPAN, BNPB, IAEA -IRMIS, ASEANTOM in data integration, so that the community can access the information system early nuclear emergency. It is required legislation as a legal basis in the development of an integrated nuclear disaster EWS, which regulates a comprehensive EWS including coordination between ministries/institutions related to planning to disseminate information continuously to the public to ensure the safety of the community so that Indonesia's security is created.

References

1. K. Kelleher, at Nucl. radiol. Emergency Anime. Whipped. Preparation System. Response Recovery. Springer, Berlin, Heidelberg, 35–52 (2021)
2. IAEA, Nuclear or Radiological Emergency Preparedness and Response. IAEA Safety Standard Series No. GSR Part 7 (International Atomic Energy Agency, Vienna (2015)
3. A. Dewi, Fukushima Accident for Improving the National Nuclear Preparedness Performance System, (Proceedings of the Nuclear Safety Seminar, 286-296 (2013)
4. T. Heryanto, Evacuation Analysis of People Affected by Radiation Hazards Due to Nuclear Accidents, Defense University, (2015)
5. A. Y. Pratiwi, The Role of the IAEA in the Development of Nuclear Power for Non-Peaceful Purposes, Calyptra **3** (2014)
6. RI, Law Number 24 Year 2007 on Disaster Management (Government of the Republic of Indonesia, Jakarta, 2007)
7. S. Maarif, Thoughts and Disaster Management in Indonesia (National Agency for Disaster Management, Jakarta, 2012)
8. RI, Law of the Republic of Indonesia No. 10 of 1997 concerning Nuclear Energy (Government of the Republic of Indonesia, Jakarta, 1997)
9. A. Dewi, BNPB Preparedness in Facing Nuclear Disaster Threats in Indonesia, Defense University, (2020)
10. BATAN, Research and Development Results of BATAN in the Energy Sector, (2019)
11. UTI, Batan and World Nuclear Association Hold World Nuclear Spotlight Seminar, (2018)
12. Bapeten, Regulation of the Head of the Nuclear Energy Supervisory Agency Number 1 of 2010 concerning Preparedness and Response to Nuclear Emergencies (Bapeten, Jakarta, 2010)

13. RI, Government Regulation of the Republic of Indonesia No. 54 of 2012 concerning the Safety and Security of Nuclear Installations (Government of the Republic of Indonesia, Jakarta, n.d.)
14. RI, Presidential Instruction No. 4 of 2019 concerning Capacity Building in Preventing, Detecting, and Responding to Disease Outbreaks, Global Pandemics, and Nuclear, Biological, and Chemical Emergencies (Government of the Republic of Indonesia, Jakarta, 2019)
15. J. Asshiddiqie, Regarding Laws in Indonesia (Raja Grafindo Persada, Jakarta, 2011)
16. D. Apriliani and I. D. K. K. Widana, Proposals for the Implementation of Nuclear Preparedness Programs in the Utilization of Radiation Facilities and Radioactive Substances, (Proceedings of Nuclear Safety, 2019), 371-379 (2019)

Lesson learn earthquake disaster in Pidie Jaya Aceh in 2016 to realize a disaster-resistant community

Najib Rahman[1*], *Heridadi*[1], *Lasmono*[1], *Mulya* Imansyah[1], and *Fauzi* Bahar[1]

[1]Disaster Management Study Program, National Security Faculty, Republic of Indonesia Defense University. IPSC Sentul Area, Bogor, West Java, Indonesia

Abstract. Indirectly, to protect and achieve prosperity, national security is needed as the freedom for the state, society, and citizens from all the threats. Natural disasters is one of non-military threat that is very real in the territory of the United State of the Republic of Indonesia. Casualties, economic, social, environmental damage, and psychological impacts caused by the disaster are real conditions occur as a result of this non-military threat. The geographical position allows the threat of natural disasters due to the movement of tectonic plates and volcanic activity. This is a natural process always occurs in the cycle of life on earth. After an earthquake 9,1 magnitude followed a giant tsunami wave that devastated the city of Banda Aceh on December 26, 2004, an earthquake 6,5 magnitude occurred again which rocked the Pidie Jaya Regency area, Aceh Province on Wednesday, December 7, 2016, at 05.03.36 am. This research aims to study the management disaster during Pidie earthquake especially in emergency response. The results of this research are to make the earthquake in Pidie Jaya District in 2016 as a Lesson Learned for all parties, especially in Pidie Jaya District and in general, for all Indonesian people to form disaster-resilient communities who have awareness of the potential disasters that will be faced to contribute to the condition of National Security.

1 Introduction

The objective of the nation and the state of Indonesia is stated in the Preamble to the 1945 Constitution of the Republic of Indonesia, namely to protect the entire Indonesian nation and all spilled Indonesian blood and to advance the general welfare. Indirectly, to protect and achieve prosperity, national security is needed as a form of freedom for the state, society, and citizens from all forms of threats. Natural disasters are one of the threats that exist in Indonesia. This is supported by Indonesia's geographical position which is located on three earth plates, namely Indo-Australia, Eurasia, and the Pacific. The three earth plates are always moving and pressing against each other, causing an accumulation of collision energy which will eventually be released in the form of an earthquake [1]. This geographical position allows the threat of natural disasters due to tectonic plate movements and volcanic activity. Natural disasters that can be caused include earthquakes, tsunamis, and volcanic eruptions.

Indonesia is an active seismic area where during the years 1976-2006 there were 4,486 earthquakes with a magnitude of more than 6.0 on the Richter scale. Research by the Meteorology and Geophysics Agency (BMKG) since 1991-2009 there have been 27 destructive earthquakes and 13 earthquakes causing tsunamis. When averaged and rounded, Indonesia experiences earthquakes twice and tsunami once a year. In 2009 there was an earthquake that damaged Papua, Tasikmalaya, Padang and Ujung Kulon [2].

One of the provinces in Indonesia with quite intensive tectonic and volcanic activity is Aceh Province. One of the districts that experienced a devastating earthquake in 2016 is Pidie Jaya Regency. Pidie Jaya Regency is located at geographic coordinates 04°06 '- 04°47' North Latitude, 95°56 '- 96°30'E, with the northern borders directly adjacent to the Malacca Strait, to the east bordering Samalanga District, Bireun Regency. The south is directly adjacent to Tangse, Mane, and Geumpang Districts, Pidie Regency, in the west it is directly adjacent to Glumpang Tiga, Glumpang Baro, and Kembang Tanjong Districts, Pidie Regency. Pidie Jaya Regency has an area of ?? 1,162.84 Km^2 and a population of approximately 165,000 people [3]

Topographically, Pidie Jaya Regency is at an altitude of 0 masl to 2300 masl with a slope of land between 0 to 40%. The area of Jangkabuya Subdistrict as a whole is a lowland area between 0 masl to 20 masl, Bandar Dua District is located at 10 masl s.dpl. 2300 masl while the Subdistricts of Ulim, Meurah Dua, Meureudu, Trienggadeng, Pante Raja, and Bandar Baru are located at 0 masl to 2,300 masl stretching from the Malacca Strait Coast to the Peak of Gunong Peuet Sagoe in the Bukit Barisan Cluster[4]. Overall, Pidie Jaya Regency is prone

* Corresponding author : najibrahman1106@gmail.com

to flooding and erosion. From the slope classification, Pidie Jaya Regency is a highland area that has a slope class area of up to 40%. After an earthquake measuring 9.1 magnitudes with a giant tsunami wave that devastated the city of Banda Aceh, Aceh Province on December 26, 2004. On Wednesday, December 7, 2016, at 05.03.036 WIB, an earthquake measuring 6.5 magnitudes occurred again which rocked the Pidie Regency area. Jaya, Aceh Province. This earthquake has resulted in significant casualties. Some victims suffered minor injuries to the point where the victim died. This earthquake also resulted in infrastructure damage in the form of buildings that were slightly damaged and collapsed [5]

The earthquake in Pidie Jaya was a natural event whose impact was not only visible on damage to building structures but also had an impact on the survival of the victims. According to data from BNPB, 102 victims died and 93 of them have been identified, and at least 650 people suffered serious and minor injuries. Not less than 85 thousand people packed the evacuation sites scattered in several sub-districts in the Pidie Jaya area. Non-material impacts such as health, mental and emotional stability (trauma), livelihoods, education, and others are considered commensurate with or maybe more significant than the impact of damage to buildings. The value of damage that is material (tangible) may be calculated immediately and followed by rehabilitation and reconstruction immediately. However, non-material impacts require more special handling and take into account the phased aspects of the time. Therefore, it is necessary to further study post-earthquake rehabilitation and recovery programs that are suitable and provide long-term stability impacts on victims as a whole from economic, social, environmental, health, education, and other aspects [6].

The purpose of this research is to provide an overview and make the Earthquake incident in Pidie Jaya Regency a lesson for all Indonesian people to create a disaster-resilient community to support National Security. The Earthquake Disaster that hit Pidie Jaya Regency in 2016 was an event that could be used as a learning for the entire community on how to process and analyze the disaster cycle that occurred and how to manage disasters more effectively and efficiently in the future.

2 Method

The methodology used in this research is a qualitative method with data collection techniques, namely interviews, observation, and document study. Qualitative research is a research procedure that produces descriptive data consisting of written and spoken words from sources [7]. Data collection techniques in this study were interviews, observation, and document study. In-depth interview techniques to obtain primary data from sources. Research sources came from related agencies, namely BPBD Pidie Jaya Regency, Syiah Kuala University Academics, Pidie Jaya District Social Service, Non-Governmental Organization (NGO) Disaster Risk Reduction Forum (FPRB) Pidie Jaya, Military District Command (KODIM) Pidie Jaya Regency, Pidie Jaya District Health Office, Pidie Jaya Regency Public Works (PU) Office and Pidie Jaya Regency Environmental Service (DLH). Documentation techniques to collect secondary data so that it supports the data obtained in the interview technique.

The data analysis in this study used the Miles and Huberman model. Data analysis based on Miles and Huberman is data reduction, display data, and conclusion drawing and verification [8].

3 Result and Discussion

Pidie Jaya is a district located in Aceh province with the capital city of Meureudu. Pidie Jaya Regency consists of 8 districts, 34 Mukim, and 222 villages. The formation of Pidie Jaya Regency is based on Law Number 7 of 2007 on January 2, 2007 [9]. Based on its geographic location, Pidie Jaya Regency is depicted on the following map:

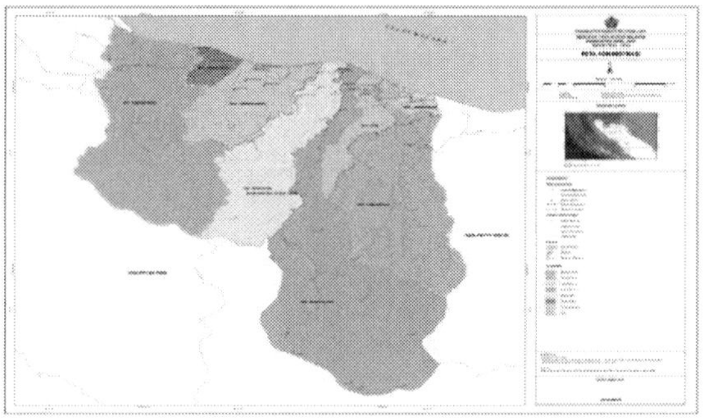

Fig. 1. Administrative Map of Pidie Jaya Regency [3]

Disaster Management Command System in the earthquake emergency response phase in Pidie Jaya district in 2016 in a disaster emergency, the Governor / Regent / Mayor appointed a disaster incidence commander at the suggestion of the Head of BPBD. Following the Regent's Decree Number 383 of 2016, the Deputy Regent was appointed as the Task Force for Emergency Response of the Pidie Jaya Earthquake. Furthermore, in the form of clusters (Sub Task Force) in disaster management such as the Refugee and Social Cluster, Field Cleaning, Health Clusters, and others [10].

Disaster Management Command System in the emergency response phase of the earthquake in Pidie Jaya district in 2016 in a disaster emergency, the Governor / Regent / Mayor appointed a disaster incident commander (Incidence Commander) at the suggestion of the Head of BPBD. Following the Regent's Decree Number 383 of 2016, the Deputy Regent was appointed as the Task Force for Emergency Response of the Pidie Jaya Earthquake. Furthermore, in the form of clusters (Sub Task Force) in disaster management such as the Refugee and Social Cluster, Field Cleaning, Health Clusters, and others.

Equations should be centered and should be numbered with the number on the right-hand side.

Table 1 Data on Damage and Loss of Pidie Jaya Disaster in 2016 [3]

No	Indicator	Description	Data Source
1	Victim Died	97 People	Rehab and Reconsent Reaction Plan for the 2017-2019 earthquake
2	Broken House	Several damaged houses: 3,205 and Moderately damaged houses: 3,269 Total 6,474 Housing Unit	Rehab and Reconsent Reaction Plan for the 2017-2019 earthquake
3	Material Loss	Rp. 511,776,697,410	Rehab and Reconsent Reaction Plan for the 2017-2019 earthquake
4	Damage Public Facilities	National Roads and Bridges Network, Provincial Roads and Bridges Network, District Roads and Bridges Network, Clean Water Network, irrigation Network, Dams, Beach Protection Buildings, Office Buildings, Hospitals, Public Health Center, Markets Facilities and services	Rehab and Reconsent Reaction Plan for the 2017-2019 earthquake

Institutions involved in the First Emergency Response period include the Indonesian Ministry of Health (Directorate of Mental Health), PDSKJI Aceh (Psychiatrist at Tamiang Hospital, Biruen, North Aceh, Central Aceh, Mental Hospital, Aceh Jaya, Pidie), CMHN Trienggadeng, CMHN Banda Aceh, CMHN Pidie, Aceh Besar Health Office, Pidie Health Office, North Aceh Health Office, Directorate General P2P Ministry of Health, Aceh HIMPSI, IPK ACEH, East Kalimantan HIMPSI, IMHA, IPKJI, UIN Psychology Students, Unsyiah, P3KJB, PKM Bandar Dua, Aceh Police, PPNI Aceh, Puskesmas Keumala, Marzoeki Mahdi Hospital Bogor, RSJ Dr. Muwardi Solo, RSJ West Java, Soeharto Heerdjan Jakarta, PDSKJI Community Psychiatry and Crisis Center, Faculty of Psychology UI, UGM Disaster Team who went directly to the field [11].

Volunteers Helping the Pidie Jaya Earthquake, namely a volunteer team totaling 1,174 people, consisting of 186 government and non-government organizations including Dompet Dhuafa Aceh, Rumah Zakat, PKPU, KARINA, MDMC, WVI, nYEU, MSF, PMI, UGM Disaster Team with expertise as follows: (1) Specialist doctors: 58 people; (2) General practitioners: 241; (3) Psychiatrist / Psychologist: 50; (4) Paramedics: 408; (5) Other supporters: 417. Based on the origin of volunteers; (1) Countries: 2 Countries (Malaysia and Singapore); (2) Provinces: 13 Provinces (Aceh, North Sumatra, West Sumatra, Riau, Riau Islands, Bangka Belitung, DKI Jakarta, DIY, West Java, Central Java, East Java, East Kalimantan, South Sulawesi); (3) Regencies: 23 districts / cities in Aceh, and 16 from cities from other provinces (Medan City, Langkat, Binjai, Tebing Tinggi, Padang, Pekan Baru, Batam, Bangka Belitung, Jakarta, Bandung , Yogyakarta, Semarang, Solo, Surabaya, Bontang and Makassar); (4) Under the coordination of the Head of the Aceh Health Service; the Aceh Health Service Sub-Cluster Team took turns to the location and the permanent team was the UPTD P2KK Team located at the Pidie Jaya Health Post. These teams are assigned and function to provide health services for disaster sites and assist Pidie Jaya District Health officers, who are victims of the disaster, either directly or indirectly [12]. The following is data on damage to health facilities at the time of the earthquake in Pidie Jaya Regency Pidie Jaya;

Table 2. Damage Data to Health Facilities [3]

No	Health facility	Type Of Damage			Total
		Severe	Moderate	Light	
1	Public Health Center		9		9
2	Hospitals	1			1
3	Doctor's Houses			28	28
4	Post	3	10	4	17
5	Post-Dpctor's	14	37	29	80
6	Public Health Office			1	1
7	Pharmacy Warehouses			1	1
	Total	18	56	63	137

On December 7, 2016, an earthquake disaster occurred in Pidie Jaya and resulted in achievements in the following stages: Emergency Response 7-20, Rehabilitation-Reconstruction Action Plan 2017-2019, Presidential Instruction for Acceleration of Rehabilitation-Reconstruction Action Plan, Ground-breaking of earthquake victims' houses on April 21, 2018, in 2019 the physical recovery has been completed but non-physical is still unknown. In Pidie Jaya the delay in recovery was caused by the impact of the disaster on the fiscal, namely revenues and expenditures. One of the factors of socio-economic recovery is always being late because disaster recovery actors do not understand the character of the relationship for each period. Different stages of disaster management require different resources [13].

Based on the results of interviews and direct in-depth studies with the Head of the Public Works and People's Permahan through a zoom meeting that the Public Works and Public Housing (PUPR) Office of Pidie Jaya Regency in carrying out the disaster management for the 2016 Pidie Jaya earthquake experienced many obstacles including heavy equipment. Regarding heavy equipment, the local government has 4 heavy equipment to clean debris after the earthquake. The Public Works and Public Housing

(PUPR) Office also requested assistance from academics to record damage and conduct rapid assessments of the earthquake disaster in Pidie Jaya [14].

The main points of post-disaster rehabilitation and reconstruction policies for the earthquake in Pidie Jaya, Pidie, and Bireuen: Facilities for building communities, principles of sustainable development, and disaster risk reduction, good governance and prioritizing the aspirations of disaster-affected communities, participatory approaches under local cultural characteristics, Earthquake-friendly housing technical standards, the openness of information, accountable distribution of funds and accountability, effective coordination and cooperation between parties across sectors, Considering the impact of damage and budget availability, the post-disaster rehabilitation, and reconstruction plan will cover the 2016 - 2018 fiscal year [15].

Problems that arise during the Emergency Period, such as the evacuation of bodies, looting, mass graves, misappropriation of aid, education, and opening of markets. Problems that arise during Rehabilitation and Reconstruction, such as relocation, land conflicts, uncooperative government officials, and rogue contractors. Good restoration is highly recommended for an indirect recovery stage from relocation to permanent buildings. There needs to be a smooth stage from emergency tents, shelters, to the construction of permanent houses. It is hoped that problems that arise during the emergency and rehabilitation and reconstruction period can be handled perfectly and create a Build Back Better. Physical recovery is easier, but socio-economic recovery will take longer because it requires a more comprehensive approach. Not only relying on the government, but universities with available resources have an important role to play in dealing with disasters.

There are 5 (five) focuses on the integration of spatial planning with disaster risk reduction, namely: (1). Integration of disaster risk studies into urban planning including special attention to settlements that are prone to disasters; (2) Prioritizing disaster risk considerations for vital infrastructure development activities; (3) Development and use of monitoring tools to measure aspects of risk reduction obtained on a spatial planning policy; (4) Integration of disaster risk studies into rural development planning, especially in mountainous and coastal areas; (5) Revision or development of building codes and reconstruction and rehabilitation practices at the national and local levels [16].

The pattern of spatial use in protected areas is aimed at realizing environmental sustainability, increasing the carrying capacity of the environment, and maintaining the balance of the ecosystem between areas to support the sustainable development process. Based on these considerations, the protected area utilization plan is: (1). Directing the function of the protected area which includes plans for spatial use in areas that protect the areas under it, nature reserves, local protected areas, and disaster areas; (2). Maintain water catchment areas or areas that have a hydrological function to ensure the availability of water sources; (3). Controlling the use of space outside the forest area so that it continues to function in protection [17].

The development of the spatial use area in the cultivation area aims to maintain the quality of the carrying capacity of the Pidie Jaya district in the planning area to create jobs, create harmony with the spatial structure plan being developed. The cultivation area includes residential areas, food crop agricultural areas, annual crops/plantations, livestock areas, fisheries areas, and tourism areas.

The development of disaster-prone areas is an effort to maintain balance against natural conditions that are prone to change. These changes occur due to floods, landslides, abrasion and earthquakes, and other changes. These disaster-prone areas must be protected from human settlements through various plans, namely forming borders and green open areas to minimize the impact of changes in these conditions. Disaster-prone areas are coastal areas of the sea and river banks (DAS).

The Concept of Flood Disaster Prone Area Arrangement, Includes: (1). Arrangement of Upstream Watershed Areas; (2). Management of Downstream Watershed Areas. The concept of Earthquake Disaster Disaster Management Arrangement includes: (1). Recommendations for RTH for Multi-functional City Disaster Evacuation Sites; (2) Disaster Evacuation Route System; (3). Evaluation of the New Development Plan at KRB Earthquake; (4). Space Allocation Alternative and Space Utilization Intensity Limitation; (5). Concept of Building Period; (6). Building Code in Earthquake Prone Areas [18].

The concept of Tsunami Disaster Prone Area Arrangement includes; (a). Structural Mitigation Development; (b) Management of Mangrove Areas; (c) Conservation of Coral Reefs. The concept of Landslide Hazard Area Arrangement, including; (1) Development of Structural Mitigation; (2) Management and Control of Protected Forest Utilization and Forest and Land Rehabilitation in Landslide KRB; (3) Soil and Water Conservation.

Disaster Risk Reduction in Pidie Jaya Regency is carried out by considering sustainability aspects and participation of all parties. This effort is carried out with a strong commitment by prioritizing actions that must be prioritized. Disaster Mitigation is carried out to reduce the risks and impacts caused by disasters on communities living in disaster-prone areas through: (a). Planning and implementation of spatial planning based on disaster risk analysis; (b) Regulation, development, infrastructure development, and building layout; (c). Providing conventional and modern education, training, and counseling.

The success of disaster risk reduction efforts in an area cannot be separated from the quality of the environment and enforcement of spatial planning, which can be seen from: (1) Availability of information on disaster-prone areas/areas; (2). The implementation of the licensing process for development activities is guided by the Disaster-based District / District Spatial Plan and the fulfillment of environmental quality. Control and enforcement of disaster-resistant building structures, including: (1). Supervision and control of land use in areas

https://doi.org/10.1051/e3sconf/202133104004

that are vulnerable to disasters; (2). Availability of a disaster evacuation route/space; (3). Increase public awareness of maintaining good spatial and environmental quality.

The following is an overview of disaster-prone areas in the Pidie Jaya district:

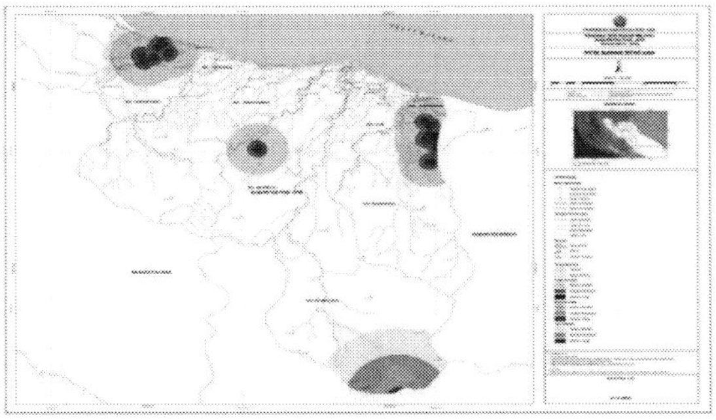

Fig. 2. Map of Disaster Hazard in Pidie Jaya Regency [3]

Based on Fig. 2, a tabulation can be made about the sub-districts in Pidie Jaya Regency which have the highest number of natural disaster threats which csan be seen in Table 3. It can be seen that Meureudu, Meurah Dua, and Bandar Baru sub-districts have three natural disaster threats and Bandar dua sub-districts which have two threats and the rest of each region has one threat of disaster, namely Tsunami.

After the 2004 Aceh earthquake and tsunami, parts of Aceh, namely Pidie Jaya district, experienced another devastating disaster in 2016, namely an earthquake. This incident gave psychological trauma to all members of the community. Many residents lost their parents, children, siblings, friends, and friends. This incident also resulted in the loss of property, environmental damage, and loss of hope for the people of Pidie Jaya Regency. Therefore the importance of vigilance and preparedness of all components of society and the nation is facing the conditions of a disaster that will occur and how to reduce the impact of the resulting risks to the minimum. The lesson (Lesson Learn) regarding the Pidie Jaya earthquake has made the community even more resilient to disasters in the future. The emergence of an awareness of disasters has made people work hand in hand to face and build disaster preparedness. The formation of disaster-resilient villages, disaster prepared schools is a description of awareness and evidence of community learning about disasters that occur to form disaster-resilient communities that will contribute to National Security [19].

4 Conclusion And Suggestion

Lessons learned from the 2016 Earthquake disaster in Pidie Jaya Regency were: (a). Initiatives must be initiated from higher education institutions because they have the resources to make initial efforts; (b). Provide recommendations to the government regarding the results of rapid assessments; (c). Various Stakeholders can always accompany the results of the rapid assessment up to the rehabilitation and reconstruction stage; (d). Disaster evacuation management must be faster and more systematic; (e) Providing more comfortable and attractive evacuation tents for earthquake victims such as disaster victims in Japan; (f). The 4000 Pidie Jaya refugees lacked facilities such as public kitchens, only 2 (two) were provided so that the distribution of consumption was often late. The bathroom which was only provided with 4 pieces caused illness in the evacuation victims. In the event of a disaster, there are always lessons to be learned for system improvement and management of what will be faced in the future.

Quick response, quick assessments, and good coordination must be carried out immediately because an action plan will emerge. Quick assessment from the socio-economic aspect: (a). Coordination between institutions and to the community: establishment of posts, contingency plans, consolidation at the provincial level; (b). Logistics channeling mechanism to the community: distribution points must be recorded by involving students and other components of society; (c). The decision to evacuate the community: there needs to be coordination and an appeal to rely on the community to evacuate themselves; (d). Capacity for the number of refugees and production capacity: mobilization must be accompanied by mobilization of resources; (e). Health and hospital service aspects: services must always be ready, either from hospital facilities or from the trauma-healing aspect of post-disaster. Processes, fast and precise responses are the keys to success in handling the earthquake disaster that occurred in the Pidie Jaya district. Careful coordination and good preparedness will form a community resilient to disasters and will ultimately contribute to a stable condition of National Security.

References

1 Center for Volcanology and Geological Mitigation., Earthquake and Tsunami. Geological Agency, Ministry of Energy and Mineral Resources (2015).

2 Sunarjo, M. T. G., S. Pribadi, Popular Edition Earthquakes. Meteorology Climatology and Geophysics Council (2012).

3 BPBD of Piedie Jaya Regency, General Condition of Pidie Jaya Regency (2020).

4 Akbar, K. N., *Earthquake Geographic Information System Damaged In Indonesia In The Center Of Vulanology And Web-Based Geological Disaster Mitigation.* (2011).

5 S. Maarif, R. Parmono, R. A. Kinseng, E. Sunarti, Contestation of Knowledge and Meaning of Natural Disaster Threats: A Case Study of Mount Merapi Disaster Threats. Journal of Dialogue and Disaster Management, **3** (1), 1-13, (2012).

6 D. Hidayati, Community preparedness: A new paradigm for natural disaster management. Journal of Indonesian Population, **3** (1), 69-84, (2008).

7 L. J. Moleong, *Qualitative Research Methodology.* PT. Rosdakarya youth, (2010).

8 Sugiyono, *Quantitative Approach Business Research Methods, Qualitative and R & D*. Alfabeta (Bandung, 2010).

9 National Center for Earthquake Studies Team, Pidie Jaya Earthquake Study in Aceh Province, Indonesia, December, 7th 2016 (M6.5). Research and Development Center for Housing and Settlements Bdan Research and Development Ministry of Public Works and Public Housing (2017).

10 M. Fadri, M. Zuhri, Effectiveness of Regional Disaster Management Agency Functions in Earthquake Management in Pidie Jaya Regency. Student Scientific Journal in the Field of State Law, **2** (2), 336-346 (2018).

11 K. Wikantika, P. A. D. Prastiwi, R. Vitriana, A. B. Harto, Identification of Post-Earthquake Damage Using the Object-Based Image Analyzer (OBIA) Method (Case Study: Pidie Jaya, Aceh) (2017).

12 Pidie Jaya District Health Office, Health Sector Condition Data (2020).

13 Q. Qadariah, A. V. Simanjuntak, M. Umar, Analysis of Focal Mechanisms Using Waveform Inversion; Case Study of Pidie Jaya Earthquake on december 7th 2016. Journal of Aceh Physics Society, **7** (3), 127-132, (2018).

14 A. Rahma, Implementing disaster risk reduction (DRR) programs through formal education. Journal of VARIDIKA, **30** (1), 1-11, (2018).

15 H. Susetyo, S. H., L, S. I, M., Towards a comprehensive security paradigm with a human security perspective in Indonesia's national security policy, Esa Unggul University (2008).

16 S. Mukhtar, National security: between theory and practice in indonesia 1. Sociae Polites, 127-137, (2017).

17 K. Anggoro, National security, national defense, and public order. In Seminar on National Law Development VIII. Denpasar: National Law Development Agency, Ministry of Justice and Human Rights of the Republic of Indonesia (2003).

18 K. Munadi, D. Yuliansyah, Multi Threat Information System for Natural Disasters in Aceh. In the National Informatics Seminar (2011).

19 B. Darmono, Indonesian National Security System and Concept. National Journal of Resilience, **15** (1), 1-42, (2010)

https://doi.org/10.1051/e3sconf/202133104005

Analysis of capacity community level in various potential disasters at dinoyo and sukun sub-district, Malang city

Mukhamad Fathoni[1*], *Turniningtyas* Ayu R[2], *Fadly* Usman[2], *Sunaryo*[2], *Lestari* Sumi[2], *Ah* Yusuf[3], and *Christrijogo* Sumartono[4]

[1]Department of Nursing, Faculty of Medicine, Universitas Brawijaya, Malang and Ph.D. in Nursing student, Faculty of Nursing, Airlangga University
[2]Universitas Brawijaya
[3]Faculty of Nursing, Airlangga University
[4]Head of School Disaster, Airlangga University

Abstract. The natural disaster in pandemic covid-19 will still ongoing in East Java, Indonesia. Apart from the virus pandemic, there are still several potential disasters in that sub-district, namely floods and landslides. Many disasters have occurred resulting in the need for people living in the area to anticipate, especially if the three disasters occur simultaneously. In Malang, this case of sub-district apparatus and the BPBD Malang City has not been able to optimally help restore the condition of the community due to the complexity of conditions that occur in the field. Therefore, the community needs to know their capacity to be able to anticipate if it is two or three will happen. Capacity is a combination of all the forces that exist in a community, social or organizational group that can reduce the impact of disaster risk or impact (UN-ISDR, 2004). The capacity assessment in this study identifies the capitals owned by the community. These capitals include natural, human, physical, financial, and social capital. The five capitals will later be described in the pentagon assets so that it will be known which capacities in the capital are the strengths and weaknesses of the people of Dinoyo and Sukun sub-district

1 Introduction

Disasters can occur at any time in any part of the world with a dramatic impact on individuals, families, and communities, causing a threat to the quality of life (International Council of Nurse (ICN) & World Health Organization (WHO), 2009). Disasters arise when a threat meets vulnerable people who have low or no capacity to respond to the threat. The combination of the two causes disruption of people's lives.

Disasters have a very big influence on humans and the surrounding environment such as mass death, disability, hunger, poverty, and infrastructure destruction (Mizam, 2012). According to Indonesian Disaster Data and Information, in 2016, there were 2,342 disasters, an increase of 35% when compared to the number of disasters in 2015 that occurred 1,582 disasters. If averaged out, it means that every day there are about five disasters that hit Indonesia. The geographical location of Malang City which is surrounded by Mount Semeru, Mount Arjuna, and Kawi makes this area also an earthquake-prone area. Meanwhile, the potential for flooding and landslides around the watershed is also large, seeing that the city of Malang is passed by four major rivers, namely the Brantas River, the Amprong River, the Metro River, and the Bango River. Disaster incidence data in Malang City in 2019 at least 217 incidents (data from PUSDALOPS BPBD Malang City) which can be categorized as disasters. The disaster occurred in five sub-districts which were grouped into several types, namely landslides, flooding / momentary inundation, tornadoes and the last one was the COVID-19 outbreak which is still ongoing today. The impact of the disaster can be seen directly from the damage and losses. Meanwhile, in the social aspect, it can be seen from the disruption of access, impaired function, and increased risk. The people affected by the disaster not only experienced the impact of physical damage to house buildings but also suffered social, economic, cultural, and psychological losses.

Floods and landslides have an impact on decreasing the quality of clean water; damaged social facilities such as meeting halls, road access, places of worship, schools; family sanitation so that the cost of fulfilling their daily needs will increase. Moreover, if treatment is not done immediately, there will be an increased risk such as an increase in disease outbreaks. Then in the economic sector, people's businesses suffer damage from both business premises, business products, and customers.

* Corresponding author: mfathony@ub.ac.id

Loss of access to work for families will result in income losses.

To handle post-disaster and minimize the risk increase due to disasters, BPBD Malang City conducts Post-Disaster Needs Assessment Activities. These activities include surveys of the social, economic, cultural, and psychological needs of the post-disaster community. Then the survey results will become material for a Focus Group Discussion (FGD) which will later be used as the basis for the preparation of the Post-Disaster Recovery Needs Assessment Results Document (Jitupasna).

2 Method

2.1 Sampling Technique

Sampling in this study using the purposive sampling technique. The purposive sampling technique is a purposive sampling technique. The sample is intended for respondents who will be given a questionnaire. Sampling will be addressed to all participants in Sukun and Dinoyo Villages who attend the meeting. The analytical tool used in determining the sample in each kelurahan is the Linear Time Function, which is an analysis based on the estimation of time constraints. This is used considering that it is still in the state of the COVID 19 pandemic, so space and time are limited. The formula is used as follows:

$$n = \frac{T - t_0}{t_i} \quad (1)$$

Note:

n = amount of minimum sample
T = avaliable time (24 days x 2 hour/day = 48 hour)
t_0 = time to take sample (12 days x 2 hour/day = 24 hour)
t_I = time to need for people to filling quesioner (0,25 hour)

$$n = \frac{48 - 24}{0{,}25}$$

n = 96 sample

The sample required is 96, with a proportion of 30 samples from the Dinoyo Sub-district and 66 samples from the Sukun Sub-district.

2.2 Data Collection Technique

Data collection is a standard for obtaining the required data. The data collection technique is in the form of primary data collection which is obtained from distributing questionnaires.

2.3 Capacity Level Analysis

The calculation capacity level of Sukun and Dinoyo Sub-district people was carried out by distributing questionnaires according to the sample proportions. The questionnaire contains questions related to capacity indicators and the classification of each parameter to be assessed in Table 1.

Table 1 Indicators of Capacity Community [1,4,6]

Capacity	Indicator
Natural Capital	1. Average condition of water quality
Social Capital	1. The average kinship between neighbors 2. Average public confidence in circulating disaster information
Human Capital	1. Average knowledge of signs of disaster 2. Average people get education disaster
Financial Capital	1. Average people's income 2. Average saving ownership 3. Average people receive assistance
Physic Capital	1. Average road and bridge condition

Before determining the level of capacity the data is processed first, because the unit of analysis desired is RW, while the data obtained is at the level of the head of the family (KK). Then it needs to be changed according to equation 2

$$Data\ for\ each\ subdistrict = \frac{\sum data\ KK}{amount\ of\ people} \quad (2)$$

Next, determine the level of capacity through the following steps:

1. The final score (equation 2) for each capital is converted to a value of 1,2, and 3 for low, medium, and high
2. Calculate conversion result with equation 3.

$$L_{pentagon} = \left(\tfrac{1}{2}\ ab.\sin 72^0\right) + \left(\tfrac{1}{2}\ bc.\sin 72^0\right) + \left(\tfrac{1}{2}\ cd.\sin 72^0\right) + \left(\tfrac{1}{2}\ de.\sin 72^0\right) + \left(\tfrac{1}{2}\ ea.\sin 72^0\right) \quad (3)$$

3. Do it with the other four capitals to get a complete pentagon image as shown in **Fig 1**

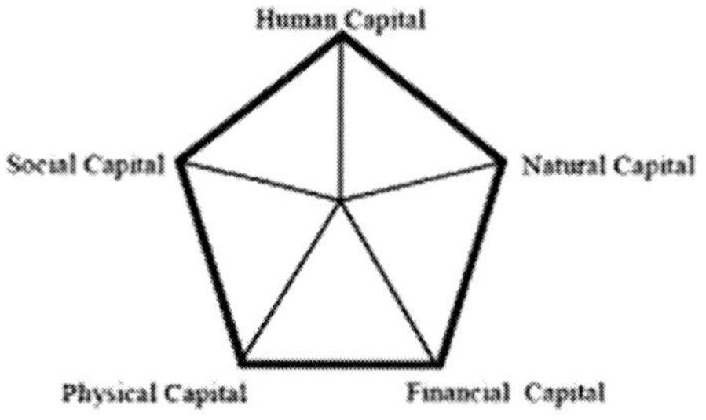

Fig 1 Pentagon Asset [4]

3 Result and Discussion

3.1 Natural Capital

The natural capital used in this study is the condition of drinking water quality. Drinking water is a basic need of urban communities. Good quality drinking water is a sign that the environment in the region is still preserved beautifully.

In Fig 2, it is known that the quality of drinking water in the two villages is quite good. The number of respondents who stated that it was quite good was 53 respondents. Good quality drinking water can be consumed by the community and indirectly affects public health. Consuming good drinking water will be able to increase the body's resilience in facing various disasters that have the potential to occur.

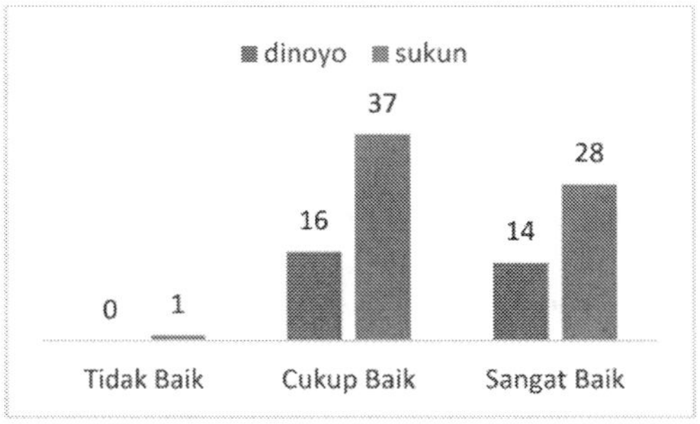

Fig 2 Water Quality

3.2 Social Capital

Factors that affect social capital in Dinoyo and breadfruit are a kinship between neighbors and the level of public confidence in the delivery of information. Kinship is seen based on how close the person is to the neighbor because the neighbor is the first person to help in the event of a disaster.

According to Fig 3, it is known that the average value for kinship is in the range of 2 to 3. Number 2 means quite good and number 3 shows good information. So that the good kinship in Dinoyo and Sukun Kelurahan is pretty good to good.

Then on the factor of public trust in the disaster information presented, the people of Dinoyo and Sukun Subdistrict have an average score of 2 to 3 again. Score 2 means that information from the Subdistrict government is trusted, while score 3 means that news delivery that people trust comes from BPBD Malang City. Dinoyo Subdistrict is more likely to trust the information conveyed by BPBD, while Sukun Subdistrict has more trust in Subdistrict government.

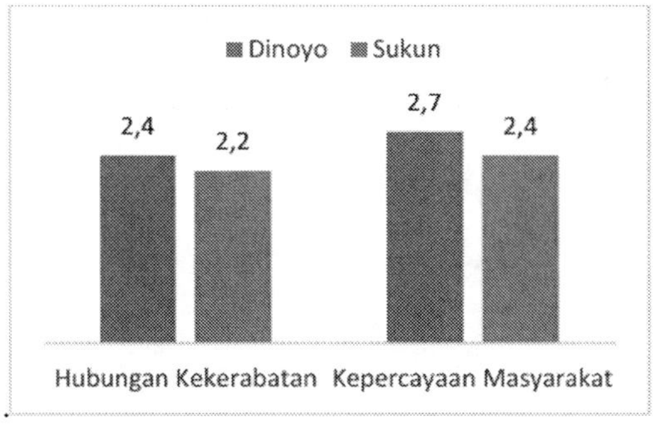

Fig 3 Kinship and Community Trust Relationship

3.3 Human Capital

Human capital is measured through the knowledge of respondents to be able to recognize signs of disasters, including floods, landslides, and the COVID-19 pandemic as well as participation in counseling on various disasters.

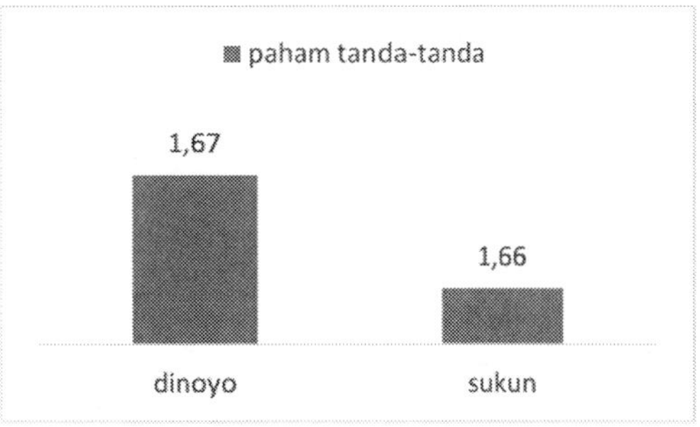

Fig 4 Public Understanding in Knowing the Signs of Disaster

Public understanding in knowing the signs of a disaster is still lacking. The value of this factor shows the number 1. Understanding to be able to know the signs is very important for the community to be able to avoid the disaster that will occur.

Then in the extension participation component, the community of the two sub-districts has on average attended counseling. There are 90% of respondents in the Dinoyo Village who stated that they had attended counseling, while for the respondents in the Sukun Village it was 71%. Respondents thought that the education related to this disaster was quite useful for them. In addition to increasing knowledge, it can also anticipate potential disasters that occur.

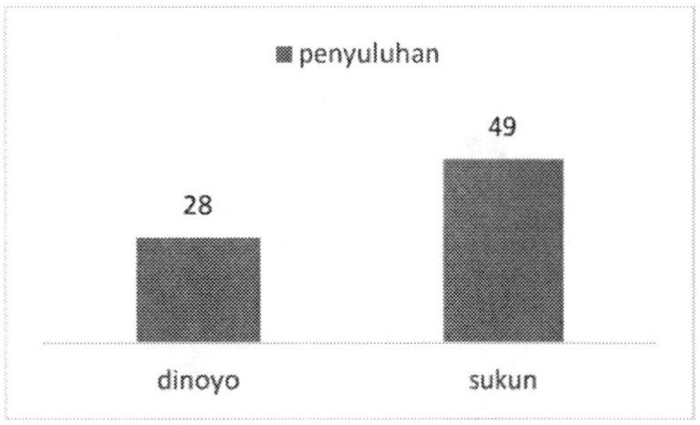

Fig 5 Community Attended Disaster Education

3.4 Financial Capital

Financial capital is the ability of the community to be able to recover from disasters, including floods, landslides, and COVID-19 pandemic, through their assets. The forms of these assets are in the form of income, ownership of savings, and assistance. Fig 6 shows the average income in the two subdistricts. The income of the people of the Sukun Subdistrict is greater than Dinoyo Subdistrict. The average income of Sukun residents is IDR 2,072,879, while in the Dinoyo community it is only IDR 1,774,333. High income indicates that the community can survive in the event of a disaster because it can spend money on necessary needs.

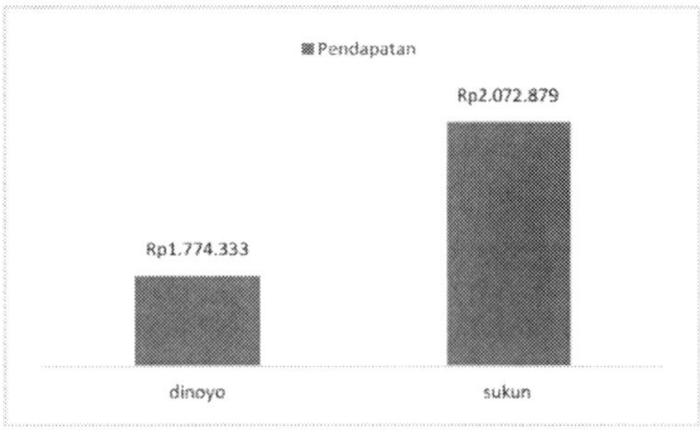

Fig 6 Average Income

Apart from income, factors affecting financial capital are ownership of savings and assistance. Fig 7 shows that the ownership of savings in the two urban villages is quite small if you look at the various types of financial institutions that are very large in Malang City. However, when compared to the Dinoyo people, 43% of respondents have savings. The amount of savings ranges from hundreds of thousands of rupiah. Ownership of savings for the community is very important to be able to survive in the current conditions. Because savings have liquidity, money can be withdrawn at any time.

Furthermore, in the aid component, the value of assistance received by respondents was still low. Only 45% of the respondents in the Sukun Subdistrict received assistance, while in Dinoyo Subdistrict it was even smaller. The funds disbursed by the central government to help the community during this pandemic were many and varied. Generally, people get assistance from the government, tough subdistricts, and banks. These assistance programs vary, some in the form of "Kartu Pra Kerja", "Bantuan Langsung Tunai (BLT)", Social Security and others. The form of assistance received was generally in the form of money, but there were also necessities.

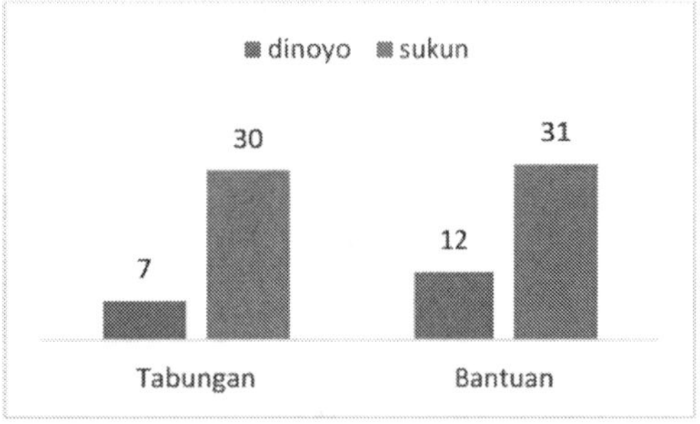

Fig 7 Savings ownership and assistance

3.5 Physic Capital

Aspects that are used to calculate the physical capital are the condition of roads and bridges. Good road and bridge conditions will affect the ease of accessibility. Easy access will increase evacuation mobilization when a disaster occurs. The road and bridge conditions in Dinoyo and Sukun Subdistrict are quite good. This is indicated by a value indicating that it is at a value of 2.

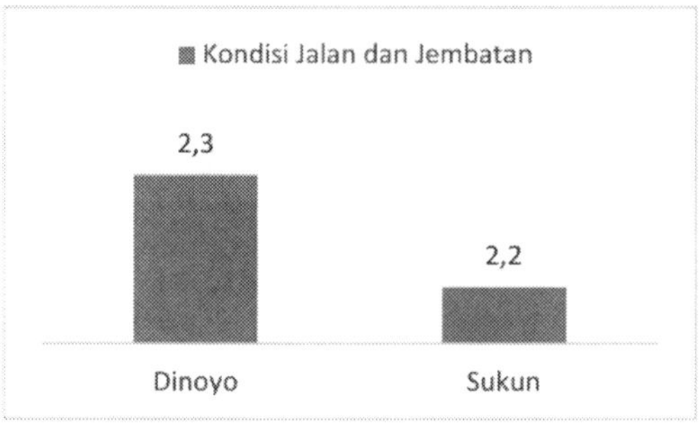

Fig 8 Condition of Road and Bridge

3.6 Capacity Community Level

Rated capacity is calculated based on the value of each stock using a composite index. Categories of the low, medium, and high which are composite index results are then translated into scoring form with a value of 1, 2, and 3. These values are used to create radar charts. Radar diagram is what is known as asset pentagon (Fig 9 and Fig 10)

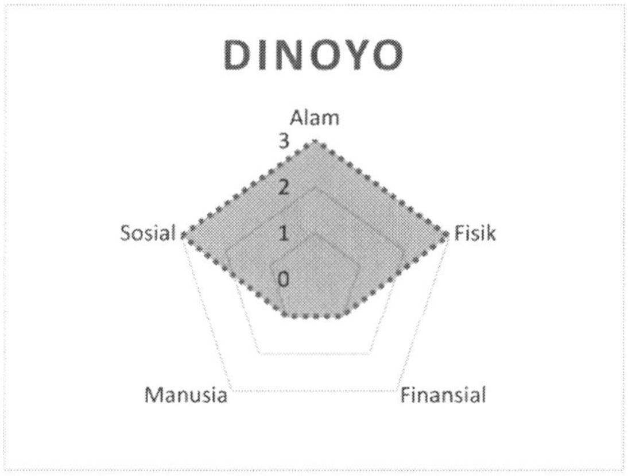

Fig 9 Capacity Communtiy in Dinoyo Subdistrict

Fig 10 Capacity Communtiy in Sukun Subdistrict

The illustration results in Fig 9 and Fig 10 show the strengths and weaknesses of the capacities or abilities in the two kelurahan. Dinoyo Kelurahan has excellence in three capacities. These three capacities include Natural, social and physical capital. Meanwhile, the Sukun sub-district is dominant in human and financial capital.

4. Conclusion and Recommendation

4.1 Conclusion

The conclusions obtained from the results of the above discussion include:

1 Dinoyo Subdistrict has advantages in social, natural, and physical capacities
2 Sukun Subdistrict has advantages in human and financial capacity

4.2 Recommendation

1 Required to increase the scope of the study area if the unit of analysis used was the village.
2 It is hoped that the results of the capacity level can become a reference for calculating disaster risk and disaster needs.

In the preparation of this study, there was support from various parties. Researchers specifically thank LPPM UB, Resilient Subdistrict Residents in Sukun and Dinoyo for taking research respondents, Permits and Facilitation by the Government of Lowokwaru District, Sukun District, Dinoyo Subdistrict and Sukun Subdistrict, Malang City BPBD, and Malang City Government who are communities which have helped both data and personnel support.

References

1. Department for International Development, Sustainable Livelihoods Guidance Sheets, Brighton: DFID (1999)
2. International Council of Nurse (ICN) & World Health Organization (WHO), ICN Framework of Disaster Nursing Competencies, WHO/WPRO. Geneva (2009)
3. A. K. Mizam, The Role of Health Workers in Disaster Management, Husada Media Health Scientific Journal, **1**(1) (2012)
4. S. Saragih, J. Lassa, A. Ramli, Sustainable Livelihoods Framework (2007)
5. UN-ISDR (United Nations secretariat of the International Strategy for Disaster Reduction). Living with risk: A global review of disaster reduction initiatives. Geneva: UN Publications (2004)
6. R. Wijayanti, M. Baiquni, R. Harini, Asset-Based Community Sustainable Livelihood Strategy in Pusur Sub-watershed, Bengawan Solo Watershed. Region and Environment, 133-152 (2016)

https://doi.org/10.1051/e3sconf/202133104006

Case study: mapping the tsunami-hazard zone in the Pasir Jambak sub-district, in Padang, Indonesia

Leli Honesti[1] and *Meli* Muchlian[1*]

[1]Department of Civil Engineering, Faculty of Engineering, Padang Institute of Technology, Indonesia

Abstract. A tsunami hazard is an adverse event that causes damage to properties and loss of life. The problem in assessing a tsunami risk zone for a small area is significant, as available tsunami inundation zone data does not give detailed information for tsunami inundation and run-up in every nested grid. Hence, this study aims to establish a tsunami risk map in the Pasir Jambak sub-district, Padang, Indonesia. The map was carried out in every nested grid point of the area and on a large scale (1:5,000). The TUNAMI N3 program was used for the simulation of the tsunami inundation. A tsunami assessment was made through simulations in nine scenarios of fault parameter data for Sipora block earthquakes. The result of the study provides a tsunami inundation map. Furthermore, this tsunami inundation map can be used for communities, local authorities, government, and others for many studies, and decision-makers can come up with mitigation plans for a small study area.

1 Introduction

Many countries in this world are located within zones where the tectonic plates are active. Most of these coastal areas are vulnerable to tsunamis. In several parts of the different regions in Indonesia, such as North Sumatra, West Sumatra, south of Java, north coast of Sulawesi, Maluku, and north coast of Papua, the risks for tsunamis are very high [1–9]. The effects of a tsunami in coastal areas in some countries are devastating, where tsunami disaster changes within the hazard zone in response to some parameters, especially reflecting the character of the urban environment [10,11]. Tsunami waves mainly affect low-lying coastal areas. However, when tsunami waves become extremely large in height, they savagely attack coastlines, causing devastating property damage and loss of life. When a tsunami occurs, the forces release hydrodynamic pressure, erosion (scouring), buoyancy, as well as the effects of objects and debris carried by the water [5,7].

If an area is very close to the earthquake epicenter, residents may only have a short time to evacuate when a tsunami occurs. Numerical simulations can estimate the arrival time of the tsunami waves on the coast. Therefore, tsunami simulation can be used to estimate tsunami run-up, arrival time, and inundation. This information can be used to determine the tsunami hazard level in an area [3,12–15]. Here, numerical models are used to predict physical tsunami characteristics. This is usually a set of mathematical formulas, called a tsunami model, to predict the tsunami waves and their coastal impact. Most computer methods were developed to simulate tsunami inundation. Tsunami prediction methods are the synthesis of earthquakes, wave evolution, and flood inundation models. Several previous tsunami simulations are presently in use, including MOST, TUNAMI, SELFE, Comcot [16–21].

Previous studies related to tsunami disasters have been carried out, but tsunami hazards that provide detailed information about tsunami waves on the coast, tsunami inundation, and run-up are still rare [7]. The problem in assessing the tsunami risk zone for a small area is significant, as available tsunami inundation zone data do not give detailed information for tsunami inundation and run-up in every nested grid [19–22]. Hence, the objective of this study is to establish a detailed tsunami inundation map in a small area. The study area is in Pasir Jambak Sub-district, in Padang, Indonesia.

Padang is the capital city of West Sumatra province, Indonesia. It has an area of 694.96 km^2 with the sea as the borders and surrounding hills that are 1,853 meters above the average sea level. The city, located between latitudes 0° 44' 00" - 1° 08' 35" S and longitudes 100° 05' 05" - 100° 34' 09" E, is among the most seismically active zones on the western coast of Sumatra. It has experienced earthquakes followed by tsunamis with at least 5 m of run-up heights between Mw 8.7 and 9.0 of earthquake magnitude and is a 200 year-return-period tsunami [4]. Pasir Jambak sub-district is located between latitude 0.8575 – 0.8608 S and longitude 100.3283 – 100.3450 E. The study area lies along the Padang coast and faces the Mentawai islands. The coastal areas are low-lying areas. Those zones consist of sandy beaches. The maximum tsunami inundation generally occurs on sandy shores [23].

* Corresponding author: melimuchlian@itp.ac.id

Those areas have large communities and different types of dwellings, with 12.5% of the whole of Padang city's population.

Based on the tsunamis that occurred in 1797 and 1833, the tsunami arrival times were around 30 minutes with a 5 m tsunami height on the coast of West Sumatra. The tsunami wave height along Muaro Padang was also about 5 m [1,2]. In 2006, the tsunami run-up simulations for West Sumatra were made by Jose C. Borrero et al. [21]. The simulation was published by the National Academy of Science in the USA in 2006. For this, they used the MOST (Method of Splitting Tsunami) numerical simulation.

Developing tsunami generation and propagation models is a very challenging task as it involves many facets: for example the formation of different wave types and the impingement of these waves on the coast. There have been several computational issues worthy of consideration. To determine tsunami generation, propagation, inundation, and run-up, most tsunami simulations have been equipped with all the dimensions and parameters. However, the influence of a tsunami height and the arrival time of tsunami wave propagation are different, since a tsunami source depth is not proportional with tsunami height [24,25].

A tsunami disaster can be designed by different simulations and development. Some parameters, such as the various predictions, approximations, domains, and coastal features, topographic and bathymetric features, play important roles in accurate tsunami hazard assessment in a coastal area. Therefore, it is important to include all the major factors for tsunami wave simulation, namely the tsunamigenic earthquake sources, earthquake parameters, tsunami wave propagation, tsunami inundation, and run-up. Many researchers have also modeled tsunami hazard zones based on the maximum inundation extent and depth.

2 Methodology

To obtain detailed attributes of tsunami inundation based on tsunami simulations, this study used the simulation results from Sipora Blocks in the Mentawai islands to acquire the water element of tsunami wave inundation at every grid point of the study area site. The TUNAMI N3 program was used for the simulation of the tsunami inundation. The tsunami simulation result is used to establish a tsunami risk zone. To get detailed and informative simulations, it would have to be carried out in nine varying scenarios and three different earthquake depths. The epicenter is located in the west of Sipora island, with magnitudes of Mw 8.0:8.5:9.0 and depths of 10:25:35 km. The tsunami generation source was estimated from the Sipora block seismological fault model. The input datasets in tsunami wave's numerical simulation, consisting of some fault parameters in the Sipora block are recorded in Table 1. The selection of this area is based on the history of 200 years of earthquakes, which took place in 1797 [1,2], [4].

The Google Earth and Map Source software were used in determining the domain areas. The bathymetry and topography datasets are the basic data required for tsunami simulation. To provide topography data, this research used the datasets from the SRTM (Shuttle Radar Topography Mission) of 90 m DEMs, with 90 m resolution or three seconds for earth ground surface, while the SRTM grids of topography elevation data were used to derive the topography data. Meanwhile, the GEBCO (General Bathymetric Chart of the Oceans) was used to obtain the bathymetry data files in the areas observed, with 30 arc-second (1 km) for the data. For this study, the GEBCO and the SRTM datasets were used for both the gridded bathymetry and the topography data. Global Mapper Software was used to process the topography data files and coastline parameters. In the next process, the topographic and bathymetric datasets were interpolated by using the Surfer software. Next, the numerical datasets were graphically shown by using the Transform software so that they could be presented in degree coordinates and UTM (Universal Transverse Mercator) coordinates and saved in (.txt) format.

The method of nested grids is a combination of linear simulation and the nonlinear approach [12,16]. In these tsunami simulations, the computational areas of study were divided into four domains, (A, B, C, and D-domains). A-domain grid is presented in Fig. 1. The nested grids of the research study with these four domains have four different spatial grid sizes. Every domain contains the bathymetric and topographic datasets in different grids.

Fig. 1. A domain grid data

The inundation and run-up were simulated for several scenarios. In this study, various parameters of tsunami simulation were presented. There were bathymetry, topography, earthquake magnitude, earthquake source, and fault rupture. The determination of earthquake source parameters was based on historical tsunamis and earthquake-induced tsunamis generated by active faults in the Sumatran megathrust. It is based on tsunami height data, which had occurred in the case study area.

The TUNAMI Number 3 (TUNAMI-N3) Imamura program was used in this study to run the linear and the nonlinear (with four different nested grid sizes) simulations of the tsunami wave inundation within the near-field. These need some input datasets to derive the tsunami wave simulation, such as the domain of observation area, the hypocentre and epicenter of earthquakes, the fault parameters, the running time, and the reference point of the tide level. The initial model of tsunami waves took reference from the vertical deformation of the ocean floor. The faults position in the framework of the tsunami wave inundation simulation was based on the reference point from the left-under angle (0, 0).

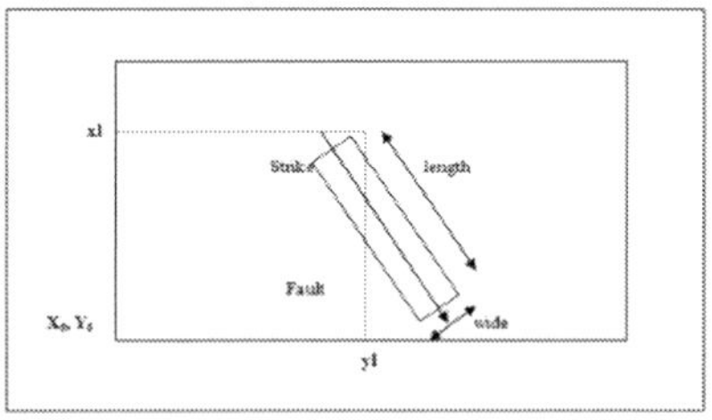

Fig. 2. Fault model in the framework of tsunami wave generation

The variables x1 and y1 of the earthquake epicenter were stated in the nested grid boundaries. The faults model in the framework of tsunami wave inundation can be seen in Fig. 2. This simulation investigated the tsunami tide level, tsunami inundation, tsunami propagation, simulations verification, and validation, as well as the tsunami hazard zones. This should be more informative for users, readers, and decision-makers. An inundation map was provided to present the tsunami simulation results in the fields.

2.1 Tsunami tide level

A graphic of tsunami tide level description displays the relation between the heights and the tsunami arrival times in four nested grid points of the areas observed. The graphic datasets of the seawater level were analyzed based on a tide level reference point within a long period. A long time was needed to simulate an initial tsunami wave to reach the tide level reference point in the observed areas. In addition, the maximum tsunami heights in the areas observed can be analyzed from the graphic.

Table 1. Fault parameter data for Sipora block earthquake

Scenario	Xgrid	Ygrid	M	HH	D	L	W	TH	DL	RD
Scenario 1	126	107	8.0	10	3.23	111	55	130	100	80
Scenario 2	126	107	8.5	10	9.09	313	156	130	100	80
Scenario 3	126	107	9.0	10	10.20	352.09	176.04	130	100	80
Scenario 4	126	107	8.0	25	3.23	111	55	130	100	80
Scenario 5	126	107	8.5	25	9.09	313	156	130	100	80
Scenario 6	126	107	9.0	25	10.20	352.09	176.04	130	100	80
Scenario 7	126	107	8.0	35	3.225	111	55	130	100	80
Scenario 8	126	107	8.5	35	9.09	313	156	130	100	80
Scenario 9	126	107	9.0	35	10.20	352.09	176.04	130	100	80

2.2 Tsunami inundation

The tsunami inundation was derived using the maximum tsunami run-up simulation data. Image processing software was used to show the numerical data in the graphic form before and after the run-up in the areas observed. By comparing the two graphics in every nested grid, the inundation levels for each nested grid were obtained. Then, all the data were interpreted in a digital inundation map by using the ArcView GIS 3.3 software, in which the size of the grid has been adjusted.

2.3 Tsunami propagation

The tsunami height datasets output program in every nested grid was 10,800 seconds, with every scene file of 40 seconds produced in 270 pieces of data. The running data was converted into an image format. Next, those files were compiled in one folder to produce an animation video. This visualization has made it easy to develop a pattern of tsunami wave propagation.

2.4 Verification of simulations

Every tide level model was verified by historical tsunami databases. In addition, if a simulation result was not based on the verification and validation of computer simulation of the historical databases, it could be simulated again by changing the earthquake parameters and/or the tide level reference point and/or the domain quality measurement.

2.5 Tsunami hazard zone

The result of tsunami simulation is the height of tsunami run-up and inundation on land. The simulation result is presented by zoning the area in the tsunami hazard zone map.

3 Results and discussion

Tsunami scenarios were obtained by conducting various tsunami simulations. The simulation result of a 5 m run-up height was used to provide the inundation data in every nested grid point. Based on the maximum heights and

https://doi.org/10.1051/e3sconf/202133104006

travel time, a tsunami generated from a potential tsunami source along the megathrust to the coast of the case study, it can be seen that bigger earthquake magnitudes generate a tsunami wave that is higher and has a shorter travel time. From the result of nine simulations of tsunami waves, it can be concluded that the highest results for tsunami obtained was 10 km depth with a magnitude of (Mw) 9.0. Tsunami wave heights, here, were taken from the maximum tsunami height.

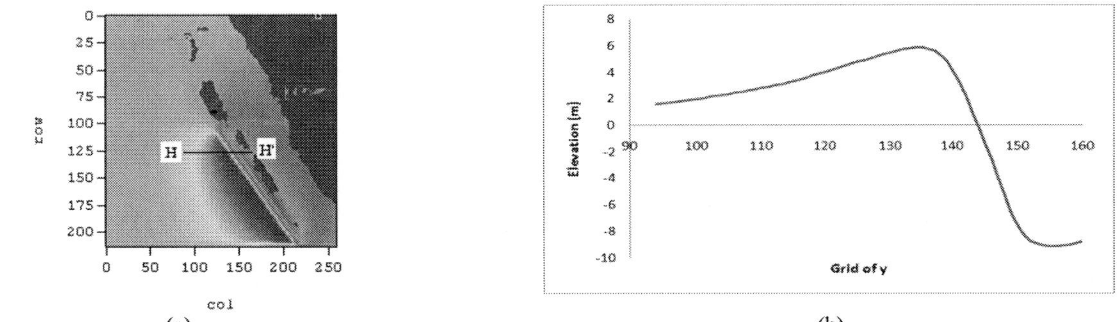

(a) (b)

Fig. 3. Tsunami height of beginning (a) and H-H' Crossection Sipora block (b)

(a) (b)

Fig. 4. Model (a) and contour (b) of tsunami inundation 34 minutes after an earthquake in A-domain

(a) (b)

Fig. 5. Model (a) and contour (b) of tsunami inundation 90 minutes after an earthquake in D-domain in Sipora block

The initial condition of a tsunami wave is shown in Fig. 3(a). Thus, the H-H' cross-section is the fluctuation of tsunami height at the tsunami generation point. The positive value is (+3.21 m), and the negative value is (-4.78 m), given the initial maximum tsunami height is 7.99 m. Apart from that, the waveform can be seen in Fig. 3(b).

The tsunami propagation can be seen in Fig. 4. Here, in 34 minutes, the seawater levels decreased in all observed area points. Water level fluctuation is shown in blue-colored contour in the event when the waves decrease and red-colored contour when the waves increase. Fig. 5 shows that the tsunami hits the observation areas in 90 minutes.

 https://doi.org/10.1051/e3sconf/202133104006

Table 2. Tide level position

Scenario	Pasir Jambak sub-district	
9	arrival time	tsunami wave height (h_{max})
	36 minute	5.70 m

Table 2 shows the results of tsunami arrival time and the tsunami wave height in the Pasir Jambak sub-district. Tide level fluctuations were caused by a tsunami sourced from the Sipora block in the Mentawai islands. The initial tsunami wave is indicated by a decrease in the tide level approximately 35 minutes after the earthquake. Before reaching the observation areas, tsunami waves are first hampered by the west of the Mentawai islands, directly facing the tsunami source.

Fig. 6 is the graphic showing the tide level elevation of the tsunami in the observation areas. The height of the tsunami wave is 5.70 m in the Pasir Jambak sub-district. This is because 80% of the initial tsunami wave has been hampered by the Mentawai islands, while some others diffracted through the gap between the islands. The reduction of tsunami energy is generally beneficial for the observation area, but it will still be so dangerous for the area that is directly stricken by the waves.

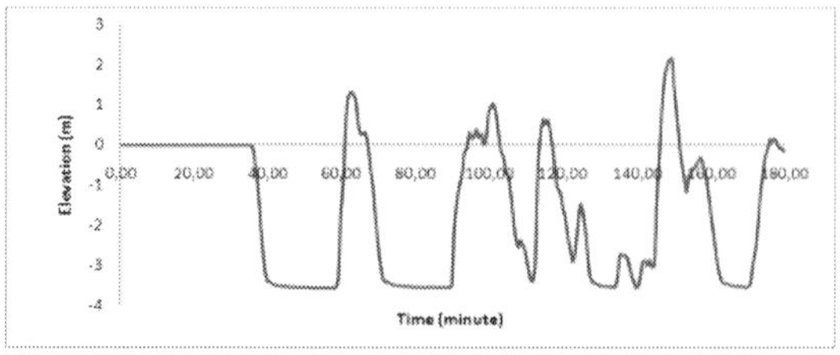

Fig. 6. Tide level from Sipora block source

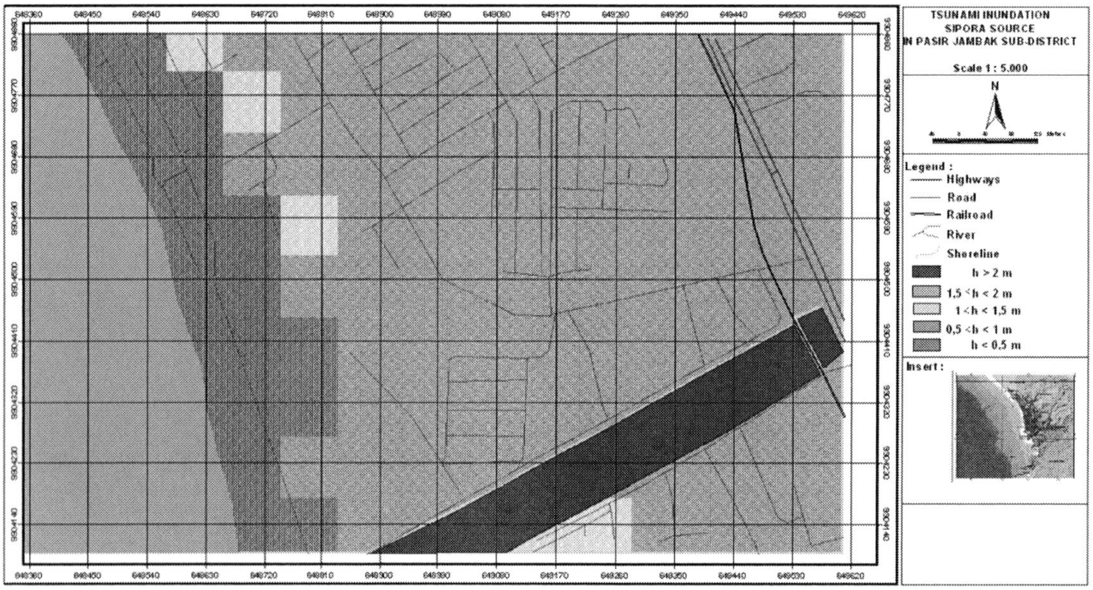

Fig. 7. Tsunami inundation map from Sipora block source in Pasir Jambak sub-district

Along the coastal area of the Pasir Jambak sub-district, the average water heights are more than 2 m, but the tsunami inundation is only 200 m from the coastline. Tsunami inundation in the Pasir Jambak sub-district in Fig 7 describes the inundation map with a red color code distributed near shore. The most vulnerable fields with red color are concentrated in the seashore. There are approximately 18 nested grids or around 154,800 m^2 of an area that can be considered the most vulnerable. A red code in the sub-district is more deeply inundated. The yellow and grey coded areas can be seen in only three nested grids or around a comparable area of 24,300 m^2, and in two nested grids or approximately 16,200 m^2 for the orange color code. More than three-quarters of the area of study is the green coded zone.

The tsunami inundation map was made on a large scale (1:5,000), and the map has detailed information for tsunami inundation and run-up in every grid. It can also be used to provide an informative tsunami hazard zone so that it plays an important role for users and decision-makers in mitigation.

4 Conclusion

The study of the tsunami hazard zone is based on a multi-scenario approach of tsunami parameters, which integrates simulated tsunami scenarios. The simulations were performed in three scenarios of magnitudes (8; 8.5 and 9.0) and three scenarios of earthquake depths (10; 25 and 35 km). Tsunami run-up and inundation simulations are shown in large-scale maps (1:5,000). The simulations have been carried out based on detailed data, attributes, and parameters, model development in the field studies. The development results of tsunami simulations were elaborated to assess the tsunami hazard zone in every nested grid point.

Tsunamis beneath the Mentawai islands of West Sumatra, Indonesia, were verified with historical tsunamis in the coastal area of Padang, based on the earthquakes and tsunamis in 1797 and 1833 so that there is a variety in the earthquake parameter and tsunami simulations with tsunami run-up are approximately 5 meters. The tsunami risk zone map includes more detailed information for tsunami inundation and run-up in each grid, a small scale, and a small area. Furthermore, this map can be used for the communities, local authorities, government as well as others for many studies.

References

1 J. du Puy, Tijdsch Neerland's Indie, **7** (1845)
2 J. du Puy, Tijdsch Neerland's Indie, **9** (1847)
3 D. Mardiatno, Tsunami Assessment Using Scenario-Based Approach, Geomorphological Analysis, and Geographic Information System, A Case Study in South Coastal Areas of Java Island-Indonesia, University of Innsbruck, (2008)
4 D. H. Natawidjaja, K. Sieh, M. Chlieh, J. Galetzka, B. W. Suwargadi, H. Cheng, R. L. Edwards, J. P. Avouac, and S. N. Ward, J. Geophys. Res. Solid Earth, **111** (2006)
5 K. Satake, Y. Nishimura, P. S. Putra, A. R. Gusman, H. Sunendar, Y. Fujii, Y. Tanioka, H. Latief, and E. Yulianto, Pure Appl. Geophys, **170** (2013)
6 K. Satake, Geosci. Lett. **1** (2014)
7 B. A. D. Van Veen, D. Vatvani, and F. Zijl, Cont. Shelf Res, **79** (2014)
8 Y. Fujii and K. Satake, Geophys. Res. Lett, **33** (2006)
9 Y. Tanioka, H. Latief, H. Sunendar, A. R. Gusman, and S. Koshimura, J. Disaster Res, **7** (2012)
10 Papathoma and D. Dominey-Howes, Nat. Hazards Earth Syst. Sci, **3** (2003)
11 M. Papathoma, D. Dominey-Howes, Y. Zong, and D. Smith, Nat. Hazards Earth Syst. Sci, **3** (2003)
12 F. Imamura, A. C. Yalçiner, and G. Ozyurt, Tsunami Model. Man, (2006)
13 E. Jelinek, R. Eckert, S., Zeug, G., and Krausmann, Tsunami Vulnerability and Risk Analysis Applied to the City of Alexandria, Egypt, Italy (2009)
14 E. L. Lekkas, E. Andreadakis, I. Kostaki, and E. Kapourani, Bull. Seismol. Soc. Am, **103** (2013)
15 UNDP, Developing a Disaster Risk Profile for Maldives (2006)
16 IOC and C. Goto, IOC Man. Guid, **35** (1997)
17 V. V Titov and F. I. Gonzalez, NOAA Tech. Memo. ERL PMEL-112 (1997)
18 Y. J. Zhang, A. Baptista, and O. Health, in Coast. Zone 07, Portland, Oregon, 1–6 (2007)
19 R. L. Priest, G. R., Witter, R. C., Zhang, Y. J., Wang, K., Goldfinger, C., Stimely, L. L., English, J. T., Pickner, S. G., Hughes, K. L.B., Wille, T. E., and Smith, Tsunami Inundation Scenarious For Oregon (2013)
20 O. Q. Gutiérrez, I. Aniel-Quiroga, M. Gonzalez, and L. Guillou, Coast. Eng. Proc, **1** (2014)
21 J. C. Borrero, K. Sieh, M. Chlieh, and C. E. Synolakis, Proc. Natl. Acad. Sci. U. S. A, 103 (2006)
22 A. P. Patel, V. M., Dholakia, M. B., and Singh, Int. J. Eng. Sci. Innov. Technol. IJESIT 2, **130** (2013)
23 L. Honesti, M. Z. Abd Majid, M. Muchlian, and N. Djali, J. Teknol, **69** (2014)
24 Y. Yudhicara, Indones. J. Geosci, (2008)
25 L. Honesti, M. Z. A. Majid, N. Djali, and M. Muchlian, J. Teknol, **72** (2015)

Analysis of Padang City community preparedness to face the earthquake and tsunami disaster

Diva Syandriaji[1] and *Junaidi*[1*]

[1]Department of Civil Engineering, Andalas University, Padang, Indonesia.

Abstract. Padang, West Sumatra is located in the collision area of two tectonic plates, namely Indo-Australia and Eurasia. Which is marked by the presence of a tectonic earthquake center in the Mentawai islands and surroundings. Realising the high risk of disasters, the people of Padang city must be ready and prepared for the possibility of an earthquake and tsunami. To reduce the risk of disaster, the use of "self-help" and "mutual-help" parameter might be the best way to help the people to know "what should they do?" and "how can they do it?" to protect themselves from disasters. This research aims to analyse the Padang city community's preparedness to face the Earthquake and Tsunami disaster by using "self-help" and "mutual-help" parameters. A questionnaire consisting of 20 questions was used to collect the data. The sample size of this research is 400. The data was collected from 4 different areas (Random area, Pondok area, Purus area, and Ulak Karang area). The study found that the Padang city community might not be prepared to face the earthquake and tsunami disaster in the future in terms of self-help and mutual help as most of the items mentioned in the questionnaire have not been applied. The government is urged to develop appropriate policies regarding further disaster risk reduction

1 Introduction

Indonesia is known as a disaster-prone country because this country is surrounded by the Pacific Ring of Fire and is above three continental plate collisions, namely Indo-Australia, Eurasia, and the Pacific plate [1]. Most Indonesian provinces and their coastal areas are identified as "disaster-high risk," yet more than half of the total population live and are dependent on these areas [2]. West Sumatera is one of the disaster high-risk Province in Indonesia. This province is located in the collision area of two tectonic plates, namely Indo-Australia and Eurasia, and marked by the presence of a tectonic earthquake center in the Mentawai islands and surroundings. Padang city, the capital city of West Sumatra located near the Mentawai island predicted to face a major Earthquake and Tsunami in the future. Past paleogeodetic, paleotsunami, and geodetic investigations [3] indicate that the Mentawai segment of the Sunda subduction zone can host a large tsunamigenic event with magnitude > 8.5 and a recurrence period of about 200 years. The last major tsunamigenic earthquakes events in this area were in 1797 and 1833 [4], while two recent events, Mw 8.4 and Mw 7.9, occurred near Bengkulu on 12 and 13 September 2007. On 30 September 2009, an earthquake with magnitude 7.6 occurred on the Sumatran subduction, right in front of the Pariaman city. Ismail, F.A., et al.[5], stated that a segment about 300 km long between the Nias segment and Bengkulu segment, called seismic gap has the potential to trigger a big earthquake in the future.

Besides, by realising the fact that Earthquake and Tsunami are unpredictable, the people of Padang City must be ready and prepared to face the risk of the earthquake and tsunami disasters. By 2019, the government and Regional Disaster Management Agency (BPBD) of Padang City already took various initiatives to reduce the risk of the disasters. The government of Padang build 3 shelters for evacuation in Ulak Karang, Parupuk Tabing, and Tabing with the capacity of about 5.000 people of each shelter and also with 58 buildings that can be used as a temporary evacuation site. They also set up about 30 small billboards consisting of information about disaster preparedness, made posters about disaster preparedness, conducted counseling to 150 schools located in the prone zone, and made tsunami-safe zone signs on 22 roads in Padang.

The Padang city government has taken many disaster risk prevention measures. Then how about the community? Did they already prepare themselves to face the disaster?. To reduce the risk of disaster, the importance of "Self-help" and "mutual-help" can be the best way to help the people to know "what did they not do?" and "how can they do it?" to protect themselves from disasters. Therefore, this research will use the "self-help" and "mutual-help" parameters to analyse the Padang city

* Corresponding author: junaidi@eng.unand.ac.id

community's preparedness to face the Earthquake and Tsunami disaster.

2 Methodology

2.1 Questionnaire

Table 1. List of 20 questions in the questionnaire.

Self-help questions	
Q1.	You already know about the fault map, earthquake, and tsunami potential map (including information on intensity, liquefaction, and tsunami risk) in the area where you live.
Q2.	You have secured your house against earthquakes.
Q3.	You have put the furniture in your home in a safe place to prevent it from falling during an earthquake.
Q4.	You and your family do not sleep in places where furniture, shelves, broken lamps, or glass can fall on you in the event of an earthquake.
Q5.	You can protect yourself; especially your head, at the moment of the earthquake.
Q6.	You can evacuate to a higher place without hesitation in case of a severe earthquake while visiting or living close to the sea, even if nobody else does
Q7.	You have read and understood the signs or information about disaster preparedness provided by the government on every road in your neighbourhood.
Q8.	You know the way to the nearest shelter where you are or live when there is a big earthquake with the tsunami potential.
Q9.	You believe you can save yourself in a situation where a large earthquake and tsunami occur.
Q10.	You have secured enough power for your mobile phone to utilize it in preparation for disasters.
Q11.	You and your family have enough food to survive for about a week without gas, electric power, and tap water if a disaster occurs.
Q12.	You have enough lighting reserves for about a week during a power failure
Q13.	You and your family had discussed what you should do after the earthquake, how to contact each other if you were in separate places when the earthquake happened, and so on.
Mutual-help questions	
Q14.	You usually take a part in community activities such as neighbourhood groups, children's groups, cleaning projects, and athletic meets.
Q15.	You have discussed what to do in preparation for natural disasters (to prevent and mitigate disasters), with your neighbours or in your community.
Q16.	You have built a good relationship with your neighbours and understood who needs help in the time of evacuation.
Q17.	Your neighborhood has a disaster preparedness training agency
Q18.	You are willing to take part in disaster drills in your community.
Q19.	You already know disaster mitigation and want/have shared your knowledge with the people in your neighbourhood.
Q20.	You believe that the people in your community can face disasters together and help one another.

A questionnaire was used in collecting data for this study, either offline or online. The questionnaire consists of 20 questions and also suggestions or comments from the community about the management and mitigation system in Padang city. In this questionnaire, respondents were asked to answer the questions using three choices (Yes, Maybe, and No). The list of questions shown in Table 1

These 20 questions are inspired by the questions from Disaster Reduction Class Application developed by Prof. Akiyoshi Takagi from Gifu University, which focused on "self-help" and "mutual help".

2.2 Data Sampling

The sampling method used in this research is purposive sampling. The sample size is determined by using the Yamane [6] formula below :

$$n = \frac{N}{1+N(e)^2} \quad (1)$$

Where,
n = sample size
N = population size
e = level of precision (0.05)

As the population of Padang was 939,112 persons (BPS, 2018), the sample size would be about 400 respondents. These respondents were divided into two survey methods. The first is by direct questionnaire for 250 respondents (Pondok area, Purus area, and Ulak Karang are) and the second is an online questionnaire for 150 respondents (Random area). Then, the respondent's information such as gender, age, educational background (started from didn't pass elementary school until college graduates), and job background are obtained randomly to see the variety of the respondent.

2.3 Data Analysis

Each answer in the three-point Likert scale was given a score shown in Table 2. [7]

Table 2. The score is assigned to the Likert response.

No.	Answer	Code	Score
1	No	1	0
2	Maybe	2	2
3	Yes	3	5

The prepared or not prepared respondents were calculated based on the final score of each respondent. First of all, we estimate the highest total score and lowest total score and then calculate the range. The range was divided by the number of categories to obtain an interval for each category. The value of scoring criteria used to determine the criteria of the respondents by using the formula below:

Total score = 5 x 20
= 100

Final Score = Σ *score* $(q1+q2+\ldots+q20)$ (2)

Then, Determine the scoring on objective criteria, whether the respondents are "prepare" or "not prepare"

Range (R) = highest score – lowest score

Category = 2, "prepare" and "not prepare"
Interval (I) = Range (R) / Category (C)
Scoring Criteria = Highest score – Interval

Those with a total score ≥ 50 would be expected to be prepared and those with a score < 50 might be not prepared. It is also possible to check every response for each question, whether the respondents are prepared to face the disaster or not using formula (3).

Final Score of Question-n = (Σ score Q-n)/(Total respondent) ....(3)

n = Number of the question

If the final score for each question was ≥ 2.5, we expected the respondent's preparation for this item already been satisfied. On the other hand, if the final score was < 2.5, expected as not prepared.

3 Result and Discussions

3.1 Result

This study was based on a questionnaire survey conducted offline and online with 400 respondents in total. The online survey collected 150 data from people who live in Padang (a Random area in Padang), and the direct survey collected 94 data in the Purus, 85 data in Pondok, and 71 data in Ulak Karang. The 400 respondents used in this study were collected randomly and came from different backgrounds. The respondents are Male and Female ranging from ages 17 to 78 years. From collected data, the gender variety of the respondents in this research, as shown in Fig.1. are balanced between Male and Female respondents. Male respondents are 200, and female respondents are 200 persons.

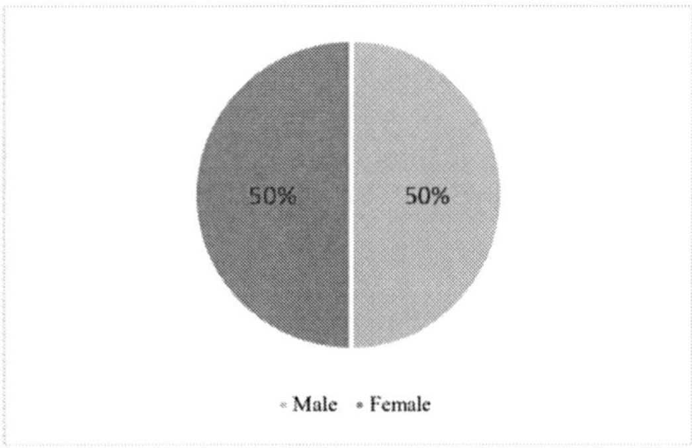

Fig. 1. Gender Variety of 400 Respondents.

For the educational background, as shown in Fig. 2., most of the respondent's educational background in this study was senior high school graduates (198 respondents) and college graduates (161 respondents). Then the rest is junior high school graduates (32 respondents), elementary graduates (7 respondents), and did not pass elementary school (2 respondents). For the job background, as shown in Fig. 3., most of the respondents in this research are Private Employee/Entrepreneur with 148 respondents and Student/College Student with 113 respondents. Then the rest is other jobs (49 respondents), housewives (44 respondents), PNS/Ex PNS (41 respondents), Farmer (3 respondents), and POLRI/TNI/ Ex POLRI&TNI (2 respondents). By knowing the variety of 400 respondents, the author then analyses the preparedness of each respondent facing the earthquake and tsunami disasters. Fig. 4. shows the percentages of 400 respondents preparedness to face earthquake and tsunami disasters.

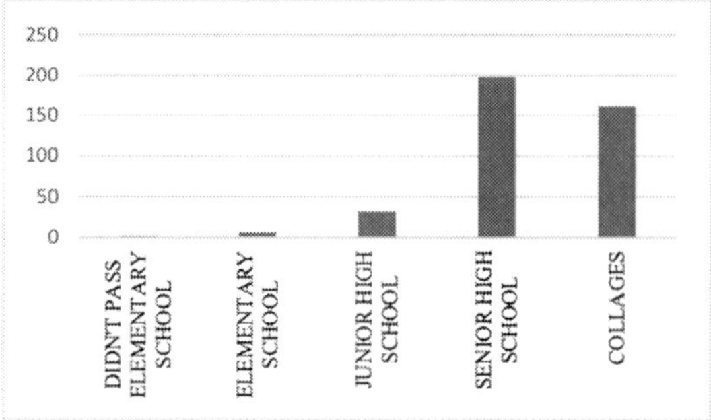

Fig. 2. Educational Background of 400 Respondents.

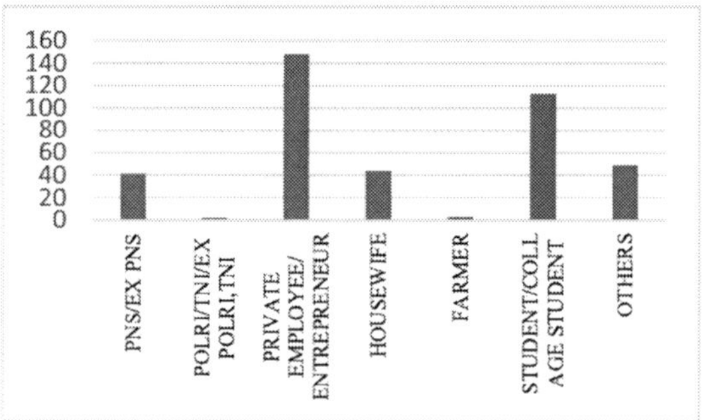

Fig. 3. Job Background of 400 Respondents.

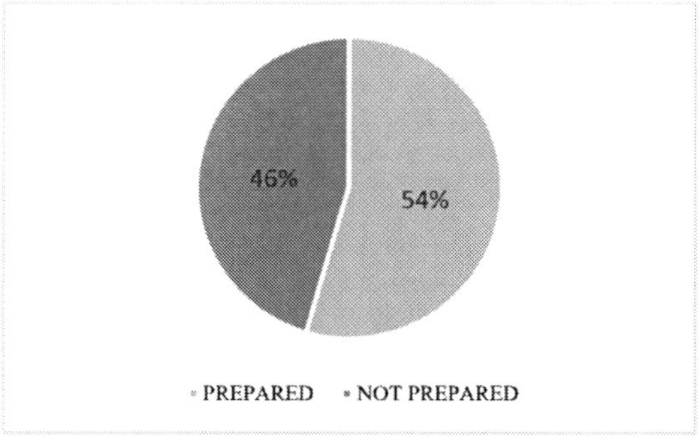

Fig. 4. Job Background of 400 Respondents.

From Fig. 4. above, it can be seen that 54% of the community are prepared to face the earthquake and tsunami, while 46% are not prepared. These percentages show a slight difference between prepared and not prepared respondents to face earthquake and tsunami disasters. The gap is only 8% of both prepared and not prepared. Then, from the result of each score on the option "Yes, Maybe, and No" in the questionnaire, we can see the community level of preparedness for each question. Fig. 5. below will show the level of preparedness from the 20 questions by 400 respondents of Padang City Community.

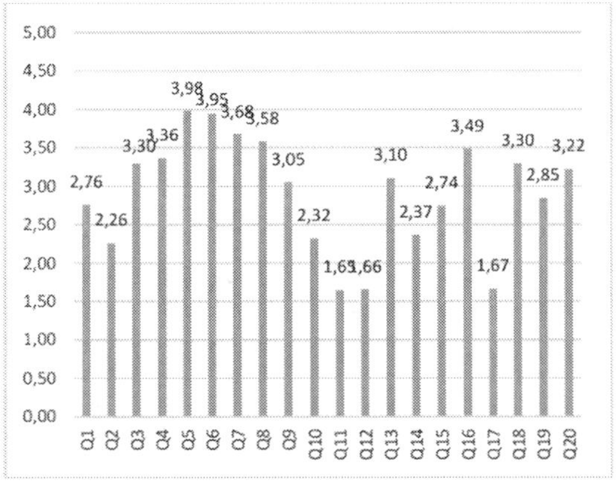

Fig. 5. Level of preparedness from the 20 questions by 400 respondents of Padang city community.

Based on Fig. 5, Padang City Community is most prepared in question number 5 "You can protect your body especially your head, at the moment of an earthquake", where the final score in this question is 3.98. Respondents mostly choose "Yes" in this question, which is means that most of the respondents already know how to protect themselves during the earthquake. However, the final score still did not reach the highest score (5 points), which is means there are still some people who hesitate and do not know how to protect themselves from the earthquake. On the other hand community of Padang city is most unprepared in question number 11 "You and your family have enough food to survive for about a week without gas, electric power, and tap water if a disaster occurs". The fact that the average score in this question is only 1.65, it seems like most of the Padang city community still lacks in the preparation of food reserves when a disaster occurs.

To see more clearly on what questions the Padang City community is more prepared or less prepared based on the questions, the author will separate the data as "Questions with the level of preparedness ≥ 2.5" and "Questions with the level of preparedness <2.5". Fig. 6. below will show the result of questions with the level of preparedness ≥ 2.5.

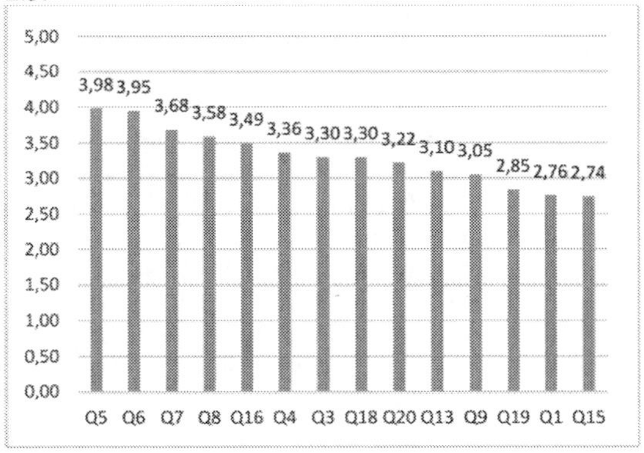

Fig. 6. Questions with Level of Preparedness ≥ 2.5.

Fig. 6. above shows that the community of Padang city is more prepared on questions number 5, 6, 7, 8, 16, 4, 3, 18, 20, 13, 9, 19, 1, and 15 in order. According to this result, most of the Padang city Community is already prepared to face the earthquake and tsunami disaster because the average score in each question is more than 2.5. However, these result also shows that not all of the community are prepared because the score on each question is still less than 5 points. It shows that some people still choose "Maybe" and "No" on those questions. Next, the questions with the level of preparedness <2.5 can be seen in Fig. 7. below:

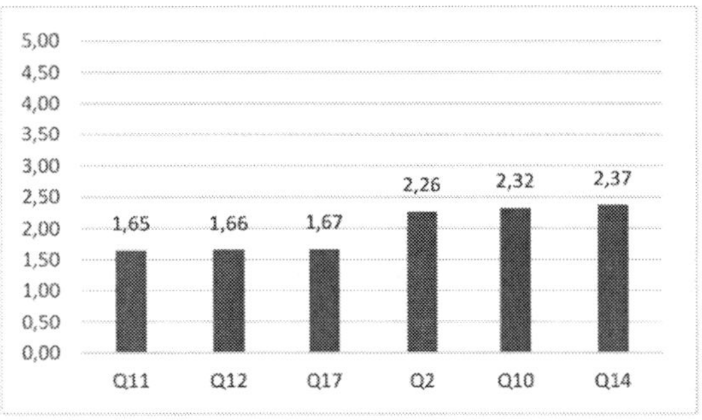

Fig. 7. Questions with Level of Preparedness < 2.5.

The result shows that the Padang city community is not prepared in Questions 11, 12, 17, 2, 10, and 14. It is because the average score in each question is less than 2.5.

3.2 Discussions

The result above shows that 400 respondents of the Padang city community, in general, are still not prepared in terms of "self-help" and "mutual-help". It is because the gap between respondents who are prepared and not prepared are only 8%, and also strengthened by the results shown in Fig. 8. below:

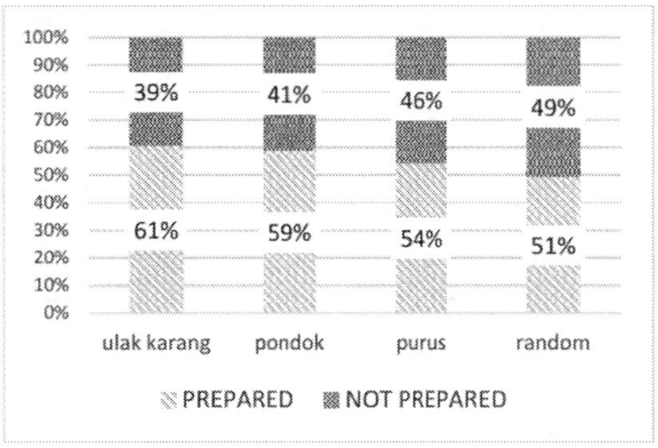

Fig. 8. Data comparison of community preparedness from Random area, Pondok area, Purus area, and Ulak Karang area.

Fig. 8. above shows that by separating the analysis into four different areas (Random area, Pondok area, Purus area, and Ulak Karang area), it can be seen that respondents from the random area in Padang city are more unprepared than the respondent from Pondok, Purus, and Ulak Karang. Respondents from Random areas in Padang city are more representative in terms of Padang city community preparedness. It is because these respondents are not only focusing on a specific area in Padang but broader. Also, based on the characteristic of each area, Pondok, Purus, and Ulak Karang used to be more prepared because of these three areas included in the High-Risk Zone (West Padang, North Padang, Nanggalo, and Part of Koto Tangah). Where most of the government programs and policies related to disaster prevention are focusing on these areas. In this study, the result of the selected areas (Pondok, Purus, and Ulak Karang) also shows that around 39%-46% of the community are still not prepared to face earthquake and tsunami disasters. Besides, based on the level of preparedness on the 20 questions, the Community of Random area had the most "Not prepared" questions than the community from Ulak Karang, Pondok, and the Purus.

Also, according to the result of "Not prepared" questions by 400 respondents of Padang city community that shown in Fig. 7, questions 11, 12, 17, 2, 10, and 14 are the important questions related to disaster preparedness. Alhadi and Sasmita [8], stated that the efforts that can improve the community preparedness are using the methods based on the local wisdom of the community, wherein community preparedness requires an active role from the communities. The role of those communities according to Alhadi and Jelita [9], starts from the smallest forms of participation in disaster preparedness which is self and family preparedness. While in a broader scope it includes community groups or community organisations. We can see those questions 11, 12, 17, 2, 10, and 14 above are included in this explanation.

This statement is also supported by LIPI-UNESCO/ISDR [10] which developed a framework for studying community preparedness to anticipate natural disasters. According to the study, There were 5th critical factors of preparedness to anticipate natural disasters, especially earthquakes and tsunami. Two of the 5th parameters of the critical factors consist of some variables related to question 11, 17, 12, 2, and 10. The first parameter mentioned knowledge and attitude towards disaster risk. One of the variables in this parameter is "an understanding of the vulnerability of physical buildings and facilities essential for disaster emergencies" which is related to question number 2 "You have secured your house against earthquakes.". Then, the third parameter mentioned plans for disaster emergency activities, at this point, several variables related to questions 11, 17, 12, and 10. The first variable is about disaster management organisations including disaster preparedness agencies, which related to question number 17 "Your neighbourhood has a disaster preparedness training agency.". The second variable is about plans for providing basic needs, including food, drinks, clothing, shelter/camps, clean water, toilets and environmental sanitation, health, and information on disasters and victims, which is related to question number 11. Then, the last variable related to question 12 and 10 about evacuation equipment and supplies as well as essential facilities for emergencies.

Also, from the past earthquake event that occurred in west Sumatera especially in Padang city in 2009, many problems happened a week after the disaster because people did not prepare themselves for the post-disaster related to questions 11, 12, 2, and 10. As reported in the Liputan 6 newspaper on 5 October 2009 [11], the distribution of aid (food, medicine, lighting, tend, and others) is not distributed evenly by the government to the community so that people end up looting. Also, many houses collapsed because most people did not secure their houses from an earthquake. If the community can provide their own basic needs, especially food and lightning reserves for post-disaster, the problem that happened in 2009 can be minimized. Unfortunately, based on the result that most of the community is still not prepared in questions 11, 12, 2, and 10, we can say that the problem in 2009 can still happen in the future.

According to question number 17 Indonesian government to involve the community participation in disaster management at the kelurahan or village level has been carried out by launching policies and programs aiming to gather local knowledge (local wisdom) about the disaster and improve the preparedness of the community/community in disaster risk areas. Desa Tangguh Bencana (DESTANA) initiated by The National Disaster Management Agency (BNPB) and Kampung Siaga Bencana (KSB) by the Minister of Social Welfare is one of the programs which involve local communities in disaster risk management.

In Padang city, one of the efforts to involve local communities in disaster management is the establishment of Kelompok Siaga Bencana (KSB) in each Kelurahan/village. Simandalahi, T. [12] stated that the Padang City Government has formed and inaugurated the KSB that consists of 2,080 members in 104 kelurahan/villages of 11 regions in Padang city. This program was initiated by The Regional Disaster Management Agency (BPBD) in 2011. Anam et.al [13] stated that the function of the KSB is an extension of the BPBD in monitoring the disaster situation in kelurahan/village. So, the local communities can report the potential and signs of disaster to BPBD through the KSB. However, Anam et.al. [13] also stated that since the formation of KSB in 2011, the presence of the KSB has not been able to achieve expectations. Alhadi and Jelita [9] also stated that since 2011, the existence of the KSB has not been seen. In 2017, the Mayor of Padang issued instructions No. 800.83/BPBD-Pdg/II/2017 concerning the Formation / Activation of KSB as one of the efforts to reactivate the Disaster Preparedness Group that already formed but did not active or vacuum. This instruction is intended for Kelurahan/Village that does not yet have the KSB to be formed immediately. However, Most of the Padang city community chose "No" on question 17, which means that the existence of KSB still not improved until now. While based the result of question number 14 also shows the Padang city community has a low interest in participating in their community organisation, this result can be one of the reasons why the KSB in Padang is still not progressing.

This study also gathered respondents' comments and hopes regarding the current risk prevention efforts undertaken by the government. The result can be seen in Fig. 9. Below:

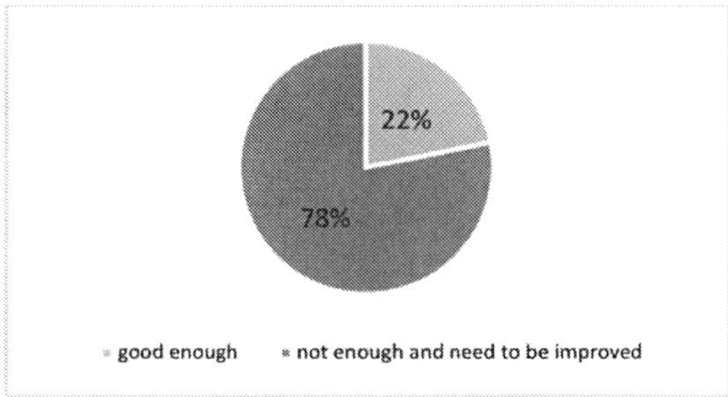

Fig. 9. Padang city community comments regarding the current risk prevention efforts undertaken by the government.

The results show that most respondents said that risk prevention efforts are not enough and still need to be improved. Most of the respondents commented that they wanted the government to increase the information, dissemination, and socialization about disaster preparedness to the community through regular training at least twice a year. They also requested that more shelters be built in high-risk areas because the existents of the shelters are still very limited and have not been maintained until now. It is also clarified by the Study of Kemal, B. M., Yosritzal & Aulia, Y. B. [14], which stated that vertical evacuation to the existing shelters is not enough because some areas are still not covered with the evacuation plan. Therefore, respondents also suggested the government should provide regulations for any building construction with more than three floors to provide access to the community during the disaster to use it as a temporary shelter. The respondents also hope that the government can provide the community with their basic needs, especially food reserved, medicine, and lightning, to face the earthquake and tsunami in the future.

From this, we can conclude that the community itself still relies heavily on government or public assistance for disaster management. The community still does not realise the importance of "Self-help" and "Mutual- help" having a big impact on disaster prevention. As Harada [15] said, the specific efforts for disaster reduction require a combination of self-help, mutual-help, and government or public assistance. Harada [15] explained that to help oneself when a disaster strikes, everyone must take individual and independent actions to save themselves and their own lives from danger. Next, as for mutual assistance, Harada [15] stated that what is very important in mutual assistance is the formation and maintenance of tight networks based on regular information sharing and collaborative relationships such as neighbourly cooperation in communities and tie-ups between organisations. Lastly, for the government or public assistance, Harada [15] explained that the role of the central government is necessary as a last resort in terms of providing guidance to the community in terms of actions and giving them a sense of reassurance, but there is a sense of distance, to a certain degree, from the natural disaster site. The problem stated by Harada [15] is that since these public institutions are public entities funded by tax money, decision-making often requires a great deal of time and procedures.

Therefore, this research can prove that the Padang city community are still not prepared to face the earthquake and tsunami disaster in the future in term of "self-help" and "mutual-help". The community needs to realise that every individual also needs self-awareness and help each other to face disasters. Then, government policies regarding risk reduction efforts also need to be improved and evaluated.

4 Conclusions

A comprehensive study on the preparedness of the vulnerable population in Padang has been conducted using a questionnaire. Preparedness assessed in this study is the application of items in terms of self-help and mutual-help among the community. The data showed that only half of the community could be declared as prepared to minimize the risk when a tsunami hit the city. Data also showed that the most prone communities, such as in Pondok, Purus, and Ulak Karang region, seem to be more prepared than those who are living in another region of the city. Furthermore, from the 20 items of preparation examined, only 70% of the items can be justified as recognized and prepared by the community. However, the level of preparedness for the items is only about 2.74-3.98 out of a 5-point scale. In terms of preparedness for a disaster, the level of preparedness should be as high as possible. Therefore, the community should increase the level of their preparedness for all items to ensure that the community prepares enough to minimize the loss due to the disaster.

5 Recommendations

Based on this study, it was recommended as follow:

1. The community should reduce their dependency on government assistance and be more reliable to self-help and mutual-help with people around them.
2. The government should educate and guide the community intensively to increase their awareness and preparedness, especially on how to rely on them-self (either self-help or mutual-help) rather than on the help of government officers more.

We would like to thank the lecturers of Civil Engineering at Andalas University and Prof. Akiyoshi Takagi from Gifu University whose provided guidance and information during this research so that this research can be completed properly.

References

1. T. Ophiyandri, A. Junaidi, A. Takagi, D. Syandriaji, *Hazard Risk Management and Mitigation System of Earthquake and Tsunami at Disaster-Prone Area.* International Journal on Advanced Science, Engineering and Information Technology, **10** (2020)
2. BNPB. Data dan Informasi Bencana Indonesia. Jakarta, Indonesia: BNPB, 2014.

3. S. S. Nalbant, S. Steacy, K. Sieh, D. Natawidjaja, and J. McCloskey, *Seismology: earthquake risk on the Sunda trench*, Nature, 435, 756–757 (2005)
4. D.H. Natawidjaja, K. Sieh, M. Chlieh, J. Galetzka, B.W. Suwargadi, H. Cheng, R.L. Edwards, J.P. Avouac, and S.N. Ward. *Source parameters of the great Sumatran megathrust earthquakes of 1797 and 1833 inferred from coral microatolls*. Journal of Geophysical Research: Solid Earth, **111**(B6), (2006)
5. F.A., Tanjung, J., Hakam, A., & Boen, T. Plastered Wire-Mesh Bandaged: An Effective Alternative Technique For Seismic Strengthening Of The Unconfined Brick Masonry Housing In Pariaman City, West Sumatera, Indonesia. International Journal of Civil Engineering and Technology (IJCIET), **8** (10), 1001–1008 (2017)
6. Yamane, Taro. Statistics, An Introductory Analysis, 2nd Ed., New York: Harper and Row. (1967).
7. A. Takagi, and Y. Azuma, *Web Application of "Disaster Reduction Class" for facilitating understanding and practice for disaster risk preparedness and reduction*. World Bosai Forum - IDRC Sendai (2017)
8. Z. Alhadi, S. Sasmita. *Padang City Community Preparedness in Facing Earthquake and Tsunami Risks Based on Local Wisdom (Study on Disaster Risk Preparedness)*. Humanus, **XIII**, .2 Th. 2014.
9. M. Jelita, Z. Alhadi, *Factors Affecting the Role of Disaster Preparedness Groups in Realizing Community Preparedness in the City of Padang*. Jurnal Ilmu Administrasi Publik, **1**, 1 (2018)
10. LIPI-UNESCO/ISDR. *Study of Community Preparedness in Anticipating Earthquake and Tsunami Disasters*. Covering Earth Sciences Indonesian Institute of Sciences. Jakarta (2006).
11. https://www.liputan6.com/news/read/246478/gempa-dan-bantuan-yang-tersendat
12. T. Simandalahi, *Knowledge of the issue of community-based disaster risk reduction in disaster preparedness groups in West Padang District, Padang City*. The Indonesian Journal of Health Science, **6** (1) (2016).
13. K. Anam, A. Mutholib, F. Setiyawan, B. A. Andini, & S. Sefniwati. *Readiness of local institutions in dealing with the tsunami disaster: Case studies of Air Manis Village and Purus Village, Padang City*. Jurnal Wilayah dan Lingkungan, **6** (1), 15-29 (2018)
14. B. M. Kemal, Yosritzal & Y. B. Aulia. Evaluation Of Tsunami Evacuation Plans In Padang, Indonesia. *International Journal of Civil Engineering and Technology (IJCIET) Technology*, **8** (10), 773-780 (2017)
15. H. Harada, *Social capital in disaster: From the great east Japan earthquake*. The Senshu Social Capital Review, **3**, 5–21, (2012)

"Monster VIPs" : disaster preparedness training for children with intellectual disabilities

Laila Isrona[1], *Yulistini*[1], *Fathma* Mardhotillah[2], *Nadyatul* Husna[2], *M.* Fauzan[2], *Izzati* Mujahidah[2], *Rafif* Helery[2], *Husna* Yetti[3*], and *Rosaria* Indah[4]

[1] Department of Medical Education, Faculty of Medicine, Andalas University, Indonesia
[2] Undergraduate medical student, Faculty of Medicine, Andalas University, Indonesia
[3] Department of Public Health and Community Medicine, Faculty of Medicine, Andalas University, Indonesia
[4] Faculty of Medicine and Research Center of Social and Cultural Sciences, Syiah Kuala University, Indonesia

Abstract. Disaster resilience is vital to everyone, including people with disabilities. However, teaching people with disabilities requires special techniques, and there is scant literature in the area. This case study designs and implements disaster preparedness training at a school for children with intellectual disabilities in Padang, Indonesia, an area prone to earthquakes and tsunamis. The training utilizes a blended learning approach called 'Monster VIPs' that combines various learning methods such as storytelling, posters, videos, puzzles, and disaster simulations to facilitate students to understand disasters responses. The training involves university teachers and disaster preparedness volunteers (Taruna Siaga Bencana/TAGANA) of the city's social services.

1 Introduction

Thousands of schools worldwide have been delivering education for around 200 million people with significant disabilities [1]. Half of those schools are dedicated to teaching children. However, there are many challenges encountered by special schools in educating these children with special needs, such as low accessibility, inadequate teacher training and support, and inflexible curriculum and material [2]. Therefore, it is essential to support efforts to increase access, train teachers, and improve special education curricula.

In the increasing number of disasters worldwide, children with disabilities are the most marginalized people who are often excluded from the disaster management framework. However, a growing body of research has started to draw attention to these children's needs regarding disaster education [3,4,5]. This case study explores a way to design and implement disaster education training for children with intellectual disabilities. It starts with a review on disaster risk knowledge in the context of special education, followed by a method section that describes the design and implementation processes, discussion, and ended with a conclusion and recommendations for future studies and actions.

1.1 Disaster Risk Knowledge and Special Education

Promoting disaster risk knowledge in formal and non-formal education is one of the top priority actions to increase disaster resilience. It is a vital strategy stated in the priority action—understanding disaster risk—described in Sendai Framework for Disaster Risk Reduction 2015-2030 established by the United Nations [6]. To increase the young generation's disaster resilience, schools around the world have been incorporating disaster risk knowledge in their existing education and training mechanisms [7,8,9]. However, special schools for children with disabilities were often overlooked by governments and other stakeholders in disaster management systems [3].

Studies show that children with disabilities are often portrayed as the 'powerless' ones in the disaster risk reduction (DRR) program [10,11,12]. The reasons are due to challenges they face to process early warning information or other disaster management phases [11] and their limited social networks. Another reason is due to their social structures and how disability is conceptualized in their society. Hence, The United Nations Convention on the Right of Persons with Disabilities (UNCRPD) has recommended several actions to protect and promote the fundamental rights of persons with disabilities [4]. In reality, many young students with disabilities are not included in disaster and mitigation management programs [10]. They were left behind in disaster education, which makes them lack improvisational skills to manage the worst situation [4]. Possible pathways towards special school education programs about DRR are essential to be explored and exercised by considering proper instructional methods that fit with the students' different abilities.

* Corresponding author: husnayetti@med.unand.ac.id

1.2 Teaching Children with Intellectual Disability

Special education schools are educational institutions prepared to handle and provide special education services for children with specific physical or mental, behavioral, and social issues. One of these issues is intellectual disability. Children with intellectual disabilities experience obstacles in cognitive, psychomotor, and social adaptation functions during their development phase. However, they have their own pace in learning.

Literature shows that instructional methods for teaching normal children are not different from those for children with intellectual disabilities, except that it will often require more time and more repetition [13]. Studies recommend several strategies, including:

a. Active learning may stimulate students' participation in learning activities with opportunities to apply information [13],
b. A structured format of learning is much more effective instead of an unstructured one [14],
c. Direct instruction is better compared to problem-solving [15], (Dagseven Emecen, 2011),
d. Teaching should use sign language or pictorial prompts to foreground verbal communication afterward [16],
e. Increasing students' interests by showing videos or pictures of people that they admire [17],
f. Reinforcers for positive behavior may help students to learn. This includes music, praise, happy face, tokens, stickers, and so on [13].

2 Methods

2.1 The Context

This case study aimed to design and implement disaster preparedness training for children with mental retardation. It took place at one of the special schools in Padang, the capital city of West Sumatra province, Indonesia. The particular school, called Sekolah Luar Biasa Perwari (SLB Perwari), has 53 students with intellectual disabilities with the assistance of eight teachers. It is located only one kilometer inland of the coastline of Purus beach connected to the Indian Ocean, where Sunda Megathrust is located. Sunda Megathrust is the most active tectonic crack in the world, the most significant source of earthquakes. This situation makes SLB Perwari prone to earthquakes and tsunamis.

West Sumatra province, in general, has a relatively high seismic activity compared to other provinces in Indonesia. Several strong earthquakes above the 5 Richter Scale have been recorded in this area, destroying several towns, including Padang. One of the most significant earthquakes occurred in 2009, which took 1.195 lives and caused significant destruction of thousands of buildings in the city [18]. West Sumatra Disaster Management Agency has predicted a big megathrust earthquake with a magnitude of 8.9 SR that will occur and cause a six-meter tsunami. It is predicted to destroy huge areas within five kilometers from the coastline of West Sumatera, including Purus beach and Padang [19].

To assess the school's need for disaster preparedness training, members of the Department of Medical Education team interviewed the principal of SLB Perwari. Based on her information, there has never been education on earthquake and tsunami disaster mitigation implemented within the school. Meanwhile, other schools in Padang are often invited by the mayor of the city to participate in earthquake and tsunami mitigation education programs. The principal was concerned that children with intellectual disabilities would not be able to save themselves because of their lack of disaster knowledge. In addition, the limited number of teachers would not be adequate to assist the students in evacuating if a disaster occurs. Children with intellectual disabilities may need extra explanations and time to evacuate due to their different intellectual and social adaptations [5].

2.2 Design Phase

Responding to the problem stated above, medical teachers in the Department of Medical Education and undergraduate medical student team of Fakultas Kedokteran Universitas Andalas aimed to design and implement disaster preparedness training for students and teachers SLB Perwari. This training's objective was to increase the children's capacities to anticipate earthquakes and tsunami.

The training should be designed well to fit the children's different abilities in processing information and fight the stigma linked to their cognitive limitation in accessing vital resources, social networks, or their disabilities to adjust emotionally during and after the disaster. By utilizing suitable instructional methods, such as those reviewed in the literature review section, the training was designed to engage the students in exciting and fun activities. After a long study and discussion, the medical teacher and student team came with the idea of blending different instructional methods. They called it 'Monster VIPS,' an acronym of 'method blending storytelling, posters, videos, puzzles, and simulations. Monster VIPS blends different methods of instructions to teach content on disaster and mitigation, emphasizing ways to save themselves when an earthquake and tsunami occur.

The purpose of this activity is to provide education and improve the knowledge of intellectually disabled students in Padang City— especially students with intellectual disabilities of SLB Perwari located in Purus beach, Ulak Karang Village, Padang Utara District, Padang City about mitigation action to save themselves during earthquake and tsunami using repetitive modification methods. It is also supposed to stimulate these students to create their participate in disaster management. They should be viewed with an active role in constructing their own life (and lives around them). The training was systematically arranged from the preparation stage, implementation stage, and evaluation stage.

2.3 Preparation phase

This phase mainly focused on preparing teaching equipment. The team also established a partnership with another community stakeholder, which is the TAGANA team (acronym of 'Taruna Siaga Bencana' or disaster preparedness and response volunteer team). After conducting a literature review on unique education teaching methods, the entire team decided to use a combination of instructional methods called 'Monster VIPS' and started to prepare the properties needed. They also planned other preparation activities, including:

2.3.1 Making hand puppets for storytelling method

The team prepared five hand puppets to perform a puppet show. These hand puppets are animating five animals made from cloth. The show delivers dialogue and simulation between those animals, including tips on what to do during and after a big earthquake. It aims to share information about earthquake and tsunami mitigation that children should understand if they are indoor and outdoor when an earthquake occurs.

2.3.2 Drawing posters

The team created two posters containing information about steps to protect themselves and evacuate when the earthquake and tsunami occur. First, the students should hide under a strong structure, such as a table, and then when the earthquake stops, they should walk out of school or other buildings while protecting their heads. They should walk to a designated gathering point. These posters were utilized to assist a trainer who tells a story about an earthquake and tsunami. Posters were made in two sizes: A5 distributed to each student and A3, which will be patched on a whiteboard in each class.

2.3.3 Video-making

The team prepared for two videos. The first video showed pictures of earthquakes and tsunamis. The other video introduced disaster responses and an educational song about self-rescue actions responding to earthquakes and tsunami.

2.3.4 Puzzle

The team provided pictures of actions to be taken when an earthquake occurs. Students would be asked to choose the proper sequential steps or actions responding to the hazard: whether they need to hide under or on a table, run to or away from the beach, stay in a building or run away to a gathering point, and so on. The puzzles aimed to facilitate the students to understand the information provided in previous methods (storytelling, posters, and videos). Hopefully, students would be able to be engaged with the puzzle as a learning game and remember the mitigation steps as they are explaining why they chose a piece of the puzzle to respond to the hazards.

2.3.5 Simulation Preparation

The team gathered tools and media needed, such as loudspeakers, sirens, and direction signs, to demonstrate the earthquake and tsunami situation and ways to evacuate.

2.4 Implementation Phase

The activities in the implementation phase aimed to visually, auditory, and kinesthetically stimulate these students to understand steps in disaster response by blending various instructional methods. The total participants of the training activities were 64, consisting of 50 students and 14 teachers. In this phase, there were three activities. Each activity took three hours and was conducted on three different days within a month time.

2.4.1 Introduction, storytelling, posters, and puzzles

The first meeting started with an introduction session between the team and students to build a good relation and understanding. The team taught the children a song about earthquakes and evacuating after teachers alarmed them by sirens and sudden. During the singing, the team demonstrates the actions that need to be taken. After the demonstrations, the team continued with storytelling using hand puppets (Fig. 1). They also provided a brief explanation of mitigation actions with posters (Fig. 2). Storytelling by using hand puppets was an innovation to keep mitigation education interesting and adjust the concentration of students. It would be easier for intellectually disabled students to absorb new information about disasters and mitigation.

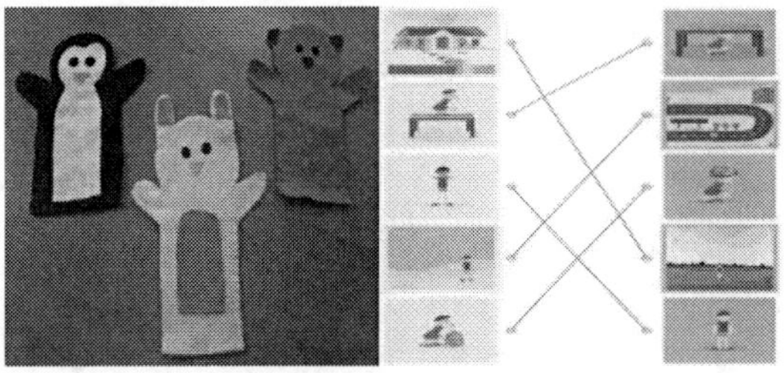

Fig. 1. Hand puppets for storytelling (left) and puzzle (right)

Fig. 2. Poster

These posters themselves were expected to help students to visualize how earthquakes and tsunamis occur and what actions should be taken in such a situation. Afterward, trainers asked the students to choose pictures, which looked like puzzles—depicting proper mitigation steps when an earthquake and tsunami occur. The use of puzzles aimed to stimulate students to understand and remember what they have learned in the previous learning activities.

2.4.2Video watching followed by simulation

The second meeting started with watching the two videos prepared by the team. The first one aimed to introduce what earthquakes are and their impacts on buildings and people. It also shows tsunami as one of the aftereffects of an earthquake occurring from the crack in the ocean floors or an erupting volcano. In sum, it visualizes the situation when the earthquake and tsunami happen. The second video shows ways to respond to an earthquake and evacuation steps if teachers or parents announce an early tsunami warning.

Afterward, the students were invited to participate in two simulations. The first one was conducted inside a classroom to simulate responses when an earthquake occurs and indoor students. Another simulation was conducted outdoors to show ways to evacuate themselves from the classroom to a gathering point in a designated open space. These simulations were organized by the TAGANA team of the Padang City Social Services.

2.4.3 Simulation

Before the third activity, the medical teacher and student and TAGANA team prepared teachers and administrators to play their roles in the disaster simulation (Fig. 3). They were given an explanation and guidance as well as opportunities to ask questions. This ensures that they understand the early warning, incident command, and communication systems as essential disaster responses. They are the ones who are responsible for the safety and security of the students during school activities. A common understanding should be established between teachers and administrators in responding to a hazard. Hence, they would effectively assist students in evacuating from the classroom to a designated gathering point in the school's front yard. Then to the nearest shelter or other designated safe location where parents would pick their children up after an earthquake or tsunami occurs.

The simulation started with an indoor activity. Trainers asked the children to sing together a song about evacuation when an earthquake and tsunami occur. An 'earthquake' announcement followed it, and students are guided to respond according to the evacuation plan explained in previous activities. On this occasion, the team utilized a route to a shelter at the Ulak Karang Cooperative Center, higher ground, and a safe location or gathering point if a 10-meter tsunami occurs.

Fig. 3. Simulation

2.5 Program Evaluation Phase

The team interviewed several teachers informally a few days after the implementation of the last training. Three themes emerged from the interviews.

First, the teachers were very thankful that community members, such as the team from the nearby university and TAGANA team, were involved in increasing their knowledge and skills in disaster preparedness. It is crucial for the safety and security of children with intellectual disability and their teachers and administrator. The training is evidence of community engagement in increasing the disaster resilience of its members.

Second, the teachers of SLB Perwari were inspired by the education approach utilized by the team. The 'Monster VIPS' approach offers alternative ways of blending different instructional methods to increase students' engagement with the content. These alternative approaches may help address various needs of the students, as some of the students came with different conditions, such as a combination of disabilities.

Third, the teachers hope that there will be a continuation of the disaster preparedness training in the future. The teachers need to upgrade their knowledge and skills to train their current and future students.

3 Discussion

This case study aimed to describe different phases in a disaster education program for children with intellectual ability in the context of a particular school prone to earthquakes and tsunamis. There were several insights obtained from the disaster education program.

3.1 The Overlooked Special Schools

The data from the preliminary study that SLB Perwari was not invited to participate in disaster preparedness programs arranged by the local government represents the reality that special education is still overlooked by education stakeholders, especially in disaster education [4,5].

This fact supports statements in previous studies that explore the same issue located in different countries in the world. This study calls for the implementation of social justice in special education schools and action to fight the stigma of 'powerlessness' of children with intellectual disabilities.

3.2 Steps In Designing An Educational Program

A good learning program should be designed by considering different steps, including identifying learning beliefs, need analysis, designing the content, media, assessment, and so on until evaluating the program [20]. This disaster preparedness training has established the designed steps from the learning belief and need analysis very well. The core of instructional designs has also been prepared accordingly by carefully thinking on the proper content, media that assisted its delivery, types of instructional methods, and the sequence of learning experiences that repeat and increase students' competencies. However, the assessment and program evaluation have not been prepared very well. Consequently, the program was inadequately evaluated, and it was not easy to spot the aspects that need improvement.

The program's implementation provides a rich experience of ways to establish good collaboration with different stakeholders Volunteers, academia, teachers, administrators, and of course, the students themselves to achieve the objective of the training, which was increasing their disaster knowledge. This rebuts the assumption that children with disabilities are powerless. They certainly have power

3.3 Blended Instructional Methods May Work

The findings of this case study show that active learning that blends pictorial (posters, videos, and puzzles), verbal (storytelling), physical prompts(simulation), reinforcements such as song and music may be effective in teaching disaster knowledge. Considering children's situations, who had a combination of disabilities, this blended learning offers multiple stimulations to their visual, auditory, and kinesthetic senses that may help them understand the core content.

4 Conclusions and Recommendations

This case study explores the design, preparation, implementation, and evaluation of disaster preparedness training for children with intellectual disabilities. Need assessment that was conducted in the form of a preliminary study revealed the exclusion of special schools in the disaster preparedness framework set by the local government. However, since building community disaster resilience is the responsibility of all parts of the community, stakeholders in disaster education should consider efforts to include people with disabilities in the resiliency framework. The medical teacher and student team and TAGANA collaboration have set a good example that can be extended to a broader audience.

This case study produces several recommendations. First, disaster education for children with disabilities should be prioritized since it is long overdue. Stakeholders should take action against the stigma of the powerlessness of people with disabilities. They are indeed robust and collaborative. Second, stakeholders may collaborate to strengthen the network of disaster preparedness. This study shows that collaborative actions between university teachers and the disaster preparedness volunteer team have produced significant results, hopefully, a good impact on the children's resilience. However, it requires a lot more effort to engage a broader audience to increase community resilience toward disasters.

References

1. The World Bank. *Disability Inclusion*. World Bank. March 19 (2021) https://www.worldbank.org/en/topic/disability
2. Theirworld. (2021, July 9). *Children with disabilities* July 9 (2021) http://theirworld.org/explainers/children-with-disabilities
3. S. Ronoh, J.C. Gaillard, J. Marlowe. *Children with disabilities and disaster preparedness: A case study of Christchurch*. Kōtuitui **10** (2), 91–102 (2015a)
4. S. Ronoh, J.C. Gaillard, J. Marlowe. . *Children with disabilities and disaster risk reduction: A review*. Int. J. Disaster Risk Sci. **6** (1), 38–48 (2015b)
5. L.M. Stough, E.M. Ducy, D. Kang, S. Lee. *Disasters, schools, and children: Disability at the intersection*. Int. J. Disaster Risk Reduct. *45*, e101447–e101447 (2020)
6. United Nations. *Sendai Framework for Disaster Risk Reduction 2015-2030*. (2015) https://www.undrr.org/publication/sendai-framework-disaster-risk-reduction-2015-2030
7. R. S. Oktari, K. Shiwaku, K. Munadi, R. Shaw. *A conceptual model of a school–community collaborative network in enhancing coastal community resilience in Banda Aceh, Indonesia*. Int. J. Disaster Risk Reduct. *12*, 300–310 (2015)
8. R. S. Oktari, K. Shiwaku, K. Munadi, R. Shaw. *Enhancing community resilience towards disaster: The contributing factors of the school-community collaborative network in the tsunami-affected area in Aceh*. Int. J. Disaster Risk Reduct. *29*, 3–12 (2018)
9. A. Sakurai, M.B.F. Bisri, T. Oda, R.S. Oktari, Y. Murayama, M. Affan. *Exploring minimum essentials for sustainable school disaster preparedness: A case of elementary schools in Banda Aceh City, Indonesia*. Int. J. Disaster Risk Reduct. 29, 73–83 (2018)
10. H. -C. Lee, H. Chen. *Implementing the Sendai Framework for disaster risk reduction 2015–2030: Disaster governance strategies for persons with disabilities in Taiwan*. Int. J. Disaster Risk Reduct. *41*, 101284 (2019)

11. E.-S. Park, D.K. Yoon, Y.-W. Choi. *Leave no one behind: Experiences of persons with a disability after the 2017 Pohang earthquake in South Korea.* Int. J. Disaster Risk Reduct. *40*, 101261 (2019)
12. S. Ronoh, J.C. Gaillard, J. Marlowe. *Children with disabilities in disability-inclusive disaster risk reduction: Focussing on school settings.* Policy Futures Educ. **15**(3), 380–388 (2017)
13. Downing, J. *Academic instruction for students with moderate and severe intellectual disabilities in inclusive classrooms.* Corwin Press (2010)
14. A.I. Siddiqui, S. Shahzadi. *Structured and Unstructured Teaching Program and Development of Mentally Retarded Children.* Pak. J. Spec. Educ. 6(2), 11 (2003)
15. D. Dagseven Emecen. *Comparison of Direct Instruction and Problem Solving Approach in Teaching Social Skills to Children with Mental Retardation.* Educ. Sci.: Theory Pract. **11**(3), 1414–1420 (2011)
16. I. Cobzaru. *Models of Intervention in the Recovery of Children with Mental Retardation and Autism Spectrum Disorders, Alternative Therapies.* Educația Plus, **28**(1), 99–109 (2021).
17. S.P. Rame, Y. Blegur, M. Selan. *Teacher's strategy in increasing the interest of children with special needs for mental retardation in participating in physical education learning at the Oelamasi State special school Kupang Regency 2019/2020.* J. Physic. Educ. Health Sport Sci. **2**(1), 17–31 (2021)
18. Kompas. (2009, October 28). *1.195 Tewas, 4,8 Triliun Kerugian Gempa Padang.* Kompas. https://edukasi.kompas.com/read/2009/10/28/11141045/~Regional~Sumatera
19. CNBC Indonesia. (2020, November 14). *Ini Fakta Potensi Gempa M 8,9 & Tsunami 10 Meter di Sumbar.* https://www.cnbcindonesia.com/tech/20201114144347-37-201847/ini-fakta-potensi-gempa-m-89-tsunami-10-meter-di-sumbar
20. N. Shambaugh, S.Magliaro. *A reflexive model for teaching instructional design.* Educ. Technol. Res. Dev. **49**(2), 69–92 (2001)

Reconstruction post disaster responsive gender in West Sumatra

Mediana desfita[1]* and *Djendrius*[2]

[1]Civil Engineering Department, Andalas University, Padang 25163, Indonesia
[2] Sociology Department, Andalas University, Padang 25163, Indonesia

Abstract. Indonesia is one of the very prone countries to disaster. The tsunami and earthquake disasters caused enormous damage to property and infrastructure as well as loss of life. An earthquake on September 30, 2009, measuring 7.9 on the Richter scale struck the west coast of Sumatra, causing loss of life and damage to infrastructure. There were around 1,115 people killed, 1,214 seriously injured and 1,688 people lightly injured. A comprehensive study of natural disaster management activities or systems should be used as learning materials to form a disaster management system. Disaster preparedness by minimizing vulnerability has been identified as a better approach to dealing with disasters than post-disaster response. Creating a culture of prevention is critical to dealing with everyday hazards and the consequences of disasters. The study is using data from a natural disaster in west Sumatra, and data from people who were involved in the process of reconstruction post disasters in west Sumatra. This study aims to obtain an accurate description of the reconstruction post-disaster and relation with responsive gender activities in West Sumatra, and identify how the gender effect on reconstruction post-disaster in the West Sumatra region.

1 Introduction

Post-disaster reconstruction activities without referring to gender aspects often make it difficult for women and children to continue their post-disaster life. Starting from the availability of shelters that are not safe and under the needs of especially women and children, the location of the shelters is far from the initial settlement and is in a position that is vulnerable to causing negative actions, because usually, its position is near the forest or mountains, far from the access. their daily activities, such as markets, schools, and health facilities. Likewise, the distribution of disaster aid has not paid attention to the special needs of this group of women and children [1].

In the context of disaster management in West Sumatra, gender considerations have received sufficient attention and are considered important to be integrated into the post-rehabilitation and disaster processes. However, in practice, there are still inconsistencies here and there. For example, gender-disaggregated data are not always available, gender analysis is not always applied in needs assessments for disaster response; assistance designs that do not always address the specific needs of each community group, aid distribution processes that do not involve/pay attention to the special conditions of women, children and vulnerable groups, temporary shelter designs that do not meet eligibility standards and the potential for increased violence against women, cases -cases have not been handled gender-based violence in refugee locations and temporary housing, [1].

Ideally, gender mainstreaming in post-disaster activities, both in the planning process, as well as in physical activities, must include gender factors in making directions and steps. This can be done if the gender factor is replicated and internalized into laws and regulations relating to post-disaster activities themselves so that the activities carried out are more integrated and support sustainable development.[1].

From all previous studies, it can be concluded that research on gender factors in post-disaster housing reconstruction activities needs to be investigated further, to reduce the level of vulnerability of women and children, as groups that are considered the most vulnerable in dealing with post-disaster conditions. This study aims to get an accurate picture of the gender position of post-disaster house reconstruction activities in West Sumatra. Knowing the effectiveness of gender-based post-disaster reconstruction activities on the matrilineal system in West Sumatra. Create a gender mainstreaming concept in post-disaster reconstruction activities in the West Sumatra region so that it can become an important part of legal guidelines and documents in post-disaster reconstruction activities in Indonesia, especially in West Sumatra. Generate a gender-friendly post-disaster shelter referral in this case to vulnerable groups of women and children.

* Corresponding author: mediana.desfita@yahoo.com

https://doi.org/10.1051/e3sconf/202133104009

Women and children are a gender group that, even under normal conditions, are very vulnerable. Especially in post-disaster conditions. So that post-disaster activities should no longer ignore gender in their implementation. Field experience shows that, 72 hours after a disaster occurs, women and children are vulnerable to violence in many forms. Meanwhile, the Huntara which was built without gender as one of the aspects reviewed resulted in a temporary shelter that did not create a sense of security for women and children. For this reason, it seems necessary to conduct research on gender mainstreaming in post-disaster reconstruction activities, which is carried out in the study area of West Sumatra, which is seen from its relationship with the order of Minangkabau cultural values, namely the Matrilineal System, which is very unique and is closely related to the position of women known as bundo kanduang. Which is the perspective of disaster is considered as a group with a high level of vulnerability, so that later there will be no post-disaster reconstruction, especially in the "Gender Blind" West Sumatra area.

In 2017, 335 natural disasters affected more than 95.6 million people, killed another 9,697 people, and cost a total of US$335 billion. This problem is not the same for all places, as Asia is the continent most vulnerable to floods and storm events, with 44% of all disaster events, 58% of total deaths, and 70% of total affected people. Nonetheless, America reported the highest economic losses, representing 88% of the total cost of the 93 disasters. China, the US, and India are the countries most affected by this incident, marked by the occurrence in each country of 25, 20, and 15 events. Considering the land area in each country. [2]

Disasters in the last decade have resulted in significant losses in human life and livelihoods, economic and social destruction of infrastructure and environmental damage has increased the need for appropriate disaster reduction with risk management strategies. A disaster is described as part of a vulnerability. Since vulnerabilities are components of disaster dependency, they must be managed and minimized to reduce disasters. Disaster reduction policies and actions, which ensure the reduction of vulnerability, need to be established and implemented to achieve a sustainable and consistent disaster plan.[3].

One of the important elements in the effort to build a natural disaster management system is to evaluate and take important lessons from activities or natural disaster management systems that have been carried out so far. Strengths and successes, as well as weaknesses and failures in natural disaster management that have been carried out, will be important lessons for building a better national system for natural disaster management in the future. Thus, a comprehensive study of natural disaster management activities or systems should be used as learning materials to form a disaster management system.[4]

Disaster preparedness by minimizing vulnerability has been identified as a better approach to dealing with disasters than post-disaster response[5], creating a culture of prevention is essential to address everyday hazards and the consequences of disasters.[6]. Disaster risk reduction is defined as a conceptual framework, which considers the possibility of minimizing vulnerability and disaster risk across society, to avoid (prevention) or to limit (mitigation and preparedness) harm, in the broad context of sustainable development.[7]. Therefore, disaster risk reduction should be included as an important component of all national and local government development strategies, policies, programs, and investments[8]. In other words, disaster reduction includes taking action in advance, addressing risk reduction, involving environmental protection, social justice, and economic growth, the three cornerstones of sustainable development, to ensure that development efforts do not increase vulnerability to hazards.[9].

In disaster management, gender is also an interesting thing to study. We recognize that there are three categories in gender issues, namely gender responsiveness, gender bias, and gender neutrality. Gender responsiveness, namely the alignment of activities/activities or circumstances/conditions in supporting the implementation of gender roles such as women in the management of the post, hygiene kits include underwear and sanitary napkins, the number of toilets for women in the post is more, including gender aspects in disaster risk studies, lactation corners at refugee posts, gender vocal points in disaster management institutional structures, and the existence of guidelines for handling gender-based violence at disaster posts. Gender bias is the impartiality of activities/activities or conditions/conditions to support the implementation of gender roles, namely open toilets without covers, separate toilets for men and women, the assumption that the head of the family is always a man, reduced reproductive health services during emergencies, women are not involved in disaster management activities because they have been represented by their husbands/fathers/brothers and there are many other examples. Gender-neutral, namely activities/activities or conditions/conditions that do not contain the gender roles of men and women, for example, disaster-prone maps,[10].

Previous research has found that gender in post-disaster reconstruction is the result of unequal relations between women and men in society, and becomes women's vulnerability.[3]. Therefore, socially constructed roles often make women more vulnerable to harm. Gender shapes the social world in which natural hazards occur,[11]. Therefore, "gender-blind" reconstruction policies and programs can only lead to increasing women's vulnerability, widening gender gaps, and creating unsustainable development in affected communities.[12]. Studies have found that failure to address gender issues in post-disaster reconstruction (PDR) policies and programs hampers opportunities for reconstruction to achieve community resilience and sustainable development.[13]. Thus, gender mainstreaming is very important to achieve sustainable development through reconstruction.

Although women are at greater risk than men in disasters, it is women who enable the community to be able to cope with disasters because their social role is very important in managing disaster management strategies,[11] and [14]. However, the ability of women to reduce hazards and prevent disasters as well as cope with and recover from the impacts of disasters has not been sufficiently calculated or developed. In the current practice of disaster management women are seen as helpless victims and their capacities, knowledge, and skills at every stage of the disaster cycle are not recognized.[15]. Gender differences in disaster mitigation have been discussed mainly in the context of vulnerability or community involvement. The absence of women in decision-making positions in emergencies results in ineffective recovery planning. Therefore, a gender perspective should be integrated into all disaster risk reduction policies and measures to reduce women's vulnerability in disasters. However, gender equality in disaster reduction requires empowering women to have increased roles in leadership, management, and decision-making positions because women are not only disaster victims but they can act as change agents in disaster reduction planning.[16].

Therefore, gender mainstreaming needs to be incorporated into policies and programs related to disaster reduction, especially because "gender shapes capacity and vulnerability to disasters.[17] as previously discussed. gender mainstreaming can promote gender equality and women's empowerment, especially if there are glaring examples of persistent discrimination against women and inequality between women and men. Gender mainstreaming can be used as an effective tool to reduce women's vulnerability, which arises due to various factors including lack of access to resources and more women entering the disaster reduction policy-making process. However, increasing gender mainstreaming is a lengthy process that requires input from many parties over a long period, including advocacy, advice and support, competency development, development of methods and tools, and vigilance in following up and evaluating progress.

The province of West Sumatra, known as the Minang realm, currently has a population of 4,528 million people. Some residents of this area, although listed as Minangkabau residents, are domiciled overseas. The Minangkabau people are the largest matrilineal ethnic group in the world and the only example for Indonesia. Its social and political organization approaches a purely matrilineal type from an anthropological point of view. However, they have also long been known as steadfast adherents of Islam, as is their market-oriented and outward-looking mental attitude. On this basis, the Minang people in Indonesia are often considered the rightful owners of the wandering tradition,[18].

The culture of the Minangkabau community is embodied in the philosophy of Tigo Tungku Sajarangan or Tigo Tali Sapilin, where the role of the three main forces is represented by the ulama, niniak mamak, bundo kanduang, and cadiak clever, which is a formal force in a matrilineal society. Socio-culturally Minangkabau women, or also known as bundo kanduang, have a unique position in Minangkabau customary law, especially in the lineage system taken from the maternal line (matrilineal). Economic resources and their use are also for women. Meanwhile, in the context of political dynamics in indigenous communities, especially in deliberations for families, the voice of bundo kanduang determines the results be achieved [18].

Women in Minangkabau have a very important position in tribes, tribes, and villages. In this case, women function as heirs of high inheritance from mamak to their nephews in the clan or tribe. Minangkabau custom places women in a very noble position. The role of women is seen in the principle of the matrilineal kinship system[18].

2 Methods

The research methodology framework of this study adopts the '*research onions*' model proposed by Saunders et al. (2019) [20]. The assumptions of the research philosophy are taken from several case studies that are used as research strategies. This study uses a mixture of qualitative and quantitative methods. In achieving its aims and objectives, a quantitative approach is used as the main method. The first data collection method (semi-structured interview) aims to identify post-disaster Reconstruction Activities with a gender-based approach. At the same time, a questionnaire survey to assess the probability and level of gender-based post-disaster reconstruction activities in the Matrilineal System in West Sumatra was identified. Content analysis was performed for qualitative analysis, while descriptive and inferential statistics were used for quantitative analysis.

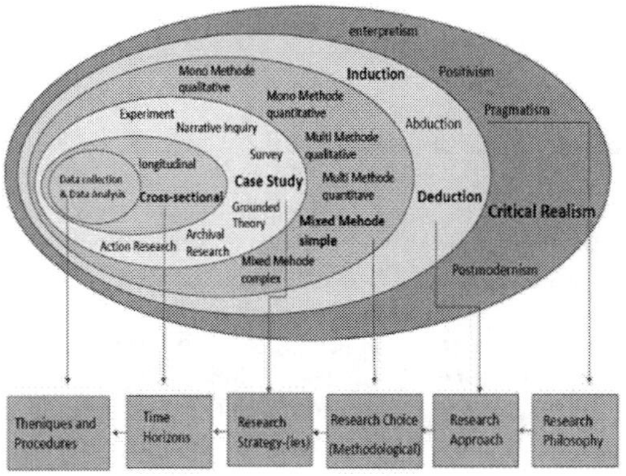

Fig 1: The research process "The Research Onion", Saunders et al. (2019)

The flow chart of the research phase can be seen in Fig 2.

The flow chart of the research phase can be seen in Figure 2.

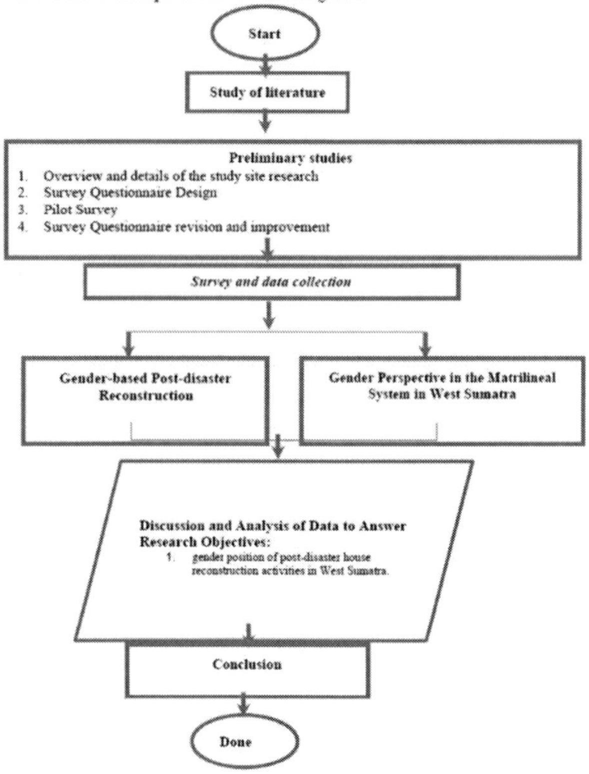

Fig 2: Flowchart of Research Stages

3 Results and Discussion

This research begins with a literature review on disasters, trends in natural disasters in the world, trends in natural disasters in Indonesia, and trends in natural disasters in the study area, namely West Sumatra. Furthermore, a literature review was conducted regarding post-disaster reconstruction activities that have occurred so far, both post-disaster reconstruction activities in the world, post-disaster reconstruction activities in Indonesia, and post-disaster reconstruction activities in the study area, namely in West Sumatra. In general, and the gender perspective concerning post-disaster reconstruction activities so far, the author here also looks for correlations between gender mainstreaming and post-disaster reconstruction activities, both in the planning stage and in the implementation stage in the field. Furthermore, the author also examines the study area, namely West Sumatra, both in terms of geographical, socio-cultural, and socio-economic conditions. Then more deeply this literature review will review the matrilineal system as a cultural order in the study area, both in terms of cultural aspects and its relation to post-disaster reconstruction activities.

The study Desk which initiated this research became the basis for the implementation of the preliminary study phase. Where at this stage, the author has introduced an overview and detailed situation of the study area related to the research to be carried out. Then the author began to design a questionnaire, as a tool in this research activity which aims to obtain the data and facts needed, to achieve the objectives of the implementation of this research. The next activity is to conduct a pilot survey to know how far this questionnaire design can answer the ability to accommodate data needs in later research. After that, revisions and improvements are made based on the results of its effectiveness in the pilot survey. At this stage, the questionnaire is considered feasible to be distributed to respondents.

The next stage is an important part of this research, namely the distribution of questionnaires that have been refined in the previous stage, which aims to collect the data needed in this study. The respondents who will be addressed in this study are people who are closely related to post-disaster reconstruction activities in West Sumatra, Indigenous Stakeholders, and the Minang Kabau Community, especially Bundo Kanduang who is the center of this research. The questionnaire as a research tool is divided into two parts, namely a questionnaire on gender-based post-disaster reconstruction, and a questionnaire on gender perspectives in the value system of the matrilineal system in the study area, namely West Sumatra. at a later stage, The relationship between these two things can be found, which is intended to answer the objectives of this study. This stage is considered complete when all data has been collected.

The next stage is the discussion and data analysis, which is carried out to get answers to the objectives of this study, namely: Identify gender position of post-disaster house reconstruction activities in West Sumatra; in this case, to vulnerable groups of women and children.

4 Conclusion

In the context of disaster management in West Sumatra, gender considerations have received sufficient attention and are considered important to be integrated into the post-rehabilitation and disaster processes. However, in practice, there are still inconsistencies here and there. For example, gender-disaggregated data are not always available, gender analysis is not always applied in disaster response needs assessments; assistance designs that do not always address the specific needs of each community group, aid distribution processes that do not involve/pay attention to the special conditions of women, children and vulnerable groups, temporary shelter designs that do not meet eligibility standards and the potential for increased violence against women, cases -cases have not been handled gender-based violence in refugee locations and temporary housing, etc. For this reason, it is necessary to conduct more significant research related to the correct integration of gender in post-disaster reconstruction activities in West Sumatra, so that the problems that occur can be overcome.

References

1. K.-S.Gender Working Paper, 1–14 (2011)
2. L. (2544), Report of the Learning Circle Association for Women, Central Sulawesi, (2020).
3. M.M. Ariyabandu and Wickramasinghe, Gender Dimensions in Disaster Management – A Guide for South Asia, Colombo: ITDG South Asia. Natural Hazards and Earth System Sciences, **18** (6), 1785–1810 (2003)
4. BNPB, Indonesian disaster risk (2016)
5. S. Childs, and M.L. Krook, Gender and politics: The state of the art. Politics, **26** (1), 18–28 (2006)
6. J. Drolet, L. Dominelli, M. Alston, R. Ersing, G. Mathbor, and H. Wu, Women rebuilding lives post-disaster: Innovative community practices for building resilience and promoting sustainable development. Gender and Development, **23** (3), 433–448 (2015)
7. E. Enarson, , "Gender and Natural Disasters", IPCRR Working Paper no.1, International Labor Organization. In Natural Disaster Risk Management: Geosciences and Social Responsibility, (2016)
8. Enarson, Elaine, Review of Women confronting natural disaster, From vulnerability to resilience. American Journal of Sociology, (2012)
9. K. Ginige, University of Huddersfield Repository Gender mainstreaming in disaster reduction: why and how? (2008)
10. isdr2002ZentelGlade2013International_Strategies_for_Disaster_Reduction.pdf. (nd).
11. H. Khatun, Livelihood strategies in disaster risk reduction in Bangladesh, In Sahni, P., and Ariyabandu, MM(eds.), Disaster Risk Reduction in South Asia, New Delhi: Prentice-Hall of India. (2018)
12. ME. McGuinness, Secretary or General? The UN Secretary-General in World Politics. Edited by Simon Chesterman. Cambridge, New York: Cambridge University Press, 2007. Pp. xiv, 280. Index. $85, £45, cloth; $31.99, £16.99, paper. American Journal of International Law, **102** (4), 930–936 (2008)
13. Saunders; K.LEwis, Philip; Thornhill, A., Research Methods for Business Students (Eighth edi), New York: Pearson, [2019]: Print edition typeset in 9.5/12 ITC Slimbach Std by Pearson CSC (2019)
14. Secretary-General, Implementation of the International Strategy for Disaster Reduction: Report of the Secretary-General to the UN General Assembly. **14**, 470–486 (2015)
15. A.B. Shamsul, Change and Continuity in Minangkabau: Local, Regional and Historical Perspective on West Sumatra. Edited by Lynn L. Thomas and Franz Von Benda-Beckmann. Monographs in International Studies Southeast Asia Series No. 71. Athens, Ohio: Ohio University Center. Journal of Southeast Asian Studies, **20** (1), 124–126 (1989)
16. i, F. . G.. Terminology of Disaster Risk Reduction (2009)
17. E. Sunarti, Evaluation of disaster management in Indonesia (Lesson learned 2006-2007). Retrieved from 202.124.205.241/handle/123456789/53265 (2007)
18. Tesco. Annual Report. Retrieved from https://www.tescoplc.com/media/392373/68336_tesco_ar_digital_interactive_250417.pdf (2017).
19. United Nations. UNISDR / 2004 / 08 26 October 2004 NATURAL DISASTERS: RICH COUNTRIES ALSO PAY THEIR TOLL. 33(October), 2759 (2004)
20. A. Yonder, S. Akçar, and P. Gopalan, Women's participation in disaster relief and recovery. In Women, Gender and Disaster: Global Issues and Initiatives (2009)
21. T. Yumarni, D. Amaratunga,Gender mainstreaming as a strategy to achieve sustainable post-disaster reconstruction. Built Environment Project and Asset Management, (2018).
22. T. Yumarni, D. Amaratunga, R. Haigh, Assessing Gender Vulnerability within Post-earthquake Reconstruction: Case Study from Indonesia. Procedia Economics and Finance, 18(September), 763–771 (2014).

Role of media in disaster preparedness: some case studies from calamity prone Odisha

Fakira Mohan Nahak[1]*

[1]Professor & HoD, University Institute of Media Studies, Chandigarh University, Mohali, Punjab, India.

Abstract. Odisha is a natural disaster prone state. Its geographical location contributes a lot for the disasters. The eastern Indian state has a 480 Kilometre long coastline. The Bay of Bengal which is the house of cyclonic storms is the major contributor for the calamities in the state. From time immemorial till today Odisha has been facing hundreds of calamities in the form of cyclones, floods and famines. It is a regular phenomenon that in the period of September to December every year Odisha faces varieties of cyclones. These affect human life, properties and agriculture to the maximum extent. After the super cyclone of 1999 the government became sensitive so also the media. In these two decades Odisha media has played a vital role in creating awareness about the disasters and helped people in displacement and rehabilitation. In recent pasts media helped the Government in reaching the "Zero Casualty" target. The role of media not only limited to this, even post-disasters it followed the condition of people and their lives. The researcher takes some case studies of different disasters and their handling by media. Also tries to find out the people's perception about media in disaster preparedness and management.

1 Introduction

Odisha is a natural disaster prone state. Its geographical location contributes a lot for the disasters. The eastern Indian state has a 480 Kilometre long coastline. The Bay of Bengal which is the house of cyclonic storms is the major contributor for the calamities in the state. From time immemorial till today Odisha has been facing hundreds of calamities in the form of cyclones, floods and famines. It is a regular phenomenon that in the period of September to December every year Odisha faces varieties of cyclones. These affect human life, properties and agriculture to the maximum extent. After the super cyclone of 1999 the government became sensitive so also the media. In these two decades Odisha media has played a vital role in creating awareness about the disasters and helped people in displacement and rehabilitation. In recent pasts media helped the Government in reaching the "Zero Casualty" target. The role of media not only limited to this, even post-disasters it followed the condition of people and their lives. The researcher takes some case studies of different disasters and their handling by media. Also tries to find out the people's perception about media in disaster preparedness and management.

2 Objectives of the study

1. To understand the role of media in disaster communication in disaster prone Odisha.
2. To know the preparedness of different media houses in order to handle the natural disasters and their coverage.
3. To know the perception of disaster affected people about media and its role in disaster communication.

3 Methodology

To measure the role of media in disaster preparedness and disaster communication, the researcher opted for a survey among the cyclone affected people of Ganjam district of Odisha using a questionnaire with a number of close ended questions. The sample size was 100 and responses are analysed. Along with this the researcher talked to different functional heads of media houses of Odisha to know their efforts in disaster communications.

3.1 Review of literature

Being in Indian eastern peninsular region Odisha has been a home for the cyclones for centuries together. The geographical location is one of the main reasons of the disasters the state encounters. Particularly when any depression in takes place in the Indian Ocean or Andaman Sea, most of the time it moves towards the Bay of Bengal and hits the eastern coast of India. As Odisha is in the same side with a long 480 Kilometres of coastline, it becomes more prone to cyclonic hits. Cyclones, severe

* Corresponding author: mohan.fakir@gmail.com

cyclones, very severe cyclones and Super cyclones are part of this cycle.

Although disasters cannot be avoided, the risk of disasters can be reduced by mitigation. Mitigation in the form of dialogue and communication has become a tool for society in dealing with life's issues. This study examines the role of government in preventing and mitigating dangers and disasters, such as natural disasters, disasters, and conflicts, through effective communication[3].

In every two years under the banner of United Nations, hundreds of countries meet together in Davos and discuss various methods and techniques to manage crisis during disasters. Researchers come with new data and its interpretation to make the disaster crisis management system more user friendly. International Disaster and Risk Conferences IDRC provide new information which helps in formulating guidelines to manage crisis and reduce risk.

Information and knowledge exchange cannot solve all problems in everyday life. There is need of appropriate and effective communication to manage life, risks, disasters and its related crisis efficiently. If the recipient is provided with correct communication, then it will help in bringing the desired significant change in the decision making and behaviour of the recipient. In return this will help him/her to mitigate risk and crisis during disaster as well as manage the post-disaster situations.

People are forced to deal with unique emergent challenges as a result of disasters. People need knowledge (about the event, support resources, activities, etc.) and the ability to comprehend and apply that information to deal with a variety of emergent hazard outcomes and demands across time in order to respond effectively. Top-down, uniform methods to disaster communication are ineffectual because of this variability[4].

Social media, Mobile applications, cell phones, Landline phones, Satellite phones (Stat phones), Two-way radio, Amateur radio (HAM radio) and word of mouth can help in communicating people during and before the disasters. Even after the disasters, this communication tools help in affected areas for rehabilitation and restoration.

It is the communication system and procedure which determines the effectiveness of the response during and after the disaster. How best may be your plan if it is not communicated effectively then it won't work. Land Mobile Radio Systems and Mass Notification Systems are some of the vital part of the disaster management and communication systems. Paton and Irons 2016, studied that a sense of community feelings can be created using the social media for disaster communication. During the disasters people face such a situation where they deal with new emergent problems. To manage that, they need concrete information about the disaster, its aftermaths, support system, resources and necessary actions. Social media based communication strategy can help in accommodating affected people from diversified backgrounds. They conclude that, development of sense of community among the affected people influence the communication quality during disaster[4].

3.2 Case studies of the role of media in disaster communication

On October 29, 1999 when the Super Cyclone hit Odisha coast, it was least communicated, ill measured and worst prepared incident of the history. Even after weeks it was not possible for the outer world to know what exactly has happened in the area. More than 30 thousand people died and thousands of homes completely devastated. Even then the weather warning system was not so developed. Only weather forecasting and warning by All India Radio and Doordarshan could not help people much. People could not gauge the severity of the incident. But after the cyclone, the gory picture of devastation could only be communicated to mass by media. Newspaper like Sambad published its special edition by handwritten designs. Reporters covered each and every part of the cyclone affected area and came up with heart wrenching photographs.

After that the communication system gradually developed in the state. Particularly the forecasting by IMD became so pinpointed that, it could be communicated to people in a proper way. During the floods of 2005, 2006 ETV Odia the first private satellite channel of the state played a vital role in disaster communication. In Odisha history, for the first time reporters ventured into the flood water and reported from the ground. Even 24 hours flood control room in the channel continued communication with the state flood control room as well as the viewers and audience through regular updates. This helped the administration in rescue operations, relief distribution, and rehabilitation.

The North Indian Ocean Cyclone Season normally occurs in between the month of May and December. The pick period is October to November. In 2013 October 12, a severe cyclonic storm named "Phailin" hit the Gopalpur on sea of Ganjam District. But by this time the state media was so active in disaster communication that, the target of the state government to achieve "Zero Casualty" could be achieved. All television channels OTV, ETV, Kanak TV, Naxatra News, MBC TV went with continuous live coverage and regular updates of the cyclonic storm and its landfall. Earlier communication to people and government's proactive initiatives compelled around one million people to leave their homes and go for safer places and cyclone shelters. Even most of the national channels (English and Hindi) sent their special correspondents to the affected areas prior to the landfall of the cyclone. The coverage became one of the most successfully handled disaster management and communication. The evacuation and disaster management brought laurels to the Chief Minister Mr. Naveen Patnaik. The United Nations felicitated the Chief Minister. The UN representative said, "Today, Odisha is a global leader in disaster management and risk reduction." So in this success story of disaster communication the role of media is clearly evident[5].

Just one year after Phailin, another cyclonic storm "Hud Hud" hit Odisha coast in the same month of October. Almost all TV channels of Odisha continued with 24 hours coverage and live bulletins. On the other hand "The Private FM Station Radio Choklate continued its communication about the cyclone with live programme

 https://doi.org/10.1051/e3sconf/202133104010

throughout the night" said one of the senior broadcasters of private FM channel Subrat Kumar Pati.

Fig.1. Disaster reporting by television media Courtesy: OTV.

In 2018, once again October became a curse for the people of Ganjam district. The severe cyclonic storm "Titili" hit the Ganjam and Gajapati districts of south Odisha. This time most of the channels continued with the same practice. 24 hours live coverage or live bulletin and continuous update about the cyclonic storm helped people in evacuation. This is not the end, the channels used their social media pages extensively for disaster communication. This time the researcher has personally listened the All India Radio, Cuttack. The AIR carried different live phone-in programmes on the cyclone "Titili" to aware the fishermen folks, farmers and affected people. This two way communication helped people in getting the expert advice as well as disaster preparedness. The kith and kin of disaster affected people staying beyond the territories of Odisha (Non-Resident Odias) get information about the disaster and overall situation in their homeland only through media. Some television channels of the state run late night shows keeping in view the timings of overseas people. In the subsequent years a number of other cyclones hit Odisha and the adjacent territories that caused a huge loss to life and property.

Cyclone Fani – 2019: Cyclonic Storm of Extreme Severity Fani was the first major cyclone of the year, as well as the fiercest tropical cyclone to impact Odisha since 1999. Fani peaked as a high-end Category 4 major hurricane on May 2, 2019, with maximum sustained wind speeds of 209-251 kmph. It made landfall in Puri, Odisha, a day later, and quickly weakened into a Category 1-equivalent 'very severe' cyclonic storm[6] . Cyclone Bulbul – 2019: Bulbul caused agricultural damage in Odisha, with 490,000 acres of crops lost across the state. It moved northward in the Bay of Bengal, eventually making landfall as a Very Severe Cyclonic Storm at Sagar Island in West Bengal on November 9, dumping torrential rain and winds of up to 135 km/h throughout southern West Bengal. Bulbul impacted 35 lakh people in West Bengal directly[6].

Cyclone Amphan – 2020: In Odisha and West Bengal, the severe tropical cyclone wreaked havoc on people and property. Storm Amphan was the first pre-monsoon super cyclone to emerge from the Bay of Bengal this century. Cyclone BOB 03 – 2020: The Indian Meteorological Department designated a depression in the Bay of Bengal as BOB 03[6]. The BOB 03 slammed the north Odisha-West Bengal coastline the day after it was identified, causing massive loss of life and property. Cyclonic Storm of Extreme Strength Yaas was a relatively strong and very severe tropical storm that made landfall in Odisha in late May 2021 and caused significant damage in West Bengal. The second cyclonic storm of the 2021 North Indian Ocean cyclone season, the second severe cyclonic storm, and the second extremely severe cyclonic storm of the year so far[7][8].

In all these cyclones, severe cyclones and very severe cyclonic storms the media played a proactive role in making people aware about the cyclonic storms and their severity. Basically almost all media informed and educated people to leave homes and go for the safe cyclone shelter centres. The contribution of all media is significant and cleanly visible during the cyclone coverage, evacuation, reliefs and rehabilitations. Not only this, the media play a vital role in the post-disaster management. Particularly media helps people in providing the pictures of affected areas. In post-disaster management media play the role of connecting link between affected people and the local and state administration. Rehabilitation of people, restoration of road transport, power supply, drinking water supply, health aids, sanitation and relief everything reported by media. There is no doubt the "Revenue and Disaster Management" department of the state works in the grass root level to assess the damages in the agriculture, livestock, housing, but media reports help in dissemination of information to the mass. Basing on media reports philanthropic and volunteer organisations extend their help to affected people.

During the disasters all media house make special efforts by deploying their special correspondents and reporters in the affected areas. Particularly television channels of the state go for special presentations with customized graphics, and data. They make all efforts to drag the attention of the viewers. During disasters the team of reporters work in a war footing manner

Fig. 2. Special coverage of disaster by the army of reporters in OTV Courtesy: OTV.

Majority of the television channels of the state of Odisha continued their live broadcasting throughout the day during the cyclones of high severity and scale. In order to be in the race and to meet the expectation of the viewers, the competition among different media houses was cleanly visible .

4 Data Analysis

The researcher has taken a random sample of 100 from the disaster affected Ganjam district of Odisha. All respondents were given a questionnaire with close ended questions. Each response was analysed thereafter to understand the people's perception about media and their role in disaster communication.

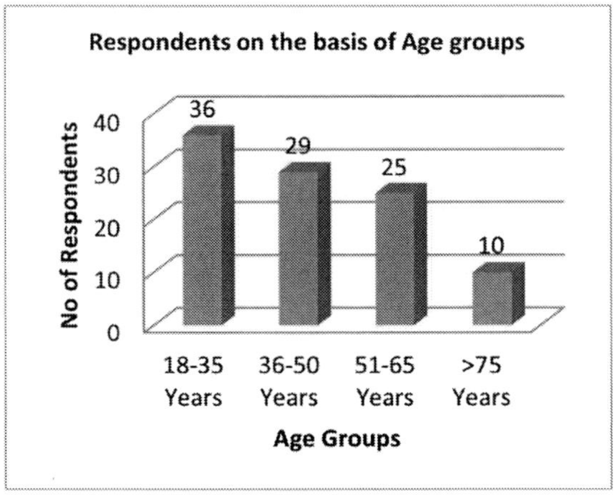

Fig. 3. Distribution of respondents based on Age groups.

From total 100 respondents 36 were from 18-35 years age group. In 36-50 years age group there were 29 respondents. 25 respondents from 51-65 years group and rest 10 were from more than 65 years age group. The age group distribution of respondents are described in figure-3.

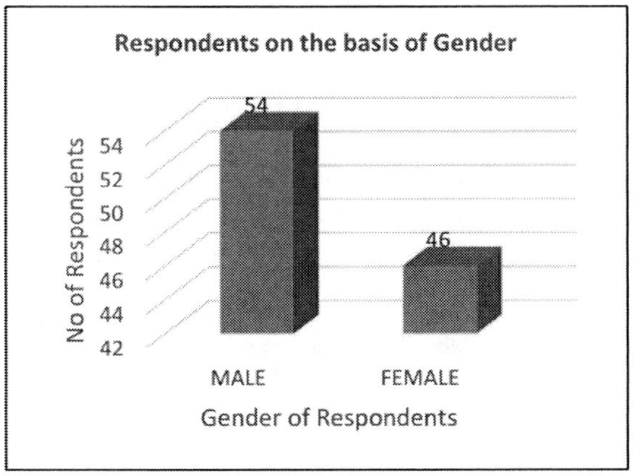

Fig. 4. Distribution of respondents based on gender.

Similarly there were 54 male and 46 female respondents in the sample. They gave their opinion about the role of media in disaster communication. Gender-based distribution is presented in the Figure-4. Figure-5 explains that, 64% respondents have a fear for natural disasters. But 36% people don't have fear for it.

In the previous figure it is explained that 36% people do not fear disaster. The reason behind this may be the development of an attitude towards disaster over a long period of time. Because Figure-6 clearly explains that, 89% respondents feel and accept that disaster has become a part of life. It has become a regular affair for them. But 11% do not feel like that.

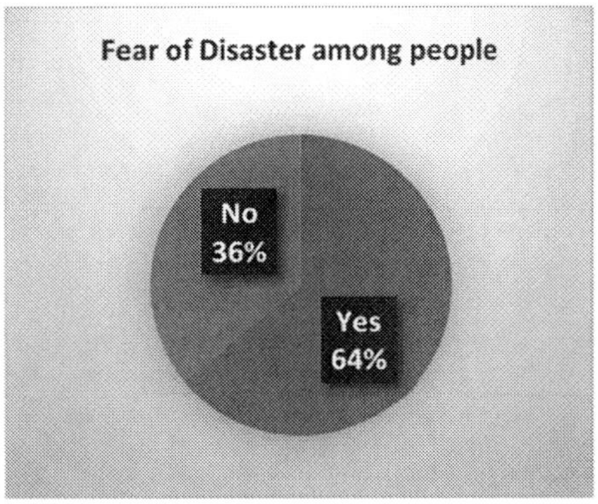

Fig. 5. Fear of disaster among respondents.

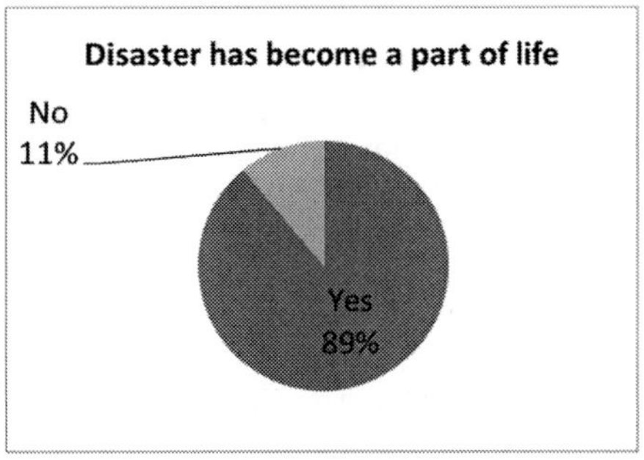

Fig. 6. Feeling of respondents about disaster becoming a part of life.

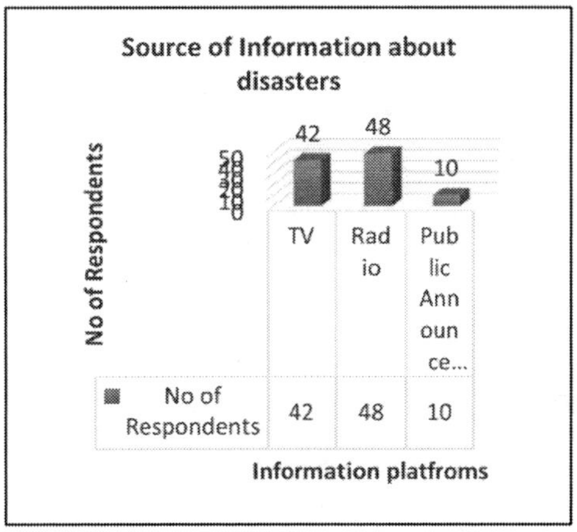

Fig.7. Respondent's source of information about disasters.

About the disaster and information related to it is normally disseminated by media. But as per the respondents 42% people depend on Television for information regarding disaster. Whereas Radio is in an advantageous position with 48% respondents depend on it. The mobility of radio gives an added edge to it over television. It is easier for people to carry a radio set anywhere. But television is fixed. On the other hand there is no need of uninterrupted power supply for radio. 10%

respondents depend on public announcement system for information about the disaster.

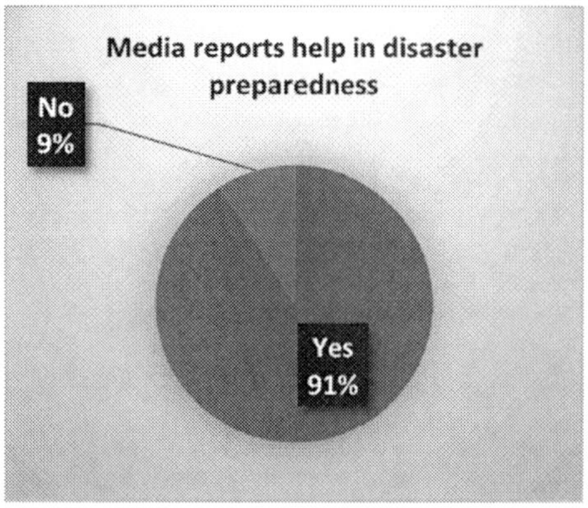

Fig. 8. Media reports help people in disaster preparedness.

As per Figure-8, 91% respondents feel that media reports help them in disaster preparedness. But 9% respondents do not feel there is any role of media in disaster preparedness. Majority of respondents make themselves alert and prepared to face the disaster. Media reports make them aware of the situation and developments. Accordingly they act to tackle the situation.

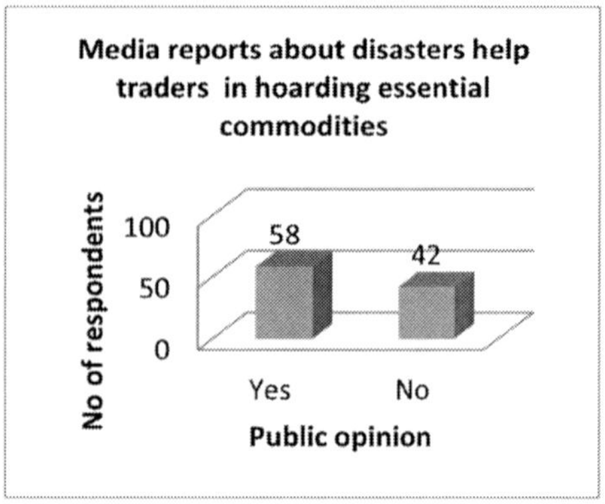

Fig. 9. Respondent's perception about media reports and illegal hoarding of commodities by traders.

Whenever there arises any disaster like situation, then it brings a golden opportunity for the illegal traders. They hoard the essential commodities and create a crisis in the consumer market. As per Figure-9, 58% respondents feel that it is the media report which help and encourage the traders for hoarding and black marketing of essential commodities. On the other hand 42% respondents do not make media responsible for this ugly situation and intentionally created crisis

Pre-emptive strategies help to an extent in saving lives during disaster. If people are stubborn, then it becomes difficult for the administration to evacuate them from their homes. This causes severe damage both property and lives. That is the reason why the tone of media and administration is always polite, because no one wants any casualty due to any disaster. As per Figure-10, majority of the respondents i.e. 82% feel that, yes they act promptly for evacuation when there is a weather warning presented in any media report. But 18% respondents are casuals. They do not act promptly after listening any warning in any media..

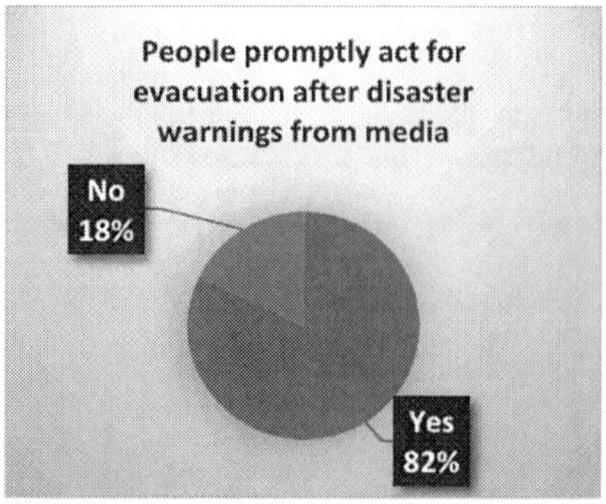

Fig. 10. Media alerts people to evacuate homes before disasters.

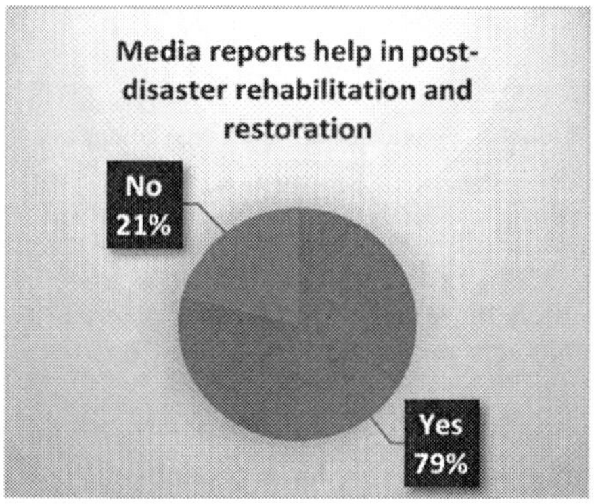

Fig. 11. Media reports help in post-disaster rehabilitation and restoration

The most important thing is that, even after disasters or natural calamities it becomes very difficult to restore and rehabilitate people. According to the Figure-11 there are around 79% respondents those who feel that, yes, media reports help a lot in disaster hit areas particularly in post-disaster rehabilitation and restoration. Whereas 21% respondents feel that, there is no role of media in post-disaster rehabilitation and restoration.

5 Findings and conclusion

This is peoples' perception about media in disaster communication. Whereas on the other side of the table is the media. Those who lead different television channels in the state say that, they act smartly as well as in a pinpointed focus for handling disasters. They make different strategies in covering disasters. Starting from deployment of reporters to movement of DSNG vans they plan everything.

From above data analysis it can be said that, people accept that media play a vital role in disaster communication. Media plays significantly alerting people, encouraging people for speedy evacuation, making people prepared for facing the crisis. Most importantly media contributes to a maximum extent in post-disaster situation for rehabilitation and restoration. Media awareness make people ready to face any exigencies during disasters.

From this study this can be concluded that, in last two decades since 1999 super cyclone, a number of natural disasters have occurred here. But the development in media sector has also played a vital role in the handling the calamities. Competitive attitude of editorial staff and dedicated work of team of reporters made it easy to handle communication during disasters. No doubt that the state administration always work in coordination with media during disasters. But it can be suggested that, there is need of more integrated approach in handling crisis during disasters. There is need of more cooperation and strategic communication between the state machinery and media, so that they work together for the benefit of people in critical conditions.

References

1. T. Lead, "National Cyclone Risk Mitigation Project (NCRMP) National Disaster Management Authority (NDMA) Consulting Services for Hazard , Risk and Vulnerability Assessment for 13 states and UT ' s in India Risk Assessment Report for Gujarat," 2016.
2. [2] PTI, "NDMA asks media to gain expertise on disaster reporting," India TV, 2013. https://www.indiatvnews.com/news/india/ndma-asks-media-to-gain-expertise-on-disaster-reporting-30477.html (accessed Sep. 13, 2021).
3. [3] Z. Khumairoh, I. D. K. K. Widana, and S. H. Sumantri, "The role of communication as the disaster risk reduction in Indonesia capital city transference policy," IOP Conf. Ser. Earth Environ. Sci., vol. 708, no. 1, 2021, doi: 10.1088/1755-1315/708/1/012101.
4. [4] D. Paton and M. Irons, "Communication, Sense of Community, and Disaster Recovery: A Facebook Case Study," Front. Commun., vol. 1, no. August, 2016, doi: 10.3389/fcomm.2016.00004.
5. [5] A. Senapati, "UN felicitates Odisha for its disaster management model during Phailin," DownToEarth, Dec. 2013.
6. [6] OB Bureau, "Odisha's Tryst With Cyclones 1737 – 2021," Odishabytes, 2021. https://odishabytes.com/odishas-tryst-with-cyclones-1737-2021/ (accessed Sep. 12, 2021).
7. [7] "Cyclone Yaas," Wikipedia. 2021, [Online]. Available: https://en.wikipedia.org/wiki/Cyclone_Yaas.
8. [8] "Explained: Cyclone Yaas to hit West Bengal, Odisha on May 26 | Here's why and how cyclones are named," Jagran English, 2021.

A literature review: model of disaster risk reduction for decision support system

Soetjipto Jojok Widodo[1*], *Hidayah* Entin[1], and *Sholikhah* Faizatus[1]

[1]Program of Civil Engineering, Jember University, 68121 Jl Kalimantan 37 Jember, Indonesia

Abstract. Disaster is a threat to human life. Many losses are caused by disasters, namely loss of life, injured people, loss of homes, and others. In addition, the frequency and intensity of disasters are also increasing every year. Therefore, research on Disaster Risk Reduction (DRR) is needed both to reduce disaster risk and to manage the disaster. The purpose of this research is to develop an appropriate DRR model in an area to assist decision-making in making policy. This research was compiled based on literature studies from various reputable journals to be used as a reference in the preparation of the right model. Then proceed with the development of a framework to model an efficient and effective DRR. The steps for making a holistic DRR model have been identified and the test design for the model has been determined, namely simulation, validation, and scenario. The recommendation given from this study is the preparation of a DSS (Decision Support System) as a tool for decision-makers to make policies regarding DRR-based regional development. This discussion will be continued in the next research by including case studies in certain urban areas.

1 Introduction

The Sendai Framework for Disaster Risk Reduction 2015-2030 report states that in the last 10 years disasters have become a threat to human life [1]. Disasters have had an impact on human welfare and safety both in the regional and global regions. Impact of the disaster: 700 thousand people lost their lives; more than 1.4 million people were injured and about 23 million were left homeless. From 2008-2012 there were 144 million people displaced by disasters, totaling more than $1.3 trillion in economic losses.

An increase in the frequency and intensity of disasters occur due to climate change. This has resulted in significant barriers to progress in sustainable development. Evidence shows that exposure to people and assets in all countries is increasing faster than vulnerability decreases, giving rise to new risks and increased disaster-related losses i.e., significant economic, social, health, cultural, and environmental impacts in the short, medium, and long term at both levels local and regional. To overcome this, the United Nations agreed on 7 targets, namely: reduction of disaster mortality, number of affected people, economic losses and disaster damage to critical infrastructure, increasing the number of countries with national and local disaster risk reduction strategies, enhancing international cooperation in support of DRR in developing countries, and increasing availability and access to early warning systems and disaster risk [2].

Based on the description above, efforts are needed for Disaster Risk Reduction, international and regional cooperation forums, development of policies and strategies as well as scientific progress and mutual learning. This will have an important impact on increasing public and institutional awareness, as well as multi-stakeholder decision-making.

2 Disaster Model

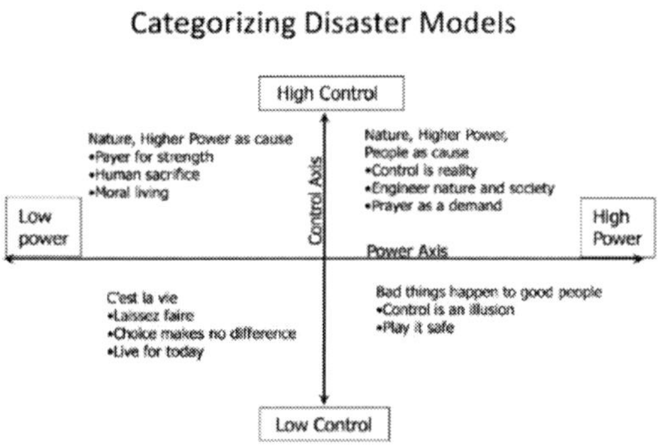

Fig. 1. A Conceptual framework for managing disaster [3].

The model is an application of theory that is a simplified representation of the real world. We need simplifying assumptions to make problems tractable [3]. Simple modeling techniques in disaster management can use quadrants as shown in Fig. 1. This model is effective

* Corresponding author: jojok.teknik@unej.ac.id

for simple systems. This model makes it easy to evaluate the effect of power and control in disaster management. Good disaster management lies in the upper right-hand quadrant.

In general, disaster management has four pillars of approach, namely:

1. **Mitigation**: actions to reduce the risk of natural disasters.
2. **Preparedness**: planning for disasters and putting in place the resources needed to deal with them when they occur.
3. **Response**: actions were taken after a disaster occurs.
4. **Recovery**: activities to rebuild and restore the community.

However, the model framework above only describes how the disaster was handled. The model still requires the type of disaster that will occur, how much expose the community and how many times occur, and so on. If this model has accurate information regarding the disaster to be faced, then this model can take effective disaster management actions to reduce its impact. Therefore, in this study, we will review several articles that have discussed disaster prediction.

3 Disaster Prediction

Disasters are very difficult to ascertain, how big they will occur when they will occur, how big the impact will be. However, several studies have tried to predict the occurrence of disasters. From various literature studies, information is obtained that several disaster events can be predicted both the events and the impacts that will be caused by exposure to the disaster.

The TOPSIS-multiple linear regression method is an application developed to predict the risk of flooding based on the database in Vietnam [4]. This method is used as an informal assessment tool that can be used by policymakers to clarify the areas most at risk and the losses incurred.

The application of Bayesian Network (BN) theory is becoming a trend in predicting the determination of various technical risks. This risk prediction has also been applied to disaster risk assessment. With BN's ability to determine parameters and correlation between variables in multivariate nonlinear equations, BN is the right consideration to be used as a risk assessment model in natural disasters [5]. Bayesian is also able to contribute to predicting the probability of flood events with good accuracy, can improve the detection of impact assessment and flood disaster management. This Bayesian capability must be supported by complete historical data [6]. The Bayesian method was also used to build a disaster chain hazard assessment model for a series of active volcanoes in China [7]. The Bayesian network combined with ArcGIS can compile a map of the chain and hazard intensity with a fairly good model performance with certain criteria.

The analysis and estimation mathematical methods have been used to predict typhoon disaster prediction and risk assessment. The Analysis method is the upper and lower limit interval estimation method, while the estimation method uses the average anomaly estimation value. Both methods were applied to the typhoon case study so that it can be used as a tool for disaster prevention and mitigation [8].

At the end of the disaster prediction sub-section, it is necessary to reaffirm the position of disaster prediction and risk assessment. This is necessary as a further step in this study. Disaster prediction is modeling objects that face danger, damage, or loss. While risk assessment models the possibility of future events that can be detrimental. In mathematical modeling, the model for disaster prediction may be an explicit function, while the model for risk assessment may be an implicit function. There are three criteria to assess whether a model is suitable for risk assessment: (i) availability of information, (ii) uncertainty of future conditions, and (iii) modeling is a comparison of the current situation with some known pattern [9]. Disaster probabilistic prediction is just a simplified risk assessment. Whereas disaster prediction can be considered as a special case of risk assessment which is likely to differ from one case to another.

4 Disaster Risk Reduction

Disaster Risk Reduction is the conceptual framework of elements considered with the possibilities to minimize vulnerabilities and disaster risks throughout a society, to avoid (prevention) or to limit (mitigation and preparedness) the adverse impacts of hazards, within the broad context of sustainable development [10]. Based on this understanding, DRR is a concept and good practice in reducing disaster risk, including reducing exposure, vulnerability, and increasing disaster preparedness. This concept must be made systematically based on analysis in formulating sustainable development policies.

Research on DRR is quite diverse because this topic connects and integrates disaster risk reduction with ecosystems through science, engineering, government, policy, and economics [11]. However, the most rapidly developing ones are the implementation of Eco-DRR and Eco-Engineering because they can provide sustainable solutions in reducing disasters with an approach to economic strengthening and social stability.

A study on DRR has been carried out to analyze policies related to code building to protect against earthquakes in Turkey. They used a probabilistic catastrophe-based cost-benefit analysis (CBA) model [12]. This model includes elements of climate change, increased hazard intensity, reduced vulnerability/ exposure to assets during the preparation of infrastructure development policies. As well as paying attention to the negative impacts of infrastructure development, namely poor land use, environmental degradation, decreased infrastructure services, overcrowding. So, this model helps them target more efficient and financially effective investments in the implementation of the DRR program before it is implemented.

Another study on the implementation of DDR in the Philippines also shows that the implementation of DDR must accommodate indigenous knowledge [13]. So far, the implementation of DDR has prioritized scientific knowledge even though the practice that has been carried out for years has naturally been proven to reduce disaster risk. Therefore, it is necessary to facilitate two forms of knowledge to further enhance the power of DDR going forward.

The simplest practice of disaster risk reduction is the study of the provision of disaster protection infrastructure. In a study on the provision of shelters for evaluation in disaster-prone zones in Padang, West Sumatra [14]. This study model the estimation of shelter needs to accommodate refugees during the evaluation process using linear regression models and binary logit. The provision of flood control infrastructure in Surakarta has also been carried out in the context of reducing disaster risk [15]. The provision of flood control infrastructure becomes ineffective if its availability is uneven and capacity decreases.

The concept of DRR in integrated disaster management is to combine the reduction of exposure and vulnerability, with capacity building for disaster preparedness in every policy/management of local and international governments. In infrastructure development, all design/planning, implementation, and implementation of the approach should incorporate DRR on disaster risk management in partnership with local authorities and stakeholders. The social success of DDR is highly dependent on implementation and approaches to adaptation to historical/current/future contexts, governance, and local gradients: urban-rural, inland-coastal landscapes, and rich-disadvantaged populations. While the technical success of DDR is highly dependent on establishing the necessary technical criteria, standards, incentives, and guidelines for participatory and interactive community resilience building across sectors, professions, and populations, it is also important for the successful implementation of this approach [11].

4.1 Hazard

The Hyogo Framework for Action defines that: Hazard is defined in Hyogo Framework for Action as: "A potentially damaging physical event, phenomenon or human activity that may cause the loss of life or injury, property damage, social and economic disruption or environmental degradation. Hazards can include latent conditions that may represent future threats and can have different origins: natural (geological, hydrometeorological, and biological) or induced by human processes (environmental degradation and technological hazards) [16].

In disaster management, it is necessary to know in advance about the potential hazards. Many studies on hazard determination have been carried out. Hazards are conditions that must be taken proactively and collected for future learning processes. Hazard data reporting techniques to be useful are were identified; sharing experiences, organizational learning, extending organizational memory, performance monitoring, and coordinating remedial actions [17]. Hazard determination using the AHP-Fuzzy method to determine hazard in geo-environmental impact [18].

4.2 Vulnerability

Vulnerability is defined as: "The conditions are determined by physical, social, economic, and environmental factors or processes, which increase the susceptibility of a community to the impact of hazards". by UN/ISDR Geneva 2004 [10]. And [16] states the types of vulnerability:

- Physical Vulnerability: infrastructure.
- Social Vulnerability: the inability of people, organizations, and societies to withstand adverse impacts to hazards
- Economic Vulnerability: the economic status of individuals, communities, and nations
- Environmental Vulnerability: Natural resource depletion and resource degradation.

Research on decision-making related to infrastructure vulnerability has been carried out. Infrastructure vulnerabilities can be referred to as Critical Infrastructure (CI). Infrastructure failure can lead to dangerous conditions for society, the economy, and even regional and national security. The infrastructure of an area can be networked and can influence each other. Therefore, to measure the vulnerability of infrastructure, we can use the Fuzzy-based DANP (FDANP) method, a hybrid method that combines the Decision Making Trial and Evaluation Laboratory (DEMATEL), Analytic Network Process (ANP), and fuzzy logic theory [19]. This model contributes to the evaluation of decision-making related to CI governance, including protection planning, and management of CI vulnerabilities, and increases the resilience of CI systems and the communities that depend on them.

4.3 Capacity

Capacity has defined a combination of all the strengths and resources available within a community, society, or organization that can reduce the level of risk or the effects of a disaster. Capacity may include physical, institutional, social, or economic means as well as skilled personal or collective attributes such as leadership and management. Capacity may also be described as capability [10].

5 Framework for Model of Regional DRR

The development of a DRR model for an area is very holistic. Many aspects that must be reviewed include how the disaster can be predicted, how much vulnerability the area has, and how much capacity the area has. In this study, a framework for making a DRR model of an area will be developed to obtain the condition of an area that is responsive to disasters by developing a DRR-based area. The concept of the model framework proposed in this study consists of: (i) disaster prediction modeling; (ii) disaster management diagram scheme; and (iii) DRR simulation.

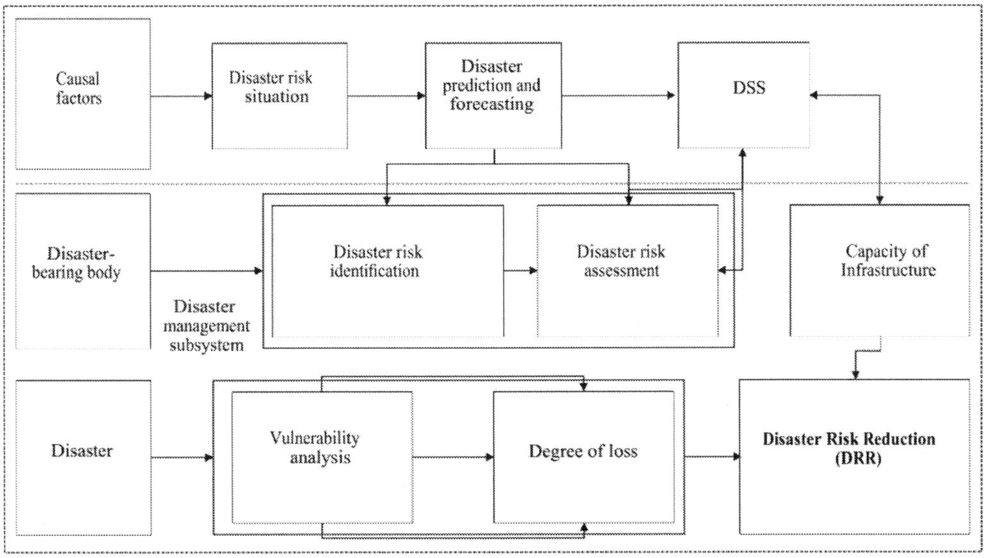

Fig. 3. Schematic disaster system

5.1 Formation of the Modeling Disaster Chain.

Disasters in an area can occur due to various kinds of disasters with various factors. Therefore, a model is needed to predict probabilistically the occurrence of disasters. This prediction is determined by the causal factors obtained from the disaster history data [20]. The chain disaster scheme can be seen in Fig 2.

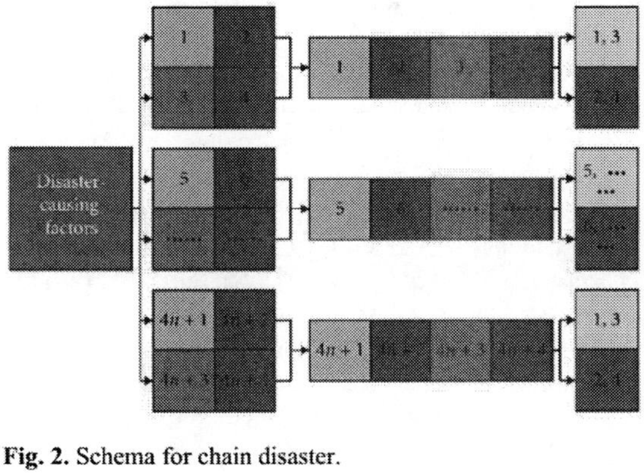

Fig. 2. Schema for chain disaster.

This disaster chain modeling can use several model options, namely bayesian or dynamic bayesian [21], markov-chain [22], artificial intelligence [23], System Dynamic [24], [25] etc.

The Bayesian model uses Bayes's theory as follows:

$$(\theta|y) = \frac{p(\theta,y)}{p(y)} = \frac{p(\theta)p(y|\theta)}{p(y)} \quad (1)$$

$$p(y) = \int p(\theta)p(y|\theta)d\theta \quad (2)$$

The formula of Markov is as follow:

$$P\left(Xt - 1 = \frac{j}{Xt} = i\right) = P\left(X1 = \frac{j}{Xo} = i\right); \quad (3)$$

The form of conditional probability transition matrix can be seen in equation (4):

$$P_{ij}^{(v)} = \begin{bmatrix} P_{00} & P_{o1} & \cdots & P_{0j} \\ P_{10} & P_{11} & \cdots & P_{1j} \\ \vdots & \vdots & \vdots & \vdots \\ P_{i0} & P_{i1} & \cdots & P_{ij} \end{bmatrix} \quad (4)$$

The System Dynamic formula is:

$$PG(t) = PG(t - dt) + TPM * dt \quad (5)$$

5.2 Diagram disaster system

In the preparation of the DRR framework for disasters in an area, it is necessary to understand and describe disasters and vulnerabilities. The risk of disaster loss depends on the vulnerability of the hazard carriers, the intensity of the hazard causes, and the disaster risk assessment method. Comprehensive urban infrastructure development must anticipate the prevention and mitigation of urban disaster systems. Therefore, to strengthen urban disaster prevention and mitigation strategies, the following schematic of comprehensive urban disaster prevention and mitigation measures, and increase the city's ability to withstand disasters (See Fig. 3).

In the first stage, the factors causing the disaster are determined as the basis for determining disaster predictions. The results of this prediction will be decided in the DSS (Decision Support System).

The next stage is the determination of disaster sub management, where at this stage a disaster risk assessment will be estimated based on risk identification data. The results of this identification will be combined in the DSS.

The last step is determining disaster vulnerability. This stage will be combined with capacity and decisions on the DSS will result in DRR action.

5.3 Simulation, Validation, and Scenario of The DRR Model

After all the stages in the preparation of the DRR model are fulfilled, the next step is to create a simulation using existing data. The simulation results will be compared with other existing data to see the validation of the model. If the model meets the validation requirements, it will continue with the model scenario.

Decision-making steps can be taken based on the results of the scenario with the smallest risk value. So that the decisions taken can provide for the management and development of infrastructure in an area.

6 Conclusion

Based on the results of the study above, it can be concluded that the preparation of DRR requires a holistic step, starting with disaster prediction, disaster impact estimation, and determination of regional infrastructure capacity.

Disaster DRR modeling can be done by selecting a model that fits the characteristics of the data, assessing future conditions, and validating using existing data to see how far the model can be accepted. The application of DDR for regional infrastructure that has big data can be modeled using Bayesian or artificial intelligence because both methods can predict cases with complex variables and data uncertainty.

A DSS is needed to simplify and speed up data processing, analysis, and decision-making. The development of this DSS system is carried out by integrating all sub-models in one analysis system.

References

1. United Nation, *Sendai Framework for Disaster Risk Reduction 2015 - 2030*. Sendai, Japan: United Nations, (2015)
2. United Nations, GLOBAL SUSTAINABLE DEVELOPMENT REPORT. United Nations, 2015.
3. D. Etkin, "Disaster Theory: Chapter 6 Disaster Models," in Butterworth-Heinemann, 193–228 (2016)
4. C. Luu, J. von Meding, M. Mojtahedi, "Analyzing Vietnam's national disaster loss database for flood risk assessment using multiple linear regression-TOPSIS," Int. J. Disaster Risk Reduct., **40**, no. March (2019)
5. M. Li, M. Hong, R. Zhang, "Improved Bayesian Network-Based Risk Model and Its Application in Disaster Risk Assessment," Int. J. Disaster Risk Sci., **9**, 2, 237–248 (2018)
6. L. Li, J. Wang, H. Leung, C. Jiang, "Assessment of catastrophic risk using bayesian network constructed from domain knowledge and spatial data," Risk Anal., **30**, 7, 1157–1175 (2010)
7. L. Han, J. Zhang, Y. Zhang, Q. Ma, S. Alu, Q. Lang, "Hazard assessment of earthquake disaster chains based on a Bayesian network model and ArcGIS," ISPRS Int. J. Geo-Information, **8**, 5 (2019)
8. L. Yang, C. Cao, D. Wu, H. Qiu, M. Lu, L. Liu, "Study on Typhoon Disaster Loss and Risk Prediction and Benefit Assessment of Disaster Prevention and Mitigation," Trop. Cyclone Res. Rev., 7, 4, 237–246 (2018)
9. C. Huang, "Differences between disaster prediction and risk assessment in natural disasters," Hum. Ecol. Risk Assess., **17**, 3, 542–553 (2011)
10. U. Nations, Living with Risk, **2** (2004)
11. A. W. Whelchel, F. G. Renaud, K. Sudmeier-Rieux, Z. Sebesvari, "Advancing ecosystems and disaster risk reduction in policy, planning, implementation, and management," Int. J. Disaster Risk Reduct., **32**, 1–3 (2018)
12. E. Michel-Kerjan et al., "Catastrophe risk models for evaluating disaster risk reduction investments in developing countries," Risk Anal., **33**, 6, 984–999 (2013)
13. M. Balay-As, J. Marlowe, J. C. Gaillard, "Deconstructing the binary between indigenous and scientific knowledge in disaster risk reduction: Approaches to high impact weather hazards," Int. J. Disaster Risk Reduct., **30**, 18–24 (2018)
14. D. Jurusan, T. Sipil, U. Andalas, J. Teknik, and U. Andalas, "Model ' Demand ' Pengguna ' Shelter ' Evakuasi Pada Zona Rawan Tsunami Di Kota Padang, 243–251 (2013)
15. A. A. S. Pramitha, R. P. Utomo, N. Miladan, "Efektivitas infrastruktur perkotaan dalam penanganan risiko banjir di Kota Surakarta," Reg. J. Pembang. Wil. dan Perenc. Partisipatif, **15**, 1, 1 (2020)
16. P. Basabe, Hyogo framework for action 2005–2015, no. January 2005 (2013)
17. J. Having, K. Bancroft, A. Rae, "Hazard reporting: How can it improve safety?," Saf. Sci., **142** (2021)
18. S. Huang, X. Li, Y. Wang, "A new model of geo-environmental impact assessment of mining: a multiple-criteria assessment method integrating Fuzzy-AHP with fuzzy synthetic ranking," Environ. Earth Sci., **66**, 275–284 (2012)
19. J. S. Chou, C. S. Ongkowijoyo, "Hybrid decision-making method for assessing interdependency and priority of critical infrastructure," Int. J. Disaster Risk Reduct., **39** (2019)
20. D. Su, X. Lei, Y. Wang, H. Wang, "Analysis of Multiobjective Scheduling Model for Emergency Management under Multi resource Combination of Disaster Chains," Complexity, 1–12 (2021)
21. J. Widodo Soetjipto, T. Joko Wahyu Adi, N. Anwar, "Dynamic Bayesian updating approach for predicting bridge condition based on Indonesia-bridge management system (I-BMS)," in MATEC Web of Conferences, Aug., **195** (2018)
22. J. Widodo Soetjipto, T. J. Wahyu Adi, N. Anwar, "*Bridge Deterioration Prediction Model Based On*

https://doi.org/10.1051/e3sconf/202133104011

Hybrid Markov-System Dynamic," MATEC Web Conf., The 6 th International Conference of Euro Asia Civil Engineering Forum (EACEF 2017), **138** (2017)

23. H. Sarvari, H. Kashefi, "Identification and prioritization of seismic risks in urban worn-out textures using fuzzy Delphi method," no. July (2021)
24. S. J. Widodo, T. J. W. Adi, N. Anwar, "System dynamics approach for bridge deterioration monitoring system," Int. J. Eng. Technol. Innov., **6** (4) (2016)
25. Y. Suponco, W. Broto, T. Joko, W. Adi, E. Suryani, J. W. Soetjipto, "Project Cost Escalation Prediction Models Based on System Dynamics Approach."

The framing of decision making support systems on increasing community resilience in disaster risk reduction efforts : a conceptual approach

Putranesia Thaha[1*] and *Febrin Anas* Ismail[2]

[1]Dept. of Quantity Surveying, Faculty of Engineering and Planning, Bung Hatta University, Indonesia
[2]Dept. of Civil Engineering, Faculty of Engineering, Andalas University, Indonesia.

Abstract. This research begins by comprehensively exploring previous research related to community resilience and what steps are used to increase community resilience in reducing disaster risk. Conceptually, it is known that the fatigue model accumulated by the time system, infrastructure system, governance system, regulatory system, and hazard system for disaster risk reduction is often associated with weakening community resilience. It is often associated with catastrophic events, which are sometimes predictable and unpredictable. In manual decision-making, people are aware of the inconsistency of subjective decisions. A decision support system hypothesizes that it will take less time to explore data to make faster and more informed decisions. As a result of this concept, it is possible to reduce the number of wrong choices when dealing with disaster risk reduction issues. In terms of disaster risk reduction, the power of decision support systems is discussed in this paper to find a framework for its effectiveness as relative decision making will differ on different dimensions of Resilience.

1 Introduction

Resilience can be described as overcoming obstacles and reducing the duration of deviations from the level of system performance and Resilience to control the difficulties encountered in life [1], [2]. Meanwhile, capability can be interpreted as a resource which is a capability, and capacity as an essential component that can affect the results or goals [3]. Furthermore, Rus, Kilar, and Koren. [4], identified Resilience as two dominant theoretical perspectives, namely socio-ecology and engineering; the socio-ecological perspective is considered a process-oriented phenomenon (a dynamic concept), while the technical perspective views Resilience as a result-or-concept (static premise).

In line with the socio-ecological philosophy above, Davies [5] said that disaster resilience could be increased when the social impact of disasters is reduced. In terms of reducing the effects of the disaster, planning for Resilience as a mitigation strategy requires wisdom from leaders in making decisions on a priority scale for at-risk populations with limited resource allocation [6]. Another study revealed that assessing Resilience from a social-ecological perspective in a community can be done using a quantitative approach [7]. There are have three fundamental problems in the Resilience of a community to reduce disaster risk, namely: preparation, response, and recovery, where the consequences of the approach affect (i) the relationship between communities, decision-makers, and planning tools (ii) system capacity civil protection (iii) economic capacity[8].

Disasters that arise in a community will undoubtedly have an impact on the community itself. The resulting consequences often cause more harm than gain, so it takes time to recover from damage, disruption, and injury. Disaster resilience refers to a community's ability to withstand, adapt, and recover from disasters as quickly as possible and with little or no outside assistance [9,10]. As disasters often have a disproportionate impact on those with the poverty of technology and financial resources [11], the idea of Resilience should be more proactive to reduce disaster risk. Disasters that arise in a community will undoubtedly have an impact on the community itself. The effects often result in more harm than good, so it takes time to recover from damage, disruption, and injury. During the recovery period, a community's ability to withstand potential hazards is determined by the extent to which it possesses the necessary and systematic resources [12]. Amaratungga and Haigh [10] also recommend the need for early thinking about disaster risk reduction by increasing the roles and responsibilities of each stakeholder in increasing community resilience.

This paper's state of the art is the influence of the decision-making system on increasing the community's

* Corresponding author: putranesia@bunghatta.ac.id

Resilience in disaster risk reduction efforts. This paper aims to determine the strength of a systematic decision-making support system in influencing community resilience in the face of disaster risk so that it can further make a more outstanding contribution to disaster risk reduction efforts in an appropriate and well-controlled manner.

1.1 The Concept of Resilience

Resilience can be defined as efforts in policy-making, framework creation, organizational/institutional, and community capacity (or non-structural actions) in disaster risk management [13]. Resilience can be described as a response to uncertainty from rapid and turbulent social change [14]. Another critical point is that the concept of Resilience recognizes community resilience through its capacity to self-regulate [15]. Resilience is a system that can be made if all subsystems and variables support the system, which can be measured by the extent to which the system can survive and absorb changes outside the system [16]. And also, according to Biesbroek, Dupuis, and Wellstead [17], Resilience has entered the public and political arena as a popular thing in viewing contemporary problems on issues of food security, climate change impacts, pollution, and natural resource management. The need to increase community resilience has become a sustainable development program (SDGs) in countries that are members of SDG 11. This is intended as an effort for Disaster Risk Reduction (DRR) in building resilient communities to reduce vulnerability to disasters Deep insight in realizing the community's unique personality, culture and understanding the mindset of the community at the individual perception level in adapting to increase their capacity to deal with disasters is needed to bridge the gap between spatial planning and disaster resilience building for disaster risk reduction in the future.

1.2 Analysis of Decision Making in Resilience

Successful decision-making in society begins with overcoming differences among community groups in decision-making [9]. As a result, decision-making must be based on uncertainty and incomplete knowledge for normative decisions based on the value of benefits (or utility) to be made by time systems, infrastructure systems, governance systems, regulatory systems, and risk systems [18]. On the other hand, Bhattacharya-Mis and Lamond [19] suggest that to increase Resilience, a realistic understanding of risk is needed to avoid external bias, where understanding risk will assist efforts in making the right decisions. In Disaster Risk Reduction, conceptually well-planned, adaptive, and sustainable planning will help policymakers avoid the worst outcome [20]. But in fact, Relative decision-making will differ on different dimensions of Resilience [21]. Due to the inconsistency of findings based on people who analyze data (subjective), a decision support system is hypothesized to reduce the amount of time needed to explore the data [22], allowing for faster and more accurate decisions. It's been shown that decision-making ability is influenced by the assessment of available resources and the decision-making power of policymakers [23]. This concept can reduce wrong decisions in overcoming problems in disaster risk reduction. The government, in this case, acts as a decision-maker that influences the six psychological constructs that will affect the performance of the construction (reconstruction) implementation, namely: culture, decision architecture, naturalistic decision making, group work, identity, and social representation [24]. Furthermore, Hettiarachchi and Weeresinghe [25] state that the primary measure for achieving disaster resilience is integrating Disaster Risk Reduction (DRR) into development at all levels.

2 Methodology

The research Methodology use “The Onion Proces” by Saunders [26].

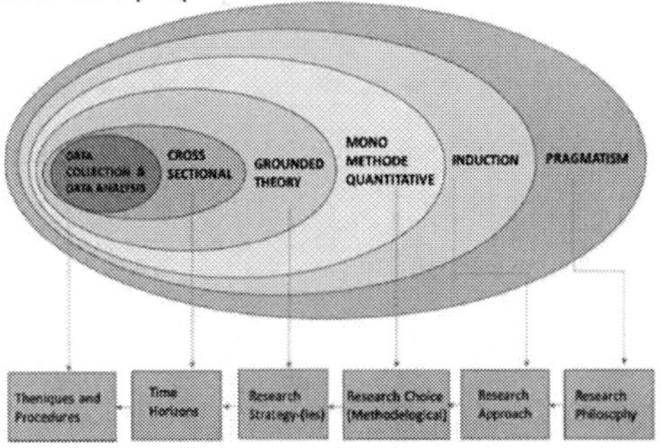

Fig.1. Developed From The Onion Proces by Saunders [26]

Fig. 1 explains that the research begins through a pragmatism approach by looking at the existing phenomena. Next, take an induction approach by exploring theories from previous researchers, which then go through a quantitative process to assess Resilience in a community from a socio-ecological perspective. Previous research was processed through a comprehensive literature review with data collection and analysis using the Feature Reporting Items for Systematic Review and Meta-Analysis (PRISMA).

3 Framing Decision Support Systems in Resilience

Recently, the concept of “resilience” has become a fundamental benchmark in various academic disciplines, including engineering research, organizational performance, ecological science, psychology, economics, climate change, disaster management, safety, security, and risk. Resilience appears to answer various problems and threats and thus attracts policymakers and researchers from multiple fields and disciplines. Policymakers are understood as people involved in organizational performance. Organizational Resilience can be understood as a learning organization that uncovers predictable and unpredictable disasters and is supported and nurtured as part of resilience goals [27]. Therefore, Resilience should be seen as an essential property of the system and not as an external quality [28]. Attempts to frame Resilience as a sustainability paradigm sometimes increase the complexity of decision making [29]. We must map the flow of information in reducing risk through collaboration, development planning and control, mitigation measures, building standards, community involvement, and effective management responses to assist all parties in making the right decisions [30], [31]. In information systems (IS), Decision Support Systems are used to support and improve managerial decision-making [32].

Learn from previous research to formulate a framework for decision-making support systems referring to disaster event data (big data) [33] and concrete steps in improving community resilience. Without the right decision support system, the amount of data will cause information overload [34]. Decision support system frameworks can be created based on hazards, resilience objectives, mitigation (ex-ante), and response (ex-post) strategies [35].

Fig. 2 explains the concept approach of Framing Decision Support Systems In Resilience

.

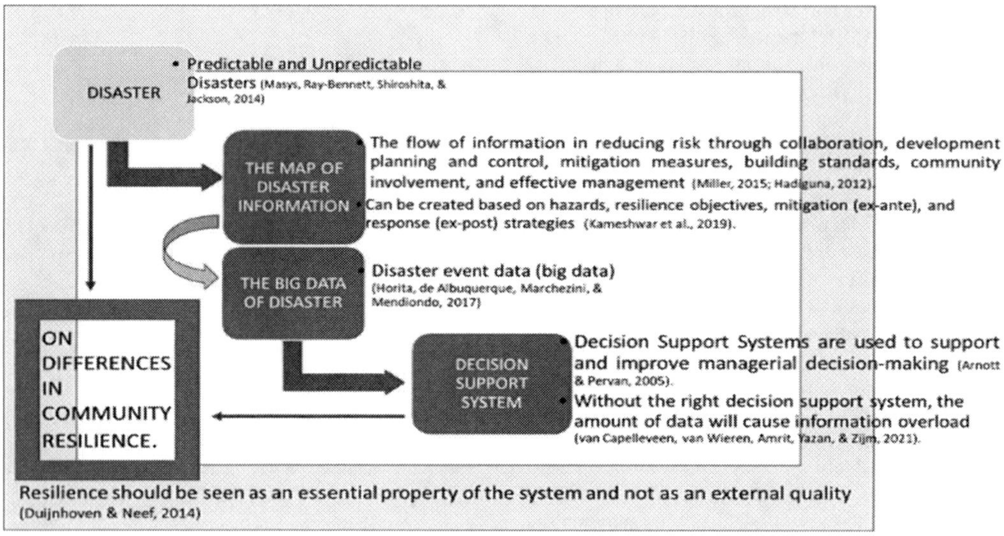

Fig.2. The Framing Decision Support Systems In Resilience [28]

4 Conclusion

As relative decision-making will differ in different dimensions of Resilience, a decision-making system will help provide effective decisions to reduce disaster risk in communities living in disaster-prone areas. In this regard, new space is obtained for further research related to decision-making. A decision support system with a good performance framework that is mixed based on differences in community resilience in disaster-prone areas can trigger disaster risk reduction of overall disaster resilience.

References

1 K. Reivich and A. Shatté, "The resilience factor: 7 essential skills for overcoming life's inevitable obstacles.," Resil. factor 7 Essent. Ski. overcoming life's inevitable Obs., 342, (2002)

2 V. Proag, "Assessing and Measuring Resilience," Procedia Econ. Financ., **18**, September, 222–229, (2014)

3 H. Lindbom, H. Tehler, K. Eriksson, and T. Aven, "The capability concept - On how to define and describe capability concerning risk, vulnerability, and resilience," Reliab. Eng. Syst. Saf., **135**,45–54, (2015)

4 K. Rus, V. Kilar, and D. Koren, "Resilience assessment of complex urban systems to natural disasters: A new literature review," Int. J. Disaster Risk Reduct.,. **31**, January, 311–330, (2018)

5 T. R. H. Davies and A. J. Davies, "Increasing communities' resilience to disasters: An impact-based approach," Int. J. Disaster Risk Reduct., **31**, May, 742–749, (2018)

6 C. E. Kontokosta and A. Malik, "The Resilience to Emergencies and Disasters Index: Applying big data to benchmark and validate neighborhood resilience capacity," Sustain. Cities Soc., **36**, June 2017, 272–285, (2018)

7 W. C. Chuang et al., "Enhancing quantitative approaches for assessing community resilience," J. Environ. Manage., **213**, 353–362, (2018)

8 M. Vona, M. Mastroberti, L. Mitidieri, and S. Tataranna, "New resilience model of communities based on numerical evaluation and observed post-seismic reconstruction process," Int. J. Disaster Risk Reduct., **28**, January, 602–609 (2018)

9 E. S. Ranjan and C. C. Abenayake, "A Study on Community's Perception on Disaster Resilience

Concept," Procedia Econ. Financ., **18**, September, 88–94, (2014)

10 C. Malalgoda, D. Amaratunga, and R. Haigh, "Challenges in Creating a Disaster Resilient Built Environment," Procedia Econ. Financ., **18**, September, 736–744, (2014)

11 D. B. Oerther, "From disaster to development: Finance provides a platform to empower technology for resilience to climate change," Procedia Eng., **159**, June, 267–271, (2016)

12 S. Platt, D. Brown, and M. Hughes, "Measuring resilience and recovery," Int. J. Disaster Risk Reduct., **19**, May, 447–460, (2016)

13 K. MacAskill and P. Guthrie, "Multiple Interpretations of Resilience in Disaster Risk Management," Procedia Econ. Financ., **18**, September, 667–674, (2014)

14 M. Iordan, M.-N. Chilian, and A. Grigorescu, "Regional Resilience in Romania – Between Realism and Aspirations," Procedia Econ. Financ., **22**, November 2014, 627–635, (2015)

15 A. J. Imperiale and F. Vanclay, "Experiencing local community resilience in action: Learning from post-disaster communities," J. Rural Stud., **47**, 204–219, (2016)

16 O. R. Danar and D. Pushpalal, "Building Community Resilience: Conceptual Framework and its Application in Post Tsunami Resettlement," Procedia Econ. Financ., **18**, September, 489–496, (2014)

17 R. Biesbroek, J. Dupuis, and A. Wellstead, "Explaining through causal mechanisms: resilience and governance of social-ecological systems," Curr. Opin. Environ. Sustain., **28**, 64–70, (2017)

18 M. H. Faber, L. Giuliani, A. Revez, S. Jayasena, J. Sparf, and J. M. Mendez, "Interdisciplinary Approach to Disaster Resilience Education and Research," Procedia Econ. Financ., **18**, September, 601–609, (2014)

19 N. Bhattacharya-Mis and J. Lamond, "Socio-economic Complexities of Flood Memory in Building Resilience: An Overview of Research," Procedia Econ. Financ., **18**, no. September, 111–119, (2014)

20 K. Yang et al., "Hurricane evacuations in the face of uncertainty: Use of integrated models to support robust, adaptive, and repeated decision-making," Int. J. Disaster Risk Reduct., **36**, October 2018, p. 101093 (2019)

21 C. W. Zobel and M. Baghersad, "Analytically comparing disaster resilience across multiple dimensions," Socioecon. Plann. Sci., **69**, (2020)

22 R. S. Renu, G. Mocko, and A. Koneru, "Use of big data and knowledge discovery to create data backbones for decision support systems," Procedia Comput. Sci., **20**, 446–453, (2013)

23 H. Lindbom, H. Hassel, H. Tehler, and C. Uhr, "Capability assessments – How to make them useful for decision-making," Int. J. Disaster Risk Reduct., vol. 31, no. May, pp. 251–259, (2018)

24 R. Turner, "Investigating how governmentality and governance influence decision making on projects," Proj. Leadersh. Soc., **1**, May, p. 100003, (2020)

25 S. S. L. Hettiarachchi and S. Weeresinghe, "Achieving Disaster Resilience through the Sri Lankan Early Warning System: Good practices of Disaster Risk Reduction and Management," Procedia Econ. Financ., **18**, September, 789–794, (2014)

26 A. Saunders ; K .LEwis, Philip; Thornhill, Research Methods for Business Students, Eighth edi. New York : Pearson, [2019]: Print edition typeset in 9.5/12 ITC Slimbach Std by Pearson CSC, (2019)

27 A. J. Masys, N. Ray-Bennett, H. Shiroshita, and P. Jackson, "High Impact/Low-Frequency Extreme Events: Enabling Reflection and Resilience in a Hyper-connected World," Procedia Econ. Financ., **18**, September, 772–779, (2014)

28 H. Duijnhoven and M. Neef, "Framing Resilience. From a Model-based Approach to a Management Process," Procedia Econ. Financ., **18**, September, 425–430, (2014)

29 G. Lizarralde, K. Chmutina, L. Bosher, and A. Dainty, "Sustainability and resilience in the built environment: The challenges of establishing a turquoise agenda in the UK," Sustain. Cities Soc., **15**, 96–104, (2015)

30 W. Miller, "What does built environment research have to do with risk mitigation, resilience, and disaster recovery?" Sustain. Cities Soc., **19**, 91–97, (2015)

31 R. A. Hadiguna, "Decision support framework for risk assessment of sustainable supply chain," Int. J. Logist. Econ. Glob.,**4**, 1/2, 35, (2012)

32 D. Arnott and G. Pervan, "A critical analysis of decision support systems research," J. Inf. Technol., **20**, 2, 67–87, (2005)

33 F. E. A. Horita, J. P. de Albuquerque, V. Marchezini, and E. M. Mendiondo, "Bridging the gap between decision-making and emerging big data sources: An application of a model-based framework to disaster management in Brazil," Decis. Support Syst., **97**, 12–22, (2017)

34 G. van Capelleveen, J. van Wieren, C. Amrit, D. M. Yazan, and H. Zijm, "Exploring recommendations for circular supply chain management through interactive visualisation," Decis. Support Syst., **140**, 113431, (2021)

35 S. Kameshwar et al., "Probabilistic decision-support framework for community resilience: Incorporating multi-hazards, infrastructure interdependencies, and resilience goals in a Bayesian network," Reliab. Eng. Syst. Saf., **191**, December, (2019)

Disaster mitigation local wisdom in the tradition of the minangkabau community

Khairil Anwar[1*]

[1]Literature of Minangkabau Department, Andalas University, Indonesia

Abstract. This study is about Minangkabau's local wisdom in disaster mitigation. Minangkabau is an ethnic and cultural group that is still alive and developing today. This ethnic group is centered in West Sumatra in the highlands of the Bukit Barisan which stretches along the island of Sumatra and develops through migrating to various regions in the world. This ethnicity is the oldest tribe on earth which is characterized by the use of the hereditary system according to the maternal or matrilineal line. His leadership made the Minangkabau ethnic have various kinds of local wisdom, especially those directly related to disaster mitigation. The Minangkabau cultural center is located in an area that has a high intensity of natural disasters by its natural topography. In responding to their natural environment, the Minangkabau community has local wisdom in the form of a superstructure that regulates infrastructure and social structure in disaster mitigation. This local wisdom is found in various literacy and traditions of the Minangkabau community. This qualitative discussion uses the perspective of cultural materialism theory. It was found that the local wisdom of disaster mitigation includes human norms and attitudes towards nature; norms before a disaster occurs; and post-disaster policies. To anticipate disasters, there are rules regarding the processing and utilization of nature, such as the use of land, hills, deserts, and swamps. In the event of a disaster, there are rules such as building *rangkiang* and filling it with food reserves, doing the *ijok* tradition, and *batangeh*.

1 Introduction

Minangkabau is an ethnic group that occupies an area centered in the central part of the Bukit Barisan on the island of Sumatra. The Minangkabau ethnicity occupies a cultural area, not an administrative area [1]. Currently, Minangkabau is known as culture, not administration. This central area is called *darek. Darek* is divided into three areas called Luhak Nan Tigo. In addition to the central area, there are development areas due to the increasing population and limited natural resources and energy in the area of origin. The area is called Ikua Darek Kapalo Rantau, and Rantau. Ikua Darek Kapalo Rantau is a transit area between darek and rantau. Meanwhile, overseas is a new local area to get a source of energy for their collective survival.

The Minangkabau cultural area is empirically located in the mountains, hills, valleys, rivers, lakes, and plains. Historically and oral tradition, this cultural area is described in tambo [2], local wisdom of Minangkabau origins, as follows:

Dari sirangkak nan badangkuang
Hinggo babuayo putiah daguak
Di sinan lagundi nan baselo
Di sinan banto nan barayun
Di sinan baringin sonsang
Dari sipisok-pisok pisau anyuik
Sampai ka pintu rajo ilia
Dari durian ditakuak rajo
Sampai ka sialang balantak basi
Dari aia babaliak mudiak
Sampai ka ombak nan badabua
Sailiran batang sikilang
Hinggo lauik nan sadidih
Rao jo Mapat Tunggua
Sarato gunuang maha lintang
Pasisia banda sapuluah
Hinggo taratak aian hitam
Sampai ka tanjuang simalidu
Pucuak jambi sambilan lurah
From crawling and snoring
Until there is a white crocodile with a chin
There's a cross-legged lagundi
There the swinging grass
There the beringin tree has hanging roots
From the Sipisok-pisok drifting knife
Up to Pintu Rajo Ilia
From Durian Ditakuak Rajo
Up to Sialang Balantak Basi

* Corresponding Author: khairilanwar@hum.gmail.ac.id

From the water back to home
Until the crashing waves
In line Batang Sikilang
Until the surging sea
Rao and Mapat Tunggua
As well as Gunuang Mahalintang
Pasisia Banda Sapuluah
Until the Taratak Aia Itam
Up to Tanjuang Simalidu
Pucuak Jambi Sambilan Lurah

In the text above, it is explained about the process of the first arrival of the Minangkabau ethnic group to the region. This ethnicity came to this area by crawling and swinging. They crawl to climb high hills with sharp slopes or high inclines and cross valleys and rivers by swinging by utilizing hanging roots from trees in these locations. The traditional historiography also describes the natural conditions of Minangkabau which consist of hills, forests, rivers, and the sea. Culturally Minangkabau [3] Initially covered several administrative areas of the provinces in Indonesia, such as West Sumatra, North Sumatra, Aceh, Bengkulu, Jambi, and Riau, and several countries on earth such as Malaysia, Singapore, Brunei, and the Philippines. At this time of course it continues to grow in various provinces and countries along with the easy access to traffic both land, air, and sea and the mobility capabilities of the Minangkabau ethnic group [4]. However, when talking about the Minangkabau ethnicity, it is more related to the administrative area of West Sumatra Province. These cultural and administrative areas can be seen in the following map (Fig. 1).

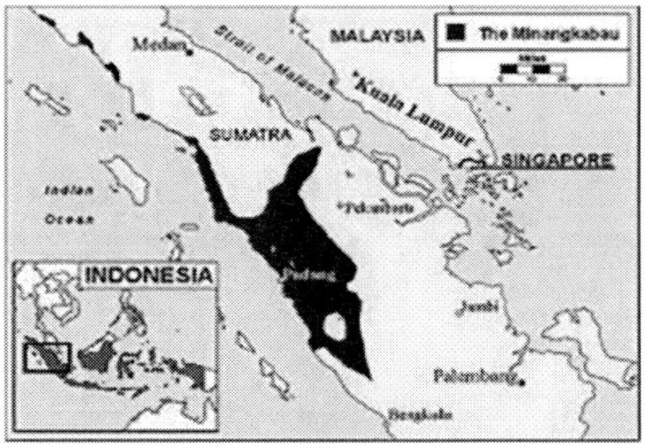

Fig. 1. Minangkabau Cultural Area Map.

Empirical facts in the form of the natural environment, traditions, and Tambo records present a fact that the Minangkabau ethnic area in West Sumatra is in an area with varied topography. The topography shows that the area has high potential and is prone to natural disasters [5]. As shown in the map in Fig. 2.

Minangkabau is the oldest ethnic group on earth. This old age is marked by the use of the hereditary system according to the maternal bloodline or matrilineal and tribal to the mother's tribe. The hereditary system was used by early humans for centuries past. His leadership makes the Minangkabau ethnic have various kinds of local wisdom that are familiar with nature. Wherever they live they always adapt to the natural situation that has existed first. The same as ancient humans who lived on the move, they came to new places by adjusting and only taking things according to their needs and not destroying nature.

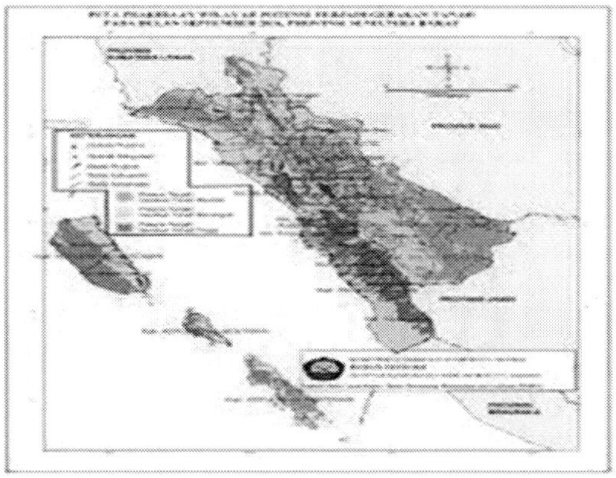

Fig. 2. Map of Disasters and Land Movements in West Sumatra

The philosophy "where the earth is stepped on there the sky is upheld" becomes a guide in the life journey of the Minangkabau ethnic group. Such a way of life so that they can be accepted everywhere they go [6]. They are not an exclusive type of ethnic group grouped in one place, by forming their ethnic identity in that place. This is directly related to their attitudes and behavior towards their natural and environmental conditions, becoming local intelligence or wisdom. A lot of their local wisdom is related to this nature, especially disaster mitigation.

How about Minangkabau local wisdom about disaster mitigation? This is what is discussed in this paper. This discussion is important to complete knowledge about disaster mitigation today, starting from the encyclopedia of knowledge and ethnic traditions that have existed for a long time and still exist today. This discussion is the result of research and studies on knowledge of oral traditions, stories, myths, and human behavior. All of that is a picture of his collective view in interacting with nature and preserving nature as a form of disaster mitigation.

2 Method and Theory

The data of this study were taken from various written data, oral traditions, oral stories, and myths as well as the traditional forms of Minangkabau human interaction with nature. The values contained in the data are called local wisdom. To explore these values, a certain method is used. This study was conducted using a qualitative method with the main objective of examining the uniqueness of the data to show Minangkabau local wisdom about disaster mitigation.

His study uses the perspective of the theory of cultural materialism formulated by Marvin Harris. Teori ini memandang bahwa bentuk, aktifitas, dan fungsi kearifan lokal mengandung tujuan dan makna. As an element of culture, it becomes a superstructure, structure, and infrastructure that is secretly [7] and hidden are recognized as norms and guidelines by the collective in their interactions.

Marvin Harris[8] applies this theory when trying to understand the cultural tradition of Hindus in India who deify the cow. He tried to dig and find out why Indian Hindus made and believed in cows as gods. The findings are that at first, cows in the community were a source of protein in the form of milk and cheese. In addition, cows are also a source of labor that helps humans cultivate large agricultural land. Meanwhile, the population is increasing and many cows are slaughtered for meat consumption. This condition worries community leaders and religious leaders. It could be, after a long time the cattle population will run out, so they will lose their source of protein and agricultural workers.

According to Harris, to maintain the cattle population, community leaders agreed in the form of a ban on cutting and eating beef. Prohibition is made the norm through suggestion and myth. The myth is embodied in various oral stories about cows as incarnations of gods who descended to earth, such as stories about the struggle for cheese with good and evil characters, and by eating cheese they become superhumans. In this regard, they perform various traditions and rituals against the cow, such as decorating the cow with clothes and various greatness attributes, giving the cows freedom to roam the village, worshiping and respecting cows, drinking cow urine and milk, and making certain days a religious day. with a cheese feast.

Myths and traditions that develop and run in a long and long time continuously from one generation to the next. It is still ongoing today as a religious belief and has become the culture of the Indian Hindu community. According to Harris, myths and oral stories are a form of the superstructure as a medium where ideas reside. The norms contained in it are the infrastructure that guides life and human interaction with humans, humans with their creators, humans with the surrounding ecology. Meanwhile, traditions and rituals are functional elements that build the social and cultural structure of Indian Hindu society.

3 Results and Discussion

Disaster mitigation implies an effort to reduce the risk and impact of disasters on communities in disaster-prone areas. Disasters can be in the form of natural disasters, man-made disasters, or a combination of both in a country or society [9]. There are four main things related to disaster mitigation, namely the availability of information and maps of disaster-prone areas; socialization and public awareness in dealing with disasters; knowing and setting actions that need to be taken and avoided if a disaster occurs at any time; structuring the area to reduce the threat of disaster[10]. Local wisdom is the intelligence of ideas, norms, and behavior of ethnic or ethnic groups that become traditions and continue from one generation to the next [7]. Local wisdom is found in various forms of oral traditions, myths, stories, and monumental works.

This disaster mitigation can be grouped into two types, namely structural and non-structural disaster mitigation [11]. Structural disaster mitigation is an effort to minimize the impact of natural disasters by building physical facilities by utilizing technology, such as building reservoirs to prevent flooding, building shelters for evacuation. Non-structural disaster mitigation is an effort to minimize the impact of disasters by making regulations and policies, such as disaster management laws, regulations on the use of spatial planning.

Local wisdom of disaster mitigation is the intelligence of ethnic groups who live in disaster-prone areas in responding to natural disasters that can occur at any time. The Minangkabau people who live in mountainous and hilly topographic areas have long had local wisdom in responding to their natural environment. There is local wisdom that is structural and non-structural. There are several forms of local wisdom of disaster mitigation in Minangkabau, such as rules for land use, building construction, building utilization rules, harvest management rules, and measures during epidemics and disasters.

Minangkabau ancestors were well aware that they lived in disaster-prone areas. Therefore, they regulate in such a way the management and use of land according to its contours [6]. The development and activities of community life are regulated by the nature of green moral or sustainable development, a development that is in favor of the future of the earth and human sustainability. The rules are contained in the collective image of the community which becomes the norm orally or in writing in the Tambo as follows:

Nan lereng tanami bambu jo tabu
Nan gurun jadikan ladang
Nan data jadikan parumahan
Nan payo jadikan ranangan itiak jo kabau
Nan bancah jadikan sawah
The slopes are planted with bamboo and sugar cane
The desert turns into fields
The flat is made into housing
The swamp is used as a place for herding ducks and buffalo
The *bancah* turns into rice fields

The traditional text above contains local wisdom that regulates land use. Land that may be used as housing or settlements is naturally flat, not leveled land. Apart from that land, it must be maintained and maintained to maintain its ecological sustainability and benefit humans. Soil with steep slopes should be planted with bamboo and sugar cane. The bamboo and sugarcane plants are symbols. The bamboo is useful for binding and holding sloped soil from possible landslides and flooding. In addition, these plants are also useful for building houses and various purposes to support human life.

Likewise, other lands must be utilized according to the available contours. Deserted land is used as fields to plant horticultural crops, wasteful land is turned into rice fields for planting rice, and swampy land is used as a breeding ground for raising buffalo and ducks. Thus, the core of land use management is to avoid the threat of natural disasters from landslides and ground movement.

In building and managing settlements, it is regulated in accordance with the autonomy and independence of the village or nagari. There are several requirements in the form of infrastructure that must exist in a nagari [1] as follows:

Nagari bakaampek suku
Barumah gadang sambilan ruang
Salanja kudo balari
Sapakiak budak maimbau
Rangkiang baririk di halamannya
Rangkiang banamo sitinjau lauik
Sabuah banamo sibayau-bayau
Ciek banamo sitangka lapa
Jo rangkiang ketek di sampiangnyo
Babalai bamusajik
Balabuah batapian
Basasok bajarami
Bapandam pakuburan
In the Nagari, there are four tribes
Having a nine-room Rumah Gadang
The breadth of the horse can run in it
The size of the building is heard screaming children calling
The rankings line up on the page
Ranking named Sitinjau Lauik
One named Sibayau-Bayau
One named Sitangka Lapa
And the little ranking beside him
Has a hall and a mosque
Has roads and edges
Has springs and rice fields
Have a public cemetery

In the text above it is explained that in a Nagari there must be infrastructure and public facilities. The building is in the form of a Rumah Gadang as a place to live, a rank for storing food ingredients, a market or public meeting room, a mosque for worship, a road as transportation access, a bank as a place of purification and communication access, springs and rice fields as a source of food and energy, and a place to live. public cemetery to bury the dead.

The infrastructure required in the Nagari is in synergy with traditional disaster mitigation. Four Rangkiang were built in the yard of the Rumah Gadang house, used to store rice after harvesting, a s shown in Fig. 3). Each Rangkiang with rice in it is intended according to the food needs of the residents and the surrounding environment. One of the rangkiang is named Sitangka Lapa [1]. Rangkiang is used to store rice that is prepared before and used after natural disasters or humanitarian disasters such as the arrival of famine or famine, helping neighbors who are stricken by natural disasters, and serving food for refugees who pass by or come for shelter.

Likewise, other infrastructures in the form of roads, halls, mosques, banks, and cemeteries are part of the Minangkabau ethnic disaster mitigation strategy. Each of these infrastructures is used as a form of preparation for anticipating and post-natural disasters. Such infrastructure, especially road access for mobilization during and after a disaster is a major requirement in the current disaster mitigation perspective [12].

Fig. 3. Rangkiang and Rumah Gadang (Source: Doc)

Fig. 4. Rangkiang in District Agam (Source: Doc)

When there is a pandemic and disease outbreak, the Minangkabau ethnic group has local wisdom to deal with it, namely Batangeh (Fig. 5). The Batangeh tradition is a way to treat members of the collective who are stricken with the disease. This tradition is carried out by placing the patient in a closed place and hot steam flowing from the results of boiling herbs in the form of spices. With this steamer method, it is hoped that the patient will recover from the plague.

Fig. 5. Batangeh and Spices (Source: Google)

In addition, the series of *Batangeh* traditions are carried out in places and spaces that are far from the public, which is called *baijok*. *Batangeh* at the *Baijok* location aims to prevent the disease outbreak from spreading to other collective members. In addition to *Batangeh*, the *Baijok* location is used as a place of isolation for patients with epidemics of infectious diseases, such as tuberculosis, cholera, smallpox, leprosy, and corona.

In addition to isolation during an epidemic disaster, this *baijok* tradition is also carried out to evacuate or flee from conflict or war areas (Fig. 6). The *baijok* tradition began during times of conflict or war with the Dutch colonialists. This tradition continued when the army attacked the central government of Indonesia against the people of West Sumatra who demanded equitable development in the early days of Indonesia's independence. People who were frightened and did not want to be caught fled to hiding places in caves or huts in the wilderness. This event is collectively called *baijok*.

Fig. 6. Corona Isolation Pondok Baijok in Agam (Source: Doc)

4 Conclusion

Minangkabau's local wisdom on disaster mitigation has been around for a long time. Following the lineage system according to the mother's line used by ancient humans adopted by the Minangkabau ethnic group means that local wisdom in disaster mitigation has long existed. This disaster mitigation has become a reference in managing and utilizing natural potentials located in disaster-prone areas.

Local wisdom of disaster mitigation is structural and non-structural. Disaster mitigation is in the form of socializing the rules on how to use and manage land to prevent natural disasters; *Rangkiang* development as a complete infrastructure for post-natural disasters; preparing rice yields for times of famine and conflict; how to isolate during a disaster or *baijok*; and treatment during the epidemic or *batangeh*.

For the presence of this paper in scientific and academic activities, thank you to the Chairperson of the Andalas University Research and Service Institute. This paper also refers to the results of previous studies that have been facilitated through research grants on Minangkabau ecological wisdom. Of course, thanks also to the 2021 ICDMM Committee who have worked hard and are willing to accept this article as part of the conference on disaster mitigation.

References

1 A. A. Navis, Nature Develops to Become a Teacher Jakarta: Grafitiperss, (1986)

2 H. D. Toeah, *Tambo Alam Minangkabau*. Pajakumbuh: Pertjetakan Limbago Pajakumbuh, (1968)

3 M. Naim, Merantau: The Migration Pattern of Minangkabau People's Migration Yogyakarta: Gajah Mada University Press (1979)

4 T. Kato, Indigenous Minangkabau and Merantau in Historical Perspective, 1st ed. Jakarta: Balai Pustaka (2005)

5 BNPB, "A Study on Natural Disaster Management in Indonesia," Orient. Consult. Co., Ltd. Asian Disaster Reduct. Cent., **2–2**, Japan International Cooperation, 189 (2009)

6 M. Nasroen, Basic Minangkabau Customary Philosophy, 2nd ed. Djakarta: Star Moon (1971)

7 K. Anwar, "Ecological Wisdom of Oral Tradition: Surface and Deep Structure Tension in Preserving the Lake Environment," *IOP Conf. Ser. Earth Environ. Sci.*, **469**, 1 (2020)

8 S. K. Sanderson, *Sosiologi Makro*, 2nd ed. Jakarta: Raja Grafindo Persada. (1995)

9 K. Kim, "Technologies for Disaster Mitigation," *Cell*, no. March, 27–30 (2009)

10 Commonwealth of Australia, "National Disaster Risk Reduction Framework," 1–23 (2018)

11 D. Triana, T. S. Hadi, and M. K. Husain, "Disaster Mitigation Through Cultural and Structural Approaches," Semin. Nas. XII "Technology Engineering. eng. and Inf. 2017", 379–384 (2017)

12 S. Aminatun, "Study of Landslide Disaster Risk Analysis as a Basis," J. Tek., **22**, 2, 372–382 (2017)

Mapping the liquefaction potential index (LPI) in Ratu Agung subdistrict, Bengkulu City, Indonesia using the shear wave velocity approach

Ceri Eliesa Suhartini[1], *Lindung* Zalbuin Mase[2], and *Muhammad* Farid[3*]

[1]Department of Geological Engineering, Faculty of Engineering, University of Gadjah Mada, Indonesia;
[2]Department of Civil Engineering, Faculty of Engineering, University of Bengkulu, Indonesia;
[3]Department of Physics, Faculty of Mathematics and Natural Sciences, University of Bengkulu, Indonesia;

Abstract. On the 4th of June, 2000 and 12th of September, 2007, Ratu Agung Sub-district, Indonesia experienced significant damage due to liquefaction after the earthquakes. Therefore, this study aims to determine the Liquefaction Potential Index in the area. Data of shear wave velocity (V_s) was collected using the Multichannel Analysis of Surface Wave (MASW) method. The measurement location was set up on a grid of 32 observations points with field investigations. Furthermore, Simplified Procedure and LPI methods were used to measure the soil liquefaction potential and vulnerability level. The results showed that the value of shear wave velocity in the Ratu Agung Sub-district ranged from 102 m/s to 442 m/s. Also, the liquefaction vulnerability levels varied from high to very high categories due to the maximum soil acceleration and conditions dominated by loose sand, as well as the influence of different geological formations in the zone. In conclusion, an empirical equation was successfully proposed to analyze the liquefaction vulnerability.

1 Introduction

Bengkulu City is vulnerable to earthquakes that trigger liquefaction [1], due to the Indo-Australian and the Eurasian Plates subduction zone. Meanwhile, the interaction movement between these plates affects the formation of the fault in the western part of Sumatra including faults of Mentawai and Semangko [2]. Geologically, Bengkulu City is dominated by alluvial materials consisting of sand, clay, and mud [3] which are potentially to undergo liquefaction [4, 5]. The trigger factor of liquefaction in Bengkulu City is the earthquake event that occurred on June 4, 2000, and September 12, 2007, with magnitudes of M_w 8.0 and M_w 8.6, respectively. Those earthquakes are later known as Bengkulu-Enggano Earthquake and Bengkulu-Mentawai Earthquake, respectively. The earthquake had triggered liquefaction that was marked by soil subsidence and structural damage as shown in Fig. 1. Several previous studies also showed that the coastal zone of Bengkulu City including Ratu Agung Sub-district is vulnerable to undergo liquefaction as well [6].

The liquefaction potential map was composed based on LPI criteria. The investigation method used by a cone penetration test, whereas the investigations using other methods, such as standard penetration test and geophysical measurement are still limited. The results showed that the soil layer in the zone of the coastal area is liquefied [7]. Therefore, it is necessary to conduct further studies on liquefaction potential and its vulnerability level in this area. Several liquefaction events had been found in Ratu Agung District.

Fig. 1. Damage due to Liquefaction in Bengkulu City after the 2007 Earthquake [8]

As an area that is identified as the most impacted area during the earthquake in 2007, the Ratu Agung area becomes an interesting study area for the local study of earthquake engineering in Bengkulu City. This paper presents the mapping of the liquefaction vulnerability of the Ratu Agung District. This study used the data from the shear wave velocity (V_s) data from geophysical

* Corresponding Author: mfarid@unib.ac.id

measurement, i.e. Multichannel Analysis of Surface Wave (MASW) method. V_s data is then used to analyze the soil liquefaction vulnerability in the Ratu Agung Sub-district. This study employed the method proposed to help for determining the liquefaction potential [9]. To quantify the liquefaction potential index in the study area, the weighted method called liquefaction potential index (LPI) is also used [10]. In general, this study could help a better understanding for local people to consider liquefaction impact in the study area.

2 Theoretical Background

2.1 Liquefaction

Liquefaction is the condition where the soil layer undergoes the loss of soil strength due to excess pore water stress arising under cyclical loading [11]. Generally, liquefaction occurs on loose sandy soils under saturated conditions. Once this soil type is receiving cyclic load, the contact between each particle suddenly disappears and the soil behaves like a liquid. It means that the bearing capacity of soil also disappears during the earthquake shaking and excess pore water pressure. The increase of pore water pressure is generally accompanied by changes in soil volume which are manifested as ground settlements, lateral spreads, and sand boils [12].

Several areas including Banda Aceh, Padang, Bengkulu, and Yogyakarta experienced sand boils, lateral spreads, as well as ground fractures, during the earthquake events that occurred in these areas. Those cause damage to buildings and grounds after earthquakes. The liquefaction phenomenon is generally found on alluvial plains, delta and swamp deposits, and brackish, as well as coastal embankments [13]. The contribution of an active fault called the Opak Fault had resulted in the earthquake and liquefaction in the Yogyakarta region. In line with those previous studies, therefore it is important to identify liquefaction in an area [14, 15].

2.2 Shear Wave Velocity

Shear wave velocity or V_s is an important parameter that can be used for various geotechnical purposes, especially for engineering practice on the aspect of soil dynamics and earthquake engineering [16]. V_s is determined from geophysical measurements, such as MASW. Soft soil has generally a lower V_s because it is not compacted. Therefore, the shear wave propagation tends to be an easy passing soft layer. For liquefaction investigation, V_s values are also used. Engineers tend to select this parameter for liquefaction investigation because the method is cheap, easy, and simple.

Before analyzing liquefaction, the corrected V_s should be firstly determined [17]. Equation 1 shows correlation for corrected V_s,

$$V_{s1} = V_s \left(\frac{P_a}{\sigma_v{}'} \right)^{0.25} \quad (1)$$

In Equation 1, it can be seen that V_{s1} is corrected V_s in the unit of m/s, P_a is as atmospheric pressure (i.e 100 kPa), and $\sigma_v{}'$is the vertical effective stress in kPa.

As elaborated in the previous section, the benefit of V_s is also able to use for soil type prediction and unit weight estimation [16]. These parameters are going to be used as the main parameter to estimate liquefaction potential [18],

$$\gamma_{sat} = 8.32 \log V_s - 1.61 \log(z) \quad (2)$$

In Equation 2, it shows that saturated unit weight is defined as γ_{sat} (kN/m^3) and z is the analyzed depth. V_s and γ_{sat} can use to predict soil type as shown by the chart presented in Fig. 2 indicate that saturated unit weight together with V_s value can be used to estimate soil type. This framework is used in this study.

2.3 Liquefaction Potential Analysis

A simple procedure method was used to analyze the liquefaction by comparing Cyclic Stress Ratio (*CSR*) to Cyclic Resistance Ratio (*CRR*) [9]. Parameters called *CSR* and *CRR* reflect the cyclic ratio of earthquake loading and the availability of soil resistance during the earthquake loading, respectively. *CSR* is estimated by using the following equation,

$$CSR = 0.65 r_d \left(\frac{a_{\max}}{g} \right) \left(\frac{\sigma_v}{\sigma_{v'}} \right) \left(\frac{1}{K_\sigma} \right) \quad (3)$$

In Equation 3, $a_{\max}$ is defined as maximum ground surface acceleration resulting during the earthquake. Both parameters, i.e σ_v and $\sigma_v{}'$are total and effective stresses at the investigated depth, and K_σ is the corrected overburden pressure factor, and r_d is the depth reduction factor. For r_d and K_σ parameters are estimated below,

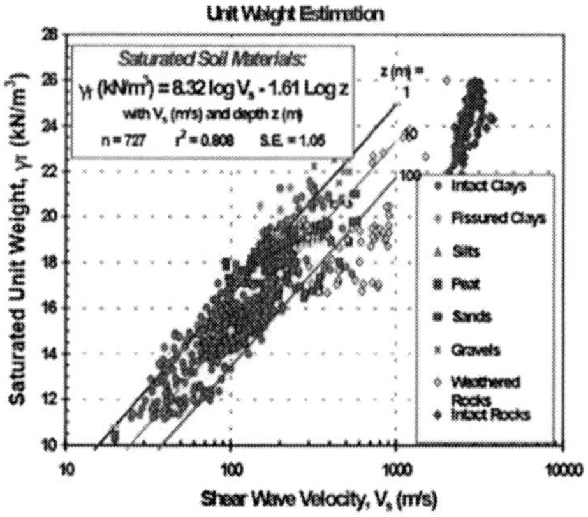

Fig. 2. The correlation between γ_{sat} against v_s [11]

$$r_d = 1 - 0.012 z \quad (4)$$

$$K_\sigma = 1 - C_\sigma \ln \left(\frac{\sigma_v{}'}{P_a} \right) \leq 1.1 \quad (5)$$

$$C_\sigma = \frac{1}{18.9 - 2.55\sqrt{(N_1)_{60}}} \leq 0.3 \quad (6)$$

$$\left(N_1\right)_{60}=\left(\frac{V_{s1}}{93.2}\right)^{\frac{1}{0.231}} \qquad (7)$$

From those equations, z is defined as analyzed depth (in m length), C_σ is the overburden pressure coeficient (dimensionless factor), and $(N_1)_{60}$ is the corrected standard penetration test (blows/ft).

CRR is estimated by using Equation 8. In this study, *CRR* is estimated based on V_s data. In Equation 8, several parameters called Magnitude Scaling Factor (*MSF*) [19] and factors for V_s correction (K_{a1} and K_{a2}) [20, 21] are introduced. For K_{a1} and K_{a2}, the detailed explanation of how to estimate those factors are presented in Andrus *et al.* [20, 21]. *MSF* factor is related to the magnitude of the earthquake (M_w). For MSF, the calculation procedure is expressed in Equation 9,

$$CRR=MSF\left(\begin{array}{l}0.022\left(\dfrac{K_{a1}V_{s1}}{100}\right)+2.8+\\ \left(\dfrac{1}{V_{s1}^{*}-\left(K_{a1}V_{s1}\right)}-\dfrac{1}{V_{s1}^{*}}\right)\end{array}\right)K_{a2} \qquad (8)$$

$$MSD=6.9\exp\left(\frac{-M_w}{4}\right)-0.058 \qquad (9)$$

Also in Equation 8, a parameter called the corrected V_s threshold based on fine content (FC) effect or V_{s1}^{*} is included. The calculation procedure how to estimate V_{s1}^{*} is expressed in these following equations,

$$V_{s1}^{*}=215 \text{ m/s} \qquad (10a)$$

$$V_{s1}^{*}=215-0.5\left(FC-5\right)\text{ m/s} \qquad (10b)$$

$$V_{s1}^{*}=200 \text{ m/s} \qquad (10c)$$

Equation 10a is used for sandy soils with FC $\leq$ 5%, Equation 10b is used for sandy soils with 5% < FC < 35% and Equation 10c is used for sandy soils with FC $\geq$ 35%.

After *CSR* and *CRR* are estimated following the procedure presented above. The factor of safety against liquefaction (*FS*) can be predicted by using the following equation,

$$FS=\frac{CRR}{CSR} \qquad (11)$$

Based on Equation 11, the liquefaction potential for a site can be determined. Liquefaction is possible to occur once *FS* is less than 1 and vice versa.

The procedure in estimating liquefaction presented in Equations 1 to 11 is used for analyzing the sandy soil's depths. However, to quantify the overall vulnerability of liquefaction, a weighting factor method can be implemented. Iwasaki *et al.* [10] introduced the liquefaction potential index (LPI) to classify liquefaction potential in an area, as expressed in Equation 12,

$$LPI=\int_0^{20} Fw\left(z\right)dz \qquad (12)$$

In Equation 12, it can be seen that LPI is estimated based on the integration procedure [21]. A parameter called F is related to *FS*. F parameter can be estimated based on these following equations,

$$F=1-FS\text{, for } FS\leq 1 \qquad (13a)$$

$$F=0\text{, for } FS>1 \qquad (13b)$$

In addition, a factor related to the depth is also introduced as expressed in this following equation,

$$w\left(z\right)=10-0.5z$$

where z is the analyzed depth.

Furthermore, LPI is estimated for each site. Iwasaki *et al.* [10] mentioned that for engineering practice, liquefaction is generally found up to 20 m depth; therefore, for calculation of LPI, a maximum depth up to 20 m depth is considered. For the category of LPI and its classification, Iwasaki *et al.* [10] mentioned that LPI = 0 means that there is no liquefaction potential, the criterion of 0 < LPI $\leq$ 5 indicates "low" liquefaction potential, the criterion of 5 < LPI $\leq$ 15 indicates "high" liquefaction potential, and LPI > 15 indicates "very high" liquefaction potential.

3 Methodology

Fig. 3 shows that this study was conducted in Ratu Agung Sub-district, Bengkulu City, Indonesia. This study was initiated by studying the geological conditions as well as soil characteristics. Primary data in the form of shear wave velocity was collected using the Multichannel Analysis of Surface Wave (MASW) method. The measurement location design was made on a grid of 32 observation points with a distance of about 400 meters to represent each geological formation. Furthermore, secondary data was collected in the form of a_{max} and soil layer conditions from previous studies. The parameters called *CSR, CRR, FS,* and *LPI* were calculated using Equations 1 to 13. Next, soil layer conditions, the liquefied depth, and liquefaction potential, as well as the liquefaction potential category were presented. For detail, Fig. 4 shows the overall framework

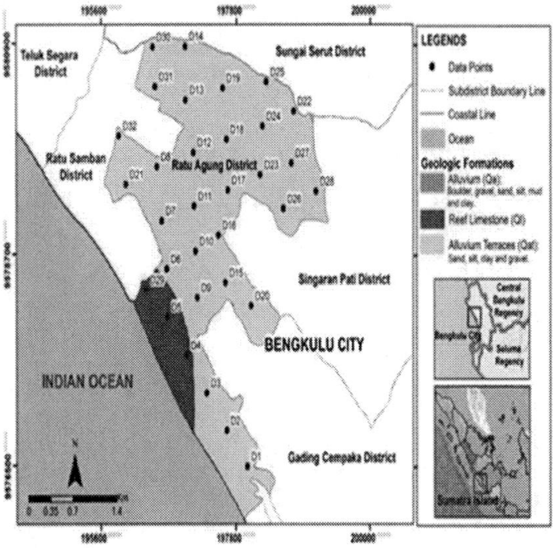

Fig. 3. Study Area, geological formation, and location of site investigation

 https://doi.org/10.1051/e3sconf/202133104014

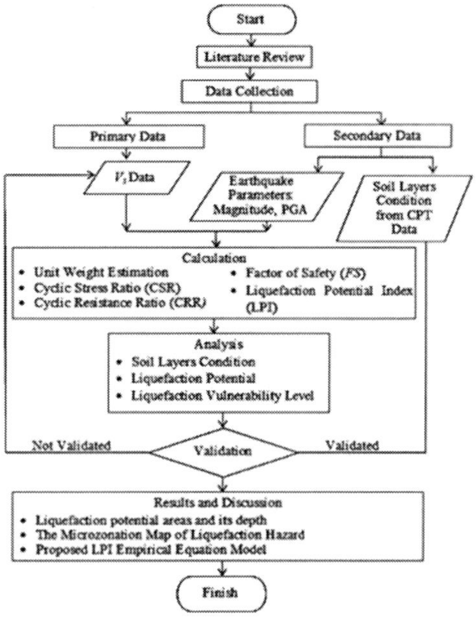

Fig. 4. A research framework for liquefaction potential analysis used in this study

4 Results and Discussion

4.1 Liquefaction Potential

In this study, the site called Site D2 is used as the representative site to represent the result in the study area. Fig. 5 presents the liquefaction potential of Site D2. It can be observed that liquefaction potentially occurs at the depth of fewer than 7.45 meters. However, the site is generally dominated by sandy soils up to 30 meters in depth. It indicates that the loose sandy soils may be found at the shallow depth, whereas at the deeper depth, the sandy soils with higher soil resistance exist. That prediction is shown by FS more than 1 for those depth ranges. The site is also categorized as Site Class D because V_{s30} of 330 m/s has lied in the range of 180 to 360 m/s. This study is generally consistent with the study conducted by Mase [22] that mentioned that the coastal area of Bengkulu City is dominated by sandy soils classified as Site Class D.

4.2 Liquefaction Potential Index (LPI)

The parameter of *FS* is obtained from the simplified procedure method. Each soil layer was then measured for its liquefaction potential index to a depth of 20 meters. Fig. 6 presents the LPI map for Ratu Agung District. Fig. 6 also shows that this sub-district has a high liquefaction potential, which is indicated by the yellow zone and the very high zone is indicated by the red color. This may be caused by several factors, such as soil types, maximum acceleration values, and groundwater elevations. Meanwhile, the potential variation is due to the differences in geological conditions in this sub-district.

The LPI values that are estimated from obtained 32 observation points spread across Ratu Agung Sub-district are observed to vary from 5.36 to 29.89. Those ranges indicate that the liquefaction potential at the site is categorized as high to very high. According to Sugalang & Buana [7], several areas namely Kebun Beler in D8, Nusa Indah in D6, D9, and D10, Sawah Lebar in D12, as well as Kebun Tebeng Village in D21, D22, D23, D24, and D25 are categorized as the area with high to very high liquefaction potential. This soil type obtained from this study is consistent with the study of Mase [23] that investigated the locations of Lempuing and Pantai Panjang closed to D2 and D27.

The results showed that all test points in the Ratu Agung Sub-district had a high liquefaction potential. This zone is dominated by a soil layer that is very susceptible to liquefaction. This sub-district has cyclical geological processes due to uplift and subsidence by tectonic or past depositional sedimentation.

Furthermore, there is maximum ground acceleration and an increase of earthquake wave vibrations because of the existence of loose sediment. This causes the increase of pore water pressure which later triggers the reduction of soil shear strength. Based on the investigation, past earthquakes that occurred on the 4th of June 2000 and 12th of September 2007 in Ratu Agung Sub-district also cause significant damage, such as lateral spreads, ground settlements, and collapsed buildings

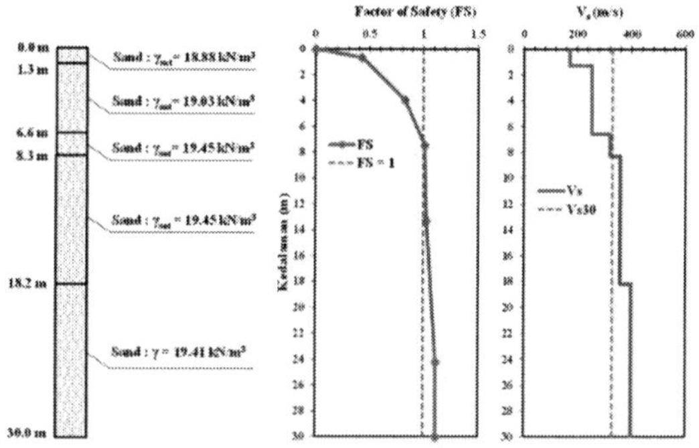

Fig. 5. Soil Layer Condition at Research Point D2

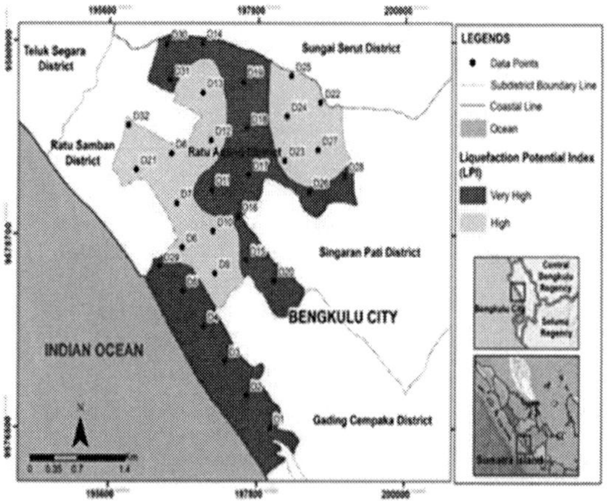

Fig. 6. Liquefaction Hazard Zoning Map in Ratu Agung Sub-district

4.3 Proposed LI Model in Ratu Agung Sub-district

The LPI value was analyzed to obtain its relationship composed of V_{s30}, A_0, f_0, and a_{max}. In this study, a multiple linear regression method is conducted to generate an LPI model for Ratu Agung District. Considering the study conducted by Mase [24], A_0 (amplification factor for the site) and f_0 (predominant frequency) parameters are known as parameters to define geo-hazard vulnerability. By adopting that previous study, the practical equation for LPI prediction could be generated. The result leads to a proposed equation that depends on internal and external factors. V_{s30}, A_0, f_0, are categorized as internal factors because they adhered to the soil characteristic, whereas a_{max} is noted as an external factor. After all, it is related to earthquake energy. The equation to estimate LPI for Ratu Agung District is expressed below,

$$LPI = 0.071V_{s30} - 1.35A_0 - 5.74f_0 + 136.64\left(\frac{a_{max}}{g}\right) \quad (14)$$

Equation 14 obtained from the regression analysis was validated to test the proposed model equation accuracy. The LPI value from Equation 14 is then compared with the calculated one. The result is accurate because it has an R^2 value of 0.97 and a Root Mean Square Error (RMSE) of 0.94. Furthermore, the Mean Absolute Percentage Error (MAPE) was 4.64% in which the value is less than 10%. According to Anggrainingsih *et al.* [25], a MAPE value less than 10% indicates a high confidence level of model accuracy. Therefore, the empirical equation proposed in this study is not only considering liquefaction factors but also presents good accuracy and acceptable result. However, it should be noted that the equation is only able to be used for the sites dominated by sandy soils.

In general, the proposed equation is used as initial justification to identify liquefaction and seismic vulnerabilities in Ratu Agung, Bengkulu City, Indonesia. According to Misliniyati *et al* [26], seismic mitigation studies are important and need to be initiated in coastal zones. In addition, Farid and Mase [27] and Mase [28] showed that these simple improvements could strengthen seismic mitigation efforts in the city.

5 Conclusion

This study presents the analysis of liquefaction potential in Ratu Agung District that is identified as one of the impacted areas during strong earthquakes in Bengkulu City, Indonesia. The geophysical measurement is conducted to obtain V_s profile that is also used as a soil parameter for liquefaction potential analysis. A mapping of LPI and an empirical equation to predict LPI are also presented in this study. Several concluding remarks can be drawn below,

1. Ratu Agung site is dominated by sandy soils at shallow depth. This sandy soil deposit could be very vulnerable to undergo liquefaction during strong earthquakes.
2. LPI map shows that the yellow and red zones in Ratu Agung Sub-district. It indicates that Ratu Agung District has a high to very high liquefaction potential.
3. The empirical equation proposed in this study could be used for engineering practice for liquefaction potential analysis in Ratu Agung Area.

References

1. L.Z. Mase, Journal of Engineering & Technological Sciences **49**, 6 (2017)
2. T. Zera, Sutrisno, A. Budiono, IOSR Journal of Applied Geology and Geophysics (IOSR-JAGG) **5** (1) (2017)
3. S. Gafoer, T.C. Amin, R. Pardede, R. Peta Geologi Lembar Bengkulu dan Sekitarnya (Pusat Penelitian dan Pengembangan Geologi Bandung, 1992)
4. M. Farid, A.I. Hadi, Telkomnika **16**, 4 (2018)
5. L.Z. Mase, S. Likitlersuang, T. Tobita, Marine Georesources & Geotechnology **37**, 7 (2019)
6. L.Z. Mase, Refrizon, Rosiana, P.W. Anggraini, Indian Geotechnical Journal **51**, 8 (2021)
7. Sugalang, T.W. Buana, Buletin Geologi Tata Lingkungan **22**, 2 (2012)
8. F. Nugroho, J. Tanjung, M. Zaidir, M. (2018). *Observasi Kerusakan Struktur Rangka Beton Bertulang Eksisting Pasca Gempa Palu 2018*. In Proceedings of the 5th Andalas Civil Engineering Conference, 28 November 2018, Padang, Indonesia (2018)
9. I.M. Idriss, R.W. Boulanger, Soil liquefaction during earthquakes (Earthquake Engineering Research Institute, Oakland, 2008)
10. T. Iwasaki, K. Tokida, F. Tatsuoka, *Soil Liquefaction Potential Evaluation with Use of the Simplified Procedure*. In Proceeding of the 1st International Conferences on Recent Advances in Geotechnical Earthquake Engineering and Soil Dynamics, 26 April to 3 May, St Louis, Missouri (1981)
11. L.Z. Mase, S, Likitlersuang, T. Tobita, Journal of Earthquake Engineering (2020) (online)
12. M. Naghizadehrokni, M. Hassanlo, M. Ramzani, International Journal of Geomate, **12**, 29 (2017)
13. E. Soebowo, A. Tohari, D. Sarah, K. Sugianti, Procedings of Pemaparan Hasil Pusat Penelitian Geoteknologi LIPI **1** (2014)
14. L.Z. Mase, International Journal of Technology **8,** 4 (2017)
15. L.Z. Mase, T.F. Fathani, A.D. Adi, ASEAN Engineering Journal 11, 3 (2021)
16. J.-S. L'Heureux, M. Long, Journal of Geotechnical and Geoenvironmental Engineering **143**, 6 (2017)
17. R. Kayen, R.E.S. Moss, E.M. Thompson, R.B. Seed, K.O. Cetin, A.D. Kiureghian, Y. Tanaka, K. Tokimatsu, Journal of Geotechnical and Geoenvironmental Engineering **139**, 3 (2013)
18. E.O. Shelley, V. Mussio, M. Rodríguez, J.G.A. Chang, Geofisica Internacional **54**, 1 (2015)
19. I.M. Idriss, R.W. Boulanger, Soil dynamics and earthquake engineering **26**, 2-4 (2006)

20. R.D. Andrus, K.H. Stokoe, C.H. Juang, Earthquake Spectra **20**, 2 (2004)
21. R.D. Andrus, K.H. Stokoe II, Journal of Geotechnical and Geoenvironmental Engineering (ASCE) 126, 11 (2000)
22. L.Z. Mase, Geotechnical and Geological Engineering **38**, 5 (2020)
23. L.Z. Mase, International Journal of Technology **9**, 5 (2018)
24. L.Z. Mase, Civil Engineering Dimension **21**, 2 (2019)
25. R. Anggrainingsih, G.R. Aprianto, S.W. Sihwi, *Time series forecasting using exponential smoothing to predict the number of website visitors of Sebelas Maret University*, in Proceedings of the 2nd *International Conference on Information Technology, Computer, and Electrical Engineering (ICITACEE)*, 16 October 2015, Semarang, Indonesia (2015)
26. R. Misliniyati, L.Z. Mase, A.J. Syahbana, E. Soebowo, (2018, December), IOP Conference Series: Earth and Environmental Science **212**, 1 (2018)
27. M. Farid, L.Z. Mase, International Journal of Geomate **18**, 69 (2020)
28. L.Z. Mase, N. Sugianto, Refrizon. Geoenvironmental Disasters **8**, 1 (2021)

E3S Web of Conferences 331, 04015 (2021) https://doi.org/10.1051/e3sconf/202133104015
ICDMM 2021

Empowering community through creative economy as a disaster risk reduction strategy in Indonesia

Ratna W.D.P[1*], *Noviansyah* Rizal[1], *Riza* B.S[1] , *Fauziyah* [2], and *Muhammad* Dimyati[3]

[1] Institut Teknologi dan Bisnis Widya Gama, Accounting Department, Lumajang East Java Province, Indonesia
[2] University of Science and Technology Beijing, Department of Computer and Communication Engineering, Beijing, China
[3] Institute of Economic Science Mandala, Accounting Department, Jember East Java Province, Indonesia

Abstract. In the 21st century, the world is faced with COVID-19 global pandemic that has had wide-ranging health, social, food, and economic negative impacts throughout the world and also in Indonesia due to the cessation of the mobility of most people and economic engines. The creative empowers the community based on innovation and creativity proven and is one of the pillars of disaster risk reduction recovering the social and economic impact of COVID-19 in the community. The method used is an ethnographic approach starting from the collection of literature sources and supported by field data. The results show that the concept of the creative economy is creative ideas developed by humans supported by the use of technology to produce innovative products and cultures, having regional imagery but are globally competitive and can ultimately provide value to the national economy. Intellectual actors included are not only academics but also include industry players, pioneers, figures in the arts, culture, and sciences who have a role in providing creative ideas, innovations in the creative industry. The result is a triplex helix model in mitigating the effect of the covid19 disaster by connecting intellectuals, businesses, and the government in a creative economic structure.

1 Introduction

The Indonesian government has set targets to support an inclusive and sustainable national economy, a full and productive workforce, and decent work for all. The government has designed and implemented policies that support sustainable tourism that enable the creation of jobs while supporting local culture and products by 2030 [1]. However, the spread of covid-19 has a disastrous impact on local and national economies around the world, many industries and policymakers are researching the impact of covid-19 on the creative and cultural industries, as well as workers in the sector [2].

The global recession caused by covid19 reduce consumption and investment in various countries. In Indonesia in the second quarter of 2020, Indonesia's economic growth was depressed to a negative 5.32% (yoy) compared to the same period last year. This contraction is more than in the previous quarter of 2.97%. Only China started showing positive growth to 3.2% in the second quarter of 2020. Whereas in the first quarter of 2020, economic growth was negative up to 6.8%.

The development of a local culture-based creative economy ecosystem, supported by competent business actors, requires support and cooperation from several parties including the government, the private sector, community, academia, as well as the media. In addition, the development of the sector can also be carried out with training to improve the welfare of business actors ([3].

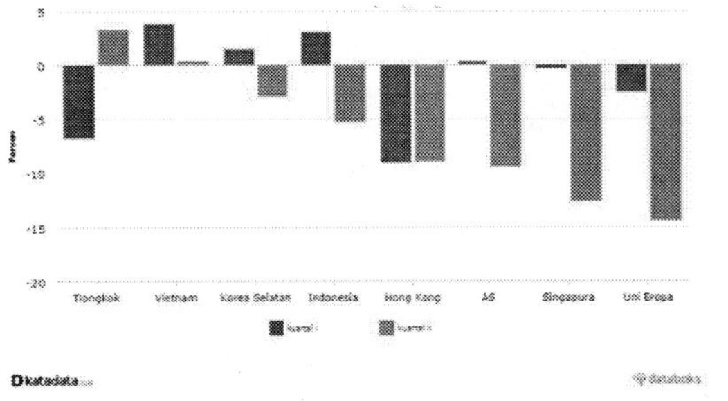

Fig. 1. 2020's Global Economic Growth (YoY)

The creative economy has been widely developed in major countries such as the US, UK [4], all countries in Australia [5], including in Indonesia [6]. Optimizing the creative economy in Indonesia has become the focus of researchers in recent years [7]. There are challenges concerning creativity and culture that are widely recognized in the academic world. However, it is often invisible to policymakers ([2]. The most populous

* Corresponding Author: pradnyataj@gmail.com

population in Indonesia is on the island of Java, along with its potential for cultural diversity and creative economic activities. Local wisdom and cultural events reflect the creativity of the community and can provide economic value for the community [8].

The creative economy has started to be discussed in Indonesia and has received proper facilities since 2006. The creative economy has contributed significantly to the economic development in Indonesia. Following the growing business world, the creative economy will also continue to develop along with the increasing creativity of business actors. In the future, it is expected that the creative industry can grow bigger and wider. The government provides various support and assistance for the development of the creative economy and development of this business. The government also holds various important events as a means of exhibiting products created by the community's creativity.

Based on the results of research between the Creative Economy Agency (Bekraf) and the Central Statistics Agency (BPS) in 2016, the creative economy contributed to the national economic growth of 922.59 billion rupiahs or 7.44 percent of the national Gross Domestic Product (GDP). Numerous empirical evidence supports that the creative industry has an impact on GDP through the creative industry sub-sector and creates a new form of governance of the cultural industry.

There are six main characteristics of the creative economy as quoted from Creative Economy Based on Local Wisdom in 2020. First is the existence of creative economy intellectual creations, emphasizing the importance of creativity. Therefore, creativity and expertise in a sector are a necessity and become one of the main characteristics of the creative economy. Second is the nature of easy to replace. Creations and innovations must continue to be developed under the development of the economic activity. The goal is to be accepted by the market and useful for the consumers. The third is the direct and indirect distribution based on company policies and consumer needs. Fourth, is the requirement of cooperation between all parties, such as between the entrepreneur and the government that regulates the policy. Fifth is based on ideas, that ideas are vital in developing creative industries and will always be related to innovation and creativity. The sixth is the nature of no limitation in product creation which means that innovation and creativity in product creation will always occur and has no definite limits.

This research focuses more on innovation and creativity to be developed, which is necessary for the growth of the creative industry in Indonesia. The result shows that the Indonesian economy can compete with other countries and will automatically become an indication that the Indonesian creative economy is one step ahead of its competitors

2 Method

This study uses an ethnographic approach. The ethnographic method [9] started with literature sources and was complemented by field data obtained through observation, participation, and in-depth interviews with selected respondents. Literature sources are obtained from researches that have been carried out starting in 2017 on culture in Banyuwangi. Informants are determined based on their involvement and role in the development of the creative industry, MSME actors, the community, and local government. The validation of the data is performed using triangulation of sources. Data analysis is carried out thoroughly and continuously starting from the process of data inventory, identification, classification, and data interpretation. Interpretation of data is carried out comprehensively by looking at the overall relationship of the data.

There are two approaches used in community development, the direct and indirect approaches [10]. The direct approach is carried out by implementing community empowerment, while the indirect approach is carried out with a process of public awareness of the importance of developing creative industries. The model developed is a participatory cycle model, which includes processes that are carried out simultaneously starting from input, process, output, impact, and result.

3 Result and Discussion

The creative industries in Banyuwangi include several creative industries spread across various regions and contribute significantly to the regional economy. One of the creative industries that are developing is the bamboo handicrafts - crafts that use natural products to be used as household products, household accessories, fashion, musical instruments, and others. The creative industry of bamboo handicrafts has shown an increasing trend, attracting the market and opening up wide business opportunities. In certain areas, such as Papring has its prosperous resources of bamboo raw materials. The bamboo craft equipped with ideas and creativity will be able to produce products that are competitive in the global market. Communities in the Papring have started their activities to develop their potential by utilizing the natural potential in their environment by producing handicrafts from bamboo. Papring area has abundant various types of bamboo such as their unique origin *Panggonane* Pring, which is the source of bamboo plants. The types of bamboo in this area include stone bamboo, apus, petung, nineg, and various other types. The bamboo crafts industry is self-sufficient in terms of raw materials since it did not require supplies from outside of the region for handicraft production.

The Triple Helix concept integrates the roles of academia, business, and government into creativity, innovation, and technology activities for the creative industry. The creativity and innovation have a close relationship that supports innovation performance. Carayannis and Campbell (2009) stated the importance of government, university, and industry policies and practices to interact intelligently, effectively, and efficiently.

The role of intellectuals is to spread and implement science, art, and technology and to form constructive values for the development of creative industries in society. Intellectuals are related to new creating activities that have the bargaining power to the market and the formation of creative people. Academics as part of intellectuals have 3 roles, namely, the role of education aimed at encouraging the birth of future creative generations that support the creation, innovation, and work in the creative industry. The role of research is carried out to provide input on creative industry development policy models and their required instruments, manage the renewal of ideas and creative processes that will produce new products, services, and technology that supports efficient working and use of resources and makes a competitive national creative industry. The role of community service is carried out in forming a society with social institutions and social orders that support the growth of the national creative industry. In realizing the importance of innovation performance in the creative industry, it is necessary to measure the performance of innovation that is appropriate for business actors in the creative industry.

Intellectual actors are not only academics but also include culturalists, artists, writers, pioneers, figures in the arts, culture, and sciences whom all have their roles in providing creative ideas, innovations in the creative industry.

The business becomes a liaison in the context of economic exchange and the transformation of creativity into economic value. The business function as actors, investors, and creators of new technology, as well as consumers. The creative industry serves as the center of excellence of creators of products, services, and creative culture, as a new market that absorbs products produced and created new jobs for creative individuals or other supporting individuals. The business also plays a role as a community and creative entrepreneur builder which function as a motor that forms a space for thinking, honing creative ideas in doing creative industry business, as well as business coaching or business management training in the creative industry.

The Government has a major role in the mechanism for providing incentive programs, controlling a conducive business climate, educative directions as well as supporting the development of creative industries for the community and the private sector. The government also acts as a catalyst and facilitator and advocate that provides stimulation, challenges, and encouragement in the development of creative industries so that innovative ideas move to a higher level of competence. This support can be in the form of the government's commitment to using its political power proportionally and by providing good public administration services in addition to financial assistance, incentives, or protection. The government also acts as a regulator that produces policies related to society, industry, institutions, intermediation, and resources and technology. The government can accelerate the development of the creative industry if they were able to make policies that create a conducive business climate for the creative industry. The government as an investor must be able to empower the state's assets to be productive within the scope of the creative industry and to be responsible for industrial infrastructure investment.

In the creative industry, the government makes a model based on creative individuals with five main pillars, namely: (1) industries involved in the production of creative industries; (2) technology as a supporter of the realization of individual creativity; (3) resources required such as natural resources and land; (4) institutions ranging from norms and values in society, industry associations and supporting communities to the protection of intellectual property; and (5) financial intermediary institutions.

4 Conclusion

The Indonesian sustainable tourism policy goals of 2030 which support that enable the creation of jobs while supporting local culture and products have their challenges with the spread of covid-19. Based on the Triple Helix concept for the creative industry, the roles of academia, business, and government in creativity, innovation, and technology activities to interact intelligently, effectively, and efficiently as required for such goal.

With its nature as being the largest archipelagic country in the world, Indonesia has the potential for a creative economy in terms of unique resources, language, and culture with the support of intellectuals such as academics, culturalists, artists, writers, pioneers, figures in the arts, culture, and sciences whom all have their roles in providing creative ideas, innovations in the creative industry. Academics intellectuals serve their roles in education, research, and community service. The business becomes a liaison in the context of economic exchange and the transformation of creativity into economic value, while the Government has its major role as major support of financial assistance, regulator, business climate, educative directions, catalyst, facilitator, and advocate of creative industries. The development of a local culture-based creative economy ecosystem, supported by competent business actors, requires support and cooperation from several parties including the government, the private sector, community, academia, as well as the media.

References

1. R. B. Sulistyan, & R. W. Paramita,. D. Business Location Planning Assistance: Preservation of Traditional Culture of Kampoeng Batara Banyuwangi. *Empowerment Society,* **4** (1), 17-21. (2021)
2. R. Comunian,& L. England,. Creative and cultural work without filters: Covid-19 and exposed precarity in the creative economy. *Cultural Trends,* **29** (2), 112-128. (2020)
3. I. Hilmiana, Soemaryani, & A. R. Prasetyo,. Strategic Partnership Model In Developing The Local Art And Culture-Based Creative Economy. *Academy of Strategic Management Journal,* **20** (1), 1-9(2021)
4. T. Kemeny, M. Nathan, , & D. O'Brien,. Creative differences? Measuring creative economy

employment in the United States and the UK. *Regional Studies,* **54** (3), 377-387. (2020)

5. S. Cunningham, M. McCutcheon, G. Hearn, & M. D. Ryan,. 'Demand' for culture and 'allied' industries: policy insights from multi-site creative economy research. *International Journal of Cultural Policy*, 1-14. (2020)
6. Y. Agustina, A. Winarno , H. Pratikto, B. S. Narmaditya, & F. Filianti,. A Creative Economy Development Strategy: The Case of Trenggalek Creative Network for Trenggalek Regency, Indonesia. *The Journal of Asian Finance, Economics and Business,* **7** (12), 1111-1122.(2020)
7. S. Madelan,. Optimizing Exports of Indonesian Creative Economy Products Towards Increasing Competitiveness. Journal of Business Economics, Communication, and Social Sciences (BECOSS), 2(3), 273-284. (2020)
8. N. Nurdiani, R. Hendarti, , & M. Tedja,. Physical Quality of Creative Economic Space on Cultural Tourism Areas in Java Island. *IOP Conference Series: Earth and Environmental Science, 452*, 012057. (2020)
9. Spradley, *Metode Etnografi*. Yogyakarta: PT Tiara J. P.Wacana. (1997)
10. N. Anoegrajekti, Asrumi, Macaryus, S., Iskandar, I., Attas, S. G., Sunarti, S., & Saddhono, K.. Optimization Pillars of Potential Culture and Creative Industry in Banyuwangi, East Java, Indonesia. *Psychology and Education,* **58** (3), 2025-2032 (2021)

https://doi.org/10.1051/e3sconf/202133104016

Networks within networks: scaling response operations in the lightning complex fires in northern california, August 2020

Louise K. Comfort[1]*

[1]Graduate School of Public and International Affairs University of Pittsburgh, Pittsburgh, PA 15260

Abstract. Wildfire risk has increased dramatically in California over more than two decades, 2000 - 2021, reflecting the intense impact of climate change on the state's environmental and ecological systems. Most urgent is the impact of wildfire at the wildland-urban-interface (WUI), and the challenge to prevent cascading disaster for regions connected via interdependent lifelines of transportation, communications, electrical power, water, sewer, and gas line distribution systems that characterize geographic regions. To what extent do large, multi-organizational, multi-jurisdictional networks of organizations learn from experience and adapt their performance in response to the dynamic conditions of an actual extreme event? This article identifies four types of networks operating in the 2020 Lightning Complex Wildfires in northern California and documents the rapid escalation of risk and costs based on a preliminary analysis of the 209 incident reports filed by California Department of Forestry and Fire Protection (CalFire) for the SCU Fire that engulfed large sections of five counties in the southeastern San Francisco Bay Region. The article concludes that Interagency Incident Management Networks provide essential intelligence to support local management of operations in the dynamic context of wildfire risk.

1 Introduction

1.1 The Challenge: Managing catastrophic hazards.

Dramatic shifts in weather patterns resulting from climate change have created more frequent and more extreme events that threaten the safety and well-being of communities across the world. Nowhere is this trend more evident than in California, USA, where drought conditions continuing over multiple years have extended wildfire season from typically dry summer months to virtually a year-round hazard, leading to catastrophic mega-wildfires [22]. Five of the 20 largest wildfires in California's history occurred within the year, 2020 [17], only to be followed by the 2021 record of 8,239 fires documented by October 2021. The largest, the Dixie Fire, ignited on July 13, 2021, consumed more than 963,000 acres, and was extinguished only by welcome rains on October 24, 2021, after burning for more than three months [24].

The dramatically altered landscape of wildfire risk in California has challenged in fundamental ways the organizational infrastructure designed for managing the rapidly changing threat of wildfire in a large state that is both ecologically and demographically diverse. While experience from previous years has demonstrated that the wildfire threat is multidisciplinary and requires a multi-organizational, multi-jurisdictional effort to mitigate the risk, the larger question is how to design and implement effective operations to respond to dynamic wildfire events in real time.

Recognizing mounting losses from the increasing frequency, scale, and scope of wildfire in the last decade, California has countered this threat by making significant investments in building capacity for mitigating wildfire risk through increased funding not only for equipment and personnel, but also for research, technologies, and training to increase 'organizational intelligence' in managing wildfire risk [3]. Doing so has extended responsibility for decision making to wider networks of public, private, and nonprofit organizations tasked with building resilience to wildfire risk statewide, but it raises a further question. *How do operational networks learn, adapt, and change in response to potentially catastrophic events in a dynamic, diverse environment?*

1.2 Policy Issue:

The question of how organizations adapt to rapidly changing, urgent conditions has drawn the attention of organizational theorists for decades (LaPorte and Consolini, 1991; Weick, 1993; Comfort, 1999; Moynihan, 2009; Berthod et al. 2016; Steelman et al., 2021) [16].

* Corresponding author: lkc@pitt.edu

Initially framed as the 'edge of chaos,' Kauffman (1993: 174) characterized this transition to innovative action as requiring "sufficient structure to hold and exchange information, but sufficient flexibility to adapt to changing conditions." From an early concept of "high reliability" organizations (Rochlin, 1987; Weick and Roberts, 1993) to a recognition of organizational structure as networks (Prietula and Carley, 1999; Comfort, 1999; Kapucu, 2006) to an emerging focus on networks undergoing change during operations (Moynihan, 2009; Berthod et al., 2016; Comfort and Zhang, 2020; Steelman et al., 2021), researchers have observed and documented organizational adaptation to changing environments. But the factors that both drive and limit organizational change, especially in networks of organizations that are mobilized for a massive, dynamic event, are not well understood, specifically under the urgent stress of escalating disaster. No single organization has the resources, personnel, equipment, and oversight to manage a catastrophic wildfire event alone. Necessarily these events are multi-organizational, multi-disciplinary, and interdependent. They require an innovative framework for managing mobilization, strategy development, and adaptive action in a dynamic field environment. These requirements exceed the capacity of most emergency response organizations, even when plans and training are in place. In California, the Standardized Emergency Management System (SEMS) is required training for all agencies engaged in emergency response operations (California Office of Emergency Services, 2015), but the size, scope, and scale of mega-wildfires strain the capacity of organizations even with advanced training and major investment in resources and personnel.

1.3 Collective investment in change.

After experiencing devastating fires in which the demand exceeded the investments made in 2019, the 2020 California state budget increased that investment to augment the number of personnel, helicopters, advanced training in anticipation of fires as the drought continued in 2020. The assumption underlying this investment is that the network of organizations tasked with responsibility for managing wildfire risk statewide would learn, incorporate new technologies for decision support into improved practice, and build collaborative relationships among multiple jurisdictions, disciplines, organizations, and communities to create an innovative approach for managing wildfire risk for the entire state. The state further relies on the Emergency Mutual Assistance Compact (EMAC) (U.S. Congress, PL 104-321, 1996) that enables one state afflicted by a serious hazard to request assistance from neighboring states, with the caveat that the receiving state would reciprocate in kind when neighboring states needed assistance. Wildfire risk, of course, does not stop at state boundaries. In the summer of 2020, all western states suffered extreme wildfire events – Washington, Oregon, Idaho, Nevada, Colorado, and Arizona – to the extent that neighboring states could not respond to most of California's requests for assistance as their resources were needed urgently to respond to wildfire outbreaks at home (Morris and Dennis, 2021: p.21). The scale and scope of wildfires in the region increased the urgency for response networks in California to learn to adapt quickly in their own context.

Inquiry into networks as learning mechanisms. To what extent do large, multi-organizational, multi-jurisdictional networks of organizations learn from experience and adapt their performance in response to the dynamic conditions of an actual event? If so, what factors contribute to network learning in large-scale, mega-disasters? Conversely, what factors inhibit the capacity of networks to adapt their performance in large, diverse, dynamic contexts of operations? These questions are particularly relevant for public policy and administrative theory in California in reference to wildfire risk. With a GDP of $3.2 trillion, California ranks as the largest economy of all 50 US states and as the fifth largest economy in the world, after Germany but ahead of India (Hughes, 2020). Maintaining that economic ranking depends on the capacity of the state to manage wildfire risk, while also maintaining the environmental resources of the state that generate economic productivity and well-being for its population.

The capacity for networks to learn returns to the earlier framing of the question in organizational theory, and the first task is to define what constitutes "sufficient structure to hold and exchange information" among the large number of organizations and jurisdictions involved in response operations in wildfire events. The second task, equally critical, is to define what is "sufficient flexibility to adapt to changing conditions" in an actual wildfire event. The inherent tension between structure and flexibility in organizational response networks is clear but needs to be redefined in the context of actual practice. Learning to calibrate organizational structure to the demand for flexibility in adapting to dynamic situations in a functional operational system is a primary challenge. In a large, diverse, multiorganizational response system, the calibration depends upon the communication flow within and between different types of networks engaged in response operations. Specifically, it means identifying the type of intelligence that is needed by emergency responders to adapt their actions to a changing operational environment, the validity and timing of that intelligence as it is shared within and between networks in the larger operational system, and the level of prior knowledge and training among a diverse group of participants to comprehend the intelligence quickly when it is shared.

1.4 Research questions:

To address specifically the challenge of calibrating structure to flexibility in practice, this paper examines networks as the primary unit of analysis in organizational response operations and focuses on four questions that explore the potential and limits of network learning in the context of actual wildfire events. The questions include:

1. What types of networks can be identified in practice in managing wildfire risk, and what role does each network play in a larger wildfire response system?

2. What factors – social and technical -- facilitate adaptive performance to changing conditions among the networks?
3. What factors – social and technical -- hinder adaptive performance to changing conditions among networks?
4. What factors operate to calibrate structure to capacity for flexibility in forging a functional operating system for managing wildfire risk?

This paper will examine these questions in the context of actual wildfire operations that occurred on the same day, August 17, 2020, in northern California largely within the San Francisco Bay Area's Metropolitan Region of nine counties, 101 cities, and 7 million residents (www.bayareacensus.ca.gov). These events were triggered by a rare, meteorological phenomenon of an intense shower of dry lightning strikes over a 24-hour period that ignited simultaneous fires in parched grasslands and wooded areas with dead trees damaged by a years-long drought. The characteristics of the wildfire events are given more detail in section 3 below. Specifically, the paper will focus on the SCU Complex Fire in the southern counties of the San Francisco Bay Area, although the full analysis in progress will address all three fires, assessing the regional impact of simultaneous fires.

The paper proceeds as follows. The next section reviews briefly the types of organizational networks relevant to the evolution of network learning. The third section describes the context of the 2020 Lightning Complex Fires that created an extraordinarily challenging environment for organizational response operations in dynamic wildfire events. The fourth section will report the data sources, collection, and methods of analysis. The fifth section will present findings, and the final section will summarize the conclusions of the study and suggest next steps for further analysis.

1.5 Theoretical Framework of Network Learning Processes

Networks have long been studied as organizational systems that form in response to urgent events. Early studies largely focused on identifying the structure of organizations engaged in response operations in past events, tracing the interactions among participating organizations, and estimating their positions of influence within the larger response system (Kapucu, 2006; Agranoff, 2007; Moynihan, 2009; Butts, 2009; Hu et al., 2015). These analyses distinguished between planned networks designed to perform specific services in emergency response operations, for example, the Incident Command System adopted among California fire departments in the late 1970's (Jensen, 2016; Nowell and Steelman, 2019) and emergent networks of volunteers who assist in recovery following community disasters (Drabek, 1981; 1985; Comfort and Zhang, 2020). The early studies established networks as primary organizational structures capable of performing complex operations embedded within larger social systems. More recent analyses of networks have focused on the processes of network adaptation in dynamic, urgent events (Carley et al., 2016; Comfort, 2019; Steelman et al., 2021).

1.5.1 Sociotechnical Networks

Recent years have seen positive change in network performance due to the integration of emerging information technologies into planned network structures for mitigation and response operations. Such networks function reasonably well for moderate events that remain within the bounds of known resources, skills, knowledge of responsibility and functions in social performance. They use technologies to increase the search, exchange, and updating of information to enhance collective cognition of risk and build the coherence needed for collective action under threat (Comfort, 2007; Comfort et al., 2020; Comfort and Rhodes, in press, 2022).

1.5.2 Operational Networks

Mobilizing operations in an actual field event involves adaptation to the local context and including local personnel and emergent groups of volunteers into structured response actions. This task means providing sufficient guidance to local personnel to enable them to provide valid information regarding local conditions, disrupted by disaster degradation and destruction, coupled with sufficient flexibility to allow the formal organizations to adapt their performance accordingly. Operational networks merge the structure of planned networks with the local knowledge of emergent networks to adjust the performance of the expanding response system dynamically as operational conditions vary. This form of adaptation relies on rapid, timely search and exchange of information between incoming response personnel and local volunteers who have knowledge and experience of local conditions (Comfort and Zhang, 2020).

1.5.3 Intelligence Networks

The fundamental challenge posed by wildfire risk in California and other states and countries lies in the sobering potential for catastrophe; that is, when the demands posed by actual hazard events exceed the scope and scale of planned operations that are constrained by tight time requirements for urgent response, limited resources and skills, and unanticipated demands for interdisciplinary knowledge. These limitations lead to high costs in loss of lives, property, livelihoods, and cascading events. Such demands overwhelm the existing operating capacity of organizational networks in emergency response operations and require a fundamental reconceptualization of the phenomena of hazard emergence, mitigation, preparedness, response, and recovery strategies. In these rapidly escalating conditions, intelligence networks become a primary resource in managing an urgent event, bringing it under control more quickly while minimizing the damage. These networks are essentially 'advice networks' (Kilduff

https://doi.org/10.1051/e3sconf/202133104016

and Krackhardt, 2008) as they do not replace existing operational personnel, but rather complement and supplement their authority to conduct field operations. Intelligence networks bring personnel with more experience, specialized knowledge, and fresh concepts in organizational performance to provide welcome back-up support to the management of operational personnel in extreme events.

1.5.4 Realigning Scale, Scope, and Time in Network Performance.

The relationships among people, jurisdictions, organizations/institutions, technologies, and risk need to be redesigned as functional networks capable of learning, redesign, and recalibration both in reciprocal response to participating actors and to the physical environment in which they operate. An initial effort to re-imagine functional networks might include systematic efforts to design sets of networks that can realign the scale, scope, and timing of their operations to fit the demands of the environment in which they are operating. The added capacity of the sub-nets to gain a systems view through visualization of operations could be provided through innovative technology in near-real time.

1.6 The Lightning Complex Fires in Northern California, August 2020.

1.6.1 The Context

On August 17, 2020, seven major wildfires were ignited in northern California by a singular meteorological event, dry lightning strikes in hot, parched grasslands and forests with trees dying from prolonged drought. More than 12,000 lightning strikes in highly combustible wildlands were recorded over two days, August 17-18, 2020, throughout northern California (CalFire, Incident Reports, 2020). Three of the largest wildfires were burning simultaneously in eight of the nine counties in the San Francisco Metropolitan Bay Area, home to seven million residents. Only the small, urban county of San Francisco escaped the wildfires.

The SCU Complex Fire spread over the three southern counties – Santa Clara, Alameda, Contra Costa and stretched over to adjacent San Joaquin and Stanislaus Counties – burning 396,622 acres and destroying 222 structures. The LNU Complex fire ignited in the northern counties of Solano, Napa, and Sonoma, spreading north to Lake County and east to Yolo County, burning 363,220 acres and destroying a larger number of structures, 1491, in a more populated area. The CZU Complex Fire ignited in San Mateo County and adjacent Santa Cruz County, destroying old growth sequoias in Big Trees State Park, and threatening the University of California, Santa Cruz campus. The next day, August 18, the Woodward Fire ignited in Marin County, burning a smaller acreage but distracting attention and resources from the larger fires burning in more heavily populated areas.

1.6.2 The Challenge

The demands for response, coordination, and control in this set of actual events were unimaginably complex under urgent threat. The California Department of Forestry and Fire Protection (CalFire), the public agency with legal responsibility for managing wildfire risk, faced an excruciatingly difficult set of interrelated and escalating tasks in leading and organizing coherent response operations to an ever-shifting dynamic of field conditions. The only approach that appeared possible was to activate simultaneously a network of networks in response operations. Yet, doing so across multiple jurisdictions that had widely varying access to resources and were constrained by health restrictions imposed by the COVID-19 pandemic presented a formidable challenge to agency personnel.

In this context, CalFire adopted a management strategy that had its origins in military operations but was adapted to the rapidly changing operational environments of extreme events. CalFire formed Interagency Incident Management Teams (IMTs) composed of experienced managers from different jurisdictions and disciplinary backgrounds with knowledge and access to resources relevant to wildfire ignition and control (Situation Reports, CalFire, 2020). A primary function of these teams is to collect, synthesize, and analyze information from multiple sources to inform action strategies for operational teams and to support the implementation of these strategies in field operations. The rationale for mobilizing these teams, especially in wildfire situations, is that timely, informed action at the local level brings the overall incident under control more quickly and more effectively, saving lives, property, and reducing overall losses. Personnel included in the IIMTs bridged organizational, jurisdictional, and disciplinary boundaries to create a systems perspective for managing the wildfire and to provide hazard intelligence and complementary analytical support to the local Incident Commander and his/her team. Importantly, the IIMTs did not replace the local Incident Command Teams, but rather acknowledged the extraordinary burden they were facing and provided an advanced level of hazard intelligence, analysis, experience, and informed access to external support for aid in coping with rapidly escalating wildfires.

2 Methods and Data.

2.1 Methods

The method most appropriate for this complex policy problem of managing mega-wildfire is to conduct an initial exploratory case study of actual events (Yin, 2016). The first task is to identify the networks that were engaged in field operations in response to the fires. Four types of networks were likely involved in response operations for each of the fire incidents, the LNU, SCU, and CZU Complex Fires. These types are: 1) planned networks that represent the formal organizational structure of the Standardized Emergency Response System (SEMS) in which all California emergency response personnel are trained; 2) emergent networks of local organizations and volunteers that responded spontaneously to each of the three fires; 3) operational networks that integrated the first two types – planned and emergent – into reasonably coherent operational networks led by CalFire; and 4) Interagency Incident Management Teams, teams of experienced personnel from federal, state, regional, and county agencies that brought advanced skills and knowledge to support local personnel in coordinating operations that involved mobilizing resources from multiple agencies, organizations, and jurisdictions under critical time constraints. This preliminary study will focus on the SCU Complex Fire in the southern counties of the San Francisco Bay Region.

2.2 Data

Access to incident data needed to distinguish the four types of networks is not easily available. Presumably, all data pertaining to wildfire operations are public and available to interested parties. In practice, incident data are held by operational agencies and are not readily released. It is possible to file a Public Records Request and specify the data one wishes to access and the reason for study, which I have done. However, the NFIRS (National Fire Incident Reporting System) data for these lightning fires is not yet available in Fall 2021, but perhaps 2022. The second option is to review all agency reports and press briefings that provide general information about the size, scale, and duration of operations, but lack specific identification of agencies, jurisdictions, private or nonprofit organizations that were involved in response operations. The third alternative is to review news reports of operations that would provide some information about community organizations as well as public agencies, but that are likely not complete. This initial exploratory study is based on data sources that are publicly available, including public reports, plans, and official briefings from CalFire and the Governor's Office of Emergency Services, State of California.

2.2.1 Data analysis

The types of investments made in preparing for the 2020 fire season can be identified from the Governor's Strike Force Report published in March 2019. These investments can be verified by the reports of new equipment purchased, number of additional personnel hired and trained, and collaborative agreements with private companies and research organizations reported in the 2020 incident action reports filed by public agencies. While these data do not specifically identify the actual networks engaged in response operations in the lightning complex fires, they do verify the types of networks that were operating and at least the number of official personnel involved in the agency networks. These reports can be used to identify the basic parameters for a system dynamics model of information flowing through the types of active networks engaged in wildfire response operations that would characterize a learning system. This model will be developed in the next stage of research.

2.2.2 Process Tracing

Given the constraints on access to public data, the most appropriate method at this stage of research is systematic use of process tracing (Bennett and Checkel, 2014). Process tracing is defined as the logical construction of an explanatory narrative for the confirmation of different types of networks and their roles not only in response operations, but also in reshaping the basic understanding of public action in response to the burgeoning threat of wildfire and related hazards stemming from climate change. This method acknowledges that the data are incomplete, but the resulting narrative, based on the most complete data available, will be used to structure the next phase of analysis on networks of learning and adaptation in anticipation of catastrophic risk environments. Process tracing assumes that research on complex, uncertain events is an iterative process and that an important outcome is to prepare an informed basis for the next stage of rigorous analysis.

3 Scaled Networks in Catastrophic Events?

Previous research has documented extensively the operation of networks in emergency response (Hu, Khosa, and Kapucu, 2015). These networks have been identified both as planned and emergent networks. Yet, any planned network needs to adapt to the operational environment in which it works, and emergent networks that provide local knowledge assist with that adaptation. Skillful leadership in both planned and emergent networks, supported by timely information from the changing hazard environment, enables the mobilization of operational networks to manage hazards within known limits of time, knowledge, skills, and resources. Large-scale, catastrophic hazards require a different order of experience and innovation, and specialized teams of Interagency Incident Management Experts dispatched to different areas of need strengthen the whole system by providing expert support at vulnerable points of performance.

In effect, in large-scale, extreme events, the operational system expands, with multiple networks serving different functions and adjusting operations to the scope and scale of the event. A systems perspective is essential to mobilize resources in knowledge, skills, and experience to manage a larger, more dangerous, catastrophic event. In doing so, previous networks have generated the capacity to recognize the skills and resources they do not have but need to counter unexpected and unknown challenges as critical information becomes available and reveals possible alternative strategies for action. In environments of uncertainty, innovations in technology that monitor change in risk conditions in real-time are critical to maintaining a valid, current profile of risk that allows informed allocation of resources, skill, and time to mitigate potential damage and prevent cascading losses.

https://doi.org/10.1051/e3sconf/202133104016

4 Preliminary Findings

The figures below show the results from a preliminary analysis of the 209 incident reports filed by Cal Fire from the SCU Fire that engulfed large sections of Santa Clara, Alameda and Contra Costa Counties, spreading into sections of Stanislaus and San Joaquin Counties. While the fire was burning largely in wildlands bordering the metropolitan cities of Oakland and San Jose, the fires came dangerously close to threatening the electrical transmission center located outside San Jose that would have compromised the electrical grid for large parts of California and neighboring states. The strategic focus for wildfires at the wildland-urban-interface (WUI) is to prevent cascading disaster for regions connected through interdependent lifelines of electrical power, communications, transportation, water, wastewater, and gas distribution systems that characterize metropolitan regions. These interdependencies transform a wildfire event in close proximity to urban centers into a potentially catastrophic event affecting millions of people and shutting down economic and social activity for the region.

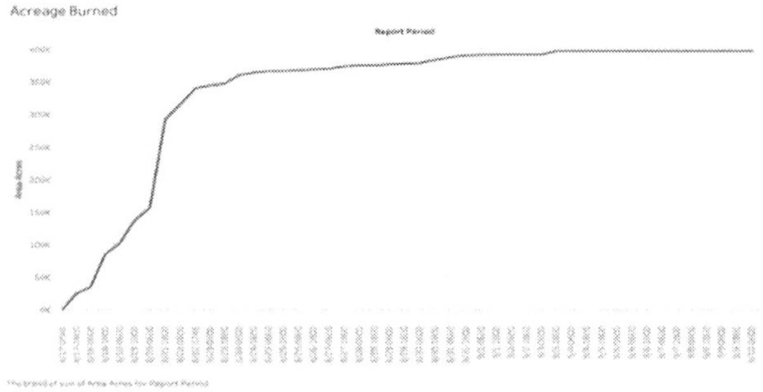

Fig. 1. Acreage burned as measured in 12-hour increments, SCU Lightning Complex Fire, 8-17-2020 to 9-10-2020.

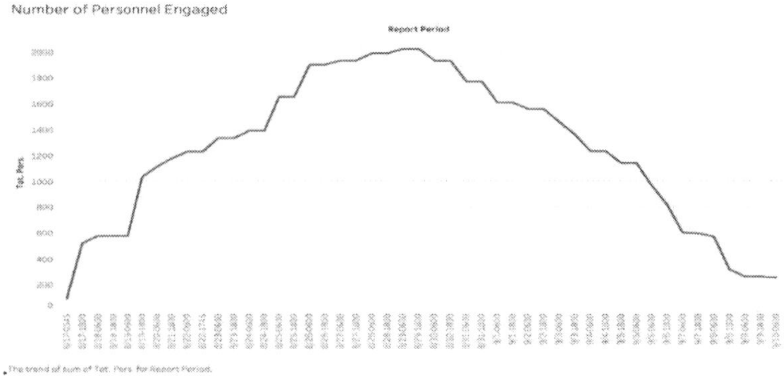

Fig. 2. Total personnel assigned to SCU Fire in 12-hour increments, 8-17-2020 – 9-10-2020.

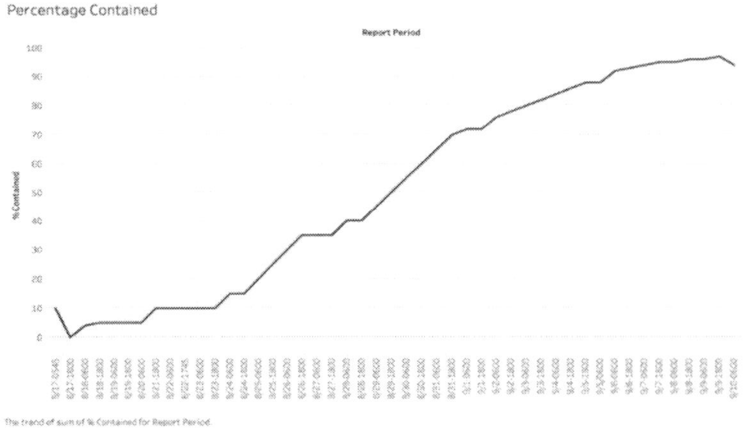

Fig. 3. Percent of SCU fire area contained by 12-hour increments, 8-17-2020 – 9-10-2020.

https://doi.org/10.1051/e3sconf/202133104016

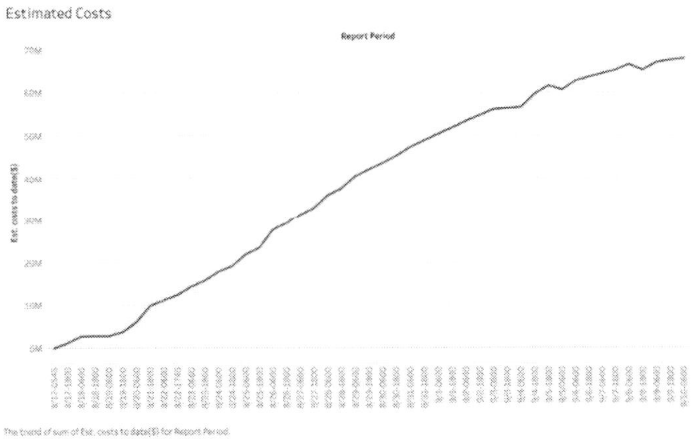

Fig. 4. Estimated costs for SCU Fire in millions of USD by 12-hour increments, 817-2020 – 9-10-2020.

As Fig. 1 illustrates, the fire escalated rapidly in the first three days, doubling in size of acreage burned by 12-hour periods, reaching nearly 350,000 acres in four days. In contrast, Fig. 2 shows the constrained mobilization of fire personnel and equipment over that same period, reaching the peak of 2,000 personnel only on August 29, 12 days into the incident.

Fig. 3 confirms the impact of constrained mobilization of resources to show that only after 8/29/2020 with full allocation of personnel and aerial equipment to fight the fire, did the percentage of contained area steadily increase to full containment 10 days later, 9-10-2020.

The final tally of costs of fire suppression in terms of personnel allocated and equipment deployed was approximately $70 million. This tally did not include costs of economic disruption and social dislocation to the communities affected.

5 Building Networks within Networks to Reduce Wildfire Risk?

The central question driving the larger study has not yet been answered fully, but preliminary analysis of data from the SCU Fire indicates the direction of that response. As local, state, and federal agencies scrambled to mobilize resources to contain the fire, the number of acres burned doubled and tripled in just four days, 8-17-20 to 8-21-20, shown in Fig. 1. In contrast, it took 12 full days, 8-17-20 to 8-29-20 to mount a fully equipped, operational response to the fire, see Fig. 2. Consequently, the delay in mobilization of resources and personnel, coupled with punishing natural conditions of continuing heat and dry winds, extended the duration of the fire for two more weeks to September 10, 2020, with expanding losses in acreage burned and mounting costs for the region, as shown in Fig. 3 and 4.

Importantly, CalFire and responding agencies adapted their strategies of fire suppression to daily changes in wind direction, temperature, humidity, and availability of resources. Yet, the sobering observation is that the massive mobilization of personnel and resources required to suppress the fire, with its consequent coordination and communication processes, was repeatedly challenged as the fire, driven by hot, strong winds, flared in new areas, and jumped firebreaks that had painstakingly been cut the day before. The detailed analysis of the four types of networks is still in progress, and the role of the Interactive Incident Management Team (IIMT) needs to be fully documented and explained, but this initial assessment reports the key role of this team in maintaining the balance between structure and flexibility in response operations to bring the fire ultimately under control (Morris and Dennis, 2021).

6 Policy Recommendations

The critical function of intelligence in managing these dynamic wildfires is demonstrated by the findings cited above from the 2020 SCU Fire. Throughout the summer and fall of 2020, the six IIMT units in California were deployed continuously, as fires ignited, burned, and threatened urban communities. In most instances, their deployment came after the fires had ignited and, given the conditions of high temperatures, dry fuel, and high winds, the operational tasks exceeded the capacity of the available teams deployed on the ground. Based on the documented experience from the SCU fire and other wildfires observed during the summer of 2020, I offer four policy recommendations:

1. Deploy IIMT units to specific regions of known wildfire risk to develop current assessments and strategic plans for minimizing risk *before* actual fires ignite.
2. Build a knowledge base of conditions in WUI areas that include current states of risk and types of resources available to minimize that risk
3. Invest in advanced communications infrastructure to facilitate the rapid exchange of information that enables collective action to reduce risk
4. Provide clear, easy access to the WUI knowledge base to inform both operational personnel and community organizations committed to wildfire mitigation

5. Engage community residents in systematic efforts to explore, comprehend, and build a sustainable future for communities at risk.

In practice, the State of California is already taking steps to build on the extraordinary events of the Lightning Complex Fires of 2020. For example, the University of California has organized a wildfire symposium to bring together researchers from all ten campuses to focus on the multidisciplinary aspects of wildfire and the conditions that contribute to long-term hazard and risk. This symposium also includes public agencies, private companies and nonprofit organizations that are engaged in activities to reduce wildfire risk. In effect, the University of California, a public research university with a mission of public service, is defining a distinct role for providing organizational intelligence for wildfire risk reduction. The University is taking an active role in collecting, aggregating, supporting, and sharing current, state-of-the-art research on wildfire risk with the broader population of California and in translating research findings into practical applications that can be adopted by organizations, communities, and households to reduce wildfire risk. The University of California system, already a network, is focusing attention and resources on a shared problem for the entire state and using these resources to bolster the capacity of public agencies to cope with a large-scale, long-term problem of wildfire risk.

Yet, the scale and severity of the dynamics of climate change driving wildfire risk in California require a statewide commitment from the broad array of public, private and nonprofit organizations and institutions to build a resilient approach to managing wildfire. Only by continuous, sustained, purposeful design and strategic action will California mitigate escalating wildfire risk.

I acknowledge, with thanks and appreciation, the Center for Information Technology Research in the Interest of Society (CITRIS), University of California, Berkeley, and my colleagues, Kenichi Soga, Bingyu Zhao, and Camille Crittenden for their thoughtful insights on the complex, interdisciplinary problem of wildfire management. I thank the CalFire Research and Development Program for assistance in gaining access to the 209 Incident Reports.

References.

1. Berthod, O.G. Hammer, Michael; M. Seitz, Gordon; Raab, Jörg; Sydow, Jörg. From High-Reliability Organizations to High-Reliability Networks: The Dynamics of Network Governance in the Face of Emergency. *Journal of Public Administration Research and Theory*, 352–371. . (2017)
2. Butts, T Carter. Revisiting the Foundations of Network Analysis. *SCIENCE*. 325, 5939, 414-416 (2009)
3. California Governor's Office of Emergency Services. *Cal OES Yellow Book: California Emergency Services Act, California Disaster Assistance Act, Emergency Compacts, California Disaster and Civil Defense Master Mutual Aid Agreement.* Sacramento,CA,https://www.caloes.ca.gov/LegalAffairsSite/Documents/Cal%20OES%20Yellow%20Book.pdf. (2015)
4. L.K Comfort. *The Dynamics of Risk: Changing Technologies and Collective Action in Seismic Events.* Princeton: Princeton University Press, Studies in Complexity. (2019)
5. Comfort, L.K., T.W. Haase, G. Ertan, and S.R. Scheinert. The Dynamics of Change Following Extreme Events: Transition, Scale, and Adaptation in Systems Under Stress. *Administration & Society* (2019)
6. L.K. Comfort, N. Kapucu, K. Ko, S. Menoni, M. Siciliano. Crisis Decision-Making on a Global Scale: Transition from Cognition to Collective Action under Threat of COVID-19. *Public Administration Review*, **80**, 4, 616–622 (2020)
7. L.K. Comfort, H. Zhang.. Operational Networks: Adaptation to Extreme Events in China. *Risk Analysis* (2020)
8. L.K Comfort,. M.L. Rhodes. Collective Cognition in Complex Systems, Chapter 2. In L.K. Comfort, and M.L. Rhodes. Global Risk Management: The Role of Cognition in Response to COVID-19. Routledge Press, (In production, March release) (2022)
9. Drabek, T. E.. *Managing multiorganizational emergency responses: Emergent search and rescue networks in natural disaster and remote area settings.* Boulder, CO: Institute of Behavioral Science, University of Colorado (1981)
10. T.E. Drabek, Managing the emergency response. *Public Administration Review,* 85-92 (1985)
11. J Gibb, A. Sune, S. Albers, 'Network learning: Episodes of interorganizational learning towards a collective performance goal'. *European Management Journal*, **35**, 15–25 (2017).
12. Hu, Q, S. Khosa, N. Kapucu. The intellectual structure of empirical social network research in public administration. *Journal of Public Administration Research and Theory*, **26** (4), 593–561 (2015)
13. Hughes, A. Ryan If California Were a Country. Bull Oak Capital. https://bulloakcapital.com/blog/if-california-were-a-country/ July 27 (2020)
14. N. Kapucu. Interagency communication network during emergencies: Boundary spanners in multi-agency coordination. *American Review of Public Administration*, **36** (2), 207–225 (2006)
15. L. Knight, 'Network learning: Exploring learning by interorganizational networks'. *Human Relations*, **55**, 427–454 (2002)
16. Moynihan, D. The network governance of crisis response: Case studies of incident command system. *Journal of Public Administration Research and Theory*, **19** (4), 895–915 (2009)
17. Morris, George III, C. Dennis.. *The 2020 Fire Siege.* Sacramento: California Department of Fire and Forestry Protection (Cal Fire) (2021)
18. B. Nowell,T. Steelman, Communication under fire: The role of embeddedness in the emergence and efficacy of disaster response communication networks. *Journal of Public Administration Research and Theory*, **25** (3), 929–952 (2015).

19. Nowell, Branda, T. Steelman.. Beyond ICS: How Should We Govern Complex Disasters in the United States? *Journal of Homeland Security and Emergency Management.* **16** (2), 2018-0067 (2019)
20. Office of the Governor, State of California.. *Wildfires and Climate Change:California's Energy Future*. Sacramento, CA: Governor Newsom's Strike Force Report. April (2019)
21. Provan, G. Keith, H.B. Milward.. A Preliminary Theory of Interorganizational Network Effectiveness: A Comparative Study of Four Community Mental Health Systems. *Administrative Science Quarterly*. **40** (1), 1-33 (1995)
22. Task Force, State of California.. *California's Wildfire and Forest Resilience Action Plan*. Sacramento, CA: Public Affairs Office, Department of Water Resources (2021)
23. U.S. Congress. *Emergency Mutual Assistance Compact (EMAC)*, Public Law 104-321.Washington, DC: 104th Congress. 110 Stat. 3877 (1996)
24. Vera, Vicente, Dixie Fire contained after burning more than 963,000 acres. ABC News. October 25. https://www.abc10.com/article/news/local/wildfire/dixie-fire-100-percent-contained/103-d2fd56f5-ca87-4bdb-8db9-68eecaa371e4 (2021)

 https://doi.org/10.1051/e3sconf/202133104017

Comparative analysis on digital surface model of urban area from Sentinel-1 SAR interferometry and aerial photogrammetry for disaster mitigation plan

Fathoni Usman[1, 2*], *Nanda*[2], *Rita* Nasmirayanti[2], and *Josapath* Tetuko Sri Sumantyo[3]

[1]Institute of Energy Infrastructure, The Uniten University, 43000 Kajang, Malaysia
[2]Department of Civil Engineering, Faculty of Engineering, Putra Indonesia University – YPTK, 25145, Padang, Indonesia
[3]Centre for Environmental Remote Sensing, Chiba University, Chiba, Japan

Abstract. This paper presents an effort to evaluate the generated digital elevation model (DEM) from an active sensor onboard satellite of Sentinel-1A and from aerial photos taken using an unmanned aerial vehicle (UAV). The objective is to compare the quality of generated DEM and review the processes for disaster mitigation and prevention plans application. The radar data acquisition used in this study is pair of SLC-type radar data. The interferogram is processed from the coherence and the phase of complex data of the pair radar imageries. Meanwhile, aerial photography was taken within the smaller urban area in Padang City. The photogrammetry process to generate the DEM was conducted using the structure from motion (SfM) technique. The quality and procedures are reviewed by comparing the DEM products with other publicly available DEM data from DEMNAS, SRTM, and AW3D. This study found that generating the DEM from Sentinel-1 interferometry SAR is a challenging process. The product is unmatched and has lower quality compared to available DEM data due to several identified factors. In contrast, high computational cost photogrammetry produced good quality DEM if sufficient ground control points (GCP) were set.

1 Introduction

The satellite era on remote sensing has given the availability of Digital Elevation Model (DEM) from its active and passive sensor on board. The application extends beyond the earthwork analysis and design for civil engineering works to various spatial analysis and environmental sciences [1, 2]. It has been used in hydrology, geology, and geomorphology research to model and predicts land deformation for disaster mitigation and many others. The accuracy and extend of uses depend on the topographic information's quality spatial resolution and temporal resolution. The detail spatial resolution ranged from 10 cm, 5 – 15 m, 30 – 90 m, and 1 km [3]. Different data types were used to generate the DEM from land surveying works, high spots from aerial photos stereography, 3D laser scanning (LIDAR), to satellite remote sensing data.

Meanwhile, data acquisition of aerial imageries has been replacing traditional land surveying products to a certain extent [4, 5]. The metropolitans and urban areas have static features on the earth's surface that predominantly shape the surface model compared to bare land. Contributions from static objects covering the landscape affect the generated DEMs. As a result, elevation refers to the height of the surface rather than the elevation of the terrain [6].

1.1 SAR Interferometry

Synthetic Aperture Radar Interferometry (InSAR) is an advanced technique to retrieve past and continuously earth surface deformation in a millimeter observation scale [7]. This technique has been applied in landslide monitoring and prediction, land subsidence due to natural and human activities in high temporal resolution [8]. However, on rapid movement and changing of surface, limited by the temporal baseline, the interpretation of the surface deformation hampered the performance [7]. The challenge has made the use of the InSAR method in the derivation of DEM less published [2].

1.2 Structure from Motion

Aerial photogrammetry has been used in complementary and, to a certain extent, replacing conventional land surveying in generating 3D surface modeling [9]. The Unmanned Aerial Vehicle (UAV), commonly called a drone, provides indispensable assistance, fast and precise, and relatively cost-saving [5]. The drone technology becomes more user-friendly to navigate and can access highly inaccessible terrain and landscape. The drone is becoming more affordable due to collaborative efforts to use the photogrammetry technology on aerial imagery and

* Corresponding author: fathoni@uniten.edu.my

 https://doi.org/10.1051/e3sconf/202133104017

its quick advancement and significant demand from retail customers.

Generating DEM using photogrammetry starts with flying a drone at a specific elevation and takes a high degree of overlapping multiple aerial photos. The flying path should cover the intended area by sticking the images to reconstruct the 3D of the surface object. The coordinates of objects are calculated using triangulation of coordinates from two coordinate systems, i.e., the camera coordinate and the world coordinate [10].

2 Method and Materials

This study evaluates the generated DEM from the interferometry technique of SAR and aerial photogrammetry. The remote sensing data of SAR cover the Padang City area. The drone that took the aerial photo has flown where the site was identified as responsive to earthquakes. Pada City is the capital of West Sumatra Province. From the two fault zones of Sumatra Fault Zone (SFZ) and Mentawai Fault Zone (MFZ), the distances are about 80 km and 120 km, respectively, to the urban area of Padang City. Fig. 1 shows the fault zones and subduction trance along Sumatra, with the circle shape representing the earthquake's location and its magnitude [11, 12].

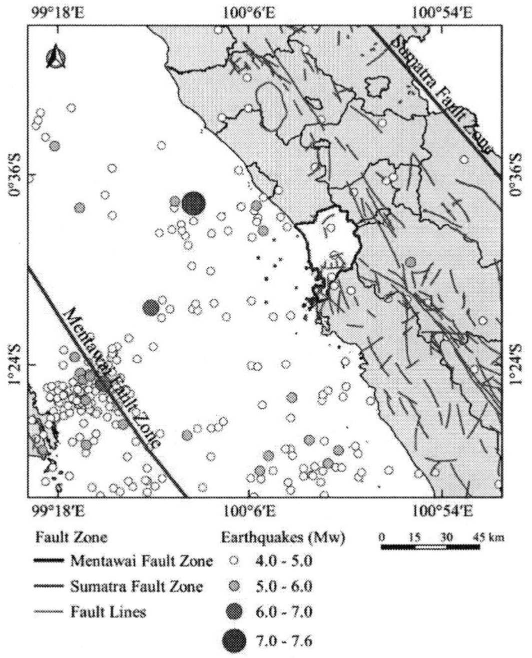

Fig. 1. Fault zones of Sumatra and Mentawai with the earthquake events that occur from August 2009 to June 2021.

The radar data are collected from the Copernicus program website that is available publicly. The data are C band Sentinel-1A mission, single look complex (SLC) type with acquisition swath mode of interferometric width (IW) in descending pass VV co-polarization (transmitting and receiving the beam in vertical polarization). Sentinel-1 was designed to measure the surface changing using interferometry. The pairs of imagery will have narrow orbital tubes [2]. The pairs of radar data in different perpendicular baselines (Bperp) and temporal baseline (Btemp) for this study are listed in **Table 1**. Those radar data will be used to study the changes in the coherence over the time of its temporal baseline.

Table 1. The information of Bperp and Btemp on pairs of SAR data in this study.

Acquisition Date	Role	Perpendicular Baseline (m)	Temporal Baseline (days)	Delta f_{DC} (Hz)
15/09/2019	Master		-	
20/03/2021	Slave	-25.35	-552	-1.60
15/09/2019	Master		-	
16/08/2020	Slave	36.30	-336	-0.23
15/10/2020	Master		-	
20/03/2021	Slave	-76.97	-156	-0.22
15/10/2020	Master		-	
02/12/2020	Slave	-19.86	-48	-0.11
23/07/2020	Master		-	
16/08/2020	Slave	-26.83	-24	-0.04
08/03/2021	Master		-	
20/03/2021	Slave	-28.41	-12	-2.77

Processing radar SLC radar data starts with selecting a pair of radar imagery to set as master and slave and reduces it into a single sub-swath with the chosen number of bursts. The orbit file is then applied and followed by backscattered geocoding. The interferogram is created using the InSAR technique, which multiplies the amplitudes and calculates the phase difference. Back geocoding and Enhanced Spectral Diversity (ESD) is used to improve the image quality. A debursted interferogram is the result of this procedure by applying the TOP SAR debursted command. The process is continued by setting the multi-look parameters and improving the image in Goldstein phase filtering. A wrapped interferogram is the finished product for the phase unwrapping process [13, 14]. The process in SeNtinel Application Platform (SNAP) 8.0 is shown in Fig. 2. The steps process the reflected wave, the backscattered from the radar sensor that emitted microwave radiation. The backscattered signal contains information on phase and amplitude. In [13] and [15], comprehensive technical tutorials for processing the SLC data from Sentinel 1 mission were explained using InSAR to generate DEM in SNAP.

Meanwhile, the basic SfM approach begins with reconstructing a sparse point cloud using feature-based matching to produce a tie point cloud. As a result of this step, multi-stereo pairs are constructed using epipolar geometry, with the collected images aligned and their relative orientation established. Outlier detection is used iteratively to remove incorrect point matches. The camera's interior and exterior orientation parameters are determined simultaneously by minimizing a global reprojection error [16]. After each tie point's space resection and intersection are resolved, a sparse point cloud with 3D coordinates can be generated in any coordinate system. Second, surveying or determining the locations and orientations of camera exposure stations are used as common georeferencing methods. A least-squares adjustment of the 3D coordinates of the sparse point cloud is made using this data along with the tie points.

Creating dense point clouds from scratch is stage three of the process. The disparities at all pixels are estimated using the previously determined epipolar geometry of the photogrammetric block from the first phase in the third stage, which uses image matching techniques. A 3D surface free of unanticipated anomalies is created using gradient-based and energy-minimization techniques, which back-project pixels onto all images and triangulate them. The SfM pipeline creates a 3D surface's raw representation as RGB-coloured dense point clouds. The process of generating DEM in Agisoft Photoscan software is presented in Fig. 3.

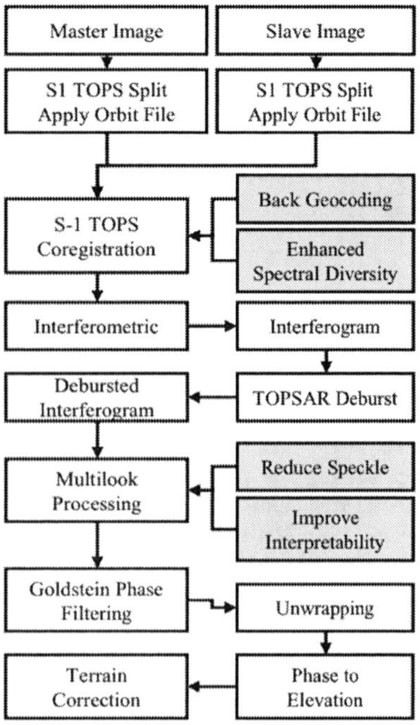

Fig. 2. The DEM generation processes in SNAP 8.0

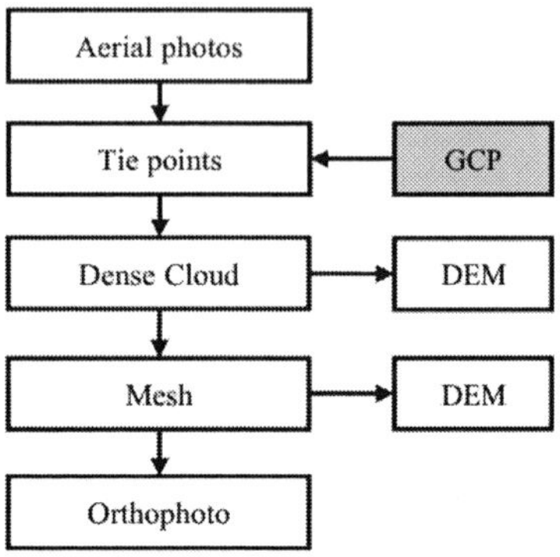

Fig. 3. The DEM generation processes in Agisoft Photoscan Professional 1.4.5

Equation 1 is used to calculate the ground sampling distance, where $\bar{D}$ is the ground sampling distance, S_w is the sensor width of the camera, H is the flight height, F_r is the focal length of the camera, and I_w is the image width of the camera sensor [10]. The precisions obtained from SfM with close-range photogrammetry principles for convergent imagery estimate the achievable coordinate accuracy, $\bar{\sigma_c}$ using Equation 2 for individual targets viewed in k images where q is a factor representing the photogrammetric network geometry, $\bar{\sigma_c}$ is the precision of image measurements, and d is the principle of convergent imagery measured the same as the lens's focal length [4].

$$\bar{D} = S_w \times H \times I_w \times \frac{100}{F_r} \quad (1)$$

$$\bar{\sigma}_c = \frac{q\bar{D}}{\sqrt{k}\,d}\bar{\sigma}_i \quad (2)$$

Three pairs of SLC radar data used for DEM generation in the urban area of Padang City are listed in Table 2. The perpendicular baseline of more than 100 m is selected, and the closest temporal baseline is 48 days. The radar data selection assisted using the Baseline search command in ASF data search. Finding a near temporal baseline with a perpendicular baseline of more than 150 m will be difficult if manually conducted in SNAP using the InSAR stack overview command for the radar data.

Table 2. The pairs of SAR data for DEM generation in this study.

Acquisition Date	Role	Perpendicular Baseline (m)	Temporal Baseline (days)
11/08/2021	Master		
28/09/2021	Slave	153	48
12/06/2021	Master		
11/08/2021	Slave	109	60
15/02/2021	Master		
12/04/2021	Slave	147	48

In preparing the data for SfM, a drone was set to fly about mid-day to minimize the shadow. There are 1329 images were taken with a flying altitude of 31.9 m. Nine ground control points (GCP) were placed (in with circle) within the coverage area of 0.126 km^{2}as shown in Fig. 4. Fig. 4 shows the number of images overlapping, with most interior regions having more than five overlapped images.

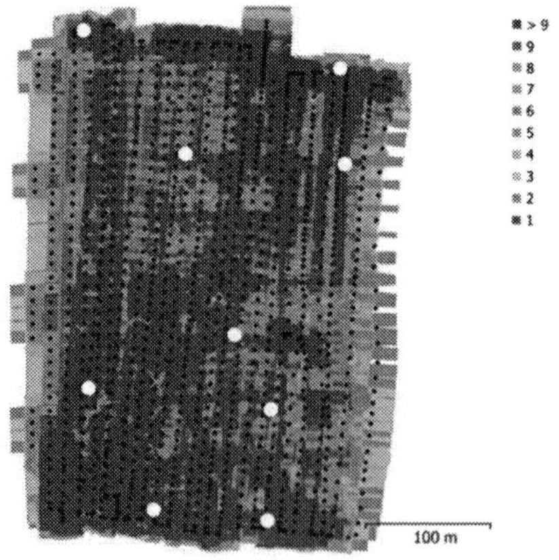

Fig. 4. The GCPs on the overlapped images within the flight coverage area.

https://doi.org/10.1051/e3sconf/202133104017

3 Discussion

This paper presents a development process of a DEM using publicly available Sentinel-1 data from the Copernicus program of the European Satellite Agency. The procedures are discussed in section 2 of this paper. The radar data of SLC that carry radiometric information have been selected with different Bperp and Btemp. The primary purpose is to evaluate the effect on the quality of coherence in developing a high-quality InSAR. It is seen that as the temporal baseline increase, the decorrelation happens to reduce the coherence. Coherence is a valuable indicator of an interferogram's quality. Fig. 5 shows the subset of coherence images on the district of Kota Padang. The box plot graph in Fig. 6 shows the reduction in coherence value on areas denser with buildings and more vegetated regions of the northern part of Kota Padang Regency.

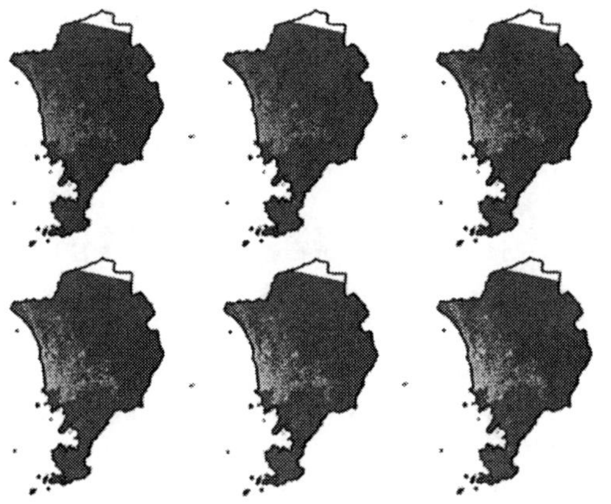

Fig. 5. The subset of coherence data on the district of Kota Padang Regency.

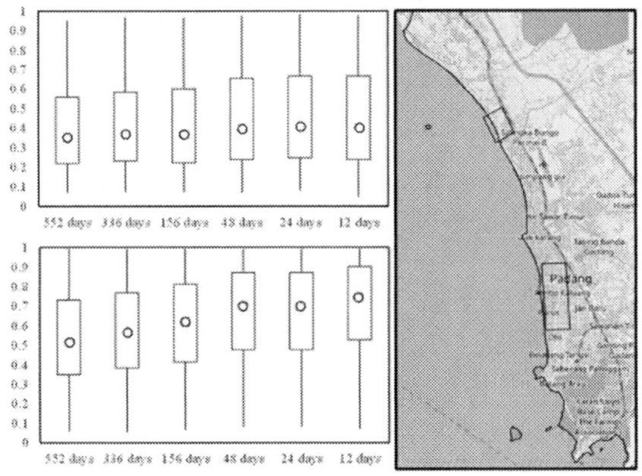

Fig. 6. Box plot of coherence value over the temporal baseline on the different areas with less vegetation on the top and denser vegetation on the bottom.

Consideration should be given to short temporal baselines, acceptable perpendicular baselines, and appropriate atmospheric conditions for data acquisition as well. Because moisture in the atmosphere creates phase delays and degrades measurement quality, the radar data suggested being collected on dry and least rainy days [17]. Between May and July, the Padang City area experienced the least rainfall. The Baseline search function on the ASF data search website did not find radar data within these three months with a short temporal baseline and a perpendicular baseline of more than 150 m. Adjustment for atmospheric delays was not used in this study but reducing the adaptive filter exponent to 0.3 in Goldstein's filtering avoided the cross-like pattern in phase on the urban area [13].

In general, the root-mean-square error (RMSE) values are incredibly high at 391.37 m in the urban area of Padang City, which is densely vegetated in its northern area and scattered vegetation in its southern urban area. Fig. 7 shows the elevation difference of the DEM generated from Sentinel-1 radar data taken on April, August, and September 2021 compared to DEM from DEMNAS, AW3D, and SRTM. The DEM of DEMNAS has a 0.27-arcsecond resolution, which is 8 m spatial resolution, and Both AW3D and SRTM have 30 m spatial resolution. Even though using the DEM generation procedures of InSAR in SNAP, the spatial resolution of Sentinel-1's DEM is 13 m, the C band wave has its limitation. The C band Sentinel-1 has the challenge to penetrate dense vegetation and irregular shape of building formations. The hillshade view of DEMs from this study is shown in Fig. 8.

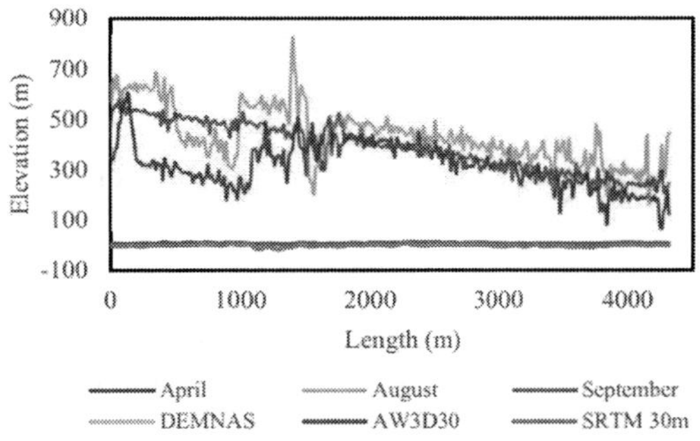

Fig. 7. The elevation difference of the DEM from Sentinel-1 from three acquisition dates.

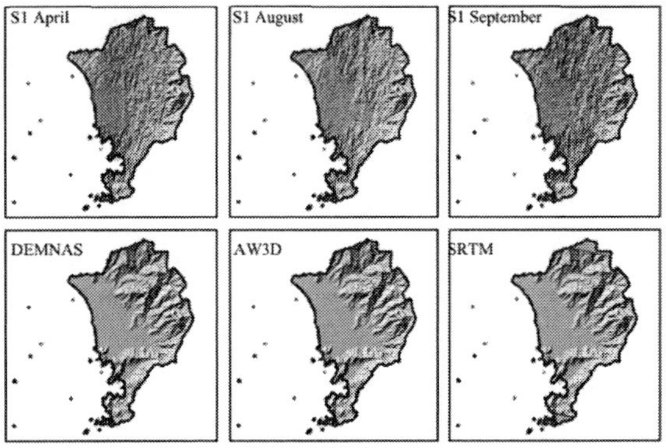

Fig. 8. The elevation difference of the DEM from Sentinel-1 (S1) from three acquisition dates in hillshade view.

On the other hand, the SfM application in generating point clouds to produce the DEM has been conducted to compare its products with and without ground control points (GCP). In this study, the GPCs tap to the DEM from DEMNAS, and the generated DEMs were taken from four products of point cloud and mesh with cleaning and without cleaning processes. Fig. 9 shows the difference between DEM products. The DEM generated without GCP and relied on the onboard positioning, and orientation only has an RMSE of 1.96 m on line A, 2.06 m on line B, and 3.56 m on line C. The placement of GCP near the perimeter of the coverage area will avoid the curvature error on the DEM products as at line B in Fig. 9. The number of GCP to quality of DEM products has a strong correlation, and it was suggested to add 2 GCPs for every hectare of land [18, 19]. Additionally, the placement of GCP close to the perimeter of the coverage area is recommended based on this study finding when a consumer-based UAV was used to take aerial images.

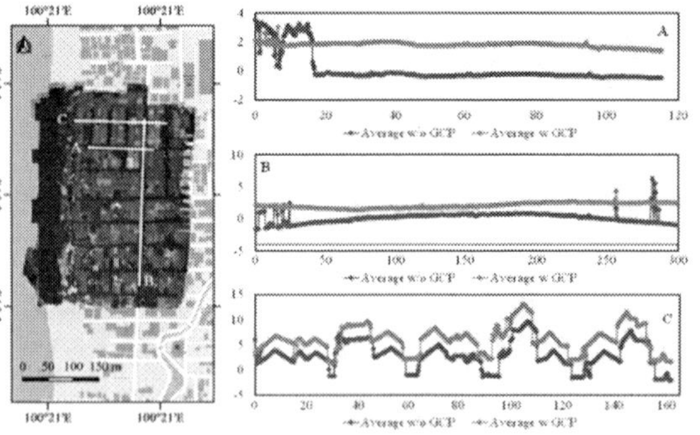

Fig. 9. The average elevation comparison between DEM products with GCPs and without GCPs.

From these two approaches of DEM derivation, the need to have updated DEM for various applications, especially for disaster mitigation plans, is still an utmost effort for urban areas where changes are still dominant due to new development and urbanization. The satellite remote sensing has extensive area coverage. While aerial photography using drones have become more affordable, they made an opportunity to provide updated DEM available. Unfortunately, derivation of DEM from Sentinel-1 radar data with its C band and available suitable baseline has limited the use. This study has identified and added some recommendations in generating DEM using both approaches.

4 Conclusion

Due to numerous usage and application, generating DEM is one of the essential information in geospatial analysis. Generating DEM from radar data is a challenging effort. This study presents an effort to generate DEM for specific purposes of a disaster mitigation plan. The Available Sentinel-1 SLC data for interferometry SAR analysis has many advantages. The Spatio-temporal of 6 days for a tandem mission and 12 days for a single mission provide a good spatial resolution. The generated DEMs were compared with the publicly available DEMs of DEMNAS, SRTM, and AW3D. The Sentinel-1 is designed for Differential SAR Interferometry to extract phase into displacement in a millimeter detail order. In generating the DEM for an urban area with vegetation canopy scattered in between buildings, it was found that lower coherence derives a low quality of generated DEM. The Sentinel-1 cannot produce a good quality DEM because of the low perpendicular baseline (Bperp) value of pair imageries. The Bperp suggested being in between 150 to 300 m which is difficult to get. The other reason for the low quality of generated DEM is Sentinel-1 uses C-band that cannot penetrate denser vegetation canopy in tropical areas. The atmospheric artifact also becomes an issue. Advance processes of developing the DEM from Sentinel-1 SLC data in SNAP Desktop software is a steep learning curve to explore.

The authors would like to thank the Centre for Environmental Remote Sensing (CEReS), Chiba University, Japan, for the CEReS Overseas Joint Research Program 2021 (CI21-111) to Universitas Putra Indonesia – YPTK Padang, Indonesia.

References

1. H. Tsunetaka, N. Hotta, Y. S. Hayakawa, and F. Imaizumi, Prog. Earth Planet. Sci. **7**, 1 (2020)
2. A. Braun, Open Geosci. **13**, 1 (2021)
3. J. Jasiewicz and T. F. Stepinski, Geomorphology, **182** (2013)
4. M. R. James and S. Robson, J. Geophys. Res. Earth Surf. **117**, 3 (2012)
5. P. Sestras, Ştefan Bilaşco, S. Roşca, B. Dudic, A. Hysa, and V. Spalević, Remote Sens. **13**, 3, (2021)
6. I. Borlaf-Mena, M. Santoro, L. Villard, O. Badea, and M. A. Tanase, Remote Sens. **12**, 18 (2020)
7. Q. Meng, F. Raspini, P. Confuorto, Y. Peng, and H. Liu, EGU Gen. Assem. (2021)
8. S. Amariena Hamim, F. Usman, and A. Kurnia Shalihat, Adv. Eng. Res. **187** (2019)
9. J. A. Navarro, N. Algeet, A. Fernández-Landa, J. Esteban, P. Rodríguez-Noriega, and M. L. Guillén-Climent, Remote Sens. **11**, 1 (2019)
10. P. Razi *et al.*, J. Phys. Conf. Ser. **1876**, 1 (2021)
11. M. Diament, Geology **20**, 3 (1992)
12. R. Hall, M. A. Cottam, and M. E. J. Wilson, The SE Asian Gateway: History and Tectonics of the Australia–Asia Collision. (Bath: The Geological Society Publishing House, 2011)
13. A. Braun, Sentinel-1 Toolbox DEM generation with Sentinel-1 Workflow and challenges, (Skywathc, 2020).
14. F. Usman, A. Syamsir, and J. Melasari, *Mapping of Earthquake-Induced Land Deformation on Urban Area Using Interferometric Synthetic Aperture Radar Data of Sentinel-1*, in Recent Advances in Earthquake Engineering, Lecture Notes in Civil Engineering **175** (2022)

15. ASF DAAC, How to Create a DEM using Sentinel-1 Data (ASF DAAC, 2018)
16. E. F. Berra and M. V. Peppa, *Advances and challenges of UAV SFM MVS photogrammetry and remote sensing: Short review*, 2020 IEEE Lat. Am. GRSS ISPRS Remote Sens. Conf. LAGIRS 2020 - Proc. (2020)
17. A. Mohammadi *et al.*, Sensors (Switzerland), **20**, 24 (2020)
18. S. Coveney and K. Roberts, Int. J. Remote Sens. **38**, 8–10 (2017)
19. L. Hashemi-Beni, J. Jones, G. Thompson, C. Johnson, and A. Gebrehiwot, Sensors (Switzerland) **18**, 11 (2018)

Development of new functional seismic brace using partially fibred carbon fiber reinforced polymer

Yukihiro Matsumoto[1*], *Takashi* Taguchi[2], *Keisuke* Shimizu[2], *Iwao* Komiya[3], and *Koji* Nakajima[3]

[1]Toyohashi University of Technology, Department of Architecture and Civil Engineering, Hibarigaoka 1-1 Tempaku-cho Toyohashi Aichi, Japan
[2] Yahagi Construction Co., Engineering Centre, Ibaragasama 1533-74 Nagakute Aichi, Japan
[3] Fukui Fibertech Co., Iwanishi 5-1 Nakahara-Cho Toyohashi Aichi, Japan

Abstract. This study proposes a new tension member with a buckling-free function using carbon fiber reinforced polymer for the seismic bracing system. Firstly, the concept of the composite brace; a partially fibred carbon fiber reinforced polymer brace, is introduced. Secondly, the material strength of the composite brace molded by modified pultrusion is investigated. Finally, cyclic loading tests to simulate the seismic loading are conducted. As a result, it was demonstrated that a seismic brace system using the composite brace can provide buckling prevention and plastic deformation performance to absorb vibration energy.

1 Introduction

Carbon fiber reinforced polymer (CFRP) have been widely applied to the airplane, automobile, civil, and building engineering field, because of its excellent mechanical performance, i.e., high strength, high elastic modulus, lightweight, and good durability. In the civil engineering field, CFRP has been applied to repair and strengthening use [1], and new composite applications were developed in recent years [2]. It is well known that CFRP can resist large tensile force by its high fiber's tensile strength. It means that the effective use of CFRP with unidirectional fibers to take advantage of material characteristics is the use of tension members. Thus, a large number of researchers have been developed a composite tension member and its connection methods [3-5]. Schmidt et al. [3] show several anchorages methods and their mechanical performance to perform the tensile capacity of FRP tendons. And it is shown that mechanical anchorage using wedges is often adopted to prevent slippage and creep. Benmokrane et al. [4] show grouted anchor applications to evaluate the tensile properties of AFRP and CFRP. Xie et al. [5] reported the cyclic loading test results of CFRP tendons with a wedge-bonded anchor.

The authors have developed tension resistible and compression non-resistible member by partially fibred CFRP (PFCFRP) to prevent a buckling deformation and plastic strain concentration at both edges of the brace by buckling shown in Fig. 1. This paper shows the adapted molding method of PFCFRP and its material strength, connection method and its strength, and cyclic test results of the PFCFRP braces to clarify the mechanical properties for the structural design.

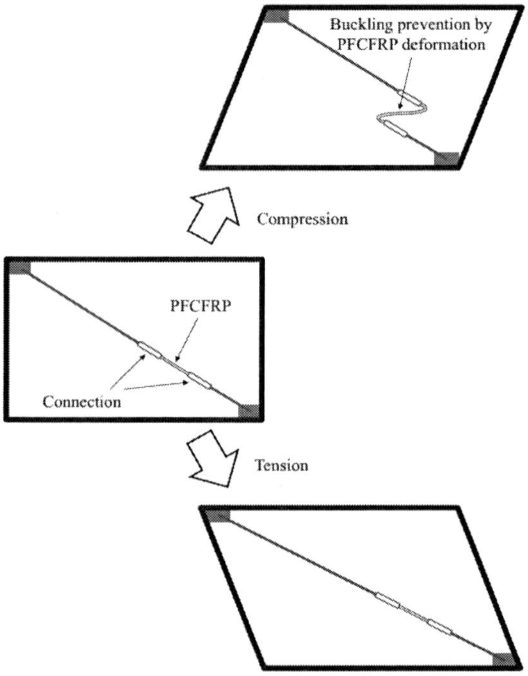

Fig. 1. Concept of new functional brace member.

2 Partially fibred CFRP

Fig. 2. shows the molding method of PFCFRP. We adopt the customized pultrusion to mold because pultrusion has some advantages; high productivity, stable properties, and

* Corresponding author: y-matsum@ace.tut.ac.jp

high fiber content. Usually, carbon fiber passes through a resin bath to become CFRP. However, the authors added a new process that bypasses the resin bath to make a partially fibred position on pultruded CFRP as shown in Fig. 2. Thus, PFCFRP can be molded without loss of productivity by applying automatic cyclic movement. Table 1 shows the properties of CFRP. We adopted a large carbon tow (PX35, commercial product of Zoltek) to make CFRPs economically. Fig. 3 shows a PFCFRP rod with a 12 mm diameter. In total 33 carbon tows having 3750 texes were installed to the PFCFRP rod.

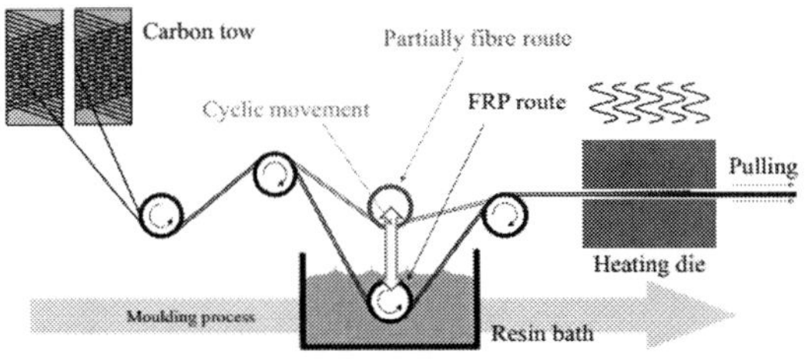

Fig. 2. The molding method of partially fibred carbon fiber reinforced polymer.

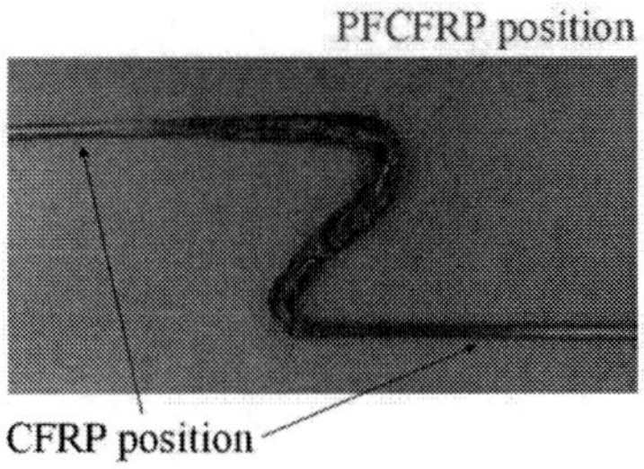

Fig. 3. Partially fibred carbon fiber reinforced polymer rod.

Table 1. CFRP properties.

Carbon tow	PX35
Elastic modulus of CF	242 GPa
Tensile strength of CF	4.137 GPa
Matrix resin	Vinylester
Fibre content	60.5 %
Diameter	12 mm

3 Experiments

3.1 Material tensile tests

This study aims to develop a new composite brace member. Thus, we have to clarify not only the material strength of PFCFRP but also the connection method and strength to apply. Therefore, we adopt a steel pipe-style turnbuckle body [6] at both edges for fixing of PFCFRP rod to simplify the connection like a sleeve connection as shown in Fig. 4. Because the thread on the side opposite to the PFCFRP rod of the turnbuckle body can be used for connection to the steel structure. Moreover, the use of a common connection method provides a possibility of replacement from existing steel turnbuckle braces to PFCFRP brace. Fig. 5 shows the experimental setup. In the case of material tests for CFRP, Fig. 5 (a), M30 steel pipe style turnbuckle body with 300 mm embedding length was adopted. An M20 turnbuckle body with 220 mm embedding length was adopted for material tests for PFCFRP shown in Fig. 5 (b). Moreover, in the case of tensile tests for connection strength, an M16 steel pipe style turnbuckle body with 100 mm embedding length and with the additional M16 bolt to simulate an actual connection condition was adopted shown in Fig. 5 (c).

Fig. 6. shows the maximum loading obtained by various tensile tests shown in Fig. 5. And nominal capacities of steel turnbuckle brace [6] and CFRP are also represented in Fig. 6. The material strength of CFRP using large carbon tow was evaluated as approximately 74% of nominal carbon fiber strength. And in the case of a fibred specimen, material strength was reduced to 54% of CFRP. However, it can be seen that the tensile strength of PFCFRP is higher than the strength of the steel turnbuckle body. Thus, the tensile strength of composite brace with 12 mm PFCFRP rod and M16 steel pipe style turnbuckle body can be evaluated by the strength of steel turnbuckle. Because the strength of steel turnbuckle body is perfectly managed to satisfy the nominal values; 83.1 kN tensile strength in the case of M16 turnbuckle body. Furthermore, both buckling prevention and plastic deformation performance can be expected using the serially connected PFCFRP brace and steel brace shown in Fig. 1.

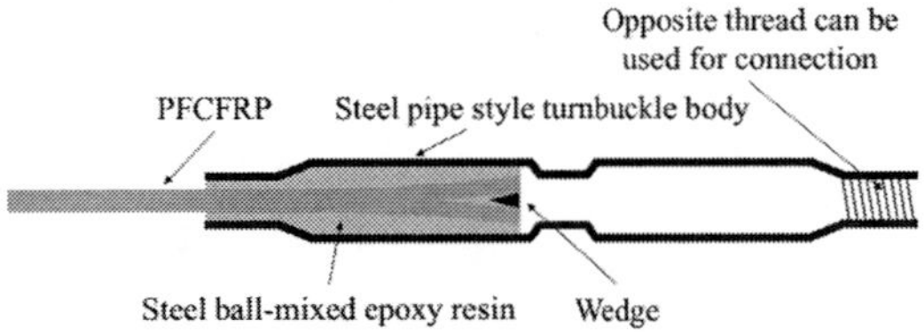

Fig. 4. Connection of PFCFRP and steel pipe-style turnbuckle body.

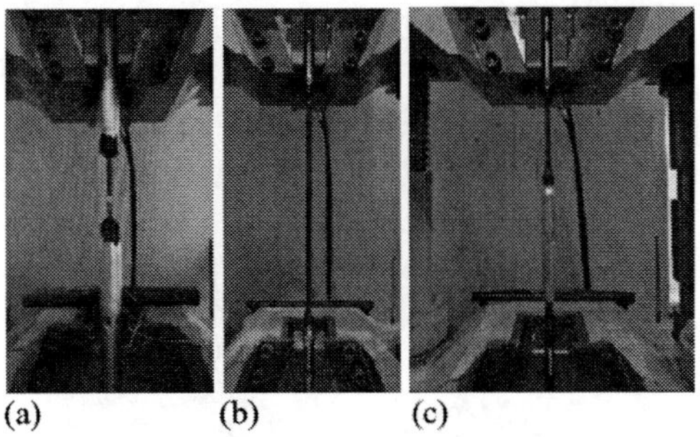

Fig. 5. Experimental setup: (a) CFRP material test; (b) PFCFRP material test; (c) Connection strength test.

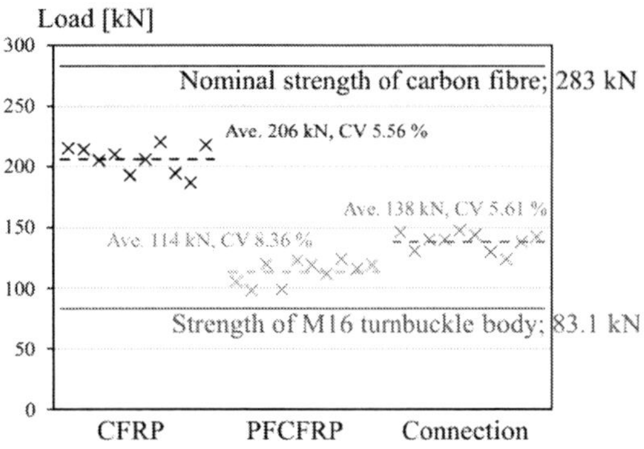

Fig. 6. Results of tensile test.

3.2 Axial cyclic loading tests

Table 2 and Fig. 7 show experimental parameter and experimental setup of longitudinal cyclic test, respectively. Fig. 8 shows deformation under compressive loading. In the case of steel brace, Fig. 8 (a), large lateral deformation occurred by buckling. This buckling deformation suggests that the steel brace will collide near structural members, finishing materials, and facilities when an earthquake occurs. On the other hand, steel and PFCFRP composite brace could maintain the shape of brace straight because PFCFRP absorbed a compressive deformation concentratedly shown in Fig. 8 (b). Fig. 9 shows axial load-deformation relations. Both steel and steel and PFCFRP composite brace performed plastic deformation. However, hardening of the M16-PF1 specimen occurred earlier than M16-S1 because the steel length of M16-PF1 is shortened by PFCFRP installation.

Table 2. Experimental parameter.

Specimen name	M16-S1	M16-PF1
Total length	3,000 mm	
PFCFRP length	N/A	750 mm
Turnbuckle size	M16	

(a) (b)

Fig. 7. Experimental setup: (a) Steel brace; (b) Steel and PFCFRP brace.

(a) (b)

Fig. 8. Deformation under compressive loading: (a) Steel brace; (b) Steel and PFCFRP brace.

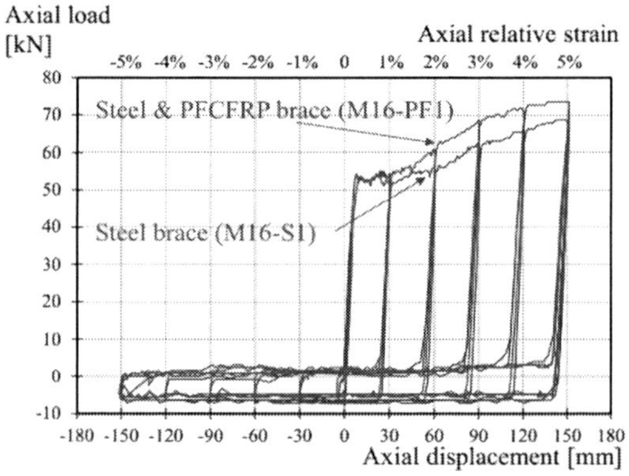

Fig. 9. Axial load-deformation relations.

4 Conclusions

This paper shows the fundamental material performance of CFRP and PFCFRP rods using large carbon tow molded by pulrusion. Then, a connection method using a steel pipe-style turnbuckle body for CFRP was proposed, and its connection strength was made clear. As the result, it was clarified that the PFCFRP rod with a 12 mm diameter has higher strength compared with the M16 turnbuckle body and turnbuckle main member. Thus, it was demonstrated that composite brace with steel and PFCFRP members can provide a function of buckling prevention and plastic deformation performance.

This research was funded by MLIT Construction Technology Research and Development Subsidy Program.

References

1. X.L. Zhao, L. Zhang, *State-of-the-art review on FRP strengthened steel structures*, Engineering Structures, **29**, 8 (2007)
2. Composites Manufacturing website, http://compositesmanufacturingmagazine.com/2017/

https://doi.org/10.1051/e3sconf/202133105001

10/ new-apple-store-Chicago-boasts-huge-composite-roof/ (2017)

3. J.W. Schmidt, A. Bennitz, B. Tãljsten, P. Goltermanm, H. Pedersen, *Mechanical anchorage of FRP tendons – A literature review*, Construction and Building Materials, **32** (2012)
4. B. Benmokrane, B. Zhang, A. Chennouf, *Tensile properties and pullout behavior of AFRP and CFRP rods for grouted anchor applications*, Construction and Building Materials, **13**, 3 (2000)
5. G.H. Xie, Y.L. Bian, Q.H. Feng, C.M. Wang, R.G. Liu, *Experimental study on wedge-bonded anchors for CFRP tendons under cyclic loading*, Construction and Building Materials, **236**, 10 (2020)
6. Japan industrial standards, JIS A 5541: Body of turnbuckle for building (2008)

https://doi.org/10.1051/e3sconf/202133105002

Masonry buildings strengthened with textile-fiber composite (TRC) layers and fiber-reinforced cementitious (FRC) layers

Teddy Boen[1*], *Hiroshi* Imai[2], *Lenny*[1], and *Sarah* E. Suryanto[1]

[1] Structural Engineer, CSI Indonesia
[2] Professor Institute of Technologists, Japan

Abstract. In June 2015, the authors published a paper titled "Brief Report of Shaking Table Test on Masonry Building Strengthened with Ferrocement Layers" [1]. The authors suggested in that paper to replace the traditional way of constructing masonry houses using the so called practical columns and beams (herein after called traditional masonry houses) with bandaging using ferrocement layers on both sides of the walls as skin facings and brick wall as core. Since then, many masonry houses bandaged with ferrocement layers are built in Indonesia. Apart from constructing new earthquake resistant houses, ferrocement bandaging is also used for retrofitting existing as well as damaged houses after earthquakes. In the past decades, continuous fiber mesh was introduced to replace the steel wire mesh in a cementitious matrix. Since the early 2000, textile-based composites were used in the field of strengthening and seismic retrofitting of masonry as well as concrete structures. Originally these new "textile fiber composite" materials are called "Textile Reinforced Concrete" (TRC) in Europe. However, in the USA, the term used is "Fiber Reinforced Cementitious Matrix systems" (FRCM). Extensive research on FRCM / TRC were conducted. A wide variety of publications on the subject matter are now available worldwide. Apart from TRC, many technical studies are published addressing fiber reinforced cement and concrete composite. The term "Fiber Reinforced Cementitious" (FRC) is used and defined as concrete and/or cementitious matrix with suitable discontinuous fibers added to it for the purposes of achieving a desired level of performance in a particular property, such as modulus elasticity, tensile strength, and ductility [2]. Lately, the use of discontinuous fibers as reinforcement for concrete and cementitious matrix FRC are introduced by many practitioners and civil engineers. Adding fibers in concrete / cementitious matrix mixer simply like adding sand or admixtures, to create a homogenous, isotropic, strong, tough, durable, and moldable structural materials [2]. In this paper the authors used the terms of TRC and FRC as defined by Naaman [2, 3], namely TRC for fiber-cement with fiber-mesh and FRC for fiber-cement with discontinuous fiber. This paper provides a simplified global analysis of the overall structure strengthened with FRCM / TRC as well as strengthened with FRC.

1 Introduction

In 2012 the authors did a global dynamic analysis of a masonry house strengthened by wrapping the walls with ferrocement layers. Subsequently a full-scale shaking table test was conducted in Tsukuba Japan for the masonry house [4] and the result of the analysis is in line with the performance of the masonry house based on the shaking table test. Subsequently in 2014, the authors did a structural analysis for the similar masonry house bandaged with ferrocement layers instead of wrapping. In June 2014, a full-scale shaking table test was conducted again in Tsukuba for the masonry house bandaged with ferrocement layers. Once again, the results of the full-scale shaking table test did match with the results of the dynamic analysis of that masonry house [1].

Based on published research, papers, and books [2, 3, 5, 6, 7], in terms of intrinsic mechanical properties, such as tensile strength, shear, strength, equivalent elastic bending stress, fracture toughness, fracture energy, ferrocement is suitable for strengthening masonry houses for earthquake resistance. Ferrocement layers of 2cm applied on both sides of masonry walls are sufficient to prevent collapse when shaken by strong earthquakes (JMA Kobe 100% 0.876g).
Advantages of ferrocement as a construction material, among others
Advantages of ferrocement as a construction material, among others:

- ferrocement is feasible in the context of social, cultural, economical, political, and technical constraints typical in developing countries.

* Corresponding author: tedboen@cbn.net.id

- ferrocement require a low level of technology and can be done by local masons with local tools / construction equipment, and local tradesmen's skills
- the construction period is relatively short so that the building can be occupied asap.
- ferrocement is suitable for self-help construction for housing; the methodology can be carried out by homeowners with minimal financial and technical assistance, and in case of retrofitting, do not require extensive reconstruction or modification of the existing building.

In spite of the advantages stated above, the constructions of ferrocement shall be strictly supervised because proper installation and adequate curing conditions for the mortar matrixes are crucial for the effectiveness of the work. Wrongly constructed ferrocement reduce the performance of ferrocement. Building earthquake resistant masonry houses bandaged with ferrocement layers have already been implemented in Indonesia as can be seen in Fig. 1. More pictures can be seen in www.teddyboen.com.

Fig. 1. Earthquake resistant masonry houses bandaged with ferrocement layers in Indonesia

Since 1980's, the use of high-performance textile fiber has gained increased attention in structural engineering applications. Textile fibers are being used instead of steel wire mesh. Currently, many countries produce textile fiber meshes with geometric and mechanical properties suitable for use in cement-based matrixes. Fiber textile meshes offer better corrosion resistance than ordinary steel wire mesh and are produced in practically any length and thus can be shaped to requirements. However, unlike wire mesh that has similar dimension and material properties, there are many different dimensions and properties of fibers around the world based on the manufactures.

The main common characteristics of Textile Reinforced Concrete (TRC) is that the ultimate tensile strain of the matrix is much smaller than the yield of reinforcement [6]. Therefore, the matrix will crack and the resistance to full separation is entirely mobilized by the reinforcement bridging the cracked surface. This does not matter for earthquake resistant masonry houses because the main aim of using ferrocement or fiber textile cement for building earthquake resistant as well as strengthening of existing masonry houses is to prevent collapse, therefore, cracks when shaken by earthquakes can be tolerated. However, the use of TRC requires careful placement and needs strict supervision. It also leads to an anisotropic building material which is not understandable for the average layman [2, 8].

For Fiber Reinforced Cementitious (FRC), there are two types of discontinuous fiber that can be used in matrix / concrete composite, i.e. natural fiber and synthetic fiber. Natural fiber reinforced concrete consists of small-diameter discontinuous, discrete natural fibers of different origin randomly distributed in a cementitious matrix [9]. It includes coconut fiber, sisal fiber, sugarcane bagasse fiber, bamboo fiber, palm fiber, jute fiber, wood fiber and some vegetable fibers. Meanwhile many varieties of synthetic fiber available with different properties that qualify as low strength, high strength or ultra-high strength, such as PVA (Poly Vinyl Alcohol), PP (Poly-Propylene), nylon which have low tensile strength and low modulus elasticity; steel, E-glass, basalt, asbestos which have high tensile strength and high modulus elasticity; carbon, poly p-phenylene benzobisoxazole (PBO), spectra which have higher tensile strength, higher modulus elasticity and lighter with specific gravity lower than one.

The low-end synthetic fibers, such as Polyvinyl Alcohol (PVA), Polypropylene (PP), nylon, are usually used in non-structural applications, particularly to control plastic shrinkage cracking of concrete at early age and the fibers contribution to the mechanical properties of the composite is ignored [2]. However, with an appropriate volume fraction of fibers, it also can be used for structural applications. The high-end synthetic fibers, such as carbon, PBO, spectra, have had limited use since their cost is among the highest as well. For those reasons above, in this paper, the author decided to use the middle level of fiber such as steel, E-glass, basalt, PVA and PP in mortar mix. The addition of a small amount of fiber in mortar / concrete mix will increase the toughness, energy absorption capacity, and ductility of the mixture [2].

Composite properties are based on volume fraction of fiber, therefore, the cost based on equal performance is more economical when using TRC/FRC. At the moment, a lot of research have been conducted in various countries, in the USA, in Europe, in Japan, etc. and the results are readily available. The authors plan to introduce ferrocement using textile fiber meshes (TRC) or discontinuous fiber (FRC) for constructing earthquake resistant masonry houses as well as retrofitting of existing traditional masonry houses in Indonesia. Unfortunately, research with regards to TRC/FRC in Indonesia is not widely available. Thus, the authors utilize available research works and material from books related to ferrocement and TRC/FRC to introduce constructing earthquake resistant masonry houses bandaged with fiber cement layers in Indonesia. The authors have no intention

https://doi.org/10.1051/e3sconf/202133105002

re-inventing the wheel by duplicating similar research already available in publications but use those researches to get the benefit for the common people. The method can also be used to strengthen existing masonry houses for earthquake resistance. So far, in ACI549.6R-20 [6] and other research papers and books, ferrocement is proven suitable for strengthening system for existing individual masonry walls, subject to in-plane and out-of-plane loads resulting from wind and earthquakes [6]. The TRC strengthening system for individual walls should evaluate the effect of strengthening as on the overall structures [6].

Similar to the strengthening system with ferrocement (using steel wire mesh) for masonry houses in Indonesia which was introduced in 2006, the author also perform a global analysis of the overall structure strengthened with TRC/FRC bandaging. For the purpose, a structural dynamic analysis similar to that performed in 2014 for a masonry house bandaged with ferrocement layers is conducted. Using textile fiber meshes (TRC) instead of wire mesh and using FRC, the authors rely on material properties and fiber types from credible research results published in scientific journals.

If the dynamic analyses show good results, the authors are of the opinion that TRC and FRC can be applied. Another method to verify the results is a full-scale shaking table test which hopefully can be conducted in the near future..

2 Structure Model

The model for analysis is taken from previous shaking table test [1], i.e. masonry structure strengthened by providing bandaging using 2cm fiber-cement layers on both sides of the walls, acting as sandwich structures. The layout of house can be seen in Fig. 2.

The model consists of four walls with size 3.6m x 2.6m with four simple roof trusses and galvanized iron-sheets roofing. The gable walls located at West and East elevations which are perpendicular to the direction of the shaking.

The fiber-cement layers are analyzed as TRC as well as FRC. Fig. 3 shows the schematic drawings for house strengthened using TRC. The fiber-mesh is installed in the middle of fiber-cement layers. Fig. 4 shows the schematic drawings for house strengthened using FRC. The discontinuous fiber is mixed together with mortar..

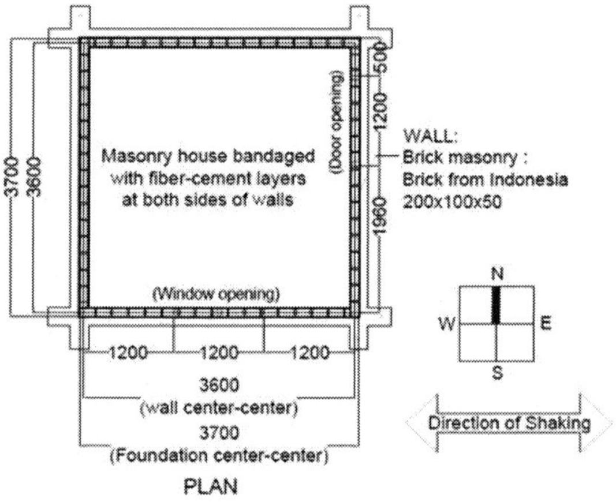

Fig. 2 Layout of house strengthened with fiber-cement la

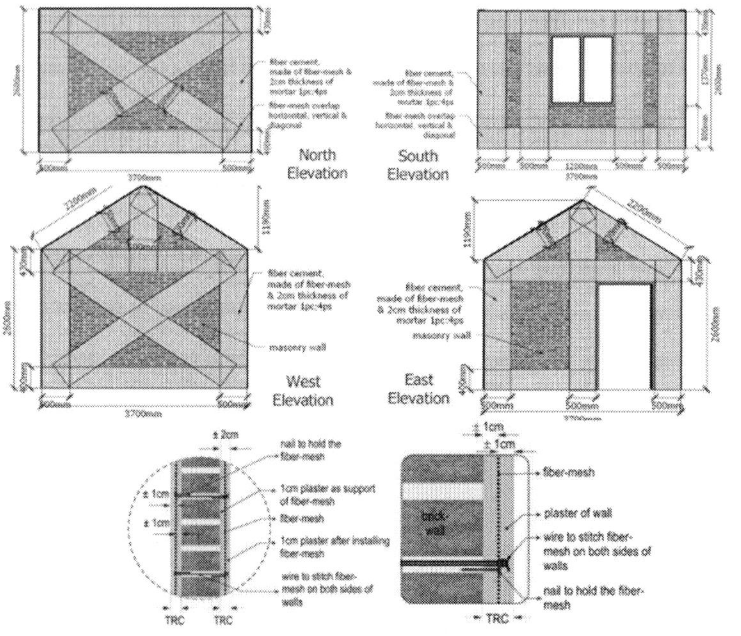

Fig 3. Schematic drawings of house strengthened with fiber-cement layers using fiber mesh (TRC)

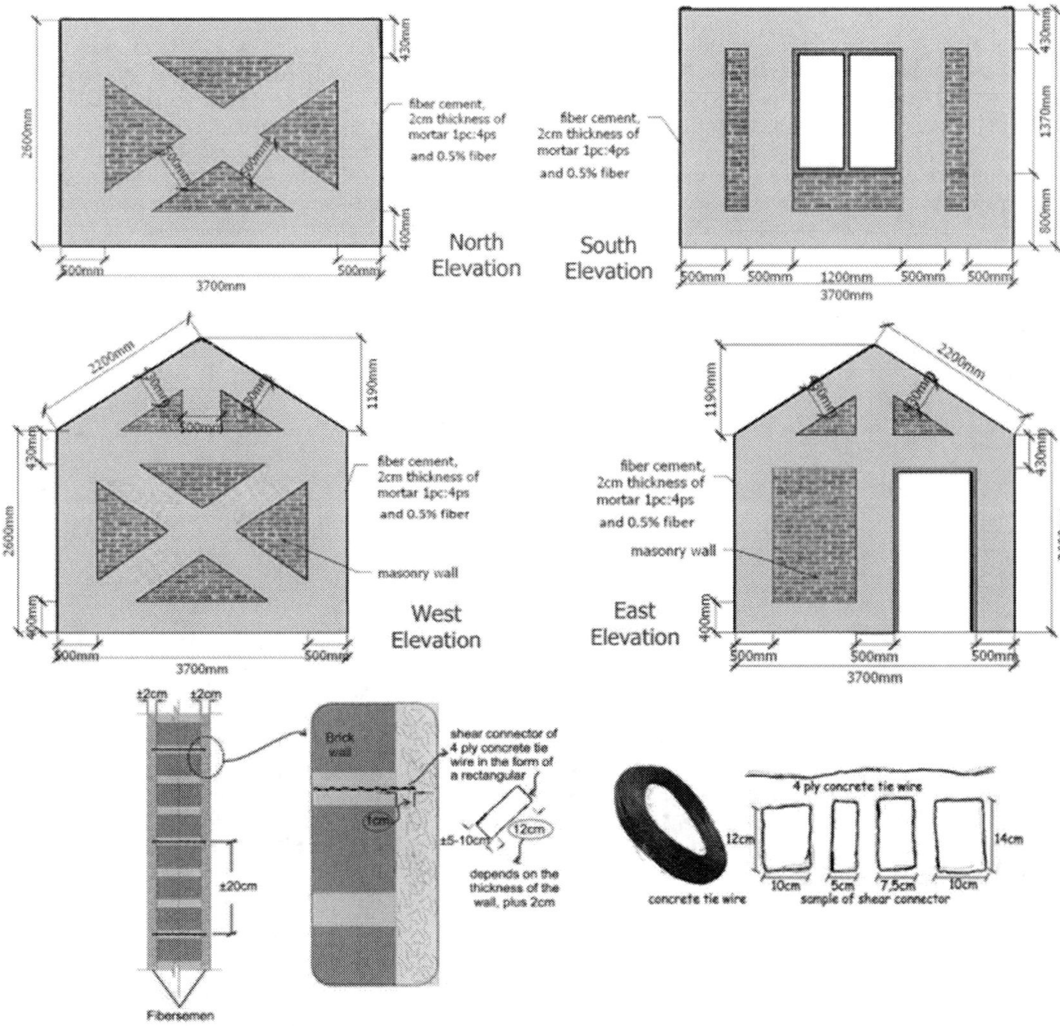

Fig. 4. Schematic drawings of house strengthened with fiber-cement layers using discontinuous fiber (FRC)

3 Material Properties

Material properties of brick-walls used for the analysis are taken from previous shaking table test in Tsukuba, 2014 [1]. The modulus elasticity of brick-walls is 5885kgf/cm2 with adhesion tensile strength of mortar (1pc:6sand) used for the brick-walls 3.5kgf/cm2. The compressive strength of mortar (1pc:4sand) for fiber-cement layer is 275.8kgf/cm2, and the modulus elasticity is 211,950kgf/cm2.

3.1. Fiber-cement using TRC

Since TRC is relatively new in Indonesia, the analysis is done for five different types of fibers, i.e. aragrid-mesh, basalt, carbon, glass, and PBO. The material properties for those fibers are taken from tests previously conducted [3, 6,7].

Table 1 below shows the material properties for 5 types of fibers that are used in the analysis, i.e. aragrid, basalt, carbon, glass, and PBO, as can be seen in Fig. 5. For this analysis purpose, the TRC only uses one layer of fiber because the masonry walls with TRC layers on both sides of the walls are acting as sandwich construction. Therefore, the TRC layer is not acting as a load-bearing single layer such as usually used for cement boards, corrugated cement sheets,pipes, cladding, shells, roofs, domes, water tanks, water channels, boats, etc. [5].

3.2. Fiber-cement using FRC

For FRC, there are 5 types of fibers used in the analysis, i.e. E-glass, basalt, steel, PVA, and PP. Cement bags and rice bags are also made from PP Woven Fiber (see Fig. 6). Therefore, those materials can also be used for low-cost housing program and not become waste. The tensile strength and modulus elasticity are taken from publications [2, 10]. Table 2 shows the material properties for 5 types of discontinuous fibers that are used in the analysis. For E-glass, basalt and steel, the volume fraction is 0.5%; however, for PVA and PP which have low modulus elasticity, the volume fraction is 0.75%-1%. The aim is to obtain the strain hardening behavior for FRC composite in tension as can be seen in Fig. 7 where maximum post-cracking stress σpc should be greater than the tensile stress at first cracking of the matrix σcc. Tensile stress at first cracking of matrix, σcc and maximum post-cracking stress, σpc can be calculated as mentioned in Chapter 9 – Modeling Tension, Fiber Reinforced Cement and Concrete Composite [2]. The tensile strength of the composite, σcc, is primarily influenced by the tensile strength of the matrix, while the post-cracking strength, σpc, is mainly dependent on the fiber reinforcing parameters and the bond at the fiber-matrix interface [2, 11, 12]. When the tensile stresses of fiber-cement are greater than σcc, multiple cracks start to develop at the mortar. Failure of FRC occurs when either the matrix fiber bonding strength is exceeded, or when the fiber breaks [11, 12], meaning when the tensile stress of FRC exceeds the maximum post-cracking stress, σpc.

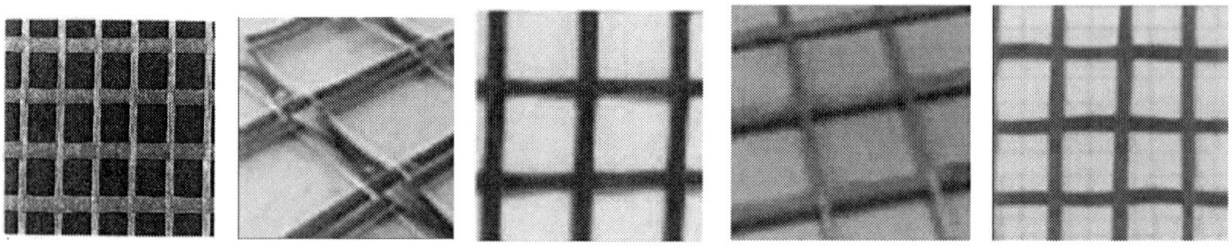

Aragrid Basalt Carbon Glass PBO

Fig. 5. Fiber-mesh used for the analysis [3, 6, 7]

Glass fiber* Basalt fiber** Steel fiber*** PVA fiber**** PP fiber

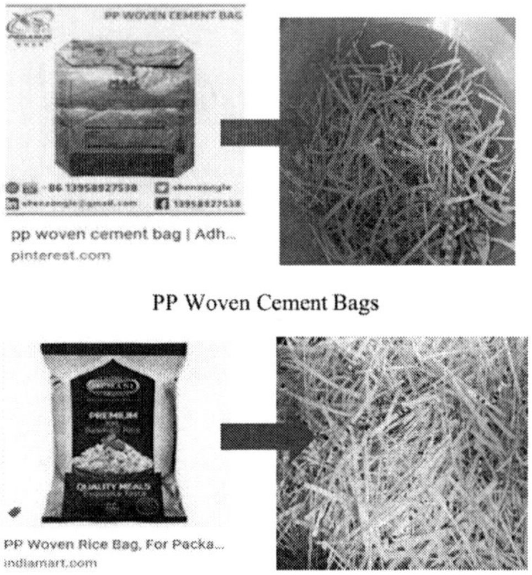

PP Woven Cement Bags

PP Woven Rice bags*******

Fig. 6. The discontinuous fiber used for the analysis

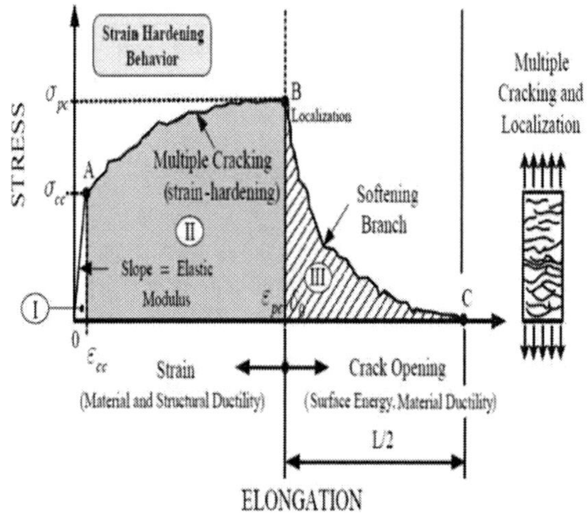

Fig. 7 Strain hardening behavior for FRC composites in tension [13]

Table 1 Material properties of fiber mesh for TRC [3, 6, 7]

Fiber Mesh:	**Units**	**Basalt**	**Carbon**	**Glass**	**Aragrid**	**PBO**
Grid spacing of fiber mesh	mm	25x25	20x20	15x15	10.5x10.5	15x15
Unit weight, W_r	g/m^2	220	170	218	154	44
Density, Y_r	g/cm^3	2.6	1.8	2.6	1.45	1.56
Tensile strength, f_{fu}	MPa	2,438	1,876	1,700	2,030	3,356
Modulus elasticity, E_f	GPa	108	219	72	45	192
Fiber-cement:						
Volume fraction of fiber, V_f	%	0.42	0.47	0.42	0.53	0.14
Volume fraction of mortar, V_m	%	99.58	99.53	99.58	99.47	99.86
Modulus elasticity of fiber-cement, E_c	kgf/cm^2	215,713	221,495	214,139	213,259	214,412
Maximum allowable tensile stress, σ_{cc}	kgf/cm^2	52.59	45.17	36.34	54.88	24.13

Table 2 Material properties of discontinuous fibers for FRC [2, 10]

Fiber:	**Units**	**E-Glass**	**Basalt**	**Steel**	**PVA**	**PP**
Tensile strength, σ_{fu}	MPa	2500	3000	2000	1240	700
Modulus elasticity along the loading, E_{f1}	GPa	76	86	200	32.5	12
Modulus elasticity ⊥ the loading, E_{f2}	GPa	76	86	200	32.5	12
Fiber-cement:						
Volume fraction of fiber, V_f	%	0.50	0.50	0.50	0.75	1,00
Volume fraction of mortar, V_m	%	99.50	99.50	99.50	99.25	99.00
Modulus elasticity along the loading, E_{c1}	GPa	21.06	21.11	21.68	20.87	20.70
Modulus elasticity ⊥ the loading, E_{c2}	GPa	20.86	20.86	20.88	20.84	20.63
Modulus elasticity of fiber composite, E_c	kgf/cm^2	213,669	213,941	216,920	212,611	210,658
Maximum allowable tensile stress at first cracking of matrix, σ_{cc}	kgf/cm^2	13.38	13.41	13.77	13.26	13.15
Maximum allowable tensile stress at post cracking of matrix, σ_{pc}	kgf/cm^2	31.86	38.23	25.48	23.70	17.84

4 Input Motions

Similar with previous shaking table test, JMA Kobe ±175mm 100% (0.876g) earthquake record is used as input motions for this analysis. JMA Kobe ± 87.5mm (2g) earthquake record that is a disastrous earthquake is also used in Fig. 8.

5 Global Analysis of Masonry House

A global analysis is performed for the masonry structure strengthened by bandaging using fiber-cement layers on both sides of the walls. The analysis use sandwich structuresanalogy and the model can be seen in Fig. 9.

The analysis is elastic-linear-time-history and utilizes commercial software. The strengthened masonry brick-walls consist of 3 layers: 2cm fiber-cement, brick-wall, and2cm fiber-cement. It must be emphasized that in an elastic design, the damages by previous earthquakes cannot be taken into consideration because all the analysis for different levels of earthquake intensity is started from the intact building condition. In non-linear analysis, previous damages can be taken into consideration, meaning the analysis can be sequential

5.1. Global Analysis of Masonry House with TRC

In TRC, fiber-cement layer is a composite of mortar and one layer of fiber-mesh. The analysis results below are for 5 different meshes used in each analysis. First, maximum tensile stress occurred in brick-wall can be seen in Fig.10. Maximum tensile stresses due to JMA Kobe 100% are almost 0.4kgf/cm^2, while stresses due to JMA Kobe 2g are around 0.75kgf/cm^2. Since the maximum allowable tensile stress of brick-wall is 3.5kgf/cm^2, all the results are still below the limit. This indicates the strength of the brick-walls can withstand JMA Kobe 100% and JMA Kobe 2g earthquakes after bandaged with fiber-cement layer on both sides of the wall.

Second, as a sandwich structure, there are two layers of fiber-cement which is at the inner and outer side of the brick-wall. Overall, those layers give similar maximum tensile stress results as shown in Fig. 11 and Fig. 12. Even though the outer fiber-cement layer seems to have slightly higher tensile stress than inner fiber-cement layer due to

JMA Kobe 100%, the results are still under the maximum allowable tensile stress (σ_{cc}) calculated in Table 1.Thus, from the analysis for 5 different meshes, fiber-cement layer on both sides of the brick-wall did not crack after shaken by JMA Kobe 100%. On the other hand, when using fiber glass mesh and PBO, maximum tensile stresses occurred due to JMA Kobe 2g surpass the σ_{cc}.

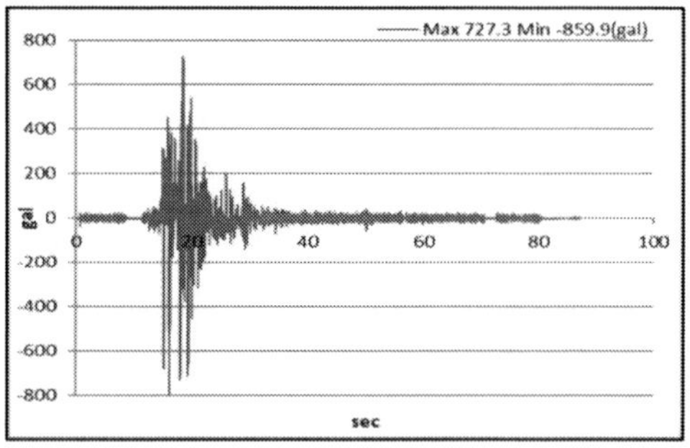

(a)

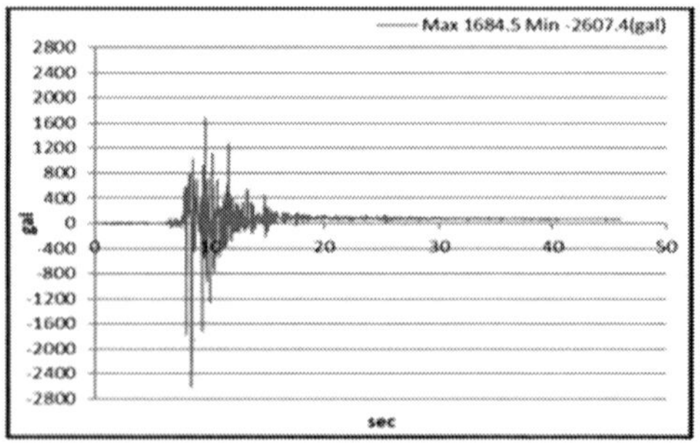

(b)

Fig. 8 Input motion of JMA Kobe NS 100% (a) and JMA Kobe 2g (b)

Fig. 9 Model 3D for analysis [14] [15]

The maximum allowable tensile stresses are about 40kgf/cm^2, while σ_{cc} for both fiber glass mesh and PBO are 36.34kgf/cm^2 and 24.13kgf/cm^2 respectively as mentioned in Table 1. Tensile stress of PBO is the highest among the other fibers, namely 3,356 MPa. However, since the volume fraction is very small (0.14%), the σ_{cc} as fiber-cement layer is the lowest. For this reason, the fiber-cement layer might crack in certain areas but will not cause collapse of the overall structure. As for three other fibers used in the analysis, namely basalt, carbon, and aragrid, all the σ_{cc} are higher as can be seen in Table 1. Therefore, those fiber-cement layers are stronger and can withstand JMA Kobe 2g.

Third, to give better clarification of the analysis results, Fig. 13 to Fig. 15 represent one of the tensile stress patterns of both brick-wall and fiber-cement layer for fiber glass mesh. The patterns are taken at certain time when maximum tensile stresses occurred in the structure. For easy reading, for low and high tensile stress that occurred in the structure, color indicators are plotted on the right side; bottom line (pink color) meaning 0kgf/cm^2; upper part indicates the maximum allowable tensile stress (blue color). Therefore, if the color in the structure is blue, the stress exceeds the limit, hence indicating possibility of crack in the structure.

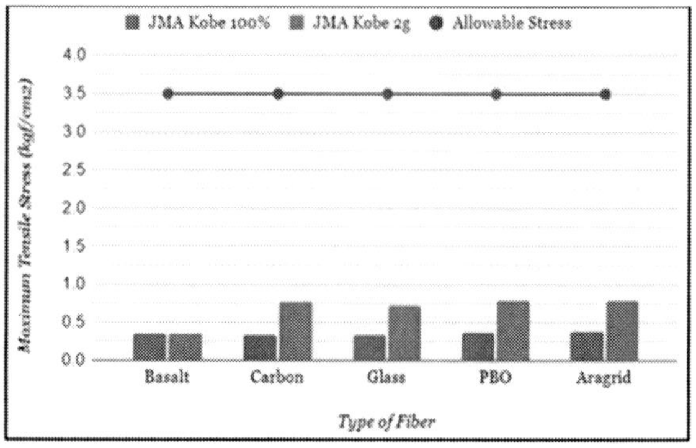

Fig. 10. The maximum tensile stress of brick-wall

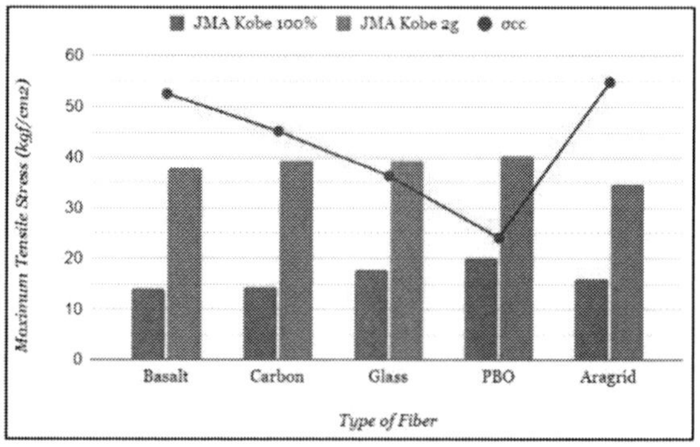

Fig. 11. The maximum tensile stress of the outer fiber-cement layer

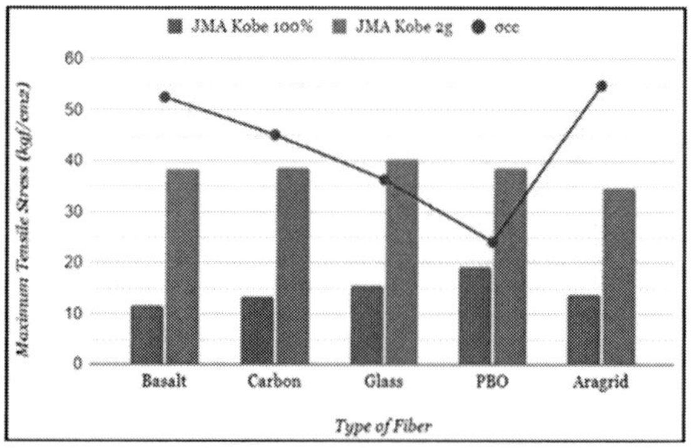

Fig. 12 Maximum tensile stress of the inner fiber-cement layer

https://doi.org/10.1051/e3sconf/202133105002

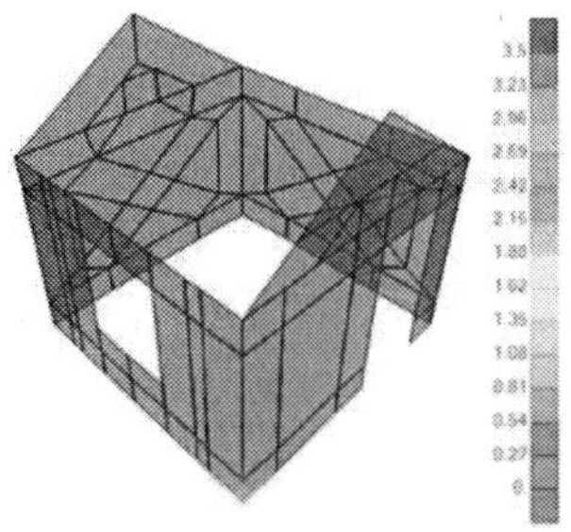

(a)

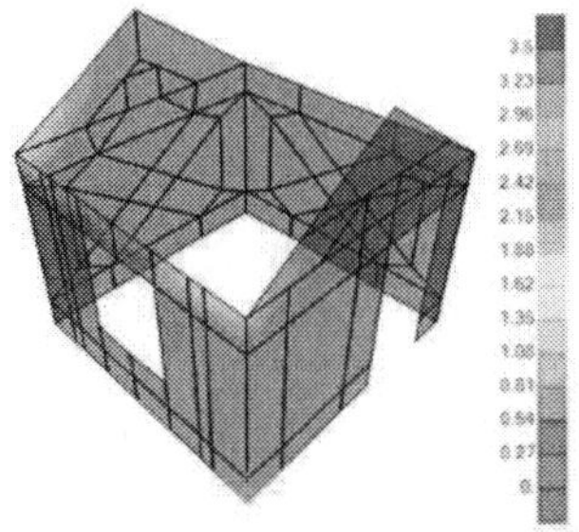

(b)

Fig. 13. Maximum tensile stress pattern (kgf/cm^2) of brick-walls due to JMA Kobe 100% (a) and JMA Kobe 2g (b) using Fiber Glass Mesh

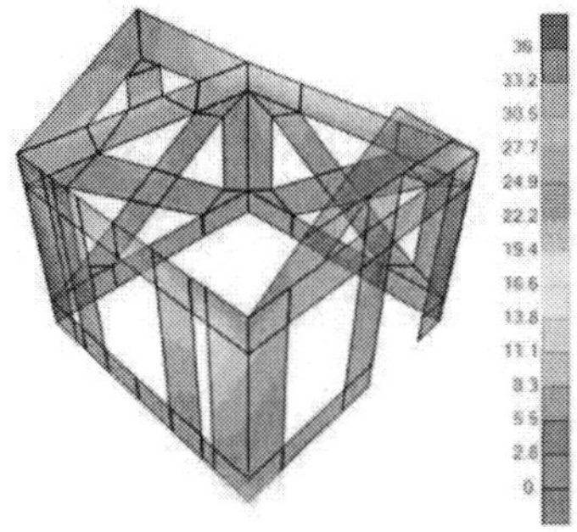

(a)

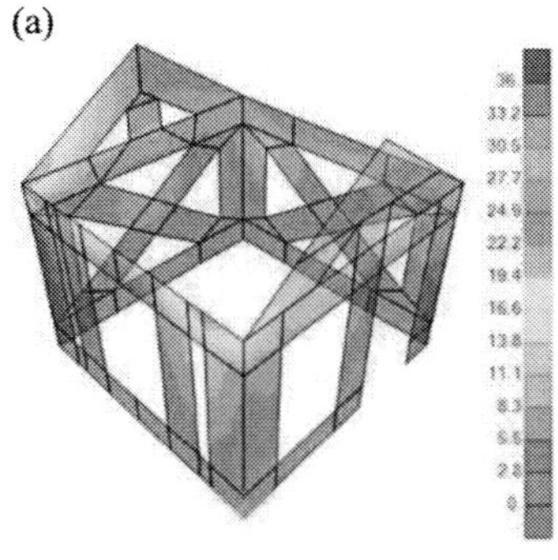

(b)

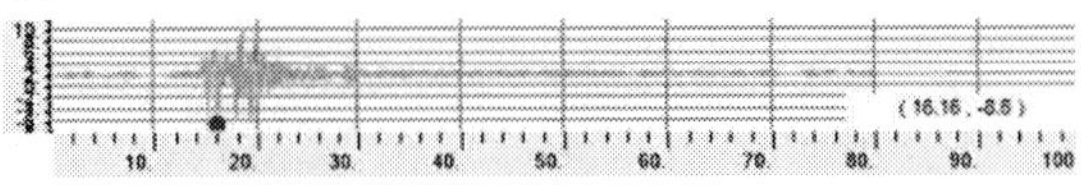

Fig. 14. Maximum tensile stress pattern (kgf/cm^2) at 16.16s ofinner (a) and outer (b) fiber-cement layers using Fiber Glass Mesh due to JMA Kobe 100%

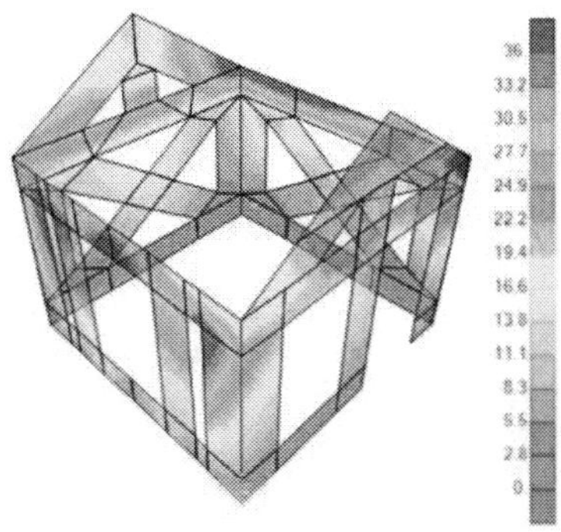

(a)

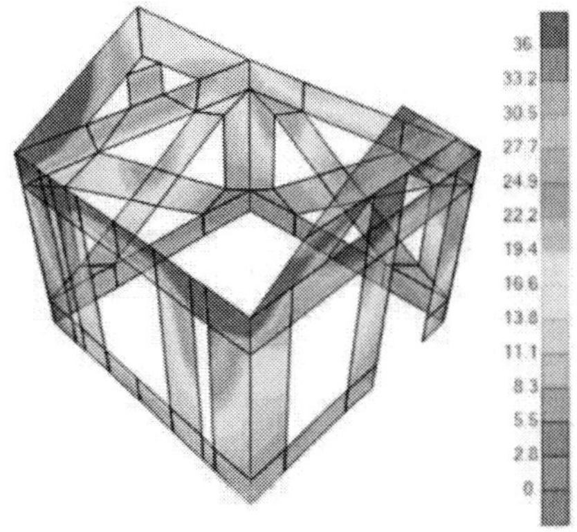

(b)

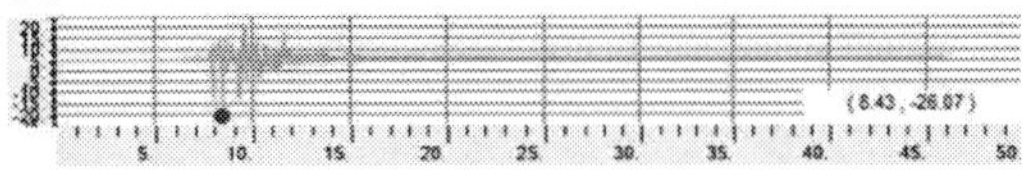

Fig. 15. Maximum tensile stress pattern (kgf/cm^2) at 8.43s of inner (a) and outer (b) fiber-cement layers using Fiber Glass Mesh dueto JMA Kobe 2g

5.2. Global Analysis of Masonry House with FRC

The fiber-cement layer is a composite of mortar and discontinuous fibers. Fig. 16 and Fig. 17 show the maximum tensile stress in fiber-cement due to JMA Kobe 100% (0.876g) and JMA Kobe 2g for five different discontinuous fibers as shown in Table 2. Since the volume fraction of fibers are relatively small (below 1%), no matter what fiber is used, the matrix (mortar) tensile strength plays an important role as a limit where the first crack in mortar occurs [2]. All tensile stresses are below the tensile stress at first cracking of the matrix, indicating there is no crack in any FRC bandage (see Fig. 16 and Fig. 17). The masonry house bandaged with FRC can survive when shaken by JMA Kobe 100% (0.876g).

When progressive cracks occur, the load will be transferred from matrix to the embedded fibers, and finally when the matrix is fully cracked, the load is mainly sustained by the fiber [2, 13, 16]. For FRC with high tensile strength and high modulus elasticity (E-glass fiber, basalt fiber, and steel fiber), the maximum tensile stress did not exceed the maximum post cracking stress (σpc), meaning when shaken by JMA Kobe 2g, substantial cracks may start to develop in the FRC walls, but the fibers are intact. For FRC with low tensile strength and low modulus

 https://doi.org/10.1051/e3sconf/202133105002

elasticity (PVA fiber and PP fiber), it seems that some of fibers in fiber-cement might be failed at several places, but not for the overall structure since Fig. 16 and Fig. 17 only show the maximum value that can occur for the overall structure at one-time (the highest stress).

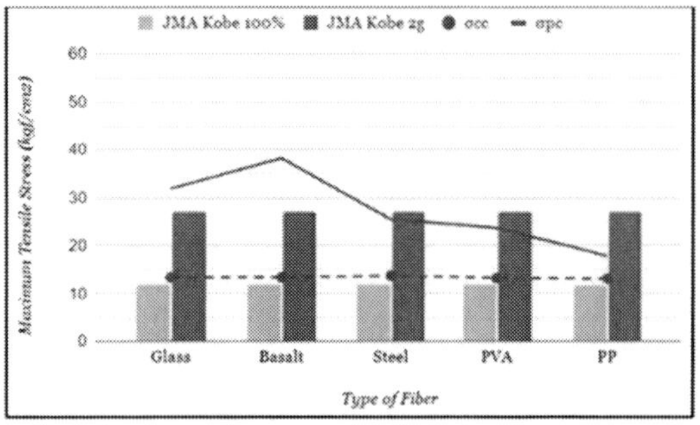

Fig. 16 Maximum tensile stress in fiber-cement of masonry wallbandaged with FRC due to JMA Kobe 100% (0.876g)

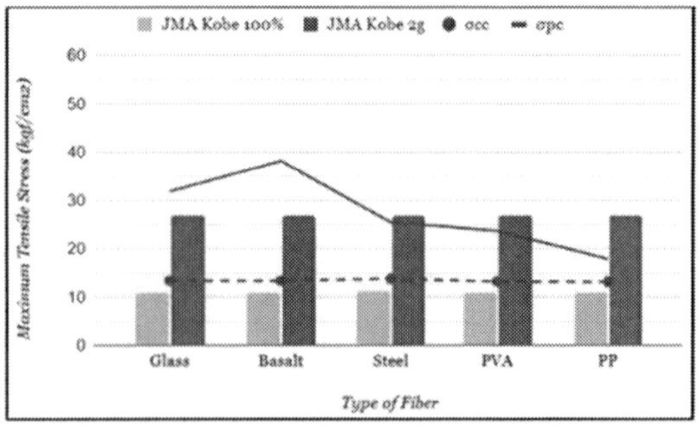

Fig. 17 Maximum tensile stress in fiber-cement of masonry wall bandaged with FRC due to JMA Kobe 2

The maximum tensile stress of brick-wall due to JMA Kobe 100% and JMA Kobe 2g is below the limit ($3.5 kgf/cm^2$) for all masonry houses bandaged with FRC with 5 different types of fiber in Fig. 18. Since all the brick-wall stresses are below the limit, it is a clear indication that fiber-cement layers (FRC) can be used similar with ferrocement layers in strengthening existing buildings as well as retrofitting earthquake damaged buildings

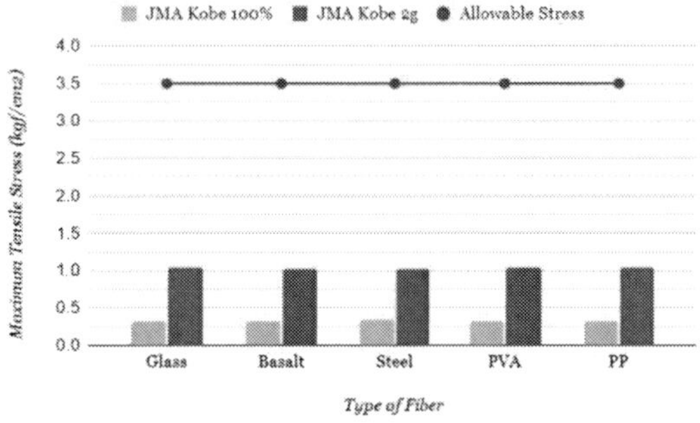

Fig. 18 Maximum tensile stress in fiber-cement of masonry wall bandaged with FRC due to JMA Kobe 2

Fig. 19 to Fig. 21 show the tensile stress patterns of FRC with E-glass fiber to give an idea what can happen in FRC layers when shaken by earthquakes. The stress patterns for basalt fiber and steel fiber are similar with that of E-glass fiber. Meanwhile, Fig. 23 to Fig. 25 show the tensile stress pattern of FRC with PP fiber that represent the low-end synthetic fibers. To improve the performance of FRC with low-end fiber, the volume fraction of fiber can be increased around 1.5%, however, it might not be necessary since earthquake with 2g magnitude most probably will not occur in the lifetime of common people's housing.

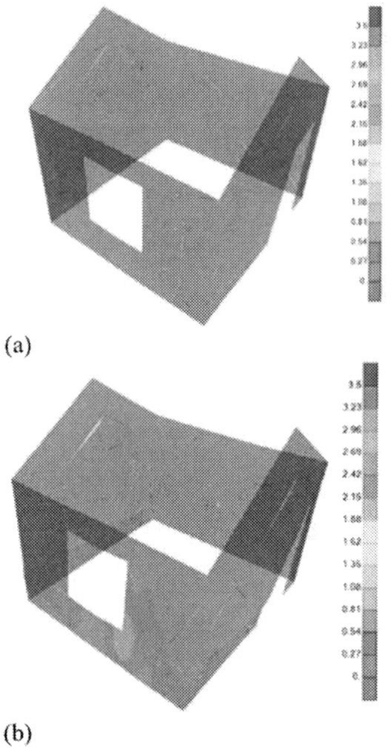

Fig. 1 Maximum tensile stress pattern (kgf/cm2) of brick-walls due to JMA Kobe 100% (a) and JMA Kobe 2g (b) using E-Glass Fiber

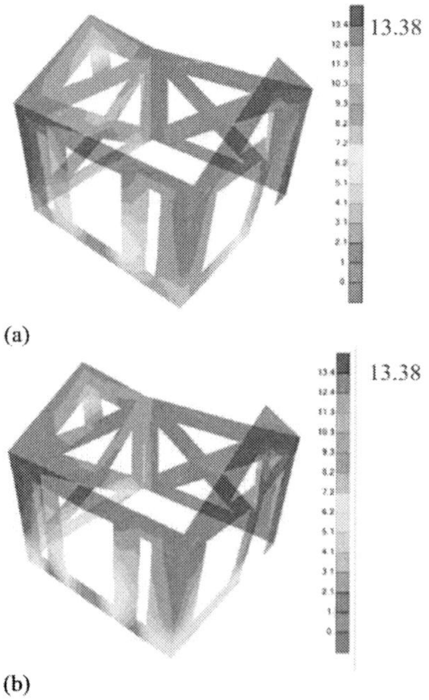

Fig. 2 Maximum tensile stress pattern (kgf/cm^2) of inner (a) and outer (b) fiber-cement layers using E-Glass Fiber due to JMA Kobe 100%; the tensile stresses are below the first cracking of matrix, σ_{cc}

 https://doi.org/10.1051/e3sconf/202133105002

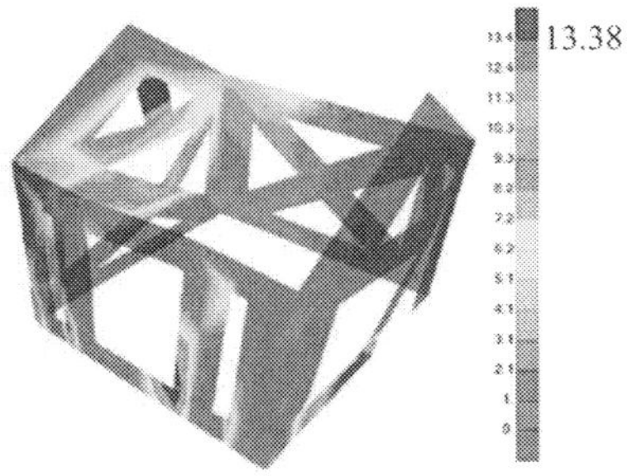

(a)

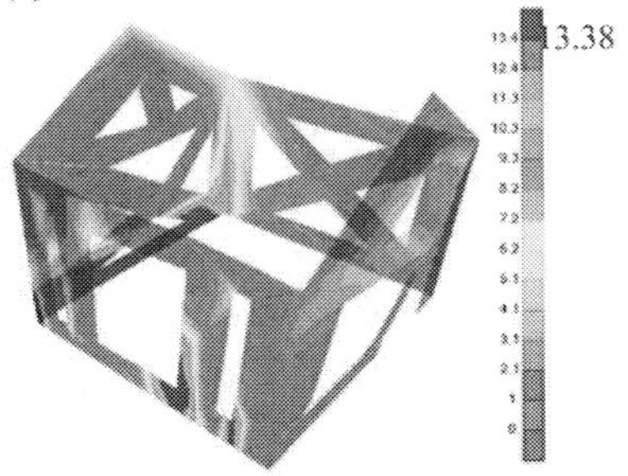

(b)

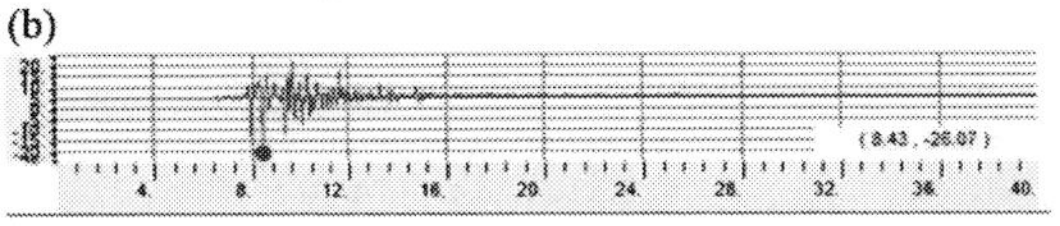

Fig. 3 Tensile stress pattern (kgf/cm^2) at 8.43s of inner (a) and outer (b) fiber-cement layers using E-Glass Fiber due to JMA Kobe 2g.

Blue color (■) indicates tensile stres4 s exceeded the tensile stress at first cracking of matrix, meaning substantial cracks start to develop in walls; however, the fiber did not fail because the stress still below the maximum post-cracking stress 31.86 kgf/cm^2, as can be seen in Fig. 4.

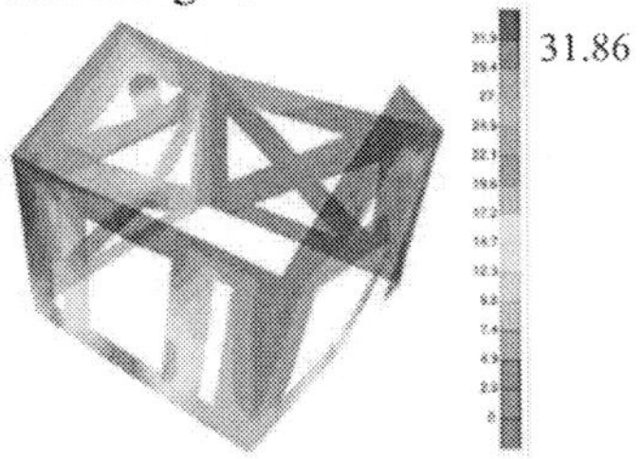

(a)

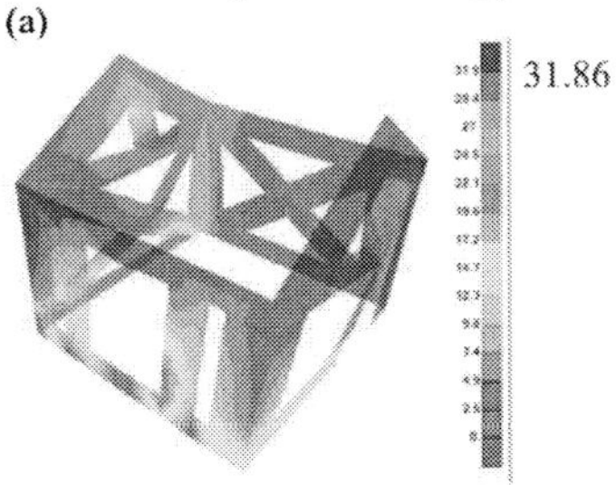

(b)

Fig. 4 Maximum tensile stress pattern (kgf/cm^2) of inner (a) and outer (b) fiber-cement layers using E-Glass Fiber due to JMA Kobe 2g; the tensile stresses are below the post cracking of matrix, σ_{pc}

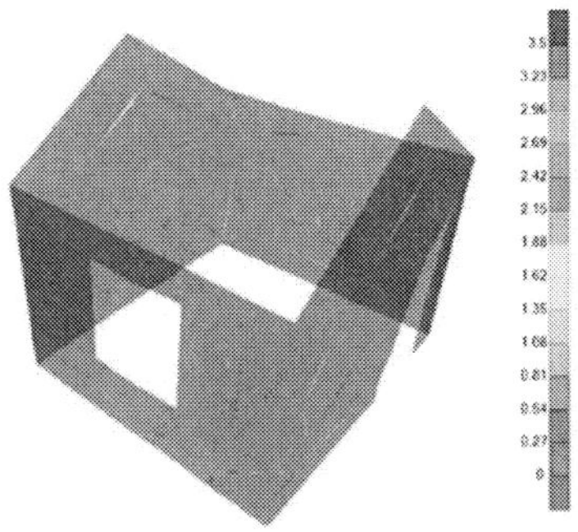

(a)

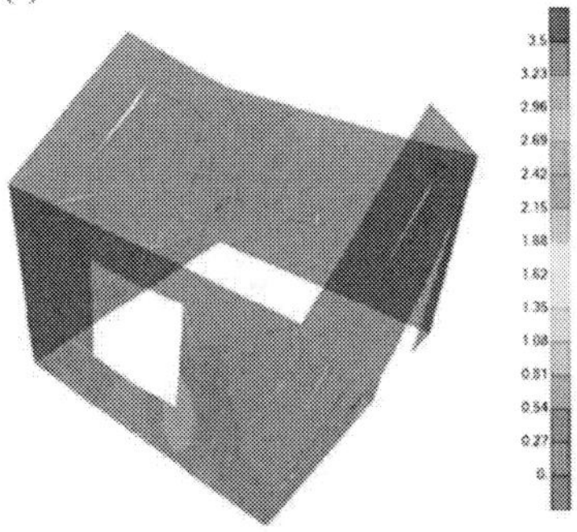

(b)

Fig. 5 Maximum tensile stress pattern (kgf/cm^2) of brick-walls due to JMA Kobe 100% (a) and JMA Kobe 2g (b) using PP Fiber

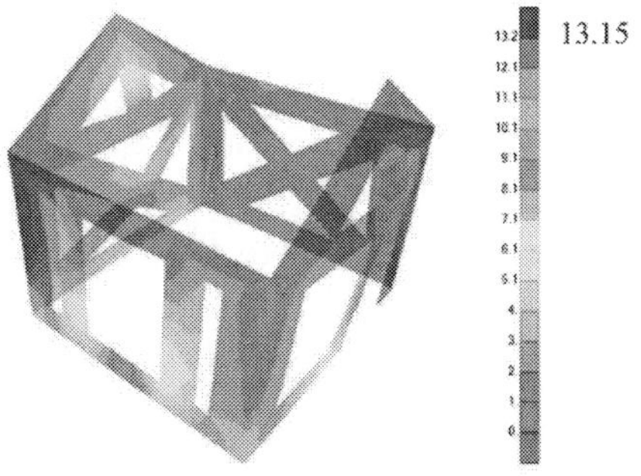

(a)

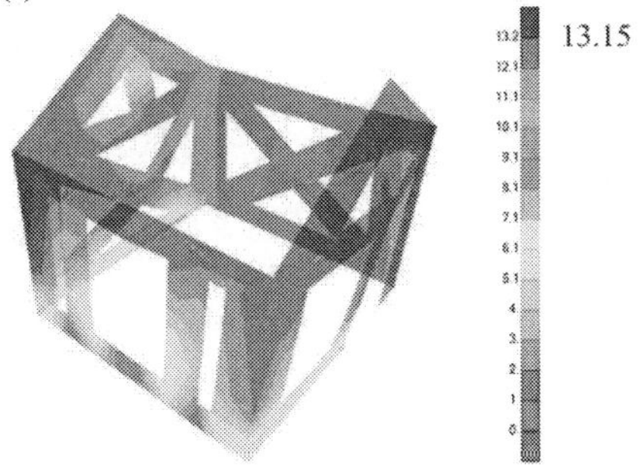

(b)

Fig. 6 Maximum tensile stress pattern (kgf/cm^2) of inner (a) and outer (b) fiber-cement layers using PP Fiber due to JMA Kobe 100%; the tensile **stresses** are below the first cracking of matrix, σ_{cc}

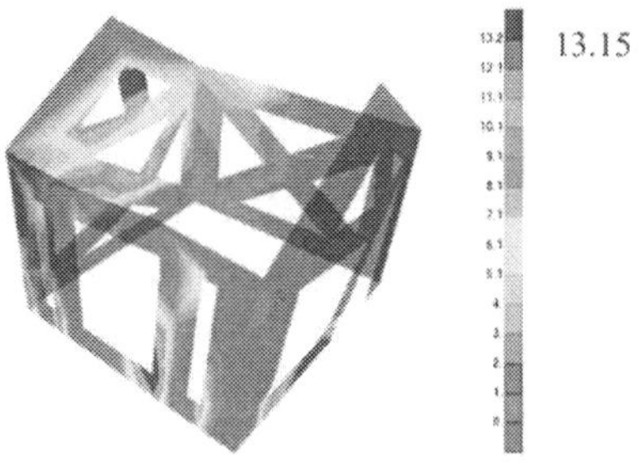

(a)

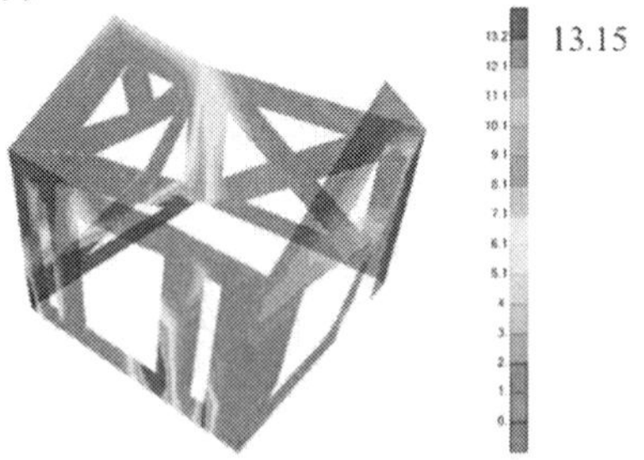

(b)

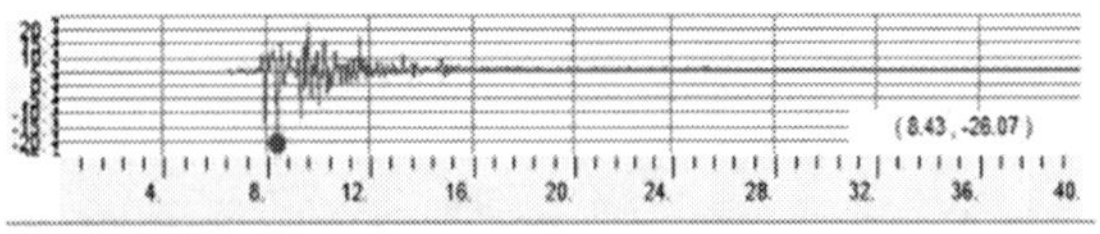

Fig. 7 Tensile stress pattern (kgf/cm^2) at 8.43s of inner (a) and outer (b) fiber-cement layers using PP Fiber due to JMA Kobe 2g

Blue color (▬) indicates tensile stress exceeded the tensile stress at first cracking of matrix, meaning substantial cracks start to develop in walls; some of PP fibers might be pulled out simultaneously since the stress exceeded the post-cracking stress 17.84 kgf/cm^2 as can be seen in Fig. 8

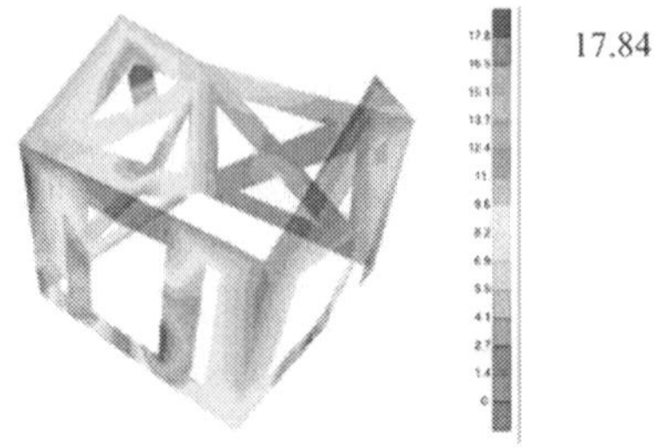

(a)

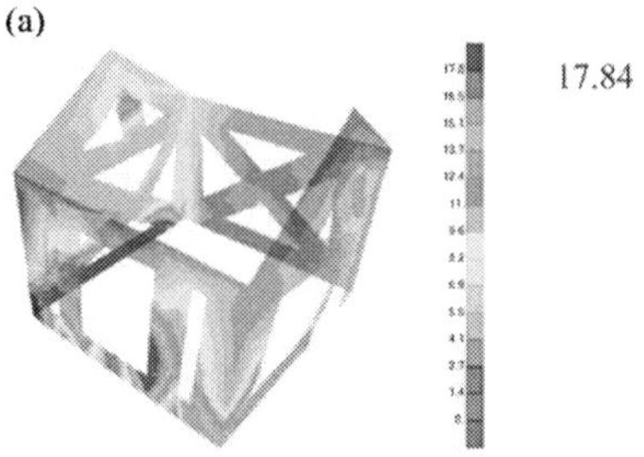

(b)

Fig. 8 Maximum tensile stress pattern (kgf/cm^2) of inner (a) and outer (b) fiber-cement layers using PP Fiber due to JMA Kobe 2g

6 Cementitious Matrix Mix

The mortar for TRC is 1cement:4sand. However, for FRC, it needs extra handling when adding the fiber to the mortar mix. The mortar should be produced first to its desired consistency, then add the fibers gradually and make sure they disperse well. The addition of fibers tends to decrease fluidity and workability; therefore, it may be necessary to hold part of the water and only add them in steps during fiber addition. After all fibers are added, the mortar mix should be mixed for 2-3 minutes. The mix of dry cement or dry sand with fibers should be avoided because it can be damaging the fibers. The fibers tend to break under the impact of mixer and the pounding of aggregates.

Problems during mixing can occur if the volume fraction of fibers used is high. The fibers cannot disperse well and forming as solid ball within the mixture. The fibers are segregating from the mix and lead to balling. If the mortar mixing procedure is well followed, the problem should not occur because the volume fraction of fibers used is small, only 0.5%-1%.

7 Conclusion

The purpose of the analysis is not to simulate the actual behavior, but to get reliable information that there is a correlation between the observed damages and the results of the analysis. The correlation is not perfect but is good enough to get an idea on how to build appropriate non-engineered constructions that can withstand earthquakes. Based on the authors' judgement, the analysis results provide justification for constructing earthquake resistant masonry houses as well as retrofitting existing masonry houses bandaged with TRC and FRC.

The type of fibers, the dimensions and shape of fibers (the yarns, yarn spacing, and direction), the presence of yarn coating or impregnation and weaving technique affect the matrix-fiber interaction and influence a specific behavior and properties of TRC system. Therefore, the fiber-mesh should be chosen carefully. In Indonesia, what fibers to be used are not yet decided. However, whatever fiber available with material properties similar to the five fibers analyzed, can safely be used. Another method to verify the results is a full-scale shaking table test which hopefully can be conducted in the near future.

References:

1. T. Boen, H. Imai, F. Ismail, T. Hanazato and Lenny, "Brief Report of Shaking Table Test on Masonry Building Strengthened with Ferrocement Layers," *Journal of Disaster Research,* **10**, 551-557 (2015)
2. A. E. Naaman, Fiber Reinforced Cement and Concrete Composite, Sarasota: Techno Press 3000 (2018)

3. A. E. Naaman, Ferrocement & Laminated Cementitious Composites, Michigan: Techno Press 3000 (2000)
4. H. Imai and A. Nakatani, "Seismic Reinforcement of Masonry Construction Experiment: Shaking Table Experiments on Full Scale Masonry Walls using Bricks made in Padang, Indonesia," National research Institute for Earth science and Disaster prevention (NIED) & Mie University, Tsukuba, Japan (2012)
5. A. E. Naaman, "Evolution in Ferrocement and Thin Reinforced Cementitious Composites," *Arab Journal Science Engineering,* 421-441, (2012)
6. ACI Committee 549.6R-20, Guide to Design and Construction of Externally Bonded Fabric-Reinforced Cementitious Matrix (FRCM) and Steel-Reinforced Grout (SRG) Systems for Repair and Strengthening Masonry Structures, Farmington Hills, MI: American Concrete Institute (2020)
7. G. de Felice, . T. D'Antino, S. de Santis, P. Meriggi and F. Roscini, "Lessons Learned on the Tensile and Bond Behavior of Fabric Reinforced Cementitious Matrix (FRCM) Composites," *Frontiers in Built Environment,* **6**, 5, 1-15 (2020)
8. T. Triantafillou, Textile Fibre Composites in Civil Engineering, Elsevier (2016)
9. M. Aziz, P. Paramasivam and S. Lee, "Natural Fibre Reinforces Concretes in Low-Cost Housing Construction," *Journal of Ferrocement,* **17**, 3, 231-240 (1987)
10. Wikipedia, "Basalt Fiber," 29 January 2021. [Online]. Available: https://en.wikipedia.org/wiki/Basalt_fiber. [Accessed 2 May 2021].
11. U. N. Gandhi, S. Goris, T. A. Osswald and Y.-Y. Song, Discontinuous Fiber-Reinforced Composites - Fundamentals and Applications, Munich: Carl Hanser Verlag (2020)
12. Cambridge University Pers, "https://in.bgu.ac.il/," 1996. [Online]. Available: https://in.bgu.ac.il/engn/mater/Documents/LaboratoryBriefings/4/Materials%20Science%20and%20Engineering%20introduction%20Chapter%2015%20Composites%207th%20ed.pdf. [Accessed 5 May 2021].
13. A. E. Naaman, "High Performance Fiber Reinforced Cement Composites: Classifications and Applications," in *CBM-CI International Workshop*, Karachi, Pakistan, November 9-11, (2007)
14. Software, "ETABS v.19 Computers & Structures Inc.," Walnut Creek, California, United States of America (2021)
15. Software, "SAP2000 v.23 Computers & Structures Inc.," Walnut Creek, California, United States of America (2021)
16. M. Leone, M. A. Aiello, A. Balsamo, F. G. Carozzi, F. Ceroni, M. Corradi, M. Gams, E. Garbin, N. Gattesco, P. Krajewski, C. Mazzotti, D. Oliveira, C. Papanicolaou, G. Ranocchiai, F. Roscini and D. Saenger, "Glass fabric reinforced cementitious matrix: Tensile properties and bond performance on masonry substrate," *Composites Part B - Elsevier,* 196-214 (2017)
17. P. K. V. R. Padalu, Y. Singha and S. Das, "Tensile properties of wire and fibre reinforced cementitious matrix composites for strengthening of masonry," *Structures - Elsevier,* **23**,164-179 (2020)

Metal cracks detection based on circular patch microstrip antenna

Hanalde Andre[1*], *Muhammad Riski* Albert[1], and *Uyung Gatot S.* Dinata[2]

[1]Electrical Engineering Department, Andalas University, Indonesia
[2]Mechanical Engineering Department, Andalas University, Indonesia

Abstract. Cracks in metal can be produced by many factors, such as external loads, physical processes, and chemical processes, such as alkali and corrosion. Structure health monitoring (SHM) is very important in maintaining the reliability of a building. Considering that the ultimate goal of a building health monitoring system is to provide sensory information that can facilitate decision making regarding the feasibility of components. Microstrip antennas have been shown to be able to detect cracking in metals according to their characteristics. In this paper will discuss the capability of a microstrip antenna with a circular patch having dual frequency operation to detect crack in metal.

1 Introduction

Metal is widely used in the construction of infrastructure as a framework to withstand the load of a building. Its role is very important in maintaining the reliability of a building. Structure health monitoring (SHM) refers to the process of implementing damage detection and characterization strategies for engineering structures. Damage is defined as a change in the material or geometric properties of a structural system, including changes in system connectivity, that affect system performance [1].

SHM is needed to ensure the robustness of a building. its application is expected to reduce losses caused by the failure of a building either due to age or disasters that occur. especially for Indonesia, which is in a disaster-prone area. Cracks can be defined as unintentional discontinuities in a structural material. In general, cracks are the result of material failure. Cracks can be produced by many factors, such as external loads, physical processes, and chemical processes, such as alkali and corrosion [2-4].

Considering that the ultimate goal of a building health monitoring system is to provide sensory information that can facilitate decision making regarding the feasibility of components, it is necessary to develop new crack sensors that can not only detect cracks but also provide quantitative information about cracks. Microstrip antenna sensor that can detect crack length and propagation with its sub-millimetre resolution [5-6]. This sensor has many advantages such as high resolution, small size, light weight, low cost, and makes use of advanced microwave detection technology to facilitate wireless detection and signal processing.

Microstrip antennas have been widely developed as sensors to detect cracks in metal. Changes in the structure of the metal will correlate with the frequency shift in the antenna characteristics [7-9]. In this article, we will discuss a circular patch microstrip antenna with an inset feed for detecting cracks in metal. frequency shift and minimum return loss are the parameters to be analyzed.

2 Methods

Microstrip antennas can be used to detect cracks that occur in metal. The characteristics of the microstrip antenna are highly correlated with the shape of the conductors in the patch and ground plane. the metal crack detection system uses a microstrip antenna using a ground plane which will be replaced with the metal to be tested.

2.1 Antenna Microstrip

Microstrip antenna generally has three parts, namely, patch, substrate and ground plane. The patch and ground plane are located at the top and bottom of the antenna, between which there is a substrate. The type of microstrip antenna used in this study has a patch with a circular shape. The substrate uses FR4 composite materials. The patch and ground plane materials used are copper conductors.

The microstrip antenna that will be used to detect cracks in metal has a circular patch shape. The feeder used is an inset feed type with an impedance of 50 ohms [10]. In Figure 1 can be seen the dimensions of the antenna. The patch antenna has a radius of 42.3 mm and the feeder has a length of 36.7 mm and a width of 3.1 mm where the

* Corresponding author: hanalde.andre@eng.unand.ac.id

feeder length of 24.78 cuts the patch antenna with a width of 1 mm on each side.

The ground plane on the opposite side of the patch and separated from the substrate has a thickness of 1.6 mm. The ground plane has dimensions of length and width of 93.2 mm and 92.2 mm, respectively

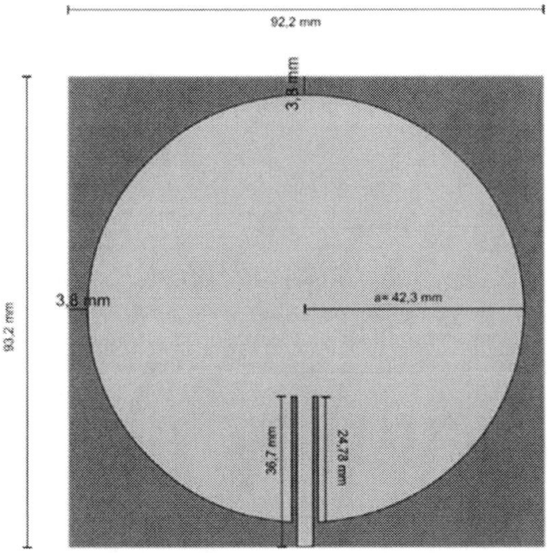

Fig. 1. Antenna Microstrip

The antenna has an operating frequency range according to its characteristics. The antenna that has been designed has two working frequency ranges. this value is obtained from the return loss. at low frequencies the antenna has a centre frequency of 1000 MHz and at high frequencies it has a value of 2803 MHz. Return loss simulation results can be seen in Fig. 2.

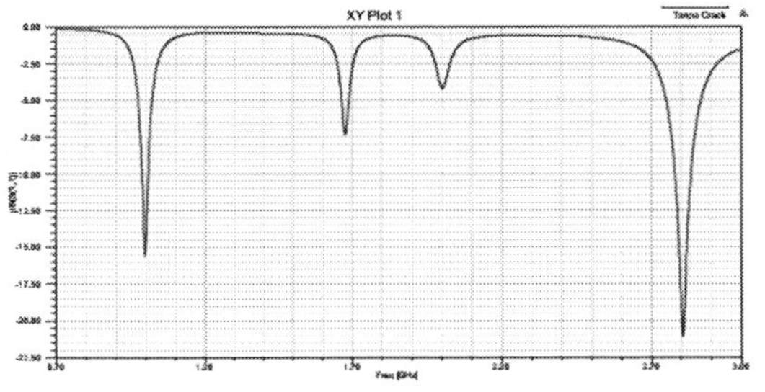

Fig. 2. Return Loss Antenna Simulation Result

2.2 Crack Detection

Cracks in metal were tested using a microstrip antenna by placing it on the ground plane. The metal tested has a width of 500 mm and a length of 93.2 mm according to the length of the antenna. Fig. 3 shows the configuration for detecting metal cracks using a microstrip antenna, where the crack position is located outside the patch.
In Fig. 4 you can see the configuration of metal crack detection, where the cracks are placed in the patch. the position of the crack at the centre of the circular patch. metal crack detection test using microstrip antenna based on simulation using the finite element method (FEM)

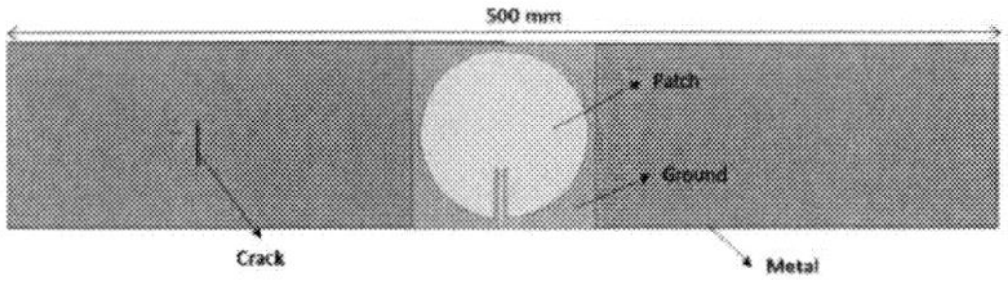

Fig. 3. Crack Detection Outside Patch

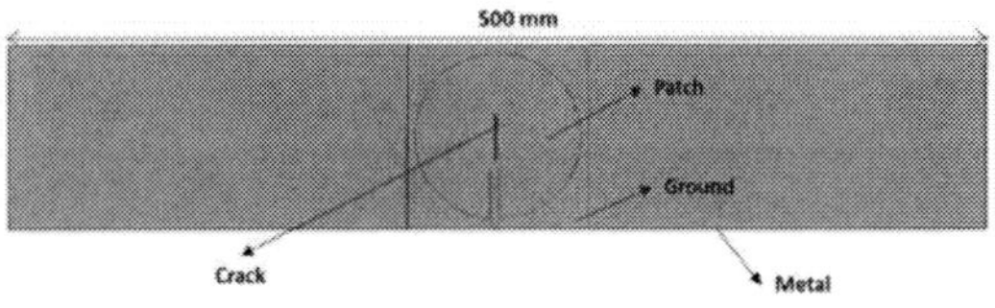

Fig. 4. Crack Detection Inside Patch

3 Results and Discussion

Cracks that occur in metal change the shape of the structure of the metal. in the test using a metal microstrip antenna placed on the ground plane of the antenna. Referring to the characteristics of the microstrip antenna, the change in structure will correlate with changes in the electrical property of the antenna.

3.1 Crack Detection Inside Patch

The test is carried out on metal cracks placed under the patch. Based on the simulation results, there is a shift in the centre frequency of the antenna operation to be greater than the operating centre frequency of the antenna without crack. in Fig. 5 can be seen the simulation results of the frequency shift caused by the crack that occurs in the patch.

The artificial crack in the test metal has a width of 1 mm with a length varying from 2 mm to 22 mm in increments of 4 mm. the results of the frequency shift of all cracks are positive from the centre frequency of the antenna without cracking for low and high frequencies. the frequency shift at high frequencies has a greater value than the frequency shift at low frequencies.

frequency shift at low frequencies based on the simulation results obtained insignificant changes in the value from 4 MHz to 6 MHz. The highest frequency shift was obtained for cracks with a length of 2 mm, 10 mm and 22 mm, while the lowest frequency shift occurred for cracks with a length of 14 mm. at high frequencies, the large frequency shift that occurs in the 2 mm crack length has the highest shift value of 19 MHz. and the lowest frequency shift is obtained at the crack length of 14 mm with a large shift of 6 MHz.
Centre frequency antenna value is obtained from the minimum return loss value. in Fig. 6 it can be seen the minimum return loss value for the antenna without cracks and the antenna on the crack test. Based on the simulation results, it is found that there is an increase in the minimum return loss value of the antenna when there is a crack in the test metal. The highest return loss value at low frequency occurs for cracks with a length of 14 mm with a return loss value of -13.4186 dB. at high frequency, the highest increase in return loss occurs in cracks with a length of 6 mm of -18.3199 dB.

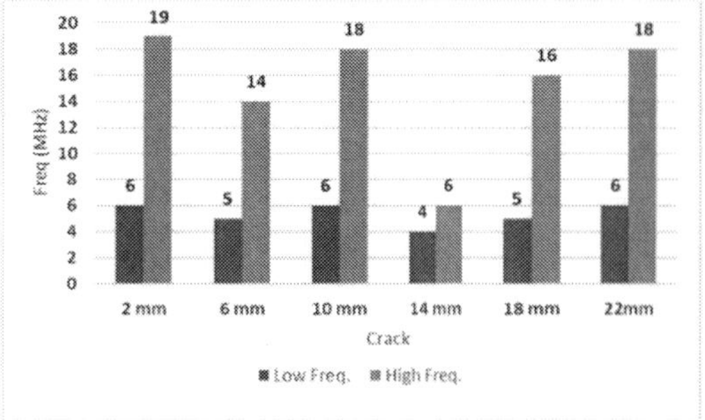

Fig. 5. Frequency Shifting Crack Detection Inside Patch

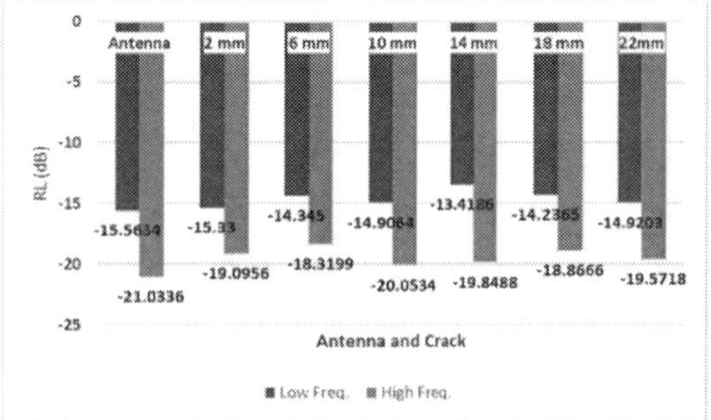

Fig.6. Minimum Return Loss Crack Detection Inside Patch

3.2 Crack Detection Outside Patch

Simulation results for metal cracks that are outside the antenna patch have a lower frequency shift than the cracks inside the patch. all crack test results for frequency shift are positive. The simulation results for the frequency shift for low and high frequencies can be seen in Fig. 7.

Frequency shifting at high frequencies has a greater value than at low frequencies. at high frequencies the frequency shift does not change much which has a value in the range of 8 MHz to 9 MHz. at low frequencies the highest shift value occurs in cracks with a length of 2 mm which have a frequency shift of 8 Mhz. The lowest frequency shift is obtained at the crack length of 10 mm with a shift value of 2 MHz.

The minimum return loss value obtained for the antenna used in the crack test on metal has a better value at low frequencies compared to the antenna without crack. The lowest minimum return loss value was obtained at the crack length of 10 mm with a value of -29,009 dB. this value is 86% lower when compared to the antenna without crack.

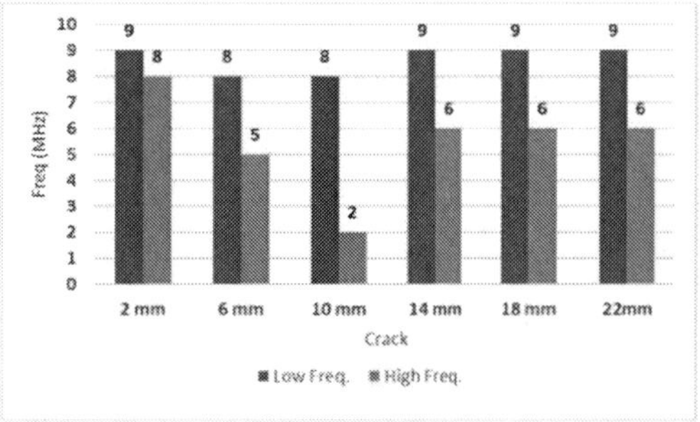

Fig. 7. Frequency Shifting Crack Detection Outside Patch

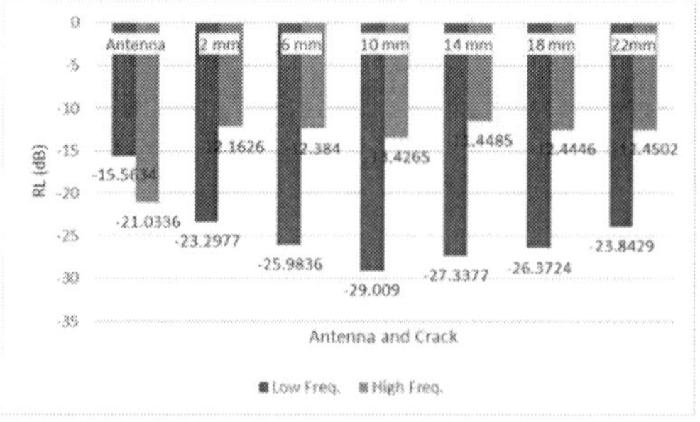

Fig. 8. Minimum Return Loss Crack Detection Outside Patch

4. Conclusions

Circular patch microstrip antenna with dual frequency at 1000 MHz and 2803 MHz has been simulated to detect cracks on metal. Based on simulation result crack position at patch antenna has better sensitivity than outside patch, especially on higher frequency. Crack with length 2 mm has maximum shifting frequency at higher frequency with value 19 MHz.

Publication of this article is supported by the Engineering Faculty of Andalas University publication grant 2021

References

1. P. Suprobo, A. Febry, Infrastructure Health Monitoring System (SHM) Development, a Necessity for Maintance and Investigation. in International Conference on Engineering and Technology Development (ICETD, 2013.
2. M. Liu, B. Li, H. Li, A crack monitoring method based on microstrip patch antenna. in 2015 Annual Reliability and Maintainability Symposium (RAMS), IEEE, (2015).
3. I. Mohammad, et al., Detecting crack orientation using patch antenna sensors. Measurement Science and Technology, **23**(1), 015102 (2011).
4. Z. Liu, et al., Crack monitoring method for an FRP-strengthened steel structure based on an antenna sensor, **17**(10), 2394 (2017).
5. P. Colombi, G. L. Fava, Sonzogni, Fatigue crack growth in CFRP-strengthened steel plates. Composits Part B: Engineering, **72**, 87-96 (2015.).
6. X. Yi, , et al., Battery-free slotted patch antenna sensor for wireless strain and crack monitoring. Smart Struct. Syst, **18**, 1217-1231 (2016).
7. M. Biswas, M. Sen, Design and development of rectangular patch antenna with superstrates for the application in portable wireless equipments and aircraft radome. Microwave and Optical Technology Letters, **56**(4), 883-893 (2014).
8. H. Andre, et al., Triple Band Circular Microstrip Antenna for Metallic Material Crack Sensing. 2018 International Conference on Applied Science and Technology (iCAST, 2018).
9. R. Fernandez, Firdaus, H. Andre, Design of wideband quadrature hybrid coupler with tapered

lines on the arms. 2018 International Conference on Applied Science and Technology (iCAST, 2018)

10. M Y. Ramesh, Kb, Design formula for inset fed microstrip patch antenna. Journal of Microwaves, Optoelectronics and Electromagnetic Applications (JMOe), **3**(3), 5-10 (2003)

Comparative study of SNI 1726:2012 and SNI 1726:2019 guidelines for response spectrum 2D method (study case: GKT II building of Bengkalis State Polytechnic)

Zev Al Jauhari[1*], *Addini* Khafifah Nur[1], *Ridho* Aidil Fitrah[2], and *Siska* Apriwelni[3]

[1] Civil Engineering Department, Bengkalis State Polytechnic, 28711 Sungai Alam, Bengkalis, Riau
[2] Civil Engineering Department, Andalas University, 25166 Limau Manis, Padang, West Sumatera
[3] Civil Engineering Department, Sumatera Institute of Technology, 35365 Way Huwi, South Lampung, Lampung

Abstract. A standard code for designing earthquake-resistant buildings in Indonesia has always improved and updated year by year. It starts 1983, SNI 1726:2002, SNI 1726:2012, and the latest one SNI 1726:2019 Standard Codes. There is a procedure scope change for designing seismic load to become wider due to the adoption of a new standard, SNI 1726:2019. This research aims to compare the building design by using the response spectrum 2D method of SNI 1726:2012 and 2019 to determine the safety of existing buildings that use the old code. The structure of the building is modelled as a classroom and office building with a height of 15.50 m in Bengkalis State Polytechnic, Riau Province, namely Gedung Kuliah Terpadu (GKT) II Building. The reviewed structure behaviors are internal forces (bending moment, shear, and axial forces) on the structural elements (beams and columns). The result is directly proportional to rising in S_{DS} and S_{D1} by 3.60% and 19.02%, respectively. By this rising, it affects the response spectra graph that has increased in SNI 1726:2019. The results prove that the change of design by using SNI 1726:2012 to 2019 has an effect, even though the calculation of the building structure forces still uses the old codes, but it meets the safe limits using designing with the SNI 1726:2019 method. This paper also compares the design of the structural elements by using SNI 2847:2019 codes with the planning of the existing building before. Overall, the building structure can be classified as able to withstand seismic force based on the new standard method.

1 Introduction

The codes for the seismic-resistant building structure have a primary role to ensure building safety. The earthquake caused building collapse has exceeded the predictions of the building regulations in Indonesia in the last decade. The unpredicted large force earthquake caused damage to severe buildings occurred in Aceh, Padang, and Yogyakarta [1]. This was influenced by the PGA increase that is greater than the value in the older code, SNI 1726:2012 [2].

In 2019, a new code on procedures for seismic resistance planning for structures of building and non-building, SNI 1726:2019, has ratified by the Government of Indonesia as a revision of SNI 1726:2012 [3]. The release of a new code has changed the procedure scope for determining and designing seismic loads to become wider. The acceleration response spectrum at the building location is one of the important scopes of information needed for designing the seismic resistance structure building [4].

An increase in Peak Ground Acceleration (PGA) may cause by the changes in the seismic hazard map that consequently increases the seismic design load. An existing building planned with older earthquake codes requires evaluation as a consequence of this seismic design load increase. Higher load demand should be investigated subjected to the performance of those buildings to determine if they need any strengthening [5].

Seismic parameters must be considered in planning and designing development in Indonesia. The ground acceleration is one of the earthquake parameters that directly affect the seismic load design because seismic waves act on the building period. The building structure must be designed as a seismic-resistant building to secure the building. It must be planned following the applicable seismic regulations in Indonesia. By modeling and analyzing the building model impacted by earthquake loads; the damage of buildings can be minimized so that planners can better design the building to construct [6].

An earthquake is a result of the sudden release of stored energy in the Earth's crust that comes from a wave in a place and spreads in all directions [7]. In SNI 1726:2012, the PGA and response spectrum acceleration for 0.2 seconds (short periods) and 1 second periods with a probability of 10% exceeding in 50 years (500 year return period), and maybe 2% exceeding in 50 years (2500

* Corresponding author: zevaljauhari@polbeng.ac.id

year return period). While in the new code, the PGA and acceleration of the response spectrum for 0.2 seconds and 1 second periods with a probability of 7% exceeded in 75 years (1000 return period years), and a 2% probability is exceeded in 50 years (2500 year return period). The 251 faults were inputted as a PSHA (Probabilistic Seismic Hazard Analysis) of the source of seismicity [8].

With the changes in the standard codes, it is necessary to find out how much the earthquake response factor changes from the old standards that affect the value of the seismic load, the deviation magnitude between the floors, and the extent to which the load difference affects the calculation of structural dimensions [9]. For this reason, a comparative analysis was carried out between SNI 1726:2012 and SNI 1726:2019. Comparison of earthquake loads applied to a 3-story building model and the force analysis results from each of these standards is calculated to determine the feasibility of the profile dimensions and reinforcement requirements [10]. In this study, a 3-story classroom and office building in Bengkalis, Riau, Indonesia which was designed using SNI 1726:2012 and SNI 2847:2013 as older Indonesian seismic code and concrete code, respectively, was chosen to be investigated.

2 Methodology and Modelling

2.1 Considered Building

The change in spectrum response in the GKT II Building of Bengkalis State Polytechnic that is still designed using SNI 1726:2012 and compared with SNI 1726:2019 was checked and evaluated in this study. After determining the spectrum response followed by the running model in SAP2000 with the dynamic method (2D response spectrum), it compares the values of the internal forces of columns and beams. Fig. 1 shows the typical structural plan view of the considered building. In general, the design and analysis were consist of modelling structure, defining material properties and frame section, then calculation and determination of loading and analysis model of the structure. Fig. 2 shows the 3D model of the considered building.

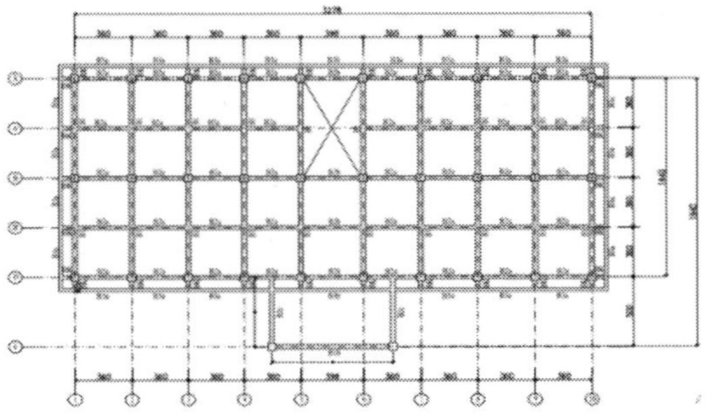

Fig. 1. Typical structural plan view of the GKT II Building.

Preliminary data such as soil investigation, as-built drawing, technical structure, and earthquake data were used to build the model. GKT II Building withstands structural loads based on SNI 2847- 2013. The building is a reinforced concrete structural system with a mini pile as the foundation. The compressive strength of concrete (f'_c) used is 25 MPa for all upper structure elements, there are, beam, slab, column, and pile. The rebar strength (f_y) for longitudinal and transversal rebars are 400 MPa and 240 MPa, respectively.

The dimension of the primary beam is typical for all structural models. Beam B1 300 mm x 600 mm is used for a 7 m span and Beam B2 250 mm x 400 mm for a 3.5 m span. The dimension of columns is 500 mm x 500 mm for all stories model. The thickness of the slab is typically 120 mm for all stories. The building is a reinforced concrete structure with a height of 3-stories building + roof (h = 15.50 m). The building was analyzed with a seismic load for soft soil (SE). The structure of the 3-story office building consists of a medium moment resisting frame system for Seismic Design Category C.

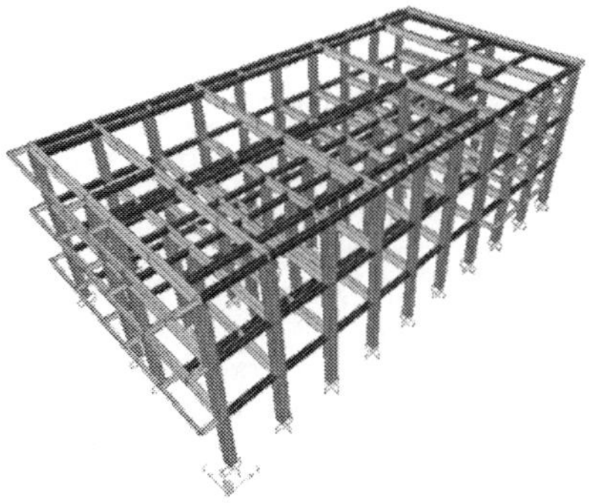

Fig. 2. 3D Model of the considered building.

2.2 Loads

There are two types of loading are used in this study, a) gravity load which includes dead and live loads, and b) lateral load (earthquake load).

2.2.1 Gravity Loads

Table 1. Type and value of loads.

Type of Load	Description	Qu (kg/m^2)
Dead Load	Slab (Reinforced Concrete)	288
Super Dead Load	Space + Ceiling + Floor Tiles + MEP	110.5
Super Dead Load for Roof	Aluminum Composite Panel + Frame	11.76
Live Load	Meeting Room, Laboratories,	400
	Offices	250
	Toilet	200
	Corridor	300
Live Load for Roof	For Slab and Water Tank	254

Gravity loads represent all tributary dead and live loads. The dead and live loads are based on PPIUG 1983 and PPURG 1987 that refer to SNI 2847:2013. The detail of gravity loads that work as distributed loads are shown in Table 1 and Fig. 3.

2.2.2 Lateral Load

This study has focused on the seismic load by using the response spectrum 2D method of SNI 1726:2012 and SNI 1726:2019 as the lateral load.

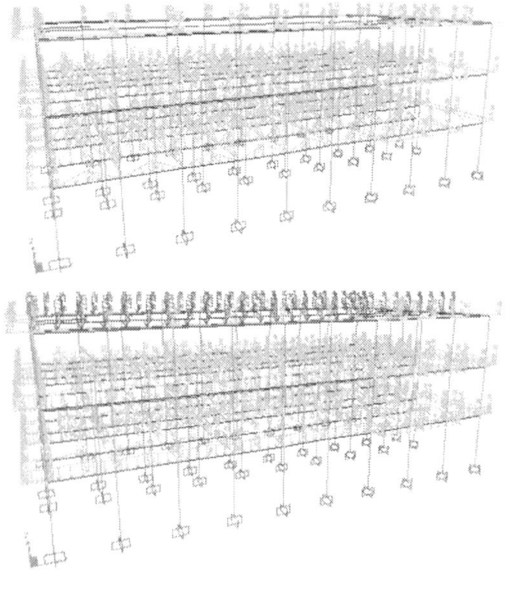

Fig. 3. Dead and live loads input on the considered building.

2.3 Load Combinations

Structural members should design to have strength for all sections at a minimum required strength calculated for the load and factored in combination according to the standards. The load combinations according to SNI 1726:2012 and 2019 are seen in Equations 1 – 6.

$$U = 1,4\ D \quad (1)$$

$$U = 1,2\ D + 1,6\ L \quad (2)$$

$$U = 1,2\ D + 1\ L + 1\ Ev + 1\ Eh \quad (3)$$

$$U = 0,9\ D - 1\ Ev + 1\ Eh \quad (4)$$

With the Ev (vertical earthquake) and Eh value (horizontal earthquake) values as follows:

$$Ev = 0,2\ S\ D \quad (5)$$

$$Eh = \rho\ Q_{EDS} \quad (6)$$

3 Results and discussions

The design spectrum response by using SNI 1726:2012 and SNI 1726:2019 will be used in the analysis comparison. To take into account the influence of internal forces on structural elements, the design response spectrum is required. The longest transverse side to be given evenly distributed load is selected to be part reviewed; it was shown in Fig. 4. Dynamic analysis for earthquake-resistant structures design is carried out. It is needed for a more accurate evaluation of the earthquake forces acting on the structure, also to determine the structural behavior due to the earthquake effects [11]. The method of spectrum response in SNI 1726:2012 and 2019 codes is used to obtain the spectrum response graphs. The data and values obtained are then inputted into SAP2000 to get the graphs of spectrum response.

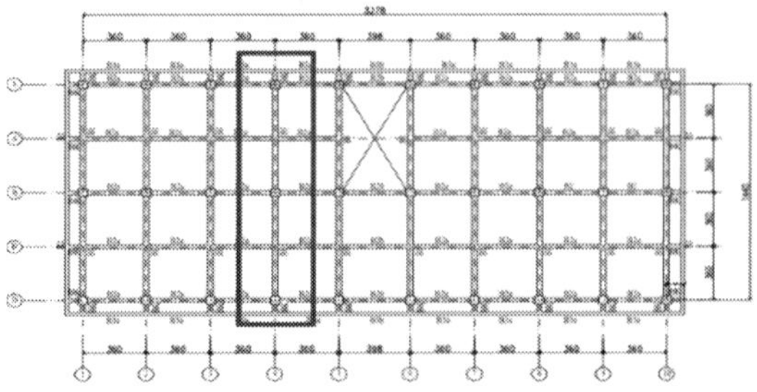

Fig. 4. Point of view on the structural building.

3.1 Response Spectrum Specified in SNI 1726:2012 and SNI 1726:2019 Codes

The steps of the design of spectrum response according to SNI 1726:2012 and SNI 1726:2019 in a way that has been adjusted to the study needs [11]:
- Look and observe the considered building location
- Determine the value of SPT and the type of soil classification
- Define the acceleration parameters of MCER (Maximum Considered Earthquake, Risk Targeted) for short periods of 0.2 seconds (S_S) and 1.0 seconds (S_1)
- Look for the amplification factor of the response parameters of the ground surface acceleration spectrum for short periods (F_a) and 1 second period (F_v).
- Define the acceleration spectrum response parameters (S_{MS} and S_{M1}) using this eq.:

$$S_{MS} = F_a \ x\ S_s \quad (7)$$

$$S_{M1} = F_v \ x\ S_1 \quad (8)$$

- Calculate the design acceleration spectrum response parameters (S_{DS} and S_{D1}) using Equations 9 and 10.

$$S_{DS} = 2/3\ x\ S_{MS} \quad (9)$$

$$S_{D1} = 2/3\ x\ S_{M1} \quad (10)$$

- Draw a response graph to images with the provisions that have been determined in SNI 1726:2012:
  a. For $T < T_0$, the S_a (design acceleration response spectrum) must be taken from Equation 11.

$$S_a = S_{DS}\ (0,4+0,6.\ (T/T_0)) \quad (11)$$

  b. For $T_0 \leq T \leq T_s$, the response spectrum of the design is $S_a = S_{DS}$.
  c. For $T > T_S$, the S_a is taken for Equation 12.

$$S_a = S_{D1}\ /\ T \quad (12)$$

- Draw a response of graph to images based on the provisions determined in SNI 1726:2019. Period provisions are as follows:
  a. For $T < T_0$, following Equation 11.
  b. For $T_0 \leq T \leq T_s$, the spectral response of the design acceleration is $S_a = S_{DS}$
  c. For $T_S < T \leq T_L$, the spectral response of the design acceleration is taken Equation 12.
  d. For $T > T_L$, the S_a is taken based on Equation 13.

$S_a = (S_{D1}.T_L) / T_2$ (13)

e. The T value is obtained by calculating the T_a (fundamental period of the structure) based on Equation 14.

$T_a = C_t \cdot h_n x$ (14)

h = the structure's height (m) above the base to the structure highest level.

f. The T_0 and T_S values can be obtained from Equations 15 and 16.

$T_0 = 0{,}2 \cdot S_{D1} / S_{DS}$ (15)

$T_S = S_{D1} / S_{DS}$ (16)

g. T_L values are obtained from SNI 1726:2019 in Fig. 20.

h. The GKT II building is categorized as IV in determining the risk category of buildings, and the earthquake priority factor is 1.50 in earthquake priority factors.

i. The seismic design categories determination is based on the acceleration response parameters category of seismic design in the 0.2 and 1 second periods

The spectrum response graphs in Fig. 5 are obtained from SNI 1726:2012 and 2019 methods.

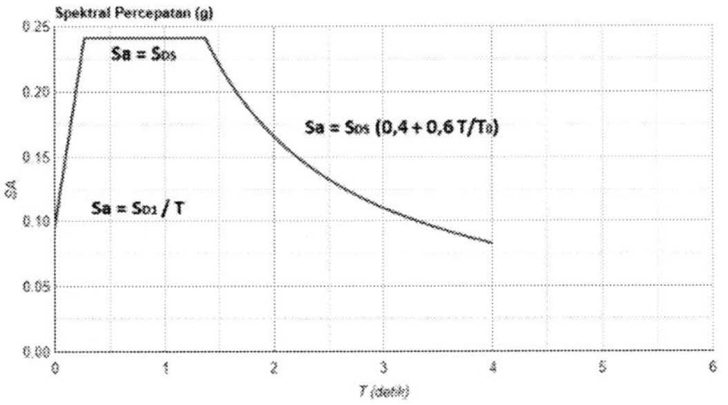

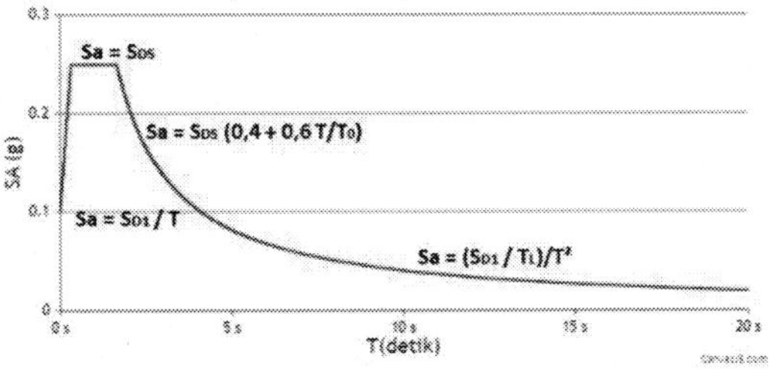

Fig. 5. Response spectrum design for Bengkalis Regency based on SNI 1726:2012 and SNI 1726:2019 (Soft Soil).

As a result of the increase in the S_{DS} value in SNI 1726:2019 which is used as a parameter for calculating vertical earthquakes, the calculation of earthquake loads in SNI 1726:2019 also increases, as seen in Table 2. The combination of loading is per the parameters contained in SNI 1726:2012 and SNI 1726:2019. Then calculated on the load combination due to vertical earthquakes to obtain 18 combinations of loading used. The combination of loading consists of dead load, live load, vertical earthquake load, and horizontal earthquake load that can affect the building. The following is a comparison of the loading combination on SNI 1726:2012 and 2019 is shown in Table 3.

Table 2. The parameter value of spectral response graph based on SNI 1726:2012 and 2019.

Parameters	SNI 1726:2012	SNI 1726:2019
S_{D1}	0,332	0,41
S_{DS}	0,241	0,25
T_0	0,275	0,33
T_S	1,377	1,64
$Sa = S_{DS} (0{,}4+0{,}6 \times T/T_0)$	0,376	0,342
$Sa = S_{DS}$	0,241	0,25
$Sa = S_{D1}/T$	0,623	0,769
T	0,533	0,533
T_L		20

Table 3. The load combinations on SNI 1726:2012 and 2019.

SNI 1726:2012	SNI 1726:2019	Percentage of difference (%)
Load combination of non- earthquake		
1.4 D	1.4 D	-
1.2 D + 1.6 L	1.2 D + 1.6 L	-
Load combination with earthquake		
1.2 D + 1 L + (E_h + E_v)	1.2 D + 1 L + (E_h + E_v)	-
1.2 D + 1 L + (1.3 Q_E + 0.2 S_{DS} D)	1.2 D + 1 L + (1.3 Q_E + 0.2 S_{DS} D)	-
0.9 D + E_h – E_v	0.9 D + E_h – E_v	-
0.9 D + 1.3 Q_E - 0.2 S_{DS} D	0.9 D + 1.3 Q_E - 0.2 S_{DS} D	-
Load combination impacted by vertical earthquake		
1.2482 D + 1 L + 1 E_h	1.25 D + 1 L + 1 E_h	0,144
0.8518 D + 1 E_h	0.85 D + 1 E_h	0,211

As seen in Table 3, entering the vertical earthquake load value can provide a different combination of the two standards. With an increase in S_{DS} by 5.52%, there is an increase in both load combinations of earthquake loading by 0.469% and 0.797%. So it affects the dead load on the building.

3.2 Structure Behaviors (Internal Forces)

With the static (dead and live) load and spectral response of SNI 1726:2012 and 2019 which have been defined in the computer software, SAP2000, the calculation and analysis can be done. Then, the values of internal forces M_u, V_u, and P_u, as shown in Fig. 6, are compared.

As can be seen in Table 4, the results of internal forces as a whole show that the new code is higher than the results of the old code in the method of spectrum response. The axial force and bending moment of columns in SNI 1726:2019 increased by 3.98% and 2.13%, respectively, from the old standard. There is also an increase in shear force values of columns about 1.52%.

The bending moment and shear force values of primary beam B1 are rising by 17.08% and 6.4% in SNI

1726:2019, respectively. While the secondary beams B2 have little rising than the B1, it is about 7.85% and 4.88% for bending moment and shear force, respectively, from the old standard. From the structural response calculation, the building structure can be categorized as still enough to resist the seismic forces based on the latest design method, even though there is an increase in the structure's internal forces. The building structure is strong enough, so it does not need to be given rehabilitation for the earthquake loads that will occur.

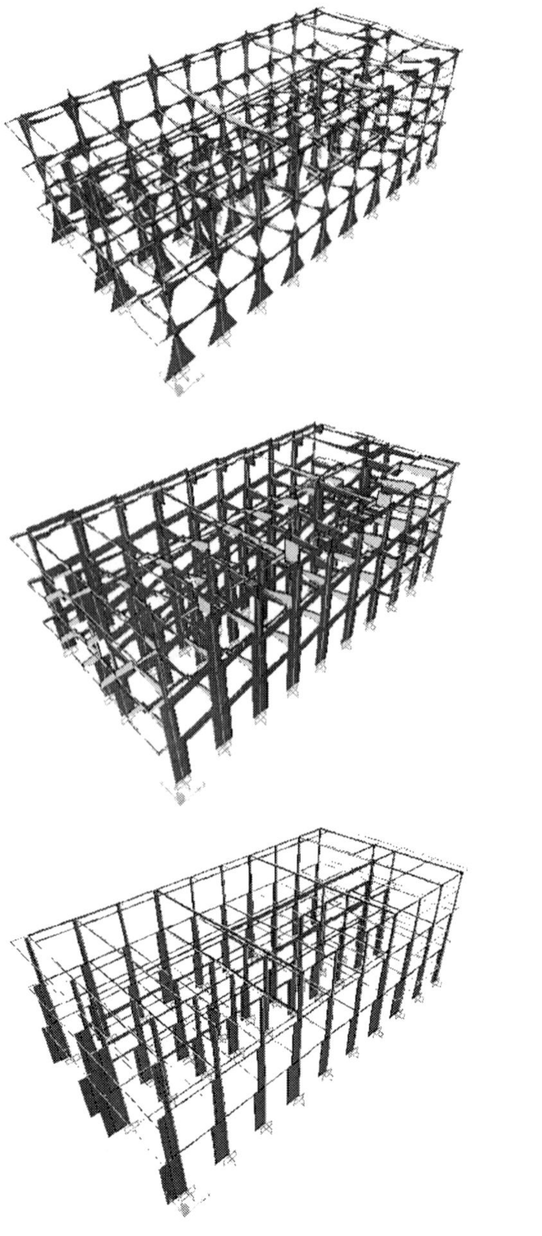

Fig. 6. Internal force (bending moment, shear, and axial forces) diagram of the building structure in Max Comb.

To obtain the reinforcement ratio (ρ), calculations were carried out using M_u data obtained in the structure model. After obtaining the amount of reinforcement at each left, right, and mid-spans; the design moment strength ($ØM_n$) calculates. Stirrups in the field area (from 1/4 L to 1/2 L from the left/ right), the value of V_u = 164,2 kN in the area L to support is greater than V_c so shear reinforcement must be provided using the maximum a_v/s equation for Beam B1. The distance between stirrups and reinforcement in the field area uses D10 - 150. The design of the Beam B1 using the requirements of the SNI 2847:2019 regulations has met the requirements set out in the standard starting from the dimensions, reinforcement design, and calculation of the design moment strength ($ØM_n$), where the details of Beam B1 can be seen in Table 5 [12].

Table 4. Comparison of beams and columns internal forces between SNI 1726:2012 and 2019.

Element	Internal Forces	SNI 2012	SNI 2019	Differ ent
K1 1st floor Ext	Pu (kg)	71403,4	72871,8	2,0%
	Vu (kg)	7415,3	7491,1	1,0%
	Mu (ton.m)	23,9	24,3	1,6%
K1 2nd floor Ext	Pu (kg)	46894,7	48363,1	3,0%
	Vu (kg)	6375,9	6618,6	3,7%
	Mu (ton.m)	13,1	13,4	2,2%
K1 3rd floor Ext	Pu (kg)	23500,8	25214,0	6,8%
	Vu (kg)	4462,73	4492,9	0,7%
	Mu (ton.m)	12,7	13,1	3,1%
K1 1st floor Int	Pu (kg)	75560,5	77028,9	1,9%
	Vu (kg)	8295,5	8360,3	0,8%
	Mu (ton.m)	26,8	27,1	1,1%
K1 2nd floor Int	Pu (kg)	47756,4	49469,6	3,5%
	Vu (kg)	7142,8	7347,8	2,8%
	Mu (ton.m)	14,5	14,9	2,7%
K1 3rd floor Int	Pu (kg)	23987,0	25700,1	6,7%
	Vu (kg)	4966,1	4970,9	0,1%
	Mu (ton.m)	14,2	14,5	2,1%
B1 1st floor Ext	Vu (kN)	155,0	164,2	5,6%
	Mu (kN.m)	151,8	208,6	27,2%
B1 2nd floor Ext	Vu (kN)	141,7	150,5	5,8%
	Mu (kN.m)	208,9	226,7	7,9%
B1 3rd floor Ext	Vu (kN)	158,8	168,1	5,5%
	Mu (kN.m)	214,8	222,1	3,3%
B1 1st floor Int	Vu (kN)	158,8	168,1	5,5%
	Mu (kN.m)	152,2	222,1	31,5%
B1 2nd floor Int	Vu (kN)	130,8	139,6	6,3%
	Mu (kN.m)	171,6	180,6	5,0%
B1 3rd floor Int	Vu (kN)	148,3	164,2	9,7%
	Mu (kN.m)	151,3	209,1	27,6%
B2 1st floor Ext	Vu (kN)	65,0	65,7	1,1%
	Mu (kN.m)	92,7	93,8	1,2%
B2 2nd floor Ext	Vu (kN)	64,1	64,3	0,3%
	Mu (kN.m)	83,7	91,2	8,2%
B2 3rd floor Ext	Vu (kN)	49,6	52,9	6,2%
	Mu (kN.m)	58,0	65,9	12,0%
B2 1st floor Int	Vu (kN)	64,1	64,4	0,5%
	Mu (kN.m)	86,1	91,3	5,7%
B2 2nd floor Int	Vu (kN)	48,3	55,7	13,3%
	Mu (kN.m)	80,5	83,7	3,8%
B2 3rd floor Int	Vu (kN)	50,7	55,7	9,0%
	Mu (kN.m)	66,6	80,6	17,4%

Table 5. Comparison of column and beams design between SNI 2847-2019 and existing building.

<table>
<tr><th colspan="2">Column K1</th></tr>
<tr><th>SNI 2487:2019</th><th>Existing Building</th></tr>
<tr><td>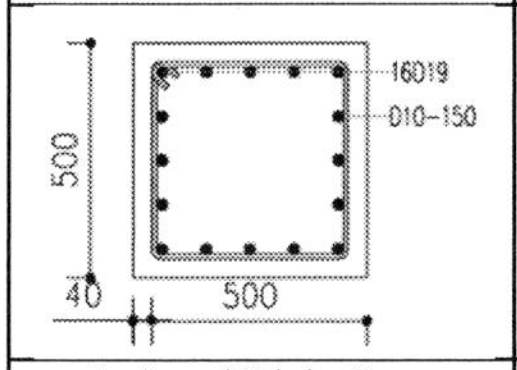</td><td>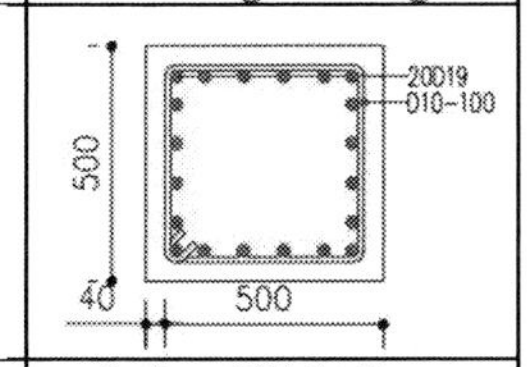</td></tr>
<tr><td>Left and Right Span</td><td>Left and Right Span</td></tr>
<tr><td>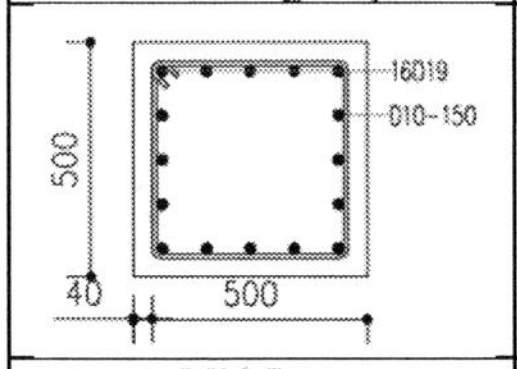</td><td>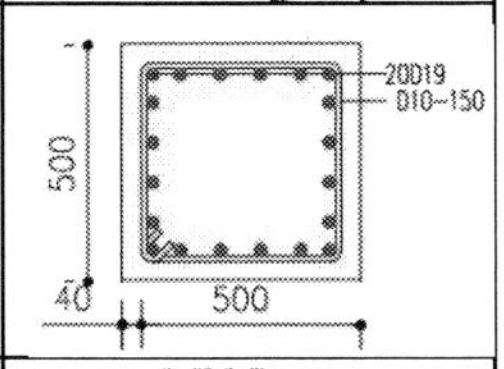</td></tr>
<tr><td>Mid Span</td><td>Mid Span</td></tr>
<tr><th colspan="2">Beam B1</th></tr>
<tr><th>SNI 2487:2019</th><th>Existing Building</th></tr>
<tr><td>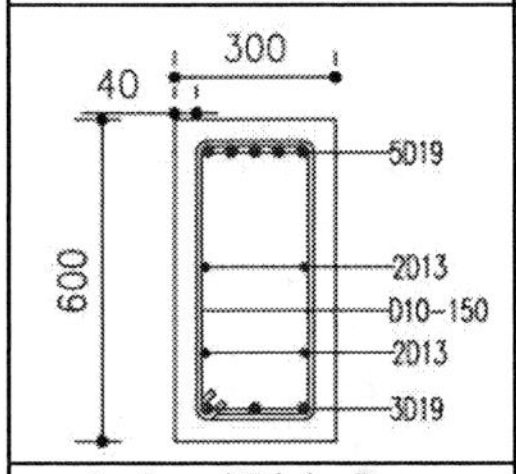</td><td>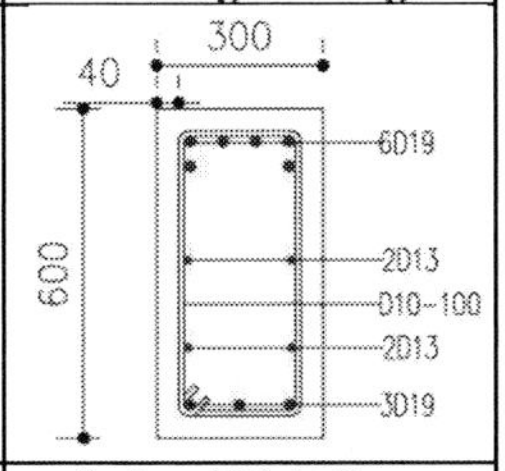</td></tr>
<tr><td>Left and Right Span</td><td>Left and Right Span</td></tr>
<tr><td>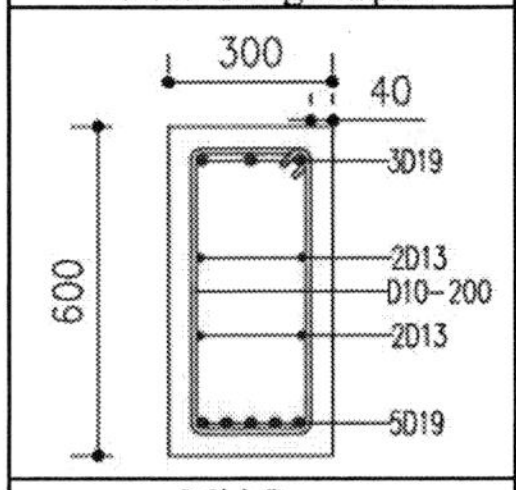</td><td>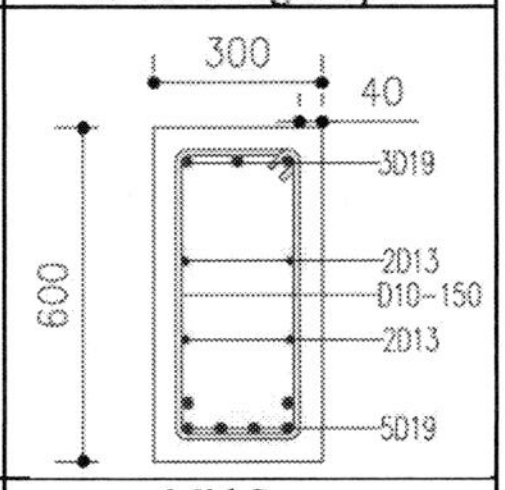</td></tr>
<tr><td>Mid Span</td><td>Mid Span</td></tr>
<tr><th colspan="2">Beam B2</th></tr>
<tr><th>SNI 2487:2019</th><th>Existing Building</th></tr>
<tr><td>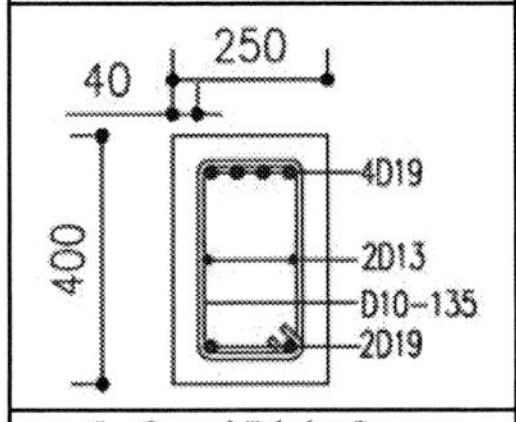</td><td>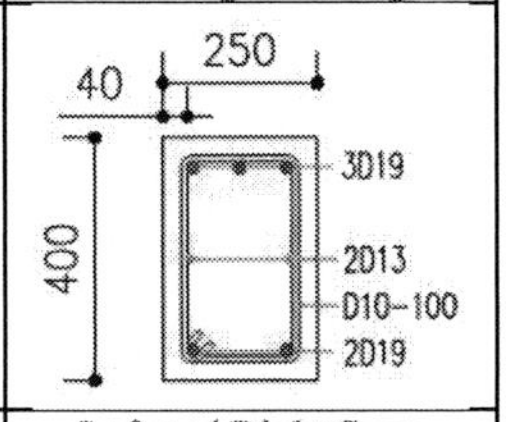</td></tr>
<tr><td>Left and Right Span</td><td>Left and Right Span</td></tr>
<tr><td>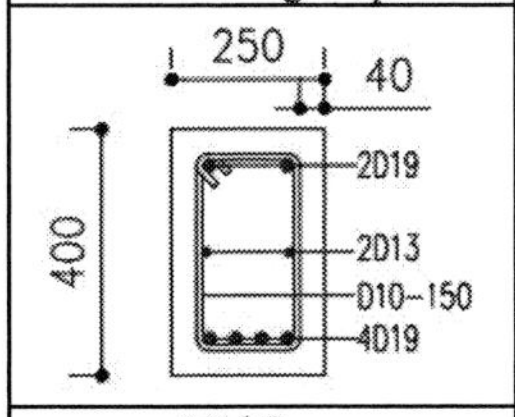</td><td>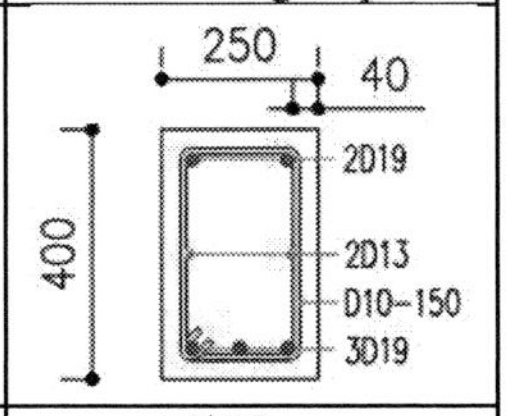</td></tr>
<tr><td>Mid Span</td><td>Mid Span</td></tr>
</table>

Based on the calculation results of the interaction diagram, the Column K1 being monitored is still safe because $M_n > M_u$; where $M_n = 573.5$ kN.m and $M_u = 244.2$ kN.m), and the strong column of the weak beam is fulfilled. For structures with Pu < 0.3 Ag fc', the total cross-sectional area of square stirrup reinforcement is taken as the largest value. After calculating Column K1 using the requirements of the SNI 2847:2019 regulation, it has fulfilled the requirements set out in the regulation starting from the dimensions, reinforcement design, and calculation of the moment strength design ($\emptyset M_n$), where the details of Column K1 can be seen in Table 5. A_s needed = $3{,}6.10^5$ mm^2 (result of SAP2000) and 16D19 flexural reinforcement is used.

Results the design of beams, columns, and slab after being redesigned with SNI 2847:2019 for dimensions already meets the requirements, but for flexural reinforcement and shear reinforcement obtained after being designed, changes are made, such as the example in Beam B2 in the initial conditions used flexural reinforcement 3D19 and after redesign used 4D19.

4 Conclusions

From the above calculation and discussion, there is an effect of changing the design plan of SNI 1726:2012 to SNI 1726:2019. In the spectrum response design, the S_{DS} and S_{D1} parameters in SNI 1726:2019 increased by 3.60% and 19.02%, respectively, from the old standard. By this rising, it affects the response spectra graph that has increased in SNI 1726:2019. For internal forces of structural elements, the axial force and bending moment of columns increased about 3.98% and 2.13%, respectively. The increase is also occurring in their shear forces by 1.52%. The bending moment and shear force values are rising by 17.08% and 6.4% for Beam B1; 7.85% and 4.88% for Beam B2, respectively.

The design of the Beam B1 using the requirements of the SNI 2847:2019 standard has met the requirements set out in the standard starting from the dimensions, reinforcement design, and calculation of the design moment strength (M_n). The Column K1 being monitored is still safe because $M_n > M_u$ and the strong column of the weak beam is fulfilled. Results the design of beams, columns, and slabs after being redesigned with SNI 2847:2019 for dimensions already meets the requirements, but for flexural reinforcement and shear reinforcement obtained after being designed, changes are made, such as Beam B2; in the initial conditions used reinforcement 3D19 and after redesign used 4D19.

The calculation of the GKT II Building structure uses the old code designing but it still meets the safe limits using designing with the new codes, SNI 1726:2019 and SNI 2847:2019 methods. Overall, the building structure can be classified as able to withstand seismic force corresponding to the latest standards.

This study was financially supported by Pusat Penelitian dan Pengabdian Masyarakat (P3M) Bengkalis State Polytechnic Indonesia through Research Grant 2021.

References

1. P. Pudjisuryadi, E.D. Mulyadi, Ferry, B. Lumantarna, Performance of an existing reinforced concrete

building designed in accordance to older Indonesian codes (PPTGIUG 1983 SK SNI T-15-1991-03): a case study for a hotel in Balikpapan, MATEC Web of Conferences **258**, 05028 (2019)

2. National Standardization Agency, SNI 1726-2012, *Procedures for Planning Earthquake Resistance for Non-Building Building Structures,* BSN, Jakarta, Indonesia (2012)
3. National Standardization Agency, SNI 1726-2019 *Procedures for Planning Earthquake Resistance for Non-Building Building Structures*, BSN, Jakarta, Indonesia (2019)
4. S. Arifah, D.J. Jaya, S. Widodo, *Comparative study of internal force in lecture buildings in Daerah Istimewa Yogyakarta between SNI 1726-2012 and 1726-2019 with the spectrum response method (2D),* Journal of Physics: Conference Series **1833** (2021) 012026
5. S.G. Makarande, V. V. Agrawal, G.D. Dhawale, A.B. Dehane, M.R. Nikhar, Analysis and Design of Multi Storeyed Building using Staad Pro and Manually for Two Seismic Zones, Iconic Research and Engineering Journals, **3,** Issue 3 (2019)
6. G.G. Kakpure, A. R. Mundhada, *Comparative Study of Static and Dynamic Seismic Analysis of Multistoried RCC Buildings by ETAB, Int. Journal of Engineering Research and Application*, **7** (5), 5, 6-10 (2017)
7. Ayuddin, Global structural analysis of high-rise hospital building using earthquake-resistant design approach, SINERGI, **24** (2), 95-108 (2020)
8. Z.A. Jauhari, Armada, F. Ananda, R. A. Fitrah, *Comparison of the design acceleration response spectra in Riau Province between SNI 1726:2019 and SNI 1726:2012 Methods, IOP Conf. Series: Earth and Environmental Science* **708** 012009 (2021)
9. R.A. Fitrah, D.I. Mazni, W. Pratiwi, Z.A. Jauhari, Seismic assessment of irregularities in steel special moment resisting frame with asymmetric-plan building (case study: Gedung D - Universitas Dharma Andalas), IOP Conf. Series: Earth and Environmental Science **708** 012007 (2021)
10. G. H. Powell, R. Allahabadi, Seismic damage prediction by deterministic methods: concepts and procedures, Earthquake Engineering & Structural Dynamics (1988)
11. D.J. Jaya, S. Arifah, S. Widodo, *Review of internal force magnitude at the 5-floor hospital in North Sulawesi Province between SNI 1726-2012 and SNI 1726-2019 with response spectrum (2D) method, IOP Conference Series: Earth and Environmental Science* (2021)
12. National Standardization Agency, SNI 2847-2019 *Structural Concrete Requirements for Buildings,* , BSN, Jakarta, Indonesia (2019)

The comparative study of analyzing the T-shaped and square-shaped concrete columns

Ade Sri Wahyuni[1*], *Mukhlis* Islam[1], and *Rizky* Arian Putra[1]

[1] Civil Engineering Department, University of Bengkulu, Indonesia

Abstract. This study aims to analyze how much changes in column shape effect have on the strength of the column structure. This calculation is comparing a square-shaped concrete column with two different axis of a T-shaped column. Each sample is compared in terms of 6 different column dimensions, however, the cross-sectional area within the three samples (Square, T-shaped axis1 and 2) has remained constant. The concrete quality (fc') is kept at 30 MPa while the reinforcement area used 1% and 5%. Column calculation analysis is using Hognestad stress-strain diagram relationship and finite difference method. The result shows, the greater the reinforcement ratio, the greater the difference of Mn value and the smaller the difference of Pn value, of T-shaped and square column for each area. The same cross-sectional area does not necessarily lead to the same behavior of a T-shaped column with a different axis

1 Introduction

Structural design plays an important role to ensure the strength, safety, and durability of the building. The column is one of the structural components of a building whose main task is to support axial loads (Pn) and nominal moments (Mn). Column design is critical in structural design as the column failure may result in the collapse of other structural components associated with it [1].

The cross-sectional dimension of the column is small relative to its overall length. Structural concrete columns are used in buildings to support the beam and floor system and transfer the vertical live loads and dead loads from the upper structure into the footings and foundations.[2]

Symmetrical reinforcement is generally used in column planning to avoid mistakes during the installation of reinforcement. This symmetrical reinforcement is also needed if there is a possibility of alternating forces on the structure, such as earthquake loads

A T-shaped column is used in the building as an alternative for maximizing room function and aesthetic issues. Using a T-shaped column will create more space in the room without the appearance of the big column. Another use of a T-shaped column is in the goods elevator with alternating doors. In this case, the T-shaped column cannot be replaced by the shear wall

2 Literature Review

Columns usually carry bending moments about both principal axis and so it is common to place reinforcing bars in all four faces. The bars in the side faces of this column of course contribute to the strength of the column [2].

Fig. 1 shows the Stress-strain diagram of the square column. The ratio of longitudinal reinforcement in a column with a rectangular cross-section using the Hognestad stress-strain relationship

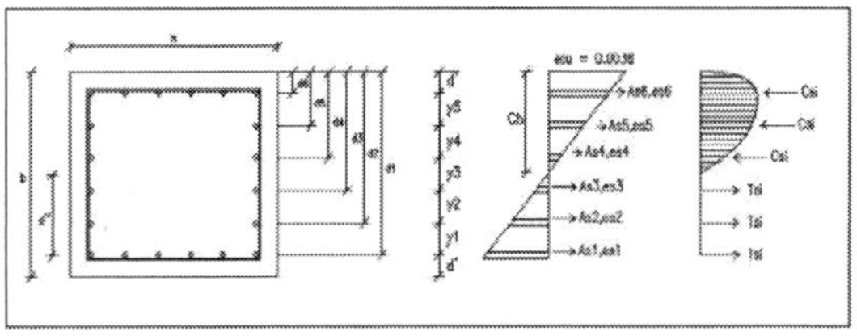

Fig. 1. Square column

When a section has bars distributed at all faces, the derivation of equations for design and analysis becomes difficult because the bars may be at various stress levels throughout the section. The analysis of such a section can be carried out using the requirements of strain compatibility and equilibrium.[3]

To take account of the bars in the side faces we simply have to consider additional layers of steel at various levels. Steel areas As_1, As_2, As_3, As_4, As_5, As_6 are located at depth d_1, d_2, d_3, d_4, d_5, d_6. The steel layers will be in tension or compression and maybe at yield or elastic, depending on the distance from the neutral axis. As the neutral axis may be placed in various positions in the section, some steel layers may be in either compression or tension [3].

* Corresponding Author: ade.sri.w@unib.ac.id

 https://doi.org/10.1051/e3sconf/202133105005

For a general bar I in the section, the strain diagram indicates that

$$\varepsilon_s = 0{,}003\frac{c - di}{c} \qquad (1)$$

Where compressive strains are positive, and tensile strains are negative. Then the stress f_{si} in bar i is given by the following relationship [2][3][4].

If $\varepsilon_{si} \geq \frac{fy}{Es}$ $\qquad f_{si} = f_y \qquad (2)$

Or if $\frac{fy}{Es} > \varepsilon_{si} > -\frac{fy}{Es}$ $\qquad f_s = \varepsilon_s.E_s \qquad (3)$

If $\varepsilon_{si} \leq -\frac{fy}{Es}$ $\qquad f_{si} = f_y \qquad (4)$

$$C_s = A_s'.f_s' \qquad (5)$$

$$T_s = A_s.f_s \qquad (6)$$

(2.1)

$$C_c = 0{,}85.f'_c.a.b \qquad (7)$$

$$Pn = C_c + C_s - T_s \qquad (8)$$

$$Mn = Pn\ .\ e \qquad (9)$$

3 Research aim

This study aims to produce the interaction diagram Pn-Mn of the square column compared with T shaped column with two different axis

Fig. 2 shows the T-shaped Column in the first axis and will be named T1 while the second axis is shown in Fig. 3 and named T2

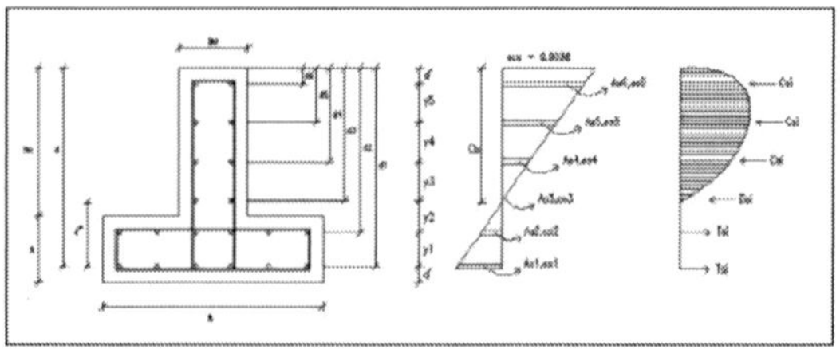

Fig. 2. T-shaped column first axis

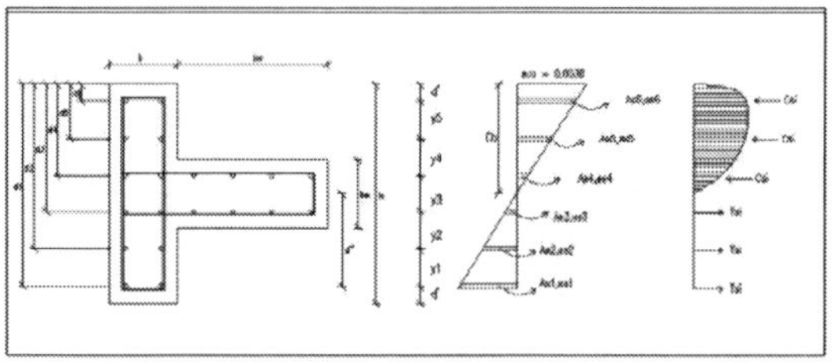

Fig. 3 T-shaped column second axis

Since there are several layers of steel, there will not be a single sharp discontinuity in the interaction diagram at the balanced failure point, rather a more curved diagram will result because not all the tension reaches the yield strength at the same time [2,3]

The foregoing approach for calculating interaction diagrams by determining the combinations for Pn and Mn at failure for various neutral axis locations can also be used for columns of shapes other than rectangular and walls, However when the neutral axis depth is small, such as in flanged walls, and the dimensions of the cross-section are large, very large tensile strains can occur in the far layers of the tension steel [2, 3]

If the maximum strength of the section is to be calculated, it is important to determine whether the tensile strains in these bars is such that the bars have entered the strain-hardening range, If the full stress-strain curve for the steel is known, the actual stresses corresponding to the strain levels can be used in the strength calculations, The additional flexural strength due to strain hardening should be taken into account when the resulting over strength could lead to an alternative brittle failure (e.g., a shear failure rather than a flexural failure. For nonsymmetrical sections or steel arrangement, two interaction curves will result, one for each sense of the eccentricity [2, 3]

4 Limitation

The limitation available in this study are:
Reinforced concrete with a square cross-section with four-face longitudinal reinforcement.
The reinforced concrete column that only experiences a combination of uniaxial bending moments
Concrete quality fc' 30 MPa
Reinforcement area 1% and 5% of the column cross-sectional area
The reinforcement quality fy 400 MPa.
E_s= 200.000 MPa

5 Methods

The flange dimension of T shaped column is 25x80, 25 x 90, 30x 80, 30x 90, 35x 80, 35x90, and the cross-sectional area of these three columns are kept constant.

The interaction diagram of Pn-Mn will be compared with two different reinforcement areas 1% and 5 %. This data analysis uses some basic assumptions commonly used in the calculation of column structure.

The iteration method is used repeatedly to get a graph at certain points. The distance between the neutral line and the edge of the compressed concrete fiber c is assumed to be 20 pieces for one value of c divided into 100 segments. The calculation will stop when it has reached 100 segments with one value of c producing one axial force and nominal moment.

The points of each value of C are combined to form the interaction diagram of the axial force (Pn) and nominal moment (Mn).

6 Results and Discussion

Any value of d_n less than d will produce a valid point on the strength line and a pair of values Pn - Mn which represent a condition of failure of the section. If very small values are chosen for d_n i.e less than the value at which pure bending failure occurs, a negative value for Pn will be obtained. This represents failure under a combination of moment and axial tension by the strength line extending down into the fourth quadrant[2, 3].

The strength line depends on the dimensions, reinforcement details, and material properties for the column cross-section and is a section property, irrespective of whether the column is short or long[2, 3].

The small compressive force increases the moment carrying capacity of the section so that the interaction diagram 'bulges' outwards at intermediate values of Pn. In such cases, the axial force Pn is reducing the tensile force in the steel, which is the primary cause of failure[2, 3]

The balance failure condition lies close to the extremity of the bulge of the interaction diagram so that the balance moment is approximately the maximum moment that the section can carry in the presence of axial compression[2, 3]

Fig.4 until Fig. 5 shows the interaction diagram of the Square column, T-shaped column first axis, and T-shaped column second axis. The square column becomes the benchmark of the axial force and moment that are allowable and can be supported by the column

The points within the inside of the red curve are also pairs of axial loads and moments that can be supported, but these pairs of loads are lower than the bearing capacity of the column, so if used, it means that the dimensions & reinforcement of the column are overdesigned.

The point outside the red curve, are pairs of axial loads and moments that exceed the bearing capacity of the column so that the column is not allowed to support these loads. Fig.4 and Fig.5 show interaction diagrams between columns with the same cross-sectional area, with the flange size of T column 250 x 800.

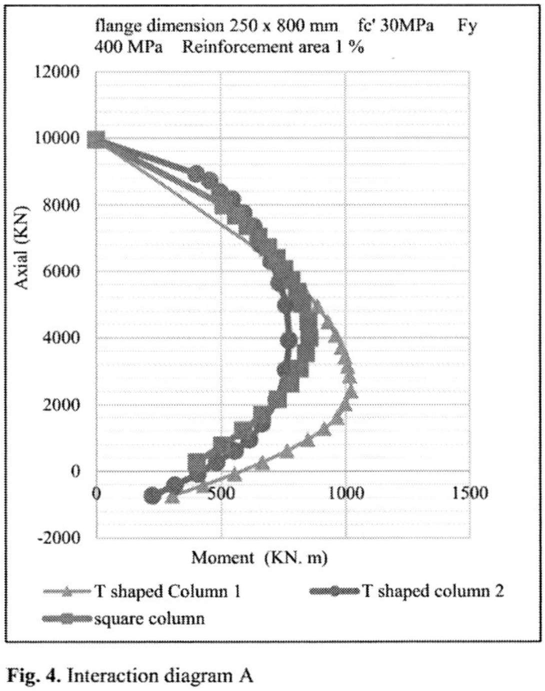

Fig. 4. Interaction diagram A

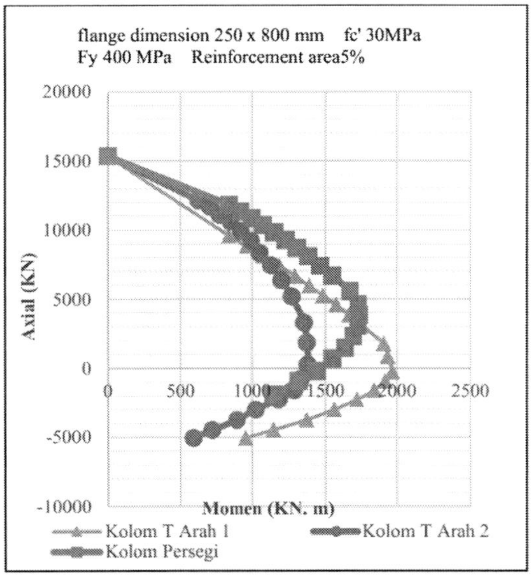

Fig. 5. Interaction diagram B

The difference between these two images is in the percentage of reinforcement area used. Fig. 1 uses 1% reinforcement area and Fig. 2 uses 5% reinforcement area

Column T2 shows that about 35% of the Pn and Mn combinations are in the safe region, while column T2 shows that only 15% of the Pn and Mn combinations are in the safe region.

When reinforcement area is increased to 5% more combinations of Pn- Mn are in a safe area. The T2-column increased to approximately 70% while the T1-column also increased from 10% to 45%.

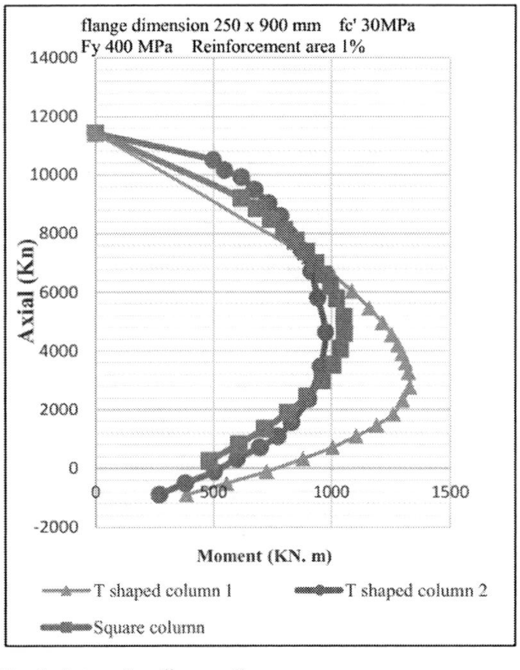

Fig. 6. Interaction diagram C

https://doi.org/10.1051/e3sconf/202133105005

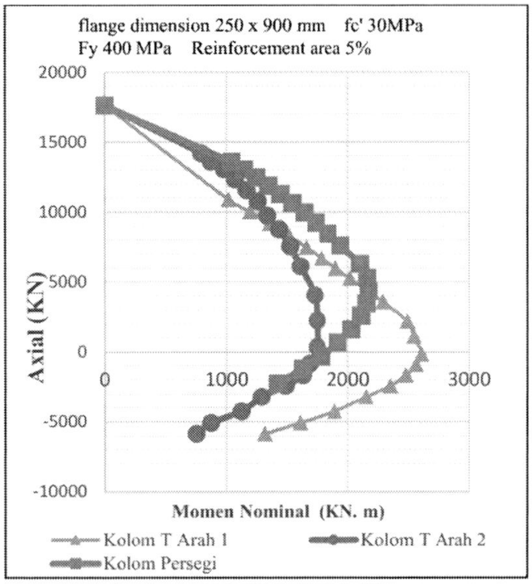

Fig. 7. Interaction diagram D

Fig. 6 and Fig. 7 show interaction diagrams between columns with the same cross-sectional area, in which the flange size of the T column is 250 x 900

Column T2 still shows that about 35% of the Pn and Mn combinations are in the safe region, while the safe region in column T1 decreases to 5% of the Pn and Mn combinations.

When reinforcement area is increased to 5% The T2-column is still approximately 70% while the T1-column is still 45%. Even though the Axial Force (Pn) and Moment increase with the bigger cross-sectional area, the percentage of the pair number of Pn-Mn which is still considered safe to hold the load is the same as the interaction diagram in Fig. 5

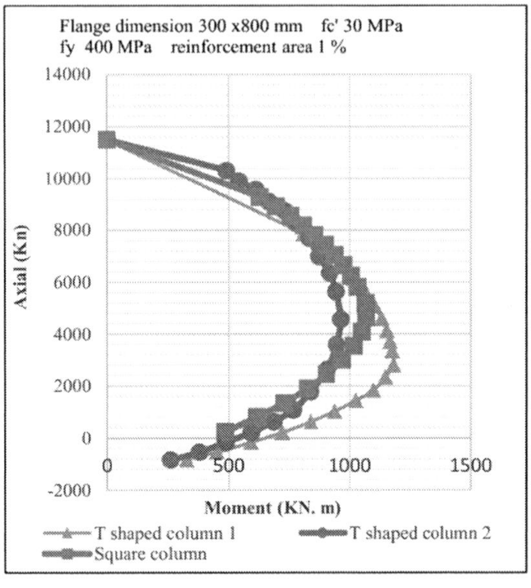

Fig. 8. Interaction diagram E

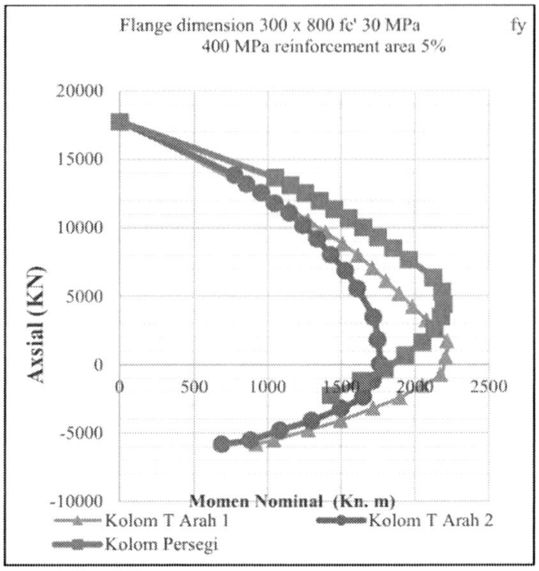

Fig. 9. Interaction diagram F

Fig. 8 and Fig. 9 show interaction diagrams between columns with the flange dimension of T column 300x800. Fig. 8 uses 1% reinforcement area and Fig. 9 uses 5% reinforcement area

Column T2 shows that about 50% of the Pn and Mn combinations are exceeding the bearing capacity, while column T2 shows that only 80% of the Pn and Mn combinations are exceeding the bearing capacity.

When the reinforcement area is increased to 5% more combinations of Pn Mn are in a safe area. The T2-column increased to approximately 60% while the T1-column also increased to 50 %.

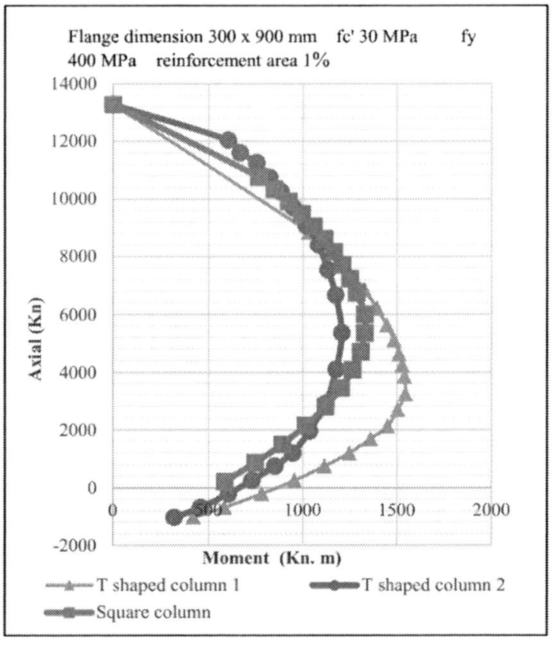

Fig. 10. Interaction diagram G

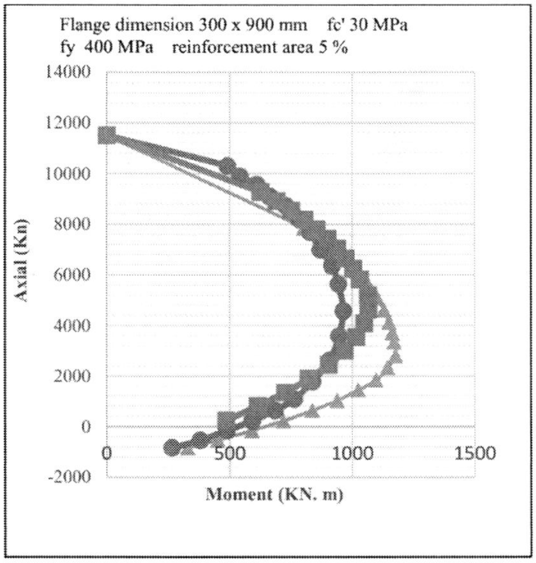

Fig.11. Interaction diagram H

Fig. 10 and Fig. 11 show interaction diagrams between columns with the size of the flange of T column 300 x 900

Column T2 shows that about 45% of the Pn and Mn combinations are in the safe region, while the safe region in column T2 slightly increases to 25% of the Pn and Mn combinations are in the safe region.

When reinforcement area is increased to 5%

The T2-column is still approximately 70% while the T1-column is still 50%.

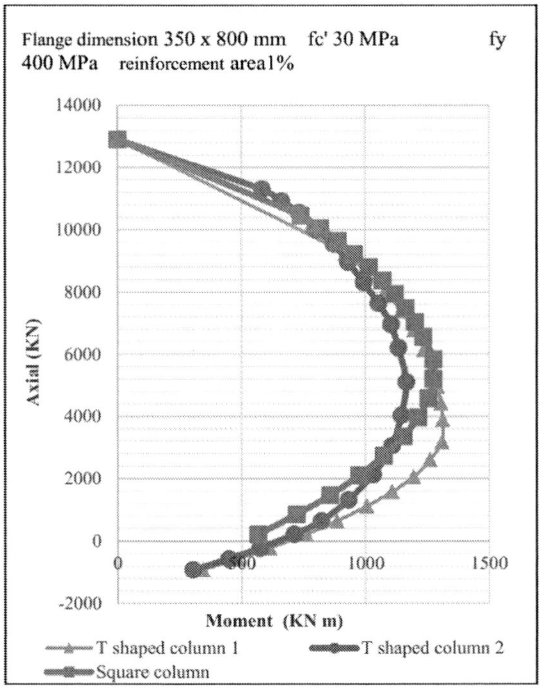

Fig. 12. Interaction diagram I

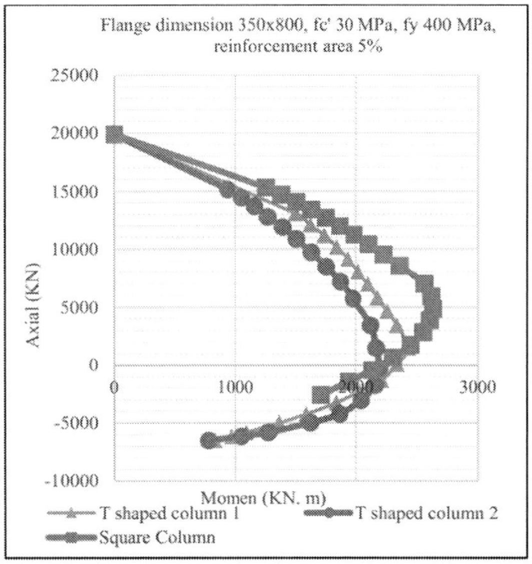

Fig. 13. Interaction diagram J

Fig. 12 and Fig. 13 show interaction diagrams between columns, with the flange size of T column 350 x 800. Fig. 12 uses 1% reinforcement area and Fig. 13 uses 5% reinforcement area

Column T2 shows that about 55% of the Pn and Mn combinations are in the safe region, while column T2 shows that only 35% of the Pn and Mn combinations are in the safe region.

When the reinforcement area is increased to 5% more combinations of PN Mn are in a safe area. The T2-column increased to approximately 60% while the T1-column also increased to 50 %.

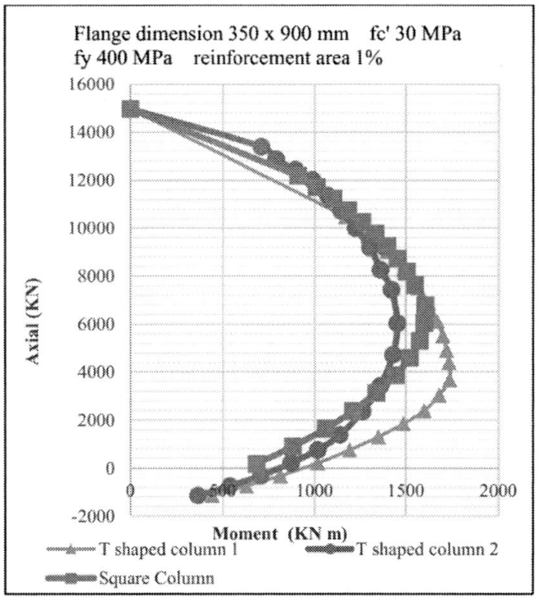

Fig. 14. Interaction diagram K

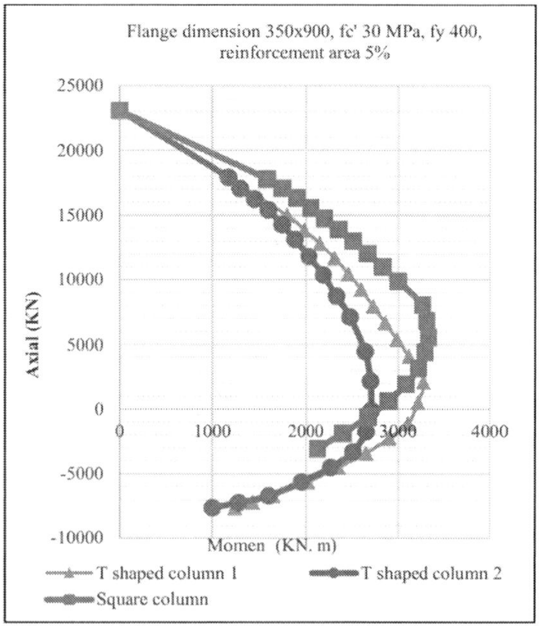

Fig.15. interaction diagram L

Fig. 14 and Fig. 15 show interaction diagrams between columns with the flange size of T column 350x900

Column T2 shows that about 50% of the Pn and Mn combinations are in the safe region, while the safe region in column T2 slightly decreases to 25% of the Pn and Mn combinations are in the safe region.

When reinforcement area is increased to 5% The T2-column is still approximately 65% while the T1-column is still 50%.

7 Conclusion

The shape of the Mn-Pn column interaction diagram is affected by an increasing amount of tensile steel.

The increase in the value of the cross-sectional area does not affect the percentage of the area that is in a safe area with a 5% reinforcement area. T1 axis ranges from 45 to 50% while the T2 axis percentage of safe areas is in the range of 60 to 70%.
As for reinforcement area, 1% for T2 ranges from 35-55%, while T1 ranges from 5-50%.

Column T with the most optimum area is on the flange dimensions 350 x 800 with the most pair of Pn-Mn which is not exceeding the bearing capacity.

The greater the reinforcement area, the bigger the cross-sectional area, the bigger the axial force and moment, T1 shows the bigger moment that exceeding the bearing capacity. However, within the same dimension but bigger reinforcement area, the T1 shows some points which allowed to hold the load.

The same cross-sectional area does not necessarily lead to the same behavior of T shaped column.

References

1. E. G. Nawy, Reinforced Concrete A Basic Approach, Bandung (1998)
2. R. F. Warner, B. V. Rangan, A. S, Hall, K. A. Faulkes, Concrete Structures, Longman (1999)
3. R. Park and T. Paulay, Reinforced Concrete Structures, A Wiley-Interscience Publication, (1933)
4. National Standardization Agency, SNI 2847 2019 - *Structural Concrete Requirements for Buildings,* BSN (2019)

The capacity analysis of wide flange steel section as beam-column elements

Mukhlis Islam[1*], *Ade Sri* Wahyuni[1], and *Jeirry* Anggara[1]

[1] Civil Engineering Department, Faculty of Engineering, University of Bengkulu. WR. Supratman Rd., Bengkulu 38371

Abstract. The design of the steel frame structure for beam, column, and beam-column elements usually takes several iterations to obtain optimum results that are safe, efficient, and economical. The problems that arise due to the number of iteration for the design process are the basis for this research. The research was conducted on 100 sections IWF and HCS which are consisted of 81 sections of IWF and 19 sections of HCS. The analysis was carried out regarding SNI 1729:2015. The results of this study are in the form of moment capacity graphs and tables of the axial capacity of steel sections as well as coefficients of m and U for beam-column design. Moment capacity graphs and tables of the axial capacity of IWF and HCS profiles produced could be used to reduce the number of iteration in designing safe and efficient steel profiles as beam, column, and beam-column elements. Analysis results show that the usage of the graphs and tables for the beam-column design process will reduce the number of iteration needed.

1 Introduction

Along with the development of technology in the field of construction, steel sections have become one of the options to be used as structural elements in building construction. This steel construction can be seen in today's buildings such as houses, factories, stadiums, hotels, and even other high-rise buildings [1].

There are various types and sizes of steel sections available in Indonesia, such as WF (Wide Flange) and HCS (Heavy Column Section). The WF section consists of I and H beam sections, wherein total there are 81 steel sections with different dimensions. The HCS section consists of 19 sections with different dimensions [2].

Various sections available on the market with different specifications make engineers carry out repeated iteration to obtain a section that is safe, efficient, and economical. The repeated iterations carried out by the engineers made the structural design process would be tedious and required a lot of effort to get an efficient steel section.

The results of this study will be design aids in the form of tables of axial capacity and nominal moments of IWF and HCS sections. These tools are then used in designing steel structures that are safe, efficient, and economical. By using this tool, there will be fewer experiments, so it doesn't take up much time and effort.

2 Literature Review

2.1 Steel Section

Steel material first appeared in the 19th century where steel is an alloy of iron and carbon. The carbon material found in steel is less than the carbon material in cast iron. The manufacture of large volume steel was first carried out by Sir Henry Bessemer, who received a patent from the British government for his findings in 1855 [1].

The first steel frame structure was the Home Insurance Company Building in Chicago [1].

2.2 Beam

Beams according to SNI 1729-2015 are nominal horizontal structural members that have the main function to withstand bending moments [3].

Beams are structural components that carry gravity loads, such as dead loads and live loads, beam structural components are a combination of compression elements and tensile elements [1].

2.3 Column

Columns according to SNI 1729-2015 are nominal vertical structural members that have the main function of resisting axial compression forces.

Compressive members are intended for structural members that carry centric compressive loads right at the center of gravity of the cross-section or column with only axial forces [4].

* Corresponding Author: islam.mukhlis@unib.ac.id

2.4 Beam-Column

Beams-columns according to SNI 1729-2015 are structural components that resist axial forces and bending moments.

Structural members sometimes the effects of the combination of axial forces and bending moments are not negligible, hence the combination of axial forces and bending moments must be considered in the design process of structural members. These structural members are often referred to as beam-column elements [1].

3 Research Methodology

The research was conducted by analyzing the cross-sectional capacity of IWF and HCS steel sections as a beam, column, and beam-column elements. This research is applied to academic research and is carried out carefully, systematically, and continuously on each IWF and HCS (heavy column section) steel profile used for structural design purposes.

The data used in the analysis process are IWF and HCS sections which consist of 81 IWF types and 19 HCS types.

3.1 Section Capacity Calculation procedure

Calculations were made for the sections of compact web and compact flange, compact web of the non-compact flange, and compact web and slender flange for beam elements, while for column elements only non-slender sections were analyzed. The steps for calculating the section's capacity are carried out with the following procedure:

3.1.1 Calculate Additional Data for Setion Capacity Calculation

a. Radius of gyration (axis-x dan axis-y)

$$rx = \sqrt{\frac{I_x}{A}} \quad (1)$$

$$rx = \sqrt{\frac{I_y}{A}} \quad (2)$$

b. Plastic section modulus (axis-x dan axis-y)

$$Zx = b \cdot tf \cdot (d - tf) + \tfrac{1}{4}\, tw\, (d - 2\, tf)^2 \quad (3)$$

$$Zy = \tfrac{1}{2}\, b^2 \cdot tf + \tfrac{1}{4}\, tw^2\, (d - 2\, tf) \quad (4)$$

c. Warping torsion constants

$$J = \frac{1}{3}(2.\, b\, .tf^3 + d'.\, tw^3) \quad (5)$$

d' is obtained by calculation of:

$$d' = h - tf \quad (6)$$

3.1.2 Column Action Calculation

The procedure is as follows:

a. Controlling of slenderness ratio of web and flange of the sections to control the effect of local buckling.
b. Calculating the critical stress of the section (F_{cr})
c. Calculating the axial capacity of the column

3.1.3 Beam Action Calculation

a. Controlling the slenderness ratio of web and flange of the sections to the effect of local buckling
b. Calculating the nominal capacity of bending moment based on lateral-torsional buckling criteria
   a) Section with compact web and compact flanges bending about its major axis, with categories of:
      - $L_b < L_p$, short span beam
      - $L_p < L_b < L_r$, medium span beam
      - $L_b > L_r$, long-span beam
   b) Section with compact web and non-compact flange bending about its major axis.

Beam-column analysis was carried out on three types of effective length factor (k). The three effective length factors consist of:

1. k=1, nominal moment capacity of the beam at span L compared to axial capacity at kL=L.
2. k=0,5, nominal moment capacity of the beam at span L compared to axial capacity at kL=0,5 L.
3. k=2, nominal moment capacity of the beam at span L compared to axial capacity at kL=2L.

3.1.4 Plotting of Nominal Bending Capacity Graphs of Beams

The nominal moment capacity graphs of the beam are formed from the relationship between the length of the unfettered beam (L_b) and the nominal moment capacity of the beam ($ØM_n$). The nominal moment capacity curve of the beam is drawn according to the mechanical properties of the steel, namely A36, A529 Gr. 50, A572 Gr. 42, A572 Gr. 50, A572 Gr. 55, A 572 Gr. 60, A572 Gr. 65, and A 913 70. The length of the curved line is limited to 30 meters. However, if the 2Lr value of the profile is less than 30 (2Lr < 30), then the length of the curve is only drawn until the curve is at the 2Lr point and if the length of the 2Lr curve exceeds 30 meters (2Lr > 30), then the nominal moment curve of the beam is limited to 30 meters.

3.1.5 Composing Table of Section Axial Capacity and Coefficient of m & U

The value of the coefficients m and U is a multiplier coefficient that is used to obtain the equivalent Pu value for each section that will be used as a beam-column element due to the effect of bending moments. Pu equivalent is the ultimate axial value which is obtained from the combination of the ultimate axial capacity P_u, the x-direction bending moment (M_x), and the y-direction bending moment (M_y) values. The equivalent Pu value is used as a determining value for the factored section axial capacity ($ØP_n$). A profile will be said to be efficient if the equivalent P_u value of the structural element is close to the P_n value of the structural element.

The table of the axial capacity of sections and m & U coefficients will be summarized in tables as the result of the analysis process. Each table will contain the axial capacity of the section, the value of the M coefficient, and

https://doi.org/10.1051/e3sconf/202133105006

the value of the U coefficient for each IWF and HCS section.

This table will contain the entire sections that have been analyzed. The table contains the axial capacity, the value of m, and the value of U from Lb = 0.5 meters to 15 meters for each steel grade.

3.2 Section Data

The data used in the analysis process are IWF and HCS sections which consist of 81 IWF sections and 19 HCS sections. The IWF sections are used in the analysis of beams, columns, and beam columns while HCS sections are used in the analysis of columns and beam-columns.

4 Results and Discussions

4.1 Bending Moment Capacity Graphs

Below are the samples of the results of the analysis of the bending moment capacity steel sections as beams

The figures of bending moment capacity for each steel grade above shows that the value of the bending moment capacity in each section increases in proportion to the quality of the steel used. The greater the quality of the steel, the greater the moment capacity. The greater the quality of steel, the cross-section of the profile closer to the limit of compactness.

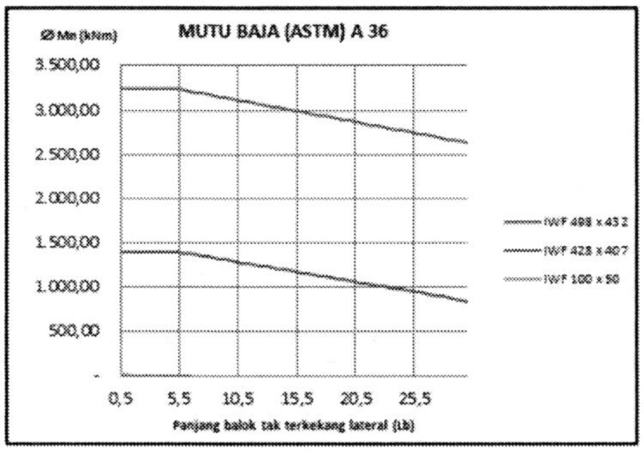

Fig. 1. Factored nominal bending capacity (ØMn) of beams using A36 steel grade

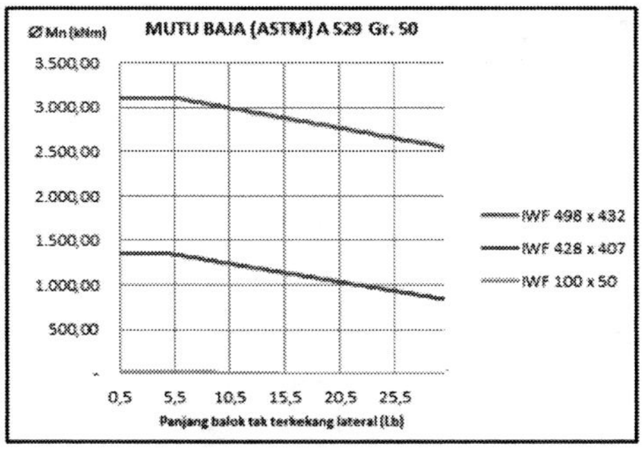

Fig. 2. Factored nominal bending capacity (ØMn) of beams using A529 Gr. 50 steel grade

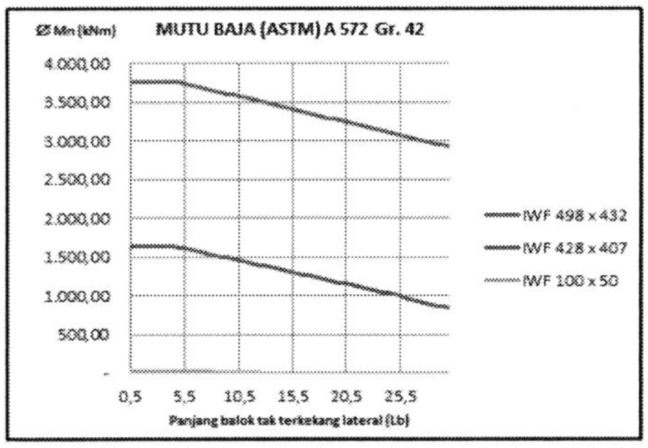

Fig. 3. Factored nominal bending capacity (ØMn) of beams using A572 Gr. 42 steel grade

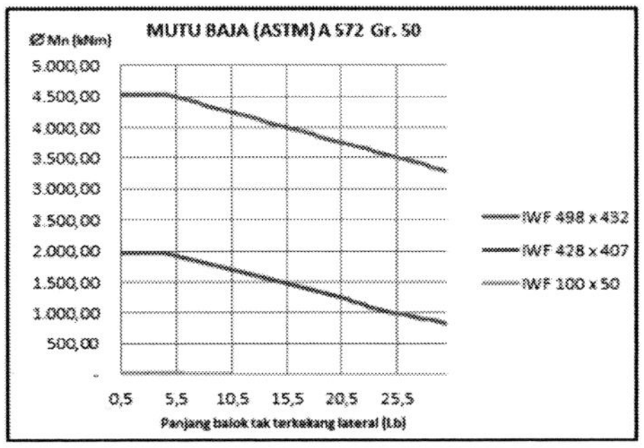

Fig. 4. Factored nominal bending capacity (ØMn) of beams using A572 Gr. 50 steel grade

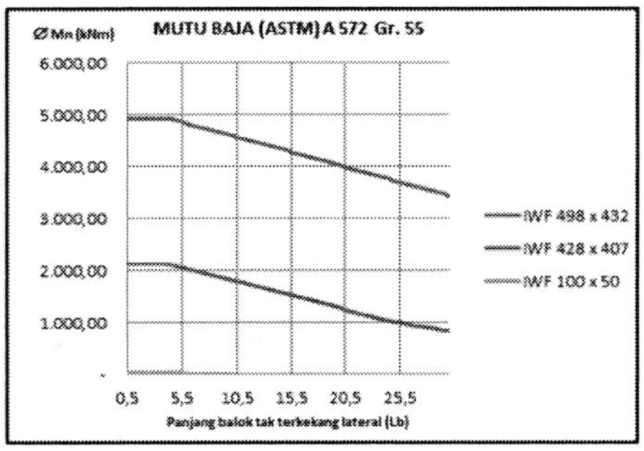

Fig. 5 Factored nominal bending capacity (ØMn) of beams using A572 Gr. 55 steel grade

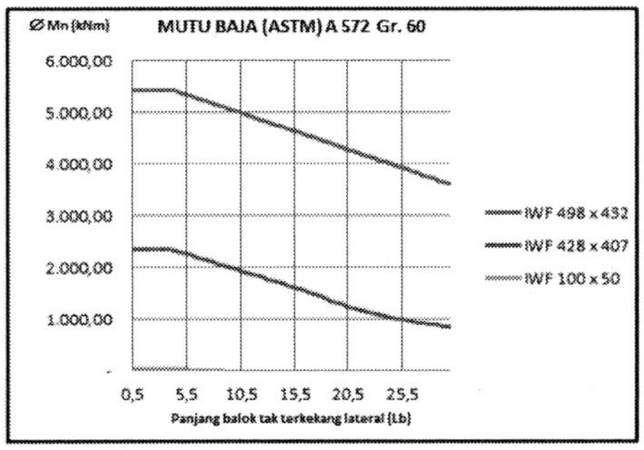

Fig. 6. Factored nominal bending capacity (ØMn) of beams using A572 Gr. 60 steel grade

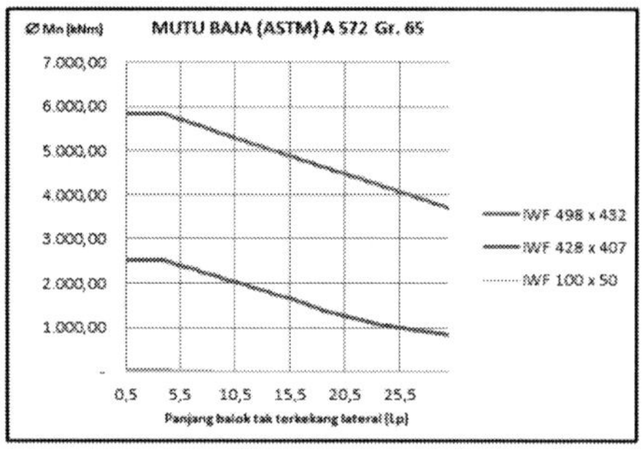

Fig. 7. Factored nominal bending capacity (ØMn) of beams using A572 Gr. 65 steel grade

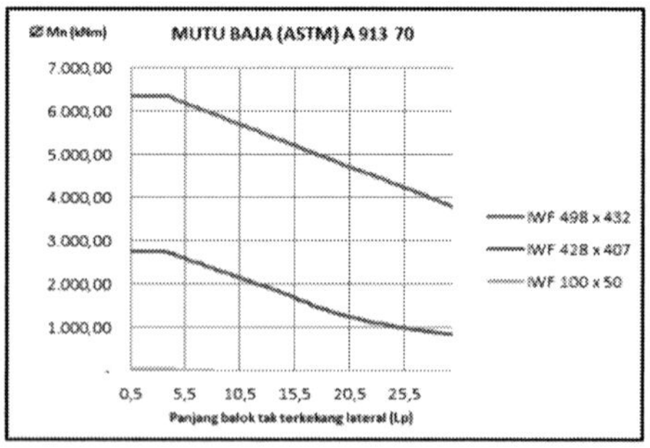

Fig. 8. Factored nominal bending capacity (ØMn) of beams using A913 Gr. 70 steel grade

4.2 Tables of Factored Axial Capacity, m & U Values

Table 1. The values of axial capacity, coefficient m, and coefficient U

kL	PROFIL IWF	(Pr / Pc) ≥ 0,2	(Pr / Pc) < 0,2
	298 x 299	m	m
0,5	2.418	7,1	15,9
1	2.382	7,0	15,7
1,5	2.328	6,8	15,3
2	2.262	6,6	14,9
2,5	2.190	6,4	14,4
3	2.118	6,2	13,9
3,5	2.048	6,0	13,5
4	1.984	5,8	13,1
4,5	1.925	5,8	13,0
5	1.873	5,7	12,9
5,5	1.827	5,7	12,9
6	1.746	5,6	12,6
U	2,2		
Lp	3,8		
Lr	13,0		

Axial capacity tables are used to determine efficient profiles for column and beam-column elements. The values of axial capacity, coefficient m, and coefficient U in the table are presented for a compact profile, with a sample of Table 1.

4.3 Table Usage Example

1. Given Pu, Mux, Muy
   P_u = 1100 KN, M_{ux} = 60 KNm, and M_{uy} = 22 KNm.
2. Determine steel specifications to be used.
   ASTM-A 36, with F_y = 250 MPa.
3. Determine effective length factor (k) and beam column element length (L). k = 1, L = 5 m
4. Determine initial values of m and U
   Determined using average of m and U values obtained from the tables
   m= 5 m^{-1}
   U= 4.
5. Calculation of Pu equivalent I.
   P_u equivalent I $= P_u + m \cdot M_{ux} + m \cdot U \cdot M_{uy}$
   P_u equivalent I = 1.100 + 5 . 50 + 5. 4 . 22
   P_u equivalent I = 1.840 kN
6. Determining IWF section with ∅ P_n value approaching P_u equivalent I value using table 2.

Table 2. IWF section with ∅ P_n value approacing P_u equivalent I

kL	PROFIL IWF	(Pr / Pc) ≥ 0,2	(Pr / Pc) < 0,2
	298 x 299	m	m
0,5	2.418	7,1	15,9
1	2.382	7,0	15,7
1,5	2.328	6,8	15,3
2	2.262	6,6	14,9
2,5	2.190	6,4	14,4
3	2.118	6,2	13,9
3,5	2.048	6,0	13,5
4	1.984	5,8	13,1
4,5	1.925	5,8	13,0
5	1.873	5,7	12,9
5,5	1.827	5,7	12,9
6	1.746	5,6	12,6
U	2,2		
Lp	3,8		
Lr	13,0		

7. IWF 298x299 is chosen
8. Calculation of P_u equivalent II
   $\frac{P_u}{\varnothing P_n} = \frac{1.100}{1.873} = 0,58 > 0,2$
   Nilai m dan U are using values obtained from the table of subsequent section:
   m = 5,7 m^{-1}
   U = 2,2
   Hence :
   P_u equivalent II $= P_u + m \cdot M_{ux} + m \cdot U \cdot M_{uy}$
   P_u equivalent II = 1.100 + 5,7 . 60 + 5,7. 2,2 . 22
   P_u equivalent II = 1.717 kN
9. Determining IWF section with ∅ P_n value approaching P_u equivalent II

Table 3. IWF section with ∅ P_n value approaching P_u equivalent II

kL	PROFIL IWF	(Pr / Pc) ≥ 0,2	(Pr / Pc) < 0,2
	294 x 302	m	m
0,5	2.347	7,8	17,5
1	2.312	7,7	17,2
1,5	2.258	7,5	16,8
2	2.192	7,3	16,3
2,5	2.120	7,0	15,8
3	2.047	6,8	15,3
3,5	1.976	6,5	14,7
4	1.910	6,3	14,2
4,5	1.850	6,1	13,8
5	1.797	6,1	13,7
5,5	1.739	6,0	13,5
6	1.641	5,8	13,1
U	2,2		
Lp	3,6		
Lr	9,1		

10. IWF 294 x 302 is chosen
11. Calculating Pu equivalent III
    $\frac{P_u}{\varnothing P_n} = \frac{1.100}{1.797} = 0{,}61 > 0{,}2$
    Nilai m dan U are using values obtained from the table of subsequent section:
    m = 6,1 m^{-1}
    U = 2,2
    Hence:
    P_u equivalent III $= P_u + m \cdot M_{ux} + m \cdot U \cdot M_{uy}$
    P_u equivalent III $= 1.100 + 6{,}1 \cdot 60 + 6{,}1 \cdot 2{,}2 \cdot 22$
    P_u equivalent III $= 1.761$ kN
12. Determining IWF section with ∅ Pn value approaching Pu equivalent III
13. IWF 294 x 302 is chosen (equal to previous trial)
14. IWF 294 X 302 is used

4.4 Testing the Efficiency of Profiles as Beam-column Elements

The efficiency of the IWF steel section as a beam-column element can be tested with the interaction formula of compression and flexural equations based on SNI 1729:2015 article (H1-1a) and article (H1-1b) according to equation 2.33 and equation 2.34:

a. for : $\frac{P_r}{P_c} \geq 0{,}2$

$$\frac{P_r}{P_c} + \frac{8}{9}\left(\frac{M_{rx}}{M_{cx}} + \frac{M_{ry}}{M_{cy}}\right) \leq 1{,}0 \quad (7)$$

b. for : $\frac{P_r}{P_c} < 0{,}2$

$$\frac{P_r}{2\,P_c} + \left(\frac{M_{cx}}{M_{rx}} + \frac{M_{cy}}{M_{ry}}\right) \leq 1{,}0 \quad (8)$$

Particularly for the example,

$\frac{P_r}{P_c} = \frac{1.100}{1.797} = 0{,}61 > 0{,}2$ hence interaction formula for beam column:

$$\frac{P_r}{P_c} + \frac{8}{9}\left(\frac{M_{rx}}{M_{cx}} + \frac{M_{ry}}{M_{cy}}\right) \leq 1{,}0$$

$$\frac{1100}{1797} + \frac{8}{9}\left(\frac{60}{268{,}320} + \frac{22}{123{,}210}\right) \leq 1{,}0$$

$0{,}96 \leq 1{,}0$ ... (OK).

The interaction coefficient is very close to 1, hence the steel section obtained is efficient.

5 Conclusions

1. The resulting factored nominal moment capacity graphs ($\varnothing M_n$) can be used as a reference in the process of designing efficient beam elements.
2. The table of factored axial capacity ($\varnothing P_n$) and the resulting m and U coefficient can be used as a reference in the design process of beam-column elements to obtain efficient sections and reduce the iteration process.

References

1. A. Setiawan, Steel Structure Design with LRFD Method, Erlangga, Jakarta (2008)
2. R. Gunawan, Steel Construction Profile Table, Yogyakarta (1987)
3. SNI 1729:2015, Specifications For Structural Steel Buildings BSN, Jakarta (2015)
4. W. Dewobroto, Behavioral Steel Structures, Analysis & Design – AISC 2010 2nd Edition (Department of Civil Engineering UPH, Tangerang 2016)

Assessment of seismic responses toward multi-story buildings of steel special moment resisting frame by considering the setback irregularities

Ridho Aidil Fitrah[1*], *Masrilayanti* Masrilayanti[1], *Gita Zakiah* Putri[2], and *Zev Al* Jauhari[3]

[1]Civil Engineering Department, Andalas University, , 25163 Limau Manis, West Sumatera, Indonesia
[2]Civil Engineering Department, Astra Manufacturing Polytechnic, 14330 Sunter, Jakarta, Indonesia
[3]Civil Engineering Department, Bengkalis State Polytechnic, 28711 Sungai Alam, Riau, Indonesia

Abstract. Setback irregularities are considered where discontinuity between adjacent stories is excessive. This irregularity caused the probability of high damage at structures subjected to strong earthquake motion. For this purpose, this study was conducted by modeling the steel special moment frame (SMF) structures using a finite element calculation program with nonlinear static analysis compared to Padang city's response spectrum. The buildings are also modeled with two types of setbacks: single and multiple setbacks. The results of this paper are discussed including the explanation of many parameters that relate to elastic and inelastic seismic responses of steel special moment frame (SMF). Based on the results, the setback irregularities, both single and multiple setbacks, the inelastic seismic responses are adequately sufficient to SNI 1726 2019 regarding drift limit. The other seismic responses are also discussed in terms of fundamental periods, inter-story drifts, story stiffness, and base shear. Referred to Indonesian Seismic Provision, SNI 1726 2019, it is found that single setback building has more adequate than multiple setbacks in terms of seismic responses. Then, the seismic assessments between these setbacks are explained to address the recommendations about future prevention toward damages and failures in steel buildings.

1 Introduction

Vertical irregularities are found in some constructions such as multi-story or high-rise buildings and industrial plants. It is considered to give better serviceability and also architectural demand. Mass, stiffness, and strength are the main parameters to determine that vertical irregularities exist in multi-story buildings. The existence of the setback irregularities is determined as the sudden change of mass, stiffness, and strength due to the discontinuity of the adjacent story. According to the seismic provision in Indonesia, SNI 1729:2019, The setback irregularities are determined as a comparison the length of bays between adjacent stories is higher than 130%.

Throughout the year, most of the researches in reinforced concrete and steel buildings considering setback irregularities has stated the seismic responses of these buildings. It was found that there are several undesirable parameters because they generated a higher inter-story drift along with building height. Humar and Wright [1] concluded the setback would cause significant inelastic deformation if the numbers of bays in some stories were smaller than its lower stories. Duan and Chandler [2] stated that the structural members located near the setback would generate severe damage, as a result, the proper and stronger design should be applied specifically to those members. Chen et. al [3] considered controlling structural responses while the larger setbacks in stories simultaneously affected local vibration. Moreover, research from Karavasilis et.al [4] revealed the parameters that need to be concerned in buildings with setbacks such as the number of stories, geometrical irregularity, and beam-column strength ratio.

On the other hand, researchers also found that there are no significant effects regarding seismic responses. Piluso [5] stated that as long as the structures are designed to have a global collapse mechanism, the setbacks irregularities do not take place as the main cause for worsening the seismic responses. Based on practical evidence, Kappos and Scott [6] also concluded that there are no significant results regarding local failures in structural members near seatbacks while Tena-Colunga [7] also analyzed that plastic deformations did not find as the cause of the setbacks irregularity's existence. Romao et.al [8] stated that the seismic responses of reinforced concrete buildings with setbacks irregularities were adequate as well as regular frame according to the

* Corresponding author: ridho@eng.unand.ac.id

Eurocode provision. Therefore, there is still a contrary issue that needs further studies to acquire the proper explanation in terms of seismic responses due to setbacks irregularities. Also, it needs further observation while seismic accelerations and intensity always increase throughout the year depending on the movement of faults around the world. It also affects the development of the design to prevent excessive seismic response toward buildings.

In Indonesia, seismic acceleration in every province had risen higher based on provision SNI 1726:2019. More restrictions are applied concerning the chosen analysis for designing structural members either steel or reinforced concrete, specifically in high-intensity earthquake regions. The studies have been done in several cases to know how it can affect seismic responses and building design [9,10]. It was found that many indicators are playing the role to make structural members get attention to adjust sufficient structural responses for instance response modification factors, and amplification factors which are determined to acquire actual elastic and inelastic seismic responses and level of building's performance. Furthermore, both horizontal and vertical irregularities in buildings are also accounted to give restrictions in analysis and design [11,12]. Then, this study is conducted to obtain seismic responses due to setback irregularities in the multi-story building of steel special moment-resisting frames. The results of seismic response discussed in this paper for getting assessment about sufficiency toward SNI 1726:2019 due to various models of setback irregularities, single and multiple setbacks.

2 Methodology

2.1 Building Modelling

To research the seismic responses of special steel moment-resisting frame buildings, three various steel buildings were modeled in finite element software with 6 stories and 24 m total height above ground level. Buildings had a regular plan view with a 6.0 m bay span. To get a seismic responses comparison, the regular building (SMRF-Reg) without setback was modeled in Fig 1. The setback irregularities were applied to be sufficiently categorized as vertical irregularities just as mentioned in SNI 1726:2019 [13]. Single and multiple setbacks were shown in Fig 1 (b) and (c). Fig 1(b) was a single setback building that has 300% setbacks between story 3 and story 4 (SMRF-Set1). Fig 1(c) was a multiple setback building that has 150% setback between story 2 and story 3 and 200% setback between story 3 and story 4 (SMRF-Set2).

Every structural member in special steel moment-resisting frame buildings was sufficiently designed based on AISC 360-16 (SNI 1729:2020) by using its specific factors such as modification response (R) is 8, amplification factor (C_d) is 5.5, and importance factor (I) is 1.5 [14]. Then, the buildings were also able to withstand toward dead load and live load respectively 2.5 kN/m^2 and 3 kN/m^2.

2.2 Seismic Analysis

Seismic analysis toward these buildings was conducted by generating a response spectrum of Padang City, West Sumatera, which included a dynamic analysis, with site class D (SD). According to SNI 1726:2019, by obtaining spectral acceleration at short period (SD_s) and one second period (SD_1) also the importance factor (I), these buildings were categorized as seismic design category of D where the special steel moment-resisting frame design is applied as the designed models. Furthermore, the response spectrum of Padang City which is used in this analysis already shown in Fig.2.

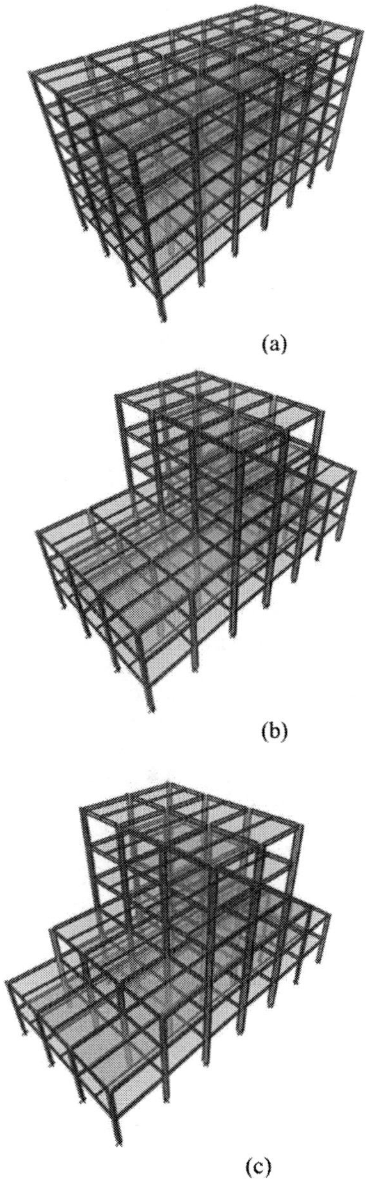

Fig. 1. Building models (a) SMRF-Reg (b) SMRF-Set1 (c) SMRF-Set2

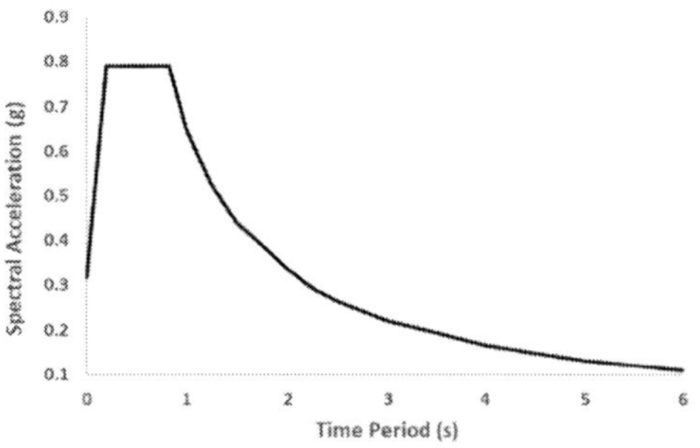

Fig. 2. Response Spectrum of Padang City (SD)

According to SNI 1726:2019, The base shears both static and dynamic analysis must have a similar comparison that meets the value 100%, which means the scale factor is needed in analysis to get the representative results toward elastic seismic response during the earthquake. The displacement and inter-story drift either X or Y direction were evaluated and discussed regarding the sufficiency toward SNI 1726:2019 drift limitation.

3 Results and Discussions

3.1 Fundamental Period

The fundamental period is one of the dynamic properties to determine how seismic force will impact building responses. Furthermore, the fundamental period is influenced by mass and stiffness which means structural configuration takes part in determining this value. The longer the fundamental period of the building, the more flexible structure will deform. It affects the value of base shear and structure's displacements.

Table 1 shows the comparison between the calculated period of the first mode through computer analysis and empirical fundamental periods according to SNI 1726:2019 in three building models. It shows that the SMRF-Reg had the biggest fundamental period among other models (SMRF-Set1 and SMRF-Set2) which these models categorized as having setback irregularities. It could be explained that floor plans were decreased due to setback irregularities while the story mass similarly decreased. There are also no torsional issues such as eccentricity between the center of rigidity and center of mass.

Table 1. Comparison of Buildings' Fundamental Period

Structure	Coeffi-cient (C_u)	Empirical Fundamental Period (T_a) (s)	First Mode Period (T_1) (s)
SMRF-Reg	1.4	$0.0724h_n^{0.8}$	1.781
SMRF-Set1			1.426
SMRF-Set2			1.478

h_n: Total of Building Height

However, as long as the center of mass and rigidity between adjacent stories tend to have a similar location, the setback irregularities would not take a significant effect on the structural modes. To emphasize the explanation, as long as the structural members were designed adequately, the setback irregularities also would not have a possibility to experience torsional either in X or Y direction.

3.2 Story Stiffness

To obtain a better comprehension of the seismic response, the story stiffness is also the parameter to determine how the flexibility of the structure. As mentioned in the previous discussion, story stiffness will determine how significant the deformation of structures is.

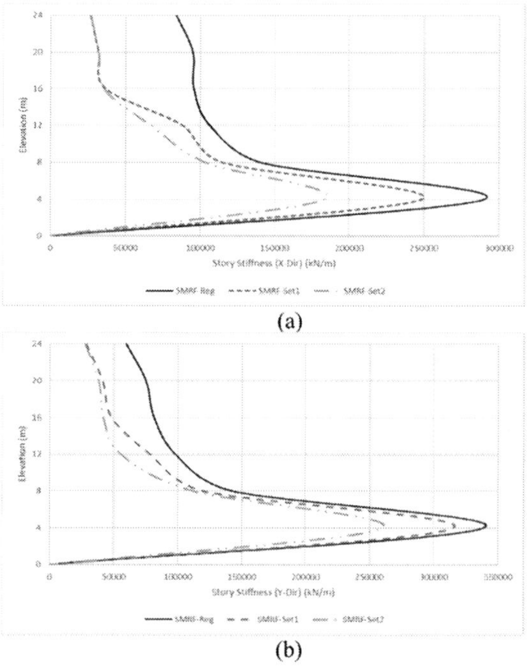

Fig. 3. Story Stiffness of Steel Buildings (a) X-direction (b) Y-direction

According to Fig.3, the graph of story stiffness of three steel structural models, either X or Y- direction, the SMRF-Reg had the largest story stiffness compared to SMRF-Set1 and SMRF-Set2. However, single and multiple setbacks were identified exist in X-direction. Then, it could be explained that the story stiffness of a single setback (SMRF-Set1) was slightly higher than the multiple setbacks (SMRF-Set2) even though the percentage of a single setback was about twice higher than multiple setbacks. Moreover, the discontinuity in the certain plane i.e setback irregularities would affect the structure to become more flexible so that the higher structural deformation in a certain direction must be larger than the regular building.

https://doi.org/10.1051/e3sconf/202133105007

3.3 Base Shear

The influence of setback irregularities on special steel moment-resisting frame (SMRF) buildings is observed according to base shear. This force is determined by the story stiffness and fundamental period. Fig.4 shows that the SMRF-Reg has a 60%-75% larger base shear force than the other setback structures (SMRF-Set1 and SMRF Set-2) either in X or Y direction. Therefore, the setback irregularities caused significant effects regarding base shear force. However, when it is compared to setback structures, it can be seen that multiple setbacks structure has slightly different base shear force than the single setback structure. Then, it can be explained that the multiple setbacks structure (SMRF-Set2) is more flexible than the single setback structure (SMRF-Set1).

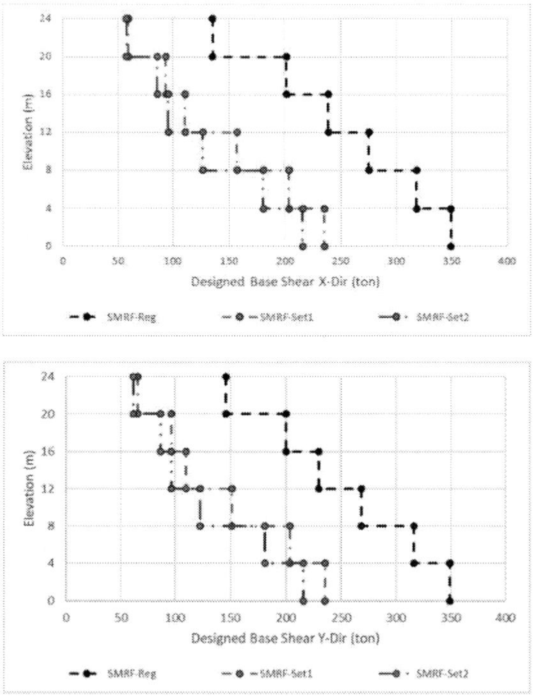

Fig. 4. Base Shear of Steel Buildings (a) X-direction (b) Y-direction

3.4 Story Displacement

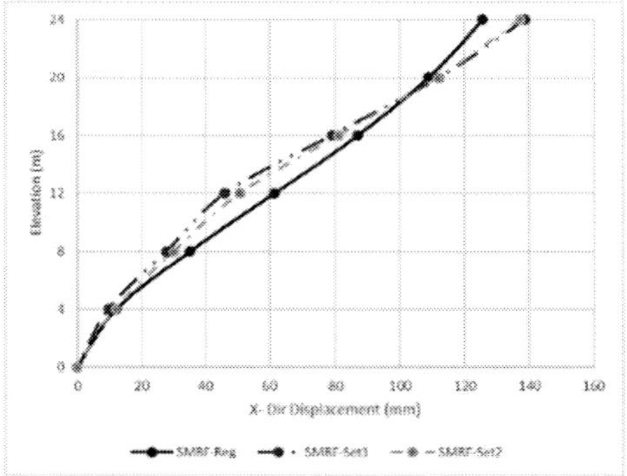

a)

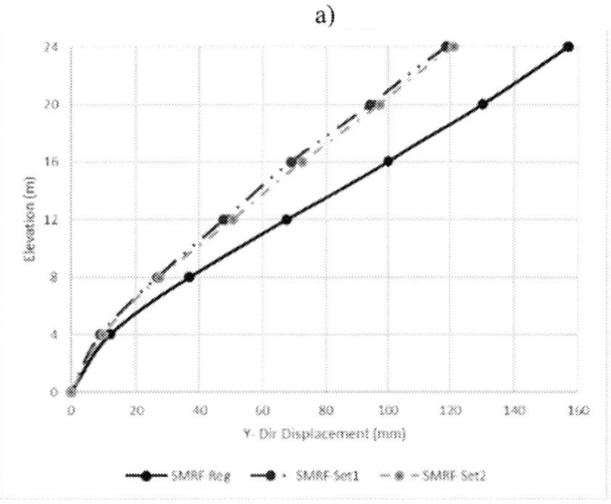

(b)

Fig. 5. Story displacement of Steel Buildings (a) X-direction (b) Y-direction

To study how the setback irregularities would affect inelastic seismic response due to seismic excitation. Then, the story displacement is discussed in Fig. 5 which represents the comparison between the story displacement in X and Y direction. The graph shows that the story displacement of regular structure (SMRF-Reg) is dominantly higher than the setback structures about 5%-25% in X-direction and 11%-33% in Y-direction. It can be annotated that the seismic weight of this building is also higher than setback structures. Therefore, the P-delta would be highly accounted to this regular frame rather than the setbacks structures.

Either in X and Y direction, the story displacements of setbacks structure (SMRF-Set1 and SMRF-Set2) tend to have similar values. Therefore, single and multiple setbacks did not influence story displacement. However, to assess the sufficiency of the elastic seismic response toward Indonesia's seismic provision SNI 1726:2019, The advanced calculation about inter-story drift is discussed below.

3.5 Inter-Story Drift

The inter-story drift is defined as the relative translational displacement between two consecutive floors. SNI 1726:2019 stated that the limitation of inter-story drift must be determined to prevent excessive inelastic responses. To obtain the comprehension of the influence of the setback irregularities regarding inter-story drift, Fig.6 shows the comparison of them in X and Y-direction. Both directions seem an adequate value for the inter-story drift. Generally, the setback irregularities did not have a significant influence. However, the buildings were designed to accommodate the load combination in serviceability and ultimate condition. This emphasizes that as long as the steel design is sufficient, so there are no issues regarding the excessive inelastic seismic responses upon the drift limit.

On the other hand, in the X-direction where the setback appeared, the inter-story drift in setback structures tends to have a bigger value rather than regular structures. Though the value is adequately sufficient to drift limit, the pattern of inter-story in setback structures (SMRF-Set1 and SMRF-Set2) did not follow a conservative pattern where the inter-story drift must be smaller while the story is increasing. For example, from story-3 to story-4 (elevation 12 m to 16 m), the inter-story drift in SMRF-Set1 was significantly changing about 78% while SMR-Set2 was significantly changing about 44%.

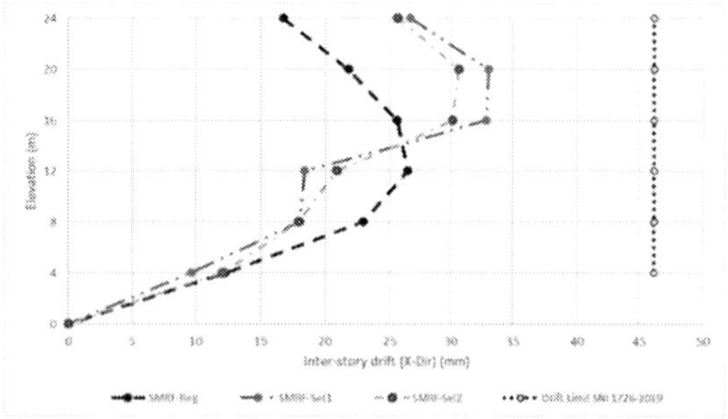

(a)

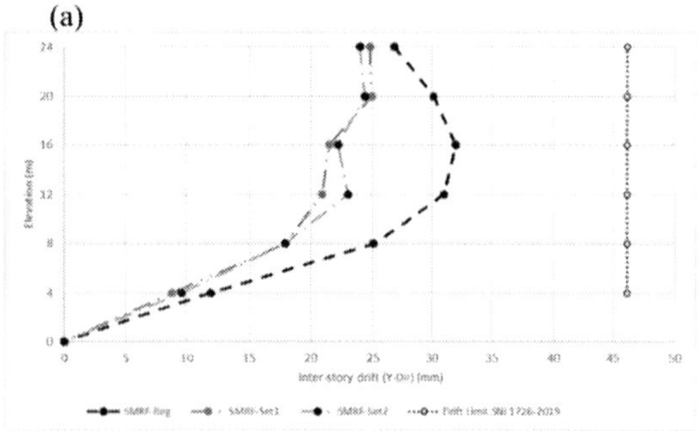

(b)

Fig. 6. Inter-story Drift of Steel Buildings (a) X-direction (b) Y-direction

The assessment can be carried out regarding the significant change between the adjacent stories. It will lead the structures would have partly plastic deformation and soft-story mechanism. Then, for the future prevention of failure in certain floors, several options can be applied to decrease the inter-story drift.

4 Conclusions

This research aimed to assess the inelastic seismic responses toward the multi-story buildings of special steel moment-resisting frames due to the existence of setback irregularities according to SNI 1726:2019. This assessment is conducted because of the increment of seismic acceleration in Padang City. To obtain the representative comparisons, the regular building was also modeled. Furthermore, these buildings were also designed to accommodate the service and ultimate loads, including the seismic load. Analysis and assessment were conducted in fundamental periods, the base shear, the story stiffness, and inter-story drift.

So, the conclusions of the assessment due to setback irregularities can be listed as follows:

1. The setback irregularities did not influence the fundamental periods and mode shapes, as long as the structural members were properly designed. Then, mode shapes 1 and 2 were translational in both X and Y-direction.
2. Story stiffness and base shear are the parameters that are affected due to the setback irregularities. It is found that these values were significantly smaller than a regular building. Multiple setbacks structure is found to be more flexible rather than single setback structure. Then, it influences the lateral deformation in a building.
3. Based on the assessment in the inter-story drift of steel buildings seems to be adequately sufficient to SNI 1726:2019. However, due to the setback irregularities, the pattern of the inter-story drift did not follow a conservative pattern. The bigger percentage of setbacks then affects the significant change between adjacent stories.
4. To prevent failure such as soft-story or partly plastic deformation because of the significant change of inter-story drift between the adjacent stories, the future design must be considered to increase the story stiffness where setbacks appeared.

References

1. J.L Humar, E.W Wright, Earthquake Engineering & Structural Dynamics, **5** (1977).
2. X.N Duan, A.M Chandler, Earthquake Engineering and Structural Dynamics, **24**, 761-77 (1995)
3. C Chen, N.T.K Lam, P Mendis, *The bifurcation behavior of vertically irregular buildings in low seismicity regions*. In Proceeding: 12th world conference on earthquake engineering. (2000).
4. T. L.Karavasilis, N. Bazeos, and D. E. Beskos, J. of Constructional Steel Research, **64** (6) (2008)
5. F. Mazzolani, V. Piluso, *Theory, and design of seismic resistant steel frames*. London, New York: FN & SPON an Imprint of Chapman & Hall (2012)
6. AJ Kappos, SG Scott, *Seismic assessment of an R/C building with setbacks using nonlinear static and dynamic analysis procedures*, In Booth, editor. Seismic design practice into the next century. (1998)
7. A Tena-Colunga, *Evaluation of the seismic response of slender, setback RC moment-resisting frame buildings designed according to the seismic guidelines of a modern seismic code*. In: 13th world conference on earthquake engineering, (2004)
8. X Romão, A Costa, R Delgado, *Seismic behavior of reinforced concrete frames with setbacks*. In: 13th world conference on earthquake engineering, (2004)
9. Z.A Jauhari, F. Ananda, R.A Fitrah, *Comparison of the design acceleration response spectra in Riau Province between SNI 1726: 2019 and SNI 1726: 2012 methods*. In Proceeding IOP Conference Series: Earth and Environmental Science, (2021)
10. R.A Fitrah, A.P Melinda, Rang Teknik Journal, **2** (2018)

 https://doi.org/10.1051/e3sconf/202133105007

11. R.A Fitrah, D.I Mazni, W Pratiwi, Z.A Jauhari, *Seismic assessment of irregularities in steel special moment resisting frame with asymmetric-plan building (case study: Gedung D-Universitas Dharma Andalas)*. In Proceeding IOP Conference Series: Earth and Environmental Science, (202
12. B. Budiono, E.B Wicaksono, Jurnal Teknik Sipil, 113-126 (2016)
13. National Standardization Agency, SNI 1726-2019, *Design Method of Earthquake Resistance for Buildings and Other Structures*, BSN, Jakarta (2019).
14. National Standardization Agency, SNI 1729: 2020, *Requirements of Steel Structure for Building*, BSN, Jakarta (2020)

Performance based evaluation of an existing continuous reinforced concrete bridge – a case study

Erwin Lim[1], *Dyah* Kusumastuti[1*], *Rildova*[1], *Mario* Asneindra[1], and *Silvester* Sandy Mulyadi[1]

[1]Department of Civil Engineering, Faculty of Civil and Environmental Engineering, Bandung Institute of Technology, West Java, Indonesia

Abstract. A performance-based evaluation was conducted on an eleven-span reinforced concrete bridge located in a high seismicity area. This bridge was designed using SNI 2833:1992 and to be checked using the updated Indonesian seismic hazard map. In addition, a non-linear time history analysis was conducted using eleven pairs of ground motions to evaluate the structural performance of the bridge pier for the anticipated service life. Elastic analysis shows that the existing bridge still satisfies the strength and serviceability requirement despite the increase in seismic demand. Meanwhile, by evaluating the average inelastic longitudinal reinforcement and concrete strain at bridge pier for selected eleven ground motions, this study suggests that the bridge is still "fully operational" and "life safety" when subjected to the lower level (100 year return period) and upper level (1000 year return period) ground motions, respectively.

1 Introduction

An existing eleven-span reinforced concrete bridge located in a high seismicity area was designed using SNI 2833:1992 [1] and constructed in early 2010. In 2016, SNI 2833 [2] related to the seismic design of the bridge was updated and in 2017, an update on the Indonesian seismic hazard map [3] was published. The bridge seismic design approach is quite different between the 1992 and 2016 versions. Therefore, it is of interest to reevaluate the bridge's structural performance using the latest nonlinear time history analysis as the tool. However, since it is an existing bridge, it would be unfair to adopt the latest design requirement and expect the old structure to meet the performance level set by the new regulation.

Federal Highway Administration [4] categorizes the performance level of an existing bridge based on the anticipated service life, bridge importance, and service life category. The performance is evaluated under two levels of ground motion, i.e. lower level (corresponding to the earthquake with a return period of 100 years) and upper level (corresponding to the earthquake with a return period of 1000 years) as shown in Table 1. Unfortunately, this document does not provide a clear approach on how to quantify the expected performance.

On the other hand, NCHRP Synthesis 440 [5] provides an in-depth discussion and background on performance-based seismic design for bridges. Moreover, it also summarizes quantitatively the acceptance criteria for some performance levels, which are available in various regulations. For instance, Table 2 shows the limit criteria for concrete strain evaluated using several regulations. However, some inadequacy was found in this document.

Table 2 suggests that AASHTO SGS, SCDOT, and Caltrans only evaluate the performance under one level of the earthquake, hence not in line with FHWA. Meanwhile, Priestley et.al., in Table 2, has suggested criteria for two-level of earthquakes, i.e. serviceability and damage levels. However, the magnitude of these earthquakes was not specified. POLA/POLB, which regulates the design of ports, has categorized the performance level into three and the magnitudes of respective earthquakes were also specified. However, the intended service life is for 50 years, not 75 years as in bridges. This problem remains until 2020 when NCHRP 949 [6] was published. This guideline fills the gap of FHWA's previous publication and provides quantitative criteria for two levels of evaluated earthquakes.

This study adopts these two guidelines [4,6] and applies them in the performance-based evaluation of the aforementioned bridge. The piers are modeled using the distributed inelasticity or commonly called as fiber model. This technique discretized the pier cross-section, the main seismic resistance element, into fibers. For each concrete and longitudinal steel reinforcement fiber, the stress-strain curve is then explicitly defined. Using this way, it can also capture implicitly the flexural stress-strain behavior of the section along the length through the integration process by assuming plane sections remain plane. Since the length to depth ratio of all piers is larger than 4 where shear behavior is less dominant, this approach is considered appropriate. Although one of the

* Corresponding author: dkusumastuti@gmail.com

 https://doi.org/10.1051/e3sconf/202133105008

drawbacks of this technique is that it cannot capture the degradation behavior, however considering that all piers satisfy the shear capacity design criteria (will be explained later), the degradation behavior may not be significant as it is controlled by the flexural response.

The performance level is then evaluated using nonlinear time history analysis under eleven pairs of ground motions. Using the average strain value obtained from eleven pairs of ground motions [7], the target performance is evaluated using the NCHRP 949 criteria as shown in Table 3

Table 1. Performance level [4]

EARTHQUAKE GROUND MOTION	BRIDGE IMPORTANCE and SERVICE LIFE CATEGORY					
	Standard			Essential		
	ASL 1	ASL 2	ASL 3	ASL 1	ASL 2	ASL 3
Lower Level Ground Motion 50 percent probability of exceedance in 75 years; return period is about 100 years.	PLO[4]	PL3	PL3	PLO[4]	PL3	PL3
Upper Level Ground Motion 7 percent probability of exceedance in 75 years; return period is about 1,000 years.	PLO[4]	PL1	PL1	PLO[4]	PL1	PL2

Notes:

1. Anticipated Service Life categories are:
   - ASL 1: 0 – 15 years
   - ASL 2: 16 – 50 years
   - ASL 3: > 50 years
2. Performance Levels are:
   - **PL0: No minimum** level of performance is recommended.
   - **PL1: Life safety.** Significant damage is sustained and service is significantly disrupted, but life safety is preserved. The bridge may need to be replaced after a large earthquake.
   - **PL2: Operational.** Damage sustained is minimal and service for emergency vehicles should be available after inspection and clearance of debris. Bridge should be reparable with or without restrictions on traffic flow.
   - **PL3: Fully Operational.** No damage is sustained and full service is available for all vehicles immediately after the earthquake. No repairs are required.
3. Spectral ordinates and peak ground accelerations may be found as described in Chapter 2.
4. Bridges assigned a Performance Level of PL0 have 15 years, or less, anticipated service life (ASL) and are candidates for replacement or rehabilitation. If the bridge is replaced or rehabilitated, the ASL category will change and so will the required Performance Level.

Table 2. Compression strain limit [5]

COMPRESSION STRAIN LIMITS—CONCRETE

Agency	Year	Ground Motion	Component and Location	Explicit Damage State?	Damage State	Strain Limit (in./in.)
AASHTO SGS	2011	7% in 75 yr (1,000-yr RP)	Column plastic hinge confined concrete	No	Transverse reinforcement fracture/core crushing	$0.004 + 1.4\frac{\rho_s f_{yh}\varepsilon_{su}}{f'_{cc}}$
			Column/pile in-ground plastic hinge confined concrete	No	Transverse reinforcement fracture/core crushing	0.020
SCDOT	2008	3% in 75 yr (1,000-yr RP)	Column plastic hinge confined concrete	No	Transverse reinforcement fracture/core crushing	$0.004 + 1.4\frac{\rho_s (f_{yh})\varepsilon_{su}}{f'_{cc}}$ Note: Omission of f_{yh} in the source may be a typographical error
Caltrans	2010	5% in 50 yr (1,000-yr RP)	Column plastic hinge confined concrete	No	Transverse reinforcement fracture/core crushing	$0.004 + 1.4\frac{\rho_s f_{yh}\varepsilon_{su}}{f'_{cc}}$
Priestley et al.	2007	Serviceability	Column plastic hinge unconfined concrete	Yes	Cover spalling	0.004
		Damage control	Column plastic hinge confined concrete	Yes	Transverse reinforcement fracture/core crushing	$0.004 + 1.4\frac{\rho_s f_{yh}\varepsilon_{su}}{f'_{cc}}$
POLA/POLB	2010/2009	50% in 50 yr (72-yr RP)	Solid concrete pile-to-deck plastic hinge extreme fiber concrete compression strain	No	Cover spalling	0.005 (cover concrete)*
		10% in 50 yr (475-yr RP)		No	Transverse reinforcement fracture/core crushing	$0.005 + 1.1\rho_s \le 0.025$ (core concrete)*
		2/3 of 2% in 50 yr (2/3 of 2, 475-yr RP)		No	NA	No limit
POLA/POLB	2010/2009	50% in 50 yr (72-yr RP)	Solid concrete pile-to-deck plastic hinge extreme fiber concrete compression strain	No	Cover spalling	0.005 (cover concrete)*
		10% in 50 yr (475-yr RP)		No	Transverse reinforcement fracture/core crushing	$0.005 + 1.1\rho_s \le 0.025$ (core concrete)*
		2/3 of 2% in 50 yr (2/3 of 2, 475-yr RP)		No	NA	No limit

Table 3. Strain limit under targeted performance level [6]

Design Parameter	Performance Level		
	PL1: Life Safety	PL2: Operational	PL3: Fully Operational
Reinf. tensile strain limit	$\varepsilon_{s\,buckling}^{bar} = 0.032 + 790\rho_s \frac{f_{yh}}{E_s} - 0.14\frac{P}{f_{ce}A_g}$	$\varepsilon_s = 0.8\,\varepsilon_{s\,buckling}^{bar}$	≤ 0.010
Concrete compressive strain limit	$\varepsilon_c = 1.4\left(0.004 + 1.4\frac{\rho_s f_{yh}\varepsilon_{su}}{f'_{cc}}\right)$	$\varepsilon_c = \left(0.004 + 1.4\frac{\rho_s f_{yh}\varepsilon_{su}}{f'_{cc}}\right)$	≤ 0.004

 https://doi.org/10.1051/e3sconf/202133105008

Table 3 tabulates the maximum strain of concrete and longitudinal reinforcement bar under three different performance levels predetermined in Table 1. The strain criteria for the reinforcement bar are governed by buckling mode, while that of concrete is limited by the crushing strain of the concrete core.

As an alternative, one can also determine the target performance based on the deformation, instead of strain level. This modeling technique is commonly known as the concentrated hinge model or lumped plasticity. Using this technique, one needs to define the backbone curve representing the force-displacement relationship of the member. Usually, this relationship is represented using a moment-rotation relationship. This relationship will need to incorporate axial and moment interaction. One of the advantages of this technique over distributed plasticity model is that it can capture the strength degradation better. Nevertheless, some assumptions are also needed, such as the determination of plastic hinge length. Also, in the development of the moment-rotation relationship, plane sections remain plane is also used.

2 Existing Bridge Information

The bridge under evaluation consists of 11 spans. The first (A1-P1) and last spans (A2-P10) connected to abutments are simply supported, while P1-P6 and P6-P10 spans are continuous as shown in Fig. 1. The typical span length is approximately 42 m, while the pier height varies from 13.5 m at P1 to 27 m at P6. The substructure was assumed to be adequate to resist all loads, and analysis was limited to the upper structures.

The typical transverse section of the bridge is shown in Fig. 2. In the longitudinal direction, the piers behave under single curvature, while in the transverse direction they behave in double curvature. The typical pier dimension is 2 m x 3.5 m, with concrete strength equivalent to 30 MPa. The piers' main longitudinal reinforcements are of D25 rebar with steel ratio ranging from 0.7 % to 1.2 % as shown in Fig. 3. Meanwhile, typical hoops and cross ties use D13 rebar.

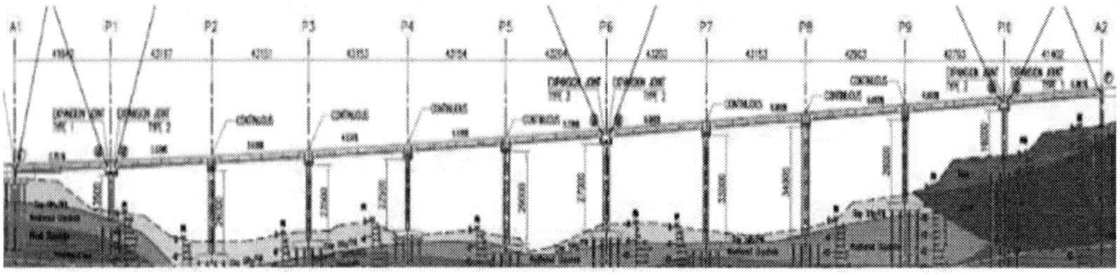

Fig. 1. Longitudinal bridge profil

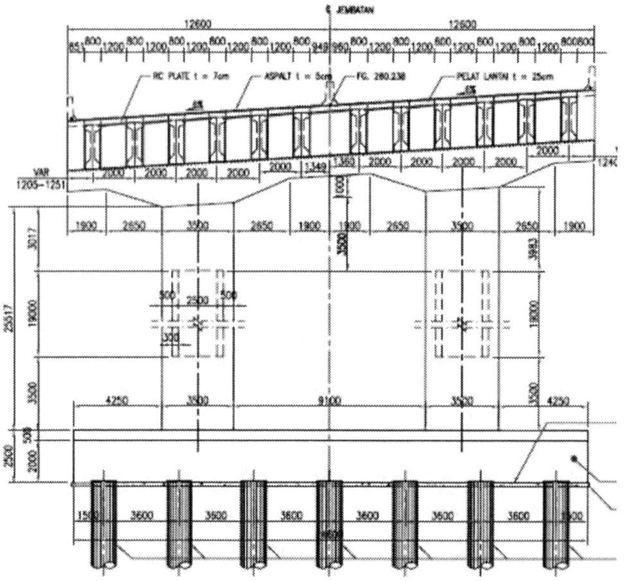

Fig. 2. Transver6se bridge profile

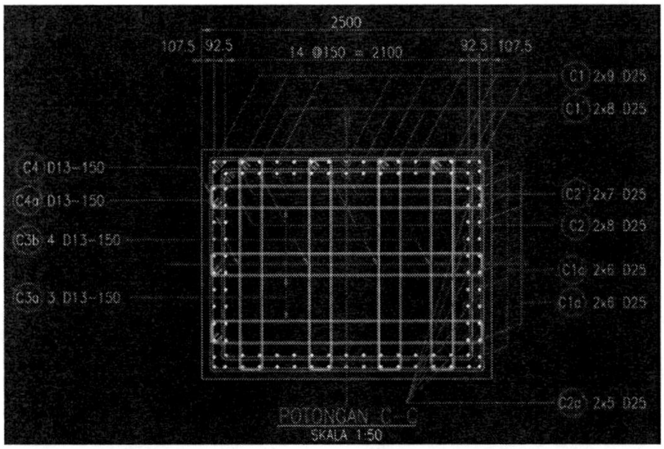

Fig. 3. Typical pier dimension and reinforcement

3 Elastic Analysis

The 3D model was developed using the properties specified in the construction drawing and is shown in Fig. 4. Based on dynamic analysis, Table 4 shows dominant mode shapes and the corresponding structural periods. A simple check on the strength adequacy based on the response spectrum in Fig. 5 analysis was carried out. The strength evaluation on piers showed that the flexural capacity of piers was sufficient as shown in the P-M diagram in Fig. 6. Moreover, it was also verified that the piers fulfilled the shear capacity design requirement. In this calculation, the shear strength is compared to the plastic shear demand due to the formation of plastic hinges in piers.

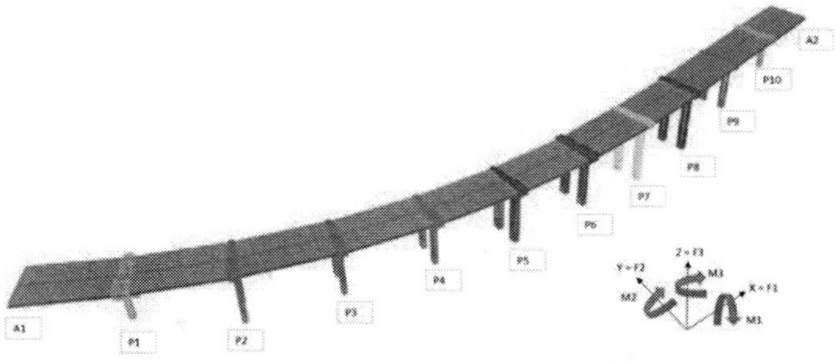

Fig. 4. Three-dimensional modeling

Table 4. Dominant mode shapes

Mode	Period	UX	UY	SumUX	SumUY	RZ	SumRZ
Unitless	Sec	Unitless	Unitless	Unitless	Unitless	Unitless	Unitless
1	0.820076	0.002544	0.43	0.002544	0.43	0.18	0.18
2	0.652082	0.12	0.2	0.13	0.63	0.22	0.4
3	0.532955	0.07068	0.08901	0.2	0.72	0.0001004	0.4
4	0.388691	0.04942	0.008963	0.25	0.73	0.14	0.54
5	0.378309	0.0003879	0.000001451	0.25	0.73	0.001505	0.54
6	0.37454	0.000001373	2.675E-07	0.25	0.73	0.000005156	0.54
7	0.356172	0.0008917	0.00004187	0.25	0.73	0.0002579	0.54
8	0.354159	0.0001242	0.000009287	0.25	0.73	0.0001441	0.54
9	0.353168	0.0006037	0.0003479	0.25	0.73	0.00001141	0.54
10	0.352934	0.0005314	0.0001004	0.25	0.73	0.0000634	0.54
11	0.335235	0.29	0.02299	0.54	0.75	0.06553	0.61
12	0.331133	0.23	0.01642	0.77	0.77	0.05516	0.66
13	0.327074	0.02939	0.002572	0.8	0.77	0.008352	0.67
14	0.307504	0.00001512	0.00008703	0.8	0.77	0.00002483	0.67
15	0.304865	1.074E-09	1.339E-08	0.8	0.77	2.711E-09	0.67
16	0.294102	0.004307	0.05487	0.81	0.83	0.00141	0.67
17	0.282639	0.0006906	0.009036	0.81	0.83	0.0003908	0.67
18	0.278257	0.00005826	0.0000103	0.81	0.83	0.000003453	0.67
19	0.27546	0.000001118	0.000301	0.81	0.83	0.0001221	0.67
20	0.27527	8.793E-07	0.0004812	0.81	0.84	0.00001895	0.67
21	0.273059	0.00003826	0.000001094	0.81	0.84	0.000009663	0.67
22	0.256783	0.0001183	0.00003244	0.81	0.84	0.00003284	0.67
23	0.239781	0.00005469	4.178E-07	0.81	0.84	0.0005385	0.67
24	0.237632	0.000002316	0.000007022	0.81	0.84	0.0003904	0.67
25	0.235973	0.000008949	0.000002656	0.81	0.84	0.0001179	0.67

 https://doi.org/10.1051/e3sconf/202133105008

The inspection on shear strength is very important because using the nonlinear time history analysis, only flexural behavior can be modeled and captured. Therefore, by ensuring the shear capacity is adequate, it can be inferred that the flexural behavior of piers can be fully developed and strength degradation due to shear behavior can be delayed. In addition, this check is also important because the pier is analyzed using distributed inelasticity model which cannot capture the degradation behavior well.

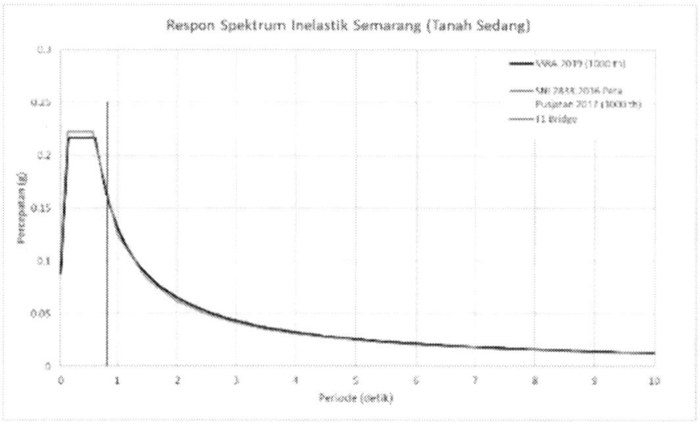

Fig. 5. Inelastic response spectrum

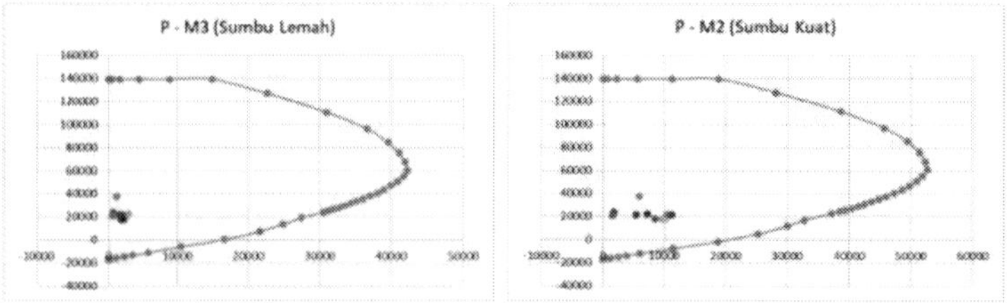

Fig. 6. Typical PM Diagram for piers

4 Non-Linear Time History Analysis

As mentioned in NCHRP 949, the criteria of performance level under two levels of an earthquake are evaluated using the strain value, both in piers' extreme concrete fiber and in longitudinal reinforcement bars. Therefore, the fiber model was generated using commercial software and is presented in Fig. 7. First, the pier section was modeled and the concrete material was discretized into 9x9 elements. The longitudinal reinforcements were modeled individually, as indicated in Fig. 7

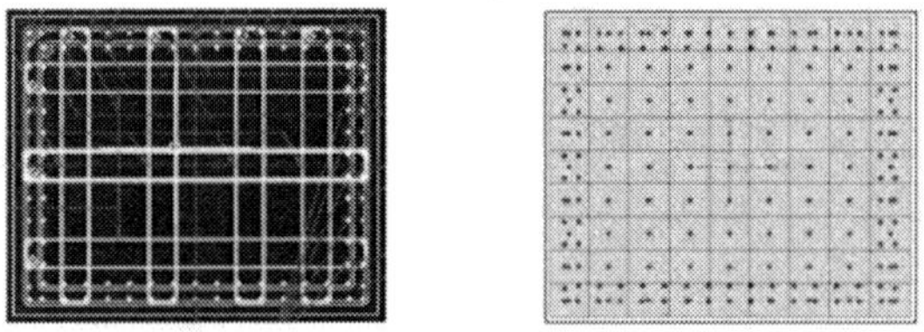

a. Pier section b. Section discretization

Fig. 7. Discretization of pier cross-section

The material constitutive equations for confined concrete and longitudinal steel reinforcement under monotonic loading followed those shown in Fig. 8. The monotonic relationships were then transformed into hysteretic relationships. For concrete material, the Takeda model was adopted, while the kinematic model was used for the reinforcement bar.

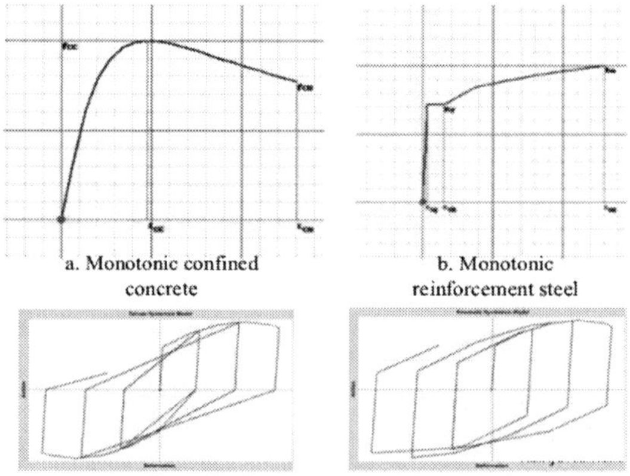

a. Monotonic confined concrete b. Monotonic reinforcement steel
c. Hysteretic confined concrete d. Hysteretic steel reinforcement

Fig. 8. Constitutive relationship for material

Eleven pairs of ground motions were developed considering all major earthquake mechanisms within a

500 km radius from the site [8], as shown in Table 5. Figure 9 also plots the acceleration histories for these ground motions. It is noteworthy to mention that two levels of the earthquake [4,6] were developed, i.e., for 100 years return period (lower level) and 1000 years return period (upper level).

After conducting nonlinear time history analyses using eleven pairs of ground motions, the stress and strain histories of extreme concrete fibers and longitudinal reinforcement bars were plotted and examined. An example of the plot is given in Fig. 10. For evaluation on the acceptance criteria, the average of the maximum value obtained for each pair of ground motions is calculated at extreme fibers. These values are then compared with the NCHRP 949's criteria shown in Table 3.

The so-called engineering design parameters values shown in Table 3 links the analysis results to decide the performance level. At the life safety level, the reinforcement strain limit of the longitudinal bar is based on its tensile limit to prevent bar fracture. Meanwhile, the concrete compressive strain limit is based on Mander's model with factor 1.4 to account for the inherited conservatism [6]. For the operational level, the tensile steel strain limit is based on statistical data, in which less than 10% of rebars are expected to buckle [9], while the concrete compression limit is based on the work of Mander [6]

Table 5. Summary of ground motions

No	Ground Motion	Time (second)	Direction	100 years rp		1000 years rp	
				Min Acc. (g)	Max Acc. (g)	Min Acc. (g)	Max Acc. (g)
1	Shallow Crustal A-GRN	19.995	A-GRN180	-0.2122	0.1585	-0.4213	0.3375
			A-GRN270	-0.1190	0.1562	-0.2203	0.3045
2	Shallow Background FDS	39.98	FDS262	-0.1936	0.1452	-0.3670	0.2644
			FDS172	-0.1715	0.1269	-0.3383	0.2953
3	Benioff Padang30sept	100	Padang30sept2009EW	-0.0283	0.0492	-0.0620	0.0939
			Padang30sept2009NS	-0.2156	0.2823	-0.5471	0.4786
4	Shallow Crustal A-HMC	17.995	A-HMC180	-0.1145	0.1078	-0.2406	0.2176
			A-HMC270	-0.1458	0.2349	-0.2703	0.4578
5	Shallow Background H-VC4	29.985	H-VC4000	-0.2052	0.1569	-0.4055	0.3031
			H-VC4090	-0.1384	0.1592	-0.2733	0.3162
6	Megathrust-HWA026	89.995	HWA026-W	-0.1459	0.2182	-0.2884	0.4230
			HWA026-N	-0.1096	0.1473	-0.2820	0.2951
7	Megathrust-MYG 013	299.99	MYG013110311146EW	-0.1059	0.1232	-0.2050	0.2386
			MYG013110311146NS	-0.1642	0.2446	-0.3484	0.4614
8	Shallow Crustal PKC	39.975	PKC360	-0.2051	0.1837	-0.4118	0.3631
			PKC090	-0.1594	0.1194	-0.3153	0.2408
9	Megathrust-HKD12	123.99	HKD1260309260450 EW	-0.1085	0.1915	-0.2335	0.3673
			HKD1260309260450 NS	-0.0913	0.1923	-0.2017	0.3737
10	Megathrust-KAU078	89.995	KAU078-W	-0.2277	0.1771	-0.3601	0.4430
			KAU078-N	-0.0866	0.1283	-0.2567	0.2195
11	Benioff KSH	32.975	KSH-T1	-0.1841	0.1481	-0.3599	0.2924
			KSH-L1	-0.1874	0.1372	-0.3648	0.2603

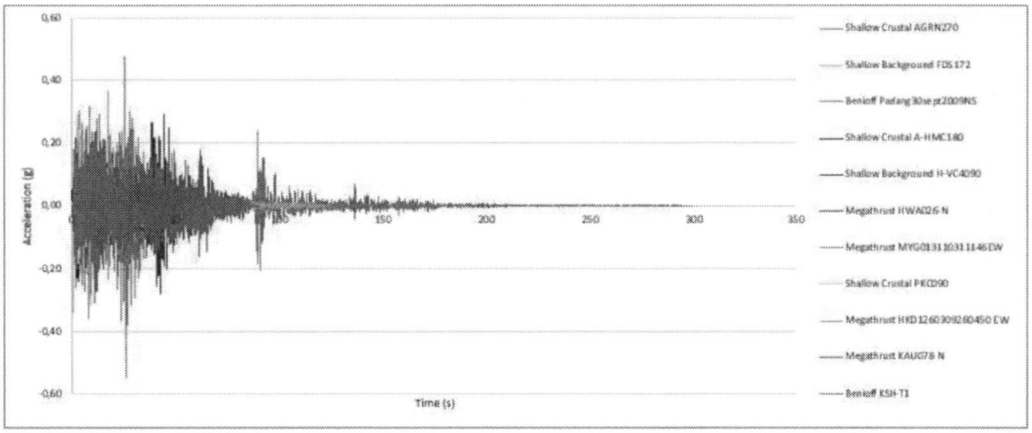

Fig. 9. Acceleration history of 11 ground motion

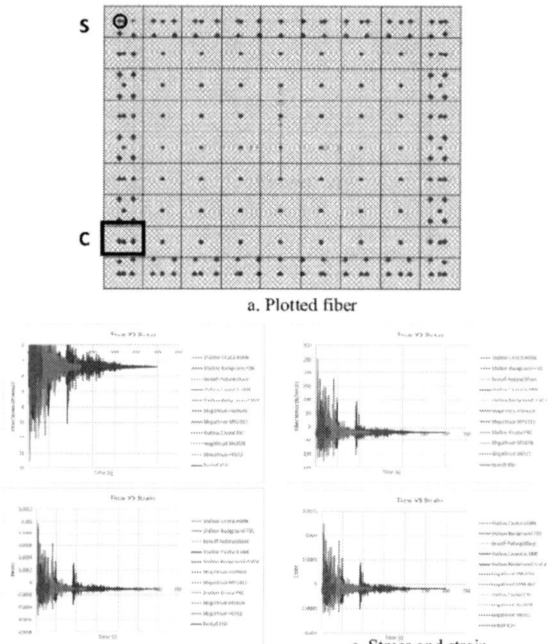

a. Plotted fiber

b. Stress and strain history of concrete

c. Stress and strain history of longitudinal

Fig. 10. Stress and strain history for extreme fiber subjected to eleven pairs of ground motio

The comparisons show that for lower-level earthquake ground motion (return period of 100 years), the attained performance level is fully operational, which complies with the FHWA requirement in Table 1. For upper-level earthquake ground motion (return period of 1000 years), the achieved performance level is life safety, which also satisfies the FHWA requirement. Therefore, the existing bridge shows adequate capacity under the current seismic load.

5 Conclusions

This study adopts the latest NCHRP 949 Guideline to quantitatively evaluate the performance level of an existing continuous reinforced concrete bridge located in a high seismic zone. This guideline fills the gap of previous guidelines in which the acceptance criteria for strain limit of concrete and longitudinal steel reinforcement under two levels of an earthquake under three different categories, depending on the anticipated service life, are well defined.

The study shows that the guideline can be applied and the results are straightforward in determining the performance level of the bridge. Nonlinear time history analysis using eleven pairs of ground motions generated specifically for the site shows that the performance level of this bridge still satisfies the requirement of FHWA, which is "fully operational" under lower level ground motion earthquakes (return period 100 years) and "life safety" under upper-level ground motion earthquake (return period 1000 years).

The authors wish to thank Prof. I Wayan Sengara, Mr. Det Komardevi, and the team for developing eleven pairs of ground motions at the site under evaluation.

References

1. National Standardization Agency, SNI 2833-1992, *Procedures for Planning Earthquake Resistance for Highway Bridges*, BSN, Jakarta (1992)
2. National Standardization Agency, SNI 2833-2016, *Earthquake Bridge Planning*, BSN, Jakarta (2016)
3. National Standardization Agency, SNI 2833-2016, *Earthquake Bridge Planning*, BSN, Jakarta (2016)
4. Federal Highway Administration, *Seismic retrofitting Manual for Highway Structures*, Virginia (2006)
5. National Academies of Sciences, Engineering, and Medicine, *NCHRP Synthesis - Performance-Based Seismic Bridge Design*, Washington DC, ***440*** (2013)
6. National Academies of Sciences, Engineering, and Medicine, *NCHRP Research Report - Proposed AASHTO Guidelines for Performance-Based Seismic Bridge Design*, Washington DC, ***949*** (2020)
7. National Standardization Agency, SNI 1726-2019, *Earthquake Resilience Planning Procedures for Building and Non-Building Structures*, BSN, Jakarta (2019)
8. W. Sengara, Det K., personal communication, Bandung (2021)

C. G. Jason, J. K. Mervyn, M. M. James," *Closure to Modified Plastic-Hinge Method for Circular RC Bridge Columns*," J. of. Structural Engineering, **143** (9) (2018).

Flexural capacity of RC beams strengthened with near-surface mounted steel bars

Rendy Thamrin[1]* and *Zaidir* Zaidir[1]

[1]Civil Engineering Department, Engineering Faculty, Andalas University, 25163, Padang, Indonesia

Abstract. An experimental study carried out to observe the flexural capacity of reinforced concrete beams strengthened with Near Surface Mounted (NSM) steel bars is presented. The test was carried out on nine concrete beams. All tested beams were subjected to two-point monotonic loads. The ratio of longitudinal reinforcement (1%, 1.4%, and 2.4%) and NSM bars (1D16 and 2D16) were used as test variables. It was found from the test result that NSM steel bars significantly increase the flexural capacity of reinforced concrete beams. However, the strengthened beams failed in brittle mode, as indicated by a sudden drop of beams capacity. An analytical study was also conducted to obtain the flexural response of the specimens through all ranges of elastic and post-peak load. The prediction of the flexural capacity of the beams compares well with the test result.

1 Introduction

In recent years, research on retrofitting reinforced concrete structures has resulted in a wide choice of methods that can be used for reinforcement. One category of the strengthening techniques is to increase the amount of reinforcement by attaching steel bars, FRP bars, steel plates, FRP plates, or FRP sheets to the surface of structural members [1]. However, a wrong choice of strengthening method may result in unexpected situations during the structure's service life. The near-surface installation (NSM) method, which is a more practical method, was introduced for strengthening reinforced concrete structures [2,3].

The idea of using the NSM method using either steel or FRP has made this method widely researched and used as a strengthening method [2-12]. Besides increasing the bending capacity of the beam structure [4] strengthening method using NSM with steel bars can also be used to improve the shear capacity of the beams [5, 10]. Another study reported that steel plates installed using the NSM method made a significant contribution to the shear strength of concrete beams [11]. This method is also proven to protect the environment where the reinforcing bars used as strengthening are covered by a material made of epoxy [6].

The purpose of this paper is to examine the flexural capacity of reinforced concrete beams strengthened with steel bars by applying the NSM method. In addition, the ratio of the appropriate steel bars used for strengthening is also checked using two different ratio values. The theoretical flexural behavior of the beam also needs to be known precisely to obtain the ultimate flexural strength of the beam. In this study, the flexural capacity of the beam was obtained using a computer program that has been developed by the author [7-12].

2 Experimental Setup

Nine concrete beams were tested to examine the flexural capacity of strengthened reinforced concrete beams using the NSM method. The test setup and loading arrangement, shown in Fig. 1, offer the equipment used during the test. The test setup consisted of beam specimens placed in simple support and loaded with two-point loads using a 500 kN capacity hydraulic jack and steel spreader beam. The load cell was used to measure the load applied from the hydraulic jack, and LVDT's were used to measure deflections of the test beams. Applied load and deflections were monitored, and the data was stored in a data acquisition system.

The dimension and types of test specimens are shown in Fig. 2. The first type is specimen strengthened with one steel bar, and the second is specimen strengthened with two steel bars. The steel bar diameter used to strengthen the beam was 16 mm with a yield strength of 476.2 MPa. The beams had a clear length of 2000 mm with a shear span length of 800 mm, as shown in Fig. 2. The beam's cross-section was 125x250 mm, and the size of tensile longitudinal reinforcing bars was 13 mm with a yield strength of 448.3 MPa. The longitudinal reinforcement ratios used were 1%, 1.5%, and 2.5%, as shown in Fig. 2. Compression reinforcement and stirrups with a diameter of 10 mm and yield stress of 355.2 MPa were used for all test specimens. Fresh concrete was supplied by a ready-mix company, and the compressive strength of the concrete cylinder at the age of 28 days was 20 MPa.

* Corresponding author:rendythamrin@eng.unand.ac.id

https://doi.org/10.1051/e3sconf/202133105009

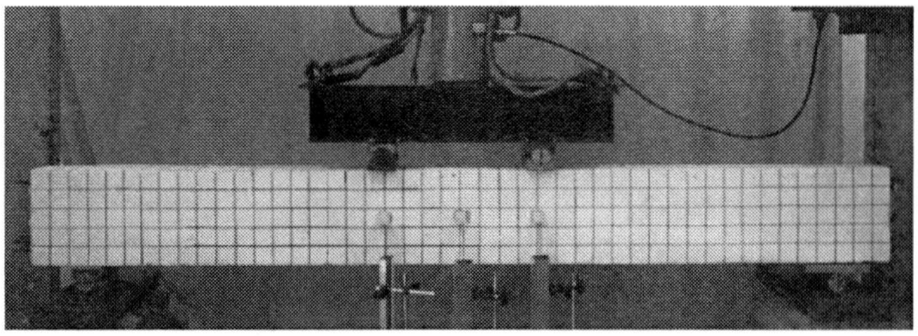

Fig. 1. Test setup and equipment used during the test.

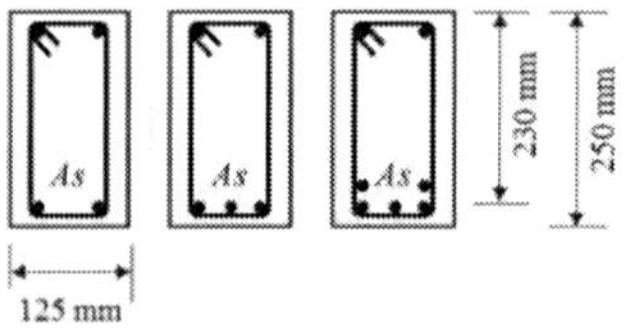

(a) Cross-sections of control beams (G6C1, G6C2, and G6C3).

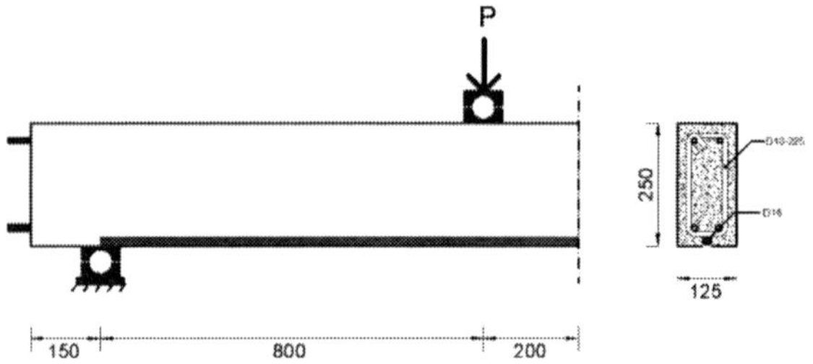

(b) Specimen strengthened with one steel bar (G6R1-1, G6R1-2, and G6R1-3).

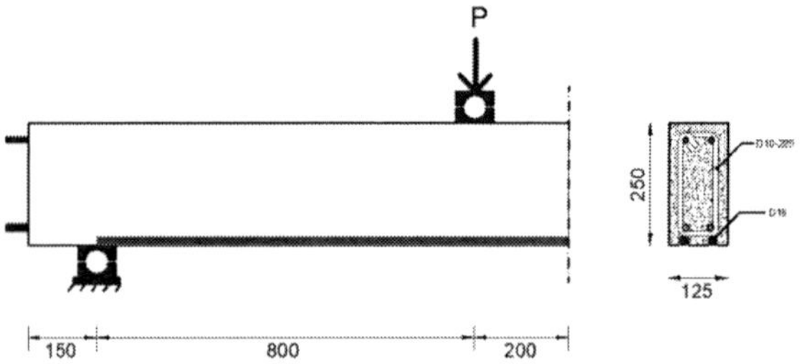

(c) Specimen strengthened with two steel bars (G6R2-1, G6R2-2, and G6R2-3).

Fig. 2. Detail of beam specimens.

The groove cuts were firstly prepared on the bottom side of the wooden formwork. The groove size was 20 mm wide and 20 mm deep. The wooden formwork was removed after 28 days, and the groove surfaces were cleaned to remove fine particles and dust. The steel bar installation process for strengthening was preceded by applying epoxy adhesive paste halfway into the grooves using a palette knife. Then the steel bar (16 mm) was placed on the top of the adhesive layer and pressed lightly along the bar. At the final stage, the grooves were again filled with epoxy adhesive paste, and then the surface was leveled.

The test started by applying an increase in load using a hydraulic jack. The applied load at each step was incremented by approximately 0.1 kN. The cracks propagation was mapped during the test, as well as the load and deflections were recorded. Loading increments continued until the specimen failed, indicated by yielding of the tensile reinforcement and crushing of the top fiber of the compression zone.

3 Analytical Study

The theoretical capacity of the reinforced concrete cross-section was obtained from a developed computer program by the author [7-13]. The calculation procedure is based on the theoretical moment-of-curvature determination [13]. In this method, the reinforced concrete section is divided into several layers of concrete and reinforcement. The computer program takes into account the nonlinearity of the concrete in the compression zone. The concrete

 https://doi.org/10.1051/e3sconf/202133105009

stress-strain curve used in the program is adopted from the model proposed by Mander et al. [14]. Meanwhile, the stress-strain curve of steel used is a bilinear model. The maximum concrete strain was assumed as 0.005. The beam deflections were calculated from the integration of the curvature along the beam length.

4 Results and Discussion

In this experimental study, the load was increased by 0.1 kN at each incremental load. Deflection at a particular position and crack propagation were recorded during the test. Initial flexural cracks developed in the constant moment region between two-point loads. When the load was increased, more flexural cracks were formed in the constant moment region and the shear span zone. Subsequently, the diagonal shear cracks were propagated in both shear-span zones when the load was increased.

All control beams (G6C1, G6C2, and G6C3) were failed in flexural failure mode. This failure mode was indicated by yielding tensile longitudinal reinforcement and crushing of concrete in the constant moment region, as shown in Fig. 3(a-c).

(a) G6C1
(b) G6C2
(c) G6C3
(d) G6R1-1
(e) G6R1-2
(f) G6R1-3
(g) G6R2-1
(h) G6R2-2
(i) G6R2-3

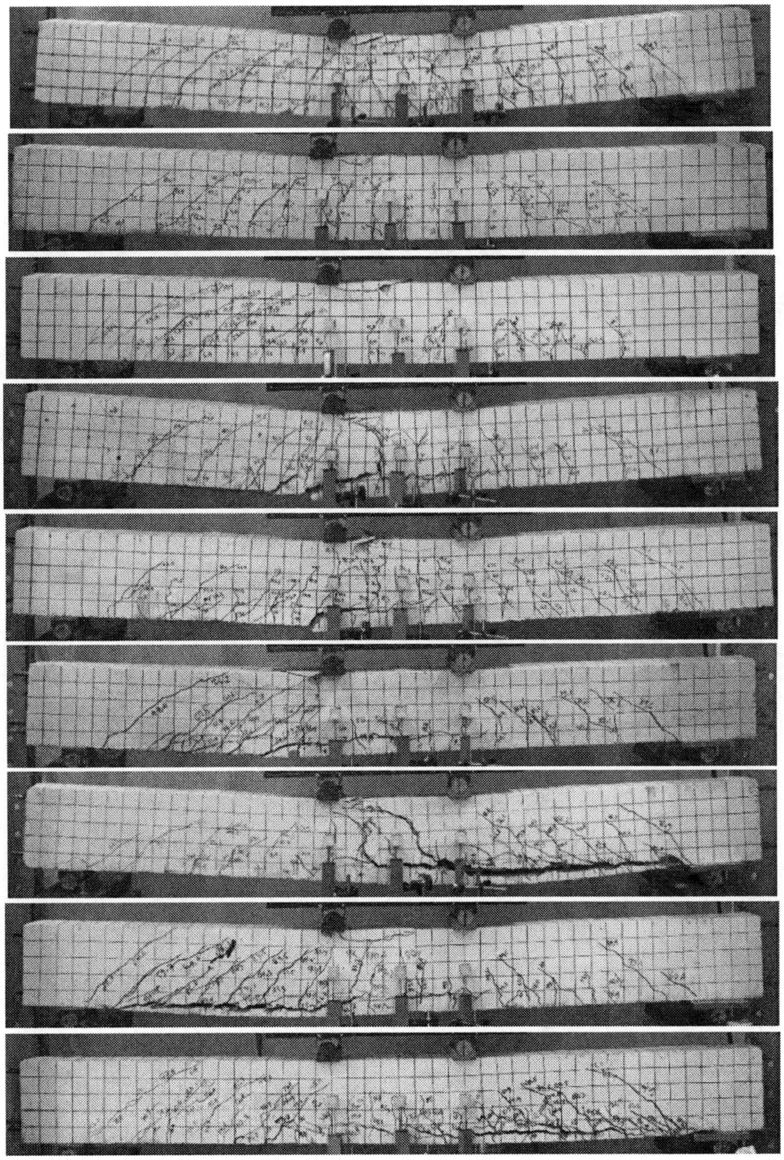

Fig. 3. Crack patterns of the beams.

However, all beams reinforced with steel bars failed immediately after the occurrence of debonding failure. Debonding failure was indicated by the loss of bond between concrete and strengthening bar. Concrete cover separation at the bottom of the beams along the beam length also indicates the debonding failure occurrence, as

https://doi.org/10.1051/e3sconf/202133105009

shown in Fig. 4. The effect of strengthening with a steel bar on the beam can be observed from the increase in the value of the first crack load on the strengthened beam.

Fig. 5 shows the load-deflection curves of the tested beams. It can be shown from Fig. 5(a) that as the ratio of reinforcement increases, the capacity of the beams increases, but the ductility decreases. Figs 5(b) and (c) show the sudden drop of the load-deflection curve after the occurrence of debonding failure. In this study, if the debonding failure occurred, the load was uninterruptedly applied until the beam reached concrete crushing at the top surface of the compression zone. It aims to see the behavior of the beam after the debonding failure. It can be seen that after the steel bar's debonding failure, the load immediately drops to the same load level as the control beam.

These curves also show that when the debonding failure occurred, both the strengthening bar and steel reinforcement did not experience yielding. Before the debonding failure, it was seen that the capacity of the strengthened beams slightly exceeded the capacity of the control beams. This proves that the capacity of the strengthened beams is higher than that of the control beams.

Fig. 6 shows a schematic plot of the stress-strain laws of concrete and steel bars used in the analytical study. The theoretical prediction of flexural capacity for each beam specimen is shown in Fig. 7. From these curves, it can be seen that the higher the ratio of longitudinal reinforcement, the more rapid and sudden debonding failure occurs.

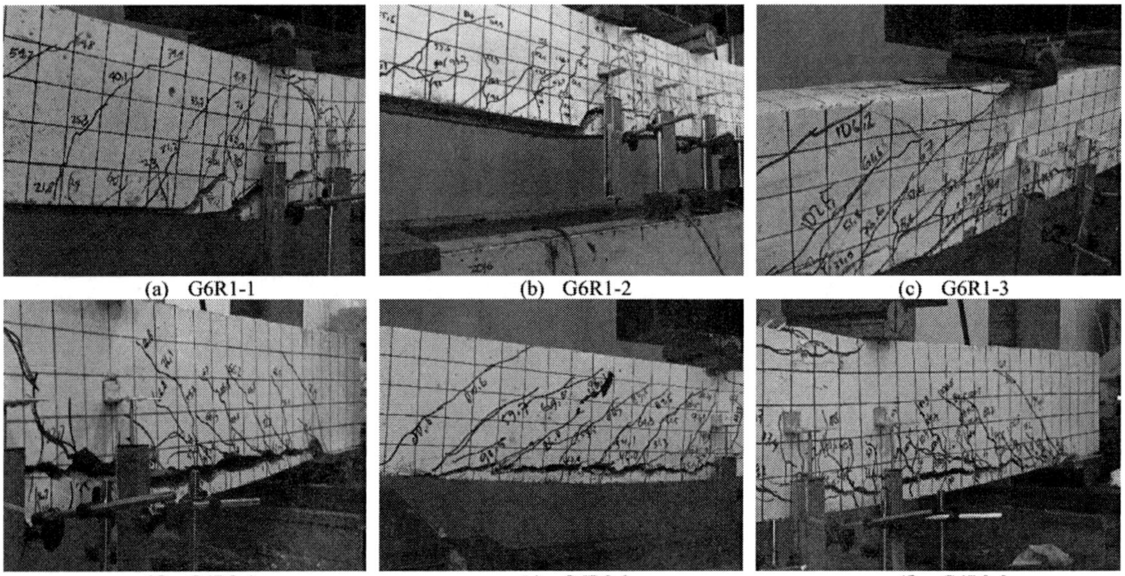

(a) G6R1-1 (b) G6R1-2 (c) G6R1-3

(d) G6R2-1 (e) G6R2-2 (f) G6R2-3

Fig. 4. Typical debonding failure of strengthened beams.

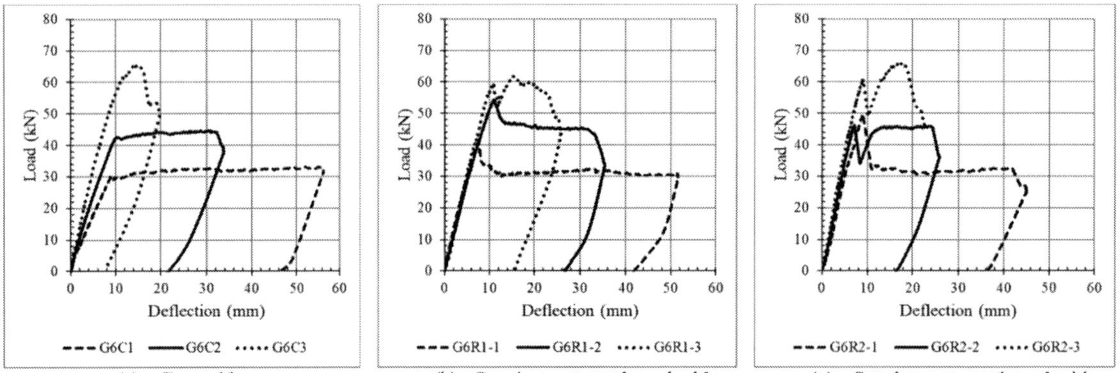

(a) Control beams (b) Specimen strengthened with one steel bar (c) Specimen strengthened with two steel bars

Fig. 5. Load-deflection curves of the tested beams.

In addition, it can be seen from these curves that the higher the number of steel bars used to strengthen, the more sudden the debonding failure will occur. In addition, it can be seen from these curves that the higher the number of steel bars used to strengthen, the more sudden the debonding failure will occur. But it can be

https://doi.org/10.1051/e3sconf/202133105009

seen that the capacity of the strengthened beams is proven to be higher than that of the control beams.

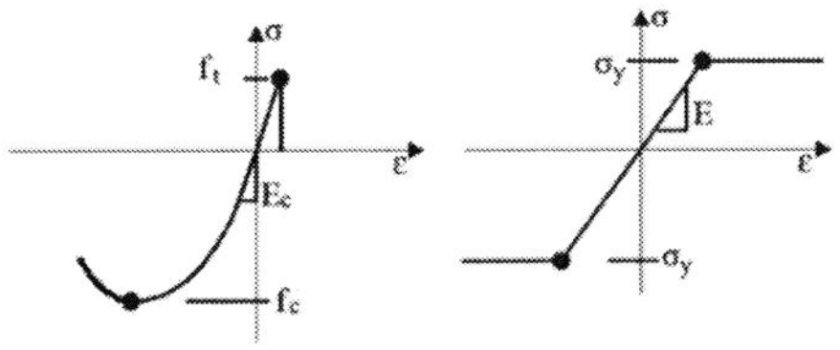

(a) Concrete (b) Steel

Fig. 6. Stress-strain laws of concrete and steel used in the analytical study.

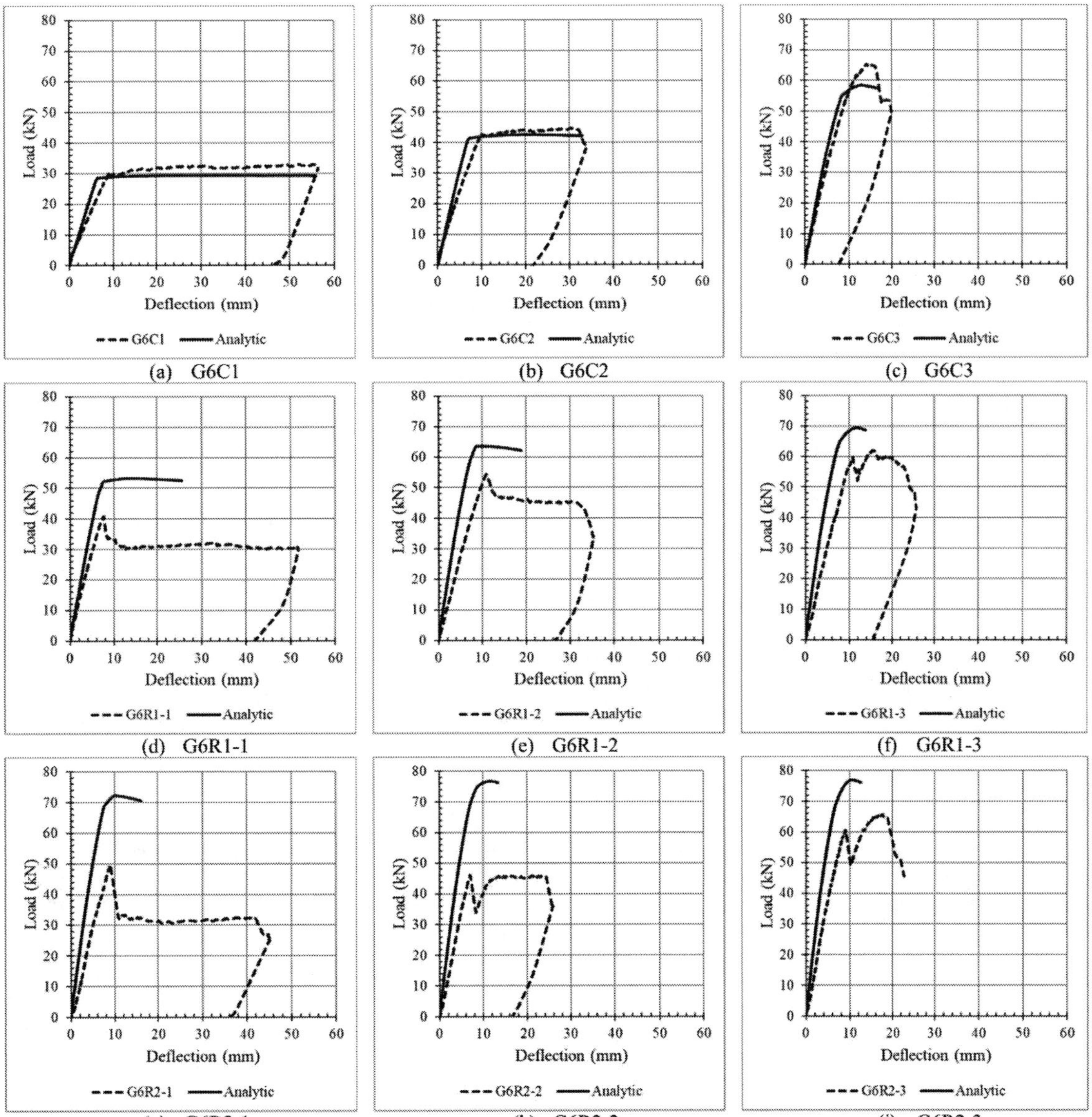

(a) G6C1 (b) G6C2 (c) G6C3

(d) G6R1-1 (e) G6R1-2 (f) G6R1-3

(g) G6R2-1 (h) G6R2-2 (i) G6R2-3

Fig. 7. Comparison between analytical and test results.

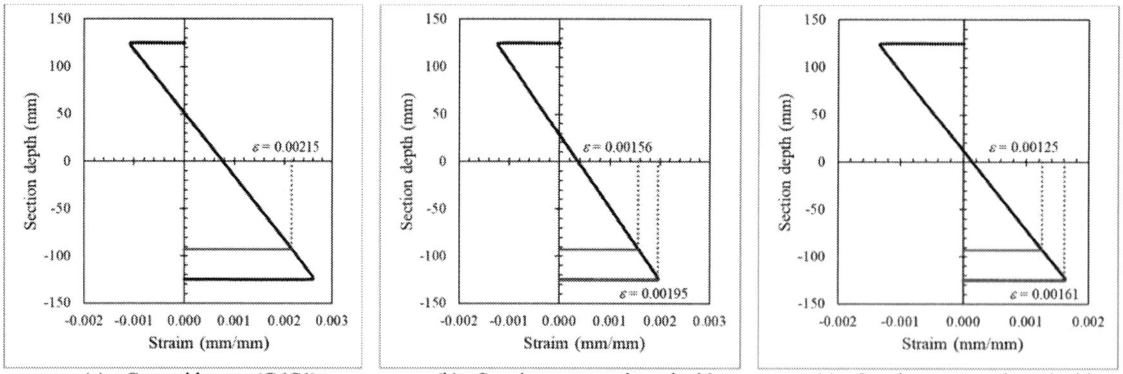

(a) Control beams (G6C1)

(b) Specimen strengthened with one steel bar (G6R1-1)

(c) Specimen strengthened with two steel bars (G6R2-1)

Fig. 8. The plot of strain distribution of the cross-section was obtained from computation results.

Fig. 8 shows the plot of strain distribution of the cross-section of specimens G6C1, G6R1-1, and G6R2-1 obtained from computation results. The plotted strain distribution diagrams have occurred at the yield stress level of the tensile reinforcement for the G6C1 specimen and the debonding failure load level for the G6R1-1 and G6R2-1 specimens. As shown in the stress distribution above, the compressive stress at the load level taken has not reached the maximum strain value assumed in the analytical study (0.005).

As is known from the yield stress values of tensile reinforcement and strengthening steel bar, the yield strain for each reinforcement is 0.0022 and 0.0024, respectively. The strain distribution shown in Fig. 8 proves that the strain in the strengthening bar and tensile reinforcement at debonding failure was not experience yielding. Since the strengthening steel bars have not yielded at the time of debonding failure, the tensile capacity of the steel bars has not been fully utilized. This may be avoided by reducing the diameter of the steel bar used as a strengthening. In addition, the strain that occurs in the reinforcement decreases with an increase in the amount of strengthening bar.

5 Conclusions

Nine reinforced concrete beams with and without strengthening were tested experimentally in this study. An analytical study was also carried out to obtain the theoretical capacity of the beams. Based on the test results, the main conclusions that can be drawn are:

1. Strengthening of reinforced concrete beams with the NSM method can increase the flexural capacity of the beam.
2. A strengthening steel bar increases the load level of the first flexural crack in the beam.
3. The strain that occurs in tensile reinforcement decreases with an increase in the amount of strengthening steel bars.
4. The area of steel bars used as strengthening must be determined as appropriately as possible to avoid premature debonding failure.

The authors express their appreciation to the Faculty of Engineering, Andalas University, for the financial support through the 2021 Faculty of Engineering Publication Grant.

References

1. BB Adhikary, H. Mutsuyoshi, Sano, M., Constr. Build. Mater., **14**, 5, 237-244 (2000)
2. L.D. Lorenzis, A. Nanni, ACI Structural J., **98**, 1, 60-68 (2001)
3. R. El-Hacha,, S.H. Rizkalla, ACI Structural J., **101**, 5, 717-726 (2004)
4. Z. Kálmán Szabó, L. Balázs, G, Periodica Polytechnica Civil Eng., **51**, 1, 33-38 (2007)
5. P Sabol, S. Priganc, Procedia Eng., **65**, 364-369 (2013)
6. A. Parvin, T. Syed Shah, Polymers. 8, 8, 298 (2016)
7. R. Thamrin, Sari, R.P., Procedia Eng., **171**, 1129-1136 (2017)
8. R. Thamrin, MATEC Web of Conf., **103**, 02012 (2017)
9. R. Thamrin, J. Eng. Sci. Technology. **13**, 3781–3794 (2018)
10. R. Thamrin, S. Haris, Zaidir, MATEC Web of Conferences 276, 01004 (2019)
11. R. Thamrin, S. Haris, Zaidir, IOP Conf. Series: Materials Science, Engineering, **771**, 012053 (2020)
12. R. Thamrin, Z. Zaidir, S. Desharma, Polymers. **13**, 16, 2738 (2021)
13. R. Park, T. Paulay, Reinforced Concrete Structures, John Wiley, New York (1975)
14. J.B. Mander, M.J.N. Priestley, and R. Park, J. of Structural Engineering, **114**, 8, 1804-1826 (198

Performance-based-plastic design method of reinforced concrete structure for operational performance level

Jati Sunaryati[1]*, *Nidiasari* Nidiasari[1], and *Rifqi* Yuliandri[1]

[1]Civil Engineering Department, Andalas University

Abstract. Under major load earthquakes, reinforced concrete structures designed according to the current codes will experience an inelastic deformation which is difficult to predict and control. Performance-based plastic design (PBPD) methodology is applied forward to design reinforced concrete structures in this study. In this method, as performance criteria, the target drift and yield mechanisms are preselected. Based on the work-energy balance principle, the design base shear is given as earthquake level and calculated as work required to push the structure as monotonically load to the target drift. The load equals the energy needed by an equivalent single degree of freedom in the same state. The plastic design is utilized to design the desired yield mechanism. The method was adopted on a 10-story reinforced concrete structure with an earthquake load in lateral forces based on SNI 1726:2019 and the Performance-Based Plastic Design (PBPD) method. Pushover analysis was carried out where the structure was pushed to obtain lateral load resistance followed by yielding gradually until plastic deformation occurred collapse From the pushover analysis, the ductility value for SNI 1726:2019 is less ductile than analytical using the Performance-Based Plastic Design (PBPD) method

1 Introduction

Important buildings such as schools, hospitals, office buildings, and even shopping centers in SNI 1726:2019 [1] are included in priority factor IV. By providing an earthquake value more significant than the planned earthquake, these important buildings are expected to have an operational performance level during a moderate earthquake. They have an immediate occupancy performance level during a large earthquake (MCE) to reduce the risk of fatalities after a building collapse when an earthquake occurs.

Performance-based seismic design (PBSD) makes it easier for the engineer to determine the selected level of structural performance. PBSD is a structural design where the design criteria are expressed in terms of the expected performance when the structure is subjected to earthquake loads. With this PSBD method, realistically, the risk of life and economic losses can be predicted if an earthquake occurs. PBSD is an iterative process in which the designer begins by setting performance objectives, followed by an initial design, analyzing the initial design whether it is following the performance targets, and finally, redesign and reassessment are carried out if necessary until the desired performance level is achieved [2][3]. The flow of this method starts from choosing the desired performance level, designing according to the performance level, and after the design is completed, the design target can become acceptance criteria through performance evaluation for the performance target level set by FEMA 356. However, the PBSD method depends primarily on an iteration process from evaluating performance to modifying design until the design structure achieves the desired performance

In recent years, we have developed a performance-based plastic design (PBPD) methodology to consider the inelastic characteristic of the structural component [4].

1.1 PBPD method

The Performance-Based Plastic Design (PBPD) method is a method with a design concept that considers the inelastic behavior of the structure with little or no iteration during the initial design. This method applies the concept of energy balance by determining the design yield mechanism in advance, as shown in Fig. 1.

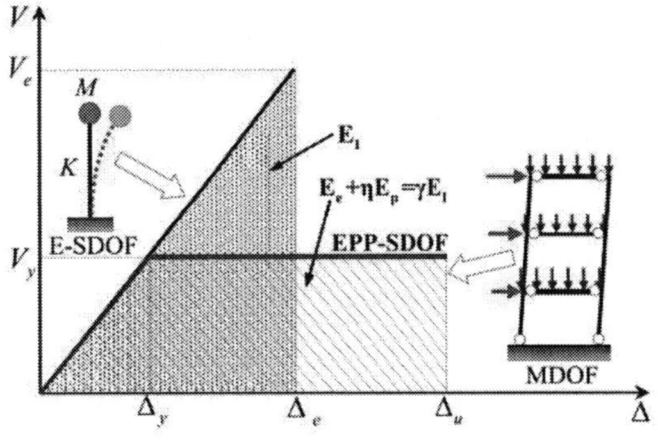

Fig. 1. Illustration of energy balance concept in PBPD [5]

* Corresponding author: jati@eng.unand.ac.id

The structural performance as planned is obtained when receiving a strong earthquake [6]. In reinforced concrete structures, it is widespread that there will be a reduction in the stiffness and strength of the structure. This reduction in strength and stiffness is also known as pinching. With the reduction in strength and stiffness in this reinforced concrete structure, a coefficient of C_2 is needed. The coefficient C_2 is a modifying factor to show the hysterical closed circuit, reduced stiffness, and reduced strength which are the main characteristics of hysteresis behavior. C_2 was chosen to modify the target drift plan [7].

1.2 Comparison of PBPD and Current Code Design Method

Indonesia already has code SNI 1726:2019 [1] regarding procedures for planning earthquake resistance for building and non-building structures to calculate earthquake loads. In this code, the design base shear is calculated by reducing the elastic strength demands to the inelastic strength demands by incorporating a seismic force-reduction factor R. This reflects the degree of inelastic response expected for design-level-ground motions as the ductility capacity of the structure.

In this code, it is also arranged how to design the structure by paying attention to the desired structure performance. For the design of earthquake-resistant structures, usually, the performance-based seismic design is used by applying a non-elastic pushover load where the lateral force is obtained from the equivalent static-load procedure [8][9]. This lateral force is also the equivalent of the shear force obtained by applying the response spectra map in Indonesia.

As mentioned before, the key of the PBPD method is the preselected target drift and yield mechanism. The lateral design forces are determined for the given seismic load and selected target drift. Therefore, R, I, and C_d are not needed.

2 Design Procedure of PBPD

The PBPD method is used to determine the first yield target of structural elements by determining the displacement target first. After that, the complete structural element design is carried out and analyzed until it reaches the ultimate limit of the entire structure [10].

A key element in the PBPD method is given hazard level to determine the design base shear [5][10][11]. It is calculated by equating the work needed to push the structure monotonically up to the target drift [10]. The work requires an equivalent elastic-plastic single degree of freedom system to achieve the same state. Assuming an idealized elastic-plastic single degree of freedom as shown in Fig. 1, the work-energy equation can be written as

$$(E_e+E_p)=\gamma\cdot\left(\frac{1}{2}M\cdot S_v^{\,2}\right)=\frac{1}{2}\gamma\cdot M\cdot\left(\frac{T}{2\pi}S_a\cdot g\right)^2 \quad (1)$$

Where E_e and E_p are the elastic and plastic components of the energy (work) needed to push the structure up to the target drift. S_v is the design pseudo-spectral velocity; S_a is the pseudo-spectral acceleration; T is the natural period, and M is the system's total mass.

Outline procedures of PBPD [8] are presented as

a. Selected the desired yield mechanism and target drift as design earthquake load.
b. Estimate the yielding drift (θ_y), the fundamental period (T) of the structure. Then determine an appropriate vertical distribution of lateral design forces of the structure.
c. Determine the elastic design spectral acceleration value, Sa
d. Calculate the design base shear (V)
e. Modify V as needed from the assumed elastic and P-Delta effect.
f. Use plastic method to design the designed yielding member.

3 Methodology

In this study, the lateral load calculation is calculated based on SNI 1726:2019 and PBPD. This lateral load will be applied to the structure at the center of mass as a nonlinear static lateral load or the pushover analysis method. In this study, what will be compared is the actual ductility value and the level of structural performance from the two lateral load calculations.

The function of the reinforced concrete structure in this study is an educational building with 10 stories, a building length of 42 m, a building width of 48 m, and a building height of 40 m. Concrete has fc' 30 MPa for columns, beams, and slabs, and fc' 35 MPa for shear walls. The dimension of the main beam is 50 cm x 80 cm and 30 cm x 50 cm for joist. There are three types of columns: 100 cm x 100 cm for the column on the first to the fourth story, 90 cm x 90 cm for the fifth to the seventh story, and 80 cm x 80 cm for the eighth to the tenth story.

Building modeling can be seen in Figs. 1 (a) and (b).

3.1 Lateral Load Using SNI 1726:2019

Based on Table 3 of SNI 1726:2019, the Education building is included in risk category IV, and according to Table 4 of SNI 1726:2019, the value of the earthquake priority factor (Ie) is 1.5. Based on the earthquake map for Padang city, the value of S_s is 1.499, and S_1 is 0.6. Based on Table 3 of SNI 1726:2019, the Education building is included in risk category IV, and according to Table 4 of SNI 1726:2019, the value of the earthquake priority factor (Ie) is 1.5. Furthermore, based on the provisions in the code (SNI 1726:2019), based on the earthquake map for Padang, the value of S_s is 1.499, and S1 is 0.6. So that the value of S_{DS} = 2/3(S_{MS}) = 2/3 x 1.499 g = 0.999 and the value of S_{D1} = 2/3(S_{M1}) = 2 /3 x 1.02 g = 0.68.

Based on Tables 8 and 9 for Education buildings with S_{DS} values = 0.999 and S_{D1} = 0.68, they are included in the seismic design category D. Based on Table 18 for moment-bearing frame structures, C_t = 0.0466 and x = 0.9 values are obtained. Based on Table 12 for the special moment-bearing reinforced concrete frame structure, the value of R = 8. The calculation of the Seismic Response

Coefficient (Cs) is 0.085. Then the base-shear force of the design earthquake is Cs x W total = 0.085 x 165354 kN.

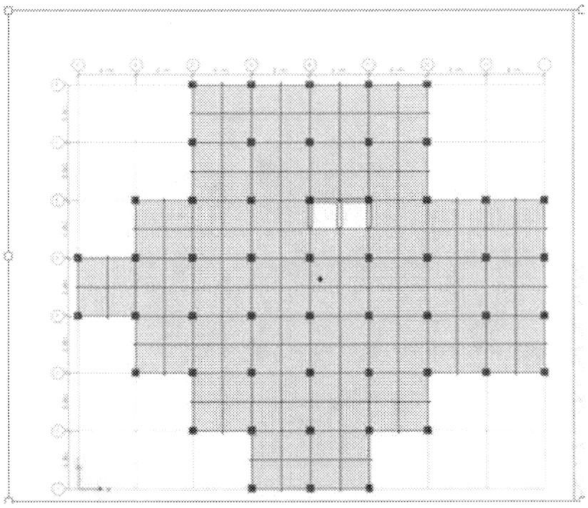

(a)

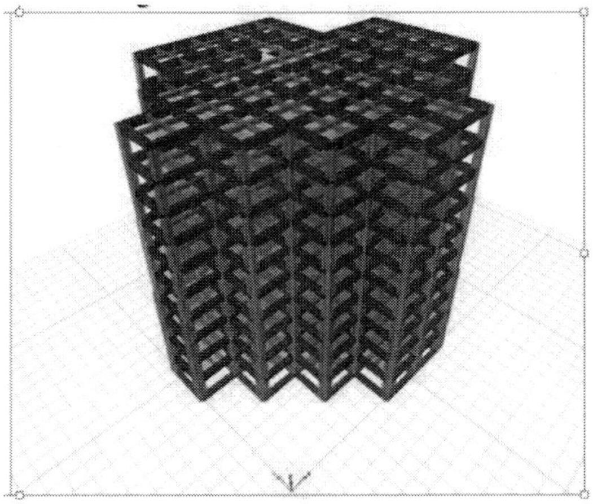

(b)

Fig. 2. Structural **modeling** (a) structure plan (b) 3D Model of the structure

Table 1. Lateral load for each story using SNI 1726:2019

Story	W_x (kN)	h_x (m)	F_i (kN)
10	12871,52	40	2480
9	16079,53	36	2671
8	16079,53	32	2263
7	16201,96	28	1889
6	17042,01	24	1600
5	16830,12	20	1222
4	17001,52	16	902
3	17890,54	12	633
2	17678,64	8	354
1	17678,64	4	133

3.2 Lateral Load Using PBPD

The fundamental period, T for the reinforced concrete frame, can be determined from the following equation, as given in ASCE 7 – 05 (2006):

$$T = C_u.T_a = C_u.C_t.h^n x \text{ if } T_{actual/model} > C_u.C_t.h_n^x \quad (2)$$

where T_a is the approximate fundamental period, C_u represents the coefficient for the upper limit on the calculated period. The value of T obtained is 1.805.

The value of the coefficient C_2 is calculated using equation (3)

$$C_2 = 1.1 - 0.045\,(T - 0.8) \quad (3)$$

The coefficient value of C_2 is obtained at 1.055.

Plastic Rotation (θ_p) is calculated using equation (4)

$$\theta_p = \theta_u / C_2 - \theta_y \quad (4)$$

The value of the drift target (θ_u) is planned at 2% for the structure's condition still in operational performance. The yield drift ratio (θ_y) can be seen in Table 3-1 (Liao, 2010). From that table, the reinforced concrete structure has a yield drift ratio of 0.5%. With the value of the C_2 coefficient of 1.055, the value of θ_p is 1.396%.

Ductility (μ_s) is calculated using equation (5)

$$\mu_s = \frac{\theta_u}{\theta_{y.C_2}} \quad (5)$$

Then the ductility value is 3.792.

Variable Factors for the Energy Balance Equation (γ) is calculated using equation (6)

$$\gamma = \frac{2\mu_s - 1}{R\mu^2} \quad (6)$$

R_μ value obtained from Table 3-2 (Liao, 2010). $R_\mu = \mu_s$, which is equal to 3.792. Thus, the value of the variable factor for the Energy Balance Equation (γ) is 0.458.

Shear Force Comparison Factor (β_i) is calculated using equation (7)

$$\beta_i = \left(\frac{\sum_{i=1}^{n} w_i \times h_i}{w_n \times h_n}\right)^{0,75.T^{-0,2}} \quad (7)$$

where w_n x h_n is the product of the weight of the structure and the height of the structure on the top floor, then the value of I for each floor is obtained in Table 2.

Design Shear Force Parameter (α) is calculated using equation (8)

$$\alpha = \left(\sum_{i=1}^{n}(\beta_i - \beta_{i+1})h_i\right)\left(\frac{w_n \times h_n}{\sum_{i=1}^{n} w_i \times h_i}\right)^{0,75.T^{-0,2}}\left(\frac{\theta_p\, 8\pi^2}{T^2 g}\right) \quad (8)$$

then the value of the Design Shear Force Parameter (α) is 1.042.

Design Earthquake Base Shear Force (V) is calculated using equation (8)

$$\frac{v}{w} = \frac{-\alpha \pm \sqrt{\alpha^2 + 4\gamma Ce^2}}{2} \quad (8)$$

then the value of the Basic Earthquake Shear Force (V) is obtained at 55074.983 KN.

Table 2. Shear force comparison factor (βi)

Story	β_i
10	1,000
9	1,652
8	2,136
7	2,521
6	2,844
5	3,097
4	3,294
3	3,445
2	3,543
1	3,592

Top-Stage Lateral Force (F_n) is calculated using equation

$$F_n = v\left(\frac{w_n \times h_n}{\sum_{i=1}^{n} w_i \times h_i}\right)^{0,75.T^{-0,2}} \quad (9)$$

then the lateral force value on the top level of the structure (F_n) is 15334 kN. Lateral force at the level I (F_i) is calculated using equation (10)

$$F_i = (\beta_i - \beta_{i+1})\, F_n \quad (10)$$

when i = n, then $\beta_{i+1} = 0$, The value of the lateral force is obtained as follows

Table 3. Lateral load per-floor using PBPD

Story	Fi (kN)
10	15334
9	10002
8	7424
7	5900
6	4955
5	3875
4	3019
3	2320
2	1501
1	743

4 Results and discussion

4.1 **Base** Shear and Lateral Force

From Table 4, it can be seen that the magnitude of the base shear force from the PBPD method is more significant than using SNI 1726:2019. The lateral force calculated using SNI 1726:2019 has increased from floors 1-9 and decreased on the 10th floor, while for the PBPD method, the value of lateral forces has increased from floors 1-10. The comparison of the values of these lateral forces is shown in Table 5.

Table 4. Base shear

	SNI 1726:2019 (kN)	PBPD (kN)
V	14146	55075

Table 5. Lateral load

Story	SNI 1726:2019 (kN)	PBPD (kN)
10	2480	15334
9	2671	10002
8	2263	7424
7	1889	5900
6	1600	4955
5	1222	3875
4	902	3019
3	633	2320
2	354	1501
1	133	743

4.2 Structural Ductility Analysis

The ductility of the structure is done by doing a pushover analysis of the structure. The amount of ductility can be determined directly by the ETABS v16 software. The ductility value for lateral forces based on SNI 1726:2019 is obtained at 3.052, while based on PBPD, the value is 3.115. Based on the value of the structure's ductility, the structure with the lateral load using the PBPD method has an actual ductility that is greater than the ductility of the lateral load of the SNI 1726:2019 method.

4.3 Structure Performance Level

The performance level of the structure is determined based on the capacity spectrum, where the process begins by analyzing the load versus displacement relationship curve by taking into account the inelastic conditions of the structure. After the performance point is obtained, then the level of structure performance is determined by referring to the provisions of ATC-40 as shown in Table 6.

Table 6. Performance level based on ATC-40

Parameter	Performance level			
	Immediate Occupancy	Damage control	Life Safety	Structural stability
Maximum total drift	0.01	0.01-0.02	0.02	0.33 Vi/Pi
Maximum inelastic total drfit	0.005	0.005 - 0.015	No limit	No limit

From Fig. 3, the performance point for structures with lateral load based on SNI 1726:2019 is marked by the intersection between the capacity (the green line) and the single demand (red line) so that the results of this pushover analysis can determine the level of structure performance

Maximum total drift is D_{max}/h = (754.242/40000) = 0.019. The value of maximum inelastic total drift is ($D - D_{yield}$)/h = (754.242 – 186.415)/40000 = 0.014. Based on ATC-40, this condition is included in the level of damage control. In the event of an earthquake, the building is still able to withstand the earthquake, and the risk of loss of life is minimal.

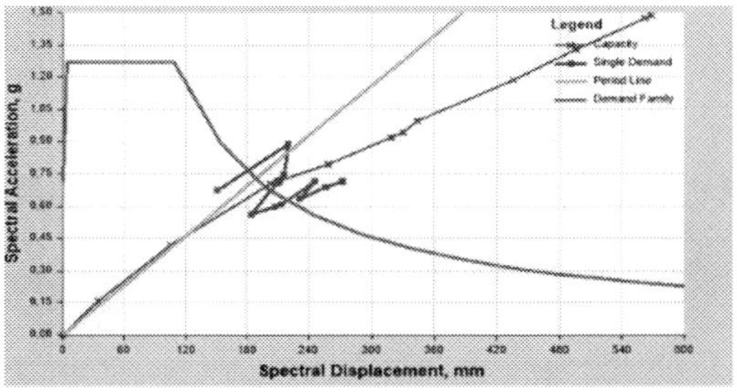

Fig. 3. Performance point structure using SNI 1726: 2019.

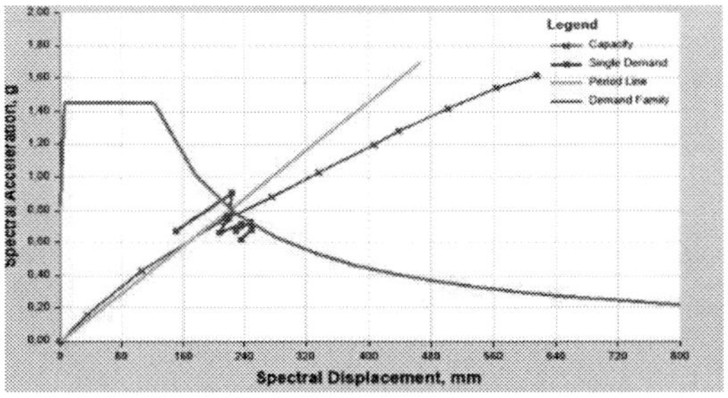

Fig. 4. Performance point structure using PBPD

Based on Fig. 4, the performance point for structures with lateral load based on the PBPD method is obtained from the intersection between capacity (green line) and single demand (red line. In the initial planning, a drifting target (θu) was set at 0.02.

Maximum total drift is D_{max}/h = (799.928/40000) = 0.02 (as same as initial drift target). The value of maximum inelastic total drift is $(D-D_{yield})/h$ = (799.928 – 197.742)/40000 = 0.015. Based on ATC-40, this condition is included in the level of life safety. In the event of an earthquake, the building is still stable, and the building can still be used if repairs are made.

5 Conclusion

Preselected target drift and yield mechanism is a key performance objective in the PBPD method. The PBPD frame model responded as intended in this research to improve their performances over the corresponding baseline frames. In addition, the basic work-energy equation can be used for seismic evaluation purposes to determine expected displacement demand for a given structure and earthquake risk.

Some conclusions that can be drawn from this research are:

a. The base shear force and lateral load on the structure will be greater if it is calculated using the PBPD method than SNI 1726: 2019.
b. By using SNI 1726:2019, the structure has a level of damage control performance. In this situation, the building is still able to withstand earthquakes when earthquakes occur, and the risk of loss of life is minimal
c. At the same time, the performance level of the structure with lateral forces calculated using the PBPD method is life safety. In an earthquake, the building is still stable, and the building can still be used if repairs are made.

Acknowledgments to Civil Engineering Department, Engineering Faculty, Andalas University which has funded this research with the contract number: 125/UN16.09.D/PL./2021

References

1. National Standardization Agency, "Sni 1726-2019," Planning Ordinance. Earthquake Resilience For Structure. Buildings and Non-Buildings, 8 (2019)
2. F. Naeim, "Performance-based seismic design of tall buildings-A USA perspective," in *Seismic Isolation, Structural Health Monitoring, and Performance-Based Seismic Design in Earthquake Engineering: Recent Developments* (2018)
3. F. Naeim, "Performance-Based Seismic Design of Tall Buildings," *Geotech. Geol. Earthq. Eng.*,17, (2010)
4. S. C. Goel, W. C. Liao, M. R. Bayat, and S. H. Chao, "Performance-based plastic design (PBPD) method for earthquake-resistant structures: An overview," *Structural Design of Tall and Special Buildings*,**19**, 1–2, (2010)
5. J. Bai, T. Y. Yang, and J. Ou, "Improved Performance-Based Plastic Design for RC Moment Resisting Frames: Development and a Comparative Case Study," *Int. J. Struct. Stab. Dyn.*,**18**, 4, (2018)
6. N. J. S. Eem Ikhsan, " Behavior of steel structure type mrf with lateral load based on sni 1726-2012 and performance based plastic design (pbpd) method " *J. Tek. Sipil*, **13**, 1 (2014)
7. Q. Zhang, E. Xiong, X. Liang, and X. Miao, "Performance-based plastic design method of high-rise steel frames," *J. Vibroengineering*, **19** (3), (2017)
8. C. P. D. Amoussou, H. Lei, Y. Halabi, and W. Alhaddad, "Performance-based seismic design methodology for tall buildings with outrigger and ladder systems," *Structures*, **34** (2021)
9. B. Rangel, A. S. Guimarães, A. Vaz Sá, and F. B. Alves, "Integrated design concept in civil engineering education," *Int. J. Eng. Educ.*, **32**, 3, (2016)
10. W. C. Liao and S. C. Goel, "Performance-based plastic design and energy-based evaluation of seismic resistant RC moment frame," *J. Mar. Sci. Technol.*,**20**, 3 (2012)
11. S. Leelataviwat, P. Doung, and N. Naiyana, "A Review on Performance-Based Plastic Design Method: Concept and Recent Developments," in *Lecture Notes in Civil Engineering*, **155** LNCE, (2021)

Analysis of the effect of various support distances to the bond-strength of plain anchor bolt in concrete

Hariyadi[1*], *Ajeng Larasati* Pradipta[2], and *Akmaluddin*[1]

[1]Department of Civil Engineering, University of Mataram, Jl. Majapahit no. 62, Mataram, NTB, Indonesia
[2]Graduated Master Study Program of Civil Engineering, University of Mataram, Jl. Majapahit no. 62, Mataram, NTB, Indonesia

Abstract. The structural components of reinforced concrete that withstand the tensile stress will be bonding stress affected by the friction between the reinforcing bar and the concrete. Anchoring depth, reinforcement diameter, and concrete quality are commonly assumed to influence bond strength; besides, the support distance of reinforcement may be one of the influencing factors. The samples in this study were cylindrical specimens with a diameter of 15 cm and 20 cm for the concrete compression strength and the bond-strength test, respectively, and double-L size 30 x 20 x 7.5 cm for the concrete shear test. Variables in this study include variations in the support distance represented by variations in the diameter of the supporting plate hole, namely 10 mm, 35 mm, 60 mm, 85 mm, and 110 mm. Plain reinforcement with a diameter of 10 mm was anchored at 150 mm depth in the centre of the concrete cylinder. The experimental results obtained 26.49 MPa and 10.86 MPa for concrete compression dan shear strength, respectively. Based on various diameters of the supporting plate hole, the hole diameter of 10 mm to 60 mm had a stable increase in bond strength and a significant decrease when a larger hole diameter was used.

1 Introduction

In reinforced concrete structures, steel reinforcement is placed into concrete to provide tensile resistance to bending loads or to assist concrete in carrying compressive loads [1]. To ensure good cooperation between concrete and steel reinforcement as a composite material, the bond strength must be seen at the interface of these materials. One subject related to the quality control of reinforced concrete structures is the bond between steel reinforcement and concrete [2]. If the steel reinforcement and concrete bond are good, there will be a slight crack when the reinforcement is pulled out of the concrete [3].

The anchoring strength in concrete specified in the standards usually assumes that the bond is uniform. However, this assumption is not entirely correct because anchoring depth affects the bond stress [4]. The strength of the deformed reinforcement ties consists of 3 parts. First is the adhesion between concrete and steel reinforcement, the second is the friction between steel reinforcement and concrete around it. The last is mechanical bonding that affects the roughness of the steel reinforcement. Mechanical bonding is called a mechanical interlock which affects the strength of the primary bond [5]. The failure mode is generated from stress, shear strength and its combinations. A typical bond failure results in reinforcing reinforcement due to the release of concrete around the surface of the reinforcement combined with an adhesive failure along with the steel reinforcement [1,6]. The mode of a connection failure in axial tensile loading follows the order of slip reinforcement, cracked concrete, steel plate bending and reinforcement breaking [7]. Normal force and shear stress were negatively and exponentially distributed in the fixed region of the anchor bolt [8]. Factors that influence the value of bond strength are anchoring depth, reinforcement diameter, and concrete quality. One aspect is thought to affect the bond strength of the anchoring in concrete, the support distance of anchors. The function of the supporting plate with a hole in bond strength testing is as a concrete barrier so that it does not lift and break during the test.

As for some studies on bond strength using plain reinforcement in normal concrete, research of bond strength using plain reinforcement with a diameter of 10 in normal concrete shows the bond strength decreases along with the increasing variation in the diameter of the supporting hole [9]. Research on bond stress on plain steel with diameters of 8, 10, 12, 16, and 19 in normal concrete shows that the strength of plain reinforcement of 8 mm, 10 mm, 12 mm diameter has increased but decreases in larger diameter [10].

1.1 Compressive Strength of Concrete

The compressive strength of concrete loads is the magnitude of the load per unit area, which causes the concrete specimens to be destroyed when loaded with a certain compressive force, which is produced by a compression testing machine [11]. The maximum compressive stress (f'c) can be formulated as Eq. 1.

* Corresponding author: hariyadi@unram.ac.id

$$f'c = \frac{P}{A} \quad (1)$$

1.2 Shear Strength Of Concrete

If the force acting on the concrete exceeds the maximum shear strength that the concrete can withstand, then a concrete crack will arise. The friction force produces shear stress between particles of one particle to another. The following formula (Eq. 2) can calculate shear strength.

$$f_{shear} = \frac{P}{bh} \quad (2)$$

1.3 Basic Concept Of Anchoring Bond

Pull-out experiments can provide a good comparison between the bond efficiency of various types of reinforcement surfaces and their length of anchor (embedment length). The ratio of the length of the distribution of plain steel reinforcement to the length of distribution of reinforcing steel bars at the same stress is a factor of two so that the average bond stress for plain steel reinforcement is obtained as Eq. 3.

$$\mu = \frac{d_b f_y}{8 l_d} \quad (3)$$

The development length of the steel reinforcement can be determined by Eq. 4 [12]:

$$l_d = \frac{d_b f_y}{2.1\sqrt{f'c}} \quad (4)$$

2 Research Methods

2.1 Testing of Concrete Compressive Strength

Compressive strength testing was conducted after the concrete was 28 days old. This study aims to obtain the compressive strength value of concrete using a Compression Testing Machine.

2.2 Testing of Concrete Shear Strength

The shear strength test was carried out after 28 days of concrete with a double L-shaped specimen with size 300 x 200 x 75 mm. The double L-shaped specimen is placed in the Compression Testing Machine. Load reading is carried out at maximum load until the sample failures.

2.3 Testing of Concrete Bond Strength

Bond strength testing is done by placing a concrete cylinder on a loading frame equipped with a hydraulic jack and load cells, then the reinforcing bar embedded in the cylinder is pulled until pulled (Pull-Out Test, Fig. 1). The variations in the diameter of the supporting holes used are 10 mm, 35 mm, 60 mm, 85 mm and 110 mm (Fig. 2). The length of the embedded reinforcement is 150 mm with a steel reinforcement diameter of 10 mm on plain reinforcing steel. This test is carried out on specimens 28 days old by pulling reinforcing steel embedded in a concrete cylinder.

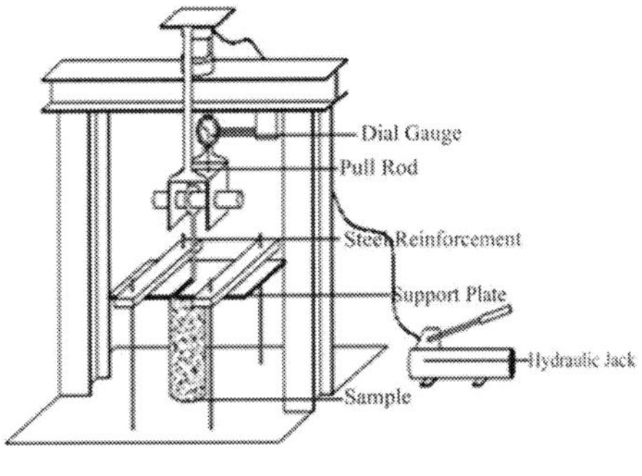

Fig. 1. Sketch of concrete bond strength test.

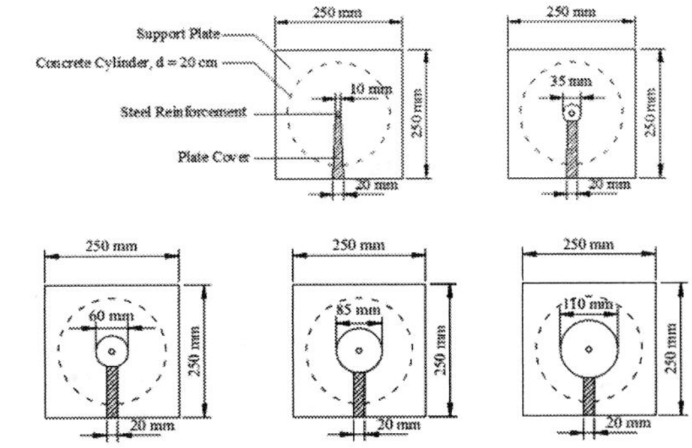

Fig. 2. Sketch of the diameter of support plate hole's variation

3 Result and Discussion

3.1 Compressive Strength of Concrete

The data obtained in this compressive strength test is the maximum load that causes the specimen to collapse. The test results obtained the value of normal concrete compressive strength of 26.492 MPa. This value indicates the compressive strength produced in all types of concrete is greater than the compressive strength required, equal to 20 MPa.

3.2 Shear Strength of Concrete

Data obtained in the shear strength test is a maximum load that causes the test object to collapse. From the test results, obtained shear strength of 10.864 MPa for normal concrete. The value of shear strength is needed so that experimental failures do not occur due to the fracture of the concrete bond before the reinforcement embedded in the test object is pulled out.

3.3 Bond Strength of Anchor Bolt

In this study, two data are obtained, namely the maximum load and the deformation value. Maximum load is the load when the steel bar is pulled from the concrete. At the same

https://doi.org/10.1051/e3sconf/202133105011

time, the deformation is the length of the shift of steel uprooted from concrete.

Test the bond strength between plain steel reinforcement with concrete in normal concrete using as many as 15 cylindrical test specimens with three pieces per variation of the supporting hole diameter. A diagram of bond strength test results with variations in the diameter of the supporting plate in a normal concrete plain reinforcement can be seen in Fig. 3.

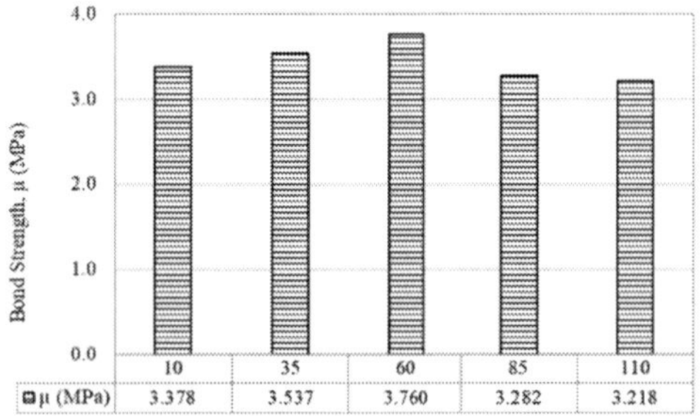

Fig. 3. Bond strength test results

Fig. 3. shows the maximum value of bond strength produced occurs at the diameter of the supporting hole size of 60 mm at 3,760 MPa, and the value of bond strength made at the minimum occurs at the diameter of the supporting hole size of 110 mm at 3,218 MPa. The diameter of a support hole of 10 mm to 60 mm shows a steady increase in bonding strength but a drastic decrease in the diameter of a support hole of 85 mm. This is due to the diameter of 85 mm and 110 mm in the presence of a short pedestal, which is only 45 mm apart so that the distribution is inadequate and the stress is not optimal. Whereas in the diameter of a supporting plate with a size of 60 mm, the bond between the reinforcement and the concrete will transfer the stress to the concrete around the reinforcement and the tension is still held by the support plate so that it reaches the maximum strength value. The graph which shows the relationship between load-deformation in normal concrete can be seen in Fig. 4.

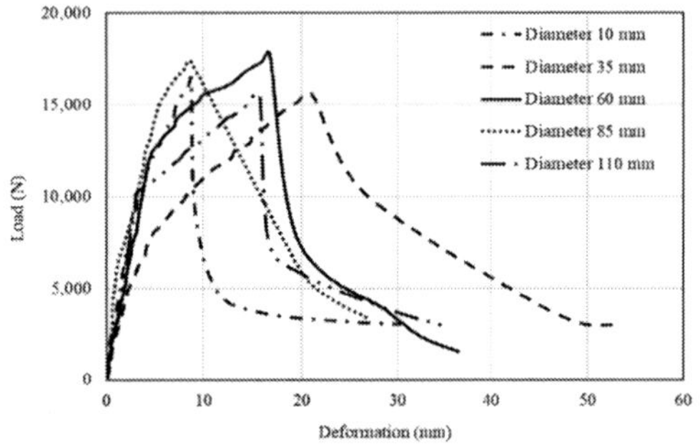

Fig. 4. The relation of load-deformation

The test results show that the maximum load value is generated by the supporting hole size of 60 mm. While the deformation obtained at 110 mm diameter of the supporting hole has a vast difference. It happens because the stress has reached a maximum at 110 mm support hole diameter and is followed by a large strain value. It shows that the level of ductility is high.

3.4 Shear Strength Relationship with The Supporting Hole Diameter

The shear strength test results are then compared with the concrete bond strength test results by making a relationship between the development length based on the shear strength and the concrete bond (Fig. 5).

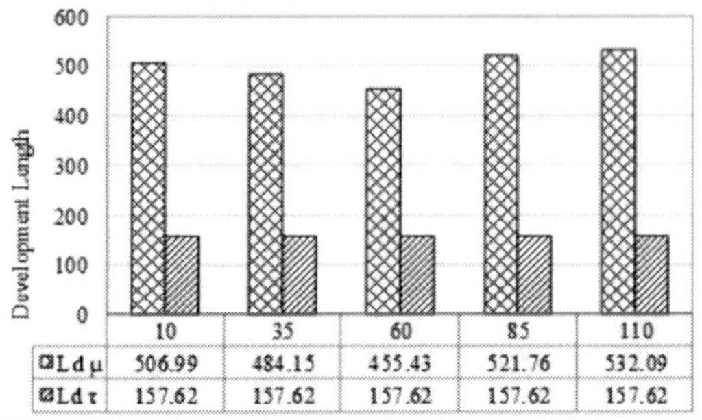

Fig. 5. Development length of bond and shear

The relationship between the development length of anchors with the shear strength for normal concrete is obtained. The development length of the bond strength is greater than the length of the shear strength. It shows that the failure of the test specimen occurred in the bond between reinforcing steel and concrete.

3.5 Analysis of the Bonding Coefficient

The diameter of plain reinforcement in normal concrete is then used to find the coefficient value of the development length of the experimental results (Fig. 6).

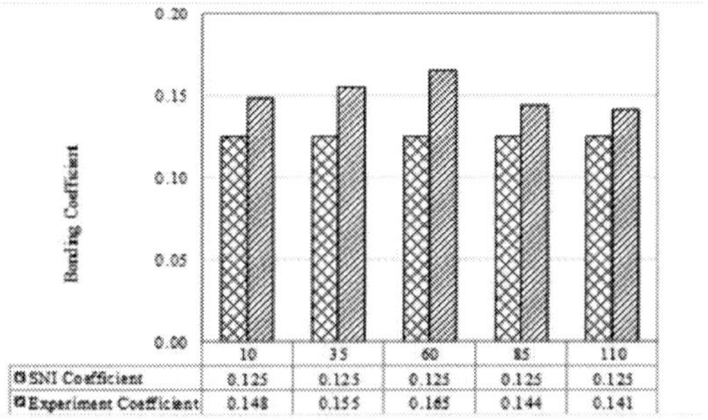

Fig. 6. The relation of diameter of support plate hole and bonding coefficient

The results shown in Fig. 6 are the bonding coefficient values that occur perfectly because the bonding coefficient values obtained from the experimental results in all variations of the diameter of the support hole are greater than the ACI coefficient value [12]. So the calculations based on experiments can be applied to the bond strength of normal concrete.

3.6 Analysis of the Development Length Coefficient

The results of testing the development length of plain reinforcement at the constant depth of anchoring are then analyzed the value of the coefficient of development length of the experimental results. This result was then compared with the value of the development length coefficient at ACI (Fig. 7) [12].

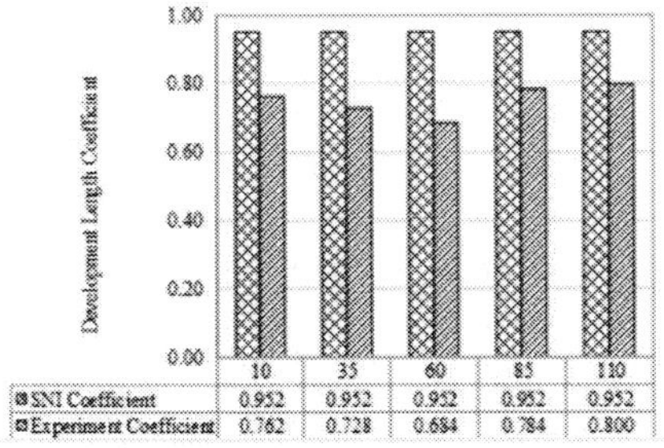

Fig. 7. The relation of diameter of hole and development length coefficient

Based on Fig. 7, it can be seen that the value of the development length coefficient obtained from experiments has a smaller value compared to the development length coefficient resulting from ACI [12].

3.7 Relationship Between Safety Factor and Supporting Plate Hole Diameter

The graph of the relationship between the safety factor and the diameter of the supporting plate hole in the normal concrete with plain reinforcement can be seen in Fig. 8.

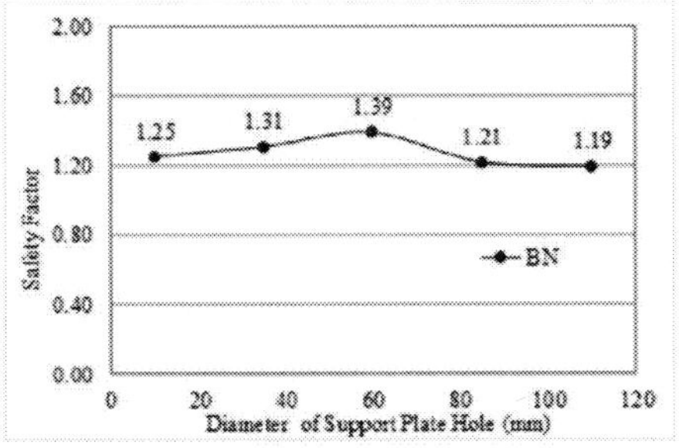

Fig. 8. The relationship between safety factor value and plate hole diameter

From Fig. 8, it can be seen that all variations in the diameter of the supporting plate hole in normal concrete with plain reinforcement have a safety factor value of more than 1. It shows that the application of anchoring depth based on ACI [12] produces a safe structure for normal concrete.

3.8 Failure Types

After the pull-out test is carried out, the next step is to observe the failure in all test specimens. The observations show that the pattern occurs in all objects, namely bonding failure, where steel is pulled out of concrete without cracking on the surface of the concrete. The bond between the steel and concrete and the restrain ability of the concrete can withstand reinforcement when the maximum load is exceeded.

4 Conclusion

Based on the results of the research, the following conclusions can be drawn:

(1) In normal concrete with plain reinforcement is obtained, the maximum bond strength at the diameter of the supporting plate is 60 mm by 3,760 MPa, and the minimum bond is obtained at 110 mm at 3,218 MPa.

(2) The relationship between bond strength and shear strength obtained correlation value based on the diameter of the supporting plate hole to the bond strength and shear strength and reinforcement. The development length of bonding distribution is greater than the development length of the shear strength distribution. It shows that the failure occurred in the bond between reinforcement and concrete.

(3) Variations in the failure of the sample obtained in normal concrete are bonding failure where the loss of bonding to the steel reinforcement.

References

1. C. W. Dolan and H.T. Hamilton, *Prestressed Concrete: Building, Design, and Construction*. Springer (2018)
2. E.P. Carvalho, E.G. Ferreira, J.C.D. Cunha, C.D.S. Rodrigues and N.D.S. Maia, Latin American Journal of Solids and Structures, 14 (11) pp 1932-1951 (2017)
3. D.L. Araújo, A.R. Danin, M.B. Melo and P.F. Rodrigues, Revista IBRACON de Estruturas e Materiais, 6 (2) pp 307-338 (2013)
4. Hariyadi, S. Munemoto and Y. Sonoda, Procedia engineering, 171 pp 926-933 (2017)
5. G. Xing, C. Zhou, T. Wu and B. Liu, Advances in Materials Science and Engineering, 2015, Article ID 604280 (2015)
6. V. Hlavička and É. Lublóy, Construction and Building Materials, 171 pp 588-597 (2018)
7. W. Guo, Z. Zhai, Z. Yu, F. Chen, Y. Gong and T. Tan, Advances in Civil Engineering, 2019, Article ID 7594132 (2019)
8. X. Xu and S. Tian, Plos One, January 17 (2020)
9. I.S. Azizah, *The Effect of Anchoring Depth Variation on Bond Strength of Plain Steel Reinforcement in Normal Concrete, Lightweight Concrete and High Quality Concrete,* Department of Civil Engineering, Faculty of Engineering, University of Mataram (in Indonesian) (2018)
10. S. Sunarmasto, GEMA TEKNIK Engineering Journal, 10 (2) pp 76-82 (in Indonesian) (2007)
11. ASTM C39/39M-04a, *Standard Test Method for Compressive Strength of Cylindrical Concrete Specimens* (2004)
12. ACI 318M-19, *Building code requirements for structural concrete (ACI 318M-19) and commentary* (2019)

 https://doi.org/10.1051/e3sconf/202133105012

Finite element modelling of monotonic pull-out test of deformed steel rebar embedded in well-confined concrete

Patria Kusumaningrum[1*], *Gigih Muslim* Prayogo[2], and *Sri* Tudjono[3]

[1]Structural Engineering Research Group, Faculty of Civil and Environmental Engineering, Bandung Institute of Technology, Indonesia
[2]Civil Engineering Master Program, Faculty of Civil and Environmental Engineering, Bandung Institute of Technology, Indonesia
[3]Civil Engineering Department, Faculty of Engineering, Diponegoro University, Semarang, Indonesia

Abstract. A finite element study carried out using LS DYNA and aimed to simulate the monotonic pull-out test of deformed steel rebar embedded in concrete is presented in this paper. Three models of the interface between deformed steel rebar and well-confined concrete, i.e. perfect bond model and two bond-slip models are observed and compared. Bond stress-slip response and rebar stress-slip response obtained numerically are validated with experimental data and empirical equations available from the literature. The full bond model overestimates the response, providing higher rebar stress. In the bond-slip models, good agreement is observed between numerical and experimental bond stress and rebar Stress–slip responses. The empirical equation of bond-slip proposed by Murcia-Delso and Shing (2014) is found to overestimate the peak bond stress.

1 Introduction

Reinforced concrete (RC) material is a composite material of concrete that has relatively low tensile strength and steel rebar to compensate for the ductility of the component with its high tensile strength and ductility. As it composes of two different materials, the interface between the concrete and steel becomes the weak part. When an RC component is loaded incrementally, stresses are gained and gradually increased all over parts of the component. As a weak link, the crack appears within this interface and stress will increase towards its capacity as the limiting value. In such a location, once the stress reaches its interface capacity, i.e., maximum bond stress, the transmitted stress between steel and concrete begins to drop to zero. At this stage, steel and concrete are unbounded, it is disconnected and no longer works as a unity. When the load is further increased, the unbounded interface grows and affects the overall structural behavior of RC components, both its load-carrying capacity and ductility/displacement.

To fully capture the accurate structural behavior of an RC component under specific loading, the bond-slip mechanism needs to be considered in the numerical model. A full bond model between steel and concrete may result in the overestimation of the structural capacity. Much past research has been made to study the bond mechanism through experimental testing and empirical analysis [1,2,3] and numerical modeling [4]. In this paper, the study is carried out numerically by finite element modeling to address this problem and to simulate the monotonic pull-out test of deformed steel rebar embedded in concrete. One of the full bond and two bond-slip models are analyzed and presented in this paper.

2 Bond Slip Mechanism

The bond-slip mechanism studied in this paper is based on experimental work done by Xu et al. (2017) [1]. The deformed rebar is well-confined concrete was monotonically loaded with a tensile force to form pull out test. The bond stress transfer at the interface between steel and concrete is represented by at least three mechanisms, i.e., adhesion, mechanical interaction, and friction.

2.1 Adhesion

Adhesion mechanism is formed during the concrete curing process by the chemical reaction of concrete, creating bonds and stresses to the rebar. This mechanism is significant to the bond stress transfer in the case of plain bars which have smooth surfaces. The failure of the adhesion stress transfer mechanism is occurred as crack initiates and propagates at the interface. In the case of deformed rebars, as studied herein, the adhesion mechanism is not significant to the stress transfer mechanism.

* Corresponding author: patria@si.itb.ac.id

 https://doi.org/10.1051/e3sconf/202133105012

2.2 Mechanical Interaction and Friction

In the case of deformed steel bars, the bond stress transfer mechanism is primarily due to the combination of the mechanical interaction and friction between deformed ribs and the concrete surrounding them. As the axial load increases, the force transfer mechanism is dominated mainly by the mechanical interaction at the deformed steel rebar ribs. At one point, microcracks start to form at the ribs in a transversal direction, perpendicular to the direction of the loading. When the axial load is further increased, cracks grow wider and hence become macro transverse cracking. The area of conical shape where transverse cracking extended within finite depth surrounding the deformed bar is called a bond zone. As the transverse cracking progresses, the concrete adjacent to the deformed bar forms inclined compression cone struts. The strut rotates, forming radial contact forces. When the concrete is insufficiently confined, it may result in splitting failure which may further spread the damage outside of the bond zone. As the concrete in this study is well confined, no splitting failure is observed at the surface.

When the stress is near to the bond stress, longitudinal cracking takes place and bond zone dilation is observed. Once the peak is reached, stress-strain softening is observed due to progressive shear failure of concrete between the deformed ribs. Fig. 1 shows the condition of the specimen post-experiment [1]. No splitting failure was observed at the surface, and the steel rebar was still in the elastic stage. The failure is herein governed by the concrete adjacent to the contact area. Hence, the bond-slip stress is deduced using the following equation:

$$\tau = \frac{F}{\pi . l_e . d} \qquad (1)$$

Fig 1. Bond Failure Pattern of Specimen Observed Experimentally [1]

3 Finite Element Modeling

3.1 Model Set-Up

The finite element modeling is carried out using explicit analysis by licensed LS DYNA software to model the monotonic pull-out test of the deformed steel rebar embedded in well-confined concrete. The set-up of the finite element model as shown in Fig. 2 follows the detail of the experimental setup for the local bond-slip test conducted by Xu et al. (2017) [1]. Well, a confined concrete specimen of size 100 mm ×100 mm ×140 mm with installed stirrups was cast together with the deformed steel rebar of 16 mm nominal diameter inserted in the center of the concrete specimen. The stirrups were added to the specimen to prevent the splitting of concrete; hence, the pull-out failure was achieved. The bonded length between rebar and concrete was 80 mm; or 5 times the nominal diameter. Displacement control is carried out to perform the monotonic pull-out test. The well confined concrete specimen was clamped by the top and bottom steel plates. Surface to surface contact algorithm is invoked to simulate the interaction between the steel plates and the concrete.

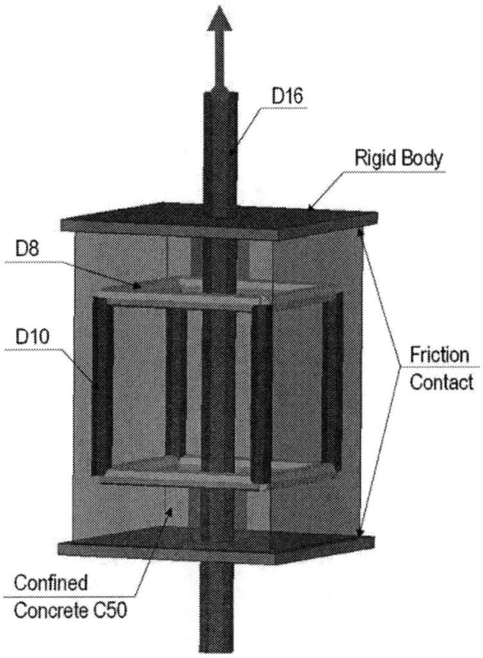

Fig 2. Pull-out Test Set Up Model

3.2 Concrete Material

There are at least four material models in LS DYNA that can be utilized to model concrete material, i.e., K&C Concrete Damage Release III (Mat 072 Rel. III), Winfrith Concrete (Mat 084), Continuous Surface Cap (CSCM, Mat 145) and Concrete Damage Plasticity (CDPM, Mat 273). Among others, Concrete Damage Plasticity Material (CDPM) is observed to be able to capture the compressive hardening and softening as well as tension stiffening for the monotonic loading condition. This material is also observed to be good at modeling the degradation of stiffness in the hysteretic curve during the loading and unloading cycles. Therefore, in this study CDPM is incorporated to model concrete material.

As for the element type, concrete is modeled as a 3D solid element with a reduced integration point. Hence, to

 https://doi.org/10.1051/e3sconf/202133105012

avoid shear locking, the hourglass control Flanagan-Belytschko stiffness form [5] with a significantly small hourglass coefficient of 0.02 is invoked. The hourglass coefficient is adjusted and optimized following the optimized mesh size from the sensitivity analysis. It is carried out by controlling the amount of hourglass energy produced, which should be less than 10% of the total internal energy. The high value of the hourglass coefficient is not recommended to be used with the stiffness form as it may stiffen the structural response.

Material regularization is carried out before FEA, and the stress-strain relations of concrete material incorporated in the FEA model are given in the following Figures. Fig. 3 shows the compressive stress-strain relation of concrete of 55 MPa compressive strength. The maximum compressive strength is obtained from the experiment [1]. The tensile behavior of concrete material is given in Fig. 4. Both the maximum tensile strength and the fracture energy of the concrete material follow the empirical equations provided by CEB FIP as given in Fig. 5.

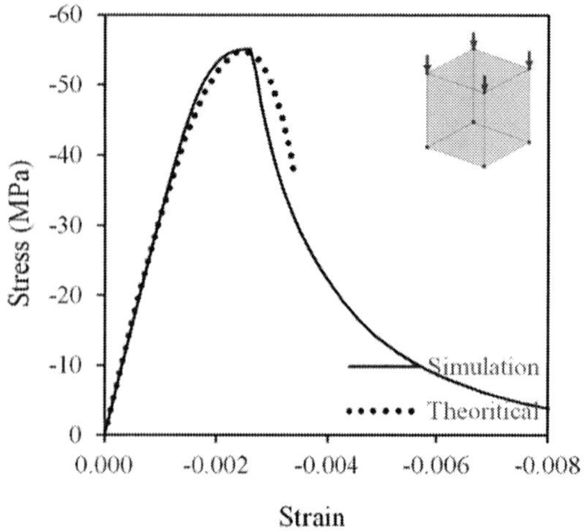

Fig 3. Compressive Stress Input of Concrete CDPM

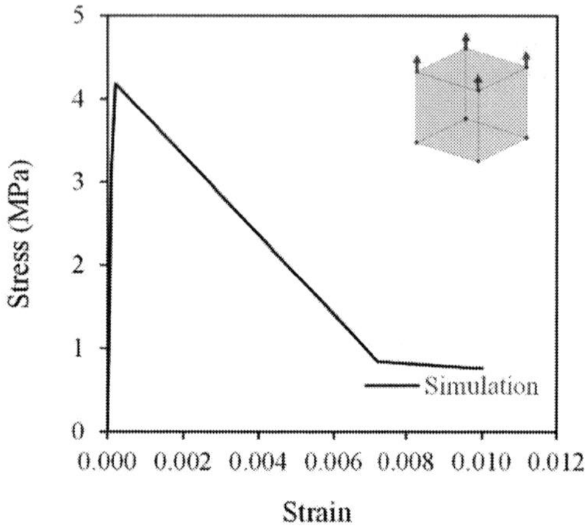

Fig 4. Tensile Stress Input of Concrete CDPM

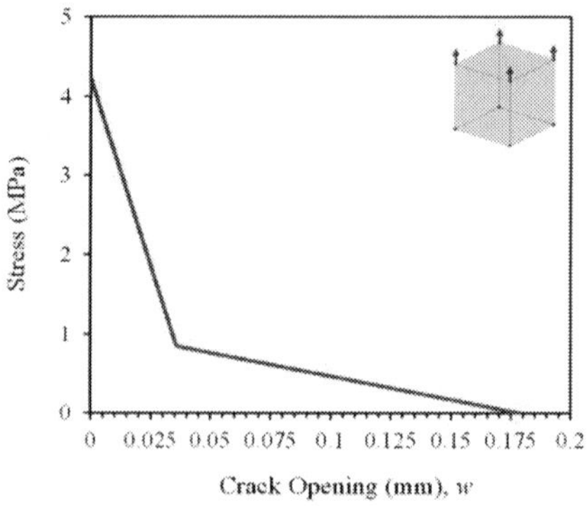

Fig 5. Fracture Energy Input of Concrete CDPM

3.3 Steel Material

To model steel material, material Piecewise Linear Plasticity (Mat 024) is incorporated into the FEA model. This material is capable to capture the elastoplastic bilinear hardening behavior of steel and incorporating isotropic hardening failure surface. The properties of steel material are given in Table 1. No element deletion is allowed in this model.

Table 1. Steel Material Properties

Material Properties	Nominal	Unit
Mass Density, ρ	7700	kg/m^3
Young's Modulus, E	2.10^5	MPa
Poisson's Ratio, υ	0,3	-
Yield Stress, f_y	363	MPa
Tangent Modulus, E_{tan}	1.10^3	MPa

The type of steel rebar element incorporated in this study is the Hughes-Liu beam element following the formula developed by Hughes and Liu in 1981 [6]. The cross-section of each beam element is divided into 4 fiber elements. Each fiber element has 1 integration point. Thus, there are 4 integration points per steel beam's cross-section are 4 pieces.

3.4 Bond Slip Interface Model

To model the debonding process between steel reinforcement and concrete, a constrained beam in solid is invoked to provide constraint relaxation along the beam in the axial direction. The debonding force is applied based on the relative motion of the steel rebar. In applying the debonding force, a user-defined function is written in dyn. f is integrated into the FEA model as given in Fig. 5. Three bond-slip models are carried out, i.e., full bond model, bond-slip model following empirical bond-slip relation by Murcia-Delso [4], and bond-slip model following empirical bond-slip relation by Murcia-Delso with adjusted maximum bond stress.

https://doi.org/10.1051/e3sconf/202133105012

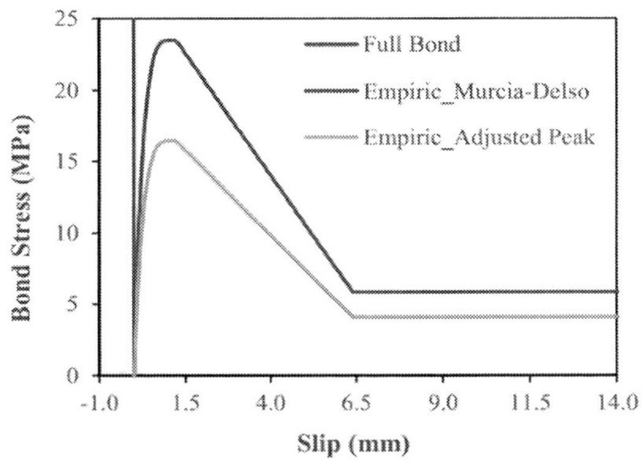

Fig 5. Three Bond-Slip models

3.4.1 Model 1: Full Bond

The first model is a full bond model. In the full bond model, constrained beam in solid is not applied as no bond-slip is allowed. The 3D solid concrete and 2D beam steel elements are sharing the nodes as the nodes are merged.

3.4.2 Model 2: Empirical Bond-Slip

In Model 2, the bond-slip is applied. The 3D solid concrete and 2D beam steel elements are constrained by using constrained beam in solid algorithm and user-defined load function as described in Section 3.4. The user-defined bond-slip relation follows the empirical equations proposed by Murcia-Delso [4] as shown in Fig. 5. The peak bond stress (τ_{max}) is given as a function of concrete compressive strength and the slip at which the peak bond stress is attained (s_{peak}) is a function of bar diameter.

$$\tau_{max} = 1.163\,(f_c')^{2/3} \quad (2)$$

$$s_{peak} = 0.07 d_b \quad (3)$$

The bond stress at softening part is cut off at friction bond resistance of 0.25 τ_{max}, when the slip occurred is beyond the clear spacing between deformed ribs (s_R). For concrete with 55 MPa compressive strength, the calculated τ_{max} is 23.49 MPa.

3.4.3 Model 3: Adjusted Empirical Peak Bond-Slip

The peak bond stress proposed by Murcia-Delso in Model 2, however, is higher than the observed bond stress of 16.43 MPa from the experiment [1]. This model is similar to Model 2 with adjustment in the peak bond stress. The peak bond stress is lowered to 16.43 MPa, yet the bond-slip interface function still follows the empirical equations proposed by Murcia-Delso in Fig. 5.

4 Numerical Results

This section presents the numerical results obtained from FEM. Validation is provided by comparing numerical results obtained with the experimental data.

Tensile damage contour plots of the three models are given in Fig. 6. Red-colored solid elements show the bond zone area which experienced tensile damage.

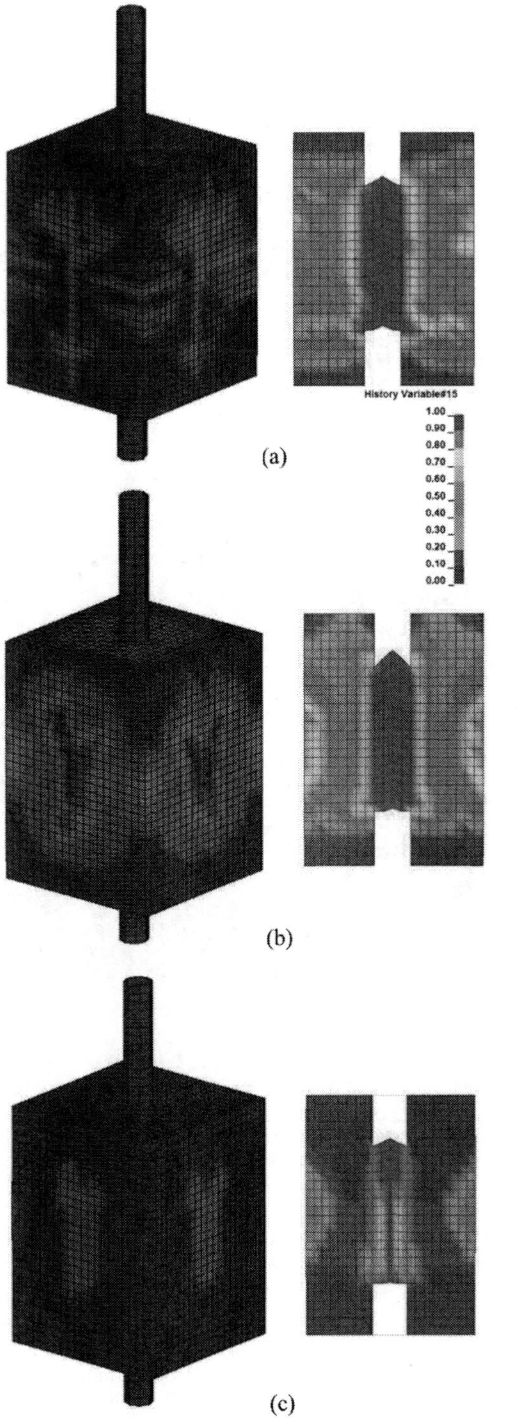

Fig. 6 Tensile Damage Contour at Final Stage of Loading of (a) Model 1: Full Bond, (b) Model 2: Empirical Bond-Slip, and (c) Model 3: Adjusted Empirical Peak Bond-Slip Specimens Subjected to Monotonic Pull-out Test

 https://doi.org/10.1051/e3sconf/202133105012

From the tensile damage contour of Model 1 and 2 as illustrated in Figs. 6(a) and (b)), the damage has spread to the concrete area adjacent to the bond zones. On the other hand, in Model 3 the bond zone is localized and has not spread. Therefore, it is observed that Model 3 which incorporates bond-slip with adjusted empirical peak bond is in good agreement with the bond failure pattern of the specimen from the experiment as shown in Fig. 1, hence, it provides more realistic measures.

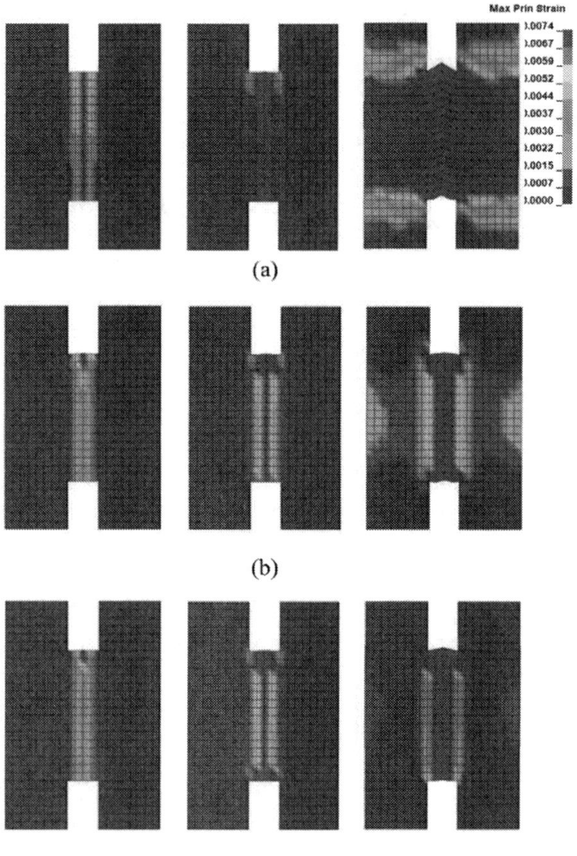

Fig. 7 Maximum Principal Strain Contour in MPa at Initial Stage (Left), Peak Bond Stress (Middle) and Final Stage (Right) of Loading of (a) Model 1: Full Bond, (b) Model 2: Empirical Bond-Slip, and (c) Model 3: Adjusted Empirical Peak Bond-Slip Specimens Subjected to Monotonic Pull-out Test

Maximum principal strains and minimum principal stresses contours of the three models are plotted and given in Figs. 7 and 8, respectively. Fig. 7 shows no splitting failure has occurred towards the transverse direction of the loading. It agrees well with the experimental result in Fig. 1. Conforming the damage observed in Fig. 6, similarly, the damage is found to be more localized around the bond zones for Model 2 and within the bond zones for Model 3 in Fig. 7. The maximum principal strain contour observed in Model 1 is, quite unrealistic as significantly large strains are observed to spread to the outer surface of the confined concrete region due to the full bond action.

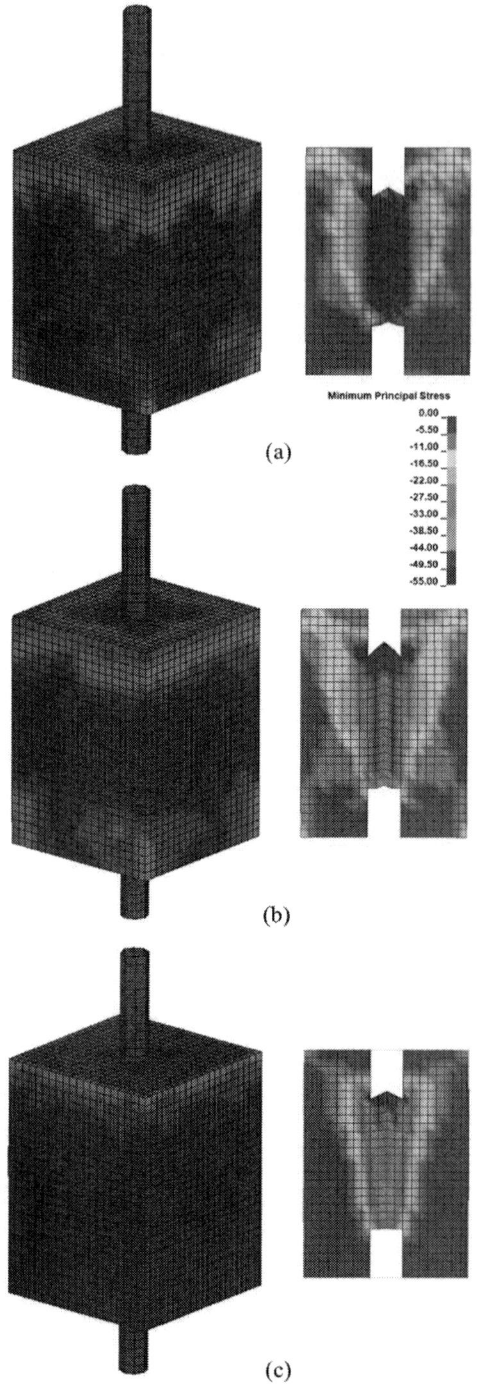

Fig. 8 Minimum Principal Stress Contour in MPa at Peak Bond Stress of (a) Model 1: Full Bond, (b) Model 2: Empirical Bond-Slip, and (c) Model 3: Adjusted Empirical Peak Bond-Slip Specimens Subjected to Monotonic Pull-out Test

In Fig. 8, the formations of inclined compression cone struts are observed. In the full bond model (Model 1), the compression struts have spread to adjacent concrete outside of the bond zone. On the other hand, the compression cone struts of Models 2 and 3 in Fig. 8 that incorporate bond slips stay within the bond zones.

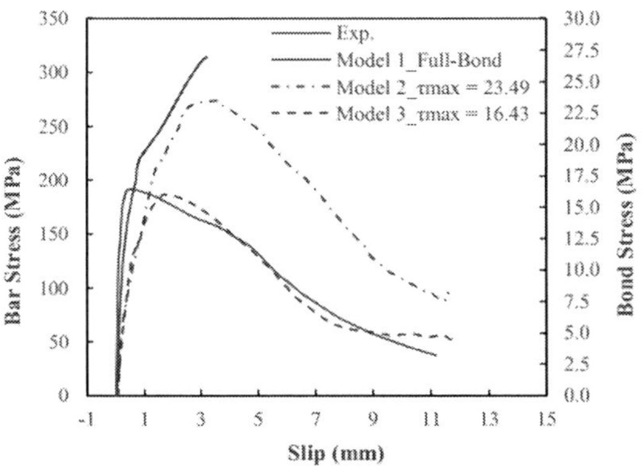

Fig. 9 Bar and Bond Stresses vs Slip

The bar and bond stresses of the three models and the experiment are plotted against the slip and given in Fig. 9. All the bar stresses are observed to have a value below the yield stress of 363 MPa. The full bond Model 1 overestimates the bar and bond stresses by 66.5%. Similarly, Model 2, the bond-slip model with Murcia-Delso empirical bond-slip relation, provides an overestimation of 44.6% when compared to the experimental result. Of the three models, Model 3, i.e., the bond-slip model with adjusted peak bond, can capture the bond-slip behavior better. The bar and bond stresses obtained from Model 3 are similar to those obtained from the experiment with only 1.2% error.

5 Conclusions

This paper investigates the behavior of the monotonic pull-out test of the deformed steel embedded into a well confined concrete specimen. Finite element analyses are carried out in three models and the results obtained are benchmarked against the experimental results available in the literature.

It is found that Model 3 of the bond slip model with adjusted peak bond provides a better estimate than the full bond of Model 1 and bond-slip Model 2 using Murcia-Delso empirical equations. Good agreement is observed between Model 3 with the adjusted empirical peak bond-slip and the experiment in terms of bond stress and rebar stress–slip response. The full bond model, as expected, provides overestimation to the response, providing higher rebar stress. In the bond-slip models, the empirical equation of bond-slip proposed by Murcia-Delso and Shing (2014) is found to overestimate the peak bond and bar stress.

References

1. S. Xu, A. Li, and H. Wang, "Bond properties for deformed steel bar in frost-damaged concrete under monotonic and reversed cyclic loading," *Construction and Building Materials*, **148**, 344-358 (2017)
2. L.J. Malvar, "Bond of reinforcement under controlled confinement," ACI Material Journal, **89**, 6, 593–601 (1992)
3. H. Sezen, J.P. Moehle, "Bond-Slip Behavior of Reinforced Concrete Members," Proceedings of the fib 2003 Symposium, Athen, Greece (2003)
4. J.Murcia-Delso, P.B. Shing, "Bond-slip model for detailed finite-element analysis of reinforced concrete structures," Journal of Structural Engineering, **141**, 4 (2014)
5. D. P. Flanagan, T. Belytschko, "A uniform strain hexahedron and quadrilateral with orthogonal hourglass control," Numerical Methods in Engineering, **17**, 5, 679-706 (1981).
6. T.J. Hughes, W.K. Liu, "Nonlinear finite element analysis of shells: Part I. Three-dimensional shells. Computer methods in applied mechanics and engineering,"**26**, 3, 331-362 (1981)

 https://doi.org/10.1051/e3sconf/202133105013

Stability analysis of concrete block retaining wall based on a scaled laboratory

Deni Irda Mazni[1,2*], *Abdul* Hakam[1], *Jafril* Tanjung[1], and *Febrin* Anas Ismail[1]

[1]Andalas University, Civil Engineering Department, Padang, Indonesia
[2]Dharma Andalas University, Program Study of Civil Engineering, Padang, Indonesia

Abstract. A concrete block retaining wall is supported by its weight. This block retaining wall structure has some advantages, including lower construction costs, a water-permeable construction that produces less water pressure behind the wall, and a more flexible construction because it can follow the ground's contours. Rankine's theory is usually used to design this block retaining wall. According to this theory, the failure pattern behind the wall forms an angle of 450 + • /2 with the horizontal plane. The laboratory tests indicate that the pattern of failure or the pattern of sand movement behind the wall is similar to the letter S. From the sandy soil of failure pattern curve, look for a functional equation approximating the pattern. The equation of the function obtained is an equation of the third-order function. An analysis of the sliding, overturning, and overall stability block retaining wall is based on this equation of the cube function. Analysis for overall stability using the method of slices, dividing the failure area by several slices. These function equation order three is needed to get the area and length failure.

1 Introduction

The purpose of retaining walls is to support the weight of sloped soil and buildings, earthquake loads, machine loads that produce vibrations, and others [1]. It is important to note that retaining walls can remain functional and stable based on the weight of the retaining wall itself and the weight of the soil above the heel of the wall.

The advantages of a concrete block retaining wall include its ease of construction; the footing is simply a gravel leveling pad, and the units are dry-stacked without the use of mortar, steel reinforcement, or grouting [2], [3]. The Concrete block retaining walls are often considered as rigid (conventional) retaining walls when designing shear and overturning stability [4–6].

The failure behavior or pattern of a block retaining wall and a gravity wall is not necessarily the same on the field. In laboratory experiments, the failure plane of sand behind the concrete block retaining wall is not a straight line, as predicted by Rankine's theory (a straight line), but is more like an S curve [7].

There have been several failures when designing retaining walls. Sharma (2011) reported on a study on the failure of retaining walls, which serves as a helpful lesson for future consideration of the stability of retaining walls [8]. The situation was discussed in detail in the report, including the retaining wall's failure mechanism. The failure of retaining walls is mainly caused by the incorrect design of retention and support systems based on experience [9-10].

As a result, a more thorough examination of this retaining wall is required, beginning with examining the failure pattern.

2 Block Retaining Wall (Concrete)

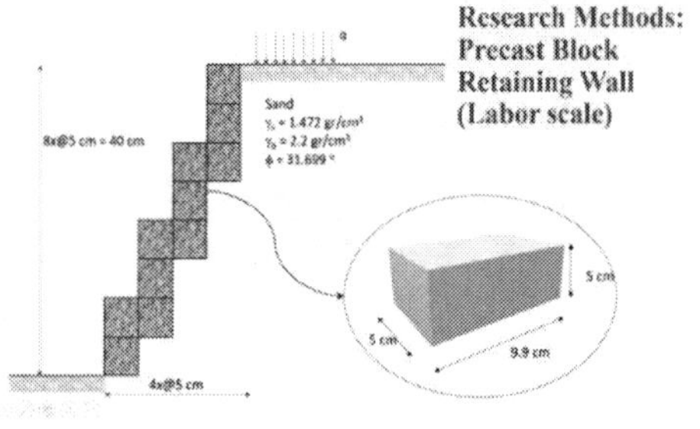

Fig. 2. Failure pattern due to static load (Rankine's theory)

In the laboratory, the model of the block retaining wall was made, as shown in Fig. 1. The block retaining wall is made of a mixture of cement, sand, and water with 9.9 (length) x 5 cm (width) x 5 cm (height). The blocks are

*Corresponding author: deniirdamazni@gmail.com

https://doi.org/10.1051/e3sconf/202133105013

arranged in a glass box with a diameter of 40 cm (length), 10 cm (width), 80 cm (height). The surface of the embankment behind the wall is level, with the soil type being sand. The total height of the wall is 40 cm, and there are 11 blocks in total. There is no adhesive or connection between the concrete blocks.

In classic theory (Rankine's theory), to calculate the overall stability of the block (segmental) retaining wall, assuming the failure pattern behind the wall resembles a straight line forming an angle of (45^0 + $\phi/2$) with the horizontal plane in Fig. 2. According to NCMA, the block retaining wall failure pattern is predicted to appear like a straight line in Fig. 3. even though the actual on the field is not always the same [11].

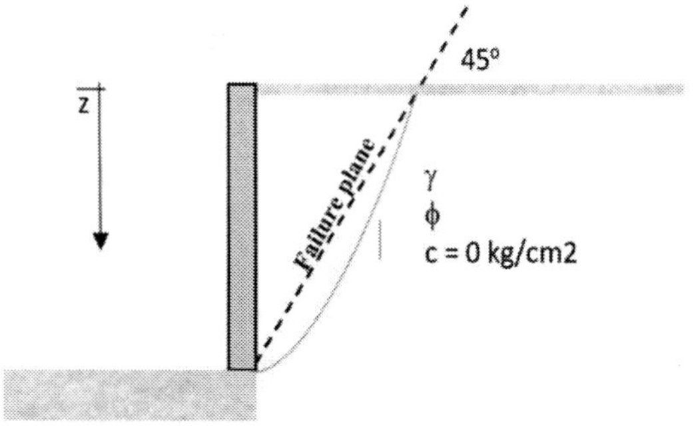

Fig. 2. Failure pattern due to static load (Rankine's theory)

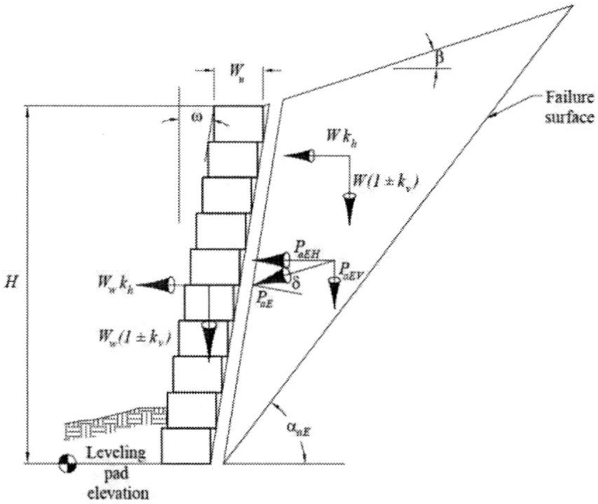

Fig. 3. NCMA, design SRW (2016)

3 Method

The static load is given gradually until the wall fails after the block retaining wall is arranged, as indicated in Fig. 1. During the testing process, videos are recorded, so the movement of sand grains can be seen. The next step is to determine the equation function of the failure pattern. This equation's function helps determine the wall's overall stability, as well as the backfill area and length. Analysis for overall stability using the slices method applies to flat, circular, or a combination of both.

With the method of slices, the soil block that has failed is divided into several slices. Vertical slicing is used to divide the slices. For each piece, the working forces are analyzed and calculated cumulatively. Furthermore, the safety factor of the assumed failure plane can be calculated by comparing the resisting and driving forces. Weight is determined by the soil's weight for each part is:

$$W = \gamma\,(\Delta A) \quad (1)$$

Next, the normal force (N) and the tangential force (T) act for each slice (α is the slope angle assumed for each slice):

$$N = W\cos\alpha \quad (2)$$
$$T = W\sin\alpha \quad (3)$$

The area of the failure (A) is calculated using the integral equation of F(h) (Equation 4). F(h) is a function of the equation for the height of the block retaining wall. The formula determines a slice's failure length (DL) in equation 5. By using equation 6, we can calculate the overall stability safety factor (c = 0).

$$area\ of\ slices\ A\,(\Delta A) = \int_a^b F(h) \quad (4)$$
$$\Delta L = \int_a^b \sqrt{1 + \left[\frac{dF(h)}{dh}\right]^2}\, dh \quad (5)$$
$$SF_{overall} = \frac{\sum_1^n (c*\Delta L_n + N_n \tan\phi)}{\sum_1^n T_n} = \frac{\sum_1^n (N_n \tan\phi)}{\sum_1^n T_n} \quad (6)$$

A safety factor of sliding and overturning is also investigated in this study. The force of lateral earth pressure, the weight of soil above the heel (Ws), and the concrete Wc should all be considered while analyzing the wall's stability. An active earth pressure condition will arise if the wall is pushed out due to the earth pressure behind it. According to Rankine, the value of earth pressure can be calculated as follows:

$$\sigma_a = \gamma\, z\, tan^2\left(45^0 - {}^{\phi}/_{2}\right) \quad (7)$$

Using the active earth pressure coefficient, Ka, as a starting point:

$$K_a = tan^2\left(45^0 - {}^{\phi}/_{2}\right) \quad (8$$

Only the weight of the soil itself contributes to the active compressive force acting behind the wall to a depth of z = h in non-cohesive soils (c = 0).

$$P_a = {}^{1}/_{2}\,\gamma\, h^2 K_a \quad (9)$$

Equation 10 is the formula for determining the value of the safety factor of sliding. The formula for the safety factor of overturning is explained in equation 11.

$$SF_{sliding} = \frac{\sum F_R}{\sum F_d} = \frac{\sum V\tan\delta + B\,c + P_p}{P_a} = \frac{\sum V\tan\delta}{P_a} \quad (10)$$
$$SF_{overturning} = \frac{\sum M_R}{P_a\,y} = \frac{\sum W\,x}{P_a\,y} \quad (11)$$

4 Result and Discussion

Table 1 shows the data on soil properties. Because the cohesion value is equal to 0 and the sieve analysis reveals 96.096% sand, the soil type is sand. Based on observations and video recordings of laboratory experiments, the movement pattern of sand grains was obtained Fig. 4. The pattern is described in the Cartesian coordinate system Fig. 5. This pattern (failure pattern) looks like the letter S.

Tabel 1. Sandy soil properties

Type of Parameter	Value
Unit Weight (γ_s)	1,472 gram/cm3
Soil Cohesion (c)	0 gram/cm2
The angle of internal friction of soil (ϕ)	31,699°
Sieve analysis of soil	96,067 % (sand)

The next step is to find the equation of the mathematical function of the pattern Fig. 6. The function obtained is the equation of polynomial order 3, which is shown in equation 12. An analysis of overall stability is performed based on the failure pattern.

Fig. 4. NCMA, design SRW (2016)

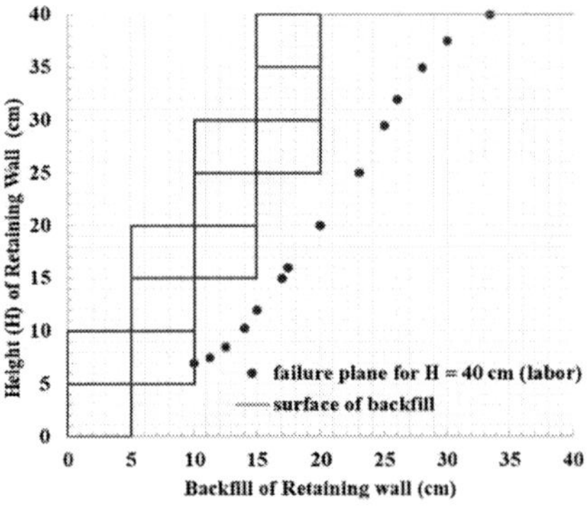

Fig. 5. Sketch of failure pattern

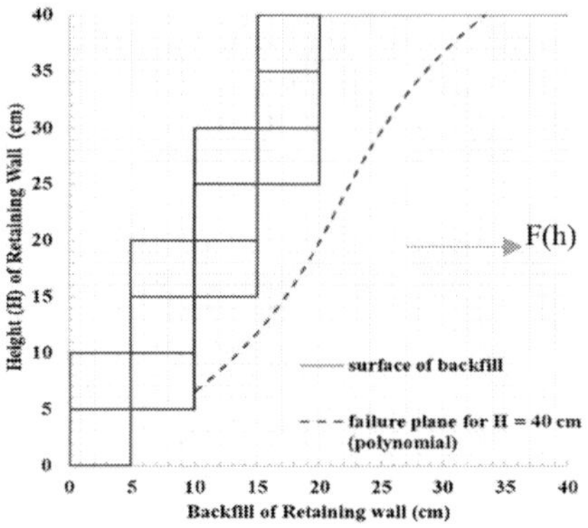

Fig. 6. The mathematical function of the failure

$$F(H) = 0.0008H^3 - 0.0558H^2 + 1.7739H + 0.5262 \quad (12)$$

4.1 Overall Stability of Block Retaining Wall

Based on the equation of function, the overall stability of the wall can be found. Fig. 7. shows the failure area divided into seven slices using the method of slices.

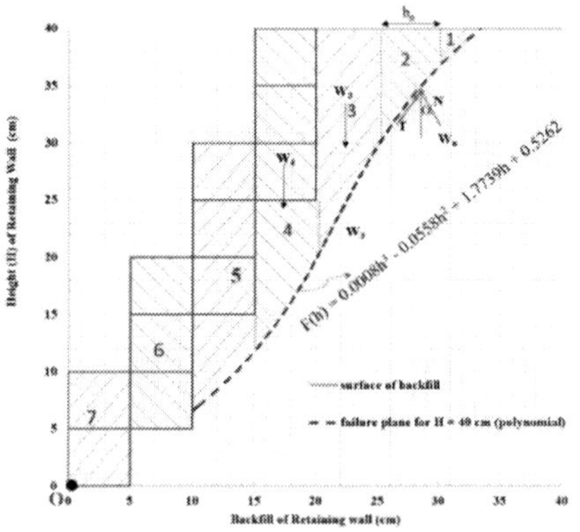

Fig. 7. Equation of failure pattern and method of slices

The area and weight of each slice are calculated using equations 1 dan 4. Additionally, the calculation is displayed in a tabular format following:

Tabel 2. Calculation of overall stability

Slice	b	Area	γ	Weight	α	sin α	cos α	W sin α	W cos α
	cm	cm^2	gr/cm^2	gr/cm^1	°			gr/cm^1	gr/cm^1
1 sand	3.40	5.42	1.47	7.98	46.00	0.72	0.69	5.74	5.55
2 sand	5.00	32.52	1.47	47.86	57.00	0.84	0.54	40.14	26.07
3 sand	5.00	75.33	1.47	110.89	62.00	0.88	0.47	97.91	52.06
4 sand	5.00	47.47	1.47	69.88	52.00	0.79	0.62	185.09	144.61
block	5.00	75.00	2.20	165.00					
5 sand	5.00	29.88	1.47	43.99	44.00	0.69	0.72	145.18	150.34
block	5.00	75.00	2.20	165.00					
6 block	5.00	75.00	2.20	165.00	0.00	0.00	1.00	0.00	165.00
7 block	5.00	50.00	2.20	110.00	0.00	0.00	1.00	0.00	110.00
							Sum	293.02	653.61

$$SF_{overall} = \frac{\sum_1^n (N_n \tan\phi)}{\sum_1^n T_n} = 1.16<1.5 \quad (13)$$

 https://doi.org/10.1051/e3sconf/202133105013

The overall stability safety factor value is 1.16, based on the preceding explanation of the slice approach (equation 6).

4.2 Sliding Stability and Overturning Stability

Fig. 2 depicts the forces involved in estimating sliding and overturning stability. The lateral earth pressure force Pa, the weight (W) of the soil and concrete, and the force's distance (x or y) from the point O (toe of a wall) are all factors to consider. Table 3 summarizes the calculations used to determine the sliding stability value.

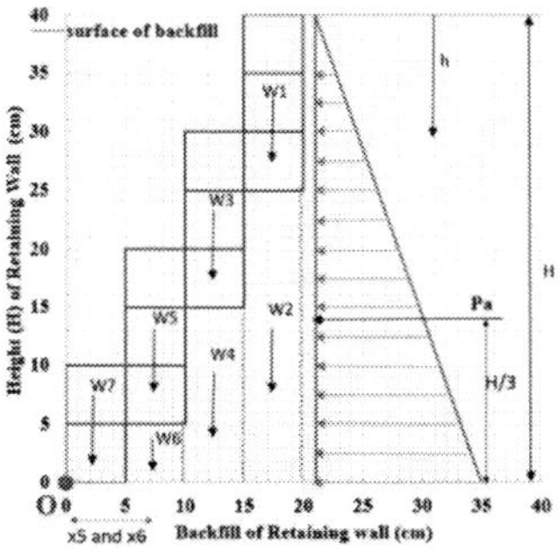

Fig.8. Forces acting on the block retaining wall

Tabel 3. Sliding stability calculation

Slice	**Area** cm^2	**γ** gr/cm^2	**W** gr/cm^1	**Distance** cm	**Moment** gr/cm^1
1	75	2.2	165	17.5	2887.50
2	125	1.47	183.75	17.5	3215.63
3	75	2.2	165	12.5	2062.50
4	75	1.47	110.25	12.5	1378.13
5	75	2.2	165	7.5	1237.50
6	25	1.47	36.75	7.5	275.63
7	75	2.2	165	2.5	412.50
Total		**ΣV =**	990.75	**ΣM$_R$ =**	11469.38
		Pa =	365.74	13.33	4875.26
		Ka =	0.311		

The safety factor value for sliding stability is:

$$SF_{sliding} = \frac{\sum V \tan\delta}{P_a} = \frac{990.75 * \tan(\frac{2}{3} * 31.699)}{365.74} = 1.05<1.5 \quad (14)$$

The safety factor value for overturning stability is:

$$SF_{overturning} = \frac{\sum W\,x}{P_a\,y} = \frac{11469.38}{4875.26} = 2.35 < 2.5 \quad (15)$$

5 Conclusion

The failure or movement of sand grains behind the block retaining wall generates an equation of a polynomial order of three functions. Order three of the function equation can be used to calculate the overall stability value. Using the integral of the equation of the power of three functions, determine the area of failure. So that calculations using these equations will get more precise results. It is necessary to redesign the block retaining wall dimensions to obtain a safe overall, sliding, and overturning stability value.

The authors are grateful for the support of this experimental work by the Indonesian Ministry of RISTEK-BRIN, Deputy for Strengthening Research and Development. Thanks to the Department of Civil Engineering and technical support staff from the Soil Mechanics Laboratory at the Faculty of Engineering, Andalas University, for their dedication and assistance before and during the test. Thank you also to Civil Engineering Study Program, Dharma Andalas University for their support in this research.

References

1. J. A. McKelvey, Lessons Learned From Failures: The Wall of Shame, Geo-Frontiers, ASCE, 3235–3244 (2011)
2. H. Brooks, J. P. Nielsen, Basics of Retaining Wall Design, 10 th. Newport Beach California: since 1992 (2013)
3. U. C. Sari, M. N. Sholeh, I. Hermanto, The stability analysis study of conventional retaining walls variation design in vertical slope, J. Phys. Conf. Ser., **1444**, 1 (2020)
4. B. Toprak, O. Sevim, I. Kalkan, Gabion Walls and Their Use, Int. J. Adv. Mech. Civ. Eng., **3**, 4, 2394–2827 (2016), [Online]. Available: http://www.iraj.in/journal/journal_file/journal_pdf/13-271-147400929156-58.pdf.
5. M. Ramli, T. J. R. Karasu, E. T. Dawood, The stability of gabion walls for earth retaining structures, Alexandria Eng. J.,**52**, 4, 705–710, (2013)
6. Nimbalkar, Pain, Ahmad, Chen, Stability Assessment of Earth Retaining Structures under Static and Seismic Conditions, Infrastructures, **4**, 2, 15 (2019)
7. D. I. Mazni, An alternative model of retaining walls on the sandy area to prevent landslides, E3S Web Conf., **156** (2020)
8. S. Sharma, Geo-Frontiers 2011 © ASCE 2011 2877, Geo-Frontiers 2011 © ASCE 2011 2877, 1993, 2877–2886, (2011) [Online]. Available:
9. A. Lim, Lesson learned from retaining wall failures: A geotechnical disaster, MATEC Web Conf., **229** (2018)
10. A. Haddad, G. Shafabakhsh, Failure of segmental retaining walls due to the insufficiency of backfill permeability, Geosynth. Civ. Environ. Eng. - Geosynth. Asia 2008 Proc. 4th Asian Reg. Conf. Geosynth., no. July, 852–856 (2008)
11. NCMA, Segmental Retaining Walls Best Practices Guide, 1st ed. 13750 Sunrise Valley Drive: NCMA Publication Number: TR308, (2016)

Assessment and retrofitting of nursing faculty building of Andalas University, Padang, Indonesia

Fauzan Fauzan[1*], *Dina* Yarmawati[2], *Seno* Willy Dharma Sipayung[1], *Willy* Kurniawan[1], *and Rozi* Adifa[1]

[1]Department of Civil Engineering, Faculty of Engineering, University of Andalas, Padang.
[2]Department of Civil and Environmental Engineering, School of Engineering, Tokyo Institute of Technology, Tokyo, Japan

Abstract. The Faculty of Nursing building is one of the administrative buildings at Andalas University. This building was designed by a planning consultant and began construction in 2013. However, during the construction period, it was found that the concrete quality was very low, that is fc'=14.32 MPa, so that the construction could not be continued because it was not in accordance with the planning quality, fc' =22.39 MPa. Therefore, it is necessary to evaluate the feasibility of the building structure before the construction is continued. In the structural analysis, the loads apllied are dead loads, live loads, and earthquake loads. The analysis was carried out using the application *(software)* ETABS. Based on the analysis results, it was found that the capacity of columns and beams of the building are not strong enough to withstand the loads acting on the structure. The inter story drift also does not meet the permit limit requirements according to the New Indonesian Earthquake Code, SNI 1726:2019. Therefore, it is necessary to retrofit (strengthen) the structure of the Nursing Faculty Building by re-designing the Detail Engineering Design (DED) on the beams and columns of the plan and jacketing the existing columns. Re-analysis results show that the retrofitted building structure has a strong enough capacity to carry loads acting on the structure and the inter story drift has met the permit limit requirements according to the SNI 1726:2019.

1 Introduction

The Faculty of Nursing building is one of the administrative buildings at Andalas University. This building was designed by a planning consultant and began construction in 2013. However, during the construction period, it was found that the concrete quality was very low, that is fc'=14.32 Mpa, so that the construction could not be continued because it was not in accordance with the planning quality, fc' =22.39 MPa [1]. The demolition of the building certainly requires a large amount of money. Therefore, it is necessary to have an action to retrofit (strengthen) the structure so that the building can carry the working load.

Therefore, it is necessary to assess of the structure and design retrofitting the structure of the Nursing Faculty Building, Andalas University.

2 Evaluation of Existing Structure

2.1 Condition of Existing Structure

The existing structure of the Nursing Faculty Building, Andalas University consists of the foundation, slab and first floor columns, as shown in Fig. 1.

Fig. 1. Existing Building Condition

From a field survey on the existing structure, it is found that the steel reinforcement had corroded (Fig. 2) and the concrete has porous and overgrown with moss (Figs. 3 and 4). From the concrete quality test results using a hammer test, the average concrete quality is fc'= 11.24 MPa, where the quality of this concrete is not in accordance with the design concrete quality.

* Corresponding author: fauzan@eng.unand.ac.id

Fig. 2. Steel Reinforcement Corrosion

Fig. 3. The Concrete Surface is Overgrown with Moss

Fig. 4. Porous Concrete and Rusty Steel Reinforcement

2.2 Structural System

The structural type of the Nursing Faculty Building, Andalas University, is a reinforced concrete structure. The building risk category is type IV because the building is classified as an educational facility building with seismic design category D, so the structural system used in the analysis is a special moment resisting reinforced concrete frame system.

2.3 Building Structure Data

Data on the structure of the Nursing Faculty Building can be seen in Table 1.

Table 1. Building data

Name Of Building	Nursing Faculty Building of Andalas University
Addres	Limau Manis, Padang
Structure Type	Concrete Reinforced
Design Concrete Quality	fc'=22,39 MPa
Eksisting Concrete Quality	fc'=11,24 MPa
Number Of Stories	2 stories with concrete slab
Building Height	8 Meter
Column Dimension	· Main Column (K1): (40x40) cm
Beam Dimension	· Main Beam (BA) : (30x50) cm
	· Secondary Beam (BA-1): (20x30) cm
Slab Thickness	12 cm
Length	2600 cm
Width	2200 cm

2.4 Modeling of Structure

Analysis of the Nursing Faculty Building structure, Andalas University, was carried out using 3D structural modeling in the ETABS program. Columns and beams of the building structure are modeled as frame elements, while the slabs are modeled as slab elements.

The results of the modeling of the building structure can be seen in Fig. 5.

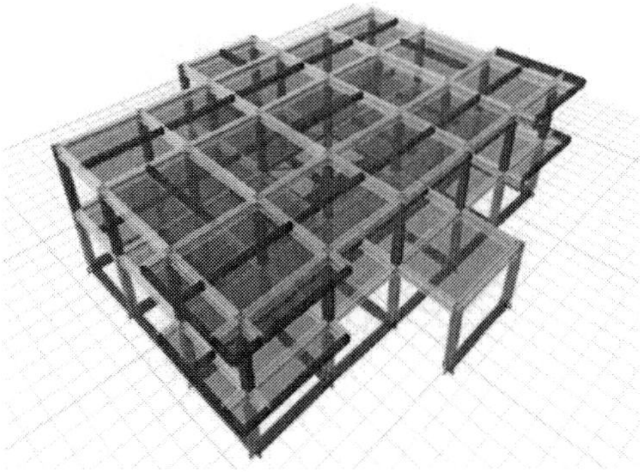

Fig. 5. Modeling of Structure

2.5 Building Loads

2.5.1 Dead Load

Based on Article 3.1.1 of SNI 1727:2020, the dead load is defined as the weight of all installed building construction materials, including walls, floors, roofs, ceilings, stairs, fixed partition walls, finishing, building cladding, and other architectural and structural components as well as service equipment. Other installed equipment includes crane weights and material conveying systems [2].

Dead loads on the Nursing Faculty Building:

a. Self-weight of structural elements is calculated directly by the ETABS structural analysis program version 2016.
b. Weight of floor covering (ceramic) 1cm thick = 1 x 24 kg/m² = 24 kg/m²
c. Ceiling weight = 20 kg/m²
d. Flooring mortar 2 cm thick floor = 2 x 21 kg/m² = 42 kg/m²
e. Mechanical Electrical and Plumbing weight = 25 kg/m²
f. Waterproofing weight= 5 kg/m²

2.5.2 Live Load

Article 4.3.1 of SNI 1727:2020 explains that the live load used in the design of buildings and other structures must be the maximum load that is expected to occur due to the occupancy and use of the building, but it must not be less than the minimum uniform load specified in Table 4.3.1 on SNI 1727:2020 [2]. The live loads acting on the building:

1. Corridor above first floor = 3.83 kN/m²
2. Office = 2.4 kN/m²
3. Meeting room = 4.79 kN/m²
4. Toilet = 2.87 kN/m²

2.5.3 Brick Wall Load

In addition to the dead load acting on the floor of the building, another load acting on the building is the brick wall. According to SNI 1727:2020, the load on the brick walls working on the building is 250 kg/m² while the wall height is 4 m, so the uniform load acting on the building with brick walls is 4m x 250 kg/m² = 1000 kg/m [2].

2.5.4 Partition Loads

Based on SNI 1727:2019 Article 4.3.2, it can be seen that the partition load is at least 72 kg/m² [2]. So that for the Nursing Faculty Building, it is assumed that the partition weight is 80 kg/m² and for the height of the building between floors is 4 meters, the partition load acting on the floor beams of the Nursing Faculty Building is 4 m x 80 kg/m² = 320 kg/m² (3.14 kN/m).

2.5.5 Earthquake Loads

Earthquake load analysis is conducted based on SNI 1726:2019 [3]. The type of earthquake load used in the design of the structure is dynamic earthquake load. The dynamic earthquake load used in building structures is the response spectrum. The earthquake spectrum response data itself is obtained from the RSA Puskim PUPR application. The Nursing Faculty Building is located in Padang city. The response spectrum data is shown in Tables 2 and 3.

Table 2. Spectra data (Source: PUSKIM PUPR RSA Application)

Variable	Value
PGA (g)	0.588
S_S (g)	1.480
S_1 (g)	0.600
C_{RS}	0.000
C_{R1}	0.000
F_{PGA}	1.000
F_A	1.000
F_V	1.700
PSA (g)	0.536
S_{MS} (g)	1.480
S_{M1} (g)	1.020
S_{DS} (g)	0.987
S_{D1} (g)	0.680
T_0	0.138
T_S	0.689

Table 3. Spectrum response period

Period (T)	Sa(g)
0.000	0.000
0.100	0.824
0.300	0.987
0.500	0.987
1.000	0.680
1.500	0.453
2.000	0.340
2.500	0.272
3.000	0.227
4.000	0.170

These tables are dynamic earthquake data obtained from the Puskim PU application. From these data, a graph of Earthquake Response Spectrum Design of Padang City was obtained, as shown in Fig. 6.

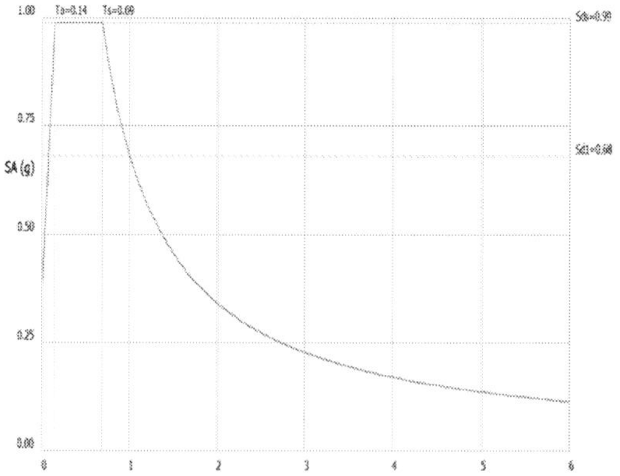

Fig. 6. Earthquake Response Spectrum Design of Padang City

2.6 Load Combination

The combination of structural loading refers to the SNI 1726:2019 [3]:

1. 1, 4D
2. 1.2 D + 1.6 L + 0.5 (Lr or R)
3. 1.2 D + 1.6 (Lr or R) + (L or 0.5 W)
4. 1.2 D + 1.0 W + L+ 0.5 (Lr or R)
5. 0.9 D+ 1.0 W
6. 1,2D + EV + EH + L
7. 0.9D - EV + EH

Table 4. Load combination

Loads Combination											
1	1.4	DL									
2	1.2	DL	+	1.6	LL						
3	1.397	DL	+	1	LL	+	1.3	EQX	+	0.39	EQY
4	1.397	DL	+	1	LL	+	1.3	EQX	-	0.39	EQY
5	1.397	DL	+	1	LL	-	1.3	EQX	+	0.39	EQY
6	1.397	DL	+	1	LL	-	1.3	EQX	-	0.39	EQY
7	1.397	DL	+	1	LL	+	0.39	EQX	+	1.3	EQY
8	1.397	DL	+	1	LL	+	0.39	EQX	-	1.3	EQY
9	1.397	DL	+	1	LL	-	0.39	EQX	+	1.3	EQY
10	1.397	DL	+	1	LL	-	0.39	EQX	-	1.3	EQY
11	0.703	DL	+	1.3	EQX	+	0.39	EQY			
12	0.703	DL	+	1.3	EQX	-	0.39	EQY			
13	0.703	DL	-	1.3	EQX	+	0.39	EQY			
14	0.703	DL	-	1.3	EQX	-	0.39	EQY			
15	0.703	DL	+	0.39	EQX	+	1.3	EQY			
16	0.703	DL	+	0.39	EQX	-	1.3	EQY			
17	0.703	DL	-	0.39	EQX	+	1.3	EQY			
18	0.703	DL	-	0.39	EQX	-	1.3	EQY			
19	Envelope										

Table 4 shows the load combinations in the structural analysis of the Nursing Faculty Building.

2.7 Inter Story Drift

Table 5. Inter story drift limit (Source: SNI 1726:2019)

Structure	Risk Category		
	I or II	III	IV
Structures, other than brick shear wall structures, 4 stories or less with interior walls, partitions, ceilings and exterior wall system which has been designed for accommodate deviations between levels.	$0{,}025h_{sx}$	$0{,}020h_{sx}$	$0{,}015h_{sx}$
Brick cantilever shear wall structure	0,010 hsx	0,010 hsx	0,010 hsx
Other brick shear wall structures	0,007 hsx	0,007 hsx	0,007 hsx
All other structures	0,020 hsx	0,015 hsx	0,010 hsx

Table 5 shows the boundary conditions for the inter story drift for the building. From the equation in the table, the calculation results of the building inter story drift are shown in Tables 6 and 7.

Table 6. Inter story drift X direction

Story	Direction	δ_e	Cd	δ_{ex}	Δ_x	Height	Δ(Limit)	Description
		mm		mm	mm	mm	mm	
2	X	31.393	5.5	115.10767	52.7817	4000	40	NOT OK
1	X	16.998	5.5	62.326	62.326	4000	40	NOT OK
0	X	0	5.5	0	0	4000	40	OK

Table 7. Inter story drift Y direction

Story	Direction	δ_e	Cd	δ_{ex}	Δ_x	Height	Δ(Limit)	Description
		mm		mm	mm	mm	mm	
2	Y	30.403	5.5	111.4777	51.5167	4000	40	NOT OK
1	Y	16.353	5.5	59.961	59.961	4000	40	NOT OK
0	Y	0	5.5	0	0	4000	40	OK

From Tables 6 and 7, it can be seen that the inter story drift that occurs in the building does not meet the permit limit requirements in SNI 1726:2019.

2.8 Cross-sectional Capacity of the Structure

2.8.1 Column Capacity

The capacity of the column is determined through the interaction diagram of the axial moment and compression of the column and the shear capacity of the column [4].

a. Momen and Axial of Column

The calculation results of the interaction diagram of the first floor and second floor columns of the building are shown in Figs. 7 and 8.

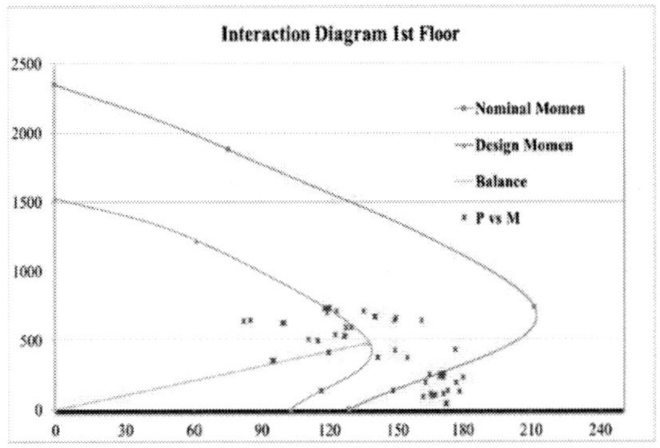

Fig. 7. First Floor Column P-M Interaction Diagram

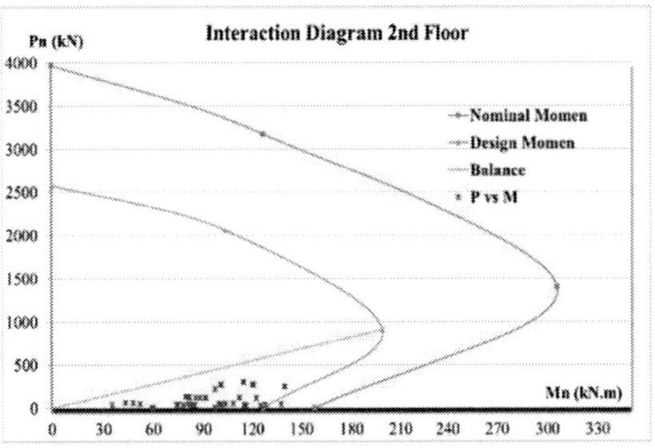

Fig. 8. Second Floor Column P-M Interaction Diagram

As seen in Figs. 7 and 8, the first and second floor columns are not strong enough to carry the working load because some axial moment and compression forces pass through the design axial compression moment line.

b. Shear Capacity of Columns

The shear capacity of the building columns is shown in Table 8. From the table, it is clearly seen that the column is able to withstand the shear forces acting on the structure.

Table 8. Shear capacity column

	Dimension (mm)	Fy	Dim. Shear Reinforced	Spasi	ΦVn	Vu	Description
1	400 x 400	310	10	100	2250.719	101.1666	OK
2	400 x 400	400	10	100	2542.005	138.0258	OK

2.8.2 Beam Capacity

a. Flexural Capacity of Beams

Table 9 shows the flexural capacity of the building beams.

Table 9. Beam flexure capacity

Type (Dimension)	Location	Main Reinforced		Nominal Flexural Capacity		Ultimate Flexural Forces		Description
		At Support	At Field	At Support	At Field	At Support	At Field	
Sloof (30X50)	Top	4D16	4D16	116.89	116.89	60.07	15.02	OK
	Bottom	4D16	4D16	116.89	116.89	30.03	30.03	
1st floor Beam (30X50)	Top	6D16	4D16	181.37	125.00	228.83	64.17	NOT OK
	Bottom	4D16	6D16	125.00	181.37	162.12	37.57	
2nd floor Beam (30X50)	Top	6D16	4D16	181.37	125.00	150.30	37.57	OK
	Bottom	4D16	6D16	125.00	181.37	93.85	82.35	
1st floor Ba-1 (20x30)	Top	5D16	3D16	74.92	49.56	63.69	15.92	NOT OK
	Bottom	3D16	5D16	49.56	74.92	81.70	98.43	
2st floor Ba-1 (20x30)	Top	5D16	3D16	74.92	49.56	64.59	16.15	NOT OK
	Bottom	3D16	5D16	49.56	74.92	44.57	85.32	

From Table 9, it can be seen that the beams on the first floor and the beams on the first and second floors are unable to withstand the loads acting on the structure.

b. Shear Capacity of Beams

The shear capacity of the building beams can be seen in Table 10. From the table, it is found that the beam is able to withstand the shear forces acting on the structure.

Table 10. Beam shear capacity

Type (Dimension)	Shear Reinforced		Nominal Shear Capacity		Ultimate Shear Forces		Description
	At Support	At Field	At Support	At Field	At Support	At Field	
Sloof (30X50)	Ø10-100	Ø10-150	366.14	269.79	25.65	55.02	OK
1st floor Beam (30X50)	Ø10-100	Ø10-150	397.86	301.52	243.54	219.82	OK
2nd floor Beam (30X50)	Ø10-100	Ø10-100	397.86	301.52	143.20	117.88	OK
1st floor Ba-1 (20x30)	Ø12-100	Ø10-150	276.25	149.92	82.27	50.23	OK
2nd floor Ba-1 (20x30)	Ø12-100	Ø10-150	276.25	149.92	71.20	52.08	OK

Based on the examination of the mass participation factor, scale factor, P-delta, and structural irregularities, all of them have met the requirements of SNI 1726:2019 [3].

From the evaluation results of the structural building performance, it was found that the structure of the columns and beams could not resist the loads acting on the structure. In addition, the inter story drift that occurred did not meet the permit limit requirements according to Indonesian building standards. Therefore, the building should be retrofitted before continuing the construction.

3 Retrofitting of the Structure

Analysis of the retrofitting/strengthening of the structure in this building is by re-design the Detail Engineering Design (DED) of the structure where the existing columns are jacketed [5-7]. The re-design is carried out on the beams, the second-floor column, while the first-floor column is retrofitted using the jacketing method.

3.1 Beam

Re-design of beams was carried out on the structure of the main and secondary beams of the building. The re-design was carried out because the beams from the initial design (DED) could not withstand the loads acting on the structure [4,8].

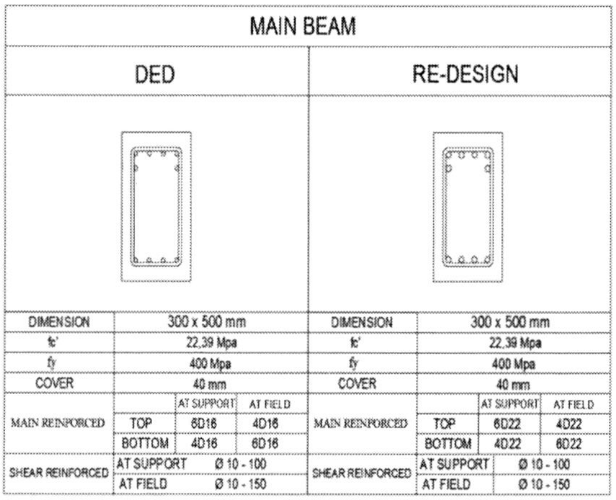

Fig. 9. Comparison of the Main Beam Details between the DED (Initial Design) and the Re-design

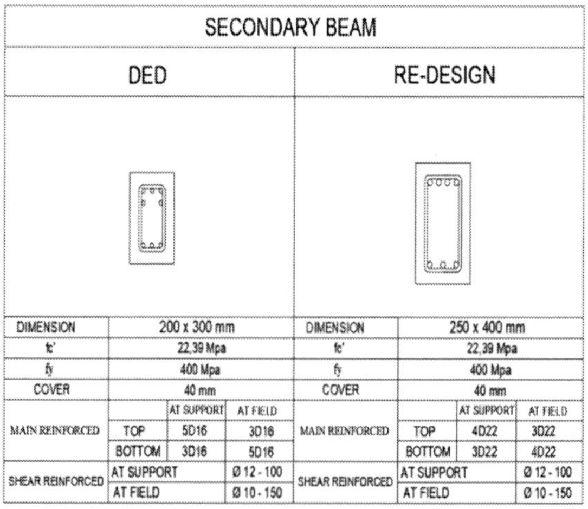

Fig. 10. Comparison of the Secondary Beam Details between the DED (Initial Design) and the Re-design

Figs. 9 and 10 show the comparison of the details of the main beam between the initial design (DED) and the Re-design.

3.1.1. Re-designed Beam Capacity

a. Flexural Capacity

The results of the beam flexural capacity analysis are shown in Table 11.

Table 11. Re-designed beam flexural capacity

Type (Dimension)	Location	Main Reinforced		Nominal Flexural Capacity		Ultimate Flexural Forces		Description
		At Support	At Field	At Support	At Field	At Support	At Field	
Sloof (30X50)	Top	4D16	4D16	116.89	116.89	60.07	15.02	OK
	Bottom	4D16	4D16	116.89	116.89	30.03	30.03	
1st floor Beam (30X50)	Top	6D22	4D22	311.90	222.56	178.07	44.52	OK
	Bottom	4D22	6D22	222.56	311.90	98.20	137.82	
2nd floor Beam (30X50)	Top	6D22	4D22	311.90	222.56	149.82	37.46	OK
	Bottom	4D22	6D22	222.56	311.90	97.08	81.66	
1st floor Ba-1 (20x30)	Top	4D22	3D22	161.97	128.06	73.67	24.92	OK
	Bottom	3D22	4D22	128.06	161.97	84.52	158.07	
2st floor Ba-1 (20x30)	Top	4D22	3D22	161.97	128.06	72.56	18.14	OK
	Bottom	3D22	4D22	128.06	161.97	47.85	91.23	

From Table 11, it can be seen that all the beam structures after the re-design have been able to withstand the loads acting on the structure.

b. Shear Capacity

Table 12 shows the results of the beam shear capacity analysis.

Table 12. Re-design beam shear capacity

Tipe (Dimensi)	Shear Reinforced		Nominal Shear Capacity		Ultimate Shear Forces		Description
	At Support	At Field	At Support	At Field	At Support	At Field	
Sloof (30X50)	Ø10-100	Ø10-150	366.14	269.79	25.65	55.02	OK
1st floor Beam (30X50)	Ø10-100	Ø10-150	397.86	301.52	240.66	190.41	OK
2nd floor Beam (30X50)	Ø10-100	Ø10-100	397.86	301.52	146.26	123.36	OK
1st floor Ba-1 (20x30)	Ø12-100	Ø10-150	396.70	221.77	93.19	34.88	OK
2nd floor Ba-1 (20x30)	Ø12-100	Ø10-150	396.70	221.77	51.36	48.95	OK

From Table 12, it is found that the re-designed beam is able to withstand shear forces due to external loads on the structure.

3.2 Second Floor Column

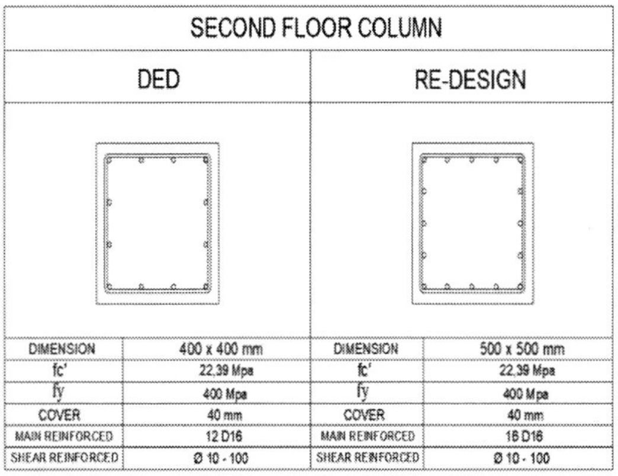

Fig. 11. Initial design (DED) and Re-design Column Details

The re-design of columns was carried out on the second floor columns of the building because the capacity of the planned columns could not withstand the working loads, so the dimensions and reinforcement of the column had to be changed [4,8].

A detailed comparison between the initial design (DED) and re-designed second floor columns can be seen in Fig. 11.

3.2.1. Column Capacity

Fig. 12 shows the second floor column P-M interaction diagram.

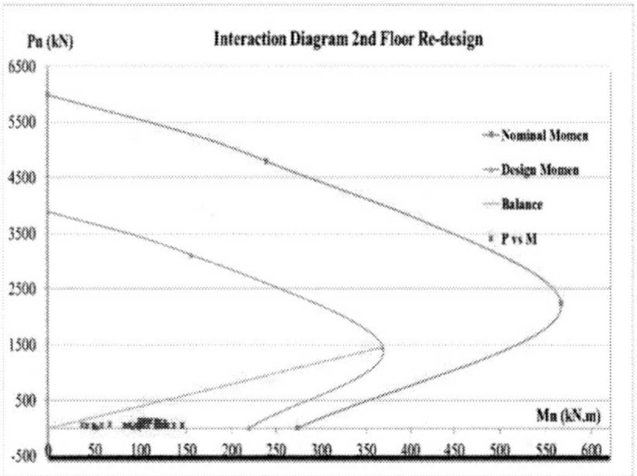

Fig. 12. Second Floor Column P-M Interaction Diagram Re-design

As seen in Fig. 12, all P-M values on the graph have been in the interaction diagram line indicating that all building columns are capable of carrying the working load.

3.3 Retrofitting of First Floor Column using Jacketing Method

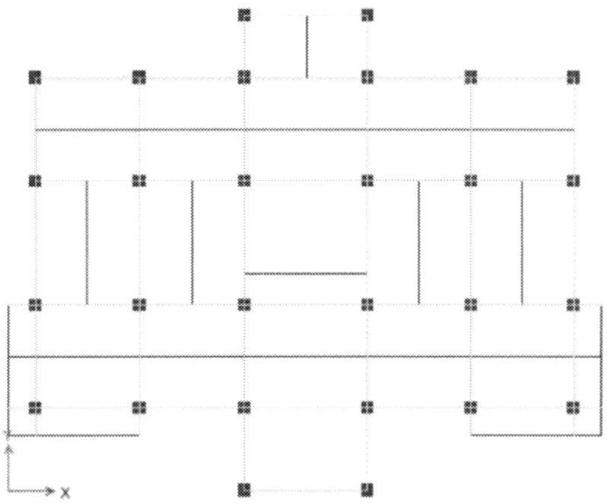

Fig. 13. Position of the Concrete Jacketing Column

Retrofit of the first floor columns is carried out using the Concrete Jacketing method on all existing columns because the concrete and steel quality in the column has decreased [9,10]. The Concrete Jacketing Column Position is shown in Fig. 13.

3.3.1. Structural Retrofitting Modeling using Concrete Jacketing Method

Retrofitting of the column using jacketing method is carried out by increasing the dimensions and adding steel reinforcement to the column with the following assumptions

- The planned jacketing column is 600 x 600 (mm).
- The column quality to be achieved is fc'=25 MPa.
- In order to reach a concrete quality of fc'=25 MPa, the quality of the jacketing concrete used as an addition to the column dimensions is a minimum of fc'= 36 MPa. Calculation of the quality of the concrete jacketing is as follows:
  - Column quality to be achieved (A)= 25 MPa
  - Existing Concrete Quality (B) = 11.24 MPa
  - Planned column area (C) = 600 x 600 = 360,000 mm^2
  - Existing column area (D) = 400 x 400 = 160,000 mm^2
  - Jacketing area (E) = 200,000 mm^2

  Quality of concrete jacketing:
  $$(D \times B) + (E \times X) = C \times A$$
  $$(160{,}000 \times 11.24) + 200{,}000 \times (X) = 360{,}000 \times 25$$
  $$X = 36 \text{ MPa}$$
- The quality and amount of added steel reinforcement are the same as the existing one, that is fy= 350 MPa with 12 D16.

From Figs. 14 and 15, it can be seen the assumption of the definition of the jacketing column in the ETABS software. Fig. 16 shows the comparison of the column cross-section between the existing column and the jacketing column.

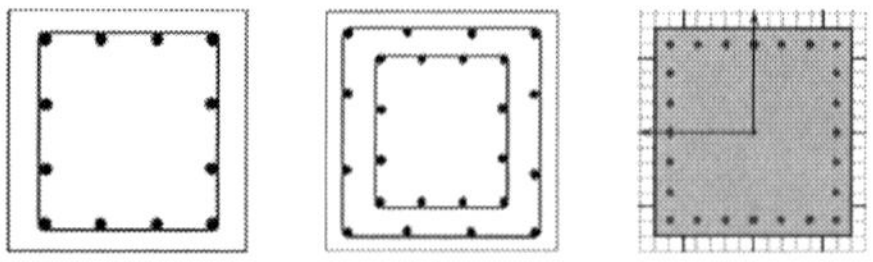

(a) Existing Column | (b) *Concrete Jacketing Coluumn* | (c) Jacketing Column with Evenly Distributed Reinforcement

Fig. 14.Conrete Jacketing Column Reinforcement Modeling

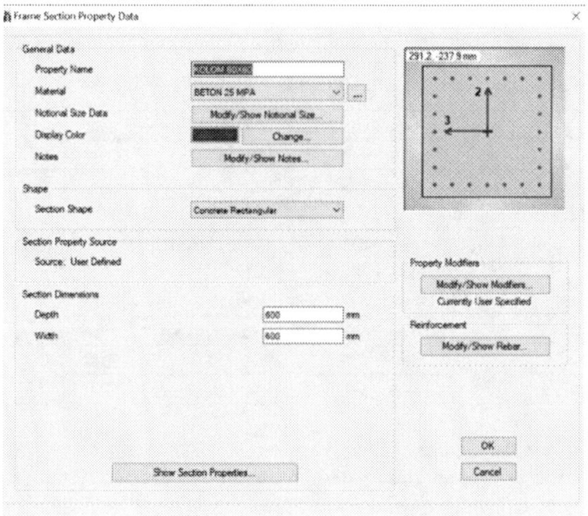

Fig. 15. Concrete Jacketed Column Modeling

EKSISTING		JACKETING	
DIMENSION	400 x 400 mm	DIMENSION	600 x 600 mm
fc'	22,39 Mpa	fc'	25 Mpa
fy	350 Mpa	fy	400 Mpa
COVER	40 mm	COVER	40 mm
MAIN REINFORCED	12 D16	MAIN REINFORCED	24 D16
SHEAR REINFORCED	Ø 10 - 100	SHEAR REINFORCED	Ø 10 - 100

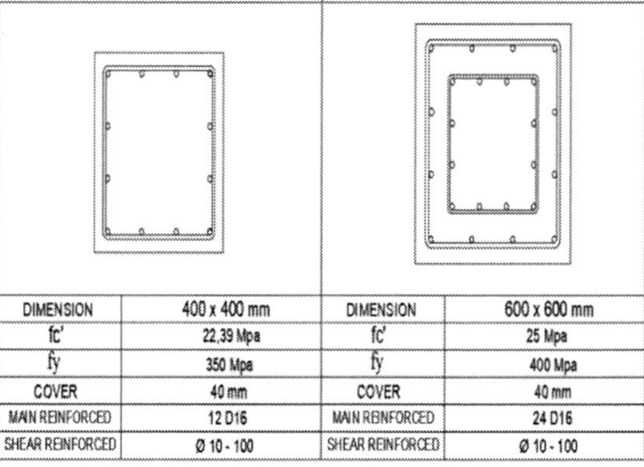

Fig. 16. The comparison of the Column Cross-section between the Existing Column and the Jacketed Column

3.3.2. Retrofitted Column Capacity

Fig. 17 shows the cross-sectional capacity of the column reinforced by the jacketing method. From the figure, it can be seen that all the P-M points on the graph have been in the interaction diagram, so that the column is able to withstand the working load.

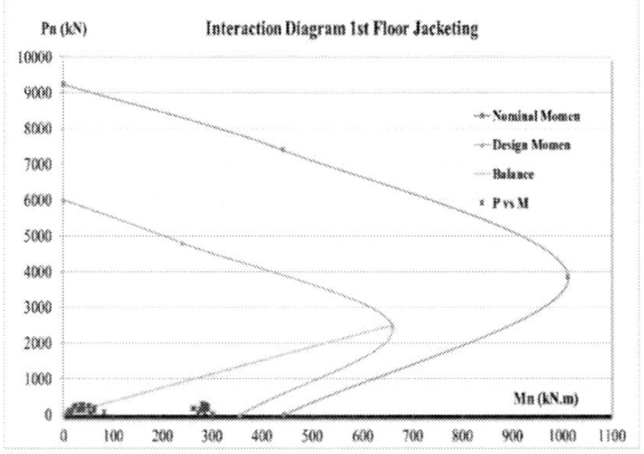

Fig. 17. P-M Interaction Diagram of Jacketed Column 600x600

3.3.3. Inter Story Drift

Retrofitting on the existing first floor column with the jacketing method also affects the building displacement, so it is necessary to check the displacement of the building structure. The calculation results of the inter story drift in the X and Y directions on the retrofitted building can be seen in Tables 13 and 14.

From Tables 13 and 14, it is clearly seen that the inter story drift that occurs have met the permit limit requirements.

Table 13. Inter story drift in X direction for retrofitted column

Story	Direction	δ_x	Cd	δ_{ex}	Δ_x	Height	Δ(Limit)	Description
		mm		mm	mm	mm	mm	
2	X	16.244	5.5	59.561	36.971	4000	40	OK
1	X	6.161	5.5	22.590	22.590	4000	40	OK
0	X	0	5.5	0.000	0.000	4000	40	OK

Table 14. Inter Story drift in Y direction for retrofitted column

Story	Direction	δ_x	Cd	δ_{ex}	Δ_x	Height	Δ(Limit)	Description
		mm		mm	mm	mm	mm	
2	Y	16.131	5.5	59.147	36.938	4000	40	OK
1	Y	6.057	5.5	22.209	22.209	4000	40	OK
0	Y	0	5.5	0	0.000	4000	40	OK

4 Conclusion

Based on the structural evaluation analysis conducted at the Nursing Faculty Building, the following conclusions can be drawn:

1. The quality of the existing concrete obtained from the hammer test results is fc'=11.24 MPa, where this value is far below the concrete quality standard for reinforced concrete building structures (minimum fc' = 17 MPa).
2. The structure of the existing first floor column, second floor plan column, and the initial design of the main and secondary beams are not able to withstand the working load.
3. The inter story drift in the existing building does not meet the permit limit requirements according to the Indonesian building standard.
4. Retrofitting of the building structure is designed by re-designing the Detail Engineering Design (DED) structure, where the existing first floor column is retrofitted by using jacketing method.
5. The retrofitted building structure has a strong enough capacity to carry the working load, and the inter story drift has met the permit limit requirements.

The authors would like to thanks Andalas University for financial support in publishing this article.

References

1. O.Dezardo, *Feasibility Evaluation and Recommendation for Strengthening the Dean Building Structure of the Faculty of Nursing, Andalas University*, (2016)
2. National Standardization Agency, *Minimum Load for Planning of Buildings and Other Structures, SNI 1727:2020*, Jakarta, BSN (2020)
3. National Standardization Agency, *Procedure for Planning Earthquake Resistance for Building Structures, SNI 1726:2019,* Jakarta, BSN (2019)
4. National Standardization Agency, *Structural Concrete Requirements for Buildings, SNI 2847:2019,* Jakarta, BSN (2019)
5. T. Boen, et al., *How to Repair a Simple Building Damaged by an Earthquake*, Jakarta (2009)
6. Central Public Works Department and Indian Building Congress, *Handbook on Seismic Retrofit of Buildings. Association of Indian Institutes of Technology,India* (2007)
7. R. Santhakumar, *Retrofit of masonry buildings*, Handb. Seism. Retrofit Build. (2007)
8. H. Khoeri, *Selecting Methods for Repair and Reinforcement of Structures Due to Earthquake (Case Study at Bank Sulteng Palu*), Journal of Construction, **12** (1) (2020)
9. J.T. Kaontole, M.D.J. Sumajouw, and R.S. Windah, *Evaluation of Reinforced Concrete Columns Capacity Reinforced by the Concrete Jacketing Method,* Civil Static, **3** (3) (2015)
10. F. S. Asmara, M. Ziddan, and A. Rosyidah, *Improving the Strength of Existing Structures by Reinforcing Structures Using CFRP & Concrete Jacketing*, Teknika, **13** (02) (2019)

https://doi.org/10.1051/e3sconf/202133105015

Assessment of building structure block D of RS BMC Padang.

Zaidir Zaidir[1] and *Wahyu* Reyza Febrian[1]

[1]Civil Engineering Department, Engineering Faculty, University of Andalas, Indonesia

Abstract. Padang city is the capital of West Sumatra province which is on the west side of the island of Sumatra, is an earthquake-prone area, it is near the confluence of the Asian Euro and Indo-Australia tectonic plates. In the last two decades, there have been two large earthquakes in West Sumatra, on March 6, 2007 (5.8 SR) and September 30, 2009 (7.6 SR). The earthquake on September 30, 2009 has caused severe damage in several areas in West Sumatra such as Padang Pariaman Regency, Padang City, Pesisir Selatan Regency, Pariaman City, Bukittinggi City, Padangpanjang City, Agam Regency, Solok City, and West Pasaman Regency. According to Satkorlak PB data, as many as 1,117 people were killed, the serious injuries reached 1,214 people, light injuries 1,688 people, and the missing 1 person. A total of 135,448 homes were severely damaged, 65,380 houses were moderately damaged, and 78,604 houses were lightly damaged. The earthquake has also caused dozens of multi-story buildings in Padang City to suffer damage, lightly, moderately, and severely damaged and some buildings were collapsed. Block D of RS BMC is one of the buildings of RS BMC Padang, which was built in early 2000. Design of the building structure of Block D of RS BMC used SNI 03-2847-1991 and SNI 03-1726-1989. Nowadays, the newest SNI is SNI 03-2847-2019 for reinforced concrete and SNI 03-1726-2019 for earthquake designs. The paper discusses the assessment of the building structure of block D of RS BMC using the new SNI 03-1726-2019 and SNI 03-2847-2019. The earthquake load used is a dynamic earthquake load using Padang city spectrum response and structural analysis using SAP2000 v.14 software. The analysis results obtained, structural elements of columns, beams, and plates are still able to resist the design loads according to the new SNI. The inter-story drift of floors, the effect of P-delta, the vertical irregularities of buildings, the concept of strong column weak beam, and mass participation are still following the newest SNI.

1 Introduction

The city of Padang is the capital of West Sumatra province which is on the west side of the island of Sumatra, is an earthquake-prone area, due to it is near the confluence of the Asian Euro and Indo Australia tectonic plates. In the last two decades, there have been two large earthquakes in West Sumatra, on March 6, 2007 (5.8 SR) and September 30, 2009 (7.6 SR). The earthquake on September 30, 2009 has caused severe damage in several areas in West Sumatra such as Padang Pariaman Regency, Padang City, Pesisir Selatan Regency, Pariaman City, Bukittinggi City, Padangpanjang City, Agam Regency, Solok City, and West Pasaman Regency. According to Satkorlak PB data, as many as 1,117 people were killed, the serious injuries reached 1,214 people, light injuries 1,688 people, and the missing 1 person. A total of 135,448 homes were severely damaged, 65,380 houses were moderately damaged, and 78,604 houses were lightly damaged. The earthquake has also caused dozens of multi-story buildings in Padang City to suffer damage, lightly, moderately, and severely damaged and some buildings were collapsed. Block D of RS BMC is one of the buildings of RS BMC Padang, which was built in the early year 2000. The design structure of this building is used SNI 03-2847-1991 for reinforced concrete and SNI 03-1726-1989 for earthquake resistant structure. Currently, the latest SNI 03-2847-2019 for reinforced concrete and SNI 03-1726-2019 for earthquake design have been published and used for earthquake-resistant building design in Indonesia. The paper discusses the assessment of the building structure of block D of RS BMC using the new SNI 03-1726-2019 and SNI 03-2847-2019. The earthquake load used is a dynamic earthquake load using Padang city spectrum response and structural analysis using SAP2000 v.14 software.

2 Building and Materials Data

2.1 Existing Condition of Block D Building

Fig.1 shows a building location of block D of RS BMC. The building located at Proklamasi Street No. 31 - 37 Padang, West Sumatra. The front and side view of the building is shown in Fig. 2 and Fig. 3. The building has 5 (five) floors with a reinforced concrete frame structure. Fig. 4 shows the typical floor dimension and column plan. The building floors has dimension with a length of 40 m and a width of 18.125 m. The beam's floor plan is shown in Fig. 5. The dimension of the column cross-section is 450 mm x 550 mm and the detailing of reinforcement is shown in Fig. 6. The beam's dimensions and their reinforcement is shown in Fig. 7.

 https://doi.org/10.1051/e3sconf/202133105015

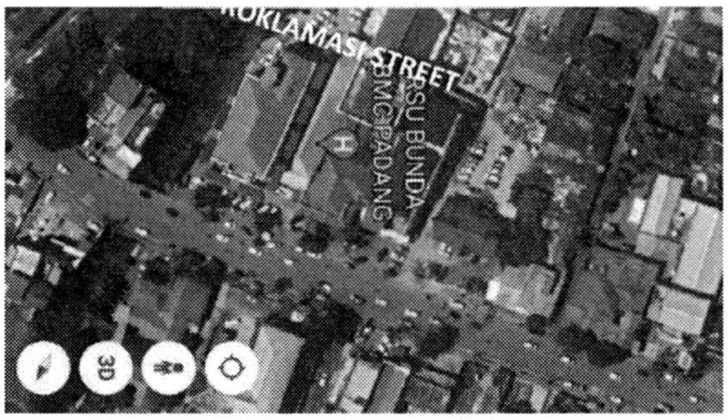

Fig. 1. Location of block D building of RS BMC Padang

Fig. 2. Front view of block D building

Fig. **3.** Side view of block D building

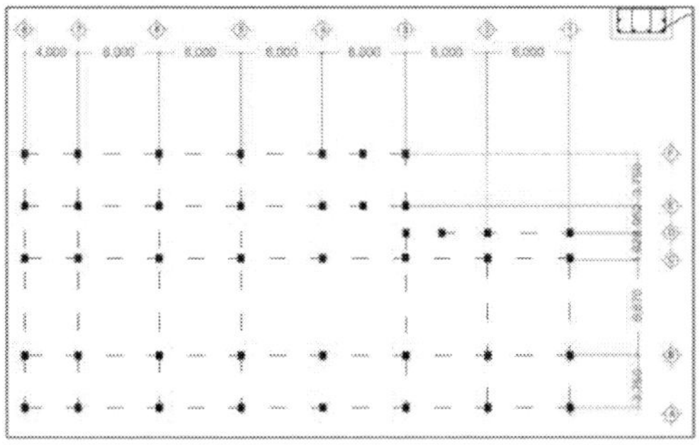

Fig. 4. Typical floor dimension and column plans

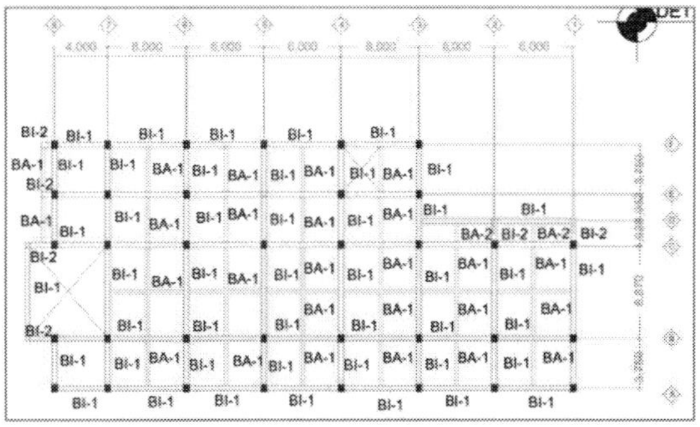

Fig. 5. Beams Floor Plan

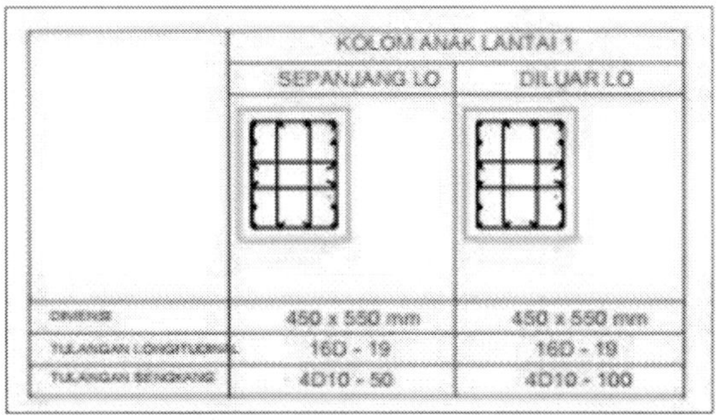

Fig. 6. Column dimensions

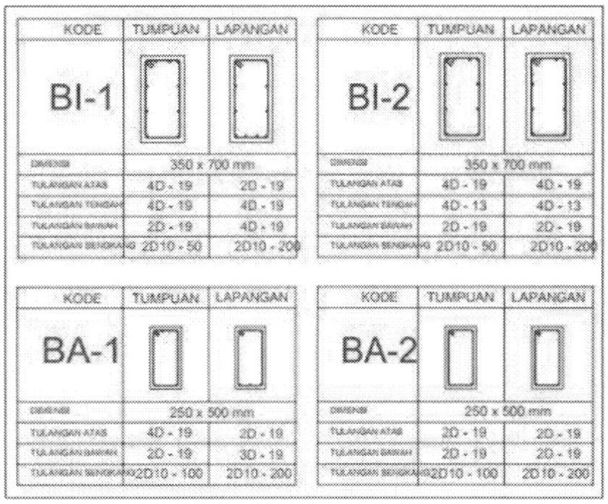

Fig. 7. Beam types and dimensions

2.2 Codes and Materials Specification

In this assessment, the newest Indonesian National Standard (SNI) for reinforced concrete design and earthquake resistant buildings design is used, namely SNI 2847-2019 Structural Concrete Requirements for Buildings and SNI 1726-2019 for Earthquake Resilience Design Procedures for Building and Non-Building Structures. The newest SNI for structural load is using SNI 1727-2018 for Minimum Load for The Design of Buildings and Other Structures.

The average concrete strength (f_c') which is collected by hammer test is 25 MPa and the reinforcing steel strength (f_y) is 390 MPa. The beam dimensions (BI1 and BI2) are 350 mm x 700 mm and 250 mm x 500 mm (BA1 and BA2). The floor thickness is 120 mm.

2.3 Structural Modelling and Loads Design

Fig 8. shows the 3D structural modelling of block D building using SAP2000 v.14. The gravitational loads are used according to SNI 1727-2018 namely, DL (dead load), LL (live load), and rain load (R).

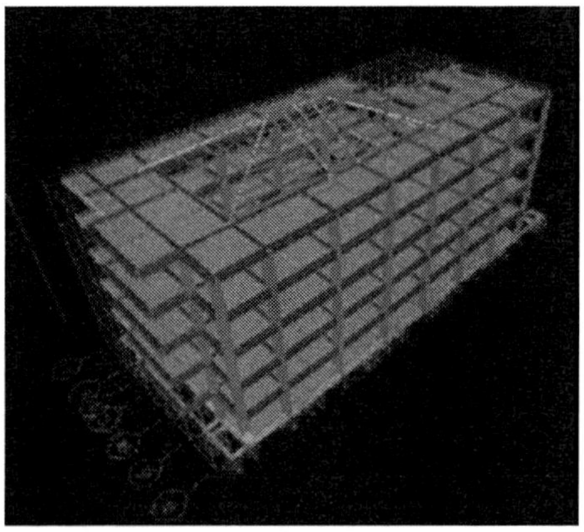

Fig. 8. The 3D Structural Modelling of block D building

For earthquake loads, according to SNI 1726-2019, the earthquake parameters which were used are :

- Hospital building: Risk Category IV
- Priority factor (Ie) = 1,5
- Site classification: SC (hard soil, very dense and soft rock)

Fig. 9 shows the earthquake response spectrum which is used for the block D building (2019 Response spectrum of PUSKIM PU)

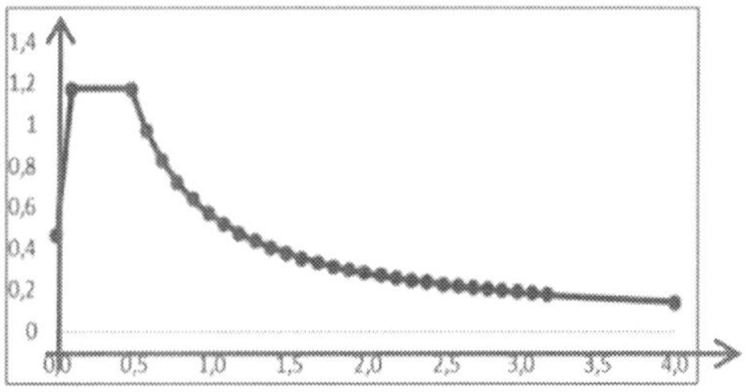

Fig. 9. Response Spectrum of Block D RS BMC Padang

The earthquake parameters which are used for a calculation were :

- S_s = 1,463 g
- S_1 = 0,600 g
- T_L = 20 sec
- F_a = 1,200
- F_v = 1,400
- S_{ms} = 1,755 g
- S_{m1} = 0,840 g
- S_{ds} = 1,170 g
- S_{d1} = 0,560 g
- T_0 = 0,096 sec
- T_s = 0,479 sec
- Earthquake load reduction (R) = 8
- System stronger factor (Ω_0) = 3
- Deflection enlargement factor (C_d) = 5,5
- Earthquake factor scale = 1839,38 mm/sec²
- Seismic Design Criteria (KDS) : D
- System Structure : SRPMK (Special Moment in Frame System)

2.4 Loads Combination

There are 26 (twenty six) load combinations which are used in this assessment as below :

1. 1,4 DL
2. 1,2 DL + 1,6 LL + 0,5 Lr
3. 1,2 DL + 1,6 LL + 0,5 R
4. 1,2 DL + 1,6 Lr + 1 LL
5. 1,2 DL + 1,6 Lr + 0,5 W
6. 1,2 DL + 1,6 R + 1 LL
7. 1,2 DL + 1,6 R + 0,5 W
8. 1,2 DL + 1 W + 1 LL + 0,5 Lr
9. 1,2 DL + 1 W + 1 LL + 0,5 R
10. 1,434 DL + 1,300 EQX + 0,390 EQY + 1 LL
11. 1,434 DL + 1,300 EQX - 0,390 EQY + 1 LL
12. 1,434 DL - 1,300 EQX + 0,390 EQY + 1 LL
13. 1,434 DL - 1,300 EQX - 0,390 EQY + 1 LL
14. 1,434 DL + 0,390 EQX + 1,300 EQY + 1 LL
15. 1,434 DL + 0,390 EQX - 1,300 EQY + 1 LL
16. 1,434 DL - 0,390 EQX + 1,300 EQY + 1 LL
17. 1,434 DL - 0,390 EQX - 1,300 EQY + 1 LL
18. 0,666 DL + 1,300 EQX + 0,390 EQY
19. 0,666 DL + 1,300 EQX - 0,390 EQY
20. 0,666 DL - 1,300 EQX + 0,390 EQY
21. 0,666 DL - 1,300 EQX - 0,390 EQY
22. 0,666 DL + 0,390 EQX + 1,300 EQY
23. 0,666 DL + 0,390 EQX - 1,300 EQY
24. 0,666 DL - 0,390 EQX + 1,300 EQY
25. 0,666 DL - 0,390 EQX - 1,300 EQY
26. ENVELOPE

where: DL = Dead Load ; LL = Live Load ; Lr = reduction of Live Load; R = Rain Load ; W = Wind Load ; EQX and EQY = Earthquake Load in X and Y direction.

3 Results and Discussions

3.1 Column Capacity (P- M Interaction Diagram)

Fig.10 (a) - (d) shows the P-M diagram interaction of building columns and Table 1 shows the column shear steel reinforcement capacities for all building floors. From figure 10 and Table 1, it can be concluded that the existing column dimension and longitudinal and shear reinforcements are still capable to resist the load's combination according to the new SNI.

3.2 Beams and Plates Capacity

Tables 2 and 3 show the moment and shear capacities of building beams. From this table, it can be concluded that the beams existing could resist the load's combination according to the new SNI. The moment capacities of plate reinforcement are shown in Table 4. From this table, it can be concluded that the existing plate and reinforcement is still can resist loads according to the new SNI.

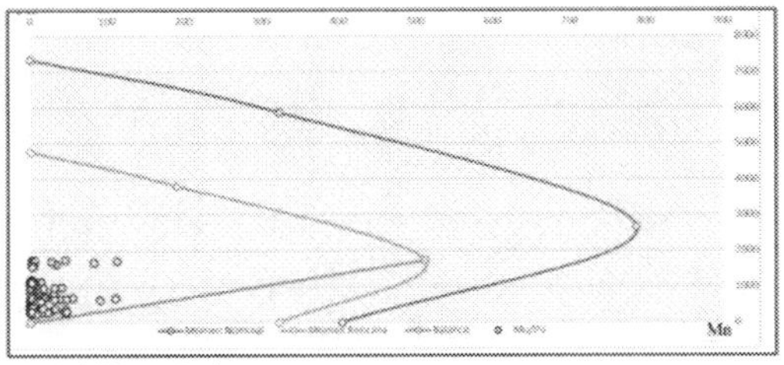

(a). 1st Floor Column

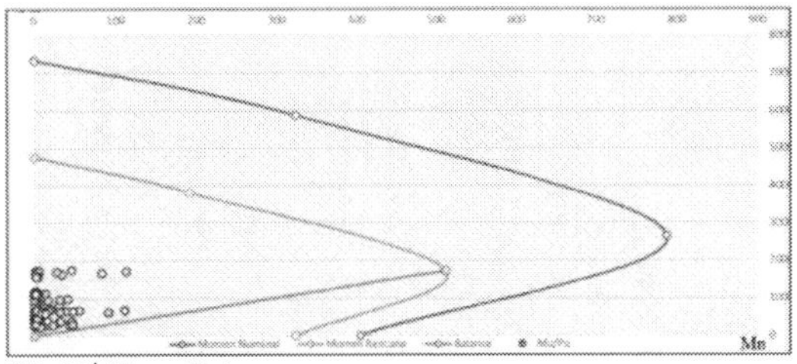

(b). 2nd Floor Column

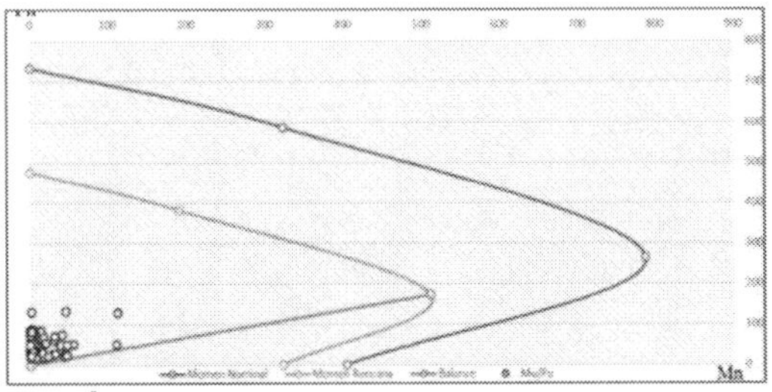

(c). 3rd Floor Column

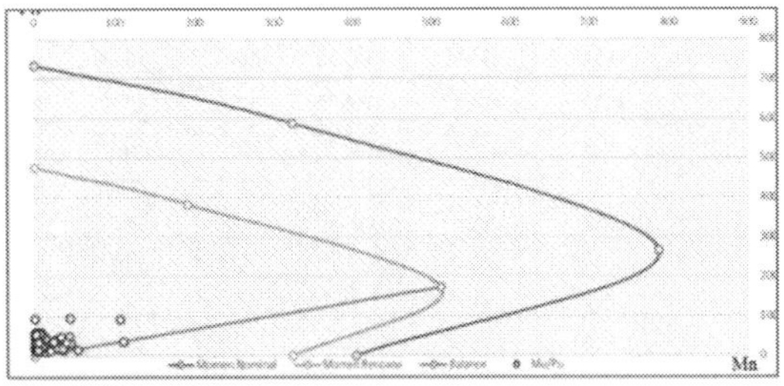

(d). 4th Floor Column

Fig. 10. P-M Diagram Interaction of Building Column

Table 1. Column Shear Reinforcement Capacity

Floor	Code	Width (mm)	Height (mm)	Main Reinforcement		Shear Reinforcement		Vr (kN)	Vu (kN)	Remarks
				Support	Middle	Support	Middle			
1	K1	450	550	16D19	16D19	4D10-50	4D10-100	311,021	69,752	OK
2	K1	450	550	16D19	16D19	4D10-50	4D10-100	311,021	61,721	OK
3	K1	450	550	16D19	16D19	4D10-50	4D10-100	311,021	65,252	OK
4	K1	450	550	16D19	16D19	4D10-50	4D10-100	311,021	62,080	OK
5	K1	450	550	16D19	16D19	4D10-50	4D10-100	311,021	104,537	OK

Table 2. Moment Capacity of Beams Reinforcement

Floor	Code	Width	Height	Comp Reinf.	Tensile Reinf.	ϕMn	Mu	Remarks
		(mm)	(mm)			(kN-m)	(kN-m)	
1	BI-1	300	700	4D-16	4D-16	325.513	235.143	OK
	BI-2	300	700	4D-16	2D-16	400.409	262.953	OK
	BA-1	250	500	4D-16	3D-16	444.642	386.046	OK
	BA-2	250	500	2D-16	2D-16	112.705	74.252	OK
2	BI-1	300	700	4D-16	4D-16	325.513	251.964	OK
	BI-2	300	700	4D-16	2D-16	400.409	263.810	OK
	BA-1	250	500	4D-16	3D-16	244.606	134.358	OK
	BA-2	250	500	2D-16	2D-16	112.705	74.852	OK
3	BI-1	300	700	4D-16	4D-16	325.513	263.282	OK
	BI-2	300	700	4D-16	2D-16	400.409	148.888	OK
	BA-1	250	500	4D-16	3D-16	244.606	87.354	OK
	BA-2	250	500	2D-16	2D-16	112.705	95.484	OK
4	BI-1	300	700	4D-16	4D-16	325.513	174.752	OK
	BA-1	250	500	4D-16	3D-16	244.606	61.993	OK
5	BI-1	300	700	4D-16	4D-16	325.513	111.986	OK
	BA-1	250	500	4D-16	3D-16	244.606	58.572	OK

https://doi.org/10.1051/e3sconf/202133105015

Table 3. Shear Capacity of Beams Reinforcement

Floor	Code	Width (mm)	Height (mm)	Shear Reinf. Support	Shear Reinf. Middle	Vr (kN)	Vu (kN)	Remarks
1	BI-1	300	700	2D10 - 50	2D10 - 200	232.606	171.421	OK
	BI-2	300	700	2D10 - 50	2D10 - 200	232.606	191.892	OK
	BA-1	250	500	2D10 - 100	2D10 - 200	147.745	134.463	OK
	BA-2	250	500	2D10 - 100	2D10 - 200	147.745	62.352	OK
2	BI-1	300	700	2D10 - 50	2D10 - 200	232.606	178.533	OK
	BI-2	300	700	2D10 - 50	2D10 - 200	232.606	188.678	OK
	BA-1	250	500	2D10 - 100	2D10 - 200	147.745	97.314	OK
	BA-2	250	500	2D10 - 100	2D10 - 200	147.745	62.102	OK
3	BI-1	300	700	2D10 - 50	2D10 - 200	232.606	172.597	OK
	BI-2	300	700	2D10 - 50	2D10 - 200	232.606	126.006	OK
	BA-1	250	500	2D10 - 100	2D10 - 200	147.745	65.051	OK
	BA-2	250	500	2D10 - 100	2D10 - 200	147.745	61.285	OK
4	BI-1	300	700	2D10 - 50	2D10 - 200	232.606	136.650	OK
	BA-1	250	500	2D10 - 100	2D10 - 200	147.745	57.995	OK
5	BI-1	300	700	2D10 - 50	2D10 - 200	232.606	97.054	OK
	BA-1	250	500	2D10 - 100	2D10 - 200	147.745	60.552	OK

Table 4. Moment Capacity of Plate Reinforcement

Floor	Dimension (m)	Reinforcement Support	Reinforcement Midspan	M_{max} Location Support kN.m/m	M_{max} Location Midspan kN.m/m	φM_n Support kN.m/m	φM_n Midspan kN.m/m	Remarks Support	Remarks Midspan
1	3,5 x 4,6	D10 - 150	D10 - 150	13.789	7.751	16.576	16.576	OK	OK
	4,25 x 4,6	D10 - 150	D10 - 150	13.420	6.884	16.576	16.576	OK	OK
	3 x 4,6	D10 - 150	D10 - 150	11.055	4.149	16.576	16.576	OK	OK
	3 x 1,5	D10 - 150	D10 - 150	3.192	2.003	16.576	16.576	OK	OK
	3 x 4	D10 - 150	D10 - 150	6.982	4.068	16.576	16.576	OK	OK
2	3,5 x 4,6	D10 - 150	D10 - 150	13.431	7.710	16.576	16.576	OK	OK
	4,25 x 4,6	D10 - 150	D10 - 150	13.322	6.938	16.576	16.576	OK	OK
	3 x 4,6	D10 - 150	D10 - 150	11.067	4.142	16.576	16.576	OK	OK
	3 x 1,5	D10 - 150	D10 - 150	3.756	2.749	16.576	16.576	OK	OK
	3 x 4	D10 - 150	D10 - 150	7.006	4.089	16.576	16.576	OK	OK
3	3,5 x 4,6	D10 - 150	D10 - 150	13.407	7.793	16.576	16.576	OK	OK
	4,25 x 4,6	D10 - 150	D10 - 150	13.240	6.945	16.576	16.576	OK	OK
	3 x 4,6	D10 - 150	D10 - 150	6.563	1.703	16.576	16.576	OK	OK
4	3,5 x 4,6	D10 - 150	D10 - 150	13.445	7.817	16.576	16.576	OK	OK
	4,25 x 4,6	D10 - 150	D10 - 150	13.186	6.975	16.576	16.576	OK	OK
	3 x 4,6	D10 - 150	D10 - 150	6.539	1.672	16.576	16.576	OK	OK
5	3,5 x 4,6	D10 - 150	D10 - 150	11.499	5.894	16.576	16.576	OK	OK
	4,25 x 4,6	D10 - 150	D10 - 150	14.022	6.837	16.576	16.576	OK	OK
	3 x 4,6	D10 - 150	D10 - 150	10.269	4.327	16.576	16.576	OK	OK

3.2 Inter story drift

Tables 5 and 6 show the calculation of building inter-story drift. The value of every floor drift (Δ_i) is less than allowable drift (Δ_{all}). It can be concluded that the building inter-story drift is qualified according to the newest SNI.

Table 5. Drift in the X direction

Floor	hsx	h	δ_e	Δ	Δ_i	Δ_{all}	Re-marks
	(m)	(mm)	(mm)	(mm)	(mm)	(mm)	
5	16	4000	26,778	98,186	23,588	30,769	OK
4	12	4000	20,345	74,598	29,740	30,769	OK
3	8	4000	12,234	44,858	29,344	30,769	OK
2	4	4000	4,231	15,514	15,514	30,769	OK
1	0	4000	0,000	0,000	0,000	30,769	OK

Table 6. Drift in the Y direction

Floor	hsy	h	δ_e	Δ	Δ_i	Δ_{all}	Re-marks
	(m)	(mm)	(mm)	(mm)	(mm)	(mm)	
5	16	4000	25,648	94,043	23,588	30,769	OK
4	12	4000	19,234	70,525	26,932	30,769	OK
3	8	4000	11,889	43,593	26,836	30,769	OK
2	4	4000	4,570	16,757	16,757	30,769	OK
1	0	4000	0,000	0,000	0,000	30,769	OK

3.2 Check of P- Δ effect

Tables 7 and 8 show a calculation of the P-Δ effect. From this table, it can be seen that the value of the stability coefficient (θ) is smaller than 1.0. According to SNI 1726:2019 article 7.8.7, if the value of the stability coefficient (θ) is smaller than 1.0, then the calculation due to the P-Δ effect could be ignored. It can be concluded that the effect of P-Δ on the structure has been stable and could be ignored.

Table 7. P-Delta Effect in the X direction

Floor	hsx	Δi	P	Vx	θ	q_{max}	Re-marks
	(m)	(mm)	(kN)	(kN)			
5	16	23,588	10.168	744	0,0055	0,091	Stable
4	12	29,740	18.257	983	0,0126	0,091	Stable
3	8	29,344	27.569	1162	0,024	0,091	Stable
2	4	15,514	34.877	1263	0,029	0,091	Stable

Table 8. P-Delta Effect in Y direction

Floor	hsx	Δi	P	Vx	θ	q_{max}	Re-marks
	(m)	(mm)	(kN)	(kN)			
5	16	23,588	10.168	776	0,0055	0,091	Stable
4	12	29,740	18.257	1049	0,0110	0,091	Stable
3	8	29,344	27.569	1256	0,0200	0,091	Stable
2	4	15,514	34.877	1263	0,029	0,091	Stable

3.4 Irregularity of Vertical Direction

Due to the shape of the building floor plan is not symmetrical for the vertical direction, it is necessary to check the irregularities of the vertical direction. Tables 9 and 10 show the irregularity of soft-level stiffness in x and y-direction. The weight irregularity calculation is shown in Table 11. From this table, the vertical and weight irregularities of the building are still following the requirements.

Table 9. Irregularity of soft-level stiffness (x-direction)

Floor	hsx	h	δe	Δ	70%	Rem	80%	Rem
	(m)	(mm)	(mm)	(mm)				
5	16	4000	26,778	98,186	0,017	Ok	0,020	Ok
4	12	4000	20,345	74,598	0,013	Ok	0,015	Ok
3	8	4000	12,234	44,858	0,008	Ok	0,009	Ok
2	4	4000	4,231	15,514	0,003	Ok	0,003	Ok
1	0	4000	0,000	0,000	0,000	Ok	0,011	Ok

Table 10. Irregularity of soft-level stiffness (y-direction)

Floor	hsy	h	δe	Δ	70%	Rem	80%	Rem
	(m)	(mm)	(mm)	(mm)				
5	16	4000	25,648	94,043	0,016	Ok	0,024	Ok
4	12	4000	19,234	70,525	0,012	Ok	0,024	Ok
3	8	4000	11,889	43,593	0,008	Ok	0,021	Ok
2	4	4000	4,570	16,757	0,003	Ok	0,017	Ok
1	0	4000	0,000	0,000	0,000	Ok	0,011	Ok

Table 11. Weight Irregularity

Floor	Weight (kg)	150% W	W < 1,5 Wt	W < 1,5 Wb
5	256,745	385,118	-	Ok
4	262,635	393,953	Ok	Ok
3	295,683	443,525	Ok	Ok
2	312,433	468,649	Ok	Ok
1	313,563	470,34	Ok	-

Table 12. Strong Column Weak Beam Calculation

Flo-or	$ØM_c$	$ØM_{bR}$	$ØM_{bL}$	ΣM_c	ΣM_b	$\Sigma M_c / \Sigma M_b$	Stan-dard	Re-ma-rk
5	363.85	157.38	157.38					
				727.70	314.76	2.31	1.20	Ok
4	363.85	157.38	157.38					
				727.70	314.76	2.31	1.20	Ok
3	363.85	157.38	157.38					
				727.70	314.76	2.31	1.20	Ok
2	363.85	157.38	157.38					
				727.70	163.82	4.44	1.20	Ok
1	363.85	81.91	81.91					

3.5 Strong Column Weak Beam Checking

Following article 18 SNI 2847:2019, it is explained that the capacity of the column must be greater than the beam, or according to the formula: $\Sigma M_{column} \geq (6/5)$ x ΣM_{beam}. Table 12 shows the strong column weak beam calculation. From the results in Table 12, it can be concluded that the column capacity is greater than 1.2 beam capacity.

3.6 Mass Participation Checking`

Following article 7.9.1.1 SNI 1726:2019, the number of natural vibrations must be 100% of the structural mass and in the period before 0.05 seconds. The calculations of building mass participation are shown in Table 13. From thi table, the mass participation in the X direction and Y direction is 100% in the 86th mode and it is still following SNI 1726:2019.

Table 13.Building mass participation calculations

Case	Mode	Period sec	UX	UY	UZ	Sum UX	Sum UY	Sum UZ	RX	RY	RZ
Modal	77	0	0,0012	0,01119	0	0,0012	0,01119	0	0,00797	0,01442	0,0025
Modal	78	0,001	0,5425	0,65	0	0,5437	0,66119	0	0,00929	0,00013	0,529
Modal	79	0,001	0,1102	0,22	0	0,6539	0,88119	0	0,01404	0,01953	0,456
Modal	80	0,005	0,00001025	0	0	0,65391	0,88119	0	0,00907	0,00255	0,0034
Modal	81	0,007	0,2405	0,001	0	0,89441	0,88219	0	0,00275	0,00041	0,234
Modal	82	0,008	0,1022	0,02	0	0,99661	0,90219	0	0,02026	0,11264	0,1544
Modal	83	0,008	0,00002273	0,03	0	0,996633	0,93219	0	0,01938	0,04759	0,0024
Modal	84	0,009	0,0004	0,00189	0	0,997033	0,93408	0	0,00131	0,00274	0,0268
Modal	85	0,01	0,002667	0,00519	0	0,9997	0,93927	0	0,00219	0,00015	0,0233
Modal	86	0,02	0,0003	0,06073	0	1	1	0	0,00875	0,00806	0,004

4 Conclusions

From the assessment which has done, the following could be concluded :

1. The analysis results of the capacity of structural elements (columns, beams, and plates) of Block D RS BMC Padang, are still able to support the design loads, according to the newest SNI (SNI 1726-2019 and SNI 2847-2019).
2. The results of checking the performance of the building structure against the inter-story drift, the P-Δ effect, and irregularities in the vertical direction, the concept of strong column weak beam, and mass participation showed that it is still following the newest standards (SNI 1726-2019)

The authors thanked the dean and the head of the civil engineering department of faculty of engineering University of Andalas for the financing provided with the publication grant scheme, contract No. 136/UN16.09.D/PL/2021.

References

1. SNI 2847-2019, "*Structural Concrete Requirements for Buildings*", BSN (2019).
2. SNI 1726-2019, "*Earthquake Resilience Design Procedures for Building and Non-Building Structures*", BSN (2019).
3. SNI 1727-2018, "*Minimum Load for The Design of Buildings and Other Structures*", BSN (2018).
4. EERI Special Earthquake Report, "*Learning from Earthquakes", The Mw 7.6 Western Sumatra Earthquake of September 30, 2009*, December 2009.
5. Fauzan, Zaidir, M.P. Laura, "*Analysis of structural failure and retrofitting of column B of SMA N 10 Padang which was damaged by the September 30, 2009 Earthquake*", Jurnal Teknika, ISSN: 0854-8471No. 34 vol 1, (2010).
6. Fauzan , F. A. Ismail, Zaidir , A. Hakam, N. Yanto, R. Ramli., "*Identification of Damage and Methods of Strengthening the Office of the Governor of West Sumatra, Proceedings of the National Seminar on Civil Engineering 1 (SeNaTS 1) 2015*", Sanur - Bali, 25 April 2015, ISBN 978-602-294-052-4,(2015).
7. Fauzan, "*Analysis of Retrofitting Implementation Methods in Simple Buildings (Case Study: SD Negeri 43 Rawang Timur, Padang)", Engineering Journal*, **8** No. 1, ISSN: 1838-2133, (2012).
8. F.A. Ismail, A. Hakam, Fauzan, J. Teknik Sipil ITB **10** (2011).
9. F.A. Ismail, et.al, *Retrofitting of Bumiminang Hotel Building in Padang*, (CECAR6, 2014)
10. P. Grundy, *The Padang Earthquake 2009 – "Lessons and Recovery,* Australian Earthquake" Eng. Soc. 2010 Conf., Perth, Western Australia, Department of Civil Engineering, Monash University.
11. G. Thermou, A.S. Elnashai, (2002), "*Performance Parameters and Criteria forAssessment and Rehabilitation, Seismic Performance Evaluation and Retrofit of Structures (SPEAR)*", European Earthquake Engineering Research Network Report, Imperial College, UK.

https://doi.org/10.1051/e3sconf/202133105015

12. Zaidir, N. Maizul., M.P., Laura, "*Evaluation of the feasibility of high-rise buildings after the 30 September 2009 earthquake, West Sumatra*", Jurnal Rekayasa Sipil, **8** No. 1, 61-73, (2012).
13. Zaidir, Fauzan, D., Angreini, "*Evaluation of the Feasibility of Building Structures Ex. PO.ANS based on SNI Earthquake 1726:2012*", Prosiding 3rd Andalas Civil Engineering National Conference, 75-89, ISBN 978-602-9081-16-9,(2016)
14. Zaidir , Fauzan, A. Hakam, F. A. Ismail, T. Boen, "*Retrofitting the Padang City Hall Building using Woven Wire*", Seminar Nasional Strategi Pengembangan Infrastruktur , Kampus ITP, Padang 27-28 Agustus 2014, ISBN : 978-602- 70570-1-2.
15. T.Boen, "*How to Repair a Simple Building Damaged by an Earthquake*", 2nd Ed., WSSI. (2010)

A study on the use of public transportation during the COVID-19 pandemic

Titi Kurniati[1*] and *Junico* Nobel Valentino[1]

[1]Civil Engineering Department, Engineering Faculty, Andalas University, Indonesia

Abstract. During the COVID-19 pandemic, public transportation occupancy has decreased significantly. In addition to the sluggish economy, the COVID-19 health protocol rules require that only 50 percent of passengers can be transported for public transportation. This study tries to identify the use of public transportation and the factors that influence this use during the adaptation period for the new habits of the COVID-19 pandemic. Data on the frequency of using public transportation before and during the pandemic was processed with the index formula. Before the pandemic period, using public transportation was often, with an index value of 60.8%. During the pandemic period, 10.1% of respondents did not do activities outside the home. Respondents who moved outside their homes (89.9%) used private vehicles (88.3%) and 11.7% public transportation for their trips. The type of public transportation often used is online transportation (motorbike and car) by 62.1%. The frequency of use of public transportation is rare (44.9%) and very rarely (40.8%); the rest are often (9.7%) and very often (4.5%). This value will produce an index value of 44.5% with a sparse interval interpretation. A hypothesis test was conducted between the respondent's characteristics (gender, age, occupation, and vehicle ownership) and the frequency of using public transportation. The result shows that the factor influencing the frequency of using public transportation is the respondent's occupation.

1 Introduction

In early 2020, a new virus, namely a new type of coronavirus (SARS-CoV-2) and its disease called Coronavirus disease 2019 (COVID-19), took the world by storm because of its widespread and speedy spread. Through the Director-General Dr. Tedros Adhanom Ghebreyesus, the World Health Organization or WHO declared the Corona COVID-19 virus a global pandemic on March 12, 2020. With the determination of the pandemic, Dr. Tedros asked each country to: activate and improve emergency response mechanisms, communicate to citizens about the risk of the coronavirus, and urge them to protect themselves, as well as find, isolate, test, and treat COVID-19 patients and trace any contact they have come into contact with [1].

Following up on this, the Indonesian government also stated that the Coronavirus problem had become a non-natural national disaster. The Central Government and Regional Governments and their staff work hand in hand to make several tactical steps to prevent the spread of the Corona COVID-19 Virus in the community. The steps taken by the government in certain areas are Large-Scale Social Restrictions (PSBB) and, after that, the implementation of new habits (new normal).

The policy, which was followed up by massive socialization to the community to work from home, a study from home, and worship from home, and the closure of tourist sites had limited the movement of people outside their homes.

Transportation is one area that is restricted in the activities of Large-Scale Social Restrictions (PSBB) and during the new normal period. Restrictions placed on public transportation are related to the number of passengers and operating hours. For traffic users who depend on public transportation, this policy results in limited access to public transportation.

MarkPlus, Inc. explains the results of a quick survey participated by 100 respondents throughout Indonesia, the majority of whom actively use public transportation. From the survey results, 66% of respondents use public motorcycles such as online motorcycle taxis, 50% minibusses, 46% city buses, 44% city trains, 31% taxis, and 1% bajaj. The survey results show a high enough fear for people to use public transportation during the COVID-19 period. 40% of respondents stated that they had never used public transportation since the pandemic, and 30% of respondents limited the intensity of their use [2].

According to the Jabodetabek Transportation Management Agency (BPTJ), public transportation in the DKI Jakarta area decreased significantly during the COVID-19 pandemic. TransJakarta services during April 2020 (until 15/4/2020) decreased by around 83,000 people per day. In fact, under normal conditions, the number of passengers reaches around 840,000 people per day. Meanwhile, on the Mass Rapid Transit (MRT)

* Corresponding author: titi@eng.unand.ac.id

service, there was a 47 percent decrease in passengers during March 2020. Usually, the daily MRT passengers reach around 85,000 people, but this has decreased to around 45,000 people [3].

Not only in Indonesia, but the impact of the COVID-19 pandemic on a significant decrease in the use of public transportation has also occurred in cities around the world such as London, New York City, Berlin, Paris, and Sydney. In Paris, for example, the decline was about 80% compared to before the pandemic [4]. In India, the pandemic has caused 19% of Indian bus operators to lose 90% of their passengers and the remaining 91% without passengers at all [5].

The purpose of this study is to identify the factors influencing the use of public transportation, as well as to analyze the factors that have the greatest influence on the use of public transportation during the adaptation of new habits to the COVID-19 pandemic.

2 Literature Review

Several studies related to the use of transportation during the pandemic in Indonesia have been published. The choice of mode during the COVID-29 pandemic in Tegal city is influenced by variables of socio-economic characteristics. To analyze, used *PLS Structural Equation Modelling (SEM)* method. The result of the analysis is that the mode choice is influenced by the vehicle ownership variable with a coefficient = 0.81. In contrast, the variables of distance, self-preservation behavior during the pandemic, and the purpose of the trip do not significantly affect people's choice of transportation mode during the pandemic [6].

The research [7] results concluded that there were changes in activity and travel due to the COVID-19 pandemic; mainly, the average travel frequency to work or school decreased by 66%. Multiple linear regression analysis was used to investigate travel-activity changes. The linear regression model between changes in travel activity and mental health variables, and socio-demographic variables met the evaluation of the feasibility of the model. However, the value of the coefficient of determination or R is relatively low. So it is necessary to consider other independent variables.

The results of the study on the factors that influence the decision to use the Bekasi Commuter Line KRL services during the COVID-19 pandemic with factor analysis are the equipment completeness variable on the health protocol factor with a contribution of 83%, the frequency of the number of departures on the availability factor with a contribution of 84%, the availability of payment methods on the convenience factor with a contribution of 86%, the new policy of PT. KCI on the convenience factor with 83% contribution, affordable cost on the cost factor with a contribution of 90%, the development and feasibility of technology on the safety factor with a contribution of 81%, and a sense of security from crime at the station on the security factor with a contribution of 86% [8].

3 Methodology

3.1 Data Collection

Data collection was carried out through online questionnaire interviews. Questionnaire questions are designed into three (3) parts. The characteristic of the questions for the three parts of the questionnaire is closed questions. Questionnaires are created in google docs forms.

The first part is a question regarding the socioeconomic characteristics of the respondents (gender, age, occupation, vehicle ownership). Part 2 contains questions regarding the movement of respondents before the pandemic. The period before the COVID-19 pandemic is before March 18, 2020. The questions are the vehicle used for movement, the type, and method of driving using a private vehicle, the purpose of the movement, the type of public transportation that is often used, and the frequency of using public transportation. Section 3 asks about the respondent's movement during the pandemic (after March 18, 2020). The questions include activities out of the house during the pandemic, the type of vehicle used, the type of public transportation that is often used, the purpose of movement, and the frequency of using public transportation.

The data collection was carried out for 1 month in September 2020. At that time, a new habit adaptation period was applied. The government's policy on public transportation is the limitation of public transport passengers with the occupation rate of 50% of capacity. Dissemination of questionnaires through social media researchers.

3.2 Data Processing

The analytical method to explain the movement of transportation users before and during the COVID-19 pandemic is a descriptive statistical method. The statistical hypotheses test at a significant level, $\alpha=5\%$, was carried out to compare the frequency of using public transportation before and during the COVID-19 pandemic. A chi-square test is carried out to determine the factors that affect the use of public transportation during the COVID-19 pandemic. Hypothesis and chi-square test using *statistical SPSS 22 software*.

For the question of frequency of use of public transportation, the answers are designed on a Likert scale with four scale choices, namely:

Score 1. Very rarely
Score 2. Rarely
Score 3. Often
Score 4. Very often

The conclusion regarding the frequency of the respondent's trip is the result of the value generated by using the Index% formula.

$$\text{Index\%} = \text{Total Score} / Y \times 100 \quad (1)$$

$$I = 100 / \text{total Likert score} \quad (2)$$

Y = highest Likert score x number of respondents

I = score interval
Criteria for interpretation of scores based on intervals:
Value 0% - 24.99% = very rarely
Value 25% - 49.99% = rarely
Value 50% - 74.99% = often
Value 75% - 100% = very often

4 Results

The survey results collected 267 responses. The personal characteristics of respondents, as shown in Table 1, the respondent is dominated by the age of 17 ≤ age < 25years with occupation as a student. Only 20% of respondents do not own a motorized vehicle.

Table 1. Personal characteristics of the respondent

Characteristics		Percentage (%)
Gender	Male	34.5
	Female	65.5
Age	Age < 17	23.6
	17 ≤ age < 25	68.5
	25 ≤ age < 33	7.1
	33 ≤ age < 41	0
	Age ≥ 41	0.7
Occupation	Student	83.1
	employees	12.4
	Businessman	1.9
	others	2.6
Vehicle ownership	Not Owning	20.6
	Motorcycle (MC)	70.0
	Car	5.2
	MC and car	4.1

Table 2. The distribution of respondent's domiciles

Province	Percentage (%)
Sumatera Barat	52.0
Kepulauan Riau	29.6
Riau	7.5
Banten	3.0
Jawa Barat	2.25
Jambi	1.5
Sumatera Utara	1.5
DKI Jakarta	0.75
DI Yogyakarta	0.37
Bengkulu	0.37
Lampung	0.37
Sumatera Selatan	0.37
Jawa Timur	0.37

The distribution of respondents' domiciles can be seen in Table 2, where most of them are domiciled in Sumatera Barat Province.

The survey results show that before the COVID-19 pandemic, 21.3% of respondents used public transportation to travel to school or campus. The more widely used public transportation is online taxis (car and motorcycles, MC). A total of 22.1% of respondents use public transportation every day (very often) (Table 3).

Table 3. Activity before the COVID-19 pandemic

Activity		Percentage (%)
the vehicle used for movement	Private vehicle	78.8
	Public transportation	21.3
The type and method of driving using a private vehicle	MC (ride alone)	67.0
	MC (delivered and picked up)	17.5
	car (drive alone)	7.5
	car (delivered +picked up)	2.3
	car and motorbike	5.7
The purpose of the movement	school/campus	62.9
	market/mall	44.6
	workplace	16.1
	others	7.9
The type of public transportation that is often used	paratransit	28.5
	bus	16.5
	MC/car online	55.1
The frequency of using public transportation.	very rarely	9.0
	rare	60.7
	often	8.2
	very often	22.1

During the pandemic, 10.1% of respondents did not do activities outside the home. Most respondents (89.9%) who move out of the house use private vehicles (79.4%) and public transportation 10.1% (Table 4).

The data processing results on the frequency of using public transportation before the COVID-19 pandemic obtained an index of 60.86%, and during the COVID-19 pandemic, it was 44.48%. Interpretation of the interval score of the 60.86% index value was often (50% - 74.99%) and 44.48% was rare (25% - 49.99%) (Tabel 5). There was a 16.4% reduction in the use of public transportation.

The null hypotheses assumption (H_0) in the hypotheses test is that there is no difference between the frequency of using public transportation before and during the COVID-19 pandemic at a significant level of α=5%. The result is $P = 0.001 < 0.05$; then the decision is to reject H_0. It was concluded that there was a significant difference between the use of public transportation before and during the COVID-19 pandemic.

Table 4. Activity during the COVID-19 pandemic

Activity		Percentage (%)
Activities out of the house	Yes No	89.9 10.1
The type of vehicle used	Private vehicle Public transportation	88.3 11.7
The type of public transportation that is often used	paratransit bus Online MC/car taxis	19.5 18.4 62.1
The purpose of the movement	school/campus market/mall workplace others	52.1 32.4 10.3 3.2
The frequency of using public transportation.	very rarely rare often very often	40.8 44.9 9.7 4.5

Table 5. The frequency of using public transportation

The frequency of using public transportation	Before	During
Mean	2.43	1.78
Standar deviation	0.933	0.799
Index%	60.86	44.48
Interpretation	often	rare

A chi-square statistical test was carried out to determine whether or not there is a relationship between socio-economic characteristics and the frequency of public transportation during the COVID-19 pandemic. Previously, it was necessary to make a crosstab analysis table between socio-economic characteristics (gender, age, occupation, and ownership of motorized vehicles) and the frequency of use. Table 6 shows a crosstab analysis between gender and frequency and table 7 between occupation and frequency.

From the results of statistical tests obtained P-value (Asymp. Sig.) = 0.537 > 0.05. Conclusion Null Hypotheses (H0) are not rejected, meaning that there is no significant relationship at a significant level of α=5% between gender and the frequency of using public transportation during the COVID-19 pandemic.

Table 6. Crosstab analysis gender vs frequency

Frequency	Gender	
	Male	Female
Very rarely	40	69
Rarely	40	80
Often	10	16
Very often	2	10

Table 7. A chi-square statistical test

	Value	df	Asymp. Sig. (2-sided)
Pearson Chi-Square	2.176	3	0.537
Likelihood Ratio	2.386	3	0.496
Linear by Linear Association	0.835	1	0.361
N of Valid Cases	267		

Table 8. Crosstab analysis occupation vs frequency

Occupation	Frequency			
	Very rarely	Rarely	Often	Very often
Student	95	100	19	8
Employees	7	17	6	3
Businessman	1	2	1	1
Others	6	1	0	0

From the results of statistical tests obtained P-value (Asymp. Sig.) = 0.035 < 0.05. Conclusion Null Hypotheses (H_0) are rejected, meaning that there is a significant relationship at a significant level of α=5% between occupation and the frequency of using public transportation during the COVID-19 pandemic.

Table 9. Summary of Chi-Square test results

Variable	P-value	Conclusion
Gender	0.537	H_0 not rejected
Age	0.546	H_0 not rejected
Occupation	0.035	H_0 rejected
Vehicle ownership	0.664	H_0 not rejected

https://doi.org/10.1051/e3sconf/202133106001

The results of the chi-square statistical test of socio-economic variables with frequency are summarized in table 9. Of the four relationships tested, only the occupation variable has a relationship with the frequency of use of public transportation during the COVID-19 pandemic at a significant level of α=5%.

5 Results

Before the pandemic period, using public transportation was often, with an index value of 60.8%. During the pandemic period, 10.1% of respondents did not do activities outside the home. Respondents who moved outside their homes (89.9%) used private vehicles (88.3%) and 11.7% public transportation for their trips. The type of public transportation often used is online transportation (motorbike and car) by 62.1%. The frequency of use of public transportation is rare (44.9%) and very rarely (40.8%); the rest are often (9.7%) and very often (4.5%). %). This value will produce an index value of 44.5% with a rare interval interpretation. There was a 16.4% reduction in the use of public transportation.

Based on the hypothesis test at a significant level of α=5%, it can be concluded that there are differences in the frequency of using public transportation before and during the COVID-19 pandemic.

The results of the chi-square statistical test of socio-economic variables with frequency concluded that only employment variables had a relationship with the frequency of use of public transportation during the COVID-19 pandemic at a significant level of α=5%

The authors would like to thank Engineering Faculty for the financial support provided through Hibah Publikasi Fakultas Teknik Tahun Anggaran 2021.

References

1. G. L. Widyaningrum, " WHO has declared COVID-19 a global pandemic. What does it mean? National Geographic Indonesia, 12 March 2020, https://nationalgeographic.grid.id/read/132059249/who-tetapkan-COVID-19-sebagai-pandemi-global-apa-maksudnya [accessed Oct. 16, 2020]
2. S. Mayasari, "Survey: Public transportation users have dropped dramatically since the COVID-19 pandemic" (in English) , kontan.go.id, 19 June 2020, https://industri.kontan.co.id/news/survei-pengguna-transportasi-umum-turun-drastis-sejak-pandemi-COVID-19 (accessed Oct. 16, 2020)
3. R. Kurniawan, "Users of Jabodetabek Public Transportation Decreased Drastically During WFH", https://otomotif.kompas.com/read/2020/04/20/154500315/pengguna-angkutan-umum-jabodetabek-menurun-drastis-selama-wfh. (accessed Oct. 16, 2020)
4. Moovit, "Moovit Public Transit Index," May 2020 https://moovitapp.com/insights/en/Moovit_Insights_Public_Transit_Index-countries. (accessed Feb. 16, 2021)
5. UITP, "Bus operations in India: what has been the impact of COVID-19?," UITP, May 27, 2020. https://www.uitp.org/news/bus-operations-india-what-has-been-impact-COVID-19 (accessed Feb. 16, 2021)
6. P. Rusmandani, R. S. Setiawan and N. Fitriani, Civil Engineering Journal / **15** No.2 119-126 (2021)
7. M. Rizki, D. Prasetyanto and A. Maulana, Environmentally Friendly Technology Journal, **5** No.2. 125-135 (2021)
8. A. M. Monica, Sukanta and Winarno Industrial Engineering Journal. **5** No.2. 17-22 (2021)

An investigation of Teratak Buluh sand as a foam mortar aggregate material

Elsa Eka Putri [1*], *Doni Rinaldi* Basri [2,3], and *Bayu Martanto* Adji[2]

[1] Civil Engineering Department, Faculty of Engineering, University of Andalas, 25163, Padang, Indonesia
[2]Civil Engineering Department, Faculty of Engineering, University of Abdurrab, 28292, Pekanbaru, Indonesia
[3]Doctoral Student, Civil Engineering Department, University of Andalas, 25163, Padang, Indonesia

Abstract. Riau Province has the largest peat soil on Sumatra island. Peat soil has a low bearing capacity and very deep hard soil depth, even in some places up to 16 meters depth. This condition makes the construction of roads on peat soils will experience significant and unavoidable degradation and damage if no special handling is conducted. One of the efforts to overcome the situation in road construction is the use of foam mortar which has been developed by the Road and Bridge Research Centre (PUSJATAN) in 2013 which discovered the Foam Mortar Light Pile technology. This foam mortar is a cement mixture that has a lower density than water. This study conducted further research on finding the ratio of Cement vs. Sand in foam mortar which has a specific gravity smaller than water. The sand comes from Teratak Buluh Village, Kampar Regency, Riau Province and Cement (PCC) comes from Padang City, West Sumatra Province. The cement and sand ratio was varied; 1C:0.7S, 1C:0.8S, 1C:0.9S, 1C:1.0S, 1C:1.1S and 1C:1.2S. Cylindrical samples with a diameter of 10 cm and a height of 20 cm were subjected to a compression test using a modified CBR tool at the 7 day curing time. From the test results, samples at a ratio of 1C: 0.7S and 1C: 0.8S can float in water, with a density of 0.77 tonne/m^3 and compressive strength of 901.20 kPa and a density of 0.83 tonne/m^3 with a compressive strength of 971.35 kPa respectively.

1 Introduction

Riau Province has the widest peat soil type on Sumatra Island, 4,044 million ha (56.1% of the total area of Sumatra island or 45% of the Riau mainland area. This peat soil has a low bearing capacity and has a very deep hard soil depth, even in some places it has a depth of up to 16 meters.

Peat soil consists of organic fragments, derived from vegetation that has changed and is chemically fossilized, the accumulation of organic matter forms peat soil [1].

Peat soil has a low bearing capacity; thus, it needs the effort to strengthen the soil to avoid excessive settlement and permanent deformation.

Road and Bridge Research Centre (PUSJATAN) in 2013 developed a new technology called Foam Mortar Lightweight Technology [2]. This foam mortar has a specific gravity that is smaller than water and can carry a minimum load of 800 kPa so that it can be used as a substitution for embankment soil on road pavements in low-bearing capacity soil.

The main purpose of this research is to determine the cement: a sand ratio that produces foam mortar under the requirements as per PUSJATAN 2013, then study the effect of the ratio of cement vs. sand on the density and compressive strength using local material, i.e. the sand comes from Teratak Buluh Village, Kampar Regency, Riau Province, and cement type is Portland Cement Concrete (PCC).

2 Literature Review

Foam mortar consists of mortar and foam agent, the mortar consists of cement, sand, water, while foam consists of foam agent and water [3].

In the direct loading test in the field, it was seen that the design of the embankment with foam mortar material resulted in lesser soil compression value than the embankment with sandstone soil material, due to the foam mortar material being lighter and more stable [4].

Research on the use of foam agents in the manufacture of lightweight concrete bricks has been carried out by adding a foaming agent in the concrete mixture to produce the lightweight concrete [5]. Sample was tested using a cube mould, with the composition of cement vs. sand of 1:0.5; 1:0.7 and 1:0.9 produce a compressive strength of 17,422 kg/cm2; 14,756 kg/cm2 and 9,778 kg/cm2 with a weight of 2.510 kg, 2.748 kg and 2.808 kg respectively [5].

* Corresponding author: elsaeka@eng.unand.ac.id

Moreover, the investigation on foam concrete has also been carried out using a mixture ratio of cement, sand, and water 1:1.5: 0.5 [6]. In this study, the foam agents were added starting from 28%, 45%, 62%, and 78%. With the addition of 28% foam agent, the density value is between 1770 kg/m, while for 78% addition the density value becomes 500 kg/m³ [6].

Using the alternative type of construction CSS-Mortar Foam can reduce the cost of flyover construction by 44.7% when compared to the PCI-Girder construction type [7].

The effect of polycarboxylate Superplasticizer (SP) content on foam concrete properties has also been investigated [8]. The results of the study show that the water-cement and Superplasticizer (SP) factors have a significant influence on the performance of foam concrete. The use of 1.35% SP improves the mechanical and transport properties of foam concrete [8].

The results of the study with the addition of ground calcium carbonate (GCC) increased the compressive strength and thermal conductivity due to the effect of filling and the addition of glass fibre (GF) contributed to the flexural strength of the concrete [9].

Meanwhile, investigation of the effect of the concentration of foaming agents and their microwave and ultrasonic treatments concerning foam stability, bulk density, microstructure, and compressive strength produced has been examined [10]. The reduction in pore size contributed to the increase in compressive strength determined in the samples prepared with lower microwave concentrations and ultrasonically treated foaming agents. The lower concentration of the foaming agent itself also has a positive effect on the stability of the foam. However, the influence is not so prominent [10].

Moreover, an experimental study of the raw material properties, possible changes, and new formations in the inner layer of porous materials provide an improved construction and operational properties of autoclaved foam concrete [11]

In this study, the ratio of cement vs. sand was varied by 1C:0.7S, 1C:0.8S, 1C:0.9S, 1C:1.0S, 1C:1.1S and 1C:1.2S. This variation was chosen to find the ratio of cement vs. sand mixture that produced the lightweight concrete which had a density of less than 1, shown by the floating of the concrete sample in the water.

3 Materials and Methods

3.1 Materials

Foam concrete can be categorized as mortar because its constituent materials consist of cement, sand, water, and foam. The foam was obtained from mixing foam agent and water in a ratio of 1:20. Foam mortar is categorized as mortar because it does not contain coarse aggregate. Meanwhile, the air trapped in the foam mortar caused by the chemical reaction was expected to reduce the specific gravity of the mortar.

The mixture of materials consists of cement, sand, and water. All materials are mixed using a laboratory mixer, with variations in the material composition according to the mix plan to obtain the desired lightweight material specifications. The mixture must be free from lumps. The sand used passed No. 10 sieve size and was retained by the No. 200 sieve size [3].

The weight of each material to form 1 m³ of mortar mixture has been calculated and the result is presented as shown in Table 1.

Table 1. Weight of Material to produce 1 m³ Foam Mortar

Materials	Cement: Sand Ratio					
	1C: 0.7S	1C: 0.8S	1C:0.9S	1C: 1.0S	1C: 1.1S	1C: 1.2S
Cement (kg)	250	250	250	250	250	250
Sand (kg)	174.9	200.12	224.89	250.05	275.02	299.98
Water (kg)	125	125	125	125	125	125
Foam (kg)	47.24	46.62	45.99	45.34	44.71	44.07

Table 1 is the weight of the material of Foam Mortar for 1 m³. Meanwhile, Table 2 is the weight of all materials for 3 samples at each composition.

Table 2. Weight of Material for Foam Mortar Composition Materials for 3 samples per mixture

Materials	Cement: Sand					
	1C: 0.7S	1C: 0.8S	1C: 0.9S	1C: 1.0S	1C: 1.1S	1C: 1.2S
Cement (kg)	2.36	2.36	2.36	2.36	2.36	2.36
Sand (kg)	1.65	1.89	2.12	2.36	2.59	2.83
Water (kg)	1.18	1.18	1.18	1.18	1.18	1.18
Foam (kg)	0.45	0.44	0.43	0.43	0.42	0.42

As can be seen in Table 1 and Table 2, the weight of the water and cement for all samples are the same. Moreover, the weight of the sand was varied according to the cement vs. ratio predetermined [3].

3.2 Methods

The weight of each foam mortar is determined based on the Circular Letter of the Minister of Public Works and People's Housing (SE Minister of PUPR) No. 44 (2015), concerning the design guidelines for a mixture of lightweight materials with foam mortar for road construction [3].

3.2.1 Volume Weight

The volume weight analysis was carried out by dividing the sample weight (tonne) by the sample volume (m³).

https://doi.org/10.1051/e3sconf/202133106002

The density was determined to find out the sample which has a density less than the water density, thus it can be used as lightweight material to replace the embankment as per the Road and Bridge Research Centre (PUSJATAN) [2].

$$Volume\ Weight = \frac{Sample\ Weight\ (tonne)}{Sample\ Volume\ (m^3)} \qquad (1)$$

The volume weight is the weight per unit volume of oven-dried soil, which is usually expressed as grams/cm^3 [12]

3.2.2 Foam Production

Mortars density was evaluated according to SE No:44/SE/M/2015 [3]. Foaming materials are liquid foam (foam agent) and water. To make foam, the foam liquid and water are mixed using a foam generator and a compressor (Fig. 1).

Fig. 1. Foam production

Samples were prepared in-cylinder mould with a diameter of 10 cm and a height of 20 cm as shown in Fig. 2.

Fig. 2. Cylinder Sample

3.2.3 Flow Test

Foam is mixed with mortar and stirred until homogeneous. A flow test was conducted based on ASTM C230-08 (2008) [5] was performed to check the workability and consistency.

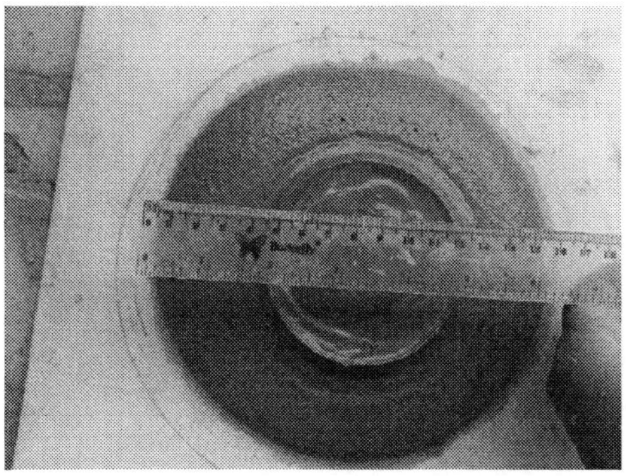

Fig. 3. Flow Test

The mortar sample was placed on a flow table and dropped 25 times within 15 seconds. After being dropped 25 times, the cone is lifted and the mortar is spread out on the flow table. The diameters for four angles of the spread mortar were measured and recorded.

Fig 3 is the measurement of the spreading mortar. The flow value is the diameter of the mortar that flows after the ring flow is removed. The flow value must be 18 cm ± 2 cm following the provisions of the SE Minister of PUPR No. 44 of 2015 [3,13].

Moreover, Fig. 4 is the cylindrical sample, which has been removed from the mould, then treated by wrapping the sample for 5 days. This is done to prevent water loss during the hardening process. So, there is no hydration due to the reaction of cement and water.

Fig. 4. Curing / Sample Maintenance

After curing for 5 days, the samples were removed and weighed. Samples were tested at the 7 days curing on a modified CBR test instrument as shown in Fig. 4.

Fig. 4. Modified CBR Equipment

3.2.4 Compressive Strength Test

A compressive strength test is done to see the strength of the sample that has been made. Whether the strength can reach 800 kPa according to the specified conditions [14].

This test was carried out with a modified CBR tool following the standards recommended by PUSJATAN. The compressive strength test with a modified CBR tool, as can be seen in Fig. 5.

Fig. 5. Compressive test of sample 1C:0.7S & 1C:0.8S

Fig. 5 shows the sample with cement vs. ratio. sand at the time of testing.

Compressive strength was determined by dividing the maximum load (Newton) by the area of the sample base (mm^2). Compressive Strength Analysis was carried out to determine the ability of the foam mortar to withstand compressive loads. The compressive strength of this foam mortar must be equal to or greater than the compressive strength of the embankment soil (>800 kPa). Equation 2 is based on SNI 1974:2011, i.e. determination of compression strength of concrete cylinder [14,15].

$$Compression\ Strength = \frac{Load\ (Newton)}{Area\ (mm^2)} \quad (2)$$

4 Results

The results for the 7 days curing at various cement vs. sand ratios have been presented in Table 3. 2 main parameters have been investigated to find the suitable lightweight foam mortar to be applied at low-bearing peat soil i.e. unit weight and compressive strength.

Table 3. Unit weight and Strength for all samples

Mixture	Sample Name	Unit weight	Strength
		tonne/m^3	kPa
1C:0.7S	A1	0.771	760.89
	A2	0.761	971.35
	A3	0.771	971.35
Average		**0.768**	**901.198**
1C:0.8S	A1	0.822	955.16
	A2	0.847	1,084.68
	A3	0.822	874.22
Average		**0.83**	**971.351**
1C:0.9S	A1	0.99	1,505.59
	A2	1.01	1,667.49
	A3	0.99	1,376.08
Average		**0.997**	**1,516.39**
1C:1S	A1	0.959	1,036.11
	A2	0.949	1,117.05
	A3	0.98	1,214.19
Average		**0.962**	**1,122.45**
1C:1.1S	A1	0.93	1,181.81
	A2	0.949	971.35
	A3	1.099	1,376.08
Average		**0.993**	**1,176.41**
1C:1.2S	A1	1.08	1,311.32
	A2	1.019	1,084.68
	A3	1.041	1,635.11
Average		**1.047**	**1,343.70**

As can be seen from Table 3, as the amount of sand in the mixture increases the volume weight of the sample also increases. Only two ratios cement vs sand that satisfy the lightweight concrete criteria specified in SE 2015, i.e. 1C:0.7S and 1C:0.8S. The 1C:0.9S or more having a density of more than 1, means the sample cannot be float when immersed in water.

Moreover, all the compressive strength values of the cement vs. sand ratio have fulfilled the required in SE 2015, with a minimum value of compressive strength of 800 kPa.

Meanwhile, the average value of the Unit Weight, as well as compressive strength, were presented in Fig. 4 and 5.

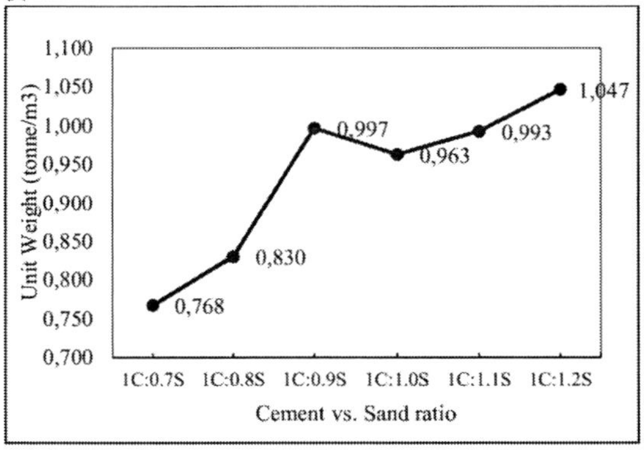

Fig. 6. Average Unit weight of sample for all various cement vs. sand

Fig. 6 shows an approximate linear transform, in which, unit weight increased with the increasing amount of sand in the mortar. It was noticeable which samples have unit weight lower than 1.

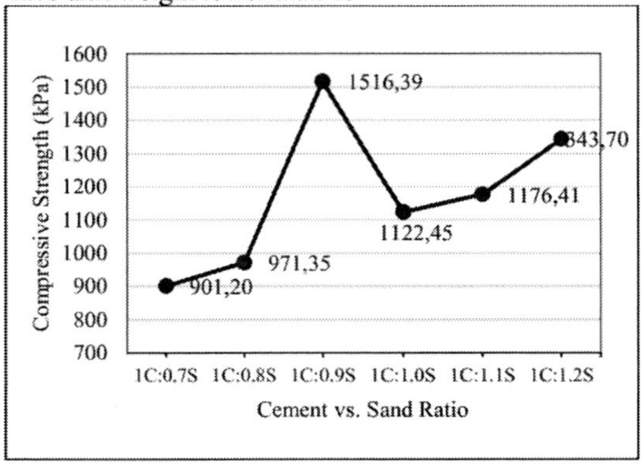

Fig. 7. Compressive Strength value cement vs. sand

As shown in Fig. 7, It shows an approximate linear regression, in which, compressive strength was increased when the sand level was increased.

It was noticeable that compressive strength in the 1C:0.9S ratio was high as well as its unit weight (Fig. 6). It seems that the higher the unit weight the higher the compressive strength.

Furthermore, as can be seen in Fig. 6, almost all samples have unit weight values below 1, only the highest sand ratio has unit weight values more than 1.

Meanwhile, the compressive strength of all samples meets the requirements, which is greater than 800kPa.

Hence, for the selection of which samples meet the requirements, the samples were directly immersed to see which cement vs. the sand ratio perfectly floats. Thus, it was found that 1C:0.7S and 1C:0.8S float perfectly, as shown in Fig. 8.

By choosing this ratio as the foundation layer, it is expected that the layer will not overburden the subgrade so that there will be no excessive settlement due to the load it carries.

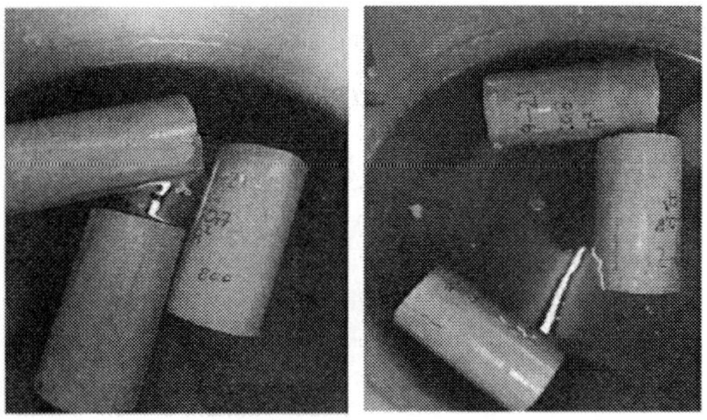

(a) (b)

Fig. 8. Floating samples in water with cement vs. sand (a) 1:0.7 (b) 1:0.8

5 Conclusions

Based on the investigation, it can be concluded that

1. The use of sand from teratak buluh, in Riau province, as a material for foam mortar, has a high compressive strength value, where the values for all cement vs. sand ratios are above 800 kPa.
2. The composition that is suitable as a lightweight foundation layer for low-bearing peat soil, is a mixture that produces a unit weight value lower than 1, namely 1C:0.7S and 1C:0.8S which floats perfectly.
3. The selected ratio 1C:0.7S and 1C:0.8S has a high compressive value, more than 800 kPa.

The authors wish to acknowledge the Faculty of Engineering, the University of Andalas for a supported grant throughout our research.

References

1. S. R. N. Panjaitan, "*A Study on the Value of Shear Strength in Muara Batang Toru Peat, North Sumatra After Experiencing Initial Compression* (Kajian terhadap nilai kuat geser tanah gambut muara Batang Toru Sumatera Utara setelah mengalami pemampatan awal)," J. Ranc. Sipil, **2**, no. 1, 2013.
2. S. Wicaksono and M. Iqbal, "*Effectiveness Of Foamed Mortar Light Weight Fill As An Approach Bridge Fill On Sheet-Pile Structures*," J. Jalan-Jembatan, **37**, no. 2, 87–101, 2020.
3. Ministry of PUPR, "*Guidelines for Specifications of Lightweight Materials with Foam Mortar for Road Construction* (Pedoman Spesifikasi Material Ringan Dengan Mortar Busa Untuk Konstruksi Jalan,)" 2015. https://binamarga.pu.go.id/uploads/files/410/pedoman-spesifikasi-material-ringan-dengan-mortar-busa-untuk-konstruksi-jalan.pdf.
4. M. A. P. Adi, Y. Lastiasih, and I. B. Mochtar, "*Alternative Planning for Road Filling with Sirtu Material and Lightweight Mortar Foam on Batang – Semarang* (Alternatif Perencanaan Timbunan Jalan dengan Material Sirtu dan Material Ringan Mortar Busa pada Jalan Tol Batang–Semarang) Seksi III

Weleri–Kendal STA 414+ 525–STA 424+ 576," J. Transp. Sist. Mater. dan Infrastruktur, **1**, no. 2, 72–77, 2019.

5. E. F. Simbolon, "*Use of Foam Agent in Making Concrete Brick* (Penggunaan Foam Agent Dalam Pembuatan Bata Beton Ringan)." Medan: Universitas Sumatera Utara, 2015.
6. S. Sulaiman, "*Water permeability and carbonation on foamed concrete*." Universiti Tun Hussein Onn Malaysia, 2011.
7. H. T. Santoso, "*Use of CSS-Mortar Foam as an Alternative for Selection of Flyover Construction Types on Construction Costs for Road and Bridge Construction* (Penggunaan CSS-Mortar Busa Sebagai Alternatif Pemilihan Tipe Konstruksi Jalan Layang terhadap Biaya Konstruksi)," BENTANG J. Teor. dan Terap. Bid. Rekayasa Sipil, **8**, no. 2, pp. 87–98, 2020.
8. A. Al-Shwaiter, H. Awang, and M. A. Khalaf, "*The influence of superplasticiser on mechanical, transport and microstructure properties of foam concrete*," J. King Saud Univ. Sci., 2021.
9. G. Calis, S. A. Yildizel, S. Erzin, and B. A. Tayeh, "*Evaluation and optimisation of foam concrete containing ground calcium carbonate and glass fibre (experimental and modelling study)*," Case Stud. Constr. Mater, **15**, e00625, 2021.
10. E. Kuzielová, L. Pach, and M. Palou, "*Effect of activated foaming agent on the foam concrete properties*," Constr. Build. Mater., **125**, 998–1004, 2016.
11. A. Mestnikov, S. Semenov, V. Strokova, and V. Nelubova, "*Autoclave foam concrete: Structure and properties*," in AIP Conference Proceedings, 2016, **1698**, no. 1, 70010.
12. H. D. Foth, "*Fundamentals of soil science*," Soil Sci., **125**, no. 4, 272, 1978.
13. ASTM, "*Standard specification for flow table for use in tests of hydraulic cement*," ASTM Int., 1–6, 2014.
14. Standar Nasional Indonesia, "*Concrete Compression Test procedure with Cylindrical Test* (Cara Uji kuat tekan beton dengan benda uji silinder)," 1974.
15. Badan Standarisasi Nasional, "*Portland Cement Mortar Compressive Strength Test Method for Civil Works* (Metode pengujian kekuatan tekan mortar semen Portland untuk pekerjaan sipil)," SNI 03-6825-2002, Indonesia, 2002.

https://doi.org/10.1051/e3sconf/202133106003

Mitigation of traffic disaster by analyzing motorist's movement behavior based on an image captured by a drone

Yosritzal[1]* , *Purnawan*[1], *Merry* Lovita[2], and *Dilum* Dissanayake[3]

[1]Center for Transportation, Civil Engineering Department, Andalas University, Indonesia
[2]Civil Engineering Department, Andalas University, Indonesia
[3]School of Engineering, Newcastle University, UK

Abstract. Uncontrolled motorists' movement behavior could trigger a disaster for road users. A deep understanding of the motorist's movement behavior might improve our mitigation planning. This paper presents an observation of the movement of motorists from a minor road crossing a major road at an unsignalised intersection. The study's objectives are to estimate the gap and lag of motorcyclists when crossing the intersection, observe the forced gap, and predict the effect of traffic volume on the variation of the gap and lag. The case study was located at the intersection between Ahmad Yani Road as a major road and Bandar Purus Road as a minor road in Padang, Indonesia. A drone was used and placed vertically above the center of the intersection to record the movement. The study found that the average lag is 2.4 seconds at off-peak, and the average gap is 4.3 seconds; at the peak hour are 2.0 seconds and 3.1 seconds, respectively. At peak hours, more than 80% of motorcyclists are committed to the force gap/lag.

1 Introduction

Motorist movements at an intersection play an important role when it comes to traffic safety and related problems. Especially when the intersection is unsignalised, this requires meticulous attention. At unsignalised intersections, the priority sign or give way sign controls vehicle maneuver and movement. However, motorists tend to disobey the sign in some cases due to knowledge limits and personal [1]. As a result, it is expected that at the unsignalised intersection, traffic problems such as congestion and traffic accidents. The disobeying traffic rule will be a disaster for road users and must be mitigated before it grows uncontrollably.

Motorcyclists are the most vulnerable road users, as shown by traffic data from Indonesia [2]. In most traffic accidents, motorcyclists are usually involved, either as a victim or a perpetrator. Fig 1 shows traffic accidents in the first half-year of 2017, where the number of motorcyclists involved in the accident is more than 31,000. In his report [3] found that from 100% of accidents, about 89% of them involved a motorcyclist.

Compared to any city globally, Padang is among the cities with the highest percentage of motorcyclists. The 2016 traffic data shows that more than 75% of the traffic in Padang are motorcyclists [4]. Each month, the number of new motorcycles registered in Padang reaches 600 units [5]. The high number of motorcyclists is also the case at the national level, where 81.3% of the traffic in 2013 was motorcycles. The annual growth rate of motorcycles is about 7.5% [6].

Researchers should have a deep understanding of motorists to be able to mitigate traffic accidents. One of the critical situations is at an unsignalised intersection.

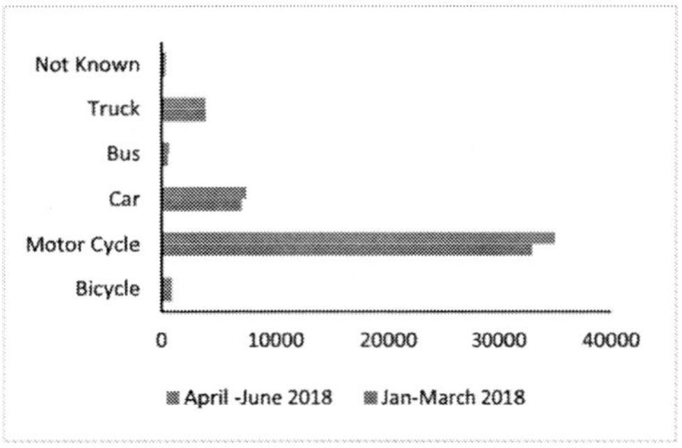

Fig. 1. The occurrence of accidents [2]

The research found that at unsignalised intersections, usually, five seconds is a critical gap for a vehicle to cross [7]. However, a study in Padang shows that the critical gap was pushed to a minimum as motorcyclists from a minor road tend to take a risk to interrupt the straight direction traffic at the major road to allow them to cross [8]. The 2016 study reported that as the gap is forced to a minimum, the straight direction vehicles must push the brake lever to avoid an accident [8].

* Corresponding author: yosritzal@eng.unand.ac.id

As the forced gap became a phenomenon in Padang, a more intensive study was conducted and reported in this paper. The study captured the movement of a motorcyclist from a minor road crossing a major road at an unsignalised intersection.

The study aims to identify the forced gap occurrence at an unsignalised intersection where motorcycles from a minor road cross a major road in Padang. The gap and lag between the passing vehicles and the incoming vehicles were observed to identify the forced gap. The terminology of a gap is defined as the interval of time when a vehicle passes a pointer to the following vehicle passes the same point [9]. Another terminology for the gap is a gap acceptance, defined as the safe gap perceived by a motorist to cross the road. Perhaps, the most relevant terminology for the interval of time for a motorist to cross a low-volume traffic road is lag. Lag is the time difference needed for an approaching vehicle from a major road to reach the crossing point at the intersection and a vehicle from a minor road to the same point. Fig 2 shows the illustration of the gap, and Fig 3 shows the illustration of the lag.

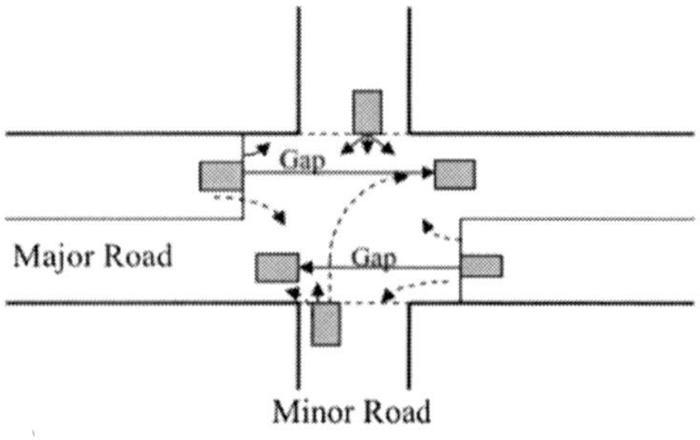

Fig. 2. Illustration of the gap [10]

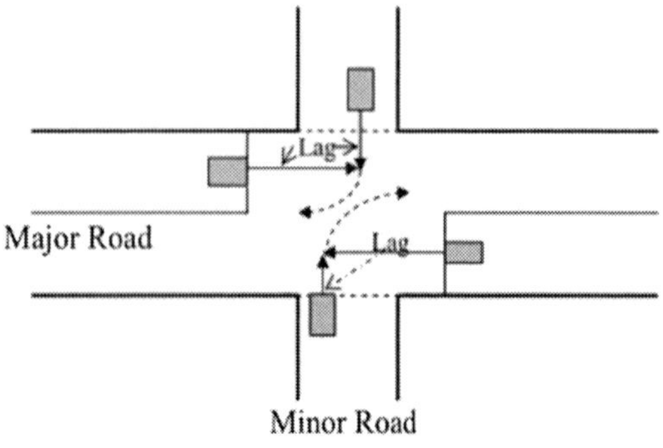

Fig. 3. Illustration of lag [10]

2 Review of Relevant Literature

2.1 Application of drone technology in engineering application and practice

A drone is a small unmanned aerial vehicle (UAV) guided by remote control or computer. Drones have cameras to take photos from different angles. The use of a drone in engineering project has some key benefits such as safer access to sites which is difficult to access by human such as a high place, working over water or steep slopes; faster and cost-effective for collecting data from a large site; and easy to integrate with GIS and CAD software [11]. The Klohn Crippen Berger has used a drone for a slope stabilization project along a river in British Columbia. The drone has also been used to inspect a landslide next to a highway.

Tkac and Mesaros [12] also highlight the use of a drone in various civil engineering fields, especially in the construction industry such as building surveys, topography mapping and land surveys, construction site inspection, equipment tracking and automating, remote monitoring, and progress reports, integration of laser scanning and aerial photogrammetry and thermal imaging recording.The use of a drone in civil engineering has many benefits[12], such as: creating real-time aerial images, overview reveals assets and challenges and also it is allowed planners to meet virtually to discuss project timing, equipment needs and challenges presented by the terrain.

2.2 The use of drone technology in transport planning

The use of drones in transportation engineering has been found in some literature, such as the American Planning Association, which uses drones to support transportation planning by air support [13]. Similarly, [14] explores the possibilities to use a drone in transportation planning, designing, and monitoring. However, the most shared articles discuss drones as air transportation to distribute logistics such as [15] and [16].

3 Methodology

In this study, some data related to the geometric of the studied intersection and traffic were collected. Manual measurement was conducted to take the geometric data, including the lane width, number of lanes, the wide of pedestrian or road shoulder. The traffic-related data was collected from a pre-recorded video of the scene from a drone. Compared to other methods, a drone might be more reliable because it has many advantages, such as allowing the camera to record the scene vertically from above at the desired height, and the recorded video can be played as many times as needed.

The drone used in this study was a quadcopter type, namely DJI Phantom 4. The DJI Phantom 4 has a 180-degree rotatable and high-resolution camera to capture live video or still images. The high capacity battery and GPS make the drone event better as it could record the live video for approximately 20 minutes each time at the exact point. The detail about the DJI Phantom 4 can be found on the DJI website [17], whilst the effective use of a drone in traffic observation was highlighted in [8].

This study was conducted in Padang, West Sumatra, Indonesia. An unsignalised intersection in a commercial area was the object of the study. The major road is A Yani road, and the minor one is Bandar Damar Road with a Bandar Purus Road at the opposite, as shown in Fig 5.

https://doi.org/10.1051/e3sconf/202133106003

Fig. 4. *Typical of* DJI *Phantom 4* [17]

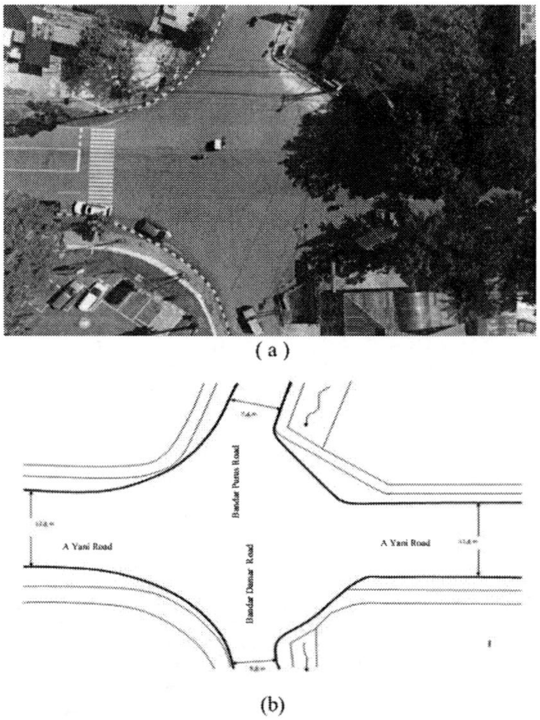

Fig. 5. (a) Image of the location from a drone and (b) two-dimensional model of the intersection.

The intersection is formed by a 12.6 m A Yani Road and a 5.9 m Bandar Damar Road with a 7.4 m Bandar Purus Road at the opposite side of the Bandar Damar Road. At the time of the observation, there was no traffic sign on both sides of the minor road. Usually, at the unsignalised intersection, there is a "Stop" sign or a "Yield" sign. Fig 6 shows the intersection from the Bandar Purus Road heads to Bandar Damar Road.

As the drone only capture the video for approximately 20 minutes, the observation will be conducted three times, the first one was at the peak hours, the second one at the medium traffic volume (medium peak hours), and the third one at the low traffic volume (off-peak hours). To know the interval time of the peak, medium, and low traffic volume, a traffic count survey was conducted to record the traffic volume at the major road from 06.00 to 18.00. Based on the recorded video, a traffic counting was

Fig. 6. Ground image of the intersection

conducted to obtain the hourly traffic volume, as shown in Fig 7. The peak hour was observed between 14.30 – 16.00 with a traffic volume was at least 2500 pcu/h. The medium traffic volume was observed between 09.00-14.00, with traffic volume was between 2000-2500 pcu/h. Traffic volume less than 2000 pcu/h was observed between 06.00-07.30. Based on the traffic variation, the drone observations were conducted two times in the off-peak time interval, two times in the medium peak interval, and two times in the maximum peak time interval.

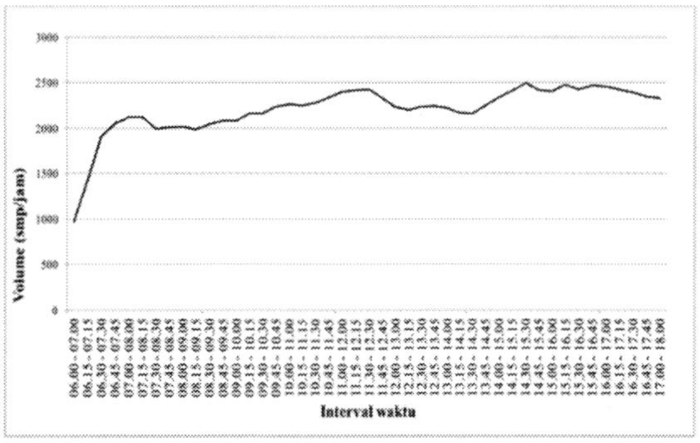

Fig. 7. Observed Traffic Volume

4 Results

4.1 Off-Peak Period Case

Low traffic volume was observed during the off-peak period, where the volume was less than 2000 pcu/h. As the traffic volume was low, and the density was low as reasonably. Considering that the speed and the density are inversely proportional, it is expected that the speed increases with decreasing the density. However, in this study, the vehicle speed was not estimated.

Fig 8(a) shows the lag in seconds at the off-peak period. On average, the lag is about 2.4 s, and the standard deviation is 0.8s. Fig 8(b) shows the gap in second at the off-peak period. The mean gap is 4.3 s, and the standard deviation is 1.1 s. Comparing the lag and the gap, we found that the lag is lower than the gap allegedly caused by the rare vehicle observed. From Fig 8, the forced gap is expected to happen when the lag is minimal. Fig 9

https://doi.org/10.1051/e3sconf/202133106003

shows the moment when the forced gap occurs at the off-peak.

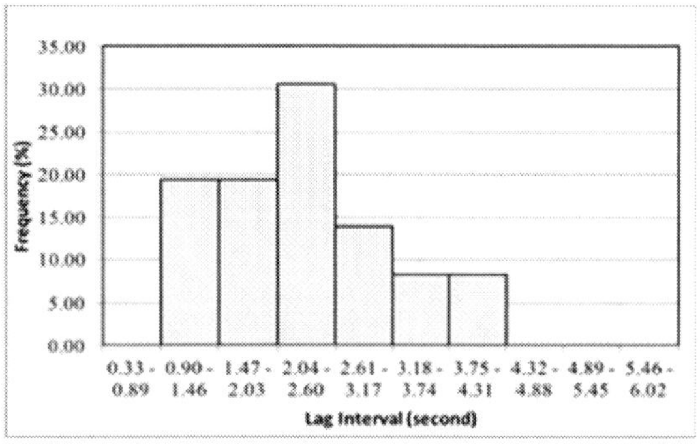

(a)

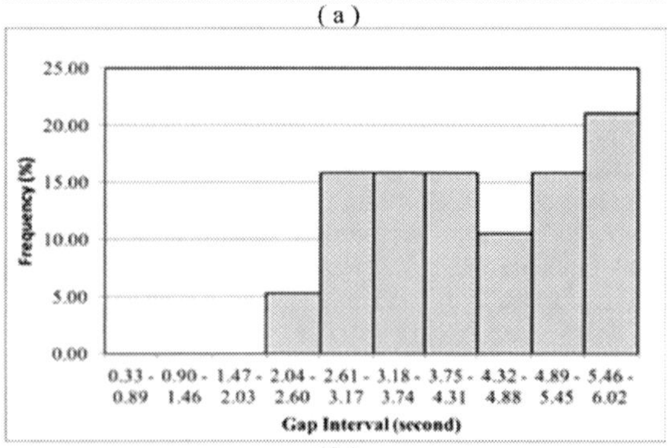

(b)

Fig. 8. The lag and gap at off-peak (a) Lag histogram and (b) Gap histogram

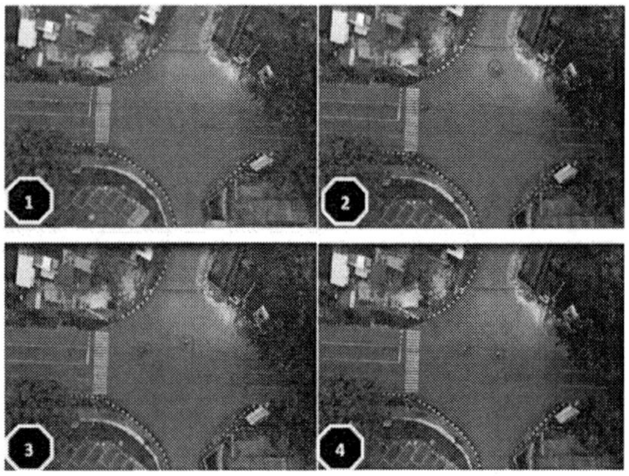

Fig. 9. Occurrence of Forced Gap during Off-Peak Period

4.2 Medium Peak Period Case

Compared to the off-peak period, the traffic volume in the medium peak period is a beat higher in the medium peak, so the density is also higher. With about 2000pcu/h to 2500 pcu/h traffic volume, the mean of lag was recorded at 2.1 s, and the standard deviation was 0.8 s. The mean gap was 3.3 s, and the standard deviation was 3.3 s. Fig 10(a) shows the histogram of lag, and Fig 10(b) shows the histogram of the gap. Fig 11 shows the occurrence of the forced gap at the medium peak period where the motorist from a minor road pushed the gap, which might lead to an accident if the vehicle on the major road did not push the brake or reduce their speed.

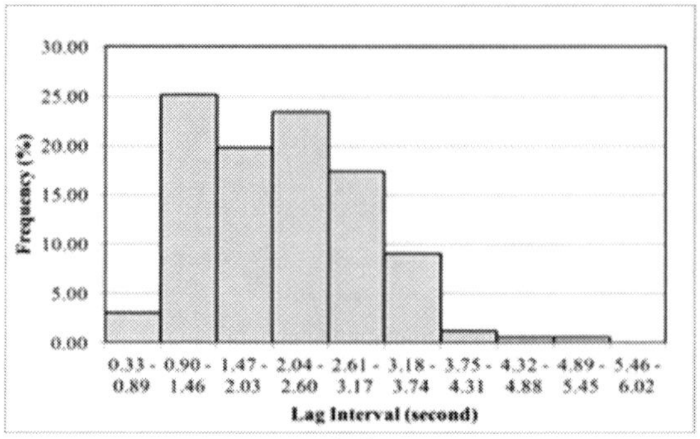

(a)

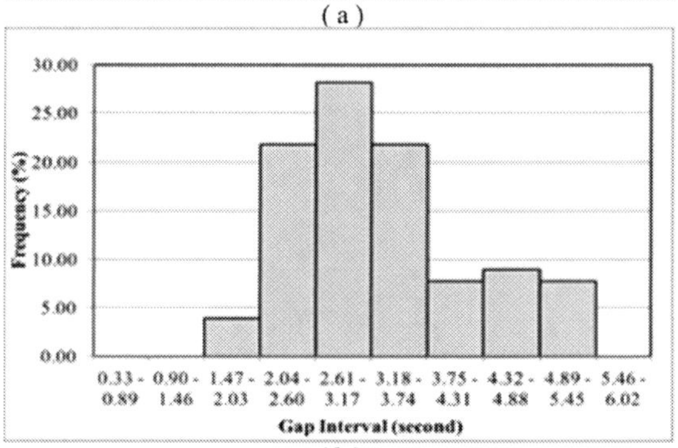

(b)

Fig. 10. The lag and gap at off-peak (a) Lag histogram and (b) Gap histogram

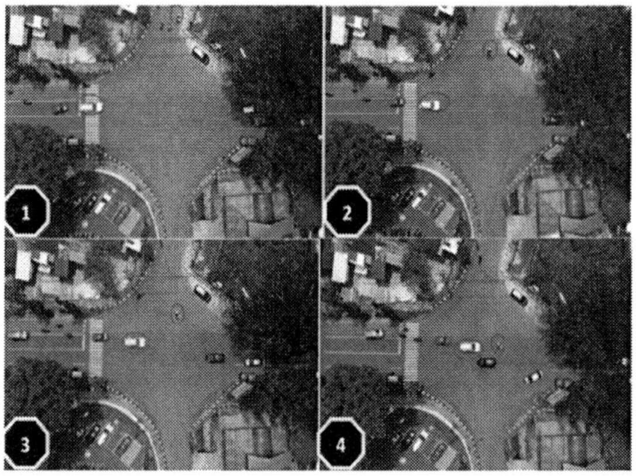

Fig. 11. A brave motorist forces the gap at Medium-peak Period

4.3 Maximum Peak Period Case

During the maximum peak period, with traffic volume increasing, the density also increases, but the speed reduces. The lag and gap also reduce as the speed reduces. The mean of lag was 2 s, and the standard deviation was 0.6 s, whilst the mean lag was 3.1 s, and the standard deviation was 0.7 s, as shown in Fig. 11(a) and (b). It was observed that in the maximum peak period, the lag and gap were shorter, and the motorists risky their life by crossing the road without waiting for a safe lag or gap, as shown in Fig 12.

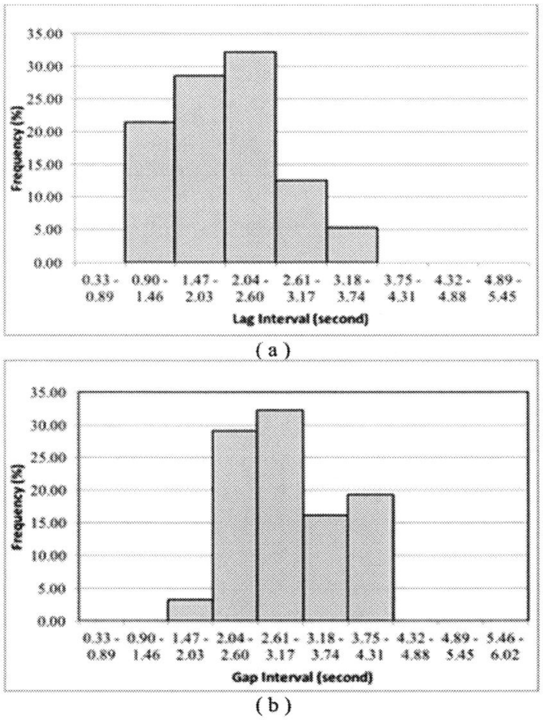

Fig. 11. Lag and Gap observed during Maximum-Peak Period

Fig. 12. Risky movement of a motorcyclist at Maximum-Peak Period

4.4 Discussion

First of all, we would like to highlight that the use of drones in data collection in this study has helped us obtain more accurate data than using a camera recorder at the roadside. The accurate data is very promising as the technology of drones proliferates. When this study was conducted, the drone's battery only lasted for about 20 minutes to keep the drone in the air. It is expected that in the future, the operating time of drone might be improved as the battery last longer.

Secondly, we will discuss the results of the study. Data in sections 4.1 to 4.3 indicated that the lag and the gap were shorter with increased traffic volume. The gap and lag found in this study are shorter than found in other studies on an international basis, such as [18] in Ahmedabad. Furthermore, as the mean of gap acceptance in Padang is 3.4 s [19], we can assume that about 69.44% of motorists were forced the lag and about 5.26% tend to force the gap because the gap and lag are shorter than average. The occurrence of the forced gap by motorists increases with the increasing traffic volume, increasing traffic density, and decreasing the average speed of traffic. Minimum gap and lag occur when the maximum traffic volume indicates that the motorists tend to push the gap and lag to gain access crossing the major road. It is rarely the case for the traffic at the major road to voluntarily access traffic from a minor road to cross at the peak hour. Concern should be raised for traffic safety in this situation. When the forced gap becomes a trend among motorists, it will be a disaster for all traffic users. The life of the motorists relies on the awareness and reaction of other drivers. An accident could be avoided if the other drivers were aware and immediately pushed the brake when a motorist forced the gap. Unfortunately, it is not safe at all as the driver behind might not have a chance to push their brake to hit the vehicle in the front. Another possibility is when the motorist forced the gap, and the other driver has no chance to push the brake, they might swerve to the left or the right with the risk of hitting or being hit by other vehicles.

With safety as our priority, the authority needs to reduce the forced gap and lag at the unsignalised intersection. On the other hand, the interest in crossing vehicles from the minor road is also essential to accommodate. Public advertisements regarding traffic safety should be displayed continuously and improvement of traffic management and engineering at the intersection, and taking action of the law enforcement against those who disobey the safety rule.

5 Conclusions

The results of the motorist's movement observation at an unsignalised intersection in Padang have been presented. An essential result of the study was that motorcyclists tend to push the gap and lag when traffic at the major road is heavy. With an average lag of 2.1 s at the maximum peak period, the motorist's risk involved in an accident is high. A traffic accident is a disaster for the big family of the victim, especially when the victim is the backbone of the family. Therefore, we recommend the authority to reduce the possibility of a motorist pushing the gap by implementing education, engineering, and law enforcement policy.

Furthermore, to the best of the author's knowledge, the application of drone technology to investigate this issue is new and has never been addressed the points in detail before. This study has proven that the use of drones could be beneficial for more accurate data collection. Our future work would be analyzing the data using microsimulation software to increase the accuracy of our prediction.

The authors would like to express their great appreciation to the Civil Engineering Department of Universitas Andalas, who provide funding for the publication of this research with contract number: 134/UN16.09.D/PL/2021.

References

1 W. Santosa, Sukarno, JRS, **3 (1),** 64-76 (2013)
2 Korlantas POLRI, Statistik Laka (Statistic of Traffic Accident) (2018)
3 S.K.A. Singh, I.S. Nasution, L. Hayati, MKS, **47** (2), 105-109 (2015)
4 B.P.S, Number of Vehicle by Type in Padang, (2018)
5 Y. Arham, New Vehicle Growth in Padang Dominated by Motorcycle (2018)
6 B.P.S. Growth of Motor Vehicles by Type, 1949-2016 (2018)
7 A. Suraji. WT, **19** (1), 5-11 (2011)
8 Yosritzal, F. Auliani, *Analysis of forced gap and gap acceptance of motorcylists, in the Proceeding of 3rd ACE National Conference*, Jurusan Teknik Sipil, Universitas Andalas, 22-23 Oktober 2016. Universitas Andalas, Padang, Indonesia (2016)
9 J.L. Gattis, ST Low, JTE, **125**(3) (1998)
10 J.E. Hummer. ITE. Prentice-Hall. inc (1994)
11 Klohn.com, Using Drones in Engineering Projects, (2018)
12 M. Tkac, P. Mesaros, JoCE, **14** (1), 27-37 (2019)
13 N. Veraart, D. Reel, A. McConnell, Air support for transportation planning (2020)
14 D. Cvitanić, *"Drone applications in transportation," in Proceedings of the 5th International Conference on Smart and Sustainable Technologies (SpliTech)*, 1-4 (2020)
15 MARSH, Drones – a view into the future for the logistics sector? (2015)
16 ITF, Ready for Take-off? Integrating drones into the transportation system, International Transport Forum (2021)
17 https://www.dji.com/id/phantom-4-pro-v2/specs
18 AK Maurya, H.J. Amin, A. Kumar, TRP, **17**, 203-212 (2016)
19 RL Vika, Undergraduate Thesis, Andalas University (2013)

E3S Web of Conferences **331**, 06004 (2021) https://doi.org/10.1051/e3sconf/202133106004
ICDMM 2021

Collaborative determination of flood shelter and evacuation route in a flooded area in Rumbai Subdistrict, Pekanbaru

Bambang Sujatmoko[1*], *Rangga* Fernando[1], and *Andy* Hendri[1]

[1]Civil Engineering Department, University of Riau, 28293 Pekanbaru, Indonesia

Abstract. Floods in Pekanbaru City have often hit the region along the Siak river, including the Rumbai subdistrict. Disasters such as floods have detrimental impacts on society, including a massive loss of lives. However, several strategies can minimize the impacts of flooding, including making a plan evacuation route mapping with Geographic Information Systems (GIS). This is a planning-based analysis of data using the algorithm djikstra for result pathways for efficient and effective evacuation. The evacuation routes involve seven simulation parameter modeling, specifically flood, length path, wide roads, road conditions, road materials, presence or absence of bridges, and the road's direction. These parameters are processed using algoritma djikstra to generate the appropriate evacuation routes based on study area conditions. The analysis focuses on one evacuation route in Palas and the other six in Sri Meranti Village. The routes in Palas Village lead to the evacuation place of the Al-Jihad Mosque, while those in Sri Meranti Village heads to Al-Ikhlas Mosque, MDA Aula Rumbai, Nurul Haq Mosque, M Nurul Mosque, vacant land, and Stadium Parking Area.

1 Introduction

The high rainfall in the watershed upstream of Siak and Pekanbaru has caused floods in some areas within Pekanbaru city. According to [1], Siak riverbanks traverse some of the towns vulnerable to flooding, including subdistrict Rumbai, Rumbai Pesisir, Payung Sekaki, Senapelan, Limapuluh Kota, and Tenayan Raya, as well as the southern section in the subdistrict Tampan, Marpoyan Damai, and Bukit Raya.

The flood caused harm, such as damaging the various types of structures, including bridges, cars, buildings, sewer systems, roads, and canals, leading to significant material losses. The communities likely to be affected should be moved to safe areas. The absence of a flood evacuation line route is attributed to the communities' ignorance. There is a need to design or plan before the onset of catastrophic flooding to minimize the possible losses to anticipate these problems. A study of the area often hit by floods is necessary to minimize the impacts. This involves examining how long and wide the flood-affected area is to draw the evacuation path planning appropriately, effectively, and efficiently. The preparation of a map of the evacuation line in a flooded area at district Rumbai using GIS application is essential for the Government to take the right policy in handling flooding. This study examines the road network as an alternative to flood evacuation path and specifies the area ideal for evacuation (Flood Shelter). This research is beneficial in several ways, such as suggesting alternative evacuation routes and an effective evacuation place for affected residents. The results can be used to anticipate or minimize casualties in the event of a disaster. The study also provides information about the path that can be traversed.

* Corresponding author: b.sujatmoko@eng.unri.ac.id

2 Literature Review

2.1 Disaster Mitigation

Mitigation measures are structural and non-structural efforts taken to limit the adverse impacts of natural hazards, environmental degradation, and technology dangers. Structural mitigation includes infrastructure development Sabo, levee, or early warning detection tool. Contrastingly, non-structural include training and capacity building in the Community [2].

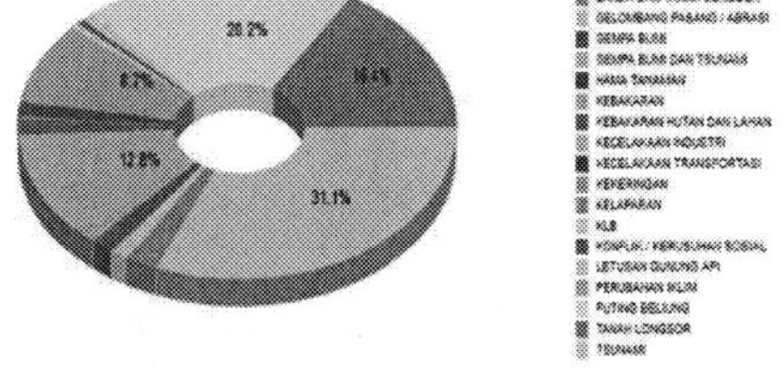

Fig.1. Amount of Disaster.

https://doi.org/10.1051/e3sconf/202133106004

Disaster mitigation refers to efforts to reduce the impact of a disaster before its occurrence, including preparedness and long-term actions to reduce risk in the affected areas. Mitigation involves several activities, such as creating disaster-prone area maps, putting up signs warning of the disaster-prone region's dangers, setting up a temporary shelter in evacuation routes, and moving the community to Flood

According to data from the national disaster mitigation agency (BNPB), 31.1% of first ranking floods have happened in Indonesia, as shown in Fig.1. The western part of the country is often affected by floods because of high rainfall compared to the central and eastern parts [3].

2.2 Modeling Simulation of River Flood

The depth of the flood simulation is retrieved based on the region's morphology in the form of contour data of DEM (Digital Elevation Model), the spatial resolution of 20 m contours. This is data in the form of imaginary lines on the ground connecting points with equal elevation or contour lines. In general, continuous are lines on the top of a map showing areas with the same height. Mapping the spread of Flood inundation in Rumbai is made through iteration using the nearest neighborhood operation. Iteration is a type of folder calculation and involves repeated mathematical operations in a row using a calculation result as input to subsequent calculations. The calculation is carried out through pixel per pixel.

2.3 Determination of the Flood Evacuation Place

Determination of Evacuation place is one of the efforts made to ensure the safety of the affected communities ([4], [5]). For the evacuation process to run smoothly, there is a need for support, including community involvement. According to Mei et al. (1995), the evacuation process can be complex when the affected communities cannot perform the evacuation. In the planning of evacuation places (flood shelters), initial steps undertaken include identifying the location of the existing conditions in the form field when the flood is not interrupted. According to research [5], flood shelter has some basic characteristics, including being flood free, Washing amenities and Latrines (PUBLIC) should match the number of refugees, the height of the building evacuation of the higher building evacuation is getting an excellent place for evacuation, spacious building to accommodate the capacity of refugees, and accessibility from the gathering point to the evacuation place.

2.4 Determination of The Evacuation Line

Evacuation is a route path that leads the community flood victims to safer places. A disaster evacuation line is a form of fundamental information needed in disaster-prone areas, such as floods. Evacuations need Government and public preparedness in flood-prone areas as listed in Act No. 24 of disaster relief [3].

Some of the principles for determining flood evacuation [5] include (1) evacuation paths designed away from the flow of the river, (2) recommended evacuation Lines do not cross the River/Bridge, (3) In the more densely populated areas, evacuation lines designed in the form of a block system limited by the flow of the River, where the movement blocks are not mixed to avoid congestion, and (4) each line of the evacuation need signs for refugees heading to a safe place.

2.5 Geographic Information Systems (GIS)

GIS can be interpreted literally as a component of hardware, software, geographical data, and human resources treated to manage data and display an information system. Understanding managing information involves taking, saving, repairing, renewing, manipulating, integrating, and analyzing data.

GIS is essential in handling and data analysis based on geographical location. It works based on integrating 4 components, including hardware, software, data management, and the user. A space scope process was conducted on GIS [6]. Spatial analysis function based on a GIS capability can also be identified from reviewing the functions that can be performed.

2.6 Djikstra's Algorithm Method

Implementing the algorithm invented by Dijkstra's shortest path is complex and challenging, hence the need to look for a more efficient one. Suppose G is a directional graph with labeled points V (G) = {v1, v2, ..., vn}, and the shortest path is searched from v1 to vn, Dijkstra's algorithm starts from the point v1. In the iteration, the algorithm will look for a one-point number weighing from point 1 to the smallest. Graf is defined as the set of pairs (V, E) written with the notation G = (V, E), which in this case is the set V of vertices, not empty nodes (vertices or nodes), while E is the set of slides (edges or arcs) connecting a pair of vertices [11].

3 Methodology

3.1 Data Collection

The first step of conducting this research is gathering data, which was obtained from (1) Administrative Data and colonization of Pekanbaru published statistics Indonesia (BPS) Pekanbaru [7], (2) Map Administration and other support maps accrued from the Spatial Plan of the city of Pekanbaru, 2014-2034 [8], (3) Data DEM obtained from Bappeda Riau Province [9], evacuation of building area Data obtained from SIMAS (Mosque information system) which can be obtained on the website https://www simas.kemenag.go.id.

3.2 Flood Modeling Simulation of Siak River

Analysis of flood simulation modeling is conducted using Quantum GIS application 2.14.1 ESSEN (*Q-GIS*) using the iteration methods (see in Fig.2).

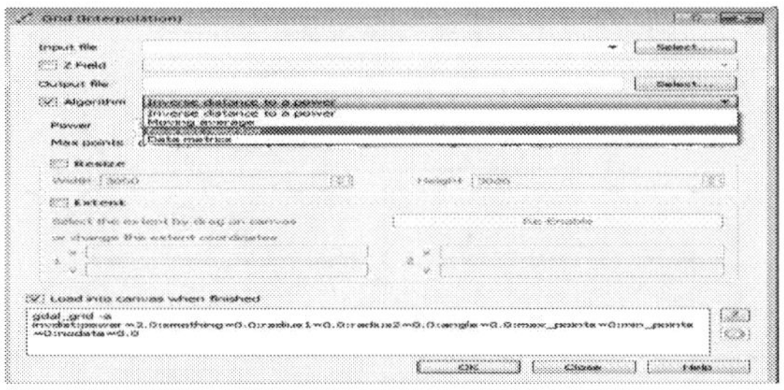

Fig.2. Display model of Q-GIS

Model simulation for the study prepared based on the maximum Flood inundation height results from the high index insecurity. In this regard, the inundation that occurred at the lowest altitude simulates a region on the DEM map. Fig.3 shows how to get the lowest height from the data map DEM contour intervals.

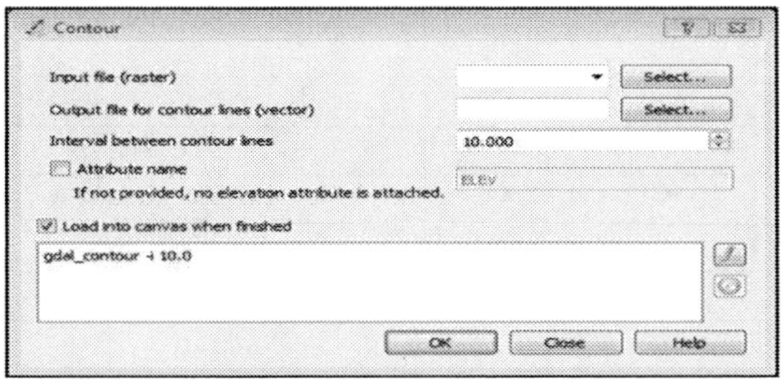

Fig.3. Display of Extract DEM of Q-GIS

3.3 Determination of The Flood Evacuation Place

The evacuation place capacity of the space required is determined based on a refugee, specifically 1 m^2 (BAPPENAS). An evacuation place is chosen through a field survey by reviewing directly buildings that can be used. The figure shows Q-GI for bringing up public facilities using layer labeling settings (see in Fig.4).

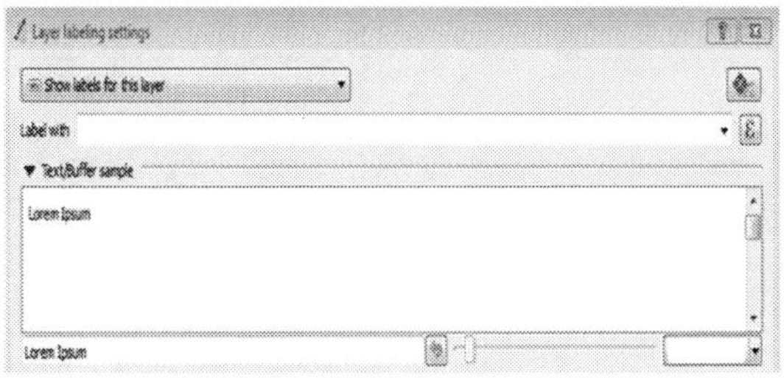

Fig.4. Display layer labeling setting Q-GIS.

3.4 Determination of The Evacuation Line

Analysis of road condition involves a survey conducted on a Subdistrict Rumbai with 6 (six) parameters, including long roads, flood simulation, the width of the road, road conditions, road surface, bridge location, and the direction of the road, as shown in Table 1.

All of the data parameters are given a score based on Table 1 before running Djikstra's algorithm method. Once all the road data network gives a score, each parameter is given different weights depending on their influence on the safety evacuation lines and speed, as shown in Table 2.

Table 1. Parameters of Evacuation Paths [5]

No.	Classification	Class	Score
1	Road Length (m)	<100	20
		>=100 - <200	30
		>=200 - <300	40
		>=300 - <400	50
		>=400	60
2	Flood Simulation	Dangerous	10000
		Save	20
3	Road Width (m)	< 2	100
		> 2 - < 5	50
		>= 5 - < 7	40
		>= 7 - <12	30
		>= 12	20
4	Road Condition	Good	20
		Enough	30
		Bad	40
5	Surface Road Material	Asphalt	20
		Conblock	30
		Cement	40
		Rock	60
		Soil	80
6	Bridge Location	No Bridge	20
		Bridge	30
7	Road Direction	Two Way dir	20
		One Way dir	30

Table 2. The Evacuation Line Parameter Weights [5]

Parameter of Evacuation Routes	Weights (%)
Flood Simulation	30
Road Length	25
Road Width	15
Road Condition	10
Surface Road Material	9
Bridge Location	7
Road Direction	4
Total	100

4 Result And Discussion

4.1 Flood Simulation Modelling of Siak River

Modeling flood inundation simulation using the DEM map involves interpolating the data, changing them into a raster map. Fig.5 shows the values of the pixels of the height data.

The flooded overflow is the scenario modeling on the maximum inundation elevation with insecurity index 0.5 m to 1.0 m. The Genesis flood maximum is at the height of 1.0 meters. The existence of the modeling of the flood area is expected to help in the process of evacuation.

Based on Table 3, the flood inundation scenario's height is greater at the low elevation area, meaning the river's area is quite flat. The biggest impact is found in scenario 1 m, covering an area of 6,777,826.16 m^2. In comparison, the smallest impact is on 0.5 m scenarios, with a broad impact of 4,234,691.31 m^2. Fig.6 shows details about the results of the simulated flood overflow with various scenarios.

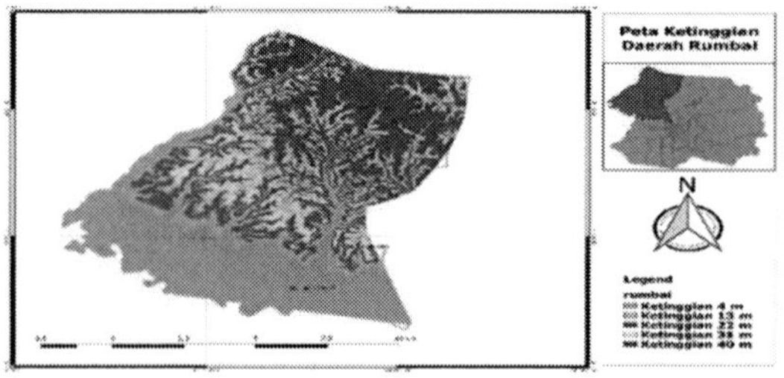

Fig.5. Raster DEM Map of Rumbai District.

Table 3. Impact of Widespread flooding from a Simulation Model

No.	The scenario of Flood Height (m)	Impact of widespread flooding (m^2)	Prosentase (%)
1	0.5	4,234,691.31	38,45
2	1.0	6,777,826.16	61,54
	Total	11,012,517.47	100%

Fig.6. Map of the Inundation Simulation.

4.2 Analysis of the Flood Evacuation Place

The main purpose of evacuation is to move people from dangerous areas. An evacuation place selected should be a location safe from flooding. In general, a secure location for an evacuation should have elevation free of flooding, specifically permanent and non-permanent buildings. Permanent buildings can be in public facilities settled on the Tassel subdistrict, mosques, schools, and other public buildings. In comparison, non-permanent involve an empty land large enough to erect a building or evacuation tents. Table 4 shows the characteristics of a place that can be used as an evacuation place.

Table 4. Characteristics of Evacuation Place

No	Building Type	Location/ Village	Amount of MCK	Building Area (m^2)	Area (m^2)	Amount of Floor	Accessibility
(1)	(2)	(3)	(4)	(5)	(6)	(7)	(8)
1	Al-Ikhlas mosque	Sri Meranti	4	432	240	2	Good
2	Aula MDA	Sri Meranti	2	145	400	1	Good
3	Nurul Haq mosque	Umban Sari	4	400	600	1	Good
4	Nurul Ukh mosque	Umban Sari	4	400	2,100	1	Bad
5	Al-Jihad mosque	Palas	4	400	500	1	Good
6	Vacant Land	Umban Sari	-	-	650	-	Good
7	Stad Parking Area	Palas	-	-	6,000	-	Good

4.3 Analysis of Evacuation Path

Table 5. The evacuation Line Route Division

Route of Evacuation	Collecting Point	Final Point	The building of Evacuation Place
1	A	D	Al-Ikhlas Msq
2	A	F	Aula MDA Rbi
3	B	G	Empty Field
4	B	E	Nurul Haq Msq
5	B	H	M Nurul Mosque
6	C	I	Al-Jihad Mosque
7	B	J	Stad Parking area

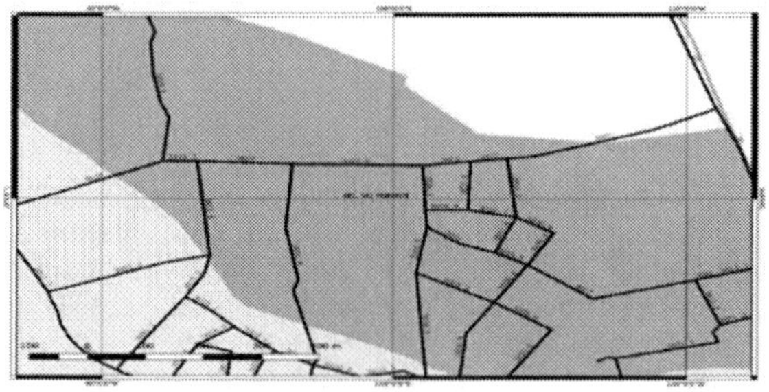

Fig.7. Weight of Road Segment

 https://doi.org/10.1051/e3sconf/202133106004

Band selection is based on evacuation scoring results using the shortest path search method (Dijkstra's Algorithm). This method will essentially choose the lowest weight value and consider a score of weights for each parameter. Fig.7 shows the results of the weighting of each road segment based on the value set.

Table 5 shows the results of the analysis of evacuation line 7 with 3 points and 7 locations.

4.3.1 Evacuation Line route 1

A is the starting point located in the village of Sri, the get-together point for the evacuation line route 1. Al-Ikhlas mosque is located at Yos Sudarso Street, the endpoint with the given symbol d. Each knot intersection or segment was given a symbol that denotes the node vertex of any route with a value or weights determined, as shown in Fig.8.

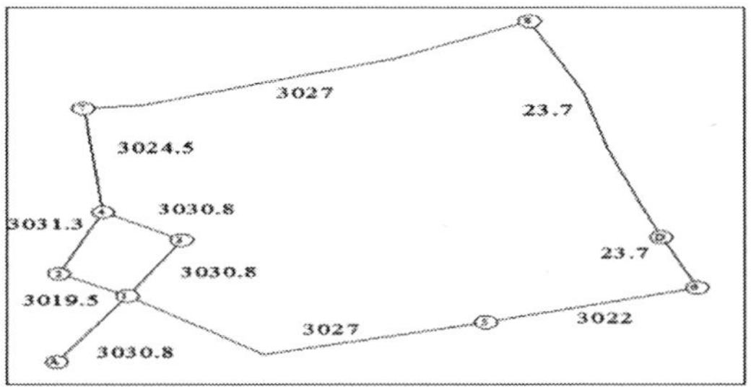

Fig.8. The determination of the Route 1

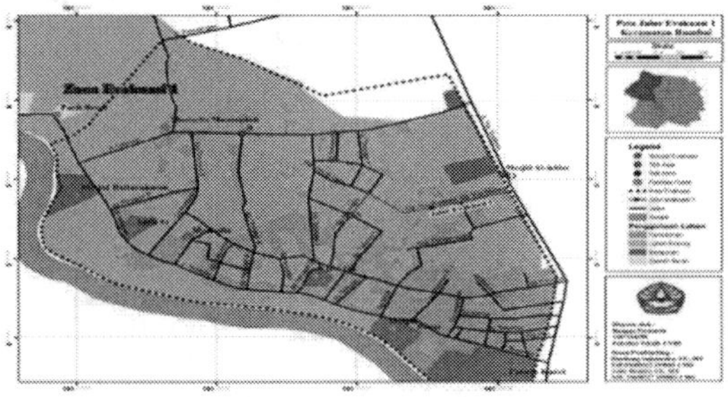

Fig.9. Evacuation Line route map 1

Table 6. The Lowest Weighting of the Route 1

From	To	Shortest Route Path	Cost
	1	A-1	3030.8
A	2	A-1-2	6050.3
A	5	A-1-5	6057.8
A	3	A-1-3	6061.6
A	4	A-1-2-4	9081.6
A	D	A-1-5-6-D	9103.5
A	7	A-1-2-4-7	12106.1

Fig.8 shows each node route is analyzed through the algorithm djikstra. Table 6 shows the shortest route search using the algorithm djikstra method. Also, the table shows the determination of the evacuation line based on the smallest weights. The line selected evacuation routes for A starting point towards endpoint D is obtained through a combination of 9103.5 with a distance of 680 meters, passing through A routing node → 1 → 5 → 6 → D (Gg. Rahmat 6 → Gg. Yos Sudarso → Yos Sudarso Street) as shown in Fig.9.

4.3.2 Evacuation Line route 2

A is the starting point located in the village of Sri. Part of it is the point of gathering for evacuation line route 2 to MDA Hall. It is located in the path of struggle, the endpoint with symbol f. Each knot intersection or segment was given a symbol that denotes the node verteks to any route with a value or weights determined, as shown in Fig.10.

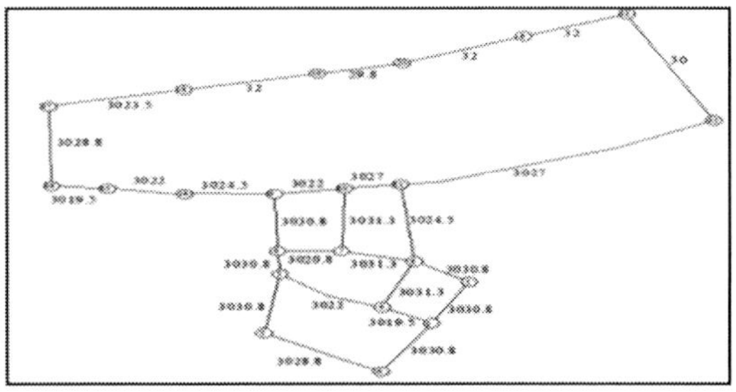

Fig.10. The determination of the Route 2

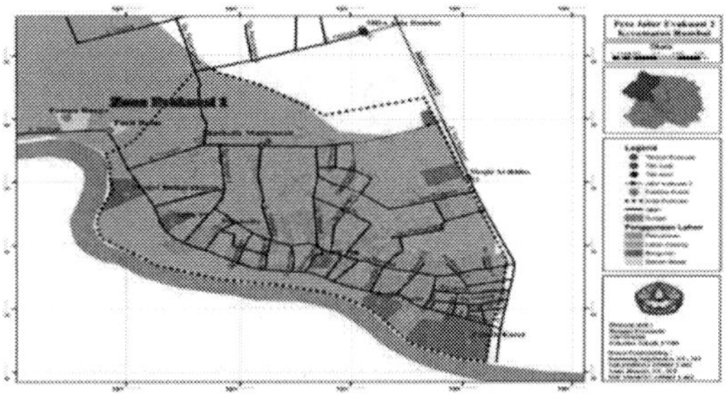

Fig.11. Evacuation Line route map 2.

According to Fig.10, each node route will be analyzed through the algorithm djikstra method. Table 7 shows the shortest route search results using this method and the determination of the evacuation line based on the smallest weights. The evacuation routes for A starting point towards the endpoint of the F line combinations are obtained with 15195.1 end weights and a distance of 1459 meters. It passes through A routing node → 1 → 4 → 6 → 11 → 12 → 13 → F (Gg. Grace 6 → Gg. Bus 1 → Bus St. → Yos Sudarso St. → Perjuangan St.), as shown in Fig.11.

Table 7. The Lowest Weighting of Route 2

From	To	Shortest Route Path	Cost
A	2	A-2	3028.8
A	1	A-1	3030.8
A	6	A-1-4-6	9081.6
A	11	A-1-4-6-11	12106.1
A	13	A-1-4-6-11-12-13	15163.1
A	F	A-1-4-6-11-12-13-F	15195.1
A	11	A-2-3-8-7-10-11	18169.5

4.3.3 Evacuation Line route 3

At the initial point B, located in the village of Sri, part of the gathering point for evacuation line Route 3, where on the goal line to evacuation is empty land, marked by an endpoint with symbol g. Each knot intersection or segment was given a symbol that denotes the node verteks to any route with a predetermined value or weights, as shown in Fig. 12.

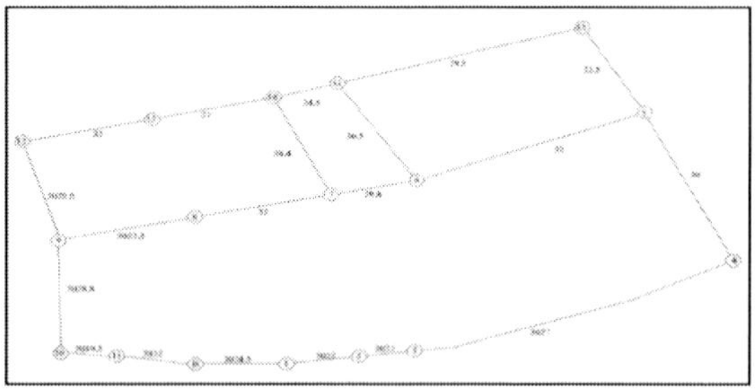

Fig.12. The determination of Route 3.

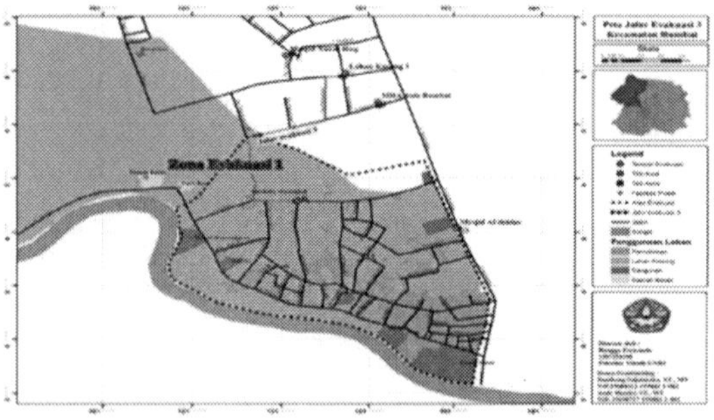

Fig.13. Evacuation Line route map 3

Table 8. The Lowest Weighting of Route 3

From	To	Shortest Route Path	Cost
B	11	B-11	3022.0
B	1	B-1	3024.5
B	9	B-11-10-9	9070.3
B	3	B-1-2-3	9073.5
B	8	B-11-10-9-8	12093.8
B	14	B-11-10-9-8-7-14	12150.3
B	G	B-11-10-9-8-7-14-G	12174.8
B	6	B-11-10-9-8-7-6	12155.6
B	G	B-1-2-3-4-5-15-G	12182.5
B	G	B-11-10-9-8-7-6-G	12186.1

In Fig.12, each node route is analyzed by the algorithm djikstra method. Table 8 shows the shortest route search obtained using this method and the determination of the evacuation line based on the smallest weights. The evacuation line starting from point B towards the endpoint G was obtained by a combination of lines with weights of the end 12174.8 and a distance of 1383 meters, passing through the nodes route B → 11 → 10 → 9 → 8 → 7 → 14 → G (Bus St. → Perjuangan St. → Gg. Struggle 2 → Utama St.) as shown in Fig.13.

4.3.4 Evacuation Line route 4

The initial point B, located in the village of Sri, is the get-together point for the evacuation line route 4. It is on the goal line to the evacuation mosque Nurul Haq located in Gg Utama, the endpoint with symbol E. Each knot intersection or segment was given a symbol that denotes the node verteks any routes with a value or weights determined as shown in Fig.14.

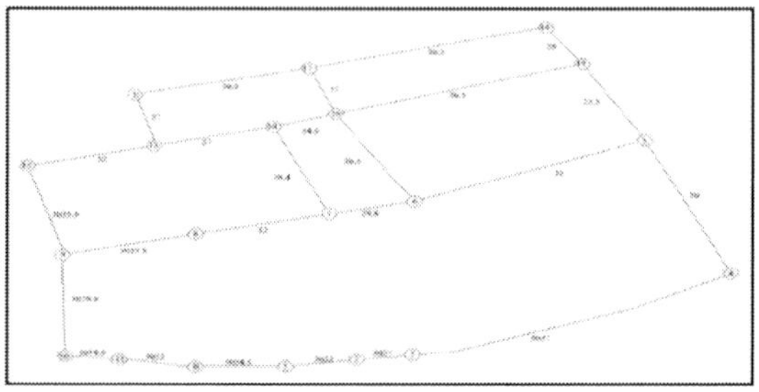

Fig.14. The determination of the Route 4

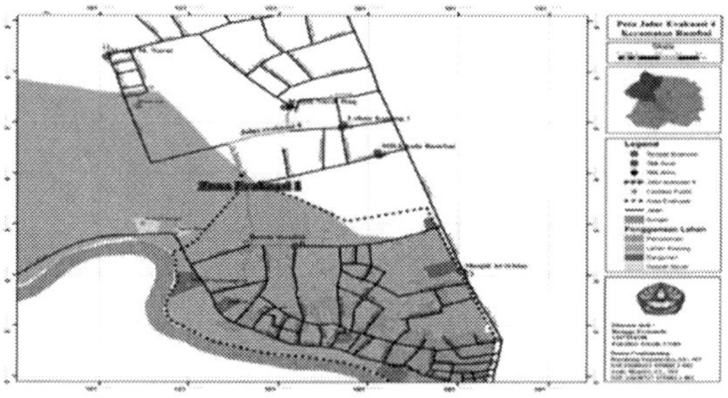

Fig.15. Evacuation Line route map 4

Table 9. The Lowest Weighting of Rute 4

From	To	Shortest Route Path	Cost
B	11	B-11	3022.0
B	1	B-1	3024.5
B	2	B-1-2	6046.5
B	9	B-11-10-9	9070.3
B	3	B-1-2-3	9073.5
B	14	B-11-10-9-8-7-14	12150.3
B	15	B-1-2-3-4-5-15	12153.0
B	E	B-11-10-9-12-13-E	12154.8
B	18	B-1-2-3-4-5-15-18	12178.0
B	6	B-11-10-9-8-7-6	12155.6

Fig.14 shows each node route is analyzed through the algorithm djikstra method. Table 9 shows the shortest route search results using this method and the determination of the evacuation line based on the smallest weights. The evacuation line selected starts from point B towards endpoint E. It is obtained through a combination line with 12154.8 end weight and a distance of 1378 meters, passing through the nodes route B → 11 → 10 → 9 → 12 → 13 → E (Bus St. → Perjuangan St. → Gg. Struggle 4 → Utama St. → Gg. 2), as shown in Fig.15.

4.3.5 Evacuation Line route 5

The initial point B, located in the village of Sri, is the get-together point for the evacuation line route 5 to M. Nurul mosque is located in Tegal Sari, which is the endpoint with the symbol h. Each knot intersection or segment was given a symbol that denotes the node verteks of any route with a value or weights determined. Fig.16 shows more details.

According to Fig.16, each node route is analyzed through the algorithm djikstra method. Table 10 shows the shortest route search results using this method and the determination of the evacuation line based on the smallest weights. The evacuation line selected starts at point B towards the endpoint H. It is obtained with the end of the line combination of 12348.2 with 1352 meters, where it passes through the nodes route B → 11 → 10 → 9 → 12 → 13 → 7 → 8 → 9 → 10 → 11 → 24 → 25 → H (Bus St. → Perjuangan St. → Gg. Struggle 4 → Utama St. → Gg. 2 → the way Achieve End → Tegal Sari), as shown in Fig. 17.

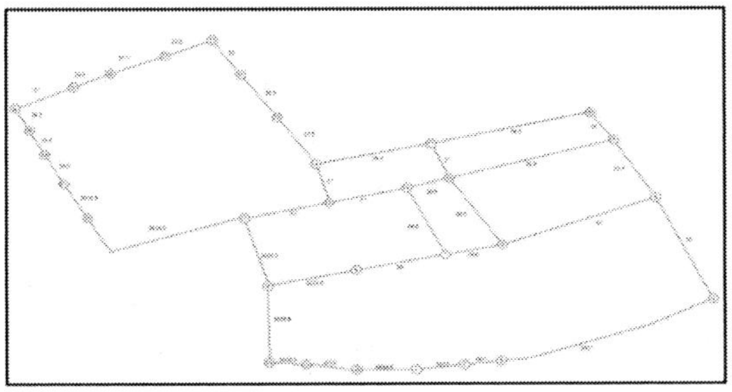

Fig.16. The determination of the Route 5

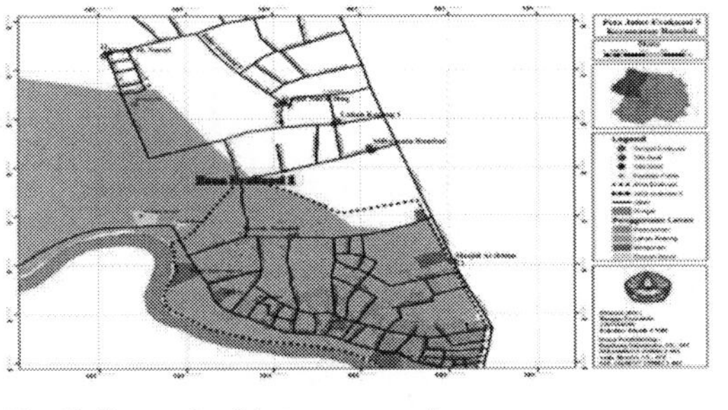

Fig.17. Evacuation Line route map 5

Table 10. The Lowest Weighting of Rute 5

From	To	Shortest Route Path	Cost
B	11	B-11	3022.0
B	1	B-1	3024.5
B	10	B-11-10	6041.5
B	9	B-11-10-9	9070.3
B	6	B-11-10-9-8-7-6	12155.6
B	14	B-11-10-9-8-7-14	12164.2
B	22	B-11-10-9-12-13-19-20-21-22	12243.8
B	25	B-11-10-9-12-13-19-20-21-22-23-24-25	12321.2
B	H	B-11-10-9-12-13-19-20-21-24-25-H	12348.2
B	26	B-11-10-9-12-26	15124.3

4.3.6 Evacuation Line route 6

The initial point C is located at village Sri Meranti, the gathering point for evacuation lane route 6 at the goal line to Al-Jihad mosque. It is located on the Siak II, the endpoint with symbol i. Each knot intersection or segment was given a symbol that denotes the node verteks to any route with a value or weights determined, as shown in Fig.18.

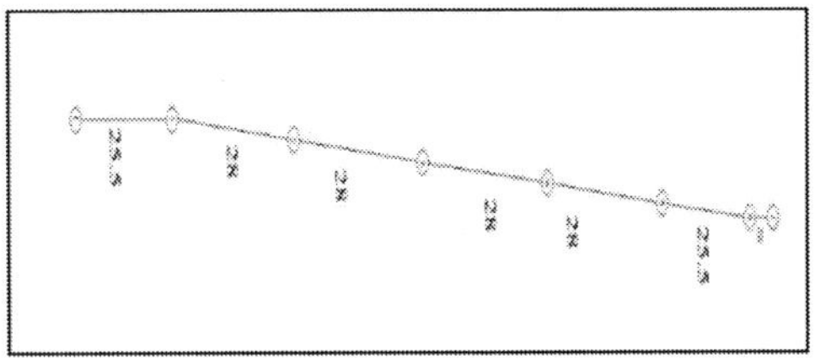

Fig.18. The determination of the Route 6

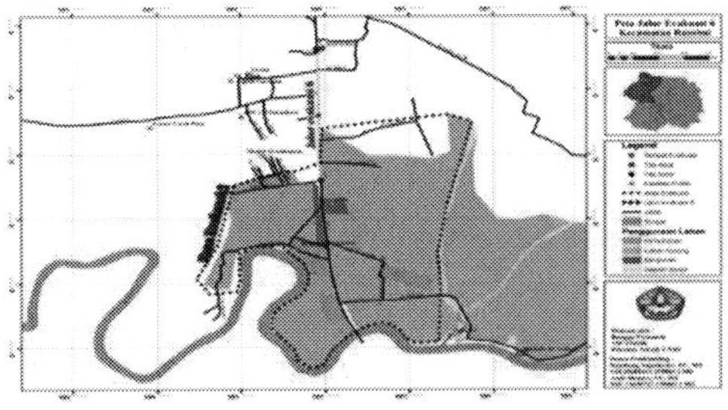

Fig.19. Evacuation Line route map 6

Fig.18 shows the evacuation line's determination directly specified because the path to the place of evacuation is not parallel. The evacuation line 6 passes through the Siak II with a distance of 1352 m, as shown in Fig.19.

4.3.7 Evacuation Line route 7

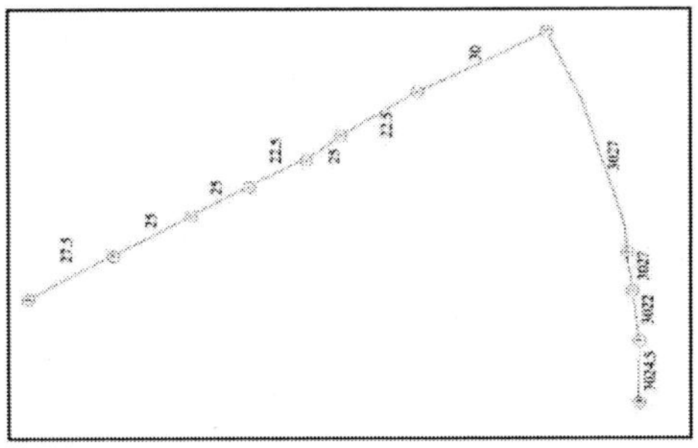

Fig. 20. The determination of the Route 7

The initial point B is located in the village of Sri, part of the gathering point for evacuation line Route 7 to the Stadium parking area. The site of evacuation tents is located at Yos Sudarso Street, the endpoint with symbol j. Evacuation Line 7 is directly determined through the Bus heading towards Yos Sudarso Street. This is because the evacuation path at point B towards the parking area

 https://doi.org/10.1051/e3sconf/202133106004

has only two road routes. One of the routes has traversed two evacuation routes to avoid the build-up of mass on the line. It passes the 7 evacuation lines leading to the Yos Sudarso because the road is wide enough. Figure 21 shows the description of the evacuation line route 7. Each node or segment intersection is given a symbol, which denotes any node verteks route with weights determined, as shown in Fig.20.

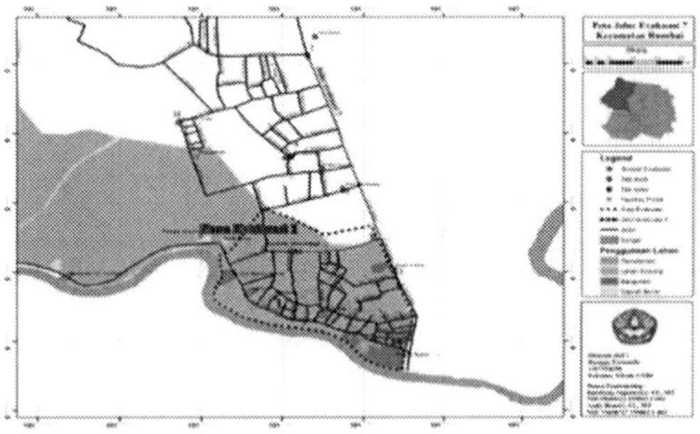

Fig. 21. Evacuation Line route map 7

5 Conclusion

Based on the analysis of the simulation of the flood, the sub-Tassel is extensively affected. The impact simulation on flood includes heights of 1, specifically with an area of 677.78 ha. The vast height was 0.5 meters (half) 423.47 ha.

A total of 7 locations can be used as a place of evacuation with 5 permanent buildings and 2 (two) land locations for permanent and non-permanent building evacuation.

References

1. N D Suprayogi, Mapping of Inner City, Flood Vulnerable Areas, 257-262 (2015)
2. T. Afrizal, Increased Risk of Flood Disaster Due to overflow of Kali Kemuning in Sampang Regency Thesis Undergraduate Program (2013).
3. BNPB Guidelines for Preparation of Disaster Management Plans (Jakarta: National Disaster Management Agency) (2016)
4. Mei and Estuningtyas W Lessons Learned From The 2010 Evacuations At Merapi Volcano J. Volcanology And Geothermal Research, Volgeo-05104, 18 (2013)
5. S. Harsini, Geographical Information System Application (Solo: UNS) 1-15 (2014)
6. P. Eddy, Geographic Information System Basic Concepts (Geodesy and Geomatics Perspective) (Bandung: Informatics) (2009)
7. Central Bureau of Statistics General Condition of Pekanbaru City (Pekanbaru: BPS) (2003)
8. Pekanbaru City Spatial Planning Revised General Plan (Pekanbaru: Riswandi) (2004)
9. Bappeda Riau Province Area of Pekanbaru City by District (Pekanbaru: Riswandi) (2012)
10. BNPB Regulation of the Head of the National Disaster Management Agency (BNPB) No. 02 of 2012 (Jakarta: BNPB Office) (2012)
11. F. Ardini, Detrmination of the Shortest Distance and Alternative Shortest Distance Using the Dijkstra Algorithm and Travel Time Estimation of the National Seminar on Applied Information & Communication Technology (2021)

Assessment of tsunami evacuation road performance

Yossyafra [1*], *Nurhuda* Fitri[1], *Rahmat* Purnama Sidhi[1], and *Marlinda*[1]

[1]Civil Engineering Department, Faculty of Engineering, Andalas University, Padang, Indonesia, 25163

Abstract. A series of disasters, from the 2004 Aceh Tsunami to the 2009 Padang Earthquake, has changed the paradigm of urban development in coastal areas in Indonesia. District or City Regional Regulations concerning Regional Spatial Planning finally oblige disaster mitigation by constructing evacuation roads. The attraction of the Tsunami Evacuation Road (TES) which was built with a relatively wide size in the city, makes the road a favourite route for motorists. The increased vehicular traffic also triggered the addition of the number of new commercial buildings along the road. The Indonesian Road Capacity Manual 1997 for urban roads is used as a reference in the analysis of the service performance of these road sections. The relatively rapid increase in traffic volume and side-road disturbances have resulted in a relative decline in road performance in a short period. The active role of the government in minimizing body side disturbances, by maximizing the implementation of traffic management and control of land-use changes are options.

1 Introduction

Since a series of disasters from the Aceh Tsunami disaster in 2004 to the West Sumatra earthquake in 2009, it has changed the paradigm of urban infrastructure development in many cities on the west coast of Sumatra, Indonesia. Municipal / Regency Regional Regulations on Regional Spatial Planning have made it mandatory to mitigate the Tsunami disaster, by constructing several evacuation roads in the city. The attractiveness of the Tsunami disaster evacuation road, which was built wide in the city, makes it the favourite route choice for motorists and drivers in fast-growing traffic activities. And in turn, triggered a sporadic increase in the number of houses and commercial buildings along the road. The Indonesian Road Capacity Manual (IRCM) 1997 in the section of urban roads is used as a reference for analysing road performance indicators.

There are several cities with a large population on the west coast of Sumatra. Many municipal and district governments along the regions understand that there is a high risk of a tsunami disaster. The government continues to make mitigation efforts, through regulation and provision of disaster-responsive and safe urban infrastructure. There are two types of evacuation options, horizontal and vertical evacuation. Vertical evacuation, by providing shelter buildings that are friendly to earthquake and tsunami disasters. And options for horizontal evacuation, by providing TES. The urban road network must be built to provide security and safety guarantees and could accommodate as much as needed when people evacuate. People can go to escape to higher locations, as a relatively safe place.

The understanding of the municipal and regency governments in mitigating the tsunami disaster is shown in the form of public policies (regional regulations), and the realization of the construction of TESs in its urban road network. The form of the policy is stated in the Regional Regulation on Regional Spatial Planning (RTRW) of each of the Municipality and Regency. Traffic growth that occurs in the evacuation route is believed to reduce the performance of the evacuation road, which is marked by frequent traffic jams, the lower travel time.

This article will provide an overview of the performance of the TES. Quantitative and qualitative methods are used in analysing.

2 Road Capacity and Tsunami Evacuation Road

2.1 Road Capacity and Degree of Saturation

Road capacity is the maximum current that passes through a point on the road per unit hour under certain conditions. The IRCM-1997 used four parameters to determine the capacity of the road, namely correction factors for road width, road lane separation, roadside disturbance, and city size [1]. The basic equation for determining capacity on urban roads is as follows:

$$C = C0 \times FCW \times FCSP \times FCSF \times FCCS \quad (1)$$

* Corresponding author: yossyafra@eng.unand.ac.id

https://doi.org/10.1051/e3sconf/202133106005

Where: C = Capacity (emp / hour), C_0 = Basic capacity (emp / hour), F_{CW} = Road width adjustment factor, F_{CSP} = Directional separation adjustment factor (only for undivided roads), F_{CSF} = Adjustment factor of side and shoulder, F_{CCS} = City size adjustment factor.

$$DS = Q / C \quad (2)$$

Where: DS = Degree of Saturation, Q = maximum traffic volume (flow) (emp / hour), C = Capacity (emp / hour).

2.2 Traffic Growth

Traffic growth predictions have been discussed in detail [2], where five types cause traffic growth. Traffic growth is likely to vary from the normal traffic growth; future traffic volumes shall be predicted by taking into account: Normal growth, Diverted Traffic, Generated/ Induced Traffic, Intermittent traffic, and Suppressed traffic. Had discussed that trade integration, and global trade, in general, has increased more than three times faster than global aggregate output .[3]. Trade gives rise to traffic. Traffic growth from the economic performance are as follows: fair cost structures e.g., a full allocation of external costs; improving efficiency using modern logistics; consensual settlement development planning; and support for regional marketing.

2.3 Tsunami Evacuation Road

Many researchers have researched TES, both on how to determine and choose evacuation routes in a city. Many considerations were taken in determining the TES [4]. The Indonesian government issued a National Standard, to determine the TES. TESs are designed practically and simply so that they are easily followed by decision-makers and stakeholders, but they are still based on scientific principles [5].

The Indonesian Standards Agency requires that in the design of TESs several things become practical considerations, namely: TESs are designed through a road alignment from the coastline; should not cross rivers or bridges; to avoid the build-up of refugees, made several parallel evacuation routes that move away from the coast; in densely populated areas, an evacuation route is designed in the form of a block system; for very low and sloping areas, which are far from high places, a safe area system in the form of buildings or artificial hills is made, which is recommended as a temporary vertical evacuation; every evacuation route needs signs to guide refugees; the direction of daily traffic is directed towards the direction of evacuation; the activities in designing the evacuation cycle need to be phased, starting with data collection, studio work, provisional design, field review, final design, printing and outreach [5].

Previous studies, such as those carried out on the Last-mile Evacuation project by the UN and the German government. The project produced a map of the travel time when tsunami evacuations hit the municipality of Padang. The project shows how much time is needed to escape from a location in the tsunami hazard zone to a safe altitude area. The project also predicts the time needed for evacuation for different periods, in the morning, afternoon, and evening [6,7].

3 Method

Many studies have been conducted to review the effectiveness of TESs, including in designing. However, of all the articles and books written about TESs, there is no writing that clearly and thoroughly states what parameters were calculated or used in designing TESs and how to evaluate the performance of existing TESs in various cities.

The main method used in this study is to review existing TESs in urban areas. Evacuation roads that have been determined by the municipality or district government are used as a sample of the assessed road, while the steps taken are:

- Reviewing several textbooks or journal articles on TESs,
- Conduct a road performance assessment by calculating the current volume and capacity of the road,

4 Analysis and Results

4.1 Road Geometric

In 2012, the Padang municipal government approved Regional Regulation number 4 of 2012, concerning the 2010-2030 Padang Municipal Spatial Plan (RTRW). The RTRW has listed six zones of the city road (Fig 1 (a)), as a TES [8]. This regulation becomes a reference in urban space development. In this study, Alai-By-pass Road, Padang was selected. This road is one of the roads that have been determined by Padang Municipality as a TES. This road is very important as a TES because it connects the western area (coastal area) from downtown Padang to the east of the city (the higher area, as a tsunami safe zone).

The technical data for the Alai-By-pass Road are as follows, are urban roads, Primary Arterial roads, road length = 4.1km, lane width (j) = 4 x 3.5 m, shoulder-width (b) = 1 m, width pavement = 14 m, raised median width (m) = 0.3 m, and pedestrian path width (t) = 1 m [9]. There are road signs and road-marks that are posted as signs indicating the road section as a TES (Fig 1. (b)). There are two types of cross-sectional sections of the TES, from the Alai-By-pass Road, one with an elevated median and the other without a road median (Fig 1. (c)).

4.2 Land Used and Building Function

Changes in land use along the TES are reviewed, aerial photographs of 2010, 2013, 2016, and 2019, respectively observed (Fig 2.). The number of buildings for the year of observation is calculated. Visually in the field visible changes in the number of buildings along with the TES of Alai-By-pass Road. Table 1. shows that for almost 10

https://doi.org/10.1051/e3sconf/202133106005

years there has been a significant increase in the number of buildings on both sides of the road (north and south side of the road). During that period, there was more than a hundred percent increase.

(a)

(b)

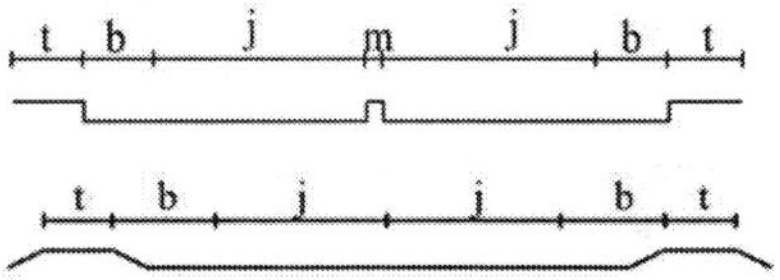

(c)

Fig. 1. (a) Map of six horizontal evacuation route sectors in Padang City [8], (b) Tsunami safety zone boundary road-mark, (c) Types of a cross-section of TES of Alai-By-pass Road.

Table 1. Number of buildings and area of vacant lots on Alai-By-pass Road, in the years 2010, 2013, 2016, 2019

Road Side	Number of the building (unit)			
	Year			
	2010	2013	2016	2019
North side	278	490	494	520
South side	161	475	491	510
Total number	439	965	985	1030
Percentage Growth		39.94%	0.69%	1.52%
	Area of vacant lots in the observation (Ha)			
Area of vacant lots	174	167.3	165.8	163.7
Reduction of Vacant Land		2.23	0.50	0.70

Based on the analysis in the GIS software, it is also obtained a change in the total area of land use from each year of observation, which is as presented in Table 1.

In terms of building function, many buildings have changed their function, such as houses that have changed into shops, restaurants, minimart, shop houses, workshops or gas stations, or vacant lots become shop houses. The main reason for changing is the high land value for the business.

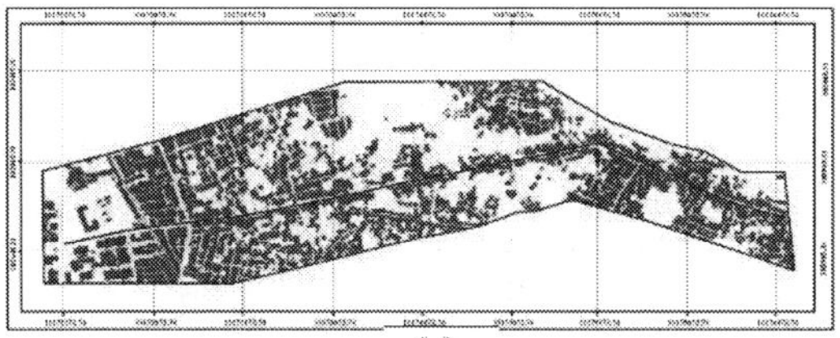

(a)

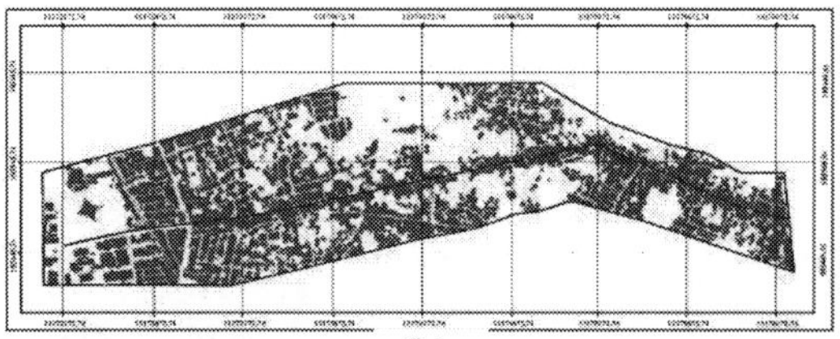

(b)

(c)

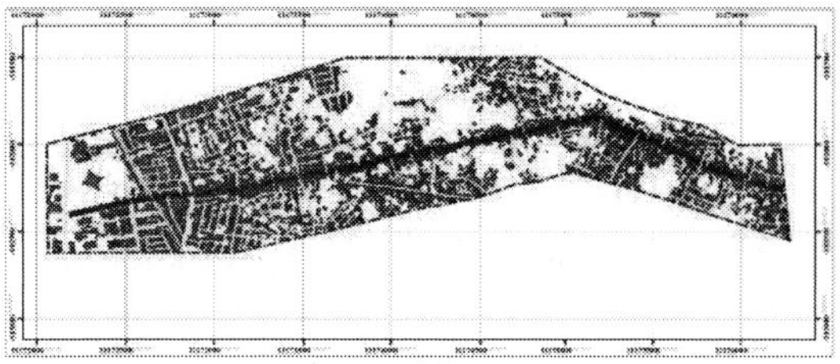

(d)

Fig. 2. Map of the area on TES of Alai-By-pass Road (a) digitizing the map in the year 2010, (b) digitizing the map in the year 2016), (c) Aerial photographs in the year 2019, (d) digitizing maps in the year 2019.

4.3 Traffic Volume

Three sets of daily traffic volume data for 2016, 2017 and 2018, on Alai-By-pass Road were obtained from previous studies [10, 11, 12]. And for 2019 data, a survey was conducted to obtain daily traffic volume. The four data sets are then compared to obtain the number of changes in traffic volume that have occurred.

Fig 3 shows the comparison of hourly daily traffic volume for the year 2016 to 2019. A collection of daily traffic data in 2016, different from other data sets in the year, because the 2016 data are only collected three morning, afternoon, and evening rush hour periods (which are taken for two continuous hours each period). It shows (1) hours and peak traffic volume in 2016 is 07.30-08.30WIB, 1612emp/hr, (2) hours and peak traffic volume in 2017, 06.45-07.45WIB, 2715emp/hr, (3) hours and peak traffic volume in 2018, 06.45-07.45WIB, 2740emp/hr, (4) hours and peak traffic volume in 2019, 07.00-08.00WIB, volume 2845emp/hr. The total peak traffic volume in both directions passing the Alai-By-pass Road has increased during the period.

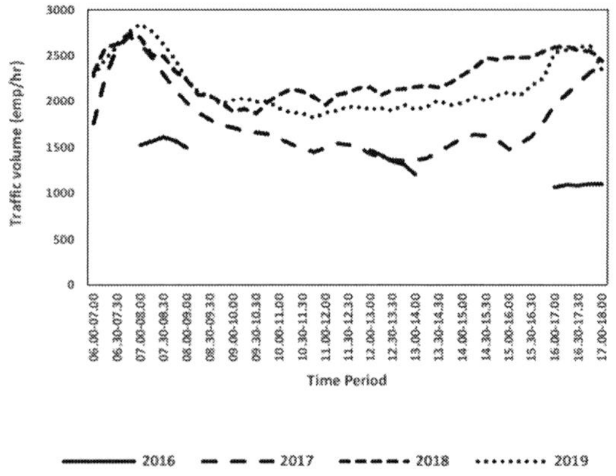

Fig. 3. Hours and peak traffic volume for year 2016 to 2019

4.4 Road Capacity and Performance

Before 2012 the Alai-By-pass Road was a road with a width of less than 5m. Over time, a major earthquake occurred on 30 September 2009, resulting in the Padang municipal government reorienting development for all sectors, including the road sector. Where several roads were widened and designated as evacuation roads. So that through central government funds, by 2012 the Alai-By-pass Road has become 12m wide. Due to a very significant road widening, this road is the first choice of drivers and motorists on their way from the west to the east (vice versa) in the city of Padang. Initially, there were relatively few houses and shops along both sides of the road. However, over time, activities along this road become congested and tend to disrupt road performance, due to vehicles parked on the road, long queues at several locations, pedestrian crossings, and school-safe zones, etc.

Using equation (1), the capacity of the Alai-By-pass Road is calculated. The road capacity is calculated based on field conditions from the four correction factors in equation (2). Then, the four correction factors were compared during the period 2016 to 2019. During that period, from the results of a review of the map and land use along the road as well as side obstacles that occurred due to activities on the constructed land. Table 2 shows the four correction factors for calculating road capacity. In 2019 there was a reduction in the value of the correction factor for the adjustment of obstacles and direction separation (some segments of the road became separated). This decrease was caused by the increasingly crowded land activities along the evacuation road (shops and minimarkets) and the increasing number of traffic disturbances on the road, resulting in a reduced road capacity value.

Table 2. Road Capacity of Alai-By-pass Road in the year 2016-2019.

Year	Factor					Road Capacity (emp/hr)
	C_O	F_{CW}	F_{CSP}	F_{CSF}	F_{CCS}	
2016	6000	1	1	0.91	0.94	5132.40
2017	6000	1	1	0.91	0.94	5132.40
2018	6000	1	1	0.86	0.94	4850.40
2019	6000	1	0.985	0.86	0.94	4777.64

Table 3. Degree of Saturation (DS) and Performance Index of TES for year 2016 to 2019.

Year	Peak hour	Volume Total (emp/ hour)	Road Cap. (emp/ hour)	DS	Perform. Level Index
2016	07.30-08.30	1612.2	4850.40	0.33	B
2017	06.45-07.45	2714.75	4850.40	0.56	C
2018	06.45-07.45	2740.30	4850.40	0.56	C
2019	07.00-08.00	2845	4777.64	0.60	C

Table 3 shows the performance level index of the TES reviewed during the study period, which is decreasing (from level B to C). The assessment that has been carried out on TES in Padang, shows that there are two reasons for the decline in the performance of TES, the first is an increase in traffic volume, and the second is an increase in traffic disturbances on the side of the road caused by an increase in land-use activities.

Maintaining the TES function as an escape route using vehicles from the Tsunami disaster in areas along the coast must be the main concern of the government. Supervision and prohibition of increased interference during TES is a priority, for TES to function.

5 Summary and Recommendation

The capacity and performance of TES have decreased over a 4-year period from 2016 to 2019.Some of the reasons for the decrease in capacity is that the level of side interference on TES has increased every year, it is seen from changes in the number and activity of buildings, as well as the level of compliance of road users who park on-street with irregular parking.

Along the Tsunami evacuation road, it is very necessary:

- o Monitoring of land-use change and enforcement of land use regulations.
- o Law enforcement against offenses by drivers and motorists who do not comply with parking rules.

The author thanks the Directorate of Research and Community Service (DRPM) of the Ministry of Research, Technology, and

Higher Education (Kemenristekdikti RI) for providing research funding through the Master Thesis Research (PTM) scheme with contract number 051/SP2H/LT/DRPM/2019. A thank you also goes to the Research and Community Service Institute (LPPM), Universitas Andalas, for helping with the administration of research and disbursing research funds, with contract #T/31/UN.16.17/PT.01.03/PKR-Kebencanaan/ 2019.

References

1. General Director of Highways, *Indonesia Road Capacity Manual*, Ministry of Public Works, Jakarta p.563 (1997)
2. G. F Koorey, T. J Mitchell, C. R. Fisk, S.F Moynihan, M.K Mara, Traffic Growth Prediction, Transfund New Zealand Research Report no.191 (2000)
3. H. Verron, B. Huckestein, G. Penn-Bressel, P. Röthke, M. Bölke, W. Hülsmann, Determining factors in traffic growth - Developments, causes, and possible future directions, Federal Environmental Agency, Germany p.64 (2005)
4. F. Ashar, D. Amaratunga, R. Haigh, The analysis of tsunami vertical shelter in Padang city, Procedia Economics & Finance 18 pp.916–923 (2014)
5. National Standardization Agency, SNI-7766:2012: *Tsunami Evacuation Route. Indonesian* National Standardization Agency, Jakarta p.19 (2012)
6. German Indonesian Tsunami Early Warning System Project, Tsunami Kit: Developing Early Warning and Community Preparedness in Indonesia, https://www.gitews.org/tsunami-kit/en/id_tsunami_evacuation_map_padang.html, (2019)

.

7. H. Taubenböck, N. Goseberg, N. Setiadi, G. Lämmel , F. Moder, M. Oczipka, H. Klüpfel, R. Wahl, T. Schlurmann, G. Strunz, J. Birkmann, K. Nagel, F. Siegert, F. Lehmann, S. Dech, A. Gress, R. Klein, "Last-Mile" preparation for a potential disaster – Interdisciplinary approach towards tsunami early warning and an evacuation information system for the coastal city of Padang, Indonesia, Nat. Hazards Earth Syst. Sci. 9 pp.1509–1528 (2009)
8. Padang City Government, Padang City Regional Regulation number 4 of 2012 concerning Padang City Spatial Planning 2010 – 2030 Padang (2012)
9. Yossyafra, Fitri N, Sidhi R P, Yosritzal, Mazni D I, Assessment of road traffic performance of the Tsunami evacuation road in Padang Municipality area based on the traffic volume simulation approach, E3S Web of Conferences 156, 04008, (2020)
10. E.H Jupri, Simulation of traffic flow modeling with Microscopic Simulator Software at the signalized arm intersection Civil Engineering University of Andalas (2016)
11. R. Fajriati, Study of the Effect of Motorcycle Proportion on Traffic Flow Variables, Civil Engineering Andalas University (2017)
12. Roselia W 2018 Traffic Flow Prediction for the Tsunami Evacuation Route Section of the Alai-Ampang-Kampung Kalawi Highway in Padang City, Civil Engineering, Andalas University.
13. National Standardization Agency, SNI 7743:2011, *Tsunami Evacuation Signs Indonesian National Standardization Agency*, Jakarta (2011)

Overview of road performance on the tsunami evacuation road during the n-COVID19 pandemic

Yossyafra [1*], *Anyta* Ramadhani[1], *Vina* Gusman[1], and *Monica* Herimarni[1]

[1]Civil Engineering Department, Faculty of Engineering, Andalas University, Padang, Indonesia, 25163

Abstract. The COVID-19 pandemic has changed the world in various sectors and human activities. Limiting human activity and mobility also has an impact on transportation and traffic. This study aims to calculate the capacity and performance of roads under normal pandemic conditions before PSBB (Large-Scale Social Restrictions) in April 2020 and New Normal in July 2002, as well as predict traffic conditions if the Tsunami disaster hits the city during both periods. Tsunami Evacuation roads in Padang City were selected for analysis. The Indonesian Road Capacity Manual 1997 on urban roads is used as a reference for analyzing road performance indicators. The results showed that; road performance during the PSBB period was better than the New Normal period. The effect of volume and side traffic disturbance factors in the New Normal period makes a significant decrease in performance. Through prediction simulations, if a Tsunami occurs in the two study periods, the analyzed roads can relatively serve evacuation movements. However, the capacity needs to be increased for normal conditions.

1 Introduction

Since several earthquakes have occurred in Padang City, the roads have become overcrowded and full of refugee vehicles who want to evacuate to a safe place. At the time of the earthquake on September 30, 2009, in Padang, the community gained experience, that in the process of evacuating from a vulnerable area (beach) to an area that is estimated to be safe from the threat of a Tsunami, there has been a long traffic jam in each evacuation road section that in the city of Padang [1]. This can result in a hazard to the community, due to the congested and congested road conditions during the evacuation. The slow evacuation time due to vehicle density requires attention from the government to review the feasibility of the existing Tsunami Evacuation Road (TER), so that it still requires a short time and there are no barrier points (congestion) during evacuation and by the standards of an evacuation route.

Padang was chosen as the location of this research because it is the city with the largest population and the highest population density along the west coast of the island of Sumatra. Padang is also one of the cities at very high risk of the current Tsunami disaster in Indonesia. Then in a decade since 2009, Padang has succeeded in building several TERs, where the TERs built to have a large width, or are seeking to have a large service capacity.

However, over time, with the decrease in the intensity of earthquakes that have the potential for a Tsunami, it seems as if the Government and Society have forgotten to maintain TER so that it has a large capacity. In addition, it is also necessary for the government to monitor the performance of TER under various conditions.

The n-COVID19 pandemic that has hit the whole world since last December 2019, also provided valuable experiences for the government and society. During this pandemic, various types of restrictions on community movement and prohibition of vehicle traffic have been imposed by the government. Various types of restrictions that have been issued by the government are intended to stop the spread of the pandemic virus in the community.

In terms of the functionality of TERs, as required, it is very necessary to know the capacity and performance of roads, in conditions before PSBB (Large-Scale Social Restrictions) and New Normal and simulate it if a Tsunami occurs during this pandemic. This article describes the road performance during evacuation in several TER locations using simulation methods.

2 Literature Review

2.1 Tsunami Evacuation Road (TER)

Evacuation is the movement of a person from a dangerous area such as a disaster hazard to a safer area. Evacuation is the act of moving, moving, and saving people from a place of danger to a safe place [2]. There are two types of evacuation, namely: (1) horizontal evacuation is evacuation that is carried out by directing refugees to areas that are outside the reach of the tsunami such as hills or highlands that are not reached by the tsunami, meanwhile (2) vertical evacuation is carried out by

* Corresponding author: yossyafra@eng.unand.ac.id

refugees to a safe height in the form of a shelter building, where the building is estimated to still be in the location of the tsunami coverage area.

TER is one type of vertical evacuation, one of the goals of the TER development is to facilitate and provide opportunities to escape from the coastline as quickly as possible to an altitude area that is safe from the threat of Tsunami. If a tsunami threatens, it is attempted so that the TER can be traversed by a large number of vehicles, to escape to a high-altitude location. According to Budi Susilo [3], TER is an important aspect to visualize the evacuation strategy that has been developed in an area. A good TER is that it can provide a safe path and there are not many barrier points (congestion) when residents are evacuated to a safe place even if only to the final evacuation place. It is required that the TER must be able to transport the entire population to be evacuated in a short time to a sheltering area or a clinic, hospital. So, the TER condition must meet the requirements in terms of quality, road material, road width, and distance from threatened areas to safe areas when evacuation occurs. An atmosphere of panic would certainly tend to occur during the evacuation so that more TER is needed in an area that is threatened with disaster.

Sea Defense Consultant [4], states that the width of the evacuation road can be used in urban areas with minimum width requirements:

- Primary Arterial Road: minimum width 10m,
- Secondary Arterial Road: minimum width 8m,
- Secondary Collector Road: minimum width 8m,
- Secondary Local Road: minimum width 4m,
- Residential road: minimum width is 4m.

The requirements of the evacuation route [5], are:

- Safety, which means that it must be completely safe from objects that pose a danger to oneself during evacuation.
- Short distance, which will be used must be a distance that is possible from the place of residence to a safe place quickly.
- Feasibility, which is selected at the time of evacuation must be feasible so that it does not hinder the evacuation process.

In 2012, the Padang City Government has issued Regional Regulation No. 4 of 2012 concerning the 2010-2030 Padang City Spatial Planning (RTRW). The document has listed six-city zones (Figure 1) [6].

2.2 Traffic Management Monitoring Stage

The sequential stages are known in Traffic Management and Engineering activities on urban roads are Planning, Organizing, Actuating, Monitoring, and Controlling. Each of these stages is a very important and mutually influential activity, that if one of the previous stages in its implementation is not or is not done well, it will adversely affect the results of the next stage.

Monitoring stages in traffic management can be done by analyzing road performance under conditions of implementation of a policy. The policy issued can be in the form of engineering in the form of (1) technical activities on roads (such as changes from two-way roads to one-way roads, prohibitions with road signs and markings) that want to improve their performance or can be in the form of public policies that regulate the general movement of people in general. comprehensive, such as (2) a public policy of limiting entry and exit of an area (city, country, etc.) that is hit by a pandemic, because it is feared that the virus will spread to people and quickly become a pandemic due to the mobility of people's movements.

Fig. 1. Map of six horizontal evacuation route sectors in Padang

2.3 COVID 19 Pandemic and Social Restriction

Most cities in many countries around the world are currently facing the COVID-19 pandemic, which impacts all aspects of life, including political, economic, social, cultural, defense, security, public welfare, and mobility aspects [7]. COVID-19, as an infectious disease caused by the newly discovered SARS-CoV-2 virus, has become a pandemic. One of the impacts of this pandemic is the limited public mobility or community movement [8]. The effect also makes all forms of mobility, whether it's the mobility of people, goods, or services that were previously used to be done, are now completely limited due to the COVID-19 pandemic. A virologist in the United States notes that stopping travel is the best way to stop the virus from spreading [9]

Other countries have implemented various policies to contain the spread of COVID-19, including implementing lockdown policies, social distancing restrictions, restrictions on international flights, and the temporary suspension of the business sector [10].

The Indonesian government in dealing with the COVID-19 pandemic has also implemented restrictions on community mobility in stages,

- Large-Scale Social Restrictions (PSBB), where one of the things regulated is restrictions on private vehicles such as cars, a passenger restriction policy has been implemented so that each four-wheeled vehicle is only allowed half of the permitted vehicle capacity.

- PSBB Transition, socio-economic activities in stages while still, applying certain limits [11].

This study aims to determine the capacity and performance of the TER section in pandemic conditions before PSBB and New Normal and to determine the performance of TER if a tsunami is simulated. The article also shows the influence of city government policies on changes in the level of community mobility that affect TER performance during the COVID-19 pandemic.

3 Methodology

This study went through several stages, which aimed to determine the effect of various government policies in restricting people's travel on TER performance in Padang. The stage was begun with conducting a literature study on TER and reviewing several public policies issued by the government in the transportation sector in reducing the impact of the pandemic. Then the next stage was the collection of primary data on several roads that have been designated as TER.

Data collection was carried out twice, the first was during a time when the pandemic had begun to spread rapidly, but there were no travel restrictions by the government. But at that time, individually or with family or one residential area, they had limited themselves not to travel from their place of residence. Traffic data was collected at three different TERs in Padang. Traffic volume data was collected by surveying for 12 hours from 06.30-18.30EIT continuously. Traffic volume data in both directions were recorded at an interval of 15 minutes based on vehicle classification. The results of recording the number of vehicles were converted into units of pcu/hour. Data on the adequacy of signs and markings on the evacuation route, carried out by direct inventory to the TER location and asked respondents' opinions on the adequacy of markings and signs along the TER.

There are some secondary data collected; data on the number of residents in urban villages along the TER (based on data published by the Central Statistics Agency of Padang) and data on vehicle ownership in Padang [12].

Then proceed with the data processing stage, which consists of data preparation, the process of preparing data for further analysis. The Indonesian Road Capacity Manual 1997 on urban roads was used as a reference for analyzing road performance indicators [13]. And the last stage was data analysis using a simulation approach. Through prediction simulations, if a Tsunami occurs in the two study periods, the analyzed performance of the TER was conducted. Then compared the volume of traffic in the two different conditions. The next step was to simulate the traffic volume prediction if there is a tsunami threat.

4 Analysis and Results

4.1 Character and Capacity of TER

In this study, three TERs were selected for investigation. Two of the TERs (Jalan Sawahan and Jalan Sisingamangaraja) are in the CBD and the other (Jalan Pagang Raya) is about 5km from the CBD, which is also one of the important roads connecting the suburbs to the city center, which is also busy with traffic. The three TERs have met the requirements for evacuation roads (safe, short, feasible).

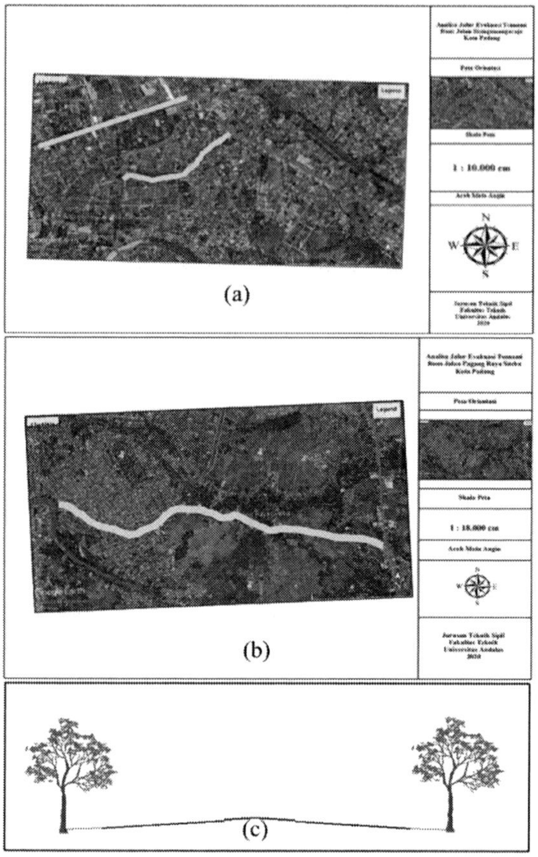

Fig. 2. (a) Location of Sawahan Road and Sisingamangaraja Road, (b) Location of Pagang Raya Road, (c) typical of the cross-section.

Table 1. The character of TER

Characteristic	Road (TER)		
	Pagang Raya (TER1)	Sisinga-mangaraja (TER2)	Sawahan (TER3)
Type	2/2 UD	2/2 UD	4/2UD
Function	secondary collector	secondary collector	secondary collector
Lane width (m)	2x3	2x3,5	4x3
Co	2900	2900	6000
F_{CW}	0.87	1	0.91
F_{CSP}	0.97	1	1
F_{CSF}	0.94	0.90	0.84
F_{CCS}	0.94	0.94	0.94
C	2162	2453	4311

- UD : Un-divided,
- C : Capacity (pcu/hour)
- Co : Basic capacity (pcu/hour)
- F_{CW} : Road width adjustment factor
- F_{CSP} : Directional separation adjustment factor
- F_{CSF} : Side drag and road adjustment factor
- F_{CCS} : City size adjustment factor

https://doi.org/10.1051/e3sconf/202133106006

Figure 2 shows (a) the location of Sawahan Road and Sisingamangaraja Road, and (b) the location of Jalan Pagang Raya, and (c) typical of the cross-section of each TER. And Table 1 shows the technical data and capacity of each TER.

4.2 Traffic Volume of TER

It has been explained in section 3, that to get road performance of TER, it was preceded by collecting traffic volume data during the PSBB and New Normal periods at each TER. The first traffic volume data were collected on Monday (April 20); Tuesday (April 21); and Wednesday (April 22) of 2020. That was a week before the government-imposed travel restrictions (PSBB) through regulations issued by the Governor of West Sumatra Province. The second collection of traffic data were collected on Monday (June 29); Wednesday (July 1); and Friday (July 3) of 2020. At that time the government had allowed the public to travel, but by complying with several health requirements (the New Normal period). Weather conditions at the time of the traffic volume survey were sunny days.

Figure 3 shows the traffic volume conditions on Jalan Pagang Raya as one of the TERs in the city of Padang, during the COVID-19 pandemic in 2020, with two types of periods, before PSBB and New Normal in 2020. In New Normal conditions (after PSBB) there is an increase in community mobility, which is indicated by an increase in the number of vehicles on the road. The difference in the average traffic volume ranges from 100 to 280pcu/hr.

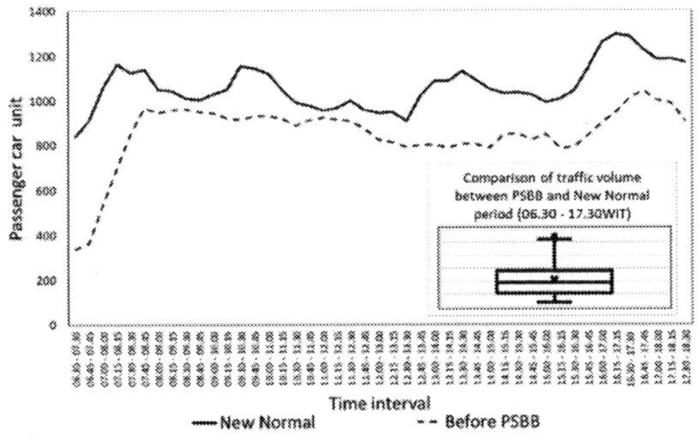

Fig. 3. Traffic volume of Pagang Raya Road in the period of before PSBB and New Normal.

Table 2. The Degree of Saturation of TER

Period	Degree of Saturation		
	TER1	TER2	TER3
Road Performance			
before PSBB period (April 2020)	0.48	0,50	0,51
New Normal periods (June and July)	0.60	0,65	0,62
Road Performance if Tsunami occur			
before PSBB period (April 2020)	0.73	0,71	1,10
New Normal periods (June and July)	0.85	0,86	1,10

Then the analysis has been continued to obtain the existing performance of each TER in the PSBB and New Normal periods. Calculations have also been continued by adding traffic predictions in the event of a Tsunami.

Table 2 shows the degree of saturation of each TER. It can be seen that on TER3 (Sawahan Road) there is a very large increase in traffic volume in the event of a Tsunami. this is because Sawahan Road is the main road from the beach or CBD, where many activity centers are busy inhabited or visited by the community. For this reason, the government's attention is needed in formulating strategies and traffic management on the TER that are directed and measurable.

4.3 Vehicle Mode selection during Tsunami Evacuation

Direct interview surveys to individuals and families in communities living around the TER area were also conducted. This survey aims to obtain a choice of what vehicle mode will be used by individuals or families for evacuation if a Tsunami disaster occurs. In each area of TER, a questionnaire from 100 respondents has been collected, so that a collection of answers from 300 respondents has been obtained, which are presented in Table 3.

Table 3. The choice of the type of evacuation and the use of vehicles by the community if a Tsunami occurs

Questions and answers	Road (TER)		
	Pagang Raya	Sisinga-mangaraja	Sawahan
Evacuation action?			
- Horizontal	60%	76%	59%
- Vertical	20%	13%	30%
- Stay at home	18%	11%	11%
If evacuating, what vehicle do you drive/ride?			
- Car	5%	7%	1%
- By-cycle	37%	48%	30%
- walk	20%	21%	28%

Table 3 shows that more than 60% of the population evacuated horizontally if a Tsunami occurred. And most of them use 2-wheeled motorized vehicles, while relatively few use 4-wheeled vehicles. It should also be noted that there are quite a several options for evacuating on foot. It is also the government's attention to providing facilities for pedestrians who are representative for evacuating the community if a Tsunami occurs.

5 Conclusion

The results of observation and analysis of three Tsunami evacuation roads (TER) in Padang Municipality indicate that:

- Under normal traffic conditions, the performance of the Tsunami evacuation road section is at Road Performance more than 0,50, meaning that traffic flow conditions can still be said to be stable and moderate traffic density.
- Especially for Sawahan Road as TER, comprehensive traffic management and engineering is needed, so that the road can serve the community to evacuate if a Tsunami occurs

- The government must conduct periodic reviews of all TER in the city of Padang, to ensure that it can become a TER that accommodates people who evacuate when a disaster occurs.

The author thanks the Faculty of Engineering, Universitas Andalas for providing research funding through the Publication of the Proceedings of the International Fund for Lecturers of the Department of Civil Engineering, with PNBP Funds in 2021 with contract No. 135/UN16.09.D/PL/2021.

References

1. Y Yossyafra, N. Fitri, N. , R.P. Sidhi, Y. Yosritzal, D.I. Mazni, E3S Web of Conferences 156, 04008, (2020)
2. National Standardization Agency of Indonesia, SNI 7766:2012, *Tsunami Evacuation Line* (2012)
3. B. Susilo, S. Yarianto,*"Analysis of Population Evacuations in Muria Nuclear Power Plant Accidents With Geographic Information SystemsApproach"*. Proceedings of the 13th National Seminar on Technology and Safety of Nuclear Power Plants and Nuclear Facilities. Jakarta, November 6, 2007. ISSN:0854 – 2910 (2007)
4. Sea Defense Consultant,*"Tsunami Evacuation Planning Guidelines SDC-R-70022"*(2007)
5. Kogami. "Knowledge*Module Facing Natural Disasters"*. Padang(2007)
6. Padang City Government,*"Padang City Regional Regulation number 4 of 2012 on Spatial Plan of Padang City Region 2010 - 2030 Padang"*(2012)
7. L. Hawryluck, "SARS Control and Psychological Effects of Quarantine", Toronto, Canadal, in Emerg Infect Dis, **11**, 1206–1212, (2004)
8. N. K. Nissa, Y. Nugraha, C. F. Finola, A. Ernesto, J. I. Kanggrawan, A. L. Suherman, Smart Systems Journal **3**, 02, 84 -94, eISSN:2622-8254, (2020)
9. J. Upton, "*Pandemics Are the Dark Side of Global Mobility*", Nautilus, no.84, (2020)
10. M. Nikola, E. Angela, H. Emily, "*Impacts of COVID-19 on Mobility*", Preliminary Analysis of Reg. Trends Urban Mobility, Slocat, 3, (2020)
11. Y. R. Damuri, H. Aswicahyono, and D. A. W. Lestari,*"Evaluation of Economic Activities and Intensity of* COVID-19 Spread in New Normal *Times: Review of Some Rapid Indicators"* [Data Update June 29, 2020], CSIS, (2020).
12. Central Statistics Agency for the City of Padang, Padang in Figures, (2020)
13. Director-General of Highways, "*Indonesia Road Capacity Manual*", Ministry of Public Works, Jakarta, 563 (1997)

https://doi.org/10.1051/e3sconf/202133106007

Study analysis of landslide vulnerability of mining area in the sub-district Lembah Gumanti, Solok regency (lubuk selasih street– surian)

Momon[1*], *Bayu Martanto* Adji[2], *Desi Widia* Kusuma[1], *Elsa* Yolarita[1], *Vivi* Ukhwatu[1] *K* Masbiran[1], and *Arian* Dodi[1]

[1]Research and Development Center, West Sumatra Province, Indonesia
[2]Department of Civil Engineering, Andalas University, Indonesia

Abstract. Open access land mining activities on the Lubuk Selasih – Surian national road have in floods and landslides that have, in losses to the community and the imposition of costs for maintenance and rehabilitation of federal roads along the mining area. This study aims to determine the existing condition of mining and analyze the vulnerability of erosions in mining areas. The data used is secondary data, using descriptive quantitative methods with photogrammetric analysis of the maps obtained. The results showed that this mining area has the characteristics of rocks that are easily crushed and tend to be unstable, as well as steep and very steep slopes. This area is also located on the active Semangko fault and volcanic mountains. The environmental carrying capacity of disaster prevention and protection ecosystem services is mainly in the shallow categories. Based on these findings, the mine site is at a very high level of vulnerability to erosions. Thus, it is necessary to mitigate mining management administratively and operationally to minimize environmental damage

1 .Introduction

The mining sector is still very important in supporting the community's economy. However, these mining activities, both near the surface or below the earth's surface, will result in disruption of the landscape and cause significant environmental impacts.

Mining activities on open access lands, which are carried out legally or illegally, in their operations mostly do not follow the rules and carry out their obligations according to what has been outlined and planned. As a result, it will harm the surrounding environment, namely the occurrence of damage and disasters such as sediment flooding and landslides.

According to ESDM (2005), landslides are the movement of slope-forming materials in the form of rocks, debris, soil, or mixed materials, moving down or out of the slope [1]. Landslides often occur due to soil movement in conditions of steep slopes, high humidity levels, and rare plants (open land). Another factor for the occurrence of landslides is seepage and geological activities such as faults, fractures, and linearization. Local environmental conditions are an interrelated component. The shape and slope of the slope, the strength of the material, the position of the groundwater level, and local drainage conditions are also closely related to the condition of slope stability [2]

On the Lubuk Selasih – Surian national road. This condition has been going on for a long time and the incident has been repeated every year, so it has caused considerable losses for the community, national road users, and destination areas such as South Solok Regency and Kerinci Regency and has caused large costs to be incurred for maintenance and repair. rehabilitation of national roads along the mining area. For this reason, it is necessary to identify landslide susceptibility in mining areas to reduce the risks and impacts of mining activities.

This study aims to determine the existing condition of mining, then analyze the vulnerability of landslides on mining land and provide mitigation directions and guidelines in preparing a recovery plan for ex-mining land damage.

2 Methods

2.1 Collecting Data

The research was carried out in June – July 2021, located in a mining area along the Lubuk Selasih –

* Corresponding author: momon.balitbang@sumbarprov.go.id

The research is quantitative. The data used in this study are secondary in the form of area/location, time of occurrence, aerial photography, records of related agencies, land cover maps, geological maps, topographic maps, hydrological maps, maps of environmental carrying capacity, and plate zone map of the mining area in Lembah Gumanti District, Solok Regency.

2.2 Processing Data

The map data was obtained from the relevant agencies: the ESDM Office, the SDA & BK Office[3], the Bina Marga Office[4], the Environment Service[5], BPBD[6], and the Road Implementation Center. An overview of the location is obtained through a literature study of journals and technical documents. The analytical method used is descriptive analysis and overlapping (overlay) using the Geographic Information System (GIS) technology to disaster vulnerability. After overlaying the map, the landslide hazard class interval is made into 4 (four) classes: low, medium, high, and very high. Next, landslide susceptibility is scored based on the criteria: slope, geology, rainfall, infiltration rate, land cover, and fault zone.

Landslide area is determined based on seven variables/criteria determining landslide-prone areas, namely (1) slope, (2) geology, (3) rainfall, (4) infiltration rate, (5) groundwater depth, (6) fault zone (earthquake), and (7) land cover [7,8,9,10,11,12], . Furthermore, to determine the level of vulnerability to landslides at the location of the mine area, it is carried out by weighting each variable according to the level of risk—variable weighting based on professional adjustment and review of previous studies.

Furthermore, to determine the landslide hazard, the researcher used the landslide-prone class interval of the Directorate of Volcanology and Geological Hazard Mitigation (DVMBG) [13]. Landslide Prone Class intervals can be seen in Table 1 below:

Table 1. Landslide Prone Class

Vulnerability Class	Score	Information
1	< 1.82	Low
2	1.82 – 2.64	Currently
3	2.64 – 3.46	Tall
4	> 3.46	Very high

3 Results and Discussion

3.1 Overview of Research Sites

Geographically, the landslide area is located in Solok Regency, where Solok Regency is located between 010 20' 27" to 010 2'39" south latitude to 1000 25' 00" and 1000 33' 43" east longitude. Solok Regency is one of the regencies in West Sumatra Province, located at an altitude area 30 Km from Padang City (Capital of West Sumatra Province).

The landslide location is on the national road where the only access for South Solok Regency is feasible to take so that if this access is cut off, South Solok Regency will be isolated. There will be a stagnant distribution of commodities or logistics.

Based on the results of the identification of mining areas along the Lubuk Selasih-Surian National Road, Lembah Gumanti District, Solok Regency, there are 7 Mining Business Permits (IUP), namely: PT. Kuansing Mineral Sejahtera, Gunawan Zulmahendra, PT. Cold water sirtu 2, Damiri, CV. PutraYLM, PT. Yuanda Putra YLM and Abdel Hanif.

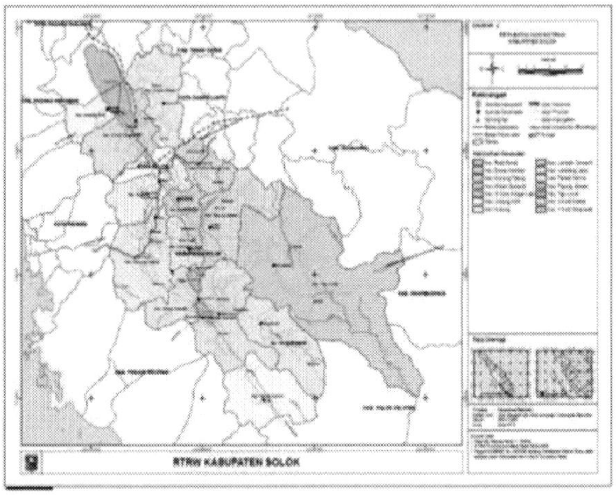

Fig. 1. Solok. District Administration Map

3.2 Landslide Site Vulnerability Analysis

3.4.1 Land Cover Analysis

The mining area is located in dry fields and sandhills. The area is functionally designated as a mining area (Perda No. 1 of 2013 concerning the RTRW of Solok Regency). However, the area has unstable rocks and has high rainfall, so that the risk of landslides is relatively high.

3.4.2 Geological Analysis

Based on data from the Provincial Energy and Mineral Resources Office, the mining area generally has a Barisan Formation (Pb) rock unit consisting of phyllite, slate, limestone, hornstone, and grow to make meta (sandstone). Rock characteristics Mining areas have rock characteristics that are scattered or easily crushed.

As it is known that the location of the mining area is on the active fault line of the Sumatra Great Fault, which causes the rock to be intensely fractured. The fractured rock will respond in the form of crack propagation so that it becomes scattered pieces.

With the characteristics of the rock being scattered, the condition becomes unstable. If there is rainwater, it will be quickly eroded by water, resulting in avalanches of material onto the national road.

3.4.3 Topographical and Morphological Analysis

To define altitude classes ranging from plains - hills - to mountains, this report uses the following relative altitude classes:
0 - 50 m: Lowland
50 – 200 m: Low hills
200 – 500 m: Hills
500 – 1000 m: High hills
> 1,000 m : Mountains

While the slope class to distinguish the surface morphology used the following slope criteria:
Slope (recommended):
0 – 8% : Flat or Slightly Sloping
8 – 15% : ramps
15 – 25% : Slightly Steep
25 – 40 % : Steep
> 40 % : Very Steep

In general, mining areas can be grouped into several morphological units in plains with a slope of 30% - 70% (EMR Department, 2020). The elevation in the mining area is generally at an altitude of 1420-1590 masl.

With a steep slope and high rainfall intensity, the material is eroded to the bottom of the valley (river). According to Martono, 2004[14], The steeper and longer slopes will increase the velocity of surface runoff, and the volume of surface water will increase so that more objects can be transported.

According to Kartasapoetra (1990)[15], soil with a slope of > 15% with high rainfall can cause landslides. Based on the expert's statement, he concluded that the current mining area in cold Aie has a high risk of landslides.

The mining area is in a potential landslide zone. Landslides have often occurred where landslides repeatedly occur four times a year. The flow of water from the top of the cliff is getting harder and harder to control so that it damages the shoulder of the road on the opposite side, which in the long term has the potential to cause the road to be cut off.

In this zone, ground movement can also occur if the rainfall is above standard. Looking at the position of the mining area, it can be concluded that the area generally has a relatively high risk of landslides.

3.4.4Hydrological Analysis

In the mining area, rainfall is included in the very high category ranging from 2,500-3,500 mm/year, and the mining area is in the upper reaches of the Batang Hari River Basin (6 tributaries of the Batang Hari to the left of the Bt. Hari River), where the catchment area is vast. The Batang Hari tributary ranges from about 13 hectares – 97 hectares with a steep slope that can debris flow and landslides.

Discharge simulation if the rainfall intensity is 15 mm/3 hours; discharge Q = 3.29 m3/s; if extreme rain > 50 mm/3 hours; discharge Q = 10.98 m3/s with a flow velocity of 5.63 m/s, the water level in the river is 0.36 m; The velocity of debris flow is very dangerous for the National Road in the mining area which is also an area prone to landslides.

Especially in the Mining Area CV. Putra YLM with an IUP area of 15 ha has changed the natural river flow pattern and impacted bringing sedimentation on the Lb National Road. Selasih - Surian (Landor/Turban Road) still has the potential for erosion where the riverbed is very prone to be eroded again when the intensity of rain is high.

3.4.5 Environmental carrying capacity and carrying capacity analysis

The concept of ecosystem services (ecosystem services) can be used to measure the carrying capacity and capacity of the environment. Ecosystem services are grouped into provisioning services, regulating services, cultural services, and supporting services. There is a common understanding between ecosystem services and the carrying capacity and carrying capacity of the environment.

The definition of service provision reflects the concept of environmental carrying capacity and regulatory services, which have similarities with environmental capacity. Meanwhile, supporting services can mean the carrying capacity of the environment and the carrying capacity of the environment. The environment's carrying capacity and capacity with the ecosystem services approach adheres to 2 (two) assumptions. First, the higher the ecosystem services of an area, the higher the ability of the environment to absorb substances, energy, and other components that enter or are included in it (regulatory services). Second, the higher the ecosystem services of an area, the higher the ability of the environment to support human life, other living things, and the balance between the two (providing services, cultural services, and support).

To find out the carrying capacity and capacity of Ecosystem Services for prevention and protection from disasters in mining areas along the Surian-Selasih national road, see Table 2.

From Table 2, it can be seen that the carrying capacity and carrying capacity of the environment from the aspect of the classification of ecosystem services in the form of prevention and protection from disasters, most of them are in a low category, this indicates that the ability of the environment to protect ecosystem services from disaster prevention and protection is low. Meager means that the potential for disaster is high to very high. The high disaster potential reaches an area of 18.82 Ha, while the very high disaster potential is an area of 13.03 Ha, located at the Gunawan Zulmahendra mine site, PT. Cold Water Sirtu, CV. YLM's son.

Table 2. Capacity for Disaster Prevention and Protection Regulatory Services

No.	IUP name	IUP Area (Ha)	Environmental Carrying Capacity					
			Disaster Prevention and Protection Ecosystem Services					
			Currently		Low		Very low	
			Ha	%	Ha	%	Ha	%
1	PT. Prosperous Mineral Kuansing	5.01					5.01	38.44
2	Gunawan Zulmahendra	5,10	-	-	-	-	5,10	39.14
3	PT. Cold Water Sirtu	5.03	3.37	32.84	1.66	8.82	-	-
4	Damiri	11.80	6.89	67.25	1.99	10.57	2.92	22.40
5	CV. YLM's son	5.07	-	-	5.07	26.93	-	-
6	PT. Yuanda Putra YLM	5.01	-	-	5.01	26.62	-	-
7	Abdel Hanif	5.09	-	-	5.09	27.04	-	-
Total		42.11	10.26	100	18.82	100	13.03	100

3.3 Fault & Earthquake Zone

Tectonically, the position of Solok Regency is included in the Asian continental plate, which is subducted by the Indian Ocean plate on the West Coast of Sumatra. The subduction of the two plates resulted in several active faults on the mainland of Sumatra in the form of the Sumatran Great Fault, known as the Semangko fault.

The mining area location is in a volcanic mountain area. The volcanic mountain area stretches from North-South in the middle of this province, with the Semangko fault in the middle.

The Semangko active fault is the main fault that extends along the Pantai Cermin sub-district, Danau Diatas, Danau Bawah to the north to Lake Singkarak and forms a graben in Solok City, which is part of an active fault along Sumatra. The active fault line in Solok Regency crosses from Pantai Cermin District and continues to the North through Lembah Gumanti District, Lake Diatas - Danau Bawah, Lembang Jaya District, Bukit Sundi District, X Koto Singkarak District and continues to Lake Singkarak. As a result of the movement of the main active fault, other associated active faults with smaller dimensions parallel to the main fault line (Sumatra) spread in Gunung Talang District, Kubung District slightly in Hiliran Gumanti District, Tigo Lurah District, Payung Sekaki District to the north. The main earthquake path in Solok Regency is along the Sumatran fault line that crosses Solok Regency. If there is a release of energy that occurs in the Indian Ocean with a strength of > 7 SR, it will potentially produce an earthquake along the fault.

Based on the analysis results above, it illustrates that the mining area is in a landslide Hazard/Prone to Landslide area. This is also reinforced by experts, where according to Arsyad (1989)[16], landslides will occur if the following three conditions are met:

- Slope conditions are steep enough so that the soil mass can move or slide down.
- The condition of the layer below the surface of the soil mass that is somewhat impermeable and soft, which will become the sliding plane.
- The condition of sufficient water in the soil so that the layer of soil mass just above the impermeable layer becomes saturated.

According to the Directorate of Environmental Geology (1981), the factors that cause landslides include:

- Topography or steep slope
- condition of soil/rock
- rainfall or water
- Earthquake/earthquake
- State of vegetation/forest and land use

These causal factors influence each other and determine the magnitude and extent of the landslide disaster. The sensitivity of an area to landslides is also determined by the influence and relationship of these factors to one another.

Meanwhile, Djamaluddin (2003) stated that landslide events are three-dimensional (3D) geological phenomena. Landslides (soil movements) in Indonesia generally occur on steep slopes formed from unconsolidated volcanic deposits. Therefore, geomorphological, hydrological, and land cover characteristics can be used as indicators that a landslide has occurred in an area. Steep slopes that are influenced by geological structures such as faults, fractures, folds are more susceptible to landslide symptoms, especially if the direction of rock layers is in the same direction as the slope and there are active faults. Layers of rock that are interspersed between an impermeable rock and water-permeable rock create areas that have the potential as slip fields.

3.4 Scoring Determination of Vulnerability

3.4.1 The basis for Determining Landslide Prone

The determinants of landslide areas can be divided into two, namely natural and human factors. Natural factors include slope, rainfall, geology, and others, while human factors change natural conditions that result in increased landslide potential.

From the review of these various literatures (see Guillard and Zezere, 2012, Althuwaynee et al., 2012, Akgun et al., 2012, Van Westen et al., 2008, Tejakusuma,

2007, Gritzner et al., 2001), seven variables were determined. The criteria for determining landslide-prone areas are (1) slope, (2) geology, (3) rainfall, (4) infiltration rate, (5) groundwater depth, (6) fault zone (earthquake), and (7) land cover.
To determine the level of vulnerability to landslides at the mine site, the researchers used several variables by giving each of these variables:

3.4.2Landslide Prone Calculation

Based on the standard for determining the landslide hazard, the calculation of the landslide hazard model is carried out. The calculation of landslide hazard can be seen as follows:

Table 3. Landslide Hazard Scoring Results

No	Criteria	Score	Weight	Score
1	Slope	5	30%	1.5
2	Geology	5	20%	1
3	Rainfall	4	20%	0.8
4	Infiltration Rate	4	10%	0.4
5	Land Cover	5	10%	0.5
6	Fault Zone	5	10%	0.5
Amount			100%	4.7

Referring to the calculation results of Landslide potential and Landslide Prone Class Interval, the location of the mining area is currently at very high Landslide Hazard. For this reason, it is necessary to direct mitigation from stakeholders related to landslide prevention to reduce the occurrence of landslides.

3.4.3 Administrative and Operational Violations

In addition to the very high vulnerability to landslides, on the other hand, there are also violations of administration and operations in mining management.

Then the company's compliance with administrative and operational order for the management of the mining area is a must; this is a form of synergy from the government and company elements so that the implementation of mining activities can be monitored and by Standard Operating Procedures (SOP).

With the landslide disaster caused by mining activities, the Mining Inspector of the Ministry of Energy and Mineral Resources of the Republic of Indonesia has examined the completeness of the mining company's documents as well as the examination of violations of technical and environmental aspects.

3.5 Landslide Disaster Mitigation Directions

From the analysis results, it is necessary to mitigate landslides to cope with more severe damage. The directives for disaster mitigation that can be carried out on post-mining land partially or for mining land closure as a whole are as follows:

3.5.1 Land Arrangement

The form of land arrangement can be suggested as follows:

1. Reconstruction of Land Forms (Reshaping)
   - Reconstruction of landforms is carried out to obtain stable slopes with low potential for landslides and erosion and is suitable for revegetation. Reconstruction of landforms at ex-mining sites with irregular and uneven landforms (perforated, with relatively steep slopes) must be arranged in such a way as to be stable with low potential for landslides and erosion. Existing piles of material are disposed of or used for filling existing holes by taking into account the characteristics of the materials used for backfilling and the presence or absence of a drainage system that may be disrupted.
   - Reconstruction of the shape of the slope is carried out to reduce the speed of runoff, erosion, sedimentation, and landslides. The shape of the slopes in this mining area is relatively steep with a significant height difference, so it is necessary to build terraces. For each terrace formed, it is necessary to make a terrace channel so that it can drain the water in the terrace area in a controlled manner.
2. Construction of Drainage Channels
   Construction of drainage channels aims to prevent excessive rainwater inundation and drain surface runoff so that excess water is planned to flow quietly and not damage the soil surface, plants, and other soil conservation structures. The drainage channels that need to be made consist of:
   a) evasion channel
      An evasion channel is made to cut off runoff from the area or plot of land above the slope so that it does not enter the surrounding area so that runoff and erosion are reduced. The evasion channel is made to cut the slope at a slight angle (0.1 – 0.5%) with the contour line to allow water to flow. When digging the channel, the

excavated soil (fill) is used to make mounds or embankments on the lower side.

Before entering the sewer, making sediment settling pond (sediment pond) is necessary to settle the sediment particles carried by the water flow.

b)Drainage.

The drainage channel (SPA) removes water from the ex-mining area that previously flowed in the evasion channel or terrace channel. The existence of this drainage channel is necessary so that the water collected in the evasion channel or terrace channel can be drained out in a controlled manner so that it is not destructive. The presence of a sewer will cut the escape channel or terrace with varying spacing (distance between SPAs) depending on field conditions.

c)Plunge

A plunge needs to be made that reduces the velocity of surface runoff so that the water flows at a speed that is not destructive and does not carry large amounts of sediment.

3.5.2 Erosion And Sedimentation Control

This ex-mining land consists of loose material (pile), which is easily eroded. Therefore, efforts to control erosion and sedimentation need to be continued. Methods that can be done are:

1. Vegetative Method
   It is carried out using cover crops, and revegetation is part of the recovery activity, which also controls erosion and sedimentation. Ground cover plants are plants intended to cover the soil so that raindrops do not directly hit the soil. Cover crops can be planted alone or planted together with main crops. Another function that is expected from cover crops is to provide organic matter to maintain soil fertility.
2. Mechanical Method
   It is carried out in locations with serious erosion problems that are not possible if only vegetative erosion control is carried out. Several mechanical methods that can be applied are terracing, evasion channels, and control dams (Sabodam/check dam).
   a. Terrace/Terrace
   In very steep areas such as mining areas, bench terraces are recommended to reduce runoff and erosion more effectively.
   b. Diversion Ditch
   In the case of trench erosion or gully erosion, a diversion channel can be made to control it. A ditch is made on the upper slope of the area affected by ditch erosion or ravine erosion. The entry of water flow into the ditch or ravine can be prevented so that ditch erosion or ravine erosion does not continue. The end of the evasion channel is directed to a part of the land that is still densely overgrown with plants or to a safe SPA, namely a SPA overgrown with grass.
   c. Control Dam (Sabo Dam)
   In locations where trench erosion occurs due to concentrated water flow, it is necessary to build a control dam that functions as a reservoir for water and sediment. The specifications of the control dam depend on the magnitude of the erosion of the trench. If the trench is significant enough (ravine erosion), where the surface runoff is large enough, then a permanent concrete control dam (Sabo Dam) is the best alternative.

3.5.3 Water Quality Management

The process of mining rock mining materials at the mine site is carried out by digging the material and then separating the desired material by a washing process using water. The material transported in the effluent is fine-sized particles. The particles in the effluent will settle according to the law of gravity, with a particular retention time. Therefore, the problem of water quality in this ex-rock mining area is the level of turbidity. To overcome this, it is necessary to make settling ponds with dimensions tailored to the needs. Settlement ponds are made with a zigzag model so that the suspended particles have sufficient time to settle. In particular, conditions where the TSS value exceeds environmental quality standards, coagulant materials such as alum, cationic polymers,

3.5.4 Management of Domestic Waste and Hazardous Toxic Waste (B3)

Mining activities cause a large amount of solid waste to accumulate in the form of organic and inorganic waste originating from the mess, kitchen, canteen, and office activities, including Hazardous and Toxic Material (B3) waste. Organic and inorganic solid waste (plastic and the like) is disposed of at the TPS to minimize it.

In addition, mining activities that use heavy equipment will produce B3 waste in used oil, used diesel, grease, and spilled fuel. To minimize the impact of waste oil/lubricant and used diesel fuel, environmental management is carried out through;

1) Oil/lubricant, grease, and used diesel fuel are packed carefully in a unique drum, then collected in the temporary shelter provided.
2) To reduce the impact of used oil spills, using diesel fuel, and others in the workshop area and fuel station, an oil trap is created, where the oil trap is separated between the oil layer and the water layer so that both can be taken separately, then transferred to the drum that has been provided and collected to the B3 waste collection site in the appropriate workshop Government Regulation No. 101 the year 2014 regarding Management of Hazardous and Toxic Waste (B3)
3) Conduct temporary bioremediation to anticipate the spilled B3 waste.

https://doi.org/10.1051/e3sconf/202133106007

3.5.5 Waste Material Stack Management

Mining activities leave large puddles in former mining sites and produce residual material that accumulates in various locations. The step taken is leveling the land using heavy equipment so that it does not have holes and becomes flat. Furthermore, land cover is carried out with organic materials from plant remains such as straw, grass, vegetables. In addition, backfilling is also carried out with particular soil for locations that are planned to be planted with plants.

3.5.6 Reclamation of Ex-Mining Land

To restore land that has been disturbed due to mining activities, the government requires mining actors to carry out land reclamation or restoration, as stated in Law no. 4/2009 concerning Mineral and Coal Mining, Law no. 41/1999 on Forestry, Law no. 32/2009 concerning Environmental Protection and Management, and their respective derivatives. Recovery/reclamation of ex-mining land is intended to obtain land conditions that are safe, stable, and not easily eroded so that it can be reused according to its designation.

The implementation of the restoration of damaged ex-mining land must comply with the principles of environmental protection and management, which include:

- Protection of the quality of surface water, groundwater, seawater, soil quality, and air quality based on environmental damage criteria under the provisions of laws and regulations.
- Protection and restoration of biodiversity.
- Guarantee the stability and safety of overburdened piles, tailings ponds, ex-mining land, and other artificial structures.
- Utilization of ex-mining land under its designation.
- Pay attention to the social, economic, and cultural conditions of the local community.

Utilization of ex-mining land that has been arranged can be in the form of revegetation with forestry plants, plantation crops, food crops as a source of water if there are ex-dug holes with good water quality.

Determination of the type of land use, whether for food crops, plantations, agro-tourism, or others, needs to be based on, among other things, the status of ownership and the biophysical condition of the land as well as the needs of the community.

3.5.7 Revegetation with Forestry Plants

Revegetation activities are carried out through the following stages:

1. Field Preparation
   Lands laid out and have drainage channels made need to be well prepared before planting with pioneer plants or other types of plants. In general, field preparation includes planting cover crops, determining the direction of the array and installing stakes, and making plant holes.
2. Procurement of Seeds
   Seeds to be planted can be produced at the nursery/nursery or through purchase. Seedlings that are made or purchased are seeds that have quality and nutritional standards. In addition, the seeds produced and purchased are seeds suitable for planting in the area. Therefore, the selection of species to be planted must be carried out. The choice of species must consider the ecological suitability of the place where it grows. The suitability of ecologically appropriate tree species for planting in land restoration areas can be determined by: i) taking an inventory of native tree species, ii) conducting species site matching through literature studies and analysis of the suitability of growing conditions, and iii) conducting a species trial or other crop testing.
3. Planting Implementation
   Planting is carried out after land preparation is complete and at the beginning of the rainy season. Healthy seedlings with a minimum height of 50 cm are transported carefully from the nursery/nursery to the planting site. The seeds are then placed near the prepared planting holes. When planting, the compost that has been inserted into the planting hole is mixed with topsoil. Then the polybag is carefully opened first so that the media remains intact, then the seeds are inserted into the planting hole and backfilled with soil up to the height of the root neck. Seedlings that have been planted are marked with a stake that is plugged near the seedling.

3.5.8 Revegetation With Plantations And Food Crops

Revegetation with plantation crops and food crops is similar to revegetation with forest crops. The difference between the two is more to the type of plant selected. Plantation crops that can be selected are rubber, cocoa, coffee, and others, while food crops such as corn, cassava, sweet potato, and others.

If the use of ex-mining land is carried out through revegetation, then improving the quality of the soil, as discussed in the previous chapter, becomes very important to do. Another thing that needs to be considered besides the quality of the soil as a planting medium is climatic conditions. Soil quality and climatic conditions will determine the success rate of land restoration.

Soil health is also an essential factor, mainly if land restoration is used for planting food crops. Not all ex-mining areas are dangerous from the food safety aspect. However, some of them need to be considered, mainly if the geochemical aspect of the extraction method of mining materials produces or uses harmful compounds to the environment.

3.5.9 Maintenance and Monitoring

Maintenance and monitoring are activities carried out after land management and revegetation activities have been completed to ensure that the results of land restoration are as expected. Maintenance includes plant

 https://doi.org/10.1051/e3sconf/202133106007

maintenance activities and maintenance of the physical condition of the field. Monitoring needs to be carried out periodically until finally all the land that has been restored is judged to have been successful.

4 Conclusion

Based on the description in the previous chapter, it can be concluded as follows:

1. Mining areas are in rocky characteristics that are scattered or easily destroyed. Conditions will reach instability if there is high enough rainfall, causing landslides.
2. With a steep or very steep slope, the velocity of runoff will increase, and the volume of surface water will increase, resulting in a high risk of landslides. Then the location of the mine site, which is upstream of the Batang Hari River Basin, has the potential to experience debris flow.
3. Most of the Ecosystem services in the form of prevention and protection from disasters are in the low/deficient category, so the potential for disaster is high to very high.
4. The mining area is in the Landslide Hazard/Prone to Landslide area because it is located on the active Semangko Fault and volcanic mountain area.
5. The results of the calculation of the potential for landslides show the mining area Landslide Hazard is very high, so it needs mitigation to reduce damage to the surrounding environment.
6. There are violations in mine management, both administrative and operational, including administrative aspects, operational aspects, mining technical aspects, environmental protection aspects, conservation aspects, health and safety aspects, and aspects of standardization and mining services business.

Thank you to all parties who helped in the research process.

References

1. Mineral Resources Energy. Introduction to the Land Movement, Vulcanological Survey of Indonesia. Mineral Resources Energy. Jakarta. (2005)
2. C Fandeli, Z Zakaria, Slope Analysis. Unpad. Bandung. (2000)
3. Department of Natural Resources Of West Sumatra Province, Mapping Of River Flow Areas, (2021) (Unpublished).
4. Ministry Of Public Works And Spatial Planning, Republic Of Indonesia
5. West Sumatra Provincial Environmental Office, Capacity And Regional Sustainability, (2021).
6. Department of Energy and Mineral Resources, Geological Map (2021).
7. Regional Disaster Management Agency Of West Sumatra Province, Disaster Location Data, (2021).
8. C Guillard, J Zezere, Landslide Susceptibility Assessment And Validation In The Framework Of Municipal Planning In Portugal: The Case Of Lourses Municipility. Environmental Management, (2012)
9. O. F. Althuwaynee, B Pradhan, & S. Lee, Application Of An Evidential Belief Function Model In Landslide Susceptibility Mapping. Computers & Geosciences, **44**, 120-135, (2012)
10. A. Akgun, C. Kineal, & B. Pradhan, Application Of Remote Sensing Data And Gis For Landslide Risk Assessment As An Environmental Threat To Izmir City (West Turkey). Environ Monit Assess, **184**, 5453-5470. (2012)
11. C. J. Van Westen, E. Castellanos, & S. L. Kuriakose, Spatial Data For Landslide Susceptibility, Hazard, And Vulnerability Assessment: An Overview. Engineering Geology Journal, **102**, 112-131, (2008)
12. I. G. Tejakusuma, Threat of landslides. Natural, **12**, 72-75, (2007)
13. M. L. G Gritzner, W. A Marcus, R. Aspinal, & S. G. Custer, Potential Landslide Assessment Using Gis, Soil Wetness Modelling And Topographic Attributes, Payette River, Idaho. Geomophology, **37**, 149-165, (2001)
14. Directorate of Volcanology And Geological Hazard Mitigation, Determine The Landslidehazard, Jakarta, (2005)
15. Martono. (Effect of Rain Intensity and Slope Slope on Soil Loss Rate in Gray Regosol Soil. Thesis. Diponegoro University, Semarang), (2004)
16. A. G. Kartasapoetra, (Damage to Agricultural Land and Efforts to Rehabilitate It.) Bina Aksara, Jakarta, (1990)
17. S. Arsyad. (Soiland Water Conservation). Ipb Press. Bogor (1989)

 https://doi.org/10.1051/e3sconf/202133107001

Overview of 1990s deadly tsunamis in Indonesia

Admiral Musa Julius[1,2,3*] and *Daryono*[1,2]

[1]Indonesia Agency for Meteorology Climatology and Geophysics (BMKG)
[2]Indonesia Association of Geophysicist (HAGI)
[3]IOC/UNESCO Indian Ocean Tsunami Information Centre (IOTIC)

Abstract. In the 1990s there were 7 (seven) deadly tsunami events due to earthquakes in Indonesia. There is M 7.8 North-east of Flores island sea earthquake (1992), M 7.6 South of Java island sea earthquake (1994), M 6.8 South of Timor Island sea earthquake (1995), M 7.8 North-west of Sulawesi island sea earthquake (1996), M 7.9 East of Biak island sea earthquake (1996), M 7.7 Taliabu island sea earthquake (1998), and M 7.4 East of Banggai Island sea earthquake (2000). Those earthquake and tsunami disasters events affected the number of casualties and damaged houses in coastal areas. Surely those events present a few lessons learned for future disaster preparedness in Indonesia.

1 Introduction

Between 1990-2000 there were 7 (seven) earthquake events that generated deadly tsunamis in Indonesia. Those tsunamis induce losses that damage houses and the number of casualties. Those tsunami events triggered by earthquakes such as **1)** M 7.8 North-east of Flores Island Sea earthquake (1992); **2)** M 7.6 South of Java Island Sea earthquake (1994); **3)** M 6.8 South of Timor Island Sea earthquake (1995); **4)** M 7.8 North-west of Sulawesi island sea earthquake (1996); **5)** M 7.9 East of Biak island Sea earthquake (1996); **6)** M 7.7 Taliabu island sea earthquake (1998); and **7)** M 7.4 East of Banggai Island sea earthquake (2000) [4,5].

The tsunami hazard triggered the buildings to be damaged and populations to be casualties on the nearest coastal area, even in a far direction or called a distant tsunami. We find interesting information on the 1990s BMKG Tsunamis Catalogue that is one of them is an unfelt earthquake able to generate tsunami which hit the coastal zone, destroy the building and induce casualties [1,2].

In this article, we write the overview of deadly tsunamis information on Indonesia to denote the possible casualties and damages due to tsunami-genic and tsunami earthquakes in the future. We suggest this event be the lesson learned for tsunami specialists to consider the multi-hazard on certain possible areas. According to the recent unexpected multi-hazard event, we also propose to conduct the vulnerable reduction for people and buildings as an initial effort to reach disaster resilience [3,6].

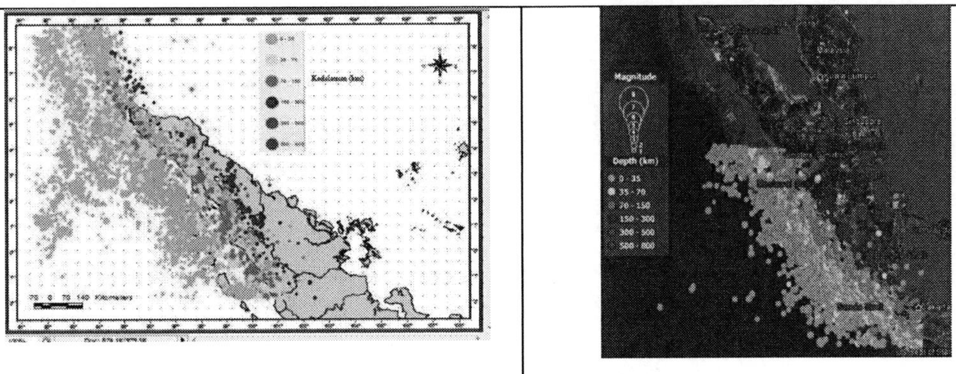

Seismicity map of Northern Sumatra	Seismicity map of Southern Sumatra

Fig. 1. Seismicity map of Sumatra 1973-2012 (USGS) [29]

* Corresponding author: admiralmusajulius@gmail.com

2 Data and Method

Since 1990-2000, we have discovered 7 deadly tsunami data in the Indonesia region. We used the data from the BMKG Earthquake and Tsunami Center online repository. We used desk literature study towards the BMKG destructive tsunami catalog, scientific reports, international tsunami bulletin, and earthquake and tsunami newsletter compilation books. The seismicity map dataset was collected from USGS (1973-2012) and processed using the ArcView program [7,8].

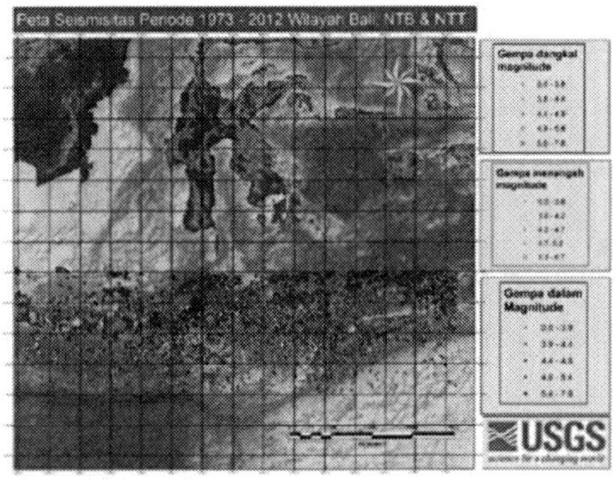

Fig. 2. Seismicity map of Bali and Nusa Tenggara Island 1973-2012 (USGS) [32]

The affected region of the M 6.8 South of Timor Island Sea earthquake (1995) nowadays has left from territorial of Republic of Indonesia to be Timor-Leste or East Timor, while the affected region of M 7.9 East of Biak Island Sea earthquake (1996) was on Irian Jaya province, nowadays has divided into 2 (two) provinces to be Papua and West Papua [9,10].

3 Result and Discussion

Most tsunamis are generated by earthquakes, we discover all tsunami events in Indonesia in the 1990s were triggered by an earthquake on the following event:

3.1 M 7.8 North-east of Flores Island Sea earthquake (1992)

December 12, 1992, North-east of Flores Island Sea earthquake (M 7.8), which caused a destructive shaking, was located in the 8.340 S - 122.490 E near Babi Island on Sikka regency at 13:29 Central Indonesian Time (Local Time). It occurred on the 20.4 km depth and was felt by people in the half of eastern part of Flores Island and its vicinities island such as in Maumere XI MMI, Ende VIII MMI, Ngada VII MMI, Larantuka VI-VII MMI, Waingapu, Sumba, and Ujungpandang IV MMI. The maximum flood of tsunami inundation can reach 300 m to the land, various run-up measured in villages 20.4 km depth and was felt by people in the half of eastern part of Flores island and its vicinities island such as in Maumere XI MMI, Ende VIII MMI, Ngada VII MMI, Larantuka VI-VII MMI, Waingapu, Sumba and Ujungpandang IV MMI. The maximum flood of tsunami inundation can reach 300 m to the land, various run-up measured in villages such as 1) Ilepadung – East Flores 11 m; 2) Babi island – Sikka 5,5 m; 3) Wuring - Sikka 2,4 - 3,2 m; 4) Mausambi – Ende 3,4 m, and 6) the highest run-up reach 25 m in Riangkroko village near Cape Batumanuk, far end of Flores island [11,12].

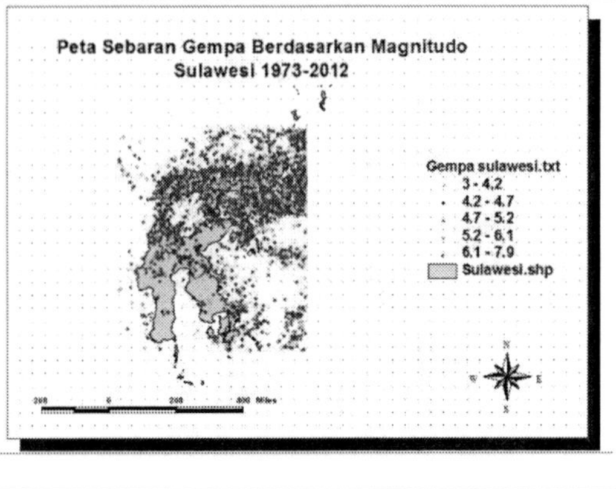

Seismicity map of Sulawesi Based on Magnitude

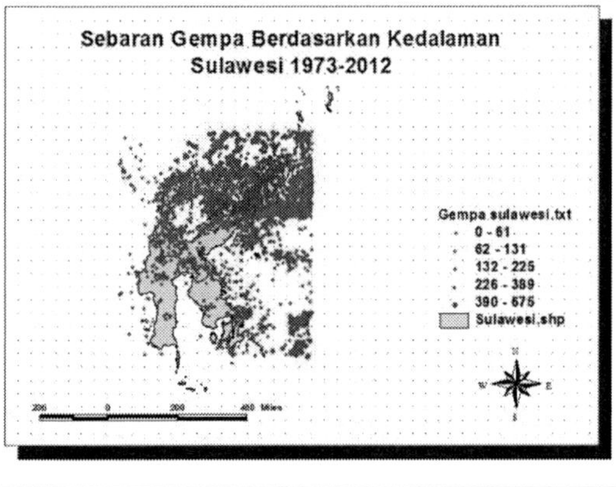

Seismicity map of Sulawesi Based on Depth

Fig. 3. Seismicity map of Sulawesi 1973-2012 (USGS) [35]

The local government noted that the total number of casualties is 1952 deaths, 2126 injuries, and 500 lost. It also noted that more than 20000 buildings including houses were damaged. This phenomenon is related to Flores's back-arc thrust tectonic activities. These earthquake-induced collateral hazards such as submarine landslides, liquefaction, and tsunami. It is strongly possible, considering most of the east of Flores Island were covered by volcanic deposits sourced from volcanic activity such as Mt. Ile Mandiri, Mt. Lewotobi, Mt. Egon, and Mt. Rokatenda. The damage also could be caused by topographical effects considering the coastal village is

 https://doi.org/10.1051/e3sconf/202133107001

located on a steep hill. This site condition can amplify the ground motion acceleration to be stronger [15,16].
Two interesting phenomena were reported from the losses on Babi island and Riangkroko village. Most of the evidence of damages and casualties in Babi island was focused on the south part, the opposite of the epicenter source of the tsunami generated. The publication of tsunami modeler answers this question that this phenomenon was due to the reflected wave which hit Flores island, the coastal zone facing the southern part of Babi island. The other interesting question also comes from the far end of east Flores island, one village reported being lost due to being swept by the giant wave. This phenomenon was long discussed due to the anomalous 25 m run-up which struck the coastal zone of Riangkroko village [19, 20].

3.2 M 7.6 South of Java island sea earthquake (1994)

On June 2, 1994, at 01:17:34 Western Indonesian Time (Local Time) a slow earthquake (M 7.2) struck south of East Java Sea. It is located 200 km south of Banyuwangi regency - East Java. It occurred at the 18 km depth and was felt by a small number of people in Banyuwangi (III-IV MMI). The local government noted that 720 buildings were damaged, 223 people die, 22 people lost, and 440 people were injured due to the tsunami hit. According to a BMG survey, the flood of tsunami inundation can reach 300 m to the land in some coastal villages [21, 22].

The seismic analysis concluded this tsunami was categorized as the tsunami-earthquake or slow slip earthquake. According to the hypocenter and focal mechanism analysis, this earthquake was generated by thrust faulting in the Indian Ocean south of the Java subduction zone or locally named Sunda trench. The field survey of run up measured in villages such as: **1)** Pancer – Banyuwangi 5,7 - 9,4 m; **2)** Lampon – Banyuwangi 9,3 m; **3)** Rajekwesi - Banyuwangi 13,9 m; **4)** Grajagan – Banyuwangi 4,1 m, and various lower run up in Jember, Tulung Agung, Blitar, Malang, Lumajang and Bali, coastal regencies near Banyuwangi. This phenomenon was focus discussed due to the slow felt shaking based on eyewitnesses or survivor storytelling in the coastal zone, and most houses were not damaged by the earthquake but tsunami waves do [23, 24].

3.3 M 6.8 South of Timor Island sea earthquake (1995)

On May 14, 1995, South of Timor Island sea earthquake (M 6.8), a major earthquake struck deep under the Timor sea, creating tsunami waves that swept houses away and damaged fishing boats in East Timor. 11 people died, 19 people were injured and 15 others were missing. The harbor was badly damaged and several office buildings and homes had gaping cracks and collapsed roofs. Five aftershocks in the hours following the quake initially prevented people from returning to their homes. This affected region has left the administration of Indonesia since 1999 to be an independent country [25, 26].

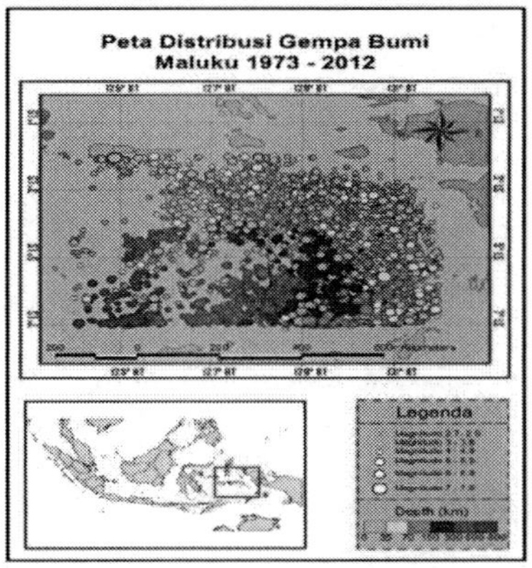

Seismicity map of Maluku

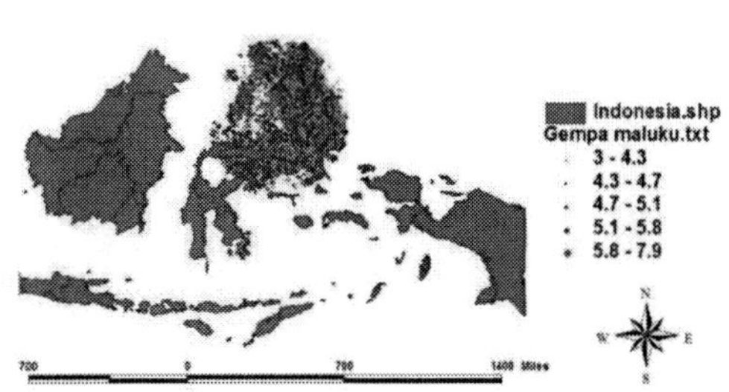

Seismicity map of Northern Maluku

Fig. 4. Seismicity map of Southern Maluku 1973-2012 (USGS) [31]

3.4 M 7.8 North-west of Sulawesi Island sea earthquake (1996)

On 1 January 1996 at 16:05 central Indonesian time, an M 7.8 earthquake struck the North-west of the Sulawesi island sea, approximately 180 km north of Palu. It generated a tsunami to the coastal zone of 100 km northeast coastal villages of Sulawesi island in Toli-toli regency. The international Tsunami Survey Team (ITST) conducting field survey of run up measured in villages such as: 1) Pagalaseang 2,28 m; 2) Munte 3,17 m; 3)

 https://doi.org/10.1051/e3sconf/202133107001

Limbosu 2,81 m; 4) Tonggolobibi 1,82 – 3,43 m, 5) Taipah 2,40 – 3,25 m; 6) Siboang 1,78 m; 7) Siwalempu 1,62 m; 8) Balukang 2,52 m; 9) Soni 1,79 m; 10) Dongko 2,39 m; and 11) Simuntu 2,00 m [27, 28].

Nine people were killed and 63 were injured in Tonggolobibi village by a 2 m high tsunami, more than 400 houses were destroyed and became unfit for living. In the coastal villages, the earthquake intensity was VI MMI. According to the tectonic setting, this earthquake is associated with the subduction interface under the North Sulawesi arm. The earthquake's source was due to a shallow-dipping thrust fault that roughly coincides with the interface between Sulawesi and the Celebes Sea subduction zone [29, 30].

3.5 M 7.9 East of Biak island Sea earthquake (1996)

On February 17, 1996, at 14:59 Eastern Indonesian Time (Local Time) an earthquake (M 7.9) struck East of the Biak island Sea. It is located 40 km east of Biak island – Irian Jaya. It occurred on the 8 km depth and was felt by people in the whole part of Biak island and its vicinities islands such as VIII-IX MMI in Biak and Supiori regency. The tsunami hit the whole coastal part of Biak island and its vicinity islands such as Yapen island, Owi island, Pai island, also the north part of Papua island coastal zones such as Manokwari and Sarmi. According to the tectonic setting, this earthquake was associated with the Carolina-Pacific plate beneath to Bismarck-Australian plate [31, 32].

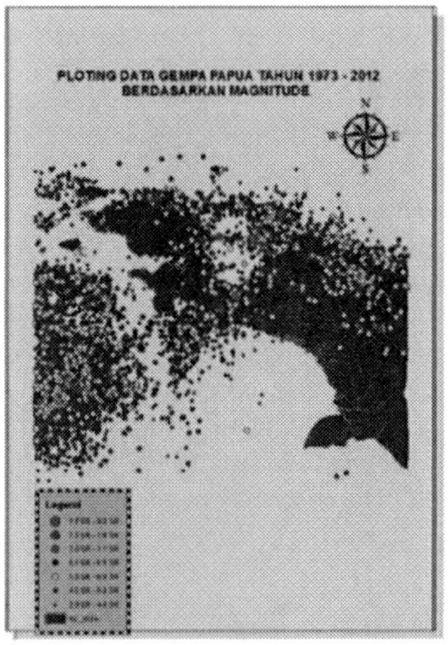

Seismicity map of Papua Based on Magnitude

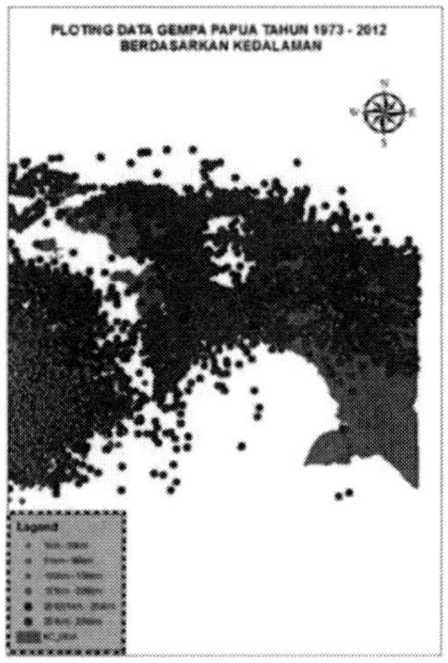

Seismicity map of Papua Based on Depth

Fig. 5. Seismicity map of Papua 1973-2012 (USGS) [33]

The interesting phenomena were discovered to refer to storytelling by eyewitnesses or survivors. The average tsunami arrival time duration approximately was 10-15 minutes since the shaking into tsunami hit in the east part of Biak island, while in the west part of Biak island were only 5 minutes, the opposite part of the earthquake source. The other interesting story is the average run-up in the eastern part was 4-6 m, while 7.7 m in the western part of Biak island. Both novel finds were long discussed due to both anomalous. However, the further approaching research mentioned this anomalous may relate to atypical sources such as submarine landslides in the near coast of the west part of Biak island [33, 34].

The affected coastal villages consist of Mansoben (Supiori), Warsa (Biak Numfor) Korem (Biak Numfor), Sauri (Biak Numfor), Farusi (Biak Numfor), and Wardo (Biak Numfor). The total amount of casualties was 110 people died, 51 people missing, 423 person injuries, more than 10.000 refugees, 5043 houses damaged. The official report on each village noted that 17 persons were dead in Saukobye, 8 persons dead in Andey, 7 persons dead in Rosayendi, and 7 persons dead in Korem [1, 35].

3.6 M 7,7 Taliabu island sea earthquake (1998)

On November 28, 1998, the M 7,7 Taliabu island sea earthquake trigger 3m tsunami waves and killed 34 persons. It is located in the 2.00 S 124.80 E.

3.7 M 7.4 East of Banggai Island sea earthquake (2000)

On May 4, 2000, at 12:21 Central Indonesian Time (Local Time) a strong earthquake (M 7.4) struck east of the Sulawesi and Banggai islands and triggered a tsunami to the Sulawesi coastal zone of Luwuk and the neighbouring islands of Peleng and Banggai. The earthquake centered in the east Sulawesi region about 100 miles (160 km) south-southeast of Gorontalo. It occurred at the 26 km

 https://doi.org/10.1051/e3sconf/202133107001

depth and was felt by several people in Banggai and Luwuk [13,14].

The tsunami waves, reportedly measuring up to six meters in height inundated two villages on Peleng Island, Kayuntayo and Uwedikum. In Luwuk (607 kilometers east of Palu, capital of central Sulawesi). Hundreds of homes on Banggai island were destroyed, and at least 54 people were reported dead, 264 injured, and more than 19.000 buildings damaged. Reports indicate that as many as 170.000 people have been made homeless [17,18].

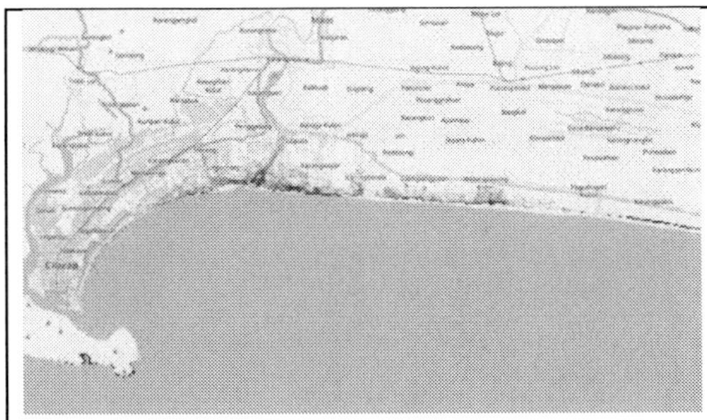

Cilacap Tsunami Inundation Map

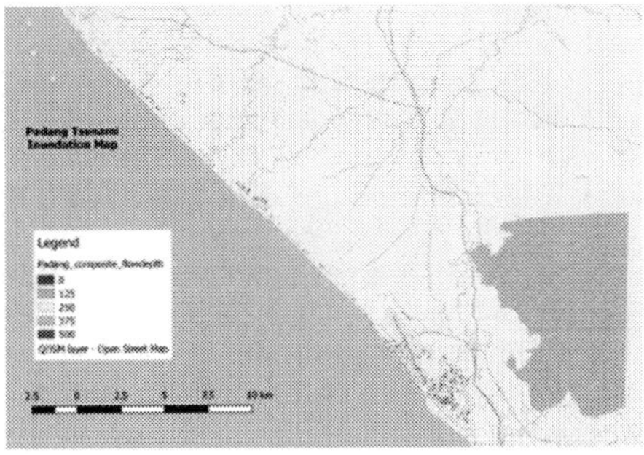

Padang Tsunami Inundation Map

Fig 6. Tsunami inundation map

3.8 Evacuation Map Based on Worst-Case Tsunami Scenario

Those 7 tsunami events have their uniqueness and will repeat themselves in the future. One of the Tsunami Ready Indicators (IOTR) Programme for Indonesia is having designated and mapped tsunami hazard zones. The primary source for mapping potential tsunami hazard zones is inundation modelling, which illustrates expected areas to be flooded by a tsunami. In this discussion, we illustrate tsunami inundation modelling for Batang Anai and Cilacap Regency Coastal Zone using the Community Model Interface for Tsunami (ComMIT).

It uses initial conditions from a pre-computed propagation database, has an easy-to-interpret graphical interface, and requires only portable hardware. We used certain worst-case scenarios for each location based on the Indian Ocean Probabilistic Tsunami Hazard Assessment (IO-PTHA) study, those are Mw 9.2 Andaman, Mw 9.2 Nicobar, Mw 9.2 Aceh, and Mw 9.2 South Sumatra earthquakes for Batang Anai, and Mw 8.8 West Java, Mw 9.0 Central Java, Mw 9.0 East Java for Cilacap tsunami hazard maps

The ComMIT server bathymetry dataset is derived from ETOPO1 one arc-minute gridded global relief model and has been interpolated from 60 arc-second to 3 arc-second to match the topography data set produced by NOAA, while the topography dataset is derived from Shuttle Radar Topography Mission (SRTM) 90m created by NASA. Multiple inundation scenarios of each coastal area were stored into a composite. These composites were converted into vector format using open-source QGIS software. Further, these inundation vector files were cleaned and the final inundation area was prepared. These inundation details overlaid on the base data (open street maps, google maps, etc.) along with the locations of safe shelters and evacuation routes. The final evacuation maps detailing the extent of inundation, evacuation routes, and locations of safe shelters were composed in QGIS. This evacuation map will help coastal residence to be more understand the way to save lives facing tsunami potential.

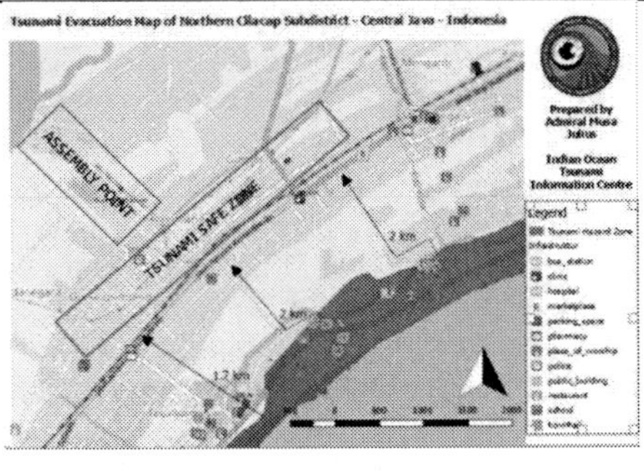

Cilacap Tsunami Evacuation Map

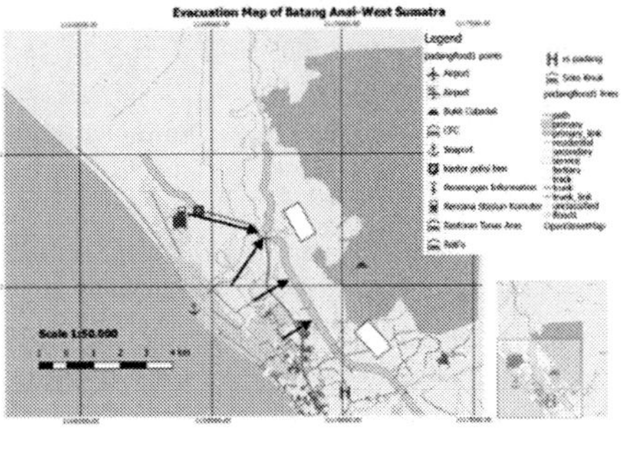

Batang Anai Tsunami Evacuation Map

Fig. 7. Tsunami evacuation map

4 Conclusion

Along the 1990s, Indonesia was hit by 7 deadly and near-field tsunami events generated by M>6.7 earthquakes in the sea of Flores (1992), Java (1994), Timor (1995), Sulawesi (1996), Biak (1996), Taliabu (1998), and Banggai (2000). Besides the tsunami, we saw the earthquake also able to generate other collateral natural hazards such as submarine landslide, rockfall, and liquefaction. We must expect more than one wave. When a tsunami struck the coastal zone, it may strike the nearest coastal zone consisting of threatened people, vulnerable buildings, and the environment. On the mitigation of tsunami risk reduction context, we can use evacuation maps, because they will be handy for the coastal disaster management authority and also used as awareness material to sensitize the administrators and people at risk.

The Authors would like to express their gratitude towards Indonesia Agency for Meteorology, Climatology, and Geophysics (BMKG), Indian Ocean Tsunami Information Centre (IOTIC) IOC/UNESCO, and Indonesia Defence University (UNHAN) for supporting this research.

References

1. R.N. Agus, I. Suardi, R. Sipayung, & D. Sianipar, Local seismicity pattern around Mt. Pandan, East Java according to February 2016 earthquake swarms activity. AIP Conference Proceedings 1987, 020034. (2018)
2. V.H. Banyunegoro, Z.A. Alatas, A. Jihad, Eridawati, and M. Umar, Probabilistic Seismic Hazard Analysis for Aceh Region. IOP Conference Series: Earth and Environmental Science, Volume 273. 012015. (2019)
3. V.H. Banyunegoro, M. Umar, and Y. Idris. Seismic microtremor experiment to determine the seismic vulnerability of North Aceh. IOP Conference Series: Materials Science and Engineering, Volume 846, 012053. (2020)
4. BMKG Earthquake and Tsunami Center. Indonesia Catalogue for Significant and Destructive Earthquakes 1821-2019. Jakarta: Indonesia Agency for Meteorology Climatology and Geophysics. In Bahasa Language. (2020)
5. BNPB Data, Information and Public Relations Center. Indonesia Disaster Data and Information. Accessed 23 November 2019. Jakarta: B Indonesia Disaster Management Agency. In Bahasa Language. (2019)
6. A.P. Cahyaningrum, A.D. Nugraha, and N.T. Puspito, Earthquake hypocenter relocation using the double-difference method in East Java and surrounding areas. AIP Conference Proceedings 1658, 030021. (2015)
7. G. Daniarsyad, and I. Suardi, Stress triggering among MW≥6.0 significant earthquakes in Manokwari Trough. AIP Conference Proceedings 1857, 020014. (2017)
8. T.H. Hududillah, A.V.H. Simanjuntak, and M. Husni, Identification of active fault using analysis of derivatives with the vertical second based on gravity anomaly data (Case study: Seulimeum fault in Sumatera fault system). AIP Conference Proceedings 1857, 030004. (2017)
9. A.M. Julius, and B. Sunardi, Earthquake Response of Storey Building in Jakarta using Accelerographs Data Analysis. AIP Conf. Proc. 1658, 040001. (2015)
10. A.M. Julius, B. Sunardi, and A. Rudyanto, Storey Building Early Monitoring Based on Rapid Seismic Response Analysis. AIP Conf. Proc. 1730, 030005. (2016)
11. L. Kesumastuti, A. Marsono, T. Yatimantoro, and S. Pribadi, Determination of the earthquake source parameters using the W-phase inversion method and its uses for tsunami modeling. AIP Conference Proceedings 1857, 090006. (2017)
12. National Center for Earthquake Study. Indonesia Map for Earthquake Source and Hazard. Jakarta: Ministry of Public Works and Resident. ISBN 978-602-5489-01-3. In Bahasa Language. (2017)
13. A. Octhav, A.M. Julius, Muzli, and A. Rudyanto, Modified of Ground Motion Prediction Equation in Indonesia, case study: South and South-East of Sulawesi at 2011-2015. AIP Conf. Proc. 1857, 020003. (2017)
14. S. H. Pandadaran, A.M. Julius, A. Widiarso, and S.E. Kurniawan, Comparison of Ground-Motion Attenuation Models for West Sumatra and Bengkulu regions based on distant Subduction Interface and Intralab Earthquake. Proceedings Joint Convention Yogyakarta 2019, HAGI-IAGI-IAFMI-IATMI (JCY 2019). ISBN 978-979-8126-37-6. (2019)
15. D.M. Puteri, A.K. Affandi, S. Sailah, N. Hudayat, and M.K. Zawawi, Analysis of peak ground acceleration (PGA) using the probabilistic seismic hazard analysis (PSHA) method for the Bengkulu earthquake of the 1900 – 2017 period. Journal of Physics: Conference Series, **1282**, 012054. (2019).
16. A. Rahman, A. Marsono. & A. Rudyanto, Rapid magnitude estimation using τC method for earthquake early warning system (Case study in Sumatra). AIP Conference Proceedings 1857, 020017. (2017)
17. A. Rahman, M.K. Zawawi, R.A. Kiat, A. Yullatifah, dan R.A.T. Putri, Statistics Report of Earthquakes in Java region. Study Program of Geophysics. Jakarta: Indonesia School for Meteorology Climatology and Geophysics. In Bahasa Language. (2012)
18. R.A. Sagala, P.P.J. Harjadi, N. Heryandoko, & D. Sianipar, Detailed seismotectonic analysis of Sumatra subduction zone revealed by high precision earthquake location. AIP Conference Proceedings 1857 (1), 020015. (2017)

19. Y.R. Serhalawan, D. Sianipar, & I. Suardi, The January 25th, 2014 Kebumen earthquake: A normal faulting in the subduction zone of Southern Java. AIP Conference Proceedings 1857 (1), 030002. (2017)
20. T. A. P. Setiadi, S. Rohadi, & N. Heryandoko, Earthquake relocation in Mollucas Sea using the teleseismic double-difference method for tectonic setting analysis. AIP Conference Proceedings **1857**, 020007. (2017).
21. R. Sipayung, M.R. Alhafiz, R.N. Agus, & D. Sianipar, Relocation of the February 2016 Mt. Pandan earthquake sequence using double-difference with waveform cross-correlation. AIP Conference Proceedings 1987 (1), 020036. (2018).
22. R. Sipayung, D. Sianipar, S. Prayoedhi, Daryono, J. Arifin, Simanjuntak, A.V.H., Umar, M., Prabu, S., Daniarsyad, G., Haryanta, Putranto, N., dan Azimi, A. Revisiting The 2018 Kalibening Earthquake Sequence in Central Java: Call for the Revision of Earthquake Hazard. (2019). IOP Conference Series: Earth and Environmental Science, **273**, 012018. (2019).
23. A.V.H. Simanjuntak, M. Umar, Qadariyah, Y. Setiawan, and Irwandi. Source Mechanism Analysis by Using Tensor Moment Inversion (Study Case: Pidie Jaya Earthquake in 2016 December 7th). IOP Conference Series: Earth and Environmental Science, **273**, 012021. (2019).
24. A.V.H. Simanjuntak, M. Husni, and M. Syirojudin, Subsurface structure identification of active fault based on magnetic anomaly data (Case study: Toru fault in Sumatera fault system). AIP Conference Proceedings 1857, 030003. (2017).
25. A.V.H. Simanjuntak, M. Umar, and F. Rahmayani, Microtremor survey to investigate seismic vulnerability around the Seulimum Fault, Aceh Besar-Indonesia. IOP Conf. Series: Materials Science and Engineering 352, 012046. (2018).
26. A.V.H. Simanjuntak, M. Umar, and R. Sipayung, Earthquake relocation using HypoDDMethod to investigate active fault system in Southeast Aceh. IOP Conf. Series: Journal of Physics: Conf. Series 1116, 032033. (2018)
27. R. M. Taruna, V. H. Banyunegoro, Earthquake Relocation Using Double Difference Method for 2D Modelling of Subducting Slab and Back Arc Thrust in West Nusa Tenggara. Jurnal Penelitian Fisika dan Aplikasinya (JPFA), v. **8**, n. 2, p. 132-143. ISSN 2477-1775. (2018).
28. R.M. Taruna, V.H. Banyunegoro, G. Daniarsyad, Peak ground acceleration at the surface for Mataram city with a return period of 2500 years using the probabilistic method. MATEC Web Conf. Volume 195, 2018:03019. (2018).
29. A. V. H. Simanjuntak, A. Rachmadan, A. Mustika, and N.A. Akuba, North Sumatra Region Earthquake Statistics Report. Geophysics Study Program. Jakarta: College of Meteorology, Climatology and Geophysics. (2012)
30. E. Nopianti, G. Daniarsyad, L. Kesumastuti, T.A. Setiadi, dan Banyunegoro. Earthquake Statistics Report for Southern Sumatra Region. Geophysics Study Program. Jakarta: STMKG. (2012)
31. A.N. Vita, B. Sihotang, Aswin, R.A. Sagala, S.Y.S. Putra, Maluku Region Earthquake Statistics Report. Geophysics Study Program. Jakarta: STMKG. (2012)
32. A.N. Rahayu, Armansyah, F. Khairina, R.A. Triputra, T. Taufik, Earthquake Statistics Report for Bali, NTB, and NTT Regions. Geophysics Study Program. Jakarta: STMKG. (2012)
33. P.K.G.A. Negara, A. Octhav, R.M. Taruna, B.A. Maulana, Y.R. Serhalawan, Papua Region Earthquake Statistics Report. Geophysics Study Program. Jakarta: STMKG. (2012)
34. A. Rahman, M.K. Zawawi, R.A. Kiat, A. Yullatifah, dan R.A.T. Putri, Java Region Earthquake Statistics Report. Geophysics Study Program. Jakarta: STMKG. (2012)
35. A.M. Julius, B.F.A. Rumy, D.S.J. Sianipar, R. Sulsaladin, and Y.H. Perdana, Sulawesi Region Earthquake Statistics Report. Geophysics Study Program. Jakarta: STMKG. (2012)

Mapping seismic exposure of building based surface peak ground acceleration: initial seismic risk assessment in Bandung Basin, West Java

Afnindar Fakhrurrozi[1*], *Anggun Mayang* Sari[1,5], *Arifan Jaya* Syahbana[1,2], *Dwi* Sarah[1], *Bambang* Setiadi[1], *Mudrik Rahmawan* Daryono[1], *Rabieahtul Abu* Bakar[3], and *Jian Cheng* Lee[4,5]

[1]Research Center for Geotechnology (P2G-LIPI), Indonesian Institute of Sciences, Bandung, Indonesia
[2]Faculty of Civil and Environmental Engineering, Bandung Institute of Technology, Bandung, Indonesia
[3]Razak Faculty of Technology and Informatics, University Technology Malaysia (UTM), Kuala Lumpur, Malaysia
[4]Institute of Earth Science, Academia Sinica, Taipei, Taiwan
[5]Integrated Research on Disaster Risk-International Centers of Excellence (IRDR-ICoE), Taipei, Taiwan

Abstract. An important key to determining the disaster risk is exposure. It is truly dynamic in space and time due to the expansion and change of the settlements. A seismic hazard is an earthquake disaster that can deliver damage to the physical environment. Hence, it is required to analyse the objects exposed by an active earthquake source in the Bandung Basin. The exposed objects that experience the most impact due to the shock of the earthquake are residential buildings. The West Bandung Regency was opted as our study area due to the high-risk zone for earthquake waves based on seismic hazard calculation. This paper tried to analyse the seismic exposure as the impact of earthquake waves around the Bandung Basin on physical buildings in the West Bandung Regency using a geospatial analytics approach. The result shows that the location area with many damaged buildings correlates with the spatial distribution of surface peak ground acceleration.

1 Introduction

During the past two decades, the number of earthquakes has increased dramatically in West Java, primarily attributed to induced seismicity from the Sunda Megathrust and the Lembang Fault, as shown in Fig. 1 [1,2,3]. The increased seismic disaster risk has affected several West Java regencies, including West Bandung Regency. Thus, scientific information related to physical buildings that are affected by seismic hazards is needed.

Peak Ground Acceleration (PGA) computation can be used to determine the seismic hazard that affects buildings[4,5]. The PGA values on the surface were obtained using Earthquake waves propagation. These values were used as an input for spatial analysis.

Spatial analysis is to map the buildings that are exposed to seismic hazards. The spatial distribution of seismic hazards and buildings that are affected by seismic waves can be achieved with a geographic information system approach [6]. Spatial analysis to estimate the spatial distribution of point samples at the unknown location required interpolation. Two spatial interpolation methods are widely used by GIS practitioners, namely Inverse Distance Weighting (IDW) and Kriging [7]. The IDW interpolation method is claimed to have a lighter performance than Kriging [7]. Hence, this study applied the IDW to map the seismic exposure of physical buildings.

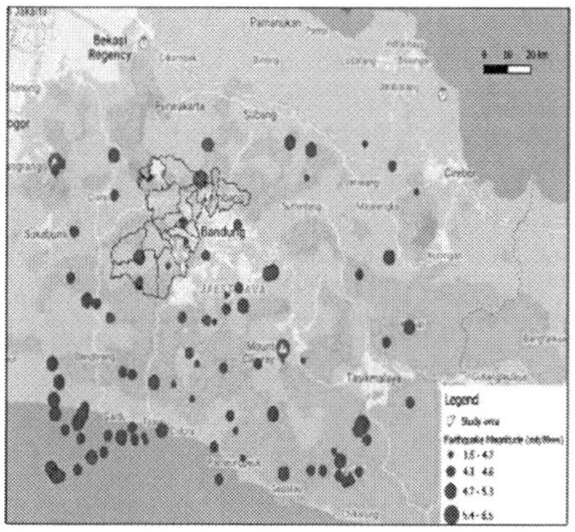

Fig. 1. Earthquake events in the last 20 years in West Java province. Mw > 3.5 (USGS)

2 Study Area

There are three sub-districts in West Bandung Regency; namely Batujajar, Cihampelas, and Cililin. Several soil field investigations, Core Penetrant Test (CPT), known as CPTu, settled on these areas, as shown in Fig. 2.

* Corresponding author: afnindar.fakhrurrozi@lipi.go.id

https://doi.org/10.1051/e3sconf/202133107002

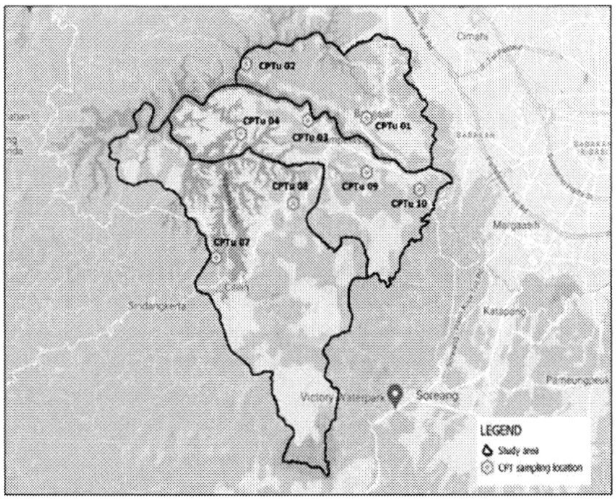

Fig. 2. The study area, along with the sampling location of CPT, is marked with an orange circle. It is consists of three sub-districts in West Bandung Regency.

CPTu-01 and CPTu-02 are located in the Batujajar sub-district. This area is a densely populated residential zone. Different from Batujajar area, Cihampelas sub-district dominated by a water area, which is Saguling dam. The physical building is located in the eastern part of this subdistrict. CPTu-03, CPTu-04, CPTu-09 and CPTu-10 were found in this subdistrict. The third is Cililin sub-district, which is the largest subdistrict in West Bandung Regency. This area consists of half land and half water, and the building infrastructures are distributed in the southern part of Cililin sub-district.

3 Methodology

The study used PGA values as the main source to map buildings that were exposed to seismic hazards. Nonlinear approach analysis was used to calculate the PGA values [8][9]. It was assumed that the shear waves had been propagated vertically in the one-dimensional layer under some soil layer conditions; homogenous, infinite extent, and horizontal motion [10]. It can be expressed in a given equation:

$$\rho \frac{\partial^2 d}{\partial t^2} + \eta \frac{\partial d}{\partial t} = \frac{\partial \tau}{\partial z} \tag{1}$$

Where ρ is the soil unit mass, d is horizontal displacement, z is the depth, t is the time, τ is the shear stress, and η is the mass damping parameter. The displacement, velocity, and acceleration are then calculated using a finite difference formulation as follows:

$$\frac{\partial \tau}{\partial z} = \frac{(\tau_i + 1) - (\tau_i - 1)}{\Delta z} \tag{2}$$

$$a_{i,n} = \frac{\partial^2 d}{\partial t^2} = \frac{\bar{v}_{i,n+\Delta n} - \bar{v}_{i,n}}{\Delta t} \tag{3}$$

Where $\bar{v}$ is motion velocity, $a_{i,n}$ is the acceleration at nodes i (i = 1, ..., n) and n-step. The PGA computation was applied to eight CPT samples. The location of each CPT sample can be seen in Fig. 2.

The PGA values were interpolated using IDW to estimate the PGA distribution. The IDW calculates the weighted averages of sample values to estimate the unknown value $\hat{z}$ given particular location x [7,11]. Mathematically, the IDW method can be represented as:

$$\hat{z}(x) = \frac{\sum_i^n w_i z_i}{\sum_i^n w_i} \tag{3}$$

$$w_i = |x - x_i|^{-\beta} \tag{4}$$

Where β is the inverse distance power and β ≥ 0, w_i is the weight value at n (n = 1, ..., n) location that is the euclidean distance, and z is the observed value.

A field survey was carried out to obtain ground truth data of physical building using the Global Earthquake Model (GEM) taxonomy framework. Moreover, stratified random sampling was employed to select sample sites. There are ten random samples placed inside a rounded buffer of 2.5km of each CPT sample. There are 80 random physical building sample data over 8 CPT points that were collected.

The parcel of building map was developed by using manual digitization due to data limitations.

4 Results

The PGA values on each sample location vary between 0.47 – 0.70, as shown in Table 1.

Table 1. The PGA computation result

ID	a_{max} $surface$
CPTu-01	0.54
CPTu-02	0.62
CPTu-03	0.54
CPTu-04	0.70
CPTu-07	0.52
CPTu-08	0.47
CPTu-09	0.51
CPTu-10	0.51

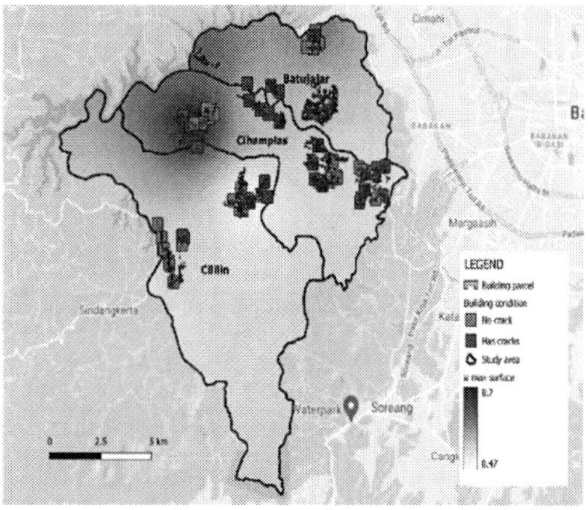

Fig. 3. The interpolated map of PGA values on the surface along with the building parcel and the building condition.

Fig. 4. A building located at CPTu-04. It has a wide crack opened and was marked inside a red box.

Fig. 5. A building located at CPTu-08 has structural damage. It was marked inside a red box.

Building condition in CPTu-04 located at Cihamplas that has been severely damaged can be seen in Fig. 4. On the one hand, the building with cracks in the Cililin sub-district can be seen in Fig. 5.

The highest value of PGA is located at Cihamplas sub-district, while the lowest value is located at Cililin sub-district. There is a strong correlation between the PGA value on the surface and the building that has cracks. It can be seen in Fig. 3. There are many buildings in Cihamplas sub-district that have cracks. The number of damaged buildings observed in Cihamplas is almost 70%. Conversely, Cililin sub-district that has the lowest value of PGA also has lesser damaged buildings, with approximately 25% of total buildings having cracks.

5 Conclusion

The seismic exposure of building maps gives us insight into the particular area prone to an earthquake disaster. It also can help to predict and estimate the building damages and losses. Our result confirms that the PGA value on the surface has a strong correlation with the number of damaged buildings. This shows that using the geospatial analytics approach can capture the seismic exposure related to the existing buildings. However, in future research, we must delve deeper into the causes of structural damage using building taxonomies classification and machine learning to automate building footprint extraction using high-resolution imageries and estimate the number of damaged buildings. Hence, we can model and quantify the number of building exposed to seismic hazards more precisely.

The authors would like to extend our appreciation to the Research Center for Geotechnology, the Indonesian Institute of Sciences, for their full support during the research. This research is funded by the Advanced Institute Seed Grant Program of the International Science Council Regional Office for Asia and the Pacific (ISC ROAP) and Integrated Research on Disaster Risk and the International Centre of Excellence Taipei (IRDR ICoE-Taipei). This research also is supported by Innovation System Research project funding (INSINAS 2021).

References

1. I. W. Sengara, Y. Munaf, and I. Susila, Int. Conf. Recent Adv. Geotech. Earthq. Eng. Soil Dyn. 1 (2001)
2. M. R. Daryono, D. H. Natawidjaja, B. Sapiie, and P. Cummins, Tectonophysics **751**, 180 (2019)
3. A. J. Syahbana, A. M. Sari, E. Soebowo, M. Irsyam, M. Asrurifak, and Hendriyawan, *Proc. 20th Southeast Asian Geotech. Conf. 3rd AGSSEA Conf. Conjunction with 22th HATTI Annu. Sci. Meet. Thema Geotech. Chall. Mega Infrastructures* (Indonesian Society for Geotechnical Engineering, Jakarta), 290–295 (2018)
4. A. Sari, A. Fakhrurrozi, and A. Syahbana, *Proc. Proc. 7th Math. Sci. Comput. Sci. Educ. Int. Semin. MSCEIS 2019, 12 Oct. 2019, Bandung, West Java, Indones.* (EAI, 2020)

5. A. M. Sari, E. Soebowo, A. Fakhrurrozi, A. J. Syahbana, and A. Tohari, Ris. Geol. Dan Pertamb. **29**, 53 (2019)
6. S. Derakhshan, M. E. Hodgson, and S. L. Cutter, Appl. Geogr. **124**, 102295 (2020)
7. A. Barbulescu, A. Bautu, and E. Bautu, Appl. Sci. **10**, (2020)
8. S. Kramer, *Geotechnical Earthquake Engineering* (Prentice-Hall, New Jersey, 1996)
9. H. Arslan and B. Siyahi, Environ. Geol. **50**, 1193 (2006)
10. J. Bardet and T. Tobita, (2016)
11. B. Hartmann, K., Krois, J., Waske, (2018)

E3S Web of Conferences **331**, 07003 (2021) https://doi.org/10.1051/e3sconf/202133107003
ICDMM 2021

Seismic hazard on West Bandung district using non-linear earthquake response analysis

Anggun Mayang Sari[1,5*], *Afnindar* Fakhrurrozi[1], *Arifan Jaya* Syahbana[1,2], *Dwi* Sarah[1], *Bambang* Setiadi[1], *Mudrik Rahmawan* Daryono[1], *Rabieahtul Abu* Bakar[3], and *Jian Cheng* Lee[4,5]

[1]Research Center for Geotechnology (P2G-LIPI), Indonesian Institute of Sciences, Bandung, Indonesia
[2]Faculty of Civil and Environmental Engineering, Bandung Institute of Technology, Bandung, Indonesia
[3]Razak Faculty of Technology and Informatics, University Technology Malaysia (UTM), Kuala Lumpur, Malaysia
[4]Institute of Earth Science, Academia Sinica, Taipei, Taiwan
[5]Integrated Research on Disaster Risk-International Centers of Excellence (IRDR-ICoE), Taipei, Taiwan

Abstract. West Bandung District is one of the government areas that grow rapidly in West Java. Here several infrastructures such as Highspeed Rail Development, Modern Residences, Tourist Attractions Area, and International Retail Company supply major contributions to economic development. Rapid construction in Bandung Basin caused the government to pay attention to risk factors, especially building structures. One of the natural hazards that must be considered is the presence of an active Lembang fault. Therefore, there is a need for a study that takes into account how the earthquake waves damaged the buildings. In this study, an analysis of the soil dynamics due to earthquake ground motion from bedrock was carried out. The analysis is carried out using the wave propagation method, which is a non-linear analysis of the soil response. The research was conducted at 12 points locations in West Bandung District. As the result, the location with the highest PGA surface value is in the Mekar Jaya area, while the lowest risk is found in the Cililin area. Furthermore, this study provides scientific information on seismic hazards to support government disaster risk reduction programs.

1 Introduction

West Bandung District grows rapidly in a few decades. As Greater Bandung is in the development stage towards a metropolitan city [1], several developments, both infrastructure and economic development, are being accelerated. The government gives full attention to development both physically and economically. One of the development aspects that need to be considered is the risk aspect due to natural hazards. Geologically, Bandung Basin is surrounded by several active fault sources such as the Lembang and Cimandiri faults [2, 4]. The source of the subduction earthquake also affects the presence of people in the Bandung Basin.

Several studies related to seismic hazards in Bandung Basin have been carried out previously [5]. One of them is a study conducted by Sari et al [5].The method used is Probability Seismic Hazard Analysis (PSHA) on a 500 km seismic source radius from a certain location in the Bandung basin. The results of the PSHA analysis are shown by the Peak Ground Acceleration at bedrock, where the area that has the highest PGA value is the West Bandung District, while the area that has the lowest value is the Bandung District (Fig. 1.).

To determine the seismic hazard that affects the building structure, it is necessary to calculate the surface PGA value. The surface PGA value is then used as a seismic load for building calculations such as foundation loading and upper structure loading. To obtain the surface PGA value, earthquake waves propagate from the bedrock to the ground surface. This study analysed the surface PGA using the soil response method. The soil profile is based on the CPTu field investigations data. Here, CPTU 01 to 12 are located in the West Bandung District (Fig. 2.)

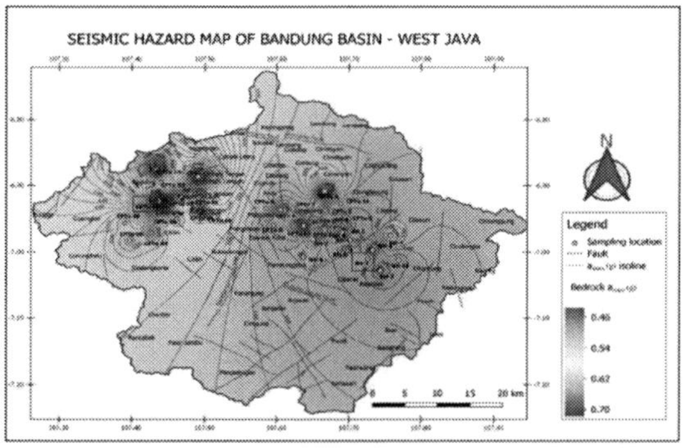

Fig. 1. Seismic Hazard Map on Bedrock located in the Bandung Basin **[5]**.

* Corresponding author: anggunmayangsari@gmail.com

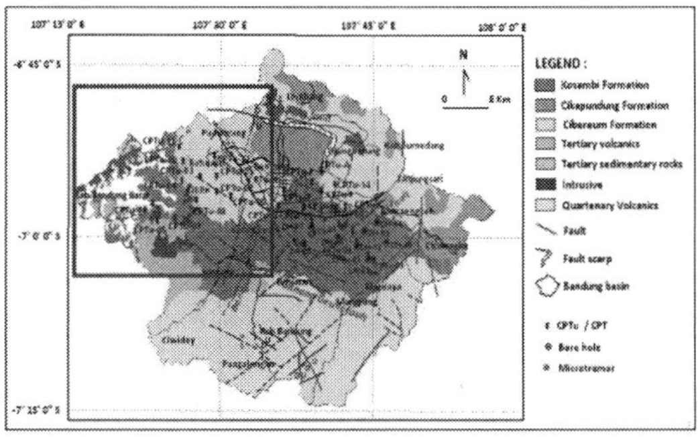

Fig. 2. Geological Map of Bandung Basin. The 12 point CPTu location in West Bandung District is marked with the red box.

2 Methods

Wave Propagation analysis can be defined as the seismic wave propagation from the bedrock to the earth's surface. Several studies on wave propagation have been carried out [6, 8]. In this study, an analysis was carried out using the non-linear earthquake response analysis method. The analysis was performed using a computer program or software, namely NERA. NERA stands for Nonlinear Earthquake Response Analysis, non-linear 1-D modelling. It produces relative velocity, amplification, and stress-strain on the surface by calculating material models, including viscoelastic models, linear equivalents, and material models. These methods require ground motion in the bedrock as a motion modelling input. The equations used by NERA are shown in Fig. 3.

Given : Material constants:

ρ_i: unit mass of soil layer i

ρ_N: density of bedrock

v_s: shear wave velocity of bedrock

(1) Initialization

$n=1,\ \tilde{v}_{i,n}=0,\ a_{i,n}=0,\ d_{i,n}=0,\ \tau_{i,0}=0,\ \gamma_{i,0}=0,\ i=1,\dots,N$ and $V_{i,0}=0,\ a_{i,0}=0$

(2) Calculate strain, strain increment and stress $(i=1,\cdots,N-1)$

$$\gamma_{i,n}=\frac{d_{i+1,n}-d_{i,n}}{\Delta z_i},\ \Delta\gamma_{i,n}=\gamma_{i,n}-\gamma_{i,n-1} \text{ and } \tau_{i,n}=IM(\tau_{i,n-1},\Delta\gamma_{i,n})$$

(3) Calculate input velocity from prescribed acceleration $a_{i,n}$

$$V_{I,n}=V_{I,n-1}+\tfrac{1}{2}(a_{I,n}+a_{I,n-1})\Delta t$$

(4) Calculate predicted velocity

$$\tilde{v}_{N,n+1}=\frac{\tilde{v}_{N,n}(\Delta z_{N-1}-v_s\Delta t)+4v_sV_{I,n}\Delta t-2\tau_{N-1,n}\Delta t/\rho_N}{\Delta z_{N-1}+v_s\Delta t} \quad \text{at node } N \text{ (bottom)}$$

$$\tilde{v}_{i,n+1}=\tilde{v}_{i,n}+2\frac{\tau_{i,n}-\tau_{i-1,n}}{\Delta z_i+\Delta z_{i-1}}\frac{\Delta t}{\rho_i} \quad \text{at node } i=2,\dots,N-1$$

$$\tilde{v}_{1,n+1}=\tilde{v}_{1,n}+\frac{2\tau_{1,n}}{\Delta z_1}\frac{\Delta t}{\rho_1} \quad \text{at node 1 (surface)}$$

(5) Calculate displacement, velocity and acceleration $(i=1,\cdots,N)$

$$d_{i,n+1}=d_{i,n}+\tilde{v}_{i,n+1}\Delta t$$

$$v_{i,n}=\frac{1}{2}(\tilde{v}_{i,n+1}+\tilde{v}_{i,n})$$

$$a_{i,n}=\frac{1}{\Delta t}(\tilde{v}_{i,n+1}-\tilde{v}_{i,n})$$

(6) $n \leftarrow n+1$ go to (2)

Fig. 3. The algorithms that were used in NERA ver.2001(Bardet and Tobitha, 2001).

The first stage in the NERA modelling is to identify earthquake ground motion on bedrock as input motion. The locations that already have a seismic record, i.e. accelerometer data or seismograph record on bedrock, can directly use the seismic motion data on programming. For the West Bandung District case, due to a lack of seismic record, the ground motion synthetic needs to be developed. This synthetic ground motion was created based on PSHA analysis using seismic sources at a 500 km radius from a certain point location in West Bandung District. The calculation resulted in a PGA value per second. This result was then used as the input motion in the NERA software (Fig. 4).

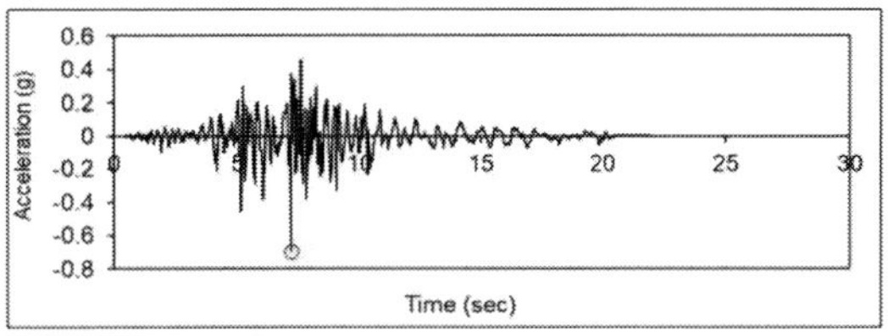

Fig. 4. Ground Motion Synthetic on CPTu 04 Location in West Bandung District.

The next step is to input soil profile data. The soil profiles data includes soil type, the thickness of the layer, total unit weight, and shear wave velocity. These data were obtained from the CPTu field investigation. The value of the shear modulus also needs to be considered. The shear modulus value is adjusted to the soil materials, which are silt, clay, and sand. The results of running data using NERA can be seen in Fig. 5.

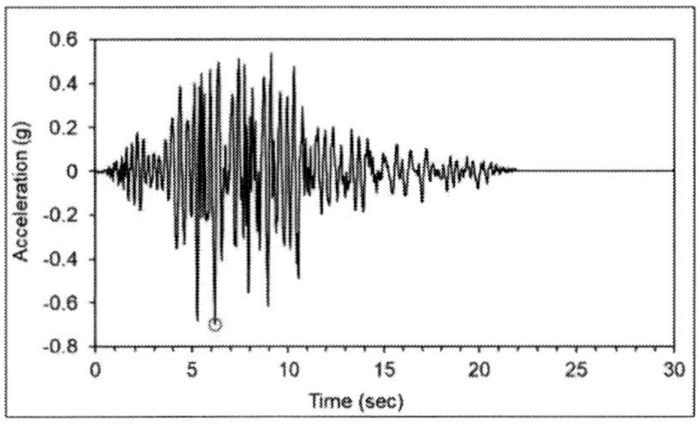

Fig. 5. Acceleration on CPTu 04 using NERA in West Bandung District.

The amplification factor is defined by the ratio between the seismic wave acceleration value in the bedrock and the acceleration value on the surface. The amplification factor equation in the F_{PGA} based on ASCE / SEI 7.10 is given by:

$$F_{PGA}=\frac{PGA_S}{PGA} \quad (1)$$

Where PGA_S is the PGA value on the surface (g), PGA is the PGA value in the bedrock (g), and F_{PGA} is the amplification factor.

3 Results and Discussion

3.1 Soil Layer Profiles

Soil layers data were obtained from CPTu field investigations. Here, 12 CPTu locations are distributed among several subdistricts in West Bandung District. The geological cross-sections were divided into three parts,

https://doi.org/10.1051/e3sconf/202133107003

which are cross-section A, cross-section B, and cross-section C. Cross-section A consists of CPTu-01, CPTu-08, CPTu-07, and CPTu-05. The cross-section can be seen in Fig. 6. Soil layers stratigraphy from the surface is clay, organic soil, silt, and sand. Most of the soil profiles consist of clay and silt materials. The length of the cross-section is about 15 km and the depth of soil varies from 10 m until 25 m.

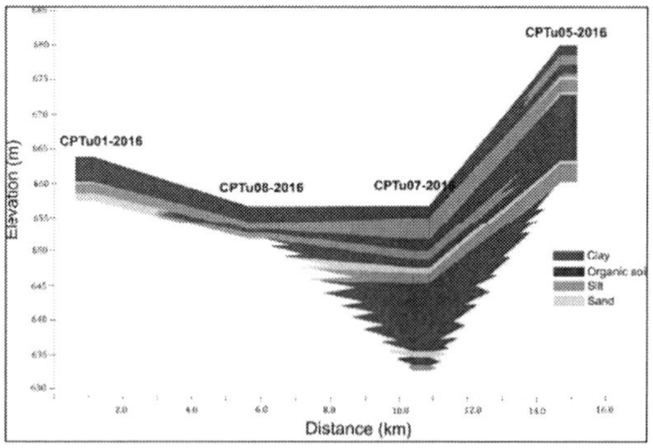

Fig. 6. Cross-section A on CPTu-01, CPTu-08, CPTu-07, and CPTu-05.

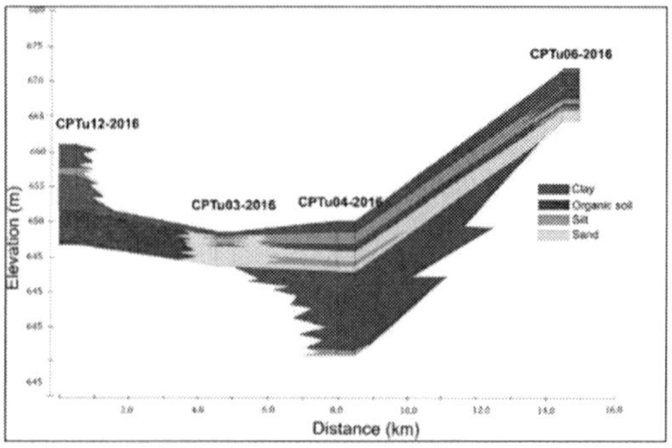

Fig. 7. Cross-section B on CPTu-12, CPTu-03, CPTu-04, and CPTu-06.

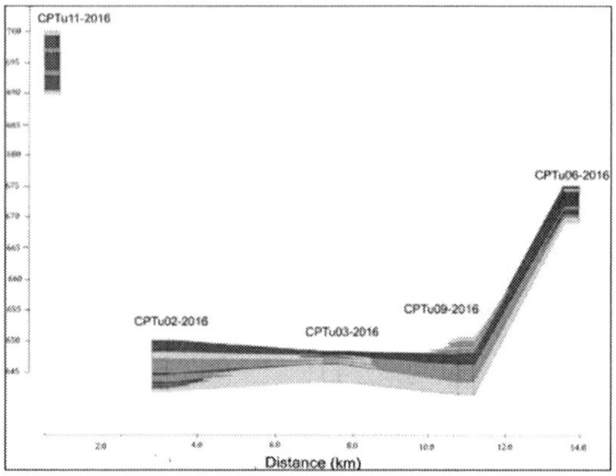

Fig. 8. Cross-section C on CPTu-11, CPTu-02, CPTu-03, CPTu-09, and CPTu-06.

Cross-section B is located on CPTu-12, CPTu-03, CPTu-04, and CPTu-06. The distance of CPTu-12 to CPTu-06 is around 15 km, while the deep of the soil layer is 8 m until 19 m. Clay material dominated the soil layer, while sand layers were found at 7 m thickness on CPTu-03. CPTu-04 has some variation of soil profiles, consisting of clay, silt, and sand, while the clay material is also found 15 m above the hard soil layer. The organic soil is also found in CPTu-12, about 5 m thickness.

The length of cross-section C is 14 km. This cross-section C consists of CPTu-11, CPTu-02, CPTu-03, CPTu-09, and CPTu-06. The depth of layers varies from 5 m to 10 m. Sand and silt materials are mostly found in CPTu-02, CPTu-03, and CPTu-9, while the clay layer is found in CPTu-11. The organic soil is also found in CPTu-06.

3.2 Peak Ground Acceleration on Surface

Peak Ground Acceleration on Surface (PGA_S) determined by using seismic wave propagation analysis. The earthquake waves propagated from the bedrock ground motion to the surface. Soil parameters; volumetric weight, stress-strain, groundwater level, damping modulus value, and shear wave velocity value, which are calculated in the propagation process. Here, NERA software is used to calculate non-linear earthquake soil response. One of the results can be seen in Fig. 9.

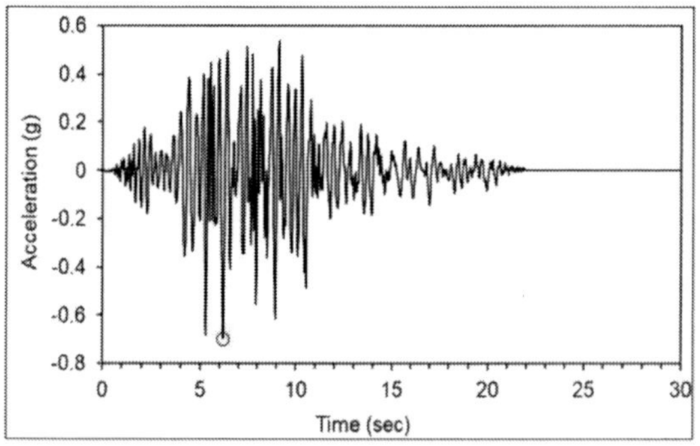

Fig. 9. Maximum PGA on the surface at CPTu 04 in West Bandung District.

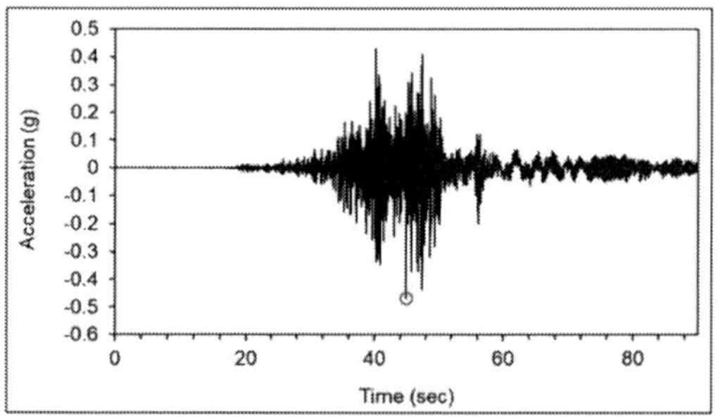

Fig. 10. Minimum PGA on the surface at CPTu 08 in West Bandung District.

The PGA surface in West Bandung District varies from 0.45g to 0.70g. The highest value was found in CPTu-04, while the lowest value was found in CPTu-05. CPTu-04 is located in Mekar Jaya. Here, the soil profile is mostly dominated by clay materials. Sand and silt are laid down on the top of the clay material. The thickness of the clay layer is about 11.3 m. This condition influenced the PGA value on the surface. As the wave propagation is very high in clay materials [6].

The minimum PGA_S value was determined on CPTu-08, on the Cililin area (Fig. 10). Here, the soil profile consists of sand materials, about 5 meters in thickness. The clay soil was found at 2 meters depth from the surface. These conditions influence the soil response on the location, where the phase of the seismic wave travels from stiffer to softer materials leading to energy dissipation. Then, the wave propagation will be fully reflected [7].

3.3 Amplification Factors

From the PGA surface calculation and PGA bedrock analysis (sari et al, 2020), then the amplification factors are calculated using equation (2). The result of these amplification factors is shown in Table.1.

Table 1. Amplification Factors (F_{PGA})on West Bandung District

Location	F_{PGA}	Location	F_{PGA}	Location	F_{PGA}
CPTu-01	0.94	CPTu-05	0.87	CPTu-09	1.00
CPTu-02	1.03	CPTu-06	1.02	CPTu-10	0.94
CPTu-03	1.04	CPTu-07	1.02	CPTu-11	0.90
CPTu-04	1.00	CPTu-08	0.89	CPTu-12	0.90

The FPGA Site Class Coefficient table in ASCE/SEI 7.10 is used to determine the Validation of Amplification Factors (FPGA) value, which is described as follows:

Table 2. F_{PGA} using Site Class Coefficient

Site Class	Mapped Maximum Considered Geometric Mean (MCE_G) Peak Ground Acceleration, PGA				
	PGA≤0.1	PGA=0.2	PGA=0.3	PGA=0.4	PGA ≥0.5
A	0.8	0.8	0.8	0.8	0.8
B	1.0	1.0	1.0	1.0	1.0
C	1.2	1.2	1.1	1.0	1.0
D	1.6	1.4	1.2	1.1	1.0
E	2.5	1.7	1.2	0.9	0.9
F	See section 11.4.7				

The Site Class type is defined by using CPTu correlation for each point location, while the PGA value is obtained from ground motion analysis. The F_{PGA} value in table 1 is then compared to the F_{PGA} of Site Class Coefficient value as shown as below:

Table 3. F_{PGA} using NERA versus F_{PGA} using Site Class Coeff.

Location	F_{PGA}	F_{PGA} (SC Coefficient)
CPTu-01	0.94	0.90
CPTu-02	1.03	1.00
CPTu-03	1.04	1.00
CPTu-04	1.00	1.00
CPTu-05	0.87	0.90
CPTu-06	1.02	1.00
CPTu-07	1.02	1.00
CPTu-08	0.89	0.90
CPTu-09	1.00	1.00
CPTu-10	0.94	0.90
CPTu-11	0.90	0.90
CPTu-12	0.90	0.90

Based on table 3, the amplification factor (FPGA) value is close to the FPGA value from the site class coefficient analysis. The results of surface PGA using soil CPTu data showed a more detailed result than PGA using site class coefficient. It means these results can be used for further calculation, such as seismic loads for structural building design.

4 Conclusions

To conclude, this study performed seismic hazards on the surface by analysing the soil response due to dynamic impact. The geological condition affected the value of surface acceleration. The Mekar Jaya area is located on Kosambi formation where this formation is composed of sedimentary rock. This condition caused the seismic wave to propagate high on these fine materials. In contrast, the Cililin area settled on the Cibereum formation. This formation is older than Kosambi Formation, which has more coarse materials than Kosambi form. These conditions produce the minimum value of wave propagation. Hence, the geological conditions are more or less influenced the acceleration calculation.

As previously determined, peak ground acceleration on clayey soil produced a high risk of ground shaking, whereas sandy soils tend to resist seismic waves. This condition needs special attention, especially the structure of the building, to minimize failures. Furthermore, this information is needed to support disaster mitigation programs, especially for natural disasters preparedness.

The authors would like to extend our appreciation to the Research Center for Geotechnology, the Indonesian Institute of Sciences for their full supports during the research. This work is based on the research supported wholly or in part by the Advanced Institute Seed Grant Program, which is implemented by the International Science Council Regional Office for Asia and the Pacific (ISC ROAP) and Integrated Research on Disaster

 https://doi.org/10.1051/e3sconf/202133107003

Risk and the International Centre of Excellence Taipei (IRDR ICoE-Taipei). This work is also supported by Innovation System Research project funding (INSINAS 2021).

References

1. Government of the Republic of Indonesia, Presidential Regulation. Number 45 of 2018 concerning Urban Spatial Plans for the Bandung Basin. (2018)
2. M. R. Daryono, D. H. Natawidjaja, B. Sapiie, and P. Cummins, Tectonophysics **751**, 180 (2019)
3. A. J. Syahbana, A. M. Sari, E. Soebowo, M. Irsyam, M. Asrurifak, and Hendriyawan, *Proc. 20th Southeast Asian Geotech. Conf. 3rd AGSSEA Conf. Conjunction with 22th HATTI Annu. Sci. Meet. Thema Geotech. Chall. Mega Infrastructures* (Indonesian Society for Geotechnical Engineering, Jakarta, 290–295 (2018)
4. A. M. Sari, A. Fakhrurrozi, and A. J. Syahbana, *Proceedings of the 7th Mathematics, Science, and Computer Science Education International Seminar, MSCEIS 2019, 12 October 2019, Bandung, West Java, Indonesia* (2020)
5. A. M. Sari and A. Fakhrurrozi, Ris. Geol. Dan Pertamb. **30**, 215 (2020)
6. G. Seidalinov and M. Taiebat, in *Comput. Methods Recent Adv. Geomech.* (2015)
7. F. Gobuzi, L. I. Bengoa, and E. Vaˇno, Predict. Wave Propag. Soils Using Semi-Analytic Method Thesis, (2017)
8. J.P. Bardet and T. Tobita. *Manual Program. The University of Southern California. Department of Civil Engineering.* (2001)
9. W.D. Iwan. J. App. Mech., **34**, 612-617 (1967)
10. Z. Mroz. J. Mech. Phys. Solids, **15**, 163-175. (1967)

Model system dynamics potential impact of disasters due to the re-period of the October 25th, 2010 earthquake in the Mentawai Islands

Aprilyanto[1*], *I Dewa* Ketut Kerta Widana[1], *Ady* Subiyanto[1], and *Hafizh* Surya Islami[1]

[1]Disaster Management Study Program, National Security Faculty, Republic of Indonesia Defense University. IPSC Sentul Area, Bogor, West Java, Indonesia

Abstract. Geographically, the Mentawai Islands Regency has a territorial boundary with the sea. The north side is the Siberut Strait, the south side is bordered by the Indian Ocean, the east side is bordered by the Mentawai Strait, and the west side is bordered by the Indian Ocean. The Mentawai Islands Regency area has a high level of seismicity which makes it prone to earthquakes and tsunamis because it has an earthquake return period. Population and economic growth in the Mentawai Islands Regency continue to increase, resulting in a higher level of threat due to earthquakes and tsunamis to people and buildings. By using earthquake return period modeling based on secondary data and population and building growth modeling using system dynamics, Pentahelix can implement disaster risk mitigation in the Mentawai Islands Regency to reduce the risk of casualties and material losses. Based on the prediction that the Mentawai earthquake return period on October 25, 2010, will occur for 24 years to 57 years or around 2034 to 2067, and the results of system dynamics modeling with Powersim Studio 10 software, the number of vulnerable people affected is 24,764 people up to 42,944 people and potential losses. housing sector between 144.73 billion to 250.98 billion.

1 Introduction

Mentawai Islands Regency with Tua Pejat as its capital located on the west side of the West Coast of Sumatra Island. While geographically it is located at 0°55'00"–3°21'00" South Latitude and 90°35'00"- 100°32'00" East Longitude with an altitude of 0 - 4.00 m above sea level. The Mentawai Islands Regency is bounded: to the north by the Siberut Strait, to the south by the Indian Ocean, to the west by the Indian Ocean, and to the east by the Mentawai Strait [1].

Based on the Indonesian Disaster Information and Database (DIBI) issued by the National Disaster Management Agency (BNPB), the Mentawai Islands Regency has a Disaster Risk Index with a score of 197.20 or is included in the high-risk category for disasters, one of which is the earthquake and tsunami disaster [2]. The Mentawai Islands Regency has a high level of earthquake activity located on the island of Sumatra. Along the west coast of the island of Sumatra, especially the Mentawai Islands is an area that often experiences earthquakes with shallow depths. This is because the area is located on one of the peripheries of the active plate where there is an interaction between the Eurasian Plate and the Indo-Australian plate. This interaction produces a pattern of subduction or subduction and the Mentawai fault zone [3]. As a result of the earthquake, the Mentawai Islands Regency area also has a potential tsunami threat, 3 (three) Villages/Kelurahan have a high tsunami hazard level and 11 (eleven) Villages have a medium tsunami hazard level [4].

The unique feature of the Mentawai Islands area is characterized by the presence of a seismic gap zone which is a historical earthquake area in 1797 with rupture zones throughout the Mentawai Islands and 1833 which occurred in the southern part of the Mentawai [5]. The high coupling condition in the Mentawai gap zone has not been released by significant earthquakes around this area such as the 2007 earthquake, the Padang earthquake magnitude 7.9 in 2009, and the Mentawai earthquake magnitude 7.8 in 2010 [6].

Several earthquakes with magnitude > 6 also occurred in the eastern area of Siberut Island with an upward fault mechanism associated with the Mentawai back thrust (MBT) so that the effect of these earthquakes on the Mentawai segment needs to be investigated further. The Padang earthquake in 2009 was an intraslab earthquake with an oblique mechanism [7]. The depth of the relocation earthquake was 87.6 km and approached the relocation result which was 90.0 km [8]. The strike-slip movement of the oblique Indo-Australian plate is accommodated by the Sumatran fault and the Mentawai fault in the eastern area of the Mentawai archipelago [7,9–11]. Meanwhile, the interpretation shows that the existence of a back thrust structure in the eastern Mentawai is the result of the analysis of reflection seismic

* Corresponding Author: aprilyanto80@gmail.com

 https://doi.org/10.1051/e3sconf/202133107004

data and bathymetric data. Back thrust with a length of more than 300 km has a strike almost parallel to the Sunda subduction and a westward slope [9].

The results of the relocation in this study showed that several earthquakes around the MBT structure continued to a depth of 20-30 km and several earthquakes had an upward fault mechanism with a dip of about 60° which was following the interpretation results [9].

Until now, the time of the earthquake cannot be predicted, but on the other hand, several studies can explain that the return period of earthquakes can be estimated. The estimated return period of an earthquake with a magnitude of 5 – 8 SR in the Mentawai Islands region has a different return period, which is around 0.4633 to 150.5033 years with details [3] in Table 1.

Table 1. Calculation of Earthquake Return Period

Magnitude (SR)	Return Period (Years)
5,0	0,463
5,5	1,214
6,0	3,185
6,5	8,351
7,0	21,895
7,5	57,404
8,0	150,503

Table 1. Explains that the October 2010 earthquake with a magnitude of 7.2 on the Richter scale, will have a return period of 22 years to 57 years. The Mentawai Islands Regency is an area that should be wary of because there is stored energy (energy of expectation) that can be released at any time in the form of a large earthquake. The last largest earthquake occurred on October 25, 2010, measuring 7.2 on the Richter Scale, centered in South Pagai [3]. The comparison of the location of the analysis of the location of the earthquake on October 25, 2021, BMKG and several international institutions is shown in 0

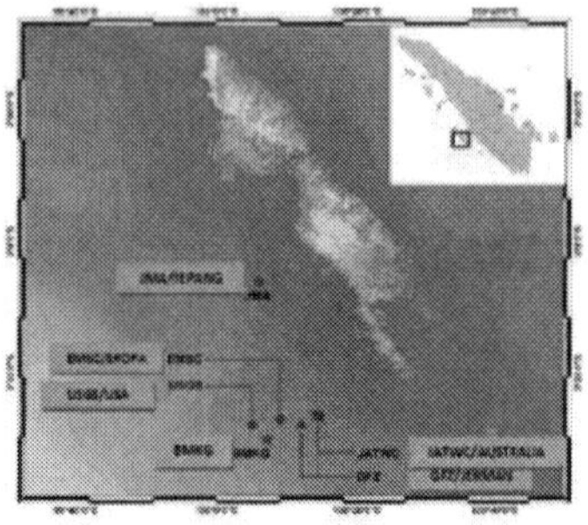

Fig. 1. Comparison of Main Earthquake Analysis Results from BMKG and Several International Institutions [12].

The Mentawai earthquake, October 25, 2010, was a repetition of the same event that had occurred in 1797 with a magnitude of magnitude 8.7-8.9 and within the range of its outbreak, an earthquake of magnitude 8.9 - 9.1 occurred in 1833 [13,14]. Table 1 shows that the return period for the Mentawai earthquake on October 25, 2010, is estimated to be around 22 years to 57 years or from 2032 to 2067.

Based on data and information from the BNPB Command Post and the Center for Operational Control for Disaster Management in West Sumatra as of November 22, 2010, the earthquake and tsunami caused 509 deaths, 17 people were injured, and 11,425 people evacuated, which are spread over refugee points in South Sipora District, South Pagai District, North Pagai District, and Sikakap District. The number of houses damaged as many as 1296 units with details of damage to 879 housing units heavily damaged, 116 housing units moderately damaged, and 274 housing units lightly damaged, with a total loss of 348.92 billion while the loss to the housing sector was around 115.82 billion (33.2%) [15]

Disaster risk assessment is an approach to show the potential negative impacts that may arise as a result of an existing potential disaster. This potential negative impact describes the potential number of lives, property losses, and environmental damage exposed to potential disasters. In conducting a disaster risk study, the functional approach of the three parameters forming the risk or risk of disaster, namely Hazard or threat, Vulnearibility or vulnerability, and Capacity or capacity related to disaster [16].

$$R_{isk} = H_{azard} X \frac{V_{ulnearibility}}{C_{avacity}} \quad (1)$$

R_{isk} : *Disaster Risk* (2)
H_{azard} : *Disaster threat* (3)
$V_{ulnearibility}$: *Disaster Vulnerability* (4)
$C_{avacity}$: *Disaster Management Capacity* (5)

Disaster risk in the Mentawai Islands Regency can increase, one of which is due to the threat of an earthquake return period and an increasing population which also increases the number of vulnerable people and population density. In law number 24 of 2007 article 55 paragraph 2 mentions vulnerable groups, namely infants, toddlers, children, pregnant and lactating mothers, people with disabilities, and the elderly. To reduce the risk that occurs, the government of the Mentawai Islands Regency by increasing the capacity, for example by structural and non-structural mitigation.

The data on the population of the Mentawai Islands in the BPS document of the Mentawai Islands Regency, in 2010 was 76,173 people and increased to 92,021 people in 2019. Population growth in the population of the Archipelago Regency has a crude birth rate of 1.17, a crude death rate of 0.22%, a migration rate entry of 1.20%, and an out-migration rate of 0.43%. The age grouping of the population can be done based on the age structure, grouped into 3 (three) major groups, namely: Young people, namely the population aged under 15 years, or the age group 0-14 years. The productive age population is the population aged 15-64 years. The elderly population, namely the population aged 65 years and over, follows the WHO provisions. The percentage of the population in the Mentawai Islands Regency under the age of 15 years is 25.9% and the elderly is 6.61% [18].

Data on the population of the Mentawai Islands Regency from 2010 to 2019 is in Table 2.

Table 2. The population of Mentawai Islands Regency from 2010 to 2019.

Year	Total Population
2010	76.173
2011	77.078
2012	78.511
2013	81.840
2014	83.603
2015	85.295
2016	86.981
2017	88.692
2018	90.373
2019	92.021

Source: [17]

The population of the Mentawai Islands Regency in 2010 was 76,173 people until 2019 as many as 92,021 and the percentage of vulnerable population in 2018 was 32.51%. The urgency of this study is the use of vulnerable population parameters in determining the purpose of this study to simulate the number of vulnerable residents and housing that could potentially be affected if there was a return period of the October 25, 2010 earthquake using system dynamics modeling.

2 Method

The data used in this study is based on the secondary data. Secondary data is a source of research data obtained by researchers indirectly through intermediary media. This means that the researcher acts as a second party because it is not obtained directly. The secondary data has been collected including the population of Mentawai regency from 2010 to 2019 (Mentawai Islands Regency BPS) and data from the earthquake return period from previous research. Data analysis method modeling system dynamics population growth of Mentawai Islands Regency is used to determine the number of vulnerable people affected and the amount of potential loss to the affected housing sector in the event of an earthquake return period. The software used to model the system dynamics of the potential threat of the Mentawai Islands earthquake return period is Powersim Studio 10.

The assumptions used in model making and model simulation are as follows: (1). The average number of people per house is 4 people (2). The percentage of heavily damaged houses is 5% (3). The percentage of moderately damaged houses is 3% (4). The percentage of lightly damaged houses is 2% (5). The value of the loss of a large damaged house is Rp. 100.000.000 per unit (6). The value of the loss of a moderately damaged house is Rp. 60.000.000 per unit (7). The value of the loss of lightly damaged houses is Rp. 40.000.000 per unit.

3 Result and discussion

3.1 Simulation and model accuracy

After the problem is understood and modeled into the Causal Loop Diagram (CLD), the next step is to simulate the system that has been poured into the CLD diagram model, input into the Stock Flow Diagram (SFD) using Powersim Studio 10 software. Total the population used in this model is the total population of the Mentawai Islands Regency in 2010-2019, used to determine the development of the population in 2020-2068. After simulating the model of the relationship between population growth and the potential for earthquake and tsunami disasters, Stock Flow Diagram (Fig. 2), it is necessary to test the accuracy of the simulation model to determine the level of model validity. The accuracy test is carried out using a simple error test, namely the Average Mean Error (AME) of the population. During the accuracy testing period is adjusted to the actual data, which is in the period 2010 to 2019. Fig. 2. Flowchart or Stock Flow Diagram (SFD) Simulation results of population growth in the Mentawai Islands Regency from 2010 to 2019 are in Table 3.

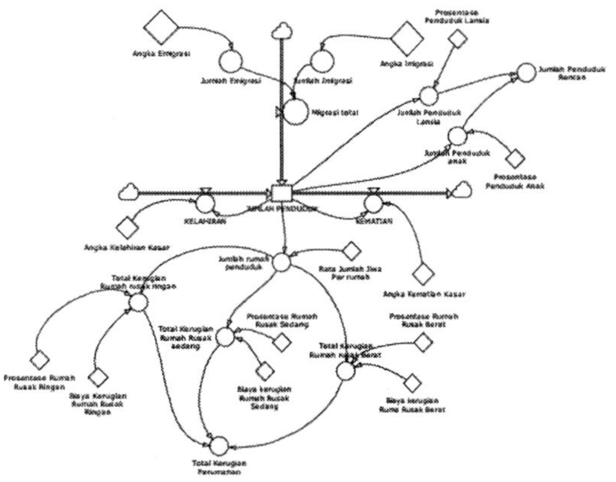

Fig. 2. Flowchart or Stock Flow Diagram (SFD)

Model validation is an attempt to conclude whether the system model is valid in representating the reality being studied, which lead to convincing conclusions. Validation in this study uses the AME value as an indicator that the model made is valid. The AME value is obtained through the equation:

$$AME = \frac{|Ds - De|}{De} x100\% \quad (6)$$

Information :
Ds = Average between simulation results (7)
De = Average between reference results (8)

A model is considered valid if it has a maximum value of AME 0.5%. To get the AME value, reference data and simulation data are needed which are shown in Table 3.

https://doi.org/10.1051/e3sconf/202133107004

Table 3. Reference Data and Simulation Data on the Population of Mentawai Islands Regency from 2010 to 2019

Year	Total population (Reference Data)	Total population (Simulation Data)
2010	76.173	76.173
2011	77.078	77.483
2012	78.511	78.806
2013	81.840	80.141
2014	83.603	82.850
2015	85.295	84.101
2016	86.981	84.223
2017	88.692	85.610
2018	90.373	87.010
2019	92.021	88.423
Average	**84.057**	**82.221**

The reference data were obtained from the population data of the Mentawai Islands Regency from 2010 to 2019 issued by the Mentawai Islands Regency BPS, while the simulation data was obtained from a dynamic model using the Powersim Studio 10 program. actual and simulated have the same value as many as 76,173 people. The actual data on the population in the period from 2010 to 2019 has an average of 84,057 people, this number does not show much difference and is almost the same as the average from the simulation results, which is 82,221 people.

$$AME = \frac{|82.221-84.057|}{84.057} x100\% = 2{,}18\% \qquad (9)$$

Based on the actual and simulation data, the results of model validation with an Average Mean Error (AME) between the two resulted in a figure of 2.18% so that the model was declared valid and could be used for further analysis.

3.2 Forecast of vulnerable population and total housing sector losses

Estimates of the number of vulnerable people and the total loss of the housing sector with a population growth model simulation are made until 2068 using a dynamic system model. From the results, it can be seen in Table 4 the predictions of the population generated by the years 2032 to 2068.

Table 4 shows that if the return period of the earthquake occurs in the range of 24 years to 57 years or occurs in 2034 to 2067, the potential for vulnerable populations to be affected is in the range of 24,764 people to 42,944 people, for the estimated loss to the housing sector between 144.73 billion up to 250.98 billion.

The potential disaster risk for housing sector losses due to the earthquake return period in the Mentawai Islands Regency increased from 115.82 billion in 2010 to 144.73 billion to 250.98 billion. Therefore, it is necessary to take mitigation measures against earthquake and tsunami disasters by implementing structural and non-structural mitigation.

Table 4. Simulation of the Number of Vulnerable Populations and Total Housing Loss for the return period in 2032 to 2068

Year	Total population	Vulnerable Population	Housing Losses
	(Person)	*(Person)*	*(Rp)*
2032	76.173	24.764	144.728.700.000
2036	81.489	26.492	154.828.827.361
2040	87.010	28.287	165.318.263.502
2044	92.743	30.151	176.212.014.304
2048	98.698	32.087	187.525.664.052
2052	104.882	34.097	199.275.397.724
2056	111.304	36.185	211.478.024.148
2060	117.974	38.353	224.151.000.046
2064	124.901	40.605	237.312.455.010
2068	132.095	42.944	250.981.217.436

Structural mitigation such as the construction of Giant Walls, Earthquake Resistant Houses, Green Belts, construction of evacuation routes, and others. Meanwhile, non-structural mitigation such as RTRW policies with a disaster perspective, prohibition of building houses in earthquake and tsunami-prone zones, increasing disaster knowledge through education and training programs.

4 Conclusion

The October 2010 earthquake disaster has a potential return period between 24 years to 57 years or will occur in the range of 2034 to 2057.

The results of modeling the potential return period of the October 2010 earthquake will threaten the population vulnerable to 24,764 people up to 42,944 people and potential housing losses between 144.73 billion to 250.98 billion.

To reduce the impact of disasters due to the return period of the earthquake and tsunami are doing by conducting mitigating both structural and non-structural.

References

1. N. A. Akhirianto, J. Sains Dan Teknol. Mitigasi Bencana **11**, 39 (2019)
2. BNPB, *Indonesia Disaster Risk Index (IRBI) 2018* (National Disaster Management Agency (BNPB), Jakarta, 2018)
3. R. Fidia, D. Pujiastuti, and A. Sabarani, J. Fis. Unand **7**, 84 (2018)
4. BNPB, *Tsunami Prone Villages/Villages Catalog (High and Medium Hazard Classes)* (National Disaster Management Agency (BNPB), Jakarta, 2021)

https://doi.org/10.1051/e3sconf/202133107004

5. C. Damayanti, A. K. Yamko, C. J. Souisa, W. Barends, and I. L. P. T. Naroly, J. Geosains Dan Remote Sens. **1**, 105 (2020)
6. A. D. Chandra, J. Fis. Dan Apl. **11**, 64 (2015)
7. J. McCloskey, D. Lange, F. Tilmann, S. S. Nalbant, A. F. Bell, D. H. Natawidjaja, and A. Rietbrock, Nat. Geosci. **3**, 70 (2010)
8. J. D. Pesicek, C. H. Thurber, H. Zhang, H. R. DeShon, E. R. Engdahl, and S. Widiyantoro, J. Geophys. Res. **115**, B10303 (2010)
9. R. B. S. Yadav, D. Shanker, S. Chopra, and A. P. Singh, Nat. Hazards **54**, 985 (2010)
10. R. Shannon, S. McDowell, and J. McCloskey, in *EGU Gen. Assem. Conf. Abstr.* (AA(the University of Ulster, School of Environmental Sciences, Coleraine, Ireland), AB(the University of Ulster, School of Environmental Sciences, Coleraine, Ireland), AC(the University of Ulster, School of Environmental Sciences, Coleraine, Ireland), 2009), 9777
11. M. Diament, H. Harjono, K. Karta, C. Deplus, D. Dahrin, M. T. Zen Jr., M. Gérard, O. Lassal, A. Martin, and J. Malod, Geology **20**, 259 (1992)
12. BMKG, *Earthquake Popular Edition* (Meteorology Climatology and Geophysics Council (BMKG), Jakarta, 2012)
13. S. Burgess and A. Briggs, Econ. Educ. Rev. **29**, 639 (2010)
14. H.-P. Plag, Nat. Hazards **72**, 1275 (2014)
15. BNPB, *Action Plan for Post-Disaster Rehabilitation and Reconstruction, and Acceleration of Development in the Mentawai Islands Region of West Sumatra Province in 2011-2013* (National Disaster Management Agency (BNPB), Jakarta, 2010)
16. National Board for Disaster Management, *Indonesia Disaster Risk Index 2020* (National Board for Disaster Management (BNPB), Jakarta, 2021)
17. BPS, *Mentawai Islands Regency in Figures 2021* (Central Bureau of Statistics of the Republic of Indonesia (BPS), Mentawai, 2021)
18. West Sumatra Provincial Government, *Population Development Profile of West Sumatra Province in 2018* (Department of Population Control, Family Planning, Population and Civil Registration of West Sumatra Province, Padang, 2018)

Earthquake and tsunami potential levels in Sulawesi (lesson learned earthquake West Sulawesi)

Rio Khoirudin Apriyadi[1*], *Sobar* Sutisna[1], *Lasmono*[1], and *Riskina* Tri Januarti[1]

[1]Disaster Management Study Program, National Security Faculty, Republic of Indonesia Defense University. IPSC Sentul Area, Bogor, West Java, Indonesia

Abstract. Earthquake and tsunami disasters always bring negative impacts and losses for humans. The high frequency of earthquake disasters in Sulawesi and the massive impact of destruction and damage to infrastructure, loss of homes and property, and death of people are partly due to the lack of knowledge of this potential disaster by the surrounding community. Therefore, it is necessary to study the potential for earthquake and tsunami disasters on Sulawesi Island, and Lesson Learned about the Earthquake Disaster that has occurred in past. This study uses a qualitative method with a descriptive analysis design of secondary data obtained through a comprehensive literature review. The results showed that Sulawesi Island has a high potential for earthquake disasters, which is in the historical records of disasters, some of these earthquakes were accompanied by tsunami waves. The lesson that can be drawn from the earthquake in West Sulawesi is the potential for aftershocks after the opening earthquake and the main earthquake. In addition, the urgency of building earthquake-resistant housing structural mitigation is the key to safety during an earthquake. The post-earthquake recovery program in West Sulawesi requires further studies related to soil classification surveys, soil dominant periods, and soil seismic vulnerability index, as part of rebuilding a better and safer post-disaster area.

1 Introduction

The earthquake disaster in early 2021 that occurred in Mamuju Regency and Mejene Regency shocked the people of Indonesia, amid the non-natural disaster the Covid-19 pandemic. On Thursday, January 14, 2021, Mamuju Regency and Majene Regency experienced a large earthquake with M 5.9 on the Richter scale at a depth of 10 km whose epicenter was on land 4 km northwest of Majene. The next earthquake shook the area again with a magnitude of M 6.2 at a depth of 10 km at a location 6 km northeast of Majene [1].

This main earthquake was followed by a series of aftershocks with magnitude of between 1 SR and 6.2 SR (Fig. 1).

Based on data from the National Disaster Management Agency (BNPB), as of January 21, the earthquake had killed 51 people (8 people in Majene Regency and 34 people in Mamuju Regency), as many as 556 people were injured (189 seriously injured and 367 people injured). people were lightly injured), and 1500 people were evacuated. Losses due to this earthquake are estimated to reach Rp. 829.1 billion [1].

The massive impact and the high frequency of earthquakes on Sulawesi Island are not new in the world of research. Tectonically, Sulawesi Island has more than 45 sources of active fault earthquakes due to collisions between the Eurasian, Indo-Australian, Pacific, and Philippine Sea plates [3]. However, the earthquake that occurred during the Covid-19 pandemic directly increased the risk of spreading Covid-19 in the disaster cluster. This is not accompanied by an increase in capacity and knowledge about disaster risk on the island of Sulawesi.

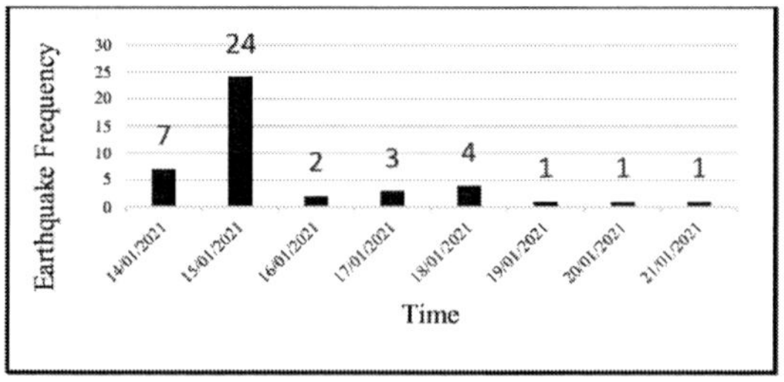

Fig. 1. Frequency of Destructive Earthquakes in Majene and Mamuju Regencies in the Period of January 14, 2021, to January 21, 2021.

This paper, besides aiming to describe the potential for earthquake and tsunami disaster on Sulawesi Island, but also describes lessons that can be drawn from the

* Corresponding Author: riokapriyadi@gmail.com

https://doi.org/10.1051/e3sconf/202133107005

earthquake disaster in Mamuju Regency and Mejene Regency, West Sulawesi in early 2021.

The significance of this research is it is important to carry out immediate measurements in the prone earthquakes area such as on the island of Sulawesi. The results of this analysis are used as database and lesson learnt in the implementation of disaster risk reduction policies in Indonesia.

2 Method

We use a qualitative method with a descriptive analysis design as an analytical tool in answering research problems. The data in this study were obtained through a literature study.

We use Shakemap Repository data obtained from the Meteorology, Climatology and Geophysics Agency (BMKG), Indonesia earthquake source and hazard maps from the National Earthquake Study Center (PUSGEN), a catalog of destructive earthquakes in Indonesia by the Center for Volcanology and Geological Hazard Mitigation (PVMBG), The National Tsunami Hazard Study for Indonesia by the Meteorology, Climatology and Geophysics Agency (BMKG), the Indonesia Disaster Risk Index by BNPB, and other documents related to the potential for earthquakes and tsunamis in Sulawesi, as well as a Situation Report on Earthquake Management in West Sulawesi [2–5].

3 Result and discussion

3.1 Earthquake Potential on Sulawesi Island

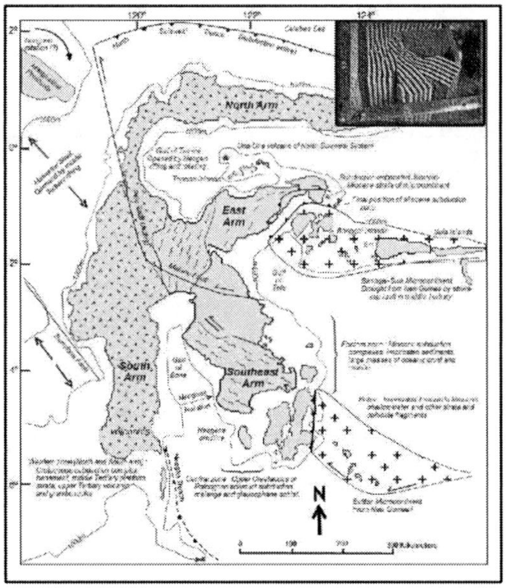

Fig. 2. Sulawesi Tectonic Dynamics Due to Intercontinental Push [10–12].

Sulawesi Island has the potential for earthquakes due to the complex and complex tectonic setting. Several studies reveal that some of the tectonic dynamics of the formation of this island due to the thrust between the Eurasian, Indo-Australian, Pacific, and Philippine Sea plates make the appearance of this island similar to the letter K (Figure 2), resulting in this island having many sources of active fault earthquakes on this island (Figure 3) [6–9].

Figure 2 explains that the formation of Sulawesi Island has a complicated fault structure consisting of faults that are still actively moving and no longer moving. The fault segment that forms the island of Sulawesi can be seen in Fig 3.

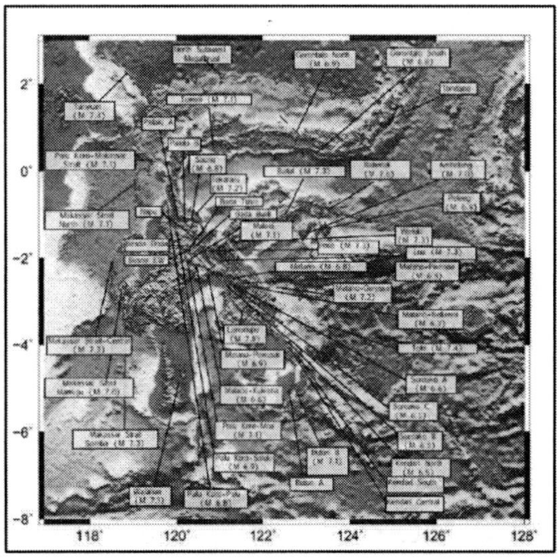

Fig. 3. Active Fault Segment on Sulawesi Island [3].

Figure 3 illustrates that Sulawesi Island has more than 45 active fault segments which are the source of earthquakes on this island [3]. Details of the active fault segments summarized in the 2017 National Earthquake map above can be seen in Table 1.

Table 1. Sulawesi Island Active Fault Segment

Structure Name	Section	Max Length (km)
Palukoro Fault	Makasar Str	130
Palukoro Fault	Palu	31
Palukoro Fault	Saluki	44
Palukoro Fault	Moa	66
Matano Fault	Kuleana	22
Matano Fault	Pewusai	46
Matano Fault	Matano	35
Matano Fault	Pamsoa	44
Matano Fault	Ballawai	26
Matano Fault	Geressa	80
Soroako Thrust	A	24
Soroako Thrust	B	20
Soroako Thrust	C	16
Batui Thrust	-	84
Balantak	-	149
Makasar Strait Thrust	North	100
Makasar Strait Thrust	Central	170

Structure Name	Section	Max Length (km)
Makasar Strait Thrust	Mamuju	40
Makasar Strait Thrust	Somba	80
Buton	A	29
Buton	B	60
Palolo	A	29
Palolo	B	24
Bada Valley Fault	Tuare WSW	16
Bada Valley Fault	Bulili NNE	26
Besoa Valley Fault	EW	7
Besoa Valley Fault	Doda NW	7
Peleng	-	44
Ambelang	-	50
Sausu	-	31
Tokararu	-	80
Tondano	-	30
Poso	-	85
Loa	-	86
Weluki	-	72
Tomini	-	59
Lawanopo	-	130
Tolo Thrust	-	120
Gorontalo	North	74
Gorontalo	South	70
Walanae	-	130
Kendari Fault	North	24
Kendari Fault	Central	11
Kendari Fault	South	10
North	-	550
Megathrust	-	0
Napu	-	11
Tarakan	-	206
Maleei	-	70

The number of active fault segments on Sulawesi Island is directly proportional to the high Earthquake Disaster Risk Index in 2020 issued by the National Disaster Management Agency (BNPB) which states that all provinces on Sulawesi Island, which include Gorontalo Province, West Sulawesi Province, South Sulawesi Province, Southeast Sulawesi Province, Central Sulawesi Province, and North Sulawesi Province have an Earthquake Disaster Risk Index with high criteria. Another interesting fact is that several districts or cities on Sulawesi Island have the highest earthquake risk index in Indonesia, such as the Bangai Laut Regency, the North Morowali Regency, and the West Muna Regency with a score of 36.0 (High-Risk Class) [5,13].

3.2 Potential Tsunami on Sulawesi Island

Based on the Indonesian Disaster Data and Information Management Database which was developed using the software and methodologies DesInventar and DesConsultar by the National Disaster Management Agency (BNPB), states that districts or cities on Sulawesi Island have a variety of Tsunami Disaster Risk Indexes, ranging from moderate to high-risk classes. with height [5].

Several studies have simulated the probability of a large tsunami with a run-up of more than 3 meters, with a probability of occurrence and a return period of 100 years on Sulawesi Island [14], as can be seen in Table 2.

Table 2. Annual Tsunami Probability and Annual Tsunami Run-up Every 100 Years Return Period on Sulawesi Island

Districts	Annual Tsunami Probability Tsunami	Tsunami Run-up Every 100 Years
Bitung City	1,1%	3,2
Sangihe Islands	1,0%	2,9
North Minahasa	1,0%	2,9
Talaud Islands	0,9%	2,8
Manado City	0,9%	2,7
Buol	0,9%	2,7
South Minahasa	0,8%	2,7
Minahasa	0,8%	2,5
Toli-Toli	0,7%	2,4
Bolaang Mengondow	0,5%	2,1
Konawe	0,5%	1,9
Muna	0,3%	1,8
Islands Banggai	0,3%	1,5
Majene	0,3%	1,2
Mamuju	0,2%	1,6
South Konawe	0,2%	1,5
Banggai	0,2%	1,3
Palu	0,2%	1,7
Boalemo	0,2%	1,0
Bone Bolango	0,2%	1,1
Donggala	0,2%	1,7
North Mamuju	0,2%	1,6
Kendari City	0,2%	1,4
Morowali	0,2%	1,3
Wakatobi	0,2%	1,6
Buton	0,1%	1,3
Polmas	0,1%	1,0
Pinrang	0,1%	1,0
Tojo Una - Una	0,1%	0,7
Poso	0,1%	0,7
Barru	0,1%	0,8
Bantaeng	< 0,1%	0,7
Jeneponto	< 0,1%	0,7
Parigi Moutong	< 0,1%	0,6
Pare-Pare	< 0,1%	0,8
Bombana	< 0,1%	0,8
Bone	< 0,1%	0,6
Bulukumba	< 0,1%	0,7
Luwu	< 0,1%	0,3
East Luwu	< 0,1%	0,2
North Luwu	< 0,1%	0,2
Maros	< 0,1%	0,6
Palopo	< 0,1%	0,2
Pangkajene Islands	< 0,1%	0,6
Selayar	< 0,1%	0,6
Sinjai	< 0,1%	0,7
Takalar	< 0,1%	0,6
Ujung Pandang	< 0,1%	0,6
Wajo	< 0,1%	0,3
Kolaka	< 0,1%	0,4
North Kolaka	< 0,1%	0,3
Bitung City	1,1%	3,2
Sangihe Islands	1,0%	2,9
North Minahasa	1,0%	2,9

https://doi.org/10.1051/e3sconf/202133107005

Table 2 shows that districts or cities located within 5 kilometers of the coast on Sulawesi Island will experience a tsunami warning with a run-up of more than 3 meters in any year.

Indonesia's disaster history records that in the period 1820 to 2021, West Sulawesi at least had an earthquake accompanied by a tsunami in 1820, 1969, and 1976 [1], as illustrated in Table 3.

Table 3. Indonesia's disaster history record period 1820 to 2021

Time	Disaster Type
1820	Earthquake with Tsunami
Feb 23, 1969	Earthquake with Tsunami
Apr 11, 1976	Earthquake with Tsunami
Jan 8, 1984	Earthquake
Jan 14-15 2021	Earthquake

3.3 Lesson Learned West Sulawesi Earthquake

An earthquake in early 2021 occurred in Mamuju Regency and Majene Regency, West Sulawesi Province, is strongly suspected trigger by Mamuju Rising Fault (Mamuju Thrust). The results of the analysis of the source mechanism show that the earthquake has a thrust fault mechanism. This upward fault mechanism is similar to the 2018 Lombok earthquake generator, the fault plane forms the slope of the fault plane to the mainland. The Naik Mamuju Fault has a targeted magnitude of 7.0 with a fault shear rate of 2 mm/year, so this active fault must be watched out for because it can trigger strong earthquakes.

Characteristics of the type of earthquake that occurred in West Sulawesi is type of earthquake opening (foreshocks), main earthquake (mainshock), and aftershocks (aftershocks). In general, this means that whenever a major earthquake occurs, there will be a main earthquake and aftershocks, as happened on January 14-15 2021 in Majene and Mamuju, West Sulawesi [15,16], as can be seen in Figures 4 and 5.

Figures 4 and 5 provide lessons learned that aftershocks will occur along with a large earthquake (M > 3 SR). Therefore, awareness of the characteristics of the opening earthquake, main earthquake, and aftershocks needs to be increased. Aftershocks will still occur as usual after a strong earthquake occurs, for that the public is asked to be aware of the possibility of aftershocks of significant strength.

The next lesson learned that we can take from the West Sulawesi earthquake is that we have not made much progress in structural mitigation, such as earthquake-resistant houses. Earthquake-resistant houses are the key to safety during an earthquake. Without any real effort to build earthquake-resistant houses, our community will always be victims when a strong earthquake occurs.

The third lesson learned is related to the post-earthquake rehabilitation and reconstruction in Mamuju Regency and Majene Regency, West Sulawesi Province. Post-disaster rehabilitation and reconstruction activities are programs to restore the lives and livelihoods of communities affected by the disaster, because before the construction of permanent housing for post-earthquake victims in Majene Regency and Mamuju Regency, West Sulawesi Province, a survey of land classification, soil dominant period and soil seismic vulnerability index, as part of rebuilding a better and safer post-disaster area.

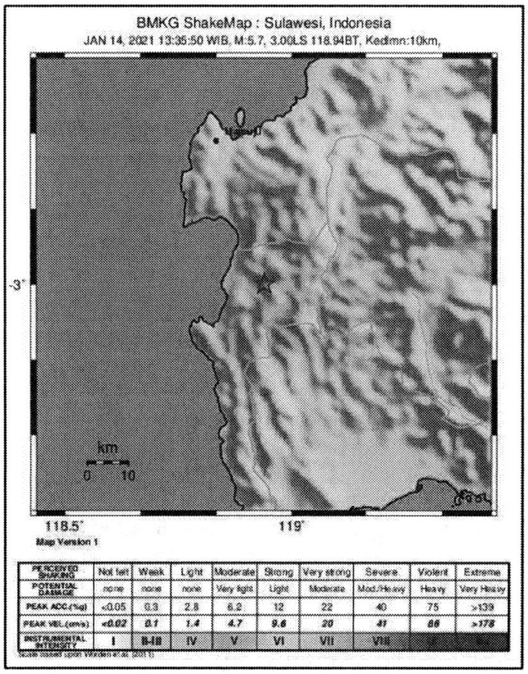

Fig 4. ShakeMap Opening M 5.9 SR

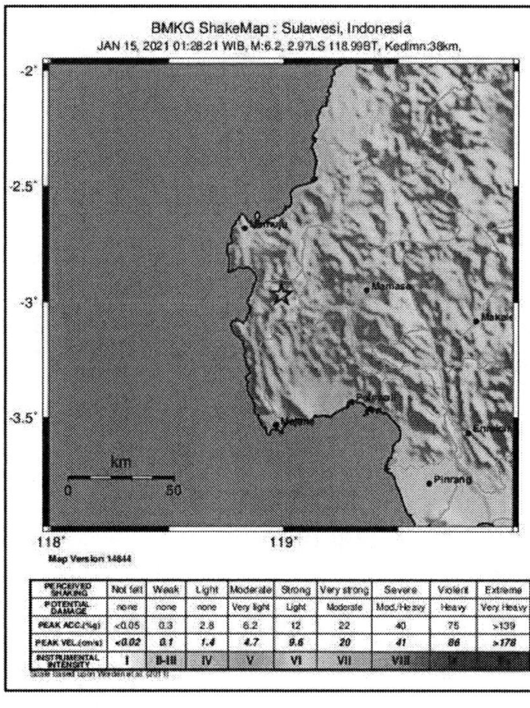

Fig 5. ShakeMap M 6.2 Main Earthquake

4 Conclusion

In conclusion, this study finds that Sulawesi Island has the potential for destructive earthquakes with a high-risk index in all districts or cities spread across the island. The high potential for earthquakes that occur on the island of Sulawesi is the result of the thrust between the plates, resulting in this island having many sources of active fault earthquakes. The source of an active fault earthquake on the island of Sulawesi, besides causing a high frequency of earthquakes, also has the potential for a tsunami disaster to occur in areas located within 5 kilometers of the coast on the island of Sulawesi with a run-up of more than 3 meters.

The history of disasters on Sulawesi Island, such as the earthquake that occurred in West Sulawesi, warns us that structural mitigation, such as earthquake-resistant houses, is the key to safety during an earthquake. Without any structural mitigation efforts such as building earthquake-resistant houses, it is predicted that there will be a spike in fatalities due to the earthquake on Sulawesi Island.

Characteristics of the earthquake that occurred on the island of Sulawesi is a type of earthquake with the type of opening earthquake, main earthquake, and aftershocks. Therefore, whenever a destructive earthquake occurs, it is necessary to watch out for a major earthquake and aftershocks with greater strength.

The lesson that can be taken is the need for future studies related to soil classification surveys, soil dominant periods, and soil seismic vulnerability index as part of the program to restore the lives and livelihoods of disaster-affected communities on Sulawesi Island that can reduce disaster risk in the community, better development and safer.

References

1. D. Hartono, R. Khoirudin Apriyadi, T. Winugroho, A. Aprilyanto, S. Hadi Sumantri, W. Wilopo, and H. Surya Islami, PENDIPA J. Sci. Educ. **5**, 218 (2021)
2. BMKG, *ShakeMap Repository* (Meteorology Climatology and Geophysics Council (BMKG), Jakarta, 2021)
3. Tim-PSGN, *2017* Earthquake Source and Hazard Map Book (Center for Housing and Settlement Research and Development Ministry of PUPR, Bandung, 2017).
4. Supartoyo, surono, and e. T. Putranto, *catalog of destroying earthquakes in indonesia years 1612 – 2014 catalog of destroying earthquakes in indonesia years 1612 – 2014 catalog of daming earthquakes in indonesia years 1612 – 2014* (center for vulcanology and geological disaster mitigation (PVMBG, Jakarta, 2014)
5. National Board for Disaster Management, *Indonesia Disaster Risk Index 2020* (National Board for Disaster Management BNPB, Jakarta, 2021)
6. I. Watkinson, R. Hall, M. Cottam, I. Sevastjanova, S. Suggate, I. Gunawan, J. Pownall, J. Hennig-Breitfeld, F. Ferdian, D. Gold, S. Zimmermann, A. Rudyawan, and E. Advokaat, Ber. Sedimentol. **23**, 21 (2012)
7. R. Hall, Tectonophysics **570–571**, 1 (2012)
8. T. Kiyota, H. Furuichi, R. F. Hidayat, N. Tada, and H. Nawir, Soils Found. **60**, 722 (2020)
9. H. Hazarika, D. Rohit, S. M. K. Pasha, T. Maeda, I. Masyhur, A. Arsyad, and S. Nurdin, Soils Found. **61**, 239 (2021)
10. T. R. Charlton, Nature **319**, 394 (1986)
11. P. Tapponnier and P. Molnar, Nature **264**, 319 (1976)
12. P. Tapponnier, G. Peltzer, A. Y. Le Dain, R. Armijo, and P. Cobbold, Geology **10**, 611 (1982)
13. BNPB, *Tsunami Prone Villages/Villages Catalog (High and Medium Hazard Classes)* (National Disaster Management Agency (BNPB), Jakarta, 2021)
14. N. Horspool, I. R. Pranantyo, J. Griffin, H. Latief, D. Natawidjaja, W. Kongko, A. Cipta, Bustamam, S. D. Anugrah, and H. K. Thio, *National Tsunami Hazard Study for Indonesia* (National Disaster Management Agency (BNPB), Jakarta, 2013)
15. P. Supendi, M. Ramdhan, Priyobudi, D. Sianipar, A. Wibowo, M. Gunawan, S. Rohadi, N. Florida, Daryono, B. Prayitno, J. Murjaya, D. Karnawati, I. Meilano, N. Rawlinson, S. Widiyantoro, A. Nugraha, G. Marliyani, K. Palgunadi, and E. Elsera, Earth Planets Sp. **73**, (2021)
16. P. Supendi, A. Nugraha, S. Widiyantoro, J. Pesicek, C. Thurber, C. I. Abdullah, Daryono, S. Wiyono, H. Ash Shiddiqi, and S. Rosalia, Geophys. J. Int. **221**, (2020)

Monitoring and analysis of seismic data during the 2018 sunda strait tsunami

Wahyu Kurniawan[1,2*] , *Daryono*[1,2], *IDK* Kerta[1], *Bayu* Pranata[2,3], and *Tri* Winugroho[1]

[1] Disaster Management Study Program, National Security Faculty, Republic of Indonesia Defense University. IPSC Sentul Area, Bogor, West Java, Indonesia
[2]Indonesian Agency for Meteorological, Climatotlogical, and Geophysics, Jakarta, Indonesia
[3] Earth science Study Program, Institute of Technology Bandung, Indonesia

Abstract. The tsunami of Sunda Strait occurred on December 22, 2018, at 21:03 West Indonesia Time (zone). An eruption of Mount Anak Krakatau caused an eruption that triggered a landslide on the slopes of Mount Anak Krakatau covering an area of 64 hectares that hit the coastal area of western Banten and southern Lampung and resulted in 437 deaths, 14.059 people were injured, and 33.721 people were displaced. Before the tsunami, signal transmissions (gaps) at the Lava seismograph station installed on the body of Mount Anak Krakatau experienced broken so that Mount Anak Krakatau Observation Post could not record volcanic earthquake signals since December 22, 2018, at 21.03 West Indonesia Time (zone). Given these facts, proper monitoring and analysis were required to monitor and analyze the source of ground vibrations originating from the eruption of Mount Anak Krakatau. Therefore, this study aims to confirm the eruptive activity of Mount Anak Krakatau based on seismic monitoring and analysis sourced from the BMKG's seismic sensor network. The method the author uses is by monitoring the seismic signal recorded by the seismometer and analyzing the seismic signal using the Seiscomp3 software. By the results of monitoring and analysis of seismic data, it was found that the location of the center of the ground shaking was on Mount Anak Krakatau with a magnitude of 3.4, and a depth of 1 km. To anticipate similar tsunami events in the future, it is very necessary to have a tsunami early warning system originating from volcanic activity and volcanic body avalanches.

1 Introduction

Indonesia is located in an area prone to destructive earthquakes, because of the presence of four active world plates in Indonesian territory, namely the Eurasian plate, the Indo-Australian plate, the Pacific plate, and the Philippine sea plate [1]. Several areas on islands that are directly opposite the subduction zone between these plates, such as the western part of Sumatra Island, southern Java Island, Nusa Tenggara, northern Papua, and Sulawesi and Maluku are areas that are very prone to tsunamis. In addition, the distribution of volcanoes that extends from Aceh to North Sulawesi for approximately 7,000 km is signing Indonesia as a ring of fire, impacting as one of the largest number countries of active volcanoes in the world [2]. There is a close relationship between volcanic density and the angle of subduction, the speed of oceanic plate motion, tectonic pressure behind the arc, and the slope of the heading of subduction [3].

Especially in the Sunda Strait area, there is a Megathrust subduction route and the presence of the Anak Krakatau Volcano makes this area very prone to tsunamis, both tsunamis caused by earthquakes and caused by the eruption of volcanic.

From year 416 to 2018, the large strength of a tectonic earthquake was occurred on August 24, 1757, in the Java Sea (northeast of the Sunda Strait) with a magnitude of 7.5, and December 16, in 1963. It occurred in Banten with a magnitude of 6.5 [4]. On the other hand, a height of tsunami up to 30 meters with a range of tsunami wave propagation of 800 km occurred on August 27, 1883, it was caused by the collapse of the body of Mount Krakatau, and a pyroclastic flow that fell into the sea which resulted in 36,000 deaths [5]. The fatalities amount in the eruption of Mount Krakatau, one of which is because the volcanic area is a fertile agricultural area, and the object of mass natural tourism visits so that it tends to be crowded by humans [6]. The distribution of the direction of the tsunami waves caused by the eruption of Mount Krakatau in 1883 can be seen in Fig. 1. After the paroxysmal eruption in 1883 of the Krakatau Volcanic Complex, a young volcano appeared in the caldera, namely the Anak Krakatau Volcano.

* Corresponding Author: wahyubintaro1985@gmail.com

Anak Krakatau Volcano, located in the Sunda Strait, is a type A stratovolcano.

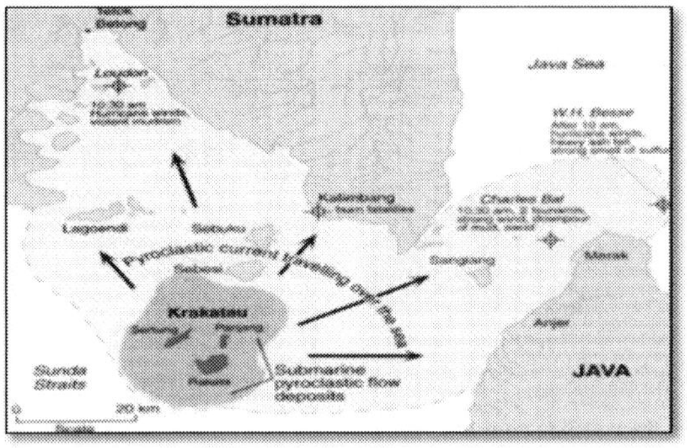

Fig. 1. Map of the distribution of the direction of the tsunami waves caused by the eruption of Mount Krakatau in 1883. [7].

Indications of the presence of an embryo of Mount Anak Krakatau have been known since the appearance of early eruption symptoms were observed from December 29, 1927, to January 5, 1928, at a geographical position of 1050 25' 27" east and 60 6' 6" South Latitude, as an underwater volcano at a depth of 28 m below sea level [8]. The body of Anak Krakatau has emerged to the sea surface since 1929. Till now Mount Anak Krakatau is in the construction phase (building its body to be big). In 2016 the eruption occurred on June 20, 2016 meanwhile the eruption occurred on February 19, 2017, in the shape of a Strombolian eruption [9].

The types and types of eruptions of Mount Anak Krakatau are phreatomagmatic, and the threat of danger can be in the form of volcanic debris that causes the tsunami.

A network of 6 seismic stations namely KRA1, KRA2, KRA3, SRTG, PULO, and LAVA stations do observations for volcanic earthquakes on Mount Anak Krakatau were monitored through the locations of the KRA1, KRA2, and KRA3 seismic stations are on Anakkrakatau Island. Meanwhile, the SRTG seismic station on Sertung Island and the PULO seismic station in the Mount Pulosari area, Pandeglang Regency, Banten. and the LAVA seismic station in the northeast of Anak Krakatau Island. The types of volcanic earthquakes recorded on Mount Anak Krakatau are eruptive earthquakes, gusts earthquakes, continuous tremors, harmonic tremors, shallow volcanic earthquakes (VB), and deep volcanic earthquakes (VA). Meanwhile, the recorded tectonic earthquakes are consist of local tectonic earthquakes and distant tectonic earthquakes [10].

On December 22, 2018, at 21.03 WIB, The LAVA Seismic Station of the Mount Anak Krakatau Observation Post did not record volcanic earthquakes. The earthquake measuring instrument installed on the body of Mount Anak Krakatau was reporting that there was an incident of loss of signal transmission from t. At 21:27 West Indonesia Time (Zone), the Meteorology, Climatology, and Geophysics Agency (BMKG) gave a press release stating that there had been a tsunami on the coasts of Banten and Lampung caused by landslides from the eruption of Mount Anak Krakatau. The flank collapse on the 64-hectare Anak Krakatau wall has caused a tsunami in the Sunda Strait (Tim Pengusul. 2019). This tsunami has resulted in 437 deaths, 14,059 people were injured, and 33,721 people are displaced. The result of the avalanche turned out to have changed the morphology of Anak Krakatau can be seen in Fig. 2.

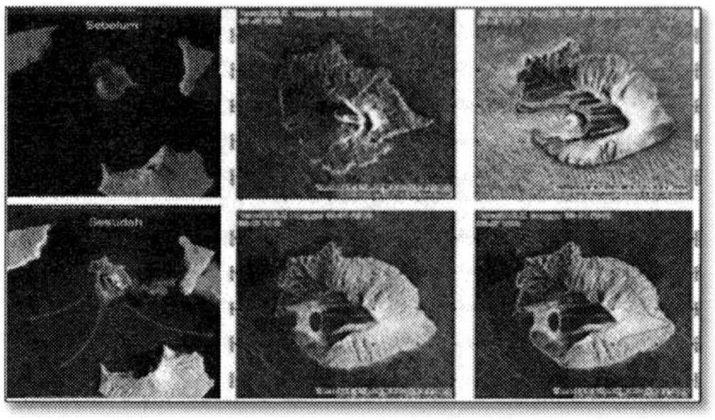

Fig. 2. Geomorphological site changes of Mount Anak Krakatau before and after the Sunda Strait tsunami.

Changes in the shape of Mount Anak Krakatau can be seen from the traces of its activities which are most likely triggered by the activity of Mount Anak Krakatau which erupts periodically and continuously, the collapse of the slopes in the western part that occurred on December 22, 2018, which resulted in waves that damaged the surrounding area, The force of the eruption had removed the peak of Mount Anak Krakatau, and Mount Anak Krakatau had begun to form again. These changes are related to the loss of mountain peaks. The upper volcanic cone of Mount Anak Krakatau disappeared due to the eruption and left a large crater wall and the highest peak of 158,635 masl at coordinates (547224,919; 9325617,886) UTM [11]. This crater wall opens to the southwest which can be seen in Fig. 3.

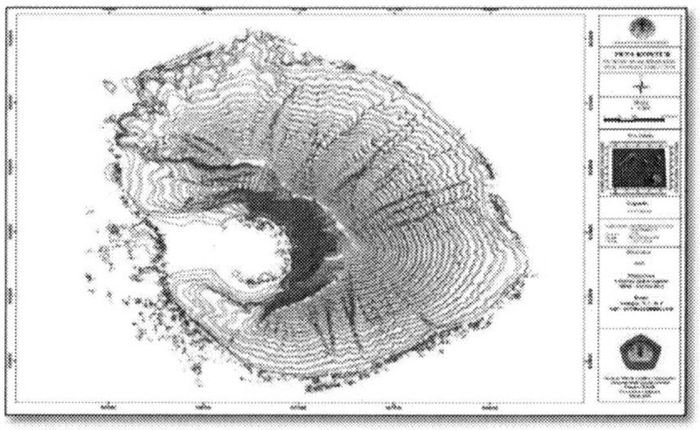

Fig. 3. Condition of Mount Anak Krakatau

The landslide may have changed the mountain passage and opened a new portion that drains magma directly into the sea surface. It is likely to trigger the meeting of hot magma with water, but over time there are also traces of seawater that stop entering the magma system so that Mount Anak Krakatau returns to normal. This kind of morphological change has always occurred on Mount Anak Krakatau throughout the history of its activities.

https://doi.org/10.1051/e3sconf/202133107006

The Sunda Strait area is a National Strategic Area. The Sunda Strait Strategic Area includes land, island, and sea areas located within Lampung Province, Banten Province, and other areas determined based on a development plan [12]. The development of a national strategic area of the Sunda Strait must pay attention to its geological aspects, both geological resource factors that support human sustainability to sustain life, and geological obstacles/obstacles that can cause human survival, vulnerability to human survival., when viewed from a geological point of view, the mandate is clear.

There are various facts regarding the 2018 Sunda Strait tsunami, so one way to monitor volcanic earthquake activity at the Anak Krakatau Volcano is by monitoring seismic data using a seismic sensor network belonging to the Meteorology, Climatology and Geophysics Agency (BMKG). In 2021, the Meteorology, Climatology and Geophysics Agency (BMKG) has 411 seismic network stations, it was making an alternative for monitoring volcanic earthquake activity at Anak Krakatau Volcano in the Sunda Strait and can analyze non-tectonic seismic data using Seiscomp3 software to obtain fast and accurate analysis results. The purpose of this study is to monitor and analyze seismic data during the 2018 Sunda Strait tsunami.

2 Method

This study uses descriptive qualitative research methods. The data obtained by the researchers were through observation, documentation, archival records, and physical devices. Data is reduced and analyzed and presented in the form of tables and pictures then conclusions are drawn. The monitoring process and anatomizing the Sunda Strait tsunami uses the Seiscomp3 software specifically monitors & analyzes seismic data. Seiscomp3 software was originally a development of GFZ and earthquake GmbH that aims to detect seismic movements quickly and measure the magnitude of earthquakes that have the potential to trigger a tsunami in Indonesia [13]. SeisComP3 is a software package as a data acquisition system, data archiving system, quality control, and interactive analysis in real-time. SeisComP is designed with the simplicity of mind, not only for development and usage but also prioritizes speed in performing data processing automatically and manually (Tim Pengkaji, 2021). Seiscomp3 is related to the Indonesian Tsunami Early Warning System (InaTEWS) seismic sensor station network. Seiscomp3 software has 3 components, these are data acquisition, data processing, and data analysis which can be seen in Fig. 4 [14].

The flow of the seismic data monitoring and analysis system begins with the data acquisition process (data acquisition) obtained from all seismic stations and seismic data recorded on seismic sensors throughout Indonesia which are sent via satellite. In this acquisition component, seiscomp3 uses the seedlink and Arclink protocols which aim to access seismic recording data (waveform) from seismic stations (remote sites) and then archive the data. At this time, Seedlink became the de-facto standard regarding the exchange of seismic data (waveform) known as SEED.

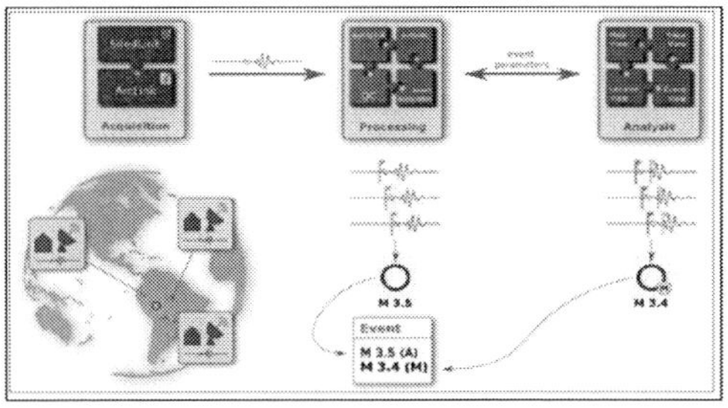

Fig. 4. Components of the Seiscomp3 software

In the data processing component, Seiscomp3 has automatic processing features that allow real-time determination of earthquake parameters such as automatic picking of p waves, automatic location, and automatic magnitude. automatic). This feature is also very useful as initial information on earthquake parameters before being analyzed (reviewed) manually by the operator.

In the data analysis component, Seiscomp3 provides features that allow operators to interact with the Seiscomp3 system easily (user friendly) to conduct reviews and manual analysis punctually. In this component, Seiscomp3 provides four GUIs, including:

1. Scmv (SC3 Map View), serves to display the condition and distribution of sensors in the form of a map which can be seen in Fig. 5.

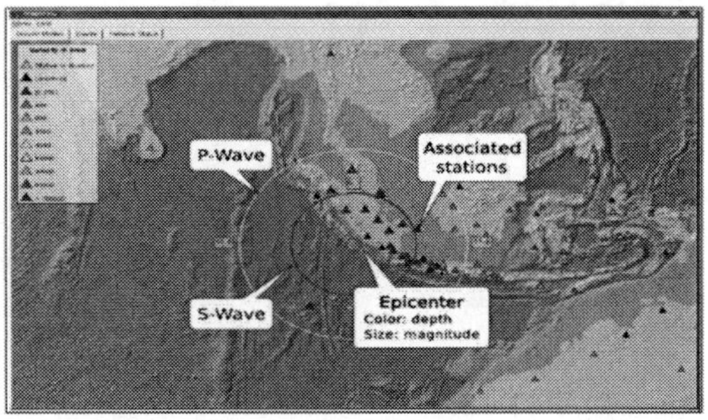

Fig. 5. GUI SCMV display.

2. Scrttv (SC3 Real-Time Trace View), serves to display seismic signals directly (real-time) which can be seen in Fig. 6.

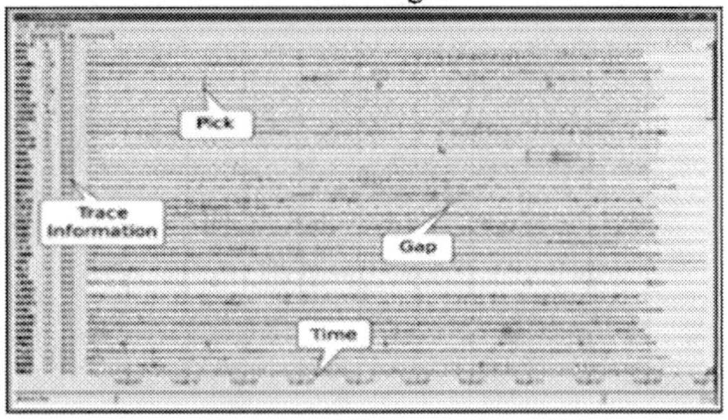

Fig. 6. GUI scrrtv display.

3. Scolv (SC3 Origin Locator View), serves to display the processing results and the window functions to perform manual picking which can be seen in Fig.7.

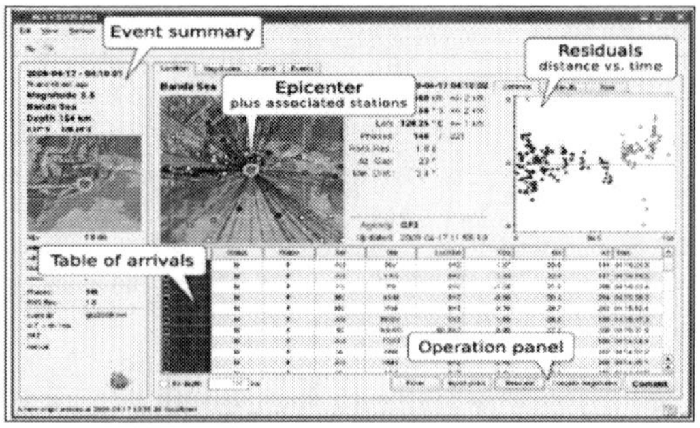

Fig. 7. GUI display of scolv.

4. Scesv (SC3 Event Summary View), serves to display the results (output) of earthquake parameters which can be seen in Fig. 8.

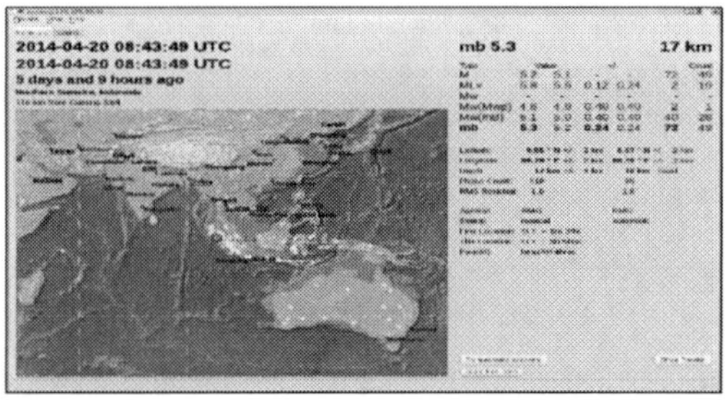

Fig. 8. GUI scesv display.

Seismic stations must obtain a minimum of 3 seismic data when manually analyzing an earthquake event. The more seismic data analyzed, the better the analysis results. Data from three seismic stations eliminates doubts because there is a meeting point of the three circles, namely the epicenter [15]. The seismic data picking process as manually, starting with picking the primary wave (p) on the vertical component and continuing to pick the secondary wave (s) on the horizontal component for each seismic station. then, the earthquake parameters are obtained such as the time of occurrence, the location of the earthquake center, namely latitude and longitude, depth, and magnitude. Constraints faced when analyzing seismic data manually are the difficulty of analyzing seismic data if there is noise (data frequency disturbance) so that it is difficult to find the best position for picking primary and secondary waves, the lack of seismic data recorded on seismic stations specifically for areas that not reachable by seismic station network.

The method that operates on the seiscomp3 software is the Geiger method. The data used in the Geiger method is the arrival time of seismic waves. The Geiger method assumes that the earth consists of a homogeneous isotropic flat layer so that the arrival time of seismic waves can be calculated due to reflection and refraction for each layer of the earth. The method used is to give the initial value of the hypocenter, then calculate the wave propagation time for each seismic station used. The basis of the calculation using the Geiger method is to find the smallest residual (r) which is influenced by the partial derivative of time to the position on each Cartesian coordinate axis (x, y, z) and the difference between the observation time (tobs) and the calculation travel time (thit) which is denoted (Δt) [16] which is shown in the following Fig 9.

$$ri = \frac{\partial T}{\partial xi}\Delta x + \frac{\partial T}{\partial yi}\Delta y + \frac{\partial T}{\partial zi}\Delta z + \Delta t$$

Fig. 9. The equation of the smallest residual in the Geiger method.

3 Result and discussion

Seismic activity that occurred on December 22, 2018, at 20:55 WIB can be monitored by the BMKG seismic network originating from 13 seismic stations around the Sunda Strait area. Seismic stations that can record seismic activity can be seen in Table 1.

Table 1. BMKG Seismic Stations

No.	Station Name	Station Location
1	CGJI	Cigeulis, Banten Province
2	SBJI	Serang, Banten Province
3	BLSI	Bandar Lampung, Lampung Province
4	TNG	Tangerang, Banten Province
5	TNGI	Tangerang, Banten Province
6	SKJI	Sukabumi, West Java province
7	LWLI	Liwa, Lampung province
8	CNJI	Cacaban, West Java province
9	MDSI	Muara Dua, South Sumatera Province
10	LEM	Lembang, West Java province
11	BBJI	Bungbulang Garut, West Java province
12	JCJI	Jatiwangi, West Java province
13	PMBI	Palembang, South Sumatera Province

Seismic data originating from 13 BMKG seismic stations were then analyzed using Seiscomp3 software by selecting the primary wave (p wave) and secondary wave (s wave) correctly on each signal for each seismic station by first using a data filter of 0.4 Hz to 1 Hz which aims to make the seismic data visible. For more details, see Fig. 10.

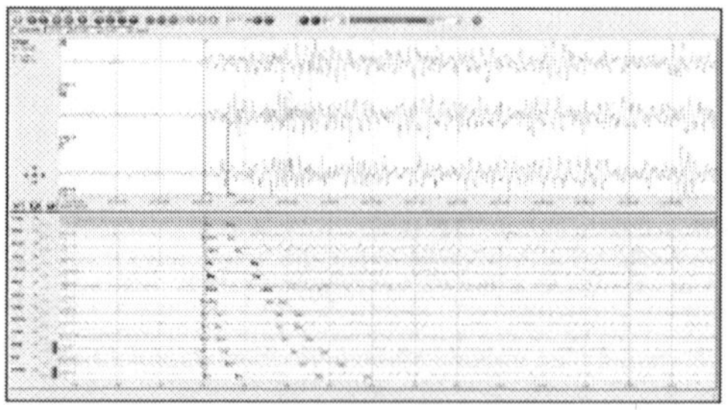

Fig. 10. Non-tectonic seismic wave recording

After selecting and analyzing the primary waves (p waves) and secondary waves (s waves), the parameters obtained are that the vibration center is on Mount Anak Krakatau, with coordinates 105.46 East Longitude – 6.10 South Latitude, 1 km depth, 3.4 MLv magnitude. The MLv magnitude was chosen because this event is seismic data in a local range and the seismic data is analyzed only on the vertical component (Z component). The residual rms value obtained is 0.5. The residual rms value of 0.5 obtained during manual analysis is still at the fair threshold value, which does not exceed 1 (one). The rms value only depends on the velocity model used in the analysis software that determines the location of the earthquake, the quality of the determination of the arrival time of the wave, and the procedure used [17]. The results of the analysis using the Seiscomp3 software can be seen in Figs 11 and 12.

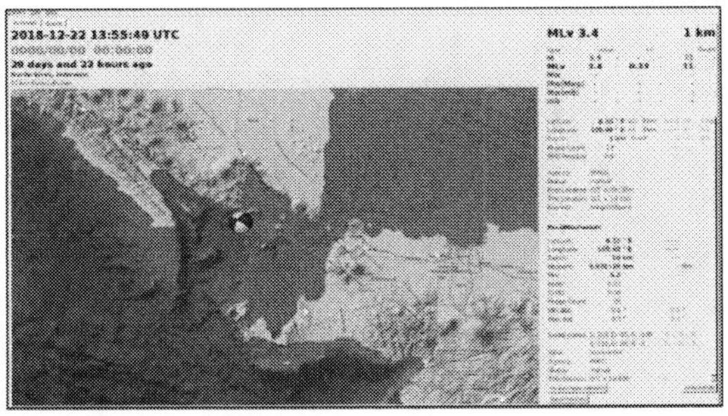

Fig. 11. Results of non-tectonic seismic data analysis using Seiscomp3

In Fig. 11 above can be seen that the position of the vibration center is right on Mount Anak Krakatau, has a vibration depth of only 1 km and the focal mechanism obtained is horizontal (strike-slip). Therefore, the focal mechanism can be summarized that the seismic records recorded by 13 BMKG seismic stations are non-tectonic seismic caused by the volcanic activity of Mount Anak Krakatau.

In Fig. 12 above can be seen the position of the vibration center is right on Mount Anak Krakatau, which is symbolized by a red circle, and the position of the BMKG seismic stations which record non-tectonic seismic records symbolized by a green triangle.

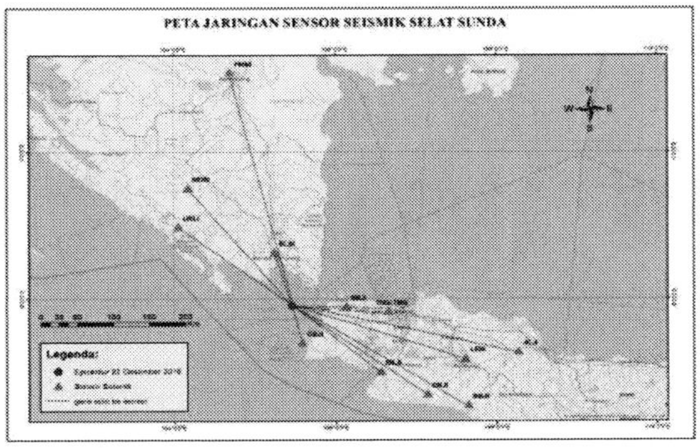

Fig. 12. Center for non-tectonic seismic vibrations from BMKG seismic network monitoring

The success of BMKG seismic stations in recording non-tectonic seismic activity is used as a way to quickly and accurately detect non-tectonic seismic activity on Mount Anak Krakatau.

The tsunami in the coastal areas of Banten and Lampung occurred after a vibration originating in Mount Anak Krakatau caused an avalanche of the body of Mount Anak Krakatau which then avalanche material fell into the sea, and has resulted in 437 deaths, 14,059 people were injured, and 33,721 people have been displaced. Based on the experience of the Sunda Strait tsunami, the Banten Provincial Government and the Lampung Provincial Government must increase their efforts to mitigate the tsunami structurally and non-structurally to increase public understanding of tsunamis to reduce the risk of a tsunami disaster.

By strength of a vibration that only has a magnitude of 3.4, the tsunami early warning system in the Indonesia Tsunami Early Warning System (InaTEWS) cannot model tsunamis in the Sunda Strait region because it is only designed to model tsunamis caused by tectonic earthquakes with large strength with a wide area coverage. wider. Therefore, a tsunami early warning system is needed that originates from the activities of the Anak Krakatau Volcano, and as a result of the avalanche of the body of the Anak Krakatau Volcano and from other tsunami generating sources.

4 Conclusion

The activity of Non-tectonic seismic is originating from the eruption of Mount Anak Krakatau can be monitored by the BMKG seismic sensor station network, and the non-tectonic seismic data can be analyzed using the Seiscomp3 software. The output of the analysis found that the center of the vibration was on Mount Anak Krakatau, with coordinates 105.46 East Longitude – 6.10 South Latitude, and a depth of 1 km, and had strength equivalent to the power of a tectonic earthquake with a magnitude of 3.4. For anticipating similar tsunami events in the future, an early warning system for tsunamis originating from volcanic activity and volcanic avalanches was required urgently, one of which is to increase the existing network of non-tectonic seismic stations.

References

1. S. Rohadi, *Study of seismotectonics as an indicator of potential earthquakes in indonesian region j. Meteorol. And Geophysics,* **10,** 111–20 (2009)
2. Fawzy N A B H M and I. Massinai, Identification of mount merapi volcanic earthquake focus mechanism (2018)
3. D. Shimozuru and N. K, *Volcanic spacing and subduction, Arc Volcanics: Physics and Tectonics,* (Tokyo: Terra Scientific Publishing, 1983)
4 Tsunami catalog year, 416-2018 (Jakarta: BMKG)
5. J. Beget, *Volcanic tsunamis* (Academic Press, 2000)

6. I. Pratomo, Classification of Indonesian active volcanoes, case studies of several volcanic eruptions in the history of J. Geol. Indonesia, **I,** 209–27 (2006)
7. Valentine, G.A, and R. Fisher, *Pyroclastic surges and blasts* (Academic Press, 2000)
8. C. E. Stehn, *The Geology and Volcanism of the Krakatau group* (Batavia, 1929)
9. PVMBG, Latest Developments on Mount Anak Krakatau's Activity, PVMBG: The height of the mountain is only 110 meters (2018)
10. Preparation Team for the 2019 Geological Agency Geological Dynamics of the Sunda Strait in Sustainable Development (Bandung: Geological Agency of the Ministry of Energy and Mineral Resources)
11. A. Armijon, E. Rahmadi, I. Susanti and F. Murdapa, Study on the Update of the Lampung Bay Coastal Tsunami Immersion Model Due to the Effect (2019)
12. Anon, Presidential Decree No.86 concerning the Development of Strategic Areas and Infrastructure of the Sunda Strait (2011)
13 J. Clinton, Seismic Monitoring and Alert In Switzerland: Realand Near-RealTime Products Of The Swiss National Seismic Networks *Second Eur. Conf. Earthq. Eng. Seismol* (Istanbul, 2014)
14 Anon, InaTEWS, *Processing System Study Report* (Jakarta: BMKG Deputy for Geophysics,2020)
15 A. Wijaya, Effect of additional seismic stations (stpi, tspi, and iwpi) on the analysis of determining earthquake parameters in the case study of earthquakes in the west papua region in 2019–2020, *Biolearning J.* 7 10–3 (2020.)
16 K. Nishi, *Hypocenter Calculation Software GAD (Geiger's method with Adaptive Damping,* 2005*)*
17 S.Husen, E. Kissling, N. Deichmann, S. Wiemer, D. Giardini, & M. Baer, Probabilistic earthquake location in complex three-dimensional velocity models *Appl. to Switzerland, J.Geophys. Res* (2003)

https://doi.org/10.1051/e3sconf/202133107007

Determination of tsunami run-up and golden time in the megathrust subduction zone of the sunda strait segment

Iko Sarikanti Ponangsera[1*], *Anwar* Kurniadi[1], *Deffi* Ayu Puspitosari[1], and *Dedy* Hartono[1]

[1] Disaster Management Study Program, National Security Faculty, Republic of Indonesia Defense University, IPSC Sentul Area, Bogor, West Java, Indonesia

Abstract. The Indonesian Disaster Data and Information Management Database state that the tsunami in the Megathrust Subduction Zone of the Sunda Strait Segment is a disaster threat with high impact and loss of life every time it occurs. Therefore, determining the run-up and golden time of the tsunami in the Megathrust Subduction Zone of the Sunda Strait Segment is necessary as part of the mitigation of seismic activity that has the potential for a tsunami. The design of this study is qualitative research with primary data to determine the run-up and golden time of the tsunami in the Megathrust Subduction Zone of the Sunda Strait Segment obtained from the Indonesia Tsunami Warning System (InaTEWS) at BMKG. The results of the study inform that seismic activity in the Megathrust Subduction Zone of the Sunda Strait Segment can cause a tsunami with a run-up of 5.99 meters and a golden time of 40 minutes 19 seconds in Banten and South Lampung, the run-up of 3.83 meters, and a golden time of 1 hour 1 minute 10 seconds in Pacitan, East Java, run up 3.28 meters and golden time 33 minutes 15 seconds in Pangandaran, West Java.

1 Introduction

One of the tasks of the state is to protect the entire nation and the homeland of Indonesia from various kinds of threats. Threats are every effort or activity that intends to threaten and endanger state sovereignty, territorial integrity, and the safety of all nations, both from within and outside the country [1,2].

The characteristics of threats are divided into 2, namely military threats and non-military threats [3–5]. The shift in the characteristics of military threats to non-military ones, such as an increase in the number of incidents and impacts due to disasters, is one of the directions of Indonesia's attention in strengthening state protection for the human security of its citizens [4].

Indonesia's geography, which is located between 4 active tectonic plates of Eurasia, Indo-Australia, Pacific, and the Philippines [6,7] makes Indonesia has a high risk of earthquakes and tsunamis [8,9].

The movement of plates in various directions causes a subduction zone of the Sumatran plate to the Java segment [10,11]. The subduction zone due to plate movement is commonly known as Megathrust. Indonesia itself has 13 megathrusts, namely, Aceh-Andaman, Nias-Simelue, Batu, Mentawai Siberut, Mentawai-Pagai, Enggano, Sunda Strait, West Java-Central Java, East Java, Sumba, North Sulawesi, the Philippines, and Papua [12–14]. The 13 megathrusts can be seen in Fig. 1.

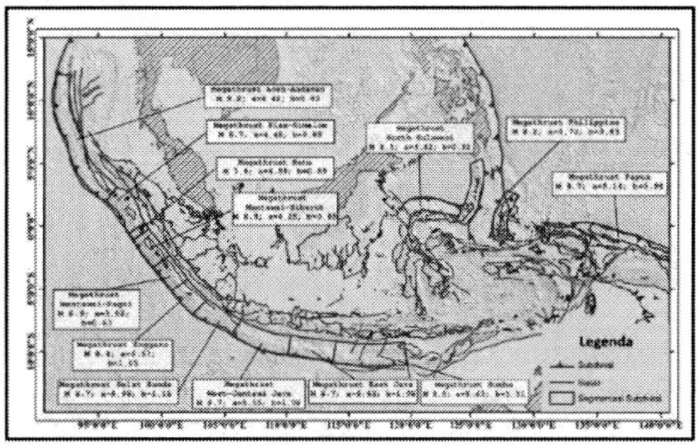

Fig. 1. Segmentation Map and Maximum Subduction Magnitude of Indonesia [14].

The 13 megathrusts in Indonesia have the potential for earthquakes with their epicenters around the subduction zone of the plate. Deformation and subduction movements of tectonic plates are the reason for the formation of active faults on the seabed and land. This is what causes the potential for tectonic earthquakes to often occur in Indonesia [15].

In addition, the movement of tectonic plates also causes the formation of active volcanoes throughout Indonesia, or commonly known as the Ring of Fire. Indonesia's geography is what makes it a high-risk index for volcanic eruptions, earthquakes, and tsunamis [14].

* Corresponding author: ikoponangsera@yahoo.co.id

https://doi.org/10.1051/e3sconf/202133107007

The historical distribution of earthquakes in Indonesia can be seen in Fig. 2.

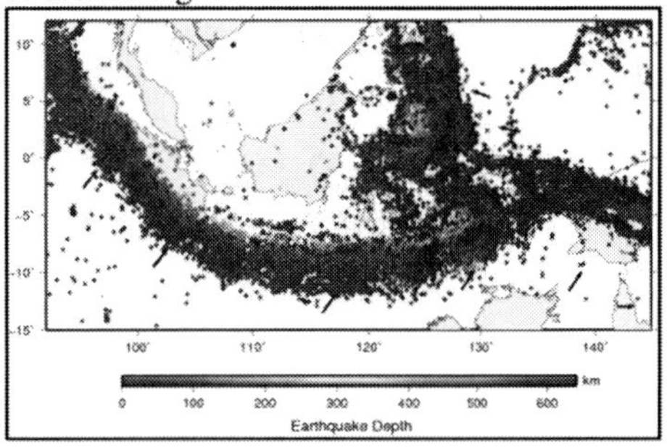

Fig. 2. Indonesia Earthquake Event Map [14].

Seismic activity in the Megathrust Subduction Zone of the Sunda Strait Segment is one of the meeting activities of the Indo-Australian and Eurasian plates which has the potential for a tsunami in the south of the islands of Sumatra and Java [10,14,16,17], as shown in Fig. 3.

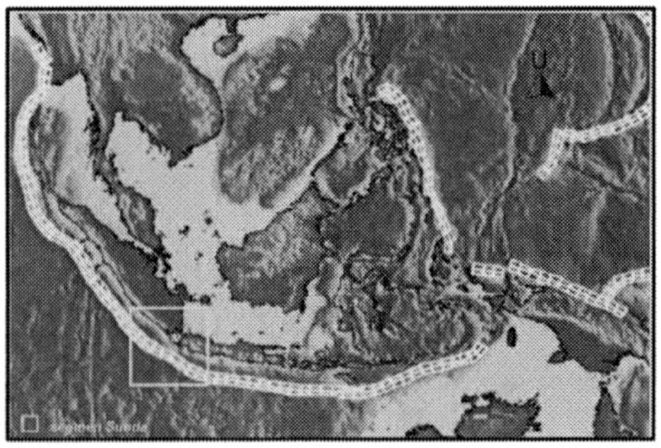

Fig. 3. Location of tsunami sources due to earthquakes associated with subduction zones (series of segments following the pattern of subduction zones) [18].

Fig. 3 explains that the Megathrust Subduction Zone of the Sunda Strait Segment is one of the tsunamigenic zones in Indonesia that can potentially cause a tsunami. Therefore, research to determine the run-up and golden time of the tsunami in the Megathrust Subduction Zone of the Sunda Strait Segment are necessary as part of disaster mitigation efforts to protect the Indonesian people from the threat of a potential tsunami disaster and as a form of national security.

2 Study Method

The design of this study is qualitative research with primary data to determine the run-up and golden time of the tsunami in the Megathrust Subduction Zone of the Sunda Strait Segment obtained from the Indonesia Tsunami Warning System (InaTEWS) at BMKG.

The guideline for modeling tsunami potential in this study uses the Tsunami Observation and Simulation Terminal software at BMKG. The time of occurrence of the earthquake (origin time) for tsunami modeling in the Megathrust Subduction Zone of the Sunda Strait Segment is 19:34:02 WIB with a strength of M 9.2. The modeling locations are at 8.23° N and 104.36° East with an epicenter of 20 km, as shown in Fig. 4.

3 Result and Discussion

The results of data processing get output in the form of an animation modeling the occurrence of a tsunami in the Megathrust Subduction Zone of the Sunda Strait Segment.

Fig. 4. Tsunami Modeling Parameters in the Megathrust Subduction Zone of the Sunda Strait Segment.

The tsunami modeling scenario in the Tsunami Observation and Simulation Terminal software uses the EasyWave2 model with parameters M 8.7 SR and epicenter 10 km. This was chosen because the spread and propagation of tsunami waves can be seen within 6 hours with the location of the tsunami wave propagation spreading to various areas. The tsunami modeling wave propagation in the Megathrust Subduction Zone of the Sunda Strait Segment can be seen in Fig. 5.

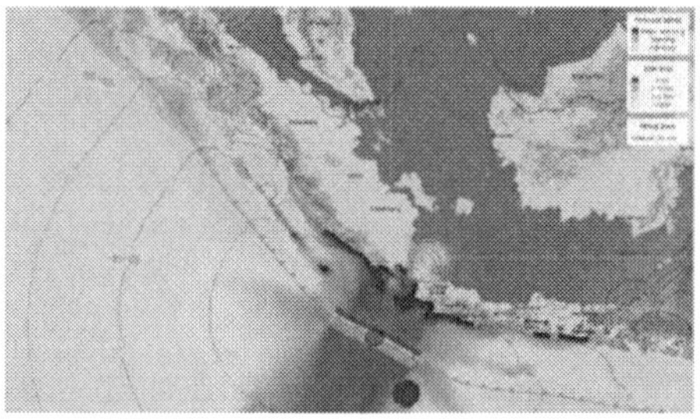

Fig. 5. Tsunami Modeling Wave Propagation in the Megathrust Subduction Zone of the Sunda Strait Segment.

Based on the modeling using the TOAST software, information on the status of tsunami early warnings in various areas with the potential for a tsunami to occur due to an earthquake is generated in the Megathrust Subduction Zone of the Sunda Strait Segment. The tsunami early warning status uses 3 levels, namely alert, alert, and alert. The alert level in this modeling is given if the estimated tsunami wave height reaches > 3 meters, the alert warning level in this model is given if the estimated tsunami wave height reaches 0.5 meters to 3 meters, the alert warning level in this model is given if the predicted wave height is tsunami reaches <0.5 meters [19,20]. Information on potential tsunami areas can be seen in Table 1.

Table 1. Potential areas affected by the tsunami.

Tsunami Modeling Time	Golden Time	Tsunami Early Warning Level	Potential Tsunami Affected Areas
19:54:45 WIB	00:05:10	Awas	Pandeglang, Banten
19:54:45 WIB	00:09:30	Awas	Lebak, Banten
19:54:45 WIB	00:45:49	Awas	Serang, Banten
19:54:45 WIB	00:59:55	Siaga	Kota Cilegon, Banten
19:54:45 WIB	00:28:55	Awas	Kep. Sebuku, Lampung Selatan
19:54:45 WIB	00:21:04	Awas	Kep. Krakatau, Lampung Selatan
19:54:45 WIB	00:20:45	Awas	Kep. Legundi, Lampung Selatan
19:54:45 WIB	00:50:15	Awas	Kalianda, Lampung Selatan
19:54:45 WIB	00:50:45	Siaga	Bakauheni, Lampung Selatan
19:54:45 WIB	01:08:00	Siaga	Panjang, Kota Bandar Lampung
19:54:45 WIB	00:17:04	Awas	Pulau Tabuan, Tanggamus
19:54:45 WIB	01:54:45	Siaga	Lampung Timur
19:54:45 WIB	00:18:00	Awas	Pesisir Tengah, Lampung Barat
19:54:45 WIB	00:16:25	Awas	Pesisir Selatan, Lampung Barat
19:54:45 WIB	00:08:30	Awas	Pelabuhan Ratu, Sukabumi, Jawa Barat
19:54:45 WIB	00:13:49	Awas	Garut, Jawa Barat
19:54:45 WIB	00:11:00	Awas	Sindangbarang, Cianjur, Jawa Barat
19:54:45 WIB	00:19:55	Awas	Tasikmalaya, Jawa Barat
19:54:45 WIB	00:25:04	Siaga	Ciamis, Jawa Barat
19:54:45 WIB	00:50:40	Awas	Wonogiri, Jawa Tengah
19:54:45 WIB	00:45:40	Siaga	Purworejo, Jawa Tengah
19:54:45 WIB	00:42:34	Siaga	Kebumen, Jawa Tengah
19:54:45 WIB	00:40:45	Siaga	Cilacap, Jawa Tengah
19:54:45 WIB	01:00:45	Awas	Malang, Jawa Timur
19:54:45 WIB	00:58:49	Awas	Tulungagung, Jawa Timur
19:54:45 WIB	00:59:49	Awas	Blitar, Jawa Timur
19:54:45 WIB	00:53:10	Awas	Pacitan, Jawa Timur
19:54:45 WIB	01:04:34	Siaga	Pulau Nusabarung, Jawa Timur

Table 1 explains that earthquakes in the Megathrust Subduction Zone of the Sunda Strait Segment can have the potential for a tsunami with a tsunami early warning level in the alert category in parts of Banten Province, such as Pandeglang, Lebak, and Serang.

In Lampung Province, seismic activity in the Megathrust Subduction Zone of the Sunda Strait Segment has the potential for a tsunami with the alert category in Kec. Rajabasa, Kalianda, Tanggamus, to the Central Coast and the West Coast.

Seismic activity in the Megathrust Subduction Zone of the Sunda Strait Segment also has the potential for tsunamis in areas in southern Java, such as West Java, Central Java, and East Java.

The shortest tsunami arrival time difference is in the Pandeglang area, Banten with a golden time of 5 minutes 10 seconds, Lebak with a golden time of 9 minutes 30 seconds. The golden time is the time difference to evacuate or save oneself from the threat of a potential tsunami after an earthquake in the Megathrust Subduction Zone of the Sunda Strait Segment.

In addition, the TOAST software's tsunami modeling output in the Megathrust Subduction Zone of the Sunda Strait Segment also produces estimates of tsunami run-up and golden time, as can be seen in Table 2.

Table 2. Estimated Run up and Golden Time Tsunami.

Tsunami Modeling Time	Run up tsunami	Golden Time	Potential Tsunami Affected Areas
12:54:45 WIB	5,99 meter	00:40:19	Pandeglang, Banten
12:54:45 WIB	5,99 meter	00:40:19	Kec. Rajabasa, Lampung Selatan
12:54:45 WIB	3,83 meter	00:56:49	Pacitan, Jawa Timur
12:54:45 WIB	3,71 meter	01:01:10	Tulung Agung, Jawa Timur
12:54:45 WIB	3,28 meter	00:33:15	Pangandaran, Jawa Barat
12:54:45 WIB	3,01 meter	00:39:45	Cilacap, Jawa Tengah

Based on Table 2, tsunami modeling in the Megathrust Subduction Zone of the Sunda Strait Segment resulted in a tsunami run-up of 5.99 meters and a golden time of 40 minutes 19 seconds in Banten and South Lampung, a run-up of 3.83 meters, and a golden time of 1 hour 1 minute 10 seconds. in Pacitan, East Java, the run-up was 3.28 meters and a golden time of 33 minutes 15 seconds in Pangandaran, West Java, and a run-up of 3.01 meters and a golden time of 39 minutes 45 seconds in Cilacap, Central Java.

4 Conclusion

The results of the study prove that seismic activity in the Megathrust Subduction Zone of the Sunda Strait Segment can have the potential for a tsunami with an alert status in the provinces of Banten, Lampung, West Java, Tangah Java, and East Java.

https://doi.org/10.1051/e3sconf/202133107007

The golden time or the difference in time for evacuation and saving from the threat of a potential tsunami after an earthquake in the Megathrust Subduction Zone of the Sunda Strait Segment varies from region to region. The fastest golden time is 5 minutes 10 seconds in Pandeglang, Banten.

The results of this study can be used as the basis for disaster mitigation policies due to seismic activity in the Megathrust Subduction Zone of the Sunda Strait Segment.

References

1. B. Lesmana, Y. Rusfiana, R. Gunawan, Peperangan Asimetrik **4**, (2018).
2. L. Suryatni, J. Ilm. Huk. Dirgant. **10**, (2020).
3. I. S. Ponangsera, R. K. Apriyadi, D. Hartono, and W. Wilopo, PENDIPA J. Sci. Educ. **5**, 277 (2021)
4. S. Yulianto, R. K. Apriyadi, A. Aprilyanto, T. Winugroho, I. S. Ponangsera, and W. Wilopo, PENDIPA J. Sci. Educ. **5**, 180 (2021)
5. J. Fitriyani, R. K. Apriyadi, T. Winugroho, D. Hartono, I. D. K. K. Widana, and W. Wilopo, PENDIPA J. Sci. Educ. **5**, 322 (2021)
6. R. K. Apriyadi and R. Amelia, PENDIPA J. Sci. Educ. **5**, 56 (2020)
7. A. Aprilyanto, R. K. Apriyadi, T. Winugroho, I. D. K. K. Widana, and W. Wilopo, PENDIPA J. Sci. Educ. **5**, 284 (2021)
8. M. Yusfania, I. M. Anjasmara, A. L. Hadi, and B. Pramudya, in *AIP Conf. Proc.* (AIP Publishing LLC, 2017), p. 40008
9. Z. Zakaria and S. Sidarto, J. Geol. Dan Sumberd. Miner. **16**, 115 (2015)
10. Y. Fauzi, W. Kongko, and K. S. Brotopuspito, in *IOP Conf. Ser. Earth Environ. Sci.* (IOP Publishing, 2020), p. 12051
11. R. McCaffrey, Annu. Rev. Earth Planet. Sci. **37**, 345 (2009)
12. S. Widiyantoro, E. Gunawan, A. Muhari, N. Rawlinson, J. Mori, N. R. Hanifa, S. Susilo, P. Supendi, H. A. Shiddiqi, and A. D. Nugraha, Sci. Rep. **10**, 1 (2020)
13. A. Muhammad, K. Goda, N. A. Alexander, W. Kongko, and A. Muhari, Nat. Hazards Earth Syst. Sci. **17**, 2245 (2017)
14. Tim-PSGN, *Buku Peta Sumber Dan Bahaya Gempa Indonesia Tahun 2017* (Pusat Litbang Perumahan dan Permukiman Kementerian PUPR, Bandung, 2017)
15. BNPB, *Indeks Risiko Bencana Indonesia Tahun 2020* (Badan Nasional Penanggulangan Bencana, Jakarta, 2021)
16. W. Widiyanto and S.-C. Hsiao, in *IOP Conf. Ser. Mater. Sci. Eng.* (IOP Publishing, 2020), p. 12034
17. W. Setyonegoro, S. Rohadi, B. Sunardi, Y. H. Perdana, T. Kurniawan, and U. Haryoko, in *43rd Annu. Sci. Meet. Himpun. Ahli Geofis. Indones. Semarang* (2018)
18. Y. Yudhicara, K. Budiono, Indones. J. Geosci. **3**, 241 (2008)
19. W. Kurniawan, E. Bangun, B. Prakoso, PENDIPA J. Sci. Educ. **5**, 204 (2021)
20. R. K. Apriyadi, W. Kurniawan, S. Yulianto, S. Syamsunasir, I. D. K. K. Widana, A. Subiyanto, F. Bahar, D. C. Kuncoro, PENDIPA J. Sci. Educ. **6**, 1 (2022)

Accelerating information of tsunami disaster using early warning system (EWS) Devices: smart solution for communication, navigation and surveillance due to tidal sea level

Nurhadi Santosa[1]*, *Nadjadji* Anwar[2], *Bangun* Muljo Sukojo[3], and *Wesam* Al Madhoun[4]

[1]Doctoral Student of Civil Engineering Department, ITS Surabaya.
[2]Professor of Civil Engineering Department, ITS Surabaya
[3]Professor of Geomatics Engineering Department, ITS Surabaya
[4]Dean and Doctor of Engineering Faculty Gaza University

Abstract. Oneof the important aspects in reducing Tsunami Risk is understanding Natural Hazards. The loss of INA TEWS equipment worth hundreds of billions of rupiah and the occurrence of the Tsunami in Aceh and other parts of Indonesia are warnings that our society needs the help of effective early warning tools at affordable prices. Using this device will make it easier for the community to monitor natural events from the field and reduce disaster risk. Currently the government has purchased hundreds of units of JRC-UNESCO products, namely IDSL (Inexpensive Device for Sea Level), but this device takes 21 minutes of information to reach the community. Therefore, a new device that has been prepared is needed. (FIDELA EARLY WARNING SYSTEM/FEWS) is a tool that involves the Equipment sector, Operations sector and Community sector. The results shows that this device is that the time for sending information to the public is only 5 minutes.

1 Introduction

Head of the data, Information and Public Relations Center of the National Disaster Management Agency (BNPB) Sutopo Purwo Nugroho revealed that the death toll was in Pandeglang, Serang, South Lampung, Pesawaran, and the Tanggamus area [3]. The large number of victims due to the tsunami was a result of the vulnerability of early detection. The latest update from Wartakotalive obtained from the BNPB Data and Information Center at the disaster site revealed that 281 people died, 1,1016 people were injured, 57 people were missing, 11687 people were displaced as shown in Fig. 1.

The same system, assisted by Germany which has been installed in West Sumatra since 2005, has also long been declared inoperative. Based on this experience, the use of tsunami early warning system technology in Indonesia must at least consider two important aspects, namely: low cost (installation and maintenance) and prevention of vandalism, in addition to technical aspects such as measurement accuracy and speed of warning delivery. early [1].

Based on data from the "Tsunami" Marine Knowledge Series, Directorate General of Marine Spatial Planning, Ministry of Marine Affairs and Fisheries, from 1964 to 2010 there were 23 tsunamis. The 23 Tsunamis were

1. Flores Tsunami in 1961
   The Flores Tsunami was caused by an earthquake at the epicenter 8.2 latitude, 122 east longitudes. There is no record of the scale of the earthquake that caused the tsunami, including the height of the tsunami and the number of fatalities.
2. Sumatra Tsunami in 1964
   The tsunami was caused by an earthquake at the epicenter of 5.8 latitude, 95.6 east longitude, with a total of 110 people killed and 479 people injured.
3. Seram Tsunami – Maluku in 1965
   This tsunami was caused by an earthquake at the epicenter of 2.4 latitude, 126 east longitude. The magnitude scale of the earthquake was 7.5 Sr, the tsunami height was 4 meters, and 71 people died.
4. Tinabung Tsunami – Sulawesi in 1967
   The tsuami occurred as a result of an earthquake with an epicenter of 3.7 latitude, 119.3 east longitude with a magnitude of 5.8 that killed 58 people.
5. Tambu Tsunami – Sulawesi in 1968
   The Tambu tsunami was caused by an earthquake with a magnitude of 7.4 on the Richter scale with an epicenter of 0.7 latitude, 119.7 east longitude. The earthquake caused a tsunami with a height of 10 meters and killed 200 people.
6. Majene Tsunami – Sulawesi in 1969

* Corresponding author: nurhadisantosa2000@gmail.com

This tsunami was triggered by an earthquake measuring 6.9 Sr at the epicenter 3.1 latitude, 118.8 east longitude with a tsunami height of 10 meters which killed 64 people.

7. Sumba Tsunami in 1977
   Triggered by an earthquake measuring 8 on the Richter scale with an epicenter of 11.1 South Latitude, 118.5 East Longitude. The earthquake triggered a tsunami as high as 15 meters and 189 people died.
8. NTT Tsunami, Flores, Atauro Island in 1977
   The NTT tsunami was triggered by an earthquake at the epicenter of 8 South Latitude, 125.3 East Longitude. There is no record of the magnitude scale of the earthquake that triggered this tsunami, the height of the tsunami, and the number of victims.
9. NTB Tsunami, Bali, Lombok, Sumbawa in1979
   This tsunami was caused by an earthquake that occurred at 8.4 latitude, 115.9 east longitude, but there is no information on the scale of the earthquake and the height that occurred. However, the Flores tsunami killed 27 people and injured 200.
10. Larantuka Tsunami in 1982
    The tsunami was caused by an earthquake at the epicenter of 8.4 South Latitude, 123 East Longitude with an earthquake scale of 5.9. The Larantuka tsunami killed 13 people but it is not known how high the tsunami was.
11. NTT Tsunami, East Flores, Pantar Island in 1987
    The NTT tsunami in 1987 was triggered by an earthquake at the epicenter of 8.4 latitude and 124.3 east longitude. The scale of the earthquake and its height are unknown but the NTT tsunami killed 83 people and injured 108 people.
12. NTT Tsunami, Alor Island in 1989
    The NTT tsunami in 1989 was caused by an earthquake at the epicenter. The tsunami natural disaster killed 7 people. The author did not find any reference indicating the magnitude scale of the earthquake and tsunami height that occurred on Alor Island in 1989.
13. Flores Tsunami in 1992
    The Flores tsunami was the worst tragedy before the Aceh tsunami in 2004. This tsunami was triggered by an earthquake centered at 8.5 latitude, 121.9 east longitude with an earthquake scale of 7.5. The tsunami was recorded as high as 26 meters and killed 2,100 people in Flores.
14. Banyuwangi Tsunami, East Java in 1994
    The Banyuwangi tsunami in 1994 was caused by an earthquake with a scale of 6.8 at the epicenter of 10.7 latitude, 113.1 east longitude. The tsunami generated by this earthquake was 14 meters high and killed as many as 208 people.
15. Palu Tsunami, Central Sulawesi in 1996
    The Palu Tsunami was triggered by an earthquake with a magnitude of 7.7 which caused a tsunami with a height of 6 meters and killed 68 people.
16. Biak Tsunami, Papua in 1996
    The Biak Tsunami in 1996 was caused by an earthquake with a magnitude of 8 at the epicenter 0.5 latitude, 136 east longitudes. The Biak tsunami was estimated to be 12 meters high and killed 160 people.
17. Taliabu Tsunami, Maluku in 1998
    The Taliabu Tsunami in 1998 was estimated to be 3 meters high and killed 34 people. The Taliabu tsunami was caused by an earthquake with a magnitude of 7.7 at the epicenter of 2 South Latitude, 124.8 East Longitude.
18. Banggai Tsunami, Central Sulawesi in 2000
    There is no information on the scale of the earthquake and the height of the Banggai tsunami in 2000. However, the Banggai tsunami killed 4 people and the tsunami was caused by an earthquake at the epicenter of 1.1 latitude and 123.5 east longitude.
19. Aceh Tsunami in 2004
    The Aceh Tsunami is the largest and the most destructive in the 21st century which killed approximately 250,000 people. The Aceh tsunami was caused by an earthquake with a magnitude of 9.2 Mw and triggered a tsunami as high as 34.5 meters on several beaches in Aceh. The epicenter of the earthquake that triggered the Aceh Tsunami was at the epicenter of 3.3 north latitude and 95.8 east longitude.
20. Nias Tsunami in 2005
    The Nias tsunami was triggered by an earthquake with a magnitude of 8 at the epicenter of 2 North Latitude, 97.8 East Longitude. The tsunami caused by this earthquake was 3.5 meters high and no one died.
21. Pangandaran Tsunami in 2006
    The Pangandaran tsunami in 2006 was triggered by an earthquake with a magnitude of 7.7 at the epicenter of 3.3 latitude and 107.9 east longitude. The 8.25-meter high Pangandaran tsunami killed 668 people.
22. Bengkulu Tsunami in 2007
    The Bengkulu tsunami in 2007 with a height of 3.8 meters was caused by an earthquake with a scale of 8.4 at the epicenter at 3.7 latitude and 101 east longitudes. The Bengkulu tsunami did not cause any casualties.
23. Mentawai Tsunami in 2010
    The Mentawai tsunami in 2010 was caused by an earthquake with a scale of 7.2 at the epicenter of 3.6 latitude and 99.93 east longitude. The Mentawai Tsunami killed 413 people and injured 448 people. The height of the Mentawai Tsunami is estimated to be around 7 meters.

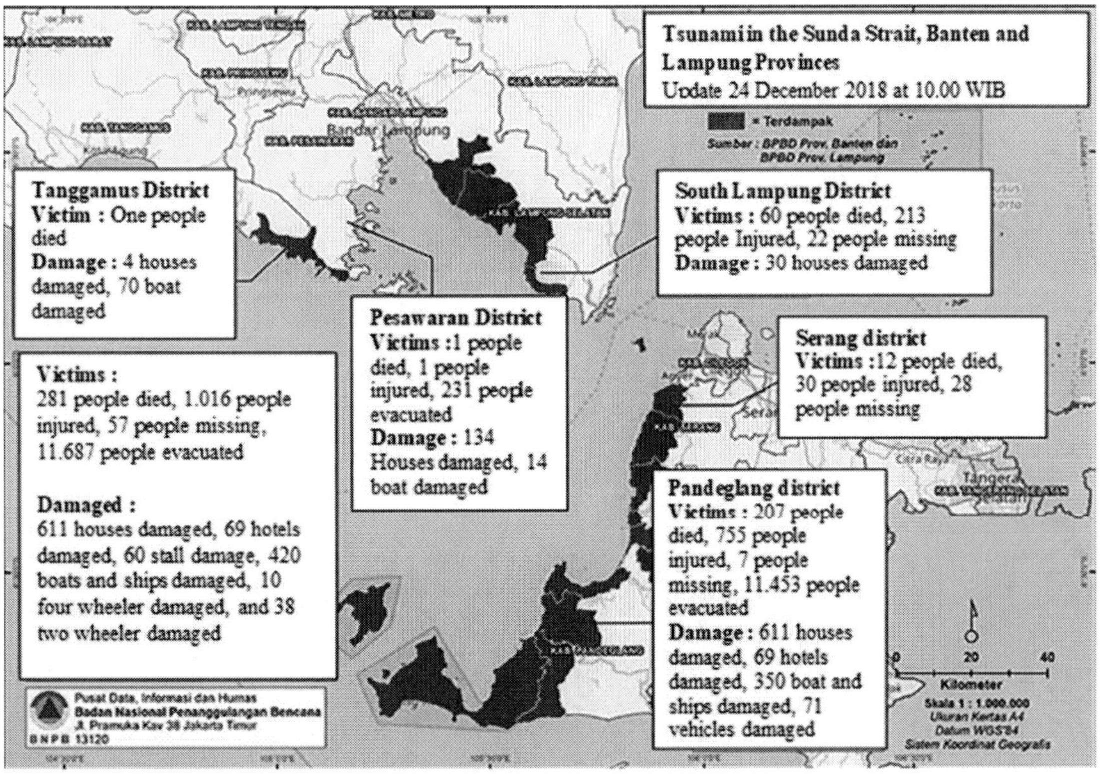

Fig. 1. Map of the Sunda Strait and its surroundings, as well as the tsunami disaster area marked in blue

Fig. 2. Out-of validity-date Early warning system devices.

In addition to the above tools, the Joint Research Centre (JRC) in collaboration with the Indonesian Tsunami Society, the Marine Research Center of the Ministry of Marine Affairs and Fisheries (KKP) and the Indonesian Meteorology, Climatology and Geophysics Agency (BMKG) has also installed equipment. This collaboration has been carried out since December 2018 in order to design and implement a new Emergency System.

This new system will adopt the Inexpensive Device for Sea Level Monitoring (IDSL) instrument developed by JRC to monitor in real time at sea level and provide communication via email, SMS alerts, CCTV images and information about potential Tsunami events (see Fig. 3).

Fig. 3. On the left is the BIG Tide Gauge installation and on the right is IDSL. Both are in Sebesi Harbor.

Example of time calculation analysis when using the IDSL system:

a. There was a landslide on Mount Anak Krakatau
b. 1 minute was received at Sabesi Island IDSL.
c. 5 minutes of data was received and processed in the central TAD Server.
d. 5 minutes to make a decision for the authorities.
e. 5 minutes to contact the relevant agencies.

f. 5 minutes sounding the siren.
g. Total 21 minutes lost.
h. Sirens can only reach a maximum of 2.5 km and are installed at only a few points from the need for a beach length of several hundred km.

In the implementation of this smart device system, GAP ANALYSIS is carried out as follows:

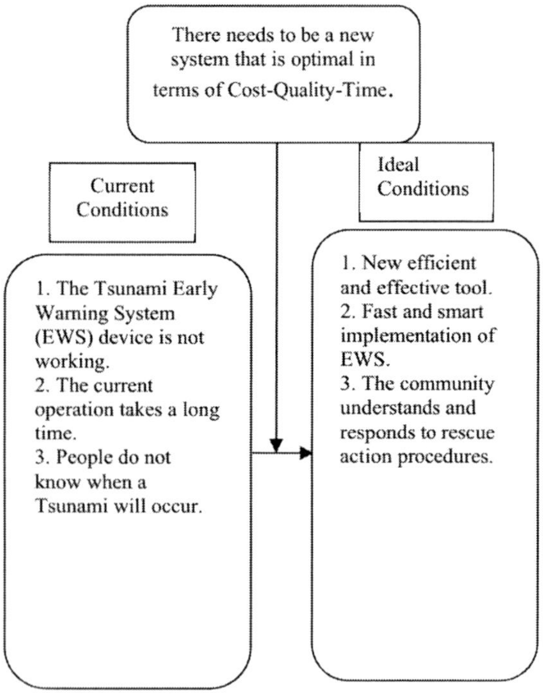

Fig. 4. Gap Analysis

Currently the government has purchased hundreds of units IDSL (Inexpensive Device for Sea Level), but this device takes 21 minutes of information to reach the community. Therefore, a new device is needed. The device is a system tool that involves the Equipment sector, Operations sector and Community sector. The expected result of this device is that the time for sending information to the public is only a few minutes

2 Methodology

2.1 Materials

Tsunami events occur when tides occur abnormally. Tsunamis are sea waves that are generated as a result of impulsive disturbance on the ocean floor. This impulsive disturbance occurs as a result of sudden changes in the geological structure of the seabed in either a vertical or horizontal direction [8] This in monitoring Tsunami is much data needed, in this case only looking at data from abnormally tidal.

To ensure this abnormal tidal event also needs to be proven in real live and real time using three parts of equipment, namely Transmitter, Main Receiver and Receiver. Transmitter consists of Tidal sensor pipe, 2-unit camera CCTV (one to monitor Tidal and other to monitor Coastal situation), radio communication, lightning rod, software and hardware to communicate between Transmitter and Main Receiver. These communication use internet facilities.

Main Receiver consists of LCD Monitor, Main Receiver software and Hardware, siren, speaker, microphone (to give a message and announcement from the Authority to many receivers contacted), lightning rod. Main internet is put on here to communicate and to spread internet network to receivers and transmitters. The Device are used for long range communication with up to 30 km line of sight. This Main Receiver received warning information from Transmitter that automatically send warning siren to Main Receiver, while the Authority make decision to turn on all of sirens in Receiver areas.

Receiver consists of unit receiver software and hardware, siren, speaker, lightning rod. These devices are placed in remote or rural place with supply internet from Main Receiver and power supply that can use from solar panel or State Electricity (PLN).

2.2 Method

In general, Accelerating Information of Tsunami Disaster Using Early Warning System (EWS) Devices: Smart Solution for Communication, Navigation and Surveillance Due to Tidal Sea Level at sea in this implementation consists of 3 aspects that were carried out, namely:

- Tools.
- Operational.
- Community.

The three optimization variables are in terms of cost, quality, and quality of data calculations.
Implementation Stages

3 Result and Discussion

To obtain warning information on changes in sea level, the following steps are needed, namely measuring sea level accurately, as close as possible to the source of the tsunami later. Fast data transmission of measurement location and presence of Software to analyze in real time and detect level changes, and provide warning and recheck data (CCTV images). Then the authorities decided to issue a warning to the beach location and activation of the siren system.

Sea level measurements are placed as close as possible to the source of the Tsunami so that the Government should also install at least every sub-district along the Coastal Coast that is prone to Tsunami and add loudspeakers in each community group. This can be done if the proposed cost for the device is very affordable for the parties concerned.

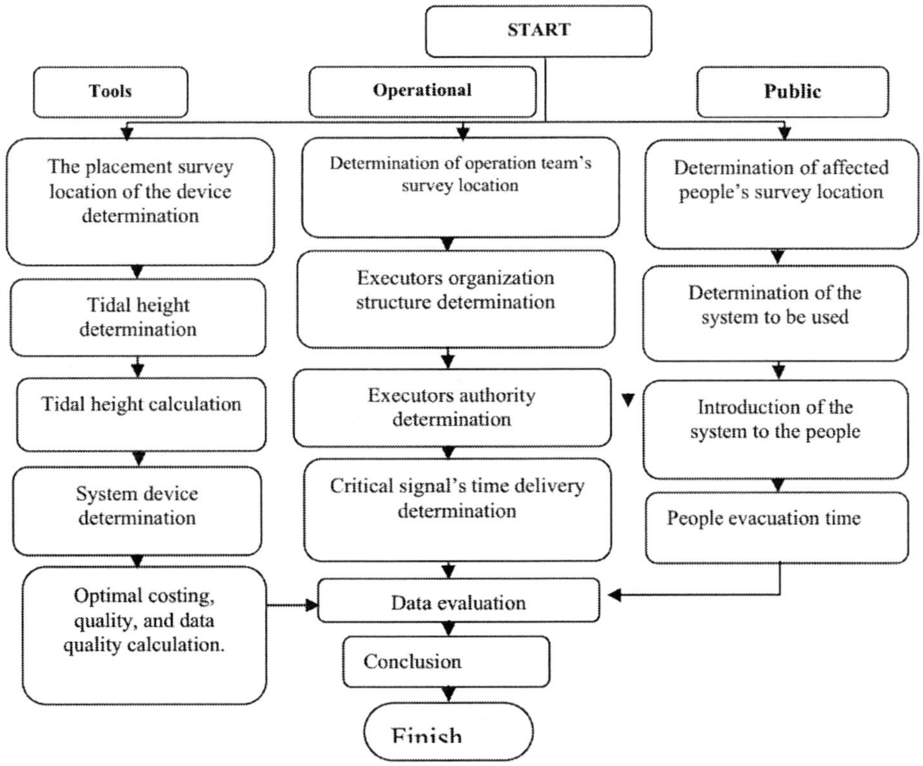

Fig. 5. Implementation Stages

Data transmission from the Measurement location must be sent rapidly. Of course, this cannot be done through a long bureaucratic path because it can hinder decision makers in providing information to the public.

Today's software is essential for processing data in real time and detecting level changes and also providing Alert signals to connected points.

In monitoring tsunami events, apart from tidal information, real information is needed in the form of re-examination through CCTV, so that there are clear signs of an impending tsunami.

Given the current tsunami incident, the decision of the authorities to issue warnings to coastal locations needs to be ready 24 hours to monitor tsunami events in their environment.

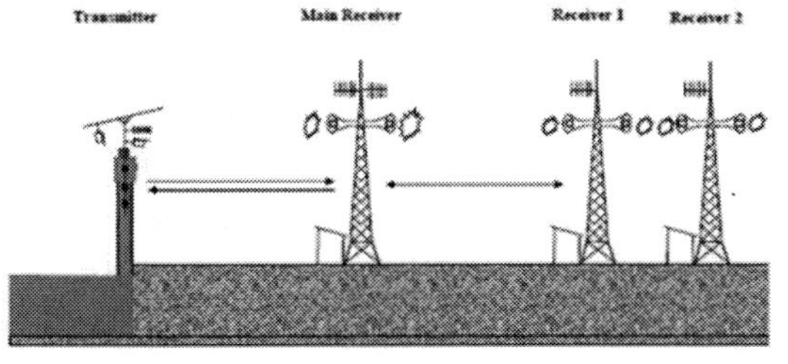

Fig. 6. Connections between the Main Receiver Transmitter and multiple receivers in remote areas.

The activation of sirens is the main thing in this early warning system and needs to be distributed evenly to remote areas prone to tsunamis, if necessary, using mosque loudspeakers that can be installed in this device system is a fairly efficient step as shown in Fig. 6.

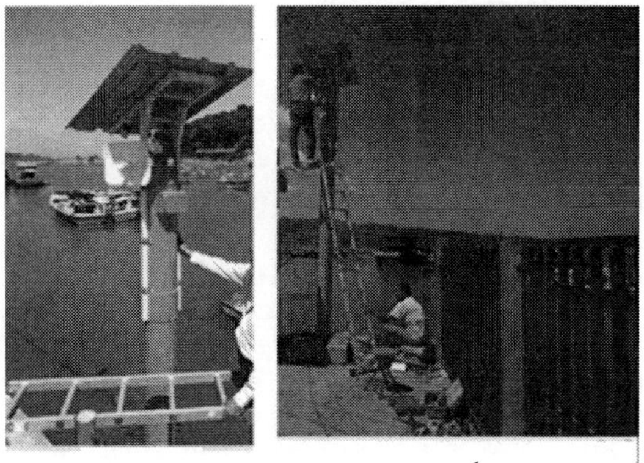

a. b.

Fig. 7. Fidela Tsunami Early Warning System in Tamperan Harbour Pacitan.

Transmitter is a Tidal Monitoring System Device that can provide real time information in the form of routine and abnormal tidal events and can provide visualization in the form of images or videos in real time using internet facilities.

The Main Receiver functions to receive data from the Transmitter and is read by the relevant agency Personnel Equipment to sound an Early Warning Signal to all Receivers in all remote tsunami-prone areas.

Receivers in the form of loudspeakers and sirens are widely available at residential corner points or at Village Halls, and others.

The image is taken from a system device that has been created based on a patent submitted by the ITS university DUDI Partner team, namely LENUS AEROSPACE INDONESIA with the FIDELA TSUNAMI EARLY WARNING SYSTEM (FTEWS) device which is currently operating at Tamperan Beach, Pacitan Fig. 7. with inexpensive devices around $1.600 for Server. This device can cut red tape to only 5 minutes compared to IDSL which takes 20 minutes.

4 Conclusion

The discussion shows that time wasted when the Tsunami incident been captured by receiving stations on the ground and transmitted to the community in an incomplete way because the recipients of siren information were only in the nearest location (maximum 2,5 km) to the siren. Consequently, there is a need for a device that can cut bureaucratic lines and increase sirens to remote areas at an affordable cost for the community. Therefore, it is recommended to use Fidela Early Warning System devices than can cut red tape to only 5 minutes and this are inexpensive devices.

In the future, to improve the system to be integrated in the System Equipment which involves the Equipment sector, Operations sector and the community sector. The expected result of this device system is that the delivery time of information to the community is only a few minutes and how they take care about it.

References

1. T. Solihuddin, H. L. Salim, S. Husrin, A. Daulat, D. Purbani. The Impact of the Sunda Strait Tsunami in Banten Province and its Mitigation Efforts. *Segara Journal* , 15-28 (2020)
2. Directorate General of Marine Spatial Management, Ministry of Marine Affairs and Fisheries, Marine Knowledge Series. Tsunami, (2017)
3. Center for Information and Public Relations of the National Disaster Management Agency. *Tsunami of the Sunda Strait, Banten and Lampung Provinces,* (2018)
4. Earthquake and Tsunami Center Deputy for Geophysics, *Katalog Tsunami Indonesia* , (2019)
5. A. Alessandro, *The Inexpensive Device for Sea Level Measurements*. Italy: EC-Joint Research Centre, (2015)
6. G. Alessandro, B. Danielle, Matteo.. *IDSL Sea Level Measurement Devices*. JRC Technical Report. (2016)
7. Santosa, Nurhadi. *Fidela Tsunami Early Warning System (Ftews).* Lenus Aerospace Indonesia, Presentation (Power Point Format) on Geofisika Institute Teknologi Sepuluh Nopember Surabaya. (2019)
8. BMKG. *Tsunami Early Warning Service Guidebook for InaTEWS.* (BMKG, Ed.) Indonesia. (2012)

Assessment on earthquake resistance spectral design load criteria for buildings and infrastructures in Indonesia

I Wayan Sengara[1,2]*and *Fahmi* Aldiamar[3]

[1]Professor, Civil and Environmental Engineering Faculty, Institut Teknologi Bandung, Bandung, Indonesia
[2]President, Indonesian Earthquake Engineers Association
[3]Geotechnical Division, Institute of Road and Bridge Research and Development, Ministry of Public Works, Bandung, Indonesia

Abstract. General assessment on earthquake resistance spectral design load criteria for buildings and infrastructures associated with the recent development of Indonesian seismic hazard maps is presented in this paper. The assessment is directed toward general identification of their associated risks for input to policy formulation of disaster risk reduction management plans or strategies. Indonesian seismic hazard maps haveevolved for the last three decades. This is originated from an early development map before 2002, where a seismic hazard map particularly for buildings (1983) was developed adopting the early process of probabilisticseismic hazard analysis (PSHA) for 200 years return period (RP). Further, a 2002 version seismic hazard maphas been developed in the form of peak ground acceleration (PGA) for 500 years RP. Spectral design criteriafor buildings and bridges have been later developed by updating PSHA involving new seismic source zones, ground-motion predictive equations, and various earthquake RP, accommodating seismic codes for buildings(2500 years RP), for bridges (1000 years RP) and dams involving various RP up to 10,000 years RP correspond to its design level. The spectral accelerations also have included PGA, short (0.2s) period, and 1-s period. The latest update hazard maps (2017) have been developed and adopted for seismic codes for buildings, bridges, dams, and other related infrastructures. The increase in spectral design load criteria is identified to assess the general risk of existing constructions, considering the results of several recent building damage surveys. Adoption of new seismic codes based on the most recent hazard maps along with its enforcement is expected to contribute to seismic disaster risk reduction in Indonesia.

1 Introduction

Since Indonesia is located within the four tectonic plates, i.e., Eurasian, Indian-Australian, Pacific, and Philippine plates, many areas in Indonesia are highly potential for seismic hazards. The most recent major earthquakes that hit Indonesia in the last two decades havecaused significant fatalities, major damages, and losses. It has been highlighted by many researchers, associations, and institutions that earthquake resistance design and construction of buildings, dams, bridges, and other infrastructures in Indonesia needs to become a priority and it is a crucial factor for earthquake disaster risk reduction.

It infers the demands to enhance the building codes and guidelines continuously, considering that structural design under earthquake shall be the vital factor of the future building performance. One of the significant efforts must be pointed to the development of the seismichazard map to be adopted in building and infrastructurecodes adopting quality improvement of new seismic source data and using the most current seismic hazard analysis methodology. The codes shall later be legally referred to and embedded within the local regulations.

Damage occurring during earthquakes needs to be identified and extensively studied, and further research needs to be conducted, ultimately leading to the development of improved building codes. Evidence from past earthquakes indicates adoption and enforcement of the latest model building codes is one ofthe most effective seismic mitigation strategies availableto communities.

Many disciplines are involved in earthquake disaster risk reduction. All groups shall work as collaborative efforts to provide the level of protection needed by communities. The activities are illustrated by followingthe ten-circle of disaster risk-chain management, as shown in Fig. 1. The chain consists of several activitycomponents from research in earthquake hazards and earthquake engineering, seismic hazard mapping, seismic building codes formulation, dissemination of the building codes, building design, building codes compliance/enforcement,

* Corresponding author: wayansenggara@si.itb.ac.id

https://doi.org/10.1051/e3sconf/202133107009

construction supervision for the quality assurance and quality control, research on building vulnerability, risk assessment process, disaster risk-reduction management planning, and back to research in hazards and earthquake engineering. The implementation of its risk reduction activities should beintegrated and linked with the efforts of preventing or minimizing the potential break of any chain.

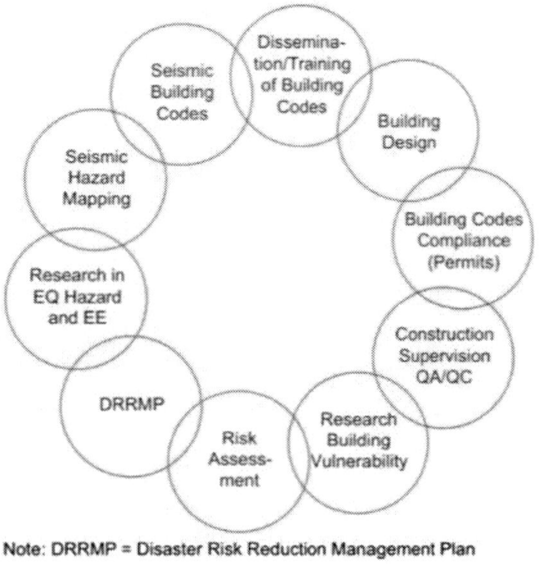

Fig. 1. Ten-Circle of Disaster Risk Chain Management for Seismic Mitigation Strategies.

This paper presents a general assessment of earthquake resistance spectral design load criteria for buildings and Infrastructures in Indonesia. In particular, the paper presents the enhancement of new seismic hazard maps adopted in new codes through the developmentof the most recent development of international building codes, to be adopted for new constructions. General identification is also made on previous hazard maps adopted for existing buildings and several infrastructures (bridge and dam). Some post-disaster building and bridge surveys are also presented in this paper, with information on the development of building vulnerability models, and the need for the development of infrastructure vulnerability models as an input in the seismic risk-reduction process for formulation of risk-reduction management plans. A particular emphasis is made on the compliance of codes as one of the main risk-reduction efforts for both central and local government officials.

2 Development of Indonesian Seismic Hazard and Mapping

In general, an earthquake is a natural phenomenon that comprises uncertainty, and consequently, it is very complex in nature to predict when and where an earthquake will occur and with what strength. The effectof an earthquake at the location of a structure is in the form of the ground-motion acceleration that will exert inertial load on the structures. The earthquake and its load are random variables as a natural phenomenon. Therefore, seismic hazard analysis and its mapping are predicted through either or both deterministic and probabilistic analysis, involving seismic sources and their seismic parameters in combination with ground-motion predictive equations.

First Indonesian hazard maps had been developed asa result of a study by Beca Carter in bilateral cooperation between Indonesia and New Zealand (Beca Carter Hollings and Ferner, 1978). This earthquake map divides Indonesia into six earthquake zones. The map was developed adopting a probabilistic approach with an earthquake return period of 200 years. Later on in the year 2002, there was an update of the Indonesian seismic hazard map particularly developed for building codes of 2002. This map is a compilation of four earthquake maps resulting from PSHA of four different research teams from Institut Teknologi Bandung (Firmansyah and Irsyam, 1999), Ministry of Public Works (Najoan etal., 1999), Geological Research Center (Kertapati, 1999), and Consultants (Shah and Boen, 1996). The hazard map was developed for reference baserock (S_B) adopting a 10% probability of exceedance (PE) in 50years (equivalent to an earthquake return period of 475 years).

After the 2002 Indonesian seismic building codes [12], many studies have been conducted toward the improvement of Indonesian seismic hazardmaps. The improvement was directed towards the adoption of the most recent data and current state of knowledge in probabilistic and deterministic seismic hazard methodology, through the use of the most recent ground-motion predictive equations (GMPEs). History on development of Indonesian seismic hazard map and elaboration on previous researches on seismic hazard analysis until 2008 is presented in Surahman et al., 2008.In 2008, during the First International Conference on Earthquake Engineering and Disaster Mitigation[25] in Jakarta, round-table discussion among Indonesian seismic hazard experts and earthquakeengineers towards the unification of several Indonesianseismic hazard maps developed by different institutionswas originated. Later on, in 2009, a national team(Team-9) was then established for the Revision of SeismicHazard Maps of Indonesia. The probabilistic seismic hazard analysis methodology is adopted for the development of the map. This team successfully developed 2010 Indonesianhazard maps [22] through the development of hazard curves for various earthquake return periods.

The 2010 hazard map was developed adopting the current seismic source zones, seismicity data and recent GMPEs, and PSHA methodology at that time. These 2010 hazard maps provide various earthquake return periods. Also, there are three seismic design maps prepared, for response spectral at T=0s (PGA), short (T=0.2s), and long (T=1s) periods regarding ASCE-SEI-7-10.

The 2010 hazard map was based on PSHA that hasbeen conducted considering many past large subductionsand shallow crustal earthquakes, such as 2004 Aceh subduction interface (M_w= 9.3), 2005 Nias subduction interface (M_w=8.7), 2006 shallow crustal Yogyakarta (M_w=6.3), 2006 Pangandaran subduction interface (M_w=7.2), 2007 Indramayu subduction intraplate

https://doi.org/10.1051/e3sconf/202133107009

(M_w=7.5), 2007 Bengkulu subduction interface (M_w=8.4), 2009 West Sumatra subduction intraplate(M_w= 7.6), 2009 West Java subduction interface (M_w= 7.0). The development of the 2010 hazard map was supported by the Indonesian Ministry of Public Work and National Disaster Management Agency (BNPB) through Australia-Indonesia Facility for Disaster Reduction (AIFDR).

Further in 2017, New 2017 Indonesian seismic hazard maps [22] have been produced and published. The New 2017 hazard maps are presented in the form of peak ground acceleration (PGA) for 50, 100, 200, 500, 1000, 2500, 5000, and 10,000-year return periods. In addition, spectral response accelerations (Sa) for short period (0.2 sec) and a 1-second period for 2% PE in 50 years (equivalent to 2,475 years return period) have been produced. In this case, 84th percentile deterministic seismic hazard analysis (DSHA) as the cap of the PSHA results has also been adopted.

The 2017 seismic hazard maps are updated by considering quality improvement with updating, adding, finding, and identifying new seismic sources of geological, geodetic, seismology & instrumentation, new GMPEs. New information regarding the identification of active faults with a significant number.The information covers both active faults (251 active faults) that have not been appropriately quantifiedduring the preparation of the 2010 map (previously only 81 faults) that are quantified better based on field research through trenching, carbon dating, epicenter relocation, and strain analysis with geodetic data. Another update was the use of a more complete and more accurate earthquake catalogue by updating theearthquake catalogue until 2016 that relocated with 3D-speed models, and a comprehensive earthquake background source.

The GMPE that also updated for the development of the Indonesian seismic hazard map for 2017, such asfor shallow crustal fault and shallow background sources models has been used the GMPE the latest studythat is: Boore-Atkinson NGA West-2 (2013), Campbell-Bozorgnia NGA West-2 (2013) and Chiou-Youngs NGA West-2 (2013), which is an update from GPME 2008 by its researchers, respectively. In addition, thenew GMPE such as BC Hydro (Abrahamson et al., 2014) was also used for the subduction source model. The above changes which are input parameters for seismic hazard analysis (SHA) have affected the hazard value, especially in areas where there are additional faults. Thisnew hazard map has also been analyzed using two software namely PSHA from USGS and OpenQuake from Global Earthquake Model (GEM), where the hazard maps previously only used USGS software.

3 Seismic Loading Criteria forBuildings and Infrastructures

3.1 Building Structures

Seismic building codes generally are intended to be applied not only by engineers, but also are used for various purposes by safety inspectors, scientists, developers, contractors, manufacturers, insurance companies, facility managers, and others for various aspects related to seismic risks. Seismic loading design criteria for a high-rise building could be simply referredfrom applicable seismic building codes specifying spectral accelerations at referenced base-rock and corresponding Site-Class to provide design spectra. The seismic building codes in general are designed to provide quality and safety for the building occupants and safeguard their assets and economic investments.

Early stage of Indonesian seismic building codes development was initiated with reference to PPTI-UG (Peraturan Perencanaan Tahan Gempa Indonesia untuk Gedung) – 1983. This code is based on the 1983 Indonesian hazard map, as shown in Fig. 2(a). Response spectra at the ground surface can be selected by taking into account local soil conditions. In PPTI- UG, local site conditions are categorized into two groups, i.e., hard and soft sites.

PPTI-UG 1983 was later renewed in 2002 with the issuance of Procedures for Earthquake Resistance Design in SNI 03-1726-2002. This new regulation was prepared regarding the 1997 Uniform Building Codes (UBC). In SNI-1726-2002, the determination of the design earthquake load was carried out adopting the 2002 Indonesian seismic hazard map, shown in Fig. 2(b). In this SNI, the maximum acceleration value and the acceleration spectra at the ground surface are determined based on the location and local site conditions. In contrast to the previous map, the site conditions in this SNI are grouped into 3 categories, i.e., hard, medium, and soft sites following UBC-1997. These categories are based on the dynamic parameters of the soil to a certain depth which is generally taken asthe top 30 m below ground surface.

After significant earthquakes hit Indonesia during the year 2000 to 2010, the Indonesia seismic design codehas been updated with SNI-1726-2012 replacing the previous SNI 03-1726-2002 which has been consideredto be out of date. New spectral loading criteria were introduced by providing ground motion values having a2% PE in 50 years (equivalent to 2475 years return period).

The significant issues of this 2012 updated code are not only changing the return period but also accountingfor the uncertainty of the collapse capacity of structures. A method, which determines the risk has been used by implementing direct integral to calculate risk-targeted ground motions (RTGM) of 1% probability of building collapse in 50 years that is derived by integrating 2% PEin 50 years hazard curves of the 2010 Indonesian seismic hazard with fragility curve of Indonesian buildings defined to have 10% probability of collapse. This follows the seismic design criteria developed for the UnitedStates of America by Luco et al. (2007), that has been adopted in ASCE-SEI-7-10. Furthermore, the SNI- 1726-2012 adopted three seismic design maps for Risk-Targeted-Maximum Considered Earthquake (MCE_R) response spectral short (T=0.2s) and long (T=1s) periods [23], as well as Geometric- Mean Maximum Considered Earthquake (MCE_G) ground-motion acceleration as referred from ASCE-SEI-7-10.

Site classification was also updated from the 2002 to 2012 building codes. There is a more specific site-classification made based on the top 30m N-SPT, Su

Undrained shear strength, and shear-wave velocity (Vs) profile from SB (soft rock), SC (hard soil), SD (medium soil), and SE (soft soil). In SNI-2012 building codes Site-Class F (SF) is introduced, in which site-specific response analysis (SSRA) shall be conducted to define the seismic design spectra criteria.

To adjust the mapped short period (S_S) and 1-second period (S_1) spectra and also the MCE_G (PGA)values for sites other than SB, site coefficients F_a, F_v, and F_{PGA} are applied to S_S, S_1, and PGA, respectively. These site coefficients have been added from the previous 2002 of simply one site coefficient concerning PGA only

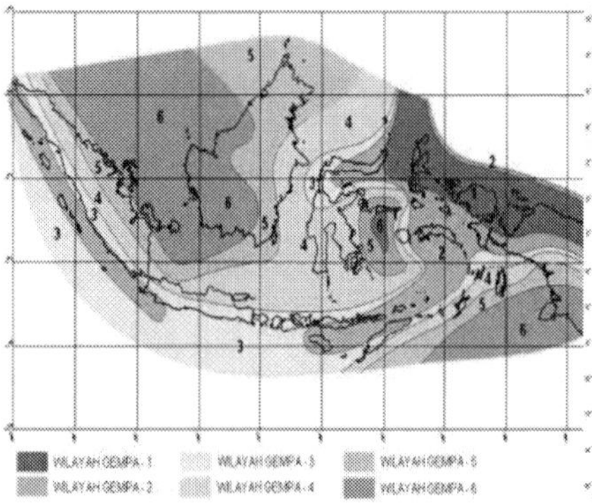

a. Indonesia seismic zone map for building (PPTI-UG 1983)

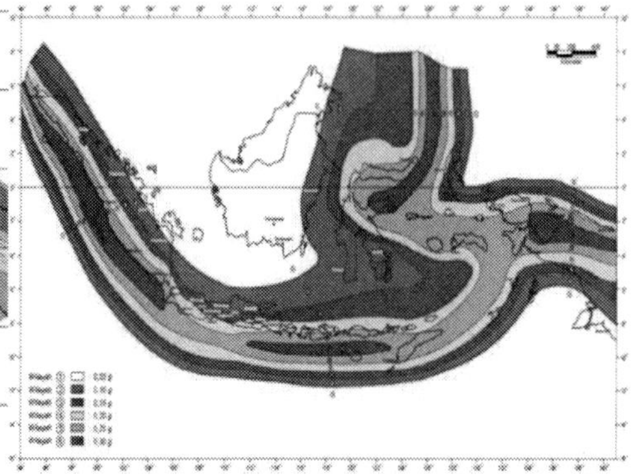

b. Indonesia seismic zone map for building [14]

Fig. 2. Indonesian Seismic Hazard Maps in Previous National Standard

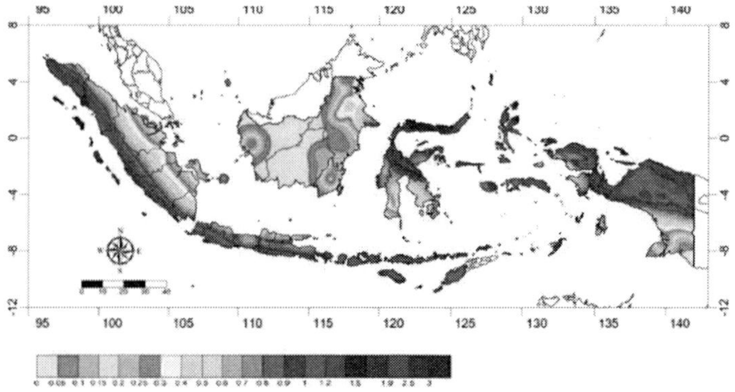

(a)

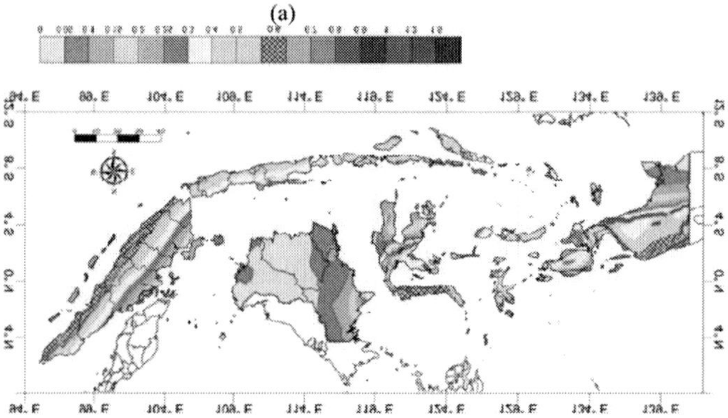

(b)

Fig. 3. (a) *Ss* Risk-Targeted Maximum Considered Earthquake (MCER) Ground Motion Parameter for 0.2 s Spectral Response Acceleration (5% of Critical Damping), Site Class B; (b) *S1* Risk-Targeted Maximum Considered Earthquake (MCER) Ground Motion Parameter for Indonesia for 1.0 s Spectral Response Acceleration (5% of Critical Damping), Site Class B. [14]

Further update of the Indonesian seismic buildingcodes 2012 was made in 2019 [14] adopting the update of the Indonesian 2017 seismichazard maps. Figure 3(a) and 3(b) show RTGM maps for Indonesia Building Code SNI-1726-2019 at 0.2s and1.0s spectral response acceleration, respectively, Site Class B for 5% of critical damping [26].

Separately from the updates to the RTGM maps, thesite coefficients used to adjust the mapped values for soil properties have also been updated for the SNI-1726- 2019. In both the SNI-1726-2012 and SNI-1726-2019, the ground motion maps are for an average shear wave velocity at small shear strains in the upper 30 meters of subsurface below a site, V_{s30}, of 760 m/s. To adjust the mapped RTGM (S_S and S_1) and MCE_G (PGA)values to other site classes, the site coefficients F_a, F_v, and F_{PGA} are applied to S_S, S_1, and PGA, respectively.

The RTGM site coefficients (F_a and F_v) have been modified as high as 20%, i.e., Fv is changed to 1.8 from 1.5 for SD Site Class at S_1 = 0.5g; Fa is changed to 1.2 from 1.0 for SC Site Class at SS ≥ 1g, and Fv is changedto 0.8 from 1.0 for SC Site Class at all values of S_1. TheMCE_G site coefficient (F_{PGA}) has similarly increased by approximately 20% for SC Site Class and PGA ≥ 0.4 g,and has increased by approximately 20% to 56% for SESite Class at PGA ≥ 0.25 g. Please note that this modification is also under consideration that the reference subsurface rock is currently specified as BC (Vs=760m/s), whereas previously, it is specified as SiteClass B.

Evaluating short-period site coefficient F_a as shownin Table 1, it is identified that the coefficient of F_a in SNI-1726-2019 has a slight difference for SE site class at Ss ≥ 1.0 than those in ASCE-7-16, for example, F_a = 1.1and 1.2 for SNI-1726-2019 and ASCE-7-16, respectively. However, similar values are shown for the coefficient of F_v and F_{PGA} in SNI-1726-2019 and ASCE-7-16. These new site amplifications are also under consideration concerning Kircher and Associates (2015).

Furthermore, the parameter T_L was introduced in SNI-1726-2019 to provide more realistic ground motions at periods T > 4 sec that would affect the design of tall buildings. Long-period transition, T_L, in this case, marks the transition between the constant velocity andconstant displacement segments of the Fourier spectrum representing a theoretical fault-rupture displacement history.

Based on several parameters that have been updated such as RTGM, spectral amplification factors, as well as T_L parameters, there is a quite significant difference between design response spectrum adopting SNI-1983, SNI-1726-2002, SNI-1726-2012, and SNI-1726-2019 parameters. As an illustration of comparison, the design response spectrum of the city of Banda Aceh is increasing for PGA, short-period (0.2s), 1-second (S_1), until reaches T_L parameters (8s). An exception is for very long period structures, such as 10s to 15s, the spectral response acceleration tends to decrease. Fig. 4 shows the difference in design response spectrum in the city of BandaAceh, using PPTGIUG (1983), SNI-1726-2002, SNI-1726-2012, and SNI-1726-2019 spectra parameters.

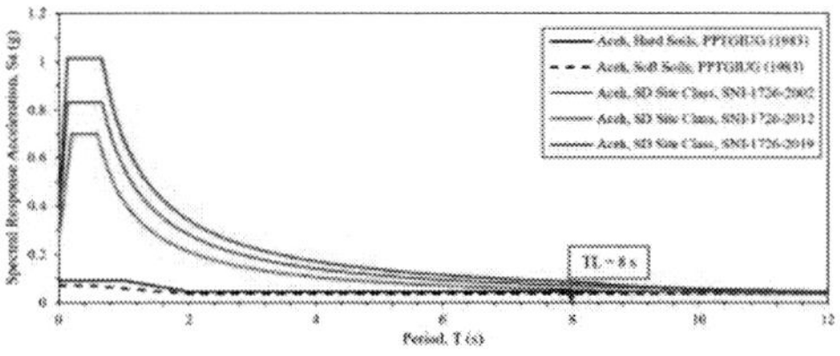

Fig. 4. Design Response Spectrum for Aceh city adoptingSNI-1726-2002, SNI-1726-2012 and SNI-1726-2019 Parameters

Table 1. Short-Period Site Coefficient, Fa (SNI-1726-2019)

Site Class	**Mapped Risk-Targeted Maximum Considered Earthquake (MCER) Spectral Response Acceleration Parameter at Short Period**					
	$S_s \leq 0,25$	$S_s = 0,5$	$S_s = 0,75$	$S_s = 1,0$	$S_s = 1,25$	$S_s \geq 1,5$
SA	0,8	0,8	0,8	0,8	0,8	0,8
SB	0,9	0,9	0,9	0,9	0,9	0,9
SC	1,3	1,3	1,2	1,2	1,2	1,2
SD	1,6	1,4	1,2	1,1	1,0	1,0
SE	2,4	1,7	1,3	1,1 (1,2)*	0,9 (1,2)*	0,8 (1,2)*

Note: *Coefficient in the parentheses ( ) is short-period site coefficient (Fa) of ASCE-7-16

Table 2. 1-Second Period Site Coefficient, Fv [14]

Site Class	**Mapped Risk-Targeted Maximum Considered Earthquake (MCER) Spectral Response Acceleration Parameter at 1-s Period**					
	$S_1 \leq 0,1$	$S_1 = 0,2$	$S_1 = 0,3$	$S_1 = 0,4$	$S_1 = 0,5$	$S_1 \geq 0,6$
SA	0,8	0,8	0,8	0,8	0,8	0,8
SB	0,8	0,8	0,8	0,8	0,8	0,8
SC	1,5	1,5	1,5	1,5	1,5	1,4
SD	2,4	2,2	2,0	1,9	1,8	1,7
SE	4,2	3,3	2,8	2,4	2,2	2,0

Table 3. Site Coefficient F_{PGA} [14]

Site Class	PGA ≤ 0,1	PGA = 0,2	PGA = 0,3	PGA = 0,4	PGA = 0,5	PGA ≥ 0,6
SA	0,8	0,8	0,8	0,8	0,8	0,8
SB	0,9	0,9	0,9	0,9	0,9	0,9
SC	1,3	1,2	1,2	1,2	1,2	1,2
SD	1,6	1,4	1,3	1,2	1,1	1,1
SE	2,4	1,9	1,6	1,4	1,2	1,1

3.2 Bridge Structures

Indonesian Government previously developed seismic hazard map and design standard for bridge,namely Standar Perencanaan Ketahanan Gempa untuk Jembatan Jalan Raya [4] in 1992 and Standar Perencanaan Ketahanan Gempa untuk Jembatan[5] in 2008. The seismic zone map for those standards were shown in Fig. 5. Afterward, updatingon seismic design standard was made in 2013 by adopting the AASHTO (2012) design criteria for 1000 years RP (7% probability of exceedance in 75 years) (Figure 6Fig. 6). This was a significant change, considering the previous national standard set to 500 years RP (10% probability of exceedance in 50 years). Other changes in the 2016 National Standard were site class, site coefficients, and design response spectrum calculation. The seismic hazard map was developedusing the methodology and seismic source parameters established by Team-9 and had been published in several seminars at Puslitbang Jalan dan Jembatan and the 18th Southeast Asian Geotechnical Conference [1].

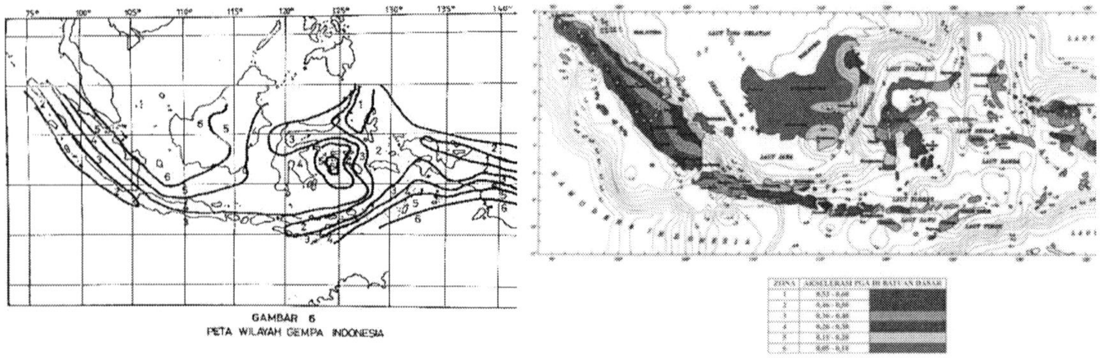

a. Indonesia seismic zone map [4]

b. Indonesia seismic zone map [5]

Fig. 5. Indonesian Seismic Hazard Maps in Previous National Standard

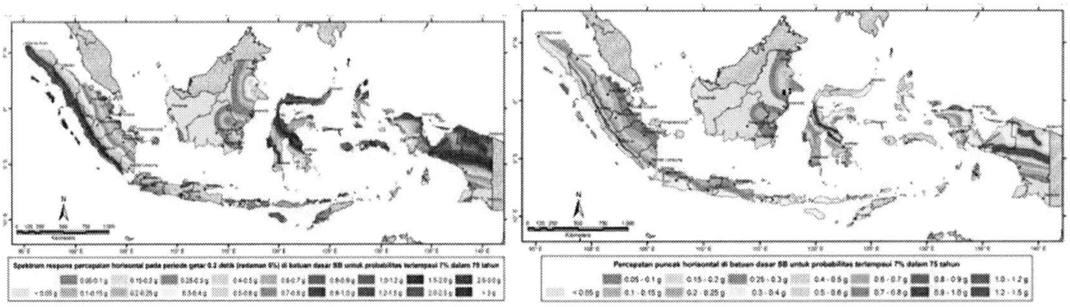

a. Peak Ground Acceleration (PGA) on Rock (SB)

b. Horizontal Spectral Response Acceleration at Period of 0.2 sec on Rock (SB)

Fig. 6. Design Horizontal Peak Ground Acceleration Coefficient (PGA) and Horizontal Response Spectral Acceleration Coefficients Maps in Latest National Standard [6]

https://doi.org/10.1051/e3sconf/202133107009

.Design response spectrum for the conventional bridge was generated using 2 procedures i.e., using seismic hazard map and site-specific seismic hazard analysis according to specific location conditions. According to National Standard, site-specific seismic hazard analysisshould be conducted for the near-fault condition or when project location less than 10km from active fault and special soil condition or determined as site class F.Design response spectrum derived from seismic hazardmap were based on 3-point values as seen in Fig. 7. Site amplification factors for each of the periods i.e., PGA0.2 second and 1.0 second are shown in Table 4 and Table 5.

A website version of the seismic hazard map for the conventional bridge to help engineers easily develop the design response spectrum can be seen at http://lini.binamarga.pu.go.id/.

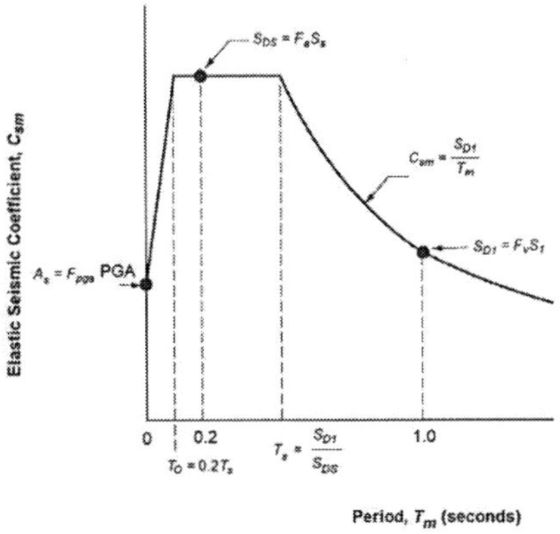

Fig. 7. Design Response Spectrum [6]

Table 4. Site Factors for PGA and Period of 0.2 sec (F_{PGA} and F_A)[6]

Site Class	PGA ≤ 0.1 S_S ≤ 0.25	PGA = 0.2 S_S = 0.5	PGA = 0.3 S_S = 0.75	PGA = 0.4 S_S = 1.0	PGA ≥ 0.5 S_S ≥ 1.25
Hard rock (SA)	0.8	0.8	0.8	0.8	0.8
Rock (SB)	1.0	1.0	1.0	1.0	1.0
Firm soil (SC)	1.2	1.2	1.1	1.0	1.0
Stiff soil (SD)	1.6	1.4	1.2	1.1	1.0
Soft soil (SE)	2.5	1.7	1.2	0.9	0.9
Specific soil (SF)	SS	SS	SS	SS	SS

Note: Linear interpolation can be used for intermediate values
S_1 = horizontal response spectral acceleration on rock (SB) at period of 0.2 sec (Figure 6b)
S_S = locations requiring site-specific geotechnical investigation and dynamic site response analysis

Table 5. Site Factors for Period of 1.0 sec (F_v)[6]

Site Class	S_1 ≤ 0.1	S_1 = 0.2	S_1 = 0.3	S_1 = 0.4	S_1 ≥ 0.5
Hard rock (SA)	0.8	0.8	0.8	0.8	0.8
Rock (SB)	1.0	1.0	1.0	1.0	1.0
Firm soil (SC)	1.7	1.6	1.5	1.4	1.3
Stiff soil (SD)	2.4	2.0	1.8	1.6	1.5
Soft soil (SE)	3.5	3.2	2.8	2.4	2.4
Specific soil (SF)	SS	SS	SS	SS	SS

Note: Linear interpolation can be used for intermediate values
S_1 = horizontal response spectral acceleration on rock (SB) at the period of 1.0 sec (Figure 6c)
S_S = locations requiring site-specific geotechnical investigation and dynamic site response analysis

3.3 Dam Structures

Seismic safety evaluation of dams is part of essential disaster risk assessment, considering the impact could be very catastrophic. International Committee on Seismic Aspects of Dam Design has representatives (dam engineers and earthquake experts) from 26 different countries from Europe, North and South America, Africa, Asia, and Australia, since 1975, following bulletins were published by ICOLD(International Commissions on Large Dams), which were mainly prepared by the seismic committee. The published bulletins consist among others are of the reviewof earthquake resistance of dams, the seismic requirement on the design of dams, inspection post-earthquakes and seismic observation, selection of seismic parameters forlarge dams, seismicity of tailings dams, and designfeatures of dams.

Beyond that, the first earthquake hazard maps for dam structure was developed by Najoan (1996). This map was adopted in guidelines for the stability analysis of the

 https://doi.org/10.1051/e3sconf/202133107009

embankment dam subjected earthquake load as shown in Figure 8 (Pd T-14-2004-A). This standard wasestablished about several international standards, i.e., USCOLD 1985, USBR 1984, USBR1987, and ICOLD 1989. Additionally, this guideline requested analysis with a particular method and specific earthquake level for evaluating the stability of the dam based on the potential risk rating as shown in Tables 6 to 8.

One (1) year after the establishment of the standard for seismic stability analysis, the Department of Public Works, Research and Development Center of Water Resources issued enhancement of Indonesia seismic zone map for dam structure, adopting Fukushima and Tanaka (1990) ground-motion predictive equation. This map was givenfor the maximum earthquake acceleration values at bedrock (SC) with return periods (RP) of 10, 20, 50, 100, 200, 500, 1000, 5000, and 10,000 years. Based on this developed map, Indonesian zonation is divided intosix (6) i.e., A, B, C, D, E, and F. The distribution was carried out by taking into account the condition of the earthquake magnitude that may occur. The map alsoprovides contour lines of the earthquake coefficients. This earthquake acceleration coefficient was required to be used as the design basis for the stability of the dams. This map was adopted in guidelines for dynamic analysis of embankment dam (Kep. Dirjen SDA 27/KPTS/D/2008,2008).

Seismic requirement for dam safety is stated in two(2) levels of safety. The upper-level earthquake is for ultimate safety termed as Safety Evaluation Earthquake(SEE). The lower level earthquake is for lower event earthquake is used to address serviceability termed as Operating Basis Earthquake (OBE). In this guideline, the earthquake load scenario associated with SEE level ground motions is corresponding to an annual probability of exceedance (PE) of about 1/3,000 to 1/10,000 (depending on risk dam level) is recommendedin defining the SEE probabilistically for high hazard dams. Meanwhile, OBE is defined probabilistically as an event with a reasonable likelihood of occurring during the life of a facility. Thus, it is typically characterized as the earthquake scenario associated with ground motions corresponding to a probability exceedance of about 1/100 to 1/200 (depending on risk dam level).

Nine years after the issuance of seismic analysis guideline for dam structure (2017), Indonesia has published successfully the Indonesian Geotechnical Code (SNI-8460-2017) as well as the new 2017 Indonesian Earthquake Hazard Map. The stability of dam structure due to earthquake also was mentioned in this standard following to reference of ICOLD Bulletin No. 148 (2016). The requirement of the dam safety criteria in this standard for SEE and OBE level earthquakes is shown in Table 9. Furthermore, the peakground acceleration (PGA) map for SEE and OBE earthquakes is corresponding to the 10,000 and 100 year return period also given in the new 2017 Indonesian Earthquake Hazard Map as depicted in Figure 9(a) and 9(b), respectively. Comparison of peak ground acceleration using three standards and maps, i.e., Pd T- 14-2004-A, Kep. Dirjen SDA 27/KPTS/D/2008 and 2017 Indonesia Seismic Hazard Map in the city of Padang are shown in Table 10.

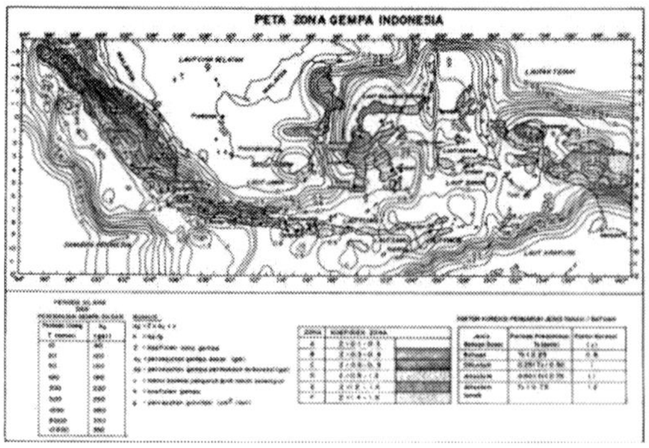

Fig. 8. Indonesia seismic zone map for dam structure (Pd T-14-2004-A)

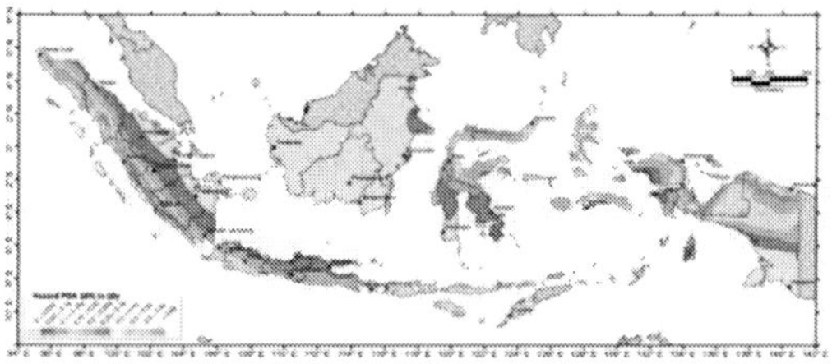

(a)

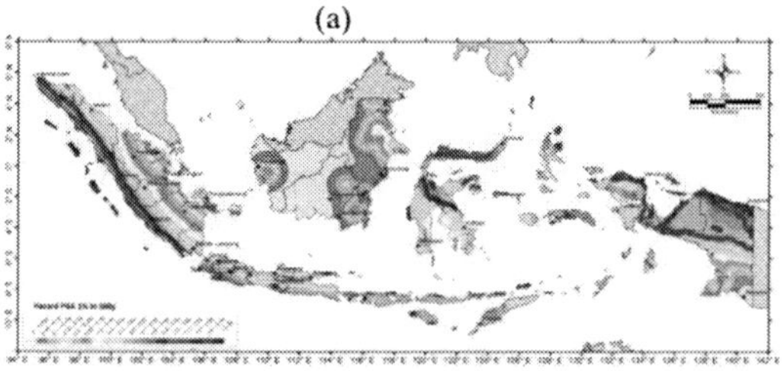

(b)

Fig. 9. Peak Ground Acceleration (PGA) Maps in the 2017 Indonesian Earthquake Hazard Map (a) 10% probability of being exceeded in 10 years (b) 1% probability of being exceeded in 100 years

Table 6. Risk Factor Criteria for Dam Safety Evaluation[10]

Risk Factor	**Extreme**	**High**	**Moderate**	**Low**
Contribution to risk (weighting points)				
Capacity (hm^3)	>120 (6)	120 – 1 (4)	1 – 0.1 (2)	< 0.1 (0)
Height (m)	>45 (6)	45 – 30 (4)	30 – 15 (2)	< 15 (0)
Evacuation Requirements (No. of persons)	>1000 (12)	1000 – 100 (8)	100 – 1 (4)	None (0)
Potential Downstream Damage	High (12)	Moderate(8)	Low (4)	None (0)

Table 7. The Risk Class of Dam based on the ComputedTotal Risk Factor [10]

Total Risk Factor	Risk Class (Risk Rating)
0-6	I (Low)
7-18	II (Moderate)
19-30	III (High)
31-36	IV (Extreme)

Table 8. Earthquake Load Criteria for Dam Design (Pd T-14-2004-A)

Risk Classes with Lifetime	OBE Level		MDE Level	
	T (year)	Analysis Method	T (year)	Analysis Method
IV N=50-100	100-200 $a_d \geq 0.1g$	Earthquake Coefficient	10,000	Earthquake Coefficientor Dynamic Analysis
III N=50-100	50-100 $a_d \geq 0.1g$		5,000	
II N=50-100	50-100 $a_d \geq 0.1g$		3,000	
I N=50-100	50-100 $a_d \geq 0.1g$		1,000	

Table 9. The Dam Safety Criteria based on SNI-8460-2017

DesignLive (year)	Probability Exceedance (%)	Return Period (year)	Safety Criteria
100	1	10,000 Safety Evaluation Earthquake (SEE)	- There is no uncontrolled flow ofwater - Maximum vertical deformation of 0.5 freeboard - Maximum horizontal deformation of 0.5 thickness of the filter
100	50	145 Operating Basis Earthquake (OBE)	- Minor damageafter the occurrence of an earthquake - No damage or loss of service must happen

Table 10. Comparison of PeakGround Acceleration at OBEand MDE Level in Padang City

Earthquake Level	Peak Ground Acceleration (g)		
	Pd T-14-2004-A[(1)]	Kep. Dirjen SDA 27 / KPTS / D / 2008[(1)]	2017 Indonesia Seismic Hazard Map[(2)]
OBE (100-year return period)	0.29	0.32	0.34
MDE (10000-year return period)	0.54	0.55	0.94

Note: [(1)] Peak ground acceleration are derived aboutAlluvium Soils
[(2)] Peak ground acceleration are derived about SD SiteClass

4 Vulnerability, Associated Risk, and its Reduction Efforts

Challenges and demands to develop an optimum construction quality of buildings and infrastructures arein the aspects of their safety and cost-effectiveness. These two aspects always need to be considered for future new constructions.

Buildings and infrastructures development in Indonesia keeps increasing in line with the economic development through a national budget plan. The development of many cities are growing relatively fast in the last decades. In addition to the existing vulnerabilities under past constructions processes and past building codes, there is a potential for vulnerabilityand seismic risk accumulation during the physical and social development of the cities and communities. The vulnerability of the new constructions could be associated with various aspects that are not well implemented well within the risk chain. With thesepotential accumulated vulnerabilities, then there is the potential of seismic disaster risks under the current conditions. These risks may be passed to generations ahead unless systematic disaster risk-reduction management plans are implemented.

General evaluation of the current conditions would suggest that the current buildings and infrastructures in Indonesia are vulnerable to various degrees as represented by the current and past construction systemto date. Many building and infrastructure damages through post-disaster surveys from several large earthquakes in Indonesia for the last 25 years have indicated vulnerabilities of many types of buildings and bridges. These vulnerabilities are considered to be among others due to a lack of knowledge on earthquakeresistance design and constructions practice as well as a lack of building enforcement by the local authorities.

4.1 Buildings Vulnerability

Several post-earthquake disaster surveys had been conducted. The surveys showed that the collapsed buildings were mostly non-engineered buildings consisting of one or two-story houses, shops, religious and poor structural quality school buildings. Most of the collapsed residential buildings were also non-engineered masonry houses. They were one story tall with unreinforced clay, brick, or block masonry in cement or lime mortar and no particular connectionframe between the timber roof and the walls. Based on the surveys and evaluations in the field, many building structures did not meet earthquake resistance requirements. In addition to the building structures, other contributing factors to the building damages werethe subsurface conditions, type of foundation and its dimensions, local topographical conditions, and the intensifying of the earthquake.

An example of a post-earthquake survey was carriedout through the post-disaster of the 30 September 2009 Padang earthquake [21]. A total of 3,896buildings were surveyed in the Padang and Pariaman region. This comprised a range of types which includedmedical facilities (108), educational buildings (460),commercial buildings (479), and residential structures (2,268). The survey works also entailed 1,700 interviews with residents. Buildings of all age categorieswere damaged and nominally engineered structures alsosuffered significant damage. Unreinforced masonry, that was not part of the structural system, was required to provide the needed resistance to seismic actions. Observations revealed poor structural configurations, poor detailing of reinforcement, and the use of low-quality construction materials. Analysis of the surveyed buildings consists of different types of structures on physical damage associated with the cost of repair (damage index) using quantity surveying information. Fig. 10 shows nine- benchmark damage models resulting from the survey andanalysis. Further detail of the surveyed buildings analysis can be referred to in Sengara et al. (2010).

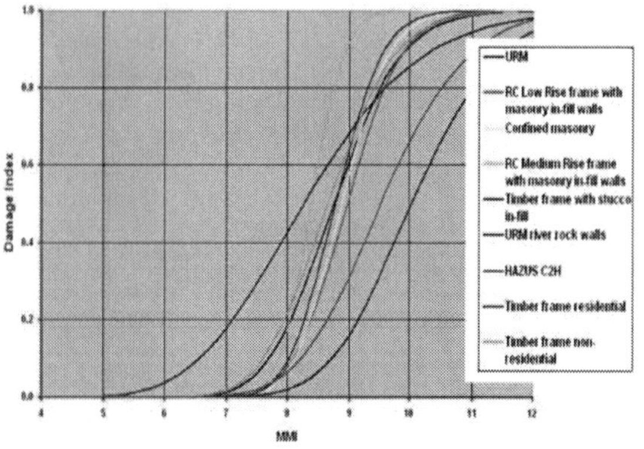

Fig. 10. Benchmark vulnerability models derived from Padang Earthquake damage observations [21].

The case of the post-Padang survey has identified that confined masonry construction in particular sufferedlower damage levels than the unreinforced masonryequivalent. The promotion of cost-effective construction practices which reduce vulnerability and the development of other structural systems with these attributes are central to reducing earthquake risk. Promotion on, enforcement of seismic building codes, training on, and community-based earthquake resistance constructions for buildings are vital to vulnerability and disaster risk reductions. The nine-benchmark models could be used as general measures of vulnerabilities of many buildings with similar types in Indonesia and identify its potential damage and its associated risks. A damage index of as high as 10 to more than 50% is anticipated for most building types for MMI8 to 9. This nine-benchmark model could be adopted for the preliminary risk-assessment process subjected to future potential earthquakes as supporting input in the formulation of disaster risk reduction management plans for cities in Indonesia.

4.2 Bridge Vulnerability

Although the performance of a bridge is based on the interaction of all its components, certain bridge components are more vulnerable to damage than othersi.e., connection, bearings, piers, columns, foundations, and abutment. Of these, connections, bearings, and inadequate seat length are the most common reasons forbridge failure and the least costly to fix. Since evaluation on bridge vulnerability post-earthquake inspection of bridges in Indonesia has not been conducted comprehensively, empirical fragility curves for the conventional bridge could be estimated using FEMA224 (1991). Damage curves for the conventional bridgesare based on ATC-13 data for FC 24, multiple simple spans, and FC 25 continuous/monolithic bridges (includes single-span bridges).

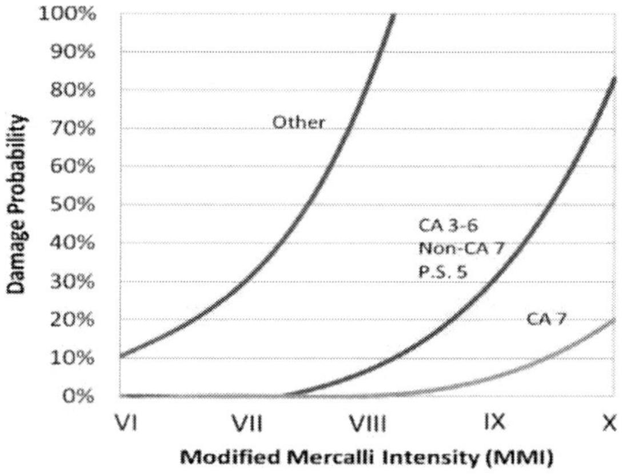

Fig. 11. Judgemental damage curves for conventional bridges[9]

Due to changes in the seismic design criteria and the design concepts it may be necessary to perform several seismic safety checks during the long economical life of existing bridges. It must be pointed out that both new and existing bridges must satisfy today's safety criteria. Therefore, a risk-based approach in which the remaining service life and the acceptable investment cost should be considered. It is essential to develop a vulnerability bridge model for Indonesia.

4.3 Associated Risks and Reduction Efforts

It has been identified and elaborated previously throughthe post-disaster survey as well as through identificationon the future potential hazards that many existing buildings and infrastructures in Indonesia is consideredto have various levels of vulnerability and associated disaster risks. The risks are accumulated through past constructions with evolving hazard maps and building codes. Additional risks would also be contributed by the performance of other components implemented within the risk-chain, previously noted.

The necessity of disaster reduction for sustainable development and social stability was recognized since several great earthquakes repeatedly occurred in Indonesia. Over the last 10 years, many research institutions, universities, including association professions in Indonesia have significantly increased their efforts towards understanding earthquake hazards and their mitigation on the impact of future large earthquakes.Some of the recent efforts to mitigate the impacts of earthquake hazards have been described briefly in this paper. The actions include updating of the seismic hazard maps of Indonesia 2010 and 2017, revision and continuous updating of building and infrastructure design codes, and development of design codes associated with geotechnical structures (SNI-8460-2017). Development of seismic micro-zonation maps for citiesin Indonesia as supporting reference on-site- classification for residential housings and other low-story buildings would be considered necessary.

In 2012, the Ministry of Public Works has practically implemented SNI 1726-2012 to replace SNI 03-1726-2002 as a national code for earthquake resistance building design. The new code for buildings follows theconcept of Maximum Considered Earthquake Geometric Mean (MCE_G) and Risk-Adjusted Maximum Considered Earthquake (MCE_R) adopted in ASCE 7–10.This has been a significant effort towards enhancement in the development of Indonesian seismic building codes from a hazard-based approach to a new concept of a uniform risk-based approach. Five years later after launching the 2012 code, the new seismic hazard maps have been enhanced (2017), and soon afterward the new maps were adopted in the SNI 1726-2019.

Updating of national standards for the design of infrastructure is also conducted for bridges, dams, and other structures. Type of infrastructure code will utilize hazard maps having different return periods. Indonesia has published successfully the Indonesian GeotechnicalCode (SNI-8460-2017) that somehow in general accommodates seismic requirements for many infrastructures. Furthermore, the new 2017 Indonesian seismic hazard maps have included several return periods that facilitate the design of many infrastructures.

Concern on the vulnerability of the existing bridge has been realized by the Directorate General of Highway. Currently,the directorate inspected a conventionalbridge located at National Road in Indonesia to evaluateeach of its conditions and prepare for maintenance and rehabilitation programs. Information of each of the bridge's conditions was stored at a web database namely, http://inspeksijembatan. pusjatan.pu.go.id. After a large earthquake, a specific survey was also conducted to identify the influence of seismic activity on the existing bridges nearby the epicenter. Mitigation scheme to retrofit existing bridges evaluated according to the damages due to large post-earthquake events. Typical damages were shown below (Fig. 12 -16).

Fig. 12. Damages of bridge deck due to the displacement of rubberbearing post-Padang Earthquake in 2009 [27]

Fig. 13. Damages of bridge deck due to compression effect in longitudinal direction post-Nusa Tenggara Barat Earthquake in 2018 [27]

Institute of Road Engineering and PT. Wiratman Chodai Indonesia is currently developing a technical guideline for inspection, evaluation, andcountermeasures of bridge structure against seismic risk. In this guideline 3 (three) concept, seismic retrofitting methods were introduced i.e., bridge element reinforcements, whole bridge reinforcements, and preventive measures against falling bridges as illustratedin Figure 14 to Figure 16.

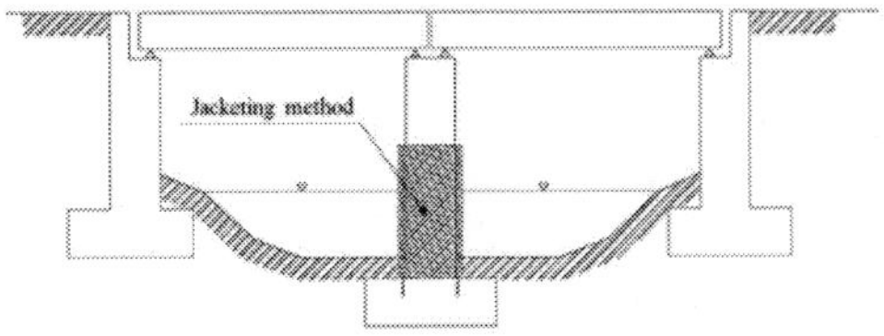

Fig. 14. Example of bridge element reinforcement (Toshiro. Y, 2019)

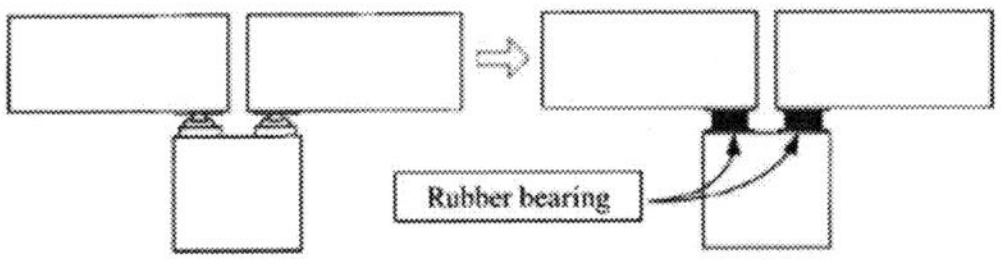

Fig. 15. Example of whole bridge reinforcement using seismic isolation (Toshiro. Y, 2019)

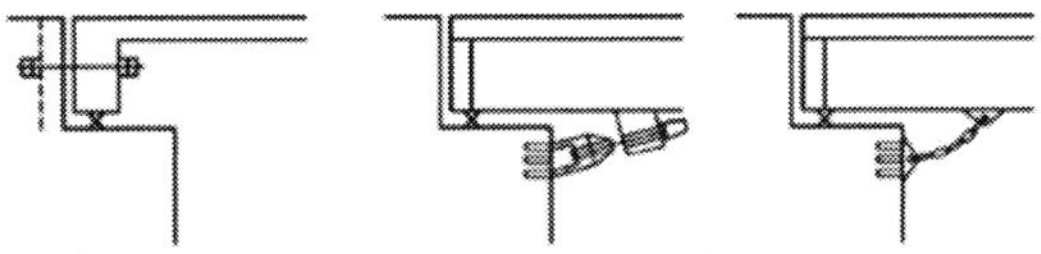

Fig. 16. Example of preventive measures against falling bridge using restrainer (Toshiro. Y, 2019)

5 Concluding Remarks

Seismic building and infrastructure (bridge and dam) spectral design load for Indonesia has evolved over the last three decades. During the development of seismic hazard mapping through the improvement of seismic source zoning and its corresponding seismic parameters,as well as the lower probability of exceedance requirement, there is a tendency that the spectral designloads requirements for buildings, bridges, dams, and other infrastructures to increase in most areas of Indonesia. Recent spectral design load requirements forbuildings and infrastructures for new construction in Indonesia are in general directed toward in-line with many International Building Codes. An increase in spectral design loads requirements would tend to expose many existing buildings and infrastructures to be at a higher level of vulnerabilities and risks.

Many building and infrastructure damages post-disaster surveys from several large earthquakes in Indonesia have indicated vulnerabilities of many types of buildings and bridges. Vulnerability building-damage models have been developed Post-Padang 2009 earthquake. The models could be adopted as preliminary bases for the risk assessment process for many types of buildings. High to very high risks of building damages need to be anticipated for many existing buildings and infrastructures under the current spectral design load in many high seismic potential areas.

It is essential to prevent or minimize building and infrastructure collapses or damages from future earthquakes through:

- Provision of proper engineering design parameters.
- Improvement of building codes and guidelines.
- Proper design, good quality construction, and building and infrastructure design enforcements.
- Systematic and good construction supervision.
- Continuous updates of seismic hazard maps and enhancement of building and infrastructures codes areexpected.
- Disaster risk assessment of buildings and infrastructures demands continuous enhancement of building vulnerability models and development of vulnerability models for infrastructures such as bridges.

Assistance by Mr. Ahmad Sulaiman for hiscontribution in editing this paper is highly appreciated. The authors are thankful to team members in the developmentof Indonesian seismic hazard maps and thecorresponding SNI for providing all the data andinformation in this paper. Other colleagues who have participated in the contribution in the supportingmaterials referred to in this paper are also appreciated.

References

1. F. Aldiamar, M. Irsyam, I W Sengara, M. Asrurifak, *Proposed seismic hazard map for conventional bridge code in Indonesia,18th Southeast Asian Geotechnical Conference,* Singapore (2013)
2. American Association of State Highway and Transportation Officials (AASHTO), Guide Specifications for LRFD Seismic Bridge Design, 2nd Edition, Washington (2012)
3. ASCE 7-10, Minimum design loads for buildings and other structures, American Society of Civil Engineers, 1 edition, ISBN-10:0784410852 (2010)
4. National Standardization Agency, SNI 03- 2833 *Earthquake Resistance Planning Procedures for Highway Bridges*, BSN, Jakarta (1992)
5. National Standardization Agency, SNI 2833:2008 *Earthquake Resistance Planning for Bridges, Indonesian National Standard*, Jakarta (2008)
6. National Standardization Agency, SNI 2833:2016 *Bridge Design Against Earthquake Load, Indonesian National Standard*, Jakarta (2016)
7. B. C. H. Ferner.. Indonesian Earthquake Study. New Zealand Bilateral Assistance Programme to Indonesia, 1-7 (1979)
8. Department of Public Works, Directorate General of Human Settlements, Directorate of Building Problems. Indonesian Earthquake Resistant Planning Regulations for Buildings (1983)
9. Federal Emergency Management Agency (FEMA), Seismic Vulnerability and Impact of Disruption of Lifelines in the Conterminous United States. Applied Technology Council, California (1991)
10. ICOLD.. Selection Seismic Parameters for Large Dams - Guidelines. Bulletin, 72 (1989)
11. ICOLD. Selection Seismic Parameters for Large Dams - Guidelines. Bulletin, 148 (2016)
12. Indonesian Standard Code, SNI 03-1726-2002, *Earthquake Resistance Design for Buildings*, National Standardization Agency (2002)
13. Indonesian Standard Code, SNI 1726-2012, *Earthquake Resistance Design for Buildings,* National Standardization Agency, 2012.
14. Indonesian Standard Code, SNI 1726-2019, Earthquake Resistance Design for Buildings, National Standardization Agency (2019)
15. M. Irsyam, I.W. Sengara, F. Aldiamar, S. Widiyantoro, W. Triyoso, D.H Natawijaya, E. Kertapati, I. Meilano, M. Asrurifak, M. Ridwan, Suhardjono, *Development of Seismic Hazards Map of Indonesia for Revision of SNI 03-1726- 2002, A Report submitted to Australia Indonesia Facility for Disaster Reduction (AIFDR) and National Agency for Disaster Management, Institute for Research and Community Services, Institut Teknologi Bandung,* (2010)
16. Irsyam et al., National Center for Earthquake Studies (PuSGeN): Map of Indonesia Earthquake Sources and Hazards in 2017, ISBN: 978-602- 5489-01-3, (2017)
17. E. Kertapati, *Probabilistic Estimates of the Seismic Ground Motion Hazard in Indonesia. Proceeding on National Conference on Earthquake Engineering, Indonesian Earthquake Engineering Association, Bandung* (1999)

18. Luco, Nicolas, Ellingwood, B.R., Hamburger, R.O., Hooper, J.D. Kimball, J.K, Kircher, C.A, Risk-Targeted versus current seismic design maps for the conterminous United States, 163- 175, in Structural Engineers Association of California 2007 Convention Proceedings, Callifornia (2007)
19. T.F. Najoan, S. Djanasoedirdja, S. Ruhijat, Map of the Earthquake Zone as a result of the Review and How to Use it as a Proposal in the Design of Earthquake Resistant Irrigation Buildings. Journal of the Research and Development Center for Irrigation, **36**, 22-36 (1996)
20. T.F. Najoan, A. Suharjoyo, A. Buditomo, S. Wibowo, Rizaldi, R.B. Nasution, Indonesia Earthquake Zone Map for Determination of Maximum Earthquake Acceleration on the Surface. Proceedings of the National Conference on Seismic Engineering, Bandung (1999)
21. I.W Sengara, M. Suarjana, D. Betham; N. Corby; M. Edwards, M. Griffith; M. Wehner; R. Weller, *The 30th September 2009 West Sumatra Earthquake, Padang Region Damage Survey, GeoScience Australia* (2010)
22. I.W. Sengara, M. Irsyam, I.D. Sidi, W. Merati,, K.S. Pribadi, M. Suarjana, M. Edwards, *Some Recent Efforts in Earthquake Hazard and Risk Analyses for Disaster Risk Reduction in Indonesia, in the 2nd International Conference on Earthquake Engineering and Disaster Mitigation (ICEEDM)*. Surabaya, Indonesia (2011)
23. I.W. Sengara, M. Irsyam, G.W. Merati, I.D Sidi, M. Suarjana, A. Mulia, M.A. Yulman, D. Hutabarat, Development MCER (Risk-Targeted Maximum Considered Earthquake) through Investigation of Hazard and Fragility Curve, Research Report – IMHERE Project B.2C FTSL, Institut Teknologi Bandung (2012)
24. I.W. Sengara, I. Masyhur, I.D. Sidi, A. Mulia, M. Asrurifak, D. Hutabarat, 2015. Development of Earthquake Risk-Targeted Ground Motion for Indonesian Earthquake Resistance Building Code SNI 1726-2012, in 12th International Conference on Application of Statistic and Probability in Civil Engineering (ICASP), July 12-15, Vancouver, Canada (2015)
25. I.W. Sengara, M.A Yulman, P. Sumiartha, A. Mulia, *Development on Risk-Based Seismic Design Criteria and Ground Motions for High-Rise Buildings in Jakarta, in the 3rd International Conference on Earthquake Engineering and Disaster Mitigation (ICEEDM).* Bali, Indonesia (2016)
26. I.W. Sengara, M. Irsyam, I.D. Sidi, A. Mulia, M. Asrurifak, D. Hutabarat, W. Partono, *New 2019 Risk-Targeted Ground Motions for Spectral Design Criteria in Indonesian Seismic Building Code, in the 4th International Conference on Earthquake Engineering and Disaster Mitigation (ICEEDM).* Padang, Indonesia (2019)
27. W. A. Riyono, Focus Group Discussion Draft Pedoman Pemeriksaan, Evaluasi dan Perkuatan Struktur Jembatan Terhadap Risiko Gempa Bumi, Presentation Materials, Institute of Road Engineering, Bandung (2019).

 https://doi.org/10.1051/e3sconf/202133107010

Distributed sensor for earthquake identification system to activate tsunami shelter finding system

Dody Ichwana Putra[1*], *Faiq* Genia Fauzul[1], *Shelvi* Ekariani[2], and *Shilva* Lioni[3]

[1] Departement of Computer Engineering, Faculty of Information and Technology, Andalas University, West Sumatera, Indonesia
[2]Departement of Mathematics, Faculty of Mathematics and Natural Sciences, Andalas University, West Sumatera, Indonesia
[3]Departement of English, Faculty of Humanities, Andalas University, West Sumatera, Indonesia

Abstract. Padang City as the capital of West Sumatera Province is at high risk of earthquakes and tsunamis due to its location between two continental plates and the Semangko Fault. Currently, there are several shelters in Padang that serve as evacuation sites as a rescue location when a tsunami hits. This paper presents an earthquake detection system that uses distributed sensors to activate the shelter search system. The proposed system will activate the shelter search system when the earthquake has medium or high magnitude intensity. The earthquake identification system is achieved by calculating the Peak Ground Acceleration (PGA) value from the p-wave and s-wave using piezoelectric and accelerometer sensors. The proposed system is a distributed node sensor placed in different shelters which communicate using MQTT protocol. To evaluate the system, system for earthquake detection using Raspberry Pi, piezoelectric sensor, accelerometer MPU-6050, and Xbee for data communication have been implemented. The result shows that the system can detect the magnitude and intensity of the earthquake.

1 Introduction

Earthquake is a natural phenomenon that can occur at any time and occur naturally due to the release of energy, elastic displacement, or the collision of the earth's plates in the fault zone all at once [1]. In addition, earthquakes can also be interpreted as tremors that are triggered at the earth's surface by seismic waves [2].

The city of Padang is the largest city on the west coast of Sumatra Island and the capital of West Sumatra Province, Indonesia. Based on data from the Central Bureau of Statistics (BPS) in 2016, Padang has a population of 902,413 people [3]. The province of West Sumatra has a high potential for earthquake disasters because it is located between the confluence of two major continental plates (the Eurasian plate and the Indo-Australian plate) and the Semangko fault [4]. Based on BMKG data, there were 130 earthquakes in the West Sumatra region during the period from 1 January 2017 to 12 March 2018, both on land and at sea [5]. Most of the earthquakes at sea occur in the Mentawai district, which has the potential to trigger a tsunami.

Currently, the city of Padang has a tsunami early warning system. This system consists of sensors off the coast that detect incoming tsunami waves. Then the sensors send data about the waves, which is converted into information so that people on the coast can be evacuated by sounding a siren [6].

After the siren is sounded, people on the beach are expected to immediately move to the nearest evacuation site. For the evacuation process to go smoothly, the community needs to know where the nearest shelter is located. The problem is that many people do not know where the nearest tsunami shelter is and only know a few evacuation locations [7]. This leads to severe congestion when the evacuation process is underway [8].

This study aims to develop a distributed sensor system that can determine the magnitude and intensity of an earthquake using accelerometer and piezoelectric sensors and the MQTT protocol for data communication to activate the tsunami shelter finding system.

2 Theory of earthquake waves

2.1 Seismic Waves

Seismic waves propagate in the earth's crust and are caused by deformation/change in structure, pressure, tension, or explosion due to the elastic nature of the earth's crust [11]. There are seismic waves that propagate inside the earth, called body waves, and those propagate at the earth's surface called surface waves.

2.2 Body Wave

* Corresponding author: dody.ichwana@it.unand.ac.id

Body waves propagate in an elastic medium and whose direction of propagation is in the earth's interior. Body waves are waves that arrive before the surface waves emanating from an earthquake. These waves have a higher frequency than surface waves. Based on the motion of particles in the medium and the direction of propagation, waves can be divided into P-waves and S-waves.

a) P-wave / primary wave
The primary wave or compression wave is a body wave that has a higher velocity than the S-wave. This wave is a longitudinal wave of particles that can propagate back and forth in the direction of propagation. These waves are created due to pressure. Since it has a high velocity, this wave arrives earlier than the S wave. This wave has a frequency of less than 4 Hz.

b) S-wave/ secondary wave
The S-wave or transverse wave is a body wave with particle motion perpendicular to the direction of propagation and arrival time after the P-wave. It has a lower velocity than the P-wave.

2.3 Earthquake Magnitude

The magnitude of the earthquake expresses the strength of an earthquake. One of the methods of calculating the value of earthquake magnitude is Peak Ground Acceleration (PGA). PGA is the maximum ground acceleration when an earthquake occurs at a location [12]. Table 1 shows the earthquake magnitude based on PGA

Table 1. Earthquake Magnitude Based on PGA

Magnitude	Peak Ground Acceleration (m/s^2)
1	<0.017
2-3	0.017-0.14
4	0.14-0.39
5	0.39-0.92
6	0.92-1.8
7	1.8-3.4
8	3.4-6.5
9	6.5-12.4
10	>12.4

2.4 Earthquake Intensity Scale

BMKG (Meteorology, Climatology, and Geophysical Agency) created the intensity scale of the earthquake to express the impact caused by an earthquake. The earthquake intensity scale (SIG-BMKG) was created by considering information about the impact of an earthquake based on the typical culture or building characteristics in Indonesia. This scale has a simpler structure and has only five levels, namely I-V, compared to the MMI (Modified Mercalli Intensity) scale, which has 12 levels. [13]. Table 2 shows Earthquake intensity based on PGA.

Table 2. Earthquake intensity based on PGA

SIG BMKG Scale	Colour	Description	MMI Scale	PGA (gal)
I	White	Not felt	I-II	<2.9
II	Green	Felt	III-V	2.9-88
III	Yellow	Slight damage	VI	89-167
IV	Orange	Moderate damage	VII-VIII	168-564
V	Red	Heavy damage	IX-XII	>564

3 System Design

3.1 Architecture System

The earthquake detection system consists of multiple sensors operating as nodes distributed in each evacuation shelter. The distributed nodes communicate wirelessly with each other using the MQTT protocol. Each node is a module consisting of a piezoelectric sensor, an accelerometer for detecting vibrations, and a Raspberry as a sensor data processor. The Xbee module is used for wireless communication between each node. The architecture of the earthquake detection system is shown in Fig. 1.

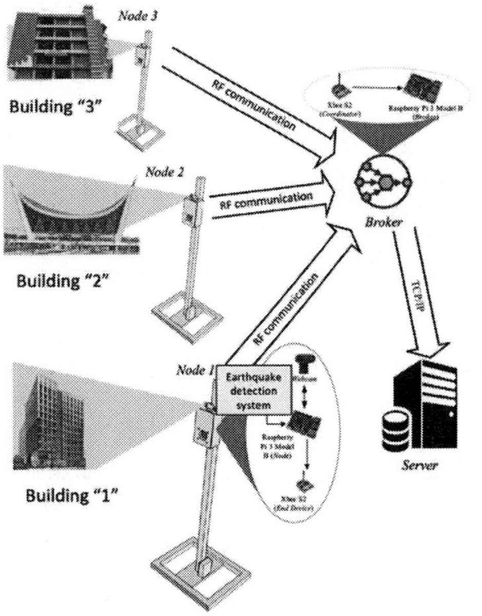

Fig. 1. System Architecture

The system will be active when the piezoelectric sensor detects the incoming vibration, the piezoelectric module amplifies the signal, and the analog output signal of the piezoelectric element is converted into a digital signal, after which the signal is sent to the microcontroller.

Besides, the microcontroller calculates the frequency of the wave. If the wave has a frequency below 4 Hz, the system activates the accelerometer to measure the S-wave vibration and sends the measurement results to the microcontroller. After that, the microcontroller calculates the peak ground acceleration (PGA) value of the S-wave and then performs the calculation process to determine the intensity and magnitude of the earthquake. If the perceived earthquake intensity is on a moderate scale or above, the system activates and proceeds to the "The tsunami shelter finding system" as an application subsystem. Next, Xbee broadcasts that an earthquake has occurred at the node to the other node. If the node also receives transmissions from other nodes, it can be concluded that an earthquake has occurred. After the shelter conditions are known, the system sends data about the location, magnitude, intensity, and state of the shelter to the coordinator/broker via Xbee. Next, the broker sends data from each node to a server that acts as a subscriber. In Fig. 2, the activity diagram to activate the subsystem of the Shelter Finding System is shown.

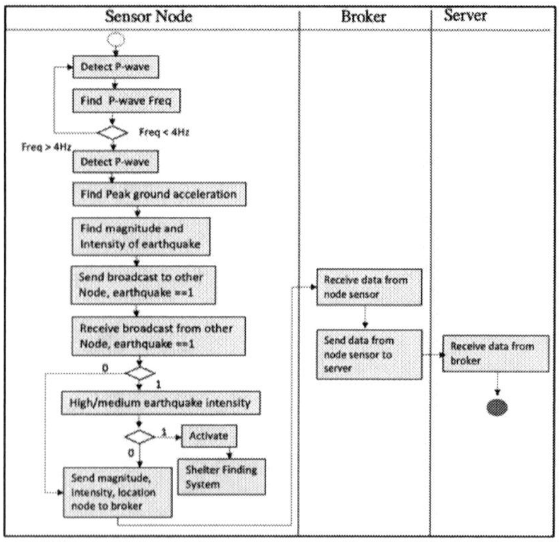

Fig. 2. Activity diagram system

3.2 Hardware and Software Design

Structure design for hardware construction is shown in Fig. 3. The descriptions are below:

1. The webcam is used to take pictures of the shelter.
2. Raspberry PI functions as a controller of the entire system, which contains instructions and program logic
3. The piezoelectric module serves to amplify the piezoelectric output (amplifier) and convert analog signals into digital signals.
4. The battery serves as a power source for the Node and coordinator.
5. Xbee serves to send data via radio frequency from the Node to the broker
6. Accelerometer functions to detect earthquake vibrations.
7. Piezoelectric is used to detect low-frequency vibrations; this system serves to detect P-wave
8. Coordinator/broker serves to receive all data from the Node (publisher) to be forwarded to the server (subscriber)
9. Xbee Coordinator functions to collect data from all Nodes.

To test the prototype wireless communication scenario using the MQTT protocol, the distance between each sensor and the broker is set to 120 meters and no obstacle between each node and the broker (line-of-sight (LOS)).

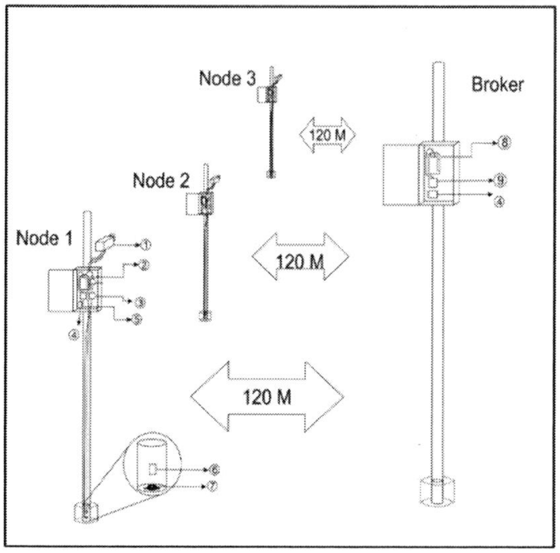

Fig. 3. Hardware design

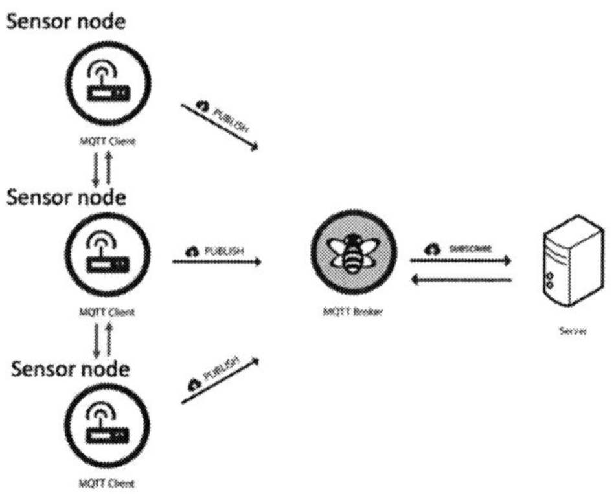

Fig. 4. Wireless communication setting using MQTT protocol

The scenario of node communications is shown in Fig.4. The scenario for earthquake detection is that the node (as publisher) sends data to the broker when it detects an earthquake, using the MQTT protocol according to the topic. The coordinator/broker is used to receive all the data from the node (publisher) and forward it to the server (subscriber). Then, the broker forwards the data to the subscriber depending on the subscriber who subscribed to the corresponding topic. When a node detects a vibration:

https://doi.org/10.1051/e3sconf/202133107010

- The node immediately sends a broadcast to other nodes that an earthquake has occurred.
- When an earthquake occurs, the node also receives an earthquake broadcast from other nodes.
- If the node does not receive any broadcast from other nodes, it can be concluded that there was no earthquake, and the detected vibrations are not from an earthquake.

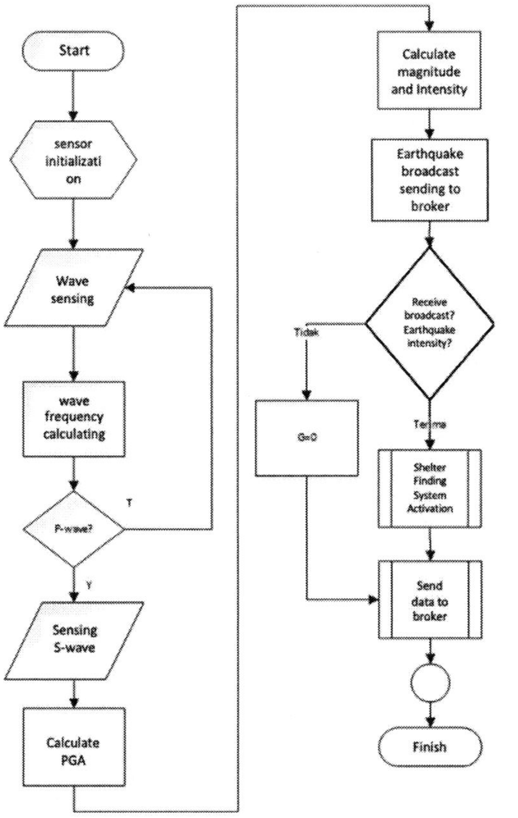

Fig. 5. Software design

The software design can be seen in Fig. 5. The system reads the vibrations using a piezoelectric sensor, and when a P-wave is detected, the accelerometer is activated to further read the S-wave. Also, the microcontroller calculates the maximum acceleration of the S-wave. The microcontroller calculates the magnitude and intensity based on the maximum acceleration of the S-wave.

3.3 P-wave and S-wave detection

The P-wave has a frequency of 4 Hz and belongs to the category of infrasonic waves. To detect the presence of these seismic waves, in this experiment, a piezoelectric sensor is used. Piezoelectric is a seismic sensor that causes low-amplitude vibrations and is very easily contaminated, so the filter is required to detect the noise signal. Low-pass filter with a program to filter vibrations with frequencies below 4 Hz. The S-wave has a higher frequency than the P-wave. The S-wave can be measured directly with an accelerometer.

The Peak Ground Acceleration (PGA) is determined from the accelerometer's measurements in terms of accelerations on the x, y, and z axes. Earthquakes occur in the x- and y-plane because earthquakes are caused by a horizontal S-wave polarization [23]. In the xy-plane, the z-axis is constant; therefore, the value on the z-axis can be neglected to measure the PGA of an earthquake. The x-axis and y-axis values obtained from the sensor can be substituted into Equation 1 to obtain the resultant value:

$$Resultan = \sqrt{x^2 + y^2} \tag{1}$$

The PGA value is obtained from the highest value of the results during an earthquake event. The algorithm of PGA mapping is shown in Fig. 6. Fig. 7 shows the block diagram of PGA detection.

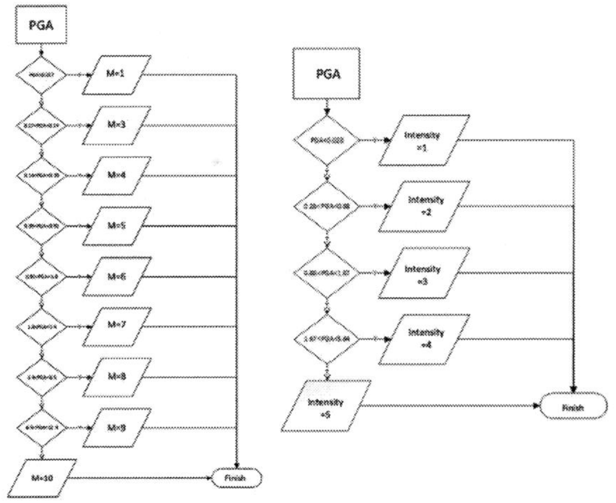

Fig. 6. Algorithm of PGA mapping

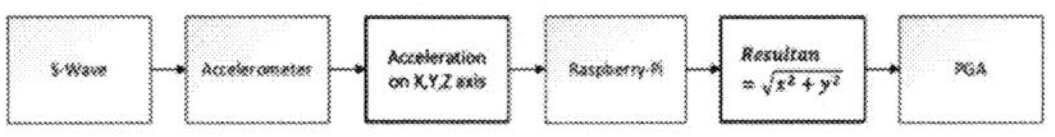

Fig. 7. Block diagram PGA detection

4 Implementation and Result

The system implementation can be seen in Fig. 8. The hardware used in this system consists of Piezoelectric, Accelerometer, Xbee, and Raspberry-Pi. This system has two input devices namely piezoelectric and accelerometer. The piezoelectric input is a P-wave from an earthquake with a wave frequency below 4 Hz, which is connected to the Raspberry-Pi through a module to facilitate communication between the microcontroller and the sensor. In addition, the accelerometer has an input for the acceleration generated by the S-wave of an earthquake.

The highest value from the input of the accelerometer is called PGA (Peak Ground Acceleration), which is used to determine the magnitude and intensity of the

https://doi.org/10.1051/e3sconf/202133107010

earthquake. In addition, Xbee communication is used to send data between the nodes and to the brokers by using radio frequencies connected to the Raspberry Pi via an Xbee adapter to facilitate the use of the Xbee module via a USB cable.

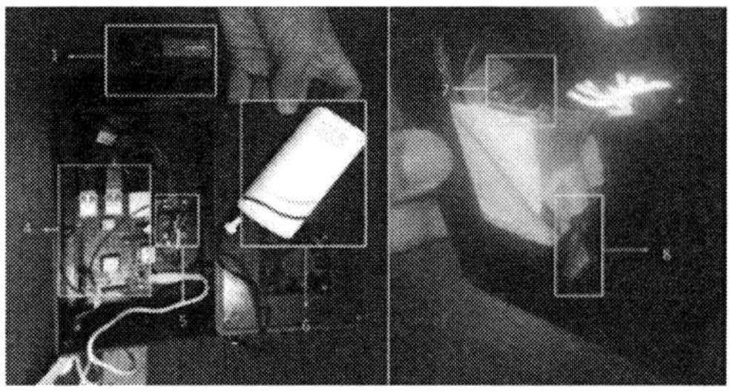

(a)

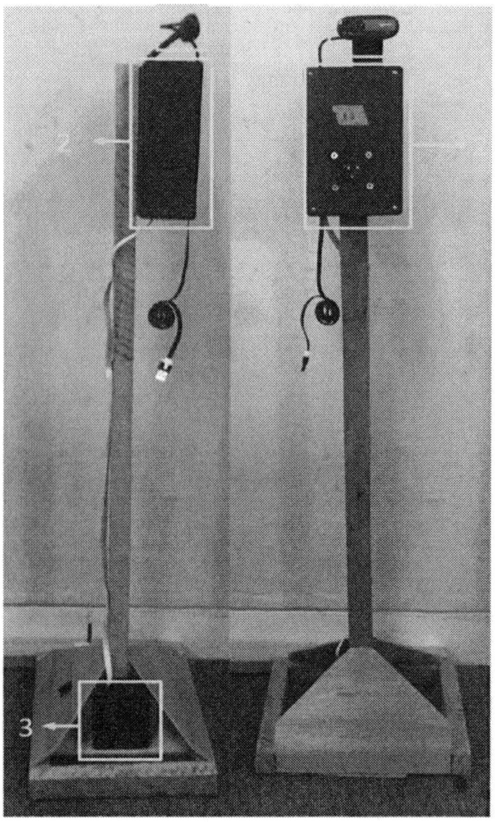

(b)

Fig. 7. Hardware implementation

Testing the S-wave detection and finding the intensity and magnitude of the earthquake aims to determine the accuracy of the PGA readings at each node to determine the intensity and magnitude of the earthquake. This test is performed by simultaneously vibrating the accelerometer at each node with the same acceleration. Table 3 shows the evaluation of S-wave detection on node 1.

Table 3. Evaluation of S-wave detection and calculating of Earthquake Magnitude and Intensity

PGA reference	*Node* 1			
	PGA	**Magnitude**	**Intensity**	***Error* PGA (%)**
1.23	1.79	6	4	44.44372
2.05	2.36	7	4	14.69156
2.46	2.13	7	4	13.75882
4.35	4.49	8	4	3.061922
5.62	5.66	8	5	0.670954
6.77	6.59	9	5	2.687554
7.00	7.44	9	5	6.145003
10.40	9.95	9	5	4.335767
12.08	11.62	9	5	3.826168
13.73	13.43	10	5	2.19724
Error mean				**9.68**
Standard deviation *Error*				**13.46**

Based on the test results for the S-wave measurements and considering magnitude and intensity, the average PGA reading error at node 1 is 9.68% with a standard deviation of 13.46%, and the average PGA reading error at node 2 is 5.02% with a standard deviation of 2.08%. The average PGA reading error at node 3 is 20.58% with a standard deviation of 19.44%. This error in PGA readings affects the earthquake intensity and magnitude results obtained from each node since the earthquake magnitude and intensity values are obtained using PGA. According to the author's analysis, one of the causes of the error in the test is that the position of the accelerometer at each node is not the same, which affects the acceleration values at each node.

Table 4. Communication time of distributed node sensor

Distance of *Node* (m)	**The transmission time of *Node* 1 (s)**	**The transmission time of *Node* 2 (s)**	**The transmission time of *Node* 3(s)**	**Mean of transmission time**	**Standard deviation**
10	1.0723	1.1556	1.1134	1.113	0.0416
15	1.0644	1.1128	1.1570	1.111	0.0463
20	1.8039	1.8035	1.8094	1.805	0.0036
25	1.8115	1.8252	1.8128	1.816	0.0075
30	1.7912	1.8217	1.8174	1.810	0.1650
35	1.8203	1.8284	1.8271	1.825	0.0043
40	1.8332	1.8397	1.8341	1.835	0.0035
45	1.9029	1.9125	1.9117	1.909	0.0053
50	1.8458	1.9114	1.9149	1.890	0.0389
55	1.9119	1.9148	1.9158	1.914	0.0020

The inter-node communication time test aims to determine the time taken by each node to transmit earthquake information to other nodes. This test is performed by measuring the time taken for the node to send an earthquake broadcast until the node receives an earthquake broadcast from another node. The tests are

 https://doi.org/10.1051/e3sconf/202133107010

performed for each node with different transmission distances. In Table 4 can be seen the time of communication of distributed node sensor. From the test results, it can be concluded that the distance can affect the transmission time between nodes to transmit the information that an earthquake has occurred. The farther the distance between nodes, the longer the data transmission time.

The average time taken by node 1 from the detection of an earthquake to the arrival of the data on the server is 135.251 seconds with a standard deviation of 0.362 seconds, the average time taken by node 2 from the detection of an earthquake to the arrival of the data on the server is 135.293 seconds with a standard deviation of 0.634 seconds, the average time taken by Node 3 from the time of detection of an earthquake to the arrival of the data on the server is 135.797 seconds with a standard deviation of 0.634 seconds. The conclusion when examining the time from the occurrence of the earthquake to the arrival of the data at the server is that the data transmission time is affected by noise at the time of data transmission from the node to the broker, as well as by the Internet connection at the time of data transmission from the broker to the server.

5 Conclusion

Based on the research conducted, the use of piezoelectric sensors to detect p-waves is quite good with an average error rate of 9.86%. The use of accelerometers to detect S-waves is good with an average error rate of 11.72%. The average time taken by the system to send the data of detected earthquake to the server is 135.25 seconds. This system is successful in activating the sub-system application of the tsunami shelter finding system after the earthquake occurred. For future development of the system, more sensitive vibration sensors such as geophones are added to detect p-waves. Adding a tool for data communication between nodes that has a longer range than Xbee.

References

1. BMKG, Meteorology, Climatology and Geophysics Agency, Available: https://inatews.bmkg.go.id/new/about_tsunami.php. [Accessed March 12, 2018]
2. D. Yona, Fundamental Oseanografi, Malang: UB Press, (2017).
3. Padang City BPS, Padang City Central Statistics Agency, Available: https://padangkota.bps.go.id/dynamictable/2017/07/14/157/sum-penresident-dan-laju-percepat-penresident-menurut-kecamatan-di-kota-padang-2010-2015- and-2016-.html. [Accessed March 12, 2018].
4. D. HR and Madzalmi, "Correlation of Magnitude of local Earthquakes with Dominant Period ," **1**, 1, (2012).
5. BMKG, "BMKG Database," Meteorology, Climatology and Geophysics Agency, Available: http://repogempa.bmkg.go.id. [Accessed March 12, 2018].
6. . R. G. Little, T. A. Birkland, W. A. Wallace and P. Herabat, "Socio-Technological Systems Integration to Support," in Proceedings of the 40th Hawaii International Conference on System Sciences - 2007, Hawaii, (2007).
7. "The people of Padang are not ready to face a potential tsunami,". Available: https://national.tempo.co/read/396774/warga-padang-belum-ready-facing-potential-tsunami. [Accessed March 2018] (2013)
8. R. Marajo, "Evacuation Route in Padang Total Jam,". Available: 10. http://www.tribunnews.com/regional/2016/03/02/jalur-evakuasi-tsunami-di-padang-macet-total. [Accessed March 2018] (2016)
9. A. Zaenudin, "Seeing How Earthquake and Tsunami Detectors Work,". Available: https://tirto.id/melihat cara-kerja-detektor-gempa-dan-tsunami-cBXd. [Accessed March 2018] (2017)
10. A. B. A. P. S. N. K. Wargantiwar, "Wireless Earthquake Alarm Design Based on MEMS Accelerometer," **4**, 128-132, (2017).
11. P. Kambe, "Wireless Sensor Networks for Earthquake Detection and Damage Mitigation System," **4**, (2016).
12. J. Douglas, "Earthquake ground motion estimation using strong-motion records: a review of equations for the estimation of peak ground acceleration and response spectral ordinates," **61**, 3, (2003).
13. BMKG, "BMKG Earthquake Intensity Scale (GIS)," Meteorology, Climatology and Geophysics Agency,. Available: https://www.bmkg.go.id/gempabumi/skala-intensitas-gempabumi.bmkg. [Accessed March 2018]
14. R. Haque, "Earthquake monitoring and Warning System," in International Conference on Advance in Electrical Engineering, Dhaka, 2015. G. Hillar, MQTT Essentials - A Lightweight IoT Protocol, Mumbai: Packt Publishing, (2017)

Hard and soft measures for earthquake and tsunami disaster mitigation

Junji Kiyono[1*]

[1]Professor, Graduate School of Engineering, Kyoto University

Abstract. A destructive earthquake struck the Kobe region on January 17, 1995, and a massive earthquake and tsunami struck eastern Japan on March 11, 2011. We present an overview of the casualty aspects of the 2011 Tohoku earthquake compared with those of the 1995 Kobe earthquake. In the Tohoku disaster, some water gates and seawalls saved some villages from the tsunami effects, though some did not. Based on these examples, we discuss the efficiency of soft and hard measures and consider their respective merits and demerits. The main causes of death in the Kobe and Tohoku EQs were, respectively, collapsing buildings and drowning in the tsunami. Although the time to death was very short in both cases, people often have more time to evacuate in the case of an interplate earthquake leading to a tsunami. Basic countermeasures against tsunamis include such hard measures as water gates, seawalls, and embankments. Soft measures need to be implemented in areas where hard measures are insufficient

1 Introduction

A massive interplate earthquake struck eastern Japan on March 11, 2011. A moment magnitude, MW, of 9.0 and maximum seismic intensity of 7 were recorded in the city of Kurihara, Miyagi Prefecture. As of May 10, 2013,the total numbers of dead, missing, and injured people were 15,883, 2,676, and 6,144, respectively. Most of the casualties were caused by the subsequent tsunami. The numbers of buildings that completely collapsed, partially collapsed, and burned down were, respectively, 126,419, 272,017, and 297 [1]. In addition, extensive parts of the infrastructure, such as road, rail, gas, water, and sewage systems, sustained considerable damage. The following is an outline of the damage.

The geotechnical effects outside the tsunami-struck area included widespread ground liquefaction in port and reclaimed land areas, earth dams, and embankments as well as slope failures. Damage caused by shaking buildings was observed throughout the Tohoku region, but there was a remarkable lack of major damage beyond the run-up area. Slope failures and disrupted sidewalks occurred along roads. Some bridges sustained damage, though most bridges were undamaged. Some coastal bridges were washed away by the tsunami. Areas of coastal local railway were destroyed at several locations by the tsunami and were out of service for several months or more. The main Shinkansen line sustained damage to its viaduct columns near the city ofSendai, though there was no significant structural damage except to some electric poles along viaducts. Inaddition to nuclear power plants, several fossil fuelpower plants went out of service because of their coastallocations. Owing to telecommunications failure, traffic congestion occurred just after the earthquake, though cell towers and services were soon restored. Gas, water, and wastewater systems were generally soon restored with few excavations being required for pipe replacement beyond the run-up area. Washout of oil tanks led to fires in several coastal areas, though several large oil tanks were observed that showed nosigns of spillage or other damage. Ports sustained heavy damage as a result of the tsunami, particularly quayside areas through ground liquefaction.

In this paper, we first present an overview of the casualty aspects of the 2011 Tohoku earthquakecompared with those of the 1995 Kobe earthquake. In the Tohoku disaster, some water gates and seawalls saved some villages from the tsunami effects, though some did not. Based on these examples, we discuss the efficiency of soft and hard measures and consider their respective merits and demerits.

2 Human Casualties of Kobe and Tohoku Earthquakes

A destructive earthquake struck the Kobe region on January 17, 1995. Most kinds of earthquake damage to buildings, houses, lifelines, railways, roadways, and harbours occurred. Figs. 1 and 2 indicate the damage to a highway viaduct, which became one of the symbolic scenes of the Kobe earthquake. Fig. 3 shows that the rate of collapsed and severely damaged houses wassignificantly different according to whether the buildings were

* Corresponding author: kiyono.kjunji.5x@kyoto-u.ac.jp

constructed before or after 1981[2] the year when the seismic code for building construction was revised. This figure indicates that severe damage leading to casualties could have been reduced if houses constructed before 1981 had been strengthenedaccording to the provisions of the revised code.

Fig. 1. A bus hangs over the edge of the collapsed highway

Fig. 2. Collapsed pillar of a highway viaduct caused byshear failure

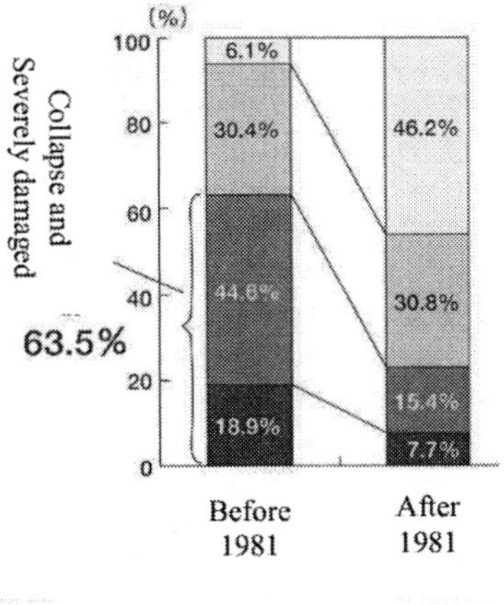

Fig. 3. Histogram representation of the proportion of structural damage according to the construction year— before and after 19812)

2.1 Causes of Death

Fig.2 shows a classification of the causes of death [2]in the 1995 Kobe earthquake (Kobe EQ). Over 80% of deaths were caused by collapsing buildings. This fact became the basis for promoting the strengthening of buildings. However, the causes of death in the 2011 Tohoku earthquake (Tohoku EQ) were quite different from the Kobe EQ, as shown in Fig. 5 [3]: more than90% of the deaths were due to drowning.

The time leading to death after being trapped in a collapsed building is classified according to the injured part of the body, as indicated in Fig. 6 [4]. In this figure, the horizontal axis represents time (hours) on a log scale, and the vertical axis is an index value between 0 (death)and 1 (survival). For example, a person survives only fora period of several minutes if they are trapped andsuffocated or are crushed by the debris (mode 1), but cansurvive for several days if they become trapped withoutsustaining any injury. In the Kobe EQ, as noted, most ofthe victims were killed by collapsed houses (Fig. 5), which falls in the category of mode 1, and those deaths occurred in buildings that predated the code revisions (Fig. 3). That is why the government promoted theintroduction of hard measures, such as the strengthening ofhouses.

In comparing the main causes of death in the two earthquakes, the difference like the disasters complicates the casualty issue. Casualties in the Kobe EQ were due to the tremor itself, whereas in the TohokuEQ the casualties were largely due to the tsunami. Both sets of casualties fall in the same category of mode 1, mentioned above. In a tsunami disaster, people often have time to prepare for evacuation, especially in the case of an interplate-type earthquake. Hard measures are effective for preventing damage caused by tremors; however, soft and hard measures should be adopted for a tsunami disaster by efficiently using the available time.

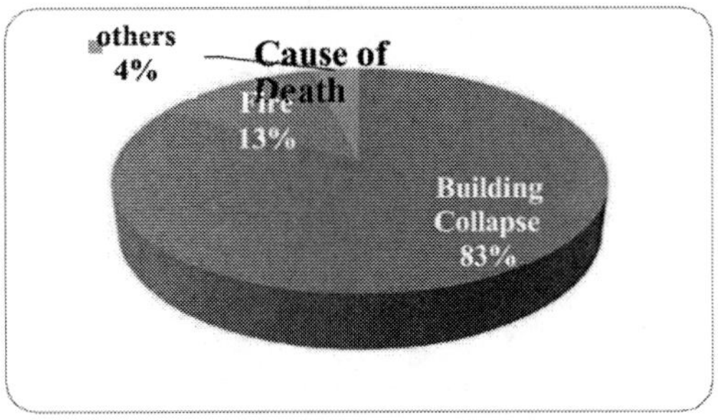

Fig. 4. Pie chart illustrating the causes of death in the 1995 Kobe earthquake [2]

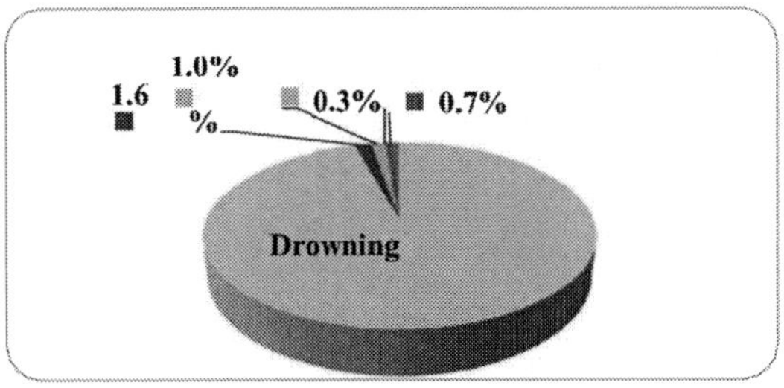

Fig. 5 Pie chart illustrating the causes of death in the 2011 Tohoku earthquake [3]

 https://doi.org/10.1051/e3sconf/202133107011

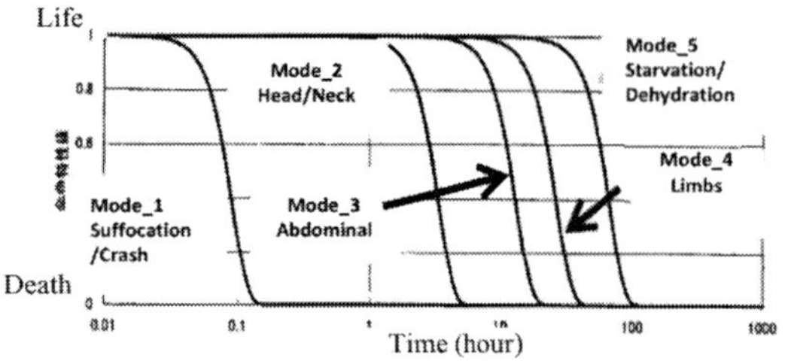

Fig. 6. Graph indicating the time leading to death after being trapped in a collapsed house [4]

- Cause damage: tall, heavy furniture
- Prevent damage: short furniture, table

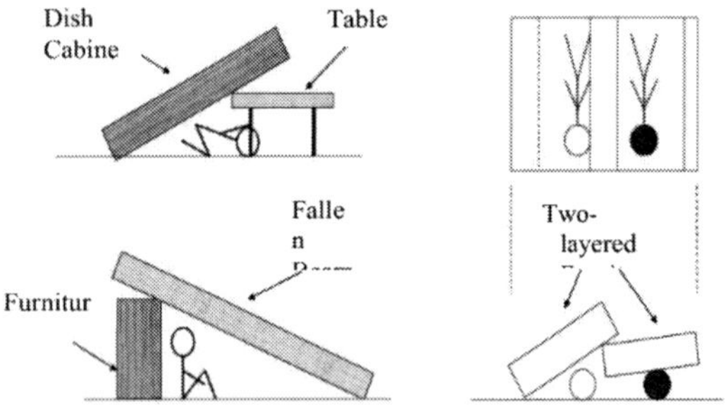

Fig. 7. Schematic representation of interior damage occurring in a house during an earthquake

In addition to the buildings themselves, household furniture is often a cause of human casualties in earthquakes. There is much anecdotal evidence to suggest that tables prevent injury from falling cabinets and other items of furniture. In general, tall, heavyfurniture items become dangerous objects, whereas tables and low furniture items provide shelter fromfalling beams, walls, and larger furniture items. In the Kobe/Tohoku EQ, it was reported that a falling piece ofdouble-layered furniture killed one person but protectedanother Fig. 7.

2.2 Rate of dead, missing, and injured persons

From the death and injury rates for the 2011 Tohoku, 1995 Kobe, 1896 Meiji Sanriku, and 1933 Showa Sanriku earthquakes [5], there are more injuries than deaths with an intraplate earthquake, such as the Kobe EQ; however, interplate earthquakes with tsunamis, such as the Tohoku, Meiji Sanriku, and Showa Sanriku earthquakes, resulting in more dead than injured people.

A famous news video taken in Banda Aceh during the 2004 Sumatra earthquake showed a tsunamiadvancing along a street. Initially, it proceeded at a walking pace; however, within tens of seconds, the wavevery quickly increased its speed. A similar situation occurred in many cities and towns along the coast in the2011 Tohoku EQ. It is possible to learn many lessons from past disasters.

3 Early warnings and Tsunami damage

3.1 Tsunami alerts and early warnings

The location and magnitude of an earthquake are determined within about 3 minutes in Japan and are immediately reported on TV and radio [6]. If the earthquake is not tsunamigenic, the message "Tsunami is not expected" is added to the report. However, if a tsunami is expected, the Japan Meteorological Agency (JMA) issues a forecast about the height of the tsunami and the arrival time of the initial wave, as shown in Fig. 8. The JMA calculates several tens of thousands of cases for potential tsunamis around the islands of Japan and in the Pacific Ocean, and the appropriate information is presented as forecasts and warnings. If the estimated height of a tsunami is greater than 3 m, a large tsunami warning is announced; if the estimated tsunami height is less than 0.5 m, the JMA announces a tsunami advisory warning.

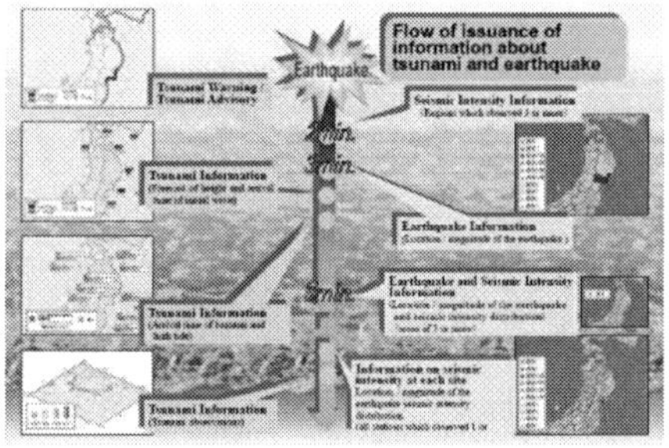

Fig. 8. Information flow-diagram relating to tsunamis and earthquakes issued by the Japan Meteorological Agency [6]

3.2 Tsunami alerts announced by the JMA in the 2011 Tohoku EQ

Table 1 presents the details of information about tsunamis released by the JMA for the Tohoku EQ. From this data, it is evident that the JMA made a tsunami announcement 4 minutes after the earthquake occurred (at 14:50). This first alert for the tsunami height by the JMA indicated 3 m, 6 m, and 3 m for Iwate, Miyagi, and Fukushima prefectures, respectively, with a magnitude of 7.9. This estimate meant that a 3-m-high tsunami had already struck the seashore of Iwate Prefecture and that 6- and 3-m-high tsunamis would strike Miyagi and Fukushima prefectures, respectively, 10 and 20 minutes later. All the tsunami heights were underestimated, but the magnitude of the earthquake was also underestimated. Twenty-eight minutes after the earthquake occurred (at 15:14), the arrival of the tsunami was confirmed in these three prefectures, and the tsunami heights were 6 m, over 10 m, and 6 m for Iwate, Miyagi, and Fukushima prefectures, respectively. The magnitude was still recorded as 7.9. After 45 minutes (at 15:31), tsunamis ofover 10 m in height, were eventually confirmed in each
Prefecture. Concerning the magnitude of the earthquake, it took an additional two days before the final value of 9.0 was released. We do not intend to criticize the difference between the estimated and actualvalues, but it should be noted that estimating a tsunami'sheight and arrival time is difficult even when high technology is employed.

Table 1. Time sequence of tsunami alerts and magnitude announcements from the JMA

Time (Mar 2011)	M	Iwate		Miyagi		Fukushima	
		arrival	height	arrival	height	arrival	height
11/14:46	Earthquake occurred						
11/14:50	7.9	Est.	3m	15:00	6m	15:10	3m
11/15:14	7.9	Conf.	6m	Conf.	>10m	Conf.	6m
11/15:31	7.9	Conf.	>10m	Conf.	>10m	Conf.	>10m
11/16:00	8.4	Magnitude modified					
11/16:09	8.4	Conf.	>10m	Conf.	>10m	Conf.	>10m
11/17:30	8.8	Magnitude modified					
11/18:47	8.8	Conf.	>10m	Conf.	>10m	Conf.	>10m
13/12:55	9.0	Magnitude modified					

3.3 Tsunami damage

Several scenes showing the damage caused by the tsunami appear in Fig 9–14. The extreme power of thetsunami is evident in the damage to buildings, harbours, lifelines, railways, roadways, industry, and agriculture.

The following sections deal with two villages in Iwate Prefecture—Fudai and Taro, as shown in Fig. 7. Each village had developed its hardware countermeasures to prevent damage from tsunamis.

Fig. 9. Damaged pipe behind a seawall

Fig. 10. Damage caused by fire after(Kita-ibaragi)

Fig. 11 Ricefield covered with seawater (Soma)

Fig. 12. Fallen pine trees (Soma)

Fig. 13. Buckled crane arm (Soma)

Fig. 14. Damaged railway station (Shinchi)

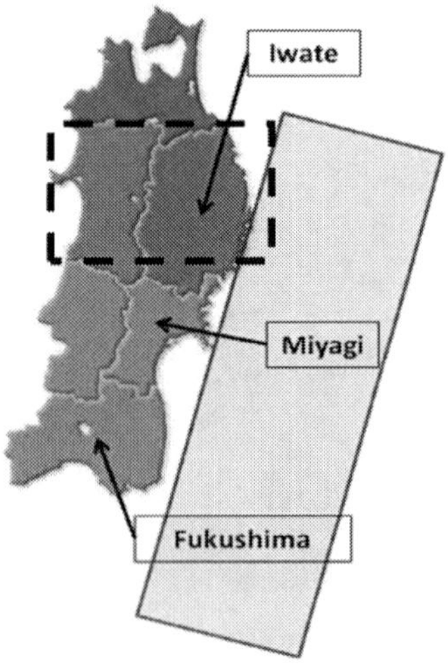

(a) Three prefectures in the Tohoku region

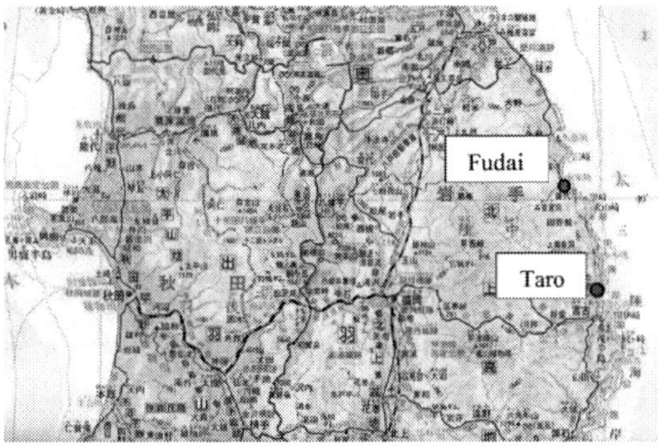

(b) Fudai and Taro

Fig. 15. Map of the Tohoku region and the location of Fudai and Taro

3.3.1 Watergate in Fudai

The Fudai watergate was constructed in 1984 at a distance of about 300 m from the coast (Figs 16 and 17).The height of the gate is TP+15.5 m and the total crest length of the four gates is 205 m. The village itself is located several hundred meters inland from the coast. Fudai experienced tsunamis associated with the 1896 Meiji Sanriku and 1933 Showa Sanriku earthquakes, which had magnitudes of 8.25 and 8.1, respectively5). In 1896, the tsunami struck the village and casualties were severe: 1,010 people were killed. In the Tohoku EQ, the tsunami height was estimated as 20 m, and so the tsunami exceeded the water gates and run-up. Fortunately, no fatalities were reported in the village after the Tohoku EQ.

3.3.2 Seawalls in Taro

The village of Taro has had an unfortunate history concerning tsunami damage. In the 1896 and 1933 earthquakes, 1,859 and 911 people were killed and tsunami heights of 15 m and 10 m, respectively, were recorded. After the 1933 Showa Sanriku tsunami, the village began constructing seawalls in three stages:1934-57; 1962-65; and 1973-78. They are depicted in Fig.11 and Fig. 8

Fig. 16 Fudai water gate (downstream side)

Fig. 17. Satellite image of the Fudai area

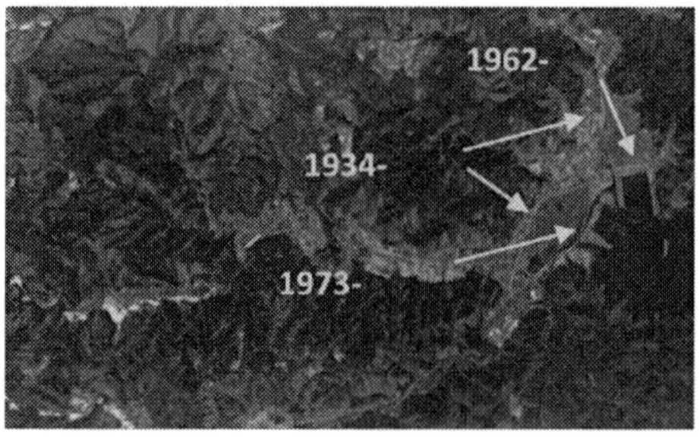

Fig. 18. Location of seawalls in taro

https://doi.org/10.1051/e3sconf/202133107011

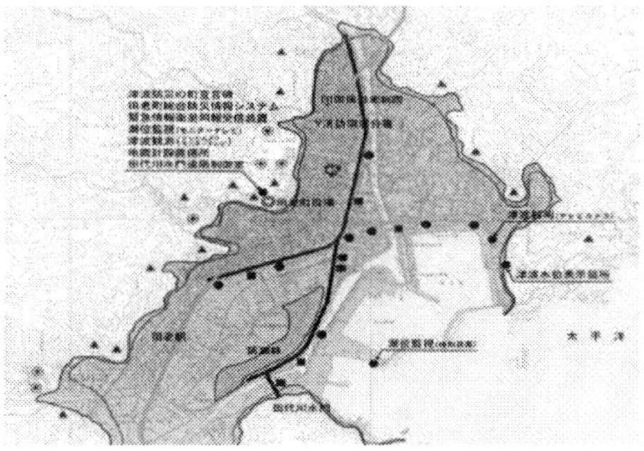

Fig. 19. Location of evacuation shelters in Taro

(a) During Tsunami

(b) After Tsunami (Fig (b) courtesy of Mr. Hatayama, from Sankei New:http://sankei.jp..msn.com)

Fig. 20. Tsunami sweeping over the 10 m high seawall at Taro

Fig. 21. The tsunami washed houses away in Taro

With the Tohoku EQ, a tsunami of over 10 m in height struck Taro, and many residents expected the seawalls to protect them. The tsunami swept over or destroyed the seawalls Fig. 20.

Fig. 21 shows the town after the tsunami. In all, 1,609 houses were destroyed in this area. The totalnumbers of dead and missing people8) for the whole ofMiyako, which comprises six districts including Taro, were 417 and 195, respectively, as of June 17, 2011 [4].

4. Tsunami Countermeasures

The basic hard countermeasures against tsunamis are such structures as water gates, seawalls, and embankments. Even though the tsunami flowed over the watergate in Fudai, the village was protected.

The term "hard" here is defined as structures constructed for the specific purpose of disaster mitigation. Conversely, "soft" indicates information, information-transmission measures, and information systems. An information-acquisition system usually requires hardware, but the main purpose of the system is the transmission of information.

One example of a successful system combining soft and hard measures is an alert system used by Japan's railways called the Urgent Earthquake Detection and Alert System (UrEDAS) [9]. This system includes data-acquisition, decision-making, and electricity shutdown processes. Japan Railways East reported that at the timeof the Tohoku EQ, five Shinkansen trains were running between Fukushima and Iwate prefectures at speeds of around 270 km/h; however, all the trains successfully decelerated and stopped quickly before the main tremor through the UrEDAS system [10]. In the 2003 Sanriku-minami earthquake, many columns of the Tohoku Shinkansen viaduct were severely damaged and retrofitted using steel plate Fig. 22. The bridge columns of the Tohoku Shinkansen sustained severe damage in the Tohoku EQ Fig. 23; however, the retrofitted columns suffered no damage at all.

Recently, the Central Disaster Prevention Council of Cabinet Office, Government of Japan [11], publisheda midterm report regarding proposals for tsunami countermeasures.

Fig. 22. Strengthened viaduct columns after 2003 Sanriku-minami-earthquake

Fig. 23. Damaged viaduct columns of the Tohoku Shinkansen line after the Tohoku earthquake (photo courtesy of Dr. Takahashi, DPRI, Kyoto Univ.)

In the report, some basic ideas for tsunami estimation were proposed. The report proposedthat two levels of tsunami should be considered: (1) thefirst level includes tsunamis that cause extensive damage, although their frequency is very low; (2) the second level includes tsunamis that cause great damage, although their frequency is higher and the tsunami height is lower than with the first level. The first level demands the construction of comprehensive countermeasures, and these cover land-use policies, erecting evacuation buildings, and setting up disaster-prevention facilities combined with key soft measures, such as residents' evacuation.

In Japan, people usually start evacuating during a disaster based on the information given by the JMA. If the tsunami height in the Tohoku EQ had been 3 m, the seawalls of Taro would have successfully protected the village from the tsunami. The accuracy of information is extremely important, but the most difficult issue concerns the combination of soft and hardmeasures. There is a tendency for people to implicitly rely on hard measures once they have been put in place.The residents in Taro put their trust in the 10-m-high seawalls, which were designed to afford protection against the 10-m-high tsunami of the 1933 Showa Sanriku earthquake, even though the village experienced a 15-m-high tsunami in the 1896 Meiji Sanriku earthquake.

Syuto (1992) [12] proposed a relationship between tsunami intensity and damage. In his paper, he pointed out that the lack of accurate data had a great effect on damage to buildings. Supporting hard measures with soft measures demands more accurate information relating to the tsunami height and arrival time than is currently available. In particular, research has to beconducted into the effects of tsunami pressure on different structures.

5 Conclusions

An outline of the destructive Tohoku EQ of March 11, 2011, was presented. From the discussion, the following results emerged.

1) The main causes of death in the Kobe and Tohoku EQs were, respectively, collapsing buildings and drowning in the tsunami. Although the time to death was very short in both cases, people often have more time to evacuate in the case of an interplate earthquake leading to a tsunami.
2) Basic countermeasures against tsunamis include such hard measures as water gates, seawalls, and embankments. Soft measures need to be implemented inareas where hard measures are insufficient.
3) Necessary elements among the soft measures that serve to support hard measures are swift operation, data accuracy and reliability, implementation of disaster plans, and achieving cooperation among residents.
4) For supporting hard measures with soft measures, information relating to tsunami height and arrival time is currently insufficient. In particular, research has to be conducted into the effects of tsunami pressure acting ondifferent structures.

References

1. National Police Agency: Damage Situation andPolice Countermeasures associate with 2011 Tohoku district – off the Pacific Ocean Earthquake, May 10 (2013)
2. Cabinet Office, Government of Japan: Disaster Management New "Bousai", 18, 14-15 (2003) (in Japanese)
3. National Police Agency: Statistics on Causes of Death (2011) (in Japanese)
4. Y. Ohta, M. Koyama, and Y Wato,: Attempt of figuring Life Span Characteristics after an Earthquake -In Case of the Hyougoken-Nanbu Earthquake, Programme, and abstracts, the Seismological Society of Japan 1999(2), 28 (1999)
5. National Astronomical Observatory of Japan: Chronological Scientific Tables (2010) (in Japanese)
6. Japan Meteorological Agency: Flow of issuance of information about tsunami and earthquake, http://www.seisvol.kishou.go.jp/eq/eng/fig/info.html
7. Japan Meteorological Agency: 2011 off the Pacific Coast of Tohoku Earthquake -Portal-, http://www.jma.go.jp/jma/en/2011_Earthquake/2011_Earthquake.html
8. City of Miyako: http://www.city.miyako.iwate. jp, July 2 (2011)
9. Y. Nakamura, Real-time Information Systems for Seismic Hazards Mitigation UrDAS, HERAS, and PIC, Quartery Report of RTRI, **37**, 3 (1996)
10. Sankei News:http://sankei.jp.msn.com/affairs/news/110529/dst11052918010008-n2.htm
11. Central Disaster Prevention Council, Cabinet Office, Government of Japan: Midterm report on tsunami and earthquake, June (2011) (in Japanese)
12. Syuto, N: Tsunami intensity and damage, Reporton Tsunami Engineering research, No.9, Tohoku University (1992) (in Japanese)

Geomagnetic anomaly associated with Fukushima earthquake on February 13th, 2021

Cipta Ramadhani[1*], *Bulkis* Kanata[1], *Abdullah* Zainuddin[1], *Rosmaliati*[1], and *Teti* Zubaidah[1]

[1]Dept Electrical Engineering, University of Mataram

Abstract. In this study, we performed research on electromagnetic anomalies related to earthquakes as early signs (precursors) that occurred in Fukushima, Japan on February 13th, 2021. The research focused on the utilization of geomagnetic field data which was derived from the Kakioka (KAK), Kanoya (KNY), and Memambetsu (MMB) observatories, particularly in the ultra-low frequency (ULF) to detect earthquake precursors. The method of electromagnetic data processing was conducted by applying a polarization ratio. In addition, we improved the methodology by splitting the ULF data (which ranged from 0.01-0.1 Hz) into 9 central frequencies and picking up the highest value from each central frequency to get the polarization ratio. The anomaly of magnetic polarization was identified 2-3 weeks before the mainshock in a narrowband frequency in the range of 0.04-0.05 Hz.

1 Introduction

Earthquakes are one of the deadliest natural disasters that occur suddenly and locally. Early signs (precursors) to detect the occurrence of an earthquake need to be conducted to reduce the hazard. Short-term earthquake forecasting gives promising results which have a time scale of about a week to a month before the event. Short-term earthquake forecasting can be potentially feasible by using geomagnetic disturbance observation instead of solely using seismometers. The weakness of using a seismometer is due to its limited sensitivity in detecting the generation of micro-fractures before an earthquake [1]. Therefore, research on electromagnetic anomalies related to earthquakes as precursors has attracted the attention of many researchers.

Geomagnetic emission in the range of ultra-low frequency provides a tremendous possibility for the detection of earthquake precursors since the emission of geomagnetic fields can propagate over the lithosphere layer of the earth with low attenuation. Thanks to the characteristic of a lithosphere layer that only damps high-frequency components instead of low-frequency components. Previous research has shown that disturbances in the geomagnetic field can be used to detect earthquakes, particularly in the ultra-low frequency (ULF) range [2]. The research began by Fraser-Smith et al [3], which reported the enhancement of geomagnetic noise at ULF 3 hours before the 1989 M7.1 Loa Prieta earthquake. Hayakawa et al [4] also reported that the anomalous variation in the vertical component of the geomagnetic field appeared 1 month before the 1995 Guam earthquake. Since then, geomagnetic earthquake forecasting studies have grown significantly.

Several studies have been conducted to monitor the activity of the earth's crust linked to electromagnetic anomalies within the limit of ULF ($f<10$ Hz) in terms of earthquake precursors. Previous studies have successfully shown that the ULF range within 0.01-0.1 Hz gave the best results for detecting electromagnetic anomalies in terms of earthquake precursors [5, 6, 7]. However, the ULF emission that can be assigned as electromagnetic anomalies has sometimes been problematic because not only is the intensity very weak (around 1nT) but it is usually difficult to understand as well due to the intense natural background of the geomagnetic field [8]. To distinguish between the emission and the background, polarization analysis has been applied to separate the vertical component of the geomagnetic field from the horizontal component [9, 10]. The Comparative Polarization signal for the H and Z components separates the ULF emission of solar wind, magnetosphere, ionosphere, or lithosphere and is derived from seismic origins [11].

The Geomagnetic Kp index is one of the most extensively used indices of geomagnetic activity. It has been used to monitor sub-auroral geomagnetic disturbance on a global scale which can indicate the occurrence of magnetic storms within the span of a day. For more than 70 years, the Kp index and derived products have shown their advantages, particularly for space weather research and service. The data set's integrity and future scientific and societal utilization are secured by a DOI [12] and the CC BY 4.0 license.

* Corresponding author: cipta.ramadhani@unram.ac.id

Table 1 The geographical positions of KNY, MMB, and KAK observatories

IAGA Code	Latitude	Longitude	Elevation(m)	Distance to EQ (km)
MMB	43.91 'N	144.19 'E	42	716
KAK	36.23 'N	140.18 'E	36	218
KNY	3.42 'N	130.88 'E	107	1217

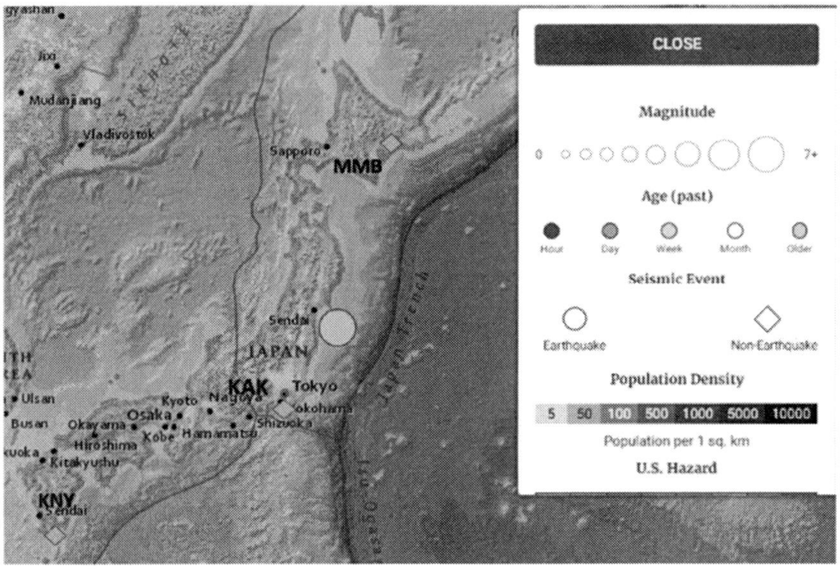

Fig. 1. The relative positions of KNY, MMB, and KAK observatories to the Earthquake (denoted by circles) that occurred on February 13th, 2021

In this paper, we tried to detect ULF geomagnetic precursors associated with the Fukushima earthquake on February 13th, 2021. The ULF data were derived from three different observatories from KAK, KNY, and MMB, which were closed from the earthquake event. The data from those observatories will be compared to each other to detect anomalies related to the earthquake precursor.

2 Methodology

2.1 Geomagnetic and Earthquake Data

The Fukushima earthquake with a magnitude of M7.1 occurred on Saturday, February 13th, 2021 at 14:07:49 (UTC) and has been reviewed by seismologists. The epicenter was located at 37.727°N, 141.775°E, at a depth of 44 km (27.3 miles). In the region near the east coast of Namie, Japan, 109 km (59 miles) ENE of Fukushima, Honshu, Japan.

The data was collected from three INTERMAGNET observatories in Japan, i.e. Kanoya (KNY), Memambetsu (MMB), and Kakioka (KAK). Table 1 shows the locations of the three observatories and their distances from the epicenter of the Fukushima earthquake, whereas Fig. 1 shows the epicenter and their locations of the Fukushima earthquake.

To reduce the influence of man-made noise, a shorter nigh local time of the geomagnetic data occurring between 00:00 to 00.03 PM (15:00 to 18:00 UTC) is used. For the analysis, a year's worth of statistical data is gathered, with seven months previous to and four months after the earthquake.

2.2 Frequency Range of ULF signal and Polarization Analysis.

2.2.1 Frequency Range of ULF Signal

In acquiring the frequency spectrum of the ULF signal, the Fast Fourier Transform (FFT) was applied with nine narrowband frequency ranges between 0.01 Hz and 0.10 Hz, which were $\Delta f = 0.01 - 0.02, 0.02 - 0.03, \ldots, 0.09 - 0.10$ Hz [13]. The ULF signal in that narrowband frequency range is acquired by a bandpass filter. Furthermore, each selected range will be observed separately with polarization ratio analysis.

2.2.2 Polarization Analysis

The polarization ratio analysis approach is according to the observation of a daily precursor parameter called $P_{Z/H}$, which is the ratio of power spectral density (PSD) of the vertical component $S_Z(f)$ to PSD of the total horizontal component $S_H(f)$. Both vertical and horizontal components were calculated using equation (1).

$$S_Z(f) = \frac{|B_Z(f)|^2}{\Delta f}$$
$$S_H(f) = \frac{|B_H(f)|^2}{\Delta f} \quad (1)$$

where the highest value of ULF signals of $B_Z(f)$ and $B_Z(f)$ is selected in each narrow band. In addition, variable Δf is the difference between $f_{up\ cut\text{-}off}$ and $f_{low\ cut\text{-}off}$.

Furthermore, both vertical and horizontal components were normalized to certain ranges using equation (2). This equation was introduced by Yusuf, et al. [13] that successfully identified possible earthquake precursors in the equatorial and mid-latitudinal regions.

$$S_{X,norm} = \left(S_X - S_{X,max}\right)\left(\frac{M-m}{S_{X,max}-S_{X,min}}\right) + M \quad (2)$$

In the equation, X can be either Z (vertical) or H (horizontal) components, $S_{x,\ norm}$, and S_x are the normalized and authentic daily values of PSD of components X, respectively. $S_{X,\ max}$, and $S_{x,\ min}$ are the maximum and minimum PSD values of component X over the observation period. The terms M and m are upper and lower boundaries of the selected range, respectively; they are written as [m, M] for the rest of this paper. Based on the normalization process introduced by Yusuf et al. [13], the most effective way to detect precursor anomalies is to use [1,2] for S_H and [1,3] for S_Z as an upper and lower boundary respectively. The combination of the range values is useful because it not only prevents any daily value of S_H from being 0, thus avoiding an infinitely high $P_{Z/H}$, but also gives S_Z a higher proportion in determining the value of the corresponding $P_{Z/H}$ because Z is the most affected component due to crustal activities.

The next step is to calculate the ratio of $S_z(f)$ and $S_H(f)$ field components using equation (3).

$$P_{Z/H}(f) = \frac{S_z(f)}{S_H(f)} \quad (3)$$

Furthermore, to consider the increase of $P_{Z/H}$ anomalous in a given day in terms of statistics, its value must be greater than two standard deviations (σ) from the mean (μ) of the entire observation period:

$$P_{Z/H} > \mu + 2\sigma \quad (4)$$

2.3 The Kp Indices of Geomagnetic Activity.

To clearly explain the magnitude of geomagnetic activity in a single day, the Kp index provides information on the occurrence of magnetic storms within the period of each day. The Kp index is computed as a consequence of the impact of solar and other cosmic events. If the Kp index ≥ 5, the range value represented a geomagnetic storm; otherwise it is classified as calm. During a single day, the magnetic field recorded at the observatory can be interfered with by a high Kp index when a geomagnetic storm occurs.

3 Result and Discussion

In this section, we present the results of the polarization ratio analysis for the earthquake that occurred in Fukushima on February 13th, 2021. Fig. 2(a) represents the horizontal component ($S_H(f)$ of the Kakioka observatory. Meanwhile, Figs. 2(b) and 2(c) represent the vertical component ($S_Z(f)$) and polarization ratio of $P_{Z/H}$, respectively. The Kp Index, which is used to determine the increase or decrease of the intensity of the magnetic field variations caused by solar activity or other phenomena, as shown in figure 2(d). The red shades denote disturbed periods where the Kp index surpassed the respective disturbed day thresholds (blue dashed lines in the figures). As previously stated, any anomalous $P_{Z/H}$ values during these periods were ignored. Furthermore, the content in figures 3 and 4 is identical to figure 2. However, they represent MMB and KNY observatories, respectively.

In all nine frequency ranges, the preliminary observation of normalized $P_{Z/H}$ was carried out. Nevertheless, they were not shown here. Instead, our focus was only on one frequency range that had an anomalous daily value and had the highest normalized value compared with the other frequency ranges.

By referring to figure 2, it was found that the range of 0.04-0.05 Hz had an apparent anomaly at the KAK observatory on 24 January 2021 (18 days before the Fukushima earthquake) with the intensity values as computed in Table 2. The normalized power spectral densities of $P_{Z/H}$ increased with a ratio of 2.3 (inside the red circle) and sigma Kp values of 10 (mean of Kp index=1.25), which means that there were no geomagnetic storms occurred throughout the anomalous period (no red shade denoted). It is acknowledged that there was an apparent second anomaly on January 11th, 2021 (31 days before the Fukushima earthquake). The normalized power spectral densities of $P_{Z/H}$ increased with a ratio of 2.16 (inside the red circle) and sigma Kp values of 18 (mean of Kp index=2.25). However, it was not considered as an earthquake precursor (it will be explained later). Both anomalous were seen higher than the respective disturbed day thresholds (blue dashed lines in the figures).

In the KNY observatory that appears in figure 3, it was anomaly signals with intensity values in the range of 0.04-0.05 Hz. Similar to the KAK observatory, it is acknowledged that there was an apparent anomaly on January 11th, 2021 with the intensity values as computed in Table 2. The normalized power spectral densities of $P_{Z/H}$ increased with a ratio of 2.39 (inside the red circle) and sigma Kp values of 18 (mean of Kp index=2.25). On the other hand, compared to the KAK observatory, it is acknowledged that there was a slight increment on January 24, 2021, that exceeded the respective thresholds (18 days before the Fukushima earthquake); nevertheless, but the excesses were minimal to meet the second criteria of being a precursor.

https://doi.org/10.1051/e3sconf/202133107012

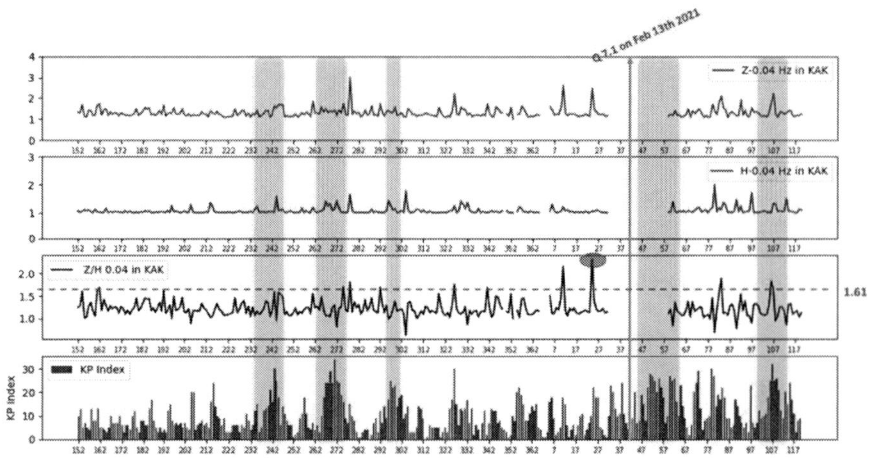

Fig. 2. Shows Power spectral densities of MMB observatory after normalization in the range of 0.04-0.05 Hz from June 1st, 2020 – May 30th, 2021, related to the M7.1 Fukushima earthquake on February 13th, 2021

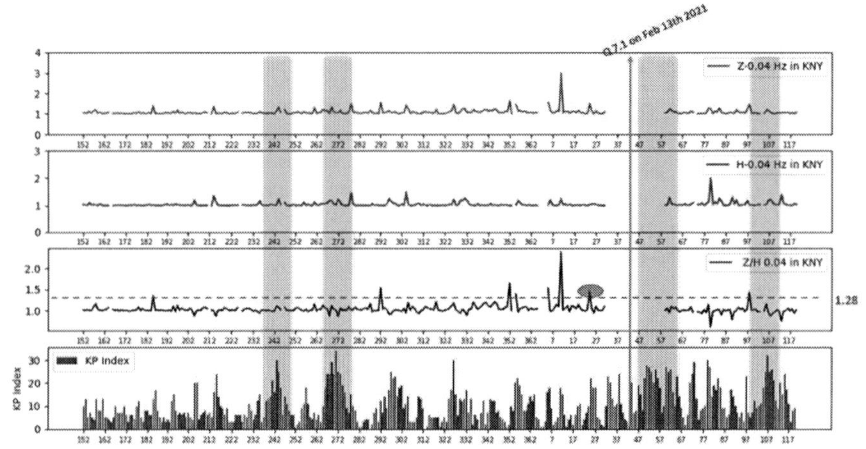

Fig. 3. Shows Power spectral densities of KNY observatory after normalization in the range of 0.04-0.05 Hz from June 1st, 2020 – May 30th, 2021, related to the M7.1 Fukushima earthquake on February 13th, 2021

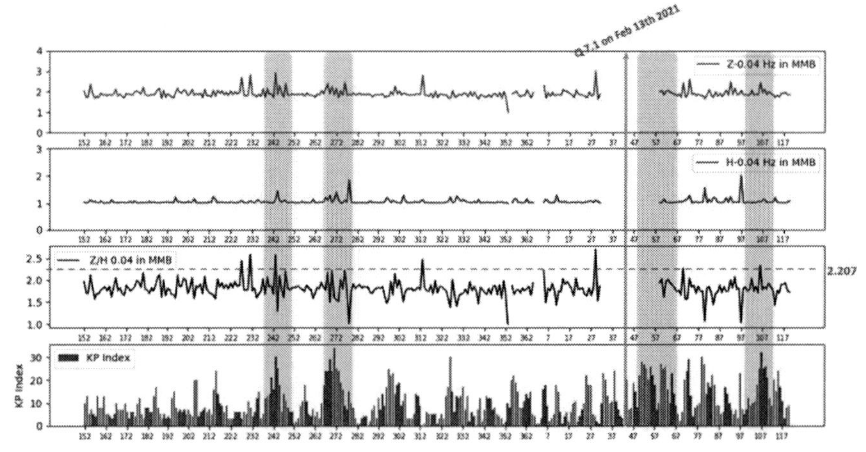

Fig. 4. Shows Power spectral densities of MMB observatory after normalization in the range of 0.04-0.05 Hz from June 1st, 2020 – May 30th, 2021, related to the M7.1 Fukushima earthquake on February 13th, 2021

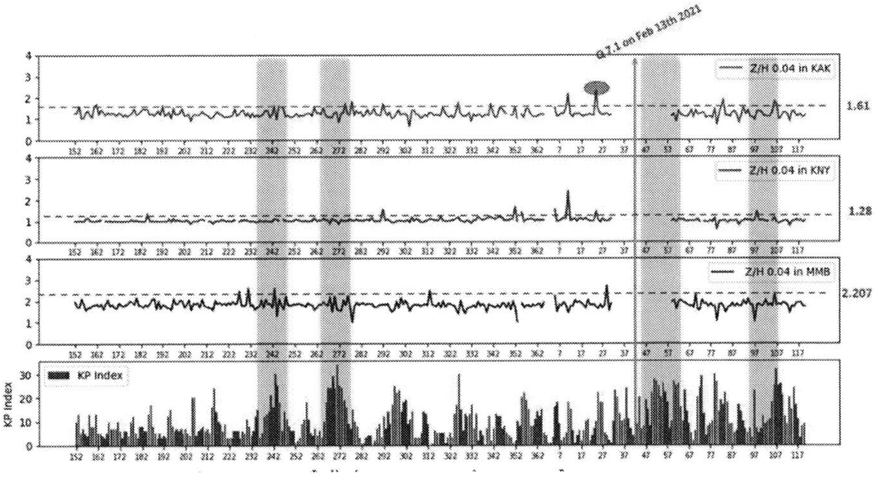

Fig. 5. Polarization ratio $P_{Z/H}$ of all observatories at the range of 0.04-0.05 Hz from June 1st, 2020 – May 30th, 2021, related to the M7.1 Fukushima earthquake on February 13th, 2021

Table 2 Characteristics of the identified anomalies preceding the main earthquakes obtained using normalized polarization ratio analysis at the range of 0.04-0.05 Hz

IAGA Code	Date of appearance	Lead time (days)	Normalized Value	Threshold	Sigma Kp
KAK	11/01/2021	31	2.162	1.61	18
	24/01/2021	18	2.322		10
KNY	11/01/2021	31	2.394	1.28	18
	24/01/2021	18	1.444		10
MMB	29/01/2021	13	2.695	2.207	22

Referring to figure 4, there were no anomalous signals at the MMB observatory on the same days that they appeared in KAK or KNY. However, it is acknowledged that there was a slight increment on January 29, 2021, with the intensity values as computed in Table 2. The normalized power spectral densities of $P_{Z/H}$ increased with a ratio of 2.6 and sigma Kp values 4 (mean of Kp index=0.2). Nevertheless, its value exceeded the respective thresholds (threshold =2.207), but the excesses were minimal to qualify as a precursor.

Figure 5 presented the results of all three observatories. Based on data, The KAK observatory provides the most visible indicators before the earthquake. This result implies that the closer an observatory is near the epicenter, the more anomalous data is recorded [14]. Based on these results, we suggest that the precursors can be noticed around three weeks (18 days) before an earthquake occurs. Based on the previous explanation, we refuse to admit the second anomaly from KAK and KNY observatories that appeared on January 11th, 2021 as an indicator before the Fukushima earthquake. Since The anomaly that appeared in both observatories did not appear in the MMB observatory. MMB observatory is closer than the KNY observatory. Therefore, if the anomaly appeared in the KNY observatory, it also appeared in the MMB observatory.

Despite the result, data for 27 days is not available from 1st February 2021 to 27th February 2021. Therefore, this research needs more evidence to make sure that the anomalies were relevant to being earthquake precursors.

4 Conclusion

This research discovered anomalous geomagnetic signals by using normalized polarization of Z/H in the narrowband frequency range of 0.04–0.05 Hz as a precursor to the Fukushima earthquake on February 13th, 2021. The anomalies are visible 2-3 weeks before the earthquake. In addition, the normalization process that was introduced by Yusof was enhanced in detecting ULF geomagnetic earthquake precursors by providing a significant proportion to the vertical component, since the vertical component is more affected by earthquake-related underground activities.

This research is funded by Universitas Mataram under Hibah Dana PNBP 2021 program entitled "Penerapan Metode Polarisasi Magnetik Untuk Analisis Prekursor Pada Kejadian Gempa Berskala Besar Di Fukushima Jepang". The data that has been presented in this paper is collected from magnetic observatories. Our appreciation is presented to the national institutes and INTERMAGNET as their effort to promote high standards of magnetic observatory practice.

References

1. M. Hayakawa, *Earthquake prediction with electromagnetic phenomena*. AIP Conference Proceedings,**1709**, (2016)
2. F. Febriani, P. Han, C. Yoshino, K. Hattori, B. Nurdiyanto, N. Effendi, I. Maulana, Suhardjono, E. Gaffar, *Ultra-low frequency (ULF) electromagnetic anomalies associated with large earthquakes in Java Island, Indonesia by using wavelet transform and detrended fluctuation analysis*. Natural Hazards and Earth System Sciences, **14 (4)**, (2014)
3. A.C. Fraser-Smith, A. Bernardi, P.R. McGill, M.E. Ladd, R.A. Helliwell, O.G. Villard, *Low-frequency magnetic feld measurements near the epicenter of the Ms 7.1 Loma Prieta Earthquake*. GEOPHYSICAL RESEARCH LETTERS, **17 (9)**, 1465–1468, (1990)
4. M. Hayakawa, R. Kawate, O.A Molchanov, K. Yumoto, *Results of ultra-low-frequency magnetic feld measurements during the Guam earthquake of 8 August 1993*, Geophys Res Lett, **23**, 241–244 (1996)
5. C.H. Chen, C.H. Lin, C.H. Wang, J.Y. Liu, T.K. Yeh, H.Y. Yen, T.W. Lin, *Potential relationships between seismo-deformation and seismo-conductivity anomalies*. Journal of Asian Earth Sciences, **114**, 327–337 (2015)
6. G. Rawat, V. Chauhan, S. Dhamodharan, *Fractal dimension variability in ULF magnetic field with reference to local earthquakes at MPGO, Ghuttu*. Geomatics, Natural Hazards, and Risk, **7 (6)**, 1937–1947,(2016)
7. K.A. Yusof, M. Abdullah, N.S.A. Hamid, S. Ahadi, *On effective ULF frequency ranges for geomagnetic earthquake precursor*. Journal of Physics: Conference Series, **1152 (1)**, (2019)
8. N.R. Zalinskiy, N.G. Kleimenova, O.V Kozyreva, S.M. Agayan, S.R. Bogoutdinov, A.A. Soloviev, *Algorithm for recognizing Pc3 geomagnetic pulsations in 1-s data from INTERMAGNET equatorial observatories*. Izvestiya, Physics of the Solid Earth, **50 (2)**, 240–248 (2014)
9. J.L. Currie, C.L. Waters, *On the use of Geomagnetic indices and ULF waves for earthquake precursor signatures*, Journal of Geophysical Research A: Space Physics, **119 (2)**, 992–1003 (2014)
10. K.A. Yusof, N.S.A. Hamid, M. Abdullah, S. Ahadi, A. Yoshikawa, *Assessment of signal processing methods for the geomagnetic precursor of the 2012 M6.9 Visayas, Philippines earthquake*, Acta Geophysica, **67 (5)**, 1297–1306 (2019)
11. K. Yumoto and The MAGDAS Group, *MAGDAS Project and Its Application For Earthquake Prediction*, Proceedings of the International Workshop on Integration of Geophysical Parameter as a Set of Large Earthquake Precursors, Research and Development Center BMG, Jakarta (2006)
12. J. Matzka, O. Bronkalla, K. Tornow, K. Elger, C. Stolle, Geomagnetic Kp index, GFZ German Research Centre for Geosciences, 11 March, Potsdam, Germany,(2021)
13. K.A. Yusof, M. Abdullah, N. Shazan, A. Hamid, S. Ahadi, A. Yoshikawa, *Normalized Polarization Ratio Analysis for ULF Precursor Detection of the 2009 M7.6 Sumatra and 2015 M6.8 Honshu Earthquakes*, Jurnal Kejuruteraan, **3 (1)**, 35-41 (2020)
14. B. Kanata, T. Zubaidah, C. Ramadhani, B. Irmawati, *Changes of the geomagnetic signals linked to Tohoku earthquake on March 11th, 2011*, International Journal of Technology, **5 (3)**, 251–258 (2014)

State of the art of seismic risk and loss assessment in structures

Riza Aryanti[1] and *Masrilayanti* Masrilayanti[1*]

[1]Andalas University, Electrical Engineering Department, 25163 Padang, West Sumatera - Indonesia

Abstract. Earthquakes are known as one of the disasters that have fatal consequences for human safety. However, inevitably, the earthquake itself is not the leading cause of the losses suffered by humans, both material and soul. The most powerful thing in human safety is infrastructure such as buildings, bridges, and houses. Therefore, an in-depth analysis of the risk factors that the infrastructure will experience in a natural disaster is needed. There is a variable seismic hazard in the Southeast Asia region, ranging from high seismic hazard allied with the underneath of the Indonesian and Philippine archipelagos to moderate and low seismic tremors associated with a sizeable stable region on the Sunda Shelf. This paper describes the history of seismic risk and loss assessment of infrastructures. The method used is by doing literature reviews of the most recent research relating to seismic risk and assessment around the world. More than fifteen research results are studied and discussed to get a deep knowledge about seismic risk and the assessment of loss due to seismic disaster.

1 Introduction

Earthquakes are known as the most dangerous natural disaster in history. Many victims and losses have occurred due to earthquakes, and because of that, research and studies have been conducted. Therefore this topic has still become trending. Along with technological advances, research on seismicity is also shifting from conventional conditions to current conditions. Much progress has been made to date, marked by reduced casualties and losses.

Engineers say that earthquakes are not dangerous. The problem comes from the structural conditions, which are not well designed and prepared. Many provisions also cause structures' failures since the provisions in many countries are not updated to the most recent research findings and technology development.

Fajfar [1], in his paper, has discussed the development of seismic provisions and possible further developments. He agrees that the essential part of seismic design and evaluation is the analysis of structures. It was begun more than a decade ago when the seismic regulations adopted 10% of the weight of structures to be considered in static analysis. This regulation was practiced for a long time in seismic codes and provisions. Eventually, by the knowledge and technology improvement, more state of the art of analysis strategies and methods were resolved and conducted, and dynamics and nonlinear analysis were considered. It is seen that from this point forward, probabilistic attention might be adopted as a preference [1]. More comprehensive evaluation could be conducted, for example, are economic efficiency and construction effectiveness.

Mejia exposes the failures that occurred in buildings during an earthquake using the study case in Popayan Earthquake in Columbia in 1983. The scale of the earthquake itself was 5.5 by an approximate depth of 9 km. It lasted for 16 seconds, and the location of the hypocentre of Papayan was 5 km. The earthquake research becomes increasingly sophisticated, and as a result, the codes and regulations too. First of all, the causes of these are because its purpose is for evaluating everything in formulas even where nature cannot be and no matter whether inaccurate assumptions support investigations. These complexities sometimes lead to losing the way because there are lots of hesitation and doubt. The result of this is the failures of the structure [2].

2 Causes of Seismic Failures

Earthquakes have occurred since this world existed. Earthquakes are the process of releasing energy due to faults in the earth's core. Earthquakes are not dangerous, but what is deadly is the debris and events that occur after the earthquake. This type of disaster is closely related to humans because it has occurred since the earth was created. Humans can shelter from thunderstorms, but they cannot protect themselves from earthquakes. Therefore, all things that are closely related to earthquakes, including seismology, basically have a long history [3][4]. Natural disasters like earthquakes cannot be prevented or predicted. Fortunately, by developing analytical and evaluation methods, structural damage can be reduced, for example, micro-zone studies, appropriate construction methods, and infrastructure that are well designed, and

* Corresponding author: masrilayanti@eng.unand.ac.id

earthquake-resistant designs. Examples of damage caused by earthquakes are ground shaking, structural damage, failure of retaining structures, and secondary hazards such as fire and liquefaction. The magnitude of the effect felt by humans depends on several things, such as the source of the earthquake, the path, the type of structure, and the conditions around it, and the readiness of the community at the scene. Meanwhile, several things that affect the intensity of soil movement include topography, groundwater, and surface hydrology [3].

In seismic hazard microzonation analysis, the methodologies that are often used are probabilistic seismic hazard assessment (PSHA) and deterministic hazard assessment (DSHA) [4]. Probabilistic Seismic Hazard Analysis has been widely used throughout the world although it has not been objectively validated. For more than 50 years, this method has been used by many parties such as government and industry, to save lives and reduce material losses. Some of the things that are often done with this method are: deciding the regulation of safety criteria for nuclear power plants, making official decisions on national hazard maps, developing building code requirements, and determining earthquake insurance rates. However, it should be noted that no method can calculate with certainty the estimated seismic hazard that will occur. Therefore, even though it has been properly analysed by experienced experts, there are still many earthquakes that cause huge loss of life and property [5].

Various studies have been carried out to find the cause of the seismic collapse in structures. In its application, a structure can be designed and analyzed with the concept of uniform hazard. But in reality, the structural failure that occurs may not experience the same collapse. Different structures are designed and analyzed using the same peak ground acceleration (PGA), but the possible risk of damage that occurs may not be the same, especially in the shape of the hazard curve and uncertainty in structural capacity. Of course, it depends on the condition of the structure, and the condition of the soil around the building [6].

Yon Buraq et a.l [7] stated in their paper that the damage caused by moderate and large earthquakes continues. They found that the cause of structural failure which then resulted in the loss of life and material was poor structural design, for example not complying with the rules of strong-weak-beam columns, short columns, imperfect supports, and wall movements which then caused damage. In addition, the quality of the materials is also poor, the quality of the workers is not good, the lack of technical services and the construction with insufficient details of the structural elements are other reasons for the damage. In addition, other causes are damage to the building from the masonry in terms of design errors which can be shown as heavy earth roofs, incorrect details of wall-to-wall and wall-to-roof joints, absence of tie beams, large openings. However, the construction of the building using poor quality local materials based on traditional rules is another reason for the failure of this building [7].

Many researchers investigate other causes of seismic failures of buildings—for example, the research conducted by Yon et al. in 2017. The researchers considered that Turkey has a high risk of earthquake disasters and evaluated the failure buildings in Istanbul. The city of Istanbul is well-known as an earthquake-prone location and has suffered a lot of damage from the large earthquakes that hit it. Some places such as the Fatih and Eminonu districts which are part of Old Istanbul are commercial centers and historical places with a large population. The researchers attempted to analyze damage data from past earthquakes along with soil amplification in the region. Likewise, the history of liquefaction and the risk of slope stability. For this purpose, a seismic microzonation map has been prepared using a geographic information system. There is a relationship between soil behavior and the damage caused by previous earthquakes. Structures in the area are damaged according to the microzonation map. This has no impact on the level of damage as slope stability is not a high risk in the area. However, it was determined that soil amplification and partial liquefaction contributed to the destruction of historical artifacts and structures [8].

The earthquake itself, of course, had a role previously associated with structural failure. Many studies have also been carried out for other structures such as bridges. Masrilayanti et al. [9] investigated the behavior of the cable-stayed bridge superstructure due to dynamic earthquake loads in the form of a time history with uniform conditions on each support. A long-span bridge was tested to obtain different reactions due to earthquake loads with uniform soil conditions. Three types of soil were used to create the response spectrum, namely hard soil, soft soil, and hard soil. The results show that the reaction of the superstructure of the bridge also varies depending on the type of soil used. [9].

To prioritize seismic risk mitigation at a large scale, rough-input methodologies are required to timely and conventionally assess the seismic risk associated with the portfolio of interest. Based on the sheer volume of information and computational effort required, it is generally only feasible to evaluate seismic vulnerability at the regional level using mechanics-based methods for a portion of the portfolio, selected according to prioritization criteria. In the literature and some guidelines, simple indices have been proposed as a convenient way to assess seismic risk, primarily by comparing the code requirements at the time of design with current seismic demands. In essence, these indices attempt to define a relative seismic risk measure that can be used for a rapid ranking to identify the part of the portfolio that merits further study. The advantage of this risk metric is that it does not require complex data, in other words, it only requires data that is easy to obtain, for example, the year when the structure was designed and the location. This makes this method easy to apply to the risk analysis industry. In addition, although traditionally, hazards and vulnerabilities can be taken into account, they can even be adjusted so that they are following what is targeted by the community. Petruzzelli et al. in their research took part in the discussion on this subject, on how to develop a large-scale seismic assessment strategy for disaster management purposes [10].

By analysing the vulnerability function, the losses that may be generated by an earthquake with a certain intensity can be described. This is part of the seismic risk assessment analysis. Research conducted by Martins et al. provides an overview of the methodology of fragility and vulnerability analysis. In this case, several things such as the spread of uncertainty, validation/verification of results, and the efficiency of the type of intensity used were also presented by the researcher. This is part of the vulnerability analysis. Martin et al. created seven modules that aim to make it easier for users to conduct vulnerability analysis, starting from choosing the earthquake intensity to be used, to model verification [11].

3 Method for Risk and Loss Assessment

It is the probability of a harmful outcome (death, injuries, property, livelihoods, economic disruption, or environmental damage) caused by natural or human-induced hazards acting in concert with vulnerable conditions.

In analysing risk, it is very important to combine all the things related to the risk aspect so that it can estimate it accurately and accurately. So, seismic risk analysis is not enough but needs to be supported by social, economic, and cultural norms as well. This risk assessment is also not the end of the analysis but can be continued with an analysis of management and allocation of appropriate and efficient resources to reduce risk to a level that is acceptable to all parties, especially the community [12].

A study conducted by Bradley et al. [13] suggested that to be able to estimate seismic losses, several things that must be analysed are a seismic hazard, structural response, damage fragility, and the consequences of damage so that in the end a seismic risk quantification is obtained. Bradley et al. [13] conducted a study related to the methodology for estimating losses of various sizes for a particular facility. Some things that can be included as inputs to this estimate are direct costs of structural and non-structural repairs, business interruption, and injury to building occupants, although the latter two are often not directly considered. The seismic loss estimation process also makes it possible to calculate the seismic risk of engineered structures, making it possible for analysts and decision-makers to communicate consistently and rationally about seismic risk acceptance and mitigation. Even from this vulnerability analysis, the interpretation of the effect of seismic performance and economic loss can be related to a function of ground motion intensity [13].

Risk can be introduced theoretically with the accompanying essential condition demonstrated in Fig. 1.

Hazard can be introduced in more ways than one. It tends to be communicated in outright or relative terms. Populace hazard can likewise be communicated as individual danger (the yearly likelihood of a solitary uncovered individual) and cultural danger (the connection between the yearly likelihood and the number of individuals that could be a financial danger). It very well may be communicated as far as Average Annual Loss, Maximum Probable Loss, or other determined from a progression of misfortune situations, each relating recurrence, and anticipated movement.

A few dangers might happen in chains: one risk causes the following. These are additionally called domino effects or concatenated hazards. These are the most dangerous sorts to dissect in a multi-peril hazard evaluation. The best methodology for investigating such peril affixes is to utilize supposed event trees.

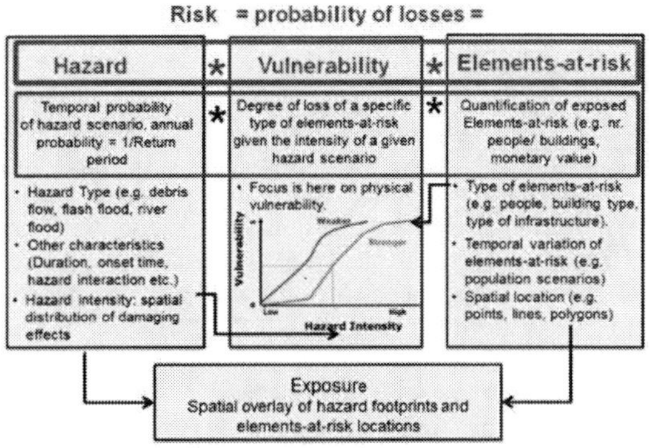

Fig. 1 Schematic portrayal of hazard as the duplication of danger, weakness, and evaluation of the uncovered components in danger [14].

An event tree is a framework that is applied to break down every one of the mixes and the related likelihood of boundaries' event that influence the framework under investigation. Every one of the dissected occasions is connected to other utilizing hubs, which is all potential conditions of the framework are considered at every hub, and each state (part of the occasion tree) is portrayed by a characterized worth of the likelihood of an event, as can be found in Fig. 2 [14].

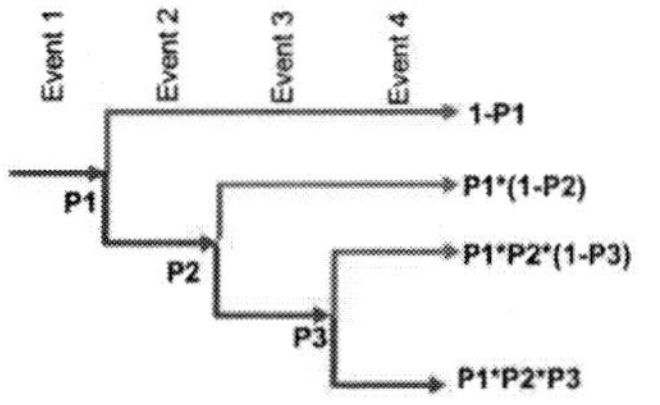

Fig. 2 Schematic portrayal of an Event-Tree examination [14].

Probabilistic Seismic Hazard Analysis (PSHA) coordinates overall conceivable quake ground movements at a site to foster a composite portrayal of the otherworldly amplitudes and dangers (yearly frequencies of exceedance) at that site. The investigation has a solid premise in studies of the quake designing and permits choices on seismic plan levels for an office to be made with regards to the tremor extents, areas, and ground movements (remembering the impacts of neighborhood site conditions for amplitudes of solid shaking) that might happen. The utilization of PSHA is normal all through the world for deciding seismic plan levels. The joint effort of two analysts during the 1960s brought about the key ideas of PSHA [15].

Intelligent use of PSHA is to decide seismic plan levels for construction regulations that reflect predictable yearly probabilities of exceedance (or proportionately the return time of a predefined level of ground movement). The initially distributed seismic zone map that included degrees of ground movement and related return periods was Esteva. The distribution depended on appraisals of repeat time frames sizes in seismic zones, constricting the ground movement to assess power in different zones, and comparing the ground movement return periods to the quake repeat spans. The accentuation of Esteva, notwithstanding, was on the general designing choice interaction for tremors, not simply on the outrageous ground movement circulation. These ideas were distributed by Esteva and in his Ph.D. theory. They incorporated the Bayesian refreshing of seismic action rates, ground movement conditions for PGA, PGV, and PGD, the induction of plan spectra from these qualities, and seismic danger bends for otherworldly ordinates that were scaled from PGA, PGV, and PGD. It was the main distribution of danger bends for otherworldly ordinates. Two essential parts of Esteva's work were especially striking. In the first place, ground movements were perceived to have critical vulnerability related to their events. This vulnerability was measured, and the presentation of the term vulnerability was 20 years in front of its broadly acknowledged use. Second, choice guidelines for the seismic plan were created utilizing underlying disappointment as the choice variable. These choice principles represented limiting future expenses and advantages by a money-related markdown rate and thought about whether designs were to be modified after quake instigated disappointment or supplanted. The choice standards likewise represented various potential degrees of harm during quake movements and noticed that plan choices rely just upon the normal pace of primary disappointment, not on its vulnerability [15].

Fajfar [2] states in his distribution that after computers opened up, i.e., in the last part of the 1960s and the 1970s, the quick advancement of techniques for seismic examination and supporting programming was seen. These days, because of gigantic advancements in processing power, mathematical strategies, and programming, there are no restrictions identified with calculation. Shockingly, information about ground movement and primary conduct has not kept up a similar speed, particularly in the inelastic reach. Additionally, it can't be anticipated that, as a rule, the fundamental abilities of architects will be better compared to previously. Today, the absence of solid information and the restricted capacities of creators address the failure point in the chain addressing the planning cycle, instead of computational apparatuses, just like the case before [2].

The seismic performance can be shown by analyzing the structure's vulnerability in resisting an earthquake motion and then developing into fragility curves. Fragility curves are found to be valuable tools for predicting the extent of probable damage. They show the probability of highway structure damage as a function of strong-motion parameters, and they allow the estimation of a level of damage probability for a known ground motion index. Karim and Yamazaki [16] conducted an analytical approach to developing the fragility curves for highway bridges based on numerical simulation. Four typical reinforced concrete bridge piers and two reinforced concrete bridge structures were considered, of which one was a non-isolated system and the other was an isolated system, and they were designed according to the seismic design code in Japan. From a sum of 250 fundamental movement records chosen from Japan, the United States, and Taiwan, nonlinear time history examinations were performed, and the damage indices for the bridge structures were obtained. Damage indices and ground motion parameters were used to construct fragility curves for the four bridge piers and the two bridge structures by assuming a log-normal distribution. It was observed that there was a critical impact on the fragility curves because of the variety of primary boundaries. The relationship between the fragility curve parameters and the over-strength ratio of the structures was also obtained by performing a linear regression analysis. The connection between the fragility curve boundaries and the over-strength proportion of the constructions was additionally acquired by playing out a direct relapse examination. In light of the noticed connection between the fragility curve boundaries and the over-strength proportion of the designs, an improved strategy was created to build the fragility curves for highway bridges using 30 non-isolated bridge models. The improved strategy might be a useful instrument to develop the fragility curves for non-isolated highway bridges in Japan, which fall inside a similar gathering and have comparable attributes [16].

Masrilayanti et al. [17] in their paper present an advantageous technique to set up the fragility curves for the cable-stayed bridge. For this reason, three spans cable-stayed bridge is surveyed utilizing a progression of seismic loads in various forces to guarantee that the construction was encountering harm in a few conditions. The fragility curve was gotten by dissecting the design utilizing Nonlinear Time History (NTHA) and Pushover Analysis. The tremor's ground movements were exposed to the extension in various intensities, which were scaled from the underlying ground movement. In the future, the construction's ductility were formed into the fragility curves as the reactions of the extension. HAZUS standard is utilized for characterizing the harms of the scaffold, which are assembled into; slight, moderate, extensive, and complete due to the seismic load. The upsides of the harm states were created to the fragility curves utilizing the probabilistic upsides of the harm states. The outcome uncovered that the fragility curve was qualified as the log-normal distribution [17].

Fig 3 describes the probabilities of a cable-stayed bridge having a different level of failure. There are four curves obtained by fragility analysis, and each of the curves describes a different percentage of prediction of the bridge performance. The blue line is for slight damage, which means the bridge will only have minor disturbance when the earthquake occurs in the various value of PGA. The orange line is for moderate damage, the red is for extensive damage, and the purple line describes the fatal damage of the structure. From the graphs, it can be predicted the damage level and the value of its percentage. For example, if the GPA is about 0.6, it can be seen from Fig.3 that the probability of its slight damage is about 98 percent, 90 percent of moderate damage, about 78 percent of extensive damage, and for complete damage, it is about 25 percent. By this analysis, the structure owner can

https://doi.org/10.1051/e3sconf/202133107013

prepare to have a better structure and avoid the loss and victim.

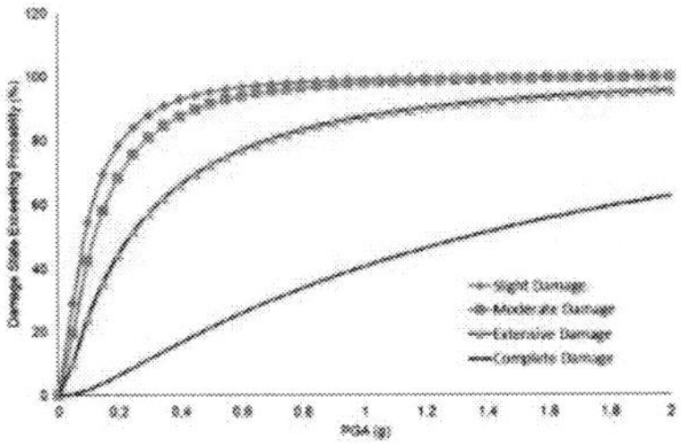

Fig. 3 Fragility curves of a cable-stayed bridge.

In the end, everything will be regulated in building and mitigation standards issued by the government. Almost all countries worldwide are very concerned about this natural disaster and have made regulations regarding earthquake maps, earthquake-resistant structures, and disaster mitigations. As with other countries, the United States has also frequently revised its standards and regulations on earthquakes by developments in science and technology, along with the advancement of the International Building Code (IBC) and other new standards. Previously, the design of earthquake-resistant structures used the assumption of a building age of 50 years and was considered capable of withstanding earthquakes with a return period of 475 years. Searer et al. [18] address the switch to a 2475-year earthquake from engineering and economic perspectives. Fenwick and Macrae [19] describe the changes in the required design strengths, stiffness levels, and capacity design provisions of Newzealand earthquake provision with particular reference to buildings where reinforced concrete moment-resisting frames provide the lateral force resistance. It is shown that basic examinations of reaction spectra and restricting between soft storeys can give deceiving ends in regards to the relative strength and solidness prerequisites except if allowance is made for some other factors.

Indonesia is known as one of the most seismically risky nations on the earth. Its enormous and weak populace focuses on solid seismic danger assessment and evaluation. The Indonesian Service of Public Works and Lodging has set up a group of tremor researchers and architects entrusted with further developing the information accessible to change the public seismic peril map in 2016. They arranged aftereffects of late dynamic issues that concentrate on utilizing land, geophysical, and geodetic perceptions, and another exhaustive seismic tremor inventory includes hypocenters moved in a three-dimensional speed model. Seismic peril examination was embraced utilizing, as of late, created ground movement forecast conditions (GMPEs) and rationale trees to incorporate epistemic vulnerability related to various decisions for GMPEs and quake repeat models. The new seismic danger maps set up the significance of emotional issues and intraslab seismicity and the subduction megathrust in deciding the degree of seismic risk, particularly in coastal, populated regions. The new Indonesian peril guides will refresh general principles for planning tremor strong structures and framework [20].

4 Conclusion

Many researchers have been conducting excellent studies to find the cause of seismic failure in structures. Experimental, simulation, and field research, has resulted in many conclusions relating to this subject. This development is in line with technological advances in all fields and areas. A lot of research has also been done to prepare structures in mitigating the disasters caused by earthquakes. Many theories, simulations, experiments, and practices have been carried out and formulated for disaster preparedness. However, further research is still needed because the danger of natural disasters will continue to threaten humankind.

Earthquakes have occurred since this world existed. Until now, it is said that there is still no technology that can determine when and how big an earthquake will occur in a specific area. However, an earthquake itself is not dangerous, but the events that occur after the earthquake must be considered harmful to human beings.

For this reason, it is crucial to carry out comprehensive and comprehensive preparations for this earthquake disaster. The latest technology has been able to predict the level of collapse in a structure, for example, by performing vulnerability and fragility analysis. With fragility analysis, the probability of structural collapse can be estimated so that the structure can be increased according to the expected level of failure at a certain intensity level.

The authors greatly thank the Civil Engineering Department for its support for this research publication by Contract Number: 131/UN16.09.D/PL/2021.

References

1. P. Fajfar, *Analysis in seismic provisions for buildings: past, present, and future: The fifth Prof. Nicholas Ambraseys lecture*, **16**, no. 7. Springer Netherlands, 2018.
2. L. G. Mejía, "A-Z of Seismic Failures," *Procedia Eng.*, **210**, pp. 433–440, 2017, DOI: doi: 10.1016/j.proeng.2017.11.098.
3. A. Ayele, K. Woldearegay, and M. Meten, "A review on the multi-criteria seismic hazard analysis of Ethiopia: with implications of infrastructural development," *Geoenvironmental Disasters*, **8**, no. 1, 2021, DOI: 10.1186/s40677-020-00175-7.
4. S. L. Kramer and S. B. Paulsen, "Practical use of geotechnical site response models," *Proceedings, Int. Work. Uncertainties Nonlinear Soil Prop. their Impact Model. Dyn. Soil Response, PEER Cent. Headquarters, Richmond, Calif.*, 1–10, 2004, [Online]. Available: http://scholar.google.com/scholar?hl=en&btnG=Search&q=intitle:Practical+Use+of+Geotechnical+Site+Response+Models#0.
5. F. Mulargia, P. B. Stark, and R. J. Geller, "Why is Probabilistic Seismic Hazard Analysis (PSHA) still used?," *Phys. Earth Planet. Inter.*, **264**, 63–75, Mar. 2017, DOI: 10.1016/J.PEPI.2016.12.002.

6. A. Kharazian, S. Molina, J. J. Galiana-Merino, and N. Agea-Medina, "Risk-targeted hazard maps for Spain," *Bull. Earthq. Eng.*, **19**, 13, pp. 5369–5389, 2021, DOI: 10.1007/s10518-021-01189-8.
7. B. Yön, E. Sayın, and O. Onat, "Earthquake and Structural Damages," in *Earthquakes - Tectonics, Hazard and Risk Mitigation*, no. Earthquakes, Intech, 2017, 319–339.
8. G. Çnce, "The relationship between the performance of soil conditions and damage following an earthquake: A case study in Istanbul, Turkey," *Nat. Hazards Earth Syst. Sci.*, 2011, DOI: 10.5194/nhess-11-1745-2011.
9. Masrilayanti, A. P. Nasution, and R. Kurniawan, "Seismic response of cable-stayed bridge subjected to single-support excitation on various soil condition," 2020, DOI: 10.1051/e3sconf/202015605003.
10. F. Petruzzelli and I. Iervolino, *NODE: a large-scale seismic risk prioritization tool for Italy based on nominal structural performance*, **19**, 7. Springer Netherlands, 2021.
11. L. Martins, V. Silva, H. Crowley, and F. Cavalieri, "Vulnerability modelers toolkit, an open-source platform for vulnerability analysis," *Bull. Earthq. Eng.*, vol. 19, **13**, 5691–5709, 2021, DOI: 10.1007/s10518-021-01187-w.
12. C. Scawthorn, "A Brief History of Seismic Risk Assessment," in *Risk Assessment, Modeling and Decision Support*, S. F. Gottlieb, Ann Bostrom, Ed. Springer, 5–81 (2006)
13. B. A. Bradley, R. P. Dhakal, M. Cubrinovski, G. A. MacRae, and D. S. Lee, "Seismic loss estimation for efficient decision making," *Bull. New Zeal. Soc. Earthq. Eng.*, **42**, 2, 96–110 (2009)
14. C.J. van Westen, "Risk assessment methods," 2011. https://www.cdema.org/virtuallibrary/index.php/charim-hbook/methodology/5-risk-assessment/5-5-risk-assessment-methods (accessed Oct. 05, 2021).
15. R. K. McGuire, "Probabilistic seismic hazard analysis: Early history," *Earthquake Engineering, and Structural Dynamics*, **37**, 3. (2008)
16. K. R. Karim and F. Yamazaki, "A simplified method of constructing fragility curves for highway bridges," *Earthq. Eng. Struct. Dyn.* (2003)
17. M. Masrilayanti, A. P. Nasution, R. Kurniawan, J. Tanjung, and S. Sarmayenti, "Fragility Curve Analysis of Medium Cable Stayed Bridge," *Civ. Environ. Eng.*, **17**, 1, pp. 209–218 (2021)
18. G. R. Searer, S. A. Freeman, and T. F. Paret, "Does it make sense from engineering and economic perspectives to design for a 2475-year earthquake?," *Spec. Pap. Geol. Soc. Am.*, **425** (2007)
19. R. Fenwick and G. Macrae, "Comparison of New Zealand standards used for seismic design of concrete buildings," *Bull. New Zeal. Soc. Earthq. Eng.*, **42**, 3, pp. 187–203 (2009)
20. M. Irsyam *et al.*, "Development of the 2017 national seismic hazard maps of Indonesia," *Earthq. Spectra*, **36**, 1_suppl (2020)

Comparing the evacuation speed through a ramp and a stair in a tsunami evacuation drill in Padang

Badrul M. Kemal[1] and Yosritzal[1]*

[1]Department of Civil Engineering, Andalas University, Padang, Indonesia

Abstract. In a vertical evacuation, there are two types of a route that could be chosen namely a ramp and a stair. A ramp is usually used to help a disabled person with a wheelchair to reach the top of a shelter and a stair is used for normal persons. However, in an evacuation, speed is very important. Therefore, a comparison between the use of stairs and ramp in terms of evacuation speed to reach the top should be evaluated. This paper presents an evaluation of the evacuation speed using stairs and ramps by normal persons. The study found that evacuating using stairs is slightly faster than using a ramp.

1 Introduction

Many authors reported that Padang City, the capital of West Sumatera Province, Indonesia, has a significant tsunami hazard potential as the city is located close to the Mentawai megathrust at the West Coast of Sumatera Island [1,4]. The Mentawai megathrust which has two segments, namely the Siberut segment and the Sipora-Pagai segment, has a strong earthquake cycle of 200-300 years. The biggest threat today is the earthquake cycle from the Siberut segment which can cause a tsunami. Fig. 1 shows several major earthquake events around the megathrust site [5].

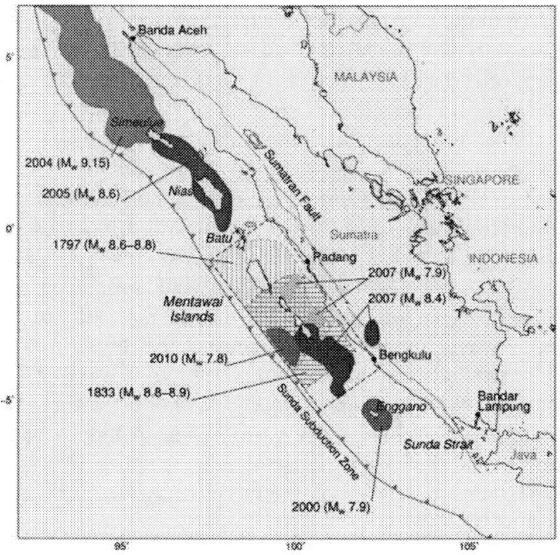

Fig. 1. Historical recent seismic activities in the Sumatra areas [5].

About half of the Padang City population are living in a tsunami-prone area with an altitude less than 5 m above sea level [6]. As the city is located near to the Ring of Fire at the West Part of Sumatera Island, the city has been hit repetitively by many big earthquakes with some of them triggering significant tsunamis to the west coastal area of Sumatera Island and nearest islands such as Mentawai and Nias Islands. McCloskey [3] reported that on February 10, 1797, and November 24, 1833, two earthquakes triggered a tsunami of 5 m and 3-4 m inundation high respectively. Furthermore, McCloskey and other scientists also predicted a huge seismic moment deficit accumulation since 1797 and 1833 at the megathrust which was predicted to trigger a 15 m tsunami inundation in Padang in the near future [3, 6, 7, 8].

The Indonesian Government with help of many countries and universities has made some progress in applying disaster mitigation systems such as developing tsunami early warning systems, increasing the capacity of tsunami evacuation routes, preparing some temporary evacuation sites and shelters, increasing the road capacity along the evacuation routes and educating people about tsunami and evacuation [9, 10]. However, many contributions are still demanded to reduce the risk and impact of the earthquake and tsunami before the predicted event becomes a reality. This paper contributes to preparing the evacuation mitigation by evaluating and comparing the evacuation speed using a ramp and a stair.

An understanding of the walking speed during an evacuation has an important role in developing evacuation plans [11,13]. However, the estimation of walking speed in the previous studies was not based on a specific observation for evacuation but based on the

* corresponding author: yosritzal@eng.unand.ac.id

observation of some short-distance crosswalks (such as [14]). Yosritzal [6] was among the first to assess the walking speed during an evacuation and found that the walking speed varies by age and gender. The study was extended to assess the role of route complexity and duration of the walking in [15]. Yosritzal found that the walking speed tends to be slower at the end section of the evacuation route because of fatigue [15]. The finding of the study triggered a question: could the evacuees reach the top floor of the tsunami shelter before the tsunami wave arrives after walking from more than 1 km away? Comparing the stairs and the ramp, which one is the fastest way to reach the top floor? This will be answered in this paper.

The presented paper aims to compare walking speed during a simulated tsunami evacuation event by using a stair and a ramp after walking more than 1 km. The observed evacuation distance of the route in the simulation was set to be equal to the estimated evacuation distance based on the effective evacuation time from [9]. The special contribution of this paper was integration between horizontal evacuation from home to a shelter and vertical evacuation from the ground of the shelter building to the top of the building either through a ramp or a stair. This paper is an extension of our previous study which observed the walking speed of evacuees during an experiment of tsunami evacuation in Padang [15].

2 Literature Review

2.1 The feature of Research Site

This study was taken place at Padang City, the capital city of West Sumatera Province, Indonesia. The city is located between 0°44' and 01° 08' South Latitude and 100°05' and 100o34' East Longitude [16]. The city's terrain is almost flat at no more than five meters above sea level, which extends about 4 km from the coast before climbing the hills further inland. The research site is shown in Fig. 2.

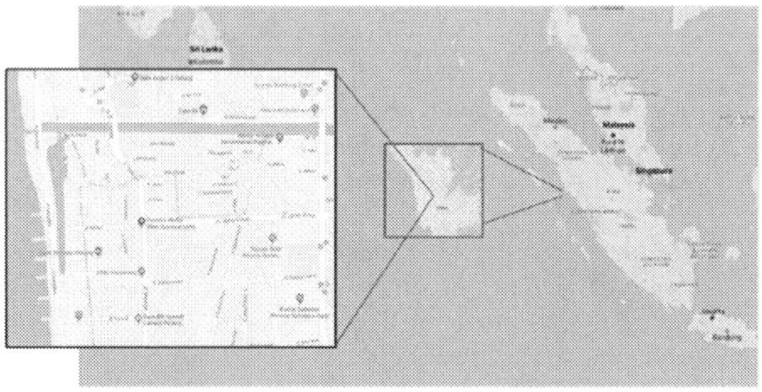

Fig. 2. Case Study Location in Padang City [6][15]

Yosritzal estimated the effective evacuation time for the people in Padang is about 17 minutes which is equivalent to around 1.4 km distance [9]. As the evacuation time is very short, it was suggested that the evacuation should be made on foot and immediately after the earthquake [17], bringing their emergency bag to the nearest shelter or the safety zones [18]. Using cars are not recommended during evacuation as it would make evacuation more difficult due to traffic jam and bottleneck [19]. This has been proven as a 7.3 SR earthquake hit Mentawai Island on March 2, 2016, majority of the evacuations were made by cars but none of them could pass through the traffic and reach the safety zone within the available evacuation time, fortunately, no tsunami occurred at the time (see [20]). This phenomenon also emerged in an agent-based model developed by Muhammad et al. (2021) and as a solution, they suggested adding more evacuation points easily accessible without a motor vehicle so that the evacuation would be more efficient.

2.2 Walking Speed Studies in Disaster Evacuation

In traffic engineering, study on walking speed is required in designing pedestrian facilities such as crosswalks and timing for traffic light operation therefore the data is commonly observed from a simulation of pedestrian movement in a corridor, sidewalk or crosswalk [22, 26]. The average walking time from those studies was used by several studies to estimate the coverage area and demand of a shelter such as in [9, 10, 27]. Abustan used average walking speed at a crosswalk as a parameter in a tsunami evacuation simulation[28]. Wood pointed that the decision to start evacuation is playing an important role in determining the minimum travel speed to evacuate from hazard zone [29].

Arguing that the normal walking speed might not be similar to the evacuation speed, Yosritzal conducted a tsunami evacuation simulation to estimate the walking speed during evacuation [60. Based on [6], Chasanah and Sakakibara used 1.4 m/s as the evacuation walking speed in the study on the volcanic eruption's evacuation case [30]. Similarly, [31] set walking speed into three categories namely slow (1.35 m/s), medium (1.4 m/s) and fast (1.51 m/s) based on [6].

In a previous study, the variation of the walking speed was observed by age and by gender [6, 28, 29]. Yosritzal suggested adding the duration of the walk and the complexity of the route as a determinant of the walking speed [15]. Yosritzal developed three scenarios in the evacuation drill which were based on the complexity of the route [15]. The complexity of the route was simplified by assuming the more turning points along the route, the more complex the route. The result shows that walking speed was not constant along the evacuation route. The walking speed tended to be slower as the distance is increasing [15] which is expected to be caused by the fatigue experienced by the evacuees.

3 Methodology

Similar to the previous study in [6] and [15], in this presented study, the walking speed was observed from an evacuation drill from a specified point inside the residential area to the rooftop of the Temporary Evacuation Site (TES) next to the West Sumatera Governor's Office, at Soedirman Road. The route of evacuation is shown in Fig. 3. Observers were placed at 6 important points along the routes to observe the time when the Evacuees pass through their position. The first four observers were placed along the evacuation road and one observer at the rooftop of the TES. The volunteers were asked to walk fast from the specified point to the specified point at the rooftop. The timing data recorded by the observers will be analyzed to achieve the objective of the study.

Fig. 3. The route of the evacuation drill (captured from Google map)

The evacuation drill was conducted on July 14, 2018. The starting point was at the residential area about 1.07 km from the TES and then climbing to the rooftop of the TES through a ramp or a stair. The volunteers were categorized into 5 groups namely: 10-20-year-old, 21-30 year old, 31–40-year-old, 41-50-year-old, and more than 50-year-old. There were three volunteers in each group. There were four scenarios in the study:

a. One by one volunteer take turns to evacuate from the original point to the destination through a specified route and climb the TES through the stair.
b. All volunteers at the same age interval group simultaneously evacuate from the original point to the destination through a specified route and climb the TES through the stair.
c. One by one by volunteers take turns to evacuate from the original point to the destination through a specified route and climb the TES through the ramp.
d. All volunteers at the same age interval group simultaneously evacuate from the original point to the destination through a specified route and climb the TES through the ramp.

The mean of observed walking speed by each age group, one by one or simultaneously, and by stair or ramp will be compared. The conclusion and recommendation will be made based on those comparisons.

The detail of the simulation segments is shown in Table 1. Segment 1 is located in the residential area and rarely to find a car or people walking as most of the people enjoy staying at home with family during the weekend. Segment 2 is located at the main road with an up to 2.5 m sidewalk on both sides of the road. Segment 3 is a busy road as many activities are available here such as a hospital, shops, and offices. A lot of cars are parked on both sides of the road. No proper sidewalk is available on this road. Segment 4 is relatively quiet on the weekend. Segment 5 is the segment where evacuees started to climb the TES either through stairs or a ramp.

Table 1. Detail Simulation Evacuation Route

Route	Road + Stairs		Road + Ramp	
	Distance (m)	Route Wide (m)	Distance (m)	Route Wide (m)
Segment I Origin (point 1) to point 2	200	4	200	4
Segment II Point 2 to point 3	457,6	7	457,6	7
Segment III Point 3 to point 4	248,1	7	248,1	7
Segment IV Point 4 to point 5 (ground)	296,9	7 and 4	320	7 and 4
Segment V Point 5 (ground) to point 5 (rooftop)	39,24	1,8	173,5	2,9

(a) Stairs

(b) Situation during the evacuation drill through stairs

https://doi.org/10.1051/e3sconf/202133107014

(c) Ramp

(d) Situation during the evacuation drill through the ramp

Fig.4. The sites and situations of evacuation drills through stairs and Ramp from the ground to the rooftop

The picture of the stairs and ramp are shown in Figure 4. In this study, the slope of the ramp and stairs were not considered as a variable in the analysis. The only variable considered was the time and the distance from the ground to the rooftop of the TES building. The dimension of the stairs and ramp were neglected as the number of samples is low. These pre-judgments are debatable but, in this study, we limit the study within the scope of those assumptions.

The evacuation drills took place in the morning and no electric bulb is on, so the light source is only from the sun. As the stairs are located inside and the ramp is in the outside of the TES building, the stairs are relatively darker than the ramp. However, the low light does not significantly influence the walking speed as the stairs and the situation around the stairs are still clearly visible by normal eyes.

4 Results

4.1 Stairs

Walking from the origin point to the destination point at the rooftop of the TES using stairs will take time as shown in Table 2.

Tabel 2. Walking Time Through Stairs by Age Interval Group in Second

Check Point	Distance (m)	Walking Time by Age Group Interval (Second)					
		11 - 20 y.o.	21 - 30 y.o.	31 - 40 y.o.	41 - 50 y.o.	> 50 y.o.	Av.
1 - 2	200.0	100	160	80	80	120	108
2 - 3	457.0	220	240	300	260	260	256
3 - 4	248.0	60	60	180	140	160	120
4 - 5	297.0	220	180	220	180	180	196
5 - 6	40.4	60	60	60	80	180	88
Total	1242.4						

Fig. 5. shows the walking speed of the evacuees from the origin point through the evacuation route to the destination point (rooftop of the TES) through the available stairs. Fig. 5. shows that the walking speed is reduced by the distances except for the case of evacuees in the age group of 21-30 and 11-20 years old. The walking speed was reduced to around 1 m/s.

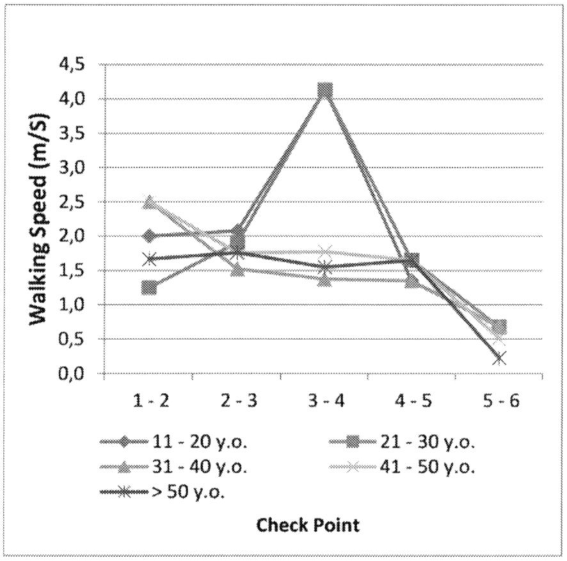

Fig. 5. Walking speed by age interval (through stairs)

4.2 Ramp

Walking from the origin point to the destination point at the rooftop of the TES using a ramp will take time as shown in Table 3. Whilst, Fig. 6. shows the walking speed of the evacuees from the origin point through the evacuation route to the destination point (rooftop of the TES) through the available stairs. Fig. 6. shows that in general, the walking speed is reduced by distances. In special cases for the climbing to the rooftop through the ramp, the speed reduced significantly to around 0.075 m/s.

Tabel 3. Walking Time Through Ramp by Age Interval Group in Second

Check Point	Distance (m)	Walking Time by Age Group Interval (Second)					
		11 - 20 y.o.	21 - 30 y.o.	31 - 40 y.o.	41 - 50 y.o.	> 50 y.o.	Av.
1 - 2	200	80	120	120	120	120	112
2 - 3	457	240	340	280	280	340	296
3 - 4	248	80	280	120	260	180	184
4 - 5	320	160	180	140	220	200	180
5 - 6	20	140	180	220	200	280	204
Total	1245						

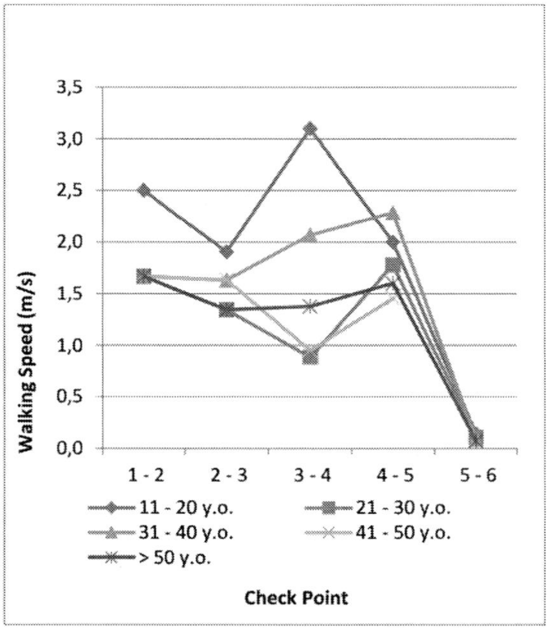

Fig. 6. Walking speed by age interval (through ramp)

5 Discussion

Figs. 5 and 6 show that the walking speed of the evacuees was reduced by the distance. However, the walking speed at segment 3 (from checkpoint 3 to 4) is more variable than at other segments. The authors expected that the variation was caused by the traffic and side frictions between parked cars, pedestrians, and unmotorised vehicles along the segment. The characteristics of the segments were not the same. Segment 1 (checkpoints 1 to 2) is a local road in the residential area which is typically low traffic and low side friction. Segment 2 (checkpoints 2 to 3) is a major road with a sidewalk on both sides of the road. Segment 4 (checkpoints 4 to 5) at the time of the evacuation drill was relatively quiet (very low traffic and pedestrian). Segment 5 (from checkpoint 5 to 6) is a segment where the evacuees climb the stair or ramp to reach the rooftop of the TES. Therefore, all of the evacuees should be aware that segment 3 is more unpredictable than other segments which potentially to prevents the evacuees to reach the safe area before the tsunami is coming. An alternative route should be considered in future studies.

Comparing data from Table 2 and Table 3, the walking time from the ground to the top is much higher using a ramp than stairs. This is because the total distance walked is shorter for the stairs than the ramp. Similarly, the speed is much higher when using the stairs than the ramp as shown in Figs 5 and 6. Therefore, it is recommended to use the stairs for those who can climb the stairs. However, for those who cannot easily climb the stair, using a ramp is recommended even though the distance is longer than using stairs.

6 Conclusion

A series of evacuation drills had been made in this study to compare the walking speed of evacuees from the origin point to the destination of evacuation which was the rooftop of the temporary evacuation site (TES). The results of the study indicated that the evacuation would be faster when using the stairs than using the ramp. However, the results should be used with caution that the slope of the stair and the ramp in this study are not the same and the time recorded by the observers are in minutes. A more detailed study is recommended to obtain more accurate data and conclusions.

The authors thank the Civil Engineering Department of Andalas University for the funding of this publication under contract number: 120/UN16.09.D/PL/2021.

References

1. J. C. Borrero, K. Sieh, M. Chlieh, and Synolakis, C. E. *P. Natl. Acad. Sci. USA*, 103. (2006).
2. R. McCaffrey, *Science*, **315**, (2007).
3. J. McCloskey, A. Antonioli, A. Piatanesi, K. Sieh, S. Steacy, S. Nalbant, M. Cocco, C. Giunchi, J. D. Huang, and P. Dunlop E & PSc. Lt., **265**, Issues 1-2 (2008)
4. T. Schlurmann, W. Kongko, N. Goserberg, , D.H. K. Natawidjaja, and Sieh, *Coastal Engineering*. (2010)
5. J. D. Griffin, I. R. Pranantyo, W. Kongko, A. Haunan, R. Robiana, V. Miller, et al. . *Indonesia: Earth Science for Disaster Risk Reduction*, eds P. R. Cummins and I. Meilano (London: Geological Society). 441 (2016)
6. Yosritzal, BM Kemal, Purnawan, H. Putra *IOP Conf. Series: E.E.S.* 140 (2018)
7. V. Cedillos, B. G. McAdoo, A. Muhari, and D. H. Natawidjaja. *Proceedings of the 9th U.S. National and 10th Canadian Conference on Earthquake Engineering* Juli 25-29 (2010)
8. A. Muhammad, K. Goda , N. A. Alexander, W. Kongko, and A. Muhari *Nat. Hazards Earth Syst. Sci., 17,* (2017)

9. Yosritzal, B. K. Kemal, and F. Siddik . *Proceeding of National Conference of Applied Sciences, Engineering, Business, and Information Technology* (2016).
10. Yosritzal,B. K. Kemal, and Y. B. Aulia IJASEIT*International Journal of Advanced Science Engineering Information Technology*, (2017)
11. E Mas, F Imamura, S. Koshimura JSCE, (2011).
12. Kim, Nam-Sook. IJST **9** (40) (2016)
13. M. Seike, N. Kawabata, M. Hasegawa, and Y. Lu. (2016)
14. J. Kiyono. and N. Mori,. *13th World Conference on Earth Engineering*, Vancouver, B.C., Canada. (2013).
15. Yosritzal, H. Putra, Kemal., E. Mas, Purnawan Adv. i.n Eng Res 193 (2020)
16. BPS (2021)
17. BPBD (2010)
18. Stanford *Final Report Stanford University* (2009).
19. E. Mas, B. Adriano, S. Koshimura, F. Imamura , J.H. Kuroiwa,F. Yamazaki ,C. Zavala , and Estrada M. *Book Chapter, Tsunami Events and Lessons Learned* **35** (2014).
20. Tribunews (2016).
21. A. Muhammad, R. De Risi, F. De Luca, N. Mori, T. Yasuda, and Goda K. SERRA, **35**, (2021).
22. M. S. Sharifi, K. Christensen , A .Chen, D. Stuart , Y. S Kim and Y. Q. Chen *Travel Behaviour and Society* 8, (2017).
23. M. Iryo-Asano and Alhajyaseen *Transportation Research Procedia* 21 (2017).
24. E. Bosina and U.Weidmann *Physica* A 468 2017.
25. Paschalidis B Politis I, Basbas S, and Lambrianidou P. *Transportation Research Part F* 42, (2016)
26. S. Chandra and A. K. Bharti, "Speed distribution curves for pedestrian during walking and crossing", *Procedia-Social and Behavioral Sciences* 104 (2013) 660-667, (2013).
27. B. M. Kemal, Yosritzal, Y. B. Aulia. *IJCIET* 2017.
28. M. S. B. H. Abustan. *Dissertation at Kyoto University* Japan, (2013).
29. N. Wood J, Jones, Schmidtlein M, Schelling J, and Frazier T. IJDRR 18 (2016).
30. F. Chasanah, H. Sakakibara Journal of Integrated Disaster Risk Management, **10** (2). (2020)
31. S. Kubisch, G. J.uth, , S. Keller, M.T. Bull, , Keller, L. and Braun, A.Ch. IJDRR 45 (2020)

Analysis of urban flood using synthetic unit hydrograph (SUH) and flood mitigation strategies along way Halim River: a case study on Seroja street, Tanjung Senang District

Dian Pratiwi[1]* *Arniza* Fitri[1], *Arlina* Phelia[1], *Nabila* Annisa Amara Adma[1], and *Kastamto*[1]

[1]Department of Civil Engineering, Faculty of Engineering and Computer Science, Indonesian Tecnocrat University, Bandar Lampung, Lampung 35132, Indonesia

Abstract. In the urban area, flooding becomes the most common disaster that has not been resolved until today. The utilization of river border area into housing and lack of absorption area becomes the trigger factor of urban flooding, as what is happening around Way Halim River on Seroja street. In this area, floods often happen during the rainy season, with the latest events recorded on January 21st, 2021. Analysis of flood intensities and discharges can be parameters for the decision-making of flood mitigation strategies. This study aims to analyze the flood discharges along Way Halim River, Seroja street by comparing the flood discharges resulting from three analysis methods of Synthetic Unit Hydrograph (SUH) including Gama I SUH, Nakayasu SUH, and Snyder SUH. Finally, suitable flood mitigation strategies were also proposed in this study based on the flood discharges and rain intensities. The results showed that Nakayasu SUH had the highest peak flood discharge than Snyder SUH and Gama I SUH. Based on the results of the investigation of land suitability; and analysis of rainfall intensities and flood discharges, the proposed flood mitigation in Seroja street is by installing biopore infiltration holes along Seroja street for storing water and reducing the risk of flooding in the area.

1 Introduction

In Indonesia, flood events are always repeated every year, especially during the rainy season. Flood occurrences can be triggered by an increase in population density, climate change, and economic development in flood-prone areas [1–6]. In the urban area, floods are the main problem that causes financial, social, and environmental losses, especially in the areas that are prone to flood events. The density of population settlements and the use of river border areas into residential areas cause flooding that is commonly called urban flooding [7].

Flood control is related to flood mitigation [8-17]. The mitigation itself differs into non-structural mitigation and structural mitigation. The mitigation that is conducted with concepts, policy, plan, and operation on integrated urban flood disaster and risk management can achieve multiple goals such as water supply, groundwater recharge, and the improvement of the urban environment [19].

Related to urban areas and flood mitigation, there are several flood mitigation strategies in urban areas with high population density, limited water catchment areas, and poor drainage systems that can be applied. These are including infiltration wells, rainwater harvesting utilizing rain barrels, and biopore hole.

Infiltration wells can reduce surface runoff, but the effectiveness of the infiltration wells depends on the density and dimensions of the well [20]. Rainwater harvesting utilizing a rain barrel is significant to reduce flood peaks in a densely populated area [21]. Biopore holes as a method of water conservation based on soil permeability, rainfall intensity, depth of groundwater level, and volume of organic waste can also be one of the flood mitigation strategies in urban areas with the dimensions of the hole can be applied in narrow residential land [22,23].

In Bandar Lampung City, where the development areas reach 69.5% of the total area (13,711.17 ha) in 2020, faces a high level of flooding vulnerability due to changing the function of water catchment areas and river borders into housing or residential areas [24] . For example, in the area of Seroja street, Tanjung Senang District, Bandar Lampung City, floods events occur after the switching functions of the Way Halim Border River area into residential and lack of water catchment area. The biggest event of urban floods in Seroja street has been recorded on July 5th, 2020, and July 21st, 2021. The

* Corresponding author: dian.pertiwi@teknoret.ac.id

floodwater level during the peak flood reached 30 to 55 cm in the area around Seroja street.

The existing conditions along way Halim river and the condition during the flood on Seroja street are shown in Figs 1 and 2.

Based on the conditions, it is necessary to analyze the intensities of rain and urban flood discharges along way Halim river. Besides, appropriate mitigation strategies in the catchment area of Way Halim river especially around Seroja street need to be proposed. Further, the information about rainfall intensities, flood discharges, and recommendations of the proper flood mitigation strategies in Seroja street can be important for the local community and local government in Bandar Lampung City.

Fig. 1. Residential conditions along Way Halim River on Seroja Street.

Fig. 2. The condition during a flood event in Seroja Street.

2 Research Method

2.1 Study Area and Data Collection

This study has been carried out along way Halim river, especially at Seroja street as shown in Fig. 3. In this study, data collections are including primary data and secondary data. Primary data was used to collect information about flood events that have been occurred around Seroja street. The secondary data are consisting of the area of the watershed, the characteristics of the watershed, and the rainfall data from 4 rainfall stations [RS) in the year 2011-2020 [Kemiling RS, Sukarame RS, Sumur Batu RS, Sumur Putri RS). Fig. 4 shows the location of the rainfall stations in Way Halim catchment.

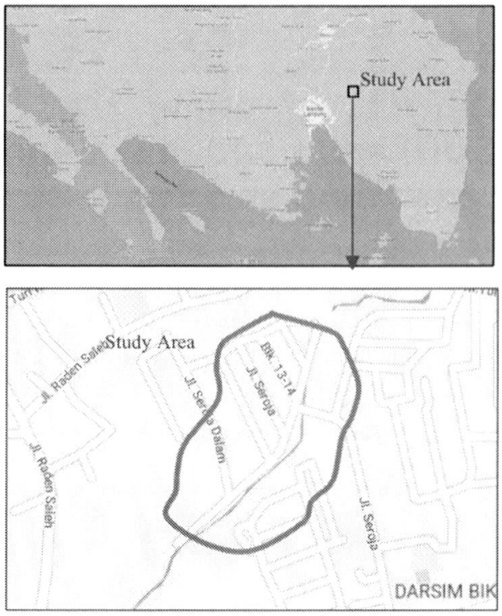

Fig. 3. Description of the study area

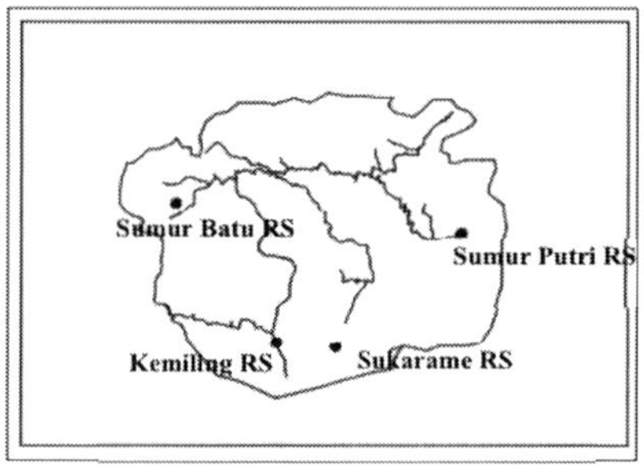

Fig. 4. Location of Rainfall Stations is Way Halim Catchment

2.2 Data Analysis Method

Data analysis involved in this study includes analysis of flood discharges, analyses of rainfall intensity, and analyses of urban flood mitigation strategies. The analysis stages include 1) rainfall area analysis using polygon Thiessen, 2) frequency distribution analysis using Log Pearson Type III, 3) statistical distribution for rainfall, rainfall intensity analysis, 4) discharge analyses using synthetic unit hydrograph [SUH), and 5) recommendation for suitable flood mitigation strategies around Seroja street, Tanjung Senang District, Bandar Lampung.

2.2.1 Rainfall Area Analysis

Rainfall measurements are made to obtain rain data which are usually only in one place, while in the analysis, it is necessary to use the average rainfall data for the watershed. To obtain the average rainfall data from a watershed with the requirement, there are at least 3 rain stations used in a watershed. Then, the average rainfall data can be obtained by using the Thiessen-Polygon method [25].

$$P_d = \sum_{i=1}^{n} a_i . P_i \quad [1]$$
$$a_i = \frac{Li}{L} \quad (2)$$

Where P_d is the average rainfall in the watershed (mm), P_i is the rainfall for each station (mm), i is the station weight factor, L_i is the area of each Thiessen-polygon (km^2) and L is the area of the watershed (km^2).

2.2.2 Frequency Distribution Analysis

Log Pearson Distribution Analysis Type III is used in the hydrological analysis as a parameter to determine rainfall intensity for a certain return period. The requirement in this distribution is that the coefficient of skewness is not equal to 0 or $Cs \neq 0$ (26).

2.2.3 Statistical Distribution for Rainfall

The rain statistical distribution used is Chi-Square and Kolmogorov–Smirnov statistic. The formula for the chi-square statistic is given by [26] :

$$X^2_{\text{hit}} = \sum_{i=1}^{k} \frac{(EF - OF)^2}{EF} \quad (3)$$

Where K is calculated by the equation $1 + 3.22.Log\ n$, OF is the observed value and EF is the expected value. For the selected frequency distribution to be acceptable, the requirements that must be met are $X^2 < X^2_{cr}$. The X^2_{cr} value can be obtained by determining the significant level with degrees of freedom.

Kolmogorov-Smirnov (D) based on the largest difference of Relationship cumulative frequency ($F_n(x)$) and Theoretical Cumulative Frequency ($F(x)$). Kolmogorov–Smirnov statistic formula is given by [26]:

$$D_n = |F_n(x) - F(x)| \quad (4)$$

2.2.4 Synthetic Unit Hydrograph (SUH)

Synthetic unit hydrograph (SUH) is a rainfall runoff model of a watershed, obtained from synthetic methods without using measured data. The SUH method used in this study including the Gama I SUH, Nakayasu SUH and Snyder SUH. With the equation given in Gama I SUH, as follows [27]:

$$B = 1{,}5518N - 0{,}14991A - 0{,}2725SIM - 0{,}0259S - 0{,}0733 \quad (5)$$
$$Qp = 0{,}1836A.0{,}5884.JN.0{,}2381.tr - 0{,}4008 \quad (6)$$
$$tb = 27{,}4132tr.0{,}1457S - 0{,}0986SN.0{,}7344RUA.0{,}25 \quad (7)$$
$$K = 0{,}5671A.0{,}1798S - 0{,}1446SF - 1{,}0897D.0{,}0452 \quad (8)$$
$$QB = 0{,}4751A.0{,}6444D.0{,}9430 \quad (9)$$

Where N is the number of train stations, A is the area of the watershed (km^2), SIM is the Symmetry Factor, S is the average river slope, B is the reduction coefficient, SN is the source frequency, RUA is the relative area upstream watershed (km^2), D is the drainage net density and QB is the base flow (m^3/s). For time to peak (tr), the related variables are river length (L), Source factor (SF), and Symmetry factor (SIM). Calculation of the peak flood discharge of HSS Gama I (Q_p) uses the JN variable, defined as several river encounters. Then, the equation for Nakayasu SUH is as follows [27] :

$$Q_p = \frac{CAR}{3{,}6(0{,}3Tp + T0{,}3)} \quad (10)$$
$$T_P = tg = 0{,}8\,tr \quad (11)$$
$$tg = 0{,}21L^{0,7} \text{ (for L<15 km)} \quad (12)$$
$$tg = 0{,}4 + 0{,}058L \text{ (for L>15km)} \quad (13)$$
$$T_{0,3} = \alpha.tg \quad (14)$$
$$Qt = (\frac{t}{T_p})^{2,4}.Q_P \quad (15)$$

Where QP is the peak flood discharge (m3/s), R is the unit rain (mm), C is the coefficient of drainage, TP is the period from early rain to time to peak of flood discharge (hours). 0.3TP is the time it takes for the discharge to decrease, from peak discharge to 30% of the peak discharge, Tg is the time of concentration (hours) and α is the hydrograph parameter (the value is between 1.5–3.0). Other equations used in Nakayasu SUH are as follows [27] :

$$Q_t = Q_P(\frac{t}{T_p})^{2,4} \quad \text{for } 0 \le t \le T_p \quad (16)$$
$$Q_t = Q_P 0{,}3^{\left[\frac{t-T_P}{T_{0,3}}\right]} \quad \text{for } T_p \le t \le (T_P + T_{0,3}) \quad (17)$$
$$Q_t = Q_P 0{,}3^{\left[\frac{t-T_P+0,5T_{0,3}}{1,5T_{0,3}}\right]} \quad \text{for } (T_P + T_{0,3}) \le t \le (T_P + T_{0,3} + 1{,}5T_{0,3}) \quad (18)$$
$$Q_t = Q_P 0{,}3^{\left[\frac{t-T_P+0,5T_{0,3}}{2T_{0,3}}\right]} \quad \text{for } t > (T_P + T_{0,3} + 1{,}5T_{0,3}) \quad (19)$$

For Snyder SUH given by the following equation:

$$t_p = C_t(L.L_c)^{0,3} \quad (20)$$
$$t_e = 0{,}278^{\frac{Cp.A}{tp}} \quad (21)$$
$$T_b = 5{,}0(T_p + \frac{t_r}{2}) \quad (22)$$

2.2.5 Flood Mitigation

After analyzing the flood discharge and rainfall intensity, appropriate structural mitigation steps were taken. Because the research area is a densely populated residential area and space requirements are very limited, it is necessary to optimize space utilization. The Biopore infiltration hole technique is one of the solutions to overcome flooding. Biopore Infiltration Hole can be a water reservoir and flood control that can be applied in narrow residential areas.

3 Result and Discussion

3.1 The Average of Annual Rainfall

Annual rainfall gave by calculating the maximum rainfall of 2011-2020 in Garuntang Sub Catchment area around Seroja street and the coefficient of polygon Thiessen. The coefficient of polygon Thiessen is given

 https://doi.org/10.1051/e3sconf/202133107015

in Tabel 1., while the average annual rainfall is given in Table 2.

Table 1. Polygon Thiessen Coefficient

No.	Rainfall Station	Thiessen Coefficient (α)
1	Kemiling RS	0,175
2	Sukarame RS	0,137
3	Sumur Batu RS	0,379
4	Sumur Putri RS	0,309
		1,000

Table 2. Annual Average Rainfall using Polygon Thiessen Method

No	Tahun	Average Rainfall (mm)
1	2020	101.7461
2	2019	61.0729
3	2018	51.3745
4	2017	89.7055
5	2016	37.1122
6	2015	40.5835
7	2014	42.2426
8	2013	71.5106
9	2012	20.3927
10	2011	196.9144

3.2 Design Rainfall and rainfall Intensity

Table 3. Statistical Parameters for Determining the Type of Distribution

Type of Distribution	Parameter	Analysis result	Information
Normal Distribution	Cs = 0 Ck = 3	Cs = 1.91 Ck = 6.50	Not accepted
Gumbel	Cs = 1.1396 Ck = 5.4002	Cs = 0.13 Ck = 38.67	Not accepted
Log Normal	Cs (InR) = 0 Ck (InR) = 0	Cs = 2.33 Ck = 7.68	Not accepted
Log Pearson Type III	Others	Cs = 0.32 Ck = 0.62	accepted

Table 3. show the value of C_s and C_k as the parameter for rainfall distribution, that has been calculated from each distribution.

Based on the statistical parameter results, it can be seen that Log Pearson Type III is accepted as the design rainfall. With 5, 25, 50, and 100 years return period.

Table 4. Design Rainfall using Log Pearson Type III

Return Period (years)	Design Rainfall (mm)
5	96.716
25	179.378
50	223.246
100	273.129

From the design rainfall, rainfall intensity can be obtained. The rainfall intensity in minute used mononobe formula in 2, 5, 25, 50 and 100 years return period given in Fig. 5.

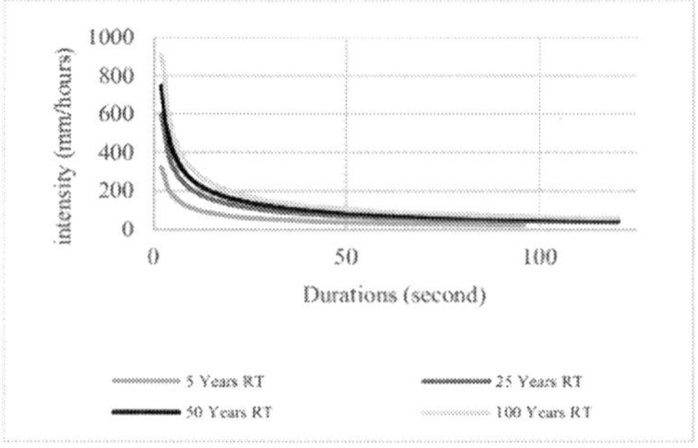

Fig. 5. IDF Curve using Mononobe Formula

IDF curve shows the intensity of Way Halim river around Seroja street. The time concentration of the IDF curve at 1.027 hours, with an intensity of 32.94 mm. The intensity is used as the parameter for the mitigation using a biopore infiltration hole.

3.3 Syntetic Unit Hydrograf (SUH)

The synthetic Unit Hydrograph (SUH) method was used to identify the flood discharges. The result from Gama I SUH, Nakayasu SUH, and Snyder SUH has given in fig (Fig 6, Fig 7, and Fig 8).

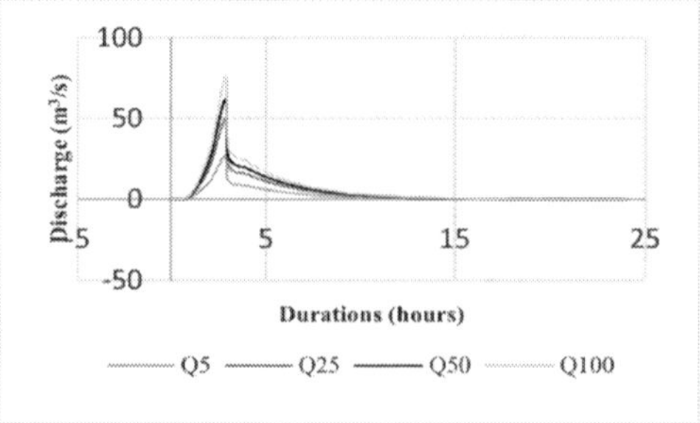

Fig. 6. Gama I SUH for different Return Period.

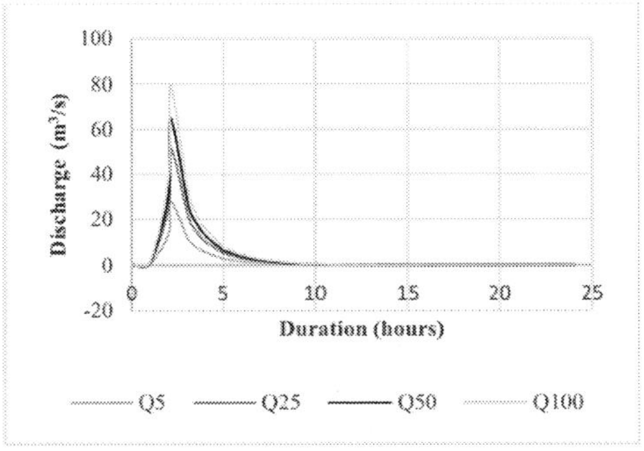

Fig. 7. Nakayasu SUH for Different Return Period

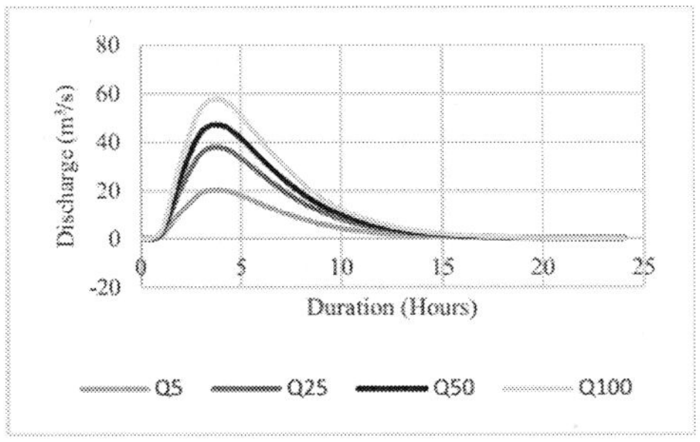

Fig. 8. Snyder SUH for Different Return Period

3.4 Biopore Hole Infiltration

Mitigation planning on Seroja street is carried out by applying biopore holes in residential areas. Biopore hole design by using a pipe with 100 mm hole diameter and 1-meter height of biopore pipe. Hole water infiltration rate assumed as 180 liter/hours. The parameter that is used in biopore hole using intensity in 5 years return period is 32.94 mm. The biopore hole infiltration plan can be seen in the following table (Table 5).

Table 5. Biopore Hole Infiltration plan in various residential areas.

Tight field area (m^2)		**Number of Biopore Hole**
Existing	76	14
	78	14
	80	15
Assumption	36	7
	45	8
	54	10
	72	13
	120	22

4 Conclusion

Urban flooding is one of the disasters that should be a concern for urban communities. By utilizing the border area of the river into housing and turning water infiltration areas into residential areas can be the triggering factor for urban flooding. During the rainy season, the peak discharge by using Nakayasu SUH for 5, 25, 50, and 100 years Return Period gave the highest value than Gama I SUH and Snyder SUH. The results were 28.041 m3/s, 52.008 m3/s, 64.727 m3/s, and 79.190 m3/s respectively. This peak discharge value is used as a recommendation for the local government in making decisions regarding mitigation that can be carried out along the Way Halim River.

Related to the mitigation that can be done by residents, several steps can be taken to reduce urban flooding. One of them is by applying biopore infiltration holes. The biopore infiltration hole can be applied to densely populated areas with limited water catchment in residential areas. Although biopore holes do not provide quick results in overcoming floods, with the function of biopore holes that can absorb rainwater runoff during the rainy season, it can be a solution step in overcoming urban flooding as well as water catchment areas that can be applied to densely populated housing.

The author expresses many thanks to Universitas Teknokrat Indonesia for supporting in the process of completing this study.

References

1. D. Pratiwi, R.O. Sinia, A. Fitri, Increasing Public Knowledge of Porous Drainage which functions as a place for rainwater infiltration, J Soc Sci Technol Community Serv.;**1**(2) (2020)
2. K.N.A. Maulud, A. Fitri, W.H.W.M. Mohtar, W.S.W.M. Jaafar, N.Z. Zuhairi, M.K.A Kamarudin, A study of spatial and water quality index during dry and rainy seasons at Kelantan River Basin, Peninsular Malaysia. Arab J Geosci.;**14** (2):1–19 (2021)
3. A. Fitri, K. Nizam, A. Maulud, F. Rossi, F. Dewantoro, N.Z. Zuhairi, *Spatial and Temporal Distribution of Dissolved Oxygen and Suspended Sediment in Kelantan River Basin. In: In 4th International Conference on Sustainable Innovation 2020–Technology, Engineering, and Agriculture (ICoSITEA 2020).* Atlantis Press.; 51–4 (2021)
4. A. Fitri, K.N.A. Maulud, D. Pratiwi, A. Phelia, F. Rossi, N.Z. Zuhairi, Trend Of Water Quality Status In Kelantan River Downstream, Peninsular Malaysia. J Rekayasa Sipil (2020)
5. A. Fitri, Z.A Hasan, A.A Ghani. *Determining the Effectiveness of Harapan Lake as Flood Retention Pond in Flood Mitigation Effort. In: In Proceedings of 2011 4th International Conference on Environmental and Computer Science (ICECS 2011).* (2011)
6. S.H. Lai, A. Fitri. Application of SWAT Hydrological Model to Upper Bernam River Basin (UBRB), Malaysia. IUP J Environ Sci; **5** (2) (2011)
7. M. Irwansyah, C. Nursaniah, L. Qadri, Alternatives for Settlement Area Management of Meureudu River watersheds towards Flood Mitigation (Case Study of Meureudu City Center, Pidie Jaya–Aceh), 63–71 (2017)

8. H. Chen, L. Yao, A. Fitri, *The influence mechanism research of inflow temperature in different time scale on the water temperature structure. In: IOP Conference Series: Earth and Environmental Science*, 12058 (2019)
9. L. Yao, X. Huang, A. Fitri, *Influence scope of local loss for pipe flow in plane sudden expansions. In: IOP Conference Series: Earth and Environmental Science*, 12056 (2019)
10. D. Pratiwi, A. Fitri, Analysis of the Potential for Tsunami Wave Propagation on the West Coast of Lampung, Indonesia. J Tek Sipil ITP. 2021;**8** (1), 29–37.
11. L. Yao, J. Li, S. Shi, A. Fitri, *Simulation of take-off angle of a ski jump energy dissipater. In: IOP Conference Series: Earth and Environmental Science*, 12057 (2019)
12. A. Fitri, R. Hashim, Motamedi S. Estimation and validation of nearshore current at the coast of Carey Island, Malaysia. Pertanika J Sci Technol **25** (3), 1009–18 (2017)
13. A. Fitri, R. Hashim, S. Abolfathi, K.N.A. Maulud, Dynamics of sediment transport and erosion-deposition patterns in the locality of a detached low-crested breakwater on a cohesive coast. Water (Switzerland), **11** (8) (2019)
14. F. Arniza, Hydro-morphodynamic responses of the detached breakwater in mangrove rehabilitation project/Arniza Fitri. University of Malaya (2018)
15. A. Fitri, L. Yao, B. Sofawi, *Evaluation of mangrove rehabilitation project at Carey Island coast, Peninsular Malaysia based on long-term geochemical changes. In: IOP Conference Series: Earth and Environmental Science,* 12055 (2019)
16. R. Hashim, C Roy, S. Shamshirband, S. Motamedi, A.Fitri, D. Petković, et al. Estimation of Wind-Driven Coastal Waves Near a Mangrove Forest Using Adaptive Neuro-Fuzzy Inference System. Water Resour Manag, **30** (7), 2391–404 (2016)
17. R. Hashim, A. Fitri, S. Motamedi, A.M. Hashim. Modeling of coastal hydrodynamic associated with coastal structures: A review. Malaysian J Sci. **32** (4), 149–54 (2013)
18. A. Fitri, R. Hashim, K. Song II, S. Motamedi. Evaluation of Morphodynamic Changes in the Vicinity of Low-Crested Breakwater on Cohesive Shore of Carey Island, Malaysia. Coast Eng J. **57** (4), 1550023 (2015)
19. T. Tingsanchali, Urban flood disaster management. Procedia Eng, 32:25–37 (2012)
20. D. I Kusumastuti, D. Jokowinarno, S.N. Khotimah, C. Dewi, F. Yuniarti, *Infiltration well to reduce the impact of land-use changes on flood peaks: a case study of Way Kuala Garuntang catchment, Bandar Lampung, Indonesia. Hydrol Earth Syst Sci Discuss.* **11** (5), 5487–513 (2014)
21. D. Jokowinarno, D.I Kusumastuti, Rainwater Harvesting for Flood Peak Reduction in Way Awi Catchment, Indonesia. Int J GEOMATE, **18** (70), 246–51 (2020)
22. N.I. Khusna, S. Amin, F.Y. Efrianinrum,A. Bashith, *The effect of using biopore on soil fertility in karst area, District of Besuki, Tulungagung Regency. IOP Conf Ser Earth Environ Sci.* **485** (1):100–4 (2020)
23. M.H. Syahruddin, Amiruddin, H. Halide, Sakka, Makhrani, *Groundwater Conservation with Hole Infiltration of Biopore Cube. IOP Conf Ser Earth Environ Sci,* **279** (1) (2019)
24. R. Wulandari, Supriatna, T. L Indra, *A simulation model for urban development in Bandar Lampung City, Lampung, Indonesia. IOP Conf Ser Earth Environ Sci*, **248**(1). (2019)
25. N. Faisal, A. Gaffar, Development of Pakistan's s New Area Weighted Rainfall Using Thiessen Polygon Method. Pakistan J Meteorol. **9** (17), 107–16 (2012)
26. M Farooq, M. Shafique, M.S. Khattak, Flood frequency analysis of river swat using Log Pearson type 3, Generalized Extreme Value, Normal, and Gumbel Max distribution methods. Arab J Geosci, **11** (9) (2018)
27. T. Marhendi, Fathurohman. *Analysis of Flood Discharges of Serayu River at Serayu Barrage using HSS Gama I and Nakayasu Models. IOP Conf Ser Mater Sci Eng.* **771**(1) (2020)

Effect of tsunami load on the elementary school building of the 23/24 Padang, Indonesia

Fauzan Fauzan[1*], *Khin Thu* Zar Htay[1,2], *Zawil* Huda[1], *Hafiz* Oktaufik[1], and *Geby* Aryo Agista[1]

[1]Department of Civil Engineering, Faculty of Engineering, University of Andalas, Padang, Indonesia
[2]Nay Pyi Taw Development Committee, NAY PYI TAW, Myanmar

Abstract. West Sumatra Province is one of the provinces in Indonesia that is vulnerable to natural disasters, especially earthquakes and tsunamis. Padang city, as the capital city of West Sumatra, is an area that is included in an area with a high level of vulnerability (High Risk Zone) to tsunamis. Therefore, the construction of public buildings such as hospitals, government offices, and school buildings must have certain technical engineering that is able to anticipate the damage and collapse of buildings due to the earthquake and tsunami. One of the public buildings as an educational facility in Padang city is the Elementary School building of the 23/24 (SD 23/24 Padang), located close to the beach. Based on the evaluation results of the Detail Engineering Design (DED) documents, it is found that this building was designed without taking into account the tsunami loads. Therefore, a building assessment should be carried out to check the capacity of the building to resist the working loads, including the tsunami loads, and to investigate the effect of the tsunami loads on the SD 23/24 Padang building. In this study, the building was analyzed using ETABS v.18 software based on the new Indonesian Seismic Code, SNI 1726-2019 for seismic load and FEMA P646-2019 for calculating tsunami loads. The results show that the SD 23/24 Padang building is strong against earthquake loads, but it doesn't have enough capacity when tsunami loads are applied, in which there are several structural elements (columns/beams) that do not have sufficient capacity to withstand the combined earthquake and tsunami loads. The effect of tsunami loads on the building structure is also discussed in this paper.

1 Introduction

West Sumatra Province is one of the provinces in Indonesia that is vulnerable to natural disasters, especially earthquakes and tsunamis [1]. This is because West Sumatra is traversed by three sources of earthquake threat : the Sumatra Fault Zone, the subduction zone that is a meeting between the India-Australia tectonic plates and the Eurasian plate, and the Mentawai Fault Zone [1,2]. Major earthquakes have occurred several times, one of which was a major earthquake on September 30, 2009, in Padang city, West Sumatra, with a magnitude of 7.6 on the Richter Scale [3]. Although it did not cause a tsunami, the devastating earthquake that hit the coast of West Sumatra has killed around 1700 people and damaged more than 200,000 buildings [3].

Padang city, as the capital city of West Sumatra, is an area that is included in the area with a high level of vulnerability (*High Risk Zone)* to tsunami [4]. Therefore, the construction of public facilities such as hospitals, government offices, sport centers, and school buildings must have certain technical engineering that is able to anticipate the damage and collapse of buildings due to the earthquake and tsunami [5]. One of the public facilities in Padang city as an educational facility is the SD 23/24 Padang building.

SD 23/24 Padang has been operating since January 1, 1976, where this school building is owned by the local government.

The SD 23/24 Padang building consists of three buildings which are three-story and two-story reinforced concrete buildings. Based on the evaluation results of the Detail Engineering Design (DED) documents, this building was designed without taking into account the tsunami loads. This building is located close to the beach, which will have an impact in the event of an earthquake with a potential tsunami. Therefore, a building assessment should be carried out to check the capacity of the building to resist the working loads, including the tsunami loads, and to investigate the effect of the tsunami loads on the SD 23/24 Padang building.

* Corresponding author: fauzan@eng.unand.ac.id

2 Building's data and analysis

2.1 Location of the building

Fig. 1 shows the location of SD 23/24 Padang on Veteran street No. 90, Padang, West Sumatra by using Google Earth, where this building is located at a distance of $\pm$ 474 M from the beach.

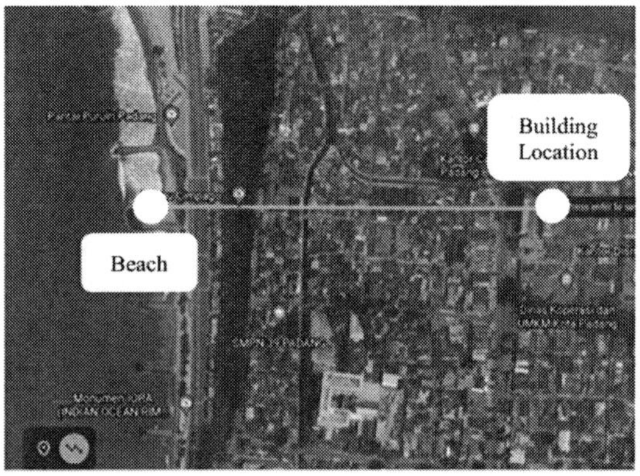

Fig. 1. The Location Of The SD 23/24 Padang Building (*Google Earth*)

The location of the building is also in the vicinity of urban areas, offices, and densely populated residential areas.

2.2 Tsunami vulnerability level

Based on the tsunami-prone zone map of Padang city shown in Fig. 2, the SD 23/24 Padang building is located in the West Padang area, an area with a high level of vulnerability (*High Risk Zone)* to tsunamis [1].

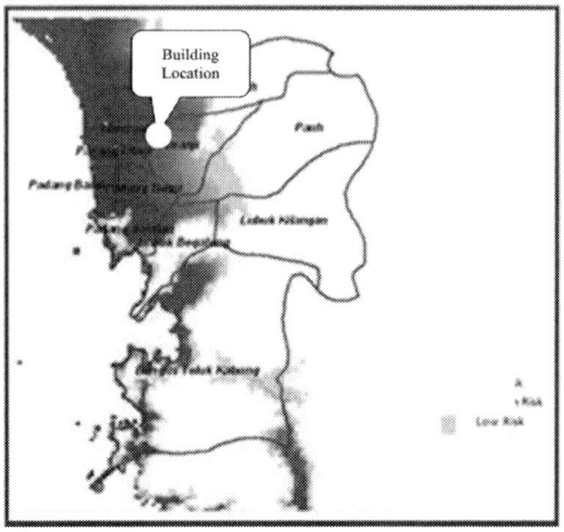

Fig. 2. Tsunami-Prone Zone Of Padang City

2.3 Building structure data

The parameters used in the structural analysis are the technical data of the building and the existing condition of the building obtained from the field survey by LPPM expert of Andalas University with the following data:

Building Name	: SD 23/24 Padang
Location	: Veteran street No.90 Ujung Gurun, Padang
Distance from the beach	: 474 m
Number of building stories	: 3 stories
Height between floors	: 4.00 m
Height above sea level	: 0.80 m

Table 1 shows the cross-sectional dimensions of columns and beams of the SD 23/24 Padang building. The plan and front view and of the building can be seen in Figs. 3 and 4.

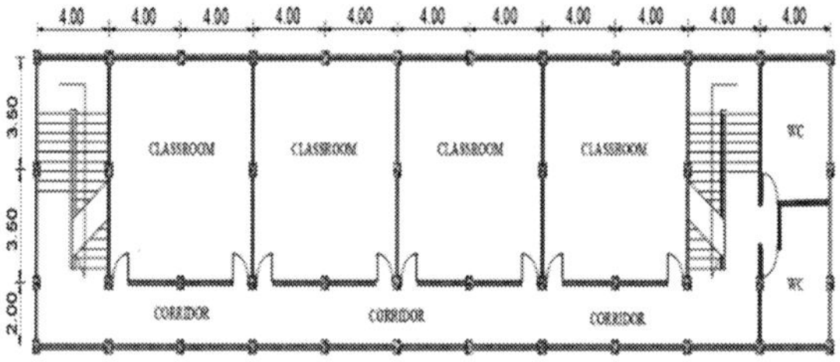

a. First and Second Floor Plans

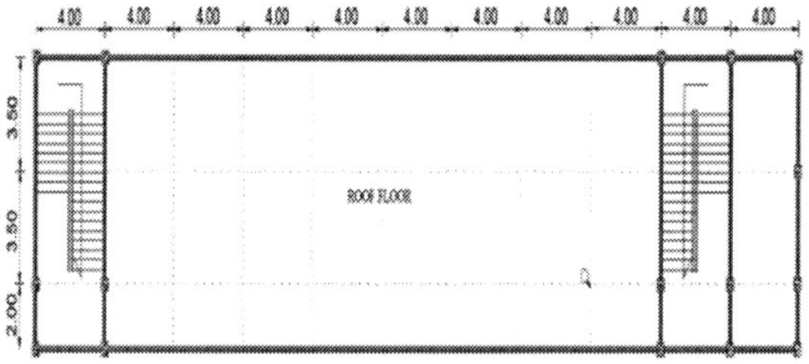

b. Third Floor Plan

Fig. 3. Plan of SD 23/24 Padang Building

Fig. 4. Front View of SD 23/24 Padang Building

Table 1. Column and beam dimensions

No	Column Code	Dimension	
		Height (mm)	Width (mm)
1	C4	450	450
2	C5	350	350
3	C6	450	350
4	CS	250	250
No	**Beam Code**	**Dimension**	
		Height (mm)	Width (mm)
1	1G1	700	300
2	1G2	550	300
3	1G3	500	450
1	2G1	700	300
2	2G2	500	300
3	2G3	450	450
1	RG1	700	300
2	RG2	500	300
3	RG3	450	350

2.4 Modeling of the building structure

The SD 23/24 Padang building structure was analyzed using software ETABS v.18 with 3D modeling. The columns and beams of the building structure are modeled as frames, the slab is modeled as a slab element, and the shear wall is modeled as a wall element. The modeling is carried out according to the current condition of the SD 23/24 Padang building. The structural analysis of SD 23/24 Padang building is conducted into two load conditions, namely:

a. Buildings are only subjected to live, dead and earthquake loads.
b. Buildings are subjected to live, dead, earthquake and tsunami loads.

All structural elements such as beams, columns, slabs, and shear walls are modeled in the ETABS v.18 structural analysis application, as shown in Fig. 5.

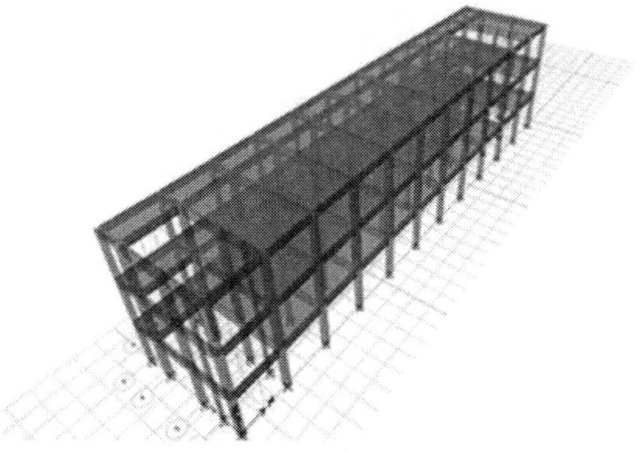

Fig. 5. Modeling of the Building Structure

2.5 Loads analysis

2.5.1 Dead load

Dead loads on the SD 23/24 Padang building are:

a. Self-weight of structural elements was calculated directly by structural analysis program ETABS v.18
b. Flooring material (1 cm thickness)
   1 x 48 kg/m^2 = 48 kg/m^2
c. Flooring mortar (2 cm thickness)
   2 x 21 kg/m^2 = 42 kg/m^2
d. Ceiling Weight
   1 x 20 kg/m2= 20 kg/m^2
   Total = 110 kg/m^2
   = 1,078 kg/m^2 $\approx$ 1,1 kN/m^2 [6]

Self-weight of building materials and components refers to Article 3.1.1, SNI 1727:2020, "Minimum load for designing buildings and other structures" [7].

2.5.2 Live load

Based on Article 4.3.1, SNI 1727:2020, the live load on the SD 23/24 Padang building is taken as 1.92 kN/m^2 for classrooms and 4.79 kN/m^2 for corridor [7].

2.5.3 Earthquake load

Analysis of the structure of the SD 23/24 Padang building uses an earthquake response analysis approach to the dynamic response spectrum. The response spectrum of the earthquake plan for the structural analysis of this building can be seen in Table 2. Earthquake loading is calculated based on the new Indonesian Seismic Code, SNI 1726:2019 (Earthquake Resistance Planning Procedures for Building and Non-Building Structures) [8]. Fig. 6 shows the response Spectrum Design of Padang City.

Table 2. Response spectrum parameters

No	Data	Variabel	Nilai
1	Risk Category Office Buildings	K	Category IV
2	Priority Factors for Earthquake	I	1.5
3	Parameters of Land Acceleration (Ss, S1)		
	a. Acceleration of MCE Spectral Response from Earthquake Map in Short Period	SS	1.500
	b. MCE Spectral Response Acceleration from Earthquake Map in 1 second Period	S1	0.600
4	Location Class/Site Classification	KS	Medium (SD)
5	Site Coefficient Factor (Fa,Fv)		
	a. Site coefficient for short period	Fa	1.000
	b. Site Coef for long period	Fv	1.700
6	Acceleration Spectrum Response Parameters (SMS and SM1)		
	a. MCE Spectral Response Acceleration Short	SMS	1.500
	b. MCE Spectral Response Acceleration Period 1 second adjusted for site class (SM1	SM1	1.020
7	Design Spectral Acceleration Parameters (SDS and SD1)		
	a. Acceleration of Spectral Response in a Short Period (SDS = 2/3 SMS	SDS	1.000
	b. Acceleration Spectral Response Over a period of 1 second (SD1 = 2/3 SM1)	SD1	0.680
8	Design Response Spectrum Time		
	a. To = 0,2 SD1/SDS	To	0.136
	B. Ts = SD1/SDS	Ts	0.680
9	Seismic Design Category	KDS	D

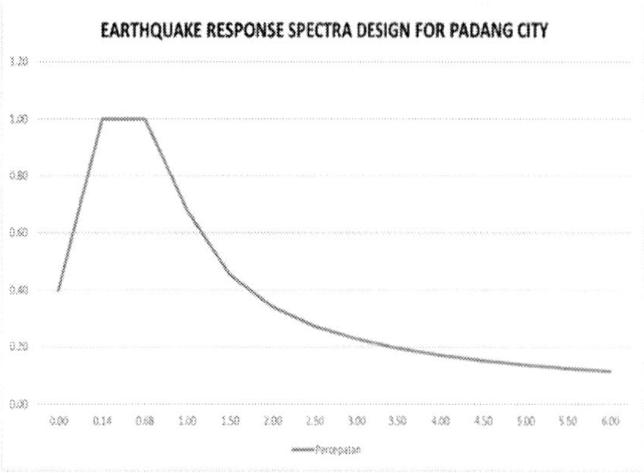

Fig. 6. Response Spectra Design of Padang City

2.5.4 Tsunami load

The calculation of the tsunami load is conducted based on the FEMA P646-2019 "Guidelines for Design of Structures for Vertical Evacuation from Tsunamis" [9]. There are several types of tsunami loads that are taken into account in the structural analysis, where the magnitude of each load value is calculated based on the predicted tsunami wave height, subgrade elevation of the design area, distance to the coast [10, 11], as shown in Fig. 7.

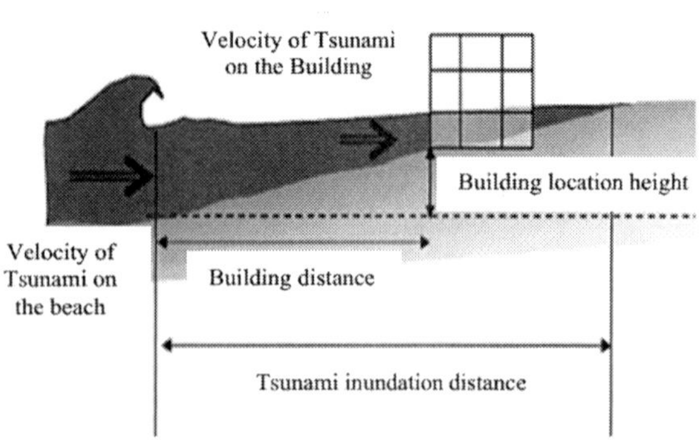

Fig. 7. Sketch of Tsunami Inundation

According to point 8.7 FEMA-2019, it is explained that not all tsunami loads affect the structural components of a building dépends on the layout of the building [9]. In this study, the tsunami loads calculated in this building are illustrated in Figs. 8 – 10.

a. Hydrostatic Force

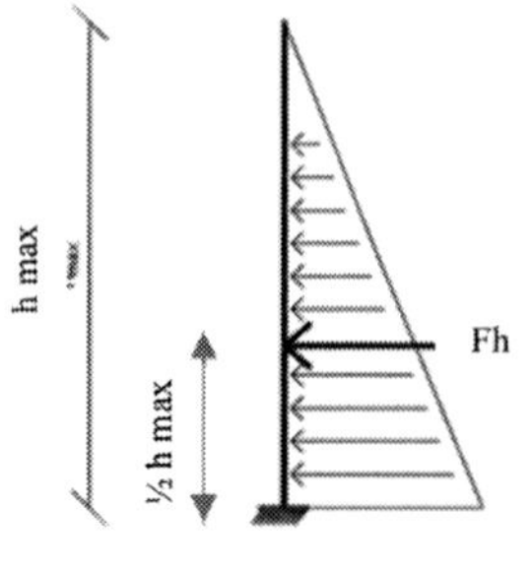

Fig. 8. Tsunami Hydrostatic Force

$$F_h = \frac{1}{2}\rho_S \; x \; g \; x \; b \; x \; h_{max}{}^2 \quad (1)$$

The fluid density including sediment, ρ_S = 1100 kg/m^3
The breadth (width) of the wall , b = 4 m
The gravitational acceleration g = 9,81 m/s^2
The maximum water height, h_{max} = 5,7 m

$F_h = 78.885\ kN$

b. Hydrodynamic Force

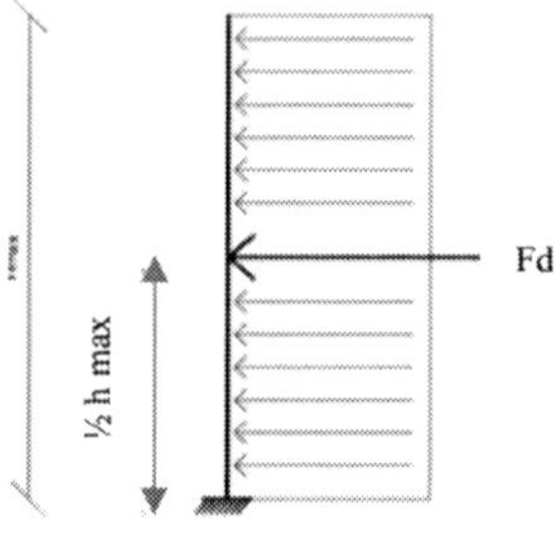

Fig. 9. Tsunami Hydrodynamic Force

$$F_d = \frac{1}{2}\rho_S \; x \; c_d \; x \; (hu^2)_{max} \quad (2)$$

The drag coefficient, c_d = 2
$F_d = 20{,}053\ kN$

c. Impulse Force

$$F_d = 1.5\ F_d \quad (3)$$
$$= 30.080\ kN$$

d. Force Due to Debris Impact

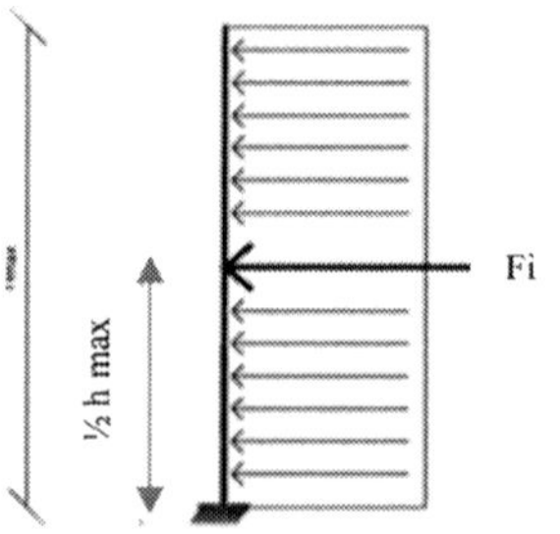

Fig. 10. The Force Due to the Impact of the Tsunami Debris

$$F_i = 1{,}3u_{max} \; x \; \sqrt{[km_d(1+c)]} \quad (4)$$

The maximum flow velocity, u_{max} = 10.575 m/s
Mass of the debris , m_d = 450 kg
Hydrodynamic mass coefficient, c = 2
The effective net combined stiffness , k = 2400000 N/m

F_i = 782.533 kN

2.6 Loads combination

Structural loads combination refers to the New Indonesian Seismic Code (SNI 1726:2019) [4]:

1. 1,4 *D*
2. 1,2 *D* + 1,6 *L* + 0,5 (*Lr* or *R*)
3. 1,2 *D* + 1,6 (*Lr* atau *R*) + (*L* or 0,5 *W*)
4. 1,2 *D* + 1,0 *W* + *L* + 0,5 (*Lr* or *R*)
5. 0,9 D+ 1,0 W
6. 1,2 *D* + *EV* + *EH* + *L*
7. 0,9 *D* - *EV* + *EH*

Combination of tsunami loads acting on the structure refers to FEMA P-646 [8]:

1. 1,2 *DL* + 1,0 *TS* +1,0 *Lref* + 0,25 *LL*
2. 0,9 *DL* + 1,0 *TS*

Where:

DL = Dead Load
LL = Live Load
Lr = Live Load on the Roof
W = Wind Load
R = Rain Load
EQX= Earthquake Load on X Direction
EQY = Earthquake Load on Y Direction
EV = 0,2SDS x D
EH = Rho x QE
Rho = 1,3 (Redundancy)
Ts = Tsunami Load
Lref = Live Load in Refuge Area

The loading combination in this analysis is shown in Table 3.

3 Results and discussion

3.1 Inter story drift

3.1.1 Inter story drift of the building

Based on article 7.8.6 of SNI 1726:2019, the determination of inter story drift (Δ) must be calculated as the difference in the deviation at the center of mass above and below the level under consideration [8]. Table 4 shows the results of inter story drift for the SD 23/24 Padang building.

From Table 4, it can be seen that the inter story drift in the X and Y directions, without and with the Tsunami loads, are still in the allowable range. The increase in maximum inter story drift due to the tsunami loads in the X direction and Y direction is 118.75% and 77.67%, respectively.

3.2 Structural sectional capacity

3.2.1 Coulomn sectional capacity

To determine the capacity of the column in carrying the load, it can be observed from the column interaction diagram [12 - 14]. The P-M interaction diagram is a diagram that describes the capability or capacity of the column based on the relationship between the moment and the axial load of the column [15, 16]. The P-M interaction diagrams of building columns without and with tsunami loads are shown in Figs. 11 and 12.

From Fig. 11, it can be seen that the column on the 1st floor of the building is strong enough to carry the working loads because the axial moment and compression forces do not outside the design axial moment line.

While when tsunami loads are added, the column on the 1st floor of the building is not strong enough to carry the working loads because there are axial moment and compression force that passes through the design axial moment line, as shown in Fig. 12.

Table 3. Load Combination

Earthquake Load Combination											
1	1.4	DL									
2	1.2	DL	+	1.6	LL						
3	1.378	DL	+	1	LL	+	1.3	EQX	+	0.39	EQY
4	1.378	DL	+	1	LL	+	1.3	EQX	-	0.39	EQY
5	1.378	DL	+	1	LL	-	1.3	EQX	+	0.39	EQY
6	1.378	DL	+	1	LL	-	1.3	EQX	-	0.39	EQY
7	1.378	DL	+	1	LL	+	0.39	EQX	+	1.3	EQY
8	1.378	DL	+	1	LL	+	0.39	EQX	-	1.3	EQY
9	1.378	DL	+	1	LL	-	0.39	EQX	+	1.3	EQY
10	1.378	DL	+	1	LL	-	0.39	EQX	-	1.3	EQY
11	0.722	DL	+	1.3	EQX	+	0.39	EQY			
12	0.722	DL	+	1.3	EQX	-	0.39	EQY			
13	0.722	DL	-	1.3	EQX	+	0.39	EQY			
14	0.722	DL	-	1.3	EQX	-	0.39	EQY			
15	0.722	DL	+	0.39	EQX	+	1.3	EQY			
16	0.722	DL	+	0.39	EQX	-	1.3	EQY			
17	0.722	DL	-	0.39	EQX	+	1.3	EQY			
18	0.722	DL	-	0.39	EQX	-	1.3	EQY			
19	Envelope										
Tsunami Load Combination											
1	1.2	DL	+	0.25	LL	+	1	Ts	+	1	Lref
2	0.9	DL	+	1	Ts						

Table 4. Inter story drift of the building on X and Y directions without and with tsunami load

Story	Displacement - X		Delta Limit	CHECK	Percentage Increase
	Without Tsunami Load	With Tsunami Load			
	Δx (mm)	Δx (mm)	mm	Δ Izin > Δx	mm
3	12.04	26.33	40.00	OK	118.75%
2	20.03	39.65	40.00	OK	97.96%
1	16.84	30.02	40.00	OK	78.26%
0	0.00	0	0	-	0.00%

Story	Displacement - Y		Delta Limit	CHECK	Percentage Increase
	Without Tsunami Load	With Tsunami Load			
	Δx (mm)	Δx (mm)	mm	Δ Izin > Δx	mm
3	9.33	16.57	40.00	OK	77.67%
2	16.83	29.22	40.00	OK	73.61%
1	16.56	26.88	40.00	OK	62.31%
0	0	0	0	-	0.00%

https://doi.org/10.1051/e3sconf/202133107016

- **Without Tsunami**

First Floor P-M Interaction Diagram

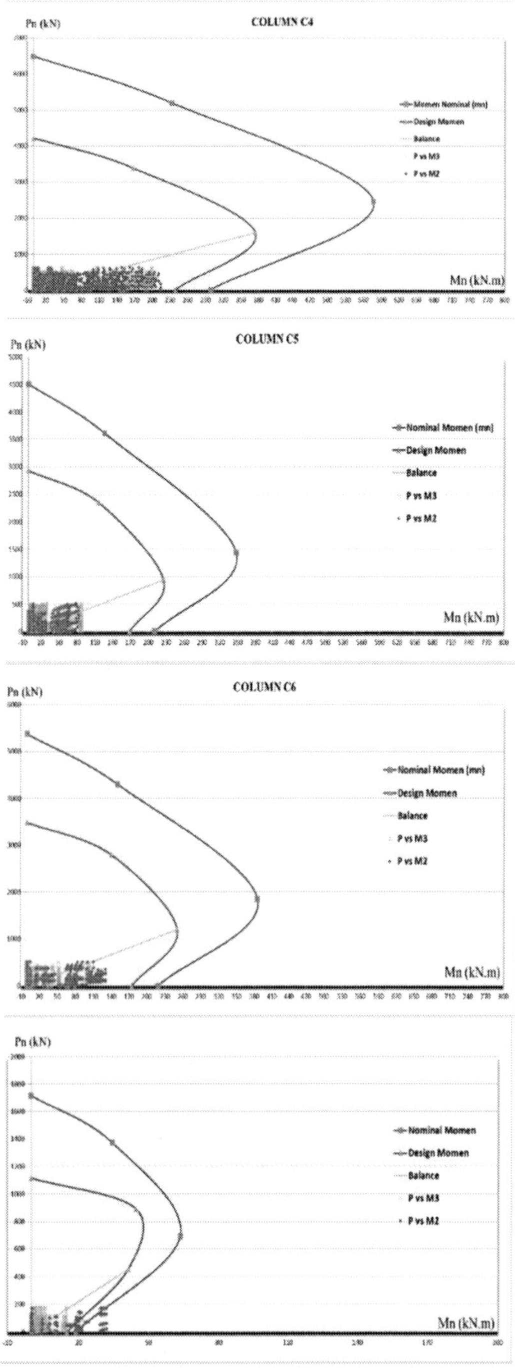

Fig. 11. Interaction Diagram of Column on 1st Floor Building Without Tsunami Load

- **With Tsunami**

First Floor P-M Interaction Diagram

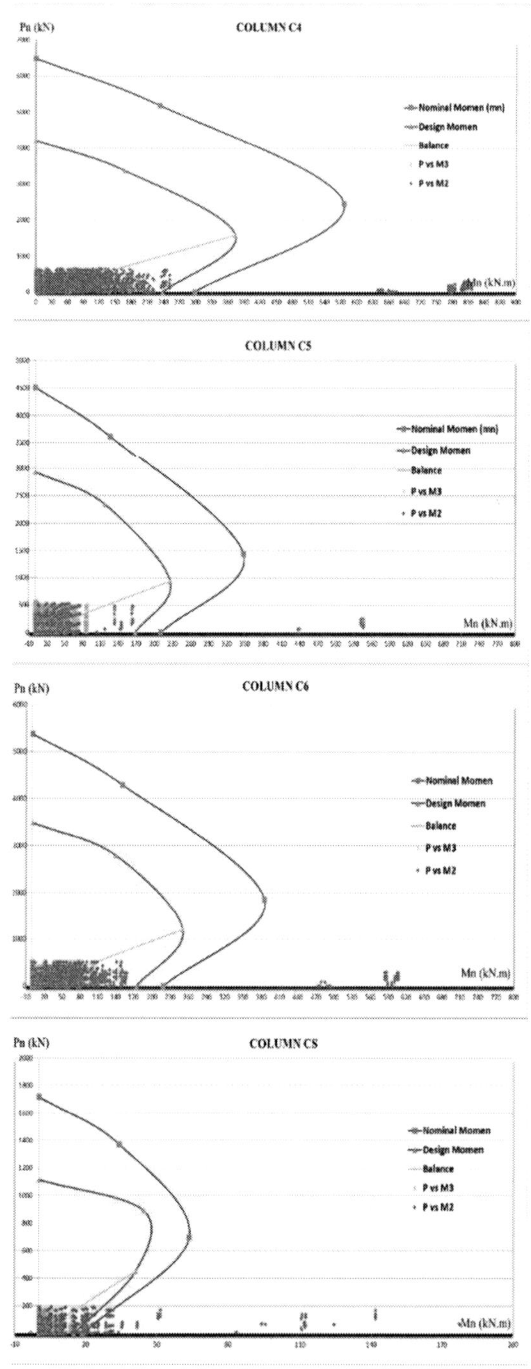

Fig. 12. Interaction Diagram of Column on 1st floor of Building with Tsunami Load

The calculation results of the column shear capacity without and with tsunami load are shown in Tables 5 and 6.

 https://doi.org/10.1051/e3sconf/202133107016

Table 5. Column shear capacity of the building without tsunami load

	Column Type	Cross-Section dimention	Stirrup Diameter	Distance between Stirrup	ΦVn	Vu	Note
		(B x H) mm	(mm)	(mm)	(kN)	(kN)	
1st Floor	C4	450 x 450	10	100	345.731	110.413	OK
	C5	350 x 350	10	100	239.129	42.900	OK
	C6	450 x 350	10	100	261.406	75.316	OK
	CS	250 x 250	10	100	146.899	19.083	OK
2nd Floor	C4	450 x 450	10	100	345.731	85.277	OK
	C5	350 x 350	10	100	239.129	33.364	OK
	C6	450 x 350	10	100	261.406	72.908	OK
	CS	250 x 250	10	100	146.899	20.662	OK
3rd Floor	C4	450 x 450	10	100	345.731	42.120	OK
	C5	350 x 350	10	100	239.129	13.277	OK
	C6	450 x 350	10	100	261.406	41.309	OK
	CS	250 x 250	10	100	146.899	11.091	OK

As seen in Table 5, the building columns are able to withstand the shear forces acting on the structure.

Table 6. Column shear capacity of the building with tsunami load

	Column Type	Cross-Section dimention	Stirrup Diameter	Distance between Stirrup	ΦVn	Vu	Note
		(B x H) mm	(mm)	(mm)	(kN)	(kN)	
1st Floor	C4	450 x 450	10	100	345.731	716.812	NOT OK
	C5	350 x 350	10	100	239.129	271.705	NOT OK
	C6	450 x 350	10	100	261.406	359.573	NOT OK
	CS	250 x 250	10	100	146.899	177.263	NOT OK
2nd Floor	C4	450 x 450	10	100	345.731	88.734	OK
	C5	350 x 350	10	100	239.129	35.755	OK
	C6	450 x 350	10	100	261.406	75.522	OK
	CS	250 x 250	10	100	146.899	22.131	OK
3rd Floor	C4	450 x 450	10	100	345.731	45.577	OK
	C5	350 x 350	10	100	239.129	16.734	OK
	C6	450 x 350	10	100	261.406	44.766	OK
	CS	250 x 250	10	100	146.899	14.548	OK

Table 7. Percentage change in shear capacity of building columns due to tsunami load

	Column Type	Vu (kN)		Percentage Increase
		Without Tsunami Load	Without Tsunami Load	
1st Floor	C4	110.413	716.812	549.21%
	C5	42.900	271.705	533.35%
	C6	75.316	359.573	377.42%
	CS	19.083	177.263	828.92%
2nd Floor	C4	85.277	88.734	4.05%
	C5	33.364	35.755	7.17%
	C6	72.908	75.522	3.59%
	CS	20.662	22.131	7.11%
3rd Floor	C4	42.120	45.577	8.21%
	C5	13.277	16.734	26.04%
	C6	41.309	44.766	8.37%
	CS	11.091	14.548	31.17%

While the columns on the 1st floor of the building are unable to withstand the shear forces acting on the structure due to the addition of the tsunami loads, as shown in Table 6.

From Table 7, it can be seen that the 1st floor column experienced an increase in the maximum shear force due to the influence of the tsunami load of 828.92%.

3.2.2 Capacity of beams

a. Bending capacity of beams

The calculation results of beam flexural capacity without and with tsunami load can be seen in **Tables 8** and **9**.

Table 8. Bending capacity of building beam without tsunami load

	BEAM TYPE	Cross-Section dimention	Note	ØMn (kN)		Mu (kN)		ØMn ≥ Mu	
				Positive	Negative	Positive	Negative	Positive	Negative
1st Floor	1G1	30 X 70	At Support	252.68	252.68	214.83	216.86	OK	OK
			At Field	252.68	252.68	138.75	134.12	OK	OK
	1G2	30 X 55	At Support	145.85	145.85	141.65	157.44	OK	NOT OK
			At Field	145.85	145.85	76.10	73.55	OK	OK
	1G3	45 X 50	At Support	130.96	130.96	130.42	134.28	OK	NOT OK
			At Field	130.96	130.96	71.70	69.17	OK	OK
2nd Floor	2G1	30 X 70	At Support	190.50	190.50	141.86	151.06	OK	OK
			At Field	190.50	190.50	95.92	98.22	OK	OK
	2G2	30 X 50	At Support	130.96	130.96	97.61	117.50	OK	OK
			At Field	130.96	130.96	55.77	50.27	OK	OK
	2G3	45 X 45	At Support	118.07	118.07	79.92	99.03	OK	OK
			At Field	118.07	118.07	50.42	43.40	OK	OK
3rd Floor	RG1	30 X 70	At Support	190.50	190.50	61.30	75.56	OK	OK
			At Field	190.50	190.50	43.74	59.33	OK	OK
	RG2	30 X 50	At Support	130.96	130.96	53.83	72.39	OK	OK
			At Field	130.96	130.96	36.09	28.08	OK	OK
	RG3	35 X 45	At Support	116.08	116.08	33.81	41.29	OK	OK
			At Field	116.08	116.08	25.80	17.81	OK	OK

Table 9. Bending capacity of building beams with tsunami load

	BEAM TYPE	Cross-Section dimention	Note	ØMn (kN)		Mu (kN)		ØMn ≥ Mu	
				Positive	Negative	Positive	Negative	Positive	Negative
1st Floor	1G1	30 X 70	At Support	252.68	252.68	227.46	229.49	OK	OK
			At Field	252.68	252.68	170.92	146.75	OK	OK
	1G2	30 X 55	At Support	145.85	145.85	148.95	164.73	OK	NOT OK
			At Field	145.85	145.85	83.39	80.84	OK	OK
	1G3	45 X 50	At Support	130.96	130.96	136.96	140.83	OK	NOT OK
			At Field	130.96	130.96	78.82	75.72	OK	OK
2nd Floor	2G1	30 X 70	At Support	190.50	190.50	147.58	156.77	OK	OK
			At Field	190.50	190.50	101.63	103.94	OK	OK
	2G2	30 X 50	At Support	130.96	130.96	101.54	121.43	OK	OK
			At Field	130.96	130.96	59.70	54.20	OK	OK
	2G3	45 X 45	At Support	118.07	118.07	83.46	102.57	OK	OK
			At Field	118.07	118.07	53.96	46.95	OK	OK
3rd Floor	RG1	30 X 70	At Support	190.50	190.50	63.20	77.47	OK	OK
			At Field	190.50	190.50	45.64	61.24	OK	OK
	RG2	30 X 50	At Support	130.96	130.96	55.14	73.70	OK	OK
			At Field	130.96	130.96	37.40	29.39	OK	OK
	RG3	35 X 45	At Support	116.08	116.08	34.97	42.45	OK	OK
			At Field	116.08	116.08	26.96	18.97	OK	OK

https://doi.org/10.1051/e3sconf/202133107016

As can be seen in Tables 8 and 9, almost all the beams in the building without and with tsunami loads are able to withstand the working loads, but some beam structural elements do not have sufficient capacity to withstand the earthquake and tsunami loads, that are the 1G2 and 1G3 beams.

Table 10. Percentage change in bending capacity of building block due to tsunami load

	BEAM TYPE	Note	Mu (kN)				Percentage Increase	
			Without Tsunami Load		WithTsunami Load			
			Positive	Negative	Positive	Negative	Positive	Negative
1st Floor	1G1	At Support	214.83	216.86	227.46	229.49	5.88%	5.83%
		At Field	138.75	134.12	170.92	146.75	23.19%	9.42%
	1G2	At Support	141.65	157.44	148.95	164.73	5.15%	4.63%
		At Field	76.10	73.55	83.39	80.84	9.58%	9.92%
	1G3	At Support	130.42	134.28	136.96	140.83	5.02%	4.88%
		At Field	71.70	69.17	78.82	75.72	9.93%	9.47%
2nd Floor	2G1	At Support	141.86	151.06	147.58	156.77	4.03%	3.78%
		At Field	95.92	98.22	101.63	103.94	5.96%	5.82%
	2G2	At Support	97.61	117.50	101.54	121.43	4.02%	3.34%
		At Field	55.77	50.27	59.70	54.20	7.04%	7.81%
	2G3	At Support	79.92	99.03	83.46	102.57	4.43%	3.58%
		At Field	50.42	43.40	53.96	46.95	7.03%	8.16%
3rd Floor	RG1	At Support	61.30	75.56	63.20	77.47	3.11%	2.52%
		At Field	43.74	59.33	45.64	61.24	4.36%	3.21%
	RG2	At Support	53.83	72.39	55.14	73.70	2.43%	1.81%
		At Field	36.09	28.08	37.40	29.39	3.63%	4.66%
	RG3	At Support	33.81	41.29	34.97	42.45	3.43%	2.81%
		At Field	25.80	17.81	26.96	18.97	4.50%	6.52%

From Table 10, it can be seen that due to the effect of the tsunami load on the building, there was no significant increase in the flexural capacity of the beams. The increase in the flexural capacity of the beams is 1.81% - 23.19%.

b. Shear Capacity of the beam

The calculation results of the shear capacity of the beams can be seen in Tables 11 and 12.

Table 11. Shear capacity of building beams without tsunami load

	BEAM TYPE	Cross-Section dimention	Note	ØVn (kN)	Vu (kN)	ØVn ≥ Vu
1st Floor	1G1	30 X 70	At Support	419.77	261.96	OK
			At Field	359.29	256.13	OK
	1G2	30 X 55	At Support	324.36	144.07	OK
			At Field	277.63	130.38	OK
	1G3	45 X 50	At Support	312.40	226.00	OK
			At Field	312.40	221.55	OK
2nd Floor	2G1	30 X 70	At Support	359.29	185.99	OK
			At Field	359.29	178.58	OK
	2G2	30 X 50	At Support	292.56	113.40	OK
			At Field	250.41	100.14	OK
	2G3	45 X 45	At Support	278.44	156.41	OK
			At Field	278.44	150.72	OK
3rd Floor	RG1	30 X 70	At Support	359.29	115.64	OK
			At Field	359.29	133.19	OK
	RG2	30 X 50	At Support	250.41	87.03	OK
			At Field	250.41	73.58	OK
	RG3	35 X 45	At Support	223.20	68.68	OK
			At Field	223.20	63.68	OK

Table 12. Shear capacity of building beams with tsunami load

	BEAM TYPE	Cross-Section dimention	Note	ØVn (kN)	Vu (kN)	ØVn ≥ Vu
1st Floor	1G1	30 X 70	At Support	419.77	385.58	OK
			At Field	359.29	377.93	NOT OK
	1G2	30 X 55	At Support	324.36	153.80	OK
			At Field	277.63	138.71	OK
	1G3	45 X 50	At Support	312.40	235.37	OK
			At Field	312.40	230.92	OK
2nd Floor	2G1	30 X 70	At Support	359.29	193.18	OK
			At Field	359.29	185.77	OK
	2G2	30 X 50	At Support	292.56	119.25	OK
			At Field	250.41	105.15	OK
	2G3	45 X 45	At Support	278.44	161.98	OK
			At Field	278.44	156.29	OK
3rd Floor	RG1	30 X 70	At Support	359.29	119.23	OK
			At Field	359.29	136.78	OK
	RG2	30 X 50	At Support	250.41	89.54	OK
			At Field	250.41	76.08	OK
	RG3	35 X 45	At Support	223.20	70.91	OK
			At Field	223.20	65.92	OK

As seen in Table 11, almost all beams in the building are able to withstand the loads acting on the structure. While one element of the beam (1G1) does not have sufficient capacity to withstand the working loads when the tsunami loads are applied (Table 12).

Table 13. Percentage change in shear capacity of building block due to tsunami load

	BEAM TYPE	Note	Vu (kN)		Percentage Increase
			Without Tsunami Load	WithTsunami Load	
1st Floor	1G1	At Support	261.96	385.58	47.19%
		At Field	256.13	377.93	47.56%
	1G2	At Support	144.07	153.80	6.75%
		At Field	130.38	138.71	6.39%
	1G3	At Support	226.00	235.37	4.15%
		At Field	221.55	230.92	4.23%
2nd Floor	2G1	At Support	185.99	193.18	3.86%
		At Field	178.58	185.77	4.02%
	2G2	At Support	113.40	119.25	5.16%
		At Field	100.14	105.15	5.00%
	2G3	At Support	156.41	161.98	3.56%
		At Field	150.72	156.29	3.69%
3rd Floor	RG1	At Support	115.64	119.23	3.11%
		At Field	133.19	136.78	2.70%
	RG2	At Support	87.03	89.54	2.88%
		At Field	73.58	76.08	3.40%
	RG3	At Support	68.68	70.91	3.25%
		At Field	63.68	65.92	3.50%

The effect of the tsunami load on the building shown in Table 13 shows that there was no significant increase in the load on the shear capacity of the beam. The load on the shear capacity of the beam increased by 2.70% - 47.56 %.

4 Conclusion

Based on this study, it can be concluded that:

1. SD 23/24 Padang building is strong against earthquake load, but it does not have enough capacity when tsunami loads are applied, in which there are several structural elements (columns/beams) that do not have sufficient capacity to withstand the combined earthquake and tsunami loads.
2. The addition of the tsunami loads on the SD 23/24 Padang building increases the inter story drift by around 78 – 119% in the X direction and 62 – 78% in the Y direction.
3. Due to the addition of the tsunami loads, there was an increase in the maximum column shear force around 828.92%.
4. In beam elements, there is a maximum increase in bending and shear forces due to the addition of tsunami loads, which are around 23.19% and 47.56%, respectively.

Acknowledgments to Civil Engineering Department, Engineering Faculty, Andalas University, which has funded this research with the contract number: 123/UN16.09.D/PL/2021.

Reference

1. National Center for Earthquake Studies, *Source and Hazard Map of the 2017 Indonesian Earthquake*, Bandung, Ministry of PUPR (2019)
2. B. Bambang et al. *Examples of Earthquake Resistant Building Designs*, Bandung: ITB (2017)
3. O. Dian, S. Manurung, *Geospatial Model of Potential Tsunami Vulnerability in Padang City*. Padang, BMKG, (2010)
4. S. U, Intan, *The Effect of the Tsunami Load on the New DPRD Building for the Province of West Sumatra*, Padang, UNAND. (2017)
5. Regional Regulation of the City of Padang Number 7 of 2015 concerning Buildings, Padang (2015)
6. Department of Settlement and Regional Infrastructure, *Indonesian Imposition of Regulations for Buildings (PPIUG) 1983,* Jakarta, BSN (2002)
7. National Standardization Agency, SNI 1727: 2020. *Minimum Load for Planning of Buildings and Other Structures*, BSN, Jakarta (2020)
8. National Standardization Agency, SNI 1726:2019 *Procedure for Planning Earthquake Resistance for Building Structures*, BSN, Jakarta (2019)
9. Applied Technology Council, *Guidelines for Design of Structures for Vertical Evacuation from Tsunamis, FEMA P646,* California, FEMA (2019)
10. I. Muhammad, *Analysis of the Effect of Earthquake and Tsunami Loads on 10-Story Reinforced Concrete Buildings Based on SNI 1727:2020 With Variations in Tsunami Heights*, Palembang, UNSRI (2020)
11. R. E. Finesa, *Vertical Evacuation Building Structure Design for Mitigation, Semarang, UNS* (2015)
12. S. Agus, *Design of Reinforced Concrete Structures (Based on SNI 2847:2013)*, Jakarta, Erlangga, (2016)
13. E. W. Peratundhika, *Planning Structure for Case Studies: Building, TPTC (Teacher Professional Training Center) Padang State University, Padang, UNAND* (2019)
14. Z. Syaifuddin, *Structure Systems in High-rise Buildings, Surabaya*, Humanities Foundation, (2011)
15. W. Sherlya, *Identification of the Resilience of Evacuation Building Structures in Lambung Village Against Tsunami Loads*, Medan, UNSYIAH (2021)
16. National Standardization Agency, *Structural Concrete Requirements for Buildings, SNI 2847:2019,* Jakarta, BSN (2019)

 https://doi.org/10.1051/e3sconf/202133108001

Designing conduit sabo dam series as a debris flow protection structure

Della Anggun Lestari[1,3*], *Teuku* Faisal Fathani[1,3], *Fikri* Faris[1,3], and *Wahyu* Wilopo[2,3]

[1]Department of Civil and Environmental Engineering, Gadjah Mada University, 55281 Yogyakarta, Indonesia
[2]Department of Geological Engineering, Gadjah Mada University, 55281 Yogyakarta, Indonesia
[3]Center of Excellence for Disaster Mitigation and Technological Innovation, Gadjah Mada University, 55281 Yogyakarta, Indonesia

Abstract. A high potential area for landslides was identified upstream of the Air Beras River on Bukit Beriti Besar which was feared to generate debris flow through the Air Beras River. One of the preventive efforts that can be conducted for debris flow is the sabo dam. This study aims to design a series of conduit sabo dams and evaluate the structure stability according to SNI 2851:2015 and Technical Standards and Guidelines for Planning and Design of Sabo Structures (JICA, 2010). By using the design discharge of debris flow of 120 m3/s, the velocity of 5.06 m/s, and the inclination of the riverbed of around 6.16°, the spillway height of the Sabo Dam 1 was determined to be 3 m, while the spillway height of Sabo Dam 1 was 2 m. The Sabo Dam 1 has a height of 10 m and a width of 30 m with a total manageable volume of 49,600 m3 while the Sabo Dam 2 has a height of 10 m and a width of 45 m with a total manageable volume of 91,500 m3. The safety factors of the sabo dam in the debris flow condition were 2.14 for overturning, 1.50 for sliding, and 5.74 for bearing capacity. Based on the result, the conduit sabo dam series effectively controls the destructive power of debris flow.

1 Introduction

Debris flow typically consists of an unsteady, non-uniform surge of the mixture of muddy water and a high concentration of sediment. It attains large mobility from the enlarged void space saturated with water or slurry. It is a secondary disaster that can be initiated by a landslide, riverbed erosion, and natural dam breaching [1-2]. It has a high destructive power that makes it one of the most catastrophic hazards in mountainous areas [3].

A landslide is a form of mass movement that is influenced by several factors such as morphology, mechanics of rock/soil, hydro-geological, and land use. Many landslide occurrences are triggered by storm rainfall and earthquake. Slope stability depends on the form of interaction between the slope constituent material and triggering factors [4]. The excess pore water pressure in the soil causes interference in slope stability. The groundwater level will reach the slip surface so that the pore structure of the soil layer will be damaged, and the soil particles lose their shear resistance, causing deep landslides [5]. The soil particles are no longer bound to each other but stay on top of the liquefied layer as an earth block. If the water in the pore of the earth block has been used up to produce the liquefied layer or the earth block has reached a flatter area, it will stop. However, the debris flow following the earth block will continue flowing down.

Referring to geological and geotechnical studies, the western area of the Air Beras River on Bukit Beriti Besar has a high potential for landslides that can trigger debris flow [6-7]. The rock lithology identified in this area has undergone alteration, resulting in brownish-white and grayish-white alterations. The debris flow in this area was predicted to flow through the Air Beras River. The upstream of this river has a characteristic of a V-shaped river valley with a narrow cross-section and steep river cliffs.

The debris flow disasters can be countered by structural and non-structural works. The structural work is intended to reduce the energy of debris flow, store sediment, or divert the flow, thereby preventing severe damage to the protected facility. In contrast, the non-structural work is aimed to increase the resilience of the environment-community and/or existing buildings against the risk of debris flow. The common structural protection works against debris flow are debris flow shed, deflection wall, and sabo dam. Several previous studies have shown that a sabo dam provides satisfactory performance in storing sediment [8-11]. Based on the construction function, sabo dams can be classified into slit sabo dams, breaker sabo dams, and barrier sabo dams. The

* Corresponding author: della.a.l@eng.ugm.ac.id

slit type can be in the form of a conduit sabo dam has been revealed as the most effective structure to trap large boulders on debris flow [12]. The debris flow in 2016 in the Air Kotok River (the southern of the upstream of the Air Beras River) on Bukit Beriti Besar showed that the debris material was a combination of boulders with a maximum diameter of 1.5 m and trees [13]. Thus, in this study, the conduit sabo dam equipped with sub sabo dam and apron was selected to mitigate debris flow and decrease impact damage.

2 Conduit sabo dam

A Sabo dam is intended to reduce flowing sediment that is moving downstream to prevent the movement of unstable sediment on the riverbed and to control the sediment runoff. A Conduit sabo dam has an additional opening in the main body to drain flowing material and the crest of the main body may be used as a road or bridge (**Fig. 1**).

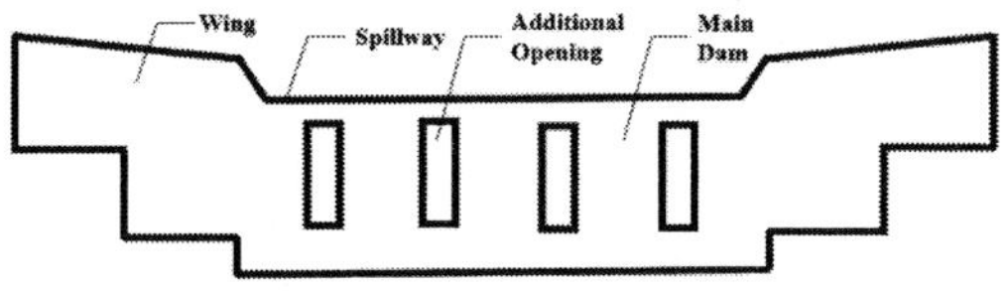

Fig. 1. The conduit sabo dam.

The structure location was selected by considering the result of the feasibility study, which included social, engineering, environmental, and economic aspects [14-15]. The location of the structure should comply with the following conditions:

a. The structural placement is carried out in a location that should satisfy the geological condition and in straight river geometries.
b. A hydraulic review is necessary when the structure is placed on a river bend to determine the runoff and scouring potential upstream and downstream of the structure.
c. Structures and their utilities are placed in an area with large storage volumes, but not in a location where there are residential and productive areas.
d. The structures are protected from the dangers of river degradation. The axis of the main body should be perpendicular to the downstream flow direction.

The use of conduit sabo dam series is intended to be able to manage more sediment volume and thus be more effective. In addition, the sabo dam series works effectively when placed between the end of the source area to the transport zone. Taking into several aspects of the foregoing, the selected location of the sabo dams was shown in **Fig. 2**.

3 Designing sabo dam

The design of the sabo dams for debris flows was in compliance with the SNI 2851:2015 on Designing Sediment Retaining Structures and Technical Standards and Guidelines for Planning and Design of Sabo Structures ratified by JICA (2010). The design work includes the design for the main sabo dam, sub sabo dam, and apron.

3.1 Main sabo dam

The dimensions of the sabo dam were designed to be favorable to catch sediment in the mixture flowing. The design is referred to SNI 2851:2015 and JICA (2010). The height of the sabo dam was determined by considering the storage volume sediment, while the width of the sabo dam was adjusted to the river width.

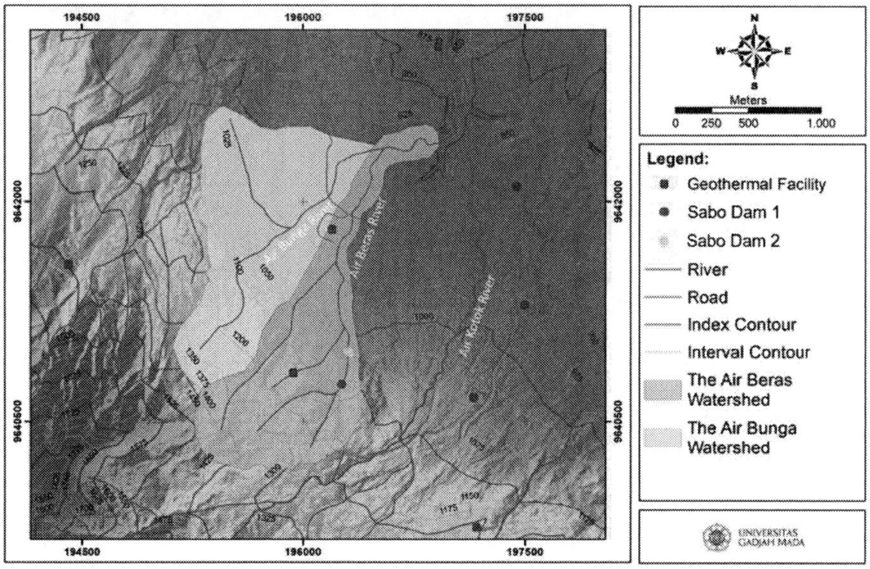

Fig. 2. The location of Sabo Dam

The manageable sediment volume that can be controlled by the sabo dam was calculated using Equation (1) and Equation (2).

$$V_p = nH^2B \quad (1)$$

$$V_c = 0.5nH^2B \quad (2)$$

where V_p is the potential storage volume (m^3); V_c is the control volume (m^3); n is the ratio of riverbed inclination ($1/I$); H is the height of sabo dam (m), and B is the width of sabo dam (m). The calculation result for the dimensions of the sabo dam and the sediment volume can be seen in Table 1.

Table 1. The dimension of the sabo dam and the sediment volume

Parameter	Sabo Dam 1	Sabo Dam 2
River width at structure location (B_{loc}, m)	33.1	61
Width of the dam ($B_{main\ dam}$, m)	30	45
Height of dam (H, m)	10	10
Potential storage volume (V_p, m^3)	33,100	61,000
Control volume (V_c, m^3)	16,500	30,500

The spillway of the sabo dam is designed to escape the stream which is perpendicular to the axis of the downstream river. The width of the crest spillway should be over 3 m and the inclination of crest spillway m_2 is 0.5. Furthermore, in the main body of the sabo dam, additional openings were also made to drain the debris flow. Therefore, if the overflow depth is h_1 (m) and the base width of the spillway is B_1, the discharge of the spillway can be calculated using Equation (3).

$$Q_{co} = (0.71 \times h_1 + 1.77 \times B_1) \times h_1^{3/2} \quad (3)$$

The dimensions of the spillway were determined by a trial-error method to obtain the optimal height. The spillway has a trapezoidal shape, while the additional opening is rectangular. Using the design discharge of debris flow of 120 m^3/s, the velocity of 5.06 m/s, and the inclination of the riverbed of around 6.16°, the dimensions of the spillway can be seen in Tabel 2, where the height of the freeboard is 0.6 m.

3.2 Sub sabo dam and apron

The sub sabo dam is a complementary building designed by considering the dimension of the river and the dimension of the main sabo dam. One of the functions of the sub sabo dam is to reduce the energy flow. Therefore, the height of the sub sabo dam was determined to be 1/3 of the main dam height and the width was adjusted to the width of the river [16]. The sub sabo dam and the main sabo dam were connected by an apron made of reinforced concrete. The apron served to avoid local scour, maintain the stability of the sabo dam, and protect the sabo dam foundation from cliff collapse. The length of the apron was calculated by using Equation (4) and the thickness of the apron with Equation (5).

$$L_a = (1.5 \text{ until } 2.0) \times (h_1 + h_3) \quad (4)$$

$$t_a = 0.2(0.3h_1 + 0.6h_3 - 1.0) \quad (5)$$

where h_1 is the height of the water flow upper spillway and h_3 is the height from the apron's surface to the crest of the main body, respectively in m. The dimension of the sub sabo dam and apron is shown in Table 3.

Table 2. The dimension of the spillway and additional opening

Parameter	Sabo Dam 1	Sabo Dam 2
Flow height (h_1, m)	1.7	1.3
Spillway height ($h_{spillway}$, m)	3.0	2.0
Spillway thickness (t_s, m)	3.0	3.0
The inclination of crest spillway (m_2)	0.5	0.5
Width of the base spillway (B_1, m)	30.0	45.0
Width of crest spillway (B_2, m)	33.0	47.0
Width of additional opening (B_{1as}, m)	1.5	1.5
Height of additional opening (h_{1as}, m)	3.0	3.0
Amount of additional opening	5	5

Table 3. The dimension of the spillway and additional opening

Parameter	Sabo Dam 1	Sabo Dam 2
Height of sub sabo dam (H_2, m)	4.00	4.00
Width of sub sabo dam (B_{sub}, m)	30.00	45.00
Length of the apron (L_a, m)	15.00	15.00
Minimum thickness of apron (t_a, m)	0.90	0.92

4 Stability of main sabo dam

The main body should satisfy the stability requirements, including the stability against overturning, sliding, and bearing capacity. Based on SNI 2851:2015, the minimum safety factor for the dam structure is summarized in Table 4.

Table 4. Requirement for the safety factor

Parameter	Safety Factor
Stability against overturning	≥ 2
Stability against sliding	≥ 1.5
Bearing Capacity	≥ 1.5

4.1 Forces acting on sabo dam

The calculation of the safety factors of these parameters started with calculating the forces acting on the structure. These forces were such as a dead load from the weight of the structure, hydrostatic pressure, earth pressure, the weight of debris flow, debris flow hydro-force, and uplift force. The diagram force can be seen in Fig. 3.

These forces will invent moment with a point of view downstream of the structure. By using the unit weight of water (γ_w) of 9.81 kN/m^3, the unit weight of concrete (γ_c) of 24 kN/m^3, the unit weight of sediment (γ_s) of 16.7 kN/m^3, and the sediment coefficient (C_e) of 0.27, the vertical force, the horizontal forces, and the moment acting on the structures can be seen in **Table 5**.

4.2 Overturning stability

The stability against overturning was evaluated by comparing the resisting moment M_r (kN.m) and the driving moment M_d (kN.m) with a base point downstream as Equation (6). The moment was obtained by multiplying the force by the moment arm length. Overturning failure can occur if the driving moment is greater than the resisting moment.

$$SF_{ot} = \frac{\sum M_r}{\sum M_d} \quad (6)$$

The stability against overturning also can be evaluated by a threshold of eccentricity. Eccentricity can occur due to the deviation of the center of resultant force from the center of mass. Eccentricity is acceptable if the resultant force acts within $b_2/3 < x < 2b_2/3$, where x is the length between the point of working resultant force at the main dam bottom and a downstream end. If the safety factor or eccentricity value does not comply with the requirements, the design of the sabo dam should be reviewed. The result is summarized in Table 6.

Based on Table 6, the resisting moment was greater than the driving moment. From the value of safety factor, it is known that the structure is qualified against overturning and fulfilling the safety requirement. In terms of eccentricity, the value of x was still within the allowed limits.

Table 6. The calc8,5ulation result of stability against overturning

Parameter	Notation	Value
Resisting Moment	ΣM_r (kN.m)	9,289.51
Driving Moment	ΣM_d (kN.m)	4,350.77
Safety Factor	SF_{ot}	2.14
Length of Resultant Force	x (m)	5.27
The base width of the main body	b_2 (m)	10.00
Eccentricity	e (m)	0.27
Threshold	$b_2/3 < x < 2b_2/3$	OK

4.3 Sliding stability

The stability against sliding was influenced by the resultant of the resisting force and the resultant of the driving force. The resultant of the resisting force is a combination of soil shear strength and resistance from the mass structure while the horizontal force is generated by the external force. The stability against sliding may be evaluated using Equation (7).

$$SF_{sl} = \frac{\sum V_{fo} \tan\phi + c \times b_2}{\sum H_{fo}} \quad (7)$$

where V_{fo} is a vertical force (kN), ϕ is internal friction angle (°), c is cohesion (kN/m^2), b_2 is the width of the main body base (m), H_{fo} is a horizontal force (kN).

The principle of it is that the resultant of the resisting force should be greater than the driving force acting on the structure and as a result, the structure can maintain its stable position. The forces acting on the structure can be seen in Table 5. Using the cohesion (c) of subgrade equal to 51.22 kN/m^2 and the internal friction (f) equal to 35.25°, the calculation result of stability against sliding is shown in Table 7.

Table 7. The calculation result of stability against sliding

Parameter	Notation	Value
Resultant of resisting force	ΣF_V (kN)	1,757.60
Resultant of the driving force	ΣF_H (kN)	1,168.93
Stability against sliding	SF_{sl}	1.50

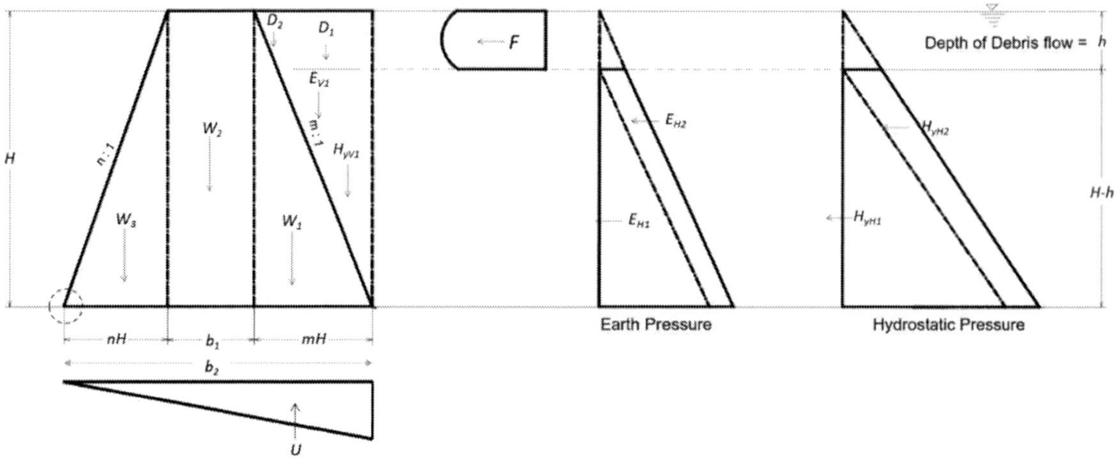

Fig. 3. The diagram force (modified from JICA, 2010).

Table 5. The calculation result of forces acting on sabo dam

Parameter	**Formula**	**Force (kN)**		**Arm Length (m)**	**Moment (kN.m)**	
		Vertical	**Horizontal**		**Vertical**	**Horizontal**
Dead Load (*W*)	$W_1 = 0.5\gamma_c mH^2$	600.00	-	6.67	4,000.00	-
	$W_2 = \gamma_c b_1 H$	720.00	-	3.50	2,520.00	-
	$W_3 = 0.5\gamma_c nH^2$	240.00	-	1.33	320.00	-
Hydrostatic Pressure (H_y)	$H_{yV1} = 0.5\gamma_w m(H-h)^2$	213.13	-	8.45	1,800.12	-
	$H_{yH1} = 0.5\gamma_w (H-h)^2$	-	426.25	3.11	-	1,324.51
	$H_{yH2} = \gamma_w h(H-h)$	-	62.00	4.66	-	288.96
Earth Pressure (*E*)	$E_{V1} = 0.5\gamma_s m(H-h)^2$	362.81	-	8.45	3,064.43	-
	$E_{H1} = 0.5\gamma_s m(H-h)^2$	-	362.81	3.11	-	1,127.39
	$E_{H2} = C_e \rho_f h(H-h)^2$	-	292.21	4.66	-	1,362.00
Weight of Debris Flow (*D*)	$D_1 = \rho_d hm(H-h)$	80.58	-	7.67	617.97	-
	$D_2 = 0.5\rho_d mh^2$	2.93	-	5.23	15.31	-
Debris Flow Hydro-Force (*F*)	$F = \frac{\rho_d}{g} h(v_d)^2$	-	25.66	9.66	-	247.90
Uplift (*U*)	$U = 0.5\gamma_w b_2 (H-h)$	-457.25	-	6.67	3,048.32	-
Total		1,762.19	1,168.93		9,289.51	4,350.77

https://doi.org/10.1051/e3sconf/202133108001

The shear strength parameters of the subgrade gave a significant effect to the resisting force. The shear strength parameters had a linear relationship with shear stability, which was the greater parameters would increase the resisting force. The subgrade had very good shear strength parameters because the constituent bedrock was a compact andesite. Based on the result in Table 7, it can be stated that the structure fulfills the safety requirement, as the value of SF_{sl} is ≥ 1.5.

4.4 Bearing capacity

The bearing capacity was evaluated with the ratio of the ultimate bearing capacity of the subgrade and the maximum normal stress acting on the structure. If q_u is the ultimate capacity of subgrade (kN/m^2) and σ_{max} is maximum normal stress (kN/m^2), a safety factor of bearing capacity is given by Equation (8).

$$SF_{bc} = \frac{q_u}{\sigma_{max}} \quad (8)$$

The ultimate bearing capacity can be calculated using the Terzaghi formula or using the recommended value by JICA (2010). The smallest value of the ultimate bearing capacity was used to obtain the conservative value in the safety factor calculation. While the normal stress was influenced by the vertical force, the base width of the main body, and the eccentricity. The calculation result of bearing capacity was summarized in Table 8.

Table 8. The calculation result of bearing capacity

Parameter	Notation	Value
Ultimate bearing capacity (Terzaghi)	q_u (kN/m^2)	6,840.38
Ultimate bearing capacity (JICA)	q_u (kN/m^2)	1,176.00
Maximum normal stress	σ_{max} (kN/m^2)	204.93
Minimum normal stress	σ_{min} (kN/m^2)	147.51
Safety factor for bearing capacity	SF_{bc}	5.74

The ultimate bearing capacity using the Terzaghi formula gave a greater value than the recommended value by JICA as the Terzaghi formula was addressed to calculate the bearing capacity on a shallow foundation with finite dimensions. The base width of the main body was 10 m; therefore, it will provide a great bearing capacity because the force is distributed on a large area while the ultimate bearing capacity was obtained by JICA recommendation with the assumption that the subgrade is classified as soft rock. Based on the calculation result in Table 8, the structure complies with the safety requirement as the safety factor is more than 1.5. The Illustration of the main sabo dam is shown in Fig. 4.

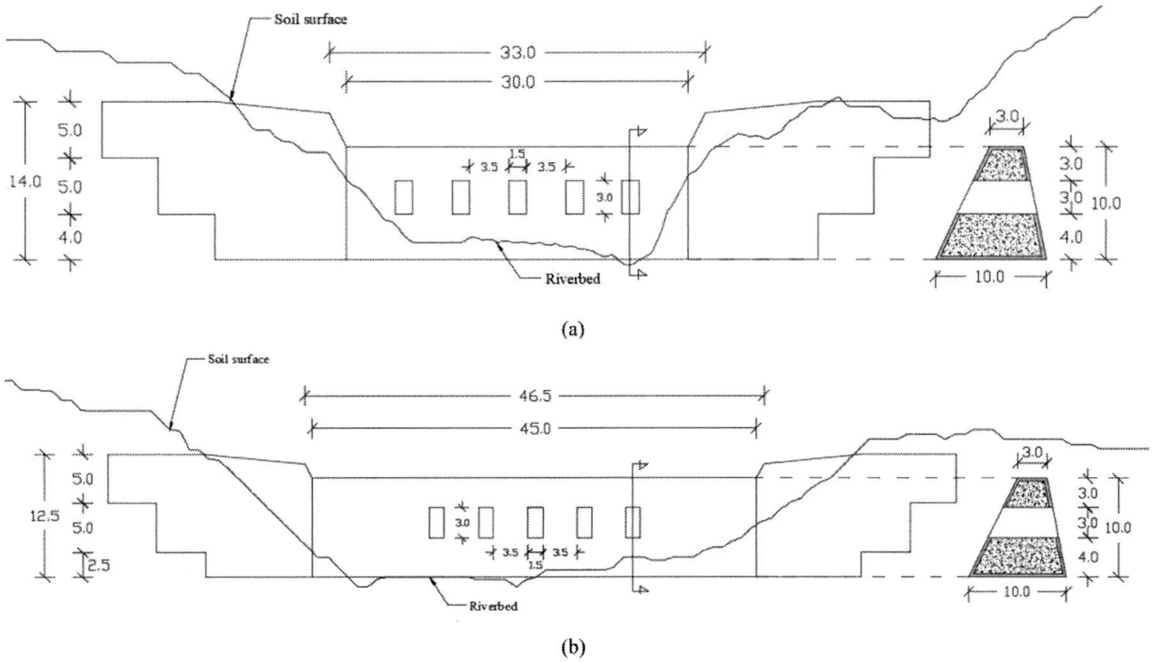

Fig. 4. The cross-section (a) Sabo Dam 1 and (b) Sabo Dam

5 Summary

The design of sabo dam in the Air Beras River was designed using two conduits sabo dams, which has a main dam height of 10 m. The width of the sabo dam was adjusted to the width of the river, i.e., 30 m for Sabo Dam 1 and 45 m for Sabo Dam 2. This conduit sabo dam series can accumulate the sediment with a total manageable volume of 141,100 m^3.

The spillway is planned to escape the flow with the discharge of 120 m^3/s. Consequently, it is determined that the spillway height is 3 m for Sabo Dam 1 and 2 m for Sabo Dam 2. In the body of the main dam, there are five additional openings, each has a height of 3 m and a width of 1.5 m.

The result of the stability evaluation for both conduits sabo dams indicates that the structures comply with the safety requirement of overturning, sliding, and bearing capacity. The safety factor of the sabo dam in the debris flow condition is 2.14 for overturning with the resisting moment equal to 9,289.51 kN.m and the driving moment equal to 4,350.77 kN.m. The safety factor for sliding is 1.50 with the resultant of resisting force of 1,757.60 kN and the resultant of driving force of 1,168.93 kN. The ultimate bearing capacity obtained is 1,176.00 kN/m^2 and the maximum normal stress is 204.93. Therefore, the safety factor for bearing capacity is 5.74.

The author would like to thank the Center for Disaster Mitigation and Technological Innovation Universitas Gadjah Mada (Gama-InaTEK) for providing field investigation data and valuable advice.

References

1. T. Takahashi, Debris Flow: Mechanics, Prediction, and Countermeasures (Taylor & Francis Group, London, UK, 2004)
2. P. Shen, L. Zhang, H. Chen, R. Fan, Geosci. Model Dev. **11**, 2841-2856, (2018)
3. S. Zhang, L.M. Zhang, H.X. Chen, Q. Yuan, H. Pan, J.Mt. Sci. **10**, 281-292
4. doi.:10.1007/s11629-012-2506-y (2013),
5. G. Samodra, D.S. Hadmoko, G.N. Wicaksono, et al, Landslides **15**, 985-993 (2018),
6. doi.:10.1007/s10346-018-0958-4
7. S. Raia, M. Alvioli, M. Rossi, R.L. Baum, J.W. Godt, F. Guzzetti, Geosci. Model Dev. **7**, 495-514, doi.:10.5194/gmd-7-495-2014 (2014)
8. T.F. Fathani, W. Wilopo, Geohazard Management Study (Faculty of Engineering UGM, Yogyakarta, Indonesia, 2019)
9. T.F. Fathani, W. Wilopo, Geohazard Management Study (Faculty of Engineering UGM, Yogyakarta, Indonesia, 2020)
10. Y. Fajarwati, T.F. Fathani, F. Faris, W. Wilopo, Inersia **16**, 105-116 (2020),
11. doi.:10.21831/inersia.v16i2.36897
12. N. Kim, H. Nakagawa, K. Kawaike, H. Zhang, J. of Japan Soc. Of Civ. Eng., Ser. B1 69, I97-I102 (2013),
13. Y.H. Zou, X.Q. Chen, Nat. Hazard Earth Syst. Sci. Discuss 3, 5777-5804
14. doi.:10.5194/nhessd-3-5777-2015 (2015),
15. J.Hübl, J. Suda, D. Proske, et al., *Debris flow impact estimation steep slopes, in Proceedings of The 11th International Symposium on Water Management and Hydraulic Engineering, Mazedonien* (2009)
16. T. Mizuyama, Int. J. of Erosion Cont. Eng. 1, 38-43, (2008)
17. T.F. Fathani, W. Wilopo, Alex. Eng. J. (under review)
18. National Standardization Agency, SNI 2851:2015 *Designing sediment retaining structures* (BSN, Jakarta, Indonesia (2015)
19. National Standardization Agency, SNI 1724:2015 *Hydrology, hydraulics, and criteria for designing river structures*, BSN, Jakarta, Indonesia (2015)
20. Japan International Cooperation Agency, Technical standards and guidelines for planning and design of sabo structures (JICA, Japan, 2010)

 https://doi.org/10.1051/e3sconf/202133108002

The effect of land uses to change on infiltration capacity and surface runoff at latung sub watershed, Padang City Indonesia

Rusli HAR[1], *Aprisal*[2*], *Werry* Darta Taifur[3], and *Teguh* Haria Aditia Putra[4]

[1]Doctoral Program in Agricultural Science, Andalas University, Indonesia
[2]Department of Soil Science, Andalas University, Indonesia
[3]Department of Economics, Andalas University, Indonesia
[4]Department of Forestry, Muhammadiyah University of Sumatera Barat, Indonesia

Abstract. Changes in land use in the Air Dingin watershed (DAS) area in Padang City, Indonesia, lead to a decrease in rainwater infiltration volume to the ground. Some land use in the Latung sub-watershed decrease in infiltration capacity with an increase in surface runoff. This research aims to determine the effect of land-use changes on infiltration capacity and surface runoff. Purposive sampling method was used in this research. The infiltration capacity was measured directly in the field using a double-ring infiltrometer, and the data was processed using the Horton model. The obtained capacity was quantitatively classified using infiltration zoning. Meanwhile, the Hydrologic Engineering Center - Hydrology Modeling System with the Synthetic Unit Hydrograph- Soil Conservation Service - Curve Number method was used to analyze the runoff discharge. The results showed that from the 13 measurement points carried out, the infiltration capacity ranges from 0.082 - 0.70 cm/minute or an average of 0.398 cm/minute, while the rainwater volume is approximately 150,000 m3/hour/km2. Therefore, the soil infiltration capacity in the Latung sub-watershed is in zone VI-B or very low. This condition had an impact on changes in runoff discharge in this area, from 87.84 m3/second in 2010 to 112.8 m3/second in 2020 or a nail of 22.13%. Based on the results, it is concluded that changes in the land led to low soil infiltration capacity, thereby leading to an increase in surface runoff. Keyword: Land use, infiltration, runoff, Watershed.

1 Introduction

Padang City has a high population growth rate and development in various sectors within the Air Dingin watershed (DAS) area, which has the Latung sub-watershed. Over the past decades, this area has continuously experienced changes in land use with the development of residential, tourist and cultivation areas [1, 3]. These changes occur in urban, cultivation, and protected areas, thereby decreasing rainwater infiltration into the ground [4]. Furthermore, these changes increase the volume of surface runoff [4, 5], erosion and sedimentation [2], inundation, and flooding in the middle and downstream parts of the watershed. This phenomenon occurs due to the inability of the river to accommodate high rainfall.

Air Dingin Watershed has several tributaries with Batang Air Dingin, which acts as the main river despite being dendritic and short in size [2]. The water source comes from the Bukit Barisan Mountains area, located between Solok Regency and east of Padang City, and empties into the Indian Ocean. The network of water and sediment supply channels to the main Batang Air Dingin river occupies a 1,200-hectare mountain range [3, 4]. One of the tributaries of this river is the Latung River which later became a part of its watershed with an area of approximately 19.2 km^2.

The Latung Sub-watershed area consists of the Koto Tangah Sub-district. This area was designated as the Government Center of Padang City on April 18, 2011, as a transition from the West Padang sub-district based on Government Regulation No. 26 of 2011. The transition made the Latung Sub-watershed experience changes in land use from green open land into residencies, government offices, and other public facilities. These changes decreased the soil's physical properties, such as infiltration capacity, thereby increasing the surface runoff [5].

In 2016 there was a major flood in the Padang city and the Koto Tangah Sub-district, located in the Latung Sub-watershed area. This tragedy became evidence that changes in land use of the Latung sub-watershed lead to increased surface runoff discharge [6]. Besides, during the dry season, this area experiences a decreased groundwater absorption. Water deficit occurs in February and April -3.26 mm/month and -43.01 mm/month, respectively. This means that water does not fill into the aquifer system in the Latung sub-watershed [1, 3, 7].

* Corresponding Author: aprisal@agr.unand.ac.id

Therefore, based on the problems above, it is necessary to examine the infiltration capacity of this area. Infiltration capacity is the vertical entry of water into an initially unsaturated soil under the influence of the soil matrix and gravity suction. The infiltration rate is usually high at the initial stage, however, with the entry of deeper water and a wet soil profile [8], the soil matrix suction decreases. The gravitational pull affects the water movement [12, 13], thereby causing the infiltration rate to decrease approaching equilibrium conditions [13, 14]. In line with this problem, Hutasoit (2009) divided the zoning of the infiltration area based on its capacity [9, 11], as shown in Table 1.

Table 1. Zoning Division of Infiltration Areas Based on Capacity

Zone	Classification of Infiltration Areas		Infiltration Rate (cm/minute)	The Volume of Water that can be Infused (m^3/hour/km^2)
I	Sangat Tinggi		> 1.5	1,050,000.00
II	Tinggi		1.25 - 1.50	825,013.00
III	Agak Tinggi		1.00 - 1.25	675,000.00
IV	Sedang		0.75 - 1.00	525,000.00
V	Rendah		0.50 - 0.75	375,000.00
	Sangat Rendah		< 0.50	
VI	1	Zona A	0.40 - 0.50	150,000.00
	2	Zona B	0.30 - 0.40	
	3	Zona C	0.20 - 0.30	
	4	Zona D	0.10 - 0.20	
	5	Zona E	< 0.10	

This research aims to examine the soil infiltration capacity based on the actual infiltration value. Furthermore, it also aims to determine the value of changes in the Latung Sub-watershed surface runoff from 2010 to 2020.

2 Methodology

2.1 Description of Research Area

The research was conducted in the Latung Sub-watershed, part of the Air Dingin Watershed of Padang City, Indonesia. This area is geographically located between 0° 49' 01.1" - 0° 50' 34" South Latitude and 100° 23' 38.3" - 100° 25' 58.8" East Longitude, at an altitude of 20 - 1,625 m asl.

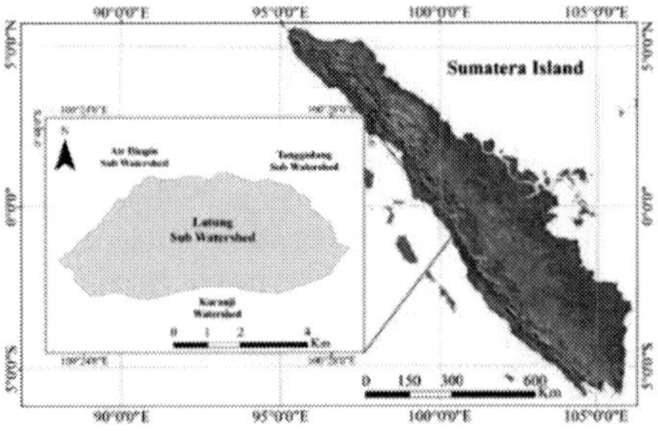

Fig.1. Map of research location

It is bordered by the Air Dingin, Batang Kuranji, and Tanggadang sub-watersheds in the north, south, and Northeast, respectively. The research area is ± 19.20 km^2 with sloping to steep topography. Based on the Schmidt Ferguson classification, the climate is classified as type A with an average rainfall of 3,411.8 mm/year. The is consists of an Inceptisol soil type, therefore the land is used for rice fields, mixed gardens, residencies, as well as secondary and primary forest. The location and condition of the research area are shown in Fig. 1.

2.2 Data Collection

The rainfall data were collected from the 3 stations closest to the Latung sub-watershed in the Batang Air Dingin watershed, namely Muaro Panjalinan, Kasang, and Bandung Koto Tuo. Meanwhile, the data on topographic, land use, and soil types were descriptively obtained from spatial and GIS and compared with survey results in the field. A double-ring infiltrometer was used to analyze the data randomly collected from 13 points in the field. This tool is made of steel with an inner circle of 30 cm, an outer circle of 60 cm, and a height of 30 cm.

2.3 Data Analysis

2.3.1 Spatial data analysis

Data on topography, land use, and soil types were used as the basis for determining the flow coefficient for the HEC-HMS modeling (HSS SCS CN method) [18]. The data were mainly used to determine the runoff coefficient (C) and the Curve Number (CN) value in the Latung sub-watershed and analyzed with ArcGIS 10.6.

2.3.2 Infiltration data analysis

Factors affecting infiltration include the influence of soil properties, cover crops, water content, rainfall, etc. The Horton method was used to calculate the infiltration capacity, which was dependent on natural parameters in the field. However, a mathematical approach or deterministic statistics is used to carry out the Horton equation assuming the data taken in the field is not accompanied by natural parameters [10-12]. This process is carried out by analyzing the trend of the data obtained from the measurement results in the field due to the influence of natural parameters. The infiltration rate is calculated based on the Horton model with equation (1).

Where f is infiltration capacity (cm/hour), f_o is the initial infiltration rate (cm/hour), f_c is the constant infiltration rate in a certain time interval t during the test (cm/hour), K is a constant, t is time (hours) and e is exponential number = 2,718. The process of determining the value of K in Equation (1) is derived as follows:

$$f = f_c + (f_0 - f_c)e^{-Kt}$$
$$f - f_c = (f_o - f_c)e^{-Kt} \quad (2)$$

The log of both sides of Equation (2) is as follows:

$$\log(f - f_c) = \log(f_o - f_c) - Kt \log e$$
$$\log(f - f_c) - \log(f_o - f_c) = -Kt \log e$$
$$t = -\frac{1}{K \log e}[\log(f - f_c) - \log(f_o - f_c)]$$

$$t = -\frac{1}{K\log e}\log(f - f_c) + \frac{1}{K\log e}\log(f_o - f_c) \quad (3)$$

To determine the unknown parameters from Equation (3), a comparison is made with a linear equation in the form: $\boldsymbol{y = mx + c}$, where y= f, $m = -\frac{1}{K\log e}$, $x = \log(f - f_c)$ and $c = \frac{1}{K\log e}\log(f_o - f_c)$.

Therefore, the value of K is determined as follows:

$$m = -\frac{1}{K\log e}$$

$$K = -\frac{1}{m\log e} = -\frac{1}{m\log 2{,}718}$$

$$K = -\frac{1}{0{,}434\, m} \quad (4)$$

Where *m* is the gradient obtained from the linear graph of the relationship between time *(t)* and *log (f – FC)*. Meanwhile, the initial infiltration value (f_o) is calculated from the regression equation generated from the graph plot of the infiltration capacity (*f*) against time (*t*). The regression equation model used to produce f_o > FC provides the largest correlation coefficient *(r)* and determination coefficient (R^2).

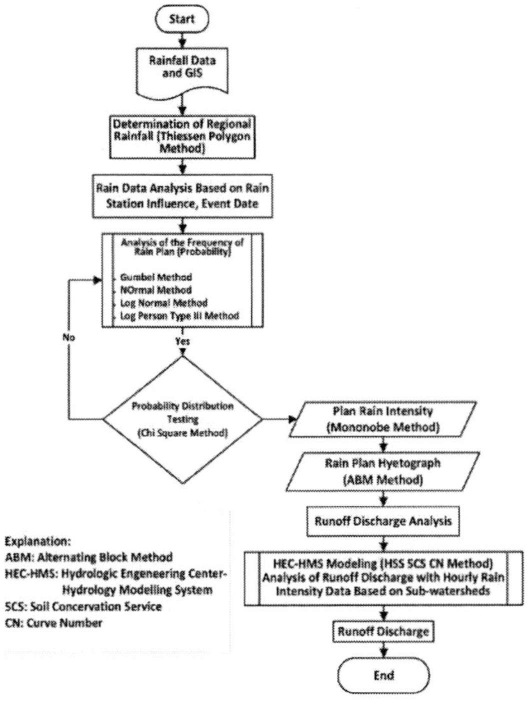

Fig. 2. Flowchart of Analysis on Rainfall and Runoff Discharge in Latung Sub-watershed

2.3.3 Analysis of rainfall and runoff discharge data

The process of examining the rainfall Influence from the 3 closest stations to the Latung sub-watershed was performed using the Thiessen Polygon method, as shown in Fig. 6. Rainfall around the Latung sub-watershed enters the tributaries and then flows and accumulates into the main river, which is the basis for calculating runoff. Furthermore, the Mononobe method was used to obtain the value of the planned rainfall intensity, which was also used to determine the peak discharge value described in the form of an IDF (Intensity Duration Frequency) curve [15,16]. The Mononobe method was an IDF curve regression equation were used to determine the rainfall data.

The runoff flow is directly generated from the analysis of the effective rainfall calculated by the Soil Conservation Service Curve Number (SCS–CN) Method [17, 18], where the rain falling on the surface is reduced by losses due to abstraction during surface runoff. Furthermore, the HEC-HMS modeling (HSS SCS CN method) [17,18] was used to analyze the runoff discharge, using the Hydrologic Engineering Center Modeling - Hydrology Modeling System (with Synthetic Unit Hydrograph - Soil Conservation Service - Curve Number) method. Meanwhile, the runoff coefficient (C) for each land cover was calculated using equation (5):

$$C_{DAS} = \frac{\sum_{i=1}^{n} C_i A_i}{\sum_{i=1}^{n} A_i} \quad (5)$$

Where C*i* is the coefficient of surface runoff of land cover I, A*i* is the area with land cover *i*. The concentration-time (tc) is calculated using the Kirpich Equation (Kamiana, 2011) (6)

$$t_c = \left(\frac{0{,}87 \times L^2}{1000 \times S}\right)^{0{,}385} \quad (6)$$

Where tc is the concentration-time (hours), L is the river length (km), and S is the gradient. In this method, the runoff discharge analysis used hourly rainfall intensity data based on the sub-watersheds in the research area. The sequence in analyzing rainfall to runoff discharge in the Latung sub-watershed is shown in Fig. 2.

3 Results and Discussion

3.1 Topography of Research Area

The topography of the Latung Sub-watershed area is shown in Fig. 3. Generally, the topographic conditions in the north bordering the Air Dingin Sub-watershed are quite steep to sloping. The Air Dingin and the Tanggadang Sub-watershed have a very steep topography in the Northeast and south. Meanwhile, in the east and west, the topography is quite steep to sloping, which becomes the basis for the calculation of runoff discharge.

3.2 Land Use

Fig. 4 shows the land use of the Latung Sub-watershed area. The figure explains that this area borders the Air Dingin and the Tanggadang Sub-watershed in the north and consists of mixed forests and gardens. To the east, it is bordered by a forest area including protected forest, while to the west, it is an area of mixed gardens and shrubs. Furthermore, the Batang Kuranji watershed is bordered to the south and is an area of mixed gardens and secondary forest. This land-use condition is also the basis for the calculation of runoff discharge.

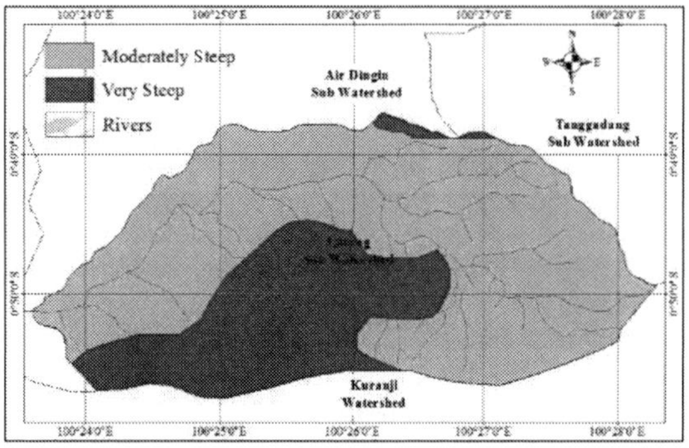

Fig. 3. Topographical Conditions of Research Area

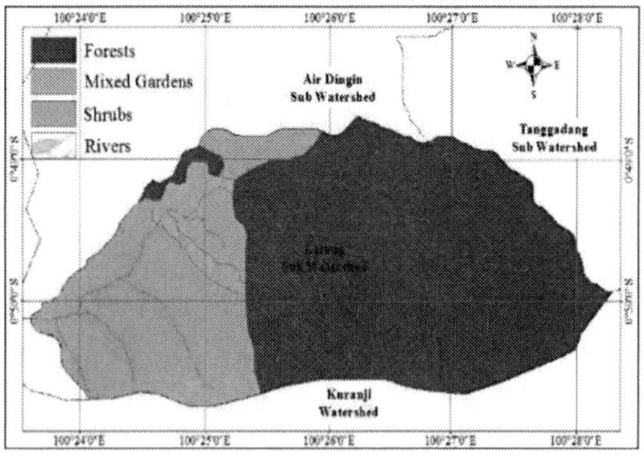

Fig. 4. Conditions of Land Use in the Latung Sub-watershed Area

3.3 Soil Type

The soil type in the Latung sub-watershed area is shown in Fig. 5, where it is mostly dominated by Humitropepts, namely Tropepts sub-order of Inceptisol rich in humus. In the western part of the Latung sub-watershed, the soil types are divided into 2, namely Dystropepts and Tropaquepts. Dystropepts are soils with base saturation below 50%, with medium to very deep depths, fine to medium texture, acidic to very acidic soil reactions, and poor in nutrients or low soil fertility. Meanwhile, Tropaquepts are a great group of the Aquepts sub-order and the Inceptisols. This soil type does not have sulfidic material at a depth below 50 cm from the mineral soil surface and has an isometric soil temperature. It is also used as a basis for determining runoff discharge in the Latung sub-watershed.

3.4 Infiltration Capacity

The analysis results on infiltration data in the Latung sub-watershed are shown in Table 2. The average initial infiltration rate in the research area was 24.85 cm/hour. Furthermore, the average constant infiltration rate was 23.54 cm/hour, with an average constant condition of 1.09 hours and an average infiltration capacity of 23.87 cm/hour, equivalent to 0.3978 cm/minute. The value of infiltration rate in Latung Sub-watershed is shown in Table 2.

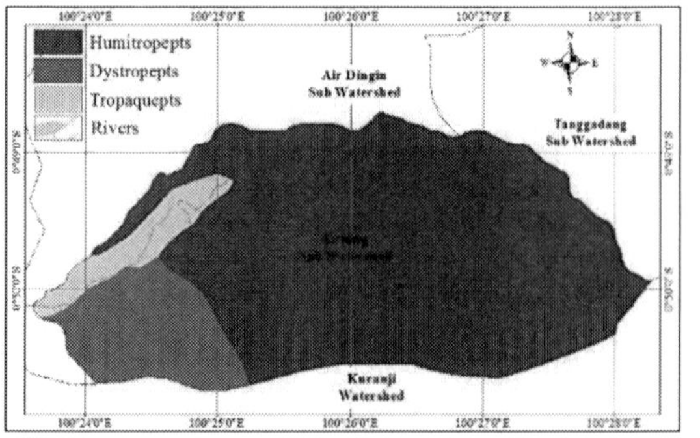

Fig. 5. Soil Types in the Latung Sub-watershed

Table 2. Infiltration Rate in Latung Sub-watershed

No	Measuring Point	Initial Infiltration Rate (fo)	Constant Infiltration Rate (fc)	Time constant (ft)	Infiltration Capacity (f)		Classification of Infiltration		The volume of water that can be absorbed	Latung sub-watershed area	Total debit absorb
		(cm/hour)	(cm/hour)	(o'clock)	cm/hour)	(cm/minute)			(m3/hour/km2)	km2	(m3/hour)
1	Ept.A.KC	16.64	16.20	1.17	16.28	0.271	VI-C	VL			
2	EPt.A.SB	23.32	22.80	1.33	22.86	0.381	VI-B	VL			
3	Ept.B.KC	28.33	24.00	0.83	24.83	0.414	VIA	VL			
4	Ept.B.SB	30.51	30.00	1.17	30.12	0.502	VIA	VL			
5	Ept.C.HT	40.16	39.00	1.50	39.00	0.65	V	L			
6	Ept.C.KC	36.62	36.00	1.67	36.03	0.60	V	L			
7	Ept.C.SB	42.65	42.00	1.50	42.00	0.70	V	L	150.00	19.20	2,880,000
8	Ept.D.HT	38.91	35.40	1.00	36.29	0.61	V	L			
9	Ept.D.KC	18.19	18.00	1.00	18.05	0.31	VI-B	VL			
10	Ept.D.SB	11.64	10.80	0.83	11.00	0.18	VI-D	VL			
11	Ept.E.HT	25.06	22.20	0.67	23.80	0.39	VI-B	VL			
12	Ept.E.KC	5.75	4.80	0.67	5.13	0.09	VI-E	VL			
13	Ept.E.SB	5.25	4.80	0.83	4.92	0.08	VI-E	VL			
Average		24.85	23.54	1.09	23.87	0.97	VI-B	VL			2,880,000

Description, Ept = Inceptisol, A, B, C, D, E = the slope of land, Kc = mix garden, SB = shrubs, HT = forest. VL = very low, L=low.

The infiltration capacity of the Latung sub-watershed in Table 2 is in zone VI-B, which is very low compared to the value in Table 1. The rainwater infiltration capacity was only 150,000 m^3/hour/km^2. Maximum and evenly distributed rain in the entire Latung sub-watershed area has a total infiltration capacity of 2.880,000 m^3 in one maximum rainy period, which is much smaller in the dry months. This shows that the infiltration capacity of the Latung Sub-watershed is in a critical condition. In the dry season, where rainfall is very low, the capacity decreases, thereby leading to a water deficit. Preliminary studies showed that the deficit of water infiltration in the Latung sub-watershed occurs in dry months, especially in February (-3.26 mm/month) and April (-43.01 mm/month) [7]. The main cause of critical infiltration is the burden on the sub-watershed. Especially in the upstream part, which keeps increasing due to changes in land use.

The infiltration capacity of the Latung sub-watershed in Table 2 is in zone VI-B, which is very low compared to the value in Table 1. The rainwater infiltration capacity was only 150,000 m^3/hour/km^2. Maximum and evenly distributed rain in the entire Latung sub-watershed area has a total infiltration capacity of 2.880,000 m^3 in one maximum rainy period, which is much smaller in the dry months. This shows that the infiltration capacity of the

Latung Sub-watershed is in a critical condition. In the dry season, where rainfall is very low, the capacity decreases, thereby leading to a water deficit. Preliminary studies showed that the deficit of water infiltration in the Latung sub-watershed occurs in dry months, especially in February (-3.26 mm/month) and April (-43.01 mm/month) [7]. The main cause of critical infiltration is the burden on the sub-watershed. Especially in the upstream part, which keeps increasing due to changes in land use.

3.5 Analysis of Rainfall and Runoff Discharge

Based on the analysis of regional rainfall using the Thiessen Polygon method, it is known that the area of the Batang Air Dingin Watershed is 131.18 km^2. Meanwhile, the Latung sub-watershed is 19.20 km^2 or 14.6% of the total area of the Batang Air Dingin watershed. Therefore, the area of rainfall influence in the Latung sub-watershed is 14.6% of the total rainfall Influence on the Batang Air Dingin watershed. Fig. 6 and Table 3 show that the total area of rainfall Influence from the 3 existing rain gauge stations on the Latung sub-watershed is 19.2 km^2.

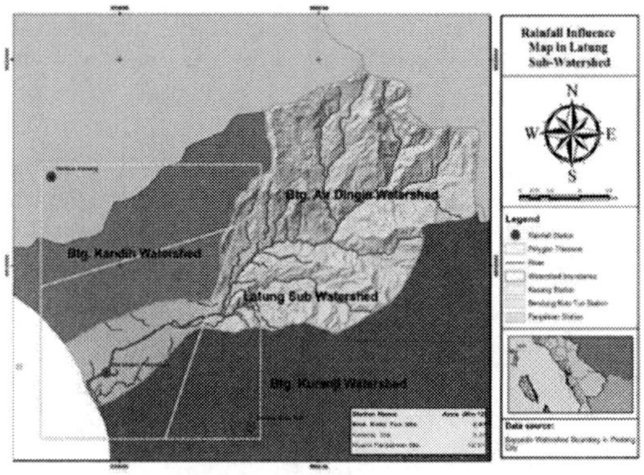

Fig. 6. Rainfall Influence on Latung Sub-watershed with Polygon Thiessen Method

Table 3. The Rainfall Influence on the Latung Sub-watershed

Station	Rainfall Influence Area (km^2)	%
Muaro Panjalinan Station	2.97	15.50
Kasang Station	3.27	17.08
Bendung Koto Tuo Station	12.92	67.42
Total	**19.16**	**100.00**

Based on the sequence of the analysis process shown in Fig. 2, the planned rainfall intensity in the Latung Sub-watershed area calculated under 5, 10, 20, 25, 50, and 100 year return periods is described in the form of an IDF (Intensity Duration Frequency) curve [15,16]. The analysis is shown in Fig. 7.

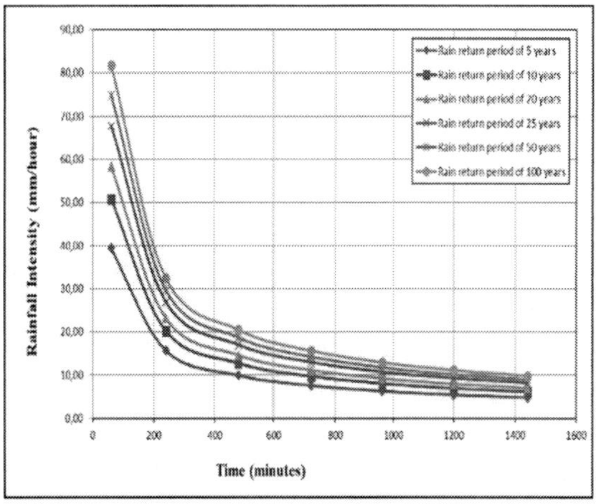

Fig. 7. IDF (Intensity Duration Frequency) Curve with Return Period in Latung Sub-watershed

3.6 Analysis of Runoff Discharge

The runoff coefficient value, Curve Number (CN), Land Use, and runoff discharge in the Latung sub-watershed are highly dependent on the type and area of land use, as shown in Figs. 3, 4, and 5. Meanwhile, the concentration-time (tc) is the time needed to drain rainwater from the farthest point to flow over the ground to a certain point in terms of the drainage area. The time is influenced by the characteristics of the infiltration area, land use, and the distance of the water path from the farthest to the consideration point. According to the existing data,

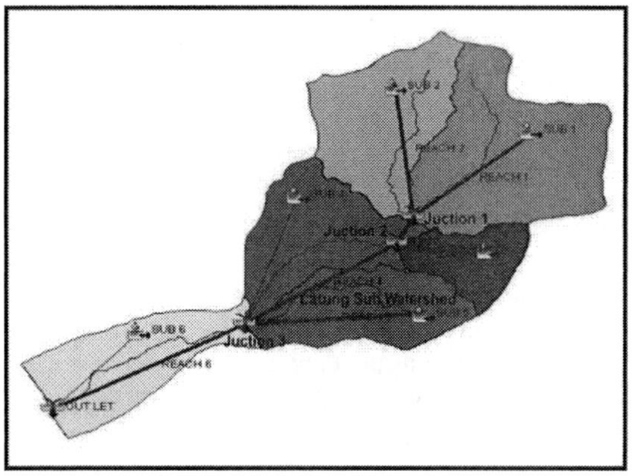

Fig. 8. Junction 3 of the Latung Sub-Watershed

The concentration-time (tc) in the Latung River calculated using equation (6) is 2.61 hours. Fig. 7 provides a detailed analysis of the runoff discharged in the Batang Air Dingin Watershed, which is divided into 6 sub-watersheds, while runoff is divided into 3 junctions. The 3rd junction is the accumulation of runoff discharge originating from the Latung sub-watershed, as shown in Fig. 8.

Table 4. Analysis Results of Runoff Discharge in Latung Sub-watershed for Return Period of 25 Years

Junction 3 (Repeat Period 25 Years)				
Time	Inflow From Junction 2	Inflow From Sub 4	Inflow From Latung Sub Watershed	Outflow From Junction 3
Hour	(m^3/s)			
0	0,0	0,0	0,0	0,0
2	1,1	3,9	1,4	6,3
4	351,5	150,1	**112,8**	614,3
6	269,5	78,0	75,1	422,7
8	170,0	47,9	45,7	263,6
10	125,9	37,4	34,4	197,6
12	104,7	31,6	28,8	165,1
14	48,7	8,0	11,4	68,1
16	16,0	1,5	3,0	20,5
18	4,5	0,2	0,7	5,5
20	1,2	0,0	0,1	1,3
22	0,3	0,0	0,0	0,3
24	0,1	0,0	0,0	0,1

The analysis results are presented seen in Table 4, while the graph of the runoff discharge hydrograph is shown in Fig. 9.

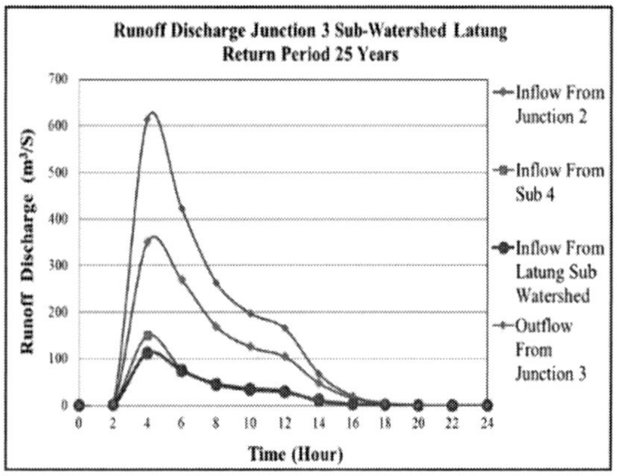

Fig. 9. Hydrograph of runoff discharge into the Latung sub-watershed

Fig. 9 and Table 4 are used to explain the preparation of the hydrograph graph for runoff entering the Latung Sub-watershed, the maximum rain duration occurred for 6 hours. The area of the Latung sub-watershed is 19.20 km^2 or 14.64% of the total area of the Batang Air Dingin watershed. Furthermore, based on the runoff discharge analysis, the value of surface runoff in 2020 calculated by the SCS CN method for a 25 year return period in the Latung sub-watershed is 112.8 m3/second, as shown in Table 4. Meanwhile, in 2010 14.64% of the design was discharged, for 87.84 m^3/second [19]. There was an increase in the runoff discharge of 25 m^3/second or 22.13% in the years before 2010. The increase in runoff discharge and the small volume of infiltration in this area were due to the flood disaster in the downstream area of the Latung Sub-watershed, namely in the Muara Penjalinan or the outlet area of the Air Dingin watershed. Therefore, efforts are needed to carry out reclamation or conservation to restore the condition of the Latung sub-watershed to maintain its function as a counterweight for the preservation of water resources.

4 Conclusion

The following conclusions were obtained from the research results:

1. The infiltration capacity in the Latung sub-watershed for 1.09 hours is 23.87 cm/hour with a relatively low infiltration capacity of 150,000 m3/hour/km^2.
2. There was an increase in runoff discharge of 25 m^3/second or 22.13% from 2010 to 2020.

The authors are grateful to the Chancellor of Andalas University for facilitating the laboratory infrastructure for soil sample analysis, which enabled the adequate conduction of this research. Furthermore, the authors are grateful to the students that helped to measure and take soil samples and analysis in the laboratory. Hopefully, this research can be one of the literary references in science development.

References

1. Aprisal, Junaidi. The Effect of Land Reclamation on Physical and Erosion Properties in Reeds Land. J. Solum **7** (2) July 2010: 118-123. ISSN: 1829-7994 (2010)
2. Aprisal, Junaidi. Prediction of Erosion and Sedimentation in Various Land Use in the Lake Limau Manis Sub Watershed in the Kuranji River Basin, Padang City. J. Solum **7** (1) January 2010:61-67. (2010)
3. R. HAR, Study on the Potential of Groundwater in the City of Padang and Its Relationship to the 2003/2004 Padang city master plan. Research and Development Agency of West Sumatra Province (2004)
4. T. Aprisal. Dyn. Of. Soil. Phys. Suboptimal Land Journal.2017 (2017).
5. R. HAR, Empowerment of Infiltration Areas as Flood Prevention Efforts, Seen from a Hydrogeological Point of View. Research and Development Agency of West Sumatra Province (2006)
6. West Sumatra Water Resources Management Office. West Sumatra Water Resources Management Profile (2017)
7. R. HAR, Study on Potential, Development, Management, and Zoning of Groundwater in Padang City, Competitive Grant 2014 (2014)
8. D. Hillel, Soil and Water Physical Principles and Processes, Academic Press New York, San Francisco, London (1978)
9. L. M. Hutasoit, Quantitative Study of Water Infiltration in North Bandung Area. LPPM-ITB Bandung (1997)
10. L. M. Hutasoit, Groundwater Surface Conditions with and without artificial infiltration in the Bandung area: Numerical Simulation Results. Indonesian Journal of Geology, **4** (3), 177-188 (2009)
11. L. M. Hutasoit, Land cover variations and their implications for baseline flow values in the Upper Citarum watershed. 2015 National Seminar on Remote Sensing (2015)

https://doi.org/10.1051/e3sconf/202133108002

12. C.W. Fetter, Contaminant Hydrogeology, Mac Millan Publishing Company, New York (1993)
13. R. A. Freeze, , and J.A. Cherry. Groundwater, Prentice-Hall, Englewood Cliffs, New Jersey (1979)
14. D. Hillel, Applications of Soil Physics. Academic Press, Inc. Harcourt Brace Jovanovich, Publishers. 1250 Sixth Avenue, Sandiego, California 92101 (1980)
15. Y. Sun, D. Wendi, D.E. Kim, and S.Y. Deriving intensity–duration–frequency (IDF) curves using downscaled in situ rainfall assimilated with remote sensing data. Official Journal of the Asia Oceania Geosciences Society (AOGS (2019).
16. F. De Paola, , M. Giugni, and M. E. Topa, et al. Intensity-Duration-Frequency (IDF) rainfall curves, for data series and climate projection in African cities. SpringerPlus 3, **133** (2014).
17. , G. J. Alves, C. R. de Mello, S. Beskow, J.A. Junqueira and M. A. Nearing. Assessment of the Soil Conservation Service–Curve Number method performance in a tropical Oxisol watershed. Journal of Soil and Water Conservation September 2019, **74** (5) 500-512(2019)
18. S. K. Mishra, Singh, and Vijay. Soil Conservation Service Curve Number (SCS-CN) Methodology. Water Science and Technology Library. eBook ISBN 978-94-017-0147-1 (2003)
19. Department of Public Works for Irrigation, West Sumatra Province. Infrastructure Development for Flood Control in Padang City. Padang City Flood Control Project (2000)

The effects of river normalization on flood risk

Nurhamidah Nurhamidah[1], *Ahmad* Junaidi[1], *Haviz* Ramadhan[1*], and *Ronny* Eka Putra[1]

[1]Department of Civil Engineering, Faculty of Engineering, University of Andalas, Padang.

Abstract. A river is a naturally occurring flow system that flows from upstream to downstream. Rivers have many benefits. Similarly, rivers have a negative impact if rainfall is high. It can lead to catastrophic flooding. Flooding often occurs along the central roads of Padang City when it rains, especially in the Maransi and Aie Pacah areas. The purpose of this study is to estimate the height of water level by using software HEC – RAS and estimate areas that have the potential for flooding in conditions before and after river normalization by using the software Arc-GIS. Data collection is carried out to achieve this goal, namely the maximum daily rainfall data, Digital Elevation Model (DEM) data obtained from the processing of aerial photo data, and cross-sectional data of the river. Then, a numerical simulation is performed to estimate the height of the water level, and then a numerical simulation is performed to determine the extent of the flood area. The results of this study are in the form of flood areas in conditions before and after river normalization with a period of 25, 50, and 100 years that can be used to see the effectiveness of river normalization work in the river.

1 Introduction

Batang Maransi River is a river located in West Sumatera Province [1], Padang City, Koto Tangah District. The land area of Padang City is ± 694.96 km² and has ± 23 streams with a total length of 155.40 km (10 major rivers and 13 small rivers). Generally, the large and small rivers in the Padang city area are not much different from the sea level height.

Batang Maransi River is one of the rivers that are prone to flooding in the rainy season. Batang Maransi River cross-section is not able to accommodate the high rainfall of Padang City, so during the rainy season in the area around Maransi and Aie Pacah, there are often floods. One of the largest floods ever occurred in the area around Batang Maransi, Koto Tangah district on March 22, 2016, where nine subdistricts in Padang City were flooded with a rain intensity of 384 mm. Especially in West Padang District, precisely at Maransi, floods entering residents' housing with a puddle height of 50-100 cm make residents required to leave their homes.

Therefore, it is necessary to reconstruct the river in Batang Maransi. The normalization of the river in an attempt to increase the capacity of the river. River normalization consists of channel improvement such as building new channels, widening or deepening the channel, channel resectioning, channel realignment, bank protection, build dikes and dams(retarding and storage dam) [2]. Currently, the Batang Maransi River is ongoing normalization of the river to overcome the flood problem in the Maransi and Aie Pacah areas, Koto Tangah Subdistrict which is one of the flood vulnerable areas. Floods can be inferred as a relatively high flow that transcends natural or artificial embankments in the river, varying magnitude by season, month to month, year to year [3]. The flooding problems damages might be the result of area cultivation and land-use changes without any control. In addition, it seems that increasing rainfall intensities in the upstream catchments of the river and a city are contributing to the problem of flooding as well. [4] Therefore, flooding in this area has become the center of attention of the government.

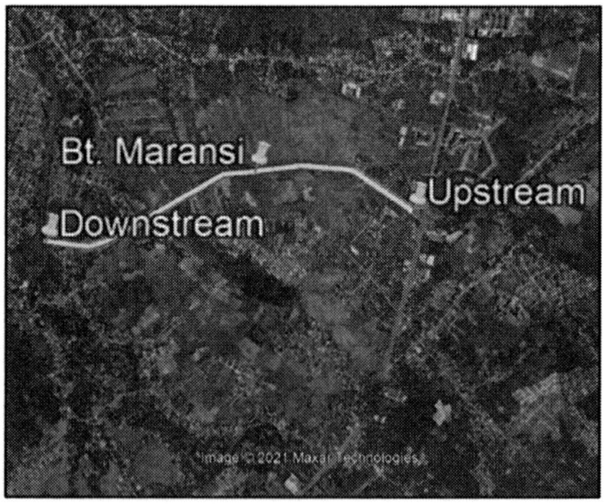

Fig. 1. Long section in Batang Maransi River.

The main objective of this study is to see the effectiveness of river normalization to reduce the risk of flooding. The specific purpose of this study is to estimate the height of the river face, then map and estimate the area of inundation of areas that have the potential flood in conditions before and after normalization in Batang Maransi River. The benefit of this study is as a reference

* Corresponding author: havizramadhan27@gmail.com

to manage flood problems as a reference in the policy-making of flood management in the area around Batang Maransi River.

2 Literature Review

The research entitled Flood Analysis Using HEC-RAS Software with a case study of sub-watershed Cisemeut downstream HM 0+000 to HM 69+00 which aims to determine the cause of the existing flooding in the region based on the capacity with discharges that occur and know the ability of the Cisimeut downstream watershed segment in accommodating river flow discharge with 50-year period. From the analysis results, hydrological analysis to get the discharge of flood plans 50-year period of 1566,439 m^3/s and then conducted a hydraulic analysis using HEC-RAS application with a steady flow and unsteady flow conditions. The result is that the capacity of the Cisimeut river in existing conditions is not able to flow the flood discharge plan, so flood control efforts are needed. Although water discharge is reduced, there is still an overflow of the river at some point. Therefore, it is necessary to repair the cross-section of the river and the elevation of the embankment to increase the capacity of the river so that the overflow can be resolved [5]

The research with the title Cilemer Watershed Flood Analysis HM 0+000 – HM 53+000 aims to identify flood-affected areas in Pandeglang Regency at HM 00+00 to HM 53+00, which can then produce recommendations in flood management. The research supporting data was obtained from the Water Resources Management Center of Banten Province. The analysis calculated the planned flood discharge based on the maximum debit data for a moment by *the* Mean Annual Flood (MAF) method and obtained the planned flood discharge data of 210,617 m^3/s. Then, flood modeling and mapping were carried out by simulating the flow using the HEC-RAS application and conducted flood modeling carried out modeling of flood-affected areas by using the Arc-GIS application. The conclusions obtained from the results of the study were river flow simulation and flood mapping using HEC-RAS and Arc-GIS applications. It was seen that 19 cross-section segments of the river (HM 35+00 to HM 53+00) of Cilemer watersheds reviewed could not accommodate the planned flood discharge [6]

The study, titled Flood Mapping Study with HEC-GeoRAS with case studies on the Tondano River to create a 2014 flood inundation map in Manado geographic information system (GIS) overlay on google earth. From the analysis, the maximum debit data was obtained on January 15, 2014, and obtained maximum debit data of 1300 m^3/s. Then, a hydraulics simulation was carried out using the HEC-GeoRAS application which appears in the form of tools in the Arc-GIS application by inputting maximum flood discharge data and carrying out flood modeling simulation with the result that the flood height is estimated at 0 meters to 6 meters, including the height of flooding in the river and the results of simulations with HEC-GeoRAS obtained flood speed is estimated at 0 m /s up to 42.9 m /s. [7]

3 Research Methodology

Research methodology is done by identifying problems according to the available data. Data collection is divided into two, namely primary and secondary. The primary data needed is cross-sectional measurement data across the river measured and by using water pass, channel roughness figure data that is reviewed based on material from the cross-section of the channel reviewed and aerial photo data of Batang Maransi River and surrounding areas taken using drones. While the secondary data needed are annual rainfall data, weir dimension data and floodgates, cross-sectional data of rivers obtained from Water Resources Management Office, West Sumatera, and Digital Elevation Model (DEM) data that can be obtained from the Geospatial Information Agency website. The methodology is carried out by conducting a hydrological analysis to obtain the value of the flood discharge plan. Then, hydraulic analysis is done, estimating the water level using software HEC-RAS. After obtaining a high water level value, in a numerical simulation to estimate the area of flood area using the software Arc-GIS. This Modeling data uses an approach with the D8 matrix to determine where the flow direction and flow accumulation are. [8]. The data is collected from various sources and it may overlap information from various data sources through various themes and layers [9]. Arc-GIS can also be used for scientific investigation, cartography, road planning and can also assist in planning in calculating responses quickly. [10]

3.1 Description of The Study Area

Batang Maransi is one of the rivers located in the city of Padang, West Sumatra, that has a role in draining the amount of water discharge flowing, starting from upstream located in Batang Lurus River and Batang Belimbing River, which is downstream of Batang Maransi. By reconstructing the river, namely dredging the Batang Maransi River and the construction of weir, the risk of flooding in the area can be minimized.

Fig. 2. Condition of Batang Maransi River.

4 Result and Discussion

4.1 Hydrology Analysis

Hydrological analysis in this study requires rainfall data for 20 years obtained Water Resources Management Office, West Sumatera. The rainfall station in need is

Gunung Sarik Rain Station. The daily rainfall data is recapitulated into maximum rainfall every month and carried out the calculation of the rainfall plan. Then, the calculation of concentration-time and intensity of rain with a period of 25, 50, and 100 years. So, the value of flood discharge plan with the rational method in the period of 25, 50, and 100 years can be seen in Table 1.

$$Q = 0{,}278 \times C \times I \times A \tag{1}$$

Description:

Q = Debit Max (m3/s)
C = Flow Coefficient
I = Rain Intensity (mm/hours)
A = Catchment Area (Km^2)

Table 1. Rational Discharge Plan Method.

Year	C	A	I_t	Q_t
		Km^2	mm/jam	m^3/s
10	0,175	17,18	70,038	58,637
25	0,175	17,18	87,862	73,559
50	0,175	17,18	102,198	85,562
100	0,175	17,18	117,387	98,277

In this study, assuming the discharge that goes into the side drain amounting to 20% of the flood discharge plan, so that the calculation of flood discharge plan period of 25 years, 50 years and 100 years after there is a discharge that goes into the side drain can be seen in Table 2.

Table 2. Discharge plan after side drain.

Year	Q_t	Q
	m^3/s	m^3/s
10	58,637	70,364
25	73,559	88,271
50	85,562	102,674
100	98,277	117,933

4.2 Hydraulic Analysis

The hydraulic analysis is performed by estimating the water level with software HEC-RAS with steady flow conditions. The results of the simulation of steady flow conditions before river normalization conditions showed that there were more flood points than after river normalization conditions.

The picture above shows the simulation results of HEC-RAS in conditions before river normalization, Q10 there are 39 points of runoff, that is Sta 0+200 to Sta 1+150, in Q25 there are 39 points of runoff, that is from Sta 0+200 to Sta. Sta 1+150, in Q50 there are 39 points of runoff, that is from Sta 0+200 to Sta 1+150, in Q100 there are 41 points of runoff, that is from Sta 0+150 to Sta 1+150.

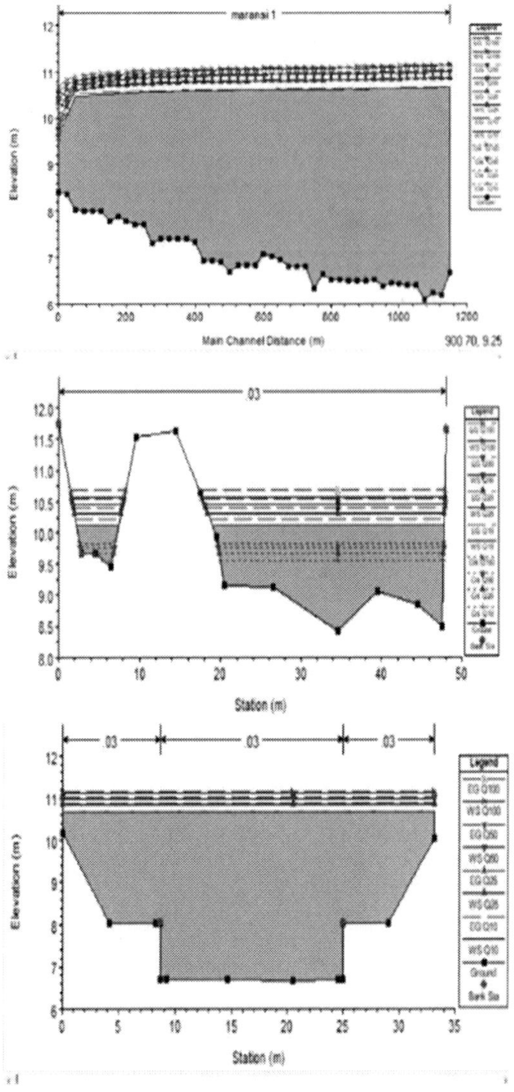

Fig. 2. Water level profile in before river normalization conditions in Batang Maransi River.

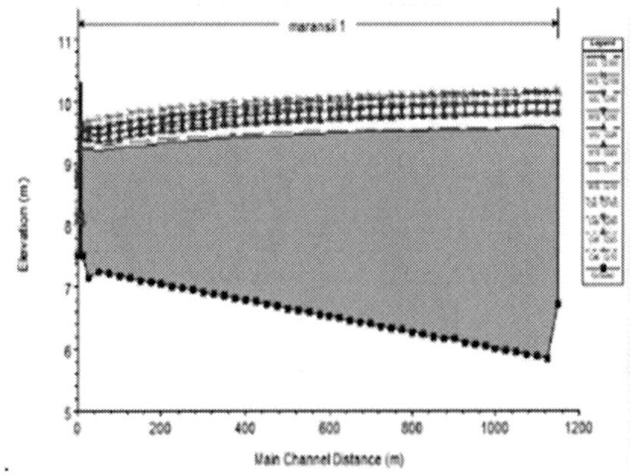

https://doi.org/10.1051/e3sconf/202133108003

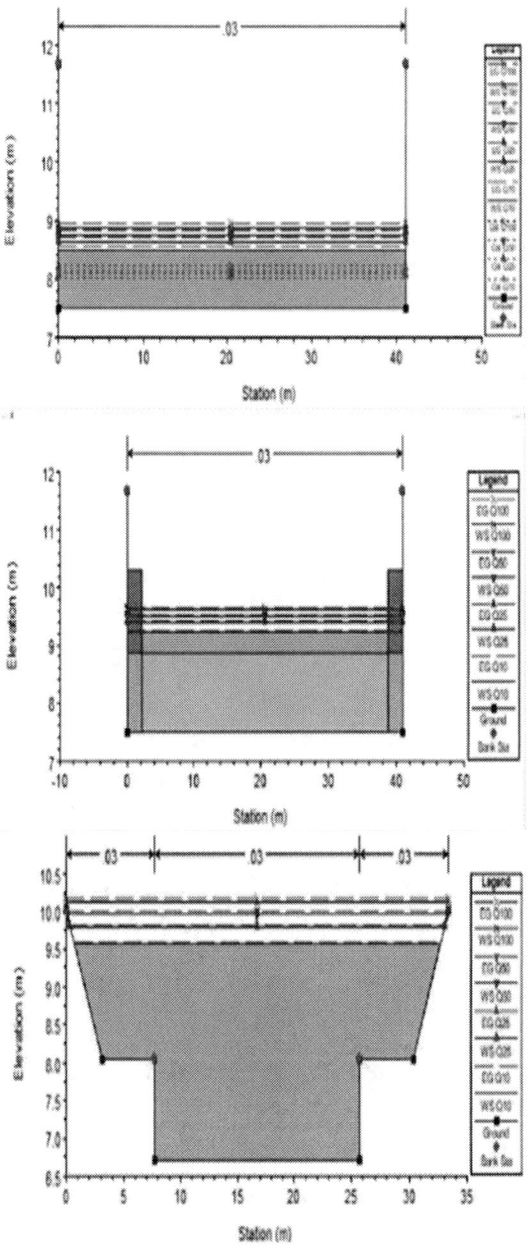

Fig. 3. Water level profile in after river normalization conditions in Batang Maransi River.

In the picture above, it is obtained that the simulation result of HEC-RAS in after river normalization conditions, the points of runoff is only found in Q100, namely at Sta 1+100 to Sta 1+150. There is a graph describing the comparison of water level elevation in conditions before and after river normalization in Fig 4.

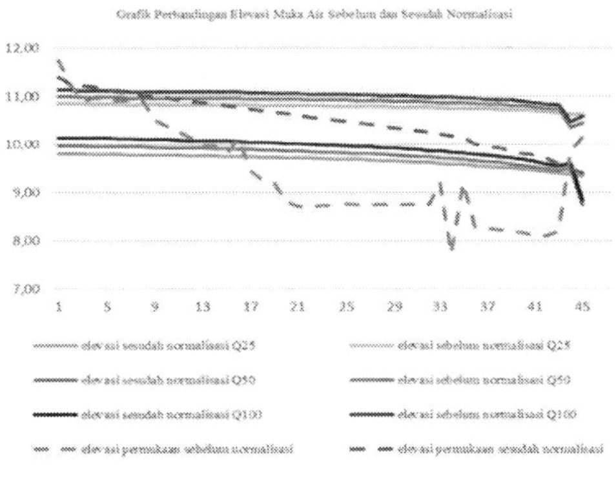

Fig. 4. Water level elevation comparison graph before and after river normalization in Batang Maransi River.

4.3 Flood Modeling Area

Based on the results of hydraulic analysis data obtained, namely the water level in Batang Maransi River, can be displayed modeling from the simulation of flood Batang Maransi River by using software Arc-GIS 10.4 version that can be seen in conditions before and after river normalization.

4.3.1 Before normalization

From the results of the flood, simulation obtained the area of flood area is 129.40 Ha for a 25-year return period, 138.38 Ha for a 50-year return period, 147.61 Ha for a 100-year return period. Simulation modeling results can be seen in Fig 4 for a 25-year return period, Fig 5 for a 50-year return period, Fig 6 for a 100-year return period.

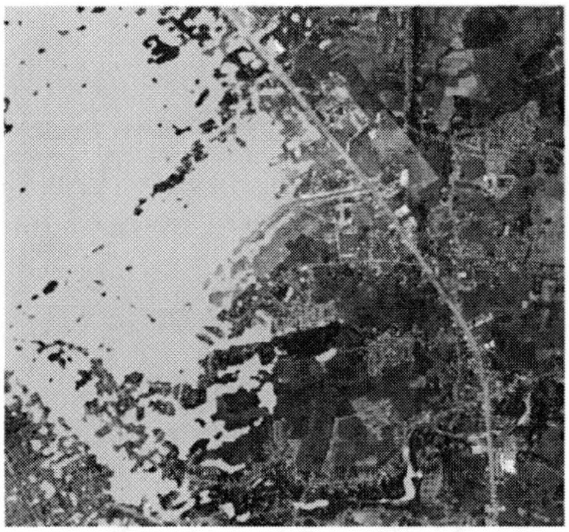

Fig. 5. Flood modeling results of conditions before river normalization with 25 years period in Batang Maransi River.

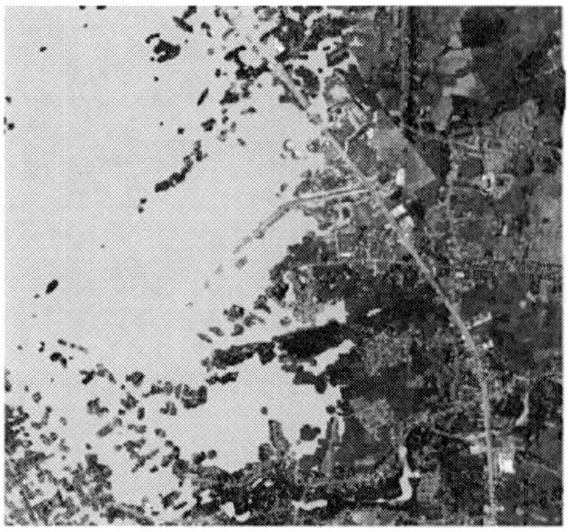

Fig. 6. Flood modeling results of conditions before river normalization with 50 years period in Batang Maransi River.

Fig. 7. Flood modeling results of conditions before river normalization with 100 years period in Batang Maransi River.

4.3.2 After river normalization

From the results of a flood, simulation obtained the area of flood area is 64.51 Ha for a 25-year return period, 71.73 Ha for a 50-year return period, 79.38 Ha for a 100-year return period. Simulation modeling results can be seen in Fig 7 for 25-year, Fig 8 for a 50-year return period, Fig 9 for a 100-year return period.

Fig. 8. Flood modeling results of conditions after river normalization with 25 years period in Batang Maransi River.

Fig. 9. Flood modeling results of conditions after river normalization with 50 years period in Batang Maransi River.

Fig. 10. Flood modelling results of conditions after river normalization with a 100 years period in Batang Maransi in Batang Maransi in Batang Maransi River.

Technically, Batang Maransi River flood control is expected to be able to manage peak discharge, their riverbeds can be used as a water reservoir, and the river environment is functional.[11]

5 Conclusion

From the research that has been done, can be concluded as follows:

1. In hydrology analysis, the debit value of the plan is used rational method. The results obtained by rational method are Q10 =58,637 m^3/s, Q25 = 73,559 m^3/s, Q50 = 85,562 m^3/s, Q100 = 98,277 m^3/s and assumed discharge that goes to the side drain of 20% of the river discharge, so that the discharge flood plan after taking into account the discharge that goes into the side drain is Q10 = 46,909 m^3/s, Q25 = 58,847 m^3/s, Q50 = 68,849 m^3/s, Q100 = 78,622 m^3/s.
2. In hydraulic analysis, the simulation results using software HEC-RAS in steady flow conditions showed the flood point that occurred in the before river

normalization condition more than after river normalization condition.

3. Before river normalization, in Q10 there were 39 flood points, namely from Sta 0+200 to Sta 1+150, in Q25 there are 39 flood points, namely from Sta 0+200 to Sta 1+150, in Q50 there are 39 flood points that are from Sta 0+200 to Sta 1+150, in Q100 there are 41 flood points that are from Sta 0+150 to Sta 1+150. In the condition after river normalization, the point of flooding only occurs in Q100, 3 flood points are at Sta 1+100 to Sta 1+150.
4. From several periods using the software Arc-GIS version, the area of flooding before river normalization is greater than after the river normalization of Batang Maransi River because it has been done river normalization.
5. The results were done by simulation using software Arc-GIS version along 1,150 km, obtained calculation of flood area that is before river normalization Q10 = 118.41 Ha, Q25 = 129.40 Ha, Q50 = 138.38 Ha, Q100 = 147.61 Ha and after river normalization Q10 = 54.71 Ha, Q25 = 64.51 Ha, Q50 = 71.73 Ha, Q100 = 79.38 Ha.
6. DEM data from Geospatial Information Agency can only be used for steep areas, whereas for relatively flat areas should be helped with DEM data from areal photos of drones.
7. From this research can be concluded that the river normalization on the Batang Maransi River proved to be effective.

This research/article's publication in supported by Department of Civil Engineering, Faculty of Engineering, Andalas University through Funding Research fiscal year 2021 with assignment agreement letter No 127/UN16.09.D/PL/2021.

References

1. Water Resources Development Profile, *The government of West Sumatra Province*, PSDA Sumatera Barat,(2018).
2. Winsemius, H. C., Aerts, J. C., Van Beek, L. P., Bierkens, M. F., Bouwman, A., Jongman, B., ... & Ward, P. J. *Global drivers of future river flood risk.* Nature Climate Change, **6**(4), 381-385. (2016).
3. Cummings, D., et al., *The tide-dominated Han River Delta, Korea: Geomorphology, sedimentology, and stratigraphic architecture,* Elsevier (2015).
4. ChowVenTe, Maidment, Mays Larry W., *Applied Hydrology, McGraw,* Hill International Edition (1988).
5. R. Wigati, Sudarsono, I.D Cahyani, *Analisis Banjir Menggunakan Software HEC-RAS 4.1(Studi Kasus sub DAS Cisimeut hilir HM 0+000 Sampai dengan HM 69+00)*, Jurnal Fondasi, **5** (2016).
6. R. Wigati, F.A. Arifin, M.D.Lestari, *Analisis Banjir Analisis Banjir Sub DAS Cilemer HM 0+00 – HM 53+00,* Jurnal Ilmiah Rekayasa Sipil, **17**(2020).
7. D. Makaesaehe, L.A.Hendratta, J.S.F. Sumarauw, *Kajian Pemetaan Banjir Dengan HEC-GEORAS Studi Kasus: Sungai Tondanao,* Jurnal Sipil Statik, **8** (2020).
8. Nurhamidah, N., A. Junaidi, and L. Anggraini, *An Immediate Review of Flood Characteristics on Delta Lowland Sumatra using D8 Model Spatial Analysis*, International Journal of Earth Sciences and Engineering, **9** (2016).
9. Saha, A. K., & Agrawal, S. *Mapping and assessment of flood risk in Prayagraj district, India: a GIS and remote sensing study.* Nanotechnology for Environmental Engineering, **5**. (2020).
10. Ntajal, J., Lamptey, B. L., Mahamadou, I. B., & Nyarko, B. K. *Flood disaster risk mapping in the lower Mono river basin in Togo, West Africa.* International journal of disaster risk reduction, **23**. (2017).
11. Fang, J., Kong, F., Fang, J., & Zhao, L. *Observed changes in hydrological extremes and flood disaster in Yangtze River Basin: spatial-temporal variability and climate change impacts.* Natural Hazards, **93**(1), 89-107. (2018)

Flood discharge analysis of Brang Ode River as a disaster mitigation effort in Kalimango Village

Imam Solihin Al-Abbas[1], *Eko* Pradjoko[1*], and *Heri* Sulistiyono[1]

[1]University of Mataram, Faculty of Engineering, Study Center of Disaster Risk Management, 83115 Mataram, Indonesia

Abstract. Flood is a hydrometeorological disaster that often occurs in West Nusa Tenggara, especially in the Brang Ode River, Kalimango Village, Alas District, Sumbawa Regency. One of the worst floods ever happened was on December 12th, 2016, which caused several villages to be inundated and houses along the river to wash away. This study aims to obtain the peak discharge from the worst flood that has ever occurred. This model is simulated using HEC-RAS 5.0.7 and QGIS for mapping the flood inundation area. Terrain data used DEMNAS. The peak discharge is obtained from the modeling results based on the flood inundation area, validated with the flood map from the DESTANA (disaster resilient village) Community of Kalimango Village. The modeling results showed that the peak flood discharge is 950 m3/s, with the inundation area 150,752.07 m2. The actual peak flood discharge can be smaller or larger than the modeling results. It may be affected by the DEMNAS raster data accuracy.

1 Introduction

Indonesia is a country that is very prone to natural disasters. A flood is an annual natural disaster that often occurs. This hydrometeorological disaster is common throughout Indonesia. Nevertheless, until now, there are still no practical steps to deal with this problem.

The heavy rain that occurred on December 12th, 2016, caused the Brang Ode River to overflow. The flood caused at least three villages to be inundated, and several houses along the river washed away. Floods caused damage to several houses and agricultural areas. In addition, floods also caused damage to the community's economy due to damage to agricultural land, access roads, and bridges.

In HEC-RAS, rivers can be modeled in one and two dimensions. One-dimensional modeling calculates water flow through a predetermined path from upstream to downstream, with the flow velocity only changing or differing in one axis direction. In two-dimensional modeling, water flow can flow in two axis directions with a changing or different flow velocity. Two-dimensional models are suitable for modeling floodwaters [1].

In dealing with disasters, especially floods, the community needs to know the risks they will face in the area where they live. Such as knowledge in disaster control which emphasizes risk reduction (mitigation). The community is the first party to deal with disaster risk so that mitigation is more effective with the active participation of the community. For this reason, a flood hazard map that can be integrated as a recommendation in spatial planning based on disaster mitigation is needed [2].

The solution is needed to see how the flood occurred and the area to be submerged based on the existing discharge. So one way of flood disaster mitigation efforts is to do the Flood Discharge Analysis of Brang Ode River as a Disaster Mitigation Effort in Kalimango Village, Alas District, Sumbawa Regency. The objective is to see the characteristics of the floods that occurred in Kalimango Village. The modeling can help the community in reducing and minimizing the impact of the flood disaster.

2 Study areas

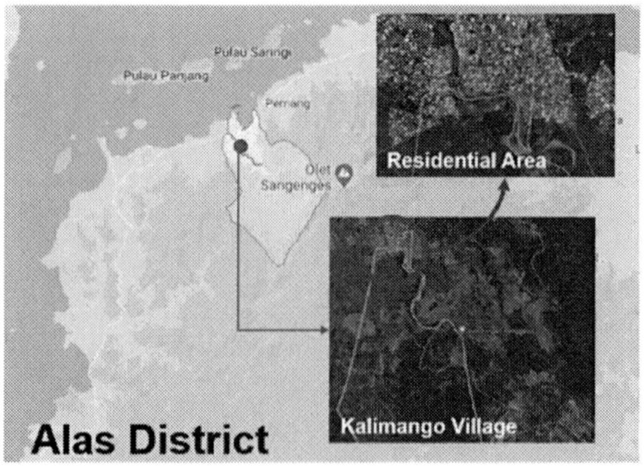

Fig. 1. Study area.

* Corresponding author: ekopradjoko@unram.ac.id

We researched in the Kalimango Village, Alas District, Sumbawa Regency. The Dalam Village bordered on the north side, the Baru Village on the west, the Marenteh Village on the south, and the Juranalas Village on the east. The Kalimango village is within the Dalap watershed coverage. This watershed is the leading supplier for the Brang Ode River and has two weirs, namely the Marente Weir and the Rengganis Weir. The Dalap watershed has an area of 97.22 km^2 and is in the category of small rivers. More details can be seen in Fig 1 below.

In the last few months during the 2020-2021 rainy season, the capacity of the Brang Ode River could still accommodate the large river water discharge caused by light to moderate rain. Based on the BMKG weather early warning report, there was light to moderate rain in the Alas District and surrounding areas on January 17-18, 2021.

Fig. 2. Brang Ode River in Cinta Bridge after rain on January 17-18, 2021.

There was an increase in the water level in the Brang Ode River during the first rain, but the river's capacity could still accommodate it. Meanwhile, during the second rain, there was a significant rise in water level. Fig 2 shows that the water level of the Brang Ode River rises until it reaches the Cinta Bridge floor slab.

3 Methodology

The data used in this study are main data and supporting data, namely data of first and second parties. The first data were obtained from interviews with local communities and observations in Kalimango Village. The second data is obtained from the relevant agencies authorized to produce the data for specific purposes. Secondary data includes DEM data obtained from the official website of the Indonesian Geospatial Information Agency and several other supporting data for the Dalap watershed area from related agencies/parties.

Data analysis was performed using HEC-RAS version 5.0.7. with input data in the form of terrain data, flow discharge, boundary conditions, river parameters, and set parameters for running the model [3]. The inundation map was then analyzed using the QGIS 3.10.

3.1 Validation map

Due to the absence of detailed records of floods that occurred in Kalimango Village, in the activity "Capacity Strengthening Program of Kalimango Village for Flood Risk Reduction," the action has been made a flood risk map of Kalimango Village. The map contains the area of flood inundation that occurred on December 12th, 2016, in Kalimango Village. The flood inundation area on the map is based on information from residents and is supported by media coverage and direct field observations. The flood inundation map can be seen in Fig 3 below.

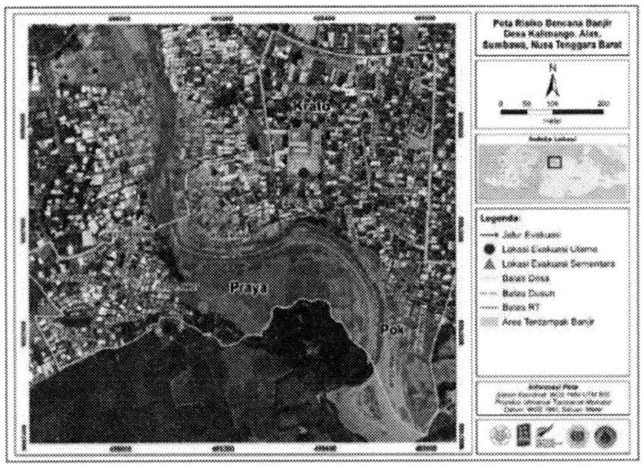

Fig. 3. Flood inundation map of Brang Ode River.

Based on the flood inundation map in Fig 3, it was found that the area of the Brang Ode River inundation in Kalimango Village is 134,468.40 m^2. The map only contains inundation that occurs in residential areas. Furthermore, the inundation area is used as a reference to validating the simulated flood inundation model. In the analysis and validation of the inundation model, the upstream inundation boundary is used as a reference for the onset of flooding in Kalimango Village. The village administrative boundary becomes the reference for the downstream inundation limit, as shown in Fig 4. The flooded overflow that occurs on the left and right banks of the river is something that will be analyzed.

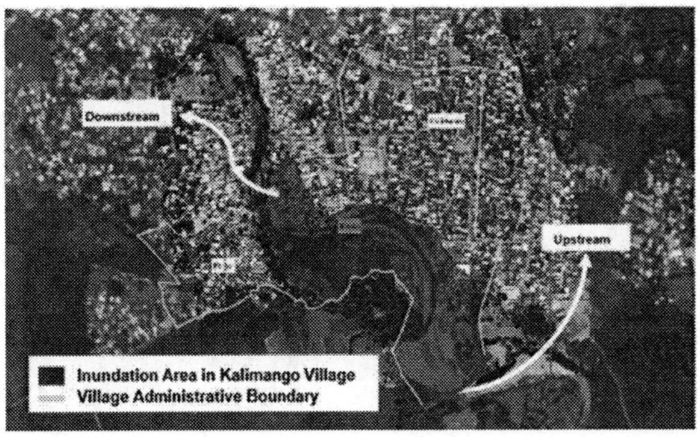

Fig. 4. Validation map.

3.2 Manning's roughness coefficient

The manning's value of the 2D modeling area is divided into 2, namely the n value for the channel and the flood plains. Because the flood plain consists of agricultural land and densely populated settlements, the value of n for the flood plain is also divided into 2. So the modeling used three values of n. The Manning's coefficient for each land-use classification can be seen in table 1 [4].

Table 1. Manning's roughness coefficients.

Land Use Type	Manning's n value
Water area	0.025
Farmland	0.04
Forest	0.16
Factory site	0.05
Residential area (low density)	0.05
Residential area (high density)	0.15
Road, vacant land	0.025

Manning's standard values for river channels obtained from kind of literature ranged from 0.03-0.04. Because there are several vegetations with different amounts of variation, it is determined that the value of the manning's roughness coefficient for the Brang Ode River is 0.05. The determination of Manning's roughness coefficient for the residential areas flood plains is based on Table 2, which is 0.1. As for agricultural land, manning's coefficient ranges from 0.035-0.04. In this case, 0.04 is used for agricultural land in Kalimango Village.

3.3 Riverbed slope

The calculation and measurement of the riverbed slope are based on changes in the slope along the river channel. Because there are several changes in certain parts of the channel slope, the river slope segment is divided based on each slope change. In this study, the slope of the riverbed channel is divided into four segments. Segment division starts from downstream to upstream.

Table 2. Bed slope of Brang Ode River.

Segments	Horizontal distance, H (m)	Elevation difference, V (m)	River slope, S_o
Segment 1	1088.10	2.8570	0.00263
Segment 2	2881.91	11.049	0.00383
Segment 3	1862.33	12.502	0.00671
Segment 4	1267.76	30.295	0.02390
Average			0.00927

From Table 2, it can be seen that the change in the slope of the river channel bed in segment 1 to segment 3 is not very significant. The average slope value of the bottom river channel is 0.00439. There is a significant change in the riverbed slope in the upstream area of the Brang Ode River, with a riverbed slope of 0.0239. The significant difference in slope changes is caused by the insufficient DEM data in the upstream part of the river. Based on this, the bottom slope value of the Brang Ode River channel is used, which is 0.0043. Since it is considered that the upstream channel is slightly steeper than the previous section, a more significant average river bed slope is used. In this study, the Brang Ode River channel bed slope of 0.005 is applied.

3.4 Terrain data

It takes at least two digital elevation model data to form a Brang Ode River. The digital elevation model data is combined in a mosaic (virtual build raster) way using QGIS 3.14. Once combined, the digital elevation model is clipped to the required area with the clip raster command. The clipped data is then used as material to create river geometry in HEC-RAS 5.0.7.

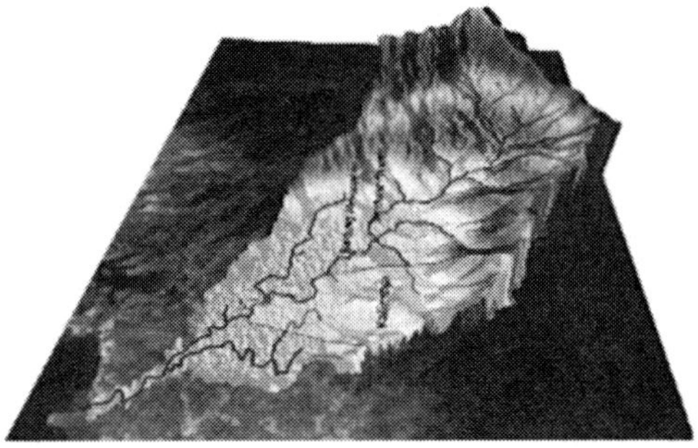

Fig. 5. Digital elevation model of Dalap Watershed

The digital elevation model data from Geospatial Information Agency is the most accurate spatial data available. The size 8 m x 8 m is quite good compared to the 30 m x 30 m digital elevation from the SRTM data. However, because DEMNAS is data from several data sources and the conversion of datums, it results in anomalies in the shape of the terrain. Such as an area that is higher or lower than the shape of the surface should be. Other landform anomalies are also seen in flow patterns and river estuaries.

3.5 Flow data (Discharge)

Given the notation Q, river flow discharge is the amount of water flowing through the cross-section of the river per unit time, which is usually expressed in cubic meters per second [5]. The discharge at a location in the river can be estimated in the following way:

- Measurement in the field (at the specified location).
- Based on discharge data from nearby stations.
- Based on rain data.

Because there is no recorded discharge data, observations and interviews with communities around flooded areas are the primary data that will be processed into flood discharge data. The data was obtained from the results of public information on the flooding that occurred. Furthermore, a simple hydrological analysis was carried out to obtain design flood discharge data by the actual flood event. The hypothetic discharge is entered to produce the inundation area as in the validation map. Discharges with similar inundation will then be analyzed.

Interviews were conducted with the community regarding the worst flooding that has occurred in recent years. It was found that the duration of the rain that occurred was for more than 2 hours. Since the first rain fell, it takes for flooding in Marente Weir is 1-2 hours. Moreover, according to information from the community, the estimated travel time for flooding from Marente Wier to Kalimango Village is about 30 minutes.

4 Results and discussion

Based on the Kalimango Village Flood Inundation map, the flood inundation area results after 2D flow modeling using HEC-RAS software. The flood inundation area generated by each discharge is then validated using the inundation area based on the "Flood Risk Map of Kalimango Village, Alas, Sumbawa, West Nusa Tenggara" in Fig 4. Three scenarios of peak discharge are used for analysis. Each of these discharges produces a different flood inundation area. After being analyzed, one scenario of peak flood discharge was selected, which resulted in the flood inundation area according to the validation map.

Based on the interview results, it was found that the length of time required by the flood from starting to rain until it reached the Marente Weir is about 1.5 hours. The flood travel time to reach Kalimango Village is about 30 minutes. There is a lot of mud content when a flood occurs. An inundation depth is about 0.5 m on the banks around the Cinta bridge.

4.1 Inundation area of scenario 1

This scenario is conducted by using a discharge amount of 800 m^3/s. A stable computational interval of 1 second is used for running the simulation and mapping out every 1 minute. The simulation does not start with an empty river condition; the hot start is used with a base flow rate of 10 m^3/s. The flood inundation simulation results can be seen in Fig 6.

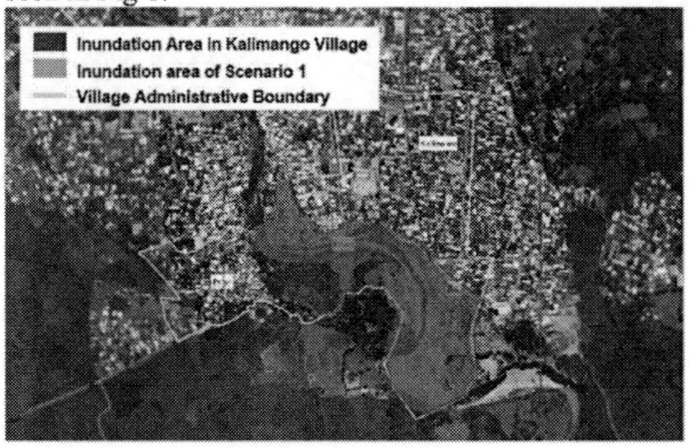

Fig. 6. Flood inundation area of scenario 1.

From the simulation results, the flood inundation in the residential area of Kalimango Village is 141,251.74 m^2. There is a leak on the right bank of the flood; if the flood area reduces it in the leak, the actual flood inundation area for scenario 1 is 122,971.59 m^2. The area is smaller than the actual flood inundation area. As shown in Fig 6, the flood inundation area of Kalimango Village is larger than the flood inundation with scenario one discharge. The flood inundation in Kalimango Village is depicted by blue polygons, while inundation with scenario one discharge is green polygons.

The average velocity of flood flow in a residential area is about 2.8 m/s. The flood travel time required from Marente Weir to reach the Kalimango Village residential area is 38 minutes. The floodwater level varies depending on the elevation of the plain and the depth of the river. At the Cinta Bridge point, the flood water level is 9.67 m, with the flood bank elevation as high as 9.4 m. The flood inundation of the Cinta Bridge floodplain is 0.27 m.

4.2 Inundation area of scenario 2

This scenario is conducted by using a discharge amount of 950 m^3/s. A stable computational interval of 1 second is used for running the simulation, and the mapping out interval is every 1 minute. The simulation does not start with an empty river condition; the hot start is used with a base flow rate of 10 m^3/s. Flood inundation simulation results can be seen in Fig 7.

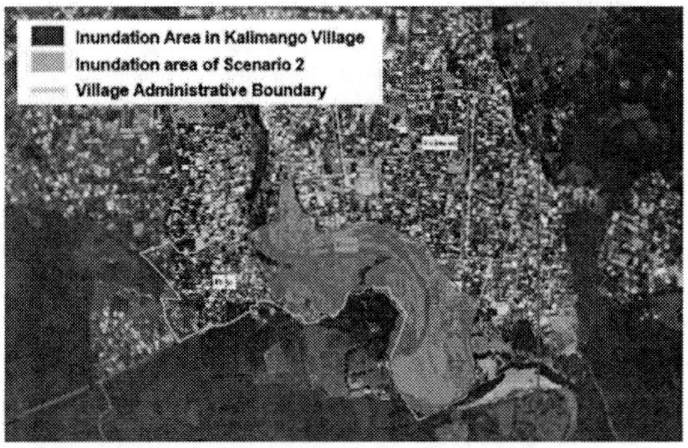

Fig. 7. Flood inundation area of scenario 2

From the simulation results, the flood inundation area in the residential area of Kalimango Village is 170.659.10 m^2. There is a leak on the right bank of the flood; if the inundation area reduces it on the leak, the actual flood inundation area for scenario 2 is 150,752.07 m^2. This area is slightly larger than the actual flood inundation area, as shown in Fig 7. The flood inundation in Kalimango Village is depicted by a blue polygon, while a yellow polygon represents the flood inundation in scenario 2.

The average velocity of flood flow in a residential area is ± 3.2 m/s. The flood travel time required from Marente Weir to reach the residential area of Kalimango Village is 32 minutes. The floodwater level varies depending on the elevation of the plain and the depth of the river. At the Cinta Bridge point, the floodwater level is 9.9 m, with the flood bank elevation as high as 9.4 m. The flood inundation of the Cinta Bridge floodplain is 0.5 m.

4.3 Inundation area of scenario 3

This scenario is conducted by using a discharge amount of 1,100 m^3/s. A stable computational interval of 1 second is used to run the simulation and map out every 1 minute. The simulation does not start with an empty river condition; the hot start is used with a base flow rate of 10 m^3/s. The simulated flood inundation can be seen in Fig 8.

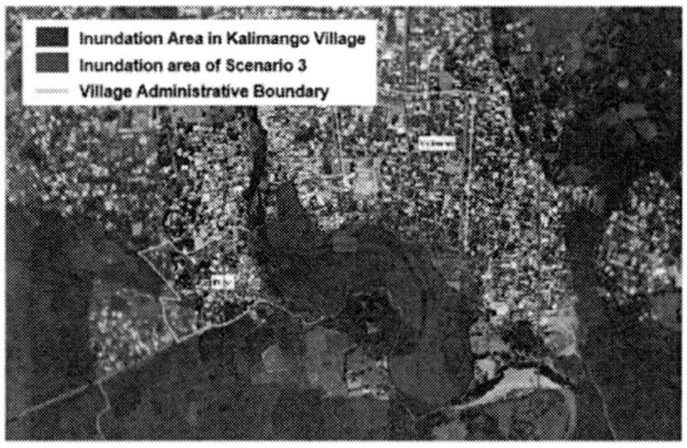

Fig. 8. Flood inundation area of scenario 3

From the simulation results, the flood inundation area in the residential area of Kalimango Village is 182,184.98 m^2. The inundation area is larger than the flood inundation area of Kalimango Village, as shown in Fig 8. In this amount of discharge, flood inundation also includes inundation on the leakage of the right bank of the river. Flood inundation in Kalimango Village is depicted with blue polygons, while flood inundation with discharge scenario three is depicted with red polygons.

The average velocity of flood flow in a residential area is ± 3.45 m/s. The flood travel time required from Marente Weir to reach the residential area of Kalimango Village is 28 minutes. The floodwater level varies depending on the elevation of the plain and the depth of the river. At the Cinta Bridge point, the floodwater level is 10.4 m, with the flood bank elevation as high as 9.4 m. The flood inundation in the Cinta Bridge floodplain point is 1 m.

4.4 Inundation validation

From the modeling (simulation) and analysis results, it was found that the error percentage value of the validation inundation area with the modeling results was 8.55% for scenario 1, 12.11% for scenario 2, and 35.49% for scenario 3. The percentage error of the inundation area in scenario 1 was close to the inundation area validation, but the resulting inundation pattern is smaller and different. For scenario 2, the percentage of inundation area error generated is slightly larger than that of the validation map inundation. However, overall, the resulting inundation pattern is similar to the validation inundation map. While in scenario 3, the percentage of error and the resulting inundation area is quite large, with a different inundation pattern from the inundation of the validation map. Based on this, scenario two was chosen as the peak discharge of the Brang Ode River in Kalimango Village.

The magnitude of the error percentage value in scenario two is caused by the poor condition of the field data used. The percentage of the inundation area error can be reduced when using fairly detailed terrain data with a small grid accuracy. The better the accuracy of the field data used, the smaller the percentage of error generated. The excellent condition of the data makes the modeled inundation area close to or even equal to the original inundation area (validation inundation).

5 Conclusions

Based on the interview results, it was found that the length of time it took for the flood from starting to rain until it reached the Marente Weir was 1.5 hours. After modeling and analysis, it was found that the appropriate flood model for the Brang Ode River in Kalimango Village is scenario 2. The travel time for flooding to reach Kalimango village is 32 minutes. The velocity of flood flow in residential areas is 3.2 m/s with a peak discharge of 950 m^3/s. So the total time required by flooding from the start of rain to flooding in Kalimango Village is 2 Hours 2 Minutes. The actual flood discharge may be larger or smaller than the modeling results. The accuracy of the DEM data influences this and also the value of the river parameters used. It is hoped that further research can develop using a good DEM and appropriate river parameters to get model results close to the actual event.

References

1. G. W. Brunner, HEC-RAS River Analysis Sistem: Hydraulic Reference Manual, US Army Corps of Engineers, Institute for Water Resources, Hydrologic Engineering Center, Davis, CA, USA (2016)
2. W. Budiarti, E. Gravitiani, Mujiyo, JTL, **18**, 242 (2017)
3. G. W. Brunner, CEIWR-HEC, HEC-RAS River Analysis System, 2D Modeling User's Manual Version 5.0. US Army Corps of Engineers, Institute for Water Resources, Hydrologic Engineering Center, Davis, CA, USA (2016)
4. A. Ryosuke, T. Ishikawa, T. Kojima, M. Tomaru, S. Maeno, Nat. Hazards Earth Syst. Sci., **17**, 1876 (2017)
5. B. Triajmodjo, Applied Hydrology, September, Yogyakarta, Indonesia (2010).

Suspended sediment transport map for disaster mitigation in the upper course of Jangkok River

Yusron Saadi[1,2*], *IB Giri* Putra[2], and *Agus* Suroso[2]

[1]Center for Disaster Risk Management, University of Mataram, Indonesia
[2]Department of Civil Engineering, University of Mataram, Indonesia

Abstract. Jangkok River, a very important river for irrigation purposes in Lombok Island Indonesia is experiencing a rapid change in the land use of its watershed. The conversion of trees into seasonal crops in the watershed causes a decrease in the carrying capacity of the watershed and reduces its ability to prevent soil erosion. Environmental damages such as erosion and sedimentation along the river are becoming increasingly difficult to control. This paper proposed an idea to identify river conditions based on suspended sediment concentration. A series of field measurements of suspended sediment concentrations and river flow discharges were carried out to obtain the relationship between these two parameters. A resulting watershed map contains information on the rate of suspended sediment transport in the form of a suspended sediment-flow discharge rating curve is designed as a sediment conservation health indicator of the watershed. The map allows the authority to observe the condition of each river order in the upper course of the Jangkok River since the nearby tributaries are most likely to have their pattern of suspended sediment transport rate. The finding from this preliminary study can facilitate the drafting and implementation of the right policy in watershed conservation management planning which leads to preventive action and better preparedness in disaster mitigation.

1 Introduction

The problem of river management is growing and the balance of nature, especially in river catchment areas is increasingly in danger due to the development of human activities. River management includes river conservation, river development, and control of water destructive power to the environment [1]. Law No. 17 of 2019 concerning Water Resources states that the destructive power of water is the power of water that is detrimental to life [2]. The destructive power of water can be in the form of floods, erosion and sedimentation, and landslides.

In the context of controlling the destructive power of water, it is also necessary to make efforts in the form of non-physical measures other than physical activities. Non-physical measures are in the form of a conservation project to maintain the existence and sustainability of the natural condition and function of the river. Its flow should always be available sufficiently both in quantity and quality to meet the needs of humans and other living organisms both in present and in the future. One thing very urgently needs to be carried out is to identify river conditions and to diagnose the problems faced by the river and its watershed. Inspection of a watershed environment is analogous to a human health checkup [3]. The so-called watershed health indicators are a set of indicators which easily and directly measured. These indicators can express the health condition in an observed watershed.

Existing health indicators mainly focused on water quality and ecological conditions. In recent years health indicators develop and accommodated parameters such as runoff ratio, denudation depth, vegetation cover ratio, turbidity, and sediment concentration [4]. The purpose of the research is to create a watershed map that contains information on one of those indicators, i.e. sediment concentration in the upper course of Jangkok River in Lombok Island. The watershed was selected based on being a very important river with the main river contributing significantly to the island development in terms of meeting irrigation needs.

Suspended sediment concentration can be used to express the state of the watershed in the upper course of the Jangkok River representing the rate of sediment transport for each river segment from tributaries to the main river. This map is expected to be used as a guide to finding out which river segments need attention according to the level or rate of suspended sediment transport. The indicator will be transformed into an equation that describes the relationship between suspended sediment concentration and river flow discharge. The suspended

* Corresponding author: y.saadi@unram.ac.id

https://doi.org/10.1051/e3sconf/202133108005

sediment-flow discharge rating curve allows identification and examination of the state of the sub-watershed that are sensitive to erosion and the sub-watershed that are relatively stable.

The production of the map will be used by the authority as a basis to determine a watershed conservation plan with treatment priorities that can be arranged by putting rivers with a high suspended sediment concentration as the initial treatment target. This is particularly important in the way to anticipate governmental limited conservation budget so that the comprehensive watershed conservation arrangement can be carried out in stages for each fiscal year.

2 Soil erosion and sediment continuity concept

Conversion of land use in the catchment area from hard crops into seasonal crops such as food crops and vegetables is immensely popular amongst people living in the nearby forest in the upper course of the Jangkok River. They did not realize the negative impact of these activities on the environment on a larger scale. The government also fails to properly address these issues as people always forward the economic reasons behind their activities. Government intervention will be considered as halting an increase in the income of people living in the watershed. This has been proven to cause uncontrolled conservation of watersheds which results in the increase of sedimentation transportation rates due to increased soil erosion. This type of erosion is considered one of the most serious environmental problems that cause soil nutrient loss and land degradation [5, 6].

A lot of research has been devoted to quantifying the effects of vegetation cover on soil erosion, e.g. Mahmoudzadeh et al. [7], Sharma [8], Han et al. [9], and Guo et al. [10] amongst others. It has been proven that vegetation can improve soil physical and chemical properties and erosion resistance [10]. Soil erosion generally flows toward river valleys and the transportation of sediment downstream leads to siltation which can reduce the cross-sectional capacity of the river. Subsequent floods harm the community and damage existing infrastructures. High sedimentation rates can also cause silting in the reservoir and reduce the effective lifetime of such a structure.

In recent years all parties gradually understood that uncontrolled land-use conversion activities not only had a local impact but also often affected the upper course and lower course of the watershed [4], threatening regional even global ecological security patterns and human development [11, 12]. Change in the form of land use may intensify soil erosion along the slope direction and contribute to an increase in river sediment transportation.

Measurement of river sediment is often considered something difficult to carry out. This activity is also receiving less attention from the authorities and is considered to have low urgency even though the magnitude and intensity of river sediment transport can be used as one of the parameters to determine the degradation level of a watershed [13]. Another limitation is that many assume the sediment transport behavior of the nearby rivers is identical and is considered to have similar patterns even though different conservation practices in the adjacent rivers and subwatershed are applied [14]. Lack of attention and the low level of urgency is the main reason that sediment measurements are usually limited in number. Each river, including its sub-watersheds, can have different characteristics in terms of sediment transport rate even though they are geographically located nearby [14]. Research conducted on three main rivers flowing into a reservoir by Saadi [15] showed that those rivers have a suspended sediment transport rate of two to fivefold different even though they are closely located. The finding indicates that the measurement of the sediment transport rate of each river in a watershed is very relevant and useful so that the respective sediment transport rates can be obtained.

The amount of sediment material transported, eroded, and deposited in rivers, especially alluvial rivers, is a function of sediment supply and sediment transport capacity by rivers. Sediment supply is obtained from soil loss within the watershed, riverbanks erosion, and river bed degradation. Sediment carrying capacity is a function of sediment size, flow rate, and geometric and hydraulic properties of the river. The equilibrium state is achieved when the carrying capacity, i.e. sediment transported by the flow is equal to the supply of sediment. In reality, it is difficult to achieve this condition, partly because the river bed stability can vary significantly even during floods of the same magnitude [16]. This is supported by the nature of the river bed which is formed from sediment particles with non-uniform grain sizes.

The concept of sediment continuity in a river reach for a certain period is expressed by the amount of sediment flowing into and deducted by the amount of sediment flowing out at the downstream end of the reach equal to the amount of sediment stored within the reach. Sediment entering the river reach can be in the form of sediment flow through the cross-section at the upstream part coupled with sediment input from both sides of the river which is the result of soil erosion from the watershed. The sediment carrying capacity by the flow determines the rate of sediment.

The stability of the river bed is strongly influenced by the nature of the flow, i.e. the flow hydrograph pattern. Continuous material transportation usually occurs in the early stages of the flood hydrograph because the riverbed material has not been well consolidated. A longer flood hydrograph increases the stability of the riverbed to a certain level due to the armoring process that takes place on the riverbed. When subsequent flood hydrograph passes the riverbed, sediment transport rate does not necessarily increase linearly even though the flood hydrograph is larger than the previous one [16].

3 Methodology

3.1 Location of the study

Water assessment and suspended sediment sampling was carried out in the upper course of Jangkok River in

Lombok Island (Fig. 1). Six measurement points were selected based on the field condition. Tembiras River was divided into upstream (T1) and downstream (T2) locations, while Jangkok River was divided into Jangkok upstream (J1) and Jangkok downstream (J2). A point each was selected in smaller tributaries called Bentoyang River (B) and Pemoto River (P). The number of samples for each location was 24 samples or 148 samples in total.

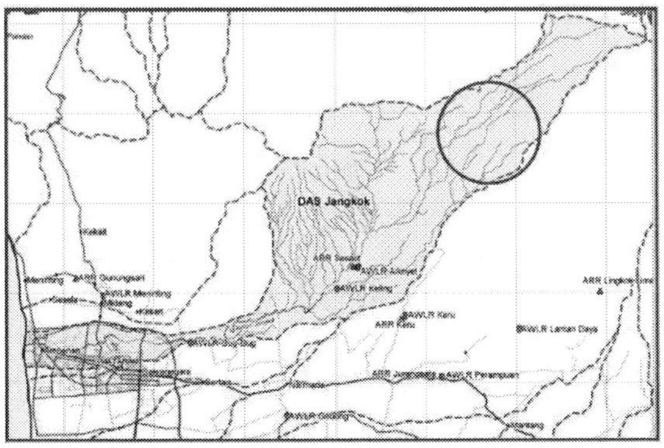

Fig. 1. Measurement location of flow discharge and suspended sediment

3.2 Flow discharge measurement

Measurement of river flow discharges was carried out at the same location with the measurement of suspended sediment to obtain an accurate relationship between suspended sediment concentration and river flow discharge. The standard procedure is to plot the daily flow discharge hydrograph and the daily sediment concentration graph. Measurement of flow discharge begins with measuring the cross-sectional area of the river. At the same point, the flow velocity measurement was carried out so that the river flow discharge can be obtained.

Flow discharge at a predetermined location was observed. The flow velocity was measured by flow meter wading set as shown in Fig. 2, whereas the water assessment was carried out between tributaries to the main river. To obtain representative data on flow velocity measurements and water assessments, measurements were made during rainy and dry seasons.

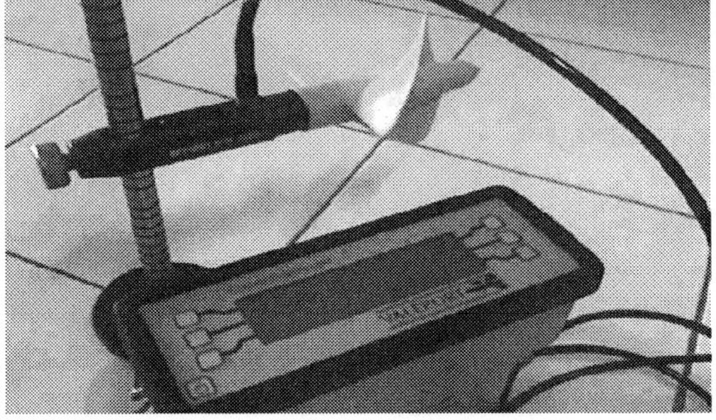

Fig. 2. Flowmeter of Valeport Model 001

3.3 Suspended sediment concentration

The measurement of suspended sediment concentration in the river is generally an adoption of a conventional method, namely the depth integration method [17]. This method measures the suspended sediment at three different water depths, i.e. 0.2h, 0.6h, and 0.8h from the water surface (h is the water depth). According to Engineer Manual 1110-2-4000, the sediment concentration in the unmeasured zone is usually estimated to be from 5-15% of the measured concentration [18]. Suspended sediment sampling was carried out after the discharge measurement was completed.

The measurement of suspended sediment concentration was carried out at the same location as the flow velocity and water assessment. Measurements were made using a suspended sediment sampler and a cylinder water sampler as shown in Fig. 3 and Fig. 4. In terms of proportional dimensions, a cylindrical measuring instrument can provide more accurate measurement results with a trapping efficiency of 0.8-1.2 for a moderate velocity [19]. Filtering and weighing samples were carried out to separate sediment contained in the water sample. Measurements of flow discharge and suspended sediment concentration were carried out as much as possible at various locations so that the existing samples represented the actual condition of the river.

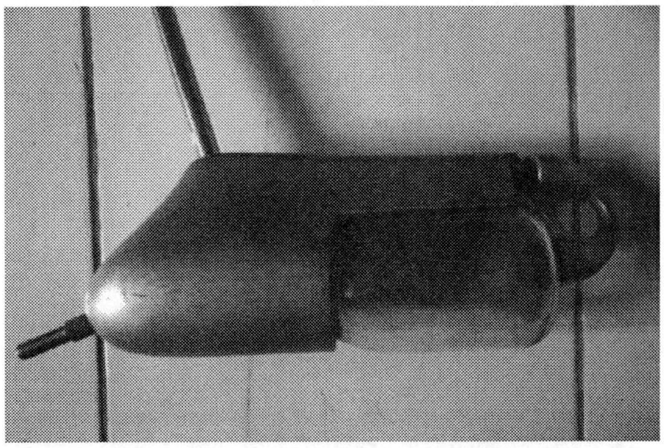

Fig. 3. Suspended sediment sampler of US DH 48 type

Fig. 4. Cylinder Water Sampler of LaMOTTE Model JT-1

In addition to field measurements, secondary data in the form of hydrological, climatological, land use, and other technical data were prepared. Watershed-related maps and additional information were also collected accordingly.

3.4 Suspended sediment-flow discharge rating curve

Processing data series that will be transformed into a mathematical equation is carried out by numerical analysis. In general, the most common data processing techniques require an understanding of the regression method whose analysis results are presented in the form of a smoothed curve (fitting curve). One of the methods is the least square method which minimizes the sum of the squares of the deviations formed by the resulting line [20]. Before the least-squares method is employed, discharge data and sediment concentration are written in the form of a logarithmic equation. In the least-squares method, the error is defined as the deviation or difference between the true value and the measured value.

The relationship between suspended sediment concentration and flow discharge is expressed by a logarithmic graph which can be mathematically expressed by an equation in the following form

$$Q_s = aQ_w^{\ b} \qquad (1)$$

where :

Q_s = suspended sediment rate (gr/lt)

Q_w = flow discharge (m^3/det)

a = coefficient

b = exponent

Based on the consideration that flow fluctuations from year to year are varied, the relationship between the measured flow discharge and the suspended sediment concentration originated from the sample is a correlation between the two factors at the time of measurement [15]. The soil erosion level may not the same for every rainfall because it depends on rainfall intensity, soil conditions, and plant growth. Certain parts of a river's reach may be more sensitive to erosion than others so that a larger sediment flow can be expected when rainfall is concentrated in the area. This shows that suspended sediment transport rate and the river flow discharge are not always directly correlated. However, the suspended sediment-flow discharge rating curve is a very useful tool for estimating suspended sediment transport rate.

3.5 Numerical analysis and suspended sediment samples test

Numerical analysis is carried out to obtain the flow discharge-suspended sediment rating curve. The analysis is conducted in two stages. The first stage is the collection of flow discharge data and suspended sediment concentration. The second stage is data processing using the least square method.

The least-squares method is employed by presenting an equation containing the relationship between suspended sediment concentration and river flow discharge (Equation 1) in the form of a logarithmic equation as follows;

$$\log Q_s = \log a + b \log Q_w \qquad (2a)$$

if $Q_s' = \log Q_s$, $Q_w' = \log Q_w$ and $a' = \log a$,

$$Q_s' = a' + bQ_w' \qquad (2b)$$

Therefore, the residual equation is

$$R = \sum_{i=1}^{n} R_i^2 = \sum_{i=1}^{n} \left(a' + bQ_{wi}' - Q_{si}'\right)^2 \qquad (3a)$$

$$\frac{\gamma R}{\gamma a'} = 2\sum_{i=1}^{n} \left(a' + bQ_{wi}' - Q_{si}'\right) = 0 \qquad (3b)$$

$$\frac{\gamma R}{\gamma b} = 2\sum_{i=1}^{n} \left(a' + bQ_{wi}' - Q_{si}'\right)Q_{wi}' = 0 \qquad (3c)$$

Sediment samples taken from the field are observed at both Geotechnical Laboratory and Hydraulics and Coastal Laboratory, Faculty of Engineering, University of Mataram. The types of tests are volumetric/ gravimetric tests and sieving analysis. Parameters resulting from these tests are specific gravity, grain size analysis (gravel, sand, silt, and clay), and the soil volume weight of wet and dry conditions.

4 Results and discussion

4.1 Suspended sediment samples

Suspended sediment measurements were carefully carried out at a location with minimum disturbance and no backwater influence (Fig. 5). This was intended to allow the suspended sediment to enter the sampler in a very smooth fashion. Following the measurement, trapped sediment was removed and stored in a bottle or a container (Fig. 6). Twenty-four samples for each location were labeled for further analysis in the laboratory.

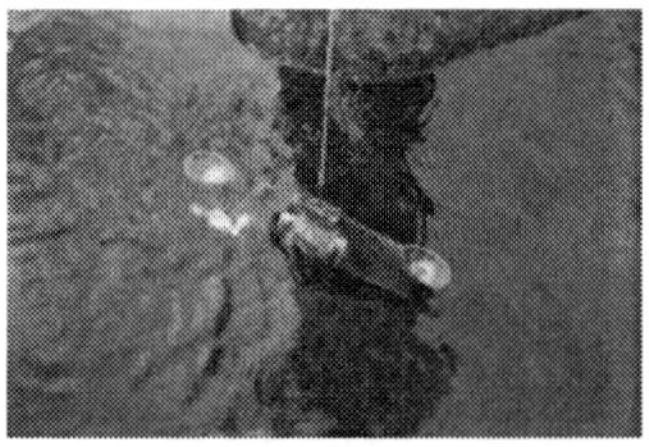

Fig. 5. Sediment sampling measurement using *Cylinder Water Sampler* (LaMOTTE Model JT-1)

Subsequent laboratory analyses such as weighing wáter and sediment followed by drying simple in the oven were conducted. After the dry weight of samples was recorded, the relationship between flow discharge and suspended sediment concentration can be obtained. The problem faced during this process was the measurement results with very little suspended sediment content, making it difficult for the drying and weighing processes. Therefore this series of work was carefully executed to obtain accurate results.

Fig. 6. Removal of sediment from US DH 48

4.2 Suspended sediment-flow discharge rating curve

The numerical analysis carried out resulted in an equation containing the relationship between suspended sediment concentration and river flow discharge. Field data from the measurement of suspended sediment concentration and flow discharge were converted into a series of data. Values of suspended sediment concentration were plotted versus flow discharge on a log-log grid (Fig. 7). Field data should be designed in such a way that samples were taken at various discharge conditions so that the relationship can be considered to represent sediment transport conditions that are close to the actual.

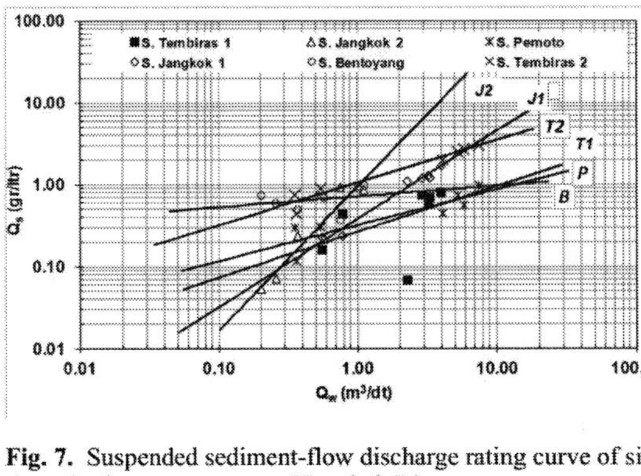

Fig. 7. Suspended sediment-flow discharge rating curve of six rivers in the upper course of Jangkok River

Sediment data measured during dry season measurement may not be adequate as the low flow was expected during this period with very limited high-flow events. A very small amount of sediment concentration may lead to an underestimated situation. On the other hand, more trapped sediment from the high-flow event during the rainy season made an overestimated calculation. Therefore, the more the data measurement taken from both seasons the more accurate the equation obtained. Measurement over a longer period than can be accommodated at present may necessitate.

Fig. 7 shows that the rating curve can be varied and has the potential to show differences in the concentration level of suspended sediment although measurement is located in adjacent sub-watersheds. It is understood that soil erosion or soil loss is not the same as sediment yield. Eroded soil may be moving and redeposited a short distance from where it was dislodged, whereas sediment yield is a portion of the eroded soil that leaves the watershed. Upon reaching the river, sediment is entrained during sporadic periods of higher flows and deposited as discharge subsides.

The rating curve can be used as a basis in estimating suspended sediment concentration and transforming it to suspended sediment transport rate at any desired discharge. As suspended sediment rate is a representation of soils loss level in the watershed the empirical relations between suspended sediment transport rate and flow discharge are beneficial to estimate watershed denudations rate and the relative stability of river courses so that the state of sediment conservation in sub-watershed can be rightly expressed. Furthermore, it is hoped that the rating curve will facilitate the assessment of the overall healthy state in the watershed when coupled with other health parameters. By understanding information about the health condition of the sub-watershed, anticipatory actions can be taken, especially related to disaster mitigation planning.

4.3 Watershed map based on suspended sediment concentration

In general, the characteristics of the sub-watersheds of Jangkok River and its tributaries have similarities, especially the average slope and vegetation cover. The difference in suspended sediment concentration shown by the suspended sediment-flow discharge rating curve is displayed on the watershed map in the form of an information box containing an equation that describes the relationship between flow discharge and suspended sediment concentration. The box also contains information on the average slope, the average flow rate, the average river width, and other related information (Fig. 8).

Determination of the river order in the upper course of the Jangkok River watershed is similar to those defined by Strahler [21]. It started from the river in a sub-watershed with no tributaries. Hence, the first order (O1) is occupied by Pemoto (P) and Bentoyang (B) river. The highest order (O4) is the Jangkok River in the downstream part (Jd). Detailed stream order for the upper course of Jangkok River is presented in Fig. 8.

Comparison between the suspended sediment-flow discharge rating curve in Fig. 7 and the stream ordering in Fig. 8 suggests that the lower order of the river has a lower suspended sediment concentration. It is expected that smaller sub-watershed such as Bentoyang (B) and Pemoto (P) rivers produce a smaller magnitude of sediment yield. Attention should be more focused on the river with a steeper line in the suspended sediment-flow discharge rating curve.

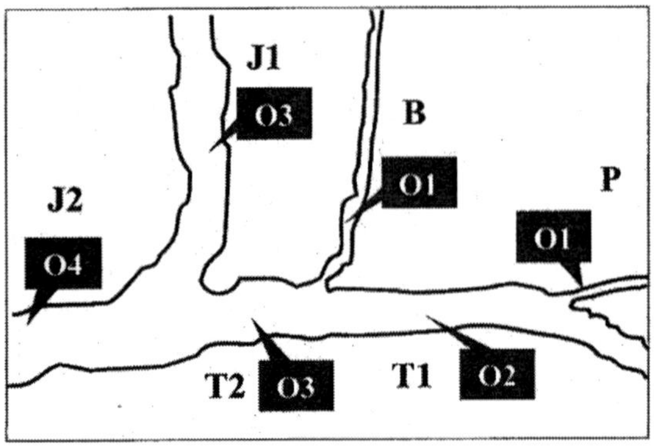

Fig. 8. Mapping of measurement locations and technical information boxes based on suspended sediment concentration

Both Jangkok upstream (J1) and Jangkok downstream (J2) experience a high increase in suspended sediment concentration as soon as flow discharge increased. Tembiras upstream (T2) has a less steep curve slope indicating a much lower increase in suspended sediment concentration. Tembiras upstream (T1) experienced a lower concentration of suspended sediment than Tembiras downstream (T2) but still higher than tributaries with the lower order, i.e. Bentoyang River (B) and Pemoto River (P).

The map in Fig. 8 is designed to provide information on watershed health assessment according to the suspended sediment concentration. Through the map, the suspended sediment rate can be estimated for any desired flow discharge in each sub-watershed. A map is a useful tool for watershed assessment whether the state of sediment rate in each sub-watershed is tolerable or intolerable magnitudes.

5 Conclusion

Measurement of suspended sediment concentration in the upper course of Jangkok River provides the preliminary and important source of information about the health state of the watershed and its sub-watershed. It is because the rivers are the principal route by which the products of watershed erosion are delivered downstream. Suspended sediment concentration and suspended sediment transport rate are the important source of estimates of watershed degradation and the stability of river courses.

A suspended sediment-flow discharge rating curve is very helpful in estimating the rate of suspended sediment transport in the upper course of the Jangkok River. The curve allows an estimation of the suspended sediment transport rate for each river order representing the health state of the watershed. The resulting watershed map containing information on sediment transport rates is a useful tool as guidance in planning watershed conservation management priorities. This is particularly important as the sediment transport behavior of each river is different even though they are located nearby.

This study is a preliminary step in broader studies of sedimentation behavior in the Jangkok River Network. It should be noted that the work focused on the investigation of suspended sediment concentration concerning the river flow discharge. Further investigation of land erosion in the Jangkok watershed or sub-watershed is necessitated and will greatly help in explaining the behavior of the river and its tributaries. This requires high measurement accuracy and representative sampling over a long period.

The result of this study allows the conservation authority to estimate the suspended sediment transport rate at the desired river flow. The level of suspended sediment concentration that will be listed on the map provides insight into the potential soil loss and subsequent sediment yield of the river network in the upper course of Jangkok River so that a conservation plan particularly related to disaster mitigation can be arranged accordingly.
Sincere thanks are conveyed to the Water Resources Information Center (BISDA) of NTB Province for the support in providing field equipment and operator allowing this preliminary research successfully conducted.

References

1. State Gazette of the Republic of Indonesia, Government Regulation of the Republic Indonesia Number 38 the Year 2011 regarding Rivers (State Gazette of the Republic of Indonesia Year 2011 Number 74, 2011)
2. State Gazette of the Republic of Indonesia, Law Number 17 the Year 2019 regarding Water Resources (State Gazette of the Republic of Indonesia Year 2019 Number 190, 2019)
3. I. Maddock, Freshwater Biology. **41**, 2 (1999)
4. P.K. Chou, B.J. Shih, C.Y. Wu, J. Chinese Inst. Eng. (2016)
5. S. Arnhold, S. Lindner, B. Lee, E. Martin, J. Kettering, T.T. Nguyen, Geoderma. s 219-220, 3 (2014)
6. S. Park, C. Oh, S. Jeon, H. Jung, C. Choi, J. Hydrology. **399**, 3 (2011)
7. A. Mahmoudzadeh, W.D. Erskine, C. Myers, Australian Forestry, **65**, 2 (2002)
8. A. Sharma, Geo-spatial Information Science, 13, 3 (2010)
9. H. Dandan, D. Jingcheng, G. Chaojun, M. Xingmin, G. Peng, G. Jianjian, Int. J. Sediment Res. (article in press)
10. Q. Guo, Z. Wang, Q. Zhang, Q. Zhang, N. Shen, B. Wu, N. Tian, J. Liu, Archives Agronomy Soil Science (2020)
11. P. Borrelli, P. Panagos, M. Marker, S. Modugno, B. Schutt, Catena. **149**, 3 (2016)
12. J.A. Martinez-Casasnovas, M.C. Ramos, G. Benites, Land Degr. Dev. **27**, 2 (2016)

13. Y. Saadi, H. Saidah, L.D.B. Irawan, *Overview of the index and class of erosion hazard in Tanggek River Basin*, in Proceeding of the 4th National Conference on Civil Engineering, 2-3 June 2010, Sanur-Bali, Indonesia (2010)

14. Y. Saadi, I.B.G. Putra, A. Suroso, Applied Mech. Materials. **845** (2016)

15. Y. Saadi, *Equation of flow discharge-sediment curve for estimating reservoir sedimentation (a case study of three rivers of Batujai Reservoir)*, in Proceeding of National Conference on Sustainable Infrastructure Development, 18 October 2007, Kuta-Bali, Indonesia [2007]

16. Y. Saadi, S.J. Tait, *The influence of time-varying antecedent flows on the stability of mixed grain size sediment deposits*, in Proceeding of 3rd International Symposium on Environmental Hydraulics, 5-8 December 2001, Tempe-Arizona, USA (2001)

.

17. National Standardization Agency, Procedure for suspended sediment sampling in the river by depth integration method based on discharge distribution, SNI 3414:2008 (BSN, Jakarta, 2008)

18. Department of the Army, Engineering, and design, Sedimentation investigations of rivers and reservoirs, Engineer Manual No. 1110-2-4000 (Department of the Army, Washington, USA, 1989)

19. L.C. van Rijn, Manual sediment transport measurements, Report S304 (Delft Hydraulics, The Netherland, 2006)

20. T.J. Akai. Applied numerical methods for engineers, (John Wiley & Sons, New York, USA, 1994)

21. A.N. Strahler, *Quantitative geomorphology of drainage basins and channel networks*, in Handbook of applied hydrology, V.T. Chow (Eds.), (McGraw-Hill, New York-USA, 1964)

Analyses of flood peak discharge in Cimadur river basin, Banten Province, Indonesia

Arniza Fitri[1], *Muhammad Shubhi* Nurul Hadie[2], *Adelia* Agustina [1*], *Dian* Pratiwi[1], *Susarman*[1], *Galuh* Pramita[1], and *Salah* H. R. Ali[3]

[1]Department of Civil Engineering, Faculty of Engineering and Computer Science, Indonesian Technocrat University, Bandar Lampung, Lampung 35132, Indonesia
[2]Department of Civil Engineering, Faculty of Engineering, Andalas University, Padang 25164, Indonesia
[3]Precision Engineering Division, National Institute for Standards, Giza (12211-136), Egypt

Abstract. Cimadur river basin is one of the most important catchment areas in Lebak District, Banten Province. For the past few years, the catchment has experienced floods during the rainy season. The big issue of flooding has been recorded recently in December 2019 which has caused damage and negative impacts to the local people and surrounding community. This study aims to analyze the possibility of flood peak discharges in the catchment area of the Cimadur river. The flood discharges are calculated for 2, 5, 10, 25, 50, and 100 years return period based on the daily rainfall data from the year 2011 to 2020. The rainfall and land use data are obtained from PT Saeba Consultant. In this study, the hydrological analyses are including 1) analyses of average annual rainfall using the Thiessen method; 2) analyses of rainfall distribution and estimation of design rainfall by considering three methods involving: Log-Normal, Log Pearson Type III, and Gumbel Type 1; and 3) analyses of flood discharges by adopting Nakayasu Synthetic Hydrograph Unit (SHU). The rainfall distribution analyses show that the Log Pearson Type III provided the best fit. Based on the flood peak discharges analyses, the results show that the flood discharges for the 5, 10, 25, and 50 years return period in the Cimadur river basin are 470.71 m^3/s, 560.16 m^3/s, 698 m^3/s, and 820.4 m^3/s, respectively.

1 Introduction

Flooding is the most frequent natural disaster that occurred in Indonesia in the past few decades, mostly triggered by extreme rain during the rainy season [1, 6]. However, the occurrences of flooding are not only natural but can be aggravated due to socio-economic factors such as urbanization and development activities which could result in larger flood-prone areas [7, 9].

Regarding flood risk reduction and prevention in a local region, estimation of flood peak discharges in a local area is becoming necessary [9]-[11]. Estimation of the flood peak discharges could be helpful in properly designing the various engineering structures in the river basin [12, 15].

Cimadur river basin is one of the most important catchment areas in Lebak District, Banten Province, Indonesia. The land-use change activities such as converting the forest into residential areas and rice fields around the river have triggered the flood occurrences in this catchment for the past few years (Fig. 1). One of the biggest flood events in the Cimadur river basin has been experienced in December 2019 which has caused damage and negative impacts to the local people and surrounding community.

Fig. 1. One of the flood events occurred in the Cimadur river basin

To reduce and control the flood events in the Lebak district, the local government and the local water resource managers are going to design the various engineering structures on the Cimadur river catchments. Due to the importance of flood peak discharges estimation in designing the engineering structures in the Cimadur river basin, therefore, this study aims to analyze the possibility of flood peak discharges in the catchment area of the

* Corresponding Author: adeliagstna@gmail.com

https://doi.org/10.1051/e3sconf/202133108006

Cimadur river. The flood discharges are calculated for 2, 5, 10, 25, 50, and 100 years return period.

2 Materials and Methods

2.1 Study Area

This study is carried out in the catchment of Cimadur river, Lebak District, Banten Province, Indonesia. Cimadur river catchment as shown in Fig. 2 is geographically located between 6°40" S to 6°55" S (latitudes) and 106°14" E to 106°25" E (longitudes). The total area of the Cimadur river catchment is approximately about 210 km^2. The catchment is influenced by 2 monsoon seasons during the year that is rainy season and the dry season. The area receives one cycle of rainfall during the rainy season which runs from April to October in the year.

The topography condition in the Cimadur river catchment is a moderate slope with an elevation between 0 and 500 m above MSL. Local people in this catchment are mostly making a living as a farmer [16].

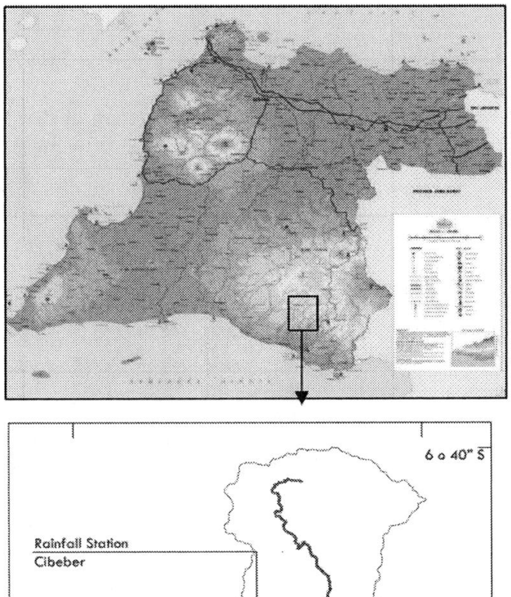

Fig. 2. Location of the study area

2.2 Data Collection

For estimation of flood peak discharges, the required data is consisting of rainfall data and land use data. There are two rainfall stations in the catchment of the Cimadur river, which are Bayah station and Cibeber station. In this study, rainfall data and land use data are obtained from PT. Saeba Konsulindo Serang, Banten. Rainfall data from both rainfall stations (Bayah station and Cibeber station) are taken for ten years period (2011 to 2020). Based on the available data, the maximum daily rainfall per year is presented in Table 1. Apart from that, Table 2 describes the land use condition in the Cimadur river basin in the year 2019 based on research done by PT Saeba Konsulindo.

Table 1. Maximum daily Rainfall per year

No	Year	Rainfall data at Bayah Station (mm), P1	Rainfall data at Cibeber Station (mm), P2
1	2011	137	75
2	2012	128	91
3	2013	185	112
4	2014	121	97
5	2015	101	71
6	2016	115	93
7	2017	120	71
8	2018	131	128
9	2019	112	201
10	2020	136	184

Table 2. Landuse Condition in Cimadur river basin in years of 2019

No	Landuse area	Area (ha)	Percentages (%)
1	forest	7284	34.65%
2	River	90	0.43%
3	residential	221	1.05%
4	ricefield	3691	17.56%
5	Mix garden	8952	42.58%
6	bush	744	3.54%
7	bare land	40	0.19%

2.3 Hydrological Analyses

In this study, hydrological analyses are carried out to estimate the flood peak discharges in the Cimadur river basin for 5, 10, 25, 50, and 100 years return period. There are several stages in hydrological analyses including analyses of average annual rainfall, analyses of rainfall frequency distribution, estimation of design rainfall, and analyses of flood peak discharges. The detailed description for those analyses are following:

2.3.1 Analysis of average annual rainfall

The average annual rainfall is analyzed by approaching the Thiessen method. Here, one rainfall station from outside of the Cimadur river basin is included to draw the polygon and to get the total areas influenced by each rainfall station. Fig. 3 shows the area division for the calculation of average annual rainfall. The calculation is further follows the equation below [17] :

$$P = \frac{A1p1+A2p2+A3p3+\cdots+Anpn}{A1+A2+A3+\cdots+An} \quad (1)$$

Where:

P = the average value of annual rainfall

P1, p2, ..., pn = maximum daily rainfall for each rainfall station (1, 2, ..., n)

A1, A2, A3, ... An = area influenced by each rainfall station (1, 2, ... , n.)

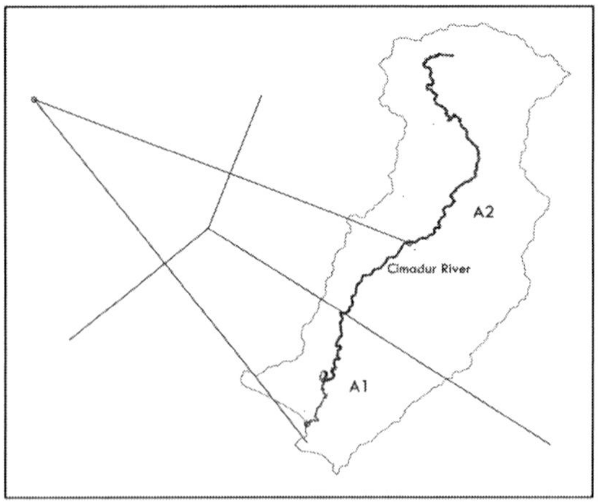

Fig. 3. Area division for each rainfall station

Based on polygon drawing, it can be described that the total area influenced by Bayah' station (A1) is 50.4 km^2 (24% from the total catchment of Cimadur), while the total area influenced by Cibeber' station (A2) is 159.6 km^2 (76% from the total catchment of Cimadur).

2.3.2 Analyses of rainfall frequency distribution and estimation of design rainfall

Concerning the estimation of design rainfall and analyses of rainfall frequency distribution, three methods are considered to be examined in this study including the Log-normal method, extreme Gumbel type 1 method, and Log person type III method.

The equation used for calculation of design rainfall in the Log-normal method is as follows [17]:

$$\text{Log } X_T = \text{Log } \bar{X} + \text{k. Sx. LogX} \quad (2)$$

Where:

Log X_T = design rainfall for T years return period

Log $\bar{X}$ = the logarithms average of rainfall data

$\bar{k}$ = coefficient factor

SxLogX = standard deviation

Further, the equation used for calculation of design rainfall in the Gumbel type 1 method is as follows [17]:

$$X_T = \bar{X} + S.K \quad (3)$$

Where:

X = design rainfall for T years return period

$\bar{X}$ = the mean value of rainfall data

K = the frequency factor for extreme Gumbel method

S = standard deviation

$$K = \frac{Ytr-Yn}{Sn} \quad (4)$$

$$Ytr = -In\{-In\frac{Tr-1}{Tr}\} \quad (5)$$

Where:

Tr = The function of time

Yn = Gumbel' reduced mean

Sn = Gumbel' reduced standard deviation

Besides, the equation used for calculation of design rainfall in the Log-Pearson type III method is as follows (15):

$$Log\ X_T = \overline{Log\ X} + ktr.S1 \quad (6)$$

Where:

Log X_T = design rainfall for T years return period

Log X = the logarithms average of annual rainfall at station

S1 = standar deviation,

Ktr = frequency coefficient based on skewness values (Cs) for T years return period (can be determined using a frequency factor table (Haan, 1977)

However, the best fit method that can be applied for design rainfall is based on the values of skewness coefficient (Cs) and kurtosis coefficient (Ck). Table 3 describes the values of Cs and Ck that must be fulfilled as a rule in choosing the fittest method for the calculation of design rainfall.

Table 3. Cs and Ck values for each distribution method

No	Type of Distribution	Rule
1	Log-Normal method	Cs = 0 Ck = 3
2	Gumbel type 1 method	Cs = 1,14 Ck = 5,4
3	Log-Pearson type III	others

The equations for Cs and Ck are following [17]:

$$Cs = \frac{n^2}{(n-1)(n-2)}\left[\frac{\sum_{i=1}^{i=n}(Xi-\bar{X})^3}{nS^3}\right] \quad ....(7)$$

$$Ck = \frac{n^3}{(n-1)(n-2)(n-3)}\left[\frac{\sum_{i=1}^{i=n}(Xi-\bar{X})^3}{nS^4}\right] \quad ....(8)$$

Where:
$\bar{X}$ = the mean value of data
n = number of data
S = standard deviation
Xi = variate value of data

2.3.3 Analysis of flood peak discharges

Flood peak discharges are analyzed based on the results of design rainfall. In this study, flood peak discharges are estimated for 5, 10, 25, 50, and 100 years return period by adopting Nakayasu Synthetic Hydrograph Unit (SHU) method (17) as follows:

$Q_{max} = C * A * Ro * (1/ (3.6 (0.3Tp+T_{0.3}))$

Where :
Tp is peak time (hour)
C is runoff coefficient
Ro is designed rainfall (mm)
A is the catchment area (km^2)
$T^{0.3}$ is the time of slowdown the discharge (until 30%)

3 Results and Discussion

3.1 Analysis Results of Average Annual Rainfall

Table 4 presents the average annual rainfall calculated using the Thiessen method.

Table 4. Analysis results of average annual rainfall

No	Year	Rainfall at Bayah Station (24%)	Rainfall at Cibeber Station (76%)	Average Annual Rainfall (mm)
1	2011	32.88	57	89.88
2	2012	30.72	69.16	99.88
3	2013	44.4	85.12	129.52
4	2014	29.04	73.72	102.76
5	2015	24.24	53.96	78.2
6	2016	27.6	70.68	98.28
7	2017	28.8	53.96	82.76
8	2018	31.44	97.28	128.72
9	2019	26.88	152.76	179.64

3.2 Estimation of Design Rainfall

Table 5 shows the values of Cs and Ck that have been calculated from each distribution method.

Table 5. The values of Cs and Ck for each distribution method

No	Type of Distribution	Rule	Analysis results	Notification
1	Log-Normal	Cs = 0	0,7860	Not acceptable
		Ck = 3	3,0005	
2	Gumbel 1	Cs = 1,14	0,7860	Not acceptable
		Ck = 5,4	3,0005	
3	Log Person III	Others	0,7860	Acceptable
			3,0005	

Based on Table 5, it can be concluded that the method of Log-Pearson type III is the most suitable fit method that can be applied to determine the design rainfall in the Cimadur river basin. The values of design rainfall for the estimated T-year return period based on the Log-Pearson type III method are presented in Table 6.

Table 6. Design rainfall using Log-Pearson Type III method

No	Years return period	The values of design Rainfall
1	2	100.20
2	5	128.92
3	10	153.42
4	25	191.19
5	50	224.70
6	100	263.30

3.3 Analysis Results of Flood Peak Discharges

Fig. 4 shows the estimation of flood peak discharges for 2, 5, 10, 25, 50, and 100 years return period by adopting Nakayasu Synthetic Hydrograph Unit (SHU) method (Fig. 4).

Based on Fig. 4, it can be demonstrated that the flood peak discharges in the catchment of Cimadur river for 2, 5, 10, 25, 50, and 100 years return periods are 398 m^3/s, 470.71 m^3/s, 560.16 m^3/s, 698 m^3/s, 820.4 m^3/s, and 1017.04 m^3/s, respectively. The flood peak discharges are found to be high even for 5 years return period of design rainfall. The results also show that the peak discharges occur after 4-hour rain continuously.

As reported by RPJMD Lebak district (14), local people in the Cimadur river basin mostly make a living as a farmer. Freshwater is always becoming a necessary thing for agriculture activities by local people in the Cimadur catchment, but the freshwater also could cause harvest failure if the farm is flooded with water (for example flooding occurrences around the agriculture areas during the rainy season).

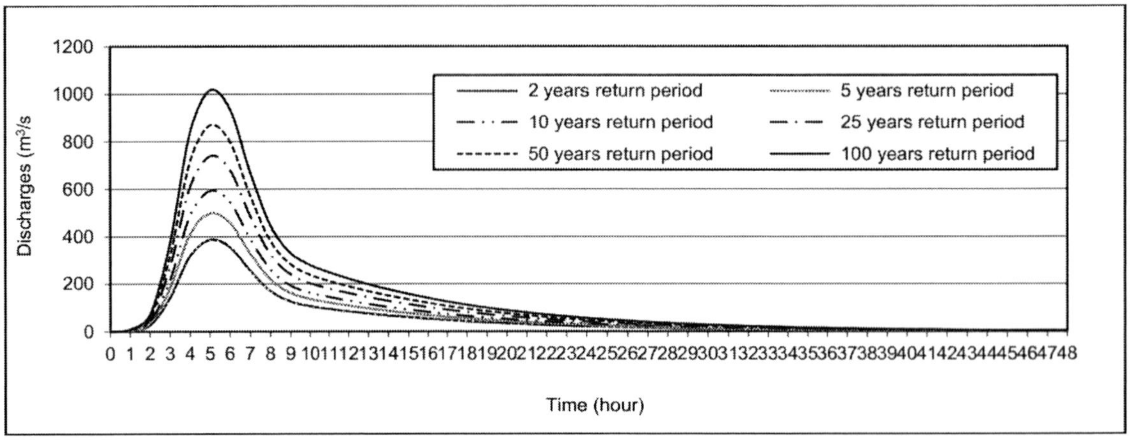

Fig. 4. Flood peaks discharge for 2 years to 100 years return period

Therefore, we have to make sure that the water supply is enough for agriculture activities during the dry season, but no flooding during the rainy season. For this, the local government must think about how to manage and store the excess water during the rainy season to be useful for the farmer during the dry season. The local government may consider constructing an engineering structure that can store and control the excess water discharge during the rainy season.

4 Conclusion

Flood is the major natural disaster in the catchment of the Cimadur river for the past few years. Converting the forest to become residential and agricultural areas has triggered more frequency of flooding events. During the rainy season, the flood peak discharges can reach 560.16 m^3/s for 10 years return period of design rainfall.

However, the agricultural areas in this district will need more water supply during the dry season. High frequency of the water during the rainy season may be utilized for water supply during the dry season. For this, purpose, the local government must think carefully about how to reduce and control the flood problem in the Cimadur catchment and also could plan how to restore the excess water during the rainy season.

The authors would like to extend a special thanks to Universitas Teknokrat Indonesia for supporting in the process of completing this study.

References

1. H. Chen, L. Yao, A. Fitri, The influence mechanism research of inflow temperature in different time scale on the water temperature structure. In: IOP Conference Series: Earth and Environmental Science. 12058 (2019)
2. R. Hashim, C. Roy, S. Shamshirband, S. Motamedi, A. Fitri, D. Petković, et al. Estimation of Wind-Driven Coastal Waves Near a Mangrove Forest Using Adaptive Neuro-Fuzzy Inference System. Water Resour Manag. **30** (7):2391–404 (2016)
3. A. Fitri, R. Hashim, S. Motamedi, Estimation and validation of nearshore current at the coast of Carey Island, Malaysia. Pertanika J Sci Technol. **25** (3):1009–18 (2017)
4. A. Fitri, R. Hashim, S. Abolfathi, K. N. A. Maulud Dynamics of sediment transport and erosion-deposition patterns in the locality of a detached low-crested breakwater on a cohesive coast. Water (Switzerland). **11** (8) (2019)
5. A. Fitri, Z. A. Hasan, A. A. Ghani, Determining the Effectiveness of Harapan Lake as Flood Retention Pond in Flood Mitigation Effort. In: In Proceedings of 2011 4th International Conference on Environmental and Computer Science (ICECS 2011). (2011)
6. S. Hutapea, Biophysical Characteristics of Deli River Watershed to Know Potential Flooding in Medan City, Indonesia. J Rangel Sci. **10**(3):316–27 (2020)
7. A. Fitri, R. Hashim, S. Abolfathi, K. N. A. Maulud Dynamics of sediment transport and erosion-deposition patterns in the locality of a detached low-crested breakwater on a cohesive coast. Water (Switzerland). **11** (8):1721 (2019)
8. L .Yao, X. Huang, A. Fitri, Influence scope of local loss for pipe flow in plane sudden expansions. In: IOP Conference Series: Earth and Environmental Science. 12056 (2019)
9. A. Fitri, K. N. Abdul Maulud, D. Pratiwi, A. Phelia, F. Rossi, N. Z. Zuhairi,Trend Of Water Quality Status In Kelantan River Downstream, Peninsular Malaysia. J Rekayasa Sipil. (2020)
10. A. Fitri, K. Nizam, A. Maulud, F. Rossi, F. Dewantoro, N. Z. Zuhairi, Spatial and Temporal Distribution of Dissolved Oxygen and Suspended Sediment in Kelantan River Basin. In: In 4th International Conference on Sustainable Innovation 2020–Technology, Engineering, and Agriculture (ICoSITEA 2020). Atlantis Press. 51–4 (2021)
11. S. H. Lai, A. Fitri, Application of SWAT Hydrological Model to Upper Bernam River Basin (UBRB), Malaysia. IUP J Environ Sci. **5**(2) (2011)

12. D. Pratiwi, R.O. Sinia, A. Fitri, Increasing Public Knowledge of Porous Drainage Functioning as a Rainwater Infiltration Place. J Soc Sci Technol Community Serv. **1**(2) (2020)
13. R. Hashim, A. Fitri, S. Motamedi, A. M. Hashim Modeling of coastal hydrodynamic associated with coastal structures: A review. Malaysian J Sci. **32**(4):149–54 (2013)
14. A. Fitri, R. Hashim, K. Song Il, Motamedi S. Evaluation of Morphodynamic Changes in the Vicinity of Low-Crested Breakwater on Cohesive Shore of Carey Island, Malaysia. Coast Eng J. **57**(04):1550023 (2015)
15. A. Fitri, L. Yao, B. Sofawi, Evaluation of mangrove rehabilitation project at Carey Island coast, Peninsular Malaysia based on long-term geochemical changes. In: IOP Conference Series: Earth and Environmental Science. 12055 (2019)
16. RPJMD. Yearly Planning Programme and Report. (2019)
17. Limantara, I L M 2018 Hydrological Engineering: Revised Edition of Andi Publisher. (2018)

Study of the effects of check dam construction on the Limau Manih river using GIS

Darwizal Daoed[1*], *Fifi* Novia Azhari[2], *Masril* Syukur[1], and *Rudy* Ferial[1]

[1]Departement of Civil Engineering, Faculty of Engineering, Andalas University, West Sumatera, Indonesia.
[2]Graduate, Departement of Civil Engineering, Faculty of Engineering, Andalas University, West Sumatera, Indonesia.

Abstract. The problem of flooding or flash floods in watersheds often occurs, both in large rivers and small rivers. The frequency of occurrence varies greatly depending on the local climate. As a result of flooding, it causes more erosion and collapse of riverbanks and washes away all kinds of materials, such as wood, sand, and stone. Sometimes it also submerges rice fields, villages, and houses downstream of the river. Likewise, in the study location of the Limau Manih river, there has been a flood with a large discharge. It was recorded twice a year, namely in July and September 2012. One of the ways to reduce energy, scour, and avalanches on riverbanks is by building a check dam or weir. A check dam is expected to reduce the energy that arises due to differences in elevation or slope of the river channel. But on the other hand, the weir will raise the water level, so that the puddle becomes wider. Therefore, it is necessary to study the impact of weir construction on the extent of inundation. The study was conducted using ArcGIS to map inundation and HEC-RAS to simulate water levels along the river. Simulations were carried out for several return periods of rainfall events. The results of the study show that the planned flood discharge in the river of Limau Manih watershed is quite large. Meanwhile, from the simulation results, the inundation area is not too significant with the increase in the return period and the planned flood discharge. Although the simulation of the weir without widening the upstream part of the weir shows a significant increase in inundation area. This is most likely due to the weir in the upstream area of the river which is rather steep, so that water flows quickly through the overflow of the weir. However, the water level in the weir is higher than without the weir, as a result, water jumps and erosion occur downstream of the weir. For this reason, it is very necessary to monitor the scour behavior in the weir and the sedimentation rate, because this area has the potential to be eroded.

1 Introduction

1.1 Background

High rainfall and short duration are very likely to cause flooding or inundation, where the water discharge will be high. Sudden flooding or flash floods or what is called Galodo by the local people of West Sumatra, Indonesia, hit the Padang City area of West Sumatra twice in 2012, 4 July 2012, and 12 September 2012. The flood washed away all kinds of materials, sand, stones, wood. and mud. In the lower reaches of Batang Kuranji in the Kuranji watershed, several locations were found to be inundated by water. The water level rises quickly, due to rain that occurs in the afternoon, and at the same time, high tide occurs in the coastal area. Flooding accompanied by high tides is a case of accelerating the occurrence of local inundation in the Padang city basin area [1,2].

Long before, flash floods also occurred in 1988 and March 16, 2008, in the Batang Kuranji and Batang Limau Manih streams. The flash flood that occurred in Batang Kuranji was caused by high rainfall and a decrease in the riverbed due to mining of rock, sand, and high velocity. Excessive natural management carried out by the community harms the preservation of nature and forests in the watershed [3]. To reduce the risk of flood events, flood and sediment controllers were built in the Batang Kuranji watershed in the Upper Segment. Check dams were built in three places along Batang Limau Manih which is a tributary of the Kuranji River in the Limau Manih sub-watershed. The Ministry of Public Works and Public Housing (PUPR) through the Sumatra River Basin Agency (BWS) V Directorate General of Natural Resources in Batang Kuranji and Batang Limau Manih and was completed in 2018 [4].

Even so, local flooding still occurs around the construction site of the Batang Kuranji check dam in the middle segment. However, it is not known with certainty whether the flood control building was due to the flood control building or the location being lower than the river embankment. Because when the discharge is large, the water level at the check dam is close to the height of the embankment but is still at a safe elevation. This means that it is still necessary to study what causes local inundation.

Based on the statement above, a study was carried out on the impact of the construction of flood and sediment controllers in Batang Limau Manih (Limauu Manih River) on the water surface area of the river and its surroundings using the geographic information system.

* Corresponding author: darwizaldaoed@gmail.com

2 Literature study

2.1 Floods

A flood is an increase in the volume of water that exceeds the capacity of the channel/river, causing an area of land to be submerged [5]. The volume of water is not a problem if it does not cause losses, victims die or are injured, does not soak settlements for a long time, does not soak rice fields, and does not interfere with community activities. On the other hand, a time-consuming inundation is called a flood disaster.

2.2 Construction of control sediment

Construction of control sediment is water structures in rivers that function as sediment retainers, gravity type, or other types [6]. Check dam can control the speed, discharge, and direction of sediment flow, accommodate sediment both permanently and temporarily and consist of a weir body including foundations, overflows that can be overflowed. sediments, wings, and complementary buildings whose crests are overflowed by the flow of water [7,8].

2.3 Catchment Area

A catchment area or rainwater catchment area is a small scale area of a watershed which means an area bounded by topographical barriers in the form of ridges or mountains that collect rainwater that falls on it and then drains it through creeks and rivers to the sea or lake [9,10]. The catchment area or water catchment area is estimated by contour line guidelines or the height of an area, the direction of flow from the conveyor channel to the main drain is usually used as a reference in the division of the catchment area.

2.4 Flood Discharge Analysis

To estimate the discharge in a watershed, where there is no discharge observation data, a design flood discharge analysis method is needed. One method that is often used is the Rational Method [2], [11], [12]. The flood discharge in this method is formulated as follows:

$Q = 1/3.6\ C\ I\ A$

Where,

Q = discharge (m^3/sec);
C = runoff coefficient,
I = intensity (mm/hour);
A = area of the watershed (km^2)

2.5 Geographic information system

Geographic Information System (GIS) is a form of information that presents information in geographic form by using a map as an interface. GIS is used as a tool to process land mapping and analyze all events that occur on earth in a computerized manner, then integrate it into database operations and statistical analysis and combine it with unique geographical analysis through mapping or using maps [13,14].

3 Methodology

The methodology applied in this research is firstly collecting climatic data, river geometry, and downloading basic digital elevation model (DEM) data and maps from google earth pro, secondly carrying out the process of forming a map of watersheds and surface runoff directions. Then do the simulation using Hec-RAS software.

The location of the weir is mapped only 1.0 km upstream and downstream of the weir, and in a transverse direction as far as 200 m perpendicular to the river.

Then rainfall analysis uses statistics and frequency for maximum daily rainfall. The results of a representative rainfall analysis are used to determine the design flood discharge. Flood discharge is calculated using the rational method.

Furthermore, the determination of the rain catchment area or watershed from the location of the weir to the upstream of the river. The area of the flow is done using ArcGIS. At the same time determine the direction of flow (flow direction). This process is carried out after interpolating the DEM data and combining it with tools in ArcGIS.

Finally, simulate the inundation surface area with conditions before and after the existence of sediment and flood control buildings using HEC-RAS. Each maximum plan debit is obtained with several return periods. The target of the simulation is to be able to describe the phenomena that will occur due to an increase in the amount of flow discharge.

4 Result and Discussion

4.1 Formation of catchment map

The flood simulation model requires input data in the form of a digital elevation model (DEM). DEM is the result of downloading through the DEMNAS website. The downloaded DEM is then processed using ArcGIS software. Where ArcGIS uses data management tools. Several DEMs are combined into a single raster and then clipped according to the Batang Kuranji watershed boundary.

In Fig. 1. DEM for Kuranji river watershed and **Fig. 2.** Flow direction for Kuranji river watershed, it can be seen that the dark blue area is an area with a higher elevation. The lighter blue color in the DEM indicates a lower area. Then the picture on the right shows the pattern of the direction of the water flow towards Batang Kuranji. In addition, it can also be seen which areas are the ridges or peaks as the topographic boundary for the formation of watersheds.

The next process is to process the flow direction map into a river network map and the distribution of catchment areas in the Batang Kuranji watershed using Spatial Analyst Tools. The results of the spatial analysis show that several catchment areas are formed, one of which is the Batang Limau Manih catchment area which is the river where the sediment control structures will be studied. The catchment area of Batang Limau Manih obtained from the results of spatial analysis in ArcGIS software is 30.0 km^2.

 https://doi.org/10.1051/e3sconf/202133108007

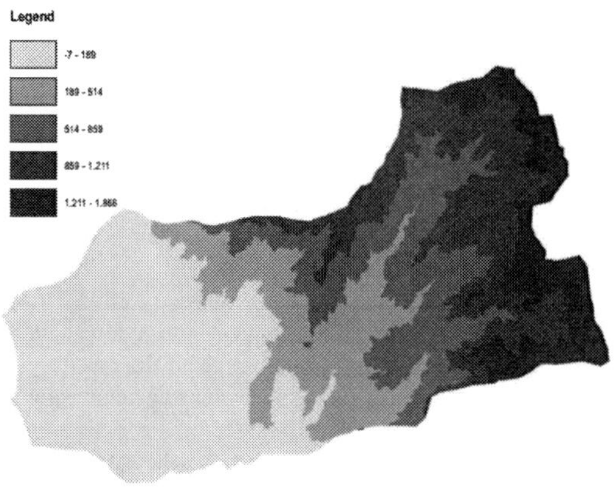

Fig. 1. DEM for Kuranji River watershed

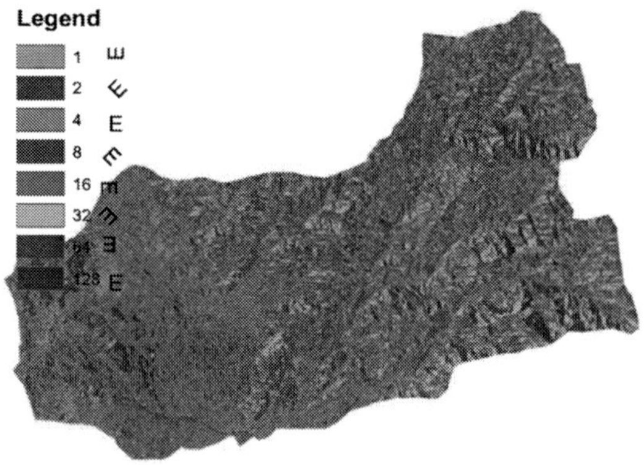

Fig. 2. Flow direction for Kuranji River watershed

The location of the Batang Limau Manih sediment control building project is in the downstream area, as shown in the following Fig. 3 Check dam location.

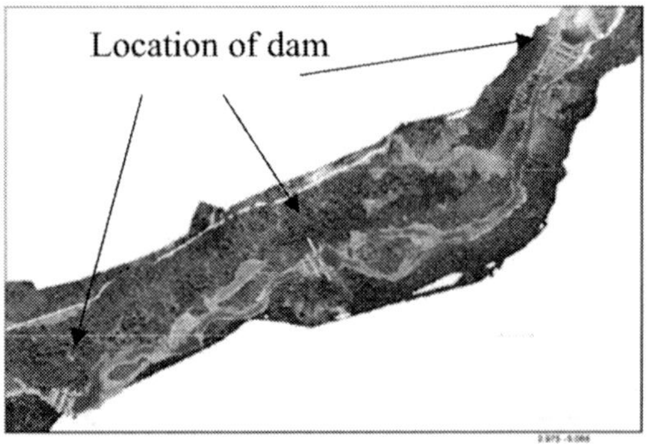

Fig. 3 Check dam location

The sediment control buildings built are in the form of three check dams, namely check dam (I, check dam II, and check dam III), Check dam (I) is at the highest elevation compared to the other two check dams.

4.2 Flood discharge

Hydrological analysis was carried out to find the value of flood discharge with a return period of 25 years, 50 years, 100 years, 500 years, and 1000 years. The data used in this hydrological analysis is the maximum rainfall data from three rain stations, namely Batu Busuak Station, Gunung Nago Station, and Ladang Padi Station. The results of the calculation of the design rainfall with the Pearson-III Log distribution and the design flood discharge are obtained as shown in **Table 1.The** watershed of Batang Limau Manih and Debit below.

Table 1.The watershed of Batang Limau Manih and Debit

Area (A)			30.0	km^2
Flow Coefficient (C)			0.65	
Time of concentration (Tc)			83.387	minute
Return period	**Tc (jam)**	**R 24 (mm)**	**I (mm/jam)**	**Q (m^3/s)**
25	1.4	195	54	294
50	1.4	203	56	305
100	1.4	209	58	314
500	1.4	215	60	323
1000	1.4	227	63	341

4.3 The surface area of water

The water surface area was carried out using the HEC-RAS 5.0.3 software. The data needed to analyze the water surface area are flood discharge, detailed map of river situation, and cross-section of the river. Using the HEC-Geo RAS tool, detailed situation maps and river cross-sections in ArcGIS can be exported to HEC-RAS.

In the analysis of the water surface area, the river cross-section data is assumed to be unchanged, except for the cross-sectional data at the check dam location station. The manning coefficient value entered is 0.03, except for the check dam cross-section, the value entered is 0.012. The water surface area at the project site is as shown in Table 2. Differences in the surface area of the water.

Table 2. Differences in the surface area of the water

TR	Discharge (m^3/s)	Without Weir (ha)	With weir (ha)	Positive/ negative (%)
25	293.6	10.3	10.2	-1.1
50	304,6	10.4	10.2	-1.7
100	314.4	10.5	10.4	-1.6
500	323.2	10.6	10.7	1.1
1000	341.0	10.8	10.9	1.1

4.4 Water surface area analysis

Analysis of the results of the water surface area is carried out through reading the inundation map obtained through several simulations. Each simulation is limited by the magnitude of the design flood discharge as input to the Arc-GIS in addition to the cross-section of the river. Simulations were carried out in two scenarios before the building and after the check dam. The picture shows a series of changes in the water surface area of the three check dams built. Analysis of the results of the water surface area is carried out through reading the inundation map obtained through several simulations. Each simulation is limited by the magnitude of the design flood

 https://doi.org/10.1051/e3sconf/202133108007

discharge as input to the Arc-GIS in addition to the cross-section of the river. Simulations were carried out in two scenarios before the building and after the check dam. The picture shows a series of changes in the water surface area of the three check dams built.

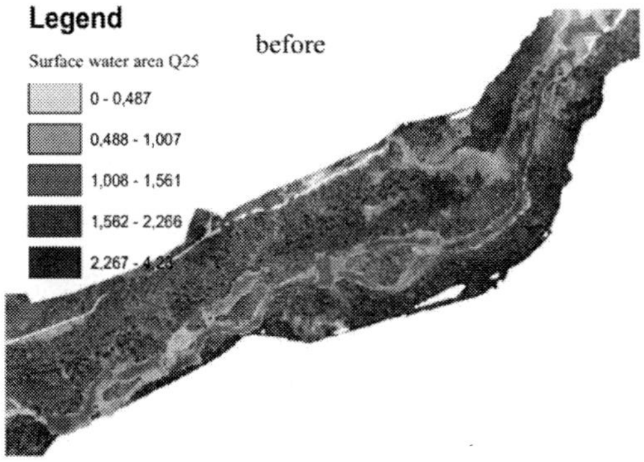

Fig. 4 Surface water area Q 25 (before)

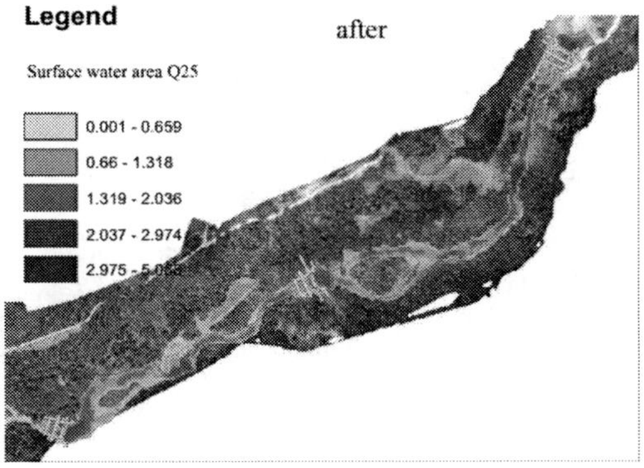

Fig. 5 Surface water area Q 25 (after)

If you look at the simulation results of the planned flood discharge with a return period of 25 years with 50 years, the change in the inundation area is not very visible. This is because the difference in discharge is not too large. However, if it is observed for changes in a discharge after 50 years, namely 100 years and 500 years, it has begun to be seen that the difference in inundation area is starting to be significant This shows the function and location of the check dam meet the requirements [8], holding sediment and escape the water [15], [16]. Please look at the following picture in fig 6,7,8,9,10 and 11.

In the picture, it can be seen that the change in the inundation area in the image is not very visible because the image scale is too large. But something is interesting, namely the change in water level upstream of the weir. The change in depth increases with the greater the water discharge. The inundation should be bigger upstream of the weir, but it is not visible in the simulation because in the geometric design the channel cross-section is widened at each upstream of the weir. In addition, the dimensions of the embankment are increased enough to accommodate a 500-year (Q500), so that water can not overflow to the side of the river [17], [18], [19].

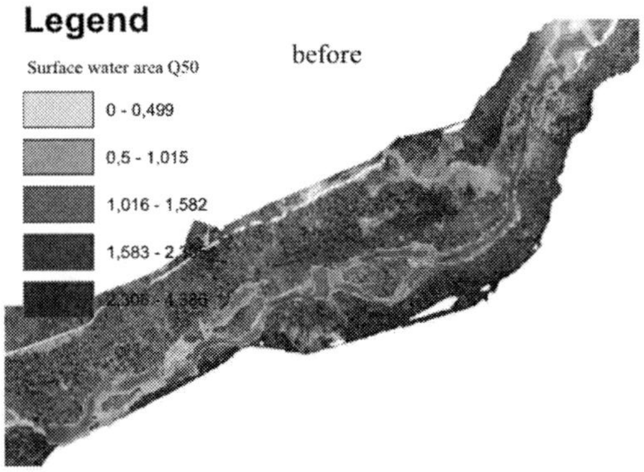

Fig. 6 Surface water area Q 50 (before)

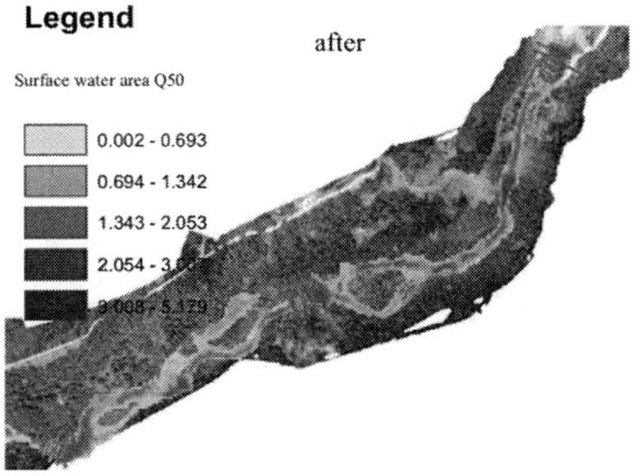

Fig. 7 Surface water area Q 50 (after)

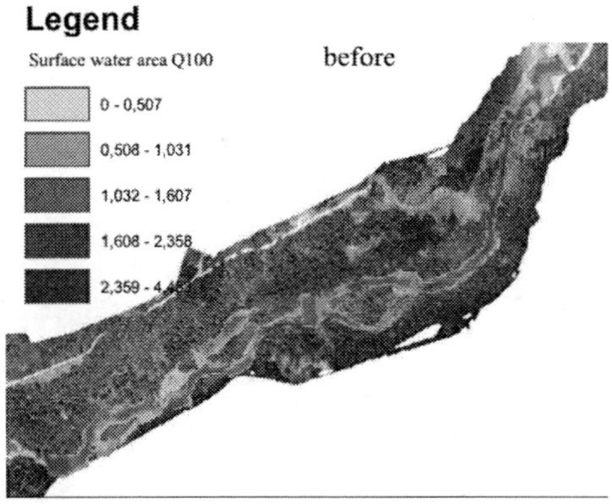

Fig. 8 Surface water area Q 100 (before)

Overall the weir can provide security against water overflows both upstream and downstream of the weir. However, due to the presence of a dam, the water level increases, where discharge in the 100-year and 500-year simulations, the dam's level is already high. So that the flow velocity after the weir is still high due to the release of potential energy and causes the downstream scour to enlarge [8], the long-term effect must be considered [20].

Physically there has been erosion downstream of the weir, especially the waterfall area. Then there is also erosion and sediment deposition along the channel after the main weir building.

https://doi.org/10.1051/e3sconf/202133108007

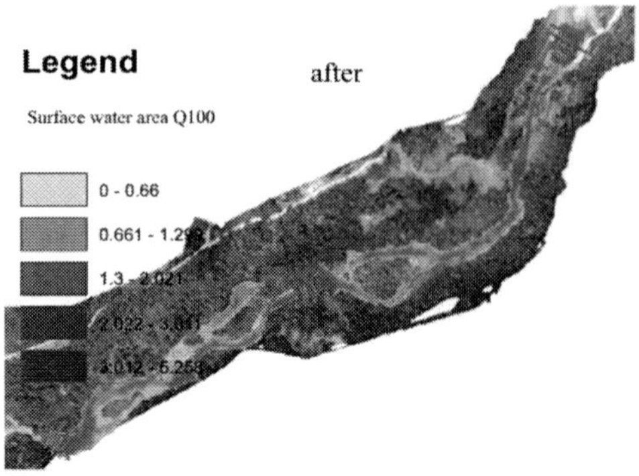

Fig. 9 Surface water area Q 100 (after)

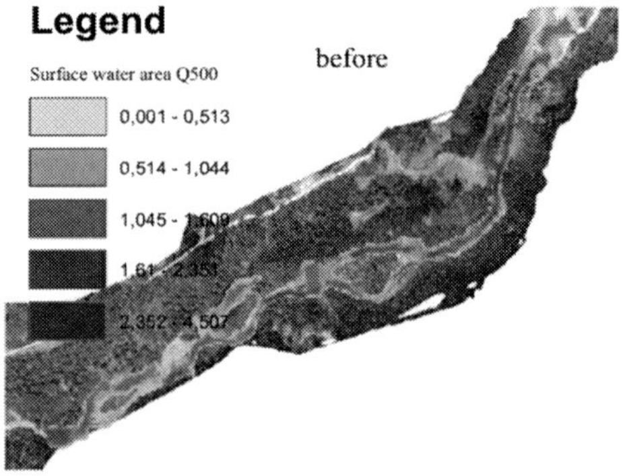

Fig. 10 Surface water area Q 500 (before)

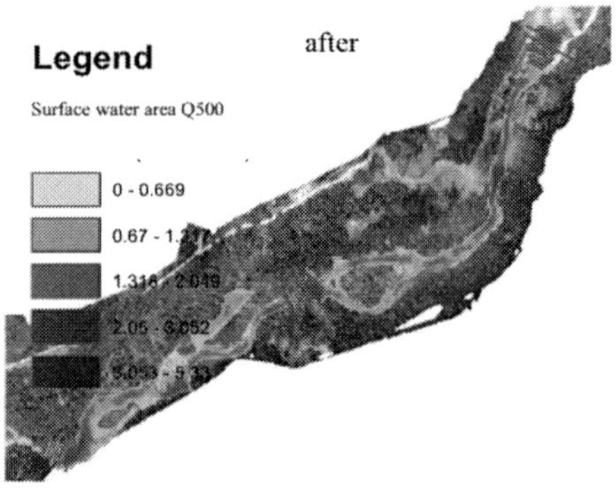

Fig. 11 Surface water area Q 500 (after)

5 Conclusion

In the impact study of the construction of the Batang Limau Manih sediment control building, it can be concluded that:

1. The location of the check dam on the Batang Limau Manih is quite good, where the water level rise upstream of the weir did not reach the next weir.

2. The water surface area due to the construction of sediment control structures does not significantly reduce or increase the water surface area that occurs. This is due to the widening of the channel cross-section upstream of the weir.

3. The construction of a check dam can affect the water level upstream and will increase the flow velocity and kinetic energy downstream of the weir.

This publication was funded by Andalas University through a publication grant from the Faculty of Engineering, Andalas University No. 122/UN16.09. D./PP/2021, and I would like to thank the Head of the Civil Engineering Department, Vice, Dean-1, Dean, and the Rector for their help.

References

1. D. Daoed, R. Bujang, Istijono, A. Hakam. Predictions of Vulnerability Flood and Flood Prone Areas in Watershed West Sumatra Province using Arc-GIS and Category Value. *International Journal of Earth Sciences and Engineering, 09 SPL No 03*(3rd International Conference on Earth Sciences and Engineering (ICEE 2016) 17th-18th June 2016), 274-279. (2016).
2. K. Subramanya, *Engineering Hydrology* (Vol. Third Edition). New Delhi: Tata McGraw-Hill Publishing Company Limited. (2008).
3. J, S. Kodoatie Robert, Flood Causes and Control Methods From an Environmental Perspective. Yogyakarta: Pustaka Pelajar. (2002).
4. Ministry of Public Works and Public Housing, (B. K. P). To Prevent Flash Floods From Recurring, Ministry of PUPR Completes Padang City Flood and Sediment Control. (2018).
5. D. Daoed, M. Syukur, & M. Rahman, Habibur.The Effect of Land Use Changes on Flow and Drainage Systems Using Geographic Information Systems (GIS) Case Study: Andalas University Campus Area, Padang. *Paper presented at the National Conference 3 (3rd ACE Engineering)*, Padang, Sumatra Barat. (2016).
6. A. Armanini, F. Dellagiacoma, & L. Ferrari, *From the check dam to the development of* functional check dams *Fluvial hydraulics of mountain regions* (pp. 331-344): Springer. (1991).
7. Ministry of Public Works and Public Housing, P. K. P.. Batang Kuranji Flood and Sediment Control Efforts in Padang City. (2015).
8. A. Armanini, & M. Larcher, Rational criterion for designing opening of slit-check dam. *Journal of Hydraulic Engineering, 127*(2), 94-104. (2001)
9. BNPB (National Disaster Management Agency). Regulation of the Head of the National Disaster Management Agency Number 8 About Standardization of Disaster Data. Jakarta. (2011).
10. D. Daoed, R. Bujang, Istijono,. M. Syukur, Evaluation of Drought Vulnerability on Watersheds in West Sumatera Province by Using Cropwat-8 and GIS. *International Journal on Advanced Science, Engineering and Information Technology, 8*(6), 2443-2449. (2018).
11. L. Utami, & Sylvia. Study of Empirical Methods for Calculating the State River Flood Discharge in the Sungai Pandan Sub-District.*Jurnal Poros Teknik.*, 103, (2016).
12. M. S. J. H. G. Kang, J.A. Song, Y.G. Chu, S.W. Her, S.W. Hwang, Park. Estimating Design Floods Based

on the Critical Storm Duration for Small Watershed. *Journal of HydroEnvironment Research, 7*. (2013).

13. M. Aziz, & S. Pujino, Desktop and Web-Based Geographic Information System. Yogyakarta: Gava Media (2006).
14. A. Padmavathy, K. G. Raj, N. Yogarajan, P. Thangavel, & M. Chandrasekhar, Check dam site selection using GIS approach. *Advances in Space Research, 13*(11), 123-127. (1993).
15. A. Romero-Díaz, F. Alonso-Sarriá, & M. Martínez-Lloris, Erosion rates obtained from check-dam sedimentation (SE Spain). A multi-method comparison. *Catena, 71*(1), 172-178. (2007).
16. A. M. Hassanli, A. E. Nameghi, & S. Beecham, Evaluation of the effect of porous check dam location on fine sediment retention (a case study). *Environmental monitoring and assessment, 152*(1), 319-326. (2009).
17. J. M. Andrew Chadwick, *Hydraulics in Civil Engineering*. London: Allen & Unwin Ltd, (1986)
18. D. Daoed, Hydraulics and Applied for Open Channels, Padang, CV Ferila. (2010).
19. C. N. Ezugwu, Sediment Deposition in Nigeria Reservoirs: Impacts and Control Measures. Innovative Systems Design and Engineering, 4 No 15(Sediment), 8. (2013).
20. K. N. Ratnam, Y. Srivastava, V. Rao, E. Amminedu, & K. Murthy, Check dam positioning by prioritization of micro-watersheds using SYI model and morphometric analysis—remote sensing and GIS perspective. *Journal of the Indian society of remote sensing, 33*(1), 25-38. (2005)

AUTHOR INDEX